ENCYCLOPEDIA

of

VIROLOGY

ENCYCLOPEDIA
of
VIROLOGY

Volume 3

Edited by

ROBERT G. WEBSTER

and

ALLAN GRANOFF

St Jude Children's Research Hospital, Memphis, USA

ACADEMIC PRESS
Harcourt Brace & Company, Publishers
London San Diego New York Boston
Sydney Tokyo Toronto

ACADEMIC PRESS LIMITED
24–28 Oval Road
London NW1 7DX

United States Edition published by
ACADEMIC PRESS INC.
San Diego, CA 92101

A catalogue record for this book is available from the British Library

ISBN 0-12-226961-6 Volume 1
0-12-226962-4 Volume 2
0-12-226963-2 Volume 3
0-12-226960-8 Set

Typeset by Alden Multimedia

Printed and bound in Great Britain by The Bath Press, Avon

Editorial Advisory Board

Guide to Use of the Encyclopedia

Structure of the Encyclopedia

The material in the Encyclopedia is arranged as a series of entries in alphabetical order. Most of the entries comprise a single article, while entries on the most common or widely studied viruses and important aspects of virology consist of two or more articles. These combined entries are arranged in a logical sequence within the entry, with a summary contents list at the beginning.

Finding Your Way Around

The Contents list

Your first point of reference will probably be the Contents list. The complete Contents list appears in all three volumes.

Dummy entries. Throughout the Contents list and the body of the Encyclopedia you will find 'dummy entries'. These occur where an article discusses two or more significant viruses together and so the dummy entry draws the reader to the correct place in the Encyclopedia.

For example:
If you wished to locate the entry which discusses Creutzfeldt-Jakob disease in the Contents list you would find the following:

Creutzfeldt–Jakob disease virus **see** Spongiform encephalopathies

Further use of the Contents list would lead you to the following:

VOLUME 3

Spongiform encephalopathies: *Mule deer, elk and bovine* Maurizio Pocchiari 1357
Creutzfeldt–Jakob disease, scrapie and related transmissible encephalopathies Laura Manuelidis and Elias E Manuelidis 1361

Scanning through the text

Alternatively if you were to try to locate the material by browsing through the text in the C section of Volume 1 you would find:

CREUTZFELDT – JAKOB DISEASE VIRUS

See Spongiform encephalopathies

This would lead you to the S section in Volume 3 to find:

SPONGIFORM ENCEPHALOPATHIES

Contents

Mule Deer, Elk and Bovine
Creutzfeldt–Jakob Disease, Scrapie and Related Transmissible Encephalopathies

---

The running headline at the top of each page helps locate the correct page.

If the reader wishes to be directed to other entries of related topics, the section before the References entitled 'See also' in most articles lists related entry titles.

For example, the entry on *Creutzfeldt–Jakob disease, scrapie and related transmissible encephalopathies* includes the following cross-references:

See also: Kuru; Nervous system viruses; Visna-Maedi viruses.

Index

The complete Index for the three volumes is provided at the back of each volume. To aid the Reader the Index is annotated with (A), (P), (B), (I) superscripts for animal, plant, bacteriophage and insect viruses respectively. Also for material which makes up a complete entry the page numbers are in bold type. Further guidance on the use of the Index is supplied with the Index.

Appendix

All viruses included in the Encyclopedia are listed in their animal, plant, bacteriophage or insect groups. Also detailed are their family grouping and any synonyms.

Contributors

A full list of Contributors is supplied at the beginning of each volume.

Contents

Editorial Advisory Board	v
Guide to Use of the Encyclopedia	vii
Contributors	xxiii
Preface	xxxix

VOLUME 1

A

Adenoviruses – *General features*	Göran Wadell	1
Molecular biology	Walter Doerfler	8
Animal adenoviruses	William C Russell	14
Malignant transformation and oncology	Geoffrey R Kitchingman	17
African swine fever virus	Maria L Salas	23
Aleutian mink disease virus **See** Parvoviruses		
Alfalfa mosaic virus and ilarviruses	J F Bol and E M J Jaspars	30
Algal viruses	James L Van Etten	35
Alpha 3 bacteriophage **See** ØX174 bacteriophage and related bacteriophages		
Amphibian herpesviruses	Allan Granoff	40
Antivirals	A Kirk Field	42
Archaebacterial bacteriophages	Pelle Stolt and Wolfram Zillig	50
Ascoviruses	Brian A Federici	58
Autoimmunity	Robert S Fujinami	63
Avian leukosis viruses	John M Coffin	66

B

Baboon herpesvirus **See** Herpesviruses – baboon and chimpanzee		
Bacillus subtilis bacteriophages – groups 1-5	H Ernest Hemphill	72
Bacterial identification – use of bacteriophages	Michael S DuBow	78
Bacteriophage ecology, evolution and speciation	Allan M Campbell	81
Bacteriophage recombination	Kenneth N Kreuzer	83
Bacteriophage taxonomy and classification	Jack Maniloff, Hans-Wolfgang Ackermann and Audrey Jarvis	93
Bacteriophage toxins and disease	Randall K Holmes and Michael P Schmitt	101
Bacteriophage transduction	Werner Arber	107
Bacteriophages as cloning vehicles	Noreen E Murray	113
Bacteriophages in industrial fermentations	Mary Ellen Sanders	116
Bacteriophages in soil	Stanley T Williams, A Martin Mortimer and Jacky Eccleston	121
Baculoviruses – *Granulosis viruses*	Norman E Crook	127
Nuclear polyhedrosis viruses	George F Rohrmann	130
Nonoccluded baculoviruses	Allan M Crawford	136
Badnaviruses	Ben E L Lockhart and Neil E Olszewski	139
BF23 bacteriophage **See** T5 bacteriophage and related bacteriophages		
Birnaviruses – animal	Hermann Becht	143
BK virus **See** JC and BK viruses		
Border disease virus **See** Bovine viral diarrhea virus and border disease virus		
Borna disease virus	Lothar Stitz and Rudolf Rott	149
Bovine herpesviruses	Michael J Studdert	155
Bovine immunodeficiency virus	Matthew A Gonda	158
Bovine leukemia virus	Kathryn Radke	166
Bovine papilloma virus **See** Shope papilloma and bovine papilloma viruses		
Bovine parvovirus **See** Parvoviruses		
Bovine spongiform encephalopathy **See** Spongiform encephalopathies		
Bovine viral diarrhea virus and border disease virus	Edward J Dubovi	175
Bromoviruses	Paul Ahlquist	181

Bunyaviruses – *General features*	Neal Nathanson and Francisco Gonzalez-Scarano	185
Replication	Richard M Elliot	192

C

Caliciviruses **See** Vesicular exanthema virus and caliciviruses of pinnipeds, cats and rabbits		
Canine distemper virus **See** Rinderpest and canine distemper viruses		
Canine parvoviruses **See** Parvoviruses		
Capilloviruses	L F Salazar	197
Caprine arthritis encephalitis virus	Gilles Quérat and Robert Vigne	199
Cardioviruses	Douglas G Scraba	205
Carlaviruses	Sergei K Zavriev	214
Carmoviruses	T J Morris and D L Hacker	218
Caulimoviruses	Robert J Shepherd	223
Cell structure and function in virus infections	Samuel Dales	226
Central European encephalitis virus **See** Encephalitis viruses		
Chandipura, Piry and Isfahan viruses	Sailen Barik and Amiya K Banerjee	233
Channel catfish virus **See** Fish herpesviruses		
Chicken herpesvirus **See** Marek's disease virus		
Chickenpox virus **See** Varicella-zoster virus		
Chikungunya, O'nyong nyong and Mayaro viruses	Charles H Calisher	236
Chilo iridescent virus **See** Tipula iridescent virus		
Chimpanzee herpesvirus **See** Herpesviruses – baboon and chimpanzee		
Chlorella viruses **See** Algal viruses		
Closteroviruses	Thierry Candresse	242
Coltiviruses **See** Orbiviruses and coltiviruses		
Comoviruses	George P Lomonossoff	249
Coronaviruses	Kathryn V Holmes	255
Cowpox virus	D Baxby and M Bennett	261
Coxsackieviruses	Odette G Gaudin	268
Creutzfeldt–Jakob Disease virus **See** Spongiform encephalopathies		
Cricket paralysis virus **See** Picornaviruses – insect		
Cryptoviruses	Robert G Milne and Cristina Marzachi	274

Cucumoviruses	Sue A Tolin	278
Cyanobacteria bacteriophages	Eugene L Martin	285
Cytomegaloviruses – *General features (human)*	Edward S Mocarski Jr	292
Molecular biology (human)	Wade Gibson	299
Animal cytomegaloviruses	Gary S Hayward	304
Murine cytomegaloviruses	John Staczek	307
Cytoplasmic polyhedrosis viruses	Serge Belloncik	312

D

Defective-interfering viruses	Laurent Roux	320
Dengue viruses	Duane J Gubler	324
Densonucleosis viruses	Jacov Tal	331
Diagnostic techniques – *Isolation and identification by microscopy*	M L Landry and G D Hsiung	335
Detection of viral antigens, nucleic acids and specific antibodies	M M M Salimans	343
Dianthoviruses	Steven A Lommel	349
Drosophila C virus **See** Picornaviruses – insect		

E

Eastern equine encephalitis virus **See** Equine encephalitis viruses		
Ebola virus **See** Marburg and Ebola viruses		
Echoviruses	Helena Kopecka	354
Ectromelia virus **See** Mousepox and rabbitpox viruses		
Encephalitis viruses – *Encephalitis viruses and related viruses causing hemorrhagic disease*	James S Porterfield	361
Tick-borne encephalitis, Wesselsbron and simian hemorrhagic fever viruses	Terje Traavik	367
Enteric viruses	Ruth F Bishop	373
Enteroviruses – *Human enteroviruses (serotypes 68–71)*	Marguerite Yin-Murphy	378
Animal enteroviruses	Elizabeth M Hoey and Samuel J Martin	384
Entomopoxviruses	Richard W Moyer	392
Epidemiology of viral diseases	Frederick A Murphy	398
Epstein–Barr virus – *General features*	Lawrence S Young	404
Molecular biology	Jeffery T Sample	410

Equine arteritis virus **See** Lactate dehydrogenase-elevating, equine arteritis and Lelystad viruses		
Equine encephalitis viruses	Diane E Griffin	416
Equine herpesviruses	Dennis J O'Callaghan and Ronald N Harty	423
Equine infectious anemia virus	Ronald C Montelaro	430
Evolution of viruses	Adrian Gibbs	436
Eye infections	John C Hierholzer	441

F

Fabaviruses	J I Cooper	451
Feline calicivirus **See** Vesicular exanthema virus and caliciviruses of pinnipeds, cats and rabbits		
Feline immunodeficiency virus	Niels C Pedersen	454
Feline leukemia and sarcoma viruses	David Onions and James Neil	459
Feline panleukopenia virus **See** Parvoviruses		
Feline sarcoma virus **See** Feline leukemia and sarcoma viruses		
Filamentous bacteriophages	Robert E Webster	464
Fish herpesviruses	Andrew J Davison	470
Fish viruses	Ken Wolf	474
Foamy viruses	Paul A Luciw, Ayalew Mergia and Philip C Loh	480
Foot and mouth disease viruses	David J Rowlands	488
Fowlpox virus	Charles C Randall and Lanelle G Gafford	496
Frog virus 3	Rakesh M Goorha and Allan Granoff	503
Furoviruses	Yukio Shirako and T Michael A Wilson	508

VOLUME 2

G

G4 bacteriophage **See** ØX174 bacteriophage and related bacteriophages		
Geminiviruses	K W Buck	517
Genetics of animal viruses	Frank Fenner	524
Giardiaviruses	Alice L Wang and Ching C Wang	532

Gibbon ape leukemia virus	Marvin S Reitz Jr	535
Goat pox virus **See** Poxviruses		
Gonometa virus **See** Picornaviruses – insect		
Goose parvovirus **See** Parvoviruses		
Granulosis viruses **See** Baculoviruses		
Guanarito virus **See** Lassa, Junin, Machupo and Guanarito viruses		

H

Hantaviruses	Connie S Schmaljohn and Joel M Dalrymple	538
Hare fibroma virus **See** Poxviruses		
Hepatitis A virus	Stanley M Lemon	546
Hepatitis B viruses – *General features (human)*	William S Robinson	554
Molecular biology (human)	Christoph Seeger	560
Avian hepatitis B viruses	William Mason and Patricia Marion	564
Hepatitis C virus	Robert H Purcell	569
Hepatitis delta virus	Michael M C Lai	574
Hepatitis E virus	Daniel W Bradley	580
Herpes simplex viruses – *General features*	Laure Aurelian	587
Molecular biology	Edward K Wagner	593
Virus glycoproteins	Gabriella Campadelli-Fiume	603
Herpesviruses – baboon and chimpanzee	S S Kalter	609
Herpesviruses saimiri and ateles	Jae U Jung and Ronald C Desrosiers	614
Herpesvirus sylvilagus	H A Rouhandeh	621
Herpesviruses 6 and 7	Koichi Yamanishi	624
History of virology – *General*	Frank Fenner	627
Polio, coxsackie, echo and other enteroviruses	Joseph L Melnick	634
Bacteriophages	Donna H Duckworth	642
Hog cholera virus	Jan T van Oirschot	649
Honey bee viruses	Leslie Bailey and Brenda V Ball	654
Hordeiviruses	R G K Donald and Andrew O Jackson	661
Host genetic resistance	David G Brownstein	664
Host-controlled modification and restriction	Detlev H Krüger	669
Human immunodeficiency viruses	Luc Montagnier and Francoís Clavel	674
Human T-cell leukemia viruses 1 – *HTLV-1*	Mitsuaki Yoshida	682
HTLV-2	Joseph D Rosenblatt and Alexander C Black	686

I

Ilarviruses **See** Alfalfa mosaic virus and ilarviruses		
Immune response – *General features*	Gordon L Ada	696
Cell mediated immune response	Peter C Doherty	703
Infectious pancreatic necrosis virus **See** Birnaviruses – animal		
Influenza viruses – *General features*	Robert G Webster	709
Molecular biology	Peter Palese and Adolfo Garciá-Sastre	715
Structure of antigens	Jonathan W Yewdell, Jack R Bennink and W Graeme Laver	722
Interference	Julius S Youngner and Patricia Whitaker-Dowling	728
Interferons – *General features*	Philip I Marcus	733
Interferons – Therapy of AIDS and cancer	Susan E Krown, Paul S Ritch and Ernest C Borden	739
Isfahan virus **See** Chandipura, Piry and Isfahan viruses		

J

Japanese encephalitis virus	Akira Igarashi	746
JC and BK viruses	Richard J Frisque	752
Junin virus **See** Lassa, Junin, Machupo and Guanarito viruses		

K

Killer virus system **See** Totiviruses and Yeast RNA viruses		
Kuru	Carlo Masullo	758
Kyasanur Forest disease virus **See** Encephalitis viruses		

L

Lactate dehydrogenase-elevating, equine arteritis and Lelystad viruses	Margo A Brinton	763
Lambda bacteriophage	Allan M Campbell	772
Lassa, Junin, Machupo and Guanarito viruses	Joseph B McCormick	776
Latency	Jack G Stevens	787
Lelystad virus **See** Lactate dehydrogenase-elevating, equine arteritis and Lelystad viruses		

Lumpy skin disease virus **See** Poxviruses		
Luteoviruses	W Allen Miller	792
Lymphocystis disease virus	Gholamreza Darai and Angela Rösen-Wolff	798
Lymphocytic choriomeningitis – *General features*	Raymond M Welsh	801
Molecular biology	Peter J Southern and Barbara J Meyer	806
Lymphoproliferative disease virus of turkeys	Arnona Gazit and Abraham Yaniv	811
Lysogeny and prophage	Max Gottesman and Amos Oppenheim	814

M

Machupo virus **See** Lassa, Junin, Machupo and Guanarito viruses		
Maize chlorotic dwarf virus	Roy E Gingery	824
Mammalian hepadnaviruses **See** Hepatitis B viruses		
Marburg and Ebola viruses	Hans-Dieter Klenk, Werner Slenczka and Heinz Feldmann	827
Marek's disease virus	L N Payne	832
Mayaro virus **See** Chikungunga, O'nyong nyong and Mayaro viruses		
Measles virus	Claes Örvell	838
Mink enteritis virus **See** Parvoviruses		
Molluscum contagiosum virus	Colin D Porter and L C Archard	848
Monkeypox virus **See** Smallpox and monkeypox viruses		
Mouse mammary tumor virus	Jaquelin Dudley	853
Mousepox and rabbitpox viruses	Frank Fenner	861
Mu and related bacteriophages	Michael S DuBow	868
Mumps virus	Bertus K Rima	876
Murine leukemia viruses	Hung Fan	883
Murine parvoviruses **See** Parvoviruses		
Murray Valley encephalitis **See** Encephalitis viruses		
Myxoma virus **See** Poxviruses		

N

N4 Bacteriophage	Lucia B Rothman-Denes	891

Necroviruses	F Meulewaeter, X Danthinne and J van Emmelo	896
Nepoviruses	M A Mayo	901
Nervous system viruses	Richard T Johnson	907
Newcastle disease virus	Peter T Emmerson	914
Nodaviruses	L Andrew Ball	919
Nonoccluded baculoviruses **See** Baculoviruses		
Norwalk and related viruses	Albert Z Kapikian and Mary K Estes	925
Nuclear polyhedrosis viruses **See** Baculoviruses		

O

O'nyong-nyong virus **See** Chikungunga, O'nyong nyong and Mayaro viruses		
Omsk hemorrhagic fever virus **See** Encephalitis viruses		
Oncogenes	Martine F Roussel	934
Orbiviruses and coltiviruses – *General features*	P P C Mertens	941
Molecular biology	Polly Roy	956
Organ system infections	Jangu E Banatvala and Felicity Nicholson	964

P

Ø6 bacteriophage	Dennis H Bamford	978
Ø29 bacteriophage	Margarita Salas	980
ØX174 bacteriophage and related bacteriophages	David T Denhardt	989
P1 bacteriophage	Jürg Meyer	997
P2, P4 bacteriophage and related bacteriophages	J Barry Egan and Ian B Dodd	1003
P22 bacteriophage	Anthony R Poteete	1009
Papillomaviruses – human – *General features*	Gérard Orth	1013
Infections of the anogenital tract	Harald zur Hausen	1021
Parainfluenza viruses – *Human parainfluenza viruses*	Allen Portner	1027
Animal parainfluenza viruses	Hiroshi Shibuta	1031
Parapoxviruses	Robert Wyler	1036
Parsnip yellow fleck virus group	A F Murant	1042

VOLUME 3

Entry	Author(s)	Page
Partitiviruses	Said A Ghabrial	1047
Parvoviruses – *General features*	John R Pattison	1052
Molecular biology	Kenneth I Berns	1057
Cats, dogs and mink	Colin R Parrish	1061
Rodents, pigs, cattle and geese	Peter Tijssen and J Bergeron	1067
Pathogenesis	Kenneth L Tyler	1076
Pea enation mosaic virus	Steven A Demler and Gustaaf A de Zoeten	1083
Persistent viral infection	Rafi Ahmed	1089
Peste des Petits ruminants **See** Rinderpest and distemper viruses		
Phocid distemper virus **See** Rinderpest and distemper viruses		
Phytoreoviruses	Donald L Nuss	1095
Picornaviruses – insect	Paul D Scotti	1100
Piry virus **See** Chandipura, Piry and Isfahan viruses		
Plant resistance to viruses	R S S Fraser	1104
Plant virus disease – economic aspects	O W Barnett	1109
Polioviruses – *General features*	Philip D Minor	1115
Molecular biology	Caroline Mirzayan and Eckard Wimmer	1119
Polydnaviruses	Don Stoltz	1133
Polyomaviruses murine – *General features*	James M Pipas	1135
Molecular biology	Walter Eckhart	1139
Potexviruses	K Andrew White, Michèle Rouleau, J B Bancroft and George A Mackie	1142
Potyviruses	John A Lindbo and William G Dougherty	1148
Poxviruses – *Rabbit, hare, squirrel and swine poxviruses*	Grant McFadden	1153
Sheep and goat poxviruses	R Paul Kitching	1160
PRD1 bacteriophage	Dennis H Bamford	1165
Prions **See** Spongiform encephalopathies		
Propagation of viruses	Steffen Faisst	1168
Pseudorabies virus	Saul Kit	1173

R

Entry	Author(s)	Page
Rabbit fibroma virus **See** Poxviruses		

Rabbit hemorrhagic disease virus **See** Vesicular exanthema virus and caliciviruses of pinnipeds, cats and rabbits		
Rabbitpox virus **See** Mousepox and rabbitpox viruses and Poxviruses		
Rabies virus	George M Baer and Noël Tordo	1180
Rabies-like viruses	Robert E Shope	1186
Reoviruses – *General features*	Bernard N Fields	1190
Molecular biology	W K Joklik	1194
Replication of viruses	V Gregory Chinchar	1203
Respiratory syncytial virus	Peter L Collins	1210
Respiratory viruses	David O White	1219
Reticuloendotheliosis viruses	Paula J Enrietto	1227
Retrotransposons of fungi	Jef D Boeke	1232
Retroviruses – type D	Maja A Sommerfelt, Sung S Rhee and Eric Hunter	1236
Rhabdoviruses – *Plant rhabdoviruses*	Dick Peters	1243
Ungrouped mammalian, bird and fish rhabdoviruses	Sailen Barik and Amiya K Banerjee	1249
Rhinoviruses	Glyn Stanway	1253
Rinderpest and distemper viruses	Tom Barrett	1260
Ross River virus	Lynn Dalgarno and Ian D Marshall	1268
Rotaviruses – *General features*	Robert D Shaw and Harry B Greenberg	1274
Molecular biology	Mary K Estes	1281
Rubella virus	Jerry S Wolinsky	1291
Russian spring summer encephalitis **See** Encephalitis viruses		

S

S13m bacteriophage **See** Ø174 bacteriophage and related bacteriophages		
San Miguel sea lion virus **See** Vesicular exanthema virus and caliciviruses of pinnipeds, cats and rabbits		
Scrapie **See** Spongiform encephalopathies		
Semliki Forest virus **See** Sindbis and Semliki Forest viruses		
Sendai virus	Morio Homma and Masato Tashiro	1299

Sheep poxvirus **See** Poxviruses		
Shope fibroma virus **See** Poxviruses		
Shope papilloma and bovine papillomaviruses	Gary L Bream and William C Phelps	1305
Sigma rhabdoviruses	Danielle Teninges	1311
Simian hemorrhagic fever virus **See** Encephalitis viruses and Lactate dehydrogenase-elevating, equine arteritis and Lelystad viruses		
Simian herpesvirus **See** Herpesviruses saimiri and ateles		
Simian immunodeficiency viruses	Hilary G Morrison and Ronald C Desrosiers	1316
Simian virus 40	Janet S Butel	1322
Sindbis and Semliki Forest viruses	Milton J Schlesinger	1330
Single-stranded RNA bacteriophages	Jan van Duin	1334
Smallpox and monkeypox viruses	Keith Dumbell	1339
Sobemoviruses	O P Sehgal	1346
SPO1 bacteriophage	Charles R Stewart	1352
Spongiform encephalopathies – *Mule deer, elk and bovine*	Maurizio Pocchiari	1357
Creutzfeldt–Jakob disease, scrapie and related transmissible encephalopathies	Laura Manuelidis and Elias E Manuelidis	1361
Squirrel fibroma virus **See** Poxviruses		
St Louis encephalitis virus **See** Encephalitis viruses		
Swine herpesvirus-1 **See** Pseudorabies virus		
Swine vesicular exanthema virus **See** Vesicular exanthema virus and caliciviruses of pinnipeds, cats and rabbits		
Swinepox virus **See** Poxviruses		

T

T1 bacteriophage	J R Christensen	1371
T4 bacteriophage and related bacteriophages	Gisela Mosig	1376
T5 bacteriophage and related bacteriophages	D James McCorquodale	1384
T7 bacteriophage	Ian J Molineux	1388
Tanapox virus **See** Yabapox and Tanapox viruses		
Taxonomy and classification – general	Claude M Fauquet	1396
Tenuiviruses	Bryce Falk	1410

Tetraviruses	Donald Hendry and Deepak Agrawal	1416
Theiler's viruses	Howard L Lipton	1423
Tick-borne encephalitis virus **See** Encephalitis viruses		
Tipula iridescent virus	James Kalmakoff	1430
Tobamoviruses	Dennis J Lewandowski and William O Dawson	1436
Tobraviruses	Alexander Mathis and Huub J M Linthorst	1442
Tombusviruses	D M Rochon	1447
Toroviruses	Marian C Horzinek	1452
Tospoviruses	Peter de Haan	1459
Totiviruses – *General features*	Said A Ghabrial	1464
Ustilago maydis viruses	Aliza Finkler and Yigal Koltin	1468
Transformation	Ron Wisdom and Inder M Verma	1472
Transplantation	Helen E Heslop, Robert A Krance and Malcolm K Brenner	1477
Transposable bacteriophages **See** Mu and related bacteriophages		
Tree shrew herpesviruses	Gholamreza Darai and Angela Rösen-Wolff	1489
Tumor viruses – human	Herbert Pfister and Bernhard Fleckenstein	1492
Turkey herpesvirus **See** Marek's disease virus		
Tymoviruses	Adrian Gibbs	1500
Ty elements **See** Retrotransposons of fungi		

U/V

Ustilago maydis viruses **See** Totiviruses		
Vaccines and immune response	Gordon L Ada	1503
Vaccinia virus	Riccardo Wittek	1507
Varicella-zoster virus – *General features*	Jeffrey I Cohen and Steven E Straus	1514
Molecular biology	William T Ruyechan and John Hay	1518
Variola virus **See** Smallpox and monkeypox viruses		
Vectors – *Animal viruses*	James Tartaglia, Russell Gettig and Enzo Paoletti	1528

Plant viruses	Thomas Hohn and Rob Goldbach	1536
Venezuelan equine encephalitis virus **See** Equine encephalitis viruses		
Vesicular exanthema virus and caliciviruses of pinnipeds, cats and rabbits	Michael J Studdert	1544
Vesicular stomatitis viruses	Stuart T Nichol	1547
Viral membranes	John Lenard	1556
Viral receptors	Horacio U Saragovi, Gordon J Sauvé and Mark I Greene	1560
Viroids	Robert A Owens	1563
Virus structure – *Atomic structure*	Ming Luo	1571
Principles of virus structure	John E Johnson and Andrew J Fisher	1573
Virus–host cell interactions	Patricia Whitaker-Dowling and Julius S Youngner	1587
Visna-Maedi viruses	Opendra Narayan	1592

W

Wesselsbron virus **See** Encephalitis viruses
West Nile encephalitis virus **See** Encephalitis viruses
Western equine encephalitis virus **See** Equine encephalitis viruses

Y

Yabapox and Tanapox viruses	H A Rouhandeh	1597
Yeast RNA viruses	Reed B Wickner	1600
Yellow fever virus	Thomas P Monath	1606

Z

Zoonoses	Thomas M Yuill	1613

Appendix: Viruses included in the Encyclopedia
Index

Contributors

HANS-WOLFGANG ACKERMANN
Department of Microbiology
Laval University
Quebec
G1K 7P4 Canada

GORDON ADA
Department of Microbiology
John Curtin School of Medical Research
The Australian National University
Canberra, ACT 2601
Australia

DEEPAK AGRAWAL
Department of Biological Sciences
Purdue University
West Lafayette
IN 47907, USA

PAUL AHLQUIST
Institute for Molecular Virology
University of Wisconsin
Madison
WI 53706-1596, USA

RAFI AHMED
Department of Microbiology and Immunology
UCLA School of Medicine
Center for Health Sciences
Los Angeles
CA 90024, USA

WERNER ARBER
Biocentrum der Universität, Basel
Abteilung Mikrobiologie
CH-4056 Basel, Switzerland

L C ARCHARD
Department of Biochemistry
Charing Cross and Westminster Medical School
London
W6 8RF, UK

LAURE AURELIAN
Virology/Immunology Laboratories
Department of Pharmacology and Experimental Therapeutics
University of Maryland
School of Medicine
Baltimore, MD 21201, USA

GEORGE M BAER
Socrates 402-601
Colonia Polanco
Mexico DF 11510, Mexico

L BAILEY (Retired)
AFRC Institute of Arable Crops Research
Rothamsted Experimental Station
Harpenden, Herts, UK

B V BALL
AFRC Institute of Arable Crops Research
Rothamsted Experimental Station
Harpenden, Herts, UK

L ANDREW BALL
Department of Microbiology
University of Alabama at Birmingham
Birmingham, AL 35294, USA

DENNIS H BAMFORD
Department of Genetics
University of Helsinki
SF-00100, Finland

JANGU E BANATVALA
Department of Virology
United Medical and Dental Schools
St Thomas' Hospital
London, SE1 7EH, UK

J B BANCROFT
Department of Biochemistry
The University of Western Ontario
London
Ontario N6A 5B7, Canada

AMIYA K BANERJEE
Department of Molecular Biology
The Cleveland Clinic Foundation
Cleveland, OH 44195, USA

SAILEN BARIK
Department of Molecular Biology
The Cleveland Clinic Foundation
Cleveland
OH 44195, USA

O W BARNETT
Department of Plant Pathology
North Carolina State University
Raleigh
NC 27695, USA

TOM BARRETT
Molecular Biology Department
AFRC Institute for Animal Health
Pirbright Laboratory
Woking
GU24 0NF, UK

D BAXBY
Department of Medical Microbiology
University of Liverpool
Liverpool L69 3BX, UK

HERMANN BECHT
Institut für Virologie
Justus-Liebig-Universität Giessen
D-6300 Giessen, Germany

SERGE BELLONCIK
Centre de Recherche en Virologie
University du Quebec
Institute Armand Frappier
Laval
Quebec H7V 1B7, Canada

M BENNETT
Department of Veterinary Clinical Science
University of Liverpool
Liverpool L69 3BX, UK

JACK R BENNINK
Laboratory of Viral Diseases
NIAID-NIH
Bethesda
MD 20892, USA

J BERGERON
Institut Armand-Frappier
Université du Québec
Laval
Québec H7N 4Z3, Canada

KENNETH I BERNS
Hearst Microbiology Research Center
Department of Microbiology
Cornell University Medical College
New York
NY 10021, USA

RUTH F BISHOP
Department of Gastroenterology
Royal Children's Hospital
Parkville
Victoria 3052, Australia

ALEXANDER C BLACK
Department of Medicine
UCLA School of Medicine, UCLA AIDS
Institute and Jonsson Comprehensive Cancer Center
Los Angeles
CA 90024, USA

JEF D BOEKE
Department of Molecular Biology and Genetics
The Johns Hopkins University School of Medicine
Baltimore
MD 21205, USA

J F BOL
Gorlaeus Laboratories
Leiden University
2333 CC Leiden, The Netherlands

ERNEST C BORDEN
Cancer Center
Medical College of Wisconsin
Milwaukee
WI 53226, USA

DANIEL W BRADLEY
Virology Laboratory Section
Hepatitis Branch
Centers For Disease Control
Atlanta
GA 30333, USA

GARY L BREAM
Division of Virology
Burroughs Wellcome Co.
Research Triangle Park
NC 27709, USA

MALCOLM BRENNER
Department of Hematology and Oncology
St Jude Children's Research Hospital
Memphis
TN 38101, USA

MARGO A BRINTON
Department of Biology
Georgia State University
Atlanta
GA 30302, USA

DAVID G BROWNSTEIN
Section of Comparative Medicine
Yale University School of Medicine
New Haven
CT 06510, USA

K W BUCK
Department of Biology
Imperial College of Science, Technology and Medicine
London, SW7 2BB, UK

JANET S BUTEL
Department of Molecular Virology
Baylor College of Medicine
Houston
TX 77030, USA

CHARLES H CALISHER
Arthropod-borne Infectious Diseases Laboratory
Colorado State University
Fort Collins
CO 80523, USA

GABRIELLA CAMPADELLI FIUME
Sezione di Microbiologia e Virologia
Dipartimento di Patologia Sperimentale
Università di Bologna
40126 Bologna, Italy

ALLAN M CAMPBELL
Department of Biological Sciences
Stanford University
Stanford
CA 94305, USA

THIERRY CANDRESSE
Station de Pathologie Végétale
Institut National de la Recherche Agronomique
33883 Villenave d'Ornon Cedex, France

V GREGORY CHINCHAR
Department of Microbiology
University of Mississippi
Medical Center
Jackson
MS 39216, USA

J R CHRISTENSEN
Department of Microbiology and Immunology
University of Rochester Medical Center
Rochester
NY 14642, USA

FRANÇOIS CLAVEL
Institut Pasteur
Viral Oncology Unit
75724 Paris Cedex 15, France

JOHN M COFFIN
Department of Molecular Biology
and Microbiology
Tufts University
Boston
MA 02111, USA

JEFFREY I COHEN
Medical Virology Section
Laboratory of Clinical Investigation
NIAID-NIH
Bethesda
MD 20892, USA

PETER L COLLINS
Laboratory of Infectious Diseases
NIAID-NIH
Bethesda
MD 20892, USA

J I COOPER
Department of Plant Virology
Natural Environment Research Council
Institute of Virology and Environmental Microbiology
Oxford
OX1 3SR, UK

ALLAN M CRAWFORD
Department of Biochemistry, Molecular Biology Unit
University of Otago
Dunedin, New Zealand

NORMAN E CROOK
Horticulture Research International
Littlehampton
West Sussex
BN17 6LP, UK

SAMUEL DALES
Department of Microbiology and Immunology
University of Western Ontario
Health Sciences Center
London, Ontario N6A 5C1, Canada

LYNN DALGARNO
Division of Biochemistry and Molecular Biology
School of Life Sciences
The Australian National University
Canberra, ACT 2601, Australia

JOEL M DALRYMPLE (Deceased)
USAMRIID
Ft Detrick
Frederick
MD 21701, USA

X DANTHINNE
Plant Genetic Systems
9000 Ghent, Belgium

GHOLAMREZA DARAI
Institut für Medizinisch Virologie
der Universität Heidelberg
6900 Heidelberg, Germany

ANDREW J DAVISON
MRC Virology Unit
Institute of Virology
Glasgow
G11 5JR, Scotland

WILLIAM O DAWSON
University of Florida
Citrus Research and Education Center
Lake Alfred
FL 33850, USA

PETER DE HAAN
Department of Biotechnology
Zaadunie BV
1600 AA Enkhuizen, The Netherlands

STEVEN A DEMLER
Department of Botany and Plant Pathology
Michigan State University
East Lansing
MI 48824, USA

DAVID T DENHARDT
Department of Biological Sciences
Rutgers University
Piscataway
NJ 08855, USA

RONALD C DESROSIERS
New England Regional Primate Research Center
Harvard Medical School
Southborough
MA 01772, USA

GUSTAAF A DE ZOETEN
Department of Botany and Plant Pathology
Michigan State University
East Lansing
MI 48824, USA

IAN B DODD
Department of Biochemistry
University of Adelaide
Adelaide
South Australia 5000, Australia

WALTER DOERFLER
Institute for Genetics
University of Cologne
D-5000 Cologne 41, Germany

PETER C DOHERTY
Department of Immunology
St Jude Children's Research Hospital
Memphis, TN 38101, USA

R G K DONALD
Department of Plant Pathology
University of California, Berkeley
College of Natural Resources
Berkeley
CA 94720, USA

WILLIAM G DOUGHERTY
Department of Microbiology
Oregon State University
Corvallis
OR 97331, USA

EDWARD J DUBOVI
Diagnostic Laboratory
College of Veterinary Medicine
Cornell University
Ithaca
New York, 14853, USA

MICHAEL S DUBOW
Department of Microbiology and Immunology
McGill University
Montreal
Quebec
H3A 2B4 Canada

DONNA H DUCKWORTH
Department of Immunology and Medical Microbiology
University of Florida
College of Medicine
Gainesville, FL 32610, USA

JACQUELIN DUDLEY
Department of Microbiology
University of Texas at Austin
TX 78712, USA

KEITH DUMBELL (Retired)
Department of Medical Microbiology
University of Cape Town Medical School
Cape 7925, South Africa

JACKY ECCLESTON
Department of Biology
University of Birmingham
Birmingham, UK

WALTER ECKHART
Arm and Hammer Center for Cancer Biology
The Salk Institute
San Diego, CA 92186, USA

J BARRY EGAN
Department of Biochemistry
The University of Adelaide
Adelaide
South Australia 5000, Australia

RICHARD M ELLIOTT
Institute of Virology
University of Glasgow
Glasgow G11 5JR, Scotland

PAULA J ENRIETTO
Department of Microbiology
State University of New York
at Stony Brook
Health Sciences Center
Stony Brook
NY 11794-5222, USA

PETER T EMMERSON
Department of Biochemistry and Genetics
Medical School
University of Newcastle upon Tyne
Newcastle upon Tyne
NE2 4HH, UK

MARY K ESTES
Division of Molecular Virology
Baylor College of Medicine
Houston
TX 77030, USA

STEFFEN FAISST
Angewandte Tumorvirologie 0610
Deutsches Krebstorschlingszentrum
im Neuenheimer Feld 242
69009 Heidelberg, Germany

BRYCE FALK
Department of Plant Pathology
College of Agricultural and Environmental Sciences
University of California
Davis
CA 95616, USA

HUNG FAN
Department of Molecular Biology and Biochemistry
University of California
Irvine
CA 92717, USA

CLAUDE M FAUQUET
1CTV Secretary
Division of Plant Biology
The Scripps Research Institute
La Jolla
CA 92057, USA

BRIAN A FEDERICI
Department of Entomology
University of California
Riverside
CA 92521, USA

HEINZ FELDMANN
Institüt für Virologie
der Philipps-Universität
3550 Marburg, Germany

FRANK FENNER
John Curtin School of Medical Research
The Australian National University
Canberra
ACT 2601, Australia

A KIRK FIELD
Hybridon Inc.
Worcester
MA 01605, USA

BERNARD N FIELDS
Department of Microbiology and Molecular Genetics
Harvard Medical School
Boston
MA 02115, USA

ALIZA FINKLER
Department of Molecular Microbiology and Biotechnology
Tel Aviv University
Ramat Aviv 69978, Israel

ANDREW J FISHER
Department of Biological Sciences
Purdue University
West Lafayette
IN 47907, USA

BERNHARD FLECKENSTEIN
Institut für Klinische und Molekulare Virologie
der Friedrich-Alexander-Universität
D-8520 Erlangen, Germany

R S S FRASER
Horticultural Research International
Littlehampton
West Sussex
BN17 6LP, UK

RICHARD J FRISQUE
Department of Molecular and Cell Biology
Pennsylvania State University
University Park
PA 16802, USA

ROBERT S FUJINAMI
Department of Neurology
University of Utah School of Medicine
Salt Lake City
UT 84132, USA

LANELLE G GAFFORD
Department of Microbiology
University of Mississippi Medical Center
Jackson
MS 39216, USA

ADOLFO GARCIÁ-SASTRE
Department of Microbiology
Mount Sinai School of Medicine
New York
NY 10029, USA

ODETTE G GAUDIN
Laboratoire di Bacteriologie-Virologie
CHU de Saint Etienne
Hopital Nord
42055 Saint Etienne CEDEX 2, France

ARNONA GAZIT
Department of Human Microbiology
Sackler School of Medicine
Tel-Aviv University
Tel-Aviv 69778, Israel

RUSSELL GETTIG
Virogenetics Corporation
Rensselaer Technology Park
Troy
NY 12180, USA

SAID A GHABRIAL
Department of Plant Pathology
University of Kentucky College of Agriculture
Lexington
KY 40546, USA

ADRIAN GIBBS
Research School of Biological Sciences
The Australian National University
ACT 2601, Australia

WADE GIBSON
Department of Pharmacology and Molecular Sciences
Johns Hopkins University School of Medicine
Baltimore, MD 21205, USA

ROY E GINGERY
US Department of Agriculture
Agricultural Research Service
Department of Plant Pathology
Ohio State University - Agricultural Research and Development Center
Wooster OH 44691, USA

ROB GOLDBACH
Department of Virology
Agricultural University
6760 EM Wageningen, The Netherlands

MATTHEW A GONDA
Laboratory of Cell and Molecular Structure
Program Resources, Inc./DynCorp
NCI-Frederick Cancer Research & Development Center
Frederick, MD 21702, USA

FRANCISCO GONZALEZ-SCARANO
Department of Neurology
University of Pennsylvania
School of Medicine
Philadelphia, PA 19104, USA

RAKESH M GOORHA
Department of Virology and Molecular Biology
St Jude Children's Research Hospital
Memphis, TN 38101, USA

MAX GOTTESMAN
Institute of Cancer Research
Columbia University
College of Physicians and Surgeons
New York
NY 10032, USA

ALLAN GRANOFF
Department of Virology and Molecular Biology
St Jude Children's Research Hospital
Memphis, TN 38101, USA

MARK I GREENE
Department of Pathology and Laboratory Medicine
University of Pennsylvania
School of Medicine
Philadelphia
PA 19104, USA

HARRY B GREENBERG
Division of Gastroenterology
Stanford University School of Medicine
Stanford
CA 94305, USA

DIANE E GRIFFIN
Department of Neurology
Johns Hopkins University School of Medicine
Baltimore, MD 21205, USA

DUANE J GUBLER
Division of Vector-Borne Infectious Diseases
Centers for Disease Control
Fort Collins
CO 80522, USA

D L HACKER
School of Biological Sciences
University of Nebraska
Lincoln
NE 68588-0118, USA

RONALD N HARTY
Department of Microbiology and Immunology
Louisiana State University Medical Center
Shreveport
Louisiana 71130, USA

JOHN HAY
Department of Microbiology
State University of New York at Buffalo
Buffalo, NY 14214, USA

GARY S HAYWARD
Department of Pharmacology and
Department of Oncology
Johns Hopkins University School of Medicine
Baltimore, MD 21205, USA

H ERNEST HEMPHILL
Department of Biology
Syracuse University
Syracuse
NY 13244, USA

DONALD HENDRY
Department of Biochemistry and Microbiology
Rhodes University
Grahamstown 6140, South Africa

HELEN E HESLOP
Department of Hematology and Oncology
St Jude Children's Research Hospital
Memphis
TN 38101, USA

JOHN C HIERHOLZER
Respiratory Virus Section
Respiratory and Enteric Viruses Branch
Centers for Disease Control
Atlanta
GA 30333, USA

ELIZABETH M HOEY
School of Biology and Biochemistry
The Queen's University of Belfast
Medical Biology Centre
Belfast
BT9 7BL, Northern Ireland

THOMAS HOHN
Friedrich Miescher Institute
CH-4002, Basel, Switzerland

KATHRYN V HOLMES
Department of Pathology
Uniformed Services University of the
Health Sciences
Bethesda, MD 20814, USA

RANDALL K HOLMES
Department of Microbiology
Uniformed Services University of the Health Sciences
Bethesda, MD 20814, USA

MORIO HOMMA
Department of Virology
Kobe University School of Medicine
Kobe and Jichi Medical School
Tochigi, Japan

MARIAN C HORZINEK
Vakgroep Infectieziekten en Immunologie
Faculteit Diergeneeskunde
Rijksuniversitateit te Utrecht
3508 TD Utrecht, The Netherlands

GUEH-DJEN HSIUNG
Virology Laboratory
Veterans Administration Medical Center
Yale University
School of Medicine
West Haven, CT 06516, USA

ERIC HUNTER
Department of Microbiology
The University of Alabama at Birmingham
Birmingham
AL 35294, USA

AKIRA IGARASHI
Department of Virology
Institute of Tropical Medicine
Nagasaki University
Nagasaki, Japan

A O JACKSON
Department of Plant Pathology
University of California Berkeley
College of Natural Resources
Berkeley
CA 94720, USA

AUDREY JARVIS
Microbial Genetics Section
New Zealand Dairy Research Institute
Palmerston North, New Zealand

E M J JASPARS
Gorlaeus Laboratories
Leiden University
Leiden, The Netherlands

JOHN E JOHNSON
Department of Biological Sciences
Purdue University
West Lafayette, IN 47907, USA

RICHARD T JOHNSON
Department of Neurology
Johns Hopkins University School of Medicine
Baltimore
MD 21287, USA

W K JOKLIK
Department of Microbiology
Duke University Medical Center
Durham
NC 27710, USA

JAE U JUNG
New England Regional Primate Research Center
Harvard Medical School
Southborough
MA 01772, USA

JAMES KALMAKOFF
Department of Microbiology
University of Otago
Dunedin, New Zealand

S S KALTER
Virus Reference Laboratory Inc.
San Antonio
TX 78229, USA

ALBERT Z KAPIKIAN
Laboratory of Infectious Diseases
National Institute of Allergy and
Infectious Diseases
National Institutes of Health
Bethesda
MD 20892, USA

SAUL KIT
Novagene Inc.
Houston
TX 77024, USA

R PAUL KITCHING
World Reference Laboratory for
Foot and Mouth Disease
Pirbright Laboratory
AFRC Institute for Animal Health
Pirbright
GU24 0NF, UK

GEOFFREY R KITCHINGMAN
Department of Virology and Molecular Biology
St Jude Children's Research Hospital
Memphis, TN 38101, USA

HANS-DIETER KLENK
Institut für Virologie
der Philipps-Universität
3550 Marburg, Germany

YIGAL KOLTIN
Department of Molecular Microbiology and Biotechnology
Tel Aviv University
Ramat Aviv 69978, Israel

HELENA KOPECKA
Unité de Virologie Moleculaire
Institut Pasteur
75724 Paris-Cedex, France

ROBERT A KRANCE
Department of Hematology and Oncology
St Jude Children's Research Hospital
Memphis
TN 38101, USA

KENNETH N KREUZER
Department of Microbiology
Duke University Medical Center
Durham, NC 27710, USA

SUSAN E KROWN
Department of Medicine
Memorial Sloan-Kettering Cancer Center
New York, NY 10021, USA

DETLEV H KRÜGER
Institute of Medical Virology
Humboldt University School of Medicine
D-10117 Berlin, Germany

MICHAEL M C LAI
Howard Hughes Medical Institute
Department of Microbiology
University of Southern California
School of Medicine
Los Angeles
CA 90033, USA

MARIE L LANDRY
Department of Laboratory Medicine
Yale University School of Medicine
New Haven, CT 06516;
Clinical Virology Laboratory
Yale–New Haven Hospital
New Haven, CT 06516;
Virology Reference Library
Veterans Administration Medical Center
West Haven, CT 06516, USA

W GRAEME LAVER
Influenza Research Unit
John Curtin School of Medical Research
The Australian National University
Canberra, ACT 2601, Australia

STANLEY M LEMON
Department of Medicine, Microbiology and Immunology
The University of North Carolina at Chapel Hill
Chapel Hill
NC 27599, USA

JOHN LENARD
Department of Physiology and Biophysics
Robert Wood Johnson Medical School
University of Medicine and Dentistry of New Jersey
Piscataway
NJ 08854, USA

ARNOLD J LEVINE
Department of Molecular Biology
Princeton University
Princeton
NJ 08544-1014, USA

DENNIS J LEWANDOWSKI
University of Florida
Citrus Research and Education Center
Lake Alfred
FL 33850, USA

JOHN A LINDBO
Department of Microbiology
Oregon State University
Corvallis
OR 97331-3804, USA

HOWARD L LIPTON
Division of Neurology
Evanston Hospital
Evanston
IL 60201, USA

HUUB J M LINTHORST
Institute of Molecular Plant Sciences
Gorlaeus Laboratories
PO Box 9502
2300 RA Leiden, The Netherlands

BEN E LOCKHART
Department of Plant Pathology
University of Minnesota
St Paul
MN 55108, USA

PHILIP C LOH
Department of Microbiology
University of Hawaii at Manoa
Honolulu
HI 96822, USA

STEVEN A LOMMEL
Department of Plant Pathology
North Carolina State University
College of Agriculture and Life Sciences
Raleigh
NC 27675, USA

GEORGE P LOMONOSSOFF
Department of Virus Research
John Innes Institute
Norwich
NR4 7UH, UK

PAUL A LUCIW
Department of Medical Pathology
University of California
Davis
CA 95616, USA

MING LUO
Center for Macromolecular Crystallography
The University of Alabama
at Birmingham
Birmingham
AL 35294, USA

ALICE LUSTIG
Department of Molecular Biology
Princeton University
Princeton
NJ 08544, USA

DMITRI K LVOV
Academy of Medical Sciences
The D.I. Ivanovsky Institute of Virology
Moscow 123098, Russia

GEORGE A MACKIE
Department of Biochemistry
The University of Western Ontario
London
Ontario N6A 5B7, Canada

JACK MANILOFF
Department of Microbiology and
Immunology
University of Rochester
Medical Center
Rochester
NY 14642, USA

ELIAS E MANUELIDIS (Deceased)
Section of Neuropathology
Yale University School of Medicine
New Haven
CT 06510, USA

LAURA MANUELIDIS
Section of Neuropathology
Yale University School of Medicine
New Haven
CT 06510, USA

PHILIP I MARCUS
Department of Molecular and Cell Biology
The University of Connecticut
Storrs
CT 06269, USA

PATRICIA MARION
Division of Infectious Diseases
Stanford University Medical Center
Stanford
CA 94305, USA

IAN D MARSHALL
Division of Biochemistry and Molecular Biology
School of Life Sciences
The Australian National University
Canberra, ACT 2601, Australia

EUGENE L MARTIN
Department of Biological Sciences
University of Nebraska
Lincoln
NE 68588, USA

SAMUEL J MARTIN
School of Biology and Biochemistry
The Queen's University of Belfast
Medical Biology Centre
Belfast
BT9 7BL, Northern Ireland

CRISTINA MARZACHI
Consiglio Nazionale delle Ricerche
Istituto di Fitovirologia Applicata
Strada delle Cacce, 73
10135 Torino, Italy

WILLIAM MASON
Institute for Cancer Research
Fox Chase Cancer Center
Philadelphia 19111, USA

CARLO MASULLO
Institute of Neurology
Catholic University School of Medicine
00168 Rome, Italy

ALEXANDER MATHIS
Institute of Molecular Plant Sciences
Gorlaeus Laboratories
2300 RA Leiden, The Netherlands

M A MAYO
Scottish Crop Research Institute
Invergowrie
Dundee
DD2 5DA, Scotland

JOSEPH B McCORMICK
Division of Parasitic Diseases
Centers for Disease Control
Atlanta
GA 30333, USA

D JAMES McCORQUODALE
Department of Biochemistry and Molecular Biology
Medical College of Ohio
Toledo
OH 436998, USA

GRANT McFADDEN
Department of Biochemistry
University of Alberta
Edmonton
Alberta
T6G 2H7, Canada

JOSEPH L MELNICK
Division of Molecular Virology
Baylor College of Medicine
Houston
Texas 77030, USA

AYALEW MERGIA
Department of Medical Pathology
University of California
Davis
CA 95616, USA

P P C MERTENS
Division of Molecular Biology
Pirbright Laboratory
AFRC Institute for Animal Health
Pirbright
Woking
Surrey, UK

F. MEULEWAETER
Laboratory of Genetics
University of Ghent
9000 Ghent, Belgium

JÜRG MEYER
Oral Microbiology
University of Basel Dental Institute
4051 Basel, Switzerland

BARBARA J MEYER
Department of Microbiology
University of Minneapolis
Minneapolis
MN 55455, USA

W ALLEN MILLER
Department of Plant Pathology
Iowa State University
Ames
IA 50011, USA

ROBERT G MILNE
Consiglio Nazionale delle Ricerche
Istituto di Fitovirologia Applicata
10135 Torino, Italy

PHILIP D MINOR
Division of Virology
National Institute for Biological Standards and Control
South Mimms
Potters Bar
Herts
EN6 3QG, UK

CAROLINE MIRZAYAN
Department of Microbiology
State University of New York at Stony Brook
Health Sciences Center
Stony Brook
NY 11794, USA

EDWARD S MOCARSKI, Jr
Department of Microbiology and Immunology
Stanford University School of Medicine
Stanford
CA 94305, USA

IAN J MOLINEUX
Department of Microbiology
The University of Texas at Austin
Austin
TX 78712, USA

THOMAS P MONATH
OraVax, Inc.
Cambridge
MA 02139, USA

LUC MONTAGNIER
Institut Pasteur
Viral Oncology Unit
F-75724 Paris Cedex 15, France

RONALD C MONTELARGO
Department of Molecular Genetics
and Biochemistry
University of Pittsburgh
Biomedical Science Tower
Pittsburgh
PA 15261, USA

T J MORRIS
School of Biological Sciences
University of Nebraska
Lincoln
NE 68588, USA

HILARY G MORRISON
New England Regional Primate Research Center
Harvard Medical School
Southborough
MA 01772, USA

A M MORTIMER
Department of Environmental and Evolutionary Biology
University of Liverpool
Liverpool
L69 3BX, UK

GISELA MOSIG
Department of Molecular Biology
Vanderbilt University
Nashville, TN 37235, USA

RICHARD W MOYER
Department of Immunology and Medical Microbiology
University of Florida
College of Medicine
Gainesville
FL 32610, USA

A F MURANT
Scottish Crop Research Institute
Invergowrie
Dundee
DD2 5DA, Scotland

FREDERICK A MURPHY
School of Veterinary Medicine
University of California
Davis
CA 95616, USA

NOREEN E MURRAY
Institute of Cell and Molecular Biology
University of Edinburgh
Edinburgh
EH9 3JR, Scotland

OPENDRA NARAYAN
Retrovirus Biology Laboratories
The Johns Hopkins University
School of Medicine
Baltimore
MD 21205, USA

NEAL NATHANSON
Department of Microbiology
University of Pennsylvania
School of Medicine
Philadelphia, PA 19104, USA

JAMES NEIL
Department of Veterinary Pathology
University of Glasgow
Glasgow, G61 1QH, Scotland

STUART T NICHOL
Special Pathogens Branch
Division of Viral and Rickettsial Diseases
National Center for Infectious Diseases
Centers for Disease Control
Atlanta, GA 30333, USA

FELICITY NICHOLSON
Department of Virology
United Medical and Dental Schools
St Thomas' Hospital
London
SE1 7EH, UK

DONALD L NUSS
Department of Molecular Oncology and Virology
Roche Institute of Molecular Biology
Nutley
NJ 07110, USA

DENNIS J O'CALLAGHAN
Department of Microbiology and Immunology
Louisiana State University Medical Center
Shreveport
Louisiana 71130, USA

NEIL E OLSZEWSKI
Department of Plant Pathology and
Plant Biology
University of Minnesota
St Paul
MN 55108, USA

DAVID ONIONS
Department of Veterinary Pathology
University of Glasgow
Glasgow, G61 1QH, Scotland

AMOS OPPENHEIM
Department of Molecular Genetics
Hebrew University
Hadassah Medical School
Jerusalem, Israel

GÉRARD ORTH
Unité des Papillomavirus, INSERM U. 190
Institut Pasteur
75015 Paris, France

CLAES ÖRVELL
Department of Virology
Central Microbiological Laboratory
of Stockholm City Council
S107 26 Stockholm, Sweden

ROBERT A OWENS
Molecular Plant Pathology Laboratory
Beltsville Agricultural Research Center
Beltsville
MD 20705, USA

PETER PALESE
Department of Microbiology
Mount Sinai School of Medicine
New York
NY 10029, USA

ENZO PAOLETTI
Virogenetics Corporation
Rensselaer Technology Park
Troy, NY 12180, USA

COLIN R PARRISH
James A Baker Institute
New York State College of Veterinary Medicine
Cornell University
Ithaca
NY 14853, USA

JOHN R PATTISON
University College London
Medical School Administration
Rayne Institute
London
WC1E 6JJ, UK

L N PAYNE
AFRC Institute for Animal Health
Compton
Nr Newbury
Berks, RG16 0NN, UK

NEILS C PEDERSEN
Department of Medicine
School of Veterinary Medicine
University of California
Davis
CA 95616, USA

DICK PETERS
Department of Virology
Agricultural University Wageningen
6700 EM Wageningen, The Netherlands

HERBERT PFISTER
Institüt für Klinische und Molekulare Virologie der Fredrich-Alexander Universität
D-8520 Erlangen, Germany

WILLIAM C PHELPS
Division of Biology
Burroughs Wellcome Co.
Research Triangle Park
NC 27709, USA

JAMES M PIPAS
Department of Biological Sciences
University of Pittsburgh
Pittsburgh
PA 15260, USA

MAURIZO POCCHIARI
Laboratory of Virology
Istituto Superiore di Sanita
00161 Rome, Italy

COLIN D PORTER
Division of Cell and Molecular Biology
Institute of Child Health
London
WC1N 1EH, UK

ALLEN PORTNER
Department of Virology and Molecular Biology
St Jude Children's Research Hospital
Memphis
TN 38101, USA

JAMES S PORTERFIELD
Formerly Reader in Bacteriology
Sir William Dunn School of Pathology
The University of Oxford
Oxford OX1 3RE, UK

ANTHONY R POTEETE
Program in Molecular Medicine
University of Massachusetts Medical Center
Worcester
MA 01605, USA

ROBERT H PURCELL
Hepatitis Viruses Section
Laboratory of Infectious Diseases
NIAID–NIH
Bethesda
MD 20892, USA

GILLES QUÉRAT
Unité INSERM 372
Campus de Luminy
13276 Marseille Cedex 9, France

KATHRYN RADKE
Avian Sciences Department
University of California
Davis
CA 95616, USA

CHARLES C RANDALL
Department of Microbiology
The University of Mississippi Medical Center
Jackson
MS 39216, USA

MARVIN S REITZ Jr
Section of Molecular Genetics
Laboratory of Tumor Cell Biology
NCI-N1H
Bethesda
MD 20892, USA

SUNG S RHEE
Department of Microbiology
University of Alabama at Birmingham
Birmingham
AL 35294, USA

BERTUS K RIMA
Division of Molecular Biology
School of Biology and Biochemistry
The Queen's University of Belfast
Medical Biology Centre
Belfast, BT9 7BL, Northern Ireland

PAUL S RITCH
Division of Cancer and Blood Diseases
Medical College of Wisconsin
Milwaukee
WI 53226, USA

WILLIAM S ROBINSON
Department of Medicine
Division of Infectious Diseases
Stanford University Medical Center
Stanford
CA 94305, USA

D M ROCHON
Agriculture Canada Research Station
Vancouver
British Columbia
Canada V6T 1X2

GEORGE F ROHRMANN
Department of Agricultural Chemistry
Oregon State University
Corvallis
OR 97331, USA

JOSEPH D ROSENBLATT
Division of Hematology
Department of Medicine
UCLA School of Medicine, UCLA AIDS Institute and Jonsson Comprehensive Cancer Center
Los Angeles
CA 90024, USA

ANGELA RÖSEN-WOLFF
Institut für Medizinisch Virologie der Universität Heidelberg
6900 Heidelberg, Germany

LUCIA B ROTHMAN-DENES
Department of Molecular Genetics and Cell Biology
University of Chicago
Chicago
IL 60637, USA

RUDOLF ROTT
Institut für Virologie
Justus-Liebig-Universität Giessen
D-6300 Giessen, Germany

H A ROUHANDEH
Department of Microbiology
Southern Illinois University at Carbondale
Carbondale
IL 62901, USA

MICHÈLE ROULEAU
Department of Biochemistry
The University of Western Ontario
London
Ontario N6A 5B7, Canada

MARTINE F ROUSSEL
Department of Tumor Cell Biology
St Jude Children's Research Hospital
Memphis
TN 38101, USA

LAURENT ROUX
Department of Genetics and Microbiology
University of Geneva Medical School
Geneva 4, Switzerland

DAVID J ROWLANDS
Molecular Science Department
The Wellcome Research Laboratories
Beckenham
Kent, BR3 3BS, UK

POLLY ROY
NERC Institute of Virology and Environmental Microbiology
Oxford OX1 3SE, UK

WILLIAM C RUSSELL
Division of Cell and Molecular Biology
School of Biological and Medical Sciences
University of St Andrews
St Andrews
Fife
KY16 9AL, Scotland

WILLIAM T RUYECHAN
Department of Microbiology
State University of New York at Buffalo
Buffalo, NY 14214, USA

MARGARITA SALAS
Centro de Biologia Molecular (CSIC-VAM), Universidad
Autonoma de Madrid
Canto Blanco
28049 Madrid, Spain

MARÍA L SALAS
Centro de Biologia Molecular (CSIC-VAM), Universidad
Autonoma de Madrid
Canto Blanco
28049 Madrid, Spain

L F SALAZAR
Department of Pathology
International Potato Center
Lima, Peru

M M M SALIMANS
Department of Virology
University Hospital Leiden
2300 AH Leiden, The Netherlands

JEFFERY T SAMPLE
Department of Virology and Molecular Biology
St Jude Children's Research Hospital
Memphis, TN 38101, USA

MARY ELLEN SANDERS
Dairy and Food Culture Technologies
Littleton
CO 80122, USA

HORACIO U SARAGOVI
Department of Pathology and Laboratory Medicine
University of Pennsylvania School of Medicine
Philadelphia
PA 19104, USA

GORDON J SAUVE
Montreal Neurological Institute
McGill University
Montreal, H3G-1Y6, Canada

MILTON J SCHLESINGER
Department of Molecular Microbiology
Washington University School of Medicine
St Louis
MO 63110, USA

CONNIE S SCHMALJOHN
Department of Molecular Virology
Virology Division
USAMRIID
Ft Detrick
Frederick
MD 21701, USA

MICHAEL P SCHMITT
Department of Microbiology
Uniformed Services University of the Health Sciences
Bethesda
MD 20814, USA

PAUL D SCOTTI
The Horticulture and Food Research Institute of New Zealand
Mt Albert Research Centre
Auckland, New Zealand

DOUGLAS G SCRABA
Department of Biochemistry
4-55 Medical Sciences Building
University of Alberta
Edmonton T6G 2H7, Canada

CHRISTOPH SEEGER
Institute for Cancer Research
Fox Chase Cancer Center
Philadelphia
PA 19111, USA

O P SEHGAL
Department of Plant Pathology
University of Missouri
Columbia
MO 65211, USA

ROBERT D SHAW
State University of New York at Stony Brook
Northport VA Medical Center
Research Service
Northport NY 11768, USA

ROBERT J SHEPHERD
Department of Plant Pathology
University of Kentucky
College of Agriculture
North Lexington
KY 40546, USA

HIROSHI SHIBUTA (Deceased)
Department of Viral Infection
Institute of Medical Science
University of Tokyo, Tokyo
Japan

YUKIO SHIRAKO
Division of Biology
California Institute of Technology
Pasadena
CA 91125, USA

ROBERT E SHOPE
Department of Epidemiology and Public Health
Yale University School of Medicine
New Haven
CT 06510, USA

WERNER SLENCZKA
Institüt für Virologie
der Philipps-Unioversität
3550 Marburg, Germany

MAJA A SOMMERFELT
Department of Microbiology
University of Alabama at Birmingham
Birmingham
AL 35294, USA

PETER J SOUTHERN
Department of Microbiology
University of Minnesota
Minneapolis
MN 55455, USA

JOHN STACZEK
Department of Microbiology and Immunology
Louisiana State University Medical Center
1501 Kings Highway
Shreveport, LA 71130, USA

GLYN STANWAY
Biology Department
Essex University
Colchester
CO4 3SQ, UK

JACK G STEVENS
Department of Microbiology and Immunology
Reed Neurological Research Center
UCLA School of Medicine
Los Angeles
CA 90024, USA

CHARLES R STEWART
Department of Biochemistry and Cell Biology
Wiess School of Natural Sciences
Rice University
Houston
TX 77251, USA

LOTHAR STITZ
Institut für Virologie
Justus-Liebig-Universität Giessen
D 6300 Giessen, Germany

PELLE STOLT
Max-Planck-Institut für Biochemie
D-8033 Martinsried bei München, Germany

DON STOLTZ
Department of Microbiology
Dalhousie University
Halifax
Nova Scotia
B3H 4H7, Canada

STEPHEN E STRAUS
Medical Virology Section
Laboratory of Clinical Investigation
NIAID-NIH
Bethesda
MD 20892, USA

MICHAEL J STUDDERT
The University of Melbourne
School of Veterinary Science
Parkville
Victoria 3052, Australia

JACOV TAL
Department of Virology
Faculty of Health Sciences
Ben-Gurion University of the Negev
Beer-Sheva 84105, Israel

JAMES TARTAGLIA
Virogenetics Corporation
Rensselaer Technology Park
Troy, NY 12180, USA

MASATO TASHIRO
Department of Virology
Kobe University School of Medicine
Kobe and Jichi Medical School
Tochigi, Japan

DANIELLE TENINGES
Centre de Génétique Moléculaire
91198 Gif sur Yvette Cedex, France

PETER TIJSSEN
Institut Armand-Frappier
Université du Québec
Laval
Québec H7N 4Z3, Canada

SUE A TOLIN
Department of Plant Pathology, Physiology and Weed Science
Virginia Polytechnic Institute and State University
Blacksburg, VA 24061-0330, USA

NOËL TORDO
Socrates 402-601
Colonia Polanco
D.F. 11510, Mexico

TERJE TRAAVIK
Department of Virology
Institute of Medical Biology
University of Tromsö
N-9037
Tromsö, Norway

KENNETH L TYLER
Neurology Service
Denver VA Medical Center
University of Colorado Health Sciences Center
Denver, CO 80220, USA

JAN VAN DUIN
Department of Chemistry
Leiden University
2300 RA Leiden, The Netherlands

J VAN EMMELO
Department of Bacteriology, Virology and Immunology
University Hospital Ghent
9000 Ghent, Belgium

JAMES L VAN ETTEN
Department of Plant Pathology
University of Nebraska
Lincoln, NE 68583, USA

JAN T VAN OIRSCHOT
Central Veterinary Institute
Edelhertweg 15
NL 8200
AB Lelystad, The Netherlands

INDER M VERMA
Molecular Biology and Virology Laboratory
The Salk Institute
San Diego
CA 92186, USA

ROBERT VIGNE
Unité INSERM 372
Campus de Luminy
13276 Marseille Cedex 9, France

GÖRAN WADELL
Department of Virology
University of Umeå
S-901 85 Umeå, Sweden

EDWARD K WAGNER
Program in Animal Virology
Department of Molecular Biology and Biochemistry
University of California
Irvine
CA 92717, USA

ALICE L WANG
Department of Pharmaceutical Chemistry
University of California
San Francisco School of Pharmacy
CA 94143, USA

CHING C WANG
Department of Pharmaceutical Chemistry
University of California
San Francisco School of Pharmacy
CA 94143, USA

ROBERT E WEBSTER
Department of Biochemistry
Duke University Medical Center
Durham
NC 27710, USA

ROBERT G WEBSTER
Department of Virology and Molecular Biology
St Jude Children's Research Hospital
Memphis
TN 38101, USA

RAYMOND M WELSH
Department of Pathology
University of Massachusetts
Medical Center
Worcester
MA 01655, USA

PATRICIA WHITAKER-DOWLING
Department of Molecular Genetics and Biochemistry
University of Pittsburgh
School of Medicine
Pittsburgh
PA 15261, USA

DAVID O WHITE
Microbiology Department
The University of Melbourne
Parkville
Victoria 3052, Australia

K ANDREW WHITE
Department of Biochemistry
The University of Western Ontario
London
Ontario N6A 5B7, Canada

REED B WICKNER
Section of Genetics of Simple Eucaryotes
NIDDK-NIH
Bethesda
MD 20892, USA

STANLEY T WILLIAMS
Department of Genetics and Microbiology
University of Liverpool
Liverpool
69 3BX, UK

T MICHAEL A WILSON
Scottish Crop Research Institute
Invergowrie
Dundee, DD2 5DA, Scotland

ECKARD WIMMER
Department of Microbiology
State University of New York at
Stony Brook
Health Sciences Center
Stony Brook
NY 11794, USA

RICCARDO WITTEK
Institut de Biologie Animale
Université de Lausanne
CH-1015 Lausanne, Switzerland

RON WISDOM
Department of Biochemistry
Vanderbilt University School of Medicine
Nashville
TN 37215, USA

KEN WOLF (Retired)
US Fish and Wildlife Service
Kearneysville
WV 25430, USA

JERRY S WOLINSKY
Department of Neurology
The University of Texas - Houston
Health Science Center
Houston
Texas 77225, USA

ROBERT WYLER
Institute of Virology
University of Zurich
CH 8057 Zurich, Switzerland

KOICHI YAMANISHI
Department of Virology
Research Institute for Microbial Diseases
Osaka University
Osaka 565, Japan

ABRAHAM YANIV
Department of Human Microbiology
Sackler School of Medicine
Tel Aviv University
Tel Aviv 69778, Israel

JONATHAN W YEWDELL
Laboratory of Viral Diseases
NIAID-NIH
Bethesda
MD 20892, USA

MARGUERITE YIN-MURPHY
Department of Microbiology
Faculty of Medicine
National University of Singapore
Singapore 0511, Republic of Singapore

MITSUAKI YOSHIDA
Department of Cellular and Molecular Biology
Institute of Medical Science
The University of Tokyo
Tokyo 108, Japan

LAWRENCE S YOUNG
CRC Laboratories
Department of Cancer Studies
The University of Birmingham
Medical School
Birmingham, B15 2TJ, UK

JULIUS S YOUNGNER
Department of Molecular Genetics and Biochemistry
University of Pittsburgh School of Medicine
Pittsburgh
PA 15261, USA

THOMAS M YUILL
Pathobiological Sciences
Institute for Environmental Studies
University of Wisconsin
Madison
WI 53705, USA

SERGEI K ZAVRIEV
Institute of Agricultural Biotechnology
Moscow 127253, Russia

WOLFRAM ZILLIG
Max-Planck-Institut für Biochemie
D-8033 Martinsried bei München, Germany

HARALD ZUR HAUSEN
Deutsches Krebsforschungs Zentrum
D-6900 Heidelberg, Germany

Preface

The goal of this Encyclopedia is to bring together the basic and practical aspects of virology in a very concise form that can provide a rapid synopsis of each area for both the professional and interested lay reader. This Encyclopedia contains the largest comprehensive reference source of current virological information available. Subjects covered include animal, insect, plant and bacterial viruses. Obviously, the large number of viruses contained in particular families could not be covered individually but examples of important members of each family can be found. Each article describing a particular virus contains the most recent taxonomic information provided by the Fifth International Committee on Taxonomy and Classification of Viruses. Articles on general subjects as they relate to particular viruses or virus diseases are also included in order to bring together various aspects of their biomedical or economic importance. This provides access to a large number of topics of direct interest not only to experimentalists and clinical virologists, but also to scientists and educators from other disciplines in health sciences and in biology in general. As well as being a primary reference source, the Encyclopedia format readily lends itself to casual reading for the lay public, particularly those articles dealing with general subjects.

The Encyclopedia is presented as three volumes with entries given in alphabetical order. The editors gave considerable thought to the organization and after taking into account the expected broad readership, opted for alphabetical listing of articles and the use of common rather than taxonomic names for viruses. Volume 1 covers Adenoviruses to Furoviruses; Volume 2 covers Geminiviruses to Parsnip yellow fleck virus group; Volume 3 covers Partitiviruses to Zoonoses. A complete list of entries is provided at the beginning of each volume and where a single article contains several different viruses or subjects, all titles are provided in the Table of Contents for these subjects. Any topic that does not have its own separate entry will be quickly located by reference to the Index. Detailed information on locating subjects is provided in the Guide to Use of the Encyclopedia.

Researchers widely acknowledged as experts in the field are authors of individual entries and because each entry has been written to be self-contained, there is some repetition between the various entries. These repetitions are not distracting or extensive and are desirable, for it helps the reader to find the information without returning to the Index. Each article has had size limitation placed on it by the editors and so could not be exhaustive. Our aim has been to provide the most pertinent information in the space allocations imposed on each contributor, thereby limiting citations to reviews rather than to specific individuals who would ordinarily be cited by name for their work. Thus, references are limited to a list of Further Reading at the end of each article. As an aid to cross-referencing, each article also contains a 'See Also' list where the reader can find other entries within the Encyclopedia relevant to that particular article. As editors, we have already found the chapters extremely valuable as a rapid source of information and as a teaching aid. Obviously, there have been important developments in virology during and after publication of these volumes. However, we feel that the information contained is based on findings confirmed by many laboratories and is unlikely to be altered fundamentally by subsequent work. To address these issues, each author has provided a brief section on the 'future' of his particular subject. Time will tell how often their predictions have been met.

Robert G Webster
Allan Granoff

PARTITIVIRUSES

Said A Ghabrial
University of Kentucky
Lexington, Kentucky, USA

History

Virues that occur naturally and multiply in fungi 'mycoviruses' were first discovered about 30 years ago. Interest in an antiviral activity associated with two *Penicillium* species led to the discovery of double-stranded RNA (dsRNA)-containing virus particles in fungi. It is now believed that viruses, mainly dsRNA viruses, are of common occurrence in fungi. The isometric dsRNA viruses with divided genomes, currently classified as belonging to the family *Partitiviridae*, were amongst the first mycoviruses to be studied and characterized. The main reason for the belated discovery of fungal viruses is that they are mostly associated with symptomless infections of their hosts. Typically, mycoviruses, including the partitiviruses, do not lyse their fungal host cells, and there is no evidence that they have an extracellular phase to their life cycles. That the partitiviruses are not infectious as purified particles is not because of any structural deficiencies, but is primarily because of their inability to penetrate fungal cell walls. Indeed, successful infections of fungal protoplasts with certain partitiviruses have been reported using purified virus preparations. For viruses with multiple dsRNA segments, the lack of conventional infectivity assays makes it extremely difficult to ascertain the number of dsRNA segments required for infectivity. The elucidation of the genome organization of these viruses thus requires that the complete nucleotide sequence of the dsRNA segments is known. To date, the partitiviruses dsRNAs have neither been cloned nor sequenced. Therefore, viruses with multiple dsRNA segments for which the present evidence (based mainly on translational analyses and/or nucleic acid hybridization data) suggests that their genomes consist of at least two unique dsRNA segments are tentatively classified as possible members of the family *Partitiviridae*. For the purpose of this article, the term partitiviruses will be applied to both members and possible members of the family *Partitiviridae*.

Taxonomy and Classification

The family *Partitiviridae* includes a single genus, *Partitivirus* (bipartite dsRNA mycovirus group), and a possible genus, *Penicillium chrysogenum* virus group. Viruses belonging to the *Partitivirus* genus have bipartite genomes comprised of two monocistronic dsRNA segments that are usually similar in size. The *Gaeumannomyces graminis* virus 019/6-A (GgV-019/6-A) is the type species of the family. Other members and possible members of the genus *Partitivirus* are: *Agaricus bisporus* virus 4 (AbV-4), *Aspergillus ochraceous* virus (AoV), *Gaeumannomyces graminis* virus T1-A (GgV-T1-A), *Penicillium stoloniferum* virus S (PsV-S), *Rhizoctonia solani* virus 717 (RsV-717), *Phialophora radicicola* virus 2-2-A (PrV-2-2-A) and *Penicillium stoloniferum* virus F (PsV-F). The *P. chrysogenum* virus group includes viruses with three or four dsRNA segments, the genome organization of which is unknown. The possible type species of the *Pencillium chrysogenum* virus group is *Penicillium chrysogenum* virus (PcV). Thermal denaturation and electron microscopic heteroduplex studies suggest that each dsRNA segment of PcV contains unique sequences. Other members and possible members of this group include *Penicillium brevicompactum* virus (PbV), *Penicillium cyaneo-fulvum* virus (Pc-fV) and *Helminthosporium victoriae* 145S virus (HvV-145S).

Particle Properties

The sedimentation coefficients $s_{20,w}$ (in Svedberg units) for members of the *Partitivirus* genus are in the range of 101S to 145S. Buoyant density in CsCl [ρCsCl (g cm^{-3})] = 1.35–1.36. With some viruses, e.g. PsV-S, purified preparations contain, in addition to the mature virions (termed L1 and L2 respectively for particles that separately encapsidate the bipartite dsRNA genome of 1.6 and 1.4 kbp), sedimenting and density components that represent empty (E) particles and intermediate components in the replication cycle. The latter include: M1 and

Table 1. Physicochemical properties of PsV-S particles

Property	Density component			
	E	M	L	H
$s_{w,20}$	66.1	87.1	101.2	112.6
ρCsCl (g cm^{-3})	1.297	M1, 1.332 M2, ND[a]	L1, 1.358 L2, 1.362	H1, 1.384 H2, 1.390
Mol. wt ($\times 10^{-6}$)	5.4	5.52	5.97	6.67
RNA (%)	0	8.7	15.5	24.4

[a] ND, not determined.

M2 particles encapsidating single-stranded (ss) RNAs, presumably the messenger sense strands of L1 and L2 dsRNAs respectively; H1 and H2 particles containing both ss and dsRNA. The physicochemical properties of these four classes of particles are summarized in Table 1.

Particle Structure and Composition

The partitiviruses have isometric particles, 30–35 nm in diameter, with icosahedral symmetry. The capsids are single-shelled and comprised of a single major polypeptide. The protein shell consists of 120 subunits of mol. wt in the range of 56 000–73 000 (for members of the *Partitivirus* genus) or 60 subunits of mol. wt of 125 000 (for the PcV group) arranged in $T = 1$ lattices. The virions of members of the *Partitivirus* genus contain two unrelated segments of dsRNA, in the size range 1.3–2.2 kbp, one encoding the capsid polypeptide and the other an unrelated polypeptide, probably the virion-associated RNA polymerase. The two segments are usually of similar size, e.g. the sizes of the two dsRNAs (kbp) of GgV-019/6-A, GgV-T1-A, AbV-4 and RsV-717 are (1.7, 1.8), (2.1, 2.2), (2.0, 2.2) and (2.0 and 2.2) respectively. PsV-F and PrV-2-2-A each contains three dsRNA segments with sizes (kbp) of: (1.5, 1.3, 0.67) and (1.9, 1.8, 1.5) respectively. The smallest segments are believed to represent satellite or defective dsRNAs. A comparison of the two serologically closely related viruses GgV-019/6-A and GgV-38-4-A reveals that whereas the former has two dsRNA components, dsRNA1 and dsRNA2, the latter has three segments dsRNA1, 2 and 3. The dsRNA1 from GgV-019/6A and the corresponding one from GgV-38-4-A show a high degree of sequence homology, as indicated by T1 oligonucleotide fingerprinting analysis and saturation hybridization assays. This is also true for dsRNA2 from the two viruses. On the other hand, dsRNA3 from GgV-38-4-A has little or no sequence homology with dsRNA1 or 2 from either virus suggesting that it is probably not required for replication, and may be regarded as satellite dsRNA.

The particles of viruses in the PcV group contain three unrelated dsRNA components with sizes in the range 2.8–3.5 kbp. Each segment is separately encapsidated, and is probably monocistronic. Some virus isolates encapsidate additional segments, which are probably satellite or defective dsRNAs.

Antigenic Properties

The partitiviruses are efficient immunogens. Members and possible members which are serologically related, e.g. PsV-S, DrV and AoV, may be strains of a virus species.

Transcription/Translation

All viruses in the *Partitiviridae* so far examined have been shown to possess virion-associated RNA polymerase activity. The RNA polymerases associated with the partitiviruses GgV 09/6-A and GgV 38-4-A are transcriptases which catalyze the synthesis and release of ssRNA copies of one of the strands of each of the virus template dsRNA molecules. Present evidence favors the idea that the *in vitro* transcription reaction occurs by a semiconservative mechanism. Thus, the released ssRNA represents a displaced strand of the parental dsRNA and the newly synthesized strand becomes part of the dsRNA duplex.

In a rabbit reticulocyte lysate system, the ssRNA transcripts of GgV-38-4-A dsRNA1, 2 and 3 have each been shown to direct the synthesis of three major polypeptides, with mol. wts of 62 000, 55 000 and 52 000 respectively. The three polypeptides are unrelated, as revealed by peptide mapping. The capsid polypeptide of the virus is encoded by dsRNA2 since the *in vitro* translation product of the ssRNA2 transcript co-migrates with the authentic virus capsid protein. The close serological relationship of GgV-38-4-A and GgV-019/6-A (the type species of the family *Partitiviridae*), coupled with the known sequence homology of dsRNA2 from the two viruses, indicates that dsRNA2 of GgV-019/6-A also encodes the capsid polypeptide. Because GgV-019/6-A has only two genome segments and requires the entire coding capacity of dsRNA2 for the capsid polypeptide of mol. wt 60 kDa, dsRNA1 probably encodes the viral RNA polymerase. The dsRNA1 from GgV-38-4-A and that from GgV-019/6-A show extensive sequence homology, suggesting that the sequence of the putative RNA polymerase sequence is highly conserved between these two strains.

Virus Multiplication

Methods for obtaining synchronous virus infection, which allow direct studies on virus multiplication, are not presently available. When full-length infectious transcripts of partitivirus dsRNAs are generated by future research, fungal protoplast infection systems may be developed so that synchronous infection can be achieved. Current knowledge on how the partitiviruses replicate their dsRNAs is derived from *in vitro* studies of virion-associated RNA polymerases and the isolation from naturally infected mycelium of particles that represent various stages in the replication cycle. The partitivirus PsV-S serves as an example. As indicated earlier, purified virus preparations are known to contain several classes of particles: the mature virions L1 and L2 which are comprised of the two genomic dsRNA segments separately encapsidated in identical capsids; a small proportion of particles M1 and M2, which contain ssRNAs corresponding to the plus-strand of the genomic dsRNAs. Furthermore, purified virus preparations still contain a large proportion of a heterogeneous population of heavy 'H' particles, which are more dense than the mature particles. The H particles represent various stages in the replication cycle which include: (1) particles containing the individual genomic dsRNAs with ssRNA tails of varying lengths; (2) particles with one molecule of dsRNA and one molecule of its ssRNA transcript; and (3) particles with two molecules of dsRNA (product or 'P' particles). In *in vitro* reactions, only the H particles exhibit RNA polymerase activity even though the capsid protein compositions of M, L and H particles are identical; all three particle types contain 120 molecules of the 56 kDa capsid polypeptide and one molecule of another polypeptide, presumably the RNA polymerase protein. The reasons for lack of activity *in vitro* by the L and M particles are not known. The *in vitro* RNA polymerase activity present in the H particles is a replicase activity directing the completion of all the intermediate stages of the replication reaction to form particles with two molecules of dsRNA. The replication reaction occurs semiconservatively, as can be demonstrated by density labeling experiments, and probably takes place in two stages. In the first stage, the newly synthesized plus-strand displaces a parental strand of the same polarity and becomes part of the duplex. In the second stage, the displaced parental strand serves as a template for the synthesis of dsRNA. It is believed that the RNA polymerase protein acts both as a transcriptase and a replicase. The transcriptase activity catalyzes the synthesis of plus-strand on parental dsRNA templates followed by the displacement of the parental plus-strand; the replicase activity catalyzes synthesis of minus-strand on the displaced parental plus-strand. There is no evidence for release of newly synthesized dsRNA from P particles or re-initiation of the replicase reaction *in vitro*.

The isolation of M particles from infected mycelium indicates that replication *in vivo* may occur asynchronously. Thus, PsV-S may replicate its dsRNA synchronously, as demonstrated *in vitro* in the H particles, or asynchronously via the M particles. It is suggested that the choice between synchronous and asynchronous pathways for PsV-S may be governed by the physiological state of the host or by early/late switches in the replication cycle.

Transmission

There are no known natural vectors and the viruses are transmitted intracellularly during cell division and sporogenesis (serial or vertical transmission), and following cell fusion between genetically compatible host strains (lateral or horizontal transmission). The hosts of the partitiviruses belong to the higher fungi classes, Basidiomycotina and Ascomycotina and their anamorphs. The hyphae are septate and grow by extension of the hyphal tip. Virus transmission during vegetative growth occurs as virus particles, which accumulate in the peripheral growth zone (the region of the mycelium where septal pores remain unplugged), are translocated with the flow of protoplasm towards the tip. New growth will continue to be virus-infected as long as the septal pores connecting the apical hyphal compartment with virus-containing compartments in the peripheral growth zone remain open.

Transmission of partitiviruses through asexual spores is highly efficient as 90–100% virus transmission into single conidial isolates of the fungal hosts can be achieved. Transmission via spores plays an important role in virus dissemination, as fungi produce many types of propagules and often in great profusion. The presence of virus particles in conidiospores has been demonstrated for a number of fungi infected with partitiviruses including PsV-S, PsV-F, PcV and PbV. The presence of virus particles in the sexual spores (basidiospores) of the cultivated mushroom *Agaricus bisporus*, possibly including the partitivirus AbV-4, has been demonstrated by electron microscopy in thin sections of basidiospores. Virus transmission via basidiospores plays an important role in the epidemiology of the LaFrance disease, a very serious disease of the cultivated mushroom believed to have a viral etiology. Unlike the basidiomycetes, the sexual spores (ascospores) of virus-infected filamentous ascomycetes, including *Gaeumannomyces graminis* (which is a host for several partitiviruses including the type species GgV-019/6-A), are virus-free. There is no evidence that virus infection of fungal spores, whether mitotic or meiotic, has any deleterious effects on their viability.

Partitiviruses are transmitted intraspecifically from one fungal strain to another via hyphal anastomosis/heterokaryosis. Fungi are known to have a potential for plasmogamy and cytoplasmic exchange during extended periods of their life cycles. Because of vegetative incompatibility, hyphal anastomosis is limited to individuals within a species or within very closely related species. Transmission of several partitiviruses, including PsV-S, PsV-F and PcV, by heterokaryosis has been demonstrated using auxotrophic and colored mutant fungal strains.

Host Range

Because of the lack of reliable conventional infectivity assays for mycoviruses, there are no known experimental host ranges for mycoviruses in general (including the partitiviruses). Due to vegetative incompatibility, natural host ranges are restricted to individuals within the species known to be naturally infected with the virus in question. The finding that different fungal species belonging to widely divergent genera may harbor identical or closely related viruses, e.g. the serologically related partitiviruses PsV-S, AoV and DrV which infect *Penicillium stoloniferum, Aspergillus foetidus* and *Diplocarpon rosae* respectively raises the question of the presence of vectors or means of transmission other than hyphal anastomosis.

Mixed infections of fungi with two or more unrelated mycoviruses are of common occurrence. Examples of mixed infections involving partitiviruses include: the bacilliform *Agaricus bisporus* virus-1 (AbV-1) and the partitivirus AbV-4; the unclassified *Gaeumannomyces graminis* virus F6-A (GgV-F6-A) and two partitiviruses, GgV-F6-B and GgV-F6-C; and the partitiviruses PsV-S and PsV-F.

Virus–Host Relationships

Ultrastructural studies carried out with a number of partitiviruses including PsV-S, PcV and PbV indicated that the viral particles accumulate in the cytoplasm of the hyphal compartments and in conidia. Particles often occur as free aggregates or enclosed in single- or double-membrane bound vesicles. Whereas the apical hyphal compartments generally contain a small number of particles, virus replication occurs at higher levels in the lower hyphal compartments, particularly in the older plugged compartments which no longer contribute to host growth.

Like other fungal viruses, the partitiviruses generally do not have deleterious effects on their hosts. Because of their intracellular mode of transmission, fungal viruses have probably evolved along with their hosts as persistent subcellular particles. Thus, latency is the rule with mycovirus infections, and viral mutations that lead to pathogenic effects and lysis of host cells are expected to be self-eliminating because of the lack of an extracellular mode of transmission. The two partitiviruses AbV-4 and HvV-145S may present exceptions to this general rule. AbV-4 may play a role in the LaFrance degenerative disease of the cultivated mushroom, as mentioned earlier. The HvV-145S is proposed as the etiological agent (or co-agent) of a degenerative disease of the plant pathogenic fungus *Helminthosporium victoriae*, the causal fungus of Victoria blight of oats (see below).

A degenerative disease of *H. victoriae*, transmissible via hyphal anastomosis, was known even before viruses were ever discovered in fungi. Diseased isolates are characterized by excessive sectoring, aerial mycelial collapse and generalized lysis. The disease of *H. victoriae* is of special interest not only because diseased isolates are hypovirulent but also because they harbor two isometric dsRNA viruses, *H. victoriae* 190S virus (HvV-190S), a member of the family *Totiviridae*, and the partitivirus HvV-145S. Several lines of evidence have been previously presented in support of a viral etiology of the disease. Because normal isolates of the fungus are either virus-free or contain only HvV-190S, and because disease severity correlates well with the concentration of HvV-145S in diseased mycelium, it has been suggested that HvV-145S alone or a mixed infection of the two viruses is the cause of the disease. HvV-145S, however, has always been found in association with HvV-190S, and no fungal isolate was ever found that contains HvV-145S alone. Recent studies have shown no sequence homology between the dsRNAs of the two viruses, and the possibility that HvV-145S is a satellite virus dependent on HvV-190S for replication or other functions cannot be ruled out.

Although evidence compiled to date indicates that mycoviruses, including the partitiviruses, are not directly involved in the synthesis of secondary metabolites, it remains possible that virus infection may affect the level of production of a metabolite or a toxin. For example, the virus-infected, diseased isolates of *H. victoriae* are known to produce reduced amounts of the pathotoxin, victorin. Because virulence expression in this plant pathogenic fungus is determined by the level of victorin produced, the diseased isolates are hypovirulent. It will be of interest to determine whether virus-encoded proteins play a role in the regulation of metabolic pathways in fungi where virus infection causes degenerative diseases.

See also: Totitiviruses; Yeast RNA viruses.

Further Reading

Bozarth RF (1979) The physico-chemical properties of mycoviruses. In: Lemke PA (ed.) *Viruses and Plasmids in Fungi*. New York: Marcel Dekker.

Buck KW (1986) Fungal virology – an overview. In: Buck KW (ed.) *Fungal Virology*. Boca Raton: CRC Press.

Buck KW and Ghabrial SA (1991) *Partitiviridae*. In: Francki RIB, Fauquet CM, Knudson DL and Brown F (eds) *Classification and Nomenclature of Viruses*, Fifth Report of the International Committee on Taxonomy of Viruses. New York: Springer Verlag.

PARVOVIRUSES

Contents

General Features
Molecular Biology
Cats, Dogs and Mink
Rodents, Pigs, Cattle and Geese

General Features

John R Pattison
University College London Medical School
London, UK

History

Members of the genus *Parvoviruses* have been discovered by a mixture of chance and the deliberate search for the causative agents of certain diseases in a variety of animals. Some of the diseases, such as that in cats due to feline parvovirus and erythema infectiosum in man due to parvovirus B19, have been known since the beginning of this century. However it was not until 1959 that Kilham and Olivier published an account of a latent virus of rats isolated in tissue culture. This was Rat virus (RV) which then became the type species of the genus *Parvovirus*. Once the distinctive size and morphology of RV had been described subsequent work investigating disease revealed a number of other parvoviruses [e.g. feline parvovirus (FPV), Aleutian mink disease virus (ADV)] and others were discovered by chance in the course of a variety of animal and cell culture experiments [e.g. minute virus of mice (MVM), Lu III].

Taxonomy and Classification

The family *Parvoviridae* consists of three genera, *Parvoviruses*, *Dependoviruses* and *Densoviruses* (formerly *Densonucleosis viruses*). Members of the genus *Parvovirus* are independent of help from other viruses for their replication and are often termed 'autonomous parvoviruses'. However even these viruses depend on certain helper functions only transiently expressed in host cells during the late S or early G-2 phase of mitosis. The dependoviruses on the other hand require such help as is only provided by co-infection with another virus. The helper viruses most frequently used and studied are adenoviruses, although herpesviruses can also act in this capacity. As a consequence these viruses are often known as adeno-associated viruses (AAVs). The densoviruses are autonomous parvoviruses of insects.

The nomenclature of parvoviruses is a mixture of descriptive names which indicate the natural host e.g. porcine parvovirus and trivial names e.g. B19 which usually refer to some incidental feature of their discovery. The following are recognized members of the genus: rat virus (RV), minute virus of mice (MVM), bovine parvovirus (BPV), porcine parvovirus (PPV), feline parvovirus (FPV), mink enteritis virus (MEV), canine parvovirus (CPV), human parvovirus B19, lapine parvovirus (LPV), Aleutian mink disease virus (ADV), goose hepatitis virus (GPV) and five viruses whose natural host is unknown (H-1, HB, RT, TVX and LuIII).

All dependoviruses with one exception have been discovered in adenovirus stocks; AAV5 was isolated from penile flat condylomata. Recognized members of the genus are the monkey viruses AAV1 and AAV4, the human viruses AAV2, AAV3 and AAV5, and the dependoviruses of cattle (BAAV), dogs (CAAV) and quail (AAAV).

Geographic and Seasonal Distribution

It is probable that parvoviruses are world-wide in their distribution although the recovery of the viruses tends to reflect the distribution of advanced systems for the health care of humans and animals. Certainly B19 virus has been found in every human population that has been studied throughout each of the five continents. The same is true of FPV and CPV in cats and dogs, respectively.

Seasonal variation has only been studied adequately for the human virus B19. Cases of infection due to this virus appear to be endemic throughout the year but there is a definite peak of incidence in the northern hemisphere in late spring or early summer.

Host Range and Virus Propagation

In most instances parvoviruses are species-specific with respect to natural infection in intact animals. This is most dramatically illustrated by the set of closely related host-range variants FPV/MEV/CPV.

Many of the autonomous parvoviruses are also species-specific for replication in cell cultures but some such as PPV and CPV will grow in cells from more than one species. Parvoviruses grow best in dividing cells therefore it is important not to inoculate confluent monolayers of resting cells but rather cultures in which division is still taking place. Adeno-associated viruses are very promiscuous with respect to the cells in which they will replicate and this has proved useful in their study.

Genetics and Evolution

All members of the family *Parvoviridae* contain single-stranded DNA genomes. Many of the viruses package only the DNA strand of negative polarity although the dependoviruses and one or two autonomous parvoviruses (e.g. B19) package both strands of the replicative form of DNA with equal efficiency.

Some minor variations of the genome of individual viruses have been revealed by endonuclease restriction analysis or by comparison of sequence data. However in most instances these variations are of no biological significance. The exception among the parvoviruses is the group of closely related host-range variants sometimes known as the FPV/MEV/CPV complex. Disease due to FPV has been known for more than a century. However in 1947–1952 there were outbreaks of enteritis in mink which were shown to be due to a virus indistinguishable from FPV. In the late 1970s disease in dogs due to a virus with considerable cross reactivity with FPV occurred. Studies of the genome of FPV and MEV show them to be very closely related with CPV being more distantly related. Some examples of FPV show genomic changes that are typical of CPV although as yet there is not a complete set of variants which show a continuous evolution from one virus to the other.

Epidemiology

Parvoviruses transmit very efficiently in both wild and captive populations. In rodents parvovirus infections are endemic and common and this is undoubtedly facilitated by the establishment of latent infection with intermittent excretion in the host and the fact that parvovirus infectivity survives well even when the virus is shed into the environment. Epidemics are not recognized in these populations and the infections are apparently sub-clinical.

With respect to the FPV/MEV/CPV complex the epidemiology was characterized by severe epizootics when the disease first occurred in an animal population followed by the occurrence of less severe disease with lower mortality as the host-parasite relationship developed.

With respect to ADV in mink it appears that asymptomatic infection is very common in feral mink. In ranch mink the artificial environment of a captive population and the strain variation in both host and parasite have led to devastating outbreaks of severe disease. Mink of the Aleutian genotype are very susceptible to infection and in them the disease has a rapid rate of progression.

In the case of the human parvovirus B19 the infection is endemic throughout the year in all populations studied but there is an annual seasonal increase in infection rates and a longer term cycle of increased incidence about every four to five years. This is typical of some other human infections which are common in childhood.

The porcine parvovirus is also remarkable for the frequency with which infection followed by latency occurs in the natural host. This is typical of the situation once infection is established in a herd although again when the virus is first introduced, the infection is at its most severe with, in the case of PPV, the reproductive failure being at its most marked.

Transmission and Tissue Tropism

The natural route of transmission of parvoviruses will vary from virus to virus. In general however the four common modes of transmission are well represented amongst the parvoviruses, namely aerosol transmission, fecal–oral transmission, venereal and transplacental transmission. In those species in which enteritis is a common result of parvovirus infection large amounts of virus are excreted in the feces and this is a very effective mode of transmission. In closed colonies of animals it is very difficult to eradicate infection without clearing the facilities for housing animals and thoroughly disinfecting them. In the case of those infections not associated with enteritis such as B19 infection the aerosol route is clearly the main mode of transmission. In many species transplacental and perinatal transmission occur with regularity. The majority of these infections seem to be asymptomatic but in each case clinical manifestations are seen in a minority. Venereal transmission seems likely in the case of boars whose semen is chronically infected with PPV.

Pathogenesis

The pathogenesis of disease due to parvoviruses represents a complex interaction of the strain of virus, the age and genetic constitution of the host and the point during the host's life at which infection occurs. Parvovirus replication will always be extensive in rapidly dividing tissues. The replication is often lytic and therefore disease due to damage of these tissues is likely to occur. Thus infections of the fetus, the intestinal epithelium and the hematopoietic system occur frequently in parvovirus infections. However, parvoviruses are not wholly pantropic but often have a predilection for cells at a particular stage of cellular differentiation and therefore some highly specific clinical syndromes do result.

Aleutian disease in mink is the best example of the importance of the strain variability in determining the severity of disease; there are highly pathogenic strains of virus and very susceptible strains of mink. With respect to the age of the animal at infection the occurrence of cerebellar ataxia in kittens and acute myocarditis in young puppies are entirely dependent on infection occurring perinatally.

At the cellular level many of the clinical manifestations of parvovirus infection are explained by the lytic properties of the virus on cells. The exception is the immune complex glomerulonephritis that occurs in mink as a consequence of ADV infection. It is also assumed that the rash and arthropathy of B19 infection in humans is immune mediated.

Clinical Features

Human parvovirus B19

Infection with B19 virus is very common in that approximately 80% of adult populations have serologic evidence of past infection. Infection occurs most commonly in childhood between the ages of 4 and 12 years and most are asymptomatic or associated with mild systemic and respiratory tract illness without any features to distinguish them from similar childhood illnesses due to other viruses. The most distinctive common illness due to B19 infection is an erythematous rash illness similar to rubella, which when it occurs in children and commences with intense erythema of the cheeks, is often called erythema infectiosum, fifth disease or slapped-cheek disease. As with rubella there may be associated joint involvement which occurs infrequently in children (about 10% of cases) but is very common in adult women (about 80% of cases). The usual manifestation is a symmetrical peripheral polyarthropathy which usually resolves within two to four weeks but may persist for months in a minority of cases. Occasional patients infected with B19 virus present with a purpuric rash.

The other common consequence of B19 infection results from the infection of early red cell precursors. The interruption of erythropoiesis occurs in all infected individuals but only becomes clinically manifest in those with pre-existing hemolytic anemia, whose red cells have a shortened life span and who are already anemic. B19 infection in these individuals causes a transient aplastic crisis which lasts for about seven days. The termination of the crisis is associated with a vigorous humoral immune response. In patients who are unable to mount such a response, e.g. those with acute lymphatic leukemia, some inherited immunodeficiency syndromes and HIV positive individuals, B19 infection persists and eventually the patient develops a chronic and relapsing anemia.

The other consequence of B19 infection occurs in pregnant women in whom the viremia that is a consistent feature of B19 infection gives ample opportunity for transplacental spread. There is evidence of intrauterine infection of the fetus in approximately one third of those pregnancies complicated by B19 infection. The majority of these are not associated with fetal damage but there is an increased risk of fetal loss and some cases of hydrops fetalis are undoubtedly associated with B19 infection. About 10% of pregnancies complicated by B19 infection will result in second trimester fetal loss and the incidence is about ten times higher than in pregnancies uncomplicated by infection. Overall about 10% of nonimmunological hydrops fetalis are due to B19 infection.

Feline panleucopenia virus

There are two distinct syndromes of infection caused by feline panleucopenia virus. In kittens infected between two weeks before birth and two weeks after birth a distinct cerebellar ataxia develops as soon as they begin to walk at about three weeks of age. If infection is acquired after the first two weeks of life then illness characterized by loss of appetite, pyrexia, lassitude, diarrhea and a leucopenia with both lymphocytes and neutrophils being reduced ensues. The mortality from this infection appears to depend largely on the general condition of the animals prior to infection.

Mink enteritis

In the mink, diarrhea with blood and mucus in the stools tends to dominate the picture although there is also vomiting, anorexia and dehydration. Lymphopenia is not so regularly found as it is in cats.

Canine parvovirus

Again there are two syndromes of infection with canine parvovirus depending upon the age of puppies when infected. Very young puppies suffer acute myocarditis with a high mortality. In older puppies the clinical picture is one of hemorrhagic enteritis paralleling that seen in the cat and the mink. Again although lymphopenia and neutropenia do occur they are not seen as regularly as they are in cats.

Aleutian disease virus

Again with Aleutian disease virus there is a distinctive clinical syndrome which results from infection of seronegative mink kits. This is an acute respiratory distress syndrome with interstitial pneumonia and hyaline membrane formation due to infection of the type 2 alveolar cells in the lung. The classical Aleutian disease occurs approximately 60 days after infection and is characterized by a plasmacytosis in many organs, hypergammaglobulinemia and developing renal failure as a consequence of immune-complex glomerulonephritis. This latter is frequently the cause of death.

Porcine parvovirus

Infection with porcine parvovirus spreads extensively in contaminated pigs. Nevertheless infections are frequently silent and virus can be recovered from apparently healthy animals. However PPV is recovered in 60–70% of dead or mummified fetuses, but from less than 1% of live piglets. It has also been shown that the persistent presence of PPV in semen may be associated with reduced fertility and the clinical syndrome of stillbirth, mummification, embryonic death and infertility has led to the use of the acronym SMEDI for the consequences of PPV infection in pigs.

Bovine parvoviruses

There appears to be infection with a transient viremia in infected animals. The bovine parvovirus has been implicated as a cause of respiratory tract illness, enteritis, fetal death and abortion.

Goose hepatitis virus

Goose hepatitis virus causes a highly contagious disease of goslings characterized by diarrhea, catarrh and hepatitis with a mortality of over 90%.

Rodent parvoviruses

There is no evidence of any link between illness and parvovirus infection in the natural hosts of RV and MVM. However there are dramatic effects of infection with these viruses when inoculated into hamsters in the perinatal period. Many animals die of acute generalized infection but others develop the characteristic osteolytic syndrome which consists of dwarfism, mongoloid-like features, missing or abnormal teeth and fragile bones. There are other clinical syndromes following experimental infection with rodent parvoviruses, namely hemorrhagic encephalopathy in newborn hamsters, rats and mice infected with RV, and cerebellar ataxia in hamsters after intracerebral inoculation with the same virus.

Adeno-associated viruses

To date there is no evidence of any link between a clinical illness and adeno-associated virus infection. There have been some studies which reveal a reduction in the number of tumors occurring in AAV infected individuals. The possible oncolytic effect is still under investigation.

Pathology and Histopathology

Human parvovirus B19

The studies of the pathology of B19 infection have been confined to examination of the bone marrow in cases of anemia and of fetal material in the cases of hydrops fetalis that have failed to recover following infection. The bone marrow in cases of aplastic crisis and the anemia associated with immunodeficiency indicates an almost total absence of red cell precursors in the early phase. It seems clear that an early but committed red cell precursor is affected. Equally in hydropic fetuses much virus is present in the nuclei of erythroid precursors in many organs such as spleen and liver. The characteristic appearance of infected cells is one in which there is a very large intranuclear inclusion which stains with conventional eosinophilic stains and probes for virus genome. The pathogenesis of the rash and arthralgia are taken to be immune-mediated because they occur at a time after infection when there is already an easily detectable immune response.

Feline parvovirus, mink enteritis virus and canine parvovirus

The most distinctive pathology seen as a consequence of infection with these viruses is in the intestinal villi. There is shortening and blunting of the villi with an intense inflammatory response resulting in the hemorrhagic enteritis with dehydration seen in the host. Virus can be seen with fluorescein labelled probes in the rapidly dividing crypt epithelial cells of the intestine of infected animals. White cell changes are most marked in the cat and lymphocytes disappear from the circulation, lymph nodes, bone marrow and thymus and it is probable that the polymorphonuclear leucocyte stem cells are also destroyed. In kittens with cerebellar ataxia the virus infects the rapidly dividing and developing cells of the external granular layer of the cerebellum and these cells cannot be replaced. In puppies with myocarditis the rapidly dividing myocardial cells in very young puppies are again a target for lytic CPV infection.

Aleutian disease virus

Aleutian mink have a lysosomal abnormality which leads to failure to destroy immune complexes after phagocytosis. This in turn leads to a persistent virus infection marked by plasmacytosis in the bone marrow, spleen, lymph nodes, liver and kidneys. Immune complexes can be demonstrated in the glomerular capillary walls and the area of fibrinoid necrosis in the arteries. In the acute respiratory distress syndrome seen in neonatal mink kits there is a very productive ADV infection of the type 2 alveolar cells.

Rodent parvoviruses

In the dramatic syndromes following experimental infection with these viruses the clinical features are a consequence of cytolytic attack on the susceptible and rapidly dividing blast cells in a variety of organs. For example the skeletal changes in the osteolytic syndrome are due to lytic infection of osteoblasts at the costochondral junction of long bones and the odontoblasts.

Immune Response

Most (perhaps all) parvovirus infections are associated with a viremia and in common with other systemic virus infections there is a prompt and readily detectable antibody response to infection. A variety of techniques such as virus neutralization, hemagglutination inhibition and various radio- and enzyme-immunoassays have been used to detect this response. Antibodies appear two to three weeks after infection and are initially a mixture of IgM, IgA and IgG. The IgM antibody response is transient being detectable for one to two months after infection. However, the IgG response is long lasting (lifelong in most cases) and is associated with resistance to reinfection. Persistently infected animals also have circulating specific antibody and in some instances the higher the antibody titer the more likely it is that the virus will be isolated from the individual animal.

One of the features of Aleutian disease is the presence of hypergammaglobulinemia and most of this circulating immunoglobulin is specific to ADV. In man, immunocompromised patients fail to produce a full repertoire of specific antibodies and this allows infection to persist.

Prevention and Control

The first vaccines to be developed were against FPV and these cell culture grown vaccines are now used routinely in veterinary practice. When disease due to CPV first emerged some cross-protection was afforded by FPV vaccines but subsequently vaccines containing cell culture grown CPV were developed. These are now used routinely and like FPV vaccines are given to young animals when maternal immunity has waned.

Vaccines made from cell culture grown PPV are also available and it is clear that these are protective. The decision about whether or not to use the vaccine in a given herd will depend upon the serological tests and the history of disease in the individual herd.

Future Perspectives

There seems little doubt that the identification of new *Parvoviruses* will continue. There are already a number which are still classified as possible members of the genus. One is the human fecal parvovirus seen in association with outbreaks of food poisoning related to shellfish. Morphologically it is a typical parvovirus. RA-1 is a virus which was recovered in cell culture experiments with human synovial tissues. However it seems more likely that this is a rodent parvovirus. Minute virus of canines (MVC), chicken parvovirus and equine parvovirus all await official inclusion in the genus. Similarly there are two dependoviruses, one isolated from horses the other from sheep which await classification.

There are likely to be some other disease associations of parvoviruses and some of the details of the pathogenesis of known diseases remain to be elucidated. Finally there are likely to be some advances in the use of vaccines for the prevention of parvoviruses and this may particularly relate to the human parvovirus B19.

See also: Adenoviruses; Aleutian mink disease virus; Densonucleosis viruses; Persistent viral infection.

Further Reading

Pattison JR (1990) Parvoviruses. In: Fields BN *et al.* (eds) *Virology* p. 17–65. New York: Raven Press.

Porterfield JS (ed.) (1982) *Parvoviridae*. In: *Andrewes' Viruses of Vertebrates* 5th edn. London: Baillière Tindall.

Siegl G (1976) The Parvoviruses. *Virol. Monogr.* 15, Springer-Verlag.

Molecular Biology

Kenneth I Berns
Cornell University Medical College
New York, New York, USA

Virion Properties

Members of the family *Parvoviridae* are among the smallest of the DNA viruses that infect eucaryotes. The virion is nonenveloped and icosahedral with a diameter of 20–26 nm. The structure of canine parvovirus has been determined by X-ray crystallography. It is rather unique; there are 60 capsomers, each a quadrilateral 'kite-shaped' wedge bounded by fivefold, threefold and twofold axes of symmetry. The 'kite-shaped' wedge is formed by a single repeating subunit. The genome is a linear single-stranded DNA of about 5 kb. Some parvoviruses encapsidate only the strand of minus polarity (antimessenger). For other members of the family a varying percentage of the virions contain the DNA strand of opposite (plus) polarity. At the extreme, 50% of adenoassociated virus (AAV) virions contain minus strands and the other half contain plus strands. The virion is stable from pH 3 to 10 at a temperature of 50°C for 30 min and is resistant to lipid solvents. Ionic detergents decrease infectivity and the virion is disrupted at alkaline pH (>11). UV radiation, formalin and β-propriolactone inactivate the virion.

The Genome

The linear single-stranded viral DNA is characterized by palindromic sequences at both termini. The terminal sequences range in length from 120 to >300 nucleotides. In some instances, the 5′-terminal sequence is an inverted repeat of the 3′ terminus (e.g. AAV and B19). In other cases the terminal sequences are unrelated. All of the 3′-terminal sequences that have been determined have an overall palindromic arrangement that is interrupted by two smaller internal palindromic sequences. When the palindromic 3′-terminal sequence is folded such that base-pairing is maximized, a T- or Y-shaped structure is formed. For those genomes in which the 5′-terminal sequence is distinct from the 3′ end, the 5′-terminal palindrome tends to form a simple hairpin structure when folded on itself.

The model of parvovirus DNA replication invokes

a single-strand displacement mechanism. The 3′-terminal palindrome hairpins to serve as a primer to initiate DNA synthesis; elongation leads to a double-stranded replicative intermediate covalently cross-linked at one end by the terminal hairpin. The linear ends are recreated by cleavage of the hairpin at the original inboard end of the palindrome, thus transferring the hairpin sequence from the parental to the progeny strand. The transfer leaves the parental strand shorter (by the length of the transferred sequence) and with a 3′ OH group which serves as a primer for repair synthesis of the parental 3′-terminal sequence using the transferred sequence as the template. The transfer reaction occurs at both ends of genomes with terminal repeats and at the 5′ ends of other parvovirus genomes. A more complex displacement reaction is used at the 3′ termini of the latter in place of the transfer. Viral structural proteins are required for sequestration of mature virion single strands from the duplex replicative intermediate.

The terminal sequences are essential in a *cis*-active manner. Any extensive deletion of these sequences inhibits DNA replication. Experiments with AAV indicate that the T conformation is essential although the sequence itself in internal positions is not so critical. On the other hand, sequence does appear to be critical at the site(s) in the hairpin that are cleaved during replication.

Viral nonstructural proteins appear to play a direct role in DNA synthesis. In the case of AAV the larger nonstructural proteins (see below) have been shown to cleave the terminal hairpin at the site predicted by the model. The assumption is that most of the proteins involved in DNA replication are of cellular origin and include DNA polymerase α and/or δ.

Gene Expression

There are two large open reading frames (ORFs) in the parvovirus genome. One in the left half of the genetic map encodes nonstructural proteins and the other, occupying the right half, encodes the structural proteins.

All parvovirus genomes have a promoter near the left end at map position 4–6. This promoter initiates transcripts encoding the nonstructural gene(s) and in the case of the human B19 virus is also responsible for the mRNAs from which the viral capsid proteins are translated. All other known parvoviruses have a second promoter at map position 38–40 for the transcripts for the coat proteins. AAV is unusual in that there is a third promoter at map position 19 which also initiates a transcript encoding two of the nonstructural *rep* genes. A third promoter has also been suggested as a possibility in the case of bovine parvovirus (BPV). All of the parvoviruses possess a common polyadenylation signal near the right end of the genome at map position 96. In addition, several parvoviruses have a second polyadenylation site near the middle of the genome [B19 and Aleutian disease of mink virus (ADV)].

The pattern of splicing is highly variable for different members of the family. However, all of the primate AAVs have similar introns, as do the autonomous rodent viruses. In some cases, alternative splice donors and acceptors are used for essentially the same intron. The significance of this variability is not always clear. In two cases though, AAV and minute virus of mice (MVM), the use of alternative splice acceptor sites is important in the translation of two different sized coat proteins with in-phase overlapping amino acid sequences from the same primary transcript.

As in any biological system parvovirus gene expression is tightly regulated. Numerous *cis*-active sequences responsive to transcriptional activators for RNA polymerase II, the enzyme responsible for parvovirus transcription, have been identified upstream of the various promoters. In addition to the cellular transactivators, viral nonstructural proteins also serve as transcriptional activators, as particularly well demonstrated for the autonomous rodent viruses and for AAV. In the cases of these viruses transactivation is clearest for the coat protein promoter at map position 38–40. For AAV, the *rep* gene product clearly transactivates its own expression as well. However, transactivation of AAV *rep* gene expression by Rep is highly dependent on the intracellular milieu. Under nonpermissive conditions, e.g. HeLa cells in the absence of an adenovirus co-infection, the Rep protein(s) appears to inhibit gene expression from the promoters at both map positions 5 and 19. These paradoxical effects on gene expression are a graphic illustration of the interaction of cellular and viral regulatory proteins. In addition to the *cis*-active sequences that are components of the several promoters, the AAV terminal repeats have been demonstrated to function as enhancers of gene expression.

In addition to regulation of gene expression in *trans*, there is evidence that AAV regulates accumulation of mRNAs from which the nonstructural proteins are translated by a mechanism involving *cis*-active sequences within the nonstructural proteins ORF. The rate of initiation of transcription is equal from all three AAV promoters but much more of the mRNA from which coat proteins are translated accumulates. Deletion of sequences from within the *rep* ORF allows an equal accumulation of the mRNAs from which Rep proteins are translated.

A different type of regulation of gene expression has been demonstrated for MVM. The prototype strain MVMp replicates in fibroblasts but not in T cells. A variant MVM has the opposite phenotype; it replicates in T cells, but not in fibroblasts. The tissue specificity appears to be at the level of transcription, but the genetic difference maps to just a few nucleotides in the middle of the structural protein ORF. Interestingly, significant levels of transcription are not seen after transfection of either the MVMp or MVMi genome into T cells, in contrast to virion infection, which demonstrates significant differences, i.e. MVMi infection leads to readily detectable transcript accumulation. Both the genetic mappings and data that result from a comparison of transfection vs infection suggest strongly that an MVM structural protein may play a role in the regulation of gene expression. Finally there is a temporal regulation of early vs late transcription of MVM and BPV, but this has not been detected for AAV.

Proteins

In common with other small viruses, the genetic information for the various parvovirus proteins is encoded in an overlapping fashion within the genome. Synthesis of different proteins from primary transcripts covering the same regions of the genome is accomplished by alternative splicing, use of alternative promoters and polyadenylation signals, use of alternative initiator codons and post-transcriptional modification.

The left side ORF encodes nonstructural proteins. To date two nonstructural proteins have been identified for the autonomous viruses and four nonstructural (*rep*) proteins have been identified for AAV. There are two promoters in the left half of the AAV genome leading to two transcripts, both of which function as mRNAs in both spliced and unspliced forms, hence there are four mRNAs and four proteins. The two autonomous nonstructural proteins are the consequence of alternative splicing. The larger autonomous viral nonstructural proteins are phosphorylated, but the AAV counterparts do not appear to be modified in this way. The two larger AAV rep proteins and the larger autonomous viral nonstructural protein function both in regulation of gene expression and directly in DNA replication. Proteins of this type from all parvoviruses appear to be potent regulators of gene expression for heterologous genes. In most cases the effect on heterologous gene expression has appeared to be inhibitory. The function(s) of the smaller nonstructural proteins is more difficult to determine. Selective inhibition of the two smaller AAV Rep proteins did not inhibit formation of the replicative DNA intermediate, but did block the sequestration (or 'peeling off') of single strands from the replicative intermediate. On the other hand, selective mutation of the MVM smaller nonstructural protein had a variable effect. The mutants were completely blocked in replication in murine fibroblasts, but suffered only a partial inhibition of replication in other cell types. This is in contrast to the complete inhibition of replication observed in all cells in the absence of functional large nonstructural proteins.

The larger nonstructural proteins bind to the termini of the parvovirus genome. Binding in the case of AAV is dependent on the hairpin structure. The large AAV Rep proteins have been demonstrated to selectively nick the hairpin at nucleotide 124 from the original end and to function as an ATP-dependent helicase. The large nonstructural protein of MVM is found covalently attached to the 5′ termini of replicative intermediates *in vivo*. This has not been observed for the larger AAV Rep proteins, but *in vitro* one of the large AAV Rep proteins does covalently attach to a 5′ end during the site-specific nicking reaction.

The ability of the larger nonstructural proteins to affect gene expression, particularly at the level of transcription, does not seem to involve direct protein–DNA interaction. It appears more likely that the effects on transcription are indirect and a consequence of interaction with various transactivators of cellular origin. The phenotypic effects of the nonstructural proteins are another manifestation of the economy of codon usage and efficiency of function demanded by such a small genome.

The parvovirus genome encodes either two or three coat proteins within the right side ORF. The coat proteins range in size from 69 to 90 kD and have overlapping amino acid sequences. Larger coat proteins are extended at the N termini. Primary differences in size are the consequence of alternative splicing of transcripts leading to use of different initiator codons. Two mechanisms account for the existence of a third coat protein for MVM and AAV. In the latter case an unusual ACG-initiator codon is used infrequently instead of the downstream AUG to produce a protein of intermediate size. For MVM the smallest coat protein is the consequence of post-translational cleavage after virion assembly.

Replication Strategy

The major taxonomic distinction between the vertebrate parvoviruses has been between the autonomous parvoviruses and the dependoviruses or AAV. The former require healthy dividing cells of the appropriate species and tissue for a productive infection. No healthy dividing cells have been discovered which are permissive for AAV. An AAV productive infection requires either co-infection with a helper virus (adeno- or herpesvirus) or in special cases, exposure of cells to toxic conditions, e.g. metabolic inhibitors, chemical carcinogens or UV irradiation. If the helper virus is appropriate (i.e. the cell is permissive for the helper), then the host range of AAV is very broad. The helper functions of adenovirus have been identified. All are general regulators of gene expression which alter the intracellular milieu and cause some genes to be expressed (e.g. adenovirus and AAV genes) and others to be inhibited. Some cellular genes are turned on which are also turned on by exposure of the cell to toxic stimuli. It is this type of intracellular milieu which seems to be permissive for AAV multiplication. In a healthy human cell in culture, the AAV virion is uncoated, but there is little detectable RNA, DNA or protein synthesis. However, under these conditions AAV DNA has a high propensity to integrate into the cellular genome. Recently it has been shown that a majority of the time the integration is site-specific for the q arm of human chromosome 19. The integrated copies of the viral DNA are integrated as a tandem repeat and are maintained for more than 100 passages in culture. The provirus can be rescued by superinfection of the latently infected cells by either adenovirus or herpesvirus.

Thus, while the autonomous viruses appear to follow a straightforward lytic cycle of replication in cell culture, AAV seems to have a strategy involving latent infection. AAV virion production appears to be the result of an event, either helper virus infection or exposure to toxic conditions, that is likely to lead to cell damage or death. The ability of the autonomous viruses to integrate to establish latent infection is not yet proven. There have been data that suggest that the human B19 virus may establish latent infection in the intact host. It is interesting to note that both AAV and B19 have inverted terminal repeats which may be important for either integration or rescue. In cell culture, recombinant plasmids containing the double-stranded form of autonomous parvovirus DNA (e.g. BPV or MVM) are infectious, so that rescue from the plasmid must occur. However, integration of the genomes of these viruses after infection has not been observed (with the possible exception of B19 as noted above).

The ability of the AAV genome to integrate in a site-specific manner with relatively high frequency has led to development of AAV as a vector for potential gene therapy. Two types of vectors have been made: (1) the AAV terminal repeats are attached to the ends of heterologous sequences up to 4 kb in length and the construct packaged using the appropriate AAV genes in *trans*; (2) the heterologous sequences are inserted into a deletion in the ORF for the structural proteins. The latter approach may be important because of the possibility of involvement of the AAV Rep protein in integration.

Future Perspectives

The members of the family *Parvoviridae* share a common structure, a linear single-stranded genome with a common genetic arrangement and seemingly similar gene products. The actual biological properties and replication cycles, however, are widely variable, ranging from active lytic infection of healthy dividing cells to a propensity for latent infection with suppression of productive infection in the same type of healthy cells. It will be the determination of the details of the mechanisms underlying these opposing outcomes that gives us the final understanding of the molecular biology of parvovirus replication.

See also: Latency; Vectors.

Further Reading

Berns KI (1990) Parvovirus replication. *Microbiol. Rev.* 54: 316.

Cotmore SF and Tattersall P (1987) The autonomously replicating parvoviruses of vertebrates. *Adv. Virus Res.* 33: 91.

Cats, Dogs and Mink

Colin R Parrish
Cornell University
Ithaca, New York, USA

History

The parvoviruses affecting carnivores have a variety of histories. Feline panleukopenia virus (FPV) has been known as an infectious agent of cats since before 1900, and was shown to be a filterable virus during the 1920s. The disease of cats was originally described by a variety of names (e.g. feline distemper, feline infectious enteritis), and during the 1930s several studies showed that those diseases were caused by the same virus. A similar viral disease of raccoons was described during the 1940s, and it appears that the raccoon parvovirus (RPV) is very similar to or identical with FPV. During the 1960s FPV was shown to be the etiologic agent of feline cerebellar ataxia.

Mink enteritis virus (MEV) was first recognized during the late 1940s as the cause of an outbreak of enteric disease amongst mink in Fort William, Ontario, Canada. Initially called Fort William disease, and later mink viral enteritis, the disease was recognized throughout the USA and Europe during the next decade. Both FPV and MEV were isolated in tissue culture cells during the 1960s, once the requirement of the viruses for rapidly dividing cells for replication was recognized.

Canine parvovirus (CPV) was first recognized as the cause of new diseases in dogs during 1978, and the virus was isolated from dogs in Europe, the USA, Australia and many other countries throughout the world during that year. Serological studies indicate that CPV was a new virus of dogs which was probably first widespread among dogs in Europe around 1974–1976, and which became globally distributed during 1978.

Minute virus of canines (MVC) was first identified during the 1960s when it was isolated from the feces of clinically normal dogs. Recent evidence indicates that MVC can cause reproductive disease in dogs, and may also be associated with some cases of canine enteritis. Disease caused by MVC appears to be rare, however, and the limited number of cultured cells permissive for the virus means that the virus is rarely isolated or identified.

Aleutian mink disease, caused by Aleutian mink disease virus (ADV), was first recognized around 1956 in ranch mink of the Aleutian genotype (see below). It was recognized later that all types of mink could develop Aleutian disease. The virus was identified during 1975, and its identity as a parvovirus confirmed during 1980.

Taxonomy and Classification

FPV, CPV, MEV and ADV are all classified among the autonomous parvoviruses in the genus *Parvovirus*, within the family *Parvoviridae*. CPV, FPV and MEV are classified as host range variants of the feline parvovirus. RPV has not been classified, but available evidence suggests that it is similar to or identical with FPV. MVC is classified as a possible parvovirus.

CPV, FPV, MEV and RPV are all very closely related antigenically, and isolates differ by < 2% in DNA sequence, while ADV is < 50% similar in DNA sequence to the other viruses. The precise relationship of MVC to the other viruses has not been established, although DNA restriction mapping analysis indicates that it is genetically distinct from CPV and FPV, and no serological relationship has been observed.

Geographic and Seasonal Distribution

CPV, FPV and the related viruses, as well as ADV, are distributed as widely as their hosts. Limited studies indicate that in most countries the viruses are endemic, although vaccination has largely controlled reported clinical disease by CPV, FPV and MEV amongst many domesticated animal populations. Serological studies indicate that MVC is widely distributed amongst dogs in the USA, but its distribution has otherwise not been extensively studied.

A seasonal distribution with greatest incidence of disease in the summer has been described for FPV in cats, most likely related to seasonal breeding patterns. The incidence of disease may be related to the population density of the suceptible host – in endemically infected areas, these are mostly young seronegative kittens or puppies. MEV disease is observed mostly during the summer, again due to the densities of susceptible kits, as mink breed and whelp once per year. Incidence of disease appears to be related to the density, immune status and size of the susceptible population.

Acute interstitial pneumonitis caused by ADV is

seen only after infection of newborn mink, and therefore is observed only in spring and summer.

Host Range and Viral Propagation

CPV, FPV, MEV and RPV

The host ranges of these viruses have been defined primarily on the observations of disease. All susceptible hosts appear to be members of the order Carnivora, the various viruses causing disease in many species of large and small cats, most members of the family *Canidae*, including wolf-like canids, Asiatic raccoon dogs (*Nyctereutes procyonoides*), Arctic foxes (*Alopex lagopus*) and probably red foxes (*Vulpes vulpes*), common raccoons (*Procyon lotor*), mink (*Mustela vison*) and ringtail coatis (*Nasua nasua*). Ferrets (*Mustela putorius*) do not appear to be naturally susceptible, but cerebellar disease can be induced by experimental infection of neonatal kits.

CPV infects wolf-like canids, as well as South American canids and raccoon dogs. As well as cats and raccoons, FPV appears to infect foxes. Although FPV and MEV can both infect mink, only MEV isolates appear to be virulent for mink, and it appears that MEV isolates are differentiated from the other viruses by their ability to cause enteric disease in mink.

MVC

MVC is known only in the domestic dog, and the host range has not been investigated.

ADV

Disease caused by ADV has been observed mostly in mink, although ferrets are also susceptible to infection and disease. Other carnivores [skunks (*Memphitis memphitis*), martens (*Martes* sp.)] may be susceptible to infection, but disease has not been described.

Propagation

CPV, FPV, MEV and RPV

Although these viruses replicate only in a limited number of tissues in the older host animal, this is at least partly due to the dependence of viral replication on cellular DNA replication. These viruses can grow in most primary cells or cell lines of feline origin. Only CPV grows well in canine cells in culture. Some cells may not propagate the viruses (e.g. the MDCK cell line appears to be insusceptible to CPV).

ADV

Wild-type ADV isolates grow poorly or not at all in most tissue culture cells, although certain viral strains have been adapted to feline tissue culture by passaging at lower temperature (~ 31.8°C). The viral strains propagated in tissue culture often do not replicate in mink. ADV isolates will replicate in mitogen-stimulated mink lymphocytes (primarily B cells) in culture.

Structure

CPV

The atomic structure of the CPV capsid has been solved by X-ray crystallography (Fig. 1). This shows that the capsid shell is assembled from 60 copies of the region of the structural proteins common to VP1 and VP2. The exterior of the capsid is 22.4 nm and 28 nm at its narrowest and widest diameters respectively.

Features observed on the capsid include a prominent 2.2 nm high by 7 nm wide 'spike' at the threefold axis of rotational symmetry. A hollow cylinder around the fivefold axis apparently allows the DNA genome or the N-terminal sequences of some copies of VP1 or VP2 to pass through in full (DNA-containing) particles. There is a canyon surrounding the fivefold cylinder and a depression or dimple on the twofold axis. DNA is seen to be associated with the interior of the protein coat – about 11 bases of single-stranded (ss)DNA with each VP2 equivalent.

Neutralizing epitopes are primarily affected by mutations of residues on the surface of the threefold spike, and a mutation which affects hemagglutination of CPV is adjacent to the twofold dimple.

ADV

Virus recovered from persistently infected mink tissues is degraded to a number of lower molecular weight forms. Particles prepared from tissue culture or from acutely infected neonatal mink contain intact proteins. Monoclonal antibody analysis shows that there are antigenic epitopes specific for the degraded or intact forms of the proteins.

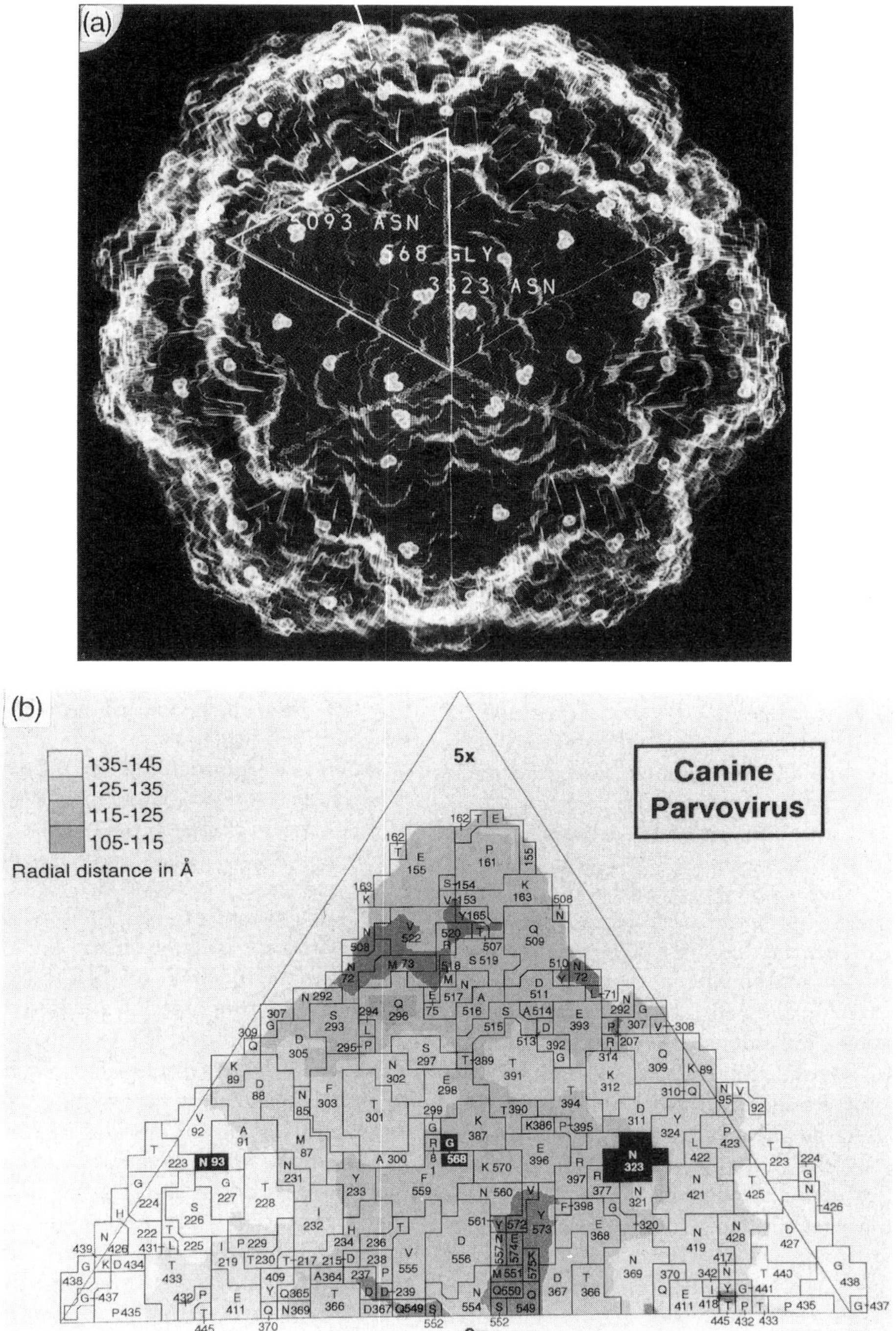

Fig. 1 (**a**) The surface-exposed differences between CPV and FPV shown on the structure of the CPV particle which was determined by X-ray crystallography, as described by Tsao J *et al.* (1991) (The three-dimensional structure of canine parvovirus and its functional implications. *Science* 251: 1456.) The particle model is shown by grid-mesh surfacing, and the view is along the threefold axis of symmetry. The particle is comprised of 60 copies of VP2 (or the very similar VP1). The icosahedral face of the virus indicated by the triangle is comprised of portions of various symmetry-related VP2 molecules. Of the three variant amino acid residues shown on the surface (residues 93, 323 and 375 in the VP2 sequence), residues 93 and 323 are contributed from either the fivefold-related VP2 protein (5093) or the threefold-related protein monomer (3323). (**b**) One icosahedral unit of the virus capsid showing a 'roadmap' of the surface-exposed amino acids of CPV. Surface-exposed residues which vary between CPV and FPV are indicated as reversed color. The contours indicate the surface features, including a 'canyon' surrounding the fivefold axis of symmetry and a 'dimple' spanning the twofold axis of the virus. Figures prepared by Jean-Yves Sgro, Institute for Molecular Virology, University of Wisconsin, Madison, WI, USA.

Genetics

CPV, FPV and the closely related viruses contain ssDNA of about 5200 bases, encapsidating the negative-sense strand (complementary to the mRNA). The ADV genome is about 4750 bases of ssDNA. The structural proteins (VP1 and VP2) are encoded towards the right hand end of the genome, while the nonstructural proteins are encoded towards the left hand end. The CPV/FPV-like viruses are thought to express genes from two promoters (at 4 and 40 genome map units), and alternative splicing gives rise to messages for VP1 and VP2, and at least one nonstructural protein (NS1). Another NS protein (NS2) may also be present which would share its N-terminal region with NS1, but would have a different C-terminal region derived by alternative mRNA splicing.

Mapping of host range differences between CPV and FPV by genetic recombination analysis shows that the canine host range of CPV is determined by a combination of surface-exposed amino acid residues within the capsid protein which also determine a CPV-specific antigenic epitope.

The ADV genome contains two promoters, and encodes two structural proteins (VP1 and VP2), as well as two or three nonstructural proteins. The structural proteins are derived from the same mRNA by differential use of translation start codons, probably by ribosome scanning. The various NS proteins are derived by differential splicing. There are two polyadenylation sites in the ADV genome, one in the middle of the genome which is used for some NS protein gene messages, while the poly(A) addition site near the 5′ end of the genome is used for the structural and some of the nonstructural gene messages.

Infectious plasmid clones which contain almost-complete genome sequences have been prepared from CPV, FPV, MEV and ADV. After transfection of susceptible cells the virus genomes resolve from the plasmids by specific nicking reactions and infectious viruses are recovered which are indistinguishable from the original viruses.

The genetic structure of MVC has not yet been examined.

Evolution

The viruses vary at relatively low rates in nature, and CPV, MEV and FPV isolates sequenced differ by less than 2% of their VP1/VP2 gene sequences over several decades. Sequence analysis shows that the FPV, MEV and RPV isolates are all closely related and comprise one phylogenetic cluster, while CPV isolates form a distinct cluster. The CPV isolates are evolving in a linear fashion over time, with a progression of sequences from CPV type-2–CPV type-2a–CPV type-2b. An estimate of the rate of DNA sequence variation of CPV since its first emergence is $\sim 1.69 \times 10^{-4}$/nucleotide per year.

Serological Relationships and Variability

By conventional serological techniques [HA inhibition (HI), serum neutralization], CPV, MEV, FPV and other closely related viruses are difficult to distinguish. Some differences can be revealed by carefully controlled HI and plaque neutralization tests, but the differences are too small to be readily used for diagnostic purposes.

However, monoclonal antibody analysis readily reveals differences between CPV isolates and the other viruses, there being at least one epitope on all CPV isolates which is not present on any of the other viruses. There is also one (sometimes two) FPV-specific epitope on FPV, MEV and RPV isolates. A virus isolated from an arctic fox appears very similar to FPV or MEV by DNA sequence analysis and monoclonal antibody typing, and it is quite distinct from CPV isolates.

Monoclonal antibody analysis has revealed variation in antigenic type among MEV isolates collected during the 1970s and 1980s in the USA and Scandinavia. The prevalence of the strains observed was not related to the year of isolation, suggesting that these strains are not replacing each other. No difference between the strains was seen in cross-protection studies of animals when vaccinated mink were challenged with the various antigenic types of virus.

Two variants of CPV have emerged since that virus was observed in 1978. The first variant strain (designated CPV type-2a) emerged around 1979, and by 1981 had replaced the original virus type (designated CPV type-2) in the USA, Denmark, Japan and Australia. A further variant strain (designated CPV type-2b) emerged around 1984, and by 1988 had become the predominant virus type isolated from dogs in the USA. The CPV-2a and CPV-2b strains each differ from the previous virus type in reactivity of one or two epitopes in the capsid.

Naturally variant strains of ADV have been defined antigenically by typing with monoclonal antibodies. The various ADV isolates differ in virulence for mink. Variation of ADV has also been shown in the DNA isolated from viruses within ADV preparations, probably representing mixed virus preparations in tissues used for inoculation. The nucleotide sequences varied by up to 5% between the three principle types of virus sequence defined (designated types 1, 2 and 3), but the significance of that variation has not been defined.

Epidemiology

CPV, FPV and MEV are shed in high titers in the feces of infected animals, and transmission is via the fecal–oral route. The viruses are highly resistant to inactivation, and may survive in the environment for up to several months. Animals that recover from infection appear to be protected for life from re-infection. Animals born to immune females are protected from infection by maternally derived immunity, and so the susceptible cohort is generally young animals with waning maternal immunity. Vaccination is able to efficiently protect animals, but effective immunization is also blocked by maternal immunity.

Spread of the viruses can be very rapid even over long distances – indeed, during 1978, CPV spread around the world within a few months, even into countries such as Australia and New Zealand with strict quarantine for dogs.

The role of antigenic variation in the epidemiology of these viruses is poorly understood. Antigenic variants of MEV all appear to co-circulate in mink. Variation of CPV has involved the sequential replacement of one viral type by the next, giving rise to the CPV-2a and CPV-2b antigenic types. The selection for the new types is presumably at least in part for the antigenic differences between the viruses, although other epidemiological advantages of the replacing strains may also be present, such as stability of the virus, or increased ability to infect or shed from dogs.

ADV is shed in the feces, urine and saliva, and as such is readily spread between animals. However, the virus can be eliminated from mink ranch populations by serological testing and subsequence culling of antibody-positive animals, indicating that infected animals do not continuously shed virus.

Pathogenicity

CPV and FPV

Natural differences in virulence of CPV, MEV, RPV and FPV in their normal hosts have not been defined. However, viruses from the various hosts may show differences in virulence when infected into a heterologous host species. For example, although FPV isolates replicate in mink they do not cause enteric disease, whereas MEV isolates cause severe clinical disease in mink. Attenuated strains of FPV, CPV and MEV have been derived by passage of the viruses in tissue culture, and ADV and CPV strains which have lost the ability to infect their natural hosts have also been derived by passaging in culture.

The viral factors that result in attenuation of the viruses have not been defined, and the pathobiological basis of the attenuated phenotype is not known. However, the attenuated CPV vaccine virus is shed at lower titers in the feces, suggesting a reduced replication in the intestinal epithelial cells.

ADV

The pathogenesis of ADV disease is at least partly dependent on the genotype of the mink infected. Aleutian mink (named after the Aleutian blue foxes, which have a similar coat color) are most susceptible to clinical disease; these mink suffer from a lysosomal abnormality similar to human Chediak–Higashi syndrome. However, normal mink also suffer from a similar disease after infection by at least some strains of ADV. Variation of the virulence of ADV strains has also been described.

Clinical Features of Infection

Dogs, cats, mink and raccoons affected by CPV, FPV, MEV or RPV respectively suffer from similar enteric diseases. In animals older than about 5 weeks at the time of infection, the first clinical sign is pyrexia between 3 and 5 days after infection, and shortly thereafter virus is shed in the feces. The animals become depressed and lethargic, and may show a panleukopenia (cats) or a relative lymphopenia (dogs). In a proportion of infections (generally fewer than 20%) the animals may develop severe enteritis with diarrhea and sometimes vomiting. Feces may be mucoid, liquid and flecked with blood. The severity of the disease depends on the condition of the animal – food deprivation prior to infection gives a more severe clinical disease. Animals may become severely dehydrated and die, possibly as a result of the dehydration and possible endotoxemia. Virus shedding stops once the immune response develops, and surviving animals recover without apparent long-term sequelae. It has been suggested that co-infections with other viral or parasitic agents may give a more severe disease due to increased replication of crypt epithelium cells of the small intestine.

A number of different syndromes have been described after infection of neonatal or fetal animals. Ataxia may be seen after FPV infection of kittens or ferrets, where the virus replicates in the external germinal cerebellum, resulting in cerebellar hypoplasia. Puppies may develop a multifocal necrosis of the myocardium as a result of CPV infection. The animals die of acute heart failure days to weeks after infection. Many of the clinical signs observed are a consequence of the heart failure, such as lethargy and dyspnea. Electrocardiographically the pups show a variety of subclinical abnormalities, and death results from ventricular fibrillation.

ADV

The various clinical syndromes recognized for ADV infection depend on the age, immune status and genotype of the mink. Infection of mink during pregnancy can result in fetal death or abortion. Kits born to ADV noninfected dams are susceptible to developing interstitial pneumonia after neonatal ADV infection, with clinical signs being observed 9–20 days after inoculation. Affected kits suffer from respiratory distress and lethargy, and most die within 24 h of the first observation of clinical signs.

Kits from immune dams or older (immunocompetent) animals develop a chronic disease which, after infection with a highly virulent strain, leads to death of a high proportion of infected animals due to immune-complex-mediated diseases. Affected animals often show few obvious clinical signs until shortly prior to death due to renal failure or to the rupture of inflamed arteries or enlarged spleens.

ADV disease is characterized by a hypergammaglobulinemia with circulating immune complexes. The IgG concentration can reach very high levels and in a proportion of affected animals there is a restricted heterogeneity of the IgG produced, suggesting a clonal expansion of some B-cell populations. Much of the antibody response is directed against the structural and nonstructural proteins of the virus.

Pathology and Histopathology

CPV and FPV diseases

The pathogenesis of these viruses in their respective hosts is determined by the requirement of the parvoviruses for dividing cells for their replication. In fetal or neonatal animals viruses may replicate in a wide variety of tissues. In older animals the viruses replicate primarily in the lymphoid tissues and the intestinal epithelium.

Initial replication after oronasal infection occurs in the tonsils and in the regional or mesenteric lymph nodes and subsequently in the thymus. Between 4 and 6 days after infection the virus infects the intestinal epithelial cells, and high titers of virus are shed in the feces. The loss of the regenerating epithelial cells results in a flattened attenuated epithelium and shortened intestinal villi. These changes may lead to loss of osmoregulation, resulting in diarrhea. Lymphoid tissues show lymphocytolysis and cellular depletion, followed by tissue regeneration and cellular repopulation in surviving animals. The spleen and bone marrow may be affected with a marked decrease in cellularity and decreases in cells of the myeloid, erythroid and megakaryocytic series. Whether these effects are due to direct viral replication or secondary effects, perhaps to endotoxemia, are unknown at present.

Feline ataxia results from viral replication in the external germinal epithelium of the developing cerebellum of newborn kittens, resulting in cerebellar hypoplasia.

In puppies which develop myocarditis there is a diffuse nonsuppurative myocarditis and often a mononuclear cell infiltration. Myocardial cells often contain Feulgen test-positive amphophilic inclusion bodies. Edema of the lungs is secondary to the acute heart failure.

ADV

In seronegative newborn mink an interstitial pneumonia results from virus infection of alveolar type II cells of the lung, resulting in decreased surfactant production and causing the observed hyaline membrane disease and respiratory distress.

In immune or older animals the virus replicates in various lymphoid tissues, particularly in follicular dendritic cells and macrophages. Probably as a result of infection of such antigen-presenting cells, and also due to the chronic infection and antigen expression by the virus, mink develop very high levels of antibody, and immune complexes are deposited in the kidney tubules and arteries.

Immune Responses

CPV, FPV and MEV

Antibodies appear to play the major role in the active immunity to and recovery from these parvovirus infections. Pups or kittens that acquire antibody from their immune dam are protected against infection until their circulating antiviral antibodies fall to very low levels. Parenterally transferred antibodies will likewise protect dogs against infection by CPV, the antibody acting to prevent systemic replication and spread of the virus. Any role for cellular immunity or secretory antibody is unknown at present. Virus may replicate locally in the gut of passively immunized animals, although at lower levels than in nonimmune animals.

ADV

Effective protective immunity against ADV infection, or immunity leading to virus clearance, has not been reported. Although ADV replication is restricted in mink in the presence of high titers of circulating anti-viral antibodies, the virus is not neutralized. The lack of neutralization is due to apparent protection of the virus by a coating of phospholipids or by viral aggregation, and ADV virus is readily neutralized by anti-viral antibody after detergent or organic solvent treatment.

Prevention and Control of Virus Infection

CPV, FPV and MEV

Vaccination efficiently protects animals against infection by CPV, FPV and MEV. Both inactivated virus and modified live virus vaccines are efficacious, although repeated doses are required to give long-lived protection with inactivated virus vaccines. Maternal immunity will prevent successful vaccination of animals. It appears that wild-type virus from the environment can infect animals with maternal antibodies at an earlier time than the same animals can be successfully vaccinated by parenteral routes. The cause of this 'window' of susceptibility to wild-type virus is not understood.

The control of virus spread between animals is difficult since the viruses are long lived in the environment, and also is shed in high titers in the feces. However, the virus may be inactivated with dilute hypochlorite solutions.

ADV

No vaccine is available against ADV, since the immune response is inextricably involved in the pathogenesis of the disease. Control measures involve the identification and culling of infected animals through serological testing, and control of re-infection by quarantine.

Future Perspectives

In future studies the basis of the host range and virulence differences between CPV, FPV and MEV will be defined, and the common features defined, at least in part, in terms of changes in the viral structure. The possible ancestors of CPV amongst the other carnivores should be defined by phylogenetic studies.

The restricted but persistent replication of ADV in mink will be examined, and the roles of the specific immune responses in the restricted replication will be defined.

See also: Immune response; Pathogenesis; Vaccines and immune response; Virus structure.

Further Reading

Parrish CR (1990) Emergence, natural history, and variation of canine, mink, and feline parvoviruses. *Adv. Virus Res.* 38: 403.

Porter DD, Larsen AE and Porter HG (1980) Aleutian disease of mink. *Adv. Immunol.* 29: 261.

Tsao J *et al.* (1991) The three-dimensional structure of canine parvovirus and its functional implications. *Science* 251: 1456.

Rodents, Pigs, Cattle and Geese

Peter Tijssen and J Bergeron
University of Québec, Institut Armand-Frappier
Québec, Canada

Introduction

Most parvoviruses were discovered by chance and their capacity to produce disease became apparent only later. A predilection for rapidly dividing cells is at the basis of their pathogenicity as illustrated, by infection of the fetus but also by their oncosuppressive activities. Parvoviruses are among the smallest viruses known and the only viruses with linear single-stranded (ss) DNA genomes.

History

Three rodent parvoviruses [minute virus of mice (MVM), H-1 (its rodent origin is sometimes questioned) and rat virus (RV)] and their variants have been isolated. RV was the first parvovirus isolated, in 1959, when Kilham and Olivier were searching for a rat papovavirus in rat embryo cell cultures. Toolan observed that neonatal hamsters developed a mongoloid-like malformation after injection of cell-free filtrates of rat-passaged transplantable human tumor Hep-1. A parvovirus was isolated and named H-1. MVM was isolated by Crawford from laboratory stocks of mouse adenovirus. These three rodent parvoviruses lack serological relationships by hemagglutinin-inhibition (HI), virus neutralization and complement fixation tests. However, several variants exist for each of these parvoviruses such X-14, LS, Krisini, 9HV, HER, H-3 variants of RV, HT variant of H-1 and MVMp and MVMi variants of MVM(RC).

Porcine parvovirus (PPV) is a major disease-causing agent in pigs and, probably, in horses. Although the major syndrome (SMEDI or *s*tillbirths, *m*ummification, *e*mbryonic *d*eath and *i*nfertility) was described by Howard Dunne and co-workers in the mid-60s, PPV was only isolated for the first time in 1975. PPV strains have been isolated from different sources and have also been associated with other syndromes (respiratory infections, dermatitis, enteritis, rhinitis, bowel edema) as well as equine abortion. Different tropisms among variant strains are also a striking feature of other parvoviruses. An example is the sudden appearance of canine parvovirus in the late 1970s probably due to mutations in a feline parvovirus. This flux in parvovirus strains and properties makes them very important, both in disease prevention and fundamental research.

Bovine parvovirus (BPV) was among the first parvoviruses known when isolated by Abinanti and Warfield in 1961. Originally, this virus was known by its acronym HADEN ('hemadsorbing enteric') which indicated the ability of infected cells to adsorb erythrocytes and the isolation of the virus from gastrointestinal tracts.

Goose parvovirus (GPV) causes a highly contagious and fatal disease of young geese. This disease emerged in the middle of the 1960s and was known under different names: goose influenza (Hungary), infectious myocarditis (Italy), hepatonephritis ascites (France), viral enteritis (former USSR), goose plaque (Netherlands) and goose hepatitis (Germany). The working group of the World Poultry Scientific Association named it Derzsy's Disease after the late Domokos Derzsy. In 1971, Schettler isolated the causative agent and showed that it is a parvovirus.

Taxonomy and Classification

All known viruses with a linear ssDNA genome belong to the *Parvoviridae* family. Traditionally, parvoviruses have been classified into three genera: (1) *Parvovirus* (autonomous parvoviruses of vertebrates), (2) *Dependovirus* (parvoviruses defined as requiring a helper virus such as adenovirus), and (3) *Densovirus* (parvoviruses of invertebrates). All viruses discussed in this entry belong to the *Parvovirus* genus. This classification needs to be revised since many exceptions exist to the original criteria. For example, some members of the *Parvovirus* genus, defined as encapsidating only negative-strands, also encapsidate positive-strands (10% for BPV, 50% for LuIII). While *Parvovirus* members are autonomous, helper viruses can increase their replication substantially. On the other hand, *Dependovirus* members can sometimes replicate autonomously. Among the four parvovirus groups discussed here, the rodent and porcine parvoviruses are very similar and will, upon reclassification, be placed in the same genus. BPV is quite different, and a definitive classification of GPV is not possible before details about its molecular biology are known.

Properties of the Virion

Parvoviruses are isometric nonenveloped particles with a diameter of 20–25 nm and a buoyant density in CsCl of about 1.40 $g\,ml^{-1}$ (full particles) or 1.30 $g\,ml^{-1}$ (empty particles). The sedimentation coefficient is about 120S for full particles and 60S for empty particles. The number of capsomers has been disputed with estimates between 12 and 42. The capsid contains three (MVM, PPV) or four (BPV) different proteins (Table 1). Data about the molecular biology of GPV are lacking. There are about 60–76 protein molecules per virion. The 60 kD structural proteins of BPV, MVM and PPV

Table 1. Viral gene products (kD)

Proteins	MVM, H-1	PPV	BPV
Nonstructural proteins	NS1 (83)	NS1 (86)	NS1A (83)
	NS2 (21)	NS2 (18)	NS1B (75)
		NS3 (12)	NS2 (28)
Capsid proteins	VP1 (86)	VP1 (84)	VP1 (80)
	VP2 (65)	VP2 (64)	VP2 (72)
	VP3 (60)	VP3 (60)	VP3 (62)
			VP4 (60)

Note. Molecular masses of NS2 and NS3 of PPV are deduced from amino acid sequences.

are proteolytic cleavage products from 64–62 kD proteins. The structural proteins are produced by a nested set of genes and have identical C-terminal sequences. The structural proteins are not glycosylated.

Parvoviruses are among the most stable viruses. Nonpurified virus has been shown to withstand 2 h at 80°C, although purified viruses are not as stable (e.g. resistant for 1 h at 56°C). Most parvoviruses are also stable within the pH range 2–11. Sonication (without heating) does not affect the virus but it is sensitive to UV irradiation. Resistance to organic solvents, such as butanol and chloroform, and enzymes, such as DNase, RNase and protease, has been noted. Capsids can also withstand many detergents, although they are dissociated in sodium dodecylsulfate (SDS) solutions.

Viral Genome and Molecular Biology of Replication

The rodent and porcine parvoviruses encapsidate virtually only the negative-strand (complementary to mRNA) although about 1% of viral DNA has positive polarity. Genome lengths reported for these viruses are 5085 (MVMi), 5149 (MVMp), 5176 (H-1), 5059 (PPV-NADL2) and 4932 (PPV-IAF6) nucleotides. Interestingly, the nonviremic strains MVMp and PPV-NADL2 differ by a repeated sequence at the 5′ end of the viral genome from those of the viremic strains. The genomic organizations are almost identical for these viruses. All have a 115-nucleotide Y-form left-hand hairpin and a cruciform right-hand hairpin (207 nucleotides for MVM, 242 for H-1 and 203 for PPV). The left-hand hairpin (3′ end) is extended (probably by DNA polymerase α or primase activity) to the right-hand hairpin and is ligated to it. It is nicked about 16 nucleotides upstream from this ligation, probably by NS1. This protein is produced after transcription of the ds intermediate. After nicking, the 3′ end can be extended, while displacing the terminal hairpin, and then folded back on to itself so that a dimeric replication form can be obtained. The hairpin transfer at the 5′ end of the negative-strand thus results in two alternative sequences, one of which is the reverse complement of the other ('flip' and 'flop'). The viral strand is, concomitant with synthesis, excised from the replicative form and encapsidated from the 5′ end in (pro)virions (containing only VP1 and VP2). The VP2 in the 1.44 g ml^{-1} particles are partially converted by proteolysis to VP3 which leads to a decrease in the density to 1.41 g ml^{-1} ('mature' particles). NS1 is still attached to the viral genome to the outside of the capsid and is often removed just before or during infection.

The BPV genome contains 5491 nucleotides. About 10% of BPV virions contain positive-strand DNA, whereas the remainder contains negative-strand DNA. The genome hairpin termini differ from those of the rodent and porcine viruses in several respects. They lack any sequence homology and are 150 (left) and 121 (right) nucleotides long. The left-hand hairpin (3′ end of negative-strand) has both flip and flop orientations, with a 10-fold excess of flip, whereas the right-hand hairpin contains both orientations in equal amounts. A kinetic model has been proposed to account for these differences between the rodent and porcine parvoviruses. No data are yet available on the genomic organization of GPV.

Transcription and Translation

Porcine and rodent parvoviruses have two genes, each coding for several products (Table 1). Their genomic organizations are almost identical. The genes are transcribed from two promoters (P_4 and P_{38}) and the transcripts coterminate near the 5′ end of the negative-strand. The P_4 transcripts code for the nonstructural proteins (left half of the transcripts). A peculiarity of this transcript is that both the nonspliced and the spliced transcripts are translated (NS1 and NS2 respectively). The N-terminal sequences of NS1 and NS2 are identical whereas the C-terminal sequences differ (overlapping ORFs). The intron removed to generate the NS2

Table 2. Properties and roles of NS proteins

NS1	NS2
Nuclear	Cytoplasmic
Phosphorylated	Phosphorylated (N terminal)
Early 1-2 h p.i.	Half-life $\approx$ 1 h
Half-life $\geqslant$6.5 h	Virulence
Cytotoxicity	
Activation of P_4 promoter	
Trans-activation of P_{38} promoter	
Binds to 5′ end of DNA	
Replication	

transcript contains an NTPase (probably ATPase), a *trans*-activation response element, which upregulates expression from P_{38}, and the P_{38} promoter elements. The roles of NS1 and NS2 are summarized in Table 2. We also observed a third nonstructural protein for PPV, NS3, which has the same N-terminal domain as NS1 and NS2 but the reading frame of the C-terminal domain overlaps the ORFs of the structural proteins. The role of this protein is unknown.

The start codon for VP1 of the various rodent and porcine parvoviruses is located almost immediately after the stop codons of NS1/NS2. There are small introns at about 40 map units in all P_{38} transcripts. The donor sites for these small introns are just upstream or just downstream of the VP1 start codon. When the upstream site is used, the start codon is removed and another start codon, almost 800 nucleotides downstream, is used resulting in VP2.

The strategy used by BPV is different: for example, there are three major open reading frames (ORFs) (left, mid and right) instead of two. Analysis of the genome suggests three promoters, P_4, P_{13} and P_{38}. Again, P_{38} is responsible for the production of capsid proteins. Unlike the rodent and porcine parvoviruses, BPV codes for two large proteins and two smaller proteins. BPV shares with all autonomous parvoviruses a glycine-rich region (a stretch of 10–20 G) near the proteolytic cleavage site to generate the smallest capsid protein. This G-run could distort the α-helix structure and make it susceptible to proteolysis. Three BPV NS proteins have been detected (Table 1). The 83 and 75 kD proteins share with NS1 of PPV and rodent parvoviruses the property that they are expressed before the capsid proteins [8 vs 14 h postinfection (p.i.)]. The NS1A and NS1B proteins of BPV may differ in the degree of phosphorylation, although both products can be detected after *in vitro* translation (suggesting primary translation products).

Gel retardation assays demonstrated that both capsid and NS proteins bind to the left end of the BPV genome and may play different roles in replication and packaging. The right end of the genome does not compete with the left end for these proteins. Cellular proteins, particularly during S-phase, also form a specific complex with the DNA.

Viral Morphogenesis

Parvoviruses lack the ability of most other DNA viruses to stimulate resting cells to DNA synthesis. The first steps during infection, i.e. absorption, penetration and uncoating in the cell, do not require viral functions but normal cellular processes. The cellular receptors are under developmental control as cell lineage determines the degree of susceptibility. Mature (containing VP1, 2 and 3) and immature particles (VP1 and 2) compete equally well for receptors. Internalization through coated pits has been observed. The mechanism of penetration of the endosomal membrane is not understood but pH shifts do not appear to play a role. It has been postulated that the virus is transported actively to the nucleus by the signal carried on the first ten amino acids of VP1. Uncoated DNA is not observed in the cytoplasm. The S-phase-dependent event appears to be the synthesis of the complementary strand from the 3′ end of the negative-strand. The resulting duplex can then serve as a template for transcription and expression. Table 2 summarizes the roles of the viral proteins. It should be noted that for MVMi and MVMp, in A9 cells, the steps up to transcription are equally efficient but that transcription and translation differ significantly. Transcription depends thus on developmentally regulated host-cell factors and is not merely a reflection of the presence of duplex DNA. An attenuation site located $\approx$ 145 nucleotides downstream from the P_4 promoter has been identified. This block could be alleviated by specific cellular factors. Moreover, the promoter for the capsid proteins is activated by NS1.

Parvoviruses interact with host nucleoli. Viral DNA replicates in the host nucleoli and nucleolar protein as well as small nuclear ribonucleoproteins are recompartmentalized. Replicative form DNA is exclusively nuclear-matrix-bound early during infection but accumulates later in the soluble fraction. The DNA–matrix interactions are, at least in part, mediated by the viral terminal proteins. ssDNA is encapsidated by the (pro)virion, as described above, usually from the 5′ to 3′ direction [occasionally (< 1%) in the other sense]. This process is probably mediated by NS1 and the virus is ready for a new round of infection after the proteolytic maturation step.

Geographic and Seasonal Distribution

All parvoviruses described here have been isolated world-wide, wherever the host is present. However, some strains may have a wider distribution than others. For example, PPV strains involved in the SMEDI syndrome have been detected throughout the world, whereas strains causing vesicular lesions have only been reported locally. Interactions of PPV with other porcine viruses (e.g. pseudorabies which is present in the USA but not in Canada) are still poorly understood. The equine strain of PPV was shown to be important in equine abortions in Manitoba (Canada) and has been detected in field samples from Québec (2500 km from Manitoba) by our laboratory (unpublished results). This equine strain has not yet been reported in other countries.

It is not known whether similar differences exist for the rodent parvoviruses. A few laboratory strains have been studied extensively whereas the epidemiology and strain distribution have received little attention. Early work showed that many strains differ in serologic or hemagglutinating properties from the laboratory strains (see below). Similarly, several BPV strains have been compared and serologic and hemagglutinating differences have been noted. Caution is warranted since comparisons are often not under identical or standardized conditions. For example, different donors of erythrocytes could be responsible for subtle variations.

Seasonal distributions are related to the age distribution of the host if infection occurs at a certain age. For example, GPV usually infects goslings younger than 1 month, although latent infections can be established in older animals. Thus, domesticated or wild geese are usually infected in spring. Similarly, BPV infects calves but latent infections may develop in older animals. The birthing pattern of the host thus contributes to the seasonality of the infections.

Host Range and Viral Propagation

In general, parvoviruses have a narrow host range, probably due to their high dependence on particular host cell functions. It is becoming increasingly clear that small genomic variations or mutations may lead to different tropisms, affecting a different tissue or host, as well as to a different virulence.

The natural host of RV and MVM are rats and mice respectively, as supported by serology and infection studies. The natural host of H-1 is disputed. Toolan considers man as the natural host of H-1 since the virus has been isolated in newborn hamsters inoculated with cell-free filtrates from human tumors. However, antibody levels to H-1 in humans are usually very low, whereas in laboratory rats it can be as high as 80%. All three rodent viruses can induce an osteolytic syndrome after perinatal inoculation of hamsters. This syndrome is characterized by dwarfism, mongoloid features, abnormal teeth and fragile bones because of infection of osteogenic tissues.

The natural host of PPV strains is the pig. The nature of the strain recovered from equine abortions needs to be established in greater detail. The possibility of cross-infection of pigs and horses by the porcine and equine parvovirus strains is still unresolved. About 40 PPV isolates have been obtained from different cell cultures (KBSH from KB cells is the prototype) and it has been suggested that these viruses were introduced with porcine trypsin used in passaging.

Cattle is the natural reservoir of BPV, but goats and possibly horses, may also be a source since many animals are seropositive. Cross-reactivity with until now unisolated viruses cannot be ruled out. Canine, monkey and human sera may also be seropositive, by immunofluorescence, but very low titers are obtained when they are assayed for neutralizing activity. Some cross-reactivity therefore seems more likely. Domesticated and wild geese as well as Muscovy ducks are susceptible to GPV. Chickens and ducks are resistant.

Parvoviruses are most often propagated in cell culture (Table 3). GPV is usually replicated in embryonated eggs. The reproduction of parvoviruses in tissue cultures can be dependent on the passage number, temperature or the nature of the strain of the same parvovirus. For instance, lymphotropic and fibrotropic strains have been described for MVM, and restricted temperature ranges for different PPV strains (some only at low temperatures, others at 37°C and still others at 39°C). PPV-NADL2 replicates most efficiently in

Table 3. Host range of parvoviruses in some cell cultures

Line	Origin	RV	H-1	MVM	PPV	KBSH	BPV	GPV
AT	Rat	+	+	+		−		
BHK-21	Hamster	+	+	+	−	−	−	−
L	Mouse	−	−	+	−	−	−	−
A9	Mouse	−	−	+				
HeLa	Human	−	+	−	+	+	−	
KB	Human	−	+	−	+	+		
NB	Human		+					
Vero	Monkey	−	−	−	−	−	−	−
PK15	Pig	−	+	−	+	+	−	
PFT	Pig				+			
MDBK	Cattle					+		

Note: MVMp is plaque-selected and will grow in A9 cells but not in whole-mouse embryonic cell cultures. MVMi, on the other hand, hardly reproduces in A9 cells but replicates in cells derived from murine lymphoma EL-4. PPV strains may also show different host cell spectra for reproduction. GPV and BPV replicate exclusively in primary embryonic cells.

PFT cells at passage 75–100 and hardly, or not at all, above 200.

Genetics

Classical genetic studies to determine cistrons and *cis*- or *trans*-acting elements have not received much attention. Rather, parvoviruses have been sequenced to predict the way the genome functions. Subsequently, these predictions were confirmed or rejected by transcription studies and by site-directed mutagenesis of genetic elements (see below).

Evolution

Rodent and porcine parvoviruses are so closely related that they have probably evolved from a common ancestor. Their relatedness with BPV, *Dependovirus* or *Densovirus* is low and decreases in this order, with one interesting exception. The amino acid sequence 10–50 of the VP1 capsid protein is also found in the structural protein of some densoviruses (≈ 90% homology), but in the opposite orientation (unpublished results).

Evolution of parvoviruses is evident even in the short time span in which they have been studied. For example, canine parvoviruses most probably evolved from feline parvoviruses. Many of the newly discovered strains with different tropisms may have evolved very recently since only a few (two to three) amino acids are involved in the tropic determinant. It can be speculated that, after a change to a new tropism through one of these amino acids, other mutations may rapidly accumulate until the optimal reproduction rate is obtained. A virulence determinant is also emerging. NS2 nonstructural protein and repetitive elements (at the 5′ end of the negative-strand) seem to be involved. For PPV, we have isolated strains having one, two or four repetitive elements (from mummified embryos almost all have one repetitive element; unpublished results).

Serologic Relationships and Variability

Almost all parvoviruses possess the ability to agglutinate erythrocytes (Table 4). Even viruses showing very close serologic relationships can be distinguished this way. Serologic relationships, as revealed by HAI, were found among RV, H-3, X-14 and HER, between H-1 and HT, and between PPV and KBSH, whereas equine PPV, HB and MVM each form a unique serologic group. Immunoprecipitation and immunofluorescence are much less specific. For example, equine PPV and porcine PPV can not be distinguished by these methods. When restriction maps of the genomes of these two viruses are compared, they are clearly very closely related (no more differences than among porcine PPV strains).

Table 4. Hemagglutination by RV, BPV, PPV and GPV

		Erythrocytes from:							
Virus	Strains	Human	Monkey	Sheep	Rabbit	Goose	Rat	Cattle	Guinea pig
RV	RV	++	++	(+)	−	−	+	−	++
	X-14	++	−	(+)	+	−	+	−	++
	H-3	++	++	(+)	(+)	(+)	++	−	++
	HER	−	−	−					++
H-1	H-1	++	++	(+)	−	(+)	++	−	++
	HT	−	(+)	−	−	−	−	−	++
MVM		++	−	−	−	−	++	−	++
PPV	NADL-8	++	++	−	−	+	++	−	++
	PPV/EPV	−	++	−	−				++
BPV	BPV-1	+		(+)	−		(+)		+
	32459	+		+	+		+		+
	BPV-CK	−	−	−	−		−		−
GPV		−		−		−	−		−

Note: sera often contain nonspecific inhibitors (treatment with kaolin required) or may contain nonspecific agglutinins (Siegl, 1976). Hemagglutination titers expressed as dilutions giving a 50% reaction: dilutions: $++ \geqslant 128$, $+ \geqslant 16$ $(+) \geqslant 4$ and − no reaction.

Transmission and Epidemiology

Parvoviruses can be transmitted horizontally by the fecal–oral route, respiratory exudates, dust, gloves, clothing and food, provided that the susceptibility of the host and the dosage of the virus are sufficiently high. They can also be transmitted vertically. Toolan noted two types of infection, one that is temporary and produces low levels of antibody which disappear within months and a second type of infection which is latent or permanent leading to high antibody titers, maintained throughout life. The latter type is exemplified by neonatal and latent infections. Similar patterns can be observed for all parvoviruses discussed in this entry. PPV- or BPV-seronegative herds are sometimes found. In most herds, most animals are seropositive but some asymptomatic long-term virus excretors may be present, mostly because of continuous replication in susceptible gut cells.

Horizontal transmission is most common but vertical transmission has been shown for BPV, GPV and PPV. Vertical transmission or the virus leads to the abortion or resorption of the fetus. A few long-term PPV excretors seem to be less efficient in horizontal transmission in age-segregated herds. However, the resistance of parvoviruses to environmental inactivation is believed to be more important than chronic carriers in the epidemiology by horizontal transmission.

The morbidity and mortality rates depend on the susceptibility of the host and the immune status of the herds. Viral infections can range from completely symptomless to mortality rates that, for GPV, may be close to 100%. Again, the strain of the parvovirus is important.

Tissue Tropism and Pathogenicity

The replication of parvoviruses depends on cellular functions, transiently expressed during late S or early G-2 phase of mitosis. Thus, cell division is an essential requirement and parvoviruses infect mitotically active tissues such as those from the fetus, intestinal epithelium and hemopoietic systems. The viruses are considered 'mitolytic', hence their oncosuppressive activities. Additional cellular factors, expressed during differentiation, are also required. Cell division is not the only determinant, and parvoviruses are therefore not truly pantropic. A tropic determinant has been identified in the capsid protein of MVM and consists of two or three amino acids. In a corresponding region of PPV structural proteins, a tropic determinant, consisting of one to three amino acids, was also found (unpublished results).

The complex set of factors involved in viral pathogenicity is still poorly defined. At a molecular level, much of the current attention is focused on the (until now unknown) role of the viral-encoded NS2 protein and noncoding genetic elements. Both seem to be involved in pathogenesis or virulence. It is not clear yet whether tropism factors can be completely separated from virulence factors. Curiously, RV, H-1 and MVMp are not pathogenic for their native hosts. Only under experimental conditions, such as intracerebral inoculation of rats with virulent RV, does a disease become apparent. RV and H-1 when isolated, were associated with tumors, and MVM has been isolated from oncogenic adenovirus stocks. All rodent parvoviruses have been shown to suppress carcinogenesis, whether spontaneous, virus-induced or carcinogen-induced. In contrast, MVMi, PPV, BPV and GPV are pathogenic. Mixed infections with other pathogens are common and a role for these parvoviruses in certain diseases is emerging.

PPV infections are often harmless (except some PPV strains causing severe dermatitis or enteritis). Pregnant animals suffer loss of embryos or fetuses when infected. Some PPV strains cause a viremia which enables PPV to cross the placenta and thus to infect the embryo or fetus. Other strains are not viremic and are harmless. Inoculation *in utero* with these nonviremic strains still causes abortion. Maternal antibodies do not cross the placenta but the fetus becomes immunocompetent after 70 days and can mount a protective immune response. Many viremic PPV strains will then not be fatal. However, some strains (Kresse, PV-7) can still cause death in late-gestation fetuses. There is a striking resemblance between MVM and PPV with respect to tropism and pathogenesis. The nonpathogenic PPV-NADL2 and MVMp are nonviremic whereas the pathogenic PPV-NADL8, PPV-Kresse and MVMi are viremic. The generalized infection of hematopoietic cells, lymphocytes and capillary endothelium leads to bilateral infarcts of the solitary renal papilli for MVMi and fetal infection for PPV. It needs hardly to be emphasized that laboratory diagnosis should distinguish the strains for reliable epidemiological surveys and herd management, instead of mere detection of virus.

Most BPV isolations are from samples from calves with diarrhea, but BPV is increasingly associated with respiratory disease and reproductive failure. Viremia can be established (in leukocytes), and, during the symptomatic phase, different targets become infected (intestinal tissue, brain, heart muscle, adrenal gland, thymus and lymph nodes). Fetal infection after viremia is particularly acute in the first trimester (particularly cerebellum). Active immunity develops in the second half of pregnancy, and a fetus infected in the third trimester usually recovers from the infection.

Clinical Features and Pathology

Except for MVMi, natural infections of rodents are clinically inapparent. Experimental infections of rodents may produce (1) an acute lethal disease in newborn animals, (2) an osteolytic syndrome, (3) cerebellar ataxia or (4) hemorrhagic encephalitis. Successive passaging of H-1 or RV in newborn hamsters can increase the pathogenicity in newborn animals to close to 100% mortality. After 4–10 days, infected animals suddenly become sluggish, gasp for breath and die. Necropsy often reveals hemorrhage of the gut and congestion of the liver. In contrast, no liver damage is observed with the RT strain, but intestines often contain a sanguineous exudate. Infected newborn rats become apathic just before dying. Intranuclear inclusions are found in cells of almost all organs of the animals.

Hamsters surviving RV or H-1 infections usually develop mongoloid features due to osteolytic activity of the parvoviruses [in rodents, osteogenic activity (osteoblasts, odontoblasts) is high]. Cerebellar ataxia can only be introduced by intracerebral inoculation of hamsters younger than 4 days (and is apparent after a month). The HER agent induces hemorrhagic encephalitis in rats after cyclophosphamide treatment. Upon necropsy, hemorrhage and necrosis are observed in brains or spinal cords of these rats.

The most common feature of an outbreak of PPV reproduction failure is the appearance of mummified fetuses. The litter may contain both mummified and stillborn fetuses. The PPV strain associated with equine abortion has not been reproduced experimentally. Interestingly, this virus also seems to be fetotropic in rabbits. Clinical features of BPV infections, mostly in calves (1–12 months), include enteritis, respiratory disease and conjunctivitis. The severity of the clinical symptoms is usually increased by co-infection with other pathogens and may only be apparent in their presence. GPV-infected goslings or Muskovy ducklings (at day 1) stop eating after about 3 days, are reluctant to move, remain near a heat source and usually die after a week. Those that survive are usually severely retarded and may become featherless. These animals may also develop a transient leg weakness. When older animals are infected, a symptomless carrier state may be obtained. Birds that succumb in the acute phase after GPV infection have characteristic lesions in the liver and heart muscle. The hyperemic liver contains small grayish–white areas. The apex of the enlarged heart can be rounded off and the myocardium may show a discoloration. Accumulation of fluid in the pericardium and abdominal cavity is often observed.

Immune Response and Prevention

A strong active immunity depends to a large degree on viral multiplication in tissues of the infected organism. The equilibrium between virus produced during latency and the production of circulating antibodies can lead to high antibody titers over long periods. A single injection with cyclophosphamide (immunosuppressive) was shown to be sufficient to convert a latent RV infection into an apparent disease. Toolan observed that infection of pregnant hamsters with H-1 did not yield any antibodies, neither did subsequent inoculations. The reason for this tolerance is not clear.

Passive immunity (maternally acquired antibodies) may protect the fetus against infection in rodents. However, immunoglobulins do not pass through the multilayer placenta (epitheliochorial) in pigs. Production of antibodies in neonatal piglets is the result of active immunity (embryos are immunocompetent from about 70 days on). After birth, pigs can contain high concentrations of PPV antibodies from colostrum while nursing seropositive dams. A similar pattern is observed for BPV; second- and third-trimester fetuses yield an IgM response that is maximum after about 10 days and is gradually replaced by IgG with maximum titers about 5 months after infection. Calves deprived of colostrum from seropositive cows may develop severe diarrhea upon infection.

Viral strains that do not produce viremia can often be used as live vaccines. The NADL2 strain of PPV, but also others such as the HT strain of PPV (no viremia), can be used. Some workers have been able to induce high antibody titers to PPV with inactivated virus. An early management method for PPV was the back-feeding of fecal material or fetal tissues from PPV-infected sows. Infections with BPV are widespread and are clinically inapparent in adults. BPV vaccines are not yet available nor has the need for vaccination been demonstrated.

Conclusion

Initially, parvoviruses attracted attention since it was felt that these small viruses would be simple and easier to understand than most other viruses. It has become clear, however, that they depend more on cellular functions which are more difficult to study. The best model may be MVM since more is known about the genetic make-up of the mouse than other vertebrates. Other parvoviruses deserve attention because of their implication in various diseases.

Further Reading

Berns KI (1990) Parvovirus replication. *Microbiol. Rev.* 54: 316.

Brownstein DG *et al.* (1992) *J. Virol.* 66: 3118.

Cotmore S and Tattersall P (1987) *Adv. Virus Res.* 33: 91.

Pattison JR (1990) Parvoviruses. In: Fields BN *et al.* (eds) *Virology*, 2nd edn, chapter 63. New York: Raven Press.

Siegl G (1976) *The Parvoviruses.* Vienna: Springer-Verlag.

Tijssen P (ed.) (1990) *Handbook of Parvoviruses*, vols I and II. Boca Raton, Fl: CRC Press.

PATHOGENESIS

Kenneth L Tyler
University of Colorado Health Sciences Center
Denver, Colorado, USA

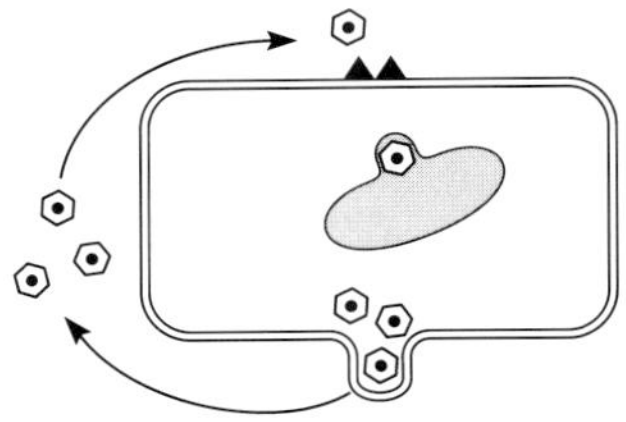

Introduction

The term *pathogenesis* is derived from two Greek words which can be translated to mean, 'the origin of disease.' The study of viral pathogenesis thus concerns itself with understanding the process by which a virus produces disease in the host. It is important to recognize that a virus can enter and replicate in a host, and even induce an immunologic response, without producing overt signs or symptoms of disease. For many viruses, the majority of infections, under normal circumstances, are asymptomatic. When illness does occur, it can be of any degree of severity. Infection may be of short duration and self-limited (acute) or may be long-term (chronic) or even persist for the life of the host. Under some circumstances chronic infection is accompanied by continued replication and shedding of virus (persistence), although in other cases virus remains in a nonreplicating or inactive state (latency) from which it may periodically re-activate. In the classic model of acute virus infection, injury to cells occurs as a direct result of the replication and release of viral particles. However, it has become increasingly recognized that viruses may also produce disease through a variety of other mechanisms. They can promote the induction of neoplasia (oncogenesis), suppress the immune system and even alter specific cellular functions without killing the target cell.

Viral pathogenesis can be analyzed in terms of a series of interactions between the virus and the host. Although the specific steps in this process may differ for individual viruses and particular hosts, the general outline remains true for most cases. A virus must survive in the environment, enter a susceptible host, multiply to increase its inoculum, and spread from the site of entry to target tissues, where it produces disease as the result of infection and injury to particular organs or populations of cells. Each of these stages will be considered separately.

Entry

The most common routes of viral entry are through the skin, respiratory tract, gastrointestinal tract, urogenital tract and conjunctiva.

Skin

Under normal circumstances the skin poses an effective barrier to the entry of viral pathogens. The dead keratinized cells of the outer skin layer (stratum corneum) do not support viral replication. Infection can be initiated when a virus enters the host through breeches in skin contiguity such as cuts, abrasions or wounds. Viruses can also be mechanically transported across the stratum corneum by an insect or animal bite or with man-made implements such as hypodermic needles. The layer of the epidermis below the stratum corneum (stratum Malpighii) contains living cells but is essentially devoid of blood vessels, lymphatics and nerves. Viruses that enter this layer, such as the papillomaviruses, typically induce local pathology (e.g. dermal warts), but only rarely disseminate to produce systemic disease. Deeper inoculation may introduce a virus into the dermis, with its luxurious supply of vessels, lymphatics and nerves, or even into the underlying subcutaneous tissue and muscle. These tissues often provide fertile ground for viral multiplication and subsequent dissemination.

The host's dermal barrier is not impervious to invasion. It is penetrated by the openings of the respiratory, alimentary and genitourinary tracts, and modified in areas such as the conjunctiva of the eye.

Respiratory tract

Viruses that initiate infection via the respiratory tract may produce symptoms secondary to local pathology (e.g. influenzal upper respiratory illness) or systemic diseases with little in the way of initial respiratory symptoms (e.g. mumps, measles). Entry is typically in the form of either droplet aerosols generated by coughing or sneezing, or by saliva exchanged in kissing or through sharing of drinking glasses, toothbrushes or other utensils. The fate of inhaled viral droplets is influenced largely by physicochemical factors including particle size, temperature and humidity. Large particles are generally trapped in the nasal turbinates and sinuses, and may initiate upper respiratory infections. Particles smaller than 5 μm can reach the alveolar airspaces to produce lower respiratory infection (pneumonia).

Host defenses against the initiation of viral infection through the respiratory tract are complex. Mucus is secreted by goblet cells, and propelled upward toward the oropharynx by the coordinated beating of ciliated epithelial cells. It helps to trap and clear foreign material. Additional antiviral action is provided by secretory immunoglobulin A (IgA) and resident phagocytic cells such as alveolar macrophages. Viral infection may be facilitated by factors that compromise mucociliary clearance including cigarette smoking, atmospheric pollutants, in-dwelling tubes or certain inherited disorders.

Gastrointestinal tract

Viruses that infect the host through the gastrointestinal (GI) tract must be able to survive gastric acidity, bile salts and a variety of proteolytic enzymes. They must also avoid inactivation by secretory IgA and the action of lymphoid cells and macrophages.

The importance of acid stability to viral survival in the alimentary tract is exemplified by the picornaviruses. Rhinoviruses are acid labile and lose infectivity at low pH. Under acidic conditions the viral outer capsid is disrupted, and viral RNA escapes, leaving noninfectious empty capsids. As expected, rhinoviruses do not produce enteric infections. By contrast, other members of the picornavirus family including polio- and coxsackieviruses are resistant to degradation under acidic conditions. They are extremely successful at initiating enteric infections, a fact recognized by their inclusion in the 'enterovirus' functional group.

Viruses that initiate infection through the enteric route must also resist inactivation by the proteolytic enzymes secreted by gastric and pancreatic cells. In fact, proteolytic digestion actually increases infectivity of several enteric viruses. For example, partial cleavage of the VP4 outer capsid protein (VP4 $\rightarrow$ VP5, VP8) by intestinal proteases enhances the infectivity of rotaviruses, which are a major cause of diarrheal illness in infants. A similar phenomenon occurs with coronaviruses (E2 peplomer glycoprotein cleavage). In the case of reoviruses, proteolytic cleavage of outer capsid proteins results in the production of infectious subvirion particles (ISVPs). ISVPs appear to play a critical role in the initiation of subsequent intestinal infection (see below). Thus, as a general principle, viruses that produce or initiate infection in the intestinal tract are not inactivated by intestinal proteolytic enzymes. In fact, quite the converse appears to be true, as in many cases proteolytic processing of viral capsid proteins seems to trigger conformational changes in the virus particle or expose new functional determinants on specific proteins which in turn facilitate specific events (receptor binding, membrane fusion, cell entry, transcriptional activation) in the viral life cycle.

Bile salts are another important factor in inhibiting viral entry through the intestinal tract. Viral envelopes are particularly susceptible to digestion by bile salts. Before direct ultrastructural visualization of virions was routinely possible, the capacity of bile salts to destroy viral infectivity was taken as prima facie evidence for the presence of a lipid envelope. With the solitary exception of coronaviruses, viruses that initiate infection through the intestinal tract are all nonenveloped.

The cellular events that underlie the initiation of systemic infection by enteric viruses are becoming better understood. Reoviruses and polioviruses initially bind to the luminal surface of specialized epithelial cells (M cells) which overlie regional aggregates of intestinal lymphoid tissue (Peyer's patches). Virions are then transported in vesicles across the M cell cytoplasm and discharged into the subepithelial lymphoid tissue where primary replication can occur in lymphoid cells and macrophages.

Genitourinary system

A number of human viruses including human immunodeficiency virus (HIV), herpes simplex and papillomaviruses are venereally transmitted. Small tears or abrasions in the epithelial lining of the rectum, urethra or vagina may occur during sexual activity and permit entry of virus. Host factors that inhibit viral entry through these routes include cervical mucus, the pH of vaginal secretions, the chemical composition and cleansing action of urine and the presence of secretory IgA.

Conjunctiva

Although viral conjunctivitis is common, and occurs either as an isolated illness or in association with certain systemic infections (e.g. measles), the conjunctiva only rarely serves as a site of entry of viruses into the host. Local infection may be initiated by direct inoculation of virus following ophthalmologic procedures (tonometry, foreign body removal) or in the process of swimming ('swimming pool conjunctivitis'). The offending viruses include adenoviruses and enteroviruses. Most such infections remain localized. Enterovirus 70 (E70) appears to be an important exception to this rule. E70 commonly produces acute hemorrhagic conjunctivitis, and on extremely rare occasions (perhaps 1 in 10 000 cases) spreads from the eye to the nervous system to produce cranial nerve palsies, myelitis or encephalitis.

Spread in the Host

For viruses that produce localized infections, the major steps in pathogenesis are entry into the host and subsequent primary replication in cells and tissues in proximity to the site of infection. Virus spreads from cell to cell in a contiguous fashion. The brunt of injury is confined to the epithelial layer, although local lymphoid tissues may also be involved. This type of circumscribed infection is typical of uncomplicated upper respiratory diseases caused by coronaviruses, rhinoviruses and influenza, and the acute diarrheal disease induced by rotaviruses. More generalized symptoms (fever, chills, myalgia, malaise, fatigue, anorexia) can accompany these infections, but are generally the result of circulating mediators induced as a result of the local infection, rather than due to systemic invasion.

The factors that restrict some viruses to local sites while allowing others to invade the host are poorly understood. They include the direction of viral release from infected cells, the distribution of viral receptors and the effects of differences between core body and epithelial surface temperature.

Release of certain enveloped viruses occurs preferentially from either the luminally facing apical surface (e.g. para- and orthomyxoviruses) or the subepithelially facing basolateral surface (e.g. rhabdoviruses) of infected epithelial cells. The pattern of release is determined by the site in the cell membrane at which viral envelope glycoproteins are inserted. This is in turn influenced by specific amino acid signal sequences within the viral protein. The potential importance of polarized release of virus in determining subsequent systemic invasion is obvious. Release of virus only toward the lumen of the respiratory or GI tract would facilitate local infection of the epithelial surface but would inhibit invasion of deep subepithelial tissues. Conversely, release of virus from the basolateral cell surface would facilitate invasion of the subepithelial mucosa, and the subsequent dissemination of virus through lymphatics, blood vessels, or nerves (see below).

Spread through the bloodstream

Direct inoculation of virus into the bloodstream is a rare event. It can occur in association with intravenous drug abuse or during transfusion of infected blood or blood products. The bite of an arthropod vector may also allow direct entry of virus into the bloodstream. More commonly, entry of virus into the host is followed by a period of primary replication in local tissues and regional lymph nodes. Important sites of primary replication include subcutaneous tissue, brown fat, skeletal muscle, endothelial cells and regional lymphatic tissue. Virus enters the bloodstream from these sites (primary viremia), and is further disseminated to reticuloendothelial organs (bone marrow, liver, spleen) and endothelial cells. Additional replication is followed by a secondary viremia, which is typically of longer duration and higher magnitude than the initial viremia. This sequence of events was originally described by Fenner and colleagues for experimental mousepox (ectromelia) infection. Distinct phases of primary and secondary viremia are often difficult to identify in human viral infections.

Virus in the bloodstream may travel free in the plasma or in association with cellular elements. For example, enteroviruses and togaviruses are frequently found in plasma, HIV is often associated with T4+ monocytes, macrophages and T cells, Epstein–Barr virus (EBV) infects B lymphocytes and Colorado tick fever virus (CTFV) infects erythrocyte precursors which subsequently mature and enter the circulation.

A variety of host defenses exist which act to clear virus from the bloodstream. In general, larger virus particles and particles coated with antibody or complement are cleared with far greater efficiency than small nonopsonized particles. Additional factors that influence clearance include the net charge of the virion particle and the composition of the viral capsid or envelope. Host factors may also influence the efficiency of clearance. Experimentally, agents such as thorotrast or silica, which decrease the phagocytic capacity of macrophages and other reticuloendothelial phagocytes, enhance the viremia of some viruses. Differences in the capacity of macrophages from immature animals to clear virus compared to their adult counterparts may account for age-related differences in susceptibility to certain viral infections (e.g. herpes simplex). Macrophages derived from different strains of mice vary in the efficiency with which they clear specific viruses (e.g. mouse hepatitis, virus, MHV), and this may explain some strain-specific differences in susceptibility to certain viruses. Finally, recent evidence has suggested that different strains of the same virus may show striking differences in their capacity to replicate in macrophages, and that this in turn may be associated with distinct patterns of organ-specific tropism and virulence.

Uptake of virus by phagocytes does not always result in their inactivation. Many viruses including HIV, lentiviruses, and certain toga-, corona-, arena- and reoviruses are capable of replicating in macrophages. Macrophages contain receptors for the Fc portion of antibody molecules, and in some cases uptake of virus is facilitated by the presence of anti-viral antibody ['antibody-mediated enhancement' (AME)]. The pathogenetic significance of AME has been clearly established in dengue infection.

As would be expected, there is a general correlation between the magnitude of viremia generated by blood-borne viruses and their capacity to invade tissues such as the central nervous system (CNS). Conversely, the failure of some attenuated viruses to generate a significant viremia may also account for their lack of invasiveness. For example, certain neurotropic bunyaviruses are fully virulent after direct intracerebral inoculation, but avirulent after peripheral inoculation because they fail to generate sufficient viremia to allow neuroinvasiveness. It is important to recognize that viremia *per se* does not automatically equate with the capacity to invade tissues from the bloodstream. This has been elegantly demonstrated by certain mutants of Semliki Forest virus (SFV) which have lost the capacity to invade the CNS while retaining the capacity to generate a viremia fully equivalent in duration and magnitude to their neuroinvasive wild-type counterparts.

The steps by which blood-borne viruses exit the bloodstream to enter tissues remain poorly understood. In some cases, the viruses appear to directly infect endothelial cells, and then are transported across these cells into the underlying parenchyma. In other cases, viruses may enter tissues inside migrating cells that are capable of emigrating across capillaries (diapedesis). Transendothelial transport of virus inside infected cells has been colorfully referred to as the 'Trojan Horse' mechanism of entry. This type of process may be important in the pathogenesis of lentivirus and HIV infections. Finally, factors that alter vascular permeability (e.g. vasogenic amines) can be shown experimentally to facilitate tissue invasion by certain viruses. This suggests that endothelial permeability may also play a role in determining tissue invasion by blood-borne viruses.

Spread through nerves

Many viruses including herpes simplex (HSV), varicella zoster (VZV), rabies and certain strains of poliovirus, reovirus and coronavirus can spread through nerves in the infected host. This pathway of spread is particularly important for viruses that invade the CNS, but theoretically also provides a route for infection of virtually any organ. Neural spread to organs other than the CNS is exemplified by the spread of rabies virus to salivary glands and VZV and HSV to the skin.

The exact mechanism(s) of neural transport of viruses have not been established, although certain basic principles have emerged. Although spread of many neurotropic viruses along nerves can occur by cell-to-cell spread through nonneural cells (e.g. Schwann cells), the pathogenetically important mode of viral spread is through the axoplasm of neurons. In the case of enveloped viruses, transport appears to involve predominantly the nucleocapsid rather than the enveloped virion. With the exception of the scrapie agent, all neurally spreading viruses appear to use fast axonal transport. This has been established by studying the kinetics of transport and through the use of selective pharmacologic inhibitors of fast and slow axonal transport. Taken as a group, neurally spreading viruses provide examples of spread through motor, sensory and autonomic nerve fibers, and in both the anterograde and retrograde direction. However, recent evidence suggests that individual strains of particular viruses (e.g. HSV) may preferentially travel in only one direction. Similarly, studies with reassortant viruses and viral mutants suggest that changes in either the viral envelope or capsid proteins (rabies, reo), or in some cases in non-structural proteins (HSV), may alter the capacity of viruses to spread through nerves, or even through specific neural pathways.

Viruses that spread within neurons also have the capacity to spread from nerve cell to nerve cell (*trans*-neuronal transport). In some cases this appears to occur specifically at synapses (*trans*-synaptic transport). The factors that influence the release of virus from presynaptic nerve terminals and facilitate their uptake postsynaptically are unknown. Fast axonal transport typically involves the microtubule-associated transport of material contained within vesicles. However, the mechanism by which viruses access this system and the form in which they are transported (intravesicular or free) has not yet been established.

Specific viral proteins play a critical role in determining whether viruses spread through the bloodstream or through nerves in the infected host, and even the specificity of the neural pathways utilized. For example, the principal pathway of spread of reovirus type 1 Lang (T1L) from muscle to CNS is through the bloodstream, and for type 3 Dearing (T3D) through nerves. The viral S1 gene, which encodes the outer capsid protein $\sigma 1$, determines this difference. As noted earlier, certain rabies virus variants with single amino acid substitutions in the envelope glycoprotein (G), seem to have lost the capacity to spread through some types of neural pathways when compared to wild-type virus.

Nonstructural proteins may also influence the efficiency with which viruses infect and spread within the nervous system. For example, HSV isolates differ in their neuroinvasiveness following footpad or corneal inoculation in mice. Most nonneuroinvasive HSV strains replicate poorly in peripheral sensory ganglia, suggesting that this, rather than an inability to be neurally transported, accounts for their lack of neuroinvasiveness. These nonneuroinvasive strains remain capable of spreading from the site of inoculation, through nerves, to the sensory ganglia, but their spread is arrested at this stage. Genetic studies of recombinant herpesviruses containing portions of the genome derived from both invasive and nonneuroinvasive viruses indicate that the viral DNA polymerase may determine the capacity of certain viruses to replicate in sensory ganglia and subsequently invade the CNS.

Thus, an increasing number of studies in several viral systems have suggested that specific viral proteins may determine both the capacity and efficiency with which viruses spread via different pathways. Neural spread is undoubtedly a complex process which requires that a virus successfully infect and replicate in neurons, be transported through the axoplasm and released from the neural cell and be capable of spreading to other nerve cells to re-initiate the process. Defects at any of these stages may inhibit 'neural spread' or result in loss of the capacity to invade the CNS. The complexity of viral transport through nerves is emphasized by the fact that different strains of virus (e.g. HSV) may spread exclusively by either anterograde or retrograde axoplasmic transport, and some viral variants (e.g. rabies) can selectively lose the capacity to spread via certain nerve fiber systems but not others.

Tropism

The capacity of a virus to selectively infect certain populations of cells in particular organs is referred to as 'tropism.' Viral tropism can depend on a variety of viral and host factors, several of which are discussed in detail in the sections which follow.

Viral receptors

Viruses must bind to target cells prior to initiating infection. Entry may be the result of the interaction of virus with a specific cellular receptor followed by receptor-mediated endocytosis. Alternatively, some viruses are capable of fusing directly with the plasma membrane (e.g. certain alphaviruses), which allows the nucleocapsid to enter the cell cytoplasm through a nonendocytosis-mediated pathway. Viruses that utilize receptor-mediated endocytosis to enter target cells may have receptors that are found on only certain types of cells (e.g. CD4 receptor for HIV), and thus may play a critical role in determining the specificity of viral infection. In other cases, the receptor appears to be ubiquitously distributed [e.g. sialic acid receptors for influenza, heparin sulfate proteoglycans for HSV, gangliosides or phospholipids for rhabdoviruses, intercellular adhesion molecule (ICAM) 1 for rhinoviruses], and other factors must account for the specificity in the pattern of viral infection.

A number of principles have emerged from studies of viral receptors. As would be expected, the cellular receptors for viruses are typically surface molecules which perform other functions in the normal host. Putative viral receptors include neurotransmitter receptors [acetylcholine receptor (AChR) for rabies, β-adrenergic receptor (βAR) for reovirus 3], complement receptors (C3d receptor for EBV), growth factor receptors [epidermal growth factor receptor (EGFR) for vaccinia], lymphocyte surface antigens (CD4 for HIV), histocompatibility antigens (SFV) and cell adhesion molecules (ICAM 1 for rhinovirus). In some cases, a cell surface molecule has been identified as a viral receptor, but its normal function in the host has not yet been established (e.g. the polio receptor is a previously unidentified member of the immunoglobulin superfamily of proteins). Preliminary characterizations of cell surface receptors for several other viruses including adenoviruses, polyomaviruses, coxsackieviruses, Theiler's virus (a picornavirus), mouse hepatitis virus (a coronavirus) and Sindbis (an alphavirus) have also been reported.

It should be emphasized that controversy surrounds some of the receptor assignments described above. In some cases (e.g. rabies, reo 3, vaccinia) it is clear that the virus can infect cells lacking the putative receptor, indicating that the particular protein is either not the viral receptor or that other mechanisms of entry into the cell (additional receptors, nonreceptor-mediated processes) exist. In the case of hepatitis B virus (HBV), the polyalbumin receptor, the IgA receptor and sialoglycoprotein have all been proposed as viral receptors. Similarly, the AChR, gangliosides and phospholipids have all been identified as receptors for rabies virus. Further studies are required to determine whether these viruses actually utilize multiple receptors or whether specific assignments are in error.

Different strains of the same virus may use different receptors, and conversely, entirely unrelated viruses may share the same receptor. For example, human rhinoviruses (HRV) 1A, 1B, 2, 49 (the minor group) bind to a receptor that is distinct from that used by HRV14 and all other rhinoviruses (the major group). Coxsackieviruses B1–6 and adenovirus 2 compete with each other for binding to certain cells, suggesting that they may share a common receptor, despite the fact that they are totally unrelated.

The presence of a viral receptor in a cell is not always sufficient to allow for infection. For example, cultured mouse cells transfected with cDNA encoding the HIV receptor, and expressing the receptor protein, remain insusceptible to HIV infection. However, some cell lines that are resistant to infection with viruses including EBV, HIV and polio become fully susceptible when they are made to express the appropriate receptor, indicating that lack of receptor can be the only barrier to susceptibility.

Viral cell attachment proteins

The interaction of a virus with its cellular receptor is typically mediated by one or more cell surface proteins. Among the proteins playing a primary role in cell attachment are envelope glycoproteins [e.g. influenza HA, E2 for coronaviruses, E2 for togaviruses, G1 for bunyaviruses, SU for retroviruses, gp120 for HIV, G for rhabdoviruses, VP1 for polyomaviruses, penton fiber protein for adenovirus, gp350/220 for EBV]. Capsid proteins play a similar role in nonenveloped viruses ($\sigma 1$ for reoviruses, VP7 for rotaviruses, VP1 for polio, large S for HBV). For HSV, more than one envelope glycoprotein (e.g. gB, gD, gH) may be involved in cell attachment.

High-resolution three-dimensional crystal structures of the influenza HA and of several picornaviruses [HRV14, poliovirus, mengovirus, encephalomyocarditis (EMC), Theiler's] has provided additional information about viral receptor-binding sites. The sialic acid-binding domain of the influenza HA lies in a small depression near the distal tip of the molecule. The receptor-binding site of picornaviruses typically forms a depression in the virion surface that has been variously described as a canyon (HRV14), a valley (polio) or a pit (mengo). Conversely, the receptor-binding site for foot-and-mouth disease virus (FMDV), a member of the aphthovirus group of picornaviruses, is located on a prominent outward-facing antigenic loop of the VP1 protein.

Tissue-specific promoters, enhancers and transcriptional activators

Although the binding of a virus to its receptor may be a necessary initiating event in most viral infections, a number of other host and viral factors influence tropism. Viruses may contain distinct genetic elements, referred to as promoters or enhancers, that may enhance the transcription of certain genes in a cell-tissue- or even species-specific manner. When mouse embryos are injected with the early region of JC virus (JCV) DNA, pathology is limited to oligodendrocytes within the CNS, despite the presence of viral genome in virtually all cells. The JCV genome contains a region that allows expression of the viral large T antigen only in oligodendrocytes and not other cells. An important role for viral enhancer elements in determining cell-type-specific gene expression has also been described for polyomaviruses, papillomaviruses and hepatitis B virus. It has recently been suggested that a cell-type-specific enhancer region may be contained within the HIV long terminal repeat (LTR), and that differences in this LTR sequence may account for differences between the neurotropism/monocyte-tropism compared to T-lymphoid tropism of some HIV isolates.

Site of entry and pathway of spread

The site of entry of virus into the host may influence its tropism. This has been clearly documented for neurally spreading viruses, whose neural spread is limited by the nature of the pathways available at the site of entry. Variations in the distribution of pathology, infectious virus or viral antigen following different sites or routes of viral inoculation have been clearly demonstrated experimentally with polio, rabies, herpes simplex, reovirus, coronaviruses and the neurotropic (NWS) influenza strain. Obviously it has been harder to document this point in human infections, although clinical studies of rabies infection and polio suggest that an identical process occurs in humans. For example, patients who developed paralytic polio after being inadvertently immunized with improperly inactivated lots of poliovirus (the 'Cutter incident') showed a preponderance of paralysis involving the inoculated limb. Similarly, with rabies infection, the site of the bite (e.g. face versus leg) influences prognosis, the initial symptomatology, the incubation period and the probability of subsequent development of clinical disease.

It has also been suggested that the site of viral entry may influence the subsequent tropism of blood-borne as well as neurally spreading viruses. This was initially suggested after clinical observations suggested that local trauma to a muscle (e.g. an injury, an injection, strenuous overexertion) increased the likelihood of this muscle becoming paralysed during a subsequent attack of paralytic polio ('provoking effect'). This effect could be reproduced experimentally if monkeys were given intramuscular injections followed by intracardiac inoculation of poliovirus type 1. The mechanism of the provoking effect has never been satisfactorily established. It was suggested that local trauma could alter the vascular permeability in the region of the spinal cord innervating the traumatized site, and the increased permeability could result in an increased likelihood that blood-borne poliovirus would localize in that segment of the spinal cord. this phenomenon does not appear to have attracted recent attention, and its existence and mechanism must be considered speculative. Nonetheless, the possibility that local factors can influence the tropism of blood-borne viruses remains intriguing.

Host Factors

It is important to recognize that host factors may play a critical role in determining the outcome and many aspects of the pathogenesis of viral infections. Although a comprehensive discussion of the role of host factors in infection is beyond the scope of this review, several of the more important ones are worthy of emphasis. Among these are age, sex, genetic background, immune status and nutritional state.

The importance of host factors in determining the outcome of viral infection is dramatically illustrated when populations of individuals are exposed to the same pathogen. Inadvertent experiments of this type have included use of hepatitis B virus contaminated yellow fever virus vaccine, and incompletely inactivated lots of poliovirus vaccine. In both cases, those vaccinated showed a wide spectrum of outcomes ranging from no obvious ill effects to severe disease (hepatitis, paralytic polio). These results occurred despite relative uniformity in the pathogen, and the dose and route of administration. Epidemics of neurotropic arthropod-borne virus (arbovirus) infection provide a less controlled illustration of the same point. Among infected individuals there are a wide variety of clinical manifestations ranging from asymptomatic seroconversion to lethal encephalitis.

The role of genetic factors in determining the outcome of viral infection has been extensively investigated using inbred strains of mice. Human genes conferring resistance or susceptibility to viral infection have not yet been identified, although one can presume that they will ultimately be shown to exist. By comparing the severity of a particular viral infection in different strains of inbred mice, it can be shown that genetic determinants of viral resistance and/or susceptibility exist for almost all groups of viruses. Genetic factors that determine susceptibility to one virus are typically unique, and differ from those involved with other viruses. In addition, there are clearly multiple mechanisms by which genetic differences lead to differences in viral susceptibility. Among those that have been characterized are differences in immune responses [cytomegalovirus (CMV), murine leukemia viruses], in the expression of viral receptors in target tissues (coronavirus) and in interferon-induced expression of antiviral proteins (influenza).

The importance of differences in the age and sex of the host in determining the outcome of viral illness can be seen in a variety of human and animal viral infections. For example, viruses such as varicella, EBV, mumps, polio and hepatitis A typically produce milder infections in children than adults, whereas the opposite is true for viruses such as rotavirus and Rous sarcoma virus (RSV). Studies of experimental viral infection suggest that age-related differences in viral susceptibility have multiple mechanisms. Among those frequently cited are the maturation of the immune system, changes in the nature and distribution of populations of mitotically active cells, or on the state of cellular differentiation.

Differences in the susceptibility of males and females to particular viral infection may be due to differences in the risk of exposure or the mechanism of viral transmission. For example, in the US, HIV infection is far more common in men than women. Similarly, the risk of transmission appears higher when the infected sexual partner is male rather than female. In some cases, sex-related differences in susceptibility cannot be accounted for by obvious epidemiologic factors such as exposure risk or mode of transmission. For example, neurologic complications of mumps infection are two to three times more common in boys than girls. Similarly, following exposure to hepatitis-B-virus-contaminated blood during hemodialysis, men are twice as likely to become chronic HB carriers as women. Striking differences in the nature of and susceptibility to viral infection also occur during pregnancy. These may be related to differences in levels of sex or steroid hormones or to pregnancy-related immunosuppression. Among the infections that are more severe during pregnancy are those caused by polio, hepatitis and herpes simplex. There is also a higher rate of re-activation of latent viruses including polyomaviruses, CMV and herpes simplex.

Exogenously administered hormones, including steroids and thyroid hormome, can worsen the course of certain experimental viral infections. Steroids are often believed to exacerbate infections due to viruses such as herpes simplex, although definitive studies on the effects of steroids on human viral infections are lacking.

Thus, a number of studies have provided evidence that a variety of host factors have play a role in determining susceptibility to and severity of viral infections.

See also: Host genetic resistance; Immune response; Latency, Nervous system viruses; Persistent viral infection; Viral receptors; Virus–host cell interactions.

Further Reading

Mims CA and White DO (1984) *Viral Pathogenesis and Immunology*. Oxford: Blackwell Scientific Publications.

Notkins AL and Oldstone MBA (1984, 1986, 1989) *Concepts in Viral Pathogenesis I, II, III*: New York: Springer-Verlag.

Tyler KL and Fields BN (1990) Pathogenesis of viral infections. In: Fields BN *et al.* (eds) Virology, 2nd edn, p. 191. New York: Raven Press.

White DO and Fenner F (1986) Pathogenesis and pathology of viral infections. In: *Medical Virology*, 3rd edn, p. 119. Orlando: Academic Press.

PEA ENATION MOSAIC VIRUS

Steven A Demler and Gustaaf A de Zoeten
Michigan State University
East Lansing, Michigan, USA

Taxonomy and Classification

Pea enation mosaic virus (PEMV) is the type and sole member of the pea enation mosaic virus group. This group is characterized by a plus-sense bipartite genome encapsidated in two isometric particles. Historically, this bipartite virion and genomic composition has caused PEMV to be affiliated with the diantho-, como- and nepovirus groups, groups with which PEMV has relatively little biological similarity. Instead, the interactions of PEMV with both its host and aphid vector are more reminiscent of the monopartite luteovirus group. In the following narrative, we will review both the similarities and differences between PEMV and the luteoviruses, and will demonstrate that PEMV is in fact a chimeric virus, one half resembling the beet western yellows–potato leafroll luteovirus subgroup and the second portion similar to members of the carmo-, tombus-, necro-, diantho- and barley yellow dwarf luteovirus groups.

Virus Structure and Composition

Virions of PEMV consist of two isometric nucleoprotein components with estimated $S_{20,w}$ values ranging from 91 to 106S for the top component and 107 to 122S for the bottom component. There is no evidence of empty viral shells in purified PEMV virion preparations. In general, the bottom component is the dominant species, although the amount of top component is highly variable and a variant strain in which the top component dominates has been reported. Particle molecular weights range from 5.6×10^6 to 5.4×10^6 for the bottom component to 4.4×10^6 to 4.7×10^6 for the top component, with an estimated RNA content of *c.* 28–33%. Particle diameters of purified preparations of PEMV virions display a bimodal diameter distribution of approximately 25 and 28 nm respectively. It has been proposed that the bottom component is composed of 180 subunits arranged in a $T = 3$ icosahedron, while the top component is composed of approximately 150 subunits lacking quasi-equivalence. The top component is also less stable (particularly under high salt conditions) than the bottom component, a possible result of the irregularity in virion composition.

Virions of aphid-nontransmissible strains of PEMV contain a single coat protein of 21 kD. In contrast, aphid-transmissible strains also contain a second minor protein of 54 kD. Repeated mechanical inoculation of aphid-transmissible PEMV strains often leads to the elimination of aphid transmissibility. Coincident with the loss of vector transmissibility is the elimination of the 54 kD protein component, an observation that has led to the suggestion that this 54 kD virion subunit is in part responsible for the aphid-transmission phenotype.

Electrophoretic examination of intact virions of aphid-nontransmissible isolates of PEMV reveals two bands, corresponding to the top and bottom components. In constrast, vector transmissible PEMV virions display a complex banding pattern attributed to incremental size differences between particles. It is believed that this multiple banding pattern reflects the incorporation of variable copies of the 54 kD subunit into the capsid infrastructure. In western blot analysis, the 54 kD protein has been demonstrated to react with antisera specific to the 21 kD monomer, suggesting that this 54 kD protein represents some form of covalent linkage between the coat protein monomer and an additional subunit.

Serology

PEMV is moderately immunogenic, with titers of up to 1:1024 attainable by standard methods. Antisera against aphid-transmissible strains of PEMV contain two antibody populations whereas antisera generated against the aphid-nontransmissible strain contain only a single population. There are no reports of serological cross-reaction between PEMV and any other virus.

Genome Structure

The genome of PEMV consists of two and sometimes three plus-sense RNAs of molecular weight 1.9×10^6, 1.4×10^6 and 0.23×10^6. The nucleotide sequence of all three species has been determined. The RNAs are nonpolyadenylated and nonaminoacylatable. A genome-linked protein of 17.5 kD is associated with virion RNA preparations, although it is not known whether all three RNAs are covalently attached to this protein. Protease treatment of virion-derived RNAs does not abolish infectivity, suggesting that VPg is nonessential for infectivity.

There has been considerable controversy as to the infectivity and particle distribution of the viral RNAs of PEMV. Part of this discrepancy is due to the inability of centrifugation and electrophoretic techniques to adequately separate the bottom and top components to homogeneity, as well as to the uncertain existence and particle affiliation of RNA3 (and its effect on particle density). The favored model suggests that the bottom component encapsidates RNA1 and the top component RNA2, with both particles mandatory for infection when assayed by mechanical inoculation. An alternative proposal argues that RNA1 and RNA3 are encapsidated in the bottom component, with RNA2 encapsidated in both the top and bottom components. In contrast to the above model, it has been reported that RNA2 alone is infective. Other conflicting models have suggested that both nucleoprotein components are individually infective or that either the top or the bottom alone is infective.

Molecular Biology: RNA1

Recent progress in interpreting the genomic strategy of PEMV is beginning to clarify the complex relationship and confusion that exists between the

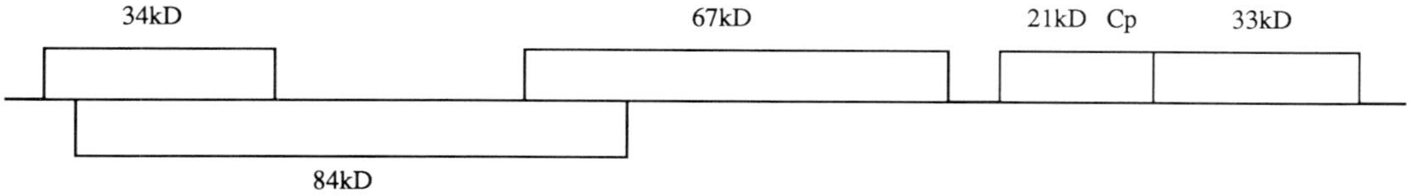

Fig. 1 Genomic organization of PEMV RNA1. Cp = coat protein.

viral RNAs. The third RNA of PEMV, which was formerly surmised to be an artifact, has been demonstrated to fulfill the criteria of a satellite RNA. This element is dispensable in PEMV infection and has been demonstrated through both hybridization and sequence analysis to lack appreciable homology to the host genome or that of its helper virus. There is no evidence of translational activity associated with this RNA, and there is no reported phenotypic consequence associated with its occurrence.

RNA1 of PEMV (5706 nucleotides) bears strong organizational and sequence similarity to the subgroup of the luteoviruses encompassing beet western yellows (BWYV), potato leafroll (PLRV) and the NY-RPV isolate of barley yellow dwarf virus (NY-RPV BYDV). RNA1 is composed of five predominant open reading frames (Fig. 1). The first reading frame encodes a protein of 34 kD of unknown function and of no known relationship with other viruses. The second open reading frame (encoding a protein of 84 kD) overlaps (out of frame) 90% of open reading frame 1. *In vitro* translation analysis of RNA1 generated products of 36 and 88 kD, suggesting that expression of the open reading frame 1 and 2 products occurs by independent translational initiation. Sequence analysis has identified a protease-like core sequence within the 84 kD protein, suggesting that proteolytic processing may be involved in the expression of the viral genome. The central region encompassing this protease motif is conserved in the luteoviruses BWYV, PLRV and NY-RPV BYDV. At this time, there is no definitive demonstration of proteolytic processing in any of these viral systems.

The third open reading frame (encoding a protein of 67 kD) overlaps (also in a unique frame) the C-terminal 23% of the open reading frame encoding the 84 kD protein. This 67 kD protein is characterized by a number of RNA polymerase and helicase-like motifs, and is postulated to represent the core of the viral RNA-dependent RNA polymerase. Sequence comparisons have established a strong relationship between this protein and those of the comparable reading frames of the luteoviruses PLRV, BWYV and NY-RPV BYDV. *In vitro* translation analysis suggests that this protein is not expressed independently from the viral RNA, but rather as a fusion of the 84 kD and 67 kD protein reading frames by translational frameshift.

The fourth open reading frame (encoding a protein of 21 kD) encodes the viral coat protein, and is immediately followed in the same reading frame by a 33 kD protein open reading frame. The amino acid sequence of both the coat and 33 kD protein bears strong sequence homology to their counterparts in the 3′-terminal reading frames described in all members of the luteoviruses. In addition, the 41-amino acid intergenic region between the opal stop codon of the coat protein and the first true start codon of the 33 kD protein contains a novel proline-rich region also conserved within the luteovirus group. By analogy with the luteoviruses, it is hypothesized that the 33 kD protein is expressed as a readthrough fusion with the 21 kD coat protein monomer. The size of this putative readthrough product (54 kD) would closely comply with the larger subunit found in aphid-transmissible isolates of PEMV. This readthrough model is also consistent with the serological recognition of both the 21 kD and 54 kD proteins with antisera specific to the 21 kD protein. At this time, there is no conclusive evidence concerning the nature of the suppression of the 54 kD protein in the aphid-nontransmissible isolate of PEMV.

Neither the coat nor the 33 kD protein are expressed in *in vitro* translation of RNA1. There is evidence from northern blot analysis of polysomal and total RNA isolated from infected tissue for an 1800-nucleotide RNA specific to the 3′-terminal region of RNA1. These data suggest that expression of these two proteins occurs from a single subgenomic messenger.

Molecular Biology: RNA2

Sequence analysis of RNA2 of PEMV (4253 nucleotides) has uncovered five open reading frames (Fig. 2). The first open reading frame encodes a protein

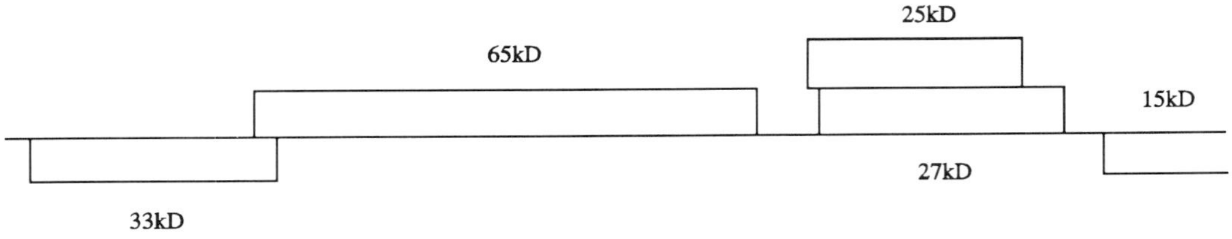

Fig. 2 Genomic organization of PEMV RNA2.

of 33 kD, which overlaps (out of frame) the first 38 amino acids of the 65 kD open reading frame 2 product. The 65 kD protein reading frame contains several sequence motifs characteristic of nucleic acid polymerases and helicases. In particular, this protein bears strong sequence homology to the putative polymerase cassettes of the diantho-, tombus-, carmo-, necro- and BYDV luteovirus groups. Following the 65 kD reading frame, there are two overlapping (out of frame) reading frames encoding products of 25 and 27 kD. These are followed by a large 703-nucleotide 3′ noncoding region. Within the noncoding region lies a 15 kD protein open reading frame at the extreme 3′ terminus which is lacking a conventional stop codon. With the exception of the 65 kD protein, none of the other proteins encoded on RNA2 bear detectable homology to any other reported viral protein. Preliminary serological screening of these three 3′-terminal reading frame products expressed *in vitro* has failed to identify a coat protein function for any of these proteins.

In vitro translation analysis of RNA2 has demonstrated only a single product of 45 kD with no evidence of the larger frameshift product postulated above. Although the 45 kD product could in principle correspond to the 33 kD protein reading frame, the size discrepancy suggests that this translational analysis requires re-examination.

Unlike other multicomponent viruses, PEMV is unique in that the 3′ and 5′ termini of the two genomic RNAs are not identical. In fact, the only appreciable similarity between any of the termini occurs between the 5′ termini of the satellite RNA and RNA2, in which 13 of the first 16 nucleotides are in alignment.

Host Range

PEMV was initially described in 1935 by Osborn to occur in *Vicia faba* in New York State. The virus occurs predominantly in northern temperate climates, but has also been reported as far south as Iran and Sicily. Economically, PEMV is considered the most significant viral disease in commercial pea production, and significant disease losses have also been described in commercial broadbean and lentil production. The virus has a narrow host range, limited mainly to the family *Leguminosae*, and includes members of the genera *Anthyllis, Astragalus, Cicer, Lathyrus, Lens, Lotus, Lupinus, Medicago, Melilotus, Phaseolus, Pisum, Glycine, Trifolium* and *Vicia*. Nonleguminous hosts include *Gomphrena globosa, Nicotiana clevelandii, Nicotiana tabacum* and members of the genus *Chenopodium* (*C. quinoa, C. amaranticolor* and *C. album*) which serve as local lesion hosts.

Transmission

Of the virus groups that utilize aphid transmission in virus dissemination, only the PEMV group and the luteovirus group are transmitted in a circulative nonpropagative manner. A notable distinction, however, is PEMV's additional capacity for effective mechanical transmission. Eight species of aphid have been reported as vectors of PEMV, with the pea aphid *Acyrthosiphon pisum* (Harris) being both the most effective and most significant in the field. There is considerable variability in the literature concerning the dynamics of vector transmission of PEMV, a reflection of variability in both the viral isolate and aphid biotype examined. In general, acquisition periods range from 15 min to 3 h, with a mandatory latent period of approximately 10–18 h, consistent with a circulative transmission pattern. Following the latent period, *A. pisum* was able to transmit PEMV following test feeding probes of less than 60 s, suggesting that inoculation of nonphloem tissues was adequate to transmit the virus. Nymphs are more efficient than adults at virus transmission. Viral particles are retained through moults, and aphids remain viruliferous in excess of 30 days. There is no concrete evidence of viral replication occurring in aphid tissues.

Extended mechanical propagation of PEMV isolates

can lead to the loss of vector transmissibility. There are no reports at this time of such aphid-nontransmissible isolates occurring under field conditions, or the reversion of these laboratory strains to aphid transmissibility.

Symptomology

Symptoms attributed to PEMV vary considerably, depending on the viral isolate, the age of the host plant and environmental conditions. Seedlings of *Pisum sativum* inoculated prior to the unfolding of the first true leaves (*c.* 7–10 days after sowing) begin to express symptoms 5–7 days after inoculation, manifested by a downward curling of the uppermost leaves. This is followed (7–10 days postinoculation) by a marked vein-clearing, and the development of both irregular chlorotic flecks and small irregular translucent lesions (often described as windows) along the leaf surface. From 10 to 14 days following inoculation, the aerial portions of the plant become severely stunted, epinastic and rugose, with continued amplification of foliar symptoms. Apical dominance is often lost, with proliferation of severely distorted axillary buds. In severe cases there is also evidence of top and bud necrosis. The diagnostic symptom of this virus, the enation, is evident on the lower leaf vein and stipule surfaces as hypertrophic undifferentiated outgrowths. These tend to form late in infection, approximately 2–3 weeks following inoculation. Although infected plants will successfully set pods, their yield, size and quality is significantly reduced. Pod symptoms consist primarily of wart-like protuberances along the outer pod surface, although pod enations have also been observed. In older plants and tolerant varieties, symptoms are often less severe, consisting mainly of the foliar mosaic symptoms with fewer growth abnormalities.

Symptoms on *Chenopodia* species consist of small chlorotic local lesions emerging 3–10 days following inoculation. The reproducibility of *Chenopodia* species as a local lesion host is highly variable and greatly influenced by environmental conditions.

Virus Epidemiology and Control

Annual and perennial leguminous host and weed species serve as overwintering reservoirs for both PEMV and its aphid vector. Aphid migration from these overwintering hosts in spring and early summer results in the dissemination of the virus into adjacent areas. Therefore, one method of reducing the incidence of PEMV is to control the occurrence of the aphid vector in both the overwintering host as well as in the secondary (crop) host. It has also been suggested that control of leguminous weed species and avoidance of perennial legumes adjacent to production fields may assist in reducing the available virus and vector reservoirs.

True resistance to PEMV has not been described for any crop species susceptible to PEMV. Tolerance to PEMV has been described in *P. sativum*, and is controlled by a single dominant gene designated *En* derived from USDA PI No. 140295 originating in Iran. Similar tolerance has also been uncovered in *Lens culinaris* accessions (PI Nos. 472547 and 472609) originating in India and Iran respectively.

PEMV, like several members of the luteovirus group, is able to serve as one partner of a disease complex with an unrelated aphid-nontransmissible virus. The bean yellow vein banding complex (BYVBV) consists of a helper-dependent association between PEMV (the helper) and BYVBV, the dependent species. This affiliation results in an increase in yield loss over that associated with PEMV infection alone. In this complex, BYVBV, which is normally mechanically transmitted, is dependent on PEMV for aphid transmission, presumably the result of a transcapsidation phenomena. Bean leafroll virus (a luteovirus) has been demonstrated to serve as a substitute helper virus for PEMV in this complex.

Cytopathology

Infections by PEMV are characterized by a rich and distinctive ultrastructural association with tissues of both the host and its aphid vector (Fig. 3). These interactions have strong parallels with those observed in luteovirus infections. Perhaps the most distinguishing characteristic of PEMV infection is its intimate replicative association with the host cell nucleus and with vesicular structures originating from the nucleus. Virions of PEMV are found throughout the nucleus as well as in the cytoplasm and vacuoles, either scattered or in loosely packed clusters. A proliferation of fibril-containing vesicular structures is evident in infected tissue, including all parenchymatic cell types and particularly in phloem tissues. In time course studies, these vesicles were shown to originate from the inner nuclear membrane

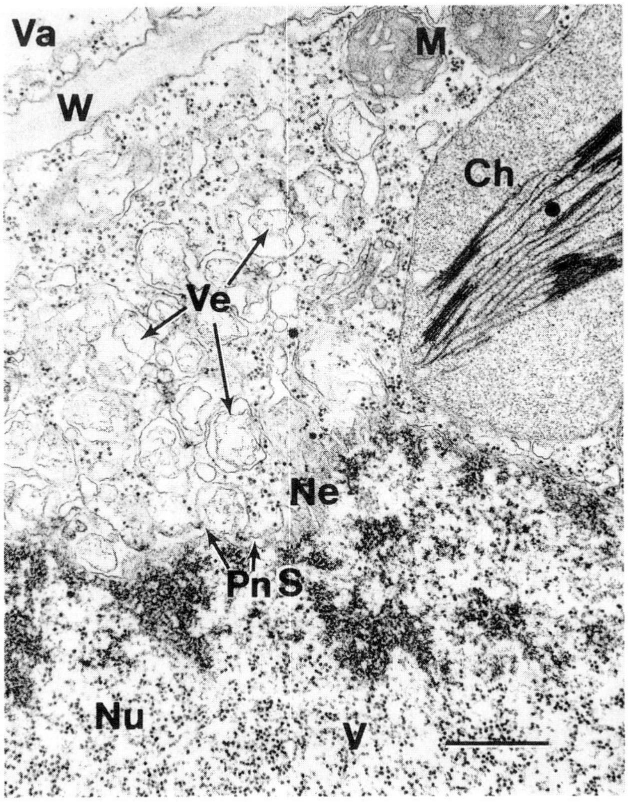

Fig. 3 A PEMV infected pea mesophyll cell. Note the cytoplasmic vesiculation (Ve) representing the replication complex of this virus. The vesicles are formed in the perinuclear space (PnS) of nuclei sustaining viral replication. In many cases, nuclei in these cells contain assembled virions (V). Ch = chloroplast; M = mitochondrion; Ne = nuclear envelope; Nu = nucleus; Va = vacuole; W = cell wall. Magnification × 31 000; scale bar = 0.5 μm.

and are deposited into the perinuclear space from which membrane-bound groups of vesicles are extruded into the cytoplasm. The internal fibrils associated with these vesicles are sensitive to digestion with DNase. Through *in situ* hybridization studies, it was demonstrated that both the nuclei and fractions enriched for these vesicular structures contained the negative-sense strand of PEMV RNA, implicating these structures as the site of viral replication. Supporting this view, PEMV-induced RNA-dependent RNA polymerase activity was also localized to both the vesicular and nuclear fractions. Indeed, isolated healthy pea nuclei were also demonstrated to support replication of PEMV RNA in *in vitro* assays. Combined, these data support the assignment of these membranous vesicles as the replicative complex of PEMV.

In addition to their role in viral replication, these vesicles are also found associated with or traversing through plasmodesmata of mesophyll, phloem parenchyma, sieve element and fully developed sieve tubes. The systemic mobility of these complexes coupled with their demonstrated polymerase activity suggests a possible role for these structures in the systemic spread of PEMV infection. In addition, a transient, electron dense dagger-like protrusion is evident associated with plasmodesmata, protruding into the cytoplasm of healthy and necrotic companion cells and sieve elements. It is not clear at this time whether this structure is of viral origin or a possible defense reaction of the host. An additional cytopathic feature characteristic of PEMV infection is the presence of elongated feather-like crystalline inclusions in epidermal cells. These anomalous inclusions are always surrounded by membranes and are rich in ribosomes and polyribosomes. They are often associated with the perinuclear space or with the vesicular membranes derived from the perinuclear space. Occasionally these structures are evident in apparently healthy tissue, and may reflect a stress-related response by the host. Their functional significance and origin are currently unknown.

A number of researchers have examined the distribution of PEMV virions within the tissues of the aphid vector of PEMV. PEMV virions were identified in the gut lumen, fat bodies, epithelium and muscle cells of the midgut, in hemocytes and in electron dense viroplasm-like structures hypothesized to be part of the lysosomal apparatus. Electron microscopic analysis of aphid salivary systems demonstrated an association of PEMV virions with the basal lamina and plasma-membrane complex of the accessory salivary gland, establishing a directional membrane-mediated shuttling of virions from the hemocoele to the stylet. In contrast, there was no evidence of a similar association of virions of the aphid-nontransmissible isolate of PEMV with these salivary structures. These observations are consistent with similar observations described for the circulatively aphid-transmitted luteoviruses.

Future Perspectives

From the discussion presented above, the PEMV system offers a number of opportunities for the analysis of virus–host, virus–vector and virus–virus interactions. Clearly, PEMV enters into a number of interactions between virus, host and aphid vector, characterized by discrete tangible structural entities.

With the enhanced understanding of the viral contribution to these interactions, the capacity to analyze them on the molecular, cellular and whole plant (or vector) level has increased.

One of the critical interactions that occurs during PEMV infection may actually be of virus–virus nature, i.e. the nature of the interaction between RNA1, RNA2 and RNA3. The presence of a second polymerase-like cassette in RNA2 raises the possibility that each of the 'genomic' RNAs may possess some level of replicative autonomy in PEMV infection. Thus, the historical confusion over the mandatory components of the PEMV genome may be a result of a chimeric viral genome composed of a luteo-like virus and a second viral genome of unknown functional competency. An understanding of the nature of the interaction between RNA1 and RNA2 will require the assessment of the independent infectivity of each RNA, the characterization of such an infection (localized, phloem-limited or systemic) and the determination of functional deficiencies in one or both RNAs that are complemented by the other member. Numerous scenarios for the interaction between RNA1 and RNA2 are feasible, including complete independence, mutualism, commensalism or absolute dependence. In addition, the small region of homology at the 5′ terminus of RNA2 and the satellite RNA raises the question of which genomic RNA (or RNAs) are responsible for the replication of the satellite element. The dissection of these interactions will provide important insight into the mechanisms leading to the development and evolution of viral diseases.

See also: Luteoviruses.

Further Reading

Demler SA and de Zoeten GA (1991) The nucleotide sequence and luteovirus-like nature of an aphid non-transmissible strain or pea enation mosaic virus. *J. Gen. Virol.* 72: 1819.

Demler SA, Rucker DG and de Zoeten GA (1993) The chimeric nature of the genome of pea enation mosaic virus: the independent replication of RNA2. *J. Gen. Virol.* 74: 1.

de Zoeten GA and Gaard G (1983) Mechanisms underlying systemic invasion of pea plants by pea enation mosaic virus. *Intervirology* 19: 85.

Harris KF *et al.* (1975) Fate of pea enation mosaic virus in PEMV-injected pea aphids. *Virology* 65: 148.

Powell CA and de Zoeten GA (1977) Replication of pea enation mosaic virus in isolated pea nuclei. *Proc. Natl. Acad. Sci. U.S.A.* 74: 2919.

PERSISTENT VIRAL INFECTION

Rafi Ahmed
School of Medicine
University of California, Los Angeles, USA

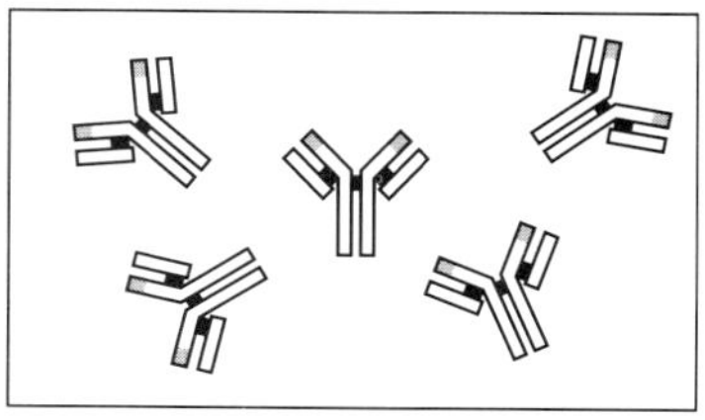

Introduction

Viral persistence and the pathological consequences associated with chronic infections now constitute major health problems. With the development of successful vaccines against several major acute viral diseases such as polio, smallpox, measles, mumps, and rubella, there has been a sharp decline in the incidence of acute viral infections. However, the number of people afflicted by viruses causing persistent infections has increased. This is most dramatically illustrated by the acquired immunodeficiency syndrome (AIDS). Also, chronic hepatitis B virus infections have steadily increased with current estimates of > 250 million virus carriers. In addition to the many known diseases associated with viral persistence (see Table 1), it is also likely that some chronic disorders of the nervous and endocrine systems that are currently of unknown etiology may have a viral origin.

A listing of viruses that can persist in humans is given in Table 1. As can be seen from this table, the ability to persist *in vivo* is not confined to a particular virus group and a variety of viruses, both DNA and RNA containing, can establish long-term infections. Persistent viral infections have traditionally been divided into two categories: (1) 'chronic (productive) infection' – a situation in which infectious virus is continuously produced and (2) 'latent infection' – a condition in which

Table 1. Persistent viral infections of humans

Virus	Site of persistence	Consequence
Human immunodeficiency virus	CD4+ T cells, macrophages, microglia	AIDS
Human T cell leukemia viruses	T cells	Leukemia
Hepatitis B virus	Liver, lymphocytes?	Cirrhosis, hepatocellular carcinoma
Hepatitis D virus	Liver	Exacerbation of chronic HBV infection
Hepatitis C virus	Liver	Cirrhosis, hepatocellular carcinoma
Polyomavirus JC	Kidney, CNS	Progressive multifocal leukoencephalopathy
Polyomavirus BK	Kidney	Hemorrhagic cystitis
Papillomaviruses	Skin, epithelial cells	Papillomas, carcinomas
Adenoviruses	Adenoids, tonsils, lymphocytes	None known
Parvovirus B19	Bone marrow	Aplastic crisis in hemolytic anemia, chronic bone marrow deficiency
Herpes simplex virus types 1 and 2	Neurons in sensory ganglia	Cold sores, genital herpes
Varicella–Zoster virus	Neurons/satellite cells in sensory ganglia	Zoster
Cytomegalovirus	Kidney, salivary gland, lymphocytes? macrophages? stromal cells?	Pneumonia, retinitis
Epstein–Barr virus	B cells, pharyngeal epithelial cells	Burkitt's lymphoma, carcinoma?
Human herpesvirus – 6	Lymphocytes	None known
Measles virus[a]	Neurons and supporting cells in the brain	Subacute sclerosing panencephalitis, measles inclusion body encephalitis
Rubella virus[a]	CNS	Progressive rubella panencephalitis, insulin-dependent diabetes mellitus (?)

[a] Measles and rubella are typically considered as viruses causing acute infections. However, in rare instances these viruses have been shown to persist in the CNS.

the viral genome is present but infectious virus is generally not produced except during intermittent episodes of reactivation. The classic example of chronic productive infection is hepatitis B virus (HBV) and of latent infection is herpes simplex virus (HSV). Although this division (i.e. latent vs productive) has been useful in classifying infections caused by different viruses - it is not absolute and as we learn more about viral persistence, it is becoming increasingly difficult to classify a particular infection as latent or productive. For example, recent studies have shown that HBV a virus known to productively infect hepatocytes can also cause a latent infection of lymphocytes. Also, there is restricted expression of HBV in hepatocellular carcinoma. Another example of both latent and productive infection is Epstein–Barr virus (EBV); EBV is latent in B cells but causes a productive infection of pharyngeal epithelial cells. Infection with HIV can either be productive or latent depending upon the cell type infected and the 'activation' stage of the cell. Thus, it is important to recognize that these categories are not absolute and that persistence involves stages of both productive and latent infection.

The consequences of viral persistence can range from the totally benign, such as with adenoviruses whose persistence is not associated with any clinical manifestation, to the almost uniformly fatal, as seen with HIV persistence and progression to AIDS. Table 1 lists the clinical manifestations associated with persistence of different viruses. It should be pointed out that persistence (with the possible exception of HIV) does not always lead to disease. For instance, although HBV carriage greatly increases the risk of developing hepatocellular carcinoma and cirrhosis is often a sequela of HBV persistence, a substantial number of HBV carriers remain asymptomatic for life. The reasons for these different outcomes are not understood – they could be due to host genetic factors, infection with different HBV variants, or environmental co-factors. The disease syndrome related to persistent infection is sometimes seen only in immunosuppressed individuals. This is the case with cytomegalovirus (CMV); in normal individuals CMV persistence is asymptomatic but in immunosuppressed people (transplant recipients, AIDS patients) CMV spreads to tissue sites it normally does not persist in, such as lung and eye, causing pneumonia and retinitis. Another example is progressive multifocal leukoencephalopathy (PML), a fatal demyelinating disease caused by infection of oligodendrocytes by polyomavirus JC. Although > 25% of normal individuals are persistently infected with JC virus, the disease PML is seen only in immunodeficient subjects and is currently the cause of death of 2–4% of AIDS patients.

Evasion of Immunity

The essence of viral persistence is evasion of antiviral immunity. In most, if not all, instances the viruses listed in Table 1 persist in immunocompetent individuals. Even HIV persistence is initially (for up to several years) in people with normal and functional immune systems. Why then, are these viruses not eliminated by the host's immune system? The precise answers to this critical issue are not known but there is now an appreciation that viruses have developed various strategies of evasion and staying one step ahead of the immune system. Before discussing the various escape

Table 2. Anti-viral immunity

Effector system	Recognition molecule	Mechanism of viral control
Antibody	Surface glycoproteins or outer-capsid proteins of virus particles	Neutralization of virus
	Viral glycoproteins expressed on membrane of infected cells	Antibody-complement mediated killing and antibody-dependent cell-mediated cytotoxicity (ADCC) of virally infected cells
CD4+ T cells	Viral peptides (10–20-mers) presented by MHC class II molecules. The peptides can be derived from any viral protein (structural-surface or internal- or non-structural). The limiting factor is the ability to bind MHC molecules	Release of anti-viral cytokines (interferon-γ, TNF) and activation of macrophages. Killing of virally infected cells (?)
CD8+ T cells	Viral peptides (9-mers) presented by MHC class I molecules of infected cells. The peptides can be derived from any viral protein, the main limiting factor being the processing and binding affinity of the peptide for MHC molecules	Killing of virally infected cells. Release of antiviral cytokines (interferon-γ, TNF) and activation of macrophages
Natural killer cells	Not known	Release of interferon-γ. Participation in ADCC. Direct lysis of virus infected cells (?)

mechanisms viruses have devised it is useful to review the basics of antiviral immunity (Table 2).

Antibodies and T cells (CD4+ and CD8+) are the two main antigen-specific effector systems for resolving viral infections. Antibodies can recognize either free virus or virally infected cells; they control virus infections by neutralizing free virus and by killing infected cells through complement-mediated cytotoxicity or antibody-dependent cell-mediated cytotoxicity (ADCC). The critical viral proteins in these above processes are surface glycoproteins or outercapsid proteins, and although antibodies against internal and nonstructural viral proteins are also made, these do not participate in viral neutralization or killing of infected cells by complement-mediated lysis or ADCC. Thus, in terms of antiviral immunity by antibody the critical recognition molecules are the viral glycoproteins expressed on membranes of infected cells and the surface glycoproteins or outercapsid proteins of virus particles. T cells play a different game altogether. Unlike antibodies that recognize viral proteins by themselves, T cells only see viral antigen in association with host MHC molecules. The antigen-specific T cell receptor recognizes short viral peptides presented by cellular MHC molecules. An important consequence of this mode of recognition is that T cells cannot recognize free virus and their antiviral activities are confined to infected cells. Thus, the T cell arm of the immune system has evolved for surveillance of infected cells whereas antibody functions primarily (though not exclusively) as defence against free virus. T cells are further subdivided into two subsets, CD4+ and CD8+ T cells. CD4+ T cells recognize viral peptides in association with MHC class II antigens, whereas CD8+ T cells recognize viral peptides plus MHC class I antigens. These peptide fragments can be derived from any viral protein, structural (either surface or internal) or non-structural. Thus, all viral proteins can be potential targets for T cell recognition. The limiting factors are the processing of the protein and the ability of the peptides to bind with MHC molecules (i.e., the affinity for various MHC antigens). How do T cells control viral infections? The primary mechanism employed by CD8+ T cells is killing of virally infected cells. These cytotoxic CD8+ T lymphocytes are highly efficient in this process and not only induce apoptotic death of the infected cell but also degrade viral nucleic acid inside the cell. Thus, both the 'factory' and the 'products' are destroyed ensuring that no infectious virus is released when the infected cells are lysed. In addition to their killing function, CD8+ T cells also control virus growth by producing antiviral cytokines such as interferon-γ and tumor necrosis factor (TNF). The primary contribution of CD4+ T cells in antiviral immunity is production of antiviral cytokines, activation/recruitment of macrophages and providing help for antibody production. Virus specific CD4+ cytotoxic T cells have been described in several systems but contribution of killing *per se* by CD4+ T cells in controlling virus infections *in vivo* is not clear. In addition to the two antigen specific effector systems (i.e., T cells and antibody), there is also evidence that natural killer (NK) cells play a role in controlling viral infections. However, it is not known what viral molecules are recognized by NK cells and how NK cells can discriminate between virally infected and normal cells.

Successful resolution of a virus infection depends upon a critical balance between viral replication/survival and the host's immune response. The various strategies viruses employ to upset this delicate balance and avoid elimination by the host's immune system are listed in Table 3. The simplest but perhaps the most effective mechanism of evasion is for the virus to become latent with minimal to no expression of viral proteins. The best example of restricted viral gene expression is latent infection of neurons by HSV. HSV gene expression is turned off except for transcription from one region of the genome and there appear to be no viral proteins expressed in the infected neurons. Under such conditions the virus essentially becomes invisible to the immune systems since immunity is directed against foreign proteins and is not programmed to distinguish between 'self' and 'foreign' nucleic acid. Thus, absolute latency is the ideal way of evading the immune system. However, this is not always feasible from the virus' point of view because often certain viral proteins are essential for replicating viral DNA. For example, during latent infection of B cells by EBV, expression of one of the EBV proteins, called EBNA-1, is necessary for propagating the EBV genome (present as an episome). A permanent state of latency is also not in the best interest of the virus in terms of transmission. Without production of infectious virus there can be no horizontal spread – an important mode of viral transmission. Thus, it is not surprising that even viruses that are highly efficient in establishing latent infections go through intermittent phases of productive infection.

Another strategy employed by viruses is infection of tissues and cell types that are not readily accessible to the immune system. A site of persistence favored

Table 3. Viral strategies for evading the immune system

Escape mechanism	Example
Restricted gene expression; virus remains latent in the cell with minimal to no expression of viral proteins	HSV in latently infected neurons, HIV in resting T cells
Infection of sites (tissues/cell types) not readily accessible to the immune system	Viral (HSV, VZV) persistence in neurons – cells that express neither MHC class I nor class II molecules and hence cannot be recognized by T cells. The presence of specialized cells like neurons and the blood–brain barrier make the CNS a privileged site of viral persistence
Antigenic variation; virus rapidly evolves and mutates antigenic sites that are critical for recognition by antibody and T cells	Lentiviruses; examples of antibody and CTL escape variants in HIV
Suppression of cellular MHC class I molecules	Adenoviruses, cytomegaloviruses, HSV
Decreased expression of cell–adhesion molecules that are required for efficient T cell–target interaction	EBV in Burkitt's lymphoma cells
Viral molecules that inferfere with the function of of antiviral cytokines	Adenovirus proteins (E3-14.7K, E3-10.4K/14.5K, and E1B-19K) and poxvirus protein T2 protect infected cells from lysis by TNF
	Adenovirus VA RNA, EBV EBER RNA, and HIV TAR RNA inhibit function of interferon-alpha/beta
	EBV protein BCRF1 (a homolog of IL-10) can block synthesis of cytokines such as IL-2 and interferon-γ

by many viruses (see Table 1) is the central nervous system (CNS). At least two factors favor viral persistence in the CNS; one, the presence of the blood–brain barrier that limits lymphocyte trafficking through the CNS, and second, the presence of specialized cells like neurons that express neither MHC class I nor class II molecules and hence, cannot be directly recognized by T cells. Another tissue where viruses tend to persist is the kidney. The human polyomaviruses BK and JC replicate in the kidney with almost life-long shedding into the urine. CMV is also found in the kidney and shed for long periods of time. It is not obvious why the immune system is less effective in eliminating microbes from the kidney. There is certainly no blood:kidney barrier and there is extensive trafficking of lymphocytes through the kidney. Although the importance of this organ in viral persistence and transmission has been appreciated for a long time, hardly any studies have been done to understand the underlying mechanisms. This neglected site of viral persistence deserves more attention.

The emergence of viral variants during persistence is a well documented phenomenon. Viruses, especially those with RNA genomes, can undergo mutation at high frequencies and under the appropriate selective pressure variants can rapidly arise. There are many examples of antibody resistant variants and mutation at sites critical for antibody recognition is a highly effective means of escape from neutralizing antibody. Similarly, it is possible to have mutations in the viral peptides recognized by T cells. Both antibody and CTL escape variants have been described in HIV. However, the evidence that persistence of HIV is due to selection of such antigenic variants is far from compelling. Nevertheless, antigenic variation is a powerful mechanism of evasion and at least in the case of another lentivirus infection (persistence of equine infectious anemia virus in horses) it is well documented that recurrent sequential episodes of disease (clinical symptoms associated with bursts of viremia) are due to the selection of antibody resistant variants.

Viruses can escape T cell recognition not only by

mutation at the epitope (peptide) seen by the T cell receptor but also by down-regulating the expression of any one of several host molecules that are necessary for efficient T cell : target interaction. In addition to the viral peptide, the interaction of virus specific T cells with infected target cells depends on the expression of host MHC class I (for CD8+ cells) or class II molecules (for CD4 cells) as well as several adhesion molecules such as ICAM-1 and LFA-3. Viruses have taken advantage of this and developed strategies for selectively inhibiting the expression of these critical host cell molecules. Reduction of MHC class I antigen expression on host cells as a consequence of viral infection has been reported for several viruses. The best documented example of viral-mediated suppression of MHC class I antigens is by the adenoviruses. One of the early proteins of adenovirus, the E3 19 000-dalton protein termed E3/19K, can bind and form a molecular complex with MHC antigens. The formation of this E3–MHC class I complex prevents the MHC antigens from being correctly processed by inhibiting their terminal glycosylation. This results in reduced cell-surface expression of class I antigen and provides an effective means of avoiding recognition by CD8+ T cells. Down regulation of the cell-surfaced adhesion molecules LFA-3 and ICAM-1 is involved in the escape of EBV positive Burkitt's lymphoma (BL) cell lines from EBV-specific CTL. Certain EBV-positive BL cell lines are not killed by MHC-matched virus-specific CTL and their resistance is not due to altered expression of MHC class I genes nor lack of viral gene expression but correlates with a reduced level of the adhesion molecules, LFA-3 and ICAM-1, on the tumor cell surface. The mechanism involved in the selective suppression of these molecules is currently not known, but this example nicely illustrates yet another viral strategy of circumventing the immune response.

Recent studies have described several viral molecules that interfere with cytokine function. Three adenovirus early proteins, E3-14.7K, E3-10.4K/14.5K, and E1B-19K, can protect virus-infected cells from lysis by TNF. The mechanism by which these adenovirus proteins counteract TNF is currently not known. The poxviruses also encode a protein called T2 that inhibits the action of TNF. The poxvirus T2 protein is a homolog of the cellular receptor for TNF and is released from infected cells. Thus, it serves as a decoy. T2 binds TNF and prevents TNF from binding its true cellular receptor and destroying virus-infected cells. Other examples of viral 'defense' molecules include the EBV protein BCRF1 that is a homolog of IL-10 and can block synthesis of IL-2 and interferon-γ. Finally, sometimes even viral RNA can function as a defense molecule; the adenovirus VA RNA, the HIV TAR RNA, and the EBV EBER RNA can inhibit the antiviral effects of interferon-alpha/beta. Interferons induce the synthesis of a protein called DAI, which in the presence of double-stranded RNA, phosphorylates initiation factor eIF-2 and prevents initiation of translation. The viral RNA's mentioned above (VA, TAR, and EBER) are molecules with extensive secondary structure and can block the interferon induced autophosphorylation of DAI and thereby interfere with the action of interferon.

Caveats and Challenges

First some caveats. Several viral strategies of evading the immune system were outlined in the preceding section. Although specific examples were used to illustrate the different means of evasion, it is unlikely that a single mechanism accounts for the persistence of a given virus. It is more likely that a combination of these strategies, plus other mechanisms that are currently not known, contribute to the persistence of virus in an otherwise immunocompetent host. Some doubts about a fashionable and widely accepted notion; despite the many elegant studies on suppression of MHC class I molecules by adenoviruses, there is surprisingly still no evidence that persistence of human adenoviruses during the natural infection is related to down-regulation of MHC class I. A word of caution also about the recent work on viral defense molecules that interfere with cytokine function and their proposed role in viral persistence. This is clearly an exciting and provocative area but most of the findings are based on *in vitro* studies and it is essential that these observations be put to the *in vivo* test. In fact, in one instance we already know that the presence of viral defense molecules is not sufficient to completely evade the host's immune response. Several molecules that interfere with various cytokines have been found in vaccinia virus and yet vaccinia virus does not cause a protracted, and certainly not a persistent infection and it is efficiently eliminated by the immune system. Although these proteins are likely to play a critical role in the pathogenesis of vaccinia virus, it is certain that, at least in this instance, they have little to do with long-term persistence within an individual host.

The challenge, of course, is to develop treatment strategies for controlling persistent viral infections. There are currently no effective means of eliminating an already established persistent viral infection of humans. It is unlikely that a single strategy will be effective against all viruses, and different regimens will have to be developed for treating the various persistent viruses. To achieve this goal it is essential to obtain a better understanding of the viral, cellular, and immunological factors involved in the persistence of a given virus. This will permit the development of new drugs that are targeted to specific viral molecules, and the production of immunotherapeutic vaccines that potentiate the critical effector systems. A marriage, or more appropriately, a ménage-à-trois, of the disciplines of virology, immunology, and chemistry should provide the answers.

See also: Adenoviruses; Cytomegaloviruses; Epstein–Barr virus; Genetics of animal viruses; Hepatitis B viruses; Herpes simplex viruses; Human immunodeficiency viruses; Immune response; Latency; Measles virus, Papillomaviruses; Polyomaviruses; Transplantation; Varicella-Zoster virus.

Further Reading

Ahmed R and Stevens JG (1990) Viral Persistence. In Fields BN *et al.* (eds) *Virology*, 2nd edn, p. 241. New York: Raven Press.

Doherty PC, Allan W, Eichelberger M and Carding SR (1992) Roles of $\alpha\beta$ and $\gamma\delta$ T cell subsets in viral immunity. *Annu. Rev. Immunol.* 10: 123.

Gooding LR (1992) Virus proteins that counteract host immune defenses. *Cell* 71: 5.

PESTE DES PETITS RUMINANTS

See Rinderpest and distemper viruses

PHOCID DISTEMPER VIRUS

See Rinderpest and distemper viruses

PHYTOREOVIRUSES

Donald L Nuss
Roche Institute of Molecular Biology
Nutley, New Jersey, USA

History

Among plant viruses, members of the family *Reoviridae* are distinguished by a segmented double-stranded (ds) RNA genome, a transmission mechanism that involves replication in the insect vector and a propensity to induce tumors in infected plant hosts. Consequently, viruses within this group have attracted attention as potential model systems for studying a range of biological and molecular interactions and processes. From an historical perspective, plant reoviruses have played important roles in several significant scientific developments. Studies with rice dwarf virus (RDV) and wound tumor virus (WTV) were crucial in confirming the once controversial hypothesis that some insect-borne plant viruses actually replicate in the vector. The technique of density gradient centrifugation was developed during attempts to purify WTV. Efforts to understand the details of plant reovirus transmission led to the establishment of a continuous cell culture line from a plant virus insect vector. In addition, several members of this group can cause considerable damage to crop plants and therefore are also of interest from an agronomic perspective.

Taxonomy and Classification

The classification scheme for the plant-infecting reoviruses continues to evolve. Until recently, the primary taxonomic considerations included the number of segments that comprised the genome, the electrophoretic mobility of the genomic segments and the identity of the insect vectors involved in virus transmission. This resulted in the construction of two genera for this group of viruses within the family *Reoviridae*, *Phytoreovirus* (12 genomic segments, leafhopper transmitted) and *Fijivirus* (10 genomic segments, planthopper transmitted). A third unclassified group was included to accommodate several plant reoviruses that did not readily fit into the two genera classification scheme. Based on additional morphological, electrophoretic and molecular data, the inclusion of a third genus was proposed by the International Committee on Taxonomy of Viruses in its fifth report released in 1991, as indicated below.

The genus *Phytoreovirus* I currently consists of three members, WTV, RDV and rice gall dwarf virus (RGDV). All members of this genus are morphologically similar, e.g. they lack capsid spikes prominent on particles of other plant reoviruses, contain 12 genomic RNA segments that exhibit similar electrophoretic profiles and are transmitted exclusively by cicadellid leafhoppers. The genus *Fijivirus* (*Phytoreovirus* II) [the name derived from the type member Fiji disease virus (FDV)] consists of three principal members, FDV, maize rough dwarf virus (MRDV) and oat sterile dwarf virus (OSDV). Several other plant-infecting reoviruses described in the literature are now considered geographical races of members of this genus. Viruses related to MRDV include rice black streaked dwarf virus (RBSDV), pangola stunt virus (PSV) and cereal tillering disease virus (CTDV). *Arrhenatherum* blue dwarf virus (ABDV) and *Lolium* enation viruses (LEV) are related to OSDV. Members of this genus have genomes consisting of 10 segments that exhibit a characteristic electrophoretic profile quite distinct from that exhibited by genomes of the members of the genus *Phytoreovirus* I and are transmitted exclusively by delphacid planthoppers. A third suggested genus, *Phytoreovirus* III, consists of the type member rice ragged stunt virus (RRSV) and the possible member *Echinochola* ragged stunt virus (ERSV). Members of this proposed genus resemble members of the genus *Fijivirus* in terms of number of genomic segments and transmission by delphacid planthoppers, but differ in virus particle morphology and the electrophoretic profile of the genomic RNA segments.

Additional support for the overall classification scheme has been provided by recent comparative sequence analyses of the terminal domains of plant reovirus genomic segments. All genomic segments of each member of the genus *Phytoreovirus* I contain the conserved terminal oligonucleotide sequences (+) 5′-GGU/CA- - - -U/CGAU-3′ (only terminal sequences of the coding strand are shown). In contrast, two members of the genus *Fijivirus* (*Phytoreovirus* II), MRDV and RBSDV, have the conserved oligonucleotide sequences (+) 5′-AAGUUUUU- - - -GUC-3′, while the consensus conserved terminal oligonucleotide sequence for members of genus *Phytoreovirus* III is (+) 5′-GUAAA- - - -GUGC-3′. Thus the nature of the genomic conserved terminal oligonucleotide sequences for the plant-infecting members of the family *Reoviridae* appear to be genus specific, providing an additional taxonomic criterion.

Structure

Plant reovirus particles consist of two protein shells surrounding a core composed of dsRNA genomic segments and associated proteins. For members of the genus *Phytoreovirus* I, the core consists of 12 dsRNA genomic segments tightly associated with three structural proteins surrounded by a shell of well-defined capsomers. This structure is further surrounded by a less well-defined outer shell consisting of two more loosely associated proteins. The complete particle is approximately 70–80 nm in diameter. RDV has now been crystallized and it is likely that X-ray diffraction analysis will soon provide a detailed view of plant reovirus particle structure. Members of the other plant reovirus genera have a slightly different morphology which is characterized by the presence of prominent spikes protruding from the inner capsid. Purified RRSV particles exhibit additional differences in that they appear to contain only one protein shell.

Several enzymatic activities are associated with the purified core structures. These include an RNA-dependent RNA polymerase responsible for mRNA synthesis, and two enzymatic activities involved in the RNA-capping reactions, an mRNA-guanyltransferase and a guanine 7-methyltransferase. An additional RNA-modifying activity, an mRNA 2′-*O*-methyltransferase, has also been demonstrated in purified WTV preparations. From an evolutionary perspective, this observation is interesting, since this enzymatic activity is not present in plant cells.

Genome Organization

The genomic RNAs of several members of the genus *Phytoreovirus* I have been subjected to extensive cDNA cloning and sequence analysis. While each of the 12 segments that comprise the viral genome is unique in terms of nucleotide sequence, all segments appear to conform to a basic organizational strategy. For example, each dsRNA segment of the WTV genome, nine of which have been completely sequenced, contains a conserved hexanucleotide sequence at one terminus and a conserved tetranucleotide sequence at the other [(+) 5′-GGUAUU——UGAU-3′, coding strand sequence shown]. It has not yet been determined whether the 5′ terminus of the coding strand is capped with an m^7G^5 ppp structure as occurs for other members of the family *Reoviridae*. The other terminus, however, appears to be a perfect base-paired duplex and lacks any homopolymeric stretches. Segment-specific regions of inverted complementarity (inverted repeats) from 6 to 14 residues in length are found adjacent to the conserved sequences. Each segment, with the exception of segment 12, contains one long open reading frame (ORF) on the strand containing the conserved 5′-terminal sequence GGUAUU——, and is flanked by a relatively short (18–63 nucleotides) 5′-noncoding region and a slightly longer (93–303 nucleotides) 3′-noncoding region. Segment 12 contains a second short ORF consisting of 40 codons located 102 nucleotides downstream from the large ORF that encodes the nonstructural protein Pns12. It is unclear whether this smaller ORF is expressed *in vivo*. *In vitro* expression of cloned WTV genomic dsRNAs has allowed the assignment of virally encoded polypeptides to their cognate genomic segments. In addition, alignment analysis of the predicted amino acid sequences of the major inner capsid protein of the three members of genus *Phytoreovirus* I have provided convincing evidence that these three viruses evolved from a common ancestor.

Only limited sequence information is currently available for members of the other plant reovirus genera. Interestingly, characterization of MRDV genomic segment S6 revealed a coding strand that contained two large nonoverlapping ORFs consisting of 363 and 310 codons and separated by a 52-nucleotide intercistronic region. This configuration is unique among the family *Reoviridae*. Only the 5′-proximal ORF was expressed in cell-free translation systems. As indicated above, the genomic segments of MRDV and RBSDV, both members of the genus *Fijivirus* (*Phytoreovirus* II), contain the same conserved terminal oligonucleotide sequences (+AAGUUUUU- - - -GUC-3′), indicating the likelihood of a common ancestor for these two viruses. It will be interesting to determine whether RBSDV segment S6 also contains two large ORFs. Sequence information for members of genus *Phytoreovirus* III is currently limited to the terminal domains. However, it is clear that the structural motif consisting of conserved terminal oligonucleotides and adjacent segment-specific inverted repeats is a common feature of the genomic segments of members of all three plant reovirus genera.

Several pieces of evidence suggest that plant reovirus particles may contain one copy of each of the 10 or 12 segments that comprise the viral genome. This evidence is based primarily on studies with WTV and includes the following observations. Genomic RNA isolated from purified virus particles contains equimolar amounts of each segment and, based on extrapolations of dilution-infectivity curves, it is predicted that an infection can be initiated by a single virus particle. The mechanisms responsible for the apparent selective packaging of one copy of each viral RNA segment into a single particle are not well understood. Molecular characterization of internally deleted defective-interfering (DI) RNAs present in some transmission-defective

WTV populations clearly established that the sequence information necessary for replication and packaging of genomic segments resides within the terminal domains of the segment. It was shown further that DI RNAs displaced only the genomic segment from which they were derived and did not interfere with the replication or packaging of unrelated genomic segments. Combined, these observations suggested that each genomic segment must contain at least two operational recognition domains: one that specifies that the RNA is viral and not cellular and a second that distinguishes one genomic RNA segment from another. It has been postulated that the initial events leading to the replication of WTV genomic segments involve the specific recognition and packaging of viral transcripts, rather than dsRNA segments, followed by subsequent synthesis of the genomic noncoding strand. It has been postulated further that the conserved terminal oligonucleotides and the segment-specific inverted repeats may serve as recognition domains, with the terminal inverted repeats serving to specify the conformational properties of the viral transcripts as a result of intramolecular base-pairing.

The kinetics of viral protein and RNA synthesis has also been measured in WTV-infected vector cells. Based on distinct changes in the level of viral gene expression, the infection was divided into an acute phase (the first 5 days after inoculation) and a persistent phase (beginning with the first cell passage). Viral specific polypeptide and viral genomic RNA accumulation was observed to increase to a maximum level during the first 5 days following inoculation and then to decrease rapidly to approximately 5% of the maximum level beginning with the first cell passage. In contrast, viral specific mRNAs were present at approximately the same level in the acute phase and in the early stage of the persistent phase of infection. Moreover, viral transcripts isolated from persistently infected cells were found to be inefficiently translated *in vitro*, reflecting the situation observed in infected cells. These results were interpreted as suggesting that a form of post-translational regulation of viral gene expression operates in vector cells persistently infected with WTV. The unavailability of cultured planthopper cell lines has precluded similar studies with members of the other plant reovirus genera.

Gene Expression

There are numerous technical difficulties associated with studying virus replication and gene expression *in planta*. In contrast, the cultured cell lines derived from leafhopper vectors provide very convenient systems for examining these processes for the plant reoviruses. All three members of the genus *Phytoreovirus* I have been shown to readily multiply in their leafhopper vector and cultured cells derived from these vectors. Infection of both the leafhopper vector and cultured vector cells results in a productive, usually asymptomatic, persistent infection. Infection of cultured vector cells by WTV has been examined in detail and was found to have the following characteristics. Infected cultures fail to exhibit any apparent cytopathology either during the initial (acute) stages of infection or after the onset of persistence. Persistently infected cells exhibit the same morphology, pattern of cellular protein synthesis and growth rate as uninfected cells and do not suffer from periodic crises. Essentially all cells in the persistently infected culture remain infected for hundreds of cell passages. Infectious virus is recoverable from persistently infected cells and can initiate a new round of acute and persistent infection.

Geographic Distribution and Economic Significance

Plant-infecting reoviruses have been isolated on every continent except Africa. Members of genus *Phytoreovirus* I have been identified only in North America and Asia, WTV in the former and RDV and RGDV in the latter. While WTV is of no economic significance and has been isolated only twice, in both cases serendipitously, RDV and RGDV are commonly found infecting rice crops and have been reported to cause severe damage under certain growing conditions and when viruliferous insect vector population levels are high. Members of the genus *Fijivirus* (*Phytoreovirus* II) are more widely distributed, found in the South Pacific, South America, Europe and Asia. Although several of the members of this genus were once responsible for serious crop losses and were expensive to control, the development of virus-resistant cultivars and efficient means of controlling vector populations have resulted in the effective control of these viral diseases. Both members of the proposed genus *Phytoreovirus* III have been found only in Asia and are so far of limited but potential economic importance.

Plant Host Range and Symptom Expression

The recorded host range of plant reoviruses, with the exception of WTV, is limited to the *Gramineae*. Severe stunting or dwarfing is the most common symptom of reovirus infection of graminaceous plants. Additional symptoms include distortion of the leaves including marginal serration, flecking or streaking, or the development of abnormally dark green coloration. These symptoms are generally accompanied by vein swelling, enations or hyperplastic growth on the abaxial leaf surface. WTV is the only reovirus known to infect dicotyledonous plants with an experimental range extending to 20 families. Other symptoms of WTV infection include stunting and leaf distortion. However, the most notable symptom is neoplastic growth of phloem tissue, the degree and duration of which are dependent on the host. For example, WTV infection of *Trifolium* spp. results in vein enlargement, while infection of *Melilotus* spp. results in the formation of massive stem and root tumors. In addition, wounding of infected plants tends to promote tumor formation. The molecular basis for tumor induction by plant reoviruses is unknown. Among the plant-infecting reoviruses, only RDV fails to induce neoplasia. RDV is also atypical in that its replication is not limited to the phloem tissue of infected plants.

Transmission

Infection of plant hosts by the plant reoviruses requires the aid of an insect vector; members of this virus group are not spread by mechanical transmission. The mode of transmission is referred to as propagative because virus multiplication occurs in the insect vector as a necessary part of the transmission process. The virus is acquired during feeding on infected plants, it sequentially infects different tissues of the vector, eventually reaching the salivary gland, and is then transmitted to uninfected plants via the salivary fluid during subsequent feeding. This process can take from 13 to 30 days to complete, depending on environmental conditions.

As indicated above, members of genus *Phytoreovirus* I are transmitted by leafhoppers, while members of the other genera are transmitted by planthoppers. Transmission of a particular plant reovirus is limited to a single or several vector species, e.g. WTV is transmitted only by *Agallia constricta* Van Duzee, *Agallia quadripunctata* Provancher and *Agalliopsis novella* (Say). In contrast to the situation in infected plants, multiplication of virus in the insect vector results in no apparent pathological consequences, although reductions in fecundity of MRDV- and RDV-infected vectors have been reported. Transovarial transmission to progeny has also been reported for several of the plant reoviruses. Continuous cell culture lines have been generated from several leafhopper species that serve as plant reovirus vectors. These cultured vector cells support virus multiplication, thus providing a valuable system for studying viral gene expression and replication as well as details of virus–vector interaction. Although the plant reoviruses do not form plaques on the vector cell monolayers, virus titers can readily be determined with the aid of a fluorescence focus assay.

Maintenance of plant reoviruses exclusively in a plant host can result in the generation of virus populations that are deficient or defective in their ability to be transmitted by the insect vector. This phenomenon has been studied in some detail for WTV. The loss of transmissibility was accompanied by a loss in the ability to infect cultured vector cells and the generation of internally deleted forms of specific genomic segments that acted as DI RNAs. In selected isolates, certain segments, e.g. segments S2 or S5, appeared to be missing. While these isolates retained the ability to induce tumors in systemically infected vegetatively propagated plants, they were completely defective in insect transmissibility. Although it has been suggested that segments S2 or S5 are required for replication in the insect vector but are dispensable for replication in the plant host, the precise basis for loss of transmissibility is unknown.

Future Perspectives

Many of the fascinating biological properties exhibited by the plant reoviruses remain poorly understood. It is anticipated that the recent progress made in determining molecular aspects of the virus group will provide new approaches for elucidating the molecular basis of such processes as virus-induced neoplasia, viral persistence, vector range and the recognition, sorting and packaging of a segmented RNA genome.

See also: Defective-interfering viruses; Persistent viral infection; Reoviruses.

Further Reading

Francki RIB and Boccardo G (1983) The Plant *Reoviridae*. In: Joklik WK (ed.) *The Reoviridae*, p. 571. New York: Plenum.
Kimura I and Omura T (1988) In: Harris KF (ed.) *Advances in Disease Vector Research*, vol 5, p. 300. New York: Springer Verlag.
Kudo H, Uyeda I and Shikata E (1991) *J. Gen. Virol.* 72: 2857.
Mizuno H *et al.* (1991) *J. Mol. Biol.* 219: 665.
Nuss DL and Dall DJ (1990) *Adv. Virus Res.* 38: 249.

PICORNAVIRUSES – INSECT

Paul D Scotti
Horticulture and Food Research Institute of New Zealand
Auckland, New Zealand

History

'Insect picornavirus' is often misused as a generic term for any small spherical virus isolated from insects. This classification is regularly, and probably incorrectly, applied to viruses which have 25–35 nm-diameter particles, but usually without considering the physicochemical properties of the virions or even determining whether the particles actually contain RNA.

The insect picornaviruses, as designated by the ICTV, are comprised of three types: cricket paralysis virus (CrPV), *Drosophila* C virus (DCV) and *Gonometa* virus (GoV). These are the only insect small RNA viruses studied with properties comparable with those of vertebrate viruses which have been formally accepted as belonging to the family *Picornaviridae*.

CrPV was discovered during the course of a mass rearing program of the Australian field cricket, *Teleogryllus commodus*, in the late 1960s. Young crickets became paralysed and died, and the disease, which spread rapidly and killed about 95% of the colony, was shown to be caused by the virus now called 'cricket paralysis virus'. Similar outbreaks in laboratory or commercial colonies of crickets have been observed elsewhere. However, CrPV is a rather ubiquitous virus with a wide host range so the name is more historical than categorical.

DCV was isolated from laboratory populations of *Drosophila melanogaster*, again in the late 1960s, during a general study of the incidence of small viruses in flies. When extracts of dead flies were injected into healthy adult *Drosophila*, the flies died within 3 days, and the virus multiplied to a higher titer than the other small spherical viruses being studied. Several years later, studies showed that CrPV and DCV were serologically related but not identical to each other.

GoV was obtained from *Gonometa podocarpi*, a lepidopteran pest of pines in Uganda, in the early 1970s. The virus was recovered at high levels from larvae and pupae and was very effective in controlling larval populations in the field. Unfortunately, the political situation in Uganda made it impossible to continue studies since no readily-available alternative host, either wild or laboratory-reared, had or has been found. Later attempts to replicate the virus in cultured cell lines were also unsuccessful. Although some crude samples of the virus still exist, it is not known whether they are still infectious, which could explain the failure to discover an alternative host for replication. Until new isolations are made, infectious virus might be considered as 'unavailable'.

Taxonomy and Classification

GoV particles, like those of the mammalian picornaviruses, are icosahedral, approximately 30 nm in diameter, and composed of four capsid proteins which enclose a single-stranded RNA genome, and whose sedimentation coefficient and density are within the range expected for picornaviruses. The small capsid protein, VP4, is not normally detected in CrPV or DCV particles since it is unstable and easily lost during purification, though it has been observed.

CrPV and DCV are usually reported as having a density of approximately 1.34 g cm^{-3} in cesium chloride but by direct comparison in the same gradient, CrPV is clearly denser (1.368 g cm^{-3} for CrPV, 1.354 g cm^{-3} for DCV and 1.346 g cm^{-3} for poliovirus) at close to neutral pH. Whether this is a significant characteristic of insect picornaviruses or simply a reflection of the various algorithms and conditions generally used in estimating virus particle density is uncertain. The particles are stable at acid pH, as are those of GoV.

The single-stranded RNA genome is infectious and is approximately the same molecular weight as poliovirus (2.8×10^6) but has a lower GC content than mammalian counterparts (around 37% compared with approximately 50%). There are some published data that suggest that CrPV and DCV possess a genome-linked protein as a primer for RNA synthesis. There is also a genetically coded poly(A) tract, so they share these genomic characteristics with the mammalian picornaviruses. The structural proteins also map at the 5′ end of the genome. The sequence of approximately 1600 bases at the 3′ end (polymerase plus noncoding region) shows no significant homology between CrPV and poliovirus, foot and mouth disease virus, or encephalomyocarditis virus (EMCV) so there is only a distant genetic relationship to the mammalian picornaviruses. Surprisingly, cDNA hybridization studies have shown no homology between CrPV and DCV, although five different strains of DCV were 60–85% homologous.

Viral Morphogenesis

In *D. melanogaster* cells, CrPV shuts down virtually all host cell protein synthesis within several hours of infection which contrasts sharply with DCV which shows a comparative inability to inhibit. The fact that CrPV eventually results in extensive cytopathic effect (CPE) while DCV does not is probably a reflection of this ability.

Although CrPV and DCV have a similar replicative strategy to the mammalian picornaviruses in that proteins are produced by post-translational cleavage of large precursor polypeptides, there are some significant differences. DCV-directed synthesis of virus-induced polypeptides is quite asymmetric, in contrast with mammalian picornaviruses. The main virus structural proteins, VP1, VP3 and VP2 and its precursor VP0, appear rapidly, yet VP4 is produced at a slower rate. Using protease inhibitors, a 200 000-mol. wt precursor of DCV proteins is seen in infected cells.

CrPV structural protein synthesis is similar to that of DCV. Processing of the precursor into viral structural proteins is found to be more rapid as compared with poliovirus, mengovirus or EMCV. The processing of precursor polypeptides is by a virus-induced protease, but *in vitro* data suggest that *Drosophila* proteases may also be involved.

Host Range and Virus Propagation

GoV was first isolated from the lasciocampid moth, *G. podocarpi*. Another virus, isolated from the lasciocampid, *Pachymetana* sp., during the same period, appeared to be identical with GoV so the host range can be assumed to cover at least two species of moths. GoV was not infectious for five European lasciocampids tested but, again, this may have been due to lack of infectivity in stored extracts.

The host range of CrPV is extensive and covers over 40 species that have been tested from five orders of insects. In fact, lack of infectivity for an insect species may be more remarkable than the ability to replicate. CrPV does not multiply in the silkworm, *Bombyx mori*, or a species of cockroach. CrPV may not replicate in coleopterans, although only a few species have been tested.

DCV has been isolated from a number of wild species of *Drosophila*. Two DCV isolates that were tested multiplied in the Mediterranean fruit fly, *Ceratitis capitata*, but not in the cricket *Gryllus bimaculatus*. CrPV, however, multiplies in this cricket species. CrPV replicates in most, but not all, *D. melanogaster* cell lines tested and in several mosquito cell lines. Attempts to infect a lepidopteran cell line from *Spodoptera frugiperda*, with either intact virions or RNA, were unsuccessful, although the virus multiplies in some other lepidopteran lines. CrPV multiplies in the lepidopteran *Lymantria dispar* cell line SCLd135 while DCV does not.

Pathology

Like their vertebrate counterparts, the insect picornaviruses can be found in crystalline arrays in the cytoplasm of cells. Particles can be found in alimentary canal cells, epidermal cells and nerve ganglia, the latter likely accounting for the paralytic effect, as well as in other tissues. In larvae infected with GoV, mitochronrdria and cell membranes show extensive degeneration and the cytoplasm contains numerous virus particles, often in groups enclosed by a membrane. Midgut cells are also affected and the fat body completely degenerates. Nuclei are unaffected.

In some cultured *D. melanogaster* cell lines, CrPV causes a distinct CPE which normally leads to cell lysis. In *D. melanogaster* cells, DCV causes cell clumping and detachment of cells from the surface of the culture vessel but without the pronounced lysis seen for CrPV. This probably reflects differences in their ability to shut off host cell protein synthesis. In *Aedes aegyptii* or *Aedes albopictus* lines, CrPV multiplies to a slight titer increase but without any obvious CPE. DCV has been detected as a persistent infection in some insect cell lines.

In insect hosts, both CrPV and DCV have frequently been found as inapparent infections, detectable only by bioassay or serology. These inapparent infections may go completely undetected until a total population collapse occurs, as happened with the discovery of CrPV.

Serological Relationships

The isolates of GoV from *G. podocarpi* and *Pachymetana* sp. could not be differentiated when tested against an antiserum prepared against the *Gonometa* isolate. There was no serological reactivity between GoV and CrPV antiserum, nor was any cross-reactivity detected between GoV and several other insect viruses. GoV is therefore quite distinct from the other two insect picornaviruses.

CrPV and DCV show serological cross-reactivity, although they are not identical. There are a number of isolates of CrPV which can be separated into three distinct serological groups by gel-diffusion tests. These generally correlate with the differences evident after analysis of the capsid polypeptide patterns by polyacrylamide gel electrophoresis. These groups can also be distinguished serologically from at least one of the DCV isolates. In gel-diffusion tests, five of the DCV isolates appeared to be serologically identical. The general serological picture is one of a complex of virus strains which are related but not identical. Since these isolates are often from totally different geographical regions and, in the case of CrPV, from a variety of hosts, it is not surprising that variability exists. Despite the serological relationships, other data indicate that CrPV and DCV are quite distinct.

One of the major difficulties in insect picornavirus classification is that a large number of isolates of 'picornalike viruses' or 'small RNA viruses' from insects have not been serologically tested to see if they are CrPV-related. Even if this is done, some viruses can easily be missed or misclassified. One of the three CrPV serotypes, typified by the isolate $CrPV_{brk}$, is quite distinct serologically, which is also reflected in its capsid polypeptide pattern. Cross-reactivity, whether by gel diffusion or immuno-osmophoresis, requires far higher concentrations of antisera to detect a reaction with a heterologous isolate. An isolate of CrPV from *Pseudoplusia includens*, called $CrPV_{ark}$ and closely related to $CrPV_{brk}$, would not have been recognized as a CrPV type if the conditions for serological testing had not been radically adjusted.

The most interesting serological data, however, have been those showing cross-reactivity between CrPV and the cardiovirus, EMCV. CrPV antigen reacted with EMCV antibody, shown to be IgG. The reciprocal gel-diffusion test, comparing CrPV and EMCV particles in reactions against CrPV antiserum, gave a spur, suggesting that the viruses are related but not identical. In tube precipitation tests, the homologous versus heterologous reactions differed from four- to 16-fold in the antiserum end points. Neutralizing antibody to EMCV, produced in guinea pigs, was enhanced when EMCV was inoculated 23 days after an initial injection with CrPV but little neutralizing activity was seen with two successive CrPV injections. The question is, therefore, should CrPV and EMCV be considered as related strains of the same virus? The late Carl Reinganum felt that these results reflected a 'structural response', i.e. a nonspecific reaction relating to similar structures rather than a taxonomic relationship. However, the suggestion that the viruses may be part of the same virus complex must be taken seriously in view of the data, although EMCV and CrPV certainly are dissimilar with respect to their sensitivities to pH, the lack of a poly(C) tract in CrPV and their densities. In comparing CrPV and DCV, a similar serological relationship is evident, yet the data on genome homology and morphogenesis support the idea that they are quite dissimilar. Clearly, this question, involving virus types with such widespread distribution and prevalence, and spanning both vertebrate and invertebrate hosts, should be explored further.

Ecology and Transmission

CrPV and DCV taken together display a wide

host range and have been isolated from a number of wild populations. In nature, most of the infections seem to be inapparent. How the viruses are transmitted from insect to insect in nature is still uncertain. Lateral transmission may occur through cannibalism. Virus could also be spread by birds feeding on infected larvae and releasing the virus in their feces. Within a population, virus could be spread by contamination of the egg surface from contaminated feces or, for example, by virus present in the soil in a pastoral environment. Experiments have shown that virus is not present within eggs so transovarial transmission could be ruled out. Virus can be transmitted *per os* from contaminated food.

Certainly, the virus can spread rapidly in closed populations. The original discovery of CrPV reflected this situation. Another strain, $CrPV_{brk}$, was isolated from a commercial cricket colony in Georgia, USA after a massive population collapse. The ability of the virus to persist in the environment was highlighted by the fact that, despite attempts to decontaminate the barn in which the crickets were reared, the disease returned and it was decided finally to burn the barn.

Natural populations of crickets have been examined for the presence of CrPV, using tissue culture bioassay to detect low levels of infection. The number of infected animals, although apparently healthy, rose steadily to 40% over several months before dropping. Presumably the infected crickets died but the study did suggest that young crickets readily acquired the virus. Transmission obviously occurs since the virus is so widespread and persists but the means by which it spreads is probably complex.

Antibodies to GoV and CrPV have been detected in vertebrates. In New Zealand, numerous cattle sera, collected from areas where CrPV-infected crickets were found, contained IgM which reacted with CrPV. IgM reactivity was also found in one horse serum and one serum from a pig. Presumably, the IgM was a response to transient but repeated exposure to virus in the soil, or on grass. A similar result was found in cattle in the UK where IgM reactivity to GoV was detected. However, GoV itself was not detected in the UK so the response may have come from exposure to a related virus. In view of the serological data linking EMCV and CrPV, it is possible that the IgM response arose from contact with a vertebrate virus.

The insect picornaviruses may have great potential for model studies, particularly when working with vertebrate types, such as poliovirus or hepatitis A virus, is difficult or even risky. CrPV can be readily assayed and replicates in insect cell systems which are generally easier to cultivate than vertebrate cell lines. Studies on the depuration of viruses from the Pacific oyster have been done using CrPV as a model rather than the often used poliovirus, and the insect virus proved to be a completely adequate substitute.

Economic Importance

The obvious use for such highly infectious and lethal viruses would be as biological agents to control pest insects. GoV was certainly effective in Uganda against a pest of pines. CrPV can devastate laboratory populations of crickets and have a major impact on business for commercial cricket farmers as was shown when $CrPV_{brk}$ was found. Strains of CrPV have been shown to be infective *per os* for the olive fruit fly, *Dacus oleae*, the Mediterranean fruit fly, *C. capitata*, and the large white or cabbage butterfly, *Pieris brassicae*, and are lethal and produce high titers in the insects. Since it would be highly desirable to affect only specific target insects, CrPV might be unsuitable given its wide host range. Although they resemble the mammalian picornaviruses, the insect picornaviruses have not been found to replicate in vertebrate cells. Given rigorous safety testing, insect picornaviruses could play a role in insect biological control programs.

See also: Cardioviruses; Coxsackieviruses; Honey bee viruses; Foot and mouth disease viruses; Polioviruses; Rhinoviruses.

Further Reading

King LA *et al.* (1985) Comparison of the genome RNA sequence homology between cricket paralysis virus and strains of *Drosophila* C virus by complementary DNA hybridization analysis. *J. Gen. Virol.* 65: 1193.

Longworth JF *et al.* (1973) Reactions between an insect picornavirus and naturally occurring IgM antibodies in several mammalian species. *Nature* (*London*) 242: 314.

Moore NF, Reavy B and Pullin JSK (1981) Processing of cricket paralysis virus induced polypeptides in *Drosophila* cells: production of high molecular weight polypeptides by treatment with iodoacetamide. *Arch. Virol.* 68: 1.

Scotti PD *et al.* (1981) The biology and ecology of strains of an insect small RNA virus complex. *Adv. Virus Res.* 26: 117.

Tinsley TW *et al.* (1984) Relationship of encephalomyocarditis virus to cricket paralysis virus of insects. *Intervirology* 21: 181.

PIRY VIRUS

See Chandipura, Piry and Isfahan viruses

PLANT RESISTANCE TO VIRUSES

RSS Fraser
Horticulture Research International
Littlehampton, West Sussex, UK

Importance of Disease Control

Plant viruses cause serious losses of yield and quality in many crops. The form of the loss can range from a mild or even symptomless infection such as with beet cryptic virus, which nevertheless is associated with an economically significant reduction in yield, to catastrophic outbreaks resulting in complete loss of marketable crop. Infection of zucchini by zucchini yellow mosaic virus is a typical example of the latter.

In contrast to the control of fungal diseases, no chemical pesticides are available for use as direct antiviral agents; thus alternative control strategies are required. Effective components can include control of viral vectors, use of virus-free seed or planting material for vegetatively propagated crops, good cultural practices to minimize transmission from infected plant debris or weed hosts, and deployment of plant resistance. Resistance is probably the most important and satisfactory of the control measures. It is important to study the mechanisms involved, to increase the effectiveness with which resistance can be used in crop protection and also because of the fundamental scientific questions raised about these types of plant–virus interaction.

Types of Resistance

Plant viruses are very diverse in their structure, mode of replication and transmission, and in their pathogenic effects on their hosts. Perhaps not surprisingly, plants have evolved a correspondingly diverse array of resistance mechanisms; some examples will be considered in this entry. However, it will first be useful to consider possible types of resistance mechanism from a more theoretical standpoint. Resistance mechanisms may be grouped into three main types depending on the complexity of the plant population involved.

Nonhost resistance operates at the level of the plant species. All members of the species show no symptoms and apparently no viral multiplication when inoculated with a nonhosted virus. This type of resistance is also referred to as nonhost immunity.

Cultivar resistance occurs in species that are normally susceptible to a particular virus but in which particular cultivars or breeding lines may show heritable resistance to that virus. This is the type of resistance used by the plant breeder in practical crop protection. A matching concept in the virus is virulence: virulent isolates of the virus are able to overcome the resistance conferred by a particular plant gene, while avirulent isolates cannot.

In *acquired resistance* (sometimes called *induced resistance*), the individual plant may have a level of resistance conferred upon it by a prior infection, a chemical treatment or genetic manipulation. This class of resistance covers a number of different mechanisms, some of which have found, or will find, great use in practical crop protection and others of which are solely of scientific interest.

Targets of the Resistance Mechanisms

It is also useful to consider which stages of the virus replicative cycle might be targets of different types of resistance mechanism. Figure 1 shows a schematic

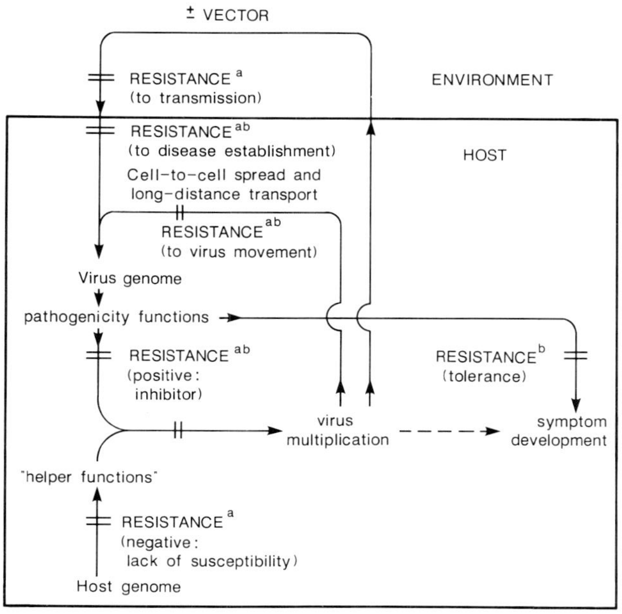

Fig. 1 The replicative cycle of a plant virus, showing various possible targets of cultivar resistance mechanisms. [a]Targets that might also be involved in the determination of host range and nonhost immunity. [b]Targets that might also be involved in various types of acquired resistance mechanism. Modified from Fraser, RSS (ed.) (1990) *Recognition and Response in Plant Virus Interactions*. Heidelberg: Springer-Verlag.

view of a replication cycle and the various possible targets of nonhost, cultivar and acquired resistance mechanisms.

When the virus penetrates a suitable plant it must first expose its genetic material for translation. Amongst other things this leads to the production of a replicase for multiplication of the viral genome and synthesis of viral structural proteins, leading to the production of progeny virus. Replication of the virus involves host 'helper functions'. These may be nonspecific, such as provision of ribosomes, amino acids, nucleoside triphosphates and an energy supply, or specific, such as a host-encoded subunit of the viral replicase.

Resistance mechanisms operating at this phase of the viral replicative cycle might be of two fundamentally different types. In positive mechanisms, the resistant host would produce an inhibitor which interfered with some stage of viral replication. In contrast, in negative mechanisms, the resistant plant would fail to produce some helper function required by the virus or would produce one that was defective from the point of view of viral replication. These two types of resistance mechanism have quite different implications for the nature of genetic control of resistance and possibly for the ability of the virus to evolve resistance-breaking behavior.

Having established viral replication in the initially infected cell, the virus or its transport form then proceeds to spread from cell to cell to establish fresh cycles of replication; it also spreads via the phloem to other parts of the plant. Cell-to-cell spread appears to involve a virus-encoded movement protein which modifies plasmodesmata to permit passage of the virus or its movement form, although the mechanism may be considerably more complex. Long-distance transport of the virus may also involve additional mechanisms for passage of the virus into and out of the sieve tubes; these mechanisms are not understood. Plant resistance can target either cell-to-cell or long distance transport.

During the course of accumulation of progeny particles and colonization of the plant, the virus usually causes formation of visible symptoms. Different viruses probably cause symptoms by a variety of mechanisms, but these are incompletely understood. Plants of a resistant cultivar may be primarily resistant to the formation of symptoms and this form of resistance is often referred to as tolerance, with the implication that viral multiplication as such is uninhibited. The presence of infectious virus can certainly be detected in tolerant plants but many studies have not measured viral concentration by quantitative methods to establish whether inhibition of multiplication occurs or not.

Progeny viruses are eventually released into the environment, by plant death and decay, mechanical handling, or into a vector such as a phloem-feeding aphid. These routes may eventually allow the virus to reach another plant and hence lead to establishment of a fresh cycle of infection. Any feature of the plant that alters its interaction with a vector to prevent selection or feeding indirectly confers resistance to any virus carried. There are many cases of viral resistance operating at this level; examples include plant color which affects host selection by insects, physical barriers such as leaf hairs or thick cuticles, and chemical barriers such as secretion by the plant of volatile insect alarm pheromones or the presence of chemical antifeedants in the sap.

Nonhost Resistance

Effectively, most plant species are nonhosts for most plant viruses: in the natural or cultivated environment they do not come into contact because of geographical separation or lack of suitable vectors. However, in the laboratory, experimental infections can often be established. Some viruses such as tobacco mosaic virus (TMV) can be shown to infect hundreds of species under these conditions. TMV and its close allies, such as tomato mosaic virus (ToMV) and cucumber green mottle mosaic virus, are also found naturally in a number of cultivated and wild species. In contrast, some other viruses have very limited host ranges. For example, bean common mosaic virus is almost entirely restricted to *Phaseolus vulgaris*. What determines host range? Why do some viruses have a much wider host range than others? What are the implications for the mechanisms of nonhost resistance?

In a series of classical papers published in the late 1960s, Bald and Tinsley postulated that plants may contain various 'susceptibility factors' and that viruses may encode various 'pathogenicity factors'. These are equivalent to the helper and pathogenicity functions shown in Fig. 1. Where there is compatibility between these two sets of functions, pathogenesis will be established. A nonhost would be such because it did not contain the required helper functions. In this context, nonhost immunity is a negative type of resistance mechanism and would be difficult or impossible to transfer to another species for use in crop protection, a possibility that has often been mooted.

An alternative mechanism proposed by Holmes was that nonhosts contain a large number of positive inhibitors of viral multiplication and that these are so common and effective that all members of the species are completely immune to large numbers of nonhosted viruses. There is no experimental evidence for this mechanism. It also leaves open the question of how the few viruses that are able to colonize a particular species can overcome this battery of putative inhibitors, while the majority of viruses that are not hosted are unable to overcome the inhibitors.

A major insight into the mechanism of nonhost resistance came with the discovery of 'subliminal' infection and subsequent experiments with protoplasts. In subliminal infection, the nonhosted virus was shown to multiply in a very small number of cells in the inoculated leaf. The implication was that these were the cells that had been directly infected during mechanical inoculation. Further work with protoplasts prepared from nonhost plants showed that they too could support viral multiplication when inoculated with a nonhosted virus. The suggestion is that the barrier to full virus colonization of the nonhost is at the stage of cell-to-cell spread and that, as the plants support a very limited amount of viral replication in the directly inoculated cells, they are not displaying complete nonhost immunity. However, comparatively few nonhost/virus combinations have been investigated at the protoplast level, and it is premature to speculate on whether all plant cells might or might not be susceptible to all or most plant viruses. There does seem to be enough evidence to suggest that inhibition of cell-to-cell spread may be one important mechanism in nonhost resistance.

Cultivar Resistance

The results of surveys of the genetics and mechanisms of resistance for a number of combinations of crops and viruses that affect them have been published. This 'database' currently contains 87 examples of resistance. The objectives were to establish the genetic basis of plant resistance to viruses, to see whether this was linked with particular types of resistance mechanism, to consider the frequency of resistance-breaking isolates, and to link these factors to theoretical models of resistance such as that advanced in Fig. 1.

Most of the examples in the survey appear to involve resistance genes that operate at a single locus. There are a few well-attested examples of resistance mechanisms which appear to involve cooperative (as opposed to additive) effects of genes at different loci, and a few examples of modifier genes.

About half of the resistances in the survey are classed as genetically dominant over the presumed susceptibility allele. About one-quarter are incompletely dominant, that is they show a clear effect of resistance allele dosage with more effective resistance in plants homozygous for the resistant allele than in those that are heterozygous. The remainder of the resistance mechanisms appear to involve recessive alleles. This distribution from dominance to recessiveness is interesting in its diversity; it contrasts with resistance to microbial pathogens which is more strongly associated with apparently dominant alleles.

In terms of mechanism, the dominant alleles tend to confer resistance which stops viral spread after the initial infection. Extremely effective examples of this are the *Tm2* and *Tm-2*^{2} genes for resistance to ToMV in tomato, which confine the invading virus to the initially inoculated cell. A more common pattern, such as with the *N* gene for TMV resistance in tobacco, is to permit a certain amount of cell-to-cell spread of the virus from the point of infection so that several hundred cells may become infected, but with further spread being inhibited a few days after inoculation. Commonly this results in the formation of a necrotic local lesion, although the evidence suggests that the necrosis itself is a secondary event and not the primary reason for cessation of viral spread. The delay before expression of the resistance probably means that the mechanism is induced by early events in pathogenesis.

The evidence suggests that the virus-encoded movement protein may be critically involved in localization resistance. In tomatoes containing the *Tm-2* or *Tm-2*^{2} resistance genes, resistance-breaking isolates of ToMV have been shown to have point mutations, which confer virulence, in their movement proteins. It is possible that the host resistance factor is a membrane protein which fails to interact appropriately with the movement protein of the avirulent strains, thereby preventing cell-to-cell spread, and that the altered movement proteins of virulent isolates are able to interact in an effective manner.

Resistance mechanisms involving necrotic local lesion formation may also operate, once induced, by blocking virus movement between cells; there is evidence that modification of the virus movement protein can alter the operation of the resistance mechanism. However, other virus-encoded components may also be involved in determining the interaction with a localizing resistance gene. In tobacco, resistance to TMV controlled by the *N*$'$ gene is effective against some strains of the virus (leading to local lesion formation), whereas other (virulent) strains of the virus spread systemically. Pioneering work by Dawson and co-workers showed that the determinant of virulence or avirulence was located in the viral coat protein gene. This does not exclude the possibility that resistance eventually operates by an interaction between the movement protein gene and a possible host receptor on the plasma membrane or in the plasmodesmata. It does suggest that the initial recognition event between host- and virus-encoded components which induces the resistance mechanism can operate at a different level and that a signal transduction chain may connect the recognition event to the eventual resistance response.

There have been reports that necrotic lesion localization of TMV in tobacco with the *N* gene is associated with synthesis of an 'inhibitor of viral replication' (IVR) and 'antiviral factor' (AVF); parallels have been drawn between the latter and human interferon. Neither has been conclusively linked with localization. IVR does have some inhibitory effects on TMV multiplication. Some of the protein components of AVF have now been shown to bear close similarity in amino acid sequence to certain of the pathogenesis-related (PR) proteins discussed below. These have as yet no known antiviral function.

Resistance controlled by alleles that show dosage dependence may be associated with mechanisms that allow viral spread through the plant but that inhibit viral multiplication or symptom development. A particularly good example is resistance to ToMV in tomato controlled by the *Tm-1* gene. In the heterozygous form, this inhibits viral multiplication by around 70%, and by around 95% in the homozygous form. Mosaic symptom formation is generally completely inhibited by either resistance genotype. The mechanism appears to be constitutive in that there is no delay between inoculation and demonstration of the resistance effect. This type of resistance is also effective in protoplasts, in contrast to the localization necrotic lesion type which is not.

Viral isolates that overcome the *Tm-1* resistance have been shown to have a mutation in the viral gene that codes for the replicase. The implication is that *Tm-1* plants may contain an inhibitor of TMV replicase activity which virulent isolates of the virus do not interact with. Alternatively, a host-encoded replicase subunit produced in resistant plants may be defective and unable to build a fully functional replicase with the virus-encoded replicase subunit of avirulent strains, but is able to do so with those of virulent strains.

A prediction from Fig. 1 is that resistance associated with recessive alleles could be of the negative type, that is the lack of a susceptibility function or the presence of a defective one. In the simplest form of the model this type of resistance could only be expressed in plants homozygous for the resistance allele. If the deleted or disabled susceptibility function was required at an early stage of the viral replicative cycle, such as in the provision of a host-encoded replicase subunit, the resistance phenotype would be expected to be near or complete immunity. This type of resistance might be very difficult for the virus to acquire virulence against, especially if it involved the absence of a host helper function. There was some supporting evidence from the survey for examples of this type of mechanism, but further investigation is required.

Acquired Resistance

When plants that respond to viral infection by formation of necrotic local lesions are inoculated with the same virus for a second time, they often show an apparent enhancement of resistance to the second infection. This is expressed as a reduction in the size and often the numbers of the lesions caused by the second infection. The effect is referred to as local acquired resistance when the same leaf is inoculated twice, and as systemic acquired resistance when the second inoculation is to upper previously untreated leaves. Viral multiplication in the smaller lesions of plants with acquired resistance may not, however, be inhibited. The apparent enhancement of resistance seems to affect the process of necrotization, rather than the localization of the virus.

These types of acquired resistance were shown many years ago to be associated with accumulation of a number of new proteins, termed PR proteins, many of which are soluble at low pH, resistant to degradation by trypsin, and secreted into the cell walls or intracellular spaces. Initially it was held that the PR proteins might be involved in the apparently enhanced resistance to viral infection. However, no direct antiviral function has been demonstrated and transgenic plants constitutively expressing PR protein genes do not show any enhanced resistance. Many of the PR proteins have now been shown to be chitinases and glucanases with potential activity against microbial pathogens. Others may have activity against insect pests. It appears that the PR proteins are part of a generalized defense mechanism against secondary invaders which is activated by a number of biotic and abiotic stresses, including necrotic viral infections.

Cross protection occurs when a plant is systemically infected by one strain of a particular virus: it may then be resistant to a second infection by a different strain. Commercially, this is used in practical crop protection by deliberately inoculating plants with mild strains of normally troublesome viruses. The mild strains multiply to low levels and cause minimal loss of yield. Cross protection is widely used in tomato against ToMV, in citrus against citrus tristeza, and in papaya against papaya ringspot. There have been many theories about the mechanisms involved, but current evidence from transgenic plants suggests that the coat protein of the protecting strain is the critical element and that part of its action in conferring resistance may be in preventing uncoating of a challenging severe strain. Cross protection can be highly effective, but there is always a risk that the mild isolate will produce mutants which cause severe losses in protected crops. It is also inefficient in that each generation of plants has to be inoculated.

A major expansion in our ability to confer resistance to plant viruses came with the development of the technology for production of DNA copies of plant RNA viruses. The increasing information about viral sequences and replication mechanisms that this led to can be used to design 'molecular spanners' to throw into the works of viral pathogenesis. These 'spanners', many of which are derived from the viral genome, are prepared as DNA sequences with appropriate promoters to ensure expression of the activity as RNA or protein. They can be given a heritable basis by direct insertion into host chromosomes using *Agrobacterium*-mediated transformation systems.

Genes for viral coat proteins have now been shown to confer resistance against more than 20 different viruses, including the economically important potyvirus group. Antisense genes can express viral information in the minus sense and thus inhibit the expression of the positive-sense viral messenger and, hence, synthesis of virus-encoded proteins. For those few viruses, such as cucumber mosaic virus, that may have their symptom expression and multiplication modified by satellite RNAs, expression of symptom-reducing isolates of the satellite from an integrated DNA copy can protect the plant against the pathogenic effects of the virus. Several other means of developing novel resistance genes from the viral genome itself have been reported, although the mechanisms are not always understood. Effective mechanisms have been found for DNA as well as RNA viruses.

Use of Resistance in Plant Breeding

Conventional breeding for resistance has had some notable successes and some unfortunate failures. The latter are largely related to the ease and speed with which the target virus has often succeeded in producing resistance-breaking isolates, and the extent to which these virulent isolates have established themselves in previously resistant crops. In the survey referred to earlier, more than half of the resistance genes had been overcome by virulent isolates and, in some cases, the resistance gene had been rendered completely ineffective. Thus, in tomato, the *Tm-1* gene for ToMV resistance was overcome within 1 year of its introduction and viral isolates able to overcome this resistance quickly became widespread. The gene had little further use in practical crop protection. In contrast, the *Tm*-2^2 gene has proved outstandingly durable and has been the mainstay for protection of the tomato crop against ToMV for many years. There have been several reports of viral isolates that can overcome this resistance, but all of these have failed to establish themselves in resistant crops; it appears that acquisition of virulence against *Tm*-2^2 may be associated with a loss of pathogenic fitness by the virus. In breeding for resistance, the breeder should attempt to assess the likely durability of his resistance gene before embarking on an extended and costly breeding program. This can be done by challenging the resistance in the breeding line with as many isolates of virus as are available, and perhaps also with nitrous acid mutants of the virus under quarantine conditions. This might indicate the need to combine available resistance genes to protect those of lower durability.

However, it is also the case that the genetic base of plant resistance to viruses is rather limited and in many crops no resistance genes are known for particular viruses that cause serious diseases. Screening wild relatives and genebank collections may reveal further natural resistance genes, although the cost of identifying these genes and incorporating them into commercial cultivars is high. In contrast, the encouraging early results with transgenic plants containing novel genes for resistance derived from viral sequences gives hope that this will be a major route to future crop improvement. As with all transgenic plants there is a need to ensure that the genetic modification does not create any environmental hazard or interfere with other aspects of plant growth and productivity. These issues are now being addressed in field trials, and, given satisfactory answers, transgenic plants containing novel genes for resistance to viral diseases should begin to make an impact in practical crop protection in the near future.

See also: Cucumoviruses; Plant virus disease – economic aspects; Potyviruses; Tobamoviruses; Vectors.

Further Reading

Bald JG and Tinsley TW (1967) A quasi-genetic model for plant virus host ranges. 1. Group reactions within taxonomic boundaries. *Virology* 31: 616.

Beachy RN, Loesch-Fries S and Tumer NE (1990) Coat protein-mediated resistance against virus infection. *Annu. Rev. Phytopathol.* 28: 451.

Dawson WO and Hilf ME (1992) Host range determinants of plant viruses. *Annu. Rev. Plant Physiology Plant Mol. Biol.* 43: 527.

Fraser RSS (1987) *Biochemistry of Virus-Infected Plants*. Chichester: Research Studies Press Ltd/John Wiley & Sons.

Fraser RSS (1990) Genetics of resistance to viruses. *Annu. Rev. Phytopathol.* 28: 179.

Wilson TMA and Davies JW (eds) (1992) *Genetic Engineering with Plant Viruses*. Boca Raton, FL: CRC Press.

PLANT VIRUS DISEASE – ECONOMIC ASPECTS

OW Barnett
North Carolina State University
Raleigh, North Carolina, USA

Introduction

Everyone has seen some of the effects of viruses on plants, perhaps without knowing it. If you are interested in art, perhaps you have seen the still-life paintings from the 17th century Dutch school which depict tulip flowers with beautiful stripes of color induced by tulip breaking virus. Perhaps the color variegations of certain ornamental plants have imparted their beauty to you. For example, in the woody ornamentals velvetleaf and flowering maple (*Abutilon* spp.), which have variegated leaf colorations, the beauty of the plants is enhanced by the viral infection, increasing their economic value.

The term 'economic aspects' can be broadly interpreted to mean not only monetary returns from crops but also the esthetic value of plants and the value of plants as members of the biosphere. Anything that affects our world will eventually have an economic effect.

There are no accurate estimates of economic impact of viral diseases on a global basis. Among plant pathogens, reduced crop production resulting from viral infections is thought to be second only to that caused by fungi. Crop production reduced from its full potential as a result of certain viral disease varies among countries, regions and even within a local area. Although reduced production on one farm can reduce income for that farmer, a farmer in another location who has a good crop may have a higher income because increased crop prices result from a lower supply.

An entire country's economy can be affected by a single viral disease. For instance, in Ghana the cacao industry provided an extremely important export commodity until cacao swollen shoot viral infections caused a decline in cocoa production by over 65% from 1936 to 1956 in the eastern region of the country.

Plant viruses infect many species of Angiosperms but only a few species of Gymnosperms, Pteridophytes and Algae. In contrast to viral infections of animals, once a plant is infected with a virus it usually remains infected for life. In most plant/virus combinations, the virus invades all parts of the plant, although shoot and root meristem regions may not be invaded by some viruses. In most plant/virus combinations, viruses are not transmitted through true seed to the next generation. To understand the economic aspects of plant viral diseases, the range of effects (symptoms) of viruses on growth and development of plants need to be summarized.

Symptoms Caused by Viral Infection

Viral infections are usually obvious when a disease condition occurs, but infections may also be latent (inapparent). The diseased condition of the plant may be expressed in many ways. Reduced growth or stunting is probably the most universal symptom of a plant infected with virus. Even latent infections may result in reduced growth. Changes in color are the most obvious symptoms of viral infection for most plant/virus combinations. Decreases in green pigments of leaves lead to the development of light green and yellow areas and result in patterns of color such as mosaic, mottling, ringspots, streaking, striping or line patterns. Necrosis or death of tissue is sometimes classified as a color deviation; for instance, necrotic streaks of plant stems is a common symptom of some viral infections. Color deviations are most obvious in leaves (e.g. tobacco infected with tobacco mosaic virus) but also occur in stems (e.g. young apple twigs infected with apple mosaic virus), flowers (e.g. tulip color breaking caused by tulip breaking virus), seed pods (e.g. edible pods of French bean infected with bean pod mottle virus), fruits (e.g. tomatoes with rings and line patterns caused by tomato spotted wilt virus) and seeds (soybean seeds may be mottled due to infection with soybean mosaic virus).

Some viral infections interfere with water relationships, which results in wilting or desiccation leading to tissue necrosis. Tabasco pepper infected with tobacco etch virus is a classic example of a viral infection that leads to wilting. Cucumber mosaic virus infection can cause wilting of cucumber plants in certain environmental conditions.

Plant morphology can be changed by viral infection. Abnormal leaf morphology commonly occurs with viral infections that cause mosaic or mottling. The leaf lamina may be crinkled, curled, blistered, narrowed or have altered margins. Abnormal growths called enations may develop from the lamina. Stems may be distorted or stem internodes shorter. Whole plants may have uncharacteristic growth patterns, such as bushy growth or a more open pattern of branching because fewer stems are produced.

Plant viral symptoms are usually not characteristic enough for identification of the virus because different viruses in the same plant can cause similar symptoms and different strains of the same virus may cause very different symptoms. Symptoms caused by viral infection may also differ with environmental conditions, plant variety and time infected. In fact, symptoms are not static and the expression of symptoms at different times after infection (symptom syndrome) must be considered when the effects of viruses on plants are studied. When viruses induce visible symptoms in plants, the initial sites of infection lead to local symptoms in 3–5 days. Upon reaching the vascular tissue, the virus is usually translocated and causes systemic symptoms. Systemic symptoms that develop first are often very severe. This acute phase is followed by a chronic phase in which symptoms are less severe or are absent even though the tissues remain infected. In some perennial plants, progressive deterioration leads to a condition called a decline. In general, viral infections cause premature senescence of plants.

Damage from Viral Infections

Viral infections of plants usually lead to symptoms of the plant disease. Symptoms result from biochemical and metabolic changes induced in the host which cause various categories of direct damage. Indirect damage from viral diseases results from procedures undertaken to maintain healthy plants. Direct damage can be categorized (Lute Bos, Research Institute for Plant Protection, Wageningen, The Netherlands) as reductions in growth, vigor, or quality and market value.

Reductions in growth

Reductions in growth due to viral infections are manifested in many ways, including crop failure, yield reductions, internode or stem length changes, reduced rooting and root growth, reduced flower and seed production and reduced seed weight.

Crop failures, while not the most frequent effect, can occur with viral diseases. Sugarbeet production in the western USA was sporadic until 1940 when varieties resistant to curly top virus were introduced. In 1925, one-third of the sugarbeet crop was destroyed by the curly top disease in the Sacramento Valley of California and all of the late plantings were destroyed in the San Joaquin Valley and southern part of the Salinas Valley. When severe infections (100%) of sugarcane with Fiji disease virus occurred in Australia, growth of the canes was reduced to only 5–10% of a healthy crop, and complete fields were destroyed by growers prior to harvest because their maintenance and harvest were not economical. In The Netherlands, fields of the lily cultivar Enchantment can be 100% infected with tulip breaking virus plus lily symptomless virus. This double infection leads to leaf and stem necrosis and eventual death of the plants.

Usually even 100% viral infection does not result in total loss of the crop. For example, the complete stock of many potato varieties were once infected with one or more viruses. The production limits that viruses imposed on the genetic potential of the cultivar could not be determined until parts of the crop could be freed of virus. Potato virus M (potato paracrinkle virus) infected the total stock of King Edward potato producing only mild symptoms, but, when virus-free potatoes became available, the virus was shown to reduce the yield of tubers by around 10%. The infected plants also produced tubers of less uniform size. Potato virus X (PVX) was known as the healthy potato virus because North American potato stocks were so generally infected, with few or no symptoms. The healthiest potatoes contained only this virus. PVX was found to reduce potato yields by 9–22%, depending upon viral strain and potato variety.

Potato crops have been grown in many countries for many years and cover large acreages. Viral damage which most agriculturists would consider insignificant is thus overall very extensive. The potato crop in the USA in 1989 was 370.5 million hundred weight valued at $2.5 billion. A 10% reduction in yield is 41.2 million hundred weight, or $278 million in reduced sales for 1 year.

Years ago, Holmes recognized that small yield reductions of crops having a high dollar value and grown world-wide are probably the most important economically. The mosaic disease of tobacco caused by tobacco mosaic virus is an example of a virus that causes readily visible symptoms. Growers may not notice symptoms on plants widely scattered in a field of healthy plants. In North Carolina and South Carolina flue-cured tobacco, this disease causes the second or third highest losses year after year because the gene for hypersensitive resistance has not been incorporated into a flue-cured cultivar having the same high quality and yield as cultivars without this gene. Ironically, tobacco mosaic virus was the first plant virus described and is probably the most studied of all plant viruses. Viruses of forage legumes reduce biomass of forage from very little up to 55%. Forages are thought of as low-value crops, but they cover large areas of land so that, from a global perspective, the reduction is large.

Other viruses cause dramatic epidemics as they spread into a crop. In Ghana, cacao swollen shoot virus infected native vegetation, but soon after cacao production began the virus spread into new plantings and caused severe economic losses. Cadang-cadang is a devastating lethal disease of coconut palms which was epidemic in the Philippines in the 1930s. Over 30 million coconut trees are estimated to have been killed by this viroid disease since it was first recognized. The loss in production for each planting site with an infected tree has been valued at $80–100 (US), based on average yield and copra prices. The economic impact becomes even more vivid when the fact is realized that 96% of the world copra production occurs on small land holdings where coconut is both a subsistence and cash crop. A similar disease, 'tenangaja,' occurs on Guam.

Reductions in growth can be related to reduction in vigor or quality. In ornamental production, tobacco ringspot virus in *Jasminum*, a woody ornamental, causes fewer cuttings to root than from uninfected plants. Cuttings that do root are of poorer quality, produce 50% less growth, and have a higher mortality rate. Dasheen mosaic virus-infected *Dieffenbachia* plants produce only 39% as many cuttings per plant as do virus-free plants. The average weight of the cuttings was not reduced by virus infection, but only plants produced from virus-free plants were marketable. Virus-free narcissi not only produce twice as many flowers but also larger ones.

Reductions in vigor

This can be manifested in many ways. Seedlings are usually larger and better able to escape disease when derived from larger seeds. Virus-infected white clover produces fewer smaller seeds which produce smaller slower growing seedlings. Infection of perennial crops may lead to reduced cold-hardiness or drought tolerance. Wheat or oat plants infected with barley yellow dwarf virus do not survive encasement in ice as well as noninfected plants. Red and white clover infected with clover yellow mosaic virus are less winter-hardy and more susceptible to drought.

Many perennial or vegetatively propagated plants develop conditions given various names such as 'degeneration,' 'running out,' and 'senility.' Many of these situations are related to the increasing prevalence of one or more viruses in the crop or stock plants. For example, from 1770 into the 1800s, potato cultivars introduced into Europe from true seed gradually became less productive and produced less vigorous growth. Eventually, this condition was found to result from mild infection of potatoes with a number of viruses.

White clover is another perennial that can be severely affected by viral infection. Individual plants infected simultaneously with as many as four viruses will not survive competition in the field. Grasses usually become predominant in a sward because the infected clover is slower growing and shaded by more vigorous uninfected clover and grasses. Infected clover is also more susceptible to root rot, and fewer and less efficient nitrogen-fixing nodules are produced. White clover–grass pastures should last 10–20 years, but, in the southeastern USA, the white clover component of pastures is often insignificant after 3–6 years. Although other factors are important, high viral incidence often precedes clover disappearance.

Early leaf fall occurs on rose bushes infected with prunus necrotic ringspot virus. Bushes lack vigor, are more sensitive to winter kill, and reestablish with difficulty after transplantation.

In general, viral infection increases susceptibility of plants to infection by fungi. For instance, sugarbeet plants infected by beet mild yellowing virus are more susceptible to infection by *Alternaria* sp. Corn leaves infected with maize dwarf mosaic virus have more lesions and earlier sporulation of *Helminthosporium maydis*. Wheat and barley infected with barley yellow dwarf virus are more susceptible to *Cladosporium* and *Verticillum* species.

In contrast, plants in which viral infection causes necrosis are often more resistant to certain fungal infections. Cucumber plants artificially inoculated with tobacco mosaic virus are resistant to anthracnose, and similarly inoculated tobacco is resistant to several fungi.

Virus-infected plants generally tolerate air pollution better than uninfected plants. For example, tobacco infected with tobacco mosaic virus had 6% less leaf damage due to ozone. Virus-infected plants may be more attractive to aphids, aphid multiplication may be enhanced, and maturation of flying aphids may be more rapid, but sometimes the longevity of aphids is less than on healthy plants.

These examples are effects of viruses on plants in relation to other stresses. In general, virus-infected plants are damaged more by additional stress factors but in some cases viral infection actually allows the plant to tolerate additional stresses. Thus virus-infected plants may be more or less vigorous depending upon other environmental and biotic factors.

Reduction of quality and market value

Quality and market value are reduced when plants are infected with viruses. Yellow summer squash from plants infected with several viruses have green markings on the fruit. Even a low percentage of green fruit means rejection of the truckload of squash at a processing plant. Snap beans for fresh market will also be rejected if they are mottled as a result of bean pod mottle virus infection which causes pod symptoms. Sweet potatoes infected with sweet potato feathery mottle virus may be harvested with apparently high-quality tubers; but after storage many tubers will have internal cork which severely reduces market quality. Tomato fruits from plants with late infections by tobacco mosaic virus develop internal necrosis which reduces marketability.

Tulip flowers with color breaking were once highly valued. Now growers realize that flower breaking results from viral infections that lead to degeneration of bulb stocks. This often cited example of a 'beneficial viral infection' does not impress modern bulb growers. Grades of flower bulbs and quality of flowers are deleteriously affected by viral infection; color deviations intensify and the size of plants and flowers and yield of bulbs are reduced.

In roses, the onset of flowering is delayed, flower size decreases, and the percentage of deformed flowers is greater from virus-infected than from healthy bushes. Pelargoniums with ringspot symptoms, which can be caused by several viruses, exhibit delayed flowering, greater flower abortion, decreased flower and floret numbers, and reduced flower stem length. Orchids infected with one of several viruses may develop necrotic streaks on the flowers about a week after opening. Acceptable flowers sometimes are shipped to wholesale houses prior to symptom appearance. Foliage ornamentals may be deformed by viral infection, for example, dasheen mosaic virus-infected *Dieffenbachia*.

Quality factors such as these are of economic importance because they affect the growers' reputation. Commercial producers must establish a good relationship with their clientele. Viral diseases can adversely affect these relationships; quality may be acceptable when shipped, but upon or shortly after receipt the quality deteriorates. Producer/client relationships then suffer.

Indirect Effects of Viral Diseases: The Cost of Maintaining Plant Health

Bawden, in a discussion of the degenerative disease of potatoes in the late 1700s, mentioned that potato viruses prevalent elsewhere were rare in the UK. In 1963 he wrote: 'This does not mean that they (potato viruses) are unimportant; they are rare only because they are controlled, at a cost to the English potato growers of about £10 000 000 a year.' Potato viruses are controlled by systems for production and certification of virus-tested seed potatoes. The virus-tested planting material allows commercial growers to produce high-quality potatoes reliably. Although planting material may be more expensive, high-quality potatoes can be produced, perhaps sold at a higher price, and with lower production costs because less land is required. Even though control is expensive, return must be greater than the cost or farmers would not have adopted the practice. Other economic aspects relate to programs that produce the virus-tested planting stock. Research scientists must monitor the crop for new viral diseases, mother potato stocks must be maintained free of virus and tested regularly, and stock increases require that inspectors monitor the process. Some growers in areas where virus spread is low will produce high-value seed potato crops. Thus, control of viral diseases means more jobs, another industry and more governmental agencies. A simple balance sheet really tells only a portion of the story about viral disease control!

When a disease problem cannot be solved by individual growers, governments may assist. Control of swollen shoot of cacao in Ghana included several approaches, but the major procedure was eradication of infected plants. The disease was recognized in 1936, the eradication program began in 1946, and by 1989 over 188.7 million infected trees had been removed. Farmers resisted eradication of infected trees, especially those without symptoms surrounding infected trees, even though compensation was paid for trees cut down and grants were made for replanting. Political concerns sometimes override control programs for this viral disease; war, farmer opposition, political intervention, lack of trained manpower and financial resources all affected the eradication program. An interesting commentary: when compulsory eradication was abandoned, eradication continued on a voluntary or consent basis, but when officials announced that the compensation program might resume, farmers discontinued voluntary eradication and less effective control resulted. Despite the control efforts, this disease is still prevalent, especially in the eastern growing region.

Zadoks contends that 'most researchers have little understanding of farmers' needs' as the farmer 'has more — and often more important — concerns than crop protection.' However, successful modern agriculturalists must make crop protection in general, and viral disease control in particular, an important management concern. Profit margins are so low that growers cannot afford to make many mistakes.

Assessment of Damage

Assessment of damage caused by a virus requires more than knowing that a plant is infected, but this is an important first step. Diagnosis (detection and identification) often requires application of serological, nucleic acid hybridization or biological assays. Mixed infections with two or more viruses commonly occur, and these plants may have symptoms similar to those of a single infection. Effects may be additive, or a synergistic reaction may occur.

Experimental measurement of damage should be carried out in several environments, with plants inoculated at different ages, in different cultivars, with different viral strains and at various plant densities. Ideally these measurements should be made in the field, but viral spread into uninfected control plants may make this impossible. Greenhouse or growth chamber experiments allow use of more defined conditions but do not mimic field conditions.

Next, viral incidence in the crop must be determined. Surveys for viral incidence may be intensive (accurate knowledge of viral incidence but of only a small area) or extensive (less accurate viral incidence but over larger areas). Field surveys may be needed several times during crop growth to determine the duration of infection or increase in incidence. Knowing how much damage an infected plant incurs and how many plants are infected allows calculation of yield reductions and even modeling of the disease for forecasting of what will happen.

The best example of predicting viral spread, viral incidence and potential loss due to viral infection and developing spray warning schedules for control of spread by vectors is for yellows disease of sugarbeet in the UK caused by a complex of aphid-borne viruses. Yield reductions can be large and are related to the length of time sugar beet plants are infected (3–5% potential sugar yield reduction per week infected). Predictions of viral incidence for the coming growing season can be made on the basis of vector numbers and winter severity. Spray warnings are usually based on aphid numbers and viral incidence.

Concluding Remarks

Viruses are very important economically because of the diseases that they cause; they are also important tools for scientific study leading to better understanding of protein structure, antibody/antigen reactions and plant functions at the molecular level. Plant viruses have also invoked curiosity, as well as inspiring poetry and art. They have been present throughout recorded history and have been a part of natural ecosystems for much longer. They are a normal part of our environment and become disruptive forces in situations where man has managed crops for human gain.

The use of resistant varieties, if available, is the most reliable means of control. However, many strains of each virus occur and resistance is often strain specific. Biotechnology offers the means of engineering plants with novel types of viral resistance which may be useful against a range of viruses. Transgenic plants with genes coding for the viral replicase, antisense RNA and satellite RNA may allow control of some viruses. Biotechnology has stimulated development of sensitive methods for identification and detection of viruses with such tools as monoclonal antibodies and DNA probes. Virology is an exciting and dynamic science which promises better control methods for viral diseases.

Further Reading

Bawden FC (1964) *Plant Viruses and Virus Diseases*, 4th edn. New York: The Ronald Press Company.

Bos L (1978) *Symptoms of Virus Diseases in Plants*, 3rd edn. Wageningen, The Netherlands: Pudoc.

Heathcote GD (1986) Virus yellows of sugar beet. In: McLean GD, Garrett RG and Ruesink WR (eds) *Plant Virus Epidemics. Monitoring, Modelling and Predicting Outbreaks*, p. 399. Sydney: Academic Press.

Matthews REF (1991) *Plant Virology*, 3rd edn. New York: Academic Press.

Zadoks JC (1985) On the conceptual basis of crop loss assessment: the threshold theory. *Annu. Rev. Phytopathol.* 23: 455.

POLIOVIRUSES

Contents

General Features
Molecular Biology

General Features

Philip D Minor
National Institute for Biological Standards and Control
Potters Bar, UK

History

A funerary stele of a Middle Kingdom Egyptian scribe from about 1500 BC bears a carving of an individual with the withered leg and foot drop deformity characteristic of paralytic poliomyelitis. If the diagnosis is correct, this is the earliest record of a virus disease. Poliomyelitis, which is caused by poliovirus, became a well-recognized disease in Europe and North America in the late nineteenth century as a result of serious epidemics which replaced the endemic pattern of disease that had occurred previously.

Poliomyelitis was transmitted to monkeys by the inoculation of extracts of feces obtained from human cases by Landsteiner and Popper in 1909. Enders and co-workers demonstrated that the virus could be grown in tissue culture in cells of nonneural origin in 1949, and this led to the development of formalin-inactivated vaccines by Salk in the mid-1950s. The difficulties of producing sufficient safe and potent vaccine led to the use of the live attenuated vaccines developed by Sabin in the late 1950s and the early 1960s. In developed countries, poliomyelitis has been controlled by vaccination, usually but not exclusively with the attenuated strains. The World Health Organization has declared its intention of eliminating the disease from the world by the year 2000. The first complete sequence of a poliovirus genomic RNA was published by Wimmer and co-workers in 1981, and the first atomic structure in 1985 by Hogle.

Taxonomy and Classification

Poliovirus is a member of the family *Picornaviridae*. Shown by electron microscopy to be icosahedral viruses about 28 nm in diameter, particles lack a lipid membrane. Current classification of the *Picornaviridae* divides them into five groups: the enteroviruses of which polioviruses is the type member; the rhinoviruses which are causative agents of the common cold; the hepatoviruses which cause hepatitis A; the cardioviruses of mice; and the aphthoviruses or foot and mouth disease viruses.

Classically the picornaviruses have been assigned to their groups largely on the basis of buoyant density and sensitivity to acid pH, although increasingly sequence comparisons will be used and may lead to some reclassification. The virion of polioviruses consists of a protein shell composed of about 60 copies each of four proteins, designated VP1, VP2, VP3 and VP4 in order of decreasing size. A myristic acid residue is attached to VP4. A lipid molecule, probably sphingosine, is found noncovalently associated with each copy of VP1 in the intact virion structure. The protein shell encases a single strand of positive-sense RNA, about 7.4 kb long. Polioviruses exist in three distinct serotypes designated type 1, type 2 and type 3. Strains may be distinguished by comparison of the sequence of small regions of the genomic RNA and are identified by serotype, name, country of isolation and year of isolation, e.g. P3/Leon/USA/1937.

Geographic and Seasonal Distribution

Polioviruses are found world-wide, although less commonly in areas where hygiene is good and inactivated rather than live attenuated vaccines are used. Isolates are thus rare in countries such as Sweden, but extremely common in developing countries or in countries where live vaccines are used extensively. Strains may be indigenous to the country of origin, imported from other areas or derived from live attenuated vaccines. It has been found that the sequence similarities between indigenous strains are surprisingly well conserved over time, and that specific genotypes (defined as those differing by less than 15% at the nucleotide level) are confined to particular geographic areas, thus making it possible to identify the probable source of epidemics. This has great value in vaccination and eradication programs, as it provides a clear measure of indigenous viral transmission and its interruption by vaccination. Viral prevalence is favored by warm wet conditions and is thus high in temperate climates in summer and autumn. In tropical climates the virus is believed to be endemic.

Host Range and Virus Propagation

Poliovirus grows naturally only in humans and chimpanzees, which may both be infected by the oral route. Old World monkeys may be infected parenterally, and certain strains of virus, notably of serotype 2, can cause a paralytic disease very similar to poliomyelitis following intracerebral inoculation of mice. Host restriction has been shown to be due to the presence or absence of a specific receptor on susceptible human and primate cells. Murine cells *in vitro* do not carry the receptor, and are not susceptible to infection despite the sensitivity of the live animal to certain strains of type 2 poliovirus. The gene for the human receptor has been isolated and transgenic mice prepared, and shown to be susceptible to infection with all three serotypes, although not readily by the oral route.

Virus may be propagated in cells of human or primate origin, such as primary monkey kidney cells, human diploid cells such as MRC5 or WI38, continuous cell lines of human origin such as HeLa or Hep2c, or continuous cell lines of simian origin such as Vero. Mouse cells transfected with the gene for the human receptor for poliovirus are also able to support virus infection.

Genetics

Polioviruses of all three serotypes rapidly mutate on replication in the human gut, accumulating point changes during an epidemic at an estimated rate of about 10 per genome per month. Deletions and insertions may be less common because of the strategy of translation of the viral proteins, which involves a single large open reading frame encoding a polyprotein which is then subjected to proteolytic processing. Mutations must therefore conserve the reading frame to give viable virus. Mutants are readily isolated *in vitro*. Recombination between viruses of different serotypes is extremely common in recipients of the live vaccines and recombination *in vitro* is well documented. It is therefore likely to occur in the wild, both between and within serotypes, although this may be difficult to prove in practice. The relative stability of the sequences associated with isolates from a given geographical region is surprising in view of the high variability of the genome.

Serologic Relationships and Variability

Polioviruses of different serotypes are antigenically distinct and immunity to one serotype is believed not to confer significant immunity to the other two. There is evidence for antibodies able to neutralize both type 1 and type 2, however, and partial denaturation of the virus exposes antigenic sites which are cross-reactive between viruses of different serotypes and other enteroviruses. Point mutations in known antigenic sites are common during infections, but the viruses remain neutralizable by polyclonal type-specific sera. Significant antigenic drift is therefore not observed, and this is best illustrated by the fact that the vaccine strains of virus in use since 1955 have remained able to induce protective immunity against wild strains. The continuing efficacy of vaccines is also evidence that no serotype 4 poliovirus exists. It is possible that the pathology associated with poliovirus is linked to its use of a particular cellular receptor site which is determined by its structural proteins. A virus whose structural proteins have drifted to such an extent that the virus is no longer antigenically recognizable as one of the three serotypes may therefore have concomitantly changed receptor site and pathology.

Epidemiology

Young children form a reservoir of infection and high frequencies of transmission are associated with poor living conditions and low socioeconomic status. Improvements in hygiene result in lesser exposure in extreme infancy, with a consequent accumulation of susceptible older individuals, leading eventually to significant epidemics, rather than the original endemic pattern.

Transmission and Tissue Tropism

Polioviruses are generally transmitted by the fecal–oral route, although transmission from the nasopharynx of infected individuals also occurs. A transient viremia may follow infection but the major site of virus growth is the gut, where replication may be expected to persist for about 5 weeks in half of infected individuals. The restriction of replication to the intestine, the central nervous system and lymphoid tissues, such as Peyer's patches and the tonsils, has been explained in the past in terms of the expression of the cellular receptor in those tissues. However, there is evidence for expression of the receptor gene and the presence of the protein on the cells of such tissues as kidneys, which are thought not to be susceptible, and the apparent tissue tropism is not understood. Virus can be excreted in stool at a level of 10^5 infectious units per gram.

Pathogenicity

Most poliovirus infections are asymptomatic, with generally no more than 1 in 100 resulting in disease. Infections with type 1 polioviruses are more likely to lead to disease than those with type 3, while infections with type 2 are the least likely to have clinical effects. The circulation of the three serotypes occurs at a similar rate, as shown by the similar age of seroconversion to each type, and it is believed that the differences in morbidity rates reflect general differences of pathogenicity between the serotypes. Within a type, however, strains may also differ widely in pathogenicity. While the molecular basis for the lack of pathogenesis of the live Sabin vaccine strains is at least partially understood for each of the three strains, the variation in the virulence of the wild strains has not been systematically examined and could have a number of different molecular explanations.

Clinical Features of Infection

Infection with poliovirus may initially be associated with slight gastrointestinal symptoms. This may be followed by fever, sore throat or influenza-like illness, from which the patient recovers within a few days. This pattern has been termed abortive poliomyelitis. Nonparalytic poliomyelitis may occur in 1–2% of infections, and is associated with the symptoms of abortive poliomyelitis, followed by invasion of the CNS leading to aseptic meningitis, often accompanied by back pain and muscle spasm. The illness lasts 2–10 days and recovery is usually complete. Paralytic poliomyelitis occurs in 0.1–2% of infections approximately 7–30 days after infections and usually begins with the symptoms of abortive poliomyelitis, progressing to flaccid paralysis resulting from lower motor neuron damage, defined as spinal, bulbar or bulbospinal, depending on the site of involvement. Paralysed patients recover wholly or to a significant degree in 10% of cases; 10% of cases are likely to be fatal, and in 80% of patients there is significant residual paralysis.

Pathology and Histopathology

Poliovirus naturally infects via the oral route and multiplies in the tonsils, lymph nodes of the neck, Peyer's patches and the small intestine. It is believed by some that infection of lymphoid tissue, such as the tonsils, occurs after replication is established in the gut and as a result of a brief viremic phase. Alternatively, the tonsils may be infected during exposure to the virus in the initial infection by mouth. That low levels of circulating antibody, including passively administered antibody, are able to prevent paralytic poliomyelitis caused by infection of the CNS leads to the view that CNS infection requires a viremic phase. However, there is also evidence from the clinical development of some cases, that infection of the CNS may occur directly from nerves associated with the intestine. The Peyer's patches are gut-associated lymphatic tissue and a major site of virus replication. Virus growth in nerve cells is restricted to motor neurons, usually the anterior horn cells of the spinal cord although in severe cases intermediate gray ganglia and posterior horn and dorsal root ganglia may be affected. Damage appears to be by replication in the neurons themselves rather than in supportive tissues, as shown by the histological location of viral genomes and antigens. In the brain, affected areas may include the reticular formation, the vestibular nuclei, the cerebellar vermis and the deep cerebellar nuclei. Polioviruses may also infect the myocardium.

Immune Response

Partial denaturation of the virus may occur by relatively mild treatments, such as heating at 56°C for 10 min or UV irradiation, and results in a change in antigenic properties from the N (or D) form found in the infectious virus to an H (or C) form. This change is also associated with the early stages of virus uncoating and is sufficiently drastic to be readily demonstrated with polyclonal sera. Individuals in the acute phase of poliomyelitis mount a humoral immune response predominantly directed against H (or C) antigen, while those in the convalescent phase have antibodies to the N (or D) form. Immune serum with antibodies specific for the N form is protective. Moreover, individuals suffering from primary immune deficiencies associated with defects in the humoral but not cellular arms of the immune response are particularly susceptible to infection and disease caused by enteroviruses in general and poliovirus in particular. The predominant protective immune response is thus believed to be humoral. The significance of cellular immunity to protection is not clear. However, infection or immunization with type 2 poliovirus is believed to be able to prime for a secondary response to type 1 and type 3 and this may be due to cross-reactive T helper cells. While an immune response may be detected in sera by 1–2 weeks postinfection, virus excretion persists for much longer periods. There is evidence for a mucosal immune response and the production of IgA antibodies. Infection with the live attenuated vaccines produces immunity to re-infection in the gut, which while real may be of short duration.

Prevention and Control of Poliomyelitis

While passive immunoglobulin confers shortlived protection from disease, strategies for the prevention and control of poliomyelitis depend on the use of vaccines either based on the formalin-killed preparations developed by Salk or the live attenuated strains developed by Sabin. No antiviral chemotherapy is available.

Vaccine viruses are grown in susceptible cells, usually primary monkey kidney cells or human diploid cells. Modern Salk-type vaccines are of far greater potency and quality than those manufactured in the 1950s and are currently prepared by concentrating, purifying and filtering the harvest to remove aggregates, then treating with 3 mM formaldehyde at 37°C for 2 weeks, before a second filtration step. The prolonged and slow inactivation method conserves the antigenic properties of the virus and the filtration steps remove aggregates which may protect virus from the formalin and therefore contain live virus particles. Such infectious aggregates are believed to have been the cause of the 'Cutter incident' of the mid-1950s in which recipients of inactivated vaccine became paralysed. It is widely believed that immunization with inactivated vaccine has little effect on the infection of the human gut by other viruses. However, circulation of wild-type viruses in several European countries appears to have been prevented by its use.

The Sabin vaccine strains were produced by passage of poliovirus isolates under a variety of regimens. The type 1 strain is designated LSc_1, the type 2 strain P2712 and the type 3 strain Leon $12a_1b$. The passage histories leading to attenuation, including *in vivo* passage in primates and *in vitro* passage under different conditions are very different for each serotype. Attenuated strains were selected on the basis of their clinical and histopathological properties when inoculated into primates by different routes, and by their growth properties and stability in the laboratory. The molecular basis of the attenuation of the three strains has been examined, and there are striking similarities between them, including the presence of comparable mutations in the 5′ noncoding portion of the genome before the single open reading frame which probably act by reducing the efficiency of initiation of translation. The viruses are usually given as a trivalent mixture, and extensive changes occur in recipients, including the loss or suppression of attenuating mutations, simple or complex intertypic recombination events in the region of the genome encoding the nonstructural proteins, point mutations in known antigenic sites and other changes of unknown significance. Despite this, the live vaccines are not only extremely effective but have eliminated poliomyelitis from many developed countries and are likely to eradicate it from the Americas in the near future. The estimated incidence of paralytic poliomyelitis due to reversion of the vaccine strain is 1 per 530 000 primary vaccinees, and 1 per 2 million of all vaccinees.

Future Perspectives

The World Health Organization has declared its intention of eradicating poliomyelitis by the year 2000, and substantial progress has been made in the Americas to this end, using the existing live vaccines. While a number of potential theoretical difficulties may be envisaged, including the question of when it is safe to stop vaccinating as the wild-type virus is replaced by viruses derived from the live vaccine strains, it is possible that the goal will be achieved. Increased understanding of the molecular and structural details of the virus and its epidemiology and evolution will help in implementing and monitoring the effects of the control program.

See also: Enteroviruses; Nervous system viruses; Pathogenesis; Viral receptors.

Further Reading

Almond JW (1987) The attenuation of poliovirus neurovirulence. *Annu. Rev. Microbiol.* 41: 153.

Molecular Biology

Caroline Mirzayan and Eckard Wimmer
State University of New York, Stony Brook
Stony Brook, New York, USA

Introduction

Polioviruses are nonenveloped icosahedral particles of 28 nm in diameter that consist of 60 copies of the structural proteins VP1 and VP3, 58–59 copies of VP2 and VP4, and one or two copies of VP0 (the uncleaved precursor of VP4 and VP2) (see Table 1). This viral capsid encloses a positive-sense single-stranded RNA genome of approximately 7500 nucleotides that is polyadenylated at its 3′ end and linked at its 5′ end to a viral polypeptide, VPg.

Polioviruses exist as three serotypes (PV1, PV2 and PV3), classified according to the ability of immune sera or monoclonal antibodies to neutralize viral infectivity. Each serotype is further subdivided into different strains that can be identified via neutralization by strain-specific antisera as well as by nucleotide and amino acid sequence. PV1 (Mahoney) was the first picornaviral genome to be sequenced in its entirety. Representative strains of the three serotypes have subsequently been sequenced and found to be highly homologous in both nucleotide and amino acid sequence. Moreover, microsequence analysis of all viral polypeptides has led to the establishment of a complete genetic map.

Earlier studies on the molecular biology of poliovirus, together with the elucidation of the primary structure and gene organization, and the development of means for genetic manipulation of its RNA genome have led to the recognition of several distinct biological processes that are representative of this class of viruses. Firstly, their positive-stranded RNA genome is of messenger-sense polarity and is thus immediately infectious upon entry into the cell. Secondly, polioviruses have evolved a cap-independent translational mechanism which, coupled to their ability to inactivate cap-dependent translation of host cellular mRNAs, selectively promotes translation of their own genome. Thirdly, while maintaining a monocistronic genome, they encode multiple polypeptides that are liberated from a single translation product, the polyprotein, by proteolysis. Limited proteolytic processing yields precursor molecules that may serve a function distinct from those of their cleavage products. Here, unique properties of poliovirus structure and proliferation will be discussed in relation to viral and host cell metabolism.

Capsid Structure and Antigenicity

Poliovirus has icosahedral symmetry and consists of 60 identical asymmetric protomers arranged around fivefold, threefold and twofold axes. Each protomer is composed of a single copy of each of the three nonidentical capsid proteins VP1, VP2 and VP3 which are the subunits of the protomers. The smallest capsid protein VP4, which may be considered as an N-terminal extension of VP2, is located on the inner surface of the virion and is relatively disordered. The crystal structure of PV1 (Mahoney) revealed remarkable similarity to that of human rhinoviruses and other spherical RNA plant viruses. Although VP1, VP2 and VP3 differ in size and amino acid sequence, they have similar tertiary structures. Each capsid protein presents a common structural motif, an eight-stranded antiparallel β-barrel core (Fig. 1a). The capsids differ in their N- and C-terminal extensions, and in the size and structure of the loops that connect the outer strands of the β barrels. The loop extensions protrude from the surface of the virion where they may become well exposed and thus represent the major antigenic sites of the virus. The folding of the β strands gives the barrel the shape of a triangular wedge where the thin end of the VP1 wedge is directed toward the fivefold axis, and the equivalent ends of VP2

Table 1. Properties of poliovirus, type 1 (Mahoney)[a]

Molecular weight of particle[b]	8.25×10^6
Molecular weight of RNA[c]	2.4×10^6
Percentage RNA of virion	29.2
Diameter	28–30 nm
Sedimentation coefficient	156S
$D_{20,w}$	1.40×10^{-7} $cm^2 s^{-1}$
V	0.685 ml g^{-1}
pH stability	3–8.5
Stable	To lipid solvents
	1% SDS, EDTA at pH 7
	4 M urea
	4 M guanidinium-HCl
	Up to 45°C in isotonic salt
	Up to 56°C in hypertonic salt, 1 M $MgCl_2$
Virions/mg	7.07×10^{13}
Virions/OD_{260} unit	9.4×10^{12}
Protein	Copies per particle
VP0	1–2
VP1	60
VP2	58–59
VP3	60
VP4	58–59
VPg-RNA	1
Ions	
K^+	4900
Na^+	900
Mg^{2+}	110
Polycations[d]	54
Lipid[e]	Sphingosine?
Carbohydrate	Not detectable

[a] Compiled from numerous publications.
[b] Calculated without water molecules and cations.
[c] Calculated without cations.
[d] 42 molecules of putrescine, 10 of spermidine, 2 of spermidine.
[e] 12 molecules of sphingosine.

and VP3 alternate around the threefold axis. The N-terminal extensions of the capsid proteins (and VP4) form an intertwined network of connections in the interior of the capsid shell which contributes largely to its stability. Of significance is the interaction of the N termini of five subunits of VP3 around the fivefold axis to form a five-stranded twisted tube of β structure, described as a 'beta annulus' or 'beta cylinder' which in turn interacts with five two-stranded β-sheets formed by the N termini of VP4. The formation of the β annulus links the protomers during the assembly of pentamers and contributes to their stability. Proteolytic processing of the P1 capsid precursor is *a priori* in this assembly pathway, since the N termini of VP3 and VP1 must be freed before pentamer formation. Additional factors may favor stability of capsids in the assembly process. One notable example is a myristic acid moiety, covalently attached to the N-terminal Gly of VP4. Myristate forms an integral part of the interior protein shell of the virion. Myristate groups are thought to stabilize the association of N termini of VP4 with the β annulus at the fivefold axis by interacting with amino acid side-chains of VP4 and VP3.

As a result of the C-terminal extensions and surface loops, the exterior of the virion is marked by protrusions, 'broad plateaus' and 'deep crevices'. One notable surface feature is a depression or 'canyon' formed at the junction of VP1 and VP3,

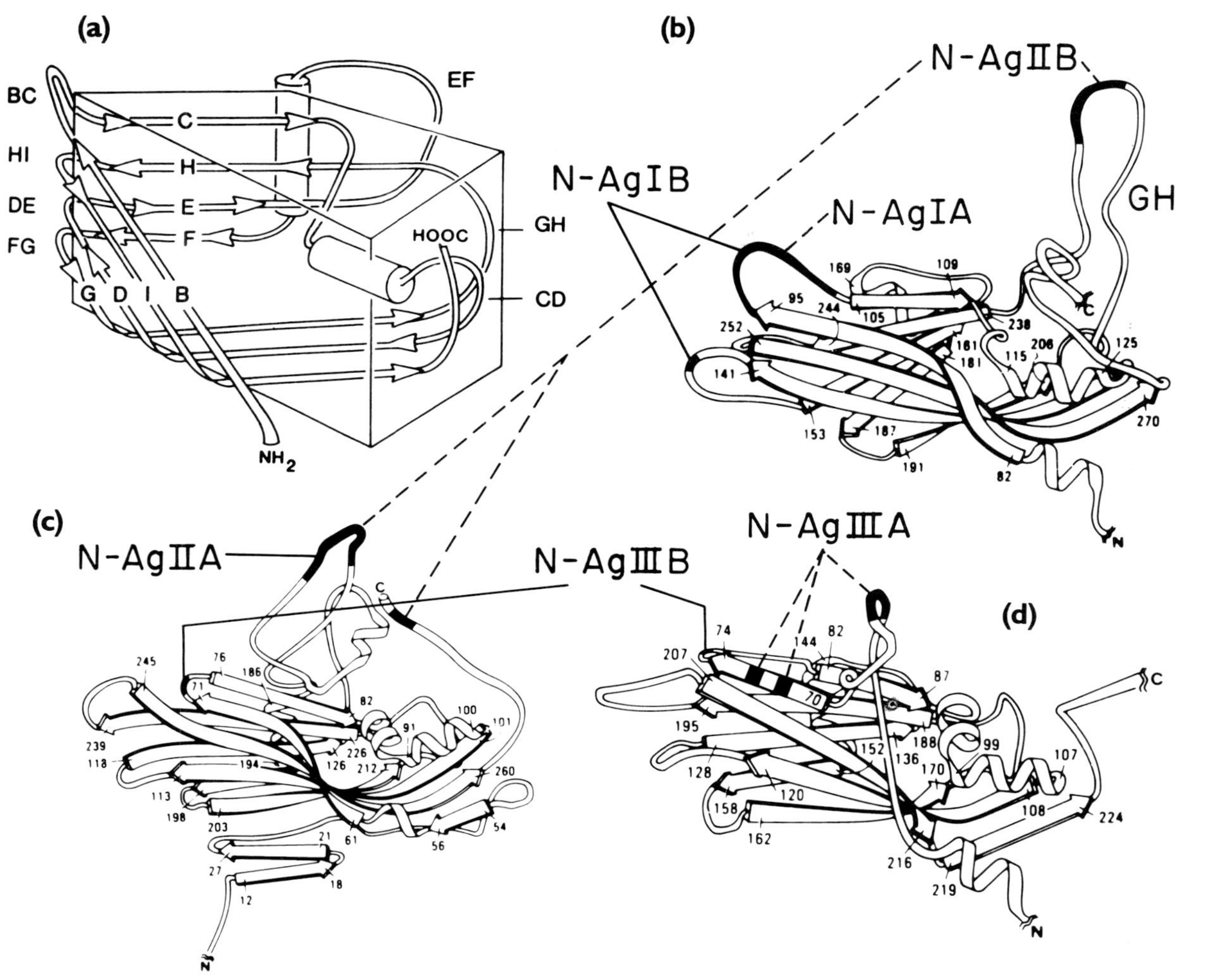

Fig. 1 Schematic representation of the poliovirus capsid proteins. A common structural motif, an eight-stranded antiparallel β-barrel core, is shared among each capsid protein (**a**). Ribbon diagrams show (**b**) VP1, (**c**) VP2 and (**d**) VP3. Four neutralization antigenic sites (N-Ags) mapped to surface loop extensions are colored black. N-AgI is a continuous sequence of amino acids mapping to positions 95–105 of VP1. N-AgII is discontinuous, being composed of amino acids 221–226 of VP1 and amino acids 164–172 and 270 of VP2. N-AgIII is composed of two independent sites, IIIA, and IIIB. N-AgIIIA consists of amino acids 58–60 and 71–73 of VP3, while N-AgIIIB consists of amino acid 72 of VP2 and 76–79 of VP3. Note that N-AgI and N-AgII may also be subdivided, but the divisions are less distinct than those in N-AgIII.

encircling the fivefold axis. The 'canyon' is thought to be the receptor binding site on the virion for human rhinovirus 14 (HRV14) and, by analogy, for polioviruses (PV) (see next section). A number of drugs known as 'WIN compounds' have been reported to inhibit attachment of HRV to the cellular receptor or uncoating, the effect depending on the type of rhinovirus. These compounds insert into a hydrophobic pocket which lies just beneath the floor of the canyon. Drug binding induces a conformational change in this pocket that inhibits virus binding to the cellular receptor. Interestingly, in PV1 (Mahoney), the hydrophobic pocket appears to be occupied by a sphingosine-like molecule. Whether this lipid substituent is a cofactor in assembly or disassembly remains to be seen.

Neutralization antigenic sites (N-Ags) have been characterized by the use of neutralizing monoclonal antibodies to select for mutant viruses resistant to neutralization. So-called neutralization 'escape mutations' have been localized to surface loop structures or adjacent β strands on the exterior of the virion. Four neutralizing antigenic sites (N-Ags I, II, IIIA and IIIB) are formed by structural elements of the capsid proteins, VP1, VP2 and VP3 (Fig. 1b,c,d). Interestingly, the three-dimensional structure of PV3 (Sabin), resolved by X ray crystallography, revealed structural differences in the N- and C-terminal extensions and in certain loops when compared with PV1 (Mahoney). The greatest difference occurs in the B–C loop of VP1 [connecting β strand B with β strand C], where the loop is considerably more exposed in PV3 (Sabin)]. The B–C loop of VP1 is a continuous sequence of amino acids (95–105) which maps to N–AgIA. The extended structure of the loop might explain the immunodominance of this antigenic site in PV2 and PV3 as opposed to PV1. Remarkably, this sequence can also carry determinants for poliovirus host range. A hybrid virus containing a six amino acid replacement in the N–AgIA site of PV1 (Mahoney) with that of PV2 (Lansing) can confer neurovirulence in mice.

Cellular Receptor

The poliovirus infectious cycle is initiated by attachment and internalization of the virus via a cellular receptor, followed by uncoating of the virus and release of the viral genome into the cytoplasm. Little is known regarding the mechanism of internalization, but the cloning and identification of the cellular receptor for poliovirus have greatly advanced knowledge of the first step of virus infection: attachment to the host cell. The cellular receptor for poliovirus is a new member of the immunoglobulin (Ig) super gene family with three distinct Ig-like domains, arranged in the order, V–C2–C2 (where V is variable and C is constant), transmembrane and a C-terminal cytoplasmic tail. All three serotypes of poliovirus compete for the same receptor, unlike all other families of picornaviruses which use different receptors. The sequence of the poliovirus receptor cDNA (PVR) predicts synthesis of a 46 kD polypeptide, but the predominant moiety observed by Western blot analyses of HeLa cell membranes and in recombinantly expressed PVR is a 67 kD protein, probably due to *N*-glycosylation of eight potential sites on the extracellular portion. Serial deletion of the Ig-like domains of the receptor molecule has demonstrated that the V domain is both necessary and sufficient for virus binding and infection, although additional sequences in the C2 domain next to the V domain may augment this interaction. Hybrid receptor molecules have been constructed whereby the PVR V domain was fused on to a truncated ICAM-1 molecule (receptor for rhinovirus major group) or the CD4 molecule (receptor for the human immunodeficiency virus, type 1). The hybrid molecules are functional poliovirus receptors, confirming the V domain as the major virus binding site. A more detailed analysis of the poliovirus:PVR interaction is presently being undertaken by site-directed mutagenesis of potential contact points.

Tissue tropism, a term defining susceptibility of specific tissue types to infection, has been generally attributed to the presence or absence of receptors. Polioviruses display restricted tissue tropism. They infect cells in the nasopharynx, Peyer's patches of the gut and the motor neurons of the spinal cord. Northern blot analyses, however, have shown ubiquitous expression of the receptor mRNA. Pending a parallel examination of protein expression, it appears that transcription of the mRNA is not sufficient for the biosynthesis of a functional receptor molecule. Results of Western blot analyses from different samples of human tissues indicate wide expression of heterologously sized proteins (30–150 kD). The exact relationship of these proteins to the 67 kD PVR is not clear. Whether receptor function depends on glycosylation and/or post-translational modification, i.e. phosphorylation or splicing, remains to be seen. Alternatively, ancillary proteins may promote receptor–virus interactions, possibly as regulatory subunits. A 100 kD protein was identified using a monoclonal antibody (MAb) that can inhibit poliovirus infection of HeLa cells by serotype 2 and to a lesser extent serotype 1. The 100 kD protein may participate as a regulatory subunit of the PVR. It is less likely that the 100 kD protein is the product of a second receptor gene not yet identified, since MAb specific for the PVR completely protect HeLa cells from poliovirus infection. The question remains whether tissues that are normally resistant to poliovirus (liver, kidney) but that express receptor-specific mRNA and protein govern infection at stages subsequent to adsorption.

Transgenic mice carrying the human PVR cDNA in the mouse genome are susceptible to all three serotypes of poliovirus and show similar clinical signs as those observed in infected monkeys. In addition, a cytological examination of infected tissues, tissue tropism and neurovirulence show similar patterns to those identified in humans and monkeys. Thus, transgenic mice might replace the monkey as the animal model for poliovirus and will greatly facilitate the study and testing of vaccines.

It has been suggested that a depression or a 'canyon' around the fivefold axis is the virus binding site for the cellular receptor. Indeed, sequences lining the canyon are found to be highly conserved, while sequence divergence is prevalent in the hypervariable loops which surround the canyon. The 'canyon hypothesis' suggests that conserved residues in the canyon allow sequence-specific binding to the receptor while excluding accessibility to antibodies elicited in an immune response. This attractive hypothesis may explain how the receptor binding site can escape host-immune surveillance in rhinoviruses. Experimental evidence strongly supports the canyon hypothesis for HRV14; it is most likely also true for polioviruses. Some poliovirus strains have been adapted to replicate in non-primate hosts. An example is the mouse-adapted PV2 (Lansing) which induces fatal poliomyelitis upon intracerebral inoculation into mice. In contrast, innoculation of PV1 (Mahoney) into mice does not result in paralysis. The molecular basis for the host range phenotype is not known. As mentioned previously, the B–C loop of VP1 may play a role in this phenotype. However, host range mutations have also been recently mapped to the N terminus of VP1 in PV1 (Mahoney). Therefore, it appears that structural determinants influencing uncoating as well as binding to a PVR-unrelated receptor play a role in mouse neurovirulence.

Although the individual steps in the process of internalization are obscure, it is believed to result from receptor-mediated endocytosis. During the period of entry into the cell and uncoating, the virus particle undergoes a conformational change. Physical characteristics associated with this structural rearrangement are: altered sedimentation rate (160S to 135S), change in antigenic properties, loss of viral capsid VP4 and extrusion of the N terminus of VP1. The 135S particle or 'A-particle' is significantly more hydrophobic than the parental derivative and, as a result, attaches to liposomes. The interaction with lipids has been attributed to the N terminal 30 amino acid residues of VP1 which are hydrophobic and may form an amphipathic helix. It is possible that the hydrophobic N terminus of VP1, in concert with the VP4–myristate moiety, make contact with the endosomal membrane and provide a pore through which the RNA is extruded into the cytoplasm.

Translation

Following uncoating of the virus in the cytoplasm, poliovirus RNA serves as mRNA to translate its single open reading frame. Initiation of translation of eucaryotic mRNA depends upon recognition of a 5′ terminal methylated cap (m^7G) by the 40S ribosomal subunit and associated factors, which are subsequently thought to scan along the 5′ nontranslated region until the first AUG triplet in a favorable context is encountered. However, poliovirus RNA displays a number of unique features that differ from the paradigm of the eucaryotic mRNA. In place of the cap structure encountered at the 5′ end of most eucaryotic mRNAs, the poliovirus genomic RNA is covalently bound to a viral polypeptide, VPg which is removed in the cytoplasm leaving a pUp residue at the 5′ end of the viral mRNA (note that poliovirus genomic RNA and mRNA are identical except for the presence or absence of VPg, respectively).

Poliovirus RNA contains an unusually long 5′ nontranslated region (NTR) of 743 nucleotides, whereas its eucaryotic counterparts are only 50–100 nucleotides in length. Moreover, the 5′ NTR contains stable secondary structures that are thought to be sufficiently stable to obstruct ribosome scanning toward the initiation codon. Finally, eight AUG codons are scattered throughout the 5′ NTR preceding the initiating AUG at nucleotide 743. At least one of these AUGs is in a correct nucleotide context for efficient translation by the criterion of Kozak yet it appears to be silent.

Given these observations, it appeared likely that polioviruses (and by analogy all picornaviruses) used an alternative method for initiation of protein synthesis to the one recognized for cap-dependent eucaryotic mRNAs. Since polioviruses translate their RNA under conditions where cap-dependent translation is inactivated (see later), investigators entertained the idea of translational initiation independently of the viral 5′ end. Evidence in support of this hypothesis was obtained by construction of dicistronic mRNAs whereby the first and second cistrons were separated by the 5′ NTR of poliovirus or EMCV. Translation of the second cistron occurred independently of the first, suggesting that ribosomes could bind to a region of the NTR placed downstream of the first cistron to direct translation of the second cistron. This region, necessary and sufficient to direct internal ribosomal binding was termed RLP for ribosome landing pad or IRES for Internal Ribosomal Entry Site. Although the function of an IRES element as internal ribosomal entry site *in vivo* was convincingly shown by expression of suitable dicistronic mRNAs in COS-1 cells, final proof for 5′-independent initiation of

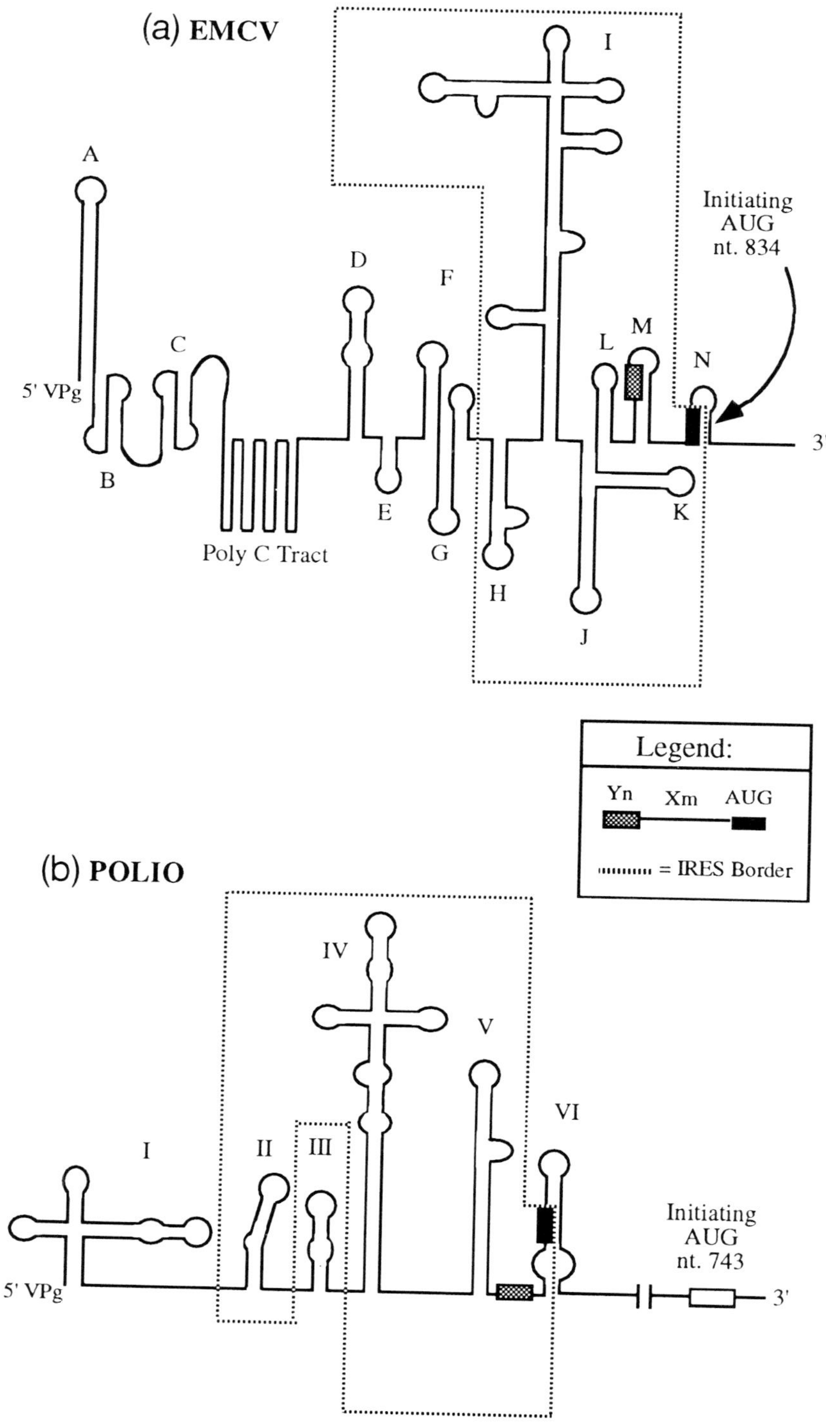

Fig. 2 5′NTRs and predicted secondary structures of representatives of Cardiovirus, (**a**) EMCV and Enterovirus (**b**) poliovirus. Dotted lines indicate the borders of the IRES elements. The Y_n–X_m–AUG motif is boxed.

translation was provided by the analysis of a dicistronic poliovirus containing two distinct IRES elements. Deletion analysis has demonstrated that large segments of the poliovirus and EMCV 5′ NTRs (around nucleotides 120–563 and 420–832 respectively) are required for IRES function. However, neither the sequences nor the structures of these two IRES elements show much homology (Fig. 2). Indeed, small deletions and point mutations have profound effects on the efficiency of translation, indicating that the structural integrity of the IRES element is of paramount importance. The identification of a large region of nucleotides defining the IRES elements argues for a super structure that promotes secondary and tertiary interactions, possibly among *cis*-acting nucleotide elements and *trans*-acting factors.

At least two *trans*-acting factors have been found that bind specifically to a region of the poliovirus 5′ NTR, in a manner which may correlate with translational efficiency. The first is a cellular protein with a molecular mass of 52 kD (p52) identified from a HeLa cell postribosomal fraction that binds to a conserved stem–loop structure (nucleotides 559–624). The second is a protein of 57 kD (p57) present in a HeLa cell ribosomal fraction that binds to a stem–loop structure (nucleotides 440–560), known as the attenuation (ATN) loop. p57 is probably the same factor originally identified from rabbit reticulocyte lysates (RRL) that binds to a stem–loop (H loop) in the encephalomyocarditis virus (EMCV) IRES. Both p52 and p57 seem to be novel cellular RNA-binding proteins not part of the complex of proteins already described in translational initiation. More recently, p57 was reported to share biochemical and antigenic properties with the cellular polypyrimidine tract-binding protein (PTB). While specific antibodies to PTB effectively inhibited polio and EMCV cap-independent translation, it had no effect on the translation of cap-dependent globin mRNA. Whether a nuclear protein that normally functions in pre-mRNA splicing has a role in the translation of cap-independent mRNAs remain to be seen.

Cis-acting sequence elements have been identified that are required for IRES function. An oligopyrimidine (Y_n; Y is pyrimidine, $n = 8-12$ nucleotides) tract preceding the initiation codon was originally observed in the 5′ NTR of foot-and-mouth disease virus (FMDV), where mutations within the Y_n tract influenced efficiency of translation. Site-directed substitutions of the Y_n tract in poliovirus defined an essential domain (nucleotides 554–573), the correct positioning of which was crucial in translation. Interestingly, in poliovirus, the Y_n tract is placed 20 nucleotides downstream of an essential, albeit noncoding AUG triplet (forming a structural unit of the general sequence Y_n-X_m-AUG where X is any nucleotide and $m = 18-20$ nucleotides). Such a Y_n-X_m-AUG motif is conserved in the 5′ NTR of all picornavirus IRES elements and it may play an important role in ribosome entry. In the poliovirus 5′ NTR the Y_n-X_m-AUG motif is separated from the downstream initiating AUG by a spacer sequence of 154 nucleotides, whereas in the EMCV IRES, the AUG of the Y_n-X_m-AUG motif is the codon from which polyprotein synthesis is initiated (see Fig. 2). The function of the spacer in poliovirus, if any, is unknown and it can be deleted without impairing virus viability in tissue culture. It has been suggested that the polypyrimidine tract in picornaviruses shows complementarity to a purine-rich sequence at the 3′ end of eucaryotic 18S rRNA. Such an interaction could promote the binding of the 40S ribosomal subunit to the mRNA. This is an attractive idea but rendered unlikely since substitutions in the Y_n tract, in contrast to HeLa cells, showed no reduction when translated in Krebs 2 cells or in RRL.

All three serotypes of poliovirus have been shown to carry specific mutations in the 5′ NTR that attenuate neurovirulence. For example, a single C_{472} to U substitution found in PV3 (Sabin), confers a strong attenuation phenotype to the virus. There is considerable evidence suggesting that attenuation, caused by mutations in the 5′ NTR, impairs the ability of the mutant RNA to initiate translation. The efficiency of translation of poliovirus RNA varies markedly between different cell types and cell-free lysates. Defective translation in deficient lysates such as RRL can be relieved or augmented by supplementation with a crude mixture of translation initiation factors (ribosomal salt wash) from permissive cell lines. This suggests that *trans*-acting factors that are of critical importance for translation of poliovirus RNA may be limiting in some cell types. Moreover, since expression of the phenotypes of attenuating mutations in the 5′ NTR is tissue-specific *in vivo* and *in vitro*, interaction with tissue-specific *trans*-acting factors is likely to be an important determinant of viral pathogenesis. As previously mentioned, p57 binds specifically to the ATN loop of poliovirus but it is not at all certain whether this relates to the phenomenon of attenuation. Finally, it should be noted that mutations other than those in the 5′ NTR strongly contribute to attenuation of the poliovirus vaccine strains.

(a)

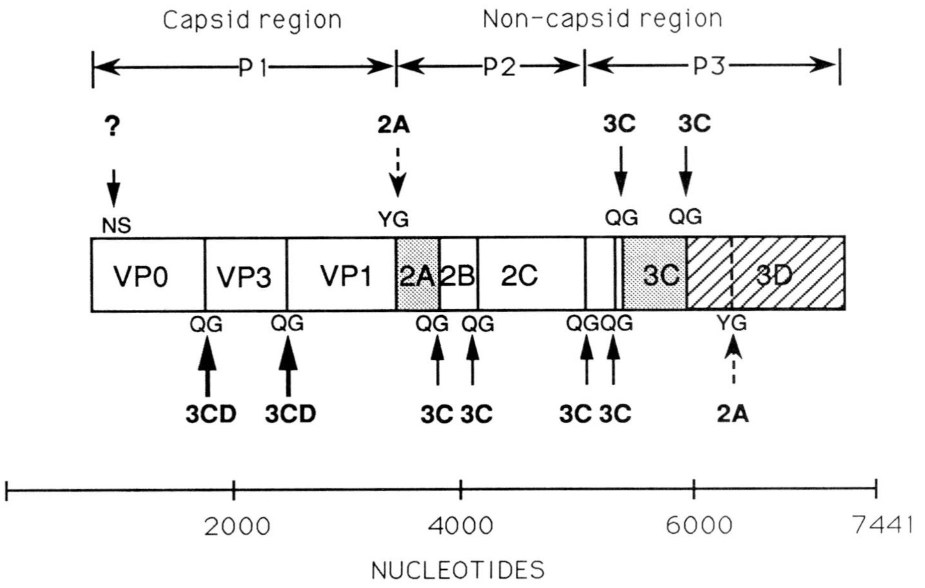

	Site	*P6*	*P5*	*P4*	*P3*	*P2*	*P1*	*P1'*	*P2'*	*P3'*	*P4'*	*P5'*
(b)	VP0/VP3	T	L	P	R	L	Q	G	L	P	V	M
	VP3/VP1	Q	K	A	L	A	Q	G	L	G	Q	M
	2A/2B	E	E	A	M	E	Q	G	I	T	N	Y
	2B/2C	P	Y	V	I	K	Q	G	D	S	W	L
	2C/3A	M	E	A	L	F	Q	G	P	L	Q	Y
	3A/3B	L	F	A	G	H	Q	G	A	Y	T	G
	3B/3C	R	T	A	K	V	Q	G	P	G	F	D
	3C/3D	Y	F	T	Q	S	Q	G	E	I	Q	W
(c)	VP1/2A	K	D	L	T	T	Y	G	F	G	H	Q
	3C'/3D'	K	L	L	D	T	Y	G	I	N	L	P

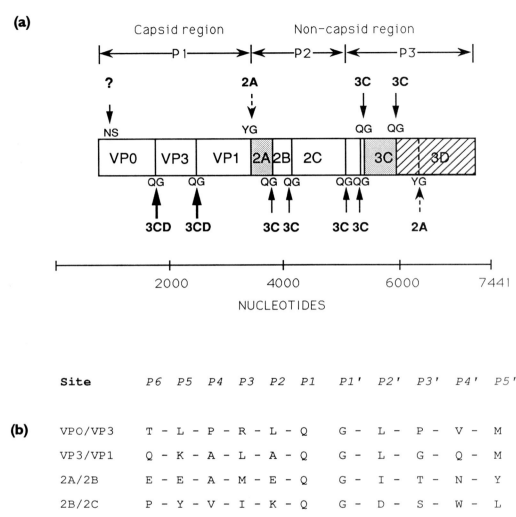

Fig. 3 Gene organization, processing scheme, and cleavage sites of the poliovirus polyprotein. (**a**) Proteolytic cleavages of the polyproteins occur between amino acid pairs indicated by standard single-letter code. Arrows above and below the polyprotein indicate sites that are cleaved in *cis*- and *trans*-reactions, respectively, by proteinases as indicated. The question mark indicates that the mechanism of cleavage at this site is not known. The positions of virus-encoded proteinases within the polyprotein are indicated by shaded boxes. Not shown is a ninth QG cleavage (within $3C^{pro}$) that yields polypeptide P3-4a in small amounts. The amino acid residues at sites cleaved (**b**) by $3C^{pro}$ and $3CD^{pro}$ and (**c**) by $2A^{pro}$ are indicated by standard single-letter code. The newly generated terminus, after cleavage of the peptide bond, is designated *P1*, preceded by the *P2* residue, etc. The newly generated N terminus is designated *P1'* followed by the *P2'* residue, etc.

Genome Organization and Proteolytic Processing

The poliovirus genome is monocistronic since it contains a single long open reading frame (ORF) which encodes a 247 kD polyprotein (Fig. 3a). A viral polyprotein is a precursor polypeptide containing distinct domains that are proteolytically processed to yield structural and nonstructural proteins. This strategy of gene expression is characteristic for most positive-strand RNA viruses and for all retroviruses. It allows maximal exploitation of genetic information under conditions of genetic austerity, the latter being dictated by the inability of RNA viruses to correct errors accumulated during genome replication. The poliovirus structural proteins encoded in the N-terminal half of the genome are four nonidentical polypeptides encoded in the order VP4-VP2-VP3-VP1, that are products of proteolytic processing of the precursor, P1. VP4 and its precursors VP0 and P1 are myristoylated at the N terminus. The downstream P2 and P3 precursors encode the nonstructural proteins. Proteolytic processing of the P2 region yields polypeptides $2A^{pro}$, 2B, 2C and that of the P3 region yields polypeptides 3A, $3B^{VPg}$, $3C^{pro}$ and $3D^{pol}$ (Fig. 3a).

Proteolysis of the polyprotein can be divided into three steps. The first step is to cleave the P1 capsid protein precursor from the nascent polyprotein. This primary cleavage event is catalyzed by the viral proteinase $2A^{pro}$ and serves to separate replicative enzymes from structural proteins. The second step is to process the noncapsid and the capsid precursors; these reactions are catalyzed by $3C^{pro}$ and $3CD^{pro}$. The third step is the processing of VP0 into VP4 and VP2 ('maturation cleavage') which takes place at the time of encapsidation of the viral RNA. The catalyst for this event has not been identified, but it is unlikely to involve any of the aforementioned viral proteinases. In poliovirus, proteolysis of the viral polypeptides occurs in a temporally controlled fashion. Moreover, precursor polypeptides liberated first may execute specific functions early in the viral life cycle, but are subsequently cleaved to mature proteins that have separate functions. Polypeptide $3CD^{pro}$ is an example of this dual function that extends the genetic content of the virus (see later).

Substrate recognition by the poliovirus proteinases is highly restricted and few polypeptides other than viral gene products are cleaved (see later). Proteinase $2A^{pro}$ cleaves at only two of the 10 Tyr-Gly dipeptides in the polyprotein. The first site is at the junction of VP1 and 2A and is cleaved in *cis*, whereas the second site (within the RNA polymerase, $3D^{pol}$) is cleaved in *trans* yielding products 3C′ and 3D′. This latter cleavage may be adventitious, since these cleavage products are not necessary for viral replication nor proliferation. Alignment of $2A^{pro}$ cleavage sites (Fig. 3c) shows the prevalence of a Thr residue in the *P2* position and a Leu in the *P4* position (the newly generated C terminus of a peptide bond is designated *P1*, followed by the *P2*, *P3*, etc. residues; the newly generated N terminus is designed *P1′*, followed by *P2′*, *P3′* etc. residues). Site-directed mutagenesis of cleavage site residues suggest that the requirements for recognition of a cleavage site by 2A in *trans* are more stringent than for recognition in *cis*.

Excision of the second virally encoded proteinase, $3C^{pro}$ and $3C^{pro}$-precursors from the polyprotein is likely to result from an intramolecular cleavage, but subsequent events catalyzed by $3C^{pro}$ occur intermolecularly (in *trans*). Cleavage by $3C^{pro}$ occurs exclusively at Gln-Gly dipeptides (a single exception occurs in a mouse-adapted W2 strain), but residues at positions flanking the scissile bond are mostly heterogeneous (Fig. 3b). Exceptions are found in the *P4* position where an Ala or other small aliphatic amino acids prevail, and to a lesser extent the *P2′* position of efficiently cleaved sites, where Pro residues prevail. Only nine of 13 Gln-Gly dipeptides in the polyprotein are cleaved (Fig. 3b), indicating that there must be additional determinants of cleavage site recognition. Oligopeptide substrates that mimic dipeptide $3C^{pro}$ cleavage sites are cleaved *in vitro* with an efficiency that parallels the efficiency of cleavage *in vivo*. Peptides corresponding to the 2B/2C site were cleaved rapidly, whereas those corresponding to the 3A/3B and 3C/3D sites were hydrolyzed slowly or not at all. The contribution of residues flanking the scissile bond to recognition by $3C^{pro}$ has been assessed by site-directed mutagenesis of polyprotein substrates, and by varying the length and composition of peptide substrates. There is a minimum substrate requirement of six or seven residues for peptide substrates to ensure cleavage; although the efficiency of cleavage increases as substrate length increases further. Experimental evidence obtained both *in vitro* and *in vivo* indicates that the *P4* position regulates the efficiency of cleavage. Thus, a peptide mimicking the 3CD site was cleaved at least 100 times more efficiently following substitution of the wild-type Thr residue by Ala. That inefficient cleavage at specific sites leads to accumulation of stable precursors *in vivo* is significant, since some precursor polypeptides such as $3CD^{pro}$ have been shown to have roles that are distinct from those played by their constituent moieties (i.e. 3C, 3D) after proteolytic scission. In addition to specificity at the primary amino acid level, additional determinants are dictated at the structural level. Experimental evidence supports that the entire P1 precursor is required for cleavage at Gln-Gly pairs within P1. Furthermore, the proper folding of the P1 β barrel structure is required for efficient cleavage at Gln-Gly pairs. Therefore, it is likely that primary substrate recognition by $3C^{pro}$ leads to the specificity for the Gln-Gly pair while differences in residues flanking this dipeptide in concert with structural determinants mediate efficiency of cleavage.

Finally, *in vitro* translation of the precursor 3CD (also 3BCD or P3) as source of proteinase catalyzes a much more efficient cleavage of P1 than when $3C^{pro}$ is translated alone. High concentrations of purified $3C^{pro}$ are needed to cleave the P1 substrate *in vitro*. Moreover, cleavage occurs only at the junction of VP3 and VP1 leading to the incomplete cleavage product, 1ABC. These data suggest that the 3D sequence of 3CD interacts with P1 in a fashion necessary for cleavage.

Both $2A^{pro}$ and $3C^{pro}$ have been classified as cysteine proteinases for their inhibition by iodoacetamide and *N*-ethylmaleimide is specific for enzymes with an active-site thiol group. Sequence alignment and tertiary structure modeling have suggested that the picornavirus proteinases are structurally related to the cellular trypsin-like serine proteinases. On this basis, an evolutionary relationship has been proposed whereby the nucleophilic Ser residue in the catalytic triad of the cellular enzyme is replaced by a Cys residue in the viral equivalent. Site-directed mutagenesis has confirmed the proposed constituents of the catalytic triad of the poliovirus proteinases as His20, Asp38 and Cys109 for $2A^{pro}$ and His40, Glu71 and Cys147 for $3C^{pro}$.

The third and final proteolytic processing step in the poliovirus replicative cycle is the cleavage of VP0 into VP4 and VP2 at an Asp-Ser site. It occurs during the final stages of virion assembly, probably after RNA encapsidation. It is not yet known what catalyzes this cleavage event, although the participation of $2A^{pro}$ or $3C^{pro}$ is unlikely since the cleavage site specificity is different from those recognized by either proteinase. It has been suggested that cleavage of VP0 may occur autocatalytically, particularly since the VP0 cleavage site is located in the interior of the capsid and may be inaccessible to viral or cellular proteinases. The involvement of Ser10 of VP2, which lies in close proximity to the carboxy end of VP4, has been suggested. In this scenario, a base from the viral RNA would act catalytically to abstract a proton from the hydroxyl group of Ser10, prompting a nucleophilic attack by the latter on to the neighboring Asn. However, this mechanism can no longer be supported since mutation of Ser10-Cys and Ser10-Ala yielded a viable virus in which the VP0 was cleaved.

Finally, it has been possible to construct a dicistronic poliovirus in which the ORF of the polyprotein has been disrupted at the P1/P2 junction by 602 nucleotides containing the EMCV IRES element. This poliovirus whose genotype is (PV)5′NTR-P1-(EMCV)IRES-(PV)P2-P3-3′NTR grows well in HeLa cells albeit with a small plaque phenotype. Interruption of the polyprotein by the EMCV IRES may serve to test which of the viral gene products function as precursors as well as end products.

Cellular substrates for the polioviral proteinases have been reported despite their unusual stringency in substrate selection. It has been shown that concomitant with poliovirus infection is rapid shut-off of host cell protein synthesis. This event is accompanied by the proteolytic cleavage of a polypeptide of 220 kD, termed p220. This protein is part of the cap-binding complex eIF-4F (consisting in addition to p220 of the cap binding protein, the initiation factor eIF-4A) that recognizes the capped 5′ end of eucaryotic mRNAs in initiation of translation. A poliovirus mutant was reported with a single amino acid insertion in $2A^{pro}$ (nucleotide 3690) that was deficient in inhibition of host cell protein synthesis. Moreover, cleavage of p220 did not occur in cells infected with this mutant. Indeed, $2A^{pro}$ synthesized *in vitro* can induce cleavage of p220 *in vitro*. However, it induces this reaction indirectly, by activating a cellular factor that degrades p220. Proteolytic activity of $2A^{pro}$ is needed for the induction of p220 cleavage since active-site mutants that reduce or abolish *cis* cleavage *in vitro* also inactivate its ability to induce p220 cleavage. Finally, the involvement of translation initiation factor, eIF3, has been proposed in p220 cleavage.

More recently, genetic and biochemical evidence have implicated $3C^{pro}$ in the inhibition of the polymerase III transcription system, specifically, in the cleavage of a TFIIIC-containing complex. Whether $3C^{pro}$ directly or indirectly cleaves TFIIIC, or a component of this complex is unclear.

RNA Replication

Poliovirus is an extremely efficient parasite. Immedi-

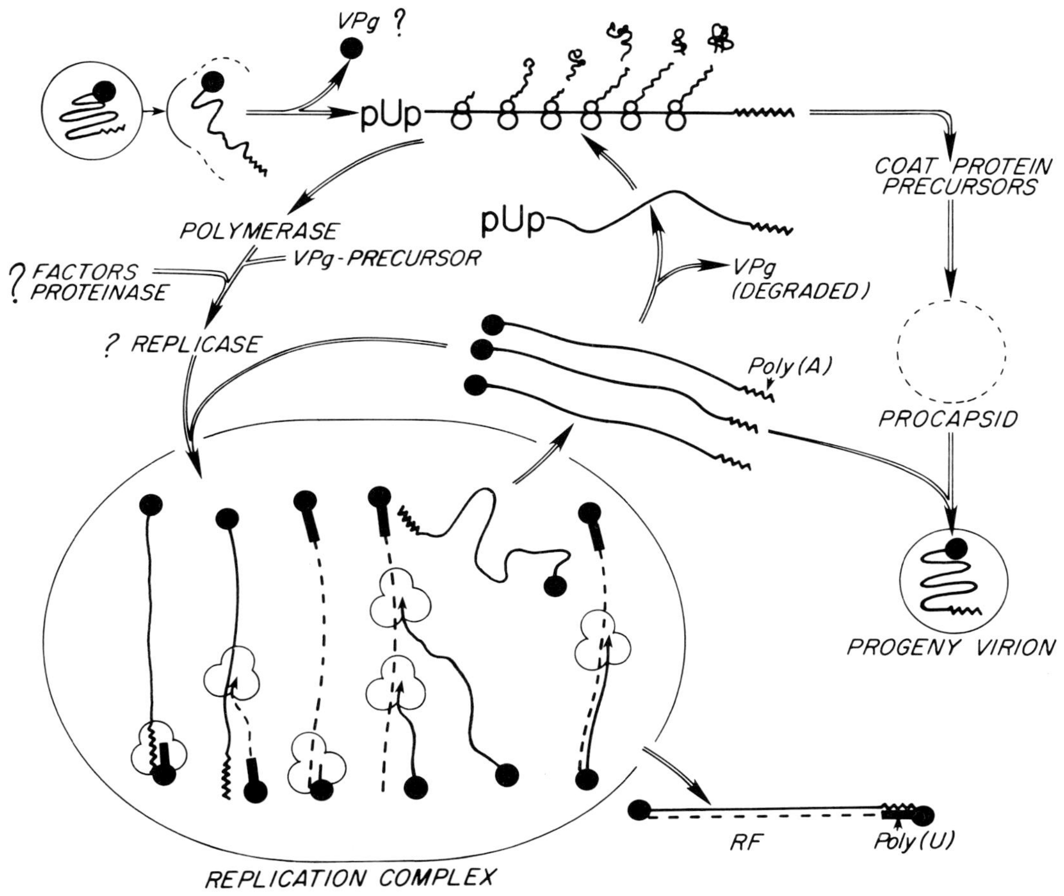

Fig. 4 Schematic representation of the poliovirus life cycle. The virion enters the cell via a cellular receptor. Following uncoating and release of the genome (solid line) into the cytoplasm, the viral RNA engages in protein synthesis. Virus-specific mRNA is identical to virion RNA in nucleotide sequence but is 'unlinked' from the 5′-terminal peptide, VPg (black dot). The RNA replication complex is membrane bound; RNA structures found associated with this complex are the RI, RF and ssRNA. The multistranded RI is usually of the type containing nascent positive strands; RI with nascent negative strands have not been observed. Whether RI molecules have a mainly single-stranded (as shown here) or a double-stranded backbone structure is unknown. Negative strands carry a VPg-linked poly(U) at their 5′ ends; the poly(U) (black bar) serves as template for the 3′-terminal poly(A) of positive-strand RNA.

ately upon entry and uncoating of the virion, its messenger-sense RNA is translated into the proteins needed for orchestrating the intracellular infectious cycle. To complete one infectious cycle takes only 6–7 h. Moreover, the intracellular events are entirely cytoplasmic and can occur in enucleated cells. The replicative cycle of poliovirus is shown schematically in Fig. 4. The steps of virion attachment to the cellular receptor, entry, uncoating, translation and proteolytic processing of the polyprotein have already been discussed. The steps that follow include RNA replication and morphogenesis. Proteins involved in replication must first copy the positive-stranded RNA into complementary negative strands that subsequently serve as templates for the synthesis of progeny positive strands. The newly replicated positive-stranded vRNAs either re-engage in multiple cycles of translation and replication or become encapsidated into virions. A number of events are characteristic of the poliovirus replicative cycle and for the purpose of this discussion will be divided into events that involve ultrastructural changes at the cellular level and events that occur at the molecular level.

There are characteristic morphological alterations that occur at the cellular level during a productive infection. Concomitant with poliovirus infection there is proliferation of membranous vesicles. Maximal viral protein synthesis occurs 2.5–3.5 h post-infection (p.i.), at which time the first signs of virus-induced vesicle formation become visible. Experiments using electron microscopy have shown that parts of the rough endoplasmic reticulum (rER) that are devoid of ribosomes begin to dilate and eventually bud off from the rER channels. This event begins at the nuclear peripheries and eventually takes over the entire cell cytoplasm. Vesicularization of the cell during infection correlates with an increase in the total phospholipid content of the infected cells, permeability of the plasma membrane, and leakage of the intracellular components that eventually result in shriveling of the cells. These are characteristic morphological changes normally associated with the cytopathic effect (c.p.e.) that can be observed at the light microscope after infection of susceptible tissue culture cell lines. However, molecular events leading to c.p.e. are obscure. Viral RNA synthesis which peaks at 3.5 h p.i. is directly associated with these membranous vesicles. Indeed, crude membrane preparations (crude replication complex, CRC) of disrupted poliovirus-infected HeLa cells can be further fractionated in a discontinuous sucrose gradient; smooth membrane fractions can be isolated that are capable of synthesizing viral RNA *in vitro*. Recent reports have demonstrated the need for a continuous phospholipid biosynthesis during poliovirus RNA synthesis, since the antibiotic, cerulenin, an inhibitor of phospholipid synthesis, blocks viral RNA synthesis. Moreover, brefeldin A, a drug inhibiting vesicle-mediated traffic between rough endoplasmic reticulum and Golgi organelles, strongly inhibits poliovirus RNA replication. The mechanism by which cerulenin or brefeldin A inhibits RNA synthesis is not understood, but it probably involves metabolism of cellular membranous components.

The interaction of viral proteins with membranous vesicles has been characterized in some detail. The results indicate association of 2BC, 2C, 3AB, $3C^{pro}$ and $3D^{pol}$ with viral-induced vesicles, albeit with differential affinities. Studies conducted with immunocytochemistry and electron microscopy have implicated 2BC and/or 2C in the induction of vesicles and/or attachment of the replication complex to the vesicular membranes. The role of membranes in RNA synthesis *in vivo* is not clear. One can imagine membranes to act as a scaffolding for the replication complex, analogous to the role of the nuclear matrix in DNA replication and transcription. Membranes may also act to concentrate the viral proteins at the site of replication. Alternatively, since the viral RNA polymerase ($3D^{pol}$) is promiscuous in its activity, membranes may serve to separate the vRNA from contaminating cellular RNAs during replication.

Three virus-specific RNA structures are found associated with the membranous vesicles. The first is single-stranded RNA (ssRNA), either virion RNA or mRNA, the latter being virion RNA from which the VPg is cleaved. ssRNA is the most abundant RNA formed after infection. The second is a replicating intermediate (RI) form which is single-stranded and partially double-stranded. RI molecules comprise a negative-stranded template with 6–8 replicative positive strands. In contrast, a positive-stranded template with nascent replicating negative strands are never seen. The third is a double-stranded RNA or replicative form (RF). RF structures are thought to be end products of RNA synthesis although they may be intermediates as well since they are infectious (note that infection by RF requires the presence of nuclei whereas ssRNA can initiate infection in enucleated cells).

RNA replication at the molecular level is poorly understood. Apart from the general scheme that positive-stranded replication proceeds via a negative-stranded intermediate, the individual steps in this pathway remain to be elucidated. Gene products encoded in the P2–P3 segment of the polyprotein, either as precursors or mature products are probably sufficient in the replication of the vRNA, although host cellular proteins may also be involved in this process. With the exception of the RNA-dependent RNA polymerase, $3D^{pol}$, and the viral proteinase, $3C^{pro}$, a biochemical characterization of nonstructural P2–P3 proteins has not been pursued rigorously due to lack of a functional assay. However, genetic studies have proposed polypeptides 2B, 2C, 3AB, and more recently, $3CD^{pro}$ to have a role in RNA synthesis.

Evidence implicating 2B and/or the precursor 2BC in viral RNA synthesis comes from the study of mutations in 2B with defined phenotypic lesions. Mutants with amino acid insertions in the 2B coding region (nucleotides 3916) have been reported to express a small plaque phenotype and a severe defect in RNA synthesis. The defect in 2B could not be complemented in *trans*, but exerted partial dominance over wild-type virus in mixed infections. The results suggest that the lesion in 2B acts in *cis*. Such a scheme is conceivable if 2B or 2BC after synthesis becomes immediately engaged in the replication complex in an irreversible manner. Alternatively, since the genome is synthesized as a polyprotein, a mutation in one genetic element could affect the subsequent role of precursor polypeptides. This may explain why it is not always possible to obtain complementation among nonstructural proteins in poliovirus. Finally, the association of 2B or 2BC with a limiting cofactor in replication could account for the partial dominance seen over wild-type virus. Host range variants of human rhinovirus type 39 (HRV39) HRV39/L have been isolated which can bypass an intercellular block in viral RNA replication in mouse L cells. The HRV39/L has an altered electrophoretic mobility in 2B and 2BC. It is possible that HRV39/L bypasses the replication block in mouse cells as a result of mutations in 2B and/or 2BC which allow its interaction with formerly nonpermissive host factors.

Selection and characterization of viral variants resistant to inhibitors of viral metabolism has led to identification of polypeptide 2C in viral RNA synthesis. One such inhibitor of poliovirus replication is guanidine hydrochloride (Gua-HCl). At millimolar concentrations, Gua-HCl can inhibit poliovirus RNA synthesis while it has no detectable effect on cellular metabolism. By growing poliovirus in the presence of inhibiting levels of Gua-HCl, it was possible to select viral variants that displayed guanidine resistance (g^r) or dependence (g^d). Characterization of natural variants identified a single mutation, Asn179-Gly, as sufficient to render a guanidine-sensitive (g^s) virus g^r, although additional point mutations in separately isolated variants could also be mapped to this polypeptide. These mutations cluster in a locus of the polypeptide with extensive conservation in amino acid sequence and antigenicity among picornaviruses. A striking conservance is noted in a region containing an NTP-binding consensus sequence that if mutated is lethal to virus viability and null in RNA replication. Furthermore, the function of 2C can be supplied in *trans* since a conditional lethal mutant of 2C displaying a strong temperature-sensitive phenotype in RNA synthesis can be rescued in a mixed infection with a virus containing a wild-type 2C gene.

Recently, the viral proteinase $3C^{pro}$ or the precursor 3CD has been implicated in RNA replication. A 4 bp insertion at nucleotide 70 in the 5′ NTR that disrupts an RNA stem–loop structure can be suppressed by mutations mapping to the viral proteinase gene $3C^{pro}$. Immunological probing has shown that $3C^{pro}$ and $3D^{pol}$ or the precursor 3CD binds to this stem–loop structure. This ribonucleoprotein complex is selectively required for positive strand synthesis, but not negative strand synthesis. Thus, the stem–loop structure may act as a recognition signal for specific binding and sequestering of viral and possibly cellular proteins engaged in positive strand RNA replication.

A virus-specific RNA polymerase activity has been identified from poliovirus-infected HeLa cells that is template and primer dependent and capable of elongating nascent RNA strands. The enzyme, $3D^{pol}$, has been characterized biochemically and can be assayed *in vitro* using polyadenylated RNA as template and oligo(U) as primer. Defects in $3D^{pol}$ which affect RNA synthesis, may or may not be complemented in *trans*, suggesting that $3D^{pol}$ has separate functional domains. However, additional mutations have to be characterized in order to devise a more complete complementation map.

A hallmark of all nascent replicating molecules is that they are covalently attached to the genome-linked protein, VPg (3B) at their 5′ end, an observation suggesting that VPg may be a primer for initiation of RNA synthesis. Using CRC isolated from poliovirus-infected cells, it is possible to synthesize VPg-linked RNA, VPg-pUpU, *in vitro*. This product can be chased into nine or more nucleotidyl residues corresponding to the viral 5′ end. VPg-pUpU can also be found in poliovirus-infected cells labeled *in vivo*. Based on this evidence, a model has been proposed in which 3AB, the precursor to VPg, attaches to membranes via a 22 amino acid hydrophobic stretch in 3A. Following membrane attachment, VPg becomes uridylylated and acts as a primer for elongation by the viral RNA polymerase ($3D^{pol}$). The viral proteinase, $3C^{pro}$, then cleaves at the junction of 3A/3B, liberating VPg-linked RNA from the membrane anchor. What catalyzes uridylylation of VPg is not known. Genetic studies have suggested the involvement of $3D^{pol}$, but this notion has not been confirmed *in vitro*. Finally, a single amino acid mutation Thr67-Ile in the hydrophobic domain of 3AB which is impaired in positive-strand RNA synthesis and *in vitro* uridylylation of VPg, provides genetic evidence for this model. An alternative model has been suggested which is primarily based on *in vitro* synthesis of negative strands. This model proposes uridylylation at the 3′ terminal poly(A) of template RNA by a host factor (uridylyl transferase), leading to the formation of a poly(A)–oligo(U) hairpin in which the oligo(U) serves as a primer for elongation by $3D^{pol}$. Subsequently, the hairpin is cleaved in a self-catalyzed transesterification reaction linking VPg to the 5′ end of the negative-stranded template.

The models presented above propose two very different roles for VPg in initiation of viral replication. It is unlikely that a small polypeptide of 22 amino acids can carry both functions selectively on positive and negative strands. Ultimately, a study of the mechanism of VPg-linkage to the 5′ end of both negative and positive strands leading to viral replication should investigate the function of additional viral proteins, i.e. 2B, 2C, 3CD (see later) and cellular proteins, if any, in RNA synthesis, the role of membranes in either providing a hydrophobic environment or an essential component for replication, a search for recognition signals at the 3′ termini of the vRNA where initiation of replication is presumably primed. Any model of initiation should corroborate the two very different 3′ termini of the poliovirus negative and positive strands; the former containing a heteropolymeric sequence, CAGUUUUAA-OH and the latter, a homopolymeric poly(A) tail. Furthermore, since there is a vast overproduction of positive strands over negative strands, there must be a regulatory step at the level of synthesis. Insight into these questions might explain why negative-stranded templates can provide multiple initiation events for positive strand replication, rather than, single run replication of negative strands from positive-stranded templates.

Polioviruses, when infecting the same host cell can undergo homologous genetic recombination with high frequency; 10–20% of viral RNA molecules can undergo recombination in a single infectious cycle. This phenomenon, that can be studied in tissue culture cells occurs readily in children who have been vaccinated with all three types of the live oral (Sabin) vaccine strains. The mechanism of recombination is not fully understood, but may be copy choice by template switching of RNA polymerase during negative strand synthesis.

Morphogenesis of viral particles from isolated components has not been achieved to date. Clearly morphogenesis involves proteolytic cleavages of P1 to yield (VP0, VP3, VP1), a complex that does not dissociate but forms pentamers $(VP0, VP3, VP1)_5$ that ultimately aggregate to form 'immature' empty capsids ('procapsids') $[(VP0, VP3, VP1)_5]_{12}$. Crystallographic studies indicate that the proteolytic cleavages of P1 lead to extensive 'migration' of newly formed polypeptide termini that, as noted earlier, result in stable interactions of subunits and protomers. It is not known at what stage the RNA associates with these structures, and how the maturation cleavage of VP0 is triggered. Biochemical and genetic analyses implicate the formation of the 'provirion', $[(VP0, VP3, VP1)_5]_{12}$.RNA, as the last step in morphogenesis, but it has not been possible to show direct conversion of provirions to infectious virus *in vitro*. The pathway of morphogenesis may thus be summarized as follows (asterisk indicates proteolytic cleavage):

$$\text{P1}^* \rightarrow \text{pentamers} \rightarrow \text{procapsid}$$

$$\text{procapsid} + \text{RNA} \rightarrow \text{provirion}^* \rightarrow \text{virion}$$

Finally, a *de novo*, cell-free synthesis of poliovirus has been achieved whereby highly efficient *in vitro* translation of poliovirus RNA can direct replication, encapsidation and the formation of infectious particles *in vitro*. Consequently, the *in vitro* translation of poliovirus RNA in a HeLa cell-free system, followed by treatment with RNase A and T1, leads to plaque formation when assayed on HeLa cell monolayers. Plaque formation was abolished when HeLa cell monolayers were treated with a monoclonal antibody directed against the PVR. Similarly, millimolar concentrations of Gua-HCl, an inhibitor of poliovirus RNA replication, and edeine, an inhibitor of translation initiation could equally abolish plaque formation. Since *de novo* RNA synthesis is a prerequisite for production of infectious particles, this system provides a unique tool for the study of RNA replication, recombination or morphogenesis *in vitro*.

See also: Genetics of animal viruses; Rhinoviruses; Virus structure.

Further Reading

Hogle JM, Chow M and Filman DJ (1985) Three-dimensional structure of poliovirus at 2.9 Å resolution. *Science* 229: 1358.

Kirkegaard K and Baltimore D (1986) The mechanism of RNA recombination in poliovirus. *Cell* 47: 433.

Kitamura N *et al.* (1981) Primary structure, gene organization and polypeptide expression of poliovirus RNA. *Nature* 291: 547.

Kräusslich H-G and Wimmer E (1988) Viral proteinases. *Annu. Rev. Biochem.* 57: 701.

Kuhn RJ and Wimmer E (1987) The replication of picornaviruses. In: Rowlands DJ, Mayo MA and Mahy BWJ (eds) *Molecular Biology of the Positive Strand RNA Viruses*. London: Academic Press.

Mendelsohn CL, Wimmer E and Racaniello VR (1989) Cellular receptor for poliovirus: molecular cloning, nucleotide sequence and expression of a new member of the immunoglobulin superfamily. *Cell*. 56: 855.

Minor PD (1991) Antigenic structure of picornaviruses. In: Racaniello VR (ed.) *Current Topics in Microbiology and Immunology: Picornaviruses*. Berlin: Springer-Verlag.

Molla A, Paul AV and Wimmer E (1991) Cell-free, de novo synthesis of poliovirus. *Science* 254: 1647.

Racaniello, VR and Baltimore D (1981) Cloned poliovirus complementary DNA is infectious in mammalian cells. *Science* 214: 916.

Wimmer E, Hellen CUT and Cao X-M (1993) Genetics of poliovirus. *Annu. Rev. Genet.* in press.

POLYDNAVIRUSES

Don Stoltz
Dalhousie University
Halifax, Nova Scotia, Canada

Taxonomy and Classification

The polydnaviruses are defined primarily on the basis of genome structure. The polydnavirus genome is segmented, consisting of a population of double-stranded circular DNA molecules of variable molecular mass (the name 'polydnavirus' is derived from *poly*disperse *DNA virus*.) Two distinct groups of the family *Polydnaviridae* are recognized: the bracoviruses, found only in the parasitoid family *Braconidae*, and the ichnoviruses, found only in the family *Ichneumonidae*. In addition to having distinct host ranges, the braco- and ichnoviruses are quite different in at least one other important parameter, morphology.

Virion Structure and Morphogenesis

Bracovirus particles consist of cylindrical nucleocapsids of uniform diameter (40 nm) but variable length (30–150 nm); these are surrounded either individually or in groups by a single unit membrane. Ichnovirus particles consist of nucleocapsids which are typically uniform in size (approximately 85 × 330 nm), having the form of a prolate ellipsoid; these are individually surrounded by two unit membranes. Morphogenesis occurs only in the wasp ovary, specifically in a region referred to as the calyx, which lies between the ovarioles and the oviducts; it begins during the late pupal stages of parasitoid development, with the *de novo* appearance within the nucleus of membranes destined to form the envelope and the inner membrane respectively of braco- and ichno-viruses. In both braconid and ichneumonid parasitoids, mature virions enter the lumen of the oviduct, where in large numbers they comprise what is generally referred to as calyx fluid, which, like typical virus suspensions at high concentration, is opalescent. Exit of mature bracovirus particles appears to involve sloughing and lysis of calyx epithelial cells; ichnovirus particles, on the other hand, bud through the plasma membrane, thereby acquiring a second envelope. It is known that the ichnoviruses bud through the nuclear envelope, but how they traverse the cytoplasm prior to budding through the plasma membrane has not as yet been determined.

The Polydnavirus Genome

The polydnavirus genome, as extracted from purified virions, consists of double-stranded circular DNA molecules of variable molecular mass. A single female wasp ovary will often contain virus sufficient for most of the virus genome to be visualized following agarose gel electrophoresis and ethidium bromide staining. Ichnovirus genomes (which have been most fully characterized) appear to consist of 20–40 different molecules, with an aggregate molecular size of 300 kbp or more. Genetic complexity may be less than this might suggest, however, since a number of genes have been observed to be present on more than one genome segment; indeed, in some cases, smaller genome segments are repeated in their entirety in one or more larger ones. There thus exist families of both genes and genome segments. Smaller repetitive motifs have also been discovered. Bracovirus genomes have not as yet been investigated to the same extent; however, preliminary observations would suggest that bracovirus genome segments are larger than those typically associated with the ichnoviruses. A rationale for the segmentation of polydnavirus genomes has not as yet been proposed.

Nothing is known concerning how polydnavirus genomes are packaged. However, it is reasonable to suspect that the relatively large ichnovirus particle would contain a complete genome; on the other hand, the variable length of bracovirus nucleocapsids may reflect packaging of DNA molecules of different sizes.

Recent work has established that there exists within the parasitoid genome stretches of DNA which are homologous to, and colinear with, each encapsidated viral DNA molecule. Common sequences, then, are represented in both linear (chromosomal) and circular (viral) form.

Life Cycle

It is now evident that all individuals of all affected species carry polydnavirus. In keeping with this observation, it now seems clear that polydnavirus genomes are chromosomally transmitted in both egg and sperm cells in accordance with simple Mendelian rules; transmission then, occurs with 100% efficiency. It also seems probable, if as yet unproven, that cognate chromosomal DNA sequences must serve as template at least for the initial stages of viral DNA replication. It is possible – although again unproven – that viral DNAs may be synthesized via an *in situ* amplification scenario followed by excision. With the recent observation that viral DNA synthesis can be hormonally induced in explanted ovaries, the replicative machinery involved may soon be elucidated.

Electron microscopy has established that virus particles are exported, initially into the lumen of the oviduct; during oviposition, both parasitoid eggs and polydnavirus particles are injected into the larvae of insects destined to serve as hosts for the next generation of parasitoids. Considerable experimentation has shown that in the absence of polydnavirus, parasitoid eggs/larvae are typically destroyed by an immune response. Polydnaviruses are in fact responsible for a variety of physiological changes which are induced in the parasitized host; in addition to immunosuppression, these include growth inhibition, hormonal perturbation, inhibition of phenol oxidase activity, and so on. Many of the observed physiological changes are assumed to be relevant to successful parasitism; this raises the very important question of how polydnaviruses mediate these myriad activities.

Following oviposition, polydnavirus particles gain entry to a variety of host tissues. Nucleocapsids reach the nuclei and become uncoated either at pores (bracoviruses) or within the nucleus itself (ichnoviruses); following this, viral DNA apparently persists within host cells throughout the natural course of parasitism (parasitoid larvae usually complete development within 1–2 weeks). Virus-specific transcription occurs in parasitized host animals, and appears to be required for successful parasitism; it should be noted that transcription occurs in the absence of viral replication (which occurs only in the parasitoid ovary). Viral gene products are assumed to be responsible for some or all of the observed changes associated with parasitism, but none of these have yet been isolated from parasitized animals. In any case, it can reasonably be concluded that polydnavirus activity in the parasitized host represents an example of genetic colonization.

The polydnavirus life cycle, then, may be envisioned as consisting of two 'arms': one of these, responsible for viral transmission and replication, is mediated by linear chromosomal DNA; the second arm, mediated by circular encapsidated DNA molecules, is responsible for genetic colonization of the parasitized host, ultimately ensuring successful parasitism.

Significance

Polydnaviruses have obvious significance in the sphere of ecology. These viruses are carried by hundreds, perhaps thousands, of species of parasitic hymenoptera, many of which parasitize more than one species of insect. Many host species are important pests, populations of which are often maintained at acceptably low densities as a consequence of parasitism by wasp species which carry polydnaviruses. It may be assumed that some of the virus-specific gene products expressed in parasitized host insects, particularly those having immunosuppressive properties, will ultimately have potential in the biopesticide industry.

What is perhaps most significant about polydnaviruses, however, is the degree to which these viruses may modify our concept of the range of possible virus–host interaction (Indeed, the polydnavirus phenomenon challenges us to come up with a more comprehensive definition for the word 'virus'!). Arising, we assume, from originally 'typical' viruses (i.e. those capable of initiating a fully productive infection), the polydnaviruses would appear to have achieved considerable success in terms of the degree to which their life cycles have become integrated with those of their hosts. Since all individuals belonging to any given affected species carry virus, the acquisition of a polydnavirus genome must presumably have preceded speciation. Perhaps in keeping with this observation, no other viruses have been shown to be required for the survival of the organisms that carry them. An investigation into the co-evolution of parasitoids and polydnaviruses – insofar as this is possible – could provide interesting reading for years to come!

Further Reading

Stoltz DB (1990) Evidence for chromosomal transmission of polydnavirus DNA. *J. Gen. Virol.* 71: 1051.

Stoltz DB (1992) The polydnavirus life cycle. In: Beckage, Federici and Thompson (eds) *Parasites and Pathogens of Insects*. New York: Academic Press, in press.

Stoltz DB and Vinson SB (1979) Viruses and parasitism in insects. *Adv. Virus Res.* 24: 125.

POLYOMAVIRUSES – MURINE

Contents

General Features
Molecular Biology

General Features

James M Pipas
University of Pittsburgh,
Pittsburgh, Pennsylvania, USA

History

Polyomavirus was discovered by Ludwig Gross in 1953 by virtue of its ability to induce tumors in inoculated newborn mice. While studying murine leukemia virus (MLV), a retrovirus that induced chronic leukemia when injected into newborn mice, Gross observed that some animals inoculated with MLV preparations developed adenocarcinomas of the parotid gland. Extracts of the tumor contained an activity, termed parotid virus, that produced the tumor in newly inoculated animals. A variety of physical techniques demonstrated that parotid virus was a different virus from MLV. Later it was shown that the virus could induce the formation of a variety of tumor types in newborn animals, and hence was given the name polyomavirus.

In 1957 Stewart, Eddy and colleagues reported that polyomavirus could be propagated in cultures of mouse embryo cells. Thus, for the first time the viral productive cycle could be analyzed in cell culture. This was followed by the development of a viral plaque assay. With these tools in hand, an exploration of the genetics of polyomavirus could begin.

Taxonomy and Classification

Murine polyomavirus is a member of the *Papovaviridae* family which consists of the papillomaviruses and the polyomaviruses. The prototype members of the polyomavirus group are simian virus 40 (SV40) and murine polyomavirus itself. Also included in this group are human viruses BK and JC, SA12 (chacma baboon), LPV (African green monkey), bovine polyomavirus, budgerigar fledgling disease virus (BFDV), hamster polyomavirus (HaPV) and KV (mice). While representatives of the family are found in species ranging from birds to humans, each virus shows a restricted host range. Members of the polyomavirus group are characterized by icosahedral virions approximately 45 nm in diameter containing the viral-encoded capsid proteins VP1, VP2 and VP3. Each virion contains a single molecule of circular duplex DNA of about 5 kb, along with the cellular histones H2A, H2B, H3 and H4. Mature virions do not contain histone H1. The viral growth cycles have been most well characterized for SV40 and polyomavirus because of well-developed cell culture systems for propagating these viruses.

Productive Infection

Polyomavirus was first propagated in primary mouse embryo fibroblast cell cultures. Established cell lines that supported productive infection and plaque formation were discovered later. The most commonly used cell line is NIH 3T3 where the virus forms plaques on monolayer cultures visible by staining with neutral red after 8 days of incubation. Infection of these cells results in a yield of approximately 300 infectious particles, or plaque forming units (PFU), per infected cell. The ratio of physical/infectious particles is about 100:1 to 1000:1 for polyomavirus. The polyomavirus infectious cycle can be divided into steps typical for animal virus productive infections. First, polyomavirus virions attach to a specific receptor on the cell surface. The receptor probably contains neuraminic acid since treatment of permissive cells with neuraminidase renders them resistant to polyomavirus infection. The mechanisms involved in penetration and uncoating are not well understood, but the result is that by approximately 4–6 h postinfection, free viral DNA is present in the nucleus of the infected cell. Soon afterwards, mRNAs representing about half of the viral coding sequences (see below) are detectable, followed by synthesis of three viral-encoded proteins, large T antigen (LT), middle T antigen (MT) and small t antigen (ST). These are collectively referred to as the early proteins, and they are encoded by the early region genes.

Early region genes are those that are expressed prior to the onset of viral DNA replication. By about 24 h postinfection, the early proteins have reached their maximum levels and viral DNA replication begins. Replication is initiated within a 65 bp segment termed the *ori* and proceeds bidirectionally around the circular genome. Viral DNA replication requires a single viral-encoded protein, LT. Much progress has been made in investigations concerning the role of the polyomavirus and SV40 LTs in initiating viral DNA replication. These results are reviewed in the following entry.

Shortly after the onset of viral DNA replication, transcription of the remainder of the viral genome, termed the late region, begins. This results in the synthesis of the three viral capsid proteins, VP1, VP2 and VP3. Late region mRNA generation is very complicated with multiple 5′ ends and numerous alternate splicing events occurring. All of the 5′ ends lie between nucleotides 5075 and 5170 on the viral genome. Both the early and late mRNA polyadenylation sites lie about 180° around the circular genome from the *ori*. Viral assembly begins shortly after the onset of late region protein synthesis. While all of the steps in assembly have not yet been elucidated it is known that purified VP1 expressed in *Escherichia coli* will self-assemble into pentamers. However, the complete assembly of infectious particles apparently requires the action of MT, which plays a role in post-translational modification of VP1.

As the infected cells die they exhibit a characteristic cytopathic effect (CPE). There does not seem to be any specific mechanism for viral release. Rather progeny virions seem to be released as the cell dies, with much of the viral material remaining attached to the cellular debris.

What has just been outlined is a typical cycle following polyomavirus infection of a permissive cell. In fact, the outcome of infection depends upon the type of cell infected. Infection of cell types that are semipermissive for polyomavirus can result in a persistent infection. This is perhaps more representative of the natural course of infection with this agent. Other cell types are nonpermissive, because either they lack a viral receptor and infection is blocked at the stage of attachment or they lack the proper machinery for allowing viral DNA replication. These latter cell types are of interest because they allow virus penetration and expression of the early viral proteins. However, since replication is blocked, these infected cells do not die, and a subset of them are transformed into neoplastic cells.

Viral Transformation

When nonpermissive cells, such as those of rat or hamster origin, are infected with polyomavirus, the early stages of infection proceed normally resulting in the synthesis of the STs, MTs and LTs. However, viral DNA does not replicate and the late region genes are not expressed. Rather, cellular DNA synthesis is stimulated and the infected cells are driven through at least one to several rounds of division. During this period, actin cables are disassembled and the cells acquire a phenotype associated with transformed cells. After a couple of days, most of the cells in the culture resume a normal appearance and viral nucleic acid and proteins can no longer be detected. This phenomenon is termed abortive transformation. During such infections a small percentage of the infected cells permanently acquire the transformed phenotype including the ability to: (1) form foci on a monolayer of normal cells; (2) proliferate in low serum concentrations; (3) grow independently of anchorage to a solid substrate; and (4) form tumors in test animals. Such cells continuously express the viral T antigens and contain viral DNA covalently integrated into the cellular chromosome. A number of genetic and biochemical experiments have lead to the conclusion that the STs, MTs and LTs are responsible for initiating and maintaining the transformed phenotype.

The rate-limiting step in neoplastic transformation appears to be the integration of viral DNA into the cellular chromosome. Polyomavirus has not evolved any specific mechanism for integration. In fact, integration is not a normal part of infection and appears to occur by cell-mediated nonhomologous recombination. Insertion occurs at random locations with respect to both the viral and cellular chromosomes. This appears to explain the phenomenon of abortive transformation. During the viral infection of nonpermissive cells, nearly all the cells are initially infected and thus express the viral T antigens. As a result the cells temporarily acquire the transformed phenotype. However, in most cells, as proliferation proceeds the viral genome is diluted out and thus nearly all progeny cells lose expression of the T antigens and revert to a normal appearance. In a minority of the infected cells, the viral DNA integrates in such a manner as to allow continuous expression of the T antigens. Since the viral genome, and thus T antigen expression, is passed on to all progeny cells, this leads to a stably transformed phenotype.

Much effort has been directed towards understanding the role each of the viral tumor antigens plays in transformation and the molecular mechanisms by which they achieve their effects. ST increases the efficiency of transformation in certain cell types or when quiescent cells are infected. LT is sufficient to immortalize primary lines. While lines expressing LT exhibit a reduced requirement for serum, they do not form foci, grow independently of anchorage or form tumors in animals. MT is the only viral gene product required to transform a number of established cell lines; however, both LT and MT are required to induce the fully transformed phenotype in primary cells. Established cell lines transformed by MT alone form foci, grow independently of anchorage and are tumorigenic. However, they do not show a reduced serum requirement.

Each of the viral T antigens are thought to act by targeting key cellular proteins that normally play a role in the regulation of proliferation. ST complexes with the cellular phosphatase 2A and alters its activity. LT complexes the retinoblastoma protein (Rb) and probably other targets. MT complexes with at least three nonreceptor tyrosine kinases, c-*src*, c-*fyn* and c-*yes*, as well as with phosphoinositide kinase.

Genome Organization

The circular double-stranded genome of polyomavirus consists of 5297 bp. By convention the nucleotide at the center of an imperfect palindrome that forms part of the origin of DNA replication is designated as 1 and numbering proceeds through the early coding sequences around the circle. Sometimes positions on the genome are indicated by map units (m.u.). This older system divides the genome into 100 m.u. using the single *Eco*RI site (nucleotide 1562) as 0 and proceeding through the remainder of the early region and then the late region and regulatory region to 100.

Regulatory region

The viral regulatory region consists of approximately 400 bp lying between the early and late coding regions (nucleotides 5086–156) and contains the *ori*, the promoters for early and late region transcription and the transcriptional enhancers. Early region transcription is initiated at position 156 with the initiation codon for all three T antigens lying at position 175. The major initiation site for the late region mRNAs is at position 5086. The *ori* and all of the known *cis*-acting sequences required for early and late transcription lie between these points. This region contains multiple binding sites for LT which is thought to recognize the consensus pentanucleotide G(A/G)GGC. There are three strong T-antigen-binding sites located upstream of the start of early region transcription. These are termed sites A (nucleotides 27–76), B (nucleotides 86–116) and C (nucleotides 121–164). Each of these contain multiple copies of the recognition pentanucleotide in various orientations and spatial arrangements. None of these sites appear to play a major role in viral DNA replication and are perhaps involved in transcriptional control. The TATA box associated with the early region promoter lies between sites B and C.

Viral DNA replication requires a stretch of nucleotides termed the core *ori* (nucleotides 45–5274) linked in *cis* to either enhancer element α or β. A stretch of DNA very rich in AT residues lies just to the late side of the core *ori*. This is followed by the β and then the α elements. The specific roles each of these sequence motifs play in viral DNA replication and transcription will be discussed in the following entry.

Early region

As mentioned above, the viral early region encodes three major proteins, ST, MT and LT. The mature mRNAs encoding each of these proteins are produced by differential splicing of a single primary transcript which starts at nucleotide 156. The polyadenylation signal for these mRNAs lies at position nucleotide 2930. All three proteins initiate at the methionine codon at position 175, and thus share a common N terminus. The mRNA for ST is generated by splicing a donor site at nucleotide 748 to an acceptor site at nucleotide 797, thus eliminating an intron of 49 nucleotides. Translation of this mRNA initiates at the ATG at nucleotide 175 and terminates at nucleotide 807. The mRNA encoding MT is generated by utilizing the same donor site (nucleotide 748) but using an acceptor at nucleotide 811. This results in loss of the termination codon used for ST and opens a new open reading frame (ORF) which terminates at position nucleotide 1499. Thus, MT shares N-terminal residues with ST, but each have unique C termini. The LT mRNA is generated by utilization of a splice donor site at nucleotide 411 and an acceptor site at nucleotide 797. This generates an ORF that shares the first 82 codons with MT and ST, but possesses a unique C terminus.

LT plays a central role in regulating productive infection. This 795-amino acid protein is the only viral product required for viral DNA replication. One way in which LT acts is to bind specific sequences located within *ori*. The consensus sequence for LT recognition is the pentanucleotide G(A/G)GGC, which occurs many times within *ori* and the viral regulatory region. LT carries an ATPase/DNA helicase function which melts and unwinds *ori* to allow initiation of DNA replication. LT also autoregulates early region transcription, at least partially by binding to sequences located to the early side of *ori*. In addition, LT transactivates the polyomavirus late promoter, as well as a number of heterologous promoters of both viral and cellular origin. LT also carries the ability to complex the cellular Rb protein as discussed above, and probably additional activities involved in transformation.

MT is a 432-amino acid protein that plays a role in viral assembly by regulating the phosphorylation of the major capsid protein VP1. In addition, MT is an oncogene which is sufficient to transform a number of established cell lines, and cooperates with LT to transform primary cells. Many of the actions of MT are mediated by its direct interaction with *src*. *fyn*, *yes* and/or phosphoinositide kinase.

The role of the 195-amino acid ST in productive infection is unclear. Under many conditions tested, polyomavirus mutants that do not produce ST are viable and can fully transform cells. Under other conditions, such as infection of quiescent cells, ST seems to increase the efficiency of transformation. ST of both polyomavirus and SV40 are found in complex with two cellular proteins, the α and β subunits of phosphatase 2A. ST seems to displace the γ subunit and serves to inhibit phosphatase activity. The SV40 ST transactivates the adenovirus E2 promoter in some cell types. Whether the polyomavirus ST carries a similar transactivation activity has not been reported.

Late region

As discussed above, the three viral capsid proteins are expressed by the late transcription unit. The major capsid protein is the 385-amino acid VP1. In infected cells, VP1 occurs as multiple species which differ in their states of phosphorylation. Some of species appear to play a role in virus adsorption and at least one is involved in virus assembly. MT seems to play a role in inducing this species. VP2 and VP3 are proteins of 319 and 204 amino acids respectively.

Additional ORFs

In addition to the major proteins discussed above, the polyomavirus genome contains several additional ORFs which could potentially encode proteins, although in most cases whether these proteins are actually made or not has not been studied. SV40 is known to express a protein termed agnoprotein from the late mRNA leader sequence. Agnoprotein is a basic protein of 61 amino acids which seems to play a role in transport of the capsid proteins to the nucleus and possibly viral assembly. Other members of the polyomavirus group, including murine polyomavirus possess a similar ORF that would encode proteins closely related to the SV40 agnoprotein. However, the observation of an agnoprotein from these other viruses has not been reported.

Genetics

The genetic analysis of polyomavirus began with the isolation of mutants that were temperature-sensitive for the ability to form plaques on a monolayer of permissive cells. In these experiments the permissive temperature was usually 32–35°C while 39–40.5°C was nonpermissive.

These mutants fell into distinct phenotypic classes which corresponded to genetic complementation groups. Simultaneous studies with SV40 led to the identification of similar classes. The *ts-a* mutants fail to accumulate progeny viral DNA at the nonpermissive temperature. All of the *ts-a* mutants carry base-pair substitutions in sequences encoding the viral LT. All temperature-sensitive mutants mapping in the T-antigen gene fall into a single complementation group and are defective for viral DNA replication at the nonpermissive temperature. The *ts-a* mutants also lose the ability to transform nonpermissive cells at the nonpermissive temperature. However, there is a class of mutants mapping to the T-antigen gene that are temperature-sensitive for viral DNA replication while retaining the ability to transform at all temperatures.

The remaining classes of temperature-sensitive mutants carry lesions that affect the structure of the viral capsid proteins. One class, exemplified by *ts*10, fails to accumulate the major viral capsid VP1 and to assemble mature virions. Another class, exemplified by *ts*3, maps to the VP3/VP3 coding sequences, and seem to exhibit a defect in the uncoating step of infection.

Other polyomavirus mutants exhibit a host-range phenotype, in that they undergo a complete productive infection in murine cells transformed by polyomavirus, but are defective in normal murine cells. The lesions in these mutants map to sequences encoding both the LT and MT, but it is their effect on MT function that is thought to result in the defect. These mutants, exemplified by *NG18*, are also defective for viral transformation.

In recent years, a large collection of polyomavirus mutants have been generated by *in vitro* mutagenesis. Most of these mutations have been targeted to the regulatory region of the virus in order to better define *cis*-acting sequences important for the regulation of DNA replication and transcription, or to the viral early protein genes. Site-directed mutagenesis of the T antigen genes has also been used to establish structure/function relations in these proteins.

Relation to Other Members of the Polyomavirus Group

To date the complete nucleotide sequences of 12 members of the polyomavirus group have been produced and the partial sequences of two others are available. These studies demonstrate that all members of this group have essentially the same genome organization and use very similar strategies for controlling genome replication and expression.

There are important differences between various members of the polyomavirus group. First of all, murine polyomavirus and HaPV both encode three early viral tumor antigens (LT, MT, ST), whereas other members of the group, such as SV40, encode only LT and ST. SV40 LT possesses a C-terminal domain, termed the host-range domain, which appears to play a role in virus assembly. Viruses that encode MT lack a host-range domain on the LT.

A second difference is that murine polyomavirus expresses each of the major capsid proteins, VP1, VP2 and VP3, from separate mRNAs. On the other hand, SV40 expresses only two late mRNAs. One encodes VP1 while the other encodes VP2 and VP3. Differential synthesis of VP2 and VP3 is controlled by the efficiency of translational initiation.

Summary

Murine polyomavirus has proved to be an excellent model for studying mechanisms of genome replication and expression in higher eucaryotes. While members of this group are not important pathogens, they have served as models for viral infection and spread in a natural population. However, most effort has been put into learning how these small DNA viruses regulate and accomplish their replication. In addition, these viruses have served as invaluable models for discerning mechanisms involved in neoplastic transformation.

See also: Persistent viral infection; Transformation.

Further Reading

Fields BN *et al.* (eds) (1990) *Virology*, 2nd edn. New York: Raven Press.

Salzman NP (ed.) (1986) *The Papovaviridae*. Vol. 1, *The Polyomaviruses*. New York: Plenum Press.

Molecular Biology

Walter Eckhart
The Salk Institute
San Diego, California, USA

Properties of the Virion

The polyoma virion is spherical in shape, with icosahedral symmetry and a diameter of about 45 nm. It contains 360 copies of a major protein, VP1, and 30–60 copies of minor proteins, VP2 and VP3, encoded by the viral DNA. The VP1 proteins are assembled as 72 pentamers, 12 of which are surrounded by five others, and 60 of which are surrounded by six others. The pentamers are roughly cylindrical, with a tapering cavity. Each pentamer appears to have one copy of VP2 or VP3 associated with it in the region of the cavity, perhaps linking the pentamers to the viral DNA.

Properties of the Genome

The polyoma genome is a covalently closed circle of double-stranded DNA about 5300 bp in size. The DNA in the virion is supercoiled. The viral DNA is associated with four cellular histones, H2A, H2B, H3 and H4, organized into nucleosomes and packaged as a 'minichromosome'.

The genome is divided into early and late regions. The early region is expressed before the onset of viral DNA replication, and encodes three proteins called T antigens. The T antigens are designated as small (ST), middle (MT) and large (LT). The late region is expressed after viral DNA replication begins, and encodes the virion proteins, VP1, VP2 and VP3. There is a single origin of DNA replication. The origin 'core' comprises about 66 bp. The early and late coding regions are located on either side of the origin of replication. There is a noncoding region of about 150 bp on the early side of the replication origin which contains early promoter sequences. A noncoding region of about 250 bp on the late side of the replication origin contains enhancer sequences.

Properties of the Protein

Polyoma LT consists of 785 amino acids. It contains two independent nuclear location signals. The protein is located in the nucleus and functions in viral DNA replication, binding to the origin of replication and unwinding DNA in the region of the origin. There are several biochemical activities associated with LT, including ATPase and helicase activities. The protein helps to 'immortalize' primary rodent cells in culture, and cooperates with certain activated oncogenes to produce neoplastic transformation. Polyoma LT forms a complex with the retinoblastoma susceptibility protein, Rb. However, unlike the simian virus 40 (SV40) LT, polyoma LT apparently does not form a complex with the tumor suppressor protein, p53.

Polyoma MT contains 421 amino acids. It is associated with the plasma membrane of the cell through a sequence of 22 uncharged and hydrophobic amino acids, bordered by positively charged amino acids, near the C terminus, which comprise a transmembrane domain. When expressed in established mouse or rat cell lines, MT induces neoplastic cell transformation. The MT protein forms a complex with pp60$^{c\text{-}src}$, the cellular counterpart of the Rous sarcoma virus transforming protein (and with other members of the src family) and elevates the protein tyrosine kinase activity of pp60$^{c\text{-}src}$. MT also forms complexes with two other cellular proteins, phosphatidylinositol-3 (PI-3) kinase and phosphatase 2A. The formation of a complex with pp60$^{c\text{-}src}$, PI-3 kinase and phosphatase 2A in each case is necessary, but not sufficient, for transformation. Modification of the function of these three cellular proteins may contribute to cell transformation by MT.

ST contains 195 amino acids. It is partitioned between the nucleus and cytoplasm. The protein is not absolutely required for viral replication, but facilitates the accumulation of viral DNA under certain conditions. Like MT, ST forms a complex with phosphatase 2A.

The major virion protein, VP1, contains 385 amino acids. It is responsible for the binding of polyoma virions to cellular receptors. Mutations in VP1 affect the plaque morphology, virion stability and hemagglutination properties of polyoma strains. The minor virion proteins, VP2 and VP3, contain 319 and 204 amino acids respectively. As noted above, they may link VP1 pentamers to DNA in the virion.

Physical Properties

Polyoma infectivity survives months at 4 °C and weeks at 37 °C. Infectivity is inactivated at 70 °C, but is not inactivated by heating to 60 °C for 30 min. The virus is resistant to ether, 2% (v/v) phenol and 50% (v/v) ethanol. Virus preparations contain several kinds of particles: infectious viruses, containing intact viral DNA, noninfectious viruses, containing deleted or rearranged viral DNA molecules, empty particles, lacking DNA, and pseudovirions, containing cellular DNA. The density of DNA-containing viral particles is 1.34 g per ml. The virus agglutinates erythrocytes of a number of species. The ratio of physical particles to infectious units under conditions of standard plaque assays is 100 : 1–1000 : 1.

Replication – Strategy of Replication of Nucleic Acid

Polyoma produces a lytic infection in mouse cells, characterized by viral replication, production of infectious progeny and lysis of the infected cell. In hamster or rat cells, the infection is generally abortive; the virus fails to replicate and the infected cells survive. In lytically infected cells, viral DNA replication begins 12–18 h after infection and continues for about 24 h until the cell dies.

Polyoma DNA replication requires the origin core and an element from the enhancer region. Replication begins at the origin core and proceeds bidirectionally, terminating about 180° away on the circular viral DNA. An initiation complex containing LT binds to the origin core of a circular covalently closed viral DNA molecule and unwinds a region of double-stranded DNA. This creates a replication 'bubble' of diverging replication forks. Nascent chains are initiated by DNA primase–DNA polymerase α. The nascent chains begin with an RNA primer, six to nine ribonucleotides long, which is extended as DNA. At the replication fork, one strand is synthesized continuously in the direction of fork movement, while the other strand is synthesized discontinuously in pieces about 135 nucleotides long in the opposite direction. As replication proceeds the torsional strain created by the unwinding of the parental strands of DNA is released by a topoisomerase activity. Bidirectional replication proceeds until the replication forks meet, at which point the daughter molecules separate.

Characterization of Transcription

The early and late regions of the polyoma genome are transcribed from opposite strands of the DNA. Transcription begins at sites on either side of the origin of replication and proceeds in opposite directions. Transcription is carried out by the cellular RNA polymerase II. Transcription of early mRNA does not require viral proteins. The early promoter resembles other eucaryotic promoters in having a TATA box about 30 bp upstream of the early mRNA initiation site. The major 5′ termini of polyoma early mRNAs are located about 20 bp upstream of the initiation codon for the early proteins. Three kinds of early mRNAs are produced by alternate splicing. Each of the mRNAs encodes one of the T antigens.

Transcription of early mRNA is regulated by sequences in the enhancer region, and by LT. Mutations in binding sites for cellular transcription factors in the enhancer region reduce transcription of early mRNA. LT suppresses early transcription, probably by binding to sequences near the early promoter and blocking RNA polymerase.

Late transcription occurs primarily following the onset of viral DNA synthesis. Polyoma late mRNAs are heterogeneous in size. The nuclei of infected cells contain large transcripts that are tandem repeats of the entire viral genome. These transcripts are processed by splicing to produce mature polyadenylated mRNAs with unique coding regions attached to a common untranslated leader sequence. The 5′ termini of the late mRNAs are heterogeneous. The leader sequence is tandemly repeated, presumably because of successive leader to leader splicing which occurs during the processing of the large primary transcripts. Alternate splicing produces individual mRNAs encoding each of the virion proteins.

Characterization of Translation

The coding regions of the T antigens overlap, producing proteins with common N-terminal regions and unique C-terminal regions, determined by alternate splicing of the early mRNAs and translation in different reading frames. The three T antigens share 79 amino acids at their N termini. ST and MT share an additional 112 amino acids which are not present in LT. The unique C-terminal regions of ST, MT and LT are 4, 230 and 706 amino acids respectively. The T antigens are synthesized in the cytoplasm. After synthesis they are partitioned to different cellular compartments: LT to the nucleus, MT to the plasma membrane and ST to the nucleus and cytoplasm.

The virion proteins, VP1, VP2 and VP3, are synthesized in the cytoplasm, but are transported to the nucleus, where assembly of infectious virions occurs. The coding regions of VP2 and VP3 overlap. The proteins have different N-terminal initiation codons, determined by alternate splicing of the late mRNAs, and a common C-terminal termination codon, so that the sequence of VP3 is entirely contained in VP2. The C-terminal coding region of VP2/3 overlaps the N-terminal coding region of VP1 for 28 nucleotides, but the amino acid sequences of the proteins are different because they are translated in different reading frames.

Post-translational Processing

Polyoma LT is phosphorylated on several serine residues in the N-terminal half of the molecule. The MT protein is also phosphorylated on serine, although the sites have not been identified, and at low levels on tyrosine-315, presumably by $pp60^{c\text{-}src}$. Tyrosine phosphorylation of MT appears to be necessary for association with PI-3 kinase because mutants lacking the tyrosine-315 phosphorylation site fail to associate with PI-3 kinase.

The virion proteins of polyoma are also modified by phosphorylation. About 15% of the VP1 molecules in virions are phosphorylated. VP2 and VP3 are phosphorylated to a lesser extent. The major phosphorylated amino acid of VP1 is threonine; phosphoserine is a minority. Phosphorylation of VP1 takes place in the cytoplasm of the cell, possibly during translation. Phosphorylation of VP1 may play a role in virus assembly. VP1 is also modified by tyrosine sulfuration and proline hydroxylation. The role of these modifications is unknown.

The N-terminal glycine residue of VP2 is acylated with myristic acid. This modification takes place cotranslationally. Mutation of the glycine residue, preventing myristoylation, results in decreased infectivity.

Assembly Site, Uptake, Release and Cytopathology

Polyoma virions are assembled in the nucleus of the infected cell. Infectious particles adsorb to susceptible cells through the interaction of VP1 with cellular receptors. Infection can be blocked by treatment of cells with sialidase, suggesting that sialic acid is involved in receptor function. Following adsorption, the virus is internalized and transported to the nucleus where replication occurs. Progeny virus particles are passively released when the infected cell dies. Each infected cell produces 10 000–100 000 virions (100–1000 plaque-forming units).

See also: Simian virus 40; Transformation; Tumor viruses – human.

Further Reading

Eckhart W (1990) Polyomavirinae and their replication. In: Fields BN *et al.* (eds) *Virology*, 2nd edn, vol. 2, pp. 1593. New York: Raven Press.
Salzman NP (ed.) (1986) *The Papovaviridae*, vol. 1: *The Polyomaviruses*. New York: Plenum Press.
Tooze J (ed.) (1980) *Molecular Biology of Tumor Viruses*, 2nd edn, part 2. Cold Spring Harbor, NY: Cold Spring Harbor Laboratory.

POTEXVIRUSES

K Andrew White, Michèle Rouleau, JB Bancroft and George A Mackie
University of Western Ontario
London, Ontario, Canada

Classification and General Characteristics

The potexviruses take their name from their prototype member, potato virus X, and constitute a group of morphologically similar flexuous filamentous viruses. The genetic material of potexviruses is a single-stranded RNA molecule of positive sense (i.e. capable of functioning as an mRNA). The infectious viral particle consists of this RNA encapsidated by 1000–1500 molecules of a single species of capsid protein. Table 1 provides an enumeration of confirmed potexviruses whose RNA sequences have

Table 1. Completely and partially sequenced members of the potexvirus group

Members	Reported particle length (nm)	Coat protein[a] (kD)	gRNA[b] (nucleotides)
Clover yellow mosaic virus (ClYMV)	540	23.5	7015
Foxtail mosaic virus (FoMV)	500	23.7	6151
Lily virus X (LVX)		21.6	
Narcissus mosaic virus (NaMV)	550	26.1	6955
Papaya mosaic virus (PapMV)	530	23.0	6656
Potato aucuba mosaic virus (PoAMV)	580	26	8000[c]
Potato virus X (strain X3) (PVX)	515	25.1	6435
Strawberry mild yellow edge-associated virus (SMYEAV)	482	25.7	6200[c]
White clover mosaic virus (strain M) (WClMV)	480	20.7	5845

[a] Molecular mass determined through sequence analysis of cloned viral cDNA.
[b] Genomic RNA (gRNA) length determined through sequencing of cloned viral cDNA.
[c] RNA length estimated by gel electrophoresis.

been determined. Although individual members of the group generally infect only a limited number of hosts, a large number of both dicotyledonous and monocotyledonous plants can be infected by potexviruses usually causing chlorotic mosaic or mottle symptoms in systemic infections. The geographic range of these viruses is a reflection of the habitat of their hosts. For the most part it would appear that the potexviruses are transmitted mechanically rather than by a vector.

Structure, Composition and Assembly

Individual potexviruses have molecular weights in the order of 35×10^6 and sediment at about 100–130S. Unlike rigid tubular viruses, the particles of potexviruses are highly hydrated and possess a fairly loose quaternary structure. The genomic RNA component (gRNA) ranges in size from 5845 to 7015 nucleotides in the members with known RNA sequence (see Table 1) and in general comprises 6.5–7.5% of the viral mass. The size of the capsid protein varies from 21×10^3 to 26×10^3, depending on the virus. Although different potexvirus coat proteins exhibit limited immunological cross-reactivity, their relatedness as determined by comparisons of their amino acid sequences (inferred from cDNA sequences) varies from 40 to 65% (identical + conserved residues). In fact, there are only ten absolutely conserved amino acids in an average of 230 residues in all known potexvirus coat proteins, a remarkable degree of diversity for functionally homologous macromolecules.

Coat protein monomers from PapMV behave as prolate ellipsoids in solution with dimensions of $3.4 \times 5.5\,nm^2$ and wrap around their cognate viral RNA in a helical fashion, with approximately nine subunits per helical turn. The coat helices of individual viruses differ mainly in the fractional departure from nine subunits per turn and in the size of the true repeating unit. This type of variation is not found among strains of rigid tubular viruses.

Several potexviruses can be reconstituted *in vitro* from their individual purified components. Of these, PapMV has been characterized most extensively. Self-assembly experiments have shown that mixtures of purified coat protein and RNA will combine in buffers of low ionic strength near pH 8.0 to form infectious particles resistant to ribonuclease A. Under these conditions, several (PapMV, ClYMV, PoAMV), but not all, potexviral RNAs are recognized by PapMV coat protein. Assembly occurs in two steps. There is a rapid initiation phase involving interactions between coat protein and 40–50 contiguous nucleotides near the 5′ terminus of the RNA in which particles of about 50 nm in length are formed in a temperature-independent manner between 1–25°C. This probably requires the organization of the coat protein into a 14S protein double disc or helix. The elongation phase in which smaller polymers of coat protein add to the initiated 50 nm particle allowing capsid growth towards the 3′ end of the genomic RNA is slower, entropically driven, and does not proceed at 1°C. In the absence of RNA, the coat proteins of most potexviruses will self-assemble at low pH to form various helical and/or stacked disc structures. Two significant differences in the *in vitro* assembly

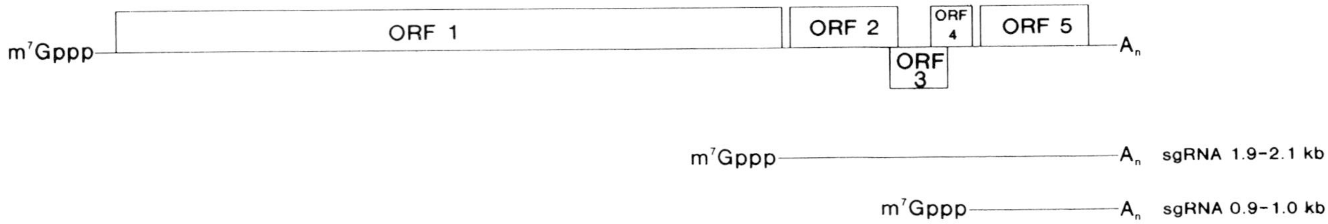

Fig. 1 General organization of a potexvirus genome. The drawing represents the organization of the genome of PVX, the type member of the potexviruses. The size of the proteins encoded by each open reading frame (ORF) is: 166 kD (ORF 1), 25 kD (ORF 2), 12 kD (ORF 3), 8 kD (ORF 4) and 25 kD (ORF 5/coat protein). For the other potexviruses the genomes of which have been sequenced, the sizes of the polypeptide products are: 147–191 kD, 24–26 kD, 11–14 kD, 6–11 kD and 21–26 kD respectively. The two most abundant sgRNAs are also depicted. A_n denotes the poly(A) tail.

of potexviruses compared to the process in tobacco mosaic virus are the unidirectionality of assembly in potexviruses and the absence of an internal initiation sequence. It is possible that a site equivalent to the latter exists for assembly of potexviruses *in vitro* since the subgenomic RNAs (sgRNAs) of some potexviruses are encapsidated.

Genome Structure and Organization

The monopartite genome of potexviruses consists of a single-stranded RNA of messenger sense bearing a cap structure (m^7GpppG) at its 5′ end and a poly(A) tail at its 3′ end. The complete nucleotide sequences of several gRNAs have been determined (Table 1) and have revealed the general organization of the potexviral genome shown in Fig. 1. An A+C-rich 5′-noncoding region varies from 80 to 107 nucleotides and leads into five principal open reading frames (ORFs) which are relatively well conserved in position and sequence. ORF 1, the largest ORF, encodes a protein that contains two amino acid sequence motifs characteristic of the conserved domains of NTP-binding helicases and RNA-dependent RNA polymerases. It is likely that this polypeptide is a component of a replicase complex which along with host factors (and possibly other viral proteins) directs the synthesis of viral RNAs. ORFs 2, 3 and 4 slightly overlap each other, constituting the 'triple gene block'. The products of these three ORFs are all necessary for infectivity in the plant host, being required for cell-to-cell movement, but are dispensible for infection of protoplasts. The predicted amino acid sequence of the product of ORF 2 suggests that it, too, possesses a helicase function. Hydrophobic segments in the products of ORF 3 and ORF 4 suggest their association with membranous structures of host cells. SMYEAV and LVX constitute two apparent exceptions to the organization of ORFs 2/3/4. The genome of SMYEAV lacks the initiation codon for ORF 2 while LVX lacks that for ORF 4. In both cases, however, the remainder of the ORF is present and is similar to that of the other potexviruses. At this time, it is unknown whether these potexviruses utilize different strategies for expressing these ORFs or are fully viable without them. ORF 5 encodes the coat protein, which has been described in the previous section. In two potexviruses, ClYMV and FoMV, ORF 5 is located within a larger ORF capable of encoding a 'readthrough' protein containing coat protein antigenic determinants. This product has only been demonstrated *in vitro*. Additional ORFs are present in some potexviral genomes embedded in either ORF 1 or ORF 5, but in a different reading frame. A 10 kD product is encoded by an ORF present within ORF 1 of FoMV and strain M of WClMV while ClYMV, PapMV and FoMV ORF 1 also contain an ORF coding for a 14–26 kD product. ORF 5 of both strains M and O of WClMV as well as of NaMV contains a smaller ORF, encoding a product of 7–10 kD. The significance of these embedded ORFs, if any, is unknown. The 3′-noncoding region varies from 43 to 138 nucleotides and precedes a poly(A) tail of variable length. A polyadenylation consensus sequence, 5′-...AAUAAA..., is present 30–60 nucleotides from the 3′ end of most sequenced potexviral RNAs.

sgRNAs are produced during potexviral infections. Like the gRNA, the sgRNAs that have been characterized are capped at their 5′ end (e.g. ClYMV) and poly(A) tailed at their 3′ end [e.g. PVX and daphne virus X (DVX)]. The 5′ end of these sgRNAs corresponds to internal regions of the gRNA and their 3′ ends are coterminal with the gRNA (Fig. 1). The two most abundant sgRNAs in infected tissues are 1.9–2.1 kb and 0.9–1.0 kb in length. Other sgRNAs of intermediate size and lower abundance have also been detected in tissues infected with either PVX or DVX. One unconventional sgRNA species, of 1.2 kb, has been found in some infections with ClYMV and can also be recovered from encapsidated viral particles. The 1.2 kb RNA has been molecularly cloned and consists of 757 residues of the 5′ end of the gRNA of ClYMV and 415 residues from the 3′ terminus joined so that an N-terminal portion of ORF 1 is fused in frame to a C-terminal portion of ORF 5. While this RNA is clearly defective, requiring a helper for its propagation, it does not alter symptom development. No other examples of defective or satellite RNAs are known within the potexviruses.

The organization of the potexvirus genome is related to that of carlaviruses as are similarities among products encoded by the various ORFs. Moreover, the products of ORFs 2 and 3 have their counterparts in sequence in the bipartite genome of furoviruses and in the tripartite genome of hordeiviruses.

Replicative Strategy and Molecular Biology

The majority of data that have contributed to the understanding of the molecular biology of the potexviruses has come from cloning and sequence determination of viral RNAs, from the analysis of viral RNAs in infected tissue, from *in vitro* translation of native and synthetic viral RNAs, and from examination of the properties of infectious synthetic transcripts. The potexviruses are believed to replicate via an RNA intermediate in which the positive-sense gRNA is first copied to produce a negative-sense RNA complement which in turn would serve as a template for synthesis of additional positive-sense RNA. The identification of double-stranded gRNA intermediates from potexvirus-infected plants supports this notion. The negative-sense RNA would, in this model, also serve as template for transcription of smaller sgRNAs which are 3′-coterminal partial copies of the gRNA (Fig. 1). The sgRNAs act as mRNAs for products encoded in the 3′-half of the gRNA which otherwise would be poorly expressed from the polycistronic genomic mRNA. With the possible exception of PVX, no sgRNA-length double-stranded RNAs have been detected in tissue from potexvirus-infected plants, supporting the notion that the full-length negative-strand must serve as the template for production of sgRNAs through internal initiation.

Several conserved sequence motifs have been found in the gRNA of most sequenced potexviruses and provide clues to the mechanism of replication. Putative promoter elements have been identified at the 5′ and 3′ ends of the gRNA as well as at internal regions immediately 5′ to the initiation site of sgRNA synthesis. The 5′ termini of all potexvirus gRNAs sequenced to date conform closely to the sequence 5′-GAAAACAAAAC..., while the six residue consensus, 5-′...ACUUAA..., occurs near their 3′ ends. These sequences may be involved in the transcription of full-length viral RNA by acting as part of a promoter element recognized by a replicase complex. An additional conserved consensus sequence (5′-...GGUUAA...) has also been identified in the gRNA of potexviruses 5′ to the site of initiation of sgRNA synthesis. This highly conserved region may represent a core element of the sgRNA promoters for the two most abundant sgRNAs. Moreover, mutations in this region of the gRNA of PVX inhibit accumulation of coat protein sgRNA.

As the RNA component of potexviruses is of the messenger sense, it can be translated by the cellular protein biosynthetic apparatus immediately following uncoating. *In vitro* translation of potexvirus gRNAs extracted from virions directs the synthesis of products with relative molecular weights corresponding to the putative replicase protein (ORF 1) encoded at their 5′ ends. In contrast, only variable amounts of coat protein can be synthesized *in vitro* from a gRNA template from some, but not all, potexviruses. This probably reflects variable degrees of *bona fide* internal initiation *in vitro* as well as fragmentation of the template. The natural template for synthesis of coat protein *in vitro* is the smaller of the two abundant sgRNAs. Indeed, efficient synthesis of coat protein *in vitro* is observed only from translation of the coat protein subgenomic mRNA. The 5′ terminus of a second abundant sgRNA (2.1 kb) maps just 5′ to ORF 2 in ClYMV, and would permit the expression of the ORF 2 product. A synthetic '2.1 kb' sgRNA transcribed from the appropriate regions of a cloned PVX cDNA can be translated *in vitro* to yield only the ORF 2 product. The mechanism of expression of the products of ORFs 3 and 4 is less clear. Additional sgRNAs, which are less abundant than the ORF 2 and coat protein subgenomic mRNAs, have been found associated with PVX and DVX. A synthetic tricistronic RNA encoding ORF 3, ORF 4 and ORF 5 of PVX, corresponding to a 1.4 kb sgRNA identified *in vivo*, can direct the synthesis of the products of both ORF 3 and ORF 4 *in vitro*. This indicates that the 1.4 kb sgRNA may serve as a bifunctional message *in vivo*. The identity of its promoter remains speculative.

The functions of the products of ORFs 2, 3 and 4 have recently been elucidated using full-length synthetic gRNAs of PVX and WClMV transcribed *in vitro* from their cloned cDNAs. RNAs containing mutations inactivating ORF 2 of PVX and WClMV, or ORF 3 or ORF 4 of WClMV, were unable to induce symptoms or spread in inoculated whole plants but were able to replicate in protoplasts. These results indicate that each of the triple-gene block proteins plays some role, not necessarily direct, in the cell-to-cell spread of the virus. Additional mutations targeted to the conserved motifs in the ORF 2 (NTPase helicase domain) and ORF 3 (hydrophobic domain) of WClMV also inactivated the virus' ability to spread *in planta*. Of the triple-gene block products, only the ORF 2 product (25 kD protein) of PVX has been localized *in planta*. It was found associated with complex lamellar structures in the cytoplasm and possibly in the nucleus of infected tobacco cells. An intact ORF 5 (coat protein) is essential for infection of tobacco and other susceptible plants with PVX.

Modification of synthetic transcripts of WClMV has also been a useful approach in determining what structural features of the gRNA are necessary for infectivity. RNA transcripts of WClMV lacking a 5′-cap structure were only 4% as infectious as capped messages. The cap structure is likely present in all the RNAs of potexviruses (including sgRNAs) and, therefore, appears to be critical for infectivity. Poly(A) tails on eucaryotic mRNAs can influence message stability and/or translation of the message. Shortening the length of the poly(A) tail of *in vitro* transcripts of WClMV decreased its infectivity. Since a portion of the poly(A) tail is apparently copied during negative-strand synthesis of potexviruses, the tail may be a necessary structure for RNA replication in addition to its conventional role. Mutation of the putative polyadenylation signal in WClMV (also found in most sequenced potexviruses) reduced the infectivity of the transcript and shortened the average length of the poly(A) tail in the progeny virus.

Potexviruses display some promise as vectors for the transient expression of nonviral genes in plants by inserting a gene of interest into a region containing the signals for sgRNA expression in a modified virus. It is too early, however, to evaluate whether potexvirus genomes exhibit sufficient plasticity to permit this strategy to succeed.

Cytopathology

Cytopathology that develops as a result of infection with different potexviruses is quite variable and ranges from undetectable to moderate. This is also true for infection of different species with one type of potexvirus. For example, infection of tobacco with PVX results in mottling or necrotic spotting, whereas infections of potato or tomato cause a mild mosaic or mosaic with slight stunting respectively. The most prevalent systemic symptoms induced by potexviruses are chlorotic mottle or mosaic patterns, but a small percentage of systemically infected hosts displays no symptoms. Stunting is also common in infected plants. Primary and systemic cytopathology are not always mutually exclusive, as both can be manifested in the same infected host. Most potexviruses also have local lesion hosts, *Gomphrena globosa* being the most common. The cytopathology induced by a potexvirus may also be altered by simultaneous infection with an additional virus. Mixed infections of PVX and potato virus Y (a potyvirus unrelated to PVX), for example, result in an increase in the severity of pathology over those observed for either virus alone.

Potexviruses can infect the majority of host tissues. Apart from tissue effects generally associated with mottling and cell death from hypersensitive reactions, the potexviruses can induce the formation of specific structures in infected cells. Potexvirus particles, packed in parallel arrays, form large aggregates termed 'inclusion bodies'. These inclusions are most commonly found in the cytoplasm of infected cells, but have also been observed in the nuclei, during NaMV and cactus virus X infections. Inclusion bodies vary in size and are generally irregularly shaped, with the exception of those induced by NaMV, tulip virus X and cactus virus X which are spindle-shaped. The laminated inclusion components (LIC) constitute a second type of inclusion, uniquely observed during PVX infections. This cytoplasmic structure is formed of thin proteinaceous sheets which are often associated with bead-like structures. LICs do not contain any viral antigen and their content and role remain obscure. A third type of inclusion, amorphous inclusions, has been observed during infection by ClYMV and discorea latent virus. These inclusions are found in the cytoplasm and vacuoles of infected cells and contain viral antigen. Their function is unknown.

Resistance and Protection

Resistance of the host to viral infection is dependent on both plant and virus strains. Three genes in various potato cultivars have been associated with viral resistance, either by hypersensitive response (*Nx* and *Nb*) or by immunity (*Rx*). Strains of PVX have correspondingly been classified into five groups (I, II, III, IV and HB) according to their ability to infect systemically a cultivar of defined genotype. Most common strains of PVX belong to group III (capable of systemic infection of an *Nb* host). It is likely that two of the three PVX strains for which complete sequences are known (the X3 and Russian strains) represent group III viruses. The nucleotide sequences of the gRNAs of these two isolates are 97.1% identical. In contrast, the nucleotide sequence of a South American isolate of PVX classified in group II differs significantly from that of the X3 strain, being only 77.6% identical, although many nucleotide changes are conservative. Thus a direct correlation between the nucleotide sequence of a particular viral strain and host susceptibility cannot be drawn easily. Nonetheless, ORF 5 and its product, the coat protein, are believed to carry the virulence determinant(s) of PVX which interact(s) with the potato *Nx* and *Rx* genes based on the behavior of chimeric viruses. Interestingly, expression of PVX coat protein in a transgenic tobacco plant can confer a delay in the onset of symptoms to subsequent infection with either PVX or PVX RNA. It is not clear whether it is the coat protein itself and/or its mRNA that confers protection.

In tobacco cultivars, the *N* gene confers resistance to viral infection. A protein factor has been identified in the sap of cultivars containing the *N* gene, termed 'inhibitor of virus replication'. This factor appears to play a role in resistance of tobacco cultivars to infection by various viruses, including PVX.

Economic Significance

Certain potexviruses can damage some crops although others possess little economic significance. Yield losses for cultivated potatoes infected with PVX are estimated at 10–20%. The availability of PVX-immune strains of potato (e.g. *Rx* cultivars) may provide partial means of protection against infection. Conversion of PVX to a resistance-breaking variant can occur with detectable frequency, however. Crop losses resulting from infection with cassava common mosaic virus can attain 30% or more. This disease may be circumvented by selecting healthy cuttings for propagation, and by roguing (culling infected plants). Potexviruses which infect bulb crops (e.g. NaMV) can reduce the longevity of the bulb and thus the value of their crops. In some cases, the presence of the virus may contribute to a desirable, but unstable, trait such as streaking.

See also: Plant resistance to viruses; Plant virus disease – economic aspects.

Further Reading

AbouHaidar MG and Erickson JW (1985) Structure and *in vitro* assembly of papaya mosaic virus. In: Davies JW (ed.) *Molecular Plant Virology*, Vol. I. Boca Raton, Fl: CRC Press.

Commonwealth Mycological Institute/Association of Applied Biologists (CMI/AAB) Descriptions of Plant Viruses, No. 200, August 1978 (Potexvirus Group) and additional numbers which describe individual potexviruses [e.g. No. 354, December 1989 (potato virus X)].

Lesemann DE and Koenig R (1977) Potexvirus (Potato virus X) Group. In: Maramorosch K (ed.) *The Atlas of Insect and Plant Viruses*. New York: Academic Press.

POTYVIRUSES

John A Lindbo and William G Dougherty
Oregon State University
Corvallis, Oregon, USA

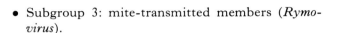

Introduction

The potato virus Y group, or *Potyviridae*, represents a large collection of plant viral pathogens. Members of this group are typically flexuous rod-shaped viruses of *c*. 700–900 nm in length and 12–15 nm in diameter. Virion particles are composed of single-stranded RNA of *c*. 9600 nucleotides, encapsidated in a helical fashion by *c*. 2000 copies of a coat protein monomer of about 30 kD. The exception is the bicomponent bymovirus subgroup. These viruses have a bipartite genome and typically have particle lengths of 200–300 nm and 500–600 nm.

All potyviridae induce the formation of characteristic cylindrical or pinwheel-shaped viral protein inclusion bodies in the cytoplasm of infected plant cells. Infections by certain group members also result in the formation of nuclear inclusion and/or amorphous cytoplasmic inclusion bodies. Most members of the potyviridae are transmitted by aphids in a nonpersistent manner, although mite, fungal, whitefly and seed transmission have been reported for a few members. Altogether, the *Potyviridae* contains more than 100 members.

Taxonomy and Classification

Several taxonomic methods have been proposed for the *Potyviridae* based on host range, ability to cross-protect, inclusion-body morphology, coat protein amino acid sequence homology, nucleotide sequence homology or serological relationships. However, with the limited number of *Potyviridae* examined and the continuum of variation observed, it has been difficult to arrive at a consensus which adequately distinguishes a 'strain' from a 'virus'. As such, controversy still remains regarding their taxonomic status. It has been suggested that the *Potyviridae* be divided into three subgroups:

- Subgroup 1: aphid-transmitted members (*Potyvirus*).
- Subgroup 2: fungus-transmitted members (*Bymovirus*).
- Subgroup 3: mite-transmitted members (*Rymovirus*).

This discussion will focus on the aphid-transmitted subgroup, the potyviruses.

Economic Importance

Several factors contribute to potyviruses being an economically important virus group. These factors include: (1) although the host range of most potyviruses is narrow, there is virtually no agricultural crop which one or more potyviruses cannot infect. (2) Symptoms on infected plants are often quite visible. Vein clearing, leaf chlorosis, necrosis, leaf distortion, stunting and distorted fruits and seeds can result in considerable crop losses in yield and/or quality. (3) Potyviruses are readily transmitted from infected to healthy plants. Aphid-transmitted potyviruses are transmitted in a nonpersistent noncirculative (stylet-borne) manner. Aphids can acquire and transmit virus with short feeding probes lasting only a few seconds. (4) Potyviruses are found in a wide range of environments. They are especially prevalent in subtropical or tropical regions where virus can overwinter in weeds, alternate crops or volunteer plants, although some potyviruses may also enter a field via contaminated seed. Tropical, subtropical and temperate climates also support insect populations, which may vector the virus.

The economic impact of the group as a whole is difficult to estimate. The extent of crop loss depends upon a complex interaction of growing conditions, cultivars, virus strains, cultural practices, time and extent of primary infection in the field, etc. Chronic losses due to viruses may often go undetected, but epidemics can cause complete crop loss. Financial investments in virus control measures such as resistance breeding programs and seed certification programs also represent one aspect of the economic impact of potyviruses.

Genome Structure

The following discussion on genome organization

(a)

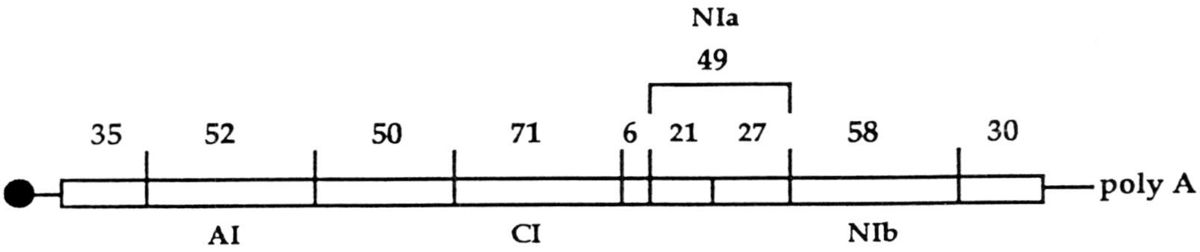

(b)

TEV GENE PRODUCT		PREDICTED AND DETERMINED FUNCTION
	35 kD	Proteolytic processing , cell-to-cell movement (N-PRO proteinase)
	52 kD	Proteolytic processing, insect transmission (helper component, HC-PRO proteinase, AI protein)
	50 kD	??
	71 kD	Replication, cell-to-cell movement (helicase, ATPase, CI protein)
	6 kD	Replication, membrane binding protein
49 kD (NIa)	21 kD	Replication (VPg)
	27 kD	Polyprotein processing, replication (NIa proteinase)
	58 kD	Replication; putative RNA-dependent RNA-polymerase (NIb protein)
	30 kD	Encapsidation and insect transmission (capsid protein, insect transmission)

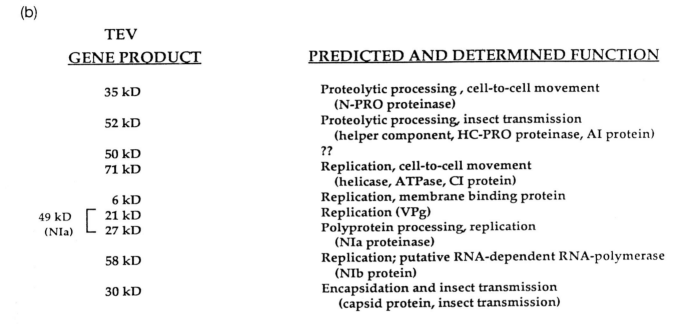

Fig. 1 Genome organization and protein products of the potyvirus tobacco etch virus (TEV). Sizes of proteins are deduced from predicted amino acid sequence derived from the TEV genomic nucleotide sequence. Other potyviruses have similar, but not identically sized, protein products. In the genome map (**a**), the solid circle represents the VPg, the thin line represents the TEV genome 5′ and 3′ untranslated sequences and the long open rectangular box denotes the open reading frame. Vertical lines are used to show locations of proteolytic cleavage of the polyprotein. Molecular masses (in kD) of the various processed TEV gene products are presented above the diagram. Proteins that aggregate to form the various inclusion bodies during infection are identified: AI and CI represent the proteins that form the amorphous cytoplasmic inclusion and cylindrical inclusion bodies respectively. The NIa polyprotein and NIb protein together make up the nuclear inclusion body co-crystal. (**b**) Predicted protein function and experimentally determined activities (in parentheses) are listed in the right hand column. The symbol, ??, denotes a protein for which no function is known. The 49 kD NIa polyprotein is processed into 21 and 27 kD proteins. The proteolytic activity is associated with the 27 kD protein, and either the 21 or the 49 kD protein can function as the VPg moiety.

and expression is based largely on experimentation with the potyviruses tobacco etch virus (TEV), tobacco vein mottling virus (TVMV) and plum pox virus (PPV). We have used the TEV system to illustrate proteolytic processing and gene expression. In the following discussion, protein sizes mentioned are for TEV; other potyviruses will not necessarily have identically sized proteins.

The RNA genome of potyviruses is a single-stranded message-sense RNA of about 9600 nucleotides. It is organized as a single large open reading frame (ORF), encompassing about 95% (>9100 nucleotides) of the RNA molecule. A viral-encoded protein (VPg) is covalently linked to the 5′-terminal nucleotide and a polyadenylate sequence is present at the 3′ terminus. At least nine gene products are encoded in this large ORF and are initially expressed as a large polyprotein which is proteolytically processed into mature viral protein products. The various TEV gene products and their identi-

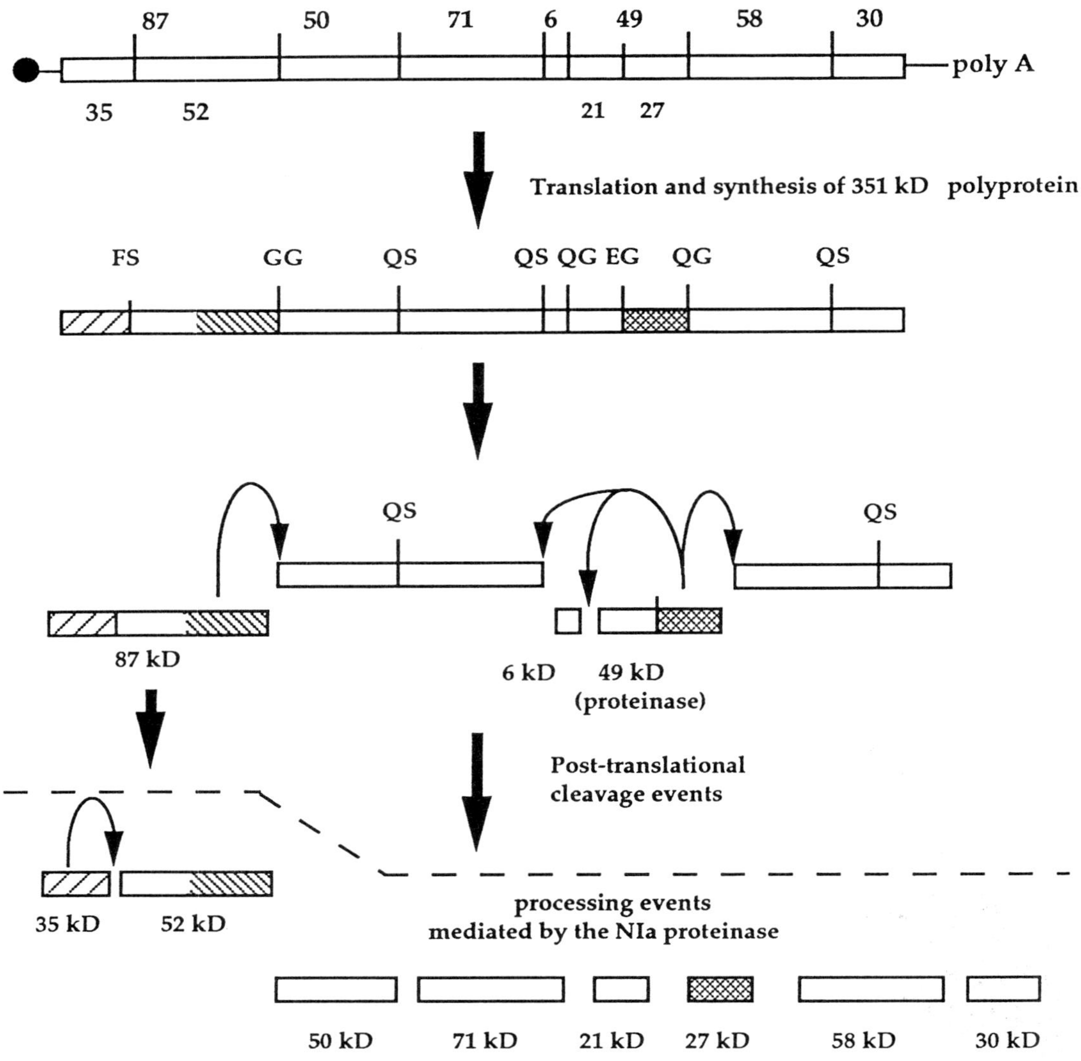

Fig. 2 Putative proteolytic processing scheme of the potyvirus tobacco etch virus (TEV). Shown is a map of the TEV genome (see Fig. 1 for details), its translation products and a proposed scheme of proteolytic processing. Proteolytic activities have been mapped to three proteins in TEV: (1) the 35 kD protein, (2) the C-terminal half of the 52 kD HC-PRO and (3) the 27 kD NIa protein. Those regions with identified proteolytic activities are shaded differently. Arrows indicate the proteinase responsible for cleavage at the dipeptide sequences (single letter amino acid code) indicated. We suggest that a series of autocatalytic events occur rapidly (above dotted line in diagram) and possibly cotranslationally. Subsequent cleavage reactions (below the dotted line) may be mono- or bimolecular. The 49 kD protein is processed at a Glu/Gly dipeptide to form a 21 and 27 kD sized protein. The 27 kD portion retains proteolytic activity.

ties/functions are presented in Fig. 1 and discussed below.

Gene Expression

Three distinct proteolytic activities have been associated with TEV gene products. The first identified proteolytic activity was ascribed to the TEV 49 kD NIa polyprotein, one of the components of the nuclear inclusion body. The 49 kD polyprotein is further processed into two proteins of *c.* 21 and 27 kD. The N-terminal 21 kD of the NIa polyprotein is the VPg of the virus. The proteolytic activity of the NIa polyprotein resides in the C-terminal 27 kD portion. The NIa proteinase of TEV has been identified as a trypsin-like cysteine proteinase similar to the 3C proteinase of picornaviruses. The proteolytic specificities associated with the 27 and 49 kD proteins appear to be identical in cell-free studies. It is uncertain whether the 49 or 27 kD form of this proteinase is responsible for *in vivo* cleavage of the polyprotein.

The TEV NIa proteinase cleaves at a highly conserved site which spans the seven amino acid sequence:

Glu–Xaa–Leu–Tyr–Xaa–Gln/Gly
Ile Glu Ser
Val

Cleavage occurs between the Gln (Glu) and Gly (Ser) residues and the positions represented by Xaa are occupied by neutral or hydrophobic amino acids. In cell-free studies, different amino acids in the Xaa positions affect the rate of cleavage. Cleavage sites composed of strictly and loosely conserved amino acids have been found in other potyviruses also.

The two other proteolytic activities described for TEV have only been demonstrated to function in a monomolecular fashion (i.e. *cis*). The 52 kD protein, referred to as HC-PRO, is a papain-like cysteine proteinase which cleaves at its C terminus to release itself from the polyprotein. Cleavage is between a Gly/Gly dipeptide. The most recently described proteolytic activity involves the 35 kD protein (N-PRO) of TEV. This proteinase, located at the N terminus of the genome-derived polyprotein, also cleaves at its C terminus to release itself from the polyprotein. Cleavage is between a Phe/Ser dipeptide.

Gene Regulation

It has been proposed that potyviruses may regulate their gene expression at a post-translational level via differential processing rates at various cleavage sites. Potyviruses may also use differential protein stabilities and/or the shunting of proteins into inclusion bodies as a means of regulating gene activity in a post-translational manner. A proposed pathway of proteolytic processing in potyviruses is presented in schematic form in Fig. 2.

Replication

Potyviruses have been placed in the super group of 'picorna-like' viruses because of their genome structure and strategy of gene expression. As such, it is likely that potyvirus replication will be similar to picornavirus replication.

Replication of picornaviruses occurs in the cytoplasm of infected cells and involves a membrane-bound replicase complex. The picornaviral proteins 2B, 2C, 3A, 3B, 3C and 3D appear to be involved in replication. Sequence homology studies have revealed regions of homology between picornavirus 2C, 3C and 3D proteins and the 71, 27 and 58 kD proteins of TEV. Structural and functional similarities have also been observed between picorna- and potyvirus proteins. On a functional level, the picornaviral 2BC protein and potyviral 71 kD protein both induce the formation of smooth-walled vesicles in the cytoplasms of infected cells. The picornaviral 3A and TEV 6 kD proteins do not show sequence homology, although they appear to be biochemically similar in that both are small proteins with hydrophobic core sequences. The 3A protein is a component of the picornavirus replicase complex. The 3B and 21 kD proteins both function as VPgs (although in a portion of TEV virions, a 49 kD polyprotein is linked to the TEV genome). The 3C and 27 kD proteins both have proteolytic activities, and the 3D and 58 kD proteins both have amino acid sequences conserved among RNA-dependent RNA polymerases. Helicase and ATPase activities have recently been demonstrated for the PPV 71 kD protein *in vitro*, implicating this protein in viral replication. A completely functional 'replicase complex' has not been isolated from potyvirus-infected tissue.

Virus–Host Relationship

The formation of large distinctive inclusion bodies is a hallmark of potyvirus-infected tissue. Potyvirus-infected cells contain cylindrical or pinwheel-shaped cytoplasmic inclusion (CI) bodies. Initially the CIs are associated with the infected cell membrane and later are found free in the cytoplasm. The CI protein has helicase and ATPase activities. These cytological and biochemical observations have led to speculation that the CI protein may function both in viral replication and cell-to-cell movement. Sequence homology studies suggest that the 35 kD proteinase may also be involved in movement of virus from cell to cell within a plant. Limited sequence homology has been detected between the TEV 35 kD protein and the TMV 30 kD movement protein.

A few potyviruses also form nuclear inclusion (NI) and/or amorphous cytoplasmic inclusion (AI) bodies. NI bodies are distinctively shaped crystals composed of two viral-encoded proteins, the 49 kD polyprotein (21 kD VPg and 27 kD proteinase) and the putative RNA-dependent RNA polymerase of *c.* 58 kD. The AI, when formed, is an aggregate of the HC-PRO protein. In addition to its proteolytic function, HC-PRO is also required for aphid transmission. The relationship between these two functions is not understood. All potyvirus-induced inclusions can be readily viewed in stained epidermal strips by light microscopy. Purification schemes to obtain inclusion bodies also have been established.

On a whole plant level, systemic symptoms usually appear within 6–8 days postinoculation (mechanical or aphid-vectored inoculum). Symptoms on systemically infected leaves often include mosaic and mottling, vein clearing, chlorosis, streaking, necrosis, leaf distortion, etc.

Tens of milligrams of virions can usually be obtained from a kilogram of infected plant tissue. Preparations of purified virus have been used to generate antibodies to potyvirus coat proteins for use in serological based detection schemes (enzyme-linked immunosorbent assay, etc.). Recent serological and structural analysis of potyvirus coat proteins has revealed that the N and C termini of the coat protein reside on the exterior of the virion. These termini greatly differ in both sequence and length among potyviruses. The internal amino acids of the coat protein, however, are relatively conserved between potyviruses. Antibodies directed toward the coat protein N or C terminus can be virus specific, whereas antibodies directed toward the internal conserved amino acid residues are usually cross-reactive with different potyviruses.

Transmission

Potyviruses are transmitted by aphids in a nonpersistent noncirculative (stylet-borne) manner. Aphids can acquire virus by probing infected tissue for only a few seconds, retain the ability to transmit virus for several hours, and can transmit virus to healthy plants with short probes lasting less than one minute. Because aphids do not retain the ability to transmit virus for extended periods of time, virus usually comes from nearby inoculum sources. A typical potyvirus is often transmissible by several different aphid species. Given the promiscuous feeding behavior of aphids and the characteristics of stylet-borne transmission, potyviruses can be spread in a field by aphids simply passing through, and not colonizing, a crop.

Aphid transmission of potyviruses involves at least two viral-encoded proteins: (1) the 'helper component' (the 52 kD HC-PRO) and (2) the viral capsid protein. Nucleotide sequence comparisons between nonaphid and aphid-transmissible virus isolates have been used to predict which amino acids in these two proteins may be important for aphid transmissibility. Infectious transcripts from a full-length cDNA clone of the potyvirus TVMV have been used to confirm the importance of these specific amino acids in both the capsid protein and HC-PRO in aphid transmissibility.

Control Schemes

In addition to cultural practices to remove primary sources of inoculum, seed certification and traditional breeding programs have been implemented to control potyviruses. Resistance phenotypes are often inherited in a recessive manner and are therefore difficult to identify. Also, resistant germplasm is often not available for every crop and viruses rapidly mutate to overcome deployed resistance genes.

Virus-resistant plants have been genetically engineered. This strategy involves deriving resistance genes from the virus itself. It has been demonstrated that transgenic host plants which express the coat protein nucleotide sequence of a particular plant virus are resistant to subsequent infection by that virus. Although the mechanism of this resistance is not entirely understood, this approach has been applied to a wide variety of plant viruses. Recently, transgenic plants expressing potyvirus coat protein RNA, but not protein, have been generated. Such plants are often immune to virus infection.

Future Perspectives

This discussion has briefly touched on eight of the nine identified gene products encoded by the TEV genome. Activities and roles in the virion replicative cycle have been described, but do not preclude other functions associated with partially processed polyproteins containing these gene products. A putative function(s) for the TEV 50 kD protein product has not been proposed.

Future research directions will most likely involve studies on the replication, movement and assembly of potyviruses and further investigation of the newly described proteolytic activities. The tools of biochemistry and molecular biology will greatly assist these efforts. For example, replication studies would benefit from the isolation or reconstitution of an active replicase complex; transgenic plants and infectious transcripts may be used to determine the proteins involved in virus movement; *in vivo* and cell-free studies may reveal the RNA and amino acid sequences necessary for potyvirus assembly; and transgenic plants, protoplasts and cell-free studies may clarify the specificities of the recently described proteolytic activities and their role in the potyvirus life cycle. A better understanding of all aspects of potyvirus biology should lead to novel virus control strategies designed to interfere with a specific step in the potyvirus replication cycle.

See also: Comoviruses; Nepoviruses; Plant virus disease – economic aspects; Polioviruses.

Further Reading

Barnett OW (ed.) (1992) Potyvirus taxonomy. *Arch. Viol.* supp. 5, Vienna: Springer Verlag.

Dougherty WG and Carrington JC (1988) Expression and function of potyviral gene products. *Annu. Rev. Phytopathol.* 26: 123.

Edwardson JR and Christie RJ (1991) *The Potyviruses*. Vols 1–4. Florida Agriculture Experimental Station Monograph no. 16.

Milne RG (ed.) (1988) *The Plant Viruses*: Vol. 4, *The Filamentous Plant Viruses*. New York: Plenum Press.

Pirone TP (1991) Viral genes and gene products that determine insect transmissibility. *Semin. Virol.* 2: 81.

Shukla DD and Ward CW (1989) Structure of potyvirus coat proteins and its application in the taxonomy of the potyvirus group. *Adv. Virus Res.* 36: 273.

POXVIRUSES

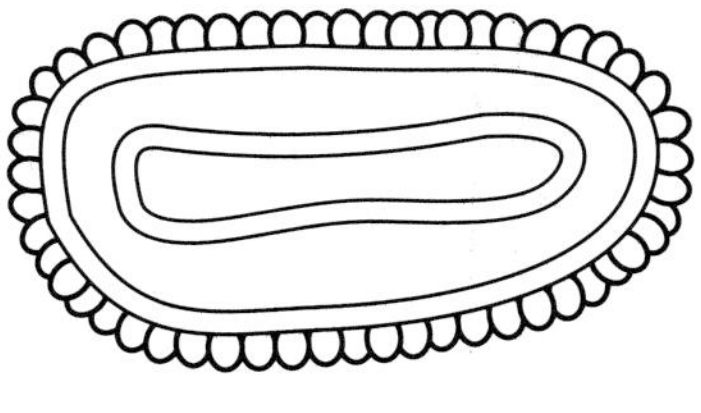

Contents

Rabbit, Hare, Squirrel and Swine Poxviruses
Sheep and Goat Poxviruses

Rabbit, Hare, Squirrel and Swine Poxviruses

Grant McFadden
University of Alberta, Edmonton
Edmonton, Alberta, Canada

History

Poxviruses of leporids and swine cause a broad range of symptoms varying from mild lesions of the skin right up to lethal systemic diseases (Table 1). The agent of myxomatosis, a virulent disease of domestic rabbits described originally by G. Sanarelli in 1896, was the first viral pathogen discovered for a laboratory animal. The close similarity of myxoma virus (MYX) with other members of the poxvirus family, such as variola and fowlpox, was first recognized by Aragão in 1927. MYX is notable because, although it causes rather benign lesions in the native *Sylvilagus* rabbit (the brush rabbit in North America and the tropical forest rabbit in South America) when introduced to the European (*Oryctolagus*) rabbit it causes an invasive disease syndrome with up to 100% mortality. MYX was the first viral agent ever introduced into the wild for the purpose of eradicating a vertebrate pest, namely the feral European rabbit population in Australia in 1950 and, 2 years later, in Europe. The resulting genetic selection of virus isolates with lesser pathogenicity and upsurgence of rabbits with greater resistance to the viral disease was studied intensively by Frank Fenner and his colleagues as a model system to investigate the ecological consequences of virus/host evolution in an outbred population.

Table 1. Members of the *Leporipoxvirus* and *Suipoxvirus* genera

Member	Abbreviation	Natural host	Major arthropod vector	Natural host disease	Disease in domesticated European rabbit (*Oryctolagus cuniculus*)
Leporipoxvirus					
Myxoma	MYX	California brush rabbit[a] S. American tapeti[b] (*Sylvilagus* sp.)	Mosquito flea	Localized benign fibroma	Systemic lethal myxomatosis
Rabbit fibroma (Shope fibroma)	SFV	N. American cottontail rabbit (*Sylvilagus floridans*)	Mosquito flea	Localized benign fibroma	Localized benign fibroma
Malignant rabbit fibroma[c]	MRV	Laboratory rabbit[d] (*Oryctolagus cuniculus*)	–	Not observed in wild	Systemic lethal syndrome similar to myxomatosis
Squirrel fibroma	SqFV	Gray squirrel (*Sciurus* sp.)	Probably mosquito	Localized or multiple fibromas	Occasional nodular dermal lesions
Hare fibroma	HFV	Wild hares (*Lepus* sp.)	Probably mosquito	Localized benign fibroma	Localized benign fibroma
Suipoxvirus					
Swinepox	SPV	Domestic pigs (*Suidae* sp.)	Hog lice	Localized cutaneous lesions	Intradermal lesions but no serial propagation

[a]Also called Marshall-Regnery myxoma.
[b]Also called Aragão's (or Brazilian) myxoma.
[c]Laboratory recombinant between MYX and SFV.
[d]MRV has been propagated only by serial inoculation of laboratory rabbits and in cultured cells.

Also of interest to the history of animal virology is that the first DNA virus associated with transmissible tumors was Shope fibroma virus (SFV), described in 1932 by Richard Shope as an infectious agent of fibroma-like hyperplasia in cottontail rabbits (*Sylvilagus floridanus*) in the Eastern USA. It is likely that the agent of 'hare sarcoma', described first in Germany in 1909, was also a poxvirus, now called hare fibroma virus (HFV). HFV remains the only leporipoxvirus to have arisen outside the Americas but its biology closely resembles that of SFV.

Very little is known about the remaining leporipoxviruses. Subcutaneous fibromatosis in gray squirrels of the Eastern USA and western gray squirrels in California, caused by poxvirus now collectively called squirrel fibroma virus (SqFV), have been observed since 1936, but their rigorous classification within the MYX-SFV family was not made until 1951 by L. Kilham. Similarly, HFV, described first in 1959 in the European hare (*Lepus europaeus*), was also shown to be a closely related poxvirus in 1961. In 1983, an outbreak of a disease resembling myxomatosis in laboratory rabbits in San Diego was caused by a novel leporipoxvirus later shown to be a genetic recombinant between SFV and a still-undefined strain of MYX. This virus, called malignant rabbit fibroma virus (MRV), has never been observed in wild rabbit populations but is of interest as an experimental model for poxvirus-induced immunosuppression and tumorigenesis.

Based on landmark experiments with pneumococcus in the 1920s, the very first example of what was believed to be genetic interaction between viruses was reported in 1936 with the discovery that heat-inactivated myxoma could be reactivated with live SFV ('Berry-Dedrick' transformation), but later work showed this to be a genome rescue phenomenon rather than true recombination.

The only known member of the *Suipoxvirus* genus, swinepox virus (SPV), has been observed sporadically in pig populations throughout the world, but is not considered a serious pathogen because infected animals usually have only moderate symptoms and recover completely.

Taxonomy and Classification

The prefix lepori- comes from Latin *lepus* or *leporis* ('hare') and sui- from Latin *sus* ('swine'), to denote the relatively restricted host range of these viruses. All of the leporipoxviruses can be shown to be closely related to each other by serology, immunodiffusion and fluorescent antibody tests although antigenic differences can be detected in strains of MYX. SPV is antigenically unique and is not known to have any closely related members. In terms of broad features, all are typical poxviruses, with characteristic brick-shaped virions containing a double-stranded DNA genome with covalently closed hairpin termini and terminal inverted repeat (TIR) sequences. Like other poxviruses, viral macromolecular synthesis takes place exclusively in the cytoplasm of infected cells.

Properties of the Virion

Like all members of the poxvirus family, the virions have a characteristic brick-shaped morphology with dimensions of approximately 250–300 nm × 250 nm × 200 nm. The leporipoxviruses are uniquely sensitive to ether and chloroform but otherwise the virions are very stable at ambient temperatures and in skin lesions. In all other respects, such as chemical composition and physical properties, the virus particles are very similar to those of vaccinia.

Properties of the Viral DNA and Protein

Detailed information about the viral DNA is available only for SFV, MYX/MRV and SPV. The leporipoxviruses have double-stranded DNA genomes of 160–163 kb, with hairpin termini and TIR sequences of 10–13 kb. SPV DNA is somewhat larger (175 kb) and the TIR is only 4–5 kb, but otherwise the genome has similar characteristics. It is believed that each virus encodes over 200 genes. Restriction cleavage maps have been deduced for these viral genomes and generally the profiles are unique for each genus member but relatively well conserved amongst substrains and variants. Viral DNAs of leporipoxviruses cross-hybridize at moderate stringencies only with other members of the genus, while SPV DNA is unique and is not known to cross-hybridize with any other poxvirus DNA. The MRV DNA genome is 95% identical to MYX, except that it encodes five genes derived from SFV plus three SFV/MYX fusion genes.

The GC content of the leporipoxviruses (40% for SFV) is higher than that of the orthopoxviruses (35% for vaccinia) but there is evidence that many of the viral genes important for replication, gene expression and viral assembly are conserved between the genera. These essential genes for viral replication are clustered near the central regions of the viral genome. In contrast, viral genes mapping near the termini show considerable variability, and are believed to encode many of the specific determinants of pathogenesis that dictate host range and disease characteristics.

The protein complexities of these viruses as determined by one-dimensional gel electrophoresis are comparable to that of vaccinia virus, although the profiles are unique for each member. In the cases where specific genes involved in viral propagation have been sequenced, such as thymidine kinase and topoisomerase, the proteins have been shown to be highly homologous to their counterparts from other poxviruses. Leporipoxvirus proteins involved in virulence and pathogenesis, such as growth factors and serine protease inhibitors, tend to be more extensively diverged from homologs in other poxviruses.

DNA Replication, Transcription and Translation

All of the major features of macromolecular synthesis by these viruses are very analogous to those deduced for the prototype poxvirus, vaccinia. DNA synthesis is restricted to cytoplasmic sites, although replication tends to be initiated somewhat more slowly than for vaccinia. The virus-encoded transcriptional apparatus is well conserved between the poxvirus genera, and many of the important regulatory signals that are used by vaccinia, such as promoters and transcription termination sequences, are also used with comparable efficiency in the leporipoxviruses. Thus, viral genes from one genus can be introduced to another by recombination or by DNA transfection technologies to generate chimeric virus constructs that maintain the correct regulation of the new genetic information. As in the case of vaccinia, transcriptional units can be of different kinetic classes (early/intermediate/late) and there is no splicing of viral mRNA.

The leporipoxviruses replicate in cytoplasmic factories that appear by microscopic analysis as eosinophilic B-type inclusion bodies. These factories, also called virosomes, can also be visualized by Feulgen, Giemsa or fluorescent antibody staining. SPV produces nuclear inclusions and vacuolations in addition to cytoplasmic bodies but these nuclear alterations are not believed to be sites of viral replication.

Molecular Mechanisms of Pathogenesis

Since these viruses are of only minor veterinary importance, recent research has focused on the elucidation of the determinants for viral virulence, particularly with respect to the cellular hyperplasia associated with viral replication in affected tissue and the mechanism(s) underlying the immune dysfunction caused by MYX/MRV in *Oryctolagus* rabbits. To date, at least two classes of leporipoxvirus gene products have been directly implicated in viral pathogenesis:

1. 'Virokines' are secreted virus-encoded proteins that are targeted to host-specific pathways outside the infected cell. For example, SFV and MYX/MRV encode growth factors related to epidermal growth factor and transforming growth factor alpha that participate in stimulating fibroblastic proliferation at primary and secondary tumors.
2. 'Viroceptors' are viral proteins that, at least in theory, could be either cell associated or secreted, but which mimic cellular receptors and function by sequestering important host cytokines that normally participate in the antiviral immune response. Leporipoxvirus-encoded receptor-like molecules have been discovered for TNF and γIFN, and may exist for other antiviral lymphokines as well.

Interference with antigen presentation by MRV/MYX is believed to play a role in circumventing T cell recognition during early stages of the virus infection. One MRV/MYX gene product responsible for evading immune clearance, designated Serpl, is believed to be an inhibitor of cellular serine proteases but its precise target remains to be identified.

Geographic and Seasonal Distribution

All three major species of *Sylvilagus* rabbits in the Americas have endemic fibroma-like poxviruses, and myxomatosis is now established in wild *Oryctolagus* rabbit populations of South America, Europe and Australia. SqFV and HFV have been reported to date only in North America and Europe, respectively. The leporipoxviruses in the wild undergo seasonal fluctuations that correlate well with increased populations of arthropod vectors in summer and autumn, most prominently mosquitoes. An exception to this is found in Britain, where the major vector of MYX is the flea, which is not as seasonally variable.

In the case of SPV, outbreaks are not tied to seasonal cycles but are generally associated with the degree of hog lice infestation.

Host Range and Virus Propagation

These viruses demonstrate a very restricted host range in terms of ability to cause disease, although viral replication can occur in cultured cells from some nonsusceptible hosts as well. In some cases viral replication in tissue culture monolayers or chicken chorioallantoic membranes produces 'foci' in which infected cells manifest minimal cytopathic effects, thus permitting macroscopic cell aggregations to develop. The extent of cytopathology is markedly influenced by both the cell type and the virus strain and in some instances the infected cells may detach from the monolayer to produce visible plaques. When viral replication is relatively slow and the toxicity to the target cell sufficiently moderate, a chronically infected carrier culture can be established in which progeny virus production persists for extended passages. Although poxviruses cannot permanently transform primary cells into an immortalized state, cells persistently infected with the fibroma-inducing leporipoxviruses assume many of the phenotypic characteristics associated with the transformed phenotype, such as novel morphology, growth in reduced serum and ability to form colonies in soft agar. It is likely that some of these phenotypic characteristics are facilitated by secreted poxviral proteins which mimic cellular mitogens, such as epidermal growth factor, and trigger neighboring cells into excessive proliferation.

In the cases of the benign leporipoxviruses and SPV, replication is restricted to dermal and subcutaneous sites, with occasional involvement of draining lymph nodes. However, MYX and MRV are unique in that they also replicate efficiently in lymphoid cells, such as macrophages, B cells and T cells.

MYX and MRV, like HIV-1, replicate in either resting or stimulated T cells, and can be readily isolated from splenocyte cultures. The molecular basis for the uniquely permissive nature of MYX and MRV replication in lymphocytes is unknown, but is unquestionably an important factor in the extreme virulence of myxomatosis.

Evolution and Genetic Variability

The deliberate release of MYX into rabbit populations of Australia, France and Britain in the early 1950s provided a unique opportunity to study the natural selection pressures exerted upon a particularly virulent virus–host interaction. An extensive literature on the ecological consequences of this eradication program exists and the rapid evolution of myxomatosis in the wild is well documented. Although the original South American MYX virus strain that was introduced left very few survivors in selected populations, within a few years attenuated viral strains with reduced virulence took over and more resistant rabbits became predominant.

In terms of the categories of viral virulence, some strains of MYX are classified as highly virulent (e.g. Moses and Lausanne), and attenuated variants exist down to relatively nonpathogenic (e.g. neuromyxoma and the Nottingham strains). Since only 5% of the MRV genome is derived from SFV, it technically can be considered to be a strain of MYX in the highly virulent category. Little is known about the extent of genetic variation in other leporipoxviruses, although different isolates of SFV show marked variations in tumorigenicity. Generally, leporids which recover from infection with one member either become resistant of undergo partial protection from infection by another member.

SPV shares some antigenic cross-reactivity with vaccinia, but neutralizing antibody does not confer cross-protection between the two.

Transmission and Tissue Tropism

The principle mode of transmission is by biting arthropod vectors, and the major inoculation route is dermal. Since these viruses do not replicate in the vector, the transmission is purely mechanical and hence virus spread can be readily accomplished by alternative routes as well. Thus, mosquitoes, fleas, blackflies, ticks, lice, mites, and even thistles and the claws of predatory birds, have all been implicated in leporipoxvirus transmission. The efficiency of transmission by arthropods is quite variable, and is related to viral titers in skin lesions as well as the size of the vector populations. There are no known respiratory or oral routes of infection with members of either genus, but in some infections, such as MYX in domestic rabbits, the disease can be transmitted by direct contact with ocular discharges or open cutaneous lesions.

The suipoxviruses and leporipoxviruses in their native hosts are specific for the epidermis or subdermis and usually do not progress to secondary sites, although draining lymph nodes can be affected. However, in the case of MYX and MRV infection of the domestic rabbit the virus can propagate efficiently in lymphocytes and migrates via infected cells through lymphatic channels to establish secondary sites of infection.

Pathogenicity

The leporipoxviruses are restricted to rabbits, squirrels and hares while swinepox is found only in domestic pigs. For SFV infection of *Sylvilagus* rabbits, tumors can last for many months before regressing, while in *Oryctolagus* rabbits recovery is usually complete within a few weeks. Only MYX/MRV manifests dramatic alterations in pathogenicity when the European rabbit is infected. For all of these viruses the immune status of the host rabbit plays a critical role. For example, in adult rabbits SFV rarely causes disease symptoms except for the primary fibroblastic lesion but in newborn or immunocompromised animals the infection can lead to invasive tumors and much higher titers of infectious virus in infected tissues. Agents such as cortisone, X rays or immunosuppressants can dramatically increase SFV tumor development, while chemical promoters like 3,4-benzopyrene or methylcholanthrene can predispose progression to invasive fibromatosis or even metastatic fibrosarcoma.

The ability to evade the host immune response, replicate in lymphocytes and efficiently spread to secondary sites is a unique property of MYX and MRV in *Oryctolagus* rabbits. The myxomatosis syndrome can be associated with multiple external signs (e.g. South American MYX) or may have relatively fewer gross symptoms (e.g. California MYX) and mortalities can range up to 100%. Supervening gram-negative bacterial infections in the respiratory tract and conjunctiva are often observed concomitantly with MRV/MYX, particularly by the adventitious pathogens *Pasteurella multocida* and *Bordetella bronchoseptica*, and contribute to the lethality of the disease.

SPV is only mildly pathogenic in pigs although it can cause a minor level of mortality, usually associated with milk feeding reduction in younger animals.

Clinical Features of Infection

The cutaneous tumors induced by the different leporipoxviruses in their natural hosts are clinically very similar to each other. The fibromas are rarely associated with any other symptoms, such as fever or appetite loss, and invariably regress as long as the animal is not otherwise immunocompromised. In the case of MRV/MYX in *Oryctolagus* rabbits, however, the symptoms rapidly become severe as the tumors fail to regress and the concomitant immunosuppression contributes to the lethal myxomatosis syndrome. The clinical features of myxomatosis are influenced by the genetic background of both the virus strain and the rabbit host. In the peracute form of the disease caused by California MYX the rabbits succumb in less than a week, and often have only minor external symptoms, such as inflammation and edema of the eyelids. Skin hemorrhages can be observed in some cases and convulsions often precede death. In the acute form caused by South American strains of MYX, the rabbits survive 1–2 weeks and develop more distinctive symptoms. The primary tumor can be either flat and diffuse or protuberant, and secondary site tumors around the nose, eyes and ears become prominent by 6–7 days, at which time purulent exudates from the nose and eyes frequently develop. The cutaneous tumors often become necrotic and a generalized immune dysfunction exacerbates the progressive secondary bacterial infestation of the respiratory tract. In the case of the more attenuated MYX isolates, such as neuromyxoma, the disease course is less severe and may be associated with little or no mortality.

The disease course of SPV in pigs is rather different, and resembles vaccinia in humans. Inoculation results in localized dermal papules, which progress on to vesicles and pustules, after which the lesions crust and scab over. The only clinical symptom is occasional minor fever and the animals recover completely within 3 weeks.

Pathology and Histopathology

The primary tumors caused by leporipoxviruses in *Sylvilagus* rabbits, squirrels and hares all closely resemble proliferant fibromas. Following inoculation, an acute inflammatory reaction occurs with infiltration of polymorphonuclear and mononuclear cells and proliferation of fibroblast-like cells of uncertain origin. The 'tumor' consists of pleomorphic cells imbedded in a matrix of intercellular fibrils of collagen. Unlike the transformed cells induced by other DNA tumor viruses, cells from poxviral tumors are not immortalized and cannot be propagated independently. Instead they appear to require secreted virus-encoded proteins in order to sustain the hyperproliferative state. Inclusion bodies characteristic of poxviral replication can be observed in the cytoplasms of epithelial and some fibroma cells. As the tumor develops, mononuclear leukocyte cuffing of adjacent vessels is observed and at the base of the tumor there is accumulation of lymphocytes, plasma cells, macrophages and neutrophils. The ratio between influx of inflammatory cells and fibroblast proliferation is variable but generally there is little or no necrosis. The speed with which immune cells clear the viral infection and reverse the hyperproliferation can range from 1–2 weeks up to 6 months, depending on both the virus and the host.

The principal difference between the benign fibroma syndrome described and the devastating disease caused by MYX/MRV in *Oryctolagus* rabbits is that the latter viruses efficiently propagate in host lymphocytes and are able to circumvent the cell-mediated immune response to the viral infection. The subcutaneous tumors consist of proliferating undifferentiated mesenchymal cells which become large and stellate with prominent nuclei ('myxoma' cells). In surrounding tissue there can be extensive proliferation of endothelial cells of the local capillaries and venules, often to the point where complete occlusion leads to extensive necrosis of the infected site. The overlying epithelial cells can show hyperplasia or degeneration, depending on the virus strain, and poxviral inclusion bodes are frequently observed in the prickle cell layer. In some MYX strains primary and secondary skin tumors can undergo extensive hemorrhage and internal lesions may be found in the stomach, intestines and heart. The virus readily migrates to secondary sites within infected immune cells and concomitant cellular proliferation can be detected in the reticulum cells of lymph nodes and spleen, as well as the conjunctival and pulmonary alveolar epithelium. The nasal mucosa and conjunctiva overlying secondary tumors undergo squamous metaplasia such that the epithelia become nonciliated and nonkeratizing. Disruption of the ciliary architecture may be one of the factors which facilitate the extensive gram-negative bacterial infections of the eyes, nose and respiratory tract. Varying degrees of inflammatory cell infiltration by polymorphonuclear heterophiles occur soon after infection but there is only a limited effective cellular immune response. The lymph nodes and spleen show evidence of aberrant T cell activation and hyperplasia and infectious virus can be isolated from all lymphoid organs except the thymus. Death is believed to be caused by a combination of tissue damage from the increasing tumor burden, generalized immunosuppression and debilitating bacterial colonization of the respiratory tract.

Little is known about SPV pathogenesis but gross features closely resemble those of the noninvasive orthopoxviruses in their native hosts.

Immune Response

The benign fibromas caused by SFV/SqFV/HFV regress due to an efficient cellular and humoral immune response. These viruses are excellent antigens and neutralizing antibody produced during recovery will also cross-react with other members of the genus. All of the leporipoxviruses are strongly cell-associated and cell-mediated immunity is probably the single most important mechanism of viral clearance. Other immune mechanisms are also activated, including interferon production, antibody-mediated cell lysis, sensitized macrophages and NK cells. Neutralizing antibody can last for many months after viral clearance and immunity is usually cross-protective to the other leporipoxviruses.

In the case of MYX/MRV in *Oryctolagus* rabbits the picture is very different. Although circulating antibody can be detected against virions, as determined by neutralization or agglutination, and against soluble antigens, as determined by complement fixation and precipitin tests, the antibody provides little protection against the disease progression. Instead, cellular immunity is severely compromised, and by day 6–7 lymphocytes (especially splenocytes) are demonstrably dysfunctional in their response to mitogens and lose the ability to secrete critical lymphokines such as interleukin-2. Unlike the case of SFV, there is a notable absence of virus-specific T cells in either the spleen or draining lymph nodes. Immune dysfunction is common in viruses that replicate in lymphocytes, but the precise levels at which MYX/MRV intervene in cellular immunity remain to be clarified. There is some evidence that these viruses interfere with the function of cell surface MHC class I molecules, which could prevent proper viral antigen presentation and hence interfere with immune recognition of infected cells. Also, several virus-specific gene products have been shown to be secreted homologs of the cellular receptors for tumor necrosis factor (TNF) and gamma-interferon (γIFN) that are believed to bind and sequester extracellular TNF and γIFN in the vicinity of virus-infected cells and thus short-circuit immune pathways dependent on TNF and γIFN.

SPV-infected pigs generally recover from the infection and become immune to secondary challenge. There are few data on the nature of this immunity, but it bears close resemblance to that of vaccinia immunization in humans.

Prevention and Control

Since these viruses are spread principally by biting arthropods, vector control is the single most effective method of disease prevention. The viruses are susceptible to standard antipoxvirus chemical agents, such as phosphonoacetic acid, arabinosyl cytosine and rifampicin, but these are of limited use in infected animals. Immunization against myxomatosis can be accomplished with live SFV or attenuated strains of MYX.

Future Perspectives

Now that DNA sequencing studies have been initiated for many different poxviruses, it is likely that more viral genes which determine the clinical characteristics of their disease will be discovered. Studies on viral gene products which stimulate fibroblastic and endothelial cells to proliferate will probably provide information on how mitogenesis is regulated by surface receptors on these target cells. The ability of MRV/MYX to replicate in lymphocytes offers an important system in which to elucidate the mechanisms of cellular tropism by these viruses and the analysis of virus-induced immunosuppression should shed light on the various immune strategies used by the host to combat viral infections in general. Finally, the restricted host ranges of the leporipoxviruses and suipoxviruses suggests the potential for the genetic manipulation of these viruses such that heterologous foreign antigen genes can be expressed for the purpose of developing novel vaccines against important pathogens of domestic leporids and swine.

See also: Cowpox virus; Epidemiology of viral diseases; Fowlpox virus, History of virology; Host genetic resistance; Immune response; Molluscum contagiosum virus; Mousepox and rabbitpox viruses; Smallpox and monkey pox viruses; Vaccinia virus; Yabapox and Tanapox viruses.

Further Reading

Buller RML and Palumbo GJ (1991) Poxvirus pathogenesis. *Microbiol Rev.* 55: 80

Fenner F and Ratcliffe FN (1965) *Myxomatosis*. Cambridge, UK: Cambridge University Press.

McFadden G (1988) Poxviruses of rabbits. In: Darai G (ed.) *Virus Diseases in Laboratory and Captive Animals*, p. 37. Boston: Martinus Nijhoff.

Strayer D (1989) Poxviruses. In: Specter S, Bendinelli M and Friedman H (eds) *Virus-induced Immunosuppression*, p. 173. New York: Plenum.

Turner PC and Moyer RW (1990) The molecular pathogenesis of poxviruses. In: Moyer RW and Turner PC (eds) *Poxviruses. Curr. Top. Microbio. Immunol.* vol. 163, p. 125. Springer-Verlag.

Upton C, Mossman K and McFadden G (1992) Encoding of a homolog of the IFN-γ receptor by myxoma virus. *Science* 258: 1369.

Sheep and Goat Poxviruses

R Paul Kitching
Institute for Animal Health
Pirbright, UK

History

Sheeppox and goatpox are malignant pox diseases of sheep and goats easily recognizable by their characteristic clinical signs, and described in the earliest texts on animal diseases. Lumpy skin disease (Neethling) of cattle (LSD), however, was first described in 1929 in Northern Rhodesia (Zambia), having apparently been absent from domestic cattle until that time. From Zambia LSD spread south to Botswana and Zimbabwe and by 1944 the disease appeared in South Africa where it caused a major epizootic, affecting over 6 million cattle. In 1957 LSD was first diagnosed in Kenya, and was thought at the time to be associated with the introduction of a flock of sheep affected with sheeppox on to the farm. Since then LSD has been present in most of the countries of sub-Saharan Africa, often associated with large epizootics followed by periods in which the disease is only rarely reported. In 1988 LSD caused a major outbreak in Egypt and in 1989 spread from Egypt to a village in Israel. This was the first time that a diagnosis of LSD outside of Africa had been supported by laboratory confirmation.

Taxonomy and Classification

The viruses that cause sheeppox, goatpox and LSD are all members of the genus *Capripoxvirus* and have morphological, physical and chemical properties similar to vaccinia virus. Originally the viruses were classified according to the species from which they were isolated, but comparisons of their genomes indicate that the distinction between them is not so clear, and that recombination events occur naturally between isolates from different species. This is reflected in the ability of some strains to cause disease in both sheep and goats and experimental results which show that all the sheep isolates examined could infect goats, and that goat isolates could infect sheep. The epidemiological relationship between sheeppox and goatpox isolates and cattle isolates is less clear, apparent in differences in the geographical distribution of sheeppox and goatpox and LSD (see later). However, some isolates recovered from sheep and goats in Kenya have genome characteristics very similar to cattle isolates. It has been proposed that confusion can be reduced by referring to the malignant pox diseases of sheep, goats and cattle, including Indian goat dermatitis and Kenyan sheep and goatpox, as capripox. It is envisaged that when sufficient isolates have been examined biochemically, no clear distinction will be possible between sheep, goat and cattle isolates, but a spectrum will emerge in which some strains have clear host preferences while others will be less well defined and will naturally infect whichever host they come into contact.

Geographic and Seasonal Distribution

Capripox of sheep and goats is enzootic in Africa

north of the equator, the Middle East and Turkey, Iran, Afghanistan, Pakistan, India, Nepal and parts of the People's Republic of China, and in 1986, Bangladesh. Sheeppox was eradicated from Britain in 1866 and from France, Spain and Portugal in 1967, 1968 and 1969 respectively. Sporadic outbreaks still occur in Europe, for instance in Italy in 1983 and Greece in 1988 and 1989.

LSD is enzootic in the sub-Saharan countries of Africa and probably now also in Egypt. The single outbreak in Israel was eradicated by slaughter of affected and in contact cattle.

There is no clear seasonality to outbreaks of capripox in sheep and goats. In enzootic areas lambs and kids are protected against infection with capripoxvirus for a variable time dependent on the immunity of the mother. However, the spread of LSD is related to the density of biting flies and consequently major enzootics have been associated with humid weather when fly activity is greatest.

Host Range and Virus Propagation

Amongst domestic species, capripoxvirus is restricted to cattle, sheep and goats. Experimentally it is possible to infect cattle, sheep or goats with isolates derived from any of these three species, although clinically the reaction following inoculation may be indiscernible. Viral genome analysis using restriction endonucleases has identified fragment size characteristics by which it is possible to classify strains into cattle, sheep or goat isolates. However, the identification of strains that have intermediate characteristics between typical sheep and goat isolates does suggest the movement of strains between these species. Analysis of some Kenyan isolates derived from sheep and goats shows very close homology with cattle LSD isolates.

The involvement of the African buffalo (*Syncerus caffer*) in the maintenance of LSD has not been clearly established. Some surveys have shown the presence of capripoxvirus antibody in buffalo, while others have failed to show its presence. Buffalo clinically affected with LSD have not been described. Experimental infection of giraffe (*Giraffe calemopardelis*), impala (*Aepyceros melampus*) and gazelle (*Gazella thomsonii*) has resulted in the development of clinical disease.

Bos indicus cattle are generally less susceptible to LSD and develop milder clinical disease than *Bos taurus*, of which the fine-skin Channel Island breeds are particularly susceptible. Similarly, breeds of sheep and goats indigenous to capripoxvirus enzootic areas appear less susceptible to severe clinical capripox than do imported European or Australian breeds.

Capripox will grow on the majority of primary and secondary cells and cell lines of ruminant origin. Primary lamb testes cells are considered the most sensitive system for isolation and growth of capripoxvirus. The virus produces a characteristic cytopathic effect (cpe) on these cells which can take up to 14 days for field isolates, but can be as short as 3 days for well-adapted strains.

Isolates of capripoxvirus derived from cattle have been adapted to grow on the chorioallantoic membrane of embryonated hens eggs, although attempts to grow isolates from sheep and goats in eggs have been unsuccessful. Capripoxvirus will not grow in any laboratory animals.

Genetics

Less is known concerning the specific genetics of capripoxvirus than is known about the orthopoxvirus genome. Studies on field isolates taken from cattle suggest that the virus is very stable, as *Hin*d III restriction endonuclease digest patterns of isolates from the 1959 Kenya outbreak of LSD are identical to those obtained from 1986 LSD isolates. However, recombination has been shown to occur between cattle and goat isolates and this could be the natural method by which the virus evolves. By analogy with the orthopoxviruses, it is also likely that sequences are deleted or repeated within the genome in the normal replicative cycle.

Comparisons between the genomes of the different poxvirus genera show a relatively low level of nucleotide sequence homology. In common with other poxviruses, capripoxviruses contain a gene coding for thymidine kinase (TK). Within the capripoxvirus genus, nucleotide divergence values suggest that the typical sheep and cattle isolates are more closely related to each other than to goat isolates. There is, however, less divergence between capripoxvirus genomes than seen in orthopoxvirus genomes, particularly in the near-terminal regions which in the orthopoxvirus is considered to account for the considerably larger host range of this genus.

The usual technique for developing capripox vaccines has been to serially passage virulent isolates in tissue culture. This has been shown to reduce the virulence of the strain, although the mechanism by which this occurs is not known.

Evolution

The capripoxviruses have evolved into specific cattle, sheep and goat lines, but as has been described above, intermediate strains exist, particularly those with cattle and goat genome characteristics. In Kenya there is evidence of movement of strains between all three species, but the absence of sheep or goat pox in LSD enzootic areas in Southern Africa, and the absence of LSD outside of Africa would suggest that host-specific strains are being maintained and presumably are continuing to evolve.

Serologic Relationships and Variability

Polyclonal sera fails to distinguish in the virus neutralization (VN) test between any of the isolates of capripoxvirus so far examined. Sheep, goats or cattle that have been infected with any of the isolates are totally resistant to challenge with any of the other isolates. On this basis it has been possible to use the same vaccine strain to protect all three species. No monoclonal antibodies have yet been developed against capripoxvirus, but it can be expected that differences will emerge between strains using these reagents.

Capripoxviruses share a precipitating antigen with parapoxviruses, but no cross-immunity has been shown between these two genera, or between capripoxvirus and any other pox virus genera.

Epidemiology

In sheeppox and goatpox enzootic areas the distribution of disease is frequently a reflection of the traditional form of husbandry. For instance, in the Yemen Arab Republic, the sheep and goat flocks kept on the grassland of the central plateau and better irrigated regions of the coastal plain move about in search of food, frequently mixing with flocks from neighboring villages at water holes, and in this situation disease is restricted to the young stock. Animals over one year of age have a solid immunity. The animals belonging to villages in the more mountainous regions, and the arid areas of the coastal plain are isolated by terrain or semi-desert from mixing with animals from other villages. It is not known what critical number of animals is required to maintain capripoxvirus within a single population, but it is over a thousand adult animals which is the approximate village sheep and goat population. In these villages disease is usually only seen following the introduction of new animals, typically from market, and generally affects animals of all age groups. The disease spreads through the village, usually within 3–6 months, and then disappears in the absence of more susceptible animals. Occasionally, even within areas of high sheep and goat density, it is possible to encounter animals that have been kept totally isolated in the confines of a domestic residence, and these may remain susceptible to infection until adult.

In Sudan, large numbers of sheep and goats are trekked from the West to the large collecting yards and markets of Omdurman, outside Khartoum. Here also it is possible to see capripox infection in adult animals. Many of the flocks originate in villages which, like in the Yemen, are isolated from their neighbors. Capripoxvirus does not persist in these villages, and as a result the animals acquire no resistance, and are fully susceptible when they first encounter disease on the long journey across Sudan. Animals being exported from countries that are free of capripoxvirus may suffer a similar fate when they arrive in a capripoxvirus enzootic area, as often seen in Australian or New Zealand sheep imported into the Middle East.

In a study of 49 outbreaks of capripox in the Yemen, only eight were reported affecting sheep and goats, the remaining 41 causing clinical disease in either sheep or goats. It is possible that both sheep and goats could have been involved in more than the eight outbreaks, but that the disease was inapparent in one species; whether, therefore, the species in which the disease was inapparent could transmit virus and become a vector for disease, has not been determined. In Kenya, capripox is frequently encountered in both sheep and goats within the same flock, and there is the possibility that the same strain of capripoxvirus could also cause LSD in cattle.

The epidemiology of sheeppox, goatpox and LSD is similar; the severity of outbreaks depend on the size of the susceptible population, the virulence of the strain of capripoxvirus, the breed affected (indigenous animals tending to be less susceptible to clinical disease than imported), and, with LSD, the presence of suitable insect vectors. Morbidity rates vary from 2 to 80%, and mortality rates may exceed 90%, particularly if the infection is in association with other disease or bad management.

Transmission and Tissue Tropism

Under natural conditions capripoxvirus is not transmitted very readily between animals, although there are circumstances when transmission appears very rapid; for example, in association with factors that damage the mucosae such as peste des petits ruminants or feeding on abrasive forage. Animals are most infectious soon after the appearance of papules and during the 10 day period before the development of significant levels of protective antibody. High titers of virus are present in the papules, and those on the mucous membranes quickly ulcerate and release virus in nasal, oral and lachrymal secretions, and into milk, urine and semen. Viremia may last up to 10 days or in some fatal cases until death. Those animals that die of acute infection before the development of clinical signs and those that develop only very mild signs or single lesions rarely transmit infection, while those that develop generalized lesions produce considerable virus and are highly infectious. Aerosol infection over a few meters only, as with other poxvirus infections, is probably the usual form of transmission. Biting flies are significant in the mechanical transmission of LSD, and *Stomoxys calcitrans* and *Biomyia fasciata* have been implicated. Experimentally, *Stomoxys calcitrans* has also been shown capable of transmitting sheeppox and goatpox.

During the recovery phase following infection, the papules on the skin become scabs. It is relatively easy to demonstrate pox virions in the scab, but difficult to isolate virus on tissue culture, probably because of the complexing of antibody and virus within the scab. Capripoxvirus is reported to remain viable in wool for 2 months and in contaminated premises for 6 months, and is reported to remain infectious in skin lesions of cattle for 4 months. The true epidemiological significance of the virus within the scab, and ultimately the environment is not clear. It has been suggested that the protein material which envelopes the virus within the type A intracytoplasmic inclusion bodies of infected cells protects the virus in the environment.

There is no evidence for the existence of animals persistently infected with capripoxvirus. Transplacental transmission of capripoxvirus may be possible in association with simultaneous pestivirus infection, as may occur with pestivirus-contaminated capripox vaccine.

Capripoxvirus can be isolated from the leukocytes during viremia, and has been isolated from lesions in the liver, urinary tract, testes, digestive tract and lungs. However, the cells of the skin and skin glands and the internal and external mucous membranes appear to be the major sites of virus replication.

Pathogenicity

There is considerable variation in the pathogenicity of strains of capripoxvirus. Nothing is known concerning the genes responsible in the capripoxvirus genome for virulence or host restriction.

Clinical Features of Infection

The incubation period of capripox infection, from contact with virus to the onset of pyrexia is approximately 12 days, although it appears frequently longer as transmission is often not immediate between infected and susceptible animals. Following experimental inoculation of virus the incubation period is approximately 7 days, and this is similar to that shown experimentally using biting flies to transmit virus.

The clinical signs of malignant disease are similar in sheep, goats and cattle. Twenty-four hours after the development of pyrexia of between 40 and 41°C, macules (2–3 cm diameter areas of congested skin) can be seen on the white skin of sheep and goats, particularly under the tail. Macules are not seen on the thicker skin of cattle, and are frequently missed on skin of pigmented sheep and goats. After a further 24 h the macules swell to become hard papules of between 0.5 and 2 cm diameter, although they may be larger in cattle. In the generalized form of capripox, papules cover the body, being concentrated particularly on the head and neck, axilla, groin and perineum, and external mucous membranes of the eyes, prepuce, vulva, anus and nose. In cattle these papules may exude serum, and there may be considerable edema of the brisket, ventral abdomen and limbs. The papules on the mucous membranes quickly ulcerate, and the secretions of rhinitis and conjunctivitis become mucopurulent. Keratitis may be associated with the conjunctivitis.

All the superficial lymph nodes, particularly the prescapular, are enlarged. Breathing may become labored as the enlarged retropharyngeal lymph nodes put pressure on the trachea. Mastitis may result from secondary infection of the lesions on the udder.

The papules do not become vesicles and then pustules, typical of orthopoxvirus infections. Instead they become necrotic, and if the animal survives the acute stage of the disease, change to scabs over a 5–10 day period from the first appearance of papules. The scabs can persist for up to a month in sheep and goats, whereas in cattle the necrotic papules which penetrate the thickness of the skin may remain as 'sitfasts' for up to a year.

Severe disease is accompanied by significant loss of condition, agalactia, possibly secondary abortion and pneumonia. Eating, drinking and walking may become painful, and death from dehydration is not uncommon. Secondary myiasis is also a major problem in tropical areas.

Pathology and Histopathology

The lesions of capripox are not restricted to the skin, but also may affect any of the internal organs, in particular the gastrointestinal tract from the mouth and tongue to the anus, and the respiratory tract. In generalized infections papules are prominent in the abomasal mucosa, trachea and lungs. Those in the lungs are approximately 2 cm in diameter, and papules may coalesce to form areas of gray consolidation.

In affected skin, there is an initial epithelial hyperplasia followed by coagulation necrosis as thrombi develop in the blood vessels supplying the papules. Histiocytes accumulate in the areas of the papules and the chromatin of the nuclei of infected cells marginates. The cells appear stellate as their boundaries become poorly defined, and many undergo hydropic degeneration with the formation of microvesicles. Intracytoplasmic inclusion bodies are present in infected cells of the dermis and also in the columnar epithelial cells of the trachea where frequently gross lesions may not be apparent. These are initially type B inclusions at the sites of virus replication, but later in infection they are replaced by type A inclusions (see earlier). The maximum titer of virus is obtained from papules approximately 6 days after their first appearance.

Immune Response

Capripoxvirus, like orthopoxvirus, is released from an infected cell within an envelope derived from modified cellular membrane. The enveloped form of the virus is more infectious than the nonenveloped form which can be obtained experimentally by freeze-thawing infected tissue culture. By analogy with orthopoxvirus, antigens on the envelope and on the tubular elements of the virion surface may stimulate protective antibodies. Animals immune to nonenveloped virus are still fully susceptible to the enveloped form. Passively transferred antibody, either colostral or experimentally inoculated will protect susceptible animals against generalized infection. However, in the vaccinated or recovered animal, there is no direct correlation between serum levels of neutralizing antibody and immunity to clinical disease. Antibody may limit the spread of capripoxvirus within the body, but it is the cell-mediated immune response which eliminates infection. In sheep, MHC-restricted cytotoxic T lymphocytes are required in the protective immune response to orthopoxvirus infection, and therefore probably also capripoxvirus infection.

Immune animals challenged with capripoxvirus by intradermal inoculation develop a delayed-type hypersensitivity reaction at the challenge site. This may be inapparent in animals with high levels of circulating antibody. It has been suggested that the very severe local response shown by some cattle at the site of vaccination against LSD may be due to an hypersensitivity reaction due to previous contact with the antigens of parapoxvirus.

There is total cross-immunity between all strains of capripoxvirus, whether derived from cattle, sheep or goats.

Prevention and Control

In temperate climates capripox can be effectively controlled by slaughter of affected animals, and movement control of all susceptible animals within a 10 km radius for 6 months. In tropical climates, particularly in humid conditions when insect activity is high, movement restrictions are not sufficient and vaccination of all susceptible animals should be considered. In outbreaks of LSD it is not considered necessary to vaccinate sheep and goats, although theoretically cattle strains of virus could infect them. Similarly in outbreaks of capripox in sheep and goats, cattle are not normally vaccinated.

Countries in which capripoxvirus is absent can maintain freedom by preventing the importation of animals from infected areas. There is always a possibility that skins from infected animals could introduce infection into a new area, although there have been no proven examples of this. The insect transmission of capripoxvirus into Israel from Egypt over a distance of between 70 and 300 km would indicate that it is impossible for countries neighboring enzootic areas totally to secure their borders.

In enzootic areas annual vaccination of susceptible animals with a live vaccine will control the disease. Calves, kids and lambs up to 6 months of age may be protected by maternal antibody, but this would only occur if the mother had recently been severely affected with capripox. Although maternal antibody will inactivate the vaccine, it is advisable to vaccinate all stock over 10 days old. No successful dead vaccines have been developed for immunization against capripoxvirus infection, other than those that give only very short-term immunity.

Future Perspectives

Capripox of sheep and goats is present in most of Africa and Asia, whereas LSD is restricted to Africa. There is no good explanation why LSD has not spread into the Middle East and India, carried by the considerable trade in live cattle. Unless there is a reservoir host in Africa which is required for the maintenance of the cattle-adapted capripoxvirus, it can be anticipated that LSD will spread out of Africa, with major economic consequences.

While considerable attention has been given to vaccinia as a vector of other viral genes for development as a recombinant vaccine, little attention has been given to capripoxvirus as a potential vector vaccine. Although its use would be restricted to the not inconsiderable capripoxvirus enzootic area, it would have the advantage of not being infectious to humans, and being a useful vaccine in its own right.

See also: Cowpox virus; Fowlpox virus; Immune response; Mousepox and rabbitpox viruses; Yaba-pox and Tanapox viruses; Vaccinia virus; Vectors.

Further Reading

Black DN, Hammond JM and Kitching RP (1986) Genomic relationship between capripoxviruses. *Virus Res.* 5: 277.
Gershon PD, Ansell DM and Black DN (1989) A comparison of the genome organization of capripoxviruses with that of the orthopoxviruses. *J. Virol.* 63: 4703.
Gershon PD, Kitching RP, Hammond JM and Black DN (1989) Poxvirus genetic recombination during natural virus transmission. *J. Gen. Virol.* 70: 485
Kitching RP, Hammond JM and Taylor WP (1987) A single vaccine for the control of capripox infection in sheep and goats. *Res. Vet. Sci.* 42: 53.
Kitching RP, Bhat PP and Black DN (1989) The characterization of African strains of capripoxvirus. *Epidemiol. Infect.* 102: 335.

PRD1 BACTERIOPHAGE

Dennis H Bamford
University of Helsinki
Helsinki, Finland

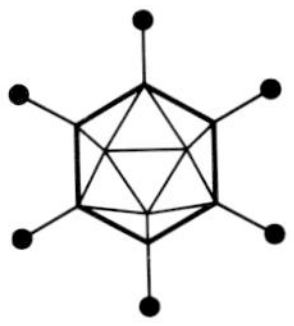

Discovery and Classification

Bacteriophage PRD1 belongs to very closely related phages isolated from different parts of the world. Other members of this family are PR3, PR4, PR5 LI7 and PR772. These viruses are lytic, infecting a vast variety of Gram-negative cells harboring a P-, N- or W-type conjugative plasmid. Among the hosts are *Escherichia coli, Salmonella typhimurium* and *Pseudomonas aeruginosa*. PRD1 is the type phage of the *Tectiviridae* family. PRD1 and PR4 are the most thoroughly studied phages in this group. Since these phages are almost identical the description of this phage system uses information from both of these viruses. The discovery of these lytic viruses (1973–1979) was associated with studies of antibiotic-resistant plasmids. In further studies it was observed that they had a lipid membrane and a linear double-stranded DNA (dsDNA) genome with 5′-covalently linked terminal proteins.

Viral Particle

The viral particle has an outer diameter of

approximately 65 nm. The outer protein layer surrounds the viral membrane. The viral genome, which resides inside this membrane vesicle. The virion is composed of approximately 70% protein, 15% DNA and 15% lipid. The outer protein layer is composed of the major coat protein P3 and the minor coat protein P5. Both of these proteins form homotrimers. Protein P5 contains a $(Gly\text{-}X\text{-}Y)_6$ motif which is found in collagens and is cleavable with collagenase. The removal of the coat proteins reveals the membrane vesicle with the rest of the viral structural proteins. These proteins function in adsorption (P2), DNA injection (P11, P14, P16, P18) and DNA packaging (P9, P20, P22). In addition to these proteins, the membrane contains several small integral membrane proteins with as yet no described function. It is most probable that proteins associated with DNA packaging and possibly some proteins needed for DNA injection are located in a specific vertex. Protein P8 is covalently linked to the 5′-terminal nucleotides of the linear dsDNA genome. The membrane composition is approximately half protein half lipid. The membrane phospholipids are composed of approximately 56% phosphatidylethanolamine, 37% phosphatidylglycerol, 5% cardiolipin and 2% neutral lipids. The fatty acid composition of the phospholipids is identical to that of the host.

Genome and Genetics

Isolated nonsense mutants form 17 genetic complementation groups. Most of these are assigned to a phage protein. In addition, at least eight gene products are known to be phage specific. The amber mutants can be complemented with cloned genomic fragments and thus the genes are located on the physical genome map determined from the restriction enzyme fragments.

Sequencing of the entire genome revealed 110 bp-long inverted terminal repeats at both ends of the 14 925 bp-long linear dsDNA (Fig. 1). The early genes for DNA replication and for transcription control are located at both ends of the genome. The genome terminal protein gene VIII and DNA polymerase gene I are at the left and two genes coding for DNA-binding proteins at the right (genes XII and XIX). The latter two are so far the only genes located in this DNA strand. Gene XV codes for the phage lytic enzyme, the synthesis of which begins before that of the late gene products. The late genes are located in the 'upper' genome strand (Fig. 1). Genes X and XVII code for nonstructural assembly factors whereas the rest of the late genes code for structural proteins. A gene designated with a roman numeral (for example I) produces an identified protein with the same arabic number (P1) whereas localized genes with as yet no identified gene product are designated with lower case letters (for example *gpc*).

Life Cycle

A schematic illustration of the PRD1 life cycle is given in Fig. 2. The phage adsorbs to the plasmid-encoded receptor. In optimal conditions, up to 50 phage particles can attach to one cell. The phage membrane contracts during the DNA injection event. The early transcription leads to the synthesis of the genome terminal protein (P8) and the DNA polymerase (P1) which can initiate the replication from each genome end. The initial reaction is the

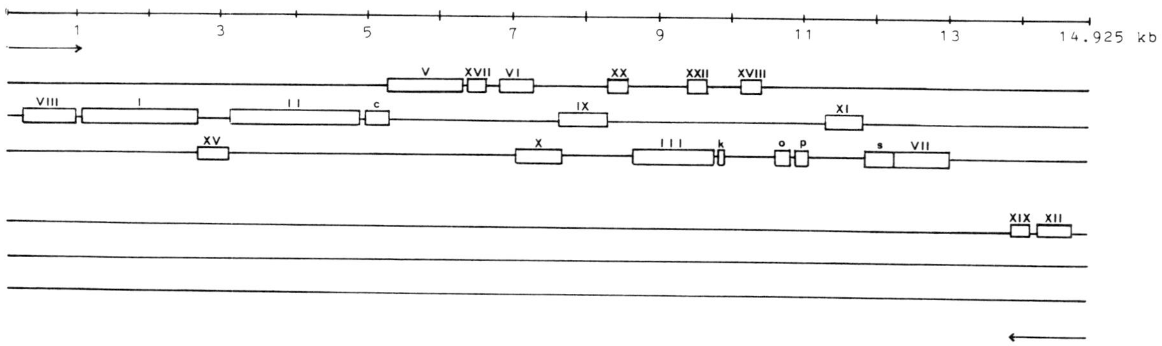

Fig. 1 Schematic presentation of the PRD1 genome. The three reading frames for both strands are shown. Genes with identified protein products are designated with roman numerals. Other localized genes are designated with lower case letters. For details see the text. (Figure by Anna-Liisa Hänninen, Department of Genetics, University of Helsinki.)

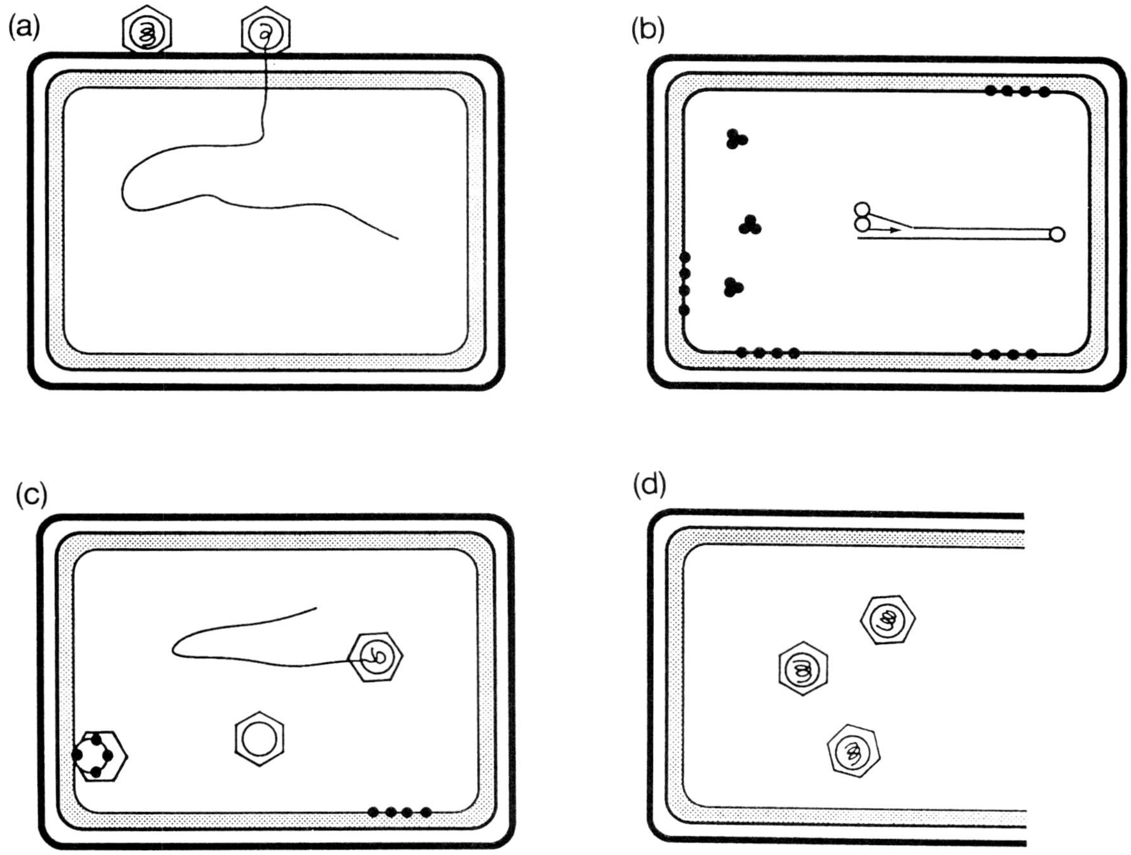

Fig. 2 Life cycle of PRD1. (**a**) Virion adsorbs to the cell surface and injects its DNA, leaving capsid and membrane behind. (**b**) DNA replication and protein synthesis begin. The major capsid protein is formed as soluble multimers, whereas membrane proteins are associated with the host membrane. (**c**) Upon translocation of the membrane and its associated proteins, empty particles are formed. DNA packaging results in the formation of mature virions. (**d**) New viral particles are released after cell lysis. (From Caldentey J, Bamford JKH and Bamford DH (1990) *J. Struct. Biol.* 104: 44, with permission.)

formation of a phosphodiester bond between Tyr-190 in the terminal protein and dGMP destined to become the 5′-terminal nucleotide of the synthesized strand. Replication initiation from both ends leads to two identical molecules. However, initiation from only one end produces one displaced single-stranded DNA (ssDNA) molecule in addition to the complete dsDNA genome. The displaced ssDNA molecule can either initiate replication using paired inverted terminal repeats or pair with another single strand with opposite polarity. The late protein synthesis leads to accumulation of the major coat protein (P3) trimers in the cytoplasm as well as to insertion of membrane proteins into the host plasma membrane.

Empty virus particles are formed in a process that is absolutely dependent at least on the major coat protein multimers and the two assembly factors P20 and P17. In this membrane translocation event, the virus-specific membrane is removed from the host plasma membrane and placed inside the virus protein coat. The empty virus particles are located in the nucleoplasm area where DNA packaging takes place. The mature virus particles are preferentially located in the cell periphery prior to cell lysis. Cell lysis liberates several hundred virus particles, 10–15% of which are empty DNA-less particles.

Future Perspectives

Bacteriophage PRD1 is the only known protein-primed DNA-replication system operating in *E. coli*. This host background allows a detailed replication host factor search. An *in vitro* replication system with purified components produces full-length PRD1 DNA. The combination of PRD1 replication components with those of *Bacillus subtilis* bacteriophage ϕ29 should reveal details of the specificity of the replication components. The PRD1 membrane system is potentially valuable in understanding membrane biogenesis and translocation.

See also: ϕ6 Bacteriophage.

Further Reading

Bamford JKH *et al.* (1991) Organization of the genome of membrane containing bacteriophage PRD1 infecting *E. coli*. *Virology* 183: 658.

Caldentey J, Bamford JKH and Bamford DH (1990) Structure and assembly of bacteriophage PRD1, an *Escherichia coli* virus with a membrane. *J. Struct. Biol.* 104: 44.

Mindich L and Bamford DH (1988) Lipid-containing bacteriophages. In: Calendar R (ed.) *The Bacteriophages*, Vol. 2, p. 475. New York: Plenum Press.

Salas M (1991) Protein-priming of DNA replication. *Annu. Rev. Biochem.* 60: 39.

Savilahti H *et al.* (1991) Overexpression, purification and characterization of *Escherichia coli* bacteriophage PRD1 DNA polymerase. *In vitro* synthesis of full-length PRD1 DNA with purified proteins. *J. Biol. Chem.* 266: 18737.

PRIONS

See Spongiform encephalopathies

PROPAGATION OF VIRUSES

Steffen Faisst
Deutsche Krebsforschungszentrum
Heidelberg, Germany

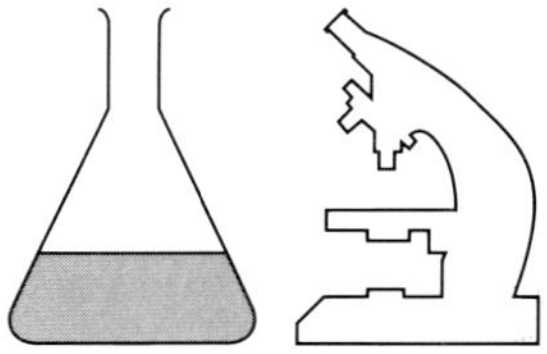

Introduction

Viruses exist in two functionally distinct forms. The viral particle, referred to as virion, represents the static extracellular form without any metabolic activity, serving as a vehicle for the viral genetic material. The second, the dynamic form, consists of the viral genetic material itself once it is uncoated in a host cell. This active form of a virus uses the host cell biosynthesis machinery and energy supplies to multiply and to generate progeny virions. Hence, virus multiplication exploits host cells in a parasitic way. This results in the disturbance of normal cellular functions in many virus-infected cells, one of the basic mechanisms of viral diseases.

Virus propagation under defined laboratory conditions provides an important experimental tool in basic research in virology, allowing studies of virus multiplication, virus–host cell interactions and viral pathogenesis. Furthermore, virus propagation provides the basis for diagnoses of viral diseases and vaccine production.

Here, the mechanisms of intracellular virus multiplication are summarized, and techniques used for virus propagation, purification and titration are described.

The Virus Multiplication Cycle

To multiply, a virus has to enter a living cell. Thereafter, the viral genome is released from the capsid, and interacts with the host cell in order to replicate and to produce viral proteins. New capsids are assembled, and the newly synthesized genomes are packaged into these capsids either concomitant with or after their assembly. This results in progeny virions, which are released from the cell in order to transfer the viral genome to new host cells.

The initial step of virus–cell contact, referred to as adsorption, is mediated by the binding of a viral protein, located at the virion surface, to a receptor at the cell surface. Cellular receptors of many viruses have been identified. Most viruses adsorb to cell surface proteins that have specific metabolic functions, and that are expressed only in a subset of differentiated cells. Rabies virus, for example, binds to the acetylcholine receptor, and accordingly adsorbs to nerve cells. Human immunodeficiency virus adsorbs to CD4 molecules of T lymphocytes and macrophages. It should also be noted that carbohydrates have been shown to act as virus receptors. For example, polyoma virus, Sendai virus and vaccinia virus adsorb to sialyloligosaccharides of glycoproteins and glycolipids, which can be found on the surface of many cell types.

After adsorption, viruses have to cross the plasma membrane in order to enter a cell. The mechanism of penetration differs from virus to virus, depending on the respective virion structure. Adsorption to cell receptors brings viruses into intimate contact with the plasma membrane. Subsequently, viruses with a membranous envelope may directly inject the viral capsid into the cytoplasm by membrane fusion. However, the virus–receptor complex can also be internalized by coated pit-mediated endocytosis and delivered to endosomes. Subsequent fusion of viral and endosome membranes leads to the release of the viral capsid into the cytoplasm.

Viruses lacking an envelope also enter cells by coated pit-mediated endocytosis, but entry of the capsid into the cytoplasm obviously cannot occur via membrane fusion. For the nonenveloped adenovirus, evidence has been presented that capsid proteins induce the lysis of endosomes, thus releasing the capsid into the cytoplasm.

Once within the cytoplasm, the viral genome has to be liberated from the capsid and transported to the appropriate intracellular site for transcription or replication. This event is referred to as uncoating.

For RNA viruses, cellular factors like proteases are thought to dismantle the virus capsids immediately after entry, releasing the viral RNA into the cytoplasm, where it is replicated and translated. Also poxviruses, which contain a double-stranded DNA genome, replicate in the cytoplasm, using a virus-encoded DNA polymerase. Only retroviruses transcribe their RNA genome in DNA, which subsequently translocate to the nucleus and integrate into the cellular genome. The integrated retroviral genome serves as a template to synthesize new RNA genomes.

For nuclear-replicating DNA viruses, the capsid is supposed to be translocated along the cytoskeleton to the nuclear pores. At the nuclear pores, the viral DNA is released into the nucleus, where it is replicated.

When the viral genome is delivered to its appropriate intracellular site, it has to interact with the host cell biosynthesis machinery in order to get amplified and transcribed into mRNAs allowing the synthesis of viral (capsid and noncapsid) proteins. Thereafter, cellular and viral factors mediate the assembly of the capsids, and the packaging of the viral genome into these capsids. These steps are referred to as capsid maturation. The assembly of capsid proteins of nonenveloped viruses occurs either in the nucleus (e.g. parvoviruses, adenoviruses and papovaviruses) or in the cytoplasm (e.g. reoviruses). Most of the nonenveloped viruses rely on host cell lysis for their egress.

The assembly of the capsids of enveloped viruses also takes place either in the cytoplasm (most RNA viruses, poxviruses), or in the nucleus (herpesviruses), but the last step in virion assembly is linked to the release of the virions. Most RNA viruses egress by budding through the plasma membrane. Herpesviruses have been shown to leave the nucleus by budding through the inner nuclear membrane. They are then released from the cells by transport through the endoplasmic reticulum. Poxviruses may be released from the cells either by budding through the Golgi apparatus, or by cell disruption, resulting in enveloped and nonenveloped particles, respectively. Both enveloped and nonenveloped particles are infectious. Once progeny virus has been released, new cells can become infected, and further multiplication cycles are initiated.

Cytopathic Effect

Different types of interaction between viruses and host cells can be observed. Many viruses kill or morphologically modify their host cells when they multiply. This is called the cytopathic effect (CPE), and the respective virus is said to be cytopathogenic.

Generally, cytopathogenic viruses code for proteins that shut off synthesis of cellular macromolecules, or that are cytotoxic. Furthermore, the capsid proteins of nonenveloped viruses seem to be implicated in cell lysis, on which these viruses may rely for their egress. Enveloped viruses may additionally insert viral proteins into the cell membrane, which also may impair the viability of the target cells. Cell death within a few days is the result of productive infection with many types of viruses, such as togaviruses, picornaviruses or autonomous parvoviruses. Infection with poxviruses, reoviruses or adenoviruses also leads to cell death, but less rapidly.

Another type of CPE is the induction of cell fusion by viruses such as paramyxoviruses, human immunodeficiency viruses and herpesviruses. Induction of cell fusion is also due to the insertion of viral proteins into the host cell membrane, and results in the formation of syncytia (giant cells with up to several hundred nuclei).

As a result of mild CPEs, a balance between cell growth and virus production can sometimes be observed (e.g. after paramyxovirus infections). Such 'carrier cultures' may be seen as the cell culture counterparts of chronic infections in animals. Several RNA viruses that are not cytopathogenic (like arenaviruses and most retroviruses) may also provoke such steady-state infections.

Host Range, Permissiveness and Susceptibility

Any cell that can be infected by a virus is said to be susceptible. However, infection of a susceptible cell does not necessarily result in a productive infection. Productive infections occur only in cells able to support a complete viral multiplication cycle. Such cells are said to be permissive. In terms of cell cultures, the spectrum of permissive cells makes up the host range of a virus. However, the host range may also define the animal species that support a productive infection.

The host range of viruses may be wide, but it may also be very limited. For example, the host range of the parvovirus H-1 comprises man, monkey, hamster and rat as animal species, and most cell types (fibroblasts, keratinocytes, lymphocytes) of human, monkey, hamster or rat origin. On the other hand, the host range of B19, another parvovirus, is restricted to man and human erythroid precursor cells.

Infections of susceptible, but nonpermissive cells do not result in virus production. Three distinct types of such nonproductive infections have been described, and are referred to as abortive, latent and restrictive infections respectively.

An abortive infection occurs in susceptible cells, which sustain some, but not all steps of the viral multiplication cycle. As stated earlier, after successful entry of a virus into a cell, the viral genome has to become uncoated, amplified and expressed. Newly synthesized genomes have to be packaged into progeny particles that need to be released from infected cells.

Virus multiplication can be blocked at all stages of this cycle in susceptible, but nonpermissive cells. For example, the polyoma virus multiplication cycle is blocked at the stage of DNA replication and transcription in embryonal carcinoma cells lacking specific DNA-binding proteins. Similarly, the multiplication of the parvoviruses H-1 and MVM is blocked at the level of DNA and RNA synthesis in some cells that are refractory to virus propagation. Furthermore, some diploid cell strains infected with parvovirus H-1 are proficient in capsid assembly, but no DNA is packaged, and no particles are released from the cells. Cellular functions implicated in these last stages of infection have not yet been identified.

It should also be stated that infection of a permissive cell with a defective virus, lacking one or several essential viral functions, results in an abortive infection. During abortive infections, the virus may exert different cytopathic effects, depending on the viral functions the host cells allow to be expressed. Hence, an abortive infection may result in cell death, if the virus was able to express its cytotoxic proteins, but may also be inapparent.

A latent infection consists of the persistence of viral genomes in infected cells for many cell generations. The viral genomes may persist in an integrated or in an episomal state or both. For example, retroviruses have been found as 'endogenous viruses' in human cells, and have been calculated to persist in these cells for approximately 40 million years. Such latent infections do not lead to host cell killing, but may nevertheless influence biologic characteristics of the cells. As an extreme example, persistence of tumor viruses is associated with cellular transformation leading to increased growth rates, abnormal cell morphology, altered cell metabolism and chromosomal aberrations.

Most viruses establishing latent infections can be rescued from their host cells, provided they have not become defective. Latent infections that can become productive are called persistent infections. To convert a latent into a productive infection, specific treatments have to be applied. For example, superinfection with adenovirus results in a productive infection of persistent adeno-associated viruses, and treatment with phorbol esters activates the multiplication cycle of persistent human immunodeficiency virus type 1 in the human macrophage cell line U1.

Finally, a rare type of nonproductive infection is the restrictive infection, that is observed in cell cultures where only a small subset of the cell population is permissive, or where the cells are only transiently permissive.

Virus Propagation in Cell Culture: Multiplicity of Infection, Defective-interfering Particles

The first cells to be cultured *in vitro* were primary cells freshly isolated from animal tissues. The growth of these cultures is restricted to five to 10 cell generations at most, which limits their value for routine virus propagation. Such primary cultures contain a variety of cell types providing a broad viral spectrum. Thus, they represent a system in which as yet uncharacterized viruses can be propagated and isolated with a high probability of success, and are therefore valuable in virus diagnosis.

Cell lines derived from cancer tissues, capable of indefinite growth in culture, and strains of diploid cells prepared from human or animal tissues (in particular from embryos), capable of growing in culture up to 100 cell generations, contain cells of a specific type and allow the propagation of viruses under defined conditions. However, some viruses, such as papillomaviruses, that require cell differentiation to multiply, cannot be propagated in conventional cell cultures. The choice of the most suitable cell line to propagate a virus will depend mainly on the host range of the virus investigated. Continuous cell lines are useful for the propagation of viruses such as adenoviruses and rhinoviruses. Other viruses, like herpesviruses or enteroviruses, may preferentially be propagated in diploid cell strains of finite life.

To obtain the best yield of progeny virus, it is important to determine the optimal conditions for the initial infection of the culture. One important parameter is the ratio of the number of cells to the number of infectious particles inoculated. This ratio is called multiplicity of infection (MOI). A multiplicity of infection of two (MOI=2) defines an infection with an average of two infectious virus particles per infected cell.

Most, if not all, viruses generate defective genomes, and, consequently, defective particles during their multiplication. This defectiveness is due to deletions in the viral genome, which render the virus incapable of expressing all the functions needed for a productive infection. These particles can arise because their defectiveness is complemented by genes of co-infecting 'wild-type' viruses. Therefore, the formation of defective particles is favored by infections at high MOIs. Adenovirus or parvovirus preparations may contain up to 1000 defective particles per infectious particle. In order to generate as few defective particles as possible, the standard protocol for the propagation of many viruses involves an initial infection at MOI 1×10^{-3} (i.e. one infectious virus per 1000 cells).

Infections are usually performed under cell culture conditions (37°C, pH 7.4) for 30 min to 1 h. The volume of the solution in which cells and viruses are brought together should be as small as possible in order to enhance the probability of host cell–virus contact. The presence of cations facilitates the adsorption process of most viruses. After infection, cell cultures are maintained at 37°C, pH 7.4, except in the case of infections with viruses such as rhinoviruses or coronaviruses, which multiply best at 33°C, the temperature encountered in the nasal mucosa. After one or several multiplication cycles, progeny virus can be purified from the infected cultures.

Virus Propagation in Whole Organisms

In initial studies in virology, the experimental tools for virus propagation and purification were whole laboratory animals or embryonated hen's eggs. Laboratory animals can be infected using the natural route of virus entry. To achieve this, the virus must be brought into contact with either the skin (e.g. papillomaviruses), the digestive tract (e.g. enteroviruses), the respiratory tract (e.g. orthomyxoviruses) or the conjunctiva (e.g. herpes simplex virus). Viral replication may be confined to the site of entry, or progeny virions may spread through the body (generally via the blood or lymphatic stream) with subsequent targeting to specific organs. Laboratory animals also can be infected by injecting the virus directly into specific organs (e.g. the brain for rabies virus). Several days or weeks after infection, the animals are killed, and cell-free extracts from the organs sustaining virus multiplication are used as a source of virus.

Furthermore, fertilized hen eggs can be infected after several days of incubation (depending on the stage of embryonic development required). Most viruses can be grown in the embryonic membranes of fertilized eggs, i.e. the yolk sac (e.g. herpesviruses), the chorion (e.g. poxviruses), the allantois (e.g. influenza virus) and the amnion (e.g. mumps virus). Although laboratory animals or embryonated eggs are still the most appropriate propagation systems for some viruses (animals for arboviruses, coxsackieviruses and rabies virus; eggs for orthomyxoviruses), they tend to be replaced by cell cultures, which are much more convenient to handle. For present-day virology, the use of animals is restricted mostly to research on viral pathogenesis and for production of vaccines.

Virus Purification

Viruses are purified from tissue (culture) homogenates, using various fractionation procedures. Different methods can be used depending on the physicochemical properties of the virions to be purified. Most virions are very sensitive to inactivation by heat, acid, alkali and lipid solvents. Accordingly, in most purification protocols, the virus is maintained at neutral pH and 4°C.

Generally, the supernatant of infected cell cultures provides a relatively clean virus suspension, but in some instances, viruses must be released by breaking up the cells by sonication, homogenization, or repeated freeze–thaw cycles. Thereafter, viruses can be concentrated and partially purified from cellular debris by adsorbing viral particles to erythrocytes, DEAE cellulose, aluminium hydroxide or ion exchange resins. Subsequently, virions are eluted with buffers of specific pH and ionic strength. Viruses may also be precipitated with ammonium sulfate.

These partially purified viruses can be further separated from contaminants by physical methods, in particular by centrifugation. Differential centrifugation, which consists of multiple centrifugation cycles at increasing speed, is used to pellet first the contaminants and then the virions. Centrifugation through a cushion of a dense sucrose solution or through a preformed sucrose gradient (rate zonal centrifugation) separates viruses from contaminants based on their size and shape, i.e. their sedimentation coefficient. Equilibrium centrifugation in cesium chloride or potassium tartrate solutions is used to purify viruses in function of their buoyant density.

Since virus purification aims at the elimination not only of cellular debris but also of defective viral material, equilibrium centrifugation may be the method of choice. This method allows the separation of infectious virions, defective particles, and empty capsids, which generally band at distinct densities (for example, infectious, defective and empty particles of the parvovirus H-1 have a density of 1.41, 1.38 and 1.32 g/ml, respectively). Before equilibrium centrifugation, it may be useful to treat the virus suspension with DNase, since contaminating cellular or viral DNA stick to some viruses and therefore would be co-purified.

Virus preparations can be concentrated by ultracentrifugation, freeze-drying or dialysis against hydrophilic agents such as polyethylene glycol. Generally, concentration will be the last step of a purification protocol, but may be useful before centrifugation steps. Most viruses are heat labile. Therefore, virus preparations should be stored as cold as possible. As a rule, the half-life of many (enveloped) viruses will be years at −196°C, months at −70°C, days at 4°C and only hours at 20°C or minutes at 37°C.

Virus Titration

There exists a variety of ways to determine the amount of virus in a preparation. Two types of methods should be distinguished: infectivity assays that measure the amount of infectious virions in a preparation; and particle assays that determine the number of all virus particles.

For lytic viruses, the most commonly used infectivity assay is the plaque assay. Monolayers of highly permissive cells are infected with serial dilutions of the virus suspension to be tested. After infection, the cells are overlayed with a semisolid agarose gel containing the culture medium, in order to restrict spread of progeny virus to cells next to the initially infected cells. Each infectious particle will give rise to a focus of infected cells, which can be seen as an area of CPE. The number of plaques in a monolayer can be easily counted after staining of the cell monolayer with a vital dye such as neutral red. Living cells stain red, and the infected areas appear as a clear plaque against a red background.

To determine the titer of noncytopathogenic viruses, a modified 'plaque' test was recently established. Prerequisites for this test are the identification of permissive cells allowing the amplification of the viral DNA by factors of more than 200, and the availability of corresponding viral DNA that can be used as a probe. After infection of the cells, at the time when DNA amplification is maximal but before release of progeny virus, the monolayer is transferred to a nitrocellulose membrane. Cells attached to the membrane are subsequently lysed by alkali treatment, and the DNA (cellular and viral) is fixed to the membrane. A single cell that was infected and the viral DNA amplified, can be detected after hybridization of the filter-bound DNA with the corresponding radioactively labelled viral DNA. From the number of infected cells detected by these methods, the number of infectious particles in a virus suspension can be calculated, and is expressed in terms of plaque-forming units (PFU) per milliliter.

Particle assays are much faster and simpler to handle. Two methods are widely used: the hemagglutination assay and particle counting by electron microscopy. Hemagglutination assays are based on the capacity of many viruses to adsorb to erythrocytes. Serial dilutions of a virus suspension are incubated with red blood cells. The viruses can bridge red blood cells, preventing them from precipitating. The virus dilution no longer capable of agglutinating erythrocytes gives a measure of the particle content of the suspension, and is expressed in terms of hemagglutinating units (HAU). This test is not very sensitive and reflects the presence of all the viral particles capable of binding to red blood cells, including defective particles and empty capsids.

Virus particles can also be counted directly by electron microscopy after negative staining. A known volume of virus suspension is deposited on a formvar-coated or carbon-coated copper grid, water and salts are removed, and viruses are subsequently negatively stained and counted. Since some virus may be lost during the washing and staining processes, it is convenient to add a marker to the virus suspension, e.g. in the form of a known concentration of latex particles. The number of latex particles counted enables the determination of particle loss during preparation and therefore allows the virus particle concentration to be determined.

See also: Cell structure and function in virus infections; Defective-interfering viruses; Replication of viruses; Viral receptors.

Further Reading

Marsh M and Helenius A (1989) Virus entry into animal cells. *Adv. Virus Res.* 36: 107.

Stevens EB and Compans RW (1988) Assembly of animal viruses at cellular membranes. *Annu. Rev. Microbiol.* 42: 489.

PSEUDORABIES VIRUS

Saul Kit
Novagene Inc.
Houston, Texas, USA

History

Aujeszky's disease (AD), also known as infectious bulbar paralysis and as pseudorabies (PR) because of its clinical similarity to rabies, was one of the first virus diseases to be recognized. In 1902, Aujeszky recovered the causative agent from infected oxen, cats and dogs, and transmitted it to rabbits and guinea pigs, in which it produced the pruritis and central nervous system (CNS) signs characteristic of naturally occurring PR disease. AD was probably present in the United States as early as 1813, but it was not until 1931 that Richard Shope established that 'mad-itch' disease of Iowa cattle was serologically identical to AD, and that a mild and usually unrecognized disease of swine was produced by the mad-itch agent, that is, PR virus (PRV). Shope showed that PRV was infectious for adult swine and was transmitted through skin abrasions to cattle pastured with infected swine. In 1933, minced rabbit brain and testicles were first used by Traub to cultivate PRV *in vitro*. In the 1940s and 1950s, established mouse cell cultures were used to serially propagate the virus. Virus isolations are now routinely made in sensitive rabbit cell cultures and propagated in a variety of vertebrate (porcine, bovine, monkey, chick, etc.) cell lines.

Taxonomy and Classification

PRV (swine herpesvirus-1) is a member of the family *Herpesviridae*, subfamily *Alphaherpesvirinae*. The genome of PRV is a linear, double-stranded, class 2 herpesvirus DNA molecule approximately 90×10^6 D in size and encodes about 80 genes. It is composed of two components, L and S. The S component consists of a short unique (U_S) sequence bracketed by large (about 10^7 D) inverted repeats. (I_R). High-frequency inversion of the U_S component, but not the L component, is observed during virus growth, so that the S genome is found in two isomeric forms. The ends of the PRV genome are unique; no terminal redundancy is present. Antigenic differences have not been detected among strains of PRV, but strain differences demonstrable by restriction endonuclease fingerprinting are common and have been used for epidemiological studies and control programs. Deletion mutations in the U_S region of the PRV genome occur 'spontaneously' following passages in chick and bovine cells.

Molecular Biology

PRV genes are expressed during lytic infection as three groups, immediate early (IE), early and late, in a coordinatedly regulated and sequentially ordered cascade. The IE genes are transcribed by the host cell RNA polymerase in the absence of *de novo* viral protein synthesis. Early gene transcription does not require viral DNA synthesis. Late genes consist of two subclasses depending on their requirement for viral DNA synthesis. Inhibition of viral DNA synthesis moderately decreases the accumulation of quasi-late gamma 1 transcripts, but totally prevents formation of gamma 2 late transcripts. Most late gene products are virion structural proteins.

The PRV genome, unlike the HSV genome, contains only one IE gene, but since this IE gene is located in the I_R region of the PRV genome, there are two copies per genome. The IE gene specifies an unspliced 5.1 kb transcript which is translated to a protein of about 180 kD. The PRV IE protein is required for continuous transcription of early and late PRV genes and for shutting off the synthesis of its own RNA. The PRV IE gene contains five to six different consensus sequences in its promoter region and is capable of *trans*-activating a variety of viral and cellular class II and class III promoters by interacting with host transcription factors and specific promoter sites.

Recent studies have shown that a latency associated transcription (LAT) unit completely overlaps the PRV IE gene. This transcription unit is at least 11.8 kb long, extends from the U_L region to the I_R region of PRV DNA (0.69 to 0.77 MU), has an antiparallel orientation to the IE transcript, and is transcriptionally active in the trigeminal ganglia of latently infected swine. Poly(A)-negative LATs of 4.5–5.5 kb and 1.0–2.0 kb have been detected in this region during latency, suggesting that precursor LAT transcripts may be spliced. The roles of the LAT transcripts in the maintenance of or reactivation from the latent state have not as yet been delineated.

Early PRV genes function in DNA metabolism and replication. These include DNA polymerase (0.18–0.29 MU), a DNA-binding protein (0.14–0.18 MU), TK, ribonucleotide reductase and probably, a helicase/primase. Replication of PRV DNA occurs in two phases. In the first phase, the viral DNA enters the cell nucleus and circularizes by ligation of the ends of the linear DNA molecules. This process does not require expression of virus functions. The first rounds of replication are associated with these circular structures. The main origin of replication is located in the region of the molecule bearing the I_R. Replication generates theta-like structures. As with HSV, a second origin of replication is also present in the middle of the U_L region of the genome. Later in the replication process, concatemers in head-to-tail alignment are generated. Cleavage of concatemeric DNA to generate mature, genome-sized linear DNA molecules requires the expression of at least nine viral gene functions and involves site-specific cleavages from signals present on both ends of the PRV genome. Both cleavage and packaging are intimately linked to capsid assembly. The molecular structure of latent PRV DNA in trigeminal ganglia has not been elucidated, but appears to be concatemeric (or circular), and is presumably generated during an aborted DNA replication cycle.

Geographic Distribution

PR is enzootic in swine in most parts of the world, except Australia, Canada and Norway. PR has been a consistent source of significant losses in swine and cattle due to increased mortality, reproductive failures and lower growth performance.

The cost of PR outbreaks has been estimated to be about $10 000 per year per outbreak, with a total annual cost in the US of from 30 to 72 million dollars. In most states in the US, PR diagnosis in a herd results in a quarantine of the swine with movement limited directly to a slaughter facility.

Host Range

Domestic and feral swine are the only identified reservoir hosts in which PR transmission is maintained. All other species are considered to be 'dead-end' hosts (i.e. the virus dies within the host without passing infection to other animals). PRV causes an acutely fatal infection in many nonporcine species, such as cattle, sheep, dogs, cats, raccoons, opossum, rats, mice and chicks. These species become infected after co-mingling with pigs or ingesting infected meat. The development of PR in carnivores may often be the first indication that the disease exists in swine on a farm. Reports of infection of humans have been rare and have not been substantiated.

Epidemiology

It seems probable that over many decades, or even centuries, commensal relationships developed between PRV and swine, so that PRV infection in swine became a largely subclinical disease under the then existing management practices. Before the late 1960s, sporadic outbreaks of PR were reported but not until the mid-1970s was PR considered a serious threat to the swine industry. Serologic surveys of swine sent to slaughter in the US revealed an increase in the PR-positive herds from 0.56% in 1974 to 8.78% in 1984. Until 1977, only 2–24 PR outbreaks in pigs were recorded in the then Federal Republic of Germany, but from 1977 onwards, their numbers increased, reaching 2000 recorded outbreaks in 1987. PRV was first isolated from New Zealand pigs on the North Island in 1976 and now threatens to infect the South Island. PR was first reported in central Japan in 1981 and has been spreading rapidly throughout that country.

The increase and spread of PR disease may be, at least in part, attributed to the emergence of more virulent PRV strains or variants. A more significant cause is the changed management within the swine industry. PR disease is most prevalent in areas that employ intensive swine housing practices and where swine are most concentrated. In Iowa in 1984/1985, only 2% of herds with fewer than 50 sows were seropositive for PRV, whereas the herd prevalence rate for herds with more than 100 sows was 55%. The incidence of PRV is highest in Europe in those areas with cold climates and intensive hog industries, such as Holland, Belgium, France and Hungary, and is less of a problem in countries with less intensive hog production.

A salient factor in the spread of PR disease is the capacity of PRV to establish latent infections in sensory ganglia of infected swine. A small percentage of these latently infected animals sporadically shed virus in saliva and nasal discharges, so that licking, biting and aerosols can cause transmission. Losses from overt disease in swine occur when nonimmune pregnant sows, or swine less than 3 months old born to nonimmune sows, are infected. When breeding sows have adequate antibody levels, overt clinical disease in their progeny may be greatly reduced, but the baby pigs may nevertheless become latently infected and later develop clinical signs of disease and/or transmit infectious virus. On farms where breeding, farrowing, nursery, growing and finishing operations are conducted separately, significant losses may occur when weaned swine from several sources are brought together into a growing–finishing facility.

Transmission and Pathogenesis

PR is usually transmitted by acutely infected pigs, which produce large amounts (10^4 PFU) of infectious PRV, or by latently infected animals, which sporadically shed small amounts of virus even though they may not show signs of disease. The primary source of virus spread is intimate nose-to-nose contact involving expelled body fluids of oronasal origin. Often, PRV is transmitted by pregnant sows to their highly susceptible newborn offspring. Latently infected pigs may be experimentally induced to excrete PRV by stress-related factors, simulated by the drug, dexamethasone. Factors influencing infection include the movement of swine (e.g. the introduction into a breeding herd of latently infected sows from outside the farm), swine and wildlife density, housing and ventilation design, access of other animals and wildlife to swine, dead pig disposal, and the source and handling of swine feed. Airborne transmission of droplets containing virus between farms can play a role in initiating disease outbreaks. The airborne transmission over a distance of 15–80 km of PRV originating in northern Germany caused the recurrent epizootics of 1984 to 1988 in both conventional and specific pathogen-free Danish herds. Contamination of feed, water, bedding and floors may likewise provide susceptible pigs with access to virus. After successful transmission, virus replication occurs in the tonsillar and pharyngeal tissues and in the olfactory epithelium. From these sites, the virus spreads to the medulla by way of the olfactory nerve system and the epineural lymph of the trigeminal and glossopharyngeal nerves, and on further replication reaches the brain. Multiplication of virus in the respiratory tract of swine facilitates its spread within the body, since macrophages and leukocytes become infected and may carry the virus to various body organs, particularly to the placenta in which it may multiply and subsequently invade the fetus. PRV can be isolated from buffy coat cells, providing further evidence of absorption on or within leukocytes. Feces and urine are not considered significant sources of infection. Most swine infected with virulent virus shed infectious virus for from 12 to 21 days. By contrast, most wildlife and domestic species excrete virus for only 1 to 2 days before death and in amounts insufficient for transmission of infection.

PRV, like other herpesviruses, is extremely labile, so that experimental virus stocks must be stored at −70°C. The virus can, however, survive in damp bedding for 140 days at 4°C and for 40 days at 37°C, for 5 weeks on shelled corn and for 3 weeks on moist meal. The virus has a very short survival time on clean concrete, green plants or well-cured hay. Heat, direct sunlight and dry conditions quickly inactivate PRV. PRV can survive in ground flesh for weeks. Dead baby pigs or infected placentas that are ingested by cats and dogs or wildlife, cause fatal infections. Dogs have become fatally infected with PRV after fighting with feral pigs. Cattle can become infected through contaminated feed or when secretions from infected pigs are introduced through existing bovine skin abrasions. Simultaneous wound contamination at the time of bite infliction also occurs, particularly in the rear limbs of cattle. To avoid fatal outbreaks of PR, it is essential that cattle on farms be isolated from swine. Iatrogenic infections, involving accidental injections of conventional, inadequately attenuated PRV vaccines into lambs and chicks, have caused fatal infections on several occasions.

Clinical Features of Infection

Young pigs infected with PRV exhibit a high morbidity, high mortality, fever, vomiting, diarrhea and CNS signs. The CNS signs include tremors, incoordination and pronounced muscle spasms, circling and intermittent convulsions accompanied by excess salivation. Death occurs by 8 days after infection. Maternal antibodies protect piglets from signs of disease, but not infection. Older (feeder) pigs show a high morbidity, variable mortality, fever, anorexia, constipation, vomiting, depression, growth arrest, listlessness, coughing and sneezing, labored breathing and CNS signs. In pregnant sows, mortality may be low, but respiratory and gastrointestinal signs are common, as are reproductive problems. Infection of sows before day 30 of gestation may result in death and resorption of the embryos. Infection after day 30 of pregnancy may terminate in abortion, or by the delivery of stillbirths and mummified fetuses. PR in cattle assumes a rapid and fatal course. Death often occurs by 2 days after the first signs of illness appear. The first sign is a decrease in milk yield followed by violent licking of parts of the body. As pruritis increases in intensity, the cow becomes frenzied, bites and gnaws at the skin and rubs its head and neck against hard objects. Rabbits, cats and dogs are among the most vulnerable of species. Subcutaneous injection of 1 PFU of PRV can be lethal to rabbits. In dogs, pruritis is accompanied by drooling of saliva and plaintive howling simulating true rabies. In cats, the disease may progress so rapidly that pruritis is not observed.

Pathology and Histopathology

Following primary oronasal infection of swine, virus replicates in the oropharynx. At 24 h after infection, viral antigens and DNA can be detected in cells of the nasal epithelium. After 48 h, areas of the epithelium become necrotic and the infection reaches the stroma killing many cells. Enveloped virus particles are detectable by electron microscopy in and around infected epithelial cells and fibroblasts. The stroma of the oronasal epithelium contains many blood vessels and neuronal axons of the trigeminal, olfactory and vegetative nerves, through which the virus can travel via the axoplasm to cranial nerve ganglia and the medulla and pons. Virus may establish latency in the nuclei of neurons, in which case production of infectious virus ceases, but the viral genome itself is maintained in a portion of the cells, or the virus may continue to spread within the CNS producing ganglioneuritis, meningoencephalitis, perivascular cuffing and focal gliosis associated with extensive necrosis of neuronal and glial cells. A common finding in severely infected swine is congestion of the nasal mucosa and pharynx, tonsillitis and rhinitis. Pulmonary congestion and edema are frequently observed in such cases. In the larger air passages, viral capsids and viruses may be seen in bronchial and bronchiolar epithelium, lymphocytes and alveolar macrophage. Swine that recover from PR may shed virus sporadically in their nasal secretions. Others from which virus cannot be isolated by conventional means may yield virus after co-cultivation of tonsillar or trigeminal ganglia with susceptible cells. That recovered swine are latently infected may also be shown by demonstrating PRV DNA or RNA in trigeminal ganglia. *In situ* hybridization techniques with polymerase chain reaction-generated probes are particularly useful for this purpose.

Immune Response

One highly sulfated, secreted glycoprotein [gX (90–98 kD)] of unknown function, and at least six structural glycoproteins are synthesized in PRV-infected cells. The PRV structural glycoproteins, i.e. gI (115–122 kD), gII, gIII (92–98 kD), gH (about 84 kD), gp50 and gp63, are homologous, respectively, to HSV-1 glycoproteins, gE, gB, gC, gH, gD and U_S7, while gX is homologous to HSV-1 glycoprotein gG. PRV gII, is a complex of three related glycoproteins that are derived from a common precursor. Two of these, gIIb (72–78 kD) and gIIc (50–58 kD), arise by proteolytic cleavage of the larger gIIa precursor (110–125 kD) and are covalently linked via disulfide bonds. gI and gp63 are noncovalently complexed to each other. Glycoproteins gp50, gII and gH are involved in virus penetration of host cells and are essential for PRV replication. Glycoproteins gI, gp63, gX and gIII, are nonessential for virus replication. Indeed, a quadruple deletion mutant lacking gI, gp63, gX and gIII is viable. The structural glycoproteins are found embedded in the nuclear and cellular membranes of infected cells as well as on the surface of mature, enveloped viruses and are major antigens that interact with the host immune system to elicit humoral and cell-mediated immune responses.

PRV glycoproteins gX, gp50, gp63 and gI map in the U_S region of the viral genome in the order given and are flanked on the left by a protein kinase gene, and on the right, by genes encoding 11 and 28 kD proteins. The gII, gIII and gH genes are found in the U_L region of the PRV genome. The gII gene maps at 0.105 to 0.130 MU. The gH gene is located immediately downstream from the PRV thymidine kinase (TK) gene [0.45 map units (MU)], while the gIII gene maps about 5 kb upstream from the PRV TK gene. gIII has a role in virus attachment to host cells through a cellular heparin-like receptor. In the absence of gIII, PRV can attach to cells by an alternative pathway not mediated by the cell receptor, and virus release, though retarded, still occurs.

Until recently, it was accepted that most viruses have only one protein or glycoprotein that is important for induction of immune responses and protection in the host. However, it is now clear that several herpesvirus glycoproteins can induce protection. Convalescent sera from animals infected with PRV have antibodies to gp50, gII and gIII, which can neutralize viruses outside cells and destroy virus-infected cells by antibody-dependent cellular cytotoxicity or antibody-dependent, complement-mediated cytolysis. In pigs, monoclonal antibodies against gI, gII, gIII and gp50 all neutralize virus, while passive immunization of mice or pigs with these monoclonal antibodies protects them from lethal virus challenge. The level of antibodies in pigs does not, however, correlate with protection against virulent virus infection, indicating that major histocompatibility complex-restricted (MHC) cytotoxic T lymphocytes and natural killer cells are also important. Unfortunately, data on these cellular immune elements are as yet limited. It is known that gI, gII and gIII are important targets, but not the only targets, for cytotoxic lymphocytes. Furthermore, experiments with bovine herpesvirus-1-infected cells suggest that glycoproteins homologous to PRV gII and gp50 may be targets for natural killer cells. The fact that the PRV envelope is composed of several glycoproteins is, in a sense, fortunate from the point of view of the host. Mutagenic alterations in a single viral glycoprotein do not abrogate the defense mechanisms of the host. Indeed, although almost all of the modified-live PRV vaccines harbor deletion mutations in viral glycoproteins, they are highly protective.

Prevention and Control of Pseudorabies

The strategy selected to control PR has depended on:

(1) the type of operation carried out by the producer (farrow-to-finish, seedstock, feeder pig producer, feeder pig finisher); (2) financial considerations; (3) availability of suitable replacements; (4) the disease profile of the herd and; (5) PRV status in the area. Slaughter of an entire herd and repopulation is considered with single isolated outbreaks in noninfected districts, but may also be necessary when disease is detected in seed stock operations. Slaughter is performed on quarantined pigs after they reach market weight of 220–260 pounds. This is done to save the meat and market value of the animals. Hence, it may take 6 months to slaughter an entire herd. The slaughter strategy is expensive and disruptive, results in the loss of valuable blood lines, and is impractical if the source of infection cannot be identified or if the re-infection risk is high. Offspring segregation has been used to produce PRV-seronegative gilt replacements from positive sow herds. Serological testing and the culling of seropositive pigs has been recommended in low-prevalence areas when the proportion of seropositive individuals in the herd was about 20% or less and when there was no evidence that virus was circulating between the sows of the breeding herd and their offspring. Seropositive pigs were then replaced with seronegative gilts. Factors impeding control through test and slaughter are the expense of repeated serologic testing, large herd size and the density at which pigs are maintained, especially if they are housed in proximity to other infected farms. Test and slaughter programs in England and Denmark have greatly decreased PR outbreaks but have not completely eliminated them.

Vaccination has been used alone or together with other strategies for over 25 years to stop PR outbreaks, to reduce virus shedding by infected pigs, to minimize clinical disease and mortality, to reduce losses from secondary bacterial infections and to reduce the incidence of latent infections. Both killed and modified-live virus (MLV) vaccines have been used, but despite frequent vaccination with conventional vaccines, the incidence of PR outbreaks in areas of intense husbandry has continued to increase, principally because it has hitherto been impossible to identify pigs latently infected with virulent PRV in a vaccinated herd. This situation has now changed with the introduction of genetically engineered, deletion-marker vaccines. Currently, 42 states of the United States have PRV eradication programs.

Conventional MLV PRV vaccines, as illustrated by the Bucharest, Bartha and NIA-4 strains, were obtained by repeatedly passing PRV in heterologous chick or bovine cells. Attenuating gI deletion mutations were thereby induced into these three vaccine strains, and, in addition, attenuating gIII point mutations were induced in the Bartha vaccine strain.

Numerous recent studies have revealed that the TK of PRV and other herpesviruses is one of the most important factors contributing to virus virulence. Therefore, irreversible deletion mutations in the TK gene have been engineered by recombinant DNA techniques to greatly increase the safety of vaccines. The Bucharest vaccine strain was the first PRV strain used by the author to develop a TK deletion mutant, but virulent Iowa, Aujeszky and NIA-3 PRV strains have also been attenuated by others through TK deletions. In addition, deletion mutations have been engineered in the gIII gene, the gI gene, or the gX gene of TK-deleted vaccines to provide diagnostic markers by which vaccinated pigs could be distinguished serologically from pigs harboring virulent PRV.

Diagnosis

The clinical signs of PR are sufficiently characteristic to permit a preliminary diagnosis which may be verified by PRV isolation, the fluorescent antibody staining of frozen tissue sections, and serum neutralization (SN) tests. Screening ELISA and latex agglutination tests are, however, more rapid and more sensitive than SN tests. With the advent of gene-deleted marker vaccines, differential blocking ELISA diagnostic tests are now being used. Marker-specific differential diagnostic tests have been developed for each of the gIII-, gI and gX-deleted vaccines. When pigs are vaccinated with the OMNIMARK vaccine, a TK-deleted gIII-deleted vaccine, for example, they develop antibodies to various PRV proteins, but not to gIII. Wild-type PRV-infected pigs develop antibodies to most PRV proteins, including gIII. Hence, a differential blocking ELISA test, which uses gIII as the antigen to coat microtiter wells and a gIII-specific monoclonal antibody conjugated to horseradish peroxidase to react with the antigen, distinguishes serologically between OMNIMARK-vaccinated and wild-type virus-infected pigs. Undiluted test sera are used with this gIII-blocking ELISA test. The specificity of the test approaches 100% and the sensitivity is about the same as that of screening ELISA and latex agglutination tests. gIII antibodies can be detected by about 7–10 days after pigs vaccinated with OMNIMARK become infected with wild-type virus and for a year or more after pigs are latently infected with wild-type PRV. SN antibody tests are often negative under these conditions. The wild-type virus-infected pigs identified by the differential diagnostic test can be removed during herd clean-up. Vaccinated swine that are negative in the differential tests are classified by the USDA as PRV-negative and their interstate and intrastate movement is permitted.

Future Perspectives

The availability of gene-deleted marker vaccines for PRV has stimulated new efforts for the control and eradication of PR disease in the US, Japan, New Zealand and the countries of the European Economic Community. It has also been recognized that latent PRV infection is prevalent in free-roaming feral swine, which may come in contact with and transmit PR disease to domestic swine and other livestock. The US Animal Health Association has therefore recommended that serological surveillance of feral swine be expanded and that procedures for their vaccination (oral?) with marker PRV vaccines be developed.

Finally, it should be emphasized that attenuated, gene-deleted marker vaccines are promising candidates for live vaccine vectors, which can confer protection against both PR and other swine diseases. Model experiments in which foreign DNA sequences have been inserted at several sites in the PRV genome have recently been carried out. In one such experiment, a live, attenuated PRV expressing the envelope glycoprotein of hog cholera virus (classical swine fever) was prepared and its efficacy in protecting swine against both PR and hog cholera investigated, with promising results. Many more such PRV vector-based subunit vaccines may be anticipated.

See also: Epidemiology of viral diseases; Herpes simplex viruses; Immune response; Latency; Vaccines and immune response; Vectors.

Further Reading

Gustafson DP (1986) Pseudorabies. In: *Diseases of Swine*, 6th edn, p. 274. Ames, Iowa: Iowa State University Press.

Kit S and Kit M (1991) In: Melnick JL (ed.) *Progress in Medical Virology*, vol. 38, p. 128. Basel: Karger.

Mettenleiter TC, Kern H and Rauh I (1990) *Virology* 179: 498.

van Oirschot JT (ed.) (1989) *Vaccination and Control of Aujeszky's Disease*. Dordrecht: Kluwer

R

RABBIT FIBROMA VIRUS

See Poxviruses

RABBIT HEMORRHAGIC DISEASE VIRUS

See Vesicular exanthema virus and caliciviruses of pinnipeds, cats and rabbits

RABBITPOX VIRUS

See Mousepox and rabbitpox viruses and Poxviruses

RABIES VIRUS

George M Baer and Noël Tordo
Laboratorios Baer
Mexico City, Mexico

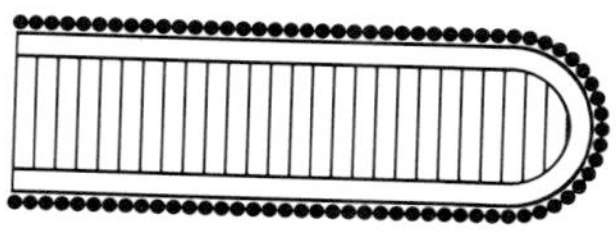

History

The dramatic clinical signs of rabies have been recorded from early times: hyperexcitability, increased tendency of animals to bite, jaw paralysis, changed facial expression, and transmission of the disease to other animals and, eventually, humans. One of the earliest references to the disease is from the third millenium BC, in the Eshnunna Code preceding the Code of Hammurabi:

> If a dog is mad and the authorities have brought it to the attention of the owner; if he does not keep it in and it bites a man and causes his death, then the owner shall pay 40 shekels of silver. If it bites a slave and causes his death he shall pay 15 shekels of silver.

Rabies in dogs and its transmission to man were well recognized by the time of Aristotle. Pliny and other writers referred to the influence of the dog-star Sirius, including the increased susceptibility of dogs to rabies during the so-called 'dog-days' of summer.

Outbreaks of rabies in dogs, foxes and wolves were reported from most European countries throughout the Middle Ages, predating the current European outbreak in foxes which began in the early 1940s. The first cases in the United States were reported in Virginia foxes in 1753. Bites by rabid wolves (animals often reported rabid in Iran, Afghanistan and the Soviet Union) still rank as the most dangerous source of the disease for humans. Mortality rates after wolf bites are well over 50%, although dogs still cause over 90% of human rabies deaths world-wide; in many developing countries the numbers of deaths still reach many thousands annually.

Rabies in vampire bats was first suggested by chronicles of Spanish conquistadores in the sixteenth and seventeenth centuries and confirmed in Brazil, Argentina and Trinidad in the early 1900s. Vampire rabies is still a major animal health problem in most Latin American countries, with hundreds of thousands of cattle deaths yearly. Insectivorous bats are a sporadic problem in North America, Latin America and Europe.

The classic studies of Pasteur beginning in 1881 showed that the central nervous system was the principal site of rabies virus replication, that the virus could be passaged in experimental animals, and that vaccines could be prepared from virus thus passaged. These studies led to his first preparing and then administering human vaccine in 1885. It is surprising that the same vaccine was not used for mass dog vaccination programs until 1921 when Umeno and Doi initiated urban control programs in Japan.

The first specific diagnostic tool, the 'Negri body', was discovered by Adelchi Negri in 1903. This rather insensitive tool was commonly used until the fluorescent antibody technique replaced it in the late 1950s.

Taxonomy and Classification

Rabies viruses belong to the family of RNA viruses *Rhabdoviridae*. These viruses have nonsegmented negative-strand RNA genomes enclosed in a lipid envelope derived from the host cell. The bullet-shaped virions are all composed of a nucleocapsid or ribonucleoprotein (RNP) core and an envelope in the form of a lipoprotein bilayer membrane closely surrounding the RNP core. The outer glycoprotein (G) coat is responsible for the induction of neutralizing (anti-G) antibodies. The virion RNP core is tightly wound into coils. Extending from the outer surface of the envelope is an array of spike-like G projections. The rabies-like viruses cross-react serologically with rabies nucleoprotein (N) antisera to varying degrees and include Lagos bat, Mokola, Duvenhage, Obodhiang and kotonkan viruses.

Geographic and Seasonal Distribution

Human rabies occurs world-wide, except in certain islands and peninsulae where the disease has never occurred (Australia and a number of small Caribbean islands) or where it has been eliminated (England, Ireland, Japan, Taiwan, Spain, Portugal). The infected countries can be divided into those that have controlled canine rabies with effective canine vaccination programs (but where wildlife rabies remains) such as all the western European countries, Canada, the US and some Latin American countries, and those countries where canine rabies is still endemic such as most of the African and Asian countries.

Genetics

Since 1981 the molecular genetics of rabies virus has followed two distinct strategies, work with viral messenger RNAs and work with the genome itself; the latter has permitted simultaneous analysis of the intergenic regions. Complete sequence is known for the PV and SAD strains of rabies and the rabies-related Mokola virus. The sequence data have helped to locate the glycosylation sites as well as the signal and transmembrane peptides on the G protein. The potential phosphorylation sites have also been characterized on the N and M1 (NS or phosphoprotein). In addition, the important regions in eliciting humoral and T cell-mediated immunity were mapped along the G, M1 and N polypeptides.

Evolution

Conservation of the elements involved in transcription and replication

A comparative analysis of rabies virus strains, rabies-related viruses, rhabdoviruses in general, and paramyxoviruses indicate that strong selective pressure has stressed the following major elements controlling the gene expression: (1) the start and stop transcription signal bordering each cistron; (2) the promoters for RNA synthesis and encapsidation at the 3′ and 5′ genomic ends, respectively; (3) the RNA-dependent RNA polymerase (L protein), by far the most conserved polypeptide; it exhibits six highly conserved domains separated by variable areas, a distribution in agreement with the notion of independent functions (RNA synthesis, capping, polyadenylation and phosphorylation) concatenated within the polypeptide.

An intermediate place of unsegmented negative-strand RNA virus evolution

Proteins other than the L polymerase are poorly conserved. The G protein, although maintaining limited sequences around the main glycosylation site, has varied a great deal, as might be expected for the polypeptide that mediates the first contact with the host. The N protein has also retained small sequence stretches most likely involved in direct interaction with the genomic backbone. The M1 phosphoproteins and M2 matrix proteins are highly varied, as are the untranslated genomic areas.

The G–L intergene is of particular interest because of its large size and its inability to encode substantial peptides. In the same genomic location a fish rhabdovirus (infectious hematopoietic necrosis virus or IHNV) encodes an mRNA of similar size, as do the paramyxoviruses for the HN hemagglutinin protein. The rabies G–L intergene is presumed to be a remnant gene, baptized ψ for pseudogene. It places rabies virus in an intermediate position in the evolution of unsegmented negative-strand RNA viruses, between the rhabdoviruses with condensed genomes (vesicular stomatitis virus, for instance, shows the dinucleotide GA between the G and L), IHNV virus and the paramyxoviruses. The L protein homology studies independently confirm that rabies virus was closer to paramyxoviruses than to VSV.

The rabies ψ pseudogene: the best thermometer of evolution

Since it is a nonprotein coding region highly susceptible to mutation, changes in the rabies ψ pseudogene are more likely to represent the natural evolution of the virus outside any external selective pressure, and this site may therefore be a suitable target for epidemiologic studies. By use of the polymerase chain reaction (PCR) technique directly from brain samples it was shown that:

1. There is up to 18% divergence in the ψ pseudogene of different vaccine strains, most of which are derived from the original Pasteur strain.
2. Wild isolates from a given geographic area are clearly related, with less than 2.5% divergence.
3. Wildlife isolates differ by approximately 15% from vaccine strains (but complete cross-protection by those vaccines is still achieved).
4. In West Africa (Ivory Coast, Cameroon, Niger and Morocco) where vaccine failures have often been noted, there is a greater (25–40%) divergence from the vaccine strains.
5. European bat isolates tend to be completely different from vaccine strains at the ψ pseudogene. At the G and N gene, those known to be important in initiating the immune response, the divergence between isolates from European bats and isolates from vaccine strains is comparable to the difference vs Mokola virus (against which rabies vaccines are ineffective).

Serologic Relationships and Variability

The lyssavirus genus includes a wide variety of rabies viruses, both laboratory-adapted and naturally occurring 'street' viruses, almost all of which can be differentiated through the use of antinucleoprotein monoclonal antibodies. Monoclonal antibody analysis of various rabies variants indicates a remarkable stability in their pattern over many years, with a bat virus from all Mexican freetail bat, for instance, giving the same pattern in the early 1980s and the early 1990s. The analysis of variants also shows that the predominant virus circulating in a given epidemiologic area or 'niche' (such as raccoon rabies in eastern US) is also found in the animals that raccoons bite such as skunks and groundhogs; bat viruses tend to give a different pattern. The common rabies vaccines prepared from 'fixed' viruses such as LEP (low egg passage Flury), HEP (high egg passage Flury), ERA, PM and PV virus strains protect against all rabies 'street' viruses.

Within the lyssavirus genus there are also 'rabies-like' viruses, originally isolated in Africa and markedly different from the rabies viruses in their NP pattern. They include a virus isolated from a bat in Nigeria (Lagos bat), from a human bitten by a bat in South Africa (Duvenhage), or from shrews in Nigeria (Mokola). Common rabies vaccines do not protect against these virus strains.

Epidemiology

Rabies is endemic world-wide, either in dogs or in wild animals, except in those limited areas of the world (mostly islands) where rabies has never existed or where it has been eliminated. Most developing countries of Asia, Africa and Latin America have many cases of rabies in dogs which result in many human antirabies vaccinations and many human deaths. In most of thse countries the approximate rate of human antirabies vaccination is 1:1000 population annually.

The picture is quite different in regions where rabies in dogs has been controlled and the disease is prevalent in a variety of wild animals, such as red foxes (western and eastern Europe and Ontario, Canada), arctic foxes (all circumpolar areas), skunks (midwest US), raccoons (eastern US), mongooses (Asia, South Africa, Cuba, Puerto Rico, the Dominican Republic, Haiti and Grenada), wolves (Iran, Afghanistan and the former USSR), vampire bats (northern Mexico to northern Argentina), insectivorous bats (Latin America, the US, Canada and, rarely, western Europe), and frugivorous bats (South America and the Caribbean). In those areas rabies is mostly transmitted within one species and rarely outside, with, for instance, rabies transmitted freely within skunk populations but only occasionally to other animals they bite, such as cows. Rabies rarely 'crosses over' and begins an outbreak in a species bitten by the original species involved.

But rabies in Latin America is almost always transmitted by one species of bat, the common vampire *Desmodus rotundus*, an animal that feeds solely on blood, mostly cattle blood. Every year there are hundreds of thousands of rabies deaths in cattle bitten by vampire bats; rare human outbreaks have also been reported, mostly in Trinidad, Brazil and Peru. Rabid insectivorous bats, on the other hand, rarely infect other species such as humans, cats and wild animals.

Transmission and Tissue Tropism

Rabies is almost always transmitted by saliva via bites or scratches. In rabid animals the submaxillary salivary glands (the most commonly involved extraneural organ) are infected about 75% of the time, with levels of virus often exceeding 10^8 infectious doses (mouse or tissue culture) per ml. After its introduction by a bite the virus stays at the local site for a variable incubation period, usually several weeks, then advances up the peripheral nerves to the central nervous system; it often reaches the brain and the salivary glands before changes in the behavior of the animal occur. There is, then, a dangerous preclinical period of virus excretion in which the animal can infect other animals or humans without exhibiting any clinical abnormality such as jaw paralysis, excitation or changes in locomotion. During the terminal centrifugal spread of the disease (again by peripheral nerves) a variety of organs are commonly infected, including salivary glands, skin, lungs, kidneys and gonads.

Pathogenicity

Rabies viruses differ in their pathogenicity, but those differences have been difficult to measure. Certain laboratory adapted viruses (i.e. 'fixed' viruses) are highly invasive when injected either intracerebrally or intramuscularly. The invasiveness (as measured by the difference in mortality after intracerebral or intramuscular injection) of two 'street' isolates from one species in one geographic area can differ by as much as a thousandfold; it is not clear what factors this difference is due to, but it may include some factor in the saliva, the number of defective-interfering particles in the sample, and the particular characteristics of the virus. The pathogenicity also depends on the species of animal bitten, some species such as foxes being exquisitely susceptible to the virus, while others, such as the opossum, are much more resistant. Humans are quite resistant to rabies, with the expected mortality in persons bitten by rabid animals yet untreated being far below 50%; in observations made in the late 1800s, before rabies treatments were initiated, it was noted that 15% of persons died after severe and multiple hand bites by rabid dogs, while 85% survived without any treatment.

Clinical Features of Infection

Rabies in humans usually begins with mild and nonspecific symptoms which lead to an initial diagnosis of a common and minor bacterial or viral infection. A specific symptom often noted during the progression of the disease is pain or paresthesia at the bite site (usually the hand or foot). The acute neurological period begins with obvious nervous dysfunction, often including hyperactivity and, later, paralysis. Fever, nuchal rigidity, muscle fasciculation, convulsions, hyperventilation and excess salivation may be seen. The majority of agitated patients ('furious rabies') develop marked anxiety or agitation, sometimes accompanied by hydrophobia and aerophobia; during periods of agitation the patient's mental state fluctuates between periods of increasingly severe agitation and periods of normal behavior or depression. This acute period ends after 2–7 days. 'Paralytic rabies', with paralysis dominating the symptoms, is seen in about 20% of patients. Coma follows a transition period which begins with apneustic breathing; death is thought to be caused by respiratory arrest.

Pathology and Histopathology

There is little gross pathology in humans or animals that die of rabies; congestion of the meningeal vessels may occasionally be noted. The most common histologic change is perivascular infiltration, especially in the brainstem (the pons and medulla) as well as the spinal cord, basal ganglia and cerebral cortex. The neuronal degeneration and other inflammatory changes are variable. Negri bodies, specific (pathognomonic) cytoplasmic inclusions, are found in approximately 75% of rabid animals; these inclusions are most notable in the Purkinje cells of the cerebellum and in the hippocampal gyri (Ammon's horn). The absence of these inclusions does *not* rule out the diagnosis of rabies.

Geographic and Seasonal Distribution

The geographic distribution of the disease is summarized in the Epidemiology section. Rabies in dogs appears to have a somewhat seasonal character, with an increase in cases during the summer months; this was early attributed to the influence of the dog-star Sirius, but actually may be due to heat cycles in female dogs which lead to fighting among male dogs and an eventual increase in subsequent rabies cases. Rabies in terrestrial wild animals has not been reported to be seasonal; bat rabies, however, is distinctly seasonal in nontropical areas.

Host Range and Virus Propagation

Rabies virus infects a very broad array of animal species, perhaps the widest range of any animal virus. Commonly infected are dogs, wolves, foxes, jackals, skunks, raccoons, raccoon dogs, vampire bats, insectivorous bats and mongooses; the animals bitten by these species develop sporadic cases of rabies (these include cats, cows, horses, badgers and woodchucks) although the virus in those species rarely becomes enzootic. Humans are mostly infected after bites by rabid dogs.

Propagation is almost always by bite. Virus appears in the saliva of infected animals either when they are symptomatic, or for days or weeks before clinical signs appear. The presymptomatic excretion of virus in saliva is a very important epidemiologic characteristic for virus survival, and a crucial piece of information for physicians who judge whether an exposure has occurred in persons bitten by rabid (or possibly rabid) animals.

Experimental hosts include mice (the animal most commonly injected for confirmatory laboratory diagnosis), hamsters and rats (often used for pathogenesis studies), dogs (pathogenesis, vaccine efficacy), raccoons and skunks (oral vaccination). Suckling (1–3 day old mice) have been used for human and animal rabies vaccine production in many countries, mainly in Latin America, since their myelin content is much lower than that of adult animals, and virus titers in these young animals tend to be very high. Embryonating eggs have been used for propagation of attenuated vaccines (LEP and HEP). Street rabies viruses can be grown on mouse neuroblastoma cells with relative ease, and those cells are the most sensitive for confirming an initial diagnosis by fluorescent antibody. Tissue culture-adapted viruses grow to high titers on BHK cells. Other cells such as VERO, human diploid, chick and duck embryo cells are also commonly used for vaccine preparation.

Immune Response

Animals or humans vaccinated with classic rabies vaccines respond with a rise in neutralizing antibodies, the level reached generally being proportional to the potency of the vaccine. Neutralizing antibodies arise in direct response to the glycoprotein (G) gene; recombinant vaccines constructed with the cDNA of rabies virus glycoprotein also give rise to neutralizing (anti-G) antibodies. Animals with neutralizing antibodies, even at low levels, almost always survive subsequent challenge with 'street' rabies virus. Recently the protective role of the N gene has been recognized in experimental animals, although its significance in vaccinated animals (or humans) is not known. The cellular immune response after vaccination involves a wide array of cells. Serum neutralizing antibodies in individuals (humans or animals) that sicken with rabies are rarely noted before the eighth day of illness; antibodies appear in the cerebrospinal fluid 1 or 2 days later.

Prevention and Control of Rabies

Rabies may be controlled at three levels: human, domestic animal and wild animal. Until widespread canine vaccination programs were initiated in the late 1940s rabies was controlled only at the human level, with hundreds of thousands of persons vaccinated world-wide for exposure to rabid animals, mostly dogs. Rabies in dogs was controlled when potent animal vaccines became available (especially after the advent of proper vaccine potency tests) along with effective dog vaccination programs (those resulting in the immunization of at least 70% of community dogs) and stray dog control. This stopped dog-to-dog rabies transmission as well as serving as a barrier between infected wild animals and human populations. Recently an additional step has been taken involving the oral rabies vaccination of wild animals (foxes, raccoons) to eliminate rabies in those populations.

Future Perspectives

Many issues remain unsolved in rabies: what causes 15% of persons bitten in the hand by rabid dogs to die, while 85% survive without treatment? Where is the virus during the long incubation periods? What makes the virus break away at the end of that period? Why do some outbreaks in wild animals fade away after a few decades while others 'simmer' for much longer periods? Is there a better and less expensive treatment for exposed humans than that now administered (antiserum – or globulin – and vaccine)?

As the examination of molecular aspects of rabies permits the unravelling of the mysteries of pathogenesis we should see more investigation of antiviral substances and the development of more effective and less lengthy human treatments. Examination of viral isolates from various areas of the world will permit examination of the viral genomes of more and more rabies viruses, and clarify whether some are so far from the 'classic' strains as to require separate vaccines for the protection of humans and animals. The future should also bring a better understanding of just how effective vaccines can be delivered to unvaccinated animals such as wild animals, or, more important, community dogs in developing countries.

See also: Epidemiology of viral diseases; Nervous system viruses; Rabies-like viruses; Vaccines and immune response; Vesicular stomatitis viruses; Zoonoses.

Further Reading

Baer GM (ed.) (1991) *The Natural History of Rabies*, 2nd edn. Boca Raton, Florida: CRC Press.

Baer GM, Bellini WJ and Fishbein DB (1990) Rhabdoviruses. In: Fields BN *et al.* (eds) *Virology*, p. 883. New York: Raven Press.

Dietzschold B *et al.* (1988) Antigenic diversity of the glycoprotein and nucleocapsid proteins of rabies and rabies-related viruses: implications for epidemiology and control of rabies. *Rev. Infect. Dis.* suppl. 4: 785.

Tordo N and Poch O (1988) Structure of rabies virus. In: Campbell JB and Charlton KM (eds) *Rabies. Developments in Veterinary Virology*, p. 25. Boston: Kluwer Academic.

RABIES-LIKE VIRUSES

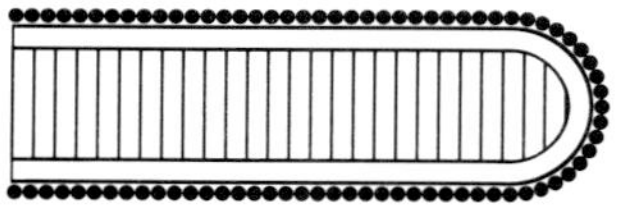

Robert E Shope
Yale University School of Medicine
New Haven, Connecticut, USA

History

Rabies virus was thought to be a single serotype without relatives until 1969. That year workers at Yale University and the Centers for Disease Control in Atlanta, Georgia discovered that two viruses from Africa were bullet-shaped like rabies, and were related serologically to rabies virus. One of these, Lagos bat virus, had been isolated in 1956 by L.R. Boulger and James Porterfield from brains of *Eidolon helvum* fruit bats captured on Lagos Island, Nigeria. The second, Mokola virus, was isolated by Graham Kemp in 1968 from the organs of shrews of the genus *Crocidura* captured in the Mokola District of Ibadan, Nigeria. Subsequently, it was discovered that the virus Obodhiang, isolated by Jack Schmidt in 1963 from *Mansonia* mosquitoes in the Sudan, and another virus isolated in 1967 by Graham Kemp and Vernon Lee from *Culicoides* midges in Ibadan, Nigeria and named kotonkan virus, were distant serologic relatives of rabies virus. The mosquito and midge viruses have not been recovered from naturally infected vertebrate animals and may be insect viruses.

The full significance of the finding in 1969 that viruses from Africa were related to rabies virus was not realized at the time. In 1971, however, Mokola virus was isolated from a fatal human case of central nervous system disease in Ibadan, Nigeria.

In 1970, a South African farmer was bitten on the lip by a bat. He later died of what was clinically thought to be rabies. The virus isolated from his brain was recognized by Courtney Meredith as a new rabies-related virus and was named Duvenhage virus.

Lagos bat virus has since been isolated from a cat and a bat in South Africa, from a bat in Central African Republic, and from a rabid dog in Ethiopia, but has not been associated with human illness. Mokola virus has been subsequently isolated from apparently rabid dogs and a cat in Zimbabwe, and from a rabid dog in Ethiopia. Both Lagos bat and Mokola viruses have thus demonstrated their potential for rabies-like pathogenicity; Mokola virus was also isolated from shrews in Cameroon and from a rodent, *Lophuromys* in the Central African Republic. The most striking finding, however, was the discovery of viruses very closely related to Duvenhage virus in bats in Germany. In 1985, a bat biologist died of this virus (in recent publications called European bat lyssavirus) infection. It is now clear that close relatives of Duvenhage virus are widely distributed in bats throughout Europe, and represent a hazard to human health.

Charles Calisher and his colleagues in 1989 tested 89 rhabdoviruses for antibody cross-reaction by immunofluorescence and found that 19 reacted with the five known rabies-related viruses or with rhabdoviruses that in turn reacted with the known rabies-related viruses. Other than rabies, Mokola, Lagos bat and Duvenhage viruses, none of these rabies relatives has been implicated as a cause of rabies-like disease in people or domestic animals. Here, the properties of the three rabies-related lyssaviruses that cause rabies in domestic animals and/or people are described, while kotonkan, Obodhiang, or the 19 other rhabdoviruses that have distant serological relationships to members of the genus *Lyssavirus* are not detailed.

Taxonomy and Classification

The rabies-related viruses belong to the genus *Lyssavirus* of the family *Rhabdoviridae* of RNA viruses. These viruses have nonsegmented negative-strand RNA genomes enclosed in a lipid envelope derived from the host cell. The particles are covered by glycoprotein (G), a layer of spikes contained in the envelope, which endows the particles of at least two of the viruses, Mokola and Lagos bat, with the ability to agglutinate avian red blood cells. Beneath the lipid envelope is a membrane protein designated NS surrounding a helical complex containing the nucleoprotein (N) and the polymerase protein (L). The viruses in the genus are bullet-shaped; Obodhiang and kotonkan resemble bovine ephemeral fever virus with mostly cone-shaped particles while the other members are indistinguishable from rabies particles. They have parallel sides, one flat end and one hemispherical end, and are approximately 180 nm in length and 65 nm in diameter.

The lyssaviruses are classified in four serotypes: serotype 1, rabies; serotype-2, Lagos bat (Lag); serotype-3, Mokola (Mok); and serotype-4, Duvenhage (Duv). These types are readily distinguished by neutralization test using polyclonal sera. Serotypes are further subtyped using a battery of anti-N monoclonal antibodies by immunofluorescence, and anti-G monoclonal antibodies by neutralization test.

Geographic and Seasonal Distribution

Rabies-related viruses are geographically limited to Africa, except for Duvenhage which is found in southern Africa and Europe. The discontinuous distribution of Duvenhage virus is not readily explained. Possibly, infected bats were transported from one region to the other, or alternatively, the distribution may not be discontinuous and the virus may exist in other parts of Africa but not yet have been detected. Bats infected with subtypes of Duvenhage virus have now been recognized throughout Europe from Spain in the West to Russia and Ukraine in the East. The rabies-related virus distribution is almost certainly a function of the range of reservoir hosts. There is no pattern of seasonality recognized.

Host Range and Virus Propagation

The reservoir hosts of Lagos bat, Mokola and Duvenhage viruses are inferred from knowledge of sources of virus in nature. Lagos bat virus has been isolated at least nine times from bats. Where the bats were identified, these were fruit-eating bats: *Eidolon helvum* in Nigeria and Senegal, *Micropterus pusillus* in the Central African Republic, and *Epomophorus wahlbergi* in South Africa. The virus has presumably spilled over into domestic animals. It caused rabies in cats in South Africa and Zimbabwe, and in a dog in Ethiopia.

The reservoir of Mokola virus is believed to be the shrew, *Crocidura* spp., from which it has been isolated in Nigeria and Cameroon. A single isolation was made from a rodent, *Lophuromys sikapusi* in Central African Republic. The virus was recovered from rabid dogs in Zimbabwe and Ethiopia, and from rabid cats in Zimbabwe indicating spill-over into the domestic animal population. A fatal human case was recorded in Nigeria.

Insectivorous bats are implicated as the reservoir host of Duvenhage virus. The virus has been isolated repeatedly from *Eptisicus serotinus* throughout Europe, and once from *Nycteris thebaica* in Zimbabwe. The bat that bit the lip of the South African farmer who died of Duvenhage virus infection in 1970 was not identified.

Lagos bat, Mokola, and Duvenhage viruses kill baby and adult mice following intracerebral inoculation. The three viruses develop plaques in Vero cells, and Duvenhage virus was readily adapted to form plaques in BHK-21 cells.

Wild-caught *Crocidura flavescens manni*, the African giant shrew, were inoculated in the laboratory with Mokola virus. Many of these became infected, including shrews exposed orally. Virus was recovered from brain, salivary glands, kidney, pancreas, lung and mouth swab. Virus was also sometimes detected in blood. Some animals became sick. They tended to save their food without eating it, to be more aggressive than uninfected animals, and to develop flaccid hind-limb paralysis. Shrews attack other animals in nature. Both sick and apparently well infected shrews transmitted Mokola virus to laboratory mice by bite.

Rhesus monkeys and beagle dogs were inoculated experimentally with Lagos bat and Mokola viruses. Animals inoculated intracerebrally invariably developed encephalitis indistinguishable from rabies and either died or were killed. Dogs inoculated intramuscularly (i.m.) survived without illness as did the majority of monkeys inoculated i.m. One of five monkeys inoculated i.m. with Lagos bat virus developed unilateral paresis, but survived. One of five monkeys inoculated i.m. with Mokola virus developed tremors and died on day 19 postinoculation. Thus Mokola and Lagos bat viruses are less pathogenic after i.m. inoculation than street rabies virus.

Mokola virus was passaged sequentially in baby mice, *Aedes albopictus* mosquito cells, and Vero cells. This passage material was found to infect *Aedes aegypti* mosquitoes by intrathoracic inoculation and was maintained by sequential passage in mosquitoes for 340 days. The virus was found in salivary glands, but in higher titer in nervous tissue of the mosquitoes. Mokola virus was transmitted transovarially in the mosquito, but infection of mice by mosquito bite was not demonstrated. Whether Mokola virus can be transmitted by arthropods in nature is still not known.

Genetics

The RNA of Mokola virus was cloned and the

cDNA sequenced. The genome is very similar in length and organization to rabies virus. The length is estimated at 12 000 nucleotides. The genes, designated by the proteins they code, are arranged as follows: 3′-N-M1-M2-G-L-5′. The 3′ and 5′ end sequences of Mokola virus are highly conserved when compared to the PV strain of rabies virus, and the Mokola end sequences appear to be complementary.

Evolution

Rabies and the other lyssaviruses are postulated to have evolved initially on the African continent. This hypothesis is based on the known African distribution of all lyssaviruses described. The nearly world-wide distribution of rabies virus can be explained by transport with people of domesticated dogs which were infected and carried extensive distances by sailing ships from Africa during the long incubation period of the virus. For this hypothesis to be credible, one also needs to postulate that rabies virus spread in the Americas from dogs to sylvatic animals such as the skunk, fox, raccoon and bat, and that further divergent evolution occurred in these hosts.

Serologic Relationships and Variability

Serologic relationships among lyssaviruses have been measured by complement fixation, neutralization and immunization challenge tests. The N protein plays a major role in complement fixation. By this method, rabies, Mokola, Lagos bat and Duvenhage viruses differ among themselves by two- to 16-fold in each direction in reciprocal tests. This degree of closeness indicates a relative conservation of the N protein antigens.

The neutralization test measures the surface G antigen. When undiluted hyperimmune sera are tested against varying dilutions of infectious virus in the baby mouse, the four lyssaviruses are very closely related. However, the same test done with constant virus dose and varying dilutions of antibody shows almost no relationship among the viruses except with undiluted antibody. When the test is done in cell culture using plaque reduction and varying dilutions of antibody, Duvenhage virus is most closely related to rabies, and Mokola virus is quite distinct.

The immunization challenge test is the most likely method to indicate the utility of a rabies vaccine to protect against challenge with other lyssaviruses. Mokola and Lagos bat cross-react minimally with rabies virus; a rabies vaccine with solid homologous protection, yields only a 1.5 log protection index against Duvenhage virus challenge. If these same relationships hold in human infection, there is some rationale for use of rabies vaccines to protect against Duvenhage infection, but the efficacy would not be expected to be optimal.

Batteries of monoclonal antibodies are used to determine the variability in reactivity by immunofluorescence of the G and N antigens. The monoclonal antibodies reactive with the G antigens have also been tested for neutralization. On the basis of reactivity with these batteries, subtypes of lyssaviruses have been recognized. Lagos bat virus has three proposed subtypes (Lagos, Central African Republic and South Africa). Mokola virus has only one, but there is some evidence that the Ethiopian Mokola virus is different from the Nigerian. Duvenhage virus has two subtypes (African and European). A single Finnish isolate in 1985 from a human being may be sufficiently different from European Duvenhage and other lyssaviruses to constitute a new serotype. The classification of the European lyssaviruses is still under active study.

Epidemiology

Lyssaviruses related to rabies are maintained in transmission cycles of terrestrial wildlife or bats. Each serotype is found in one or a relatively limited number of related species. It is believed that transmission to man or domestic animals represents spill-over from the wildlife cycle. These spill-over events are rare and occur when people intrude into wildlife habitats, such as a Finnish spelunker who was infected in 1985 with a Duvenhage-related virus, or when the wildlife enter houses, such as may have happened in Ibadan, Nigeria in 1971 when a 6-year-old girl was infected fatally with Mokola virus.

There is no evidence that epidemics or epizootics occur. Animals and people may survive infection with Duvenhage, Lagos bat and Mokola viruses. Survival of experimentally infected dogs and monkeys following Lagos bat and Mokola virus infection is documented. Mokola virus was reportedly isolated from a child that recovererd from poliomyelitis-like disease in Nigeria, although the patient did not develop antibody during convalescence and the validity of the isolation is in doubt. The seasonality, carrier rates, sex and age susceptibility are not known for animals or people.

Transmission and Tissue Tropism

Experimentally, Mokola virus is transmitted by bite of shrews, and there is anecdotal evidence that a bat transmitted Duvenhage virus to a person by bite. Aerosol transmission has not been eliminated as a possible mechanism of spread. Mokola virus was adapted to mosquitoes in the laboratory, but there appears to be no need to postulate mosquito transmission to explain maintenance in nature.

The limited data available from autopsy and infection of experimental animals indicates that Mokola, Lagos bat and Duvenhage viruses are neurotropic. Like rabies, the brain is the prominent target organ, and virus is also found in salivary glands late in infection.

Pathogenicity and Clinical Features of Infection

The range of pathogenicity of different strains of rabies-related viruses is not known. The clinical features of Mokola and Duvenhage virus infections are recorded in three patients; all presented with central nervous system disease. Duvenhage is associated with Guillain–Barre ascending paralysis and radiating pain in the arm and neck, followed later by agitation, increased respiration rate, muscle spasms, then coma and death. Diabetes insipidus with polyuria complicates the course. The child who died of Mokola virus infection had a prodromal illness of fever and vomiting, then flaccid paralysis of the limbs, progressing to deep coma before death.

Pathology and Histopathology

Specific lesions in the child who died of Mokola virus infection after a 9-day illness, are limited to the brain. There is lymphocytic perivascular cuffing, neuronal changes of chromatolysis, eosinophilic necrosis, and nuclear pyknosis most marked in the midbrain and basal ganglia. Eosinophilic cytoplasmic inclusion bodies are found clustered in degenerating neurons. There are several lytic changes in the brain of the Finnish case of Duvenhage-like virus. Rabies-reactive antigen is found in central nervous system neurons postmortem.

Immune Response

The very limited observations of patients infected with Mokola and Duvenhage viruses failed to show a specific serological response. The Finnish case of Duvenhage-like infection was followed through 23 days of illness.

Prevention and Control of Disease Caused by Rabies-related Viruses

Prevention of human infection with rabies-related viruses is based on elimination of exposure. In Europe and Africa, contact with bats should be avoided. Sick animals or those with abnormal behavior should be handled only with protective gloves. There is no vaccine available for Mokola or Lagos bat viruses. Commercial rabies vaccine and immune globulin may give limited protection against Duvenhage infection. Thus, persons with bat bites and other bat exposures should receive classic rabies postexposure treatment, but should not expect complete protection. Persons whose occupations lead to exposure to bats should receive preexposure rabies vaccination.

Vaccines prepared from Mokola and Duvenhage viruses would be useful biologicals, but with the very low attack rates, it is not likely that such vaccines will be developed commercially.

Future Perspectives

The finding in Zimbabwe and Ethiopia of dogs and cats infected with Lagos bat and Mokola viruses is cause for concern. Human exposure and disease may be prevalent and not diagnosed. Active surveillance in Africa for Lagos bat, Mokola and Duvenhage viruses is needed.

Likewise, the threat posed by European bat lyssavirus is not yet completely appreciated. The scientific community needs to maintain vigilance for possible human cases, and needs increased research on the prevalence in bats, the possible spread to species other than serotine bats, and the extent to which terrestrial wildlife and domestic animals are exposed and infected.

See also: Rabies virus; Rhabdoviruses.

Further Reading

Bourhy H, Tordo N, Lafon M and Sureau P (1989) Complete cloning and molecular organization of a rabies-related virus, Mokola virus. *J. Gen. Virol.* 70: 2063.

Mebatsion T, Cox JH and Frost JW (1992) Isolation and characterization of 115 street rabies virus isolates from Ethiopia by using monoclonal antibodies: identification of 2 isolates as Mokola and Lagos bat viruses. *J. Infect. Dis.* 166: 972.

Roine RO *et al.* (1988) Fatal encephalitis caused by a bat-borne rabies-related virus. *Brain* 111: 1505.

REOVIRUSES

Contents

General Features
Molecular Biology

General Features

Bernard N Fields
Harvard Medical School
Boston, Massachusetts, USA

History

In 1951 Stanley and colleagues in Australia isolated a virus from stools of a young aboriginal child. He proposed the name 'hepatoencephalomyelitis virus' (HEV) for the new isolate. Subsequent isolates were originally classified as ECHO 10 or were unclassified. In 1959, Albert Sabin coined the name 'reovirus' (respiratory, enteric, orphan viruses) for this group of viruses which are larger than known enteroviruses (> 75 nm), produce inclusions in monkey kidney cell culture, are pathogenic for newborn but not adult mice and hemagglutinate human Type O erythrocytes. Subsequent studies established that the reoviruses are indeed a unique group which contain a genome of segmented double-stranded RNA (dsRNA) segments and a double-capsid icosahedral shell.

Taxonomy and Classification

Reoviruses are the prototype viruses of the *Reoviridae* family of RNA viruses. The classification is based on the presence of a segmented dsRNA genome as well as a double-capsid shell built on icosahedral symmetry. The mammalian reoviruses, as described here, contain ten genome segments and share a common complement-fixing antigen. Three serotypes are distinguished based on neutralization and hemagglutination-inhibition (HI) tests. Morphologically the three serotypes are quite similar but differences in the homologous RNA segments and proteins can be seen in their mobilities on polyacrylamide gels.

Infectious Agent

Virion structure

Reovirus particles consist of an inner protein shell (core) containing the ten dsRNA segments, surrounded by an outer protein shell (outer capsid). The core contains the viral transcriptase and replicase activities and consists of three major proteins ($\lambda1$, $\lambda2$, $\sigma2$) and two minor proteins ($\lambda3$, $\mu2$). The most conspicuous feature of the core is the projection (spike) at each of the 12 vertices of the core icosahedron. The core diameter is approximately 60 nm.

The outer capsid is arranged in a $T = 131$ icosahedral lattice with 12 vertices and fivefold, threefold and twofold axes of rotational symmetry. The virion diameter is ~ 81 nm. The outer capsid consists of three proteins $\sigma1$, $\mu1/\mu1C$ and $\sigma3$. The $\sigma1$ protein is located at the 12 vertices and exists as a homomultimer. It is not yet clear if there are three or four molecules per multimer. The vertices consist of surface extension of the $\lambda2$ pentamers, forming a spike.

The $\mu1/\mu1C$ and $\sigma3$ proteins are present in equal amounts on the outer capsid ($\sim$ 600 copies each) and form the basic shell of the outer capsid except at the icosahedral vertices.

In addition to the intact virus, there are two stable subvirion forms, infectious subviral particles (ISVPs) and cores that are derived from virions following exposure to proteases under different digestion conditions. ISVPs differ from virions in that they lack $\sigma3$ and contain cleaved forms of the $\mu1$ and $\mu1C$ proteins (named δ and $\mu1\delta$). In addition, the $\sigma1$ protein remains attached to the viral particle but appears as an extended fiber. Cores lack all the external proteins ($\sigma1$, $\mu1$, $\mu1C$, δ, $\mu1\delta$ and $\sigma3$) and have wide 'spikes' consisting of the $\lambda2$ pentamers.

Host Range and Virus Propagation

Reoviruses are ubiquitous in geographic distribution and host range. Virus is often found in the environment (rivers, stagnant water, sewage). Virus and viral antibody have been found in the vast majority of mammals studied, including humans, monkeys, mice, dogs and cattle. There have been occasional isolates in insects, fish and reptiles. Reovirus serotype three is commonly found in colonies of laboratory mice and rats.

Animals display a variety of syndromes when infected with reoviruses. Natural infection in animals other than humans involves primarily the respiratory and enteric systems. Infected mice may have diarrhea, runting, oily hair, jaundice and neurologic symptoms. In nonhuman primates, cases of pneumonia and other respiratory diseases have been found. In addition, there have been occasional reports of hepatitis and extrahepatic biliary atresia, meningitis and necrosis of ependymal and choroid plexus epithelial cells. Under experimental conditions, mice have been studied most extensively and display a wide variety of systemic disease outcomes (including encephalitis, pneumonia, myocarditis and gastroenteritis).

Reoviruses cause cytopathic effects (CPE) in a wide variety of mammalian cells in culture (including L cells, HeLa cells, CV-1 cells, KB cells and many monkey cells). The cells become granular and develop cytoplasmic inclusions. The inclusions appear initially as dense granular material in the cytoplasm and ultimately coalesce and become perinuclear.

Viral infection of L cells in one-step growth curves indicates that peak yields appear around 18–24 h. Most virus remains cell-associated, and physical disruption of cells is used to completely recover infectious virus.

All three serotypes agglutinate erythrocytes via the $\sigma1$ protein (viral hemagglutinin) found on the viral outer capsid shell. Binding of the reovirus type 3 hemagglutinin to erythrocytes is to glycophorin in the erythrocyte membrane. This attachment requires the presence of sialic acid residues on the glycophorin molecule.

Effect in Host Cells

Following reovirus infection, there is an inhibition of the synthesis of cellular DNA, RNA and protein. Inhibition of cellular DNA synthesis appears to be distinct from that of cellular RNA and protein since it occurs with UV-inactivated reovirus particles (in contrast to inhibition of RNA and protein synthesis), suggesting that viral replication is not needed to inhibit cellular DNA replication. Genetic analysis indicates that the reovirus $\sigma1$ protein is the viral component responsible for inhibition of DNA synthesis while the $\sigma3$ protein is responsible for the inhibition of RNA synthesis. In neither of the above instances is the biochemical mechanism understood. Analysis of the morphology of cells infected with reoviruses shows progressive accumulation of granular material in the cytoplasm (viral factories). Ultrastructural immunohistochemical analysis of these structures indicates that cellular microtubules are coated with viral particles or protein and that the intermediate filaments are progressively disrupted.

Replication

Reoviruses bind to sialylated glycoproteins on cell surfaces. The viral attachment protein is the viral hemagglutinin. Following binding, reovirus enters cells via receptor-mediated endocytosis. Initially, virus localizes at coated pits and subsequently appears in clathrin-coated vesicles. Alternatively, ISVPs are thought to be capable of directly penetrating certain cells (L cells) at the cell surface. In endosomal vesicles and/or lysosomes, reovirus is processed to an ISVP. This process is associated with proteolytic cleavage of the primary product of the viral M2 gene (μ1 or μ1C) to three products, μ1N, δ and ϕ. The μ1N fragment is at the N terminus while ϕ is at the C terminus. All three fragments of the M2 gene product stay associated with the ISVP and in this form associate with membranes. Precisely how, or if, the ISVP crosses the membrane is not known. In any case, the next step involves initiation of primary transcription by the viral core in the cytoplasm. The conversion of ISVP to transcriptionally active cores is thought to be independent of further proteolysis and is facilitated by specific ions (Na^+, K^+, Mg^{2+} and Ca^{2+}) or particle–particle interactions.

It is not known whether transcription takes place within endosomes/lysosomes or from viral cores that have escaped into the cytoplasm. In either case, the transcriptase is fully conserved within the viral core and transcription involves utilization of the minus-strand of each of the ten dsRNA segments producing ten full-length plus-polarity mRNA molecules. The precise nature of the transcriptase is not known but is thought to include the L1 gene product (λ3). The 5′ termini of mRNA transcripts are capped by a series of core enzymes that are independent of the transcriptase. One of these capping enzymes, the guanylyltransferase, is a property of the λ2 protein, the protein that makes up the pentameric core spike.

In vitro-generated cores transcribe all ten mRNAs while *in vivo* there appears to be some regulation. Four mRNAs, the products of L1, M3, S3 and S4 genes, appear before the other six. These mRNAs have been called pre-early mRNAs. How this regulation occurs and its significance, if any, is not known.

After early mRNA synthesis, replication of the newly formed dsRNA genome takes place in the cytoplasm within nascent progeny subviral particles. The mechanism that ensures that ten unique dsRNA segments will assemble into nascent particles is unknown.

Subsequently, transcription takes place from the progeny dsRNA genome and represents 'late' mRNA. Such late mRNAs begin to appear 4–6 h after infection, reach maximum levels at 12 h and then decrease. Some of the late transcripts serve as templates for minus-strand synthesis of dsRNA by nascent progeny subviral particles.

Shortly after onset of viral infection, host-cell protein synthesis starts gradually to decrease. Newly synthesized viral proteins can be detected by 2 h postinfection and by 10 h they are the majority of proteins being synthesized. The mechanism whereby reovirus predominates in the infected cell is not known for certain. Proposed mechanisms include: a modification in the cap-dependence of the host-cell machinery such that the late viral uncapped mRNAs are preferentially translated over capped cellular mRNAs; a host factor; mRNA competition for a limiting message-discriminating factor. Since there are disagreements in different studies, no single explanation can, at present, be provided. Similarly there is regulation of the amount of different mRNAs and proteins synthesized. This is also not fully explained.

Reovirus assembly takes place in cytoplasmic inclusions that are not membrane bound. There is no clear morphogenic scheme that provides details on the steps in assembly.

Progeny virus stays largely cell associated. Reoviruses do not have an envelope or 'bud' into cell organelles or from the cell surface. It is thus likely that virus is eventually released by cell lysis. The specific steps leading to cell lysis are not known.

Genetics

The reoviruses contain segmented genomes and thus, not surprisingly, display reassortment of gene segments following infection of cells with different isolates. This phenomenon of gene segment reassortment is also detected following infection of young mice. In contrast to the ease of detecting reassortment, complementation following co-infection of cells with pairs of *ts* mutants at high temperatures is hard to detect and is inefficient. The reason for this is the capacity for such mutants to interfere with the growth of wild-type virus.

Reassortment analysis has served as a way to group mutants. Not surprisingly, there are ten 'reassortment' groups of *ts* mutants, corresponding to lesions on the ten genome segments.

As with other RNA viruses, mutation occurs at high frequency. Such mutations may have profound effects on the biologic properties of the virus and need to be considered in any genetic analysis.

Molecular Pathogenesis

Reovirus strains differ in pathogenicity, some causing more disease than others following inoculation into experimental hosts. Newborn mice have been most extensively studied and have provided the most information concerning the correlation between pathogenesis and the function of individual viral genes.

Following introduction into the gut of a newborn mouse, the virion is exposed to luminal proteases. Such proteases cleave the σ3 protein and convert the virion to a subviral form, the ISVP. This conversion is a required first step since it allows the viral particle to penetrate through specialized cells (M cells) found in the lining of the small intestine which deliver virus to Peyer's patch. Virus that has not been 'activated' in this way does not bind to M cells and thus does not initiate primary replication. In Peyer's patch, ISVPs enter macrophages where primary replication takes place. The capacity of ISVPs to bind to cell surfaces is a property of the viral hemagglutinin (σ1 protein), which is found in an extended state in the ISVP. The ISVP is subsequently internalized and reaches the cytoplasm to continue replication (see section on replication). After primary replication in Peyer's patch, virus spreads beyond the primary site by entering the bloodstream or adjacent nerves and subsequently spreading to distant sites in the body. The hemagglutinin plays a central role in determining the choice of pathway for spread and is the key protein that determines ultimate cell and tissue tropism.

Tropism has been studied most extensively in the central nervous system (CNS) and illustrates the role of the hemagglutinin most precisely. Reovirus T1 when introduced into the CNS localizes in ependymal cells (cells lining the ventricles in the brain) while reovirus serotype 3 localizes in neurons (causing encephalitis). This difference in localization is a property of the viral hemagglutinin. In addition, mutations in the gene encoding the hemagglutinin result in restricted tissue spread and attenuated disease.

It is apparent from the above comments that the outer capsid proteins (σ3, μ1/μ1C and σ1) act as a 'delivery' system which brings reoviruses into the host and host cells. Following penetration of the subviral particle into the cytoplasm, transcription is initiated by the viral core and the remainder of the viral gene products function to allow the genome to be transcribed, the RNA to replicate and the capsid to be assembled. Thus the remaining genes are part of the viral 'payload'.

Pathology

In man generally, there is no pathologic material to evaluate since reoviruses are largely associated with asymptomatic infection. Thus descriptions of the pathology of reovirus-related disease are largely described in experimental animals, such as newborn mice. When studied in animals other than mice, the results appear similar.

The highlights of pathologic findings under conditions of experimental inoculation of newborn mice are listed below.

Heart

Myocardial injury can be seen. Grossly white or yellow patches are seen that may coalesce. Antigen appears in myocytes which become swollen, fragmented and develop nuclear pyknoses and karyorrhexis.

Hepatobiliary

Hepatic injury may occur. Grossly yellow lesions appear on the surface of the liver. Microscopically, there is hepatic necrosis beginning midzonal and centrilobular and progressing to widespread zones of coagulation necrosis and hemorrhage.

Reovirus type 3 (Dearing) can produce, in mice, infection of the bile ducts leading to chronic obstructive jaundice. Whether this occurs in humans is not known.

Other organs

Reovirus can cause extensive pancreatic injury. Pituitary infection and mild thyroiditis can occur. Extensive myositis is also found.

Oily hair and runting

A classic finding in mice is the generation of a syndrome called the oily hair effect. Such neonatal mice appear to be dipped in oil. The mice are also very small ('runts'). Many animals become jaundiced and develop acholic steatorrheic stools. The syndrome is thought to be due to malabsorption produced by coexisting hepatic or biliary disease.

Clinical Features

Reoviruses are still largely the 'orphan' viruses they

were when the name was proposed in the 1950s. Although human disease is quite rare, the vast majority of humans are infected as shown by serologic analysis. The majority of infections are thus likely to be either asymptomatic or to blend into minor respiratory or gastrointestinal illness of infancy or early childhood.

In one volunteer study, in human adults using intranasal challenge, after a 1–2-day incubation period, an illness developed consisting of headache, pharyngitis, sneezing, rhinorrhea, cough and malaise. The illness occurred in the majority of volunteers, lasted 4–7 days and was self-limited. Other studies did not generate illness. Thus these studies indicate that mild self-limited illness is the major symptomatic illness likely to be present.

It has been proposed that reovirus may be responsible for certain cases of human neonatal biliary atresia. Although initial reports appeared promising, subsequent studies have not found an association, and this link between reovirus and this important human disease is currently controversial.

There have been a few scattered reports of reovirus causing human CNS disease (fatal encephalitis): a fatal systemic illness in a 10-month-old (pneumonitis, myocarditis, hepatitis and encephalitis); a 51-year-old man with fatal spinal muscular atrophy. Other rare reports include a case of fatal pneumonia and exanthematous illness in children.

It is clear that the extreme rarity of reovirus as a cause of documented severe illness suggests that reovirus should not be considered as causally related to serious illness until all other possible causes have been extensively evaluated. The overwhelming majority of individuals infected with reovirus are asymptomatic or, at most, demonstrate minor respiratory or gastrointestinal symptoms.

Diagnosis and Treatment

Diagnosis is based on isolation of virus from tissues or body fluids, detection of viral antigen in infected material and serologic studies. Most isolates of reovirus are from stool or respiratory secretions. Isolations from biopsy or autopsy material should lead to further evaluation.

A variety of cultured cell lines can be used for viral isolation. Monkey kidney cells have generally served for primary isolation. In addition to its characteristic CPE, viral antigen can be detected by immunofluorescence.

Determination of serotype can be difficult and is mainly of interest in epidemiologic evaluations. HI and neutralization testing can be used to identify the isolate and to evaluate subsequent serologic response. HI antibodies generally appear within 3 weeks and are present for life. A fourfold rise is significant.

There is no specific therapy.

Epidemiology

Surveys of populations for reovirus indicate that the virus is present in a variety of geographic and cultural settings. The majority of individuals develop antibodies against all three serotypes during childhood.

Reoviruses are widely found in virtually every species of mammal. There do not appear to be any differences between human and animal isolates. Whether animals serve as a reservoir for virus for human infection is not known. Reoviruses are frequently isolated from water supplies and sewage.

See also: Genetics of animal viruses; Pathogenesis.

Further Reading

Nibert ML, Furlong DB and Fields BN (1991) Distinct forms of reoviruses and their roles during replication in cells and host. *J. Clin. Invest.* 88: 727.

Schiff LA and Fields BN (1990) Reoviruses and their replication. In: Fields BN *et al.* (eds) *Virology*, 2nd edn, p. 1275. New York: Raven Press.

Tyler KL and Fields BN (1990) Reoviruses. In: Fields BN *et al.* (eds) *Virology*, 2nd edn, p. 1307. New York: Raven Press.

Molecular Biology

WK Joklik
Duke University Medical Center
Durham, North Carolina, USA

Introduction

Reoviruses (*Re*spiratory *e*nter*o*) are members of the genus *Orthoreovirus* of the family *Reoviridae*, which also includes five other genera, namely, the *Rotaviruses* and the *Orbiviruses*, both of which are also viruses of vertebrates, the *Cypoviruses* of insects and the *Phytoviruses* and *Fijiviruses*, both of which are plant viruses. The common characteristic features of all these genera are a basically similar particle morphology and size, and double-stranded RNA (dsRNA)-containing genomes that comprise 10, 11 or 12 genome segments.

Reoviruses are ubiquitous among mammals in which they usually do not cause overt disease. Mammalian reoviruses fall into three serologic subgroups or serotypes (GT1, GT2 and GT3), members of each of which are not neutralized by antisera against the other two.

Reoviruses have also been isolated from other vertebrates such as birds and reptiles, as well as from invertebrates, such as insects and molluscs. None of these viruses have been studied in biochemical or molecular terms except the avian reoviruses which fall into five serotypes. In contrast to mammalian reoviruses, they cause disease [the viral arthritis syndrome (VAS) and the stunted growth syndrome (transient digestive system disorder, TDSD)], as well as pericarditis/myocarditis, hepatitis and nephritis.

Most of our knowledge concerning the biochemistry and molecular biology/genetics of orthoreoviruses derives from studies of the mammalian reoviruses.

The Reovirus Particle

Reovirus particles consist of a core some 55 nm in diameter which is surrounded by an outer capsid shell that is about 12.5 nm thick (Fig. 1). Since the thickness of the core shell is about 7.5 nm, the central cavity accounts for about 12.5% of the reovirus particle volume, and the core for about 33%.

Both the outer capsid and the core shell are composed of capsomers which are arranged with icosahedral symmetry. The reovirus particle surface reveals 600 protrusions, presumably capsomers, arranged with $T = 131$ symmetry in the form of shared hexamer rings around 120 holes. The core shell also possesses $T = 131$ symmetry elements, but the arrangment of capsomers in them is difficult to discern (Fig. 2). However, cores possess 12 icosahedrally located columnar projections or spikes which are about 10 nm in diameter, possess central channels 5 nm in diameter, and project about halfway through the outer capsid shell in which they are visible as depressions or craters. The spikes, which can be removed from cores by incubating them at pH 11.4 at 4°C for 15 min, are pentamers of protein $\lambda 2$ and appear to be partially covered by 3 nm-thick 'lids' which appear to be trimers of protein $\sigma 1$ (see below).

The outer capsid shell is stable at high and physiological salt concentrations, but loses capsomers on storage at low ionic strength. It is readily digested by chymotrypsin to which the core shell is completely resistant. Virus particles and cores are readily separated by density gradient centrifugation; their sedimentation coefficients are about 630 and

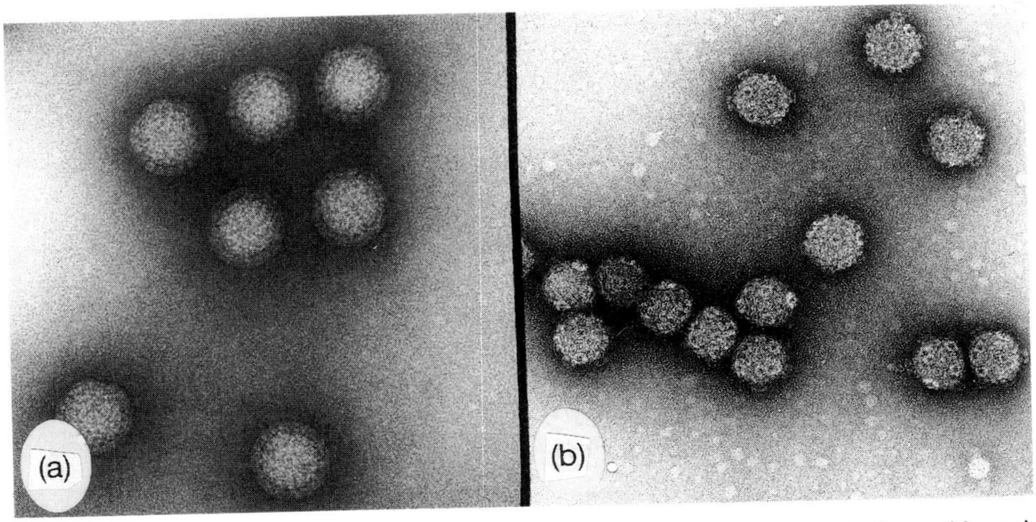

Fig. 1 (**a**) Reovirus particles. Note double-capsid shell. The arrangement of capsomers is discernible at the periphery. (**b**) Reovirus cores. Note the large spikes, located as if situated at the 12 vertices of an icosahedron. Magnification for both × 120 000. (Courtesy of Dr R.B. Luftig.)

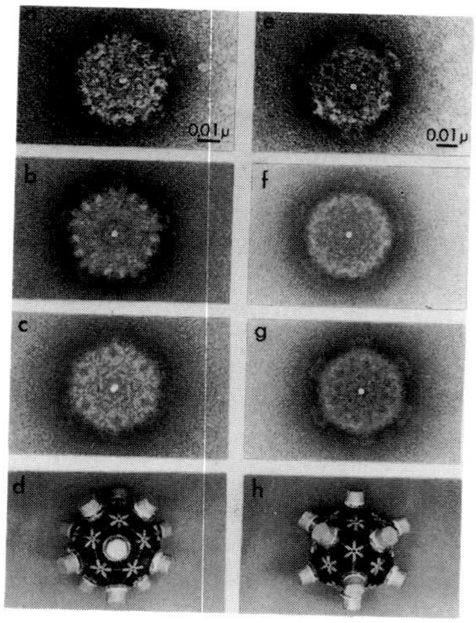

Fig. 2 Two reovirus core particles the central axis of which passes through either a presumptive fivefold (**a**) or threefold (**e**) vertex. Enhancement of the five peripheral spikes of (**a**) was achieved by an $n = 5$ rotation (**b**), but not by an $n = 6$ rotation (**c**). (**d**) Model that depicts the spike orientations when the central axis is through a fivefold vertex. Enhancement of the six peripheral spikes (**e**) was exhibited with an $n = 6$ rotation (**g**), but not an $n = 5$ rotation (**f**). (**h**) Model with the central axis through a threefold vertex. The particles were stained with 2% uranyl acetate. [From Luftig *et al.* (1972) *Virology* 48: 170.]

470S respectively, and their densities in cesium chloride density gradients are 1.36 and 1.43 g ml^{-1} respectively.

The reovirus genome

The reovirus genome consists of ten segments of dsRNA, all of which have been cloned and sequenced (Table 1). They fall into three size classes termed L, M and S, members of which are about 3900, 2250 and 1300 bp long, respectively. The total length of the reovirus genome is 23 549 bp; it is one of the largest and therefore most complex of RNA-containing genomes. Small-angle X-ray diffraction studies indicate that the RNA segments exist within the central cavity in a tight well-ordered packing arrangement with adjacent helices aligned locally parallel to each other, rather like the DNA in T even bacteriophages.

Each of the ten reovirus genome segments possesses a major open reading frame (ORF) which varies in length from 365 to 1289 codons. Most of them also contain short ORFs in other reading frames at least one of which, in S1, is translated in infected cells (see below). The 5′-untranslated regions are all short (12–32 nucleotides); those at the 3′-ends are longer, but still [with one exception, L3 (182 nucleotides)] less than 85 nucleotides long. The four 5′-terminal and five 3′-terminal

Table 1. The reovirus serotype 3 genome segments and the proteins that they encode

Genome segment			Protein				
Segment	Length (bp)	Protein	Number of amino acids	Number of molecules/particle	Percent of viral protein	Location in particle	α-helix/β-sheet ratio
L1	3854	λ3	1267	120	15	Core	0.9
L2	3916	λ2	1289	60	8	Core	0.5
L3	3896	λ1	1223	12	1.5	Core	0.9
M1	2304	μ2	687	12	1	Core	1.2
M2	2203	μ1	708	24	1.5	OCS*	0.9
		μ1C	666	600	39	OCS	
		μ1N	42	?		OCS	
M3	2235	μNS	719	–	–	–	2.5
		μNSC	678	–	–	–	
S1	1416	σ1	455	36	1.5	OCS	0.5
		σ1S	120	–	–	–	
S2	1331	σ2	418	240	10	Core	0.5
S3	1198	σNS	366	–	–	–	1.8
S4	1196	σ3	365	600	22	OCS	0.9

*OCS, outer capsid shell.

base pairs of all ten genome segments are identical (5′-GCUA- and -UCAUC-3′ for the plus-strands).

Reovirus oligonucleotides

Reovirus particles also contain about 3200 oligonucleotides of which about 2400 are 5′-G-terminated and are the products of abortive reiterative transcription catalyzed by the reovirus transcriptase (see below), and about 850 are oligoadenylates from 2 to 20 residues long. The latter may represent either an untemplated polymerase activity of the reovirus transcriptase committed to transcribe but unable to move along its template, or slippage transcription of the three genome segments (L2, M3 and S3) the 3′ ends of the minus-strands of which are CGAUUU- (see below).

Enzymes in reovirus cores

Since host cells do not contain enzymes capable of generating mRNA from dsRNA templates, reovirus particles contain the enzymes necessary for this purpose. Since reovirus mRNAs are capped at their 5′ ends, this involves possession of a transcriptase or RNA polymerase, an RNA triphosphatase to convert the 5′-terminal ppp groups generated by the transcriptase to the pp groups required by the guanylyltransferase, the guanylyltransferase itself and two methyltransferases to methylate the 7 position of the cap G and the 2′-O position of what was the 5′-terminal ribose of the uncapped RNA.

In core form the transcriptase is very stable and functions for long periods of time (more than 24 h) at elevated temperatures (up to 50°C), transcribing the genome segments many times. The transcriptase is the minor core component λ3 and the guanylyltransferase is protein λ2 (see below); but the natures of the RNA triphosphatase and the methyltransferase(s) are not known.

An oligo- or polyadenylic acid polymerase activity in the presence of ATP has also been observed repeatedly in reovirus cores. Presumably it represents an untemplated or slippage polymerase function of the transcriptase (see above).

The reovirus proteins

The various reovirus proteins are listed in Table 1. Their functions are summarized in Table 2.

All reovirus proteins except μ1 and μ2 have been isolated in native form from cells infected with vaccinia viruses into the TK gene of which the various reovirus genome segments have been cloned under the control of powerful promoters like the T7 polymerase promoter or the cowpox virus A-type inclusion body protein gene promoter.

The core shell components λ1 and σ2

The major components of the core shell are proteins λ1 and σ2, apparently in the ratio of 1:2. Protein λ1 possesses a nucleotide-binding site –TKGKSSG– at its N terminus, and a zinc finger motif centered around codon 194. Both λ1 and σ2 possess weak dsRNA-binding activity. They associate with each other *in vitro*, and in cells infected with hybrid vaccinia viruses containing the L3 and S2 genome segments, core-like particles are formed.

The spike component λ2

Protein λ2, in the form of pentamers, is the reovirus spike. It is also the reovirus guanylyltransferase. However, it possesses neither RNA triphosphatase nor methyltransferase activity; that is, it is the reovirus capping enzyme, but it neither provides the required 5′-ppG-terminated substrate nor methylates the cap.

Protein λ2 possesses affinity for λ1 and λ3, as well as for σ1.

The minor core components λ3 and μ2

Cores also contain 12–24 molecules of each of two additional proteins. One is protein λ3, which appears to be the reovirus RNA transcriptase because (1) it specifies its pH optimum and (2) it contains a highly conserved motif the core of which is the tripeptide GDD which is present in all viral RNA polymerases of plus-stranded and double-stranded RNA-containing viruses; the other is protein μ2. By itself, λ3 is a poly(C) dependent poly(G) polymerase; in conjunction with μ2 it transcribes RNA. It is likely that the λ3–μ2 complex catalyzes both the transcription of dsRNA into ssRNA (the transcriptase function that generates mRNA molecules), as well as the transcription of plus-strands into minus-strands (the replicase function which generates progeny genome segments, see below). Presumably the two forms of the enzyme differ in some modification or are associated with different cofactors.

The outer capsid shell

The reovirus outer capsid shell consists of 600 capsomers composed of two –SS– bonded μ1C and two σ3 molecules. The cleavage of μ1, which is myristoylated at its N terminus, to μ1C and μ1N, the 42-amino acid-long myristoylated N-terminal fragment, occurs when μ1 associates with σ3 (see

Table 2. Functions of the reovirus proteins

Protein	Function
λ1	Core shell component; Zn metalloprotein; nucleotide-binding site
λ2	Spike component; elicits formation of group-specific neutralizing antibodies; mRNA capping enzyme (guanylyltransferase)
λ3	Transcriptase
μ1	Myristoylated
μ1C	Tissue tropism; modulation of virulence; induction of tolerance following peroral inoculation
μ2	Component of the functional transcriptase complex (?)
μNS	Phosphoprotein; binds ssRNA; possesses affinity for elements of the cytoskeleton
σ1	Cell-attachment protein; hemagglutinin; tissue tropism; virulence; elicits formation of type-specific neutralizing antibodies; reacts with immune system
σ1S	Inhibition of host DNA synthesis (?) Role in generation of cytopathic effects (?)
σ2	Core shell component
σNS	Binds ssRNA
σ3	Possesses putative protease site and binds dsRNA; Zn metalloprotein; inhibits host RNA and protein synthesis; functions in establishment of viral RNA translation; regulation of viral RNA transcription

below), well before the incorporation of the resulting capsomers into virus particles: most μ1 in cells is free, whereas most μ1C is associated with σ3. Interestingly, most μ1N remains associated with μ1C and is also present in virus particles, as is a small amount of uncleaved μ1 (about one-twentieth of the amount of μ1C). The presence of the myristoyl group in the reovirus outer capsid shell raises interesting questions concerning its function in the assembly and structural stability of the outer capsid shell and during the uptake and entry of reovirus particles into cells (in analogy with the role of myristoyl groups on picornavirus and other viral capsid proteins).

Protein μ1C accounts for almost 40% of the reovirus protein complement. Not surprisingly, it plays a major role in specifying how reovirus particles interact with their environment. On the one hand, it controls sensitivity to chemical reagents like ethanol and phenol; on the other, it controls tissue tropism because susceptibility of μ1C to proteolytic cleavage controls activation of the transcription of mRNA as the first step of the reovirus multiplication cycle (see below). It therefore specifies the extent to which reovirus multiplies in the intestine, which in turn determines how efficiently it spreads to the central nervous system, and thus controls neurovirulence. It also plays a role in inducing serotype-specific immunologic tolerance for delayed-type hypersensitivity responses. Antibodies against μ1C neither precipitate reovirus particles nor neutralize their infectivity.

The second major component of the reovirus outer capsid shell is protein σ3 which has a variety of functions. First, it possesses, near its N terminus, a Zn-binding site ($CX_2CX_{12}HX_3H$) which partially overlaps a sequence ($CGGX_3CXH$) with strong similarity to picornavirus proteases; since σ3 is strongly implicated in the cleavage of μ1 to μ1C, this may be the catalytic site that effects this cleavage. Second, σ3 possesses affinity for dsRNA, a remarkable property for an outer capsid component. It is likely that this affinity is not for dsRNA as such, but rather for short regions in ssRNA with a dsRNA character such as hairpins (stem-loops) or intermolecular complementary regions. The dsRNA-binding domain of σ3 is located in the C-terminal portion of the molecule.

Protein σ3 has been implicated in the following functions by genome segment reassortant analysis: (1) inhibition of host cell RNA and protein synthesis; (2) the establishment and maintenance of persistent infections; and (3) modulation of the efficiency of mRNA translation, mediated apparently by its ability to bind dsRNA and so prevent the activation of the cellular protein kinase that phosphorylates the α subunit of eucaryotic initiation factor 2 (eIF-2), thereby inactivating it and inhibiting protein synthesis (see below).

The third component of the reovirus outer capsid shell is protein σ1 which exists in the form of trimers which are associated with the reovirus spikes. Normally σ1 appears to form a lid which covers the channel of these spikes (see above), without, however, preventing λ2 from reacting with antibodies against it; but mild heat causes σ1 to assume the form of 48 nm-long 4–6 nm-wide fibers topped by 9.5 nm-diameter globular heads (which contain the cell-attachment sequence) which extend from the surface of reovirus particles. Most of the N-terminal one-third of σ1 is made up of a series of about 20 tandemly arranged heptads in which the first and fourth amino acids are hydrophobic. This type of sequence causes an α-helical coiled-coil type structure in which the hydrophobic residues form the interfaces between α-helices. It is these sequences that cause the trimerization of σ1 which in turn generates the signals for its association with/insertion into the spike channels.

Although present in reovirus particles to the extent of only 36 molecules, protein σ1 plays an extremely important role in specifying the interactions of reovirus particles with their host cells and with the immune system. Protein σ1 is the cell-attachment protein; as such, it is also the reovirus hemagglutinin and specifies tissue tropism (as discussed in the previous entry). It also possesses the epitopes that elicit the formation of neutralizing antibodies, which cross-react minimally, if at all, so that it is also the type-specific antigen. It also induces delayed-type hypersensitivity, generates suppressor T cells and cytolytic T lymphocytes, and is the protein that is recognized by cytolytic T lymphocytes; and inhibits cellular DNA synthesis and specifies the extent of association of reovirus particle assembly intermediates with microtubules.

The nonstructural reovirus proteins

Reovirus encodes three nonstructural proteins. Two are encoded by the major ORFs of genome segments M3 and S3 and are produced in large amounts. The former, μNS, is produced in two forms, one of which is translated from the entire M3 ORF, whereas the other, designated μNSC because it was at first thought to be a cleavage product of μNS, is translated in the same reading frame but starting at an AUG at codon position 42. Protein μNS is a phosphoprotein, μNSC is not. Protein μNS has a very high α-helix content and in its C-terminal region shares a periodic sequence similarity pattern with various myosins. It rapidly combines with newly formed plus-stranded RNAs with which it remains associated until they are transcribed into minus-strands to form progeny genome segments (see below). It possesses affinity for elements of the cytoskeletal framework (CGK) which may play a role in morphogenesis.

Protein σNS is another ssRNA-binding protein; it also forms complexes with reovirus mRNA molecules very soon after they are transcribed. Proteins μNS and σNS possess affinity for each other as demonstrated by the fact that antibodies against either also precipitate the other (but this association may be mediated by RNA).

The third nonstructural reovirus protein is the basic protein σ1S, encoded by the minor ORF in S1, which is formed in infected cells in readily detectable amounts and is present both in the cytoplasm and in the nucleolus. It, rather than σ1, appears to play a role in inhibiting host DNA replication, and also increases the severity of the cytopathic effects (CPE) that follow infection with reovirus.

Reovirus Replication

Strategy

Reovirus multiplication exhibits two unusual features, both consequences of the fact that the reo-

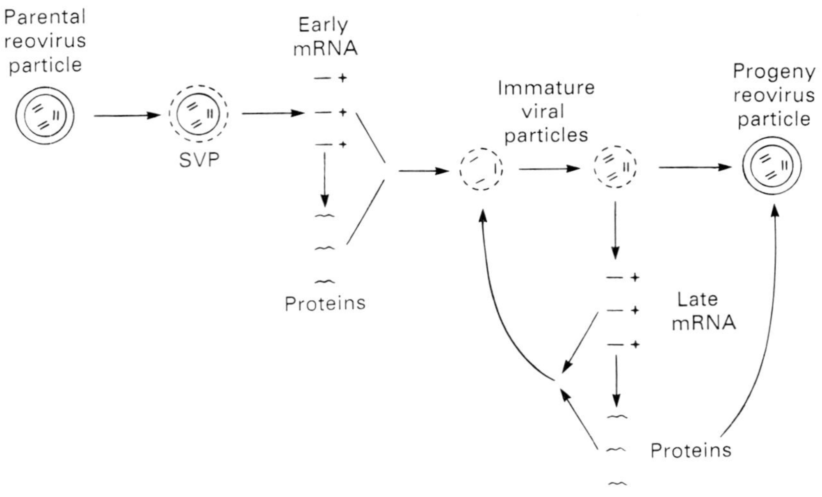

Fig. 3 The reovirus multiplication cycle. SVP, subviral particle.

virus genome is dsRNA, which means that a virus-associated polymerase is required to generate mRNA, and the fact that it is segmented. As a result, parental genomes are not uncoated to naked RNA, and a highly complex mechanism functions to assort the ten genome segments into complexes that contain one, and one only, of each (Fig. 3).

Reovirus particles adsorb to specific receptors (see below) and are internalized in endoplasmic vesicles which fuse with lysosomes within which the reovirus capsid shell is extensively degraded: a 12 kD fragment is cleaved from the C-terminal portion of μ1C to generate the 60 kD protein δ, and proteins σ3 and σ1 are lost. The resultant particles, which are cores covered by an outer shell which consists only of δ and which are known as subviral particles (SVPs), are liberated into the cytoplasm. The major functional difference between reovirus particles and SVPs is that reovirus particles are unable to transcribe their genome segments into mRNA, whereas SVPs can do so. The mRNAs are then translated into the various reovirus proteins and, after a certain interval, generally at about 3 h after infection, they begin to be transcribed into minus-strands with which they remain associated, thereby generating the progeny dsRNA-containing genome segments; and at the same time equimolar amounts of them are assorted into complexes. These complexes, or particles, then transcribe the dsRNA molecules back into plus-strands, which are again either translated and/or used as templates for the generation of further ds genome segments. This cycle represents the multiplication phase of reovirus replication. As more dsRNA-containing complexes (dsRCCs) are generated and more reovirus proteins are formed, the protein complement of the dsRCCs is modified in stepwise fashion to generate first core-like particles, and then finally complete reovirus particles. The SVPs take no part in reovirus replication but may be detected among the virus yield if the multiplicity was high and the yield low.

The nature of the reovirus receptor

There are several reovirus receptors that differ in their affinity for the cell-attachment sequences on the σ1 proteins. Protein σ1 itself possesses highest affinity for a 67 kD glycoprotein, as well as somewhat lower affinity for several other glycoproteins. The use of anti-idiotypic antibody has suggested that the β-adrenergic receptor can act as a reovirus receptor; and L cells possess either two reovirus receptors, one of which binds only serotype 3 (ST3), while the other binds both ST1 and ST3, or one receptor with higher affinity for ST3 than for ST1. However that may be, the minimal essential receptor determinant for reovirus recognition is the sialic acid residue at the ends of the carbohydrate prosthetic groups of glycoproteins; in fact, reovirus possesses affinity for sialic acid ligated to bovine serum albumin. Presumably the remainder of the carbohydrate moiety and the protein modulate affinity and specificity.

The reovirus receptor on erythrocytes is glycophorin.

All this evidence suggests a relatively nonstringent requirement for reovirus receptor recognition which is consistent with its very wide host range and its ability to bind to and infect a variety of cell types in the body.

The activation of the reovirus transcriptase

Reovirus particles do not transcribe mRNA; SVPs do so. This is not due to the fact that the transcriptase is inactive in virus particles; on the contrary, transcription initiation proceeds normally in virus particles, but transcription ceases before transcripts are more than four to six residues long and then reinitiates. The products of this abortive reiterative transcription are 5′-triphosphorylated – and also often capped – oligonucleotides, the sequence of most of which is $GCUA_n$ or $GCUAU_n$ where n is 1–4, and which remain associated, at least temporarily, with reovirus particles, in which they make up the bulk of the approximately 2400 5′-G-terminated oligonucleotides that they contain (see above).

SVPs, by contrast, are capable of transcribing full-length mRNA molecules, although for them also the majority (of the order of 90%) of transcripts are abortive transcripts. Clearly, what is activated is not the transcriptase, but rather the movement of template segments relative to the transcriptase catalytic site. Exactly what is involved in this release of movement is not known. Physicochemical evidence indicates that the conversion of reovirus particles to SVPs or cores triggers a conformational change in the dsRNA. This brings up the old question of whether the genome segments in reovirus particles are linked by protein molecules that represent the vestiges of the mechanism that assorts them. It is conceivable that the 'activation' process dissociates such bonds, thereby permitting the independent transcription of each genome segment.

Transcription and translation of reovirus mRNA

Under optimal conditions of NTP and Mg^{2+} concentration, reovirus cores transcribe all ten genome segments at the same rate, that is, in amounts inversely proportional to their sizes. In infected cells, the relative proportions of the various species of mRNAs that are formed generally differ in two respects: there is usually a deficiency in the relative amounts of the *l* size class species, and during the early phase of the multiplication cycle, before progeny dsRNA genome segments have been formed, several species of mRNA (*l*1, *m*3, *s*3 and *s*4) appear to be formed in amounts larger than the rest. The basis of this effect, which is variable – often it is the *m*2, *s*3 and *s*4 species that are formed in relatively higher proportions, and often the effect is not limited to the early period – is not known.

Whereas these effects on the relative transcription frequencies are variable and quantitatively minor (that is, the excess transcription efficiencies are no more than two- to threefold), the relative translation efficiencies of the ten species of reovirus mRNAs differ enormously. The most efficiently translated mRNA species is usually species *s*4, followed by *m*2 (relative translation efficiency 0.67), *s*2 and *s*3 (slightly less than 0.5), *m*3, *l*2 and *l*3 (0.25–0.33), *s*1 (0.1) and *l*1 and *m*1 (0.01). Many of the reasons for these differences in relative translation efficiencies are inherent in the sequences that surround and lie upstream of the initiation codons. Not only does the well-known Kozak rule apply, namely, positions −3 and +4 relative to the first nucleotide of the initiation codon must be G or A, but the nature of nucleotides at least as far upstream as position −8 also profoundly affects translation efficiency, depending on the nature of the nucleotides in positions −1 to −3. There is also an optimal length of the 5′-untranslated region (about 14 nucleotides), and there are also secondary structure constraints: the 5′-upstream sequence, including the initiation codon, must not be part of a stable stem-loop, and the 5′-cap must be accessible, that is, it also must not be part of or too close to, a stem-loop.

The fact that reovirus cores contain enzymes that catalyze the entire capping reaction indicates that capping is very important for reovirus mRNA translation. Some studies have suggested that late reovirus mRNAs are uncapped and that reovirus infection modifies the host cell translational machinery, so that late *un*capped viral mRNAs are translated preferentially over capped cellular mRNAs, or that a factor in reovirus-infected cells stimulates translation of late uncapped reovirus mRNAs. Other studies have found that reovirus-infected cells can translate both uncapped and capped mRNAs, which confirms earlier studies that indicated that possession of a cap *facilitates* translation, but is not essential. It has also been suggested that the preferential translation of reovirus mRNAs in infected cells is mediated at the level of competition between mRNAs for a limited amount of a message-discriminatory factor. The resolution of all this conflicting evidence may lie in the finding that, like infection with many other viruses, infection with reovirus activates, via the generation of dsRNA or of short sequences in ssRNA with a locally dsRNA-like character, a cellular protein kinase that phosphorylates the α-subunit of protein synthesis initiation factor eIF-2, thereby inactivating it and inhibiting protein synthesis, including the synthesis of reovirus proteins; and that protein $\sigma3$, by virtue of its affinity for dsRNA (see above), prevents this activation. Thus protein $\sigma3$ itself may be the factor that is essential for the efficient translation of late reovirus mRNAs.

The assortment of genome segments into reovirus genomes

The mechanism responsible for assorting the ten genome segments into genomes containing one of each is one of the most fascinating problems of reovirology; and very little is known about it. Since it is much easier to imagine how ssRNA molecules can be recognized by each other and by proteins than dsRNA molecules, it has always been assumed that assortment proceeds at the level of ssRNA. However, such ssRNA-containing complexes (ssRCCs) cannot be found. Rather it appears that the plus-strands associate with three viral proteins very soon after they are formed: the nonstructural protein μNS, the nonstructural protein σNS and $\sigma3$. The resultant complexes contain one molecule of RNA and 15–30 molecules of these three proteins, depending on their length; most of them contain μNS, as well as σNS and/or $\sigma3$. Presumably the binding is sufficiently reversible not to interfere with the RNAs being translated. The relative amounts of the various ssRNA species in the populations of these complexes reflects the relative frequencies with which they are transcribed (see above). Very significantly, however, even the first *double*-stranded RNA-containing complexes (dsRCCs) that can be detected (and which contain $\lambda2$ as a major component as well as, presumably, $\lambda3$) contain strictly equimolar amounts of all ten genome segments. This suggests that the generation of dsRNA genome segments and their assortment into genome sets are functionally linked and concomitant events; and it focuses attention on the RNA polymerase $\lambda3$ as a key effector of assortment.

Infectious reovirus RNA

Since the only molecular links between parental and progeny virus particles are the plus-strands transcribed by SVPs, infectious reovirus should in theory be formed in cells into which the ten species of plus-stranded RNA are introduced. Conditions have indeed been found recently under which reovirus RNA is 'infectious' in this sense. The basic system consists of lipofecting into cells all ten species of ST3 ssRNA together with rabbit reticulocyte lysates in which all ten species of ST3 ssRNA have been translated for 60 min, and infecting these cells 4 to 8 h later with ST2 reovirus. When analyzed for their virus content 24 h later, ST3 virus is found to be present in these cells to the extent of 10^3 to 10^4 plaque-forming units (PFU)/10^6 cells. If ST3 dsRNA is also lipofected at the same time, the virus yields are 100 times higher; and most of the increased amount of virus can be shown to be the progeny of the ssRNA, *not* the dsRNA, the function of which is therefore to enhance the infectivity of the former. The function of the protein-synthesizing system is fascinating: (1) it is not absolutely essential, but increases virus yields 100-fold; (2) translation must be allowed to proceed for 60 min, when an activity plateau is reached; (3) if RNA is hydrolyzed at that time, all activity is lost and cannot be restored by re-addition of RNA without permitting it to be translated (that is, the essential factor that is formed is not protein, but protein–RNA complexes); and (4) the RNA of reovirus *ts* mutants fails to yield active complexes even if translated under fully 'permissive' conditions, which indicates that the conformation of the protein in these protein–RNA complexes is extremely important and cannot be achieved by proteins that have folding problems.

This system is important in the sense that it permits the introduction into the reovirus genome of novel genetic information. It can be used to discover the nature of the assortment and encapsidation signals in reovirus RNA and the nature of the functional domains in reovirus proteins. Further, it permits the construction of reoviruses with any desired property or phenotype, and containing any desired inserted sequence information.

Effect of reovirus infection on infected cells

There is no special mechanism for the release of reovirus progeny; reovirus particles are released when cytopathic effects have progressed sufficiently for cell necrosis to result in cell lysis. In cells infected with reovirus, masses of granular material develop in areas scattered throughout the cytoplasm which eventually move toward the nucleus and coalesce, forming characteristic inclusions or 'viral factories'. These inclusions are easily identified with fluorescein-labelled antibodies against reovirus proteins; they represent the areas where viral assembly proceeds, and quasicrystalline arrays of reovirus particles are often associated with them. Microtubules appear to extend throughout these viral factories and appear to be covered with viral protein;

in particular, proteins σ1 and μNS possess affinity for elements of the CSK, which suggests that they may play a role in facilitating or mediating reovirus morphogenesis. However, microtubules *per se* are *not* essential, since colchicine does not inhibit reovirus multiplication or assembly. Like infection with all lytic viruses, reovirus infection causes progressive disruption of the CSK organization, but it is not known which reovirus protein(s) cause(s) this effect.

Reovirus infection inhibits host macromolecular biosynthesis. Since the rate at which this occurs differs in different cells infected with members of the three reovirus serotypes, it has been possible to identify which viral proteins cause such inhibition, using genome segment reassortant analysis (that is, using batteries of reovirus particles that contain some genome segments of one serotype and others of another and correlating specific effects, such as rapidity of inhibition of host protein synthesis, with the presence of a specific genome segment of one serotype, or using reassortants that contain nine genome segments of one serotype and one of another). It has been found in this way that the S1 genome segment, either through σ1 or σ1S, specifies the efficiency with which host DNA replication is inhibited; and that the S4 genome segment (protein σ3) controls the rate of host protein and RNA synthesis inhibition.

See also: Pathogenesis; Replication of viruses.

Further Reading

Joklik WK (ed.) (1983) *The Reoviridae*. New York: Plenum Press.

Joklik WK (1985) Recent progress in reovirus research. *Annu. Rev. Genet.* 19: 537.

Schiff LA and Fields BN (1991) Reoviruses and their replication. In: Fields BN *et al.* (eds) *Virology*, 2nd edn, chapter 25. New York: Raven Press.

REPLICATION OF VIRUSES

V Gregory Chinchar
University of Mississippi Medical Center
Jackson, Missisippi, USA

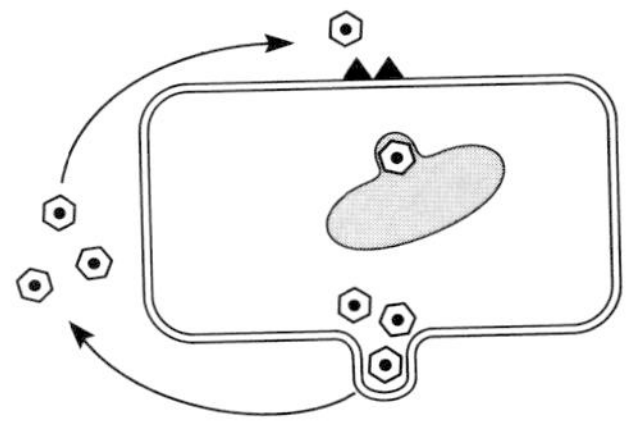

Introduction

Viruses are obligate intracellular parasites, the replication of which occurs only within living animal, plant or bacterial cells. Of more than 60 taxonomically defined virus families, 21 contain members that infect vertebrate animals, and it is these families that will be the focus of this overview. Among the smallest vertebrate viruses, the virion consists only of the viral genome and a closely associated protein coat (nucleocapsid), whereas larger viruses also possess a host-derived lipid bilayer containing one or more virus-encoded glycoproteins. In animal cells, virus replication is complete within several hours to at most a few days, and results in the synthesis of 10^3–10^5 virus particles/cell. Conceptually, virus replication can be divided into three stages: (1) early events (attachment to susceptible cells, penetration and uncoating), (2) viral biosynthetic events (replication of the viral genome, transcription and translation) and (3) virion assembly. In this overview, general strategies of animal virus replication will be considered. However, early events and mechanisms of virion assembly will be dealt with briefly so that viral biosynthetic strategies can be considered in greater detail.

Early Events

Attachment

Infection begins with the attachment of a virion, via capsid or envelope proteins, to specific cell-surface macromolecules (viral receptors). Because of the specificity of this interaction, the host range (tropism) of a given virus is determined primarily by the presence of viral receptor molecules on the cell surface. As a group, viruses utilize a variety of proteins, lipids and oligosaccharides as receptors. One class of receptors includes cellular macromolecules involved in ligand binding, endocytosis and cell recognition. For example, the receptors for poliovirus, human rhinovirus (ICAM-1), human immunodeficiency virus type 1 (CD4) and Epstein–Barr virus (CR2) are members of the immunoglobulin superfamily of proteins. In contrast, the receptor for a murine C type retrovirus is an integral membrane protein of unknown cellular function, whereas sialic acid-containing glycoproteins serve as receptors for paramyxo- and orthomyxoviruses.

Penetration and uncoating

Following attachment, the virion must enter the cell (penetrate) and release its genomic material (uncoat) in order to initiate a productive infection. The process by which many viruses accomplish this dual task is termed receptor-mediated endocytosis and is the same mechanism used by the cell to import growth factors, peptide hormones and other large molecules to which the plasma membrane is not permeable. Virions, bound to their cognate cellular receptors, are transported laterally within the plasma membrane to clathrin-coated pits and endocytosed as the clathrin-coated pit invaginates. Subsequently, the clathrin-coated vesicle fuses with an endosome and, within this acidic compartment, uncoating takes place. The acidic pH of the endosome is critical and agents that raise the intraendosomal pH (e.g. NH_4Cl, chloroquine, etc.) block virus uncoating. For enveloped viruses uncoating involves fusion of the viral envelope with the endosomal membrane followed by release of the nucleocapsid into the cytoplasm. Nonenveloped viruses also appear to utilize receptor-mediated endocytosis, although here uncoating does not involve membrane–membrane fusion. For example, following attachment of poliovirus to target cells, one of the capsid proteins (VP4) is released exposing hydrophobic residues buried inside the virion. Interaction of these residues with the endosomal membrane may provide a pore through which viral RNA is extruded into the cytoplasm. In the adenovirus system, low endosomal pH induces conformational changes in the capsid which rupture the endosomal membrane at virion–membrane contact points. Following its release into the cytoplasm, adenovirus is transported via microtubules to nuclear pores where viral DNA enters the nucleus. In contrast to the above mechanism, several viruses [e.g. paramyxoviruses, herpesviruses and human immunodeficiency virus type 1 (HIV)] do not require an acidic environment for uncoating and enter cells by fusion at the plasma membrane.

Although the presence of viral receptors is a primary determinant of infectivity, not all cells carrying the appropriate receptor are susceptible to infection. In several 'restrictive' systems, the synthesis of infectious progeny is blocked at a post-attachment step. For example, some mammalian cell lines bind influenza virus and support the synthesis of all viral macromolecules, yet do not generate infectious virions because they lack the protease required to cleave the hemagglutinin precursor (HA_0) and generate activated (i.e. fusion-competent) HA_1 and HA_2. Conversely, some cells that lack the appropriate viral receptor can nonetheless support a productive infection if the viral genome is introduced into the cell by transfection.

Synthesis of Virus-specific Macromolecules

The 21 families of vertebrate viruses, although differing in genomic make-up, virion morphology and their repertoire of viral-encoded enzymes, can be classified on the basis of replicative mechanisms. However, before examination of different viral replicative strategies, several common themes need to be addressed.

Viral transcription and genome replication

Viral nucleic acid synthesis is catalyzed by both viral and host enzymes, the relative contribution of which is determined by the type of virus and the specific molecule. Viruses with RNA genomes, except for the retroviruses, synthesize mRNA and replicate their genomes using virus-encoded RNA-dependent RNA polymerases. In contrast, retroviruses synthesize a double-stranded complementary DNA (cDNA) copy of their single-stranded RNA genome using a virion-encoded RNA-dependent DNA polymerase (reverse transcriptase). In subsequent steps, the retroviral cDNA is integrated into the host chromosome and transcribed by host-encoded DNA-dependent RNA polymerase II (pol II) to yield viral messages and genomic RNA. DNA viruses, except for poxviruses, also use host-encoded pol II to transcribe their messages. Poxviruses, because they replicate in the cytoplasm and do not have access to pol II, assemble a novel transcriptase composed of multiple poxvirus-specific (and possibly one or more host-derived) subunits. Most DNA virus families (e.g. *Poxviridae*, *Iridoviridae*, *Herpesviridae*, *Adenoviridae* and *Hepadnaviridae*) synthesize a virus-encoded DNA polymerase. However, two families (e.g. *Parvoviridae* and *Papovaviridae*) utilize host DNA polymerase.

Gene regulation

Viruses have evolved a variety of mechanisms to control gene expression and maximize efficiency. In some systems, viral gene expression is divided into temporal phases in which catalytic and regulatory proteins are synthesized early in infection, while the synthesis of structural proteins is limited to late times. Alternatively, the expression of viral genes may be controlled by differences in the transcription rate of specific genes (e.g. rhabdoviruses and paramyxoviruses), the translational efficiency of different viral messages (e.g. reovirus) or in the replication of transcriptional templates (e.g. influenza virus). Moreover, it is likely that, even within a single virus family, multiple mechanisms regulate gene expression. At the molecular level, gene expression is controlled by both *cis*- and *trans*-acting signals. In some cases, the nucleotide sequence of viral messages and transcriptional templates may be the primary factor in determining how efficiently a given sequence is translated or transcribed. For example, the differential synthesis of the various coronavirus mRNAs is thought to be controlled by interaction between *trans*-acting coronavirus leader RNA and *cis*-acting sequences located at the beginning of each gene. Furthermore, transcription and genome replication among DNA viruses (and retroviruses) is regulated by the (often) combined action of *trans*-acting viral- and host-encoded factors with *cis*-acting viral nucleotide sequences. For example, herpesvirus immediate-early gene transcription requires, aside from pol II, both host- (OTF-1) and virus-encoded (α-TIF) transcription factors. Lastly, in two systems (*Poxviridae* and *Iridoviridae*), there are hints that viral proteins may regulate viral gene expression at the translational level.

Viral protein synthesis

Viral protein synthesis is completely dependent on the cell's translational machinery (i.e. ribosomes, tRNAs, initiation factors, etc.). Reflecting that dependence, viral mRNAs, despite some prominent exceptions (e.g. picornaviruses), are similar in overall structure to host messages, i.e., they are capped and methylated at their 5′ terminus and polyadenylated at their 3′ end. Viral mRNAs are monocistronic and are translated as are other eucaryotic transcripts. However, in some systems, viral proteins are synthesized as part of a larger multiprotein precursor (polyprotein) which is cleaved to generate the final products. This mechanism overcomes the inability of eucaryotic ribosomes to translate polycistronic messages and allows one viral mRNA to code for several proteins. Finally, like their cellular counterparts, viral proteins are post-translationally modified (e.g. glycosylated, phosphorylated, etc.) using host-cell-specific pathways and enzymes.

As infection progresses, viral protein synthesis often supplants cellular translation. In some cases, this simply reflects the increased abundance of viral messages, whereas in others viral messages appear to initiate translation at a higher rate than host messages. Alternatively, virus infection may actively inhibit host translation by (1) proteolytically inactivating initiation factors required solely or preferentially by cellular messages, (2) selectively degrading host messages or (3) altering the intracellular ionic environment to favor viral over host translation. Furthermore, because infection can lead to the phosphorylation and functional inactivation of eucaryotic initiation factor 2 (eIF-2), several virus families (*Poxviridae*, *Reoviridae*, *Orthomyxoviridae*, *Adenoviridae* and *Picornaviridae*) have evolved mechanisms to block eIF-2 phosphorylation and maintain adequate levels of protein synthesis throughout infection. Virus infection also blocks host cell RNA and DNA synthesis. While transcriptional shut-off may be the direct result of inactivating specific transcription factors, the inhibition of cellular DNA synthesis is likely due to the earlier inhibition of protein synthesis.

Cytoskeleton

Aside from providing enzymes, transcription factors, protein and nucleic acid precursors, the cell also supplies the virus with an intracellular highway to facilitate infection and assembly. There is growing evidence that the transport of infecting virions to the nucleus and the transport of viral proteins into assembly sites takes place along the various fibers of the cellular cytoskeleton.

RNA Viruses

RNA-containing viruses will be discussed in the light of four basic organizational/transcriptional strategies. These strategies, depicted diagrammatically in Fig. 1, encompass (1) viruses with a message-sense genome (positive-strand viruses), (2) viruses with a genome that is complementary to

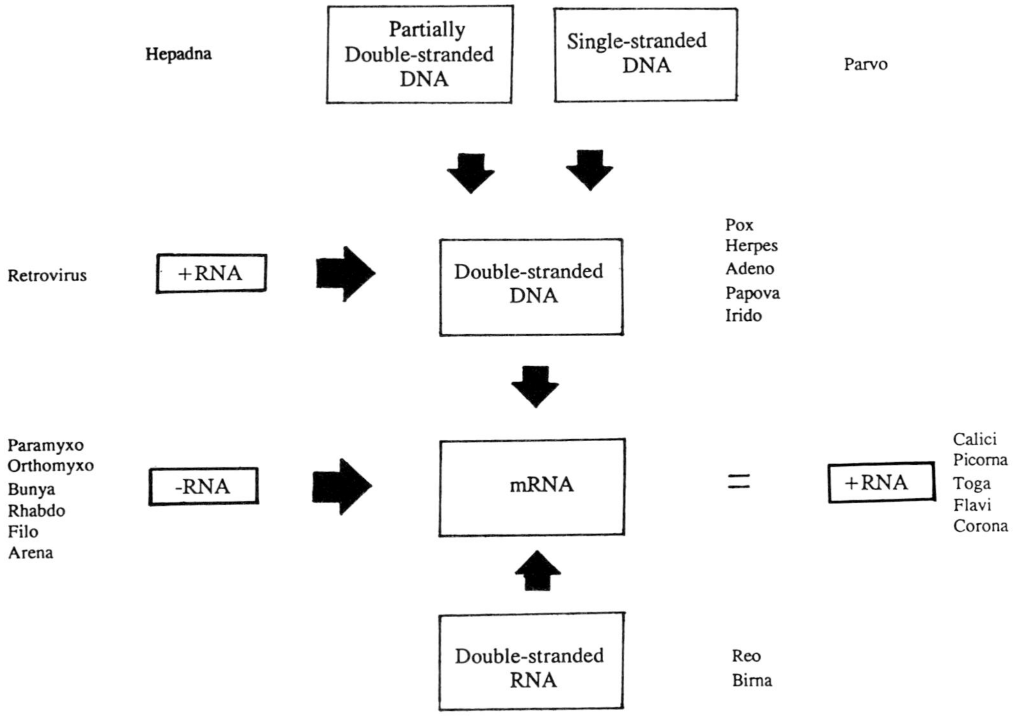

Fig. 1 Strategies for the production of viral mRNA. Based on the schema of D. Baltimore (1971); updated from *Bacteriol. Rev.* 35; 235.

mRNA (negative-strand viruses), (3) viruses that package their replicative form as genome (double-stranded RNA viruses) and (4) viruses that utilize 'reverse transcription'. Although this approach is conceptually useful, not all viruses within a class conform precisely to the prototypic replication strategy. Despite this caveat, representative examples will be cited to illustrate the replication mechanism.

Positive-strand virus families (*Picornaviridae*, *Togaviridae*, *Flaviviridae*, *Calciviridae* and *Coronaviridae*)

Positive-strand viruses contain a single-stranded message-sense RNA genome which is translated immediately following uncoating. To simplify this discussion, poliovirus (family, *Picornaviridae*; genus, *Enterovirus*) will be used as a prototype because it is the most extensively studied positive-strand virus and provides a clear view of this strategy.

The primary translation product of the poliovirus genome is a high molecular weight polypeptide (mol. wt ~ 200 000) termed the polyprotein which is cleaved to yield viral structural and catalytic proteins. Initial cleavage of the polyprotein occurs cotranslationally and is mediated by the autocatalytic activity of a virus-encoded protease, polypeptide 2A. Subsequent cleavages yield the structural (capsid) and catalytic proteins. As infection proceeds, translation of capped host mRNAs is blocked due to the virus-induced degradation of the large subunit of the mRNA cap recognition factor (eIF-4F). In contrast, viral messages, which are uncapped and possess a highly structured 5′-nontranslated region (5′-NTR), escape shut-off because the 40S ribosomal subunit is able to bind to a sequence within the 5′-NTR and thus bypass the requirements for a capped 5′-terminus and intact eIF-4F.

Following its synthesis, the viral RNA-dependent RNA polymerase catalyzes the synthesis of a full-length negative-sense copy of the genome. Subsequently, the negative-strand serves as template and directs the synthesis of multiple plus-strands. Early in infection, when the concentration of viral structural proteins is low, newly synthesized positive-strands most likely are translated and serve to amplify the synthesis of viral proteins. Later when the concentration of virion precursors is high, newly synthesized plus-strands are encapsidated to generate infectious virus particles. Interestingly, a small virus-encoded protein (VPg) is covalently linked to the 5′ end of all picornavirus RNAs except message. VPg is thought to play a role in RNA synthesis, but it is unclear whether VPg functions as a primer or in some other capacity.

Negative-strand viruses (*Paramyxoviridae*, *Rhabdoviridae*, *Filoviridae*, *Orthomyxoviridae*, *Bunyaviridae* and *Arenaviridae*)

Because the genome of the single-stranded negative-sense RNA viruses cannot be translated, the first virus-specific biosynthetic event following uncoating is the synthesis of viral mRNA by a virion-associated RNA-dependent RNA polymerase using the viral genome as template. Negative-stranded viruses can be divided into two classes: viruses with unsegmented (monopartite) genomes (i.e. paramyxo-, rhabdo- and filoviruses) and those with segmented (multipartite) genomes (i.e. the orthomyxo-, bunya- and arenaviruses). Although each class uses the negative-strand strategy, they possess unique attributes and will be dealt with separately.

The replication of unsegmented negative-strand viruses will be discussed using vesicular stomatitis virus (VSV), a rhabdovirus, as the prototype. Immediately after uncoating, the VSV genome is transcribed to yield a small nontranslated 'leader' RNA followed, in decreasing molar amounts, by five capped, methylated and polyadenylated viral mRNAs. Transcription occurs within the nucleocapsid, a structure containing the viral genome and multiple copies of three virus-encoded proteins, the nucleocapsid protein (NP), a phosphoprotein (P) and the polymerase (L). Polypeptide L, an ~ 200 000-mol. wt protein present at about 50 molecules/nucleocapsid, catalyzes initiation and elongation, as well as capping, methylation and polyadenylation. The phosphoprotein P, present in about 500 molecules/nucleocapsid, plays a variety of roles in RNA synthesis. It binds L to the nucleocapsid, maintains the solubility of free NP and may function in chain elongation. The viral transcriptase binds genomic RNA at its 3′ terminus and initiates transcription. At each intergenic junction (with the exception of the leader–NP junction), a poly(A) tract is added to the newly synthesized mRNA by repetitive copying ('stuttering') of an oligo(U) sequence present at the end of the gene. After the synthesis of the poly(A) tract, transcription terminates, releasing newly synthesized mRNA, but maintaining the transcriptase on its template. Re-initiation at the next gene downstream occurs via a conserved start sequence present at the beginning of each gene. However, because re-initiation does not take place every time, downstream genes (i.e. those coding for the envelope and polymerase proteins) are transcribed less frequently than upstream ones encoding the nucleocapsid, phosphoprotein and matrix proteins. Thus transcriptional polarity controls viral gene expression.

Viral genome replication, i.e. the synthesis of a full-length positive-sense copy of the genome and the subsequent generation of progeny negative-strands, is catalyzed by the same polymerase that directs transcription. The switch between the transcriptive and replicative modes of RNA synthesis appears to be controlled by the concentration of the nucleocapsid protein. When NP reaches a critical concentration, it binds to newly synthesized RNA within the leader sequence and allows the polymerase to read-through intergenic regions and synthesize full-length positive-strands. These, in turn, serve as templates for virion RNA synthesis.

Segmented viruses encode their genetic information in multiple molecules of negative-sense RNA. In the case of influenza A virus (*Orthomyxoviridae*), the genome is composed of eight unique segments of virion RNA. In contrast to most RNA virus families, orthomyxoviruses require a functional cell nucleus for replication. This requirement reflects the fact that the orthomyxovirus polymerase complex can neither initiate transcription *de novo* nor cap and methylate viral mRNAs. Instead, the complex 'pirates' the capped and methylated 5′ terminus from a selected set of newly synthesized host messages and uses these to prime viral transcription. Once initiated, transcription of each segment continues until an oligo(U) tract, about 22 nucleotides from the end of virion RNA, signals addition of the 3′ poly(A) tail by repetitive copying. Because of this unique method of transcription, the viral genome, and the mRNAs coded from it, are not completely complementary, but differ at both their 5′ and 3′ ends. As with the unsegmented viruses, the trigger controlling the transition from transcription to replication may be the concentration of nucleocapsid protein.

Some bunyaviruses (tripartite genome) and all arenaviruses (bipartite genome) possess another unusual feature, 'ambisense' genomic RNA. These genomes are termed 'ambisense' because nonoverlapping subgenomic messages are transcribed from the 3′ ends of both virion RNA and its full-length complement. The former message encodes the nucleocapsid protein, whereas the latter message specifies the glycoprotein precursor, a 'late' protein.

Double-stranded RNA (dsRNA) viruses (*Reoviridae*, *Birnaviridae*)

Animal viruses with dsRNA genomes are segmented

and can be viewed as a variant of the negative-sense theme in which the virion encapsidates the replicative form of the genome. Genomic dsRNA is transcribed within partially uncoated ribonuclease-resistant viral cores by the virion-associated polymerase to yield viral mRNAs. Early in infection, some progeny plus-strands function as translational templates, whereas others associate with nonstructural proteins and form complexes which are transcribed once to yield dsRNA. Newly synthesized dsRNA serves as template for the synthesis of additional viral mRNA which amplifies the replication cycle. Later, as the concentration of core and capsid proteins increases, the dsRNA–protein complex exchanges nonstructural for structural proteins and forms mature virus particles.

RNA viruses that utilize a reverse transcription strategy (*Retroviridae*)

Retroviruses replicate their genome and transcribe mRNA using a dsDNA copy of viral genomic RNA as template. This unconventional mechanism, in which single-stranded virion RNA is used as a template for dsDNA synthesis, is catalyzed by a virion-associated RNA-dependent DNA polymerase (reverse transcriptase). Following viral entry, the virion capsid is partially uncoated and a complementary DNA copy of the RNA genome is synthesized using reverse transcriptase. An endonucleolytic activity, integral to the reverse transcriptase, degrades the RNA template and second-strand DNA synthesis begins. Completion of second-strand synthesis results not only in a dsDNA copy of virion RNA, but also generates a unique structure termed the long terminal repeat (LTR). The LTR, present in two copies/genome, flanks the viral cDNA and is composed of unique sequences from the 5′ and 3′ ends of the genome and a repeat element common to both ends. Retroviral DNA is next integrated into the host chromosome, and, in this form, is termed the 'provirus'. Subsequently, the provirus is transcribed by pol II to yield full-length progeny RNA and one or more subgenomic mRNAs. The upstream LTR plays a very important role in retrovirus gene expression because it contains enhancer elements which interact with cellular and/or viral transcription factors to regulate pol II-catalyzed transcription. (The downstream LTR is not involved in viral gene expression, but may activate host oncogenes and play a role in cell transformation.) Full-length genome-sized RNA can either be packaged within virions or serve as messenger for the capsid and catalytic viral proteins. Among most retroviruses, translation of genomic RNA yields two classes of polyproteins. The majority (~ 95% of the total) encode only the capsid and core proteins. However, a minor population, resulting from bypassing a stop codon at the end of the capsid/core sequences via a frameshift mechanism, encodes the protease, reverse transcriptase and integrase in addition to the capsid proteins. Envelope glycoproteins are translated from a singly-spliced subgenomic mRNA complementary to the 5′ end of the genome, while lentiviruses, such as HIV-1, utilize doubly-spliced subgenomic mRNAs to direct the synthesis of TAT, REV and several other regulatory proteins.

TAT and REV are the two best-studied of the HIV-1 regulatory proteins. TAT is a *trans*-acting protein that binds to a sequence present at the 5′ end of all HIV-1 mRNAs and enhances HIV-1 gene expression by relieving a block in transcriptional elongation. However, other data suggest that TAT may act by increasing transcriptional initiation or enhancing HIV-1 translation. Whether these different mechanisms reflect alternative facets of TAT's activity profile or the vagaries of different experimental systems and protocols is not yet clear. REV directs the transport to the cytoplasm of unspliced genome-sized RNA and singly-spliced envelope message. Thus, REV mediates the switch between the synthesis of regulatory proteins (i.e. TAT and REV) and the generation of structural and catalytic polypeptides.

DNA Viruses

With the exception of parvoviruses and hepadnaviruses, the genomes of which are respectively single-stranded and partially double-stranded, DNA-containing animal viruses possess a dsDNA genome. However, even in the former two families, viral mRNA is ultimately transcribed from a dsDNA template using cellular DNA-dependent RNA polymerase (Fig. 1). In place of a detailed discussion of each family, broader issues of viral DNA replication will be discussed. To begin with, DNA viruses differ greatly in their genetic content ranging in size from 5 kbp (*Parvoviridae*) to greater than 120 kbp (*Herpesviridae*, *Poxviridae* and *Iridoviridae*). Thus the small DNA viruses are about as genetically complex as a typical RNA virus, whereas the larger DNA viruses encode 100 or more proteins. Not unexpectedly, the degree to which virus replication is dependent upon cellular functions reflects the genetic complexity of the virus. Thus, parvoviruses and papovaviruses require extensive host involvement to support viral biosynthetic events (including DNA synthesis), whereas other families are progressively more independent.

Among herpes-, pox- and iridoviruses, viral genes are expressed in a coordinate temporal sequence (cascade) of immediate-early (α), early (β) and late (γ) genes. Immediate-early genes code for factors required to initiate virus replication, early genes encode catalytic functions (e.g. the herpesvirus DNA polymerase is an early protein), and late genes specify viral structural proteins. Furthermore, immediate-early genes activate early and late gene transcription, while specific early and late genes downregulate immediate-early and early gene expression. Aside from specific regulatory proteins, full late gene expression also requires viral DNA synthesis, thus inhibitors of viral DNA replication block late gene expression despite the presence of functional immediate-early and early activators.

Because DNA polymerase requires a primer with an available 3′-OH to initiate DNA synthesis, all virions with a linear DNA genome have evolved specialized features that allow them to maintain their termini intact during replication. For example, adenoviruses solve the 'end-problem' by using a nucleotide-linked terminal protein to initiate DNA replication, herpesviruses replicate through a rolling circle mechanism and poxviruses and parvoviruses utilize a 'self-priming' mechanism to ensure replication of their termini. In contrast to other DNA viruses, hepadnaviruses possess a circular, partially single-stranded DNA genome that is replicated through an RNA intermediate using a virus-encoded reverse transcriptase. Upon entry into the cell, the gaps are repaired and the negative DNA strand is transcribed to yield viral mRNAs and a supergenome-length RNA. The latter is encapsidated and transcribed into complementary DNA using a protein primer and reverse transcriptase. As with retroviruses, the RNA template is degraded and second-strand DNA synthesis takes place. However, before completion of this strand, the virion is exported from the cell leaving genomic DNA partially single-stranded.

Unlike other DNA viruses, poxviruses replicate solely within the cytoplasm in morphologically distinct viral 'factories'. Reflecting their metabolic independence from the host cell, poxviruses synthesize unique DNA and RNA polymerases, and their virions contain all the proteins needed to transcribe the earliest class of viral mRNAs. Furthermore, viral transcriptional promoters and termination sequences are unique and are regulated by virus-specific factors.

Iridoviruses, occupying a taxonomic middle-ground between poxviruses and the nuclear DNA viruses, possess several distinctive features. Viral DNA replication takes place in two distinct compartments (genome-length progeny DNA is synthesized in the nucleus, followed by the synthesis of concatemeric DNA in the cytoplasm), whereas virion assembly is confined to cytoplasmic viral 'assembly sites'. Viral DNA is highly methylated with nearly 25% of cytosine residues present as methylcytosine. Methylation is catalyzed by a virus-encoded enzyme and, as in some bacteriophage systems, appears to function as part of a restriction-modification system. Surprisingly, despite the high content of methylcytosine, host RNA polymerase II has been implicated in at least the early rounds of viral transcription. However, it is not known whether unmodified pol II transcribes viral DNA late in infection or whether viral-encoded proteins modify pol II and alter its specificity.

Virus Assembly

Once sufficient stores of viral nucleic acid and protein have accumulated in the infected cell, nucleocapsid formation and virion assembly begin and continue as long as the cells are metabolically competent. Despite the large number of animal virus families, only three types of nucleocapsids are found: complex, helical and icosahedral (spherical). The nucleocapsids of poxviruses, retroviruses and hepadnaviruses do not conform to the geometric symmetry found among the helical and icosahedral viruses and are considered to be 'complex'. Little is known about the molecular mechanisms controlling their assembly and they will not be considered further. Helical nucleocapsids (which, among animal viruses, enclose only RNA genomes) form as viral proteins, bind to nascent RNA transcripts and encapsidate them. During assembly, helical nucleocapsids migrate to cellular membranes where viral glycoproteins have concentrated. There, through concerted interaction between the nucleocapsid and the cytoplasmic tails of viral glycoproteins, the nucleocapsid is enveloped by the cellular membrane in a process referred to as 'budding'. In this process, host proteins are excluded and the resulting envelope contains only virus-encoded glycoproteins. Moreover, because envelopment is not a precise process, dual infections with different strains of the same multipartite virus (e.g. influenza virus A) lead to high-frequency genetic reassortment. Although virion envelopment takes place commonly at the plasma membrane (e.g. among the *Paramyxoviridae*, *Orthomyxoviridae* and *Rhabdoviridae*), intracellular membranes (e.g. those of the Golgi, endoplasmic reticulum and, in the case of DNA viruses, the nucleus) are used by other virus families.

In contrast to helical nucleocapsids, icosahedral nucleocapsids enclose both DNA and RNA genomes. In picornavirus-infected cells, nucleocapsids form spontaneously within the cytoplasm when capsid precursors reach a critical concentration. However, icosahedral nucleocapsid assembly also occurs within the nucleus (as with most families of DNA viruses) or within defined viral assembly sites (iridoviruses). In some families nucleocapsids are not enveloped (i.e. virion = nucleocapsid), whereas in other systems nucleocapsids are enveloped as described above.

See also: Cell structure and function in virus infections; Frog virus-3.

Further Reading

Banerjee AK and Barik S (1992) Gene expression of vesicular stomatitis virus genome RNA. *Virology* 188: 417.

Marsh M and Helenius A (1989) Virus entry into animal cells. *Adv. Virus Res.* 36: 107.

Fields BN and Knipe DM (eds) (1990) *Virology*, 2nd edn. New York: Raven Press.

Joklik WK *et al.* (eds) (1992) *Zinsser Microbiology* 20th edn. Norwalk, CT: Appleton and Lange.

Porterfield JS (ed.) (1989). *Andrewes' Viruses of Vertebrates.* London: Baillière Tindall.

Schneider RJ and Shenk T (1987) Impact of virus infection on host cell protein synthesis. *Annu. Rev. Biochem.* 56: 317.

RESPIRATORY SYNCYTIAL VIRUS

Peter L Collins
National Institute of Allergy and Infectious Disease,
Bethesda, Maryland, USA

History

Respiratory syncytial virus was first isolated in 1956 from one of a group of laboratory chimpanzees with upper respiratory tract disease. Shortly thereafter, an apparently identical virus was isolated from children ill with pneumonia or croup. Serologic studies established that the virus indeed was a human pathogen and that infection of children and infants was common. Human respiratory syncytial virus (HRSV) is now recognized as a major, ubiquitous, highly infectious etiologic agent of pediatric respiratory tract disease world-wide. HRSV infects most children by 2 years of age, is the leading cause of bronchiolitis and pneumonia in infants, and can be a significant cause of disease in immunocompromised adults and the elderly. The name of the virus includes reference to its prominent cytopathic effect in tissue culture, the formation of syncytia.

There also are bovine (BRSV), caprine (CRSV) and ovine (ORSV) counterparts which are related antigenically to HRSV but are distinct (see later). BRSV is one of several respiratory tract pathogens associated with shipping fever, an important disease of cattle.

Taxonomy and Classification

HRSV is classified in the *Paramyxoviridae* family which, together with families *Rhabdoviridae* (e.g. vesicular stomatitis and rabies viruses) and *Filoviridae* (e.g. Marburg and Ebola viruses), comprise the nonsegmented negative-strand RNA viruses. These are enveloped, cytoplasmic viruses that have as genome a single strand of protein-coated RNA (vRNA) that is of negative sense and, for the paramyxoviruses, is 15.2–15.9 kb in length.

Paramyxoviridae contains four genera organized into two subfamilies (as of 1992). Subfamily *Pneumovirinae* has a single genus, *Pneumovirus*, which contains HRSV (the type species), its animal counterparts mentioned above, pneumonia virus of mice (PVM), and turkey rhinotracheitis virus (TRTV). The molecular organization of PVM corresponds closely to the HRSV model, and initial characterization of TRTV demonstrated clear similarities with regard to proteins but identified differences in gene order (see later). The second subfamily, *Paraflumorbillivirinae*, contains the more commonly known, better characterized paramyxoviruses, namely: Sendai virus and human parainfluenza viruses types 1 and 3 (genus *Respirovirus*), measles and canine distemper viruses (genus *Morbillivirus*), and mumps virus, simian virus type 5, Newcastle disease virus and human parainfluenza viruses types 2 and 4 (genus *Rubulavirus*).

Pneumoviruses differ from other paramyxoviruses in that the diameter of the helical nucleocapsid (see later) is 12–15 nm rather than 18 nm. They also differ in aspects of vRNA and protein structure as described later (see also Fig. 2). With regard to virion surface markers, pneumoviruses lack neuraminidase activity, which is present in members of *Respirovirus* and *Rubulavirus* but is absent in members of *Morbillivirus*. HRSV also lacks a hemagglutinin, which is found in all members of *Paraflumorbillivirinae*. But PVM has a hemagglutinin, and thus its absence from HRSV is not a characteristic of *Pneumovirus*.

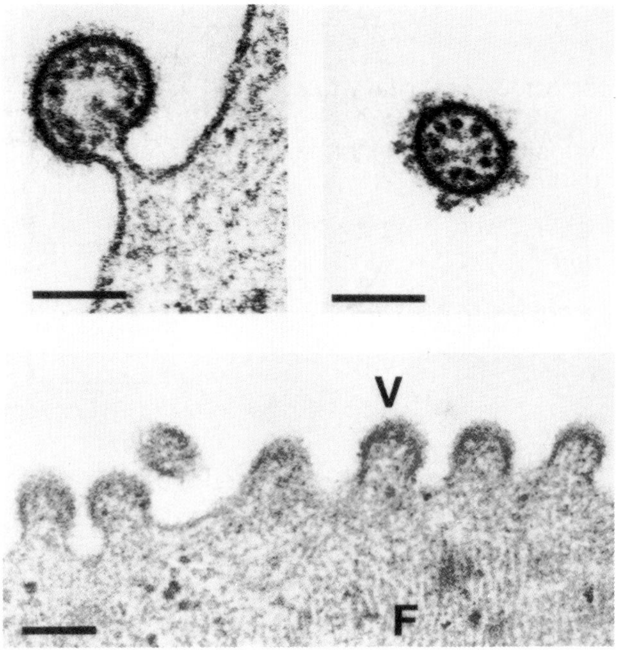

Fig. 1 Electron micrographs of thin-layer sections of: (top left) a spherical-type HRSV virion in the final stage of budding from the plasma membrane of an infected cell; (top right) a free spherical virion; and (bottom) a series of budding virions (V). Virion spikes are visible as a fringe on the exteriors of the virions, and the dense round structures in the interiors of the virions in the top two panels are probably of nucleocapsids. In the bottom panel, the letter (F) is placed in an accumulation of intracellular fibrils (running vertically in the micrograph) which commonly underlie regions of budding and probably also are nucleocapsids. Scale bar = 100 nm. Adapted from Kalica *et al.* 1973 *Archiv für die gesamte Virusforschung* 41: 248.

Virion Structure and Viral Proteins

HRSV virions grown in tissue culture appear in the electron microscope as spherical particles of 80–350 nm in diameter (Fig. 1) and filamentous particles of 60–100 nm in diameter and up to 10 μm in length. The virion contains a nucleocapsid packaged in a lipoprotein envelope which is acquired from the host cell plasma membrane during budding. The virion surface has spike-like projections of 11–20 nm at intervals of 6–10 nm. These consist of viral-encoded transmembrane glycoproteins that are involved in attachment to and penetration of host cells.

Ten HRSV proteins have been described (Fig. 2). Three are nucleocapsid-associated virion proteins: the major nucleocapsid protein N (43.4 kD) is tightly associated with vRNA, the phosphoprotein P (27.2 kD) is probably a polymerase cofactor, and the large protein L (250.2 kD) is the major polymerase subunit which probably contains the catalytic domains. By analogy with more extensively studied nonsegmented negative-strand viruses, these three proteins are thought to function together in viral RNA synthesis.

There are three transmembrane virion proteins which assembly separately into homo-oligomers that make up the membrane spikes: the fusion F glycoprotein (69 kD; polypeptide moiety 61 kD) responsible for viral penetration, the attachment G glycoprotein (90 kD; polypeptide moiety 32.5 kD), and the small hydrophobic protein SH (present in virions as a 7.5 kD unglycosylated form and as a 21–30 kD glycosylated form) of unknown function. The F contains a cleaved N terminal signal sequence and is anchored in the membrane by a C terminus-proximal membrane anchor such that the N terminal 91% of the molecule is extracellular. The F protein contains five to six potential acceptor

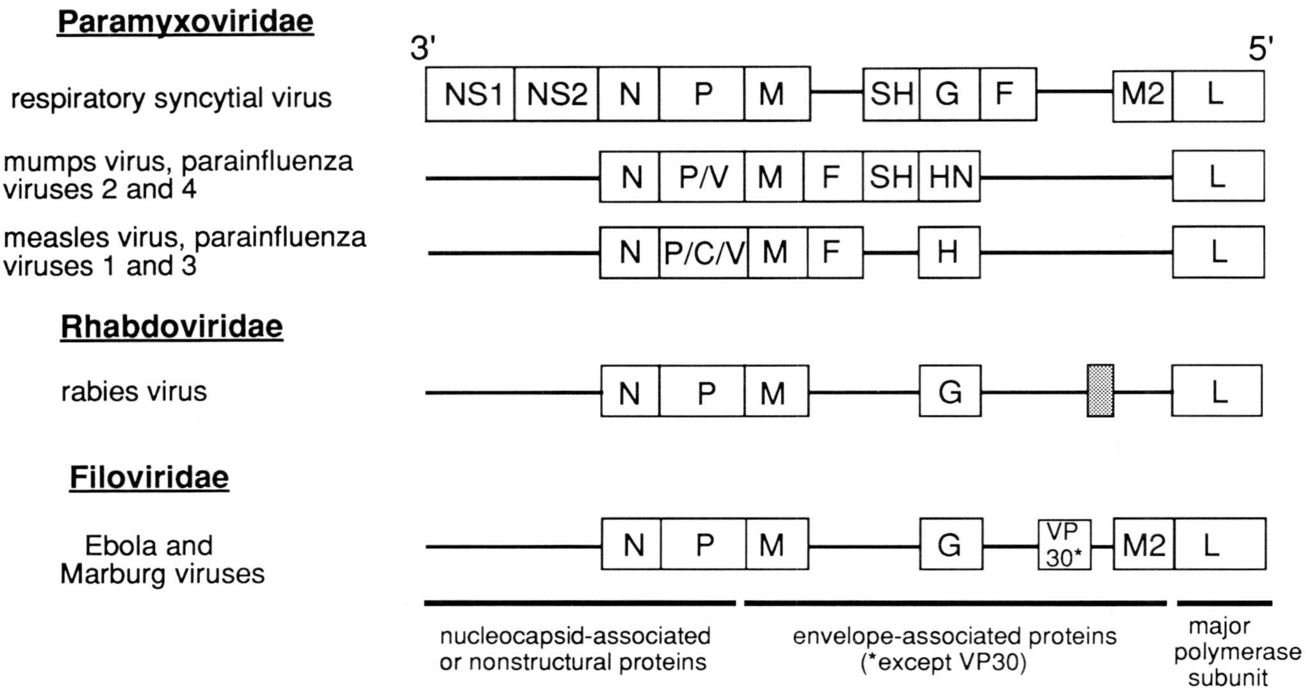

Fig. 2 Comparison of the gene maps of HRSV and selected viruses (of humans) of the three families of nonsegmented negative-strand RNA viruses. Each box represents a separate encoded mRNA, and acronyms within identify proteins. For the mRNAs which contain more than one distinct open reading frame, the acronyms of the encoded proteins are separated by slashes. When proteins of different viruses have the same name, it indicates that they are thought to have similar functions (even though in many cases there are substantial differences in protein structure and little or no apparent sequence relatedness). The one exception is the viral attachment protein, which varies in name (G, HN or H) among different viruses despite having the common function of attachment. The box sizes and spacings are not to scale and were drawn to place analogous proteins in columns when possible. The identification of the M and M2 proteins of the filoviruses is preliminary; VP30 might be a nucleocapsid-associated protein. The shaded box in the rabies virus map is an apparent relict of an unknown gene. Within *Paramyxoviridae*, RSV represents the genus *Pneumovirus*, mumps and parainfluenza viruses 2 and 4 represent *Rubulavirus*, measles represents *Morbillivirus* and parainfluenza viruses 1 and 3 represent *Respirovirus*. Adapted from Pringle CR (1991) In: Kingsbury DW (ed.) *The Paramyxoviruses*, p. 1.

sites for N-linked carbohydrate side chains. It is synthesized as a precursor, F_0, that is cleaved intracellularly by a cellular protease into two disulfide-linked subunits. These subunits, F_1 (50 kD) and F_2 (19 kD), constitute the biologically active form and are arranged as follows: NH_2-F_2-F-S-S-F_1-COOH. The F protein assembles into a tetramer, and intermonomeric contacts appear to involve the part of the molecule which, upon subsequent cleavage, becomes the F_1 subunit.

The G protein is anchored in the membrane by an N terminus-proximal hydrophobic signal/anchor sequence such that the C terminal 78% of the molecule is extracellular. The G protein, which assembles into a trimer or tetramer, contains several N-linked carbohydrate side chains (there are three to eight potential acceptor sites, depending on the virus strain). It is somewhat unusual in containing a relatively large amount of O-linked sugars which, by one determination, are present as 24 or 25 side chains (there are more than 70 serine and threonine residues as potential acceptor sites, giving G an unusually high content of these two amino acids). The G protein also has an unusually high content of proline residues. These features might cause it to have an extended nonglobular structure which would be very different from that of the highly folded attachment HN or H proteins of the members of subfamily *Paramorbillivirinae*. The G protein also is unusual in that the amino acid sequence of most of the ectodomain is highly divergent among different strains (see later). But this otherwise divergent ectodomain does contain a single conserved sequence of 13 amino acids which is thought to form an unglycosylated, disulfide-linked domain involved in receptor binding.

The SH protein is anchored in the membrane by a centrally located signal/anchor sequence such that the C terminal 36% of this short species is extracellular. The SH protein, which assembles into a homo-oligomer containing at least five monomers, is present mainly as an unglycosylated 7.5 kD form. But some molecules contain a single N-linked side chain which in turn is modified by the addition of polylactosamine, the significance of which is unknown. One interesting speculation on the function of the SH protein is that it forms ion channels, by analogy to the suggested function of the M2 protein of influenza A virus. But it is not evident how such an activity would be important for HRSV.

There are two nonglycosylated matrix proteins located between the nucleocapsid and the envelope of the virion: the M protein (28.7 kD) and the M2 protein (sometimes called the 22K protein, 22.2 kD). Their functions are unknown but, by analogy to other M proteins, probably include rendering nucleocapsids transcriptionally inactive before packaging and mediating the association between the nucleocapsid and envelope during budding. Lastly, there are two nonstructural (nonvirion) proteins of unknown function: NS1 (sometimes called 1C, 15.6 kD) and NS2 (sometimes called 1B 14.7 kD). The M2 mRNA has a second, conserved translational open reading frame which might encode an eleventh, unidentified protein.

Genome Organization, Transcription and Replication

The genome of HRSV is a single negative strand of RNA (Fig. 2). It has been sequenced in its entirety for the A2 strain and contains 15 222 nucleotides. It encodes 10 major subgenomic polyadenylated mRNAs and 10 major proteins (compared to 6–7 mRNAs and 7–8 proteins for members of *Paraflumorbillivirinae*).

The vRNA gene order is 3′NS1-NS2-N-P-M-SH-G-F-M2-L (Fig. 2). (Interestingly, TRTV has a different order for the sixth through ninth genes in the map: 3′-F-M2-SH-G-5′, suggestive of gene rearrangement.) The first nine HRSV genes are separated by intergenic regions of 1–52 nucleotides (for strain A2); these regions do not have any apparent conserved sequences. The last two genes overlap by 68 nucleotides and are expressed as separate mRNAs which each contain the overlap sequence (TRTV lacks this overlap, again suggestive of past rearrangements in genome structure). All HRSV genes initiate with a nine-nucleotide conserved gene-start sequence and terminate with a 12–13 nucleotide semiconserved gene-end/polyadenylation sequence: these serve as transcriptive signals that direct the synthesis of the individual mRNAs. The latter signal ends with a run of 4–7 U residues (vRNA sense) which are thought to encode the poly(A) tail of the mRNA by a mechanism of reiterative copying. The 3′ vRNA end contains an extragenic 'leader' region of 44 nucleotides that contains the major viral promoter, and the 5′ vRNA end contains an extragenic 'trailer' region of 155 nucleotides.

HRSV vRNA transcription (to yield subgenomic, positive-sense, polyadenylated mRNAs) and replication (to yield progeny vRNA) follow the nonsegmented negative-strand strategy and take place in the cytoplasm without apparent nuclear involvement. The viral-encoded polymerase, which is assumed to be packaged in the virion, initiates within the leader region of N protein-encapsidated vRNA. Viral genes are transcribed in their 3′ to 5′ order by a sequential stop–start mechanism that is guided by the gene-start and gene-end/polyadenylation signals and yields single-gene transcripts. A model of sequential linear transcription does not provide for transcription of the overlapping M2 and L genes; an alternative possibility is that the polymerase initiates internally for transcription of L. There are two known mechanisms that determine the relative levels of expression of the different genes. First, genes proximal to the 3′ terminal promotor are transcribed more frequently than are downstream genes due to polymerase fall-off. Second, the presence of the gene-end/polyadenylation signal of the M2 gene within the L gene (due to the gene overlap) results in premature termination for 90% of L gene transcripts, resulting in a 10-fold reduction in the synthesis of full-length L mRNA. There is no evidence of temporal regulation of gene expression.

vRNA replication occurs by an antitermination mechanism that causes the polymerase to ignore transcriptive signals and synthesize a full-length, positive-sense, encapsidated replicative intermediate. This in turn is the template for the synthesis of progeny vRNA. The synthesis of replicative intermediate RNA and progeny vRNA is thought to be tightly coupled to their encapsidation by N protein, and concurrent encapsidation might be the mechanism for antitermination.

Since vRNA is negative sense, it is not directly infectious alone. Instead, the minimum unit of infectivity is thought to be a transcriptionally active nucleocapsid. In recent studies, cDNA-encoded analogs of vRNA were made in which the 10 viral genes were deleted and replaced by a gene encoding a heterologous reporter enzyme. The vRNA analog was transfected into tissue culture cells and complemented with proteins supplied by superinfection with HRSV. Under these conditions the synthetic vRNA was encapsidated, amplified, transcribed and packaged into virions. These studies showed that the RNA signals involved in these activities are in two segments contained a total of 105 nucleotides derived from the two vRNA ends. This system for 'rescuing' vRNAs encoded by cDNA will allow detailed characterization of the functions of vRNA sequences and viral proteins and might make possible the rescue of infectious virus from full-length cDNA-encoded vRNA.

Antigenic Subgroups

HRSV is monotypic serologically, with up to fourfold differences between disparate strains in cross-neutralization *in vitro* by postinfection serum. However, more refined antigenic analyses with monoclonal antibodies (MAbs) showed that HRSV isolates can be segregated, by the presence or absence of certain epitopes, into two distinct antigenic subgroups designated A and B. Most of the dimorphism occurs in the G protein, although differences in most of the other proteins also have been noted. In studies with postinfection serum, the antigenic relatedness between the two subgroups was greater than 50% for the F protein compared with only 5% for the G protein. Overall, the amino acid sequences of all proteins are 87% or more identical between the two subgroups except for the extracellular domains of the SH and G proteins which are only 50% and 43% conserved, respectively. In contrast, within each subgroup the amount of sequence identity between strains is substantially higher (usually >96%, although in some comparisons G was <90% identical), with a correspondingly higher degree of antigenic relatedness. Divergence in the G protein both within and between subgroups also can involve extensive variability in the number and location of acceptor sites for N- and O-linked carbohydrate side chains.

Antigenic mapping studies of the F protein with murine MAbs identified four or five major antigenic sites. At least three of these were nonoverlapping and reacted with neutralizing MAbs. Many of the available F MAbs efficiently neutralize HRSV *in vitro*. In one study, one-half of the neutralizing F-specific mabs were reactive with numerous strains from both antigenic subgroups and several were reactive with BRSV, attesting to the antigenic conservation of the F protein.

In contrast, most MAbs to the G protein neutralize infectivity poorly or not at all and are not broadly reactive within or between the antigenic subgroups. Weak neutralization by MAbs with the homologous strain might be influenced by the high content of host-specified sugars. This might involve a masking of the virus-specified polypeptide chain as well as possible microheterogeneity in the placement of side chains among the many available potential acceptor sites. This latter factor might result in subpopulations of molecules, some of which might not be efficiently bound by individual MAbs. In one study, most of the tested G-specific mabs were reactive only with the fully glycosylated protein, indicating that the O-linked sugars indeed are important in antigenicity. Differences in epitopes between heterologous strains could be due to divergence both in the amino acid sequence and in the locations of carbohydrate side chains. Antigenic mapping studies with murine mabs indicated that the G protein contains numerous antigenic sites that are overlapping rather than distinct. Viral variants selected for resistance to neutralization by individual G-specific mabs were found also to lose reactivity with G-specific MAbs from most of the other antigenic sites as well as with polyclonal G-specific sera. Sequence analysis showed that these MAb-resistant mutants contained large deletions or truncations in the C-terminal region of the G protein. Taken together, these observations suggest that the G protein contains a large immunodominant region near the C terminus which could be deleted in the MAb-resistant mutants without affecting their replication *in vitro*. It is not clear whether this escape mechanism has any relevance *in vivo* since similar naturally occurring deletions in the G protein have not been described to date.

Antigenic and Sequence Relatedness of HRSV to Other Viruses

Antigenic cross-reactivity has been observed between HRSV and BRSV for the N, P, M and F proteins, and between HRSV and PVM for the N and P proteins. CRSV appears to be more closely related to BRSV than to HRSV. Several of the genes of BRSV have now been sequenced. The amount of sequence divergence between BRSV and HRSV is approximately twice that between the two HRSV subgroups. (For example, the N and F proteins are, respectively, 7% and 19% divergent between HRSV and BRSV compared with 4% and 9% divergence between the HRSV subgroups. The BRSV G protein is the most divergent relative to HRSV, being 70% divergent with HRSV compared with 47% divergence between the G proteins of the HRSV subgroups.) Some sequence information is now available for PVM and TRTV; for example, the PVM N protein and the TRTV F protein are 60 and 39% identical, respectively, to their HRSV counterparts.

Regarding sequence relationships between HRSV and other nonsegmented negative-strand viruses: the HRSV F protein has low but significant sequence relatedness with the F proteins of other members of *Paraflumorbillivirinae*, and the L protein has low but significant relatedness with the L protein of the *Paraflumorbillivirinae*, *Rhabdoviridae* and *Filoviridae* (Fig. 2). It seems clear that these two HRSV proteins have global similarity to their counterparts in the above-mentioned groups and are derived from common ancestral species. The other HRSV proteins lack obvious, unambiguous sequence relatedness with their counterparts outside of *Pneumovirinae*. Nonetheless, it is likely that functional similarity (and evolutionary relatedness) exists between these viruses for proteins such as N, P and M. Proteins such as the nonstructural species might or might not be related evolutionarily between members of *Pneumovirinae* and *Paraflumorbillivirinae*, but might have similar functions. The HRSV G protein is unusual in that it is very dissimilar structurally to the attachment proteins of the other paramyxoviruses, and it is difficult to envision a common ancestral protein. Nucleotide sequences of HRSV are not conserved with viruses beyond *Pneumovirinae*.

Virus Infection in Tissue Culture and Animals

The cellular receptor for HRSV attachment has not been identified but, by analogy to other paramyxoviruses, it might be an oligosaccharide moiety of glycoprotein or glycolipid. This type of receptor typically is abundant and relatively ubiquitous and probably would not be a major factor in tissue tropism. HRSV enters the cell by fusion at the plasma membrane.

HRSV has not been successfully propagated in eggs but can be grown in a variety of cultured cells of human, simian or bovine origin, with the HEp-2 line being the most commonly used. During infection with strain A2 at an input multiplicity of 5, the intracellular production of viral proteins and nucleic acids can be detected by 10 h postinfection, reaches maximum at 15–24 h and is quite abundant. Virus release begins by 10 h and is maximal after 20 h, syncytia become evident by 20 h, and extensive cytopathology and destruction of the monolayer occurs at 30–48 h. Host cell macromolecular synthesis does not appear to be inhibited except by the indirect effects of cytopathology. In cultured cells, most of the progeny virions remain cell-associated and are released by freeze-thawing or sonication. HRSV is very labile to inactivation during unfrozen storage or freeze-thawing: harvested tissue culture supernatants typically are adjusted to be pH 7.5 and to contain 0.1 M magnesium sulfate to stabilize infectivity. Virus yield is 2×10^5 to 5×10^7 PFU per ml depending on the strain. The virus tends to associate with membranous cell debris. The instability of the virion makes further purification and concentration difficult.

Chimpanzees are highly permissive to HRSV infection and exhibit fully developed symptoms of respiratory tract disease. The level of HRSV replication in the respiratory tract is similar to that observed in humans. HRSV also can replicate the respiratory tract of several species of monkeys as well as in hamsters, guinea pigs, infant ferrets, mice and cotton rats. But in these animals the infection is semipermissive, the titer of recoverable virus is 100 to 1000-fold lower than in the fully permissive chimpanzee, and disease either does not occur or is very greatly reduced in severity.

In animals and in humans, HRSV infection generally is restricted to the superficial layers of the respiratory tract epithelium. The reasons for this tropism have not been established. For some paramyxoviruses the ability of the host cell to cleave the F_0 precursor and thereby produce infectious virus is an important determinant of tissue tropism, but this does not appear to be the case for the readily cleaved HRSV F protein. The spread of HRSV to secondary organs has been described under conditions of immunosuppression or immunodeficiency. This indicates that the virus is capable of infecting other organs but normally is restricted by host immunity. Other factors which have not yet been investigated, such as the topography of virus budding from the respiratory epithelium, might also be involved in the predilection of HRSV to remain localized in the respiratory tract. Infection of monocytes and macrophages has been reported but is of unknown significance. HRSV inoculated intramuscularly appeared to undergo a single cycle of replication without the production of infectious virus.

Epidemiology and Clinical Factors

HRSV causes yearly epidemics that usually occur in the winter in temperate regions or in the rainy season in the tropics. Infection is associated with crowding. HRSV is an important cause of nosocomial infection. Spread involves inoculation of conjunctival or mucosal surfaces by hand or particles containing respiratory secretions. Most humans are infected by age 1 or 2 years, and the greatest incidence of serious disease is between 6 weeks and 6 months of age. The relative sparing of newborns is thought to be due to the protective effects of transplacentally derived maternal serum antibodies, an effect which quickly wanes. The higher incidence of serious disease in young infants probably reflects in part the greater susceptibility of smaller airways to obstruction by edema and secretions. The risk of hospitalization due to HRSV infection during infancy is 1 in 70 to 1 in 200. With full hospital care, mortality is very low for normal children. But infants and children with bronchopulmonary dysplasia, heart disease or immunodeficiency are at special risk for serious, life-threatening HRSV disease and mortality can be as high as 30%. In some underdeveloped countries the infant death rate from respiratory disease can exceed 2000 per 100 000 births (World Health Organization Report 1986, *Vaccine* 4: 201), and it is estimated that 20–25% of these would be due to HRSV.

HRSV has an incubation period of about 4–5 days. It causes upper respiratory tract disease with symptoms of a common cold. Progression to the lower respiratory tract causing bronchitis, bronchiolitis or pneumonia occurs with 25–40% of primary infections. Symptoms include rhinorrhea, middle ear disease, fever, coughing and wheezing. Seriously ill infants have increased coughing and wheezing, rapid respiration and hypoxemia requiring the administration of humidified oxygen. Duration of illness is 7–12 days. Virus is shed in large amounts (10^4 to 10^6 PFU per ml of nasal wash) throughout infection and sometimes during recovery.

HRSV infection can be diagnosed rapidly and efficiently by the detection of viral antigens by immunofluorescence of exfoliated cells or enzyme-linked immunoadsorbant assay of respiratory secretions. Other methods in common use include isolation of the virus in tissue culture or detection of an antibody rise by serology.

One hallmark of HRSV is the ability to infect young infants despite the presence of transplacentally derived virus-neutralizing serum IgG. This reflects the inefficiency with which serum IgG transudates into the respiratory tract and indicates the importance of local immunity in restricting virus replication. But it is not clear why HRSV is more infectious under these conditions than are other viruses of the respiratory tract. Another striking feature of HRSV is its ability to re-infect repeatedly during childhood and throughout life. Certainly the transient nature of local secretory immunity and the relative ineffectiveness of serum antibodies are factors in frequent re-infections. But it is not clear why re-infections are so much more frequent with HRSV than with other viruses of the respiratory tract. The above-mentioned antigenic variation may contribute to reducing the effectiveness of host immunity in restricting re-infection but does not appear to be a major factor. Although the virus can re-infect, serious disease is associated mostly with first or second infection. Sparing during subsequent infections presumably is due to immunological restriction of virus replication in the lower respiratory tract.

Much of the pathogenesis of HRSV is the direct result of destruction of epithelial cells by virus replication and the concomitant edema, mucus secretion and influx of lymphocytes and macrophages. Pathogenesis probably also can be influenced by additional immune factors, such as antibody-mediated or cell-mediated hypersensitivity, which remain to be elucidated. It has been suggested that maternally derived virus-specific serum antibodies in infected infants might participate in immunopathology, but this now is considered unlikely. In some individuals, airway reactivity in HRSV disease might involve an allergic-type reaction mediated by IgE. The ability of immune factors to influence RSV pathogenesis profoundly is illustrated by the enhancement of RSV disease associated with immunization with a formalin-treated RSV vaccine (see later). A long-term reduction in pulmonary function is a common sequel to serious HRSV disease, but it is not clear whether this is due to the infection or whether such individuals already had underlying pulmonary deficiencies which predisposed them to serious HRSV infection.

Immunity

Infection with HRSV induces virus-neutralizing secretory and serum antibodies and virus-specific CD4+ and CD8+ T^+ lymphocytes. The F and G proteins are the only HRSV antigens which have been shown to induce neutralizing antibodies. Of the two, the F protein appears to be substantially more immunogenic, although both antigens are important independent protective antigens. The CD8+ T lymphocytes have MHC class I-restricted, HRSV-specific cytolytic activity, whereas the CD4+ subset lacked quantitatively significant cytolytic activity (at least in the BALB/c [H-2d] mouse) and presumably are of helper or otherwise regulatory phenotypes. In the BALB/c mouse, the major target antigen for CD8+ CTL is the M2 protein, with F and N being secondary antigens. The profiles of CTL target antigens in other murine genetic backgrounds are somewhat different, as has been observed in other viral systems. HRSV-specific CTL tend to be reactive between the antigenic subgroups. Studies in humans are less complete, but CTL from (multiply infected) adults recognized most of the viral proteins, with the N protein being the predominant antigen.

The major mechanisms for resolving primary infection appear to be CTL and secretory antibodies. Studies in naive mice showed that reduction in pulmonary HRSV replication coincided with the appearance of HRSV-specific CD8+ CTL and was inhibited by prior depletion of that subset. Children deficient in cell-mediated immunity experience more serious disease and have difficulty resolving infection. Studies in calves and humans indicated that the appearance of secretory antibodies is coincident with viral clearance. The late appearance of serum antibodies suggested that they are less important in resolving the primary infection. HRSV infection in mice can be restricted by the passive transfer of either RSV-specific CTL or RSV-neutralizing antibodies, and antibodies applied directly to the respiratory tract are 160-fold more effective than those administered systemically.

Humans are easily re-infected with HRSV, but (as noted earlier) serious involvement of the lower respiratory tract generally occurs only in the first or second infections. With regard to resistance to re-infection, in humans there is a good correlation between levels of HRSV-neutralizing secretory IgA in the upper respiratory tract and resistance to challenge virus replication. However, this immunity is relatively short-lived following initial infection. Studies in experimental animals showed that high titers of neutralizing serum antibodies alone also can restrict virus replication in the lower respiratory tract. This presumably is due to transudation, albeit at a very low level, of serum antibodies to the respiratory epithelium. Thus, serum antibodies probably contribute to the long-lasting immunity that develops in the lower respiratory tract following several HRSV infections. Immunization of experimental animals with individual HRSV proteins showed that antigens which elicited predominantly either (1) neutralizing antibodies or (2) HRSV-specific CTL could induce resistance to subsequent challenge virus replication. However, the resistance induced by CTL target antigens was short-lived compared with that afforded by antigens which induced neutralizing antibodies. This suggests that, while CTL clearly have an important role in resolving infection, the long-term resistance to virus replication which develops from prior infection or immunization might involve antibodies rather than primed CTL.

Immunoprophylaxis and Treatment

It seems unlikely that an HRSV vaccine can be developed which can provide complete, long-lasting resistance to re-infection, since that is not achieved following natural infection. Nonetheless, it is clear that prior immunization can restrict viral replication in the lower respiratory tract and thus greatly reduce the incidence of serious disease. The development of a safe and effective vaccine is an important, ongoing goal that faces unique challenges. Given the young age of peak disease, vaccination would involve very young infants and underscores the importance of vaccine safety. Young infants have been shown to have reduced immune responses to HRSV infection due to immunologic immaturity. Also, immunosuppression by maternally derived antibodies can profoundly reduce the induction of antibodies and CTL, although this effect can be partly abrogated by direct immunization of the respiratory tract. A vaccine which was made from formalin-inactivated concentrated HRSV and was tested in 1966 failed to prevent natural infection and, paradoxically, primed the vaccinees for an increased frequency and the severity of HRSV disease. The lack of protective efficacy probably was due to denaturation of neutralization epitopes, and disease enhancement is hypothesized to have been due to cell-mediated delayed hypersensitivity. The elucidation of this phenomenon, and the testing of HRSV vaccines in general, is complicated by the semipermissive nature of HRSV infection in most experimental animals.

Live attenuated HRSV strains for immunization by intranasal infection are under development and may represent the most effective and safe vaccine. Live recombinant vaccinia and adenoviruses which express the F or G glycoproteins have been tested in experimental animals as prototype vaccines but, to date, have not been sufficiently immunogenic. Purified F and G glycoproteins for intramuscular immunization have been produced from cultured mammalian cells infected with HRSV or from cultured insect cells infected with recombinant baculoviruses expressing F or G. The noninfectious nature of such a subunit vaccine might be especially appropriate for immunization of the very young. But these purified antigens appear to induce antibodies with low neutralizing activity, perhaps due to denaturation during preparation, and there are indications that they induce hypersensitivity.

Ribavirin treatment of hospitalized normal or high-risk infected children is well established, although debate as to its level of efficacy has continued. The administration of aerosolized ribavirin to children who were seriously ill and supported by mechanical respirators has now been clearly shown to be associated with marked improvement and reduced hospital stay. As another (still experimental) approach, the systemic or topical application of purified HRSV-neutralizing antibodies (either donor polyclonal IgG or murine mabs) to infected experimental animals was shown to reduce viral replication rapidly. This approach is under evaluation and further development for use in humans both for treatment of infection and as a method of prophylaxis for high-risk individuals. These applications ultimately will require the 'humanization' of existing murine mabs or the production of human HRSV-neutralizing mabs.

Future Perspectives

The development of an HRSV vaccine remains as a challenging task, but our understanding of the immune mechanisms involved in protection and disease continues to increase. It will be particularly important to determine the basis of the enhanced disease which was observed following immunization with formalin-treated HRSV. Prophylaxis and treatment of disease with topically applied antibodies, if successful in humans, would be a major advance. The general features of the molecular genetics of HSRV are now known. Furthermore, a genetic system has been developed for complementing cDNA-encoded vRNA analogs with HRSV proteins in tissue culture, which will allow detailed characterization of the functions of vRNA sequences and viral proteins. The next step will be to complement full-length synthetic vRNA molecules with HRSV proteins so that nondefective virus can be produced from cDNA. Apart from its usefulness in molecular studies, this would make possible the characterization and rational design of live HRSV vaccine strains.

See also: Immune response; Parainfluenza viruses; Respiratory viruses; Sendai virus; Vaccines and immune response.

Further Reading

Chanock RM, McIntosh K, Murphy BR and Parrott RH (1989) Respiratory syncytial virus. In: Evans, AS (ed.) *Viral Infections of Humans*, p. 525. New York: Plenum.

Collins PL (1991) The molecular biology of human respiratory syncytial virus (RSV) of the Genus *Pneumovirus*. In Kingsbury DW (ed.) *The Paramyxoviruses*, p. 103. New York: Plenum.

McIntosh KM and Chanock RM (1990) Respiratory syncytial virus, In: Fields BN *et al.* (eds) *Virology*, 2nd edn, p. 1045. New York: Raven Press.

McIntosh K and Fishaut JM (1982) Immunopathologic mechanisms in lower respiratory tract disease of infants due to respiratory syncytial virus. *Prog. Med. Virol.* 26: 578. New York: Plenum; Collins PL *ibid*, p. 103; and Feldmann *et al.* (1992) *Virus Res.* 24: 1.

RESPIRATORY VIRUSES

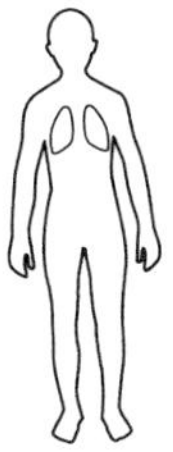

David O White
The University of Melbourne
Parkville, Victoria, Australia

Introduction

Respiratory infections are the most common afflictions of man and most of them are caused by viruses. Children contract up to half a dozen respiratory illnesses each year, adults perhaps two or three. Admittedly these are mainly trivial colds and sore throats but they account for millions of lost working hours and a significant proportion of all visits to family physicians. More serious lower respiratory tract infections tend to occur at the extremes of life, and in those with pre-existing pulmonary conditions. The most important human respiratory viruses are influenza and respiratory syncytial viruses (RSV), the former killing mainly the aged and the latter the very young. Of the estimated 5 million deaths from respiratory infections in children annually world-wide, at least one million are viral in origin.

Altogether, there are over 200 human respiratory viruses, falling mainly within six families: orthomyxoviruses, paramyxoviruses, picornaviruses, coronaviruses, adenoviruses and herpesviruses. Here those that enter the body via the respiratory route and cause disease confined largely to the respiratory tract are described. Many other 'respiratory' viruses are disseminated via the bloodstream to produce a more generalized disease, as is the case with most of the human childhood exanthems such as measles, rubella and varicella, or rinderpest and foot-and-mouth disease in cattle. Yet other viruses, entering by nonrespiratory routes can reach the lungs via systemic spread, and pneumonia may represent the final lethal event, e.g. in overwhelming infections with herpesviruses or adenoviruses in immunocompromised neonates or AIDS patients.

Epidemiology

By definition, respiratory viruses are transmitted via the respiratory route. Virions are shed from the respiratory tract of an infected human or animal, particularly during sneezing, coughing, talking or barking. A sneeze generates an aerosol comprising up to a million tiny droplets less than 10 μm in diameter that quickly evaporate to yield droplet nuclei which remain suspended in the air for several minutes. This particulate material containing virions transmits infection following inhalation by someone nearby. Larger droplets (up to 100 μm) contain more virions but fall to the ground within seconds. They are a danger to anyone directly in the line of fire. Alternatively, respiratory infections can spread by direct contact, e.g. kissing, or by transfer of nasal or oral secretions via hands to nose or mouth. At the height of a common cold such secretions are particularly copious and readily find their way on to handkerchiefs, towels, toothbrushes, eating utensils, doorknobs and so on, as well as hands.

Enveloped viruses such as orthomyxoviruses, paramyxoviruses and coronaviruses tend to be rather susceptible to inactivation by desiccation or by summer temperatures, but icosahedral viruses such as adenoviruses and picornaviruses are more stable; for example, certain outbreaks of foot-and-mouth disease have been attributed to virus carried in the wind for 100 km.

We associate respiratory infections with cold wet weather but there is no evidence that the winter incidence of the common cold, or indeed any other respiratory disease, is attributable to cold or wet *per se*. Colds are not common in Arctic or Antarctic explorers, for example. It seems more likely that the striking winter peaks of respiratory disease caused by influenza and RSV (Fig. 1) are a reflection of our predilection during that season for avoiding the invigorating outdoor climate and shutting ourselves away in ill-ventilated centrally heated buildings and vehicles, in close apposition to others of like mind. This hypothesis is supported by the observation that in the tropics, where summer and winter are replaced by 'wet' and 'dry' seasons, respiratory infections are more prevalent during the monsoonal rains when people spend more time indoors, exchanging parasites in crowded, often squalid conditions. An additional factor in the Third World is the domestic air pollution (smoke) generated by the ever-present fire, lit for cooking and warmth inside poorly ventilated huts. Outbreaks of influenza often occur in boarding schools, army camps, nursing homes, etc.;

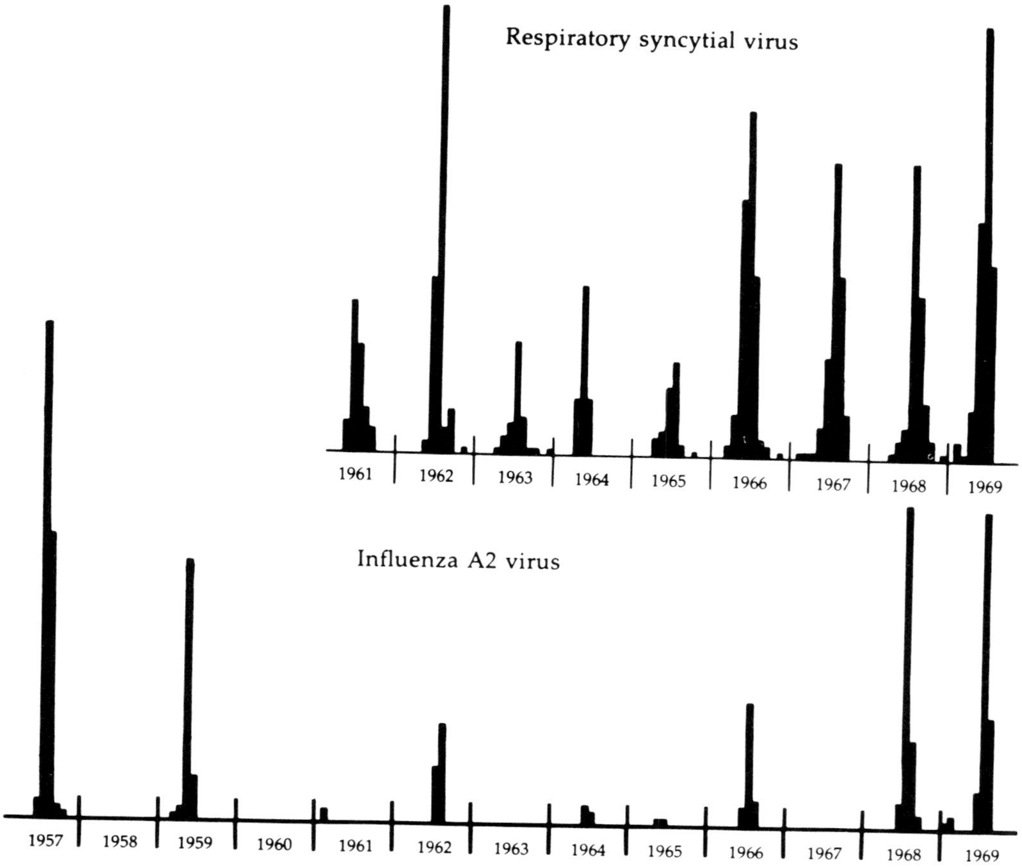

Fig. 1 Epidemic occurrence of influenza A and respiratory syncytial viruses (RSV). The histograms show the monthly isolations of these two viruses from patients admitted to the Fairfield Hospital for Infectious Diseases, Melbourne, over the periods indicated. Compare the regular winter epidemics of RSV (causing significant disease mainly in infants) with the less regular winter epidemics of influenza. There were major peaks of influenza in 1957 (first appearance of the new influenza subtype H2N2, known as 'Asian flu') and then again in 1968 (marking the emergence of the novel subtype H3N2, known as 'Hong Kong flu'). Data courtesy of Drs A.A. Ferris, F. Lewis and I.D. Gust. From Fenner, F and White, DO (1976) *Medical Virology*, 2nd edn, San Diego: Academic Press.

similarly, nosocomial spread of RSV is common in hospital nurseries. Livestock such as cattle are particularly vulnerable when crowded together in feedlots or transport vehicles, e.g. shipping fever.

Respiratory viruses spread with great facility and speed, albeit not with the explosive onset that characterizes certain 'common source' outbreaks of enteric viruses when feces contaminates food or water supplies. Firstly, respiratory diseases have a very short incubation period, usually 2–7 days. Secondly, very large numbers of virions (10^3–10^9 per ml of respiratory secretions) are shed, commencing even before symptoms develop, and peaking around the time the patient is coughing or sneezing with greatest abandon. A single infectious particle may infect a susceptible contact. Typically, a young child picks up the latest virus at school, brings it home and passes it on to the rest of the family and perhaps to the neighbors' children. Within 2 or 3 months up to half the population of a city may have contracted the infection and developed immunity to the virus. As the proportion of uninfected susceptibles in the community falls, the epidemic burns itself out (Fig. 1).

Respiratory viruses may evolve quite rapidly in the field. RNA viruses in particular display a very high rate of mutation, because their RNA polymerase is error-prone and lacks the error-correcting capability that accompanies DNA replication. Any spontaneously arising mutant that is capable of replicating in the presence of antibody against the wild-type virus will have a growth advantage. Eventually mutants emerge which contain amino acid substitutions in most or all of the immunodominant anti-

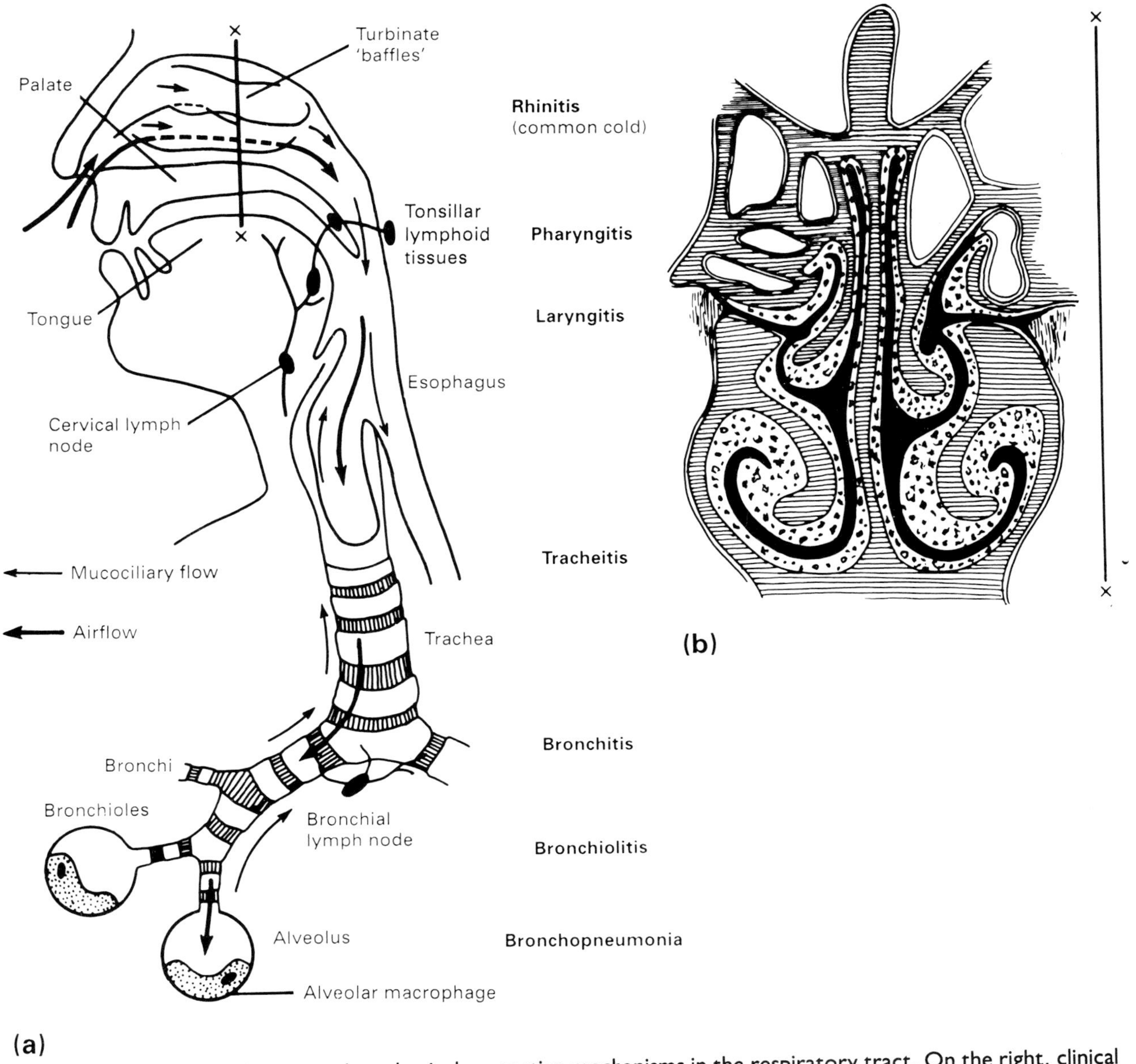

Fig. 2 (**a**) Pathways of infection and mechanical protective mechanisms in the respiratory tract. On the right, clinical syndromes produced by infection at various levels of the respiratory tract. (**b**) Section of the turbinates (X——X) showing the narrow and complicated pathway of inspired air, and thus the ease with which slight swelling 'blocks the nose'. (Modified from Mims CA and White DO (1984). *Viral Pathogenesis and Immunology*, Oxford: Blackwell Scientific Production).

genic domains on the critical surface protein of the virion. Such multiple mutants, no longer neutralizable by wild-type antibody, are designated a new strain and may initiate another epidemic. This phenomenon is known as antigenic drift. While best characterized for influenza it is presumably the mechanism that has given rise also to the dozens of known serotypes of rhinoviruses, enteroviruses, adenoviruses, and so on. A second, more dramatic type of change, known as antigenic shift, is observed in RNA viruses with segmented genomes that can undergo genetic reassortment. Novel subtypes of human influenza A viruses, capable of initiating pandemics that can infect the majority of the world's population within a year, arise every generation or so as a result of reassortment following co-infection of, say a pig, with a human and an avian influenza virus simultaneously.

Pathogenesis and Immunity

Inhaled droplets >10 μm in diameter are trapped in the turbinates of the nose (Fig. 2), whereas those measuring 5–10 μm often reach the trachea and bronchioles. Many of these particles become trapped in the layer of mucus that blankets the ciliated epithelium and are carried by ciliary action to the pharynx where they are swallowed or coughed out. Smaller particles still can be inhaled directly into the lung and some may reach the alveoli. Here, virus may be phagocytosed and destroyed by alveolar macrophages (although some viral species undergo an abortive cycle of replication and others have developed the capacity to replicate in macrophages). A few virions will succeed in attaching to susceptible epithelial cells via the appropriate ligand–receptor pairing and thereby initiate infection. Progeny virions will be released a few hours later, often by budding from the apical surface of the cell into the lumen of the respiratory tract, then initiate a second cycle of infection in adjacent or more distant cells. Some of the enveloped species of respiratory viruses are dependent upon a particular cellular protease to cleave the appropriate viral envelope glycoprotein, e.g. influenza HA, or RSV F protein, otherwise the progeny are noninfectious. Moreover, mucus contains glycoprotein inhibitors or mannose-binding lectins such as pulmonary surfactants which can neutralize the infectivity of certain viruses, e.g. influenza. Interferon, synthesized by and secreted from virus-infected cells binds to interferon-receptors on nearby uninfected cells and protects them by inhibiting viral replication.

If specific neutralizing antibodies of the IgA class are already present in the mucus coating the respiratory tract as a result of previous infection or vaccination, they will bind to the corresponding epitopes on the surface of the virion and neutralize its infectivity by blocking attachment or fusion of viral envelope with plasma membrane or endosome, thus preventing uncoating of the viral genome. However, in a primary infection, antibody synthesis does not become significant for several days, hence other mechanisms are required if the infection is to be cleared promptly. First to appear are NK (natural killer) cells, mobilized from bone marrow and activated by interferon to lyse virus-infected cells. Shortly thereafter, the relevant clones of T lymphocytes are called into play. Helper T (T_h) cells (CD4+) recognize peptides generated from endocytosed virions by proteolysis and presented in the peptide-binding groove of the class II MHC molecules on the surface of antigen-presenting cells (dendritic cells and macrophages) and are triggered to proliferate and to secrete a range of lymphokines that mediate inflammation by attracting macrophages and other leukocytes to the site, and by upregulating macrophages, B cells and T cells. Cytotoxic (T_c) cells (CD8+), on the other hand, see endogenous viral peptides generated by proteolysis of newly synthesized viral proteins and bound to class I molecules on the surface of infected cells.

Recovery from viral infection is mainly attributable to lysis of infected cells by activated T_c lymphocytes. Children with a congenital T cell deficiency may die from measles or RSV infection; conversely influenza virus-infected athymic mice may be saved by adoptive transfer of virus-specific CD8+ T cells. CD4+ T cells also contribute to the process of recovery, as do antibodies, which act by (1) neutralizing virus, (2) complement-mediated lysis of infected cells, or (3) antibody-dependent cell-mediated cytolysis (ADCC). If virus-specific memory B and T cells are present in the bronchus-associated lymphoid tissue (BALT) following an earlier infection with the same or a sufficiently-related strain of virus, anamnestic antibody and T_c responses will greatly expedite recovery.

Whereas the immune response to respiratory infection is instrumental in recovery, it can also, paradoxically, exacerbate the disease itself. CD4+ Th1 cells may induce such a strong inflammatory response ('delayed hypersensitivity' or DTH) as to cause lethal consolidation of the lung (pneumonia). Furthermore, antibodies of the IgE class can precipitate a life-threatening attack of asthma in a young infant infected with RSV. Other factors may also contribute to RSV bronchiolitis: virus infection not only enhances bronchial reactivity to antigen but also destroys the ciliated epithelial cells responsible for mucociliary clearance, thus allowing the infant's narrow bronchioles to become plugged with mucus, inflammatory cells and necrotic cell debris, while bronchoconstriction may also be triggered by vagal nerve reflexes or by release of mediators by inflammatory cells. Blockage of airways causes hypoxia and a pathophysiologic cascade that leads to acidosis and uncontrollable fluid exudation into airways.

Superinfection with bacteria, typically *Streptococcus pneumoniae, Haemophilus influenzae* or *Staphylococcus aureus* often complicates viral pneumonitis, and without chemotherapy can lead to a fatal outcome. The very young and very old are particularly at risk, as are the immunocompromised, and premature or malnourished infants.

Systemic viral infections such as measles generate a strong memory response and prolonged production of IgG antibodies, which protect against re-infections for life. In contrast, viruses that cause infection localized to the respiratory tract, with little or no viremia, e.g. RSV or rhinoviruses, induce only a relatively short-lived mucosal IgA antibody response, hence re-infections with the same or a somewhat different strain can recur repeatedly throughout life. In addition, numerous strains arising by antigenic drift may cause sequential episodes of the same disease in a single patient.

Viral Diseases of the Human Respiratory Tract

While some viruses have a predilection for one particular part of the respiratory tract, most are capable of causing disease at any level and the syndromes to be described next overlap somewhat (Fig. 3). Nevertheless, for ease of description we will designate six basic diseases of increasing severity as we descend the respiratory tract: rhinitis, pharyngitis, croup, bronchitis, bronchiolitis and pneumonia (Table 1).

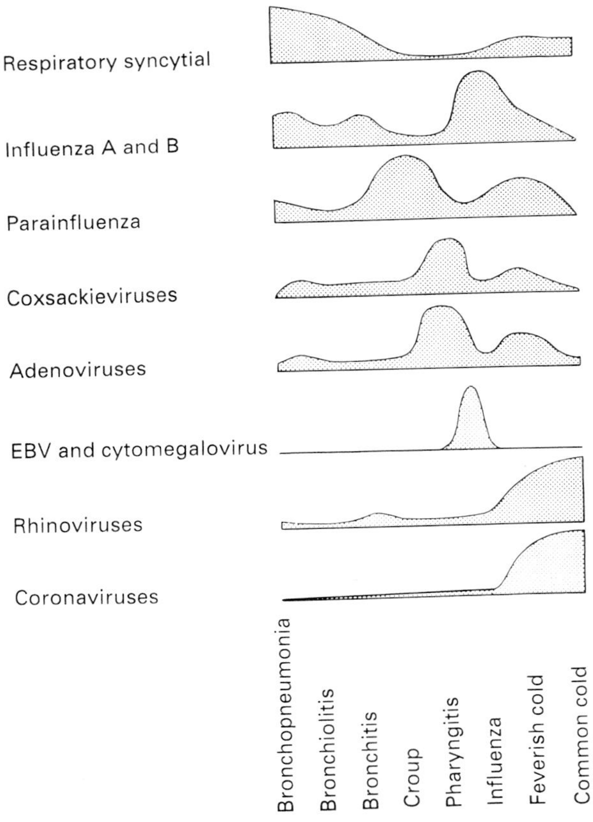

Fig. 3 Diagram showing the frequency with which particular viruses produce disease at various levels of the human respiratory tract. Data courtesy of Dr D.A.J. Tyrrell. From White DO and Fenner F (1993) *Medical Virology*, 4th edn, San Diego: Academic Press.

Rhinitis (common cold)

The classical common cold (coryza) is marked by copious watery nasal discharge and congestion, sneezing, and perhaps a mild sore throat or cough, but little or no fever. Rhinoviruses are the major cause, several serotypes being prevalent all year round and accounting for about half of all colds. Coronaviruses are responsible for about another 15%, mainly those occurring in the winter. Certain enteroviruses, particularly coxsackieviruses A21 and A24, and echoviruses 11 and 20 cause febrile colds and sore throats in the summer. In children, RSV, parainfluenza viruses and the lower-numbered adenoviruses are between them responsible for up to half of all upper respiratory tract infections ('URTI' or 'URI').

Otitis media or sinusitis sometimes complicate URI. Bacterial superinfection is generally involved, but viruses have also been recovered from the effusion. Respiratory infections with RSV, influenza, parainfluenza, adenovirus or measles viruses predispose to otitis media. Indeed, repeated viral infections can precipitate recurrent middle ear infections, leading to progressive hearing loss.

Pharyngitis

Most pharyngitis is of viral etiology. URI with any of the viruses just described can present as a sore throat, with or without cough, malaise, fever and/or cervical lymphadenopathy. Influenza, parainfluenza and rhinoviruses are common causes throughout life, but other agents are prominent in particular age groups: RSV and adenoviruses in young children; herpesviruses in adolescents and young adults. Adenoviruses, though not major pathogens overall, are estimated to be responsible for about 5% of all respiratory illnesses in young children, often presenting as pharyngoconjunctival fever. Primary infection with herpes simplex virus (HSV), if delayed until adolescence, presents as a pharyngitis and/or tonsillitis rather than as the gingivostomatitis seen principally in younger children; the characteristic vesicles, rupturing to form ulcers can be confused only with herpangina, a common type of vesicular pharyngitis caused by coxsackie A

Table 1. Human respiratory viral diseases

Disease	Virus	
	Common	Less common
Rhinitis (common cold)	Rhinoviruses Coronaviruses	RSV, parainfluenza, influenza Adenoviruses Coxsackie A21, 24; echo 11,20
Pharyngitis	Parainfluenza 1–3 Influenza Herpes simplex EB virus Coxsackie A	Rhinoviruses Adenoviruses 1–7 RSV Cytomegalovirus
Laryngotracheobronchitis (croup)	Parainfluenza Influenza	RSV
Bronchiolitis	RSV Parainfluenza 3	Influenza
Pneumonia	RSV Parainfluenza 3 Influenza	Adenoviruses 3, 7 Cytomegalovirus Measles Varicella

From White DO and Fenner F (1993) *Medical Virology*, 4th edn, San Diego: Academic Press.

viruses. Infectious mononucleosis (glandular fever) is usually marked by a very severe pharyngitis, often with a membranous exudate, together with cervical lymphadenopathy and fever; this syndrome is generally caused by EB virus in 15–25-year-olds, and less commonly by cytomegalovirus.

Laryngotracheobronchitis (croup)

Croup is one of the serious manifestations of parainfluenza and influenza virus infections. A young child presents with fever, cough, inspiratory stridor and respiratory distress, sometimes progressing to complete laryngeal obstruction and cyanosis. Parainfluenza viruses are responsible for about half of all cases, type 1 being commoner than type 2. Influenza and RSV are important causes during winter epidemics.

Bronchitis

Influenza, parainfluenza and RSV are the main viral causes of acute bronchitis. There is also evidence that chronic bronchitis, which is particularly common in smokers, may be exacerbated by acute episodes of infection with influenza viruses, rhinoviruses or coronaviruses.

Bronchiolitis

RSV is the most important respiratory pathogen during the first year or two of life, being responsible, during winter epidemics, for about half of all bronchiolitis in infants. Parainfluenza viruses (especially type 3) and influenza viruses are the other major causes of this syndrome. The disease can develop with remarkable speed. Breathing becomes rapid and labored, and is accompanied by a persistent cough, expiratory wheezing, cyanosis, a variable amount of atelectasis, and marked emphysema visible by X ray. The infant may die overnight and is one of the causes of unexplained ‘cot deaths’.

Pneumonia

Whereas viruses are relatively uncommon causes of pneumonia in adults, they are very important in young children. RSV and parainfluenza (mainly type 3) are between them responsible for 25% of all pneumonitis in infants in the first year of life. Influenza also causes a considerable number of deaths during epidemic years. Adenoviruses 3 and 7 are less common but can be severe and long-term sequelae such as obliterative bronchiolitis or bronchiectasis may permanently impair lung function. Up to

Table 2. Major respiratory viral diseases of animals

Host	Disease (virus)	Virus family
Cattle	Infectious bovine rhinotracheitis	*Herpesviridae*
Cattle	Respiratory syncytial virus	*Paramyxoviridae*
Cattle, swine	Foot-and-mouth disease	*Picornaviridae*
Horse	Equine rhinopneumonitis	*Herpesviridae*
Cat	Feline calicivirus	*Caliciviridae*
Cat	Feline rhinotracheitis	*Herpesviridae*
Dog	Canine laryngotracheitis	*Adenoviridae*
Dog	Canine distemper	*Paramyxoviridae*
Chicken	Infectious laryngotracheitis	*Herpesviridae*
Chicken	Avian infectious bronchitis	*Coronaviridae*
Birds	Newcastle disease	*Paramyxoviridae*
Birds, horses, swine	Influenza	*Orthomyxoviridae*

20% of pneumonitis in infants has been ascribed to perinatal infection with cytomegalovirus (CMV). CMV may also cause potentially lethal pneumonia in immunocompromised patients, as may measles, varicella and adenoviruses. Moreover, viral pneumonia not uncommonly develops in adults with varicella, and in military recruits involved in outbreaks of adenovirus 4 or 7, while measles is often complicated by bacterial pneumonia, especially in malnourished children in Africa and South America. In the elderly, and particularly in those with underlying pulmonary or cardiac conditions, influenza is a major cause of death, either via influenza pneumonitis or more commonly via secondary bacterial pneumonia.

Viral pneumonitis often develops insidiously following URI and the clinical picture may be atypical. The patient is generally febrile, with a cough and a degree of dyspnea, and auscultation may reveal some wheezing or moist rales. Unlike typical bacterial lobar pneumonia with its uniform consolidation, or bronchopneumonia with its streaky consolidation, viral pneumonitis is usually confined to diffuse interstitial lesions. The radiological findings are not striking; they often show little more than an increase in hilar shadows or, at most, scattered areas of consolidation.

Space does not permit a discussion of veterinary diseases, but Table 2 lists some of the most important respiratory viral diseases of farm and companion animals.

Laboratory Diagnosis

The etiology of respiratory viral infection can be established in the laboratory by identifying the virus itself, viral antigen, or the viral genome. The most appropriate specimen is generally a throat swab or, better still, mucus aspirated from the nasopharynx, taken early in the disease. Because enveloped respiratory viruses with helical nucleocapsids are notoriously labile, the specimen is kept cold and moist, transported promptly to the laboratory and processed as soon as practicable.

Enzyme immunoassay (EIA) is the method of choice for the rapid detection of antigen in respiratory secretions. Diagnostic kits, based on appropriate monoclonal antibodies for antigen-capture and detection respectively, and often incorporating a biotin–avidin readout system, are now available for all common human respiratory viruses and many of the important animal pathogens. EIA is replacing immunofluorescence (application of fluorescein-labeled monoclonal antiviral antibody to infected cells aspirated from the throat). Gene amplification by the polymerase chain reaction (PCR) followed by nucleic acid hybridization using appropriate probes is a feasible option but currently appears unlikely to supplant antigen detection as the method of choice.

Most of the known respiratory viruses can quite readily be cultivated in appropriate cell lines from the corresponding host species, e.g. diploid lung fibroblasts for human viruses. Growth of the virus is detected by cytopathic effects (CPE) and/or hemadsorption or immunofluorescence, and the virus recovered from the supernatant is then typed using any of a variety of serological techniques. However, this sequence is so time-consuming and expensive that isolation and identification of virus is today undertaken mainly by reference laboratories requiring a large supply of the virus for further characterization, for research, or for antigen or vaccine production.

'Serology', i.e. identification and quantification of antibody in the patient's serum, is also far too slow to be of value in influencing the management of the patient, hence is used principally in seroepidemiological surveys to assess the immune status of populations. IgM-capture EIA does provide a rapid diagnosis but fills only a limited role in the routine diagnostic laboratory.

Vaccines and Chemotherapy

Very successful live attenuated vaccines are in general use against certain 'respiratory' viruses like measles, mumps and rubella, which, though naturally transmitted via the respiratory route, are absolutely dependent upon viremic spread to their target organs elsewhere in the body. In contrast, it is a much more challenging assignment to develop effective vaccines against viruses whose pathogenicity is essentially confined to the respiratory tract. The major reasons for this are that (1) secretory IgA memory is relatively short-lived, and (2) numerous antigenically distinct strains or serotypes are capable of causing the same clinical syndrome. Thus, a common cold vaccine might need to contain dozens of different serotypes of rhinoviruses (± coronaviruses). An inactivated vaccine is used to protect the aged and other risk groups against the currently prevalent strains of influenza, but its composition must be updated regularly to keep abreast of antigenic drift and shift and, even so, its efficacy in the aged is only of the order of 50–80%. Live 'cold-adapted' mutants have been constructed which replicate only at the low temperature of the nose and are generally avirulent, but it has proved difficult to walk the tightrope between genetic stability of avirulence on the one side, and adequate replication (hence immunogenicity) on the other. A better understanding of the mucosal immune system is required before we can expect the development of truly safe and effective live vaccines against such viruses as influenza and RSV, for delivery by aerosol or by mouth. In the long term the greater hope may lie with antiviral chemotherapeutic agents.

Antiviral chemotherapy moved slowly during the decades following the discovery of interferon in 1957. Indeed, the only agents currently in use against respiratory viruses are (1) the ion channel blockers, amantadine and its analog rimantadine, which display activity against influenza A if given prophylactically, and (2) a nucleoside analog, ribavirin, which may be of value in severely ill infants with RSV bronchiolitis/pneumonia when administered as a small-particle aerosol. Recently, thanks to X ray crystallography, the receptor-binding sites on rhinoviruses and influenza viruses have been resolved in atomic detail, and the race is on to develop antivirals that fill these canyons hence block cell attachment, penetration and/or uncoating of the virions.

See also: Adenoviruses; Coronaviruses; Diagnostic Techniques; Influenza viruses; Parainfluenza viruses; Respiratory Syncytial virus; Rhinoviruses; Vaccines and immune response.

Further Reading

Fenner F, Gibbs EPJ, Murphy FA, Rott R, Studdert MJ and White DO (1993) *Veterinary Virology*, 2nd edn. San Diego: Academic Press.

Fields BN *et al.* (1990) *Virology*, 2nd edn. New York: Raven Press.

Mims CA and White DO (1984) *Viral Pathogenesis and Immunology*. Oxford: Blackwell Scientific.

White DO and Fenner FJ (1993) *Medical Virology*, 4th edn. San Diego: Academic Press.

RETICULOENDOTHELIOSIS VIRUSES

Paula J Enrietto
State University of New York at Stony Brook,
Stony Brook, New York, USA

History

The reticuloendotheliosis viruses (REV) constitute a group of pathogenic avian retroviruses that are distinct from and unrelated to the avian sarcoma leukemia viruses. The five members of this group, reticuloendotheliosis virus strain T (REV-T), reticuloendotheliosis-associated virus (REV-A), duck infectious anemia virus (DIAV), Trager duck spleen necrosis virus (SNV) and chick syncytial virus (CSV) share similar morphologies, a common group reactive antigen and extensive RNA sequence homology.

REV-A, DIAV, SNV and CSV are all replication competent, while REV-T, the prototype member of this group, is replication defective and requires helper virus for productive infection. REV-T was isolated from the tissue of a turkey that died of visceral reticuloendotheliosis with associated nerve lesions. REV-T was subsequently shown to cause acute malignant disease since it carries within its genome an oncogene sequence, *rel*.

Taxonomy and Classification

The reticuloendotheliosis viruses are members of the family *Retroviridae*. This family includes all viruses containing an RNA genome and an RNA-dependent DNA polymerase which converts the viral genomic RNA into proviral DNA that integrates into the host cell genome. *Retroviridae* is further subdivided with the REVs classified as *Oncovirinae* (i.e. viruses that are oncogenic and closely related nononcogenic viruses).

Retroviruses, including REVs, have common morphological, biochemical and physical properties that group them in a single virus family. The *Retroviridae* contain linear, positive-sense, single-stranded RNA that serves as the viral genome. The genome itself is composed of two identical subunits, the 5′ end containing a cap structure and a 3′ polyadenylated end. Thus, genomic retroviral RNA structurally resembles cellular mRNA. In addition, each complete virion contains a tRNA associated with the viral RNA which serves as a primer for replication.

Morphologically classified as C-type retroviruses, the REVs contain a centrally located core appearing as a sphere with an electron dense center. The membrane of the virus, derived from the plasma membrane of the host cell, contains distinct spikes made up of the viral glycoproteins. Electron microscopic investigation of REV-A showed that, although it was morphologically similar to type C retroviruses, its linkage at 5′ ends was similar to the genomic structure of mammalian retroviruses. Sequence analysis of the replication competent REVs confirmed their unrelatedness to any other avian or mammalian retrovirus, although some homology has been detected with the MAC-1 group of primate viruses. Furthermore, the RNA primer used for the reverse transcription of the REVs is $tRNA^{Pro}$, the same primer used by murine leukemia viruses (MLV). Taken together, these data suggest a closer relationship between the REVs and mammalian retroviruses than between the REVs and avian retroviruses, especially in view of the lack of a relationship between REV and avian leukemia sarcoma viruses (ALSV). This is supported by biological data that indicate lack of relationship between REV and the avian leukemia sarcoma viruses (ALSV).

Properties of the Virion

Reticuloendotheliosis virus particles resemble in all aspects classical retroviral C-type particles. As is generally characteristic of type-C viruses, no intracytoplasmic viral structures are observed until budding has started at the plasma membrane. As maturation proceeds, the core of the virus appears with an electrolucent center and the plasma membrane containing the virally encoded glycoprotein spikes, surrounds and pinches off the viral core. Finally, extracellular particles of 80–110 nm in diameter appear which have a centrally located electron dense core.

Properties of the Genome

The nondefective REVs contain positive-sense, single-stranded RNA of approximately 60S, which dissociates into two 35S subunits that are linked at the 5′ end in a manner similar to that found in the genomes of mammalian retroviruses. The genome of REV-A has a complexity of 3.9×10^6 D and, although their sequences are not identical, the four nondefective REVs show greater than 90% sequence homology. The genome structure of the nondefective REVs is similar to other nondefective avian retroviruses and contains the genes encoding three structural proteins in the following order:

5′ cap R U_5 *gag* *pol* *env* U_3 R poly(A) 3′

gag: encodes the viral structural proteins
pol: encodes the viral reverse transcriptase
env: encodes the viral envelope proteins

The genome of the replication defective member of this group, REV-T, has been studied extensively. REV-T contains an extensive deletion of replicative sequences (Fig. 1a), including portions of *gag*, most of *pol* and portions of *env*. The cellular oncogene *rel* has been transduced within *env* sequences and the resulting genome is 5.9 kb (28S).

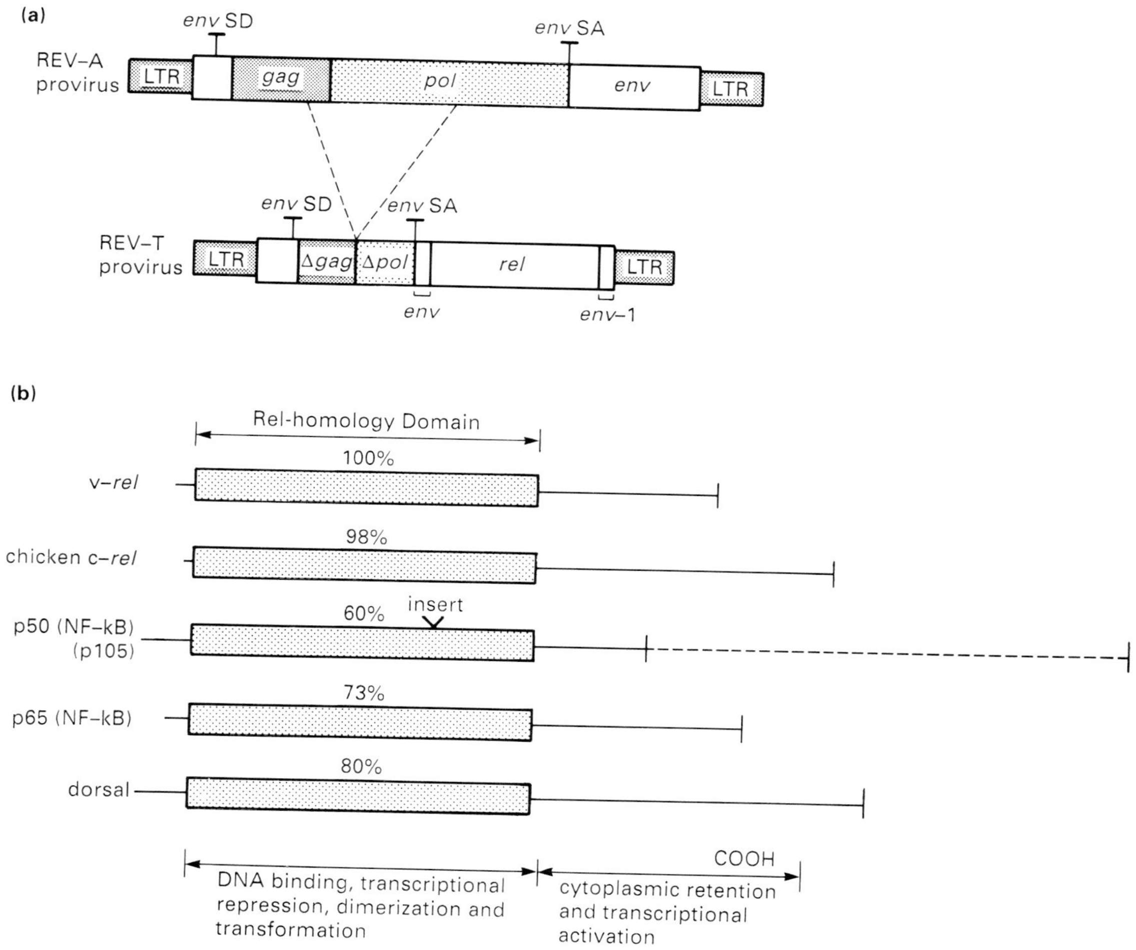

Fig. 1 **(a)** Comparison of REV-A and REV-T demonstrating the extent of the deletions in the structural genes and the position of the v-*rel* gene in REV-T. SD indicates the splice donor site and SA the splice acceptor site from which the subgenomic mRNA encoding *env* is derived in REV-A. In REV-T these splice sites are used to produce the mRNA encoding v-*rel*. **(b)** The *rel* protein family is compared. Percentages indicate degree of homology between the various members of the family. p50, which appears to be encoded as a precursor of mol. wt 105 000, and p65 are two subunits of the transcription factor NF-kB. The functional domains, determined by mutational analyses of the proteins, are indicated below the arrows and the *rel* homology region is indicated above the arrows.

Properties of Proteins

Structural protein analysis of the nondefective REVs has shown that the *gag* gene encodes proteins of 29 000 (p29), 15 000 (p15) and 13 000 (p13) D which serve as core proteins (p29 being the major component). The *env* gene encodes two proteins found in the viral envelope, which are modified by glycosylation gp71/73 and gp22. None of these proteins is antigenically related to the equivalent structural proteins from other avian retroviruses and the major core protein (p29) is very different in composition and sequence from the major core protein of the avian leukemia sarcoma viruses (ALSV), p27. However, it is related to the murine leukemia virus (MLV) core protein p30 by radioimmunological and sequence analysis. REV p29 is about 40% homologous to MLV and feline leukemia virus (FLV) p30 molecules and has greater than 50% homology to the major core protein of the endogenous retroviruses of macaques, thus reinforcing the notion that the REVs are more closely related to mammalian rather than avian viruses.

The REV reverse transcriptase is unrelated to other avian retroviral enzymes but is antigenically cross-reactive with those of other mammalian retroviruses. Like mammalian retroviral transcriptases, the REV reverse transcriptase has a divalent cation requirement for manganese. By analogy with other retroviruses, the enzymatic activity of the *pol* gene product (a single 84kD protein) includes both an RNase H reverse transcriptase and an integrase activity.

Replication Strategy

The REVs follow a classical retroviral replication strategy. Following absorption and uncoating, the reverse transcriptase, which is encoded by and contained within the virus, converts the viral RNA genome into linear, double-stranded proviral DNA which in turn integrates into the host cell genome. Only a general outline of the replication cycle will be given here.

Replication of the nucleic acids

The two major requirements for the initiation of DNA synthesis are template and primer. The template, a viral RNA genome, has the following structure:

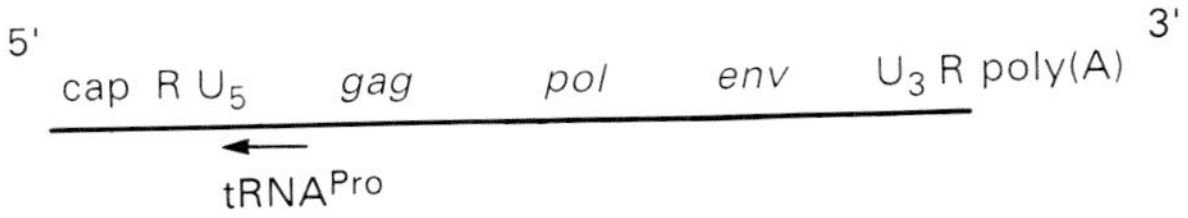

In this simple diagram, R indicates a repeated sequence present at both the 5′ and 3′ ends of the genomic RNA. U_5 and U_3 represent unique noncoding sequences at the 5′ end and 3′ end of the genome, respectively. The repeated sequences are essential for replication. The second requirement for primer is met in the REVs by a $tRNA^{Pro}$, which binds at the primer binding site approximately 200 bp from the 5′ end of the viral RNA. First strand (negative strand) synthesis proceeds from the primer toward the 5′ end of the molecule.

The template then circularizes through the R sequences and the polymerase jumps to the 3′ end of the molecule. Upon completion (or concurrent with) negative-strand synthesis, the RNase H activity within the reverse transcriptase molecule degrades the RNA template. These degradation products serve as primers for positive-strand DNA synthesis which occurs on the newly synthesized negative strand.

The final product of viral replication is a double-stranded linear proviral DNA molecule with the following structure:

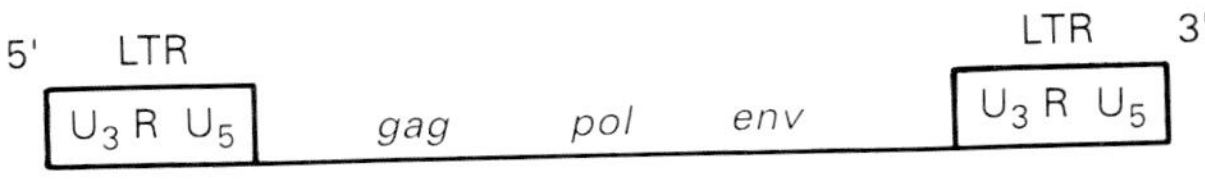

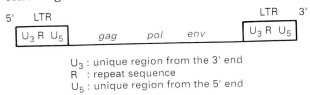

U_3 : unique region from the 3' end
R : repeat sequence
U_5 : unique region from the 5' end

The repeated sequences (U_3 R U_5) at the end of each molecule are termed long terminal repeats (LTR) and are a direct result of the replication mechanism. These sequences contain all of the signals required for transcription of the proviral DNA into viral mRNA.

Integration

Following synthesis, which probably occurs in the cytoplasm, the proviral DNA moves to the nucleus and integrates into the host cell chromosome. The integration event is random, resulting in the duplication of adjacent cellular sequences, and is mediated by the integrase function contained within reverse transcriptase.

Transcription

Following integration, the proviral DNA serves as a

template for the synthesis of viral genomic and messenger RNAs. The LTR meets all the sequence requirements for initiation of transcription and after transcription begins, the replication competent REVs are transcribed into two mRNAs. The 8.7 kb mRNA represents the genome length message, and a smaller subgenomic mRNA is also produced representing a spliced mRNA encoding only the *env* gene.

The replication defective REV-T is transcribed into a full-length genomic mRNA of 5.9 kb and a subgenomic mRNA containing v-*rel* sequences of 1.9 kb. The subgenomic mRNA is spliced through the *env* splice acceptors and donor which are still present in REV-T (Fig. 1a).

Transcription of viral proteins and post-translational processing

In the replication competent REVs, the viral structural proteins encoded by the *gag* gene are translated from the 8.7 kb mRNA. The protein product is synthesized as a 63 000 D precursor which by analogy with other avian retroviruses is most likely proteolytically cleaved to yield the viral structural proteins. In a similar fashion, the reverse transcriptase molecule is translated from genome size RNA, as a *gag–pol* fusion protein of 180 000 D, which is subsequently cleaved to give a protein of about 84 000 D. Nothing is known about further cleavage of this protein.

The *env* proteins are translated from the subgenomic mRNA into a precursor of 80 000 D, which is glycosylated and subsequently cleaved into proteins of 73 000 and 22 000 D.

The *rel* Gene

Rev(T), an acutely transforming retrovirus, contains sequences of cellular origin that are responsible for its ability to cause rapidly fatal neoplastic disease. These oncogenic sequences, termed *rel* encode a phosphoprotein of 503 amino acids, $p59^{V\text{-}rel}$. Transduction of the cellular *rel* gene occurred within *env* sequences (see Fig. 1a). Thus, translation of the subgenomic mRNA encoding v-*rel* begins with envelope sequences, generating a fusion protein with 12 amino acids of *env* at the N terminus. Additionally, v-*rel* is fused to 19 amino acids of out-of-frame envelope sequences at the C terminus. This fusion protein has been localized to both the nucleus and cytoplasm of Rev(T)-transformed avian lymphoid cells. Further biochemical studies have indicated that v-*rel* exists in multiple high-molecular-weight complexes with cellular proteins of 36, 115 and 124 kD and with $p68^{C\text{-}rel}$, the cellular homolog of v-*rel*.

Comparisons of the viral and cellular *rel* genes have indicated that v-*rel* is lacking two amino acids at the N terminus and 118 amino acids at the C terminus relative to c-*rel*. v-*rel* also contains multiple point mutations and small deletions within the body of protein. These differences between the viral and cellular homologs are most likely contributory to the oncogenic potential of $p59^{V\text{-}rel}$.

Recently, studies on the mammalian transcription factor NF-kB and the dorsal protein of *Drosophila* have lent insight into the biochemical function of the *rel* proteins. It was discovered that these proteins from divergent sources all contain strong homology over approximately 300 amino acids of N-terminal sequence (see Fig. 1b). Shortly thereafter, studies indicated that these proteins interacted with similar DNA sequences and suggested that they are involved in the control of gene transcription. In fact, v-*rel* has been demonstrated to bind to an NFkB site and repress transcription under certain circumstances. Mutagenesis studies have defined domains on the protein that appear to be required for function and localization. These domains are outlined in Fig. 1b.

Host Range and Virus Propagation

While the structural data described earlier indicate that the REVs have evolved from mammalian D-type viruses, the natural hosts of the REVs are birds. REV field isolates were derived from turkey and ducks. REVs have a broad host range in birds and in avian cell culture where each strain has been shown to replicate in all avian cells tested. In addition, the REVs have been shown to replicate in dog and rat cell lines and act as a helper virus for Moloney murine sarcoma virus.

Transmission

Transmission of retroviruses, including REVs, occurs either horizontally (via contact or aerosols) or vertically through congenital infection or genetic transmission. Congenital infection occurs when infectious particles are released by the mother which infect the egg resulting in direct infection of the offspring. Genetic transmission occurs when the viral genome is transmitted from one generation to the next as a DNA provirus that is maintained as part of the genome within the gametes.

Pathogenicity

In vivo

The viruses composing the REV group can be divided into two types on the basis of their genome structure and pathogenicity. REV-A, DIAV, SNV and CSV are all replication competent viruses and do not carry an oncogene within the viral genome. These nondefective REVs cause a wide variety of syndromes in birds including visceral reticuloendotheliosis, splenomegaly, spleen necrosis, lymphoproliferative nerve lesions, anemia and lymphomas. These lesions (excluding lymphoma) are thought to be the result of the cytotoxicity of these nondefective viruses for avian cells and, while life-threatening, are not neoplastic. It was shown subsequently, however, that CSV could induce a neoplastic disease (lymphoma) in birds. Like the avian leukemia viruses, which also have the ability to induce lymphomas, CSV proviral DNA was found integrated near the cellular oncogene c-*myc*, whose activation is thought to be critical in the induction of lymphomas. This disease was of relatively long latency (3–6 months) reflecting the selection of the clone of cells in which insertional activation of c-*myc* has occurred.

In contrast to the replication competent REVs, REV-T, which is replication defective, induces rapidly fatal (3–10 days postinfection) disease in birds. REV-T was isolated originally from an adult turkey that died from extensive visceral reticuloendotheliosis and infiltrative nerve lesions. Young chickens experimentally infected with REV-T and the helper virus REV-A develop a severe hepatosplenomegaly with either lymphoproliferative or necrotic lesions. Some lesions are composed almost entirely of large undifferentiated lymphoreticular cells, whereas others are composed of smaller lymphoid cells. The progression of the disease is rapid and the mortality rate approaches 100% when 1-day-old chicks are injected intraperitoneally.

The type of helper virus was recently demonstrated to influence the disease spectrum induced by REV-T. When cytotoxic helper virus (REV-A) was used in co-infections with REV-T, the resulting tumors were IgM negative and had markers of the T cell and myeloid lineages on their surfaces. In contrast, when CSV (a less cytotoxic helper virus) was used as a helper, the resulting tumors were B cell in origin since they were IgM positive and had rearranged immunoglobulin genes. Thus REV-T appears able to induce tumors in a variety of different cell types with the resultant tumor depending on the helper virus used for infection. What role does the helper virus play? Besides providing virion proteins in *trans* for the replication defective REV-T, the helper virus rapidly and severely depresses the cellular immune response of the infected bird. REV-A and the other nononcogenic members of this group induce or activate a suppressor cell population in the spleens of infected chickens that inhibit the proliferation of T cells. The induction of the suppressor cell population in birds infected with REV-A has been shown to correlate with the level of viremia and can be detected 3 days after infection. Although the immunosuppression induced by REV-A alone is transient, of about 4 weeks in duration, it is within this time that birds co-infected by REV-T die. Chickens that exhibit a suppressed proliferative response develop disease, whereas, in the absence of immunosuppression, birds fail to develop visible tumors. Therefore, the rapid induction of the suppressor cell population that prevents T cell proliferation may permit the unrestricted replication of REV-T-transformed lymphocytes.

The rapidly fatal disease induced by REV-T is the result of the expression of the oncogene *rel* which was transduced into the viral genome at the expense of viral structural genes (see later).

In vitro

The nondefective REVs do not transform cultured cells; instead, acute infection is characterized by extensive cytopathicity. Large amounts of unintegrated viral DNA accumulate and multiple proviral DNA copies are detected. Cell death at this stage of infection is thought to be the result of the toxic effects of large amounts of unintegrated viral DNA in the cell. Surviving cells proliferate but exhibit much lower copy numbers of unintegrated viral DNA.

REV-T is the only member of this group capable of transforming cells *in vitro*. Analysis of the ability of REV-T to transform fibroblasts and hematopoietic cells *in vitro* has been complicated by the cytotoxicity of the helper virus required for viral replication. Early studies suggested that fibroblast transformation by REV-T could be observed only after extended periods of time in culture (i.e. several weeks). However, more recent studies using a replication competent virus containing the v-*rel* gene showed that transformation of infected fibroblasts was obvious after three to four passages in culture. The phenotype of the transformed cell resembles that observed with the viral oncogenes *erb*B and *sea*. The morphology is fusiform and the cells exhibit a loss of actin cables and fibronectin on their cell surface. Perhaps the most striking result of v-*rel* expression is the extended life span of the transformed cells. Normal chicken embryo fibroblasts have a life span in culture in the range of weeks, whereas v-*rel*-transformed fibroblasts can be maintained in culture up to 6 months before going into crisis and dying.

Because the tumors induced by REV-T *in vivo* were hematopoietic in origin, several groups attempted to identify the hematopoietic target cell for transformation by REV-T. On the basis of studies using lineage-specific antibody markers, early studies suggested that the cell transformed by REV-T was of the B cell lineage. More recent studies, however, suggest that *rel*-containing viruses can transform more than one cell type including a cell that has the phenotype of an uncommitted precursor cell. As has been shown *in vivo*, the helper virus had a clear effect on the target cell population available for transformation by REV-T *in vitro*.

See also: Avian leukosis viruses; Retroviruses – type D.

Further Reading

Weiss R, Teich N, Varmus H and Coffin J (eds) (1982) *RNA Tumor Viruses: Molecular Biology of Tumor Viruses*, 2nd edn. Cold Spring Harbor, NY: Cold Spring Harbor Laboratory.

RETROTRANSPOSONS OF FUNGI

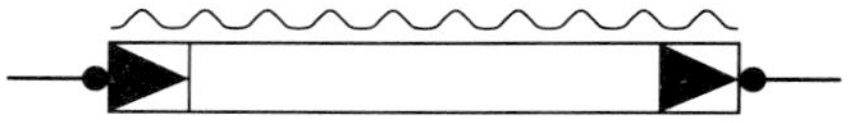

Jef D Boeke
The Johns Hopkins University
Baltimore, Maryland, USA

Introduction

Transposable elements in eucaryotes can nearly all be classified into three basic types; these are compared in Fig. 1. The first type, typified by the Ac elements of plants and the P elements of *Drosophila*, resemble bacterial transposons in that they bear short inverted repeat termini; the available evidence strongly suggests that this type of element transposes directly via a DNA intermediate. However, most eucaryotic transposons differ from the bacterial elements in that they encode a reverse transcriptase (RT) or RT-like protein. Two basic types of these 'retrotransposons' are known – the long terminal repeat (LTR)-containing types, which are structurally highly reminiscent of retroviruses, and the poly(A) types, which lack LTRs and usually (but not always) contain an oligo(A), poly(A) or similar sequence tract at their extreme 3′ end. It is important to note that both classes of retrotransposons are now known from organisms as phylogenetically distinct as fungi, trypanosomes, insects and mammals. This brief review focuses on what is currently known about fungal retrotransposons, and provides only a small and highly selective glimpse of the total picture regarding the universe of retrotransposons.

Structural Features

The structural features of the known fungal retrotransposons are summarized in Table 1. LTR-containing retrotransposons isolated from fungi resemble retroviral proviruses in structure in that they contain LTR sequence of a few hundred base pairs; these flank a central coding region that contains one or two open reading frames (ORF). As is the case with proviruses as well as DNA-based transposons, target site duplications of a fixed length flank the elements; these vary greatly in sequence and are presumably generated during the integration process by a transposon-encoded integrase function. In cases where the RNA has been

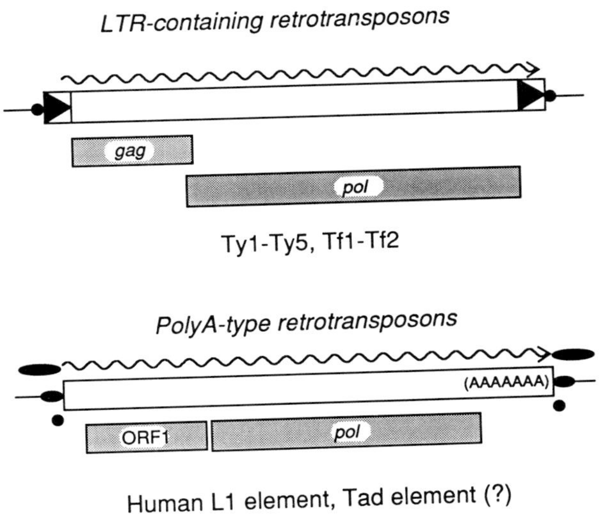

Fig. 1 Retrotransposon types in fungi. All elements found to date in the yeasts *Saccharomyces cerevisiae*, *Schizosaccharomyces pombe* and *Candida albicans* are long terminal repeat (LTR)-containing transposons (upper panel). The only elements found to date in the filamentous fungi *Neurospora crassa* and *Magnaporthe grisea* are of the 'poly(A)' type. Since no functional copies of these fungal elements have yet been sequenced, the structure of a putative functional copy of the related human L1 element is shown. Note that unlike almost all members of this class of elements, TAD does not contain the 3′ poly(A) tract.

examined, it resembles retroviral genomic RNA in that it extends from LTR to LTR and is terminally repetitious, allowing definition of U3, R and U5 regions of the LTR sequence in a manner formally analogous to that used by retrovirologists. The elements with two open reading frames clearly contain the equivalents of *gag* and *pol*, but no analog of retroviral *env*. The element with a single ORF, Tf1, apparently produces a Gag/Pol fusion protein only.

Thus far there is only one report in the literature of a fungal retrotransposon that lacks LTRs, and that is the TAD element from the filamentous fungus *Neurospora crassa* Adiopodoume strain, although a structurally similar element has now been isolated from another filamentous fungus, *Magnaporthe grisea*.

Transposition Mechanism

There is probably more known about the LTR-containing retrotransposons of fungi than about those of any other species. By far the most heavily studied elements are the Ty1 and Ty3 elements of *Saccharomyces cerevisiae*. In both cases, the overall outline of the life cycle appears quite similar. Largely by analogy, much of this outline can also be applied to the Tf1 elements of *Schizosaccharomyces pombe*. In contrast, except for the presence of a cytoplasmic transposition intermediate, and evidence for reverse transcription during its transposition, relatively little is yet known about TAD transposition. This is because these poly(A) type elements are generally less well known (and retroviral analogies are uncertain at best). Thus, the discussion which follows applies to yeast Ty elements specifically, and to LTR retrotransposons generally.

A transposon copy (usually studied in the laboratory in the form of a *GAL* promoter/Ty element fusion) produces a transcript that extends from a

Table 1. Fungal retrotransposons

Host	Element name	Type	ORFs	Target site duplication	LTR length	Primer negative-strand
Candida albicans	T_{cal}	LTR	?	5 bp	388 bp	?
Saccharomyces cerevisiae	Ty1	LTR	2	5 bp	334 bp	$tRNA_i^{Met}$
S. cerevisiae	Ty2	LTR	2	5 bp	334 bp	$tRNA_i^{Met}$
S. cerevisiae	Ty3	LTR	2	5 bp	340 bp	$tRNA_i^{Met}$?
S. cerevisiae	Ty4	LTR	2	5 bp	371 bp	$tRNA_i^{Asn}$
S. cerevisiae	Ty5	LTR	?	?	250 bp	$tRNA_i^{Met}$
Schizosaccharomyces pombe	Tf1	LTR	1	5 bp	358 bp	?
Sch. pombe	Tf2	LTR	1	5 bp	358 bp	?
Neurospora crassa	TAD	poly(A)	?	14, 17 bp	N.A.	?

point in the 5′ LTR to a different, downstream point in the 3′ LTR. Thus, a terminally redundant RNA is generated. This RNA is polyadenylated and exported to the cytoplasm, where it can have two different fates: (1) it can serve as an mRNA for TYA and TYB protein products, and/or (2) it can serve as genetic material for transposition.

Translation of Ty elements is somewhat unconventional. TYA is produced directly by conventional translation of this mRNA, whereas TYB is expressed as a TYA/TYB fusion protein (readthrough protein); this 'frameshifting' process is mediated at a special sequence within the region of overlap of TYA and TYB.

TYA...Leu Arg(4 amino acids)

......CUUAGGC......

Leu Gly...*TYB*

Both the Ty1 and Ty3 frameshifts differ from those of more conventional retroviruses in that they are +1 frameshifts rather than −1 frameshifts, and the intrinsic mechanism used to effect the frameshift is different. In the −1 frameshifts used by retroviruses, coronaviruses and the yeast killer double-stranded RNA virus, a 'slippery site' allows for a simultaneous slip of the ribosome during the translational step in which the ribosomal P and A sites are simultaneously occupied. Stem–loop structures or pseudoknots that are often found just downstream of the slippery site are though to cause ribosomal pausing and perhaps even to induce the ribosome to slip backwards. In contrast, the Ty1 element appears to use a completely different mechanism. There appears to be no requirement for any special RNA secondary structure; rather, the frameshifting appears to be sequence mediated, because a seven nucleotide sequence from the overlap region readily confers frameshifting on a heterologous reporter gene. Ribosomal pausing, thought to be required for frameshifting, is apparently caused by limiting amounts of a rare tRNA, $tRNA^{Arg}_{CCU}$. The stalled ribosome, which unlike the retroviral case has an empty A site, has the P site occupied by a specific $tRNA^{Leu}$ that recognizes all six Leu codons. The stall caused by low $tRNA^{Arg}$ levels allows time for the slippage event to occur – the $tRNA^{Leu}$ slips to an overlapping Leu codon in the +1 frame. This then exposes a Gly codon recognized by an apparently abundant tRNA in the A site, allowing translation to continue in the TYB frame. This mechanism results in an efficiency of frameshifting that varies with context, but on average is about 10–20%.

Once sufficient Ty protein products, in the form of intact TYA and TYA/TYB readthrough proteins, are produced, an assembly process that is not yet well-understood begins. What follows is the author's working model for this assembly process, based largely on interpretations of limited experiments on Ty assembly and by analogy with retroviral systems and models. Ty RNA is apparently selectively packaged, together with at least one specific tRNA, the primer tRNA, into a capsid initially consisting of a co-assemblage of unprocessed TYA and TYA/TYB proteins. Presumably, these co-assemble via TYA–TYA interactions, and are made such that the C termini of the proteins, and the RNA, reside in the internal cavity of the VLP. A protease encoded within TYB, presumably activated by a dimerization process facilitated by the high protein concentration involved in the assembly process, then cleaves the precursor TYA proteins and TYA/TYB proteins to their mature forms, which are presumed to be the physiologically relevant ones. This results in a change in the morphology of the VLPs, as well as an apparent activation of the endogenous reverse transcriptase activity.

Once reverse transcriptase is activated, a negative-strand strong stop DNA is synthesized, using a tRNA as primer. This primer was recently shown to be $tRNA^{Met}_{i}$ for Ty1. Eventually, a full-length double-stranded DNA is made, although the detailed steps of reverse transcription remain unknown for Ty1. For example, the positive-strand primer has not yet been identified, and may differ from that used by retroviruses, at least for Ty1, as there is no polypurine tract adjacent to the 3′ LTR of Ty1 as there is in the retroviruses and most other LTR-retrotransposons.

The double-stranded DNA remains associated with TYA and TYB proteins such as reverse transcriptase inside the cell. These DNA-containing VLPs (isolated as a mixture of RNA-containing and DNA-containing VLPs) have been shown to contain all of the macromolecular factors needed to carry out an *in vitro* transposition (integration) reaction. Like retroviral core particles, which have this same activity, these VLPs require only a divalent cation for activity. A persistent question about the mechanism of Ty transposition is how does the Ty DNA, apparently synthesized in the cytoplasm and presumably in the form of a VLP, get delivered to the nucleus? This is a particularly interesting question to ask in yeast, because fungi, unlike mammalian cells, undergo a 'closed mitosis' in which the nuclear membrane does not break down and re-form during each mitosis, but apparently remains intact. Unfortunately this is a difficult question to ask biochemically, because most of the VLPs in a cell lack DNA, and the number of VLPs in a cell greatly outnumbers the number of successful transposition events observed.

Once the Ty DNA and Ty integrase enter the nucleus, a concomitant cleavage of host DNA and joining to transposon ends similar to that occurring during retroviral integration occurs. Although Ty1 is relatively nonspecific in its target site selection, at least at a gross level, with many phosphodiester bonds in any given gene presenting a reasonably efficient target, Ty3 is extremely selective, and apparently inserts only at the transcription initiation sites for RNA polymerase III. Thus the integrases encoded by these elements presumably have very different 'specificity domains'.

Virus-like Particles – Evolutionary Vestige or Transposition Intermediates?

The presence of virus-like particles in Ty elements has often raised the question of the evolutionary relationship between retroviruses and LTR-containing retrotransposons. Is the virus-like particle a degenerate leftover of some decaying retrovirus? Or are LTR-containing retrotransposons a family of modern-day descendants of the precursor of the fearsome retroviruses? Presumably all retroelements descended ultimately from a 'cellular reverse transcriptase gene' as originally proposed by Temin. This gene may be ancient, and its original product may have been the molecule that archived the genetic information of the RNA world into DNA. If one examines the spectrum of modern-day retroelements from this perspective, one observes a natural progression from the simple to the very complex, as follows: starting with a simple RT gene (perhaps represented by modern-day telomerase?); to the poly(A)-type retrotransposons, which contain a second open reading frame in addition to RT; to the LTR-containing retrotransposons, which have the above plus LTRs; to the simple retroviruses, which have picked up a third open reading frame, *env*; and on to the most elaborate of all, the lentiviruses, with multiple additional regulatory reading frames in addition to the basic three. Since all LTR-containing retrotransposons studied appear to involve a VLP intermediate (that is, they have many properties of a transposition intermediate), the implication is that, as is the case in all complex biological reactions, a structure is built to ensure (1) high local concentration of numerous macromolecules required for the reaction, and (2) appropriate orientations/conformations of these macromolecules to allow the reactions to proceed appropriately. The strongest support for this idea comes from studies on Ty1, which suggest that the VLP is a direct, functional transposition intermediate. Thus, if LTR-containing retrotransposons predated retroviruses, and their VLP structure evolved in response to a selection for this organizing function, it does not stretch the imagination too far to suggest that these elements were 'preadapted' for subsequent selection for infectivity. In fact, one may see such transitional forms within the *gypsy* family of LTR-containing retrotransposons, in which some family members resemble the Tys in

Table 2. Host genes affecting Ty1

Gene name	Normal function/product	Function for Ty1	Found by
Transcriptional effect genes			
SPT3, 7, 8	Transcription factor?	Transcription initiation	Suppression of Ty- or LTR-induced mutation
SPT10, 21	Repressor and activator	Repress 3′ LTR	Suppression of Ty- or LTR-induced mutation
Post-transcriptional effect genes			
IMT1-4	Translation initiation, $tRNA_i^{Met}$	Prime reverse transcription	Intentional mutagenesis of genes to reveal interaction
(none)	$tRNA^{Arg}_{CCU}$	Low level causes ribosome stalling, frameshifting	Search for genes that interfere with transposition when overexpressed
DBR1	Debranch intron lariats (2′-5′ phosphodiesterase)	Unknown	Search for chromosomal mutations that interfere with transposition

having two ORFs, but a few elements have a third *env*-like ORF in the appropriate genomic position.

Host Functions in Retrotransposition

The Ty1 system has provided some insights into the roles of host-encoded proteins and RNAs on the retrotransposition process, and it is anticipated that many more remain to be discovered. A large number of host genes that play roles in Ty and host gene transcription have been uncovered genetically; these are called *SPT* genes because they were originally identified by mutations that *sup*pressed the effect of Ty or LTR insertions. Although some of these affect *Ty* transcription, they do not affect the production of GAL/Ty mRNA. The development of sensitive assays for transposition of GAL/Tys *in vivo* has led to the identification of a number of host factors important for transposition at a post-transcriptional level; these are reviewed in Table 2.

See also: Coronaviruses; Human immunodeficiencies viruses.

Further Reading

Belcourt MF and Farabaugh PJ (1990) Ribosomal frameshifting in the yeast retrotransposon Ty: tRNAs induce slippage on a 7 nucleotide minimal site. *Cell* 62: 339.

Boeke JD and Sandmeyer SB (1991) Yeast transposable elements. In: Broach J, Jones E and Pringle J (eds) *The Molecular and Cellular Biology of the Yeast* Saccharomyces. NY: Cold Spring Harbor Laboratory.

Eichinger DJ and Boeke JD (1988) The DNA intermediates in yeast Ty1 element transposition copurifies with virus-like particles: cell-free Ty1 transposition. *Cell* 54: 955.

Kinsey JA (1990) *Tad*, a LINE-like element of *Neurospora*, can transpose between nuclei in heterokaryons. *Genetics* 126: 317.

Levin HL and Boeke JD (1992) Demonstration of retrotransposition of the Tf1 element in fission yeast. *EMBO J* 11: 1145.

Levin HL, Weaver DC and Boeke JD (1990) Two related families of retrotransposons from *Schizosaccharomyces pombe*. *Mol. Cell. Biol.* 10: 6791.

RETROVIRUSES – TYPE D

Maja A Sommerfelt, Sung S Rhee and Eric Hunter
The University of Alabama at Birmingham,
Birmingham, Alabama, USA

History

The first type D retrovirus described was isolated in 1970 from a spontaneous mammary carcinoma of an 8-year-old rhesus macaque. This virus, named Mason-Pfizer monkey virus (M-PMV), has since become the prototype for an enlarging family of both endogenous and exogenous viruses. Although isolated from a mammary carcinoma, M-PMV is not oncogenic; instead infected primates succumb to a severe and often fatal immunosuppressive disease, distinct from that caused by the Simian immunodeficiency viruses (SIV) which are lentiviruses. Simian AIDS (SAIDS) was first defined in 1983 as a disease entity caused by type D viruses. The precise mechanism of type D virus-induced immune suppression is not known, but seems to be restricted to the genus *Macaca* (subfamily *Cercopithecinae*) of which has been reported in seven species.

Thus, approximately 12 years after the discovery of M-PMV, several additional horizontally transmitted type D viruses were isolated. These viruses were initially named Simian AIDS D-type (SAIDS-D) and the primate center of origin was appended to the name. Therefore the viruses isolated at the California and New England centers were named SAIDS-D/CA and SAIDS-D/NE respectively. Viruses isolated in the Washington (SAIDS-D/W) and Oregon (SAIDS-D/OR) primate centres, in addition to causing immune suppression, were directly associated with a severe retroperitoneal fibromatosis (RF). Based on the observation that the type D viruses could be divided into distinct neutralization groups, a new nomenclature was devised to distinguish these viruses from each other and to differentiate them from the HIV-like AIDS-inducing SIV. Members of the group are now named Simian Retrovirus Type 1 (SRV-1) corresponding to SAIDS-D/CA and SAIDS-D/NE, SRV-2 corresponding to SAIDS-D/W and SAIDS-D/OR and SRV-3 corresponding to M-PMV. Serotypes 1–3 have all been molecularly cloned and sequenced. More recently, SRV-4 and SRV-5 have been isolated from the cynomologous and rhesus monkey respectively. SRV1-5 correspond to the exogenous type D viruses. Two endogenous viruses have been described, one endogenous to the New World squirrel monkey *Saimiri sciureus* (squirrel monkey retrovirus SMRV) and the other to the Old World spectacled langur *Presbytis obscurus* (PO-1-Lu). Although the exogenous type D viruses readily infect primates, they are not endogenous to the macaque genus. Hybridization studies with tissues of different monkey species show that M-PMV and SRV-1 DNA hydridize more readily to langur DNA (in the subfamily *Colobinae*) than to rhesus DNA, suggesting that M-PMV may have been derived from a virus similar to that endogenous in the langur monkey. The host species of origin for the type D viruses has not yet been defined. There have been reports of type D viruses isolated from human permanent cell lines: HEp-2V and PMFV, but these appear to be contaminants since only certain cell stocks carried the virus.

Taxonomy and Classification

Type D retroviruses are classified in the family *Retroviridae*. These are positive-strand RNA viruses that replicate by reverse transcription to form a proviral DNA intermediate which can integrate covalently into the host chromosomal DNA. M-PMV has a morphogenesis similar to that of the type B mouse mammary tumor virus (MMTV) in that it preassembles immature particles [sometimes referred to as intracytoplasmic A-type particles (ICAPs)] within the infected cell cytoplasm. The morphology of the mature extracellular virion is different in that it has a centrally located nucleoid similar to type C retroviruses and a much less dense fringe of glycoprotein than MMTV. Being of primate origin with these differences, a new morphological class of retrovirus was designated type D.

Although M-PMV was isolated from a rhesus mammary tumor, it is evident that these viruses are not 'oncogenic' viruses but are associated with immune suppression in primates.

Properties of the Virion

Type D viruses assemble an immature capsid in the cytoplasm. These intracellular spherical particles, 60–95 nm in diameter, migrate to the plasma membrane where they acquire, during release by budding, an envelope containing virus-encoded glycoproteins. Following release, a proteolytic event termed 'maturation' results in a morphological change in the virion from an electron luscent core to an electron dense core. These particles have a diameter of 100–120 nm. Type D viruses exhibit a buoyant density of 1.16–1.17 g/ml in sucrose and 1.21 g/ml in cesium chloride. The three-dimensional structure of the virion remains to be determined.

Properties of the Genome

The genome of type D retroviruses that have been sequenced to date, is composed of two identical positive-sense, single-stranded RNAs of approximately 8 kb in length (7943 nucleotides for M-PMV). The RNA molecules are proposed to be hydrogen-bonded to each other and to a host $tRNA^{Lys}$. This diploid structure of about 5.3×10^6 D (70S) is denatured by heat (80°C for 2.5 min) or 40% formamide to 2.65×10^6 D (35S). Like eucaryotic mRNA, the retroviral RNA genome is capped at the 5′ end with a 7-methyl GTP, polyadenylated at the 3′ end and internally methylated on scattered adenosine residues.

As with any replication competent retrovirus genome, coding regions of type D viruses are flanked at both ends by 5′- and 3′-terminal sequences which are important for virus replication and regulation of gene expression. The 5′ sequences include a binding site for a host $tRNA^{Lys}$ which serves as a primer for the synthesis of negative-strand DNA by the viral reverse transcriptase enzyme, and an untranslated region preceding the coding regions. The RNA splice donor site (AAGUAAGU) for subgenomic mRNAs is located 21 nucleotides upstream of the *gag* AUG for M-PMV. A packaging signal sequence for encapsidation of genomic RNAs into virions is also present. The 3′-terminal sequences contain at least two transcriptional elements; the AUUAAA signal sequence for poly(A) addition, and the UAUAUAAG sequence corresponding to the TATA box as promoter for viral RNA synthesis.

Properties of the Proteins

Type D virus proteins are encoded by four genes arranged as 5′ *gag-pro-pol-env* 3′. The most detailed information is available for the prototype virus M-PMV. Viral genes are expressed as polyprotein precursors – Pr78 is encoded by *gag*, Pr95 by *gag-pro* and is generated by a −1 frameshifting event at the end of *gag*. The *pol* gene is similarly expressed as a Pr180 *gag-pro-pol* product that is generated from two −1 frameshifting events, one at the end of *gag* and a second at the end of *pro*. The envelope glycoprotein is translated from a spliced mRNA as a Pr86 precursor. This is comprised of the surface (SU) glycoprotein gp70 which interacts with receptors on susceptible cells, and gp22 the transmembrane glycoprotein (TM). During virus maturation the gp22 is cleaved to gp20 by the viral protease; the significance of this cleavage event in the virus life cycle remains unclear.

Immature intracytoplasmic A-type particles are composed of Pr78, Pr95 and Pr180 in an approximate 80 : 15 : 5 ratio. During maturation, the Pr70 prescursor is cleaved to yield six internal structural proteins in the order NH_2-p10(MA)-pp24/16-p12-p27(CA)-p14(NC)-p4-COOH. The 10 kD matrix protein (MA) forms the envelope-associated outer shell of the polyhedral capsid and is modified with myristic acid. The 27 kD capsid protein (CA) is the major structural component of the capsid while the 14 kD nucleic acid binding protein (NC), presumably functions in genomic RNA packaging and dimer formation. The functions of the 12 kD (p12), 4 kD (p4) proteins and the phosphorylated pp 24/16 protein remain to be determined. SMRV particles contain a major capsid protein p35 which is likely to represent a fusion protein of p12 and p27.

The protease gene of M-PMV could potentially encode a protein of 314 amino acids, fused to the truncated Gag precursor. This region is cleaved by the viral protease to two proteins of 17 kD; the N-terminal protein has dUTPase activity, while the C-terminal portion has protease activity and can be further cleaved to an 11 kD protein with protease activity. The protease is an aspartyl protease and remains inert within the cell until it is activated by an unknown mechanism following virus release. The importance of dUTPase activity in the virus life cycle is unknown.

The reverse transcriptase enzyme (80–90 kD) for all the type D retroviruses has a divalent cation preference for magnesium in contrast to manganese which is used by many of the nonhuman mammalian type C viruses. This enzyme also has RNase H and integrase activity.

Physical Properties of the Virion

Type D virions are rapidly inactivated when exposed to high temperature (56°C for 30 min) and are also sensitive to lipid solvents and detergents. The mature virions of M-PMV are easily disrupted in mild detergent (0.5% Triton X-100), whereas immature intracytoplasmic capsid particles remain stable under these conditions.

Although retroviruses are quite resistant to UV or gamma irradiation, UV irradiation of M-PMV results in inhibition of virus-induced syncytial formation in rhesus monkey embryonic lung cells.

Type D viruses are composed of approximately 60–70% protein, 30–40% lipid, 2–4% carbohydrate and 1% nucleic acid by weight.

Strategy of Replication of the Nucleic Acid

In common with other retroviruses, the type D retroviral RNA genome is reverse transcribed within the infected cell cytoplasm to form a double-stranded DNA intermediate which migrates to the nucleus and integrates covalently into the host chromosomal DNA where it exists as a provirus. Transcription of the provirus results in messenger RNA that will function to produce virus-specific polyprotein precursors, and also genomic RNA that will be packaged into assembling virions.

Characterization of Transcription

The integrated type D provirus genome is transcribed into RNA by cellular RNA polymerase II. In the cytoplasm of infected cells, two classes of viral RNA transcripts are detected. A genomic-sized mRNA (8 kb) which is either translated into polyprotein precursors or encapsidated into progeny virus particles and a 3 kb mRNA representing subgenomic spliced mRNA which is translated into Env polyproteins.

Characterization of Translation

Genomic-size RNA subsets are translated on free

polysomes into the Gag polyprotein and, as a result of ribosomal frameshifting into the Gag-Pro and Gag-Pro-Pol polyproteins. The heptanucleotide sequence motifs for the −1 frameshift has been localized in both M-PMV and SRV-1 sequences; GGGAAAC for Gag-Pro fusion and AAAUUUU for Pro-Pol fusion. Owing to the frequency of frameshifting, the majority of precursor proteins are Gag polyproteins, 10–15% are Gag-Pro polyproteins and up to 5% are Gag-Pro-Pol polyproteins.

The spliced subgenomic *env* mRNA is translated on polysomes associated with the endoplasmic reticulum. The co-translationally glycosylated product is transported via the secretory pathway of the cell to the plasma membrane where it is anchored and oriented as a type I glycoprotein.

Post-translational Processing

The Gag-containing polyproteins are co- and post-translationally modified by cellular enzymes. The N-terminal methionine residue is co-translationally removed and modified by a myristic acid addition to an N-terminal glycine residue. The polyproteins are also phosphorylated following translation, but the significance of this modification is not clear.

Following entry into the endoplasmic reticulum of the cell, Env polyproteins form an oligomeric structure (presumably a trimer), which is the transport competent form of the precursor. During transport to the cell surface through the secretory pathway, core oligosaccharide side chains on the SU domain of M-PMV Env polyproteins are processed to complex oligosaccharide chains. The single high-mannose oligosaccharide chain added to the TM domain during translation is maintained in an unmodified form. In a late Golgi compartment the Pr86 polyprotein is cleaved into the two mature glycoproteins, gp70(SU) and gp22(TM), by a furin-like cellular endopeptidase. The gp22 is further processed, by a C-terminal cleavage event, to gp20 during virus maturation by the viral aspartyl protease.

Assembly Site, Release and Cytopathology

Type D viruses are characterized by the preassembly of immature intracytoplasmic A type particles (ICAPs). These are assembled from uncleaved Gag-containing precursor proteins and migrate to the plasma membrane where they are released by budding. They acquire an outer membrane, derived from the host cell plasma membrane, containing virus-specific envelope glycoproteins during this process. The preassembly appears to occur at a site in the cytoplasm where precursors congregate in sufficient quantities to allow self assembly. A signal which directs the proteins to the intracytoplasmic assembly site has been identified in the MA protein of M-PMV.

Although the pathway of intracytoplasmic transport has not been defined, the transport of capsids to the membrane requires a specific targeting process, in addition to the presence of myristic acid which is also required for intracellular transport. During the budding process, preassembled capsids associate with the plasma membrane, presumably via the MA protein and are extruded from the cell. It has been postulated for M-PMV that there is an interaction between the cytoplasmic domain of the TM glycoprotein, gp22, and the matrix protein which directs incorporation of envelope glycoproteins into the virion. As with other retroviruses, the viral glycoproteins are not essential for virion release but are required for virus infectivity.

Type D virus infection *in vitro* is not associated with any cytopathology, instead the cells become chronically infected and continually shed virus.

Other Subjects Relevant to Virus Group

Type D viruses are capable of inducing cell fusion *in vitro*. This is mediated by the interaction of virus envelope glycoproteins expressed at the cell surface with available receptors present on adjacent uninfected cells and has provided a method of titrating the virus in culture.

Geographic Distribution

The primates that are infected with exogenous type D viruses comprise the *Macaca* (Cercopithecinae) genus of Old World monkey. At present the disease spectrum of this virus group has mainly been associated with primate research centers in the United States. The spectacled langur which harbors the endogenous virus PO-1-Lu inhabits India and Southeast Asia and the squirrel monkey harboring SMRV (Cebidae) inhabits central and southern America.

Host Range and Virus Propagation

Viruses enter cells following attachment to specific cell surface receptors by a pH-independent pathway, suggesting that they penetrate cells at the plasma membrane. Type D virus receptors appear to be constitutively expressed on human cells, and the same receptor moiety appears to be used by all the type D viruses as well as two type C viruses, RD114 and BaEV. Viruses within this group therefore exhibit receptor interference; cells chronically infected with one virus will be resistant to superinfection by a second in the group. The receptor gene has been assigned to human chromosome 19q13.1–13.3. The type D viruses are not considered to be cytopathic, instead they integrate into the cell genome and persist as a chronic infection where the cells do not die but continuously shed virus.

The *in vivo* host range of the exogenous type D viruses shows that SRVs can infect many tisssue types within the infected animal including lymph nodes, salivary gland, spleen, thymus and brain. There have been reports that the SRVs may remain latent in the brain.

The *in vitro* host range of the type D viruses is extensive, including a variety of human lymphoblastoid (H9, Raji) and nonlymphoblastoid cell lines (HeLa and HOS). Mink, bat lung, chimp lung, African green monkey (Vero), horse epithelial and dog thymus cells are also susceptible. Cells that appear to be resistant to productive infection include hamster, rat and mouse.

The endogenous viruses have a more restricted *in vitro* host range. PO-1-Lu infects only human and bat lung cells. It cannot infect langur cells indicating that it has a xenotropic host range, typical of endogenous viruses. SRMV has a broader xenotropic host range infecting cells of mink, dog, bat, rabbit, chimp, rhesus monkey and human but not marmoset, owl monkey, howler monkey or baboon.

Genetics

The type D retroviruses in common with other retroviruses contain two identical RNA genomes; this diploidy allows for genetic recombination.

SRV-1, -2 and -3 have been fully sequenced and show a high degree of both nucleotide and amino acid identity. They have slightly different restriction endonuclease patterns but are uniform within each group. The difference between the SRV genomes probably results from base pair mutations over time rather than large recombinational events. The exogenous SRVs have been described as variants of M-PMV. The major differences are found in regions encompassing the Gag phosphoproteins, the p12 proteins and the envelope glycoproteins. SRV1 and SRV-3 (M-PMV) appear to be more related than SRV-2 which is also associated with RF.

The viral genome is comprised of four genes in the order 5′-*gag-pro-pol-env*-3′, where *gag* encodes the structural components of the capsid, *pro* encodes the viral aspartyl protease, *pol* encodes the polymerase enzyme (reverse transcriptase) and integrase, and *env* encodes the surface envelope glycoproteins. The type D retroviruses lack regulatory genes found in the immunosuppressive primate lentiviruses.

Evolution

Interestingly, the type D viruses such as M-PMV show sequence homology and immunological cross-reactivity in their envelope glycoproteins to two related endogenous type C retroviruses, namely baboon endogenous virus (BaEV), endogenous to *Papio* spp., and RD114, endogenous to cats of the *Felis* spp. The *gag* and *pol* proteins of the type D viruses are unrelated.

In contrast, the genomic organization, reverse transcriptase and Gag proteins of type D viruses are related to those of the B-type MMTV and the intracisternal A-type particules (IAPs). The type D viruses may therefore have arisen as a recombinational event in which the *env* gene was acquired from a virus ancestral to the C-type BaEV and the capsid and reverse transcriptase genes were derived from a retrovirus ancestral to MMTV or the IAPs.

Serologic Relationships and Variability

There are five different serotypes each with distinct neutralization epitopes amongst the exogenous SRVs. These neutralization epitopes lie in the external envelope glycoproteins which may show variability because of adaption to different macaque species. The viruses can also be distinguished using antibodies to the smaller Gag proteins, p10 and p12. In addition to these type specific determinants, the type D retroviruses demonstrate a high level of antigenic cross-reactivity within the major capsid protein. Studies addressing sequence variability following infection have not been undertaken.

Epidemiology

Spontaneous SAIDS has been described for the genus *Macaca* in at least seven species at four primate centers in the US. These species are: *M. mulatta* (rhesus macaque), *M. arctoides* (stump tailed macaque), *M. cyclopsis* (Formosan/Taiwanese rock macaque), *M. fascicularis* (crab eating macaque) *M. nemestrina* (pigtailed macaque), *M. fuscata* (Japanese macaque) and *m. nigra* (Celebese black macaque). The disease is undoubtedly more widespread, although its distribution in primate holding facilities, zoos or in the wild has not yet been fully investigated.

Transmission and Tissue Tropism

Transmission of the exogenous SRVs requires close physical contact and is probably spread by saliva and blood through biting, grooming and fighting; an inherent way such primates establish a social hierarchy. Unlike HIV, SRV transmission does not appear to be via sexual contact. High titers of virus can be found in infected macaque saliva, allowing entry to the bloodstream of a recipient individual following trauma.

The virus can be experimentally transmitted by inoculation of primates as was done to prove the disease association of these viruses. The *in vivo* tissue tropism shows a broad host range for both lymphoid and nonlymphoid organs. Perinatal transmission and *in utero* transmission of the exogenous type D viruses appears to be infrequent.

Transmission of the endogenous viruses occurs vertically as inherited genetic elements.

Pathogenicity

SAIDS is a naturally occurring, experimentally reproducible infection, where infected rhesus monkeys show a broad spectrum of clinical and pathological abnormalities. Following infection of juvenile rhesus macaques, the animals may die within one year. The course of infection can result in either fulminating viremia and death, a relatively mild acute phase of disease resulting in a chronic carrier state, or recovery from both viremia and latent infection.

The endogenous viruses are not associated with any pathogenic effects in either the squirrel monkey or the spectacled langur.

Clinical Features of Infection

Infection of macaques with all the SRVs isolated to date is associated with a severe and often fatal immune suppression characterized by immune deficiency, wasting, chronic diarrhea, splenomegaly, neutropenia, mesenchymal proliferative disorders, noma, disseminated CMV, bacterial pneumonia, progressive generalized lymphadenopathy (PGL) and anemia. There is a decrease in both T and B cell populations and a low response of peripheral blood lymphocytes to mitogens together with an increased incidence of tumors. Virus can be isolated from persistently infected animals from saliva, peripheral lymph nodes, plasma, peripheral blood leukocytes and exfoliated cells in milk.

One feature unique to SRV-2, is that in addition to immunosuppression, this virus is also associated with retroperitoneal fibromatosis (RF), which has been compared to Kaposi's sarcoma in humans. RF is characterized by an aggressive proliferation of highly vascular fibrous tissue which generally remains localized to the peritoneum. In 25% of the cases the RF progresses to involve the entire abdominal cavity, inguinal canal and thoracic cavity. A cutaneous form of RF has also been recognized in a small number of animals, immunohistochemical studies show a factor VIII-related antigen in endothelial cells and scattered fibroblast-like cells throughout the RF lesions, similar to that described for human Kaposi's sarcoma.

The endogenous viruses SMRV and PO-1-Lu have no disease association in their natural host.

Pathology and Histopathology

Typical type D virus particles can be found in the salivary gland acinar cells (secretory cells), macrophages, lymph nodes, peripheral blood lymphocytes, thymus and spleen, but not in muscle, nor in the brain. Viral nucleic acid can be found in brain parenchyma, suggesting that the virus may remain latent in this organ. Neuropathy, however, is not a characteristic of SAIDS. Brain cells may thus present a post-transcriptional block to SRV replication. The tissue localization of SRV-1 is also dependent on the severity of the disease. In severe SAIDS, virus is present in a higher percentage of cells of the salivary gland, as well as in the sinusoidal cells lining the spleen, and the stellate cells of the thymus. If the primate presents with PGL or splenomegaly, virus antigen is not as prevalent, being limited to geminal centers of lymphoid organs and scattered salivary gland acinar cells. There have been reports that some virus may also be found in other secretory cells, e.g. sweat glands, mammary glands and pancreatic acinar cells.

The histology of RF associated with SRV-2 infection reveals thymic atrophy, follicular and paracortical atrophy of the lymph nodes and variable myeloid and lymphoid hyperplasia in the bone marrow.

Immune Response

Infection usually results in an overall decline in both B and T lymphocytes. Impairment of B cell function has been demonstrated by a diminished antibody response *in vivo* and *in vitro* to antigenic stimulation, and progressively falling levels in serum of all immunoglobulin subclasses. The complement C3 and C4 levels remain intact or are slightly elevated through the course of the disease. Antibodies are mounted, particularly to the Gag antigens. Resistance to type D infection corresponds to the presence of high levels of neutralizing antibodies directed against the envelope glycoproteins.

Prevention and Control

Effective vaccines have been generated to both SRV-1, SRV-2 and SRV-3 respectively. Juvenile macaques inoculated with formalin-inactivated SRV-1 generate neutralizing antibodies which are protective upon challenge. Vaccinia recombinants expressing the envelope glycoproteins of SRV-1, SRV-2 and SRV-3 have been generated which also protect against virus challenge, demonstrating that the envelope glycoproteins alone are sufficient to elicit a protective immune response. These vaccines have limited cross-protection; SRV-3 vaccinated animals were protected from SRV-1 challenge but no cross-immunity between the SRV-1 and SRV-2 serotypes has been observed. Control of infection *in vitro* has also been successful using AZT, which inhibits the reverse transcriptase.

Future Perspectives

Although type D virus-induced immunosuppression has been a problem at primate centers in the United States, effective screening measures are possible to identify infected primates. Polyvalent vaccines may eventually be generated to protect primates from all the different serotypes identified to date. A recent report of the isolation of a type D virus, highly related to M-PMV, from an AIDS patient suffering from lymphoma raises the possibility of human infections by type D viruses – further research is required.

See also: Feline immunodeficiency virus; Human immunodeficiency viruses; Mouse mammary tumor virus; Simian immunodeficiency viruses.

Further Reading

Fine D and Schochetman G (1978) Type D primate retroviruses: a review. *Cancer Res.* 38: 3123.

Gardner MB and Marx PA (1985) Simian acquired immunodeficiency syndrome. In: Klein G (ed.) *Advances in Viral Oncology*, vol. 5, New York: Raven Press.

Weiss RA, Teich N, Varmus H and Coffin J (eds) (1984) *RNA Tumor Viruses: Molecular Biology of Tumor Viruses*, 2nd edn, p. 146. Cold Spring Harbor, NY: Cold Spring Harbor Laboratory.

RHABDOVIRUSES

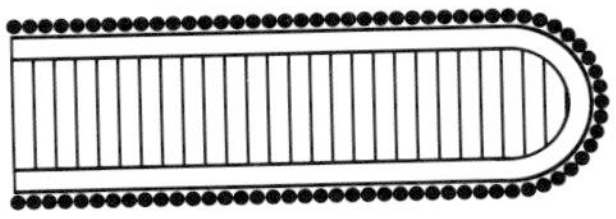

Contents

Plant Rhabdoviruses
Ungrouped Mammalian, Bird and Fish Rhabdoviruses

Plant Rhabdoviruses

Dick Peters
Agricultural University Wageningen
Wageningen, The Netherlands

Introduction

The family *Rhabdoviridae* is a large family of virus species; about half of them infect animals, the other half infect plants. The plant rhabdoviruses are included in this family by biological, morphological and molecular features which they have in common with the animal rhabdoviruses. The animal-infecting viruses are grouped into two genera, the plant rhabdoviruses into two subgroups which will probably be reclassified into three genera.

A few plant rhabdoviruses cause serious diseases in the crops they infect, but many are only discovered in a few infected plants and therefore cause diseases of less economic importance. Since the economic impact of the plant-infecting rhabdoviruses is minimal, research on these viruses is scarce. Several papers have been published which report the finding of rhabdovirus-like particles by electron microscopy in thin sections of plant cells or leaf dip preparations without any virus transmission or other data.

Taxonomy and Classification

The large number of plant-infecting rhabdoviruses, some 80 viruses are currently recognized, offer a taxonomic challenge. Many are considered to be distinct species, although several have been included as possible or probable species in the *Rhabdoviridae* on the basis of different biological (plant species infected, vector species), molecular (protein composition), serological and cytopathogenic (accumulation either in the perinuclear space or cytoplasm) features.

A classification of plant rhabdoviruses into different serogroups has not yet been proposed as serology has only been used to a very limited extent. Studies made so far show that viruses transmitted by planthoppers (*Delphacidae*) between grasses are closely or distinctly related whereas those transmitted by leafhoppers (*Jassidae*) are serologically completely distinct. Close serological relationships were found between eggplant mottled dwarf virus (EMDV), tomato vein yellowing virus and Tom-P, a rhabdovirus isolated in Portugal from tomato. These studies were made with viruses where relationships could be expected. They were selected by various criteria such as infecting the same group of host plants (*Graminae*), transmission by the same vector (*Laodelphax striatellus*) and possessing identical dimensions. Although the relationships between the different plant rhabdoviruses have to be elucidated in more detail, the limited cases in which some relationships are found strengthen the idea that plant rhabdoviruses form a group of many different species. Features such as nucleotide sequence homology between the different viruses and genome organization have not been used to study their taxonomic relations.

The International Committee on Taxonomy of Viruses distinguished two subgroups, A and B, based on the cellular location of particle assembly, kinetics of transcriptase activity and protein composition of the particles. The particles of subgroup A occur in the cytoplasm, contain one protein with low molecular weight, denoted M, and possess a transcriptase activity that can readily be detected *in vitro*. These properties resemble those of species of the genus *Vesiculovirus*. The particles of subgroup B accumulate in the perinuclear space, possess two low molecular weight proteins, often called M1 and M2, but are presumably matrix (M) and nonstructural (NS) proteins, and have low *in vitro* transcriptase activity. They share these properties with the viruses of the genus *Lyssavirus*.

Serological relationships between rhabdoviruses infecting animals and plants have not been reported.

Particle Morphology

The rhabdoviruses that infect plants often have a bacilliform morphology, whereas those that infect animals are usually bullet-shaped. The particles contain five proteins: a large protein (L); a glycoprotein (G); a nucleocapsid protein (N), which encapsidates the single-stranded RNA of negative-sense polarity; a matrix proteins (M); and a nonstructural protein NS, which is sometimes designated P. The M and NS proteins are usually denoted M1 and M2 in those plant rhabdoviruses that replicate in the nucleus and accumulate in the perinuclear space. M1 and M2 proteins are designated according to their electrophoretic mobility on polyacrylamide gels and not to their position in the virus particles or their function in replication.

The difference between animal and plant rhabdoviruses in morphology is not fundamental, but is of a qualitative nature. Purified preparations of plant rhabdoviruses with a bacilliform appearance *in situ* often contain a high number of bullet-shaped particles, indicating that the bacilliform shape is not a very stable form.

The form of the rhabdoviruses may be determined by either the fate of the membrane at the end where the particle has been closed after the budding process or the way they detach from the membrane after envelopment of the nucleocapsid. The accepted form of the nucleocapsid in the viral particle is a hollow bullet-shaped structure: one of the ends is blunt, whereas the other forms a half hemisphere. After envelopment of the nucleocapsid, the particle may detach by closing the membrane under formation of an envelope continuity on top of the blunt end of the nucleocapsid. Owing to the fragility of the envelope at this end, because it covers an empty space, the membrane may readily collapse and protrude subsequently into the cavity of the nucleocapsid. This process results in a bullet-shaped form characteristic of most animal rhabdoviruses but found less frequently for plant rhabdoviruses (Fig. 1).

Alternatively, the bullet-shaped particles can also arise after abrupt detachment from the membrane without the formation of a membrane continuity after ensheathment of the nucleocapsid. Also, in cells infected by animal rhabdoviruses, particles are often observed that remain attached to the membrane after ensheathment of the nucleocapsid. These particles will have a bullet-shaped form after disruption of the cell.

Plant-infecting rhabdoviruses vary more in size than do animal rhabdoviruses. Their length varies from 200 to 350 nm and their width between 35 and 95 nm. The volume of most plant rhabdovirus nucleocapsids is often a factor of 1.2–1.8 larger than that of vesicular stomatitis virus (VSV), and some are even a factor of 2 larger. Since the volume is a three-dimensional parameter and the length of the genome a one-dimensional parameter, the genomes of most plant rhabdoviruses will be 1.1 to 1.3 times larger than the genome of VSV. The genome of *Sonchus* yellow net virus (SYNV) (13 000 nucleotides) is 1.18 longer than that of VSV (11 000 nucleotides). In comparison to animal viruses, plant rhabdoviruses may have a larger genome, as they possess an extra gene which encodes a protein required to transport the virus from cell to cell in plants. These estimates on the nucleocapsid volumes and size of the plant rhabdovirus genomes do not support the belief of some virologists that plant rhabdoviruses are formed by two bullet-shaped nucleocapsids formed by blunt end aggregation.

Defective-interfering (DI) particles are common in animal rhabdovirus preparations. They arise in cell cultures upon application of an inoculum with a high multiplicity of infection. These DI particles, the genomes of which lack part of the complete genome, are dependent for replication on wild-type virus. The DI particles have a distinct replicative advantage over the complete genome and accumulate therefore at the expense of wild-type virus. Only two cases of putative DI particles have been reported in plant rhabdoviruses. Repeated transfer of potato yellow dwarf virus (PYDV) under conditions in which symptoms developed rapidly in tobacco resulted in a decreased recovery of the virus by purification. Particles were found which sedimented at a lower rate than the normal particles. When this variant form was mixed with normal particles, the infectivity of the latter was drastically reduced, suggesting that the variant fraction contained DI particles.

In the second instance, defective particles of SYNV found in *Nicotiana edwarsonii* plants inoculated with sap extracted from chronically infected calyx tissue. The infected plants exhibited a chlorotic mottling, instead of the normal vein-clearing symptoms. The purified short particles were not infectious. Plants inoculated with a mixture of these short and normal particles developed mottling symptoms and yielded predominantly short particles. The RNA of these short particles was 77% of the size of the standard virus. Since the DI particles of SYNV arose during a chronic infection of calyx tissue, the ability to generate DI particles is not solely a consequence of repeated transfers at high multiplicities of infection.

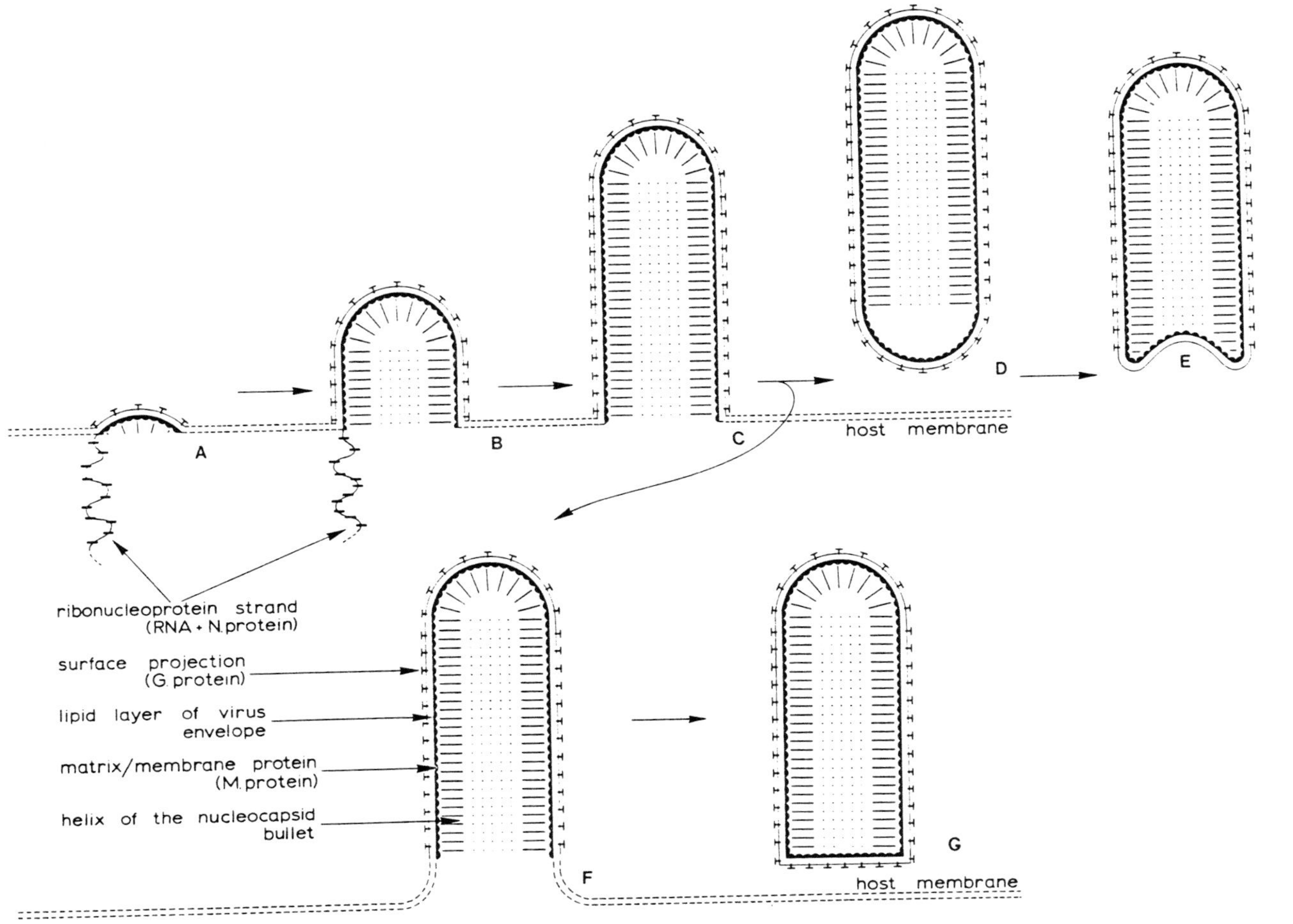

Fig. 1 Schematic representation of the maturation of the rhabdoviruses in cells. **A–C**, formation of the wound nucleocapsid and its envelopment; **D**, mature virus particle formed by membrance closure; **E**, particle with collapsed and inward protruding membrane continuity; **F**, to membrane attached particle; **G**, detached particle by membrane disruption.

Genomic Organization

Only the genome of SYNV has been the subject of detailed molecular studies. The genes are arranged in the same order as those in VSV, the order of which is N, NS, M, G and L in the 3′ to 5′ direction preceded by a leader sequence. However, the genome of SYNV contains an extra gene, denoted SC4. This is located between the M2 (NS in VSV) and the M1 (M in VSV) gene. A discrete mRNA transcribed from this gene has been detected. Its putative translation product has not yet been found.

Transcription of a rhabdovirus genome is preceded by the synthesis of a leader RNA. The leaders carry functions that regulate transcription and replication. The leader of SYNV differs considerably in length from those encountered in VSV, rabies (RV) and spring viremia carp virus (SVCV). Two motifs on these leaders are very similar but their locations vary greatly (Fig. 2). The hexanucleotide 3′-UUGGU, thought to initiate VSV RNA synthesis, is encountered at position 143 of the SYNV genome, whereas this motif can be found between position 15 and 20 in VSV, RV and SVCV.

A second motif, a U-rich 3′-end region, is about 30 nucleotides long in SYNV but only 20 nucleotides long in VSV, RV and SVCV. The already mentioned hexanucleotide is then part of this motif in the animal-infecting rhabdoviruses.

In addition to the homology in gene arrangement and the two similar motifs in the leader sequence, a nucleotide sequence with the common structure 3′-AUACUUUUUUNAUUGUCNNUAG-5′ occurs at the intergenic regions and in the flanking region

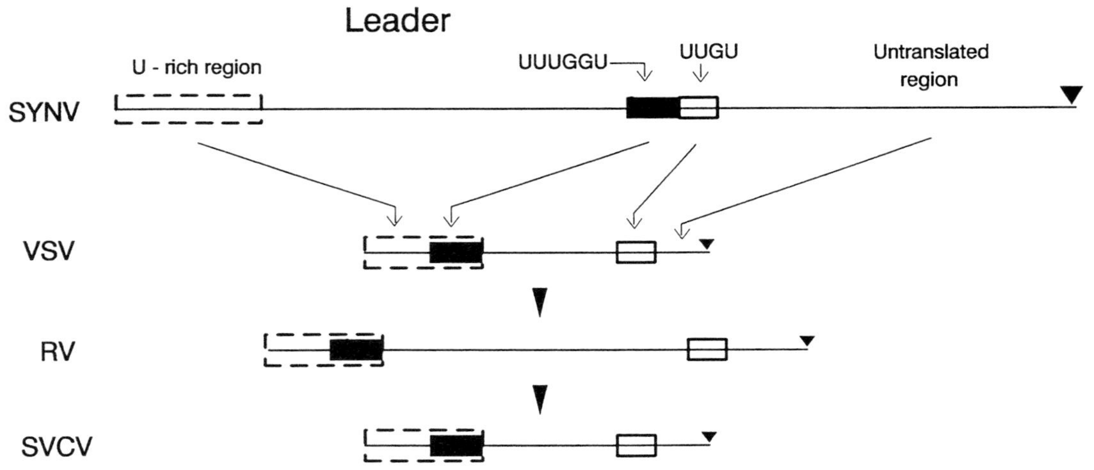

Fig. 2 The location of regulatory sequences in the 3′ ends of the genome of *Sonchus* yellow vein virus (SYNV), vesicular stomatitis virus (VSV), rabies virus (RV) and spring viremia carp virus (SVCV).

in front of the N protein gene at the 3′end. These sequences are strikingly similar in all rhabdoviruses analyzed. The underlined dinucleotide, present in all intergenic regions but different in composition in the various viruses, is lost during transcription.

A tetranucleotide UUGU, a template for the four complementary nucleotides, that follow the cap structure of all mRNAs of VSV, the N protein mRNA of RV and all mRNAs of SYNV and SVCV. This sequence can be found several nucleotides apart from the hexanucleotide UUUGGU in VSV, RV and SVCV. They are neighbors in SYNV using an overlapping U. The strong homology at these junctions indicates that the regulatory functions in the genomes of different rhabdoviruses are extremely well conserved during their evolution in the two different kingdoms. Thus, the processes that use these sequences did not evolve in different directions when the plant and animal viruses diverted from their common ancestor.

Ultrastructure and Replication

Although all animal rhabdoviruses replicate and assemble in the cytoplasm, plant rhabdoviruses differ markedly in their morphogenesis and site of accumulation. Electron microscopy has permitted detailed studies of the cytopathology of infected plants. These studies show that plant rhabdoviruses can be divided into at least four groups depending on the site of nucleocapsid formation and assembly of viral particles, and on the cytopathogenic structures encountered in the infected cells.

A large group including viruses such as SYNV, PYDV, sowthistle yellow vein virus (SYVV) and EMDV mature in association with the inner nuclear membrane and accumulate in the perinuclear space. Extensive aggregation in the perinuclear space can lead to the formation of invaginations filled with viral particles in the cytoplasm and or nucleus. A second group of viruses, including lettuce necrotic yellow vein virus (LNYV) and broccoli necrotic yellows virus, appear to mature in association with the endoplasmic reticulum and accumulate in vesicles of the endoplasmic reticulum. A third group of viruses, of which barley yellow striated mosaic virus (BYSMV) and Northern cereal mosaic virus are examples, mature in association with membrane-bound granular structures, called viroplasms. The particles accumulate after budding from the membranes surrounding the viroplasms in the vacuolike spaces. Finally, a fourth group, represented by coffee ring spot virus, can be recognized. They accumulate in the nucleus and their nucleocapsids are arranged as spokes in wheel-like structures surrounded by membranes.

For the study of replication of plant rhabdoviruses, protoplast suspensions are preferred above whole plants. Such suspensions have, however, not been extensively used in the study of plant rhabdoviruses. Infection of *Vigna unguiculata* protoplasts with SYNV and *Festuca* leaf streak virus (FLSV) has been shown to be a useful system for the study of replication of plant rhabdoviruses. SYNV was detected 11–12 h after inoculation and increased in titer until 30 h after inoculation. In total, 97% of the protoplasts became infected using polyethylene glycol during the inoculation. Numerous particles could be detected by electron microscopy in the perinuclear space. Direct evidence for the involvement of the nucleus in SYNV assembly was obtained by incubating the protoplasts with tunicamycin, a specific inhibitor of glycosylation. Nucleocapsids in the form of coiled structures accumulated inside the nucleus of the periphery, while granular material almost filled the nucleus. FLSV, which replicates in the cytoplasm of its natural host, was also found to replicate in the cytoplasm of *V. unguiculata* protoplasts.

Besides demonstrating that protoplasts have a large potential in studies on the replication of plant rhabdoviruses, the results also show that replication occurs in protoplasts from plant species that are not naturally susceptible to the virus studied.

Geographic Distribution and Host Range

Plant rhabdoviruses have been reported from most parts of the world. Some viruses are fairly widespread, such as maize mosaic virus (MMV) which has been reported in tropical and subtropical countries, but many seem to have a restricted distribution.

Their natural host range is usually rather small, especially those infecting dicotyledonous plant species. Some viruses found in gramineous crops species. Some viruses found in gramineous crops occur naturally in several grass species, which may function as reservoir and maintenance hosts. Some distinct viruses can be encountered in one and the same host, e.g. *Sonchus oleraceus* which occurs world-wide as a weed. SYVV has been discovered in this weed in California as well as at some locations in Europe. In Florida, this weed is infected by SNYV, while it hosts LNYV in Australia, and possibly also in France. In Argentina, a virus, called *Sonchus* virus, occurs in this weed.

Ecology and Pathology

Plant rhabdoviruses induce a variety of symptoms which often result in a lower production of leaf material or in a reduction in the yield of fruits or seeds. Perhaps the most serious disease caused by a plant rhabdovirus is that caused by MMV. This virus has been known to cause serious diseases in maize for more than 50 years. Yield losses of 100% have been reported. Major losses can occur when susceptible maize cultivars are planted in areas where its vector, *Peregrinus maidis*, transmits the virus while migrating from grasses, known as reservoirs, to maize. The infected plants initially develop yellow stripes between the leaf veins, followed by a yellowing and subsequently necrotization of the leaves. Early infected plants are stunted with shortened internodes and produce deformed cobs.

LNYV causes serious outbreaks of disease in lettuce in Australia. The infected plants become chlorotic and develop a flattened appearance during the acute phase of the disease. Upon necrotization of the infected leaves, many plants will die. The surviving plants, which are stunted, fail to produce a marketable head. The virus is mainly transmitted by the aphid *Hyperomyzus lactucae*. The spread of LNYV seems to be a complex phenomenon. Infected *Sonchus oleraceus* plants, a common weed with the common name sowthistle, form a reservoir for this virus in Australia. On development of large populations of *H. lactucae* on this host, which does not show symptoms, winged aphids are produced which migrate to other plants in search for suitable host plants. Migrants, unable to find a suitable host after some time, may probe lettuce plants and use them for a short time as hosts, long enough to infect them. Major disease outbreaks occur about a month after the beginning of extensive flights of *H. lactucae*. Spread of this virus may be controlled by eradicating sowthistle plants in and around lettuce fields.

A third disease of economic importance, rice transitory yellowing (caused by RYTV), causes serious losses in rice in the central and southern region of Taiwan, where two rice crops are grown in one season. After infection, some leaves, near the distal half of the plant, turn yellow, plants become stunted and tillering is reduced. Yield losses depend on the age of the plant at the time of infection. The virus is transmitted by three *Nephotettix* species.

In the past, PYDV caused serious diseases in the northeastern part of the United States. Infected potato plants developed leaf chlorosis and other symptoms like stem necrosis, stunting and reduced tuber production. The tubers that were produced became necrotic and had difficulties in sprouting. The ability of perennial weeds to serve as hosts provides the potential for regular outbreaks of the disease. Serious crop losses occurred in the 1930s in dry years, when the vector, *Aceratagallia constricta*, migrated from natural vegetation to potato fields. Since that time the incidence of the disease has decreased and major losses have not occurred since the 1940s. This decline can be explained by the advent of wide-spread use of insecticides, use of tolerant varieties and planting of certified seed.

Transmission

The biological cycles and spread of most plant-infecting rhabdoviruses, like those of animal rhabdoviruses, are poorly understood. Spread of plant rhabdoviruses depends heavily on transmission by insects. To overcome the natural barrier around the plant cell, the cell wall, plant rhabdoviruses, like other plant viruses, can only infect a host after an initial introduction into a cell by a vector (or eventually by mechanical abrasion). As these viruses remain localized in the cell after replication, vectors have to ingest and transmit them to healthy plants.

Plant rhabdoviruses are transmitted by arthropods sucking sap from plants. Of the some 30 species for which the vector is known, all, except two, are transmitted by aphids (*Aphididae*), leafhoppers (*Jassidae*) or planthoppers (*Delphicidae*). Beet leaf curl virus (BLCV) is transmitted by the lacebug *Piesma quadratum* and coffee ringspot virus by the mite *Brevipalpus phoenicis*. Viruses that infect grasses are all transmitted by leaf- or planthoppers. Some viruses that infect dicotyledonous plant species can also be transmitted by sap inoculation techniques to hosts that are not naturally infected. Thus the artificial host range is wider than the natural host range.

Most insects vector of plant rhabdoviruses ingest food from the phloem or xylem. With this food they acquire the virus. The longer they feed, the higher the chance that they acquire an infectious dose of virus. After a latent period, which may last several days and during which the virus will replicate in the vector, the virus will be transmitted to healthy plants. The ability to inoculate is generally retained until death of the vector. Efficiency of transmission declines with age, but this is possibly related to changes in feeding behavior rather than to reduced virus concentration in the salivary glands.

Each individual virus is usually vectored by one or only a few insect species. The species belong then, in the latter instance, to one and the same family or genus. Insects that vector plant rhabdoviruses generally accept only a restricted number of plant species as hosts. The long feeding periods required to ingest an infectious dose demand that the vectors colonize the plants susceptible to the virus transmitted. This means that they have to accept as a host the plant on which they can complete one or more life cycles, including egg and nymphal stages. This explains why plant rhabdoviruses often only flourish in small ecological and rather specific niches. In these narrow biocycles the plant functions as reservoir and maintenance host.

The kinetics of the transmission can be expressed in terms such as: the acquisition access period $(AAP)_{50}$, which is the period in which half of the vectors acquire an infectious dose; the latent period $(LP)_{50}$, which is the period after which half of the vectors are able to transmit; and the inoculation access period $(IAP)_{50}$, which is the period required by 50% of the viruliferous aphids to infect a plant. The AAP_{50} and the IAP_{50} have hardly been studied for plant rhabdoviruses. Most studies have been limited to determining the minimum period that a vector spends on a plant to acquire an infectious does or to infect a plant. These periods appear to be a few minutes. The periods AAP_{50} and IAP_{50} will be longer; the first may last a day whereas the latter will require some hours. The latent period of *Coriander* feathery red-vein virus in its vector, the aphid *Hyadaphis foeniculi*, is 5–7 days. Longer periods have been reported for BYSMV in *Laodelphix striatellus*. Several plant rhabdoviruses replicate in their vector, as has been shown for BLCV, northern cereal mosaic virus and SYVV using serial transmission techniques, for SYVV using primary aphid cell cultures of *Hyperomyzus lactucae* and for PYDV using secondary cultures of *A. constricta* cells.

Transovarial transmission has been demonstrated for SYVV and LNYV in the aphid *H. lactucae*. Of the leafhopper-borne viruses, those that infect eggs have only been reported for BYSMV and winter wheat mosaic virus. Ecologically, transovarial passage could provide a potential virus reservoir even if only a small proportion of the progeny transmit the virus.

See also: Defective-interfering viruses; Evolution of viruses; Plant virus disease – economic aspects; Rabies virus; Vesicular stomatitis viruses; Vectors.

Further Reading

Ismail ID and Milner JJ (1988) Isolation of defective interfering particles of Sonchus yellow net virus from chronically infected plants. *J. Gen. Virol.* 69: 999.

Jackson AO, Francki RIB and Zuidema D (1987) Biology, structure, and replication of plant rhabdoviruses. In: Wagner RR (ed.) *The Rhabdoviruses*, p. 427. New York: Plenum Press
Peters D (1981) Plant rhabdovirus group. CMI/AAB Descriptions of Plant Viruses No. 244, 6p.
Peters D (1991) Divergent evolution of *Rhabdoviridae* and *Bunyaviridae* in plants and animals. *Semin. Virol.* 2: 27.

Ungrouped Mammalian, Bird and Fish Rhabdoviruses

Sailen Barik and Amiya K Banerjee
The Cleveland Clinic Foundation
Cleveland, Ohio, USA

Introduction

Although viruses belonging to the family *Rhabdoviridae* are promiscuous infective agents that multiply in a variety of hosts throughout the evolutionary tree (invertebrates, vertebrates and plants), detailed information is available for only a few. However, most viruses do share antigenic reactivity with some others, thus forming an antigenic group (e.g. vesicular stomatitis group, bovine ephemeral fever group, Hart Park group, etc.). This chapter deals with those few rhabdoviruses that are not serologically related to other viruses or viral isolates and hence remain antigenically 'ungrouped'. The hosts (and their viruses) include: bat (Gossas, Oita-296), rodent (Klamath), bird (Navarro) and fish (viral hemorrhagic septicemia, infectious hematopoietic necrosis, spring viremia and pike fry disease). Except for the fish ones, very little is known about these viruses. All virions are bullet-shaped. The mammalian and bird rhabdoviruses grow in most common laboratory cell lines, e.g. BHK-21, Vero (CL) etc. In addition, like other rhabdoviruses, they readily infect newborn or weanling mice when injected intracerebrally, the death of the animal being the eventual outcome. No human disease is known for any of these viruses.

Gossas Virus

This virus was isolated in 1984 by P. Bres from the salivary glands of an adult bat (*Tadarida sp.*) caught in Dakar, Senegal (Africa). No obvious pathology of the infection was seen nor has neutralizing antibody been detected in either bat or human sera. However, in laboratory infections, antibody can be detected in both mice and rabbits. The virus fails to cross-react antigenically with more than 150 viruses tested, including Semliki Forest, Sindbis, West Nile, Dakar bat (a flavivirus), Le Dantec, herpes, blue tongue, Uukuniemi and Mount Elgon bat viruses. The virus grows in a variety of cell lines, such as from chick embryo, BHK, Vero, porcine kidney (PS) and rhesus monkey (LLC-MK2). Oita rhabdovirus was isolated by A. Oya in Japan from bats; no other information is available on it.

Klamath Virus

This was originally isolated in 1965 by Harold N. Johnson from a 3-month old (immature) meadow vole (*Microtus montanus*) collected in Klamath Falls, Oregon. The rodent eventually died, presumably as a result of the viral infection. The virus was subsequently detected in two regions in Alaska: Dot Lake (red-backed mice, *Clathrionomys rutilus*) and University of Alaska (meadow vole, *Microtus oeconomus*) in 1976. It is a bullet-shaped virus (167 × 80 nm) with an envelope containing projections of various lengths (average 8 nm) typical of a rhabdovirus. Lung as well as brain of intracerebrally inoculated newborn mice contain infectious virus. The virus also grows well in chicken embryonated eggs inoculated via the yolk sac. The majority of laboratory cell lines are susceptible to infection. The cytoplasm of infected cells contains nucleocapsids, and budding virions are found around the cisternae of endoplasmic reticulum. In complement fixation tests, Klamath does not react with antisera against 154 viruses tested.

Navarro Virus

One of the five avian rhabdoviruses reported, ungrouped Navarro virus, was isolated in 1984 by the Cali Virus Laboratory (Cali, Colombia) from the spleen of an adult wild turkey vulture (*Cathartes aura*) shot in Navarro, Colombia. The known properties of the virus are essentially similar to those of Klamath virus, described above.

Fish Rhabdoviruses

The four reported rhabdoviruses of fish are classified

into two divisions reflecting their host origin: salmonid and nonsalmonid. Viral hemorrhagic septicemia virus (VHSV or Egtved virus) and infectious hematopoietic necrosis virus (IHNV) and related viruses are examples of salmonid fish rhabdoviruses while spring viremia of carp virus (SVCV) and pike fry virus (PFV) belong to the non-salmonid group. These four classes of virus do not antigenically cross-react and remain essentially ungrouped. Because of their highly infectious nature, the diseases they cause rapidly reach epidemic proportions, resulting in great financial loss to pisciculture farms. Unlike other rhabdoviral diseases, those due to fish viruses occur at a characteristically low temperature (12–15°C), probably reflecting an adaptation to the colder aquatic habitats. In the laboratory, these viruses can be grown in cultures of poikilothermic cell lines, such as FHM (fathead minnow), RTG-2 or STE-137, and also in a human diploid cell line (WI-38) and in BHK-21. The optimum temperature for the stability of the virions and for viral growth in all cell lines is typically about 15–18°C. Clinical symptoms and pathology at autopsy are very similar for all fish viral diseases. The morphology of virion (bullet shape) and properties of the RNA genome (nonsegmented, linear, single-stranded, negative or anti-message sense) of all fish rhabdoviruses are typical of other rhabdoviruses. The immune response to these viruses remains unknown. Like vesicular stomatitis virus (VSV), both PFV and SVCV virions contain protein kinase activity. Specific features of fish rhabdoviruses follow.

Viral Hemorrhagic Septicemia Virus (VHSV) (Egtved Virus)

History and host range

Originally described in 1938 as 'Nierenschwellung' by Schaperclaus, the viral disease was rediscovered among rainbow trouts (*Salmo gairdnerii*) in the Egtved region of Jutland (Denmark) in 1950. In Europe, the disease has been described by half a dozen other names, such as 'Bauchwassersucht der Forellen', 'Anemie infectieuse', 'L'anemie pernicieuse des truites' etc. The name 'viral hemorrhagic septicemia' was internationally accepted at the first symposium of fish diseases in Turin, Italy in 1962. Brown trout (*Salmo trutta*) and brook trout (*Salvelinus fontinalis*) are considered immune to the disease; however, brown trout, whitefish (*Coregonus* sp.) and grayling (*Thymallus thymallus*) can be infected experimentally.

A viral isolate called 'Strain No. 23-75', isolated from brown trout in an outbreak of heavy mortality, appears to be a variant of VHSV that has acquired a broader host range. In experimental infections, the strain killed 95% of rainbow trout fry and 50% of brown trout. In control infections with VHSV, only rainbow trout were affected.

Pathology, histopathology and clinical features of infection

Affected trout appear quite black especially on the head and abdomen and have intense exophthalmia of both eyes (sometimes with protruding eyeballs giving a 'popeye' effect due to the hemorrhages in the connective tissues of the eye pit), distended abdomen and severe anemia. The gills become pale pink or greyish white. Dropsy may be found in some, and the hemorrhages seen in the acute form at the base of pectoral fins and at the lateral line are less frequent or may be absent. In the terminal stage (2–3 weeks after infection) of an epizootic, a small percentage of fish show neurological and motor disorders, such as spiral swimming at the bottom of the pond, tilted swimming and darting through and out of water. Death occurs within several days.

The most specific gross pathological changes are the scattered hemorrhages, particularly in periocular connective tissue, mouth cavity, skeletal muscle, perivascular adipose tissue, air bladder, intestine and sex organs. They are more intense in the acute form and tend to disappear in the chronic stage. The liver is hyperemic and wine-red in color in the acute stage; in the chronic stage, it becomes pale with petechiae. Histology of liver shows necrotic foci with hepatocytes, including cytoplasmic vacuoles, karyolysis and pyknosis. The gross color changes of the kidney and the histology of the nephrons are similar to those of liver. In addition, detachment of epithelium of uriniferous tubules and glomerular edema is observed. Lymphoid tissue shows a decreased number of lymphoid cells and an increased percentage of mononucleated and immature erythrocytes. Similar reactions can be seen in the spleen.

Immunology and diagnosis

The commonly used anti-VHSV antibody is produced by intravenous or subcutaneous inoculation of rabbits followed by long immunization periods of 5–18 months. The antibody is active in precipitation, neutralization of viral infectivity and complement fixation assays. Fluorescein isothiocyanate (FITC)-conjugated antibody is also used for histological detection of viral antigens. Neutralization studies of 76 natural isolates of VHSV from Danish, Norwegian and Swedish rainbow trouts showed that 72 of them were essentially identical to the standard F1 strain, but four (three Danish, one Norwegian) differed, suggesting that they may represent separate serotypes. However, in cell cultures all were stained by FITC-conjugated anti-F1 antibody.

A preliminary diagnosis of the disease is made on the basis of clinical and histological examination. Conclusive diagnosis and proof of VHSV etiology is obtained by either actual isolation of the virus from tissue homogenates followed by antibody typing or staining of frozen sections of tissue with specific fluorescent antibody (FA). Commonly used tissues are those of kidney, spleen and liver. Although these methods work well in the stages of epizootic when mortality is high, they often fail in the final phases when motor-neuronal symptoms predominate.

Transmission and prophylaxis

Despite failures to detect VHSV in symptom-free fish, it is almost certain that the fish serve as carriers of the virus, the most frequent cause being exchange of infected fish between farms. Infection spreads primarily through the water contaminated by excretions of carrier fish. Experimental transmission can be achieved by injecting infected tissue homogenates intraperitoneally or by brushing them on the gills, and through physical contacts between fish, but not by feeding infectious material. VHSV is not transmitted through eyed eggs spawned by infected mothers; even if small amounts of virus are present in the initial stages, they are presumably washed away by the time the eggs reach the eyed stage. Gulls and other piscivorous birds are also suspected transmission agents, and in Denmark all fish ponds are required to be covered with plastic wire to eliminate this hazard.

Destruction of all infected fish and disinfection of infected equipment constitute the only method presently available for prevention of VHS epizootics. A solution of 200 p.p.m. available chlorine is a commonly used and effective disinfectant. The hatchery may then be restocked with disease-free trout or with eyed eggs.

Interferon production

Experimental infection of rainbow trouts by VHSV results in the production of interferon which reaches a maximal level (2750 units per ml of serum) around 3 days postinfection. Physicochemical properties of fish interferon are comparable with those of other animal interferons, e.g. a molecular mass of ~ 26 kD (by gel filtration), sedimentation coefficient 2.5S (sucrose gradient) and an isoelectric point of 4.5–6.2. It has been suggested that interferon production may play a role in rendering the trout resistant to virus infection at temperatures above 15°C. Autointerference observed in FHM cells infected with VHSV at a high multiplicity of infection (MOI) could be due to the induction of defective-interfering (DI) particles or T particles (see the entry on Vesicular stomatitis viruses); its relevance in nature remains unknown.

Molecular aspects

The VHSV genome codes for the large protein L (~ 150 kD), glycoprotein G (74 kD) and nucleocapsid protein N (41 kD), similar to those of other rhabdoviruses. Notable differences are the absence of a P phosphoprotein (NS) gene and the presence of two matrix proteins, M1 and M2 (as opposed to one in VSV), about 22 and 19 kD respectively. In this respect, VSHV resembles rabies virus. The status of phosphorylation of N protein remains uncertain; specifically, it is not clear whether the highly phosphorylated species comigrating with N protein in sodium dodecyl sulfate–polyacrylamide-gel electrophoresis (SDS–PAGE) is phosphorylated N protein itself or a variant of a P-like protein. All proteins are structural, and their distribution in the virion is identical to that of VSV. RNA-dependent RNA polymerase (transcriptase) activity of all fish rhabdoviruses including VHSV have a lower temperature optimum (15–22°C) than the 30–32°C for VSV. Also unlike other rhabdoviruses, the VHSV and IHNV transcriptases are stimulated by Mn^{2+} rather than Mg^{2+}.

Infectious Hematopoietic Necrosis Virus (IHNV) and Related Viruses

History

IHNV, Oregon sockeye salmon disease virus (OSDV) and Sacramento River Chinook disease virus (SRCDV) are antigenically similar and produce diseases with nearly identical symptoms. IHN has been described in salmon and trout by a number of workers. In 1946, an outbreak of a disease of high mortality in juvenile (fingerling) sockeye salmon (*Oncorhyncus nerka*) occurred in two hatcheries in the drainage region of the Columbia river in northwestern Washington. About the same time, similar casualties were reported from a number of hatcheries in Oregon; hence the name. The SRCD was initially limited to chinook salmon (*Oncorhyncus tschawytscha*) in one locality on the bank of the Sacramento river in northern California, but was later also reported in sockeye and spring chinook salmon from Washington. All three diseases appear to be cold-dependent: epizootics appear only at temperatures below 13°C and disappear above 15°C or so, supporting the notion that fish rhabdoviruses (and their transcription machinery, see VHSV above) are adapted to low temperatures.

Clinical features and pathology

An epizootic of IHNV usually begins with a sudden rise in mortality. Moribund fish show dark coloring, loss of appetite, anemia, exophthalmia, distension of the abdomen with ascites, general viremia and fecal casts. Petechial hemorrhages are found at the base of the fins or on the fat or mesenteries surrounding the viscera. Histological examination shows extensive necrosis of the hematopoietic tissues in the anterior kidney and spleen. Severe necrosis is also found in liver, pancreas and granular cells in the wall of the alimentary tract.

Moribund fish infected with SRCDV do not feed, and have symptoms similar to those of OSDV epizootics. The gills turn pale and red blotches appear on the skin due to subcutaneous hemorrhage. Necrotic lesions in kidney, pancreas, spleen and adrenal cortex, vacuoles in pancreas and massive vascular damage in the head are visible. Viral particles are found mostly in interstitial spaces of diseased organs and occasionally in cytoplasmic vacuoles.

Transmission

IHNV is transmitted through water, by feeding on infected carcasses (cannibalism) or by exposure to eggs or fry from infected brood fish, gills and gastrointestinal tract being the most probable route of entry. In additon, transmission from adult carriers to fry constitute a major pathway. Virus is frequently demonstrated in ovarian or seminal fluid of carrier fish, female carriers outnumbering males. Unlike in VHS, eggs represent a major vector for IHNV.

As in VHSV, high MOI of cells in culture produces autointerference and liberates shorter virions, likely to be DI particles.

Molecular aspects

The morphology and protein composition of the virus are essentially identical to that of VHSV, including the existence of two M proteins, M1 and M2, and the absence of a P protein. The M1 protein of IHNV is phosphorylated while that of VHSV is not. The gene order in the genome is 5′-L-G-M2-M1-N-3′, and the messages of the genes are well characterized. Also, the G–L intergenic region of IHNV is large, a situation similar to the rabies virus. However, while the rabies intergenic region cannot code for any protein, that of IHNV contains an intact gene which codes for a 12 kD nonstructural protein (named NV for nonviral) of unknown function. This constitutes the first instance of a rhabdoviral nonstructural protein. Properties of the RNA transcriptase of both salmonid viruses are similar.

Spring Viremia of Carp Virus (SVCV)

History, symptoms, pathology, transmission and molecular aspects

The infectious dropsy of carp (*Cyprinus carpio*) was reported in Europe as early as 1930. It was not until 1950 that the viral origin of the disease was correctly identified. The virus is also called *Rhabdovirus carpio* for obvious reasons. Although slightly different in seasonal variation, pathogenesis, timing and geographic distribution, the swim bladder inflammation (SBI) disease of the cyprinids (aerocystitis) is due to a rhabdovirus which is serologically identical to SVCV. The symptoms of SVC vary depending on the form of the disease: acute, chronic, asymptomatic or latent. In an overt disease, central nervous system and peripheral nerves are initially affected. The fish become hyperactive, their abdomen distended, gills pale and scales protrude where ruptured and ulcerated dermal vesicles (carp erythrodermatitis) appear. Kidney and spleen are enlarged and contain the highest titer of virus; intestines show hemorrhagic necrosis. Some have protruding eyes. Moribund carp become dark, respire slowly and lie on their side. Progress of the infection with systemic viremia is as follows: peak virema appears on the 6th day postinfection and again on the 9th and 10th, excretion of virus in feces and mucus occurs on the 11th, and finally death on the 20th. SVC has also been described in 'big-head' fish in the former USSR. In experimental infections, fingerling carp, pike fry and the larvae and fry of carp are also found susceptible. SVC is a primarily European (former USSR, Yugoslavia, Czechoslovakia, France, Austria and Germany) disease.

SVCV transcriptase is greatly stimulated by *S*-adenosyl-L-methionine (SAM) *in vitro*.

Pike Fry (Rhabdo)virus (PFV)

PFV is responsible for two diseases of fry of the northern pike (*Esox lucius* L.): 'head disease', first seen in Nieuw-Vennep hatchery in the Netherlands in 1959, is characterized by a swelling or lump on the head; and the 'red disease', first observed in 1956, is characterized by large areas of the body and tail becoming swollen and reddish in color. Both diseases have a high mortality rate. The hydrocephalus associated with the 'head disease' makes the fish lose their equilibrium and swim erratically near the surface of the water. Characteristic features are poor growth, abnormal amounts of cerebral fluids, petechial hemorrhage in brain, spinal cord, spleen and pancreas, and degenerative necrotic changes in kidney tubules. 'Red disease' in pike fry causes pale gills, hemorrhages in the trunk and in muscle connective tissue, and red swollen areas above pelvic fins. Upon autopsy, virus is found in the hematopoietic tissues of the kidney, but not in skin, muscle or nervous system. The causative agent of the grass carp (*Ctenopharyngodon idella* Val.) rhabdovirus infection, isolated in 1974, appears to be a variant of PFV.

Of all fish viruses, PFV is the most similar to VSV at the molecular level, since it codes for the same profile of proteins, N, P, M, G and L.

See also: Defective-interfering viruses; Fish viruses; Interferon; Rabies virus; Vesicular stomatitis viruses.

Further Reading

Pilcher KS and Fryer JL (1980) The viral diseases of fish: A review through 1978. Part I: Diseases of proven viral etiology. *CRC Crit. Rev. Microbiol.* 7: 287.

Wolf KE (1988) *Fish Viruses and Fish Viral Diseases*. Ithaca, NY: Cornell University Press.

RHINOVIRUSES

Glyn Stanway
University of Essex
Colchester, UK

History

Rhinoviruses are the major causative agents of the familiar, mild, upper respiratory tract infection usually known as the common cold and are among the most frequent human pathogens. References to colds are found from earliest times and both Hippocrates and Pliny the Younger discussed possible therapies. Although long suspected, their infectious nature was formally demonstrated in the early 1900s by Kruse, Foster and Dochez who transmitted colds from donors to recipients using filtered nasal secretions. It was not until the advent of organ and tissue culture that rhinoviruses were first isolated, initially in the laboratories of Pelon and Price. When culture conditions were modified, notably by using temperature and pH similar to those found in the nose, the number of identified rhinovirus serotypes grew rapidly.

Taxonomy and Classification

Rhinoviruses include human, bovine and equine serotypes. Human rhinoviruses (HRVs) comprise by far the largest group with over 100 known serotypes (1A, 1B and 2 to 100) and will form the major focus for this discussion. There are two serotypes each of bovine and equine rhinoviruses. All are classified as picornaviruses; HRVs and bovine serotypes assigned to the rhinovirus genus and equine rhinoviruses unassigned.

Serotypes are defined in terms of neutralization by antisera and absence of cross-reactivity. Several attempts have been made to further divide the large number of HRVs, the most useful being on the basis of receptor tropism. Around 90% of HRVs (the major receptor group) use the membrane protein, intercellular adhesion molecule 1 (ICAM-1) as their cell receptor, while the other 10% (the minor group) use a second, unidentified receptor. A single serotype (HRV-87) apparently has a distinct receptor.

Properties of the Rhinovirus Virion

Electron microscopy of rhinoviruses reveals the typical picornavirus appearance of spherical, largely featureless particles approximately 30 nm in diameter. The particles are nonenveloped, of icosahedral symmetry and are made up of the virus RNA surrounded by 60 copies each of four capsid proteins VP1–4. An intermediate in capsid assembly is the protomer, made up of one copy of VP1, VP2 and VP0, a precursor in which VP4 and VP2 are covalently linked. Five of these protomers, arranged symmetrically about a fivefold axis, give another important intermediate, the pentamer which will form one of the 12 corners of the icosahedron. Twelve pentamers come together to form the protein coat. Cleavage of VP0 is the final step of assembly and this may be related to encapsidation of the nucleic acid and to stabilization of the mature particle. One or two copies of VP0 remain uncleaved but the significance of this is not known.

Our understanding of picornaviruses, particularly HRVs, has been aided greatly by work done in the laboratory of Rossmann where the three-dimensional structures of two HRVs have been solved to high resolution. These studies have shown that VP1, VP2 and VP3 are structurally similar, each being composed of an eight-stranded β barrel, and differing mainly in the loops and elaborations which join or project from the β sheets which make up this core structure. The same basic structure for the capsid proteins is seen in all picornaviruses, as is their relative position in the virus particle. At the surface, the area surrounding the fivefold axis of the pentamer is composed primarily of VP1, while VP2 and VP3 are arranged side by side at a greater distance from this axis. VP4 has an extended configuration and is located internally, underneath the other proteins. The pentamer is stabilized by interactions involving the N and C termini of VP1 and VP3, while adjacent pentamers are held together largely by hydrogen bonds between parts of VP2 and VP3. The loops between the β sheets of VP1–VP3 are the location of the regions of antigenic importance and they form much of the surface of the virus. The amino acid sequences of these regions are highly variable between serotypes and this is presumably the origin of the many HRV serotypes.

Properties of the Rhinovirus Genome

In common with other picornaviruses, the rhinovirus genetic information is carried by a single-stranded RNA molecule, the 5′ terminus of which is linked covalently to a small virus-encoded protein, VPg. The RNA has positive polarity and functions directly as a message to encode the synthesis of a single polyprotein which is then cleaved by proteases to give the final products. In the case of HRVs, which are the only rhinoviruses yet sequenced, the genomic RNA is approximately 7200 nucleotides in length and is made up of a 600 nucleotide 5′ untranslated region (UTR), an open reading frame of approximately 2200 codons, a short 3′ UTR and a poly(A) tract. At least seven HRV serotypes have been sequenced completely (HRV-1B, -2, -9, -14, -15, -85 and -89) and partial data are available for several more. The serotypes are closely related, except for HRV-14 which shows more similarity to the enteroviruses than do the other HRVs. All HRVs have a characteristic nucleotide composition with a preponderance of A and U, particularly in the third position of codons.

Properties of Rhinovirus Proteins

The HRV proteins are numbered 1A(VP4), 1B(VP2), 1C(VP3), 1D(VP1), 2A, 2B, 2C, 3A, 3B(VPg), 3C, 3D, according to their location in the initial polyprotein. There are 11 proteins but in the infected cell, several precursors, for example 3AB and 3CD, have a significant half-life and may be functionally active. The capsid proteins were discussed earlier, and much of our knowledge of the nonstructural proteins is inferred from other picornaviruses. The largest nonstructural proteins, 2C and 3D, together with 3AB, are known to be involved in RNA replication. 2C has amino acid sequence motifs seen in nucleotide binding proteins and, by analogy with other positive-stranded virus proteins, may have a helicase function. 3D has been shown to be the RNA-dependent RNA polymerase and contains motifs (YGDD for example) which are seen in other polymerases. 2A and 3C are cysteine proteases involved in processing the polyprotein and both are structurally homologous to the trypsin group of proteases. 2A is also involved in host cell protein synthesis shut off. The other proteins are not well understood but 2B may include determinants of host range.

Physical Properties

Rhinoviruses have a buoyant density in CsCl of 1.38–1.42. The characteristic feature which distinguishes HRVs from the similar pathogens, enteroviruses, is their lability below pH 6.0. In contrast, they are relatively thermostable, surviving for days at 20–37°C and this may be an important factor in their spread. As they do not have a lipid envelope, they are resistant to organic solvents such as ether and are unaffected by the detergent sodium deoxycholate. Alcohol/phenol disinfectants are effective virucidal agents.

Replication: Strategy and Early Events

At the molecular level, the replication of HRVs resembles that of the enteroviruses. Since one enterovirus, poliovirus, has been studied in great detail, HRV replication will be considered only briefly except where HRV-specific features are found. Little is known about bovine and equine rhinovirus replication.

In common with other viruses, the replication of HRVs requires several steps, including attachment, penetration, uncoating, protein synthesis, RNA replication and assembly. The extent of our knowledge of these steps varies considerably and the study of the first step, attachment, has probably advanced the furthest. Structural analysis by Rossmann has revealed the presence of deep depressions, often termed 'canyons', running at a constant radius around each of the fivefold axes and these contain the sites at which most, if not all, HRVs interact with their cellular receptor. As the canyons are rather narrow, they exclude antibodies and it is thought that this is a mechanism for protecting the virus receptor-binding site from the immune system. ICAM-1, used by the major HRV receptor group, is an immunoglobulin-like molecule which is sufficiently narrow to penetrate the canyon.

Replication: Translation

HRVs have a positive-sense RNA genome and the first step in macromolecular synthesis is translation, to enable the production of virus-specific proteins. The HRV 5′ UTR is very similar to that of poliovirus, in terms of primary nucleotide sequence and predicted secondary structure and translation probably proceeds in the same manner, i.e. following internal ribosomal entry within the 5′ UTR. Similarly, HRVs also shut off host cell protein synthesis by inactivating the cap binding complex and their cap-independent, internal ribosome entry allows them to circumvent this inactivation.

Post-translational Processing

Synthesis of a single virus polyprotein necessitates cleavage to give the mature proteins and this is brought about by at least two proteolytic activities encoded by the virus RNA. The 3C protease carries out the majority of cleavages, usually at peptide bonds between the dipeptides QG, QS, QA, QT or QM. The amino acids P, A, T or V are often found close to the cleavage site and these, with other features, probably help in its definition. The 2A protease performs the first cleavage event of processing, liberation of the capsid protein precursor, P1. 2A usually cleaves between AG, VG or YG residues. Details of the cleavage of VP0 to VP4 and VP2, which does not occur until the new virus particles are assembled, are not yet known.

Replication: RNA Synthesis

The nature of the genetic material requires that an RNA-dependent RNA polymerase is present and this must be virally encoded as no host enzyme exists. Once the first rounds of translation and processing are complete, the synthesized polymerase (3D), together with other virus products and host proteins, forms a membrane-associated complex.

This uses the genomic-sense RNA as a template for producing negative-sense copies which in turn act as templates for genomic-sense RNA synthesis. Some of these act as messages, while some are packaged into virus particles. The details of the process are not yet elucidated and little specific work has been done on HRVs.

Assembly, Release and Cytopathology

Assembly of HRV particles is via the protomers and pentamers already described and takes place in association with membranes. The determinants important in RNA packaging and the details of this process are not known. The gross effect on the cell (the cytopathic effect) probably stems from the inhibition of protein and RNA synthesis, together with the accumulation of virus components. These lead to degeneration of the cell and release of the progeny virions.

Geographic and Seasonal Distribution

HRVs infect individuals living under all climatic conditions, from arctic to tropical. Infections occur throughout the year, but in temperate countries there is an increased occurrence of HRV colds in September and less markedly in May. These may be correlated with summer/autumn and spring/summer changes in the weather. There is probably also a world-wide distribution of equine rhinoviruses.

Host Range and Virus Propagation

HRVs exhibit a restricted host range, a major factor probably being the lack of receptors in nonhuman organisms. They cannot be transmitted to any commonly used laboratory animals, and, although some serotypes can infect chimpanzees and gibbons, no symptoms are observed. In contrast, equine rhinoviruses can infect other species, including man.

Tissue cultures useful for HRV isolation are mainly of human origin and include fetal kidney and tonsil as well as continuous cell lines. Human diploid lung fibroblasts are commonly used for primary isolation, while MRC-5 and HRV-sensitive strains of HeLa (e.g. Ohio-HeLa) give good results. Several tissue culture systems need to be used to maximize HRV recovery from clinical samples. Equine rhinoviruses grow, with difficulty, in cells of equine origin.

Genetics

HRVs are typical picornaviruses in terms of structure and genome organization. The genetic information is carried on one piece of RNA, precluding reassortment and the major evolutionary mechanism is the accumulation of point mutations. Recombination has been demonstrated in some picornaviruses and may play a role in HRV evolution, although there is no direct evidence for this.

Evolution

HRVs are closely related to another genus of human pathogenic picornaviruses, the enteroviruses, and it is certain that they have a common ancestor. Little is known about evolution within the HRVs on a contemporary time-scale, although it has been shown that antigenic variants can be isolated during an outbreak. Some serotypes, for example HRV-36 and HRV-89, are very closely related at the RNA sequence level, suggesting that they have diverged relatively recently. The existence of intertypes, with antigenic properties intermediate between two serotypes, strengthens the conclusion that some of the currently known HRVs have a recent common ancestor.

Serological Relationships and Variability

A major factor in the incidence of HRV infection is the large number of serotypes; additionally, several of these can co-circulate within the community. Although there is evidence of an increase in the frequency with which higher numbered and therefore more recently identified serotypes are isolated, many lower numbered HRVs are still prevalent. Thus, many serotypes seem to coexist within the human population. The multiplicity of serotypes is in contrast to two other medically important picornaviruses, the polioviruses (three serotypes) and hepatitis A (a single serotype).

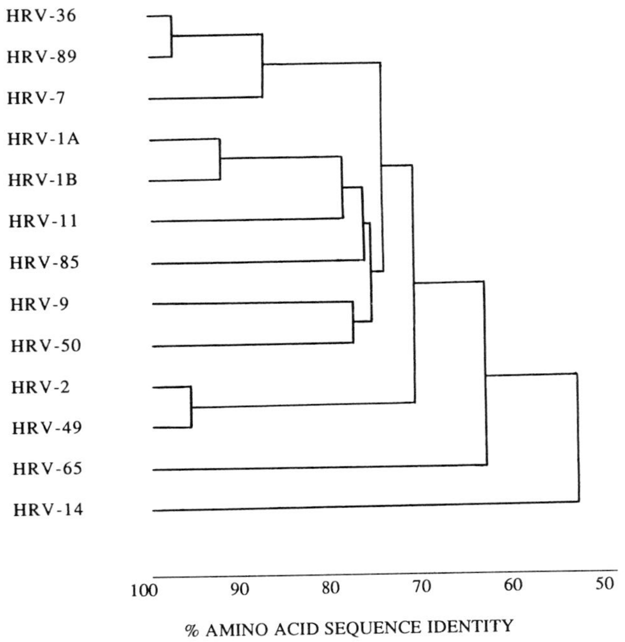

Fig. 1 The relationship between several HRVs in a 150 amino acid section of the VP2 protein, including one of the major antigenic sites. The comparisons show the degree of amino acid sequence identity and incorporate published sequences and data generated by PCR amplification and sequencing of previously unstudied serotypes (data from Gama and Stanway).

Although there are many HRV serotypes, several can be shown to fall into groups on the basis of low-level immunological cross-reactivity with hyperimmune serum. Whether this plays any role in protection from heterologous serotypes is not known. Recent results, obtained using PCR followed by sequencing as a rapid method of sequence accumulation, give a fuller indication of the molecular relationships within the HRVs (see Fig. 1). These data show that some HRVs are particularly closely related at the amino acid sequence level (HRV-36 and HRV-89; HRV-2 and HRV-49 for example) and correlate with the known antigenic cross-reactivity between these serotypes.

Epidemiology

HRVs are a major cause of morbidity and economic loss. They have been implicated in 10–40% of cases of acute respiratory disease, a category accounting for around half of all acute illnesses in the developed world. Thus, although the common cold is less severe than many other respiratory diseases, HRVs are responsible for a significant proportion of working days lost in industry, commerce and education.

HRV infections are most common in young children and babies and the infection rate decreases with age, possibly due to prior exposure to a growing number of serotypes. Estimates of the average incidence of HRV colds have varied, but it is probably at least 0.5 per adult per year. It is agreed that the rate in babies and infants is approximately three times that in adults. Once an individual is infected, the virus often spreads to other members of the family, the home being a major site of transmission. Young children are frequent causes of introduction into the family unit and other young children and the mother the most frequent recipients. The mother's infection rate presumably reflects greater exposure to the infected individual, although other factors may be involved as it has been shown that susceptibility to colds varies with the stage of the menstrual cycle. Schools and preschool groups also facilitate HRV transmission, though to a lesser extent.

Transmission and Tissue Tropism

Two routes may be important in HRV spread: direct contact and airborne transmission. People with colds contaminate their hands and environmental surfaces with virus from nasal secretions. The hands of uninfected individuals can then become contaminated by direct contact with the person with a cold or by touching the contaminated surface. The virus can enter the body when the hand is used to rub an eye or pick the nose, normal features of human behavior. In experiments which excluded this route, some transmission still occurred, suggesting that contaminated airborne particles or aerosols may also be important. Transmission in the family context correlates with time of exposure to the infected individual, severity of their symptoms and titer of HRV in nasal secretions.

Once introduced into the body, the primary and major site of infection is the epithelial surface of the nasal mucosa. HRVs thus show a restricted tissue tropism which may be correlated with their optimum growth temperature (33°C) and which largely limits their clinical manifestation to common cold-like symptoms. Sometimes though, HRVs infect other tissue, particularly the lower respiratory tract, the maxilliary sinus and the middle ear.

Pathogenicity

HRV infections are usually trivial and not life-threatening. Extensive work has been performed on their etiology and pathogenesis, notably at the Common Cold Unit, Salisbury, England, using human volunteers. These studies show that an HRV infection can be initiated by less than one $TCID_{50}$ (50% tissue culture infectious dose) of virus, if it is administered to the nasopharynx. Virus shedding is often detected after 24 h and reaches a maximum after 2–3 days, coinciding with the onset of symptoms. Virus titers thereafter fall rapidly but may remain detectable for 3 weeks.

Clinical Features of Infection

HRV infection is accompanied typically by the symptoms of a common cold. These vary with the individual and possibly the particular HRV, but usually include nasal discharge and obstruction, often with sneezing, coughing and sore throat. Fever and malaise are less commonly seen than in infections with other respiratory viruses, but gastrointestinal disorders are not uncommon, particularly in children.

Although limited largely to the upper respiratory tract, HRV infections are believed to predispose some individuals to bacterial sinusitis and otitis media. In addition, HRVs can precipitate asthmatic attacks in susceptible children and these children also experience more HRV colds than their non-asthmatic counterparts. Up to 40% of exacerbations of chronic bronchitis may be due to HRV infections.

Pathology and Histopathology

Physical examination of patients with an HRV infection usually reveals nasal obstruction and discharge, the nasal mucosa being pale and edematous. Elevated levels of bradykinin found locally, possibly account for this edema. The detailed histopathology of HRV infections is not well documented. Nasal mucosa biopsies reveal few histological abnormalities although shed, virus-containing, columnar epithelial cells can be detected in nasal secretions, suggesting that the epithelial surface of the nasal mucosa is primarily involved. Infection of bovine tracheal organ cultures with bovine rhinovirus leads to the shedding of large numbers of ciliated epithelial cells, leaving a smooth epithelial surface.

Immune Response

HRV infection stimulates the production of type-specific IgA, IgG and IgM antibodies in up to 90% of individuals. These are detectable within 2–3 weeks in both serum and nasal secretions and their levels rise for 5–6 weeks. Most immunoglobulin in nasal secretions is IgA and it is thought that this is a major factor in protection against re-infection (or at least amelioration of disease symptoms) by the homotypic HRV. Serum and secretory antibody persist for several years after infection, although their levels decline. As antibody appears late in the infection, it probably plays little part in recovery but it may be involved in final virus clearance. Other mechanisms, including interferon involvement, may therefore be involved in the recovery process. There has been little work on the cellular immunology of HRV infection.

Prevention and Control

At present there is no means of protecting against or treating common colds produced by HRVs. The large number of virus serotypes apparently precludes conventional vaccines and few attempts have been made to pursue this approach. Most effort has been expended on the development of chemical antivirus agents but of the many shown to have anti-HRV activity *in vitro*, none has, to date, proved clinically useful. A more promising agent is interferon, produced in large amounts by recombinant DNA means. High, intranasal doses, initiated several days before virus challenge, have proved to be effective in preventing illness.

However, side-effects limit long-term use and as symptoms are only reduced if treatment is commenced before virus infection, interferon has limited applicability. It may prove useful in the family context when it is important to prevent virus spread to specific individuals, e.g. asthmatics and bronchitics. In these cases, it may also be possible to use what we know about the properties of HRVs and their mode of transmission to limit spread. Interrupting transmission by avoiding direct contact with the infected person and by frequent hand-washing is sensible.

Furthermore, experiments have been performed in which paper tissues impregnated with virucidal agents (mild acid to exploit HRV lability at low pH) were used for frequent nose-blowing and hand-wiping by infected and uninfected individuals kept together under confined conditions for prolonged periods. The tissues were effective in preventing infection. One semi-empirical approach is the regular topical application of warm, moist air to the upper respiratory tract. Its beneficial effect may be due to the temperature increase in the nose, making it less conducive to HRV replication, although the stimulation of host mechanisms may also be involved. In contrast, the widely publicized notion that large amounts of vitamin C have some protective effect has been shown to be fallacious.

Future Perspectives

The past few years have seen major advances in the study of HRVs, particularly the determination of three-dimensional structures and the identification of a rhinovirus receptor. Several nucleotide sequences have been determined, revealing conserved regions which may, in the future, be exploited by PCR-based systems for rapid HRV detection. These would be an improvement on the current, cumbersome procedures for detection and identification and may give a more complete picture of the role of HRVs in human disease.

The serotype diversity of HRVs requires that if infections are to be controlled, broadly reactive prophylactic or therapeutic approaches must be devised. The economic and social significance of the common cold continues to stimulate research and our knowledge of the structure of the virus particle, the determinants of antigenicity, the cellular receptor for most HRVs and other advances made in the past few years will contribute greatly to the development of these rational approaches. It still seems unlikely that a vaccine can be produced, although the ability to pin-point areas of antigenic importance means that it may become possible to construct molecules which can mimic the antigenicity of several serotypes and thus reduce the complexity of the antigenic diversity problem. Even so, several components to the vaccine would be necessary and there may be a reluctance to use such a vaccine, with possible side-effects, against what is usually a mild pathogen.

The anti-HRV drug route may prove more feasible and the development of a group of agents which bind within a hydrophobic pocket in the HRV capsid, thereby blocking virus attachment to cells and/or uncoating, is encouraging. They have high specificity and efficacy against HRVs *in vitro* and seem to lack toxicity. As the pocket is well conserved between HRVs, the drugs are potentially broadly reactive and it is possible that one, or a small number, of agents would be effective against all HRV serotypes. However, even such highly specific reagents are likely to suffer from problems which may limit their usefulness, for example, in common with other RNA viruses, HRVs have a high mutation rate and mutants resistant to these 'pocket-binding' drugs can appear rapidly *in vitro*. If this occurs in an infected individual and the mutants are pathogenic, little benefit would be derived from the drug. Furthermore, maximal benefit from other agents tested is possible only if they are administered before infection. Such agents are not truly 'cures for the common cold' but rather prophylactics effective only if used over long periods of time. Whether it is worthwhile for most people to undergo prolonged medication to reduce the number of colds they suffer is open to question. Thus, despite improvements in understanding, HRV infections may continue to be a familiar feature of our lives.

See also: Antivirals; Enteroviruses; Interferons; Polioviruses; Respiratory viruses; Vaccines and immune response; Virus structure.

Further Reading

Couch RB (1990) Rhinoviruses. In: Fields BN *et al.* (eds) *Virology*, 2nd edn, p. 607. New York: Raven Press.

Gwaltney JM Jr (1989) In: Evans AS (ed.) *Viral Infections of Humans: Epidemiology and Control*, p. 593. New York: Plenum Medical Book Co.

Racaniello VR (ed.) (1990) *Picornaviruses. Current Topics in Microbiology and Immunology*, vol. 161. Berlin: Springer-Verlag.

RINDERPEST AND DISTEMPER VIRUSES

Tom Barrett
AFRC Institute of Animal Health
Pirbright, UK

History

Rinderpest (also known as cattle plague) is one of the oldest known plagues of domestic livestock with recognizable descriptions dating back to the fourth century AD. Rinderpest was first shown to be a filtrable agent in 1902 and transmission is by direct contact of susceptible animals with the secretions or excretions of infected animals. It is of ancient Asiatic origin but in more recent times devastating plagues of rinderpest swept across Europe in the eighteenth and nineteenth centuries.

In 1711 the disease entered Europe through Venice and had spread as far as Britain by 1714. The economic effect of the subsequent European plagues was so drastic that it led to the establishment of the first veterinary schools to deal specifically with the problems of animal health. The first was located at Lyon in France in 1762 and shortly afterwards other European countries followed France's example and set up their own schools. A vigorous slaughter and quarantine policy succeeded in controlling the disease and by the beginning of the twentieth century Europe was free of rinderpest. Subsequently periodic introductions occurred through importation of live infected cattle. The last serious outbreak in domestic cattle occurred in Belgium in 1920 and was caused by infected zebu cattle in transit from India to Brazil.

Following the 1920 outbreak in Europe, the Office International des Epizooties (O.I.E.) was set up in Paris to deal with the problems of animal health in relation to international trade. The last known case reported in Europe was in an imported zoo animal in Rome in 1949. Isolated cases of rinderpest have occurred in Brazil (1920) and Australia (1924), again in association with importation of live infected cattle.

In 1889 a catastrophic outbreak of rinderpest occurred in Africa and was caused by the importation of infected cattle from India to feed Italian soldiers engaged in a military campaign in Abyssinia (Ethiopia). The subsequent panzootic spread to nearly all parts of the continent and wiped out over 90% of domestic cattle and wild buffalo. Many other wildlife species were also severely affected. Descriptions at the time stated that the East African Plains were so littered with dead carcasses that the vultures were unable to clear carrion. Previously rinderpest was only seen in Africa, in Egypt and in parts of Senegal, where it was periodically introduced from Europe or the Middle East.

A similar plague in small ruminants (peste des petits ruminants or PPR), virus was first described in West Africa in 1942 by Gargadennec and Lalanne. The disease is also known as Kata in West Africa. At first it was thought to be a variant of rinderpest virus adapted to grow in sheep and goats. However, it was subsequently shown to be an immunologically distinct virus with a separate epizootiology in areas where both viruses are enzootic. A disease of small ruminants, which was almost certainly PPR, was first described in Senegal in 1871.

Canine distemper virus is also a disease with a long history. Edward Jenner studied its neurological symptoms but it was Carré in 1906 who first showed that it was caused by a virus. In French the virus is known as la maladie de Carré. In the summer of 1988 a large number of harbor seals (*Phoca vitulina*) died in the Baltic Sea and on the North Sea coasts of Northern Europe. The epizootic was eventually shown to have been caused by a morbillivirus, at first thought to be canine distemper. Monoclonal antibody analyses and nucleic acid hybridization showed that it was a new morbillivirus distinct from canine distemper and it is now named phocid distemper virus (PDV).

A disease with similar clinical signs which caused mass mortality in Siberian seals (*Phoca sibirica*) in Lake Baikal occurred in the winter of 1987. There was no apparent epidemiological connection between the two outbreaks and subsequent work showed that, in contrast to the European situation, this epizootic was caused by a virus indistinguishable from canine distemper. The source of virus in this outbreak is likely to have been lakeside dogs which were suffering from canine distemper at that time.

Taxonomy and Classification

The viruses are classified as the *Morbillivirus* genus

within the *Paramyxoviridae* family. They are antigenically closely related to human measles virus and are large enveloped viruses with a negative-strand RNA genome of about 16 kb. The virus particles are pleomorphic with an average diameter of around 200 nm. Measles virus is the only member of the group which has been shown to hemagglutinate red blood cells reproducibly. No neuraminidase activity has been demonstrated in any morbillivirus.

Geographic Distribution

Rinderpest is enzootic on the Indian subcontinent, in the Middle East and in Eastern Africa. Sporadic outbreaks occur in countries bordering the enzootic regions. Peste des petits ruminants is enzootic in parts of West Africa with occasional outbreaks in Eastern Africa and the Middle East. Recently its presence in Southern India has been confirmed using specific cDNA hybridization probes.

Canine distemper has a world-wide distribution but is not found in very hot, arid regions. The development of an attenuated vaccine for canine distemper virus in the 1950s greatly reduced the incidence of disease in domestic dogs. However, many wildlife species are susceptible to the disease and can act as reservoirs of infection.

The origin of the European seal morbillivirus is unknown but it appears to be enzootic in Arctic waters since sera collected from Greenland seals dating back to the early eighties have been shown to be positive for morbillivirus antibodies. Morbilliviruses have been isolated from porpoises in the North Sea and Irish Sea and more recently from dolphins in the Mediterranean. There is serological evidence of a morbillivirus infection in North American dolphins.

Host Range and Virus Propagation

All species of the order *Artiodactyla* are susceptible to infection with rinderpest virus but some species are more susceptible than others. In the case of cattle, the Asian species are more resistant than the European. The opposite is the case in pigs. In domestic animals PPR only causes clinical disease in sheep and goats; goats being particularly susceptible. It can also infect wild ruminants, as was illustrated recently by an outbreak of PPR in a zoo in the United Arab Emirates, but its full host range is unknown. In that outbreak gazelle, ibex and gemsbok were involved. Some large ruminants and pigs can be infected subclinically but they are dead-end hosts and do not transmit the disease.

Canine distemper can infect most mammalian carnivores but in some cases the disease can be subclinical, as in Felidae (cats, lions, tigers). It causes disease in all members of the Canidae (dog, wolf, fox) as well as Mustelidae (ferret, weasle, mink), Procyonidae (raccoon, panda) and collared peccaries (*Tayassu tajacu*). The outbreak of canine distemper virus in Siberian seals has extended its host range to aquatic mammals.

Phocid distemper is known to infect several species of seal ranging from the harp seals of the North Atlantic to the gray and harbor seals which have a more southerly distribution. In contrast to the harbor seals which died in large numbers, very few gray seals (*Halichoerus grypus*) succumbed to the infection.

The morbilliviruses can be isolated in a variety of cell types. Primary bovine kidney cells are usually used to isolate field strains of rinderpest and primary lamb kidney cells for PPR. More recently a marmoset lymphoblastoid cell line (B95a) and a *Theileria parva*-transformed bovine lymphocyte cell line have been reported to be suitable for rinderpest virus isolation. Virus can be best isolated from tissues such as mucosal lesions, lymph nodes or by co-cultivation of washed buffy coat from infected animals with susceptible cells such as bovine kidney. Cytopathic effects are usually evident 3–12 days after infection and the control cells with antirinderpest antiserum should remain healthy.

CDV is usually isolated by co-cultivation of lymphocytes from infected animals with mitogen-stimulated canine or ferret lymphocytes and can then usually be adapted to grow in MDCK or Vero cells. Lung tissue is also a good source of virus for CDV and PDV isolation. Primary seal kidney cells were used initially to isolate PDV but the virus can also be adapted to grow in Vero cells. Typical cytopathic effects such as cell elongation, cell rounding, the formation of stellate cells and syncytia can be observed 3–12 days post infection. Several blind passages may be necessary before cytopathic changes are observed in the cells.

Genetics

The viruses are made up of six structural virus proteins and two proteins (proteins C and V), termed nonstructural proteins, which have been identified in infected cells but not so far in the virus particles. Figure 1 is a diagrammatic representation of the virus and its RNA genome. Measles and CDV virus have been completely sequenced; the measles virion RNA is 15 892 nucleotides long and CDV 15 616. The virion RNA consists of a short 3′ leader RNA followed by the coding regions of the six structural protein genes and ending in a short 5′ leader RNA. There are defined stop–start sequence motifs and short intergenic regions between each gene. The gene order for canine distemper is identical to that of measles. Studies to date on the other members of the group show them to be identical in genetic configuration to measles and canine distemper.

The virus envelope consists of two virus-coded glycoproteins, the hemagglutinin (H protein) and fusion proteins (F protein), embedded in a lipid membrane derived from the host cell during virus budding. A nonglycosylated matrix protein (M protein) interacts both with the cytoplasmic domains of the envelope glycoproteins and with the nucleocapsids formed within the host cell during genome replication. The M protein is thought to be essential for virus morphogenesis and budding. The newly synthesized virion RNA is surrounded and protected in the cytoplasm by the nucleocapsid (N protein) protein which, in association with the virus polymerase (L protein) and phosphoprotein (P protein), forms the ribonucleoprotein complex.

The two nonstructural proteins found in infected cells are derived from messenger RNAs transcribed from the P protein gene. The C nonstructural protein is translated from an alternate reading frame in the phosphoprotein mRNA beginning at the second AUG codon. The V protein is translated from an mRNA which is not an exact copy of the P gene sequence but from an edited mRNA with an insertion of an extra G residue at position 751. This editing (or alternative transcription occurs in) about 50% of the mRNAs transcribed from the P gene in the case of measles and there is evidence that it occurs in all other morbilliviruses. Translation of this mRNA produces a chimeric protein consisting of the N terminus of the P protein with a new C terminus, rich in cysteine residues, derived from sequences in the third reading frame. This

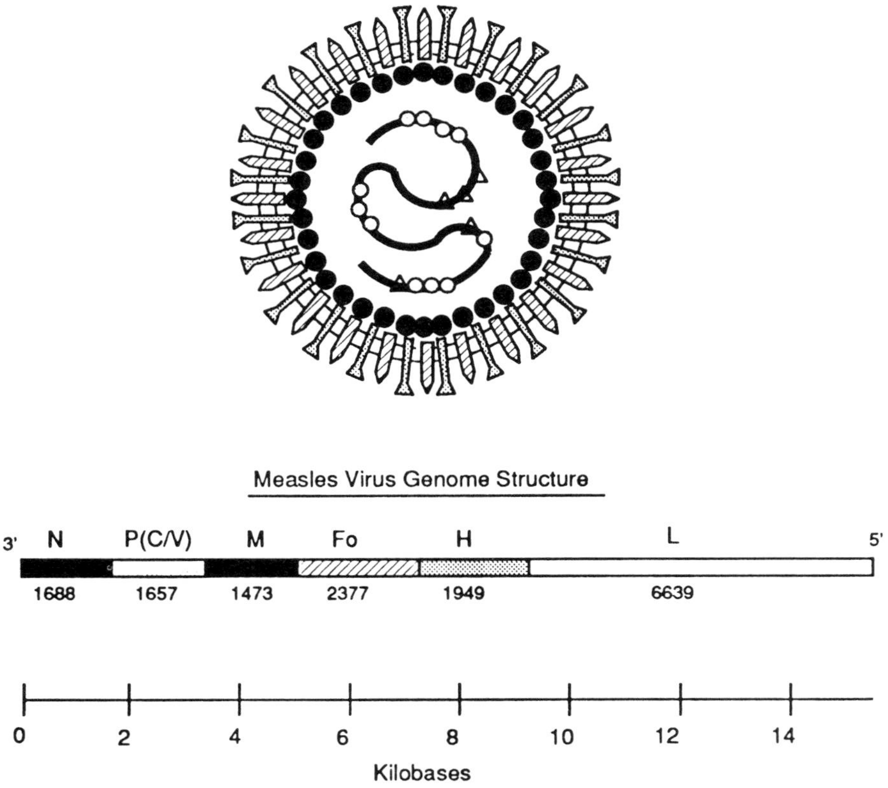

Fig. 1 Diagrammatic representation of a typical morbillivirus showing the outer envelope with the two projecting surface glycoproteins and the inner layer of matrix protein surrounding the helical nucleocapsid containing the virion RNA. The arrangement of the genes along the virion RNA is shown below.

mRNA is also capable of translating the C protein since its coding region is located in front of the editing sequence position. The functions of these proteins are unknown but they most likely play a part in transcription and replication of the genome RNA.

Evolution

Monoclonal antibody studies and sequence analysis indicate that measles, rinderpest and PPR viruses are antigenically closely related and that phocid distemper is most closely related to canine distemper. Sequence analysis of PDV showed that the nucleic acid identity is only about 70% when compared to canine distemper virus. The two viruses are, therefore, as different as rinderpest and measles and must have had a long period during which they diverged from a common ancestor. The new dolphin and porpoise morbilliviruses are antigenically more related to rinderpest and PPRV than to CDV or PDV but they form a quite distinct group when sequence data are compared. The dolphin and porpoise viruses are very closely related to each other and are possibly just different strains of the same virus (see Fig. 2).

The exact evolutionary relationship between the different morbilliviruses is unknown but rinderpest has been proposed as the 'archevirus' of the group. Monoclonal antibody reactivities support this conclusion since rinderpest reacts with a wider range of monoclonals produced against other morbilliviruses. In addition, the lifelong immunity following vaccination or recovery from infection, the lack of a persistent carrier state and the fragility of the virus particle dictate that the morbilliviruses have a large pool of susceptible hosts to maintain the virus. Such populations would most likely have been large herds of cattle or other ruminants before man settled to form sufficiently large populations (estimated as 300 000) to maintain measles virus. When cattle were domesticated they may have passed a morbillivirus infection to humans which evolved into measles virus. However, canine distemper infects a wide range of carnivores and is maintained by a large reservoir of virus in wildlife species. This virus, unlike rinderpest virus, can in rare cases maintain a persistent infection from which the disease can reactivate as old dog distemper. It is therefore possible that canine distemper is the progenitor of the group and that predators passed the disease to other nonpredator species which then evolved into distinct viruses. More extensive sequencing studies may resolve this problem.

Serological Relationships and Variability

All viruses in the group are antigenically related, the fusion and nucleocapsid proteins being the most

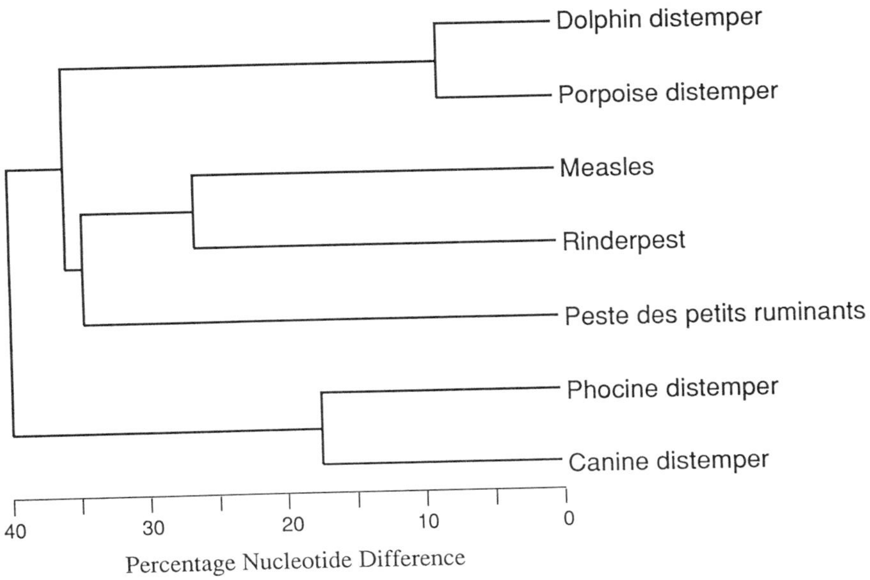

Fig. 2 A computer generated tree showing the relationship between the morbilliviruses based on partial sequence analysis of the P protein gene [from Barrett *et al.* (1993) *Virology*, 193: 1010].

highly conserved across the group. In fact immunity induced to these proteins may be responsible for the strong cross-protection seen after vaccination with heterologous virus. Rinderpest vaccine is routinely used to vaccinate against PPR and inactivated canine distemper virus vaccine has been shown to protect seals against infection with phocid distemper virus. It has been shown experimentally that measles virus can protect dogs against distemper and that distemper virus can protect humans against measles. This has no epidemiological significance since the viruses do not naturally cross-infect. The H protein, responsible for attachment to the host cell receptor, is the least cross-reactive. In immunoprecipitation reactions only the H protein fails to cross-precipitate with heterologous antisera, although some one way precipitations are seen, e.g. measles antiserum will precipitate canine distemper H but not vice versa. Strain variations within each virus group can be readily demonstrated using monoclonal antibodies but these variations do not result in different serotypes for each virus.

Epizootiology

Traditionally rinderpest outbreaks follow wars and civil disturbance where there is unrestricted movement of people and troops with live food animals which can carry the virus. Recent outbreaks in Lebanon, the Middle East and Sri Lanka follow this pattern. The outbreak in Sri Lanka was seen after a 40 year span free of the disease and the likely source was live goats brought from India with the troops and traded locally. More recently, rinderpest has reappeared in Turkey and this may be a consequence of the Gulf War.

Rinderpest and PPR are normally introduced into an area by importation of live animals from an enzootic area. Transmission by infected materials such as meat is very rare and considered to be a low risk. The most dangerous sources of virus are subclinically infected animals. Sheep, goats and possibly other ruminants can be subclinically infected with rinderpest and pass the infection to cattle. Subclinically infected pigs act as a source of virus for cattle; only Asian breeds of pigs and warthogs show clinical signs when infected with rinderpest although all are susceptible to virus infection. Wild ruminants vary greatly in their response to rinderpest infection with species such as buffalo and eland being highly susceptible and others such as the hippopotamus and Thompson's gazelle being highly resistant. Another factor which may be important in the maintenance of rinderpest is the presence of mild strains which cause subclinical infections in some enzootic areas. These can persist unnoticed for many years in the cattle population and then flare up clinically when animals are put under stress. The role of wildlife species in maintaining the disease is unclear.

There is no good evidence that wild ruminants act as a reservoir of infection for domestic animals but they may be important in helping to spread disease once an outbreak occurs. To date the evidence suggests that domestic animals are generally the source of infection for wildlife species and the disease has been eradicated successfully from South Africa and Tanzania, despite the presence of considerable numbers of susceptible wild animals. The epizootiology of PPR and the role wildlife plays in its maintenance has not been studied in any detail.

CDV is enzootic in wild animal populations and it remains a problem in poor urban areas where there are many stray dogs and vaccination is not rigorously carried out. The virus is also an important factor in the ecology of wild animal populations. The last free-living population of black-footed ferrets in Wyoming was almost wiped out by a canine distemper infection. As noted previously, canine distemper and a closely related phocid distemper virus have recently been isolated from several species of aquatic mammals. Such a serious disease could have a devastating effect on small populations of rare sea mammals, such as that of the monk seals (*Monachus monachus*) in the Mediterranean or the Siberian seals in Lake Baikal, and so it is important to monitor the populations of such species for signs of infection. The epizootiology of PDV is poorly understood. Analysis of historic sera from Canadian harp seals (*Phoca groenlandica*) and arctic ringed seals (*Phoca hispida*) indicated that they were infected with a morbillivirus several years before the appearance of PDV along the coasts of Northern Europe. It has been suggested that climatic changes caused the harp seals to migrate further south, thus passing the infection to other seal species. However, there is not sufficient evidence to be confident of the true source of infection of the European seals.

Transmission and Tissue Tropism

Morbillivirus transmission is by direct contact with secretions or excretions of infected animals. The morbilliviruses are highly contagious: all discharges can carry the virus. However, since the virus is extremely sensitive to environmental factors such as heat, sunlight and chemical inactivation it requires close contact with an infected animal for successful transmission. It is therefore relatively easy to control by regulating animal movements in conjunction with a strict quarantine and slaughter policy where necessary. Even without the availability of vaccines rinderpest was successfully eliminated from Europe by these means.

The morbilliviruses are highly lymphotropic and cause a transient immunosuppression in infected animals. The more virulent strains of the virus also have a strong tropism for epithelial tissues and this helps the spread of the disease by contact since high titers of virus are then excreted. Mild strains do not replicate so readily in epithelial surfaces and so are more difficult to transmit by contact.

Pathogenicity

Although there is only one serotype of each virus, differences between isolates can be shown using monoclonal antibodies and they also vary in their ability to cause disease in infected animals. In the case of rinderpest, extreme variation in pathogenicity has been reported, ranging from the mild strains enzootic in East Africa to highly virulent strains, such as those prevalent in the Middle East and India, which can cause 100% mortality in susceptible hosts. Passage of rinderpest in wild ruminants attenuates it for domestic cattle but it gradually regains virulence as it spreads through a herd, presumably by selection of pathogenic variants. There are no reports of nervous signs or meningitis following rinderpest or PPR infections.

Variation in the severity of clinical disease and neurotropism have also been reported with different strains of CDV which are known as 'biotypes'. Nothing is known about the determinants of pathogenicity in the morbilliviruses but secondary bacterial infections can greatly influence the outcome of the disease.

Clinical Features of Infection

Typical rinderpest is an acute febrile disease with mortality reaching close to 100% with some of the most virulent strains of rinderpest virus such as the Saudi/81 strain. The course of the disease is divided into five stages: after a short (1) incubation period of 3–5 days the (2) prodromal phase is seen in which there is a rapid rise in temperature. This is followed by the (3) mucosal phase in which severe mouth lesions are seen and there is a copious nasal and ocular mucopurulent discharge. The affected animals become depressed and anorexic and on post-mortem many lesions are seen throughout the digestive and lymphatic system. This is followed by the (4) diarrheal phase where there is severe bloody diarrhea, the animal is prostrated and dies from dehydration and weakness. In nonfatal cases there follows the (5) convalescent phase which may take many weeks and during which pregnant animals may abort. Some strains such as those prevalent in Eastern Africa are generally mild and there may be no clinical signs in many cases even though the animals seroconvert. With the mild strains the incubation period can extend up to 15 days.

Clinically, PPR closely resembles rinderpest. There is an incubation period of 4–5 days then a rise in temperature followed by nasal and ocular discharges which can be very severe. Mucosal lesions and severe diarrhea then appear and the animals, in fatal cases, die of dehydration within 6 days of the onset of fever. Bronchopneumonia is a frequent complication.

The incubation period for CDV is usually about one week. There are often two temperature peaks, the second corresponding with other signs such as nasal discharge, conjunctivitis and anorexia. As in the case of rinderpest, respiratory and gastrointestinal lesions follow and severe leukopenia is also seen. In some cases CNS signs, convulsions and seizures, are seen in association with the normal clinical signs, or they can also follow a subclinical infection. Recovered dogs frequently show persistent nervous 'ticks' or involuntary movements of one or more legs. Old dog encephalitis is a rare disease thought to be caused by CDV persistence in the brain and nervous tissue. Seals infected with PDV show clinical signs, including CNS lesions, reminiscent of CDV infection in dogs and this gave the first clue as to the nature of the etiological agent.

Pathology and Histopathology

Severe morbillivirus infections are accompanied by a marked leukopenia leading to a deficiency in the immune system. This often results in the activation of latent or other concurrent infections. In addition, secondary bacterial infections are common.

These factors may complicate both the clinical and pathomorphological findings. In the case of rinderpest leukopenia is most marked during the erosive mucosal phase. Histologically the virus shows a tropism for lymphoid and epithelial cells. All lymphoid organs are affected with the greatest damage occurring in the mesenteric lymph nodes and gut associated lymphoid tissue; severe destruction of the B and T-cell areas is seen. Intracytoplasmic and intranuclear eosinophilic inclusion bodies are commonly found in morbillivirus infected animals.

Epithelial tissues of the upper respiratory, urogenital and alimentary tracts are also infected by the highly pathogenic, and highly contagious strains of the virus. In acute rinderpest and PPR there is extensive erosive inflammation of mucosal surfaces of the digestive and upper respiratory tracts with cellular necrosis and the formation of syncytia. Mild strains induce less extensive mucosal lesions and this may account for the reduced ability of these strains to transmit by contact. In the cecum, colon and rectum of animals infected with RPV and PPRV so-called zebra or tiger stripes are commonly found. The stripes are caused by greatly distended capillaries packed with erythrocytes. Similar pathological changes are seen in infections with the other animal morbilliviruses. In acute CDV and PDV infections a serous inflammation of intestinal and respiratory surfaces, associated with interstitial pneumonia, occurs. A nonsuppurative encephalitis is also seen.

Dehydration in acute morbillivirus infections due to profuse diarrhea causes changes in hematology and blood chemistry. As a consequence total serum proteins and serum chloride levels drop. There is an apparent increase in erythrocyte count and the packed cell volume increases by 40–65%. At death the blood is dark, thick and slow to coagulate.

Immune Response

There is a strong cell-mediated component in the response to morbillivirus infection. In the case of measles virus, individuals with a genetic defect leading to agammaglobulinemia can quite easily overcome measles infection but immunosuppressed individuals are at extreme risk from the disease. Animals with any detectable neutralizing antibody are considered immune to rinderpest and immunity following morbillivirus infection is lifelong. Studies using ISCOM vaccines containing purified H or F proteins of CDV and with vaccinia recombinants expressing either the H or F protein of rinderpest have shown that either antigen can confer immunity to clinical disease. Similar observations were made in the case of measles virus but, in addition, it was shown that a vaccinia recombinant expressing the N protein could protect mice from lethal infection.

Prevention and Control

Rapid and accurate differential diagnosis is the key to success in controlling a morbillivirus outbreak. Morbilliviruses are extremely fragile; they are sensitive to sunlight, high temperature, low and high pH and chemicals which can destroy their outer lipid-containing envelope. Outbreaks of these viruses are therefore easily controllable by proper quarantine and hygienic measures. There is only one serotype of each virus and, once recovered from the infection, the animal is immune for life. There is no evidence for a persistent or carrier state in the recovered animals. Consequently vaccination is a very effective means of controlling these diseases.

Diagnosis

Identification of the morbilliviruses is generally based on the species in which the disease is seen, e.g. measles in humans, rinderpest in cattle, canine distemper in dogs, etc. However, this is not always accurate since the viruses can sometimes infect unusual hosts and so diagnosis can not be based solely on clinical signs. Differential diagnosis between rinderpest and bovine virus diarrhea is also not easy based on clinical signs alone and PPR can be confused with pasteurellosis and other microbial pleuropneumonias. Therefore laboratory diagnosis, based on either virus isolation or specific antigen detection, is essential to confirm the presence of the disease. General morbillivirus-specific antigen detection tests have been developed for field use but they cannot distinguish between the viruses of rinderpest and PPR. Differential neutralization tests can be used to classify virus isolates but these are time-consuming. Hybridization techniques using cDNA clones specific for each virus are more rapid and can be used to identify virus in postmortem samples (see Fig. 3). Recently the polymerase chain reaction has also been used for specific morbillivirus diagnosis. Competitive ELISA systems using morbillivirus species-specific monoclonal antibodies are now available for serological analysis which can distinguish between infections caused by the different viruses.

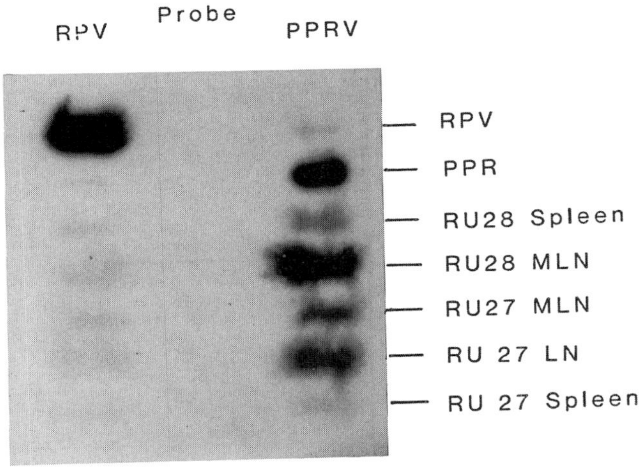

Fig. 3 Examples of RNA extracted from the tissues of suspected PPR (Peste des Petits Ruminants) virus infected goats from Ethiopia hybridized with probes specific for rinderpest and PPR. The control PPR and rinderpest RNAs are in the first and second row, respectively.

Vaccination

During the 1930s attenuated rinderpest vaccines were developed by passage of the virus in non-natural hosts e.g. rabbits and embryonated chicken eggs (lapinized/avianized) or goats (caprinized). In Japan a lapinized/avianized vaccine was developed which was used extensively to control the disease in Asia. In India and Africa the caprinized virus was used. However, the latter virus was not completely attenuated and it may have been responsible for the circulation of rinderpest in small ruminants in India. In the early 1960s the Plowright tissue culture attenuated vaccine was introduced which was completely safe and relatively easy to produce. In the 1960s an internationally funded rinderpest eradication campaign (Joint Programme 15 or JP 15) was carried out in Africa using the tissue culture attenuated vaccine which almost succeeded in clearing the disease from Africa. However, political instability, lack of funds to continue vaccination and disease surveillance and the persistence of mild strains of the disease resulted in a resurgence of disease in the 1980s. Quite often countries are reluctant to report cases of rinderpest or PPR because of the repercussions for trade in live animals and this makes the task of controlling the outbreaks more difficult. Currently a series of internationally funded vaccination campaigns are underway in Africa (Pan African Rinderpest Campaign or PARC), West Asia (WAREC) and South Asia (SAREC) in an attempt to eradicate the disease. Generally rinderpest vaccine is used to control the spread of PPR but an homologous vaccine has now been developed and is being field tested in West Africa. The role that wildlife species play in maintaining rinderpest and PPR needs to be more clearly understood before it is possible to say for certain that vaccination can eliminate these diseases.

The tissue culture vaccine strain of rinderpest virus is extremely safe with no clinical signs following vaccination in domestic animals. In addition, the virus does not replicate at epithelial surfaces and cannot be transmitted by contact. Immunity following vaccination is complete and lifelong. The vaccine is, however, very heat labile and so is expensive and difficult to use in the hot climates where rinderpest is enzootic. The establishment of an effective cold-chain and follow-up seromonitoring to determine the level of herd immunity are essential prerequisites for a successful vaccination campaign.

Canine distemper vaccines are not attenuated for all species and in some, such as the giant panda, they cause quite severe infections. There are two widely used vaccines for CDV. The Onderstepoort strain was attenuated by growth in avian cells and the Rockborn strain in canine tissue culture cells. Immunity lasts for several years following vaccination with either vaccine. A less effective subunit vaccine must be used for nondomestic species, particularly in the case of valuable zoo animals. Since there is a large wildlife reservoir of CDV and the vaccines are not attenuated for many wild animal species, it may be impossible to eradicate the disease completely.

Future Perspectives

The most pressing need is for a rapid, easy to use diagnostic kit for diagnosis of the economically important morbillivirus diseases in the field. Preferably such a test would be able to distinguish between rinderpest and PPR. There is also a need to improve the heat stability of the rinderpest/PPR vaccines for easier and more effective use in hot climates where these diseases are endemic. Research is continuing to select more heat stable clones of the tissue culture virus and to improve methods of freeze-drying and storage. Other research into the development of poxvirus recombinants may yield a more stable vaccine for use in hot regions.

See also: Epidemiology of viral diseases; Host genetic resistance; Immune response; Measles virus; Pathogenesis; Vaccines and immune response.

Further Reading

Appel M (1987) Canine distemper virus. In: Appel MJ (ed.) *Virus Infections of Carnivores*, p. 133. Amsterdam: Elsevier.

Losos GJ (1986) Peste des petits ruminants. In: *Infectious Tropical Diseases of Domestic Animals*, p 549. Harlow: Longman.

Plowright W (1982) The effects of rinderpest and rinderpest control on wildlife in Africa. In: MA Edwards and U McDonnell (eds) *Animal Disease in Relation to Animal Conservation*, p.1. (Symposia of the Zoological Society of London no. 50). London: Academic Press.

Scott GR (1981) Rinderpest and peste des petits ruminants. In: EPJ Gibbs (ed), *Virus Disease of Food Animals* Vol 2, p. 401. London: Academic Press.

Scott GR (1985) Rinderpest in the 1980s. In: Paudey R (ed.) *Infections and Immunity in Farm Animals* (Progress in Veterinary Microbiology and Immunology 1) p. 145. Basel: S. Karger.

Visser IKG, Van Bressem M-F, Barrett T and Osterhaus ADME (1993) Morbilliviruses infections in aquatic mammals. *Vet. Res.*, **24**: 169.

ROSS RIVER VIRUS

Lynn Dalgarno and Ian D Marshall
The Australian National University
Canberra, Australia

History

The first reports of a disease probably caused by Ross River virus (RRV) appeared in 1928 describing epidemics of transient arthritis and rash in two Murrumbidgee River towns on the semiarid inland plains of New South Wales (NSW), southeastern Australia. During World War II, epidemics of arthritis with rash were described in troops serving in the tropical regions of Australia and on islands to the immediate north. Most of these outbreaks were differentially diagnosed against a background of endemic dengue fever, and published reports of several large series of cases allowed the description of a syndrome adequate for clinical diagnosis of the disease, at least during epidemics. Several attempts failed to isolate or define the nature of the causative agent. These wartime reports provided the terms 'epidemic polyarthritis', and 'epidemic polyarthritis with rash' to characterize the disease; although these are still widely used, the noncommittal 'Ross River virus disease' is gaining currency. As fever is usually absent or unremarkable, 'Ross River fever' is inappropriate.

The first large, adequately documented community epidemic occurred in the Murray Valley of southeastern Australia in 1956, but again two groups of investigators failed to isolate the causative agent. However, following consideration of the epidemiology and nature of the disease, and subsequent serological testing of convalescent sera, it was concluded that a mosquito-borne Group A arbovirus was the most likely candidate, and for several years the Malaysian virus Bebaru was used as a surrogate diagnostic and survey antigen. Eventually, in 1963, the specific alphavirus was isolated from a pool of *Aedes vigilax* mosquitoes collected during dengue investigations near the Ross River at Townsville, coastal north Queensland. The first human isolate was in 1971 from the serum of a mildly febrile 7-year-old aboriginal boy, but, as is usual before puberty, characteristic signs and symptoms did not develop. The final incrimination of RRV did not occur until 1979, when the virus was introduced to Fiji, presumably by a viremic tourist from Australia. RRV was isolated without difficulty in newborn mice inoculated with the acute stage serum of the indicator case. Numerous isolations were subsequently made in mice, mosquitoes and cell cultures during the ensuing series of virgin soil epidemics which, over the next 2 years, extended across the South Pacific from New Caledonia to the Cook Islands. RRV has since been isolated from sera of Australian patients.

Over recent years it has become apparent that a small proportion of clinically well defined cases of viral polyarthritis with rash are due to etiological agents other than RRV. There is now no doubt that another alphavirus, Barmah Forest virus (BFV), is one such agent. BFV was first isolated in 1974 from a pool of *Culex annulirostris* mosquitoes trapped during investigations of a widespread epidemic of Murray Valley encephalitis. It has now been isolated from a range of mosquito species in diverse tropical and temperate ecosystems on the Australian mainland, and is classified as the only member, so far, of a seventh alphavirus serocomplex. As with RRV, its primary cycle involves mammals rather than birds. Antibody prevalence in humans is considerably lower than for RRV. The virus was isolated from the serum of a mildly ill young patient, and there have been a number of cases, thoroughly documented both clinically and serologically, which exhibited virtually the full range of signs and symptoms associated with a fulsome RRV infection. An episode of 'epidemic polyarthritis' in a remote mining community near Darwin in 1992 was due to the concurrent circulation of both RRV and BFV.

Taxonomy and Classification

RRV is a mosquito-borne arbovirus belonging to the genus *Alphavirus* of the *Togaviridae* family. It has a single-stranded, positive-polarity RNA genome.

Properties of the Virion

Purified virions have two glycosylated envelope proteins: E1, the hemagglutinin (mol. wt 52 000) and E2, the neutralizing antigen (mol. wt 49 000). The nucleocapsid protein, C, is of mol. wt 32 000. RRV is sensitive to chloroform, ether, detergents, ultraviolet irradiation and low pH. The infectivity titer of RRV T48 is virtually unaffected by incubation in cell growth medium at 50°C for 45 min. Under the same conditions, a mutant of RRV T48 with a deletion of seven amino acids (residues 55–61) in the E2 glycoprotein is thermolabile, showing a three log unit loss of infectivity.

Properties of the Genome

The RRV genome is a single-stranded, positive-sense RNA molecule of 11 851 nucleotides (T48 strain) excluding the poly(A) tail. The 5′ two-thirds encodes the NSPs; the 3′ one-third of the genomic RNA is not translated from genomic RNA itself but is expressed as a subgenomic (26S) mRNA molecule which is transcribed from full-length negative strands. In the genome, a 5′ noncoding region of 78 nucleotides is followed by an open reading frame (ORF) of 7440 nucleotides which is interrupted after 5586 nucleotides by a UGA termination codon (as is found in SIN and Middelburg RNAs but not SFV). By analogy with SIN, the 5′ two-thirds of the genome encodes two polyproteins. One is NSP1-2-3 (1862 amino acids), and the second is produced by readthrough of the 'leaky' UGA codon to generate NSP1-2-3-4. Four in-phase stop codons, three of which are in a region corresponding to the 5′ noncoding sequence of the 26S subgenomic RNA, ensure termination of NSP translation. The 3′ one-third of the genomic RNA has an ORF of 3762 nucleotides which encodes the polyprotein C-E3-E2-6K-E1. For the T48 strain the 3′ noncoding region is 524 nucleotides in length. Four closely related sequence blocks, 48–58 nucleotides in length, are repeated in the 3′ noncoding region. The degree of homology between these blocks is striking, but they show no homology with corresponding sequence blocks for alphaviruses other than the closely related Getah virus. RRV strains show marked diversity in the 3′ noncoding region; deletions, insertions, sequence rearrangements and single nucleotide substitutions are found; for example, the length of the 3′ noncoding region in Nelson Bay strain NB5092 is ~348 nucleotides.

The RRV genome has three regions which are strongly conserved in sequence between alphaviruses. These are: (1) a tract of 23 nucleotides next to the 3′ poly(A) tail; (2) 21 nucleotides at the 3′ terminus of the NSP4 gene; and (3) 50 nucleotides near the 5′ end of the NSP1 gene. The RRV genome also contains a moderately conserved sequence element close to its 5′ end. All four sequence elements appear to have roles in the regulation of viral RNA replication. The 23 nucleotide tract is believed to be a promoter for negative-strand synthesis; the complement of the 21 nucleotide element may be recognized, in the negative strand, by the 26S RNA transcriptase.

A full-length cDNA clone of RRV (T48) has been constructed which can be transcribed *in vitro* to produce infectious RNA. Chimeric viruses have been constructed in which the 5′ and 3′ noncoding regions of the RRV and SIN genomes have been exchanged. Virus chimeras containing heterologous 5′ noncoding regions show host-dependent defects in growth; exchange of the 3′ noncoding regions gives rise to virus which grows surprisingly well.

Properties of Virus Proteins

The subgenomic (26S) RNA is translated into the polyprotein precursor of the structural proteins. The 26S RNA, which is co-terminal with the 3′ terminus of the genomic RNA, comprises a 5′ noncoding region of 48 nucleotides, an ORF and a 3′ noncoding region (see earlier). Sequence data predict that C, E3, E2, the '6K' protein and E1 are 270, 64, 422, 60 and 438 residues in length respectively, assuming no post-translational trimming. At the amino acid sequence level the polyprotein is 75 and 48% homologous with the corresponding SFV and SIN polyproteins respectively. The capsid protein is highly basic in its N-terminal half, consistent with a role in interacting with genomic RNA. The N-terminal 10 amino acids of E3 are hydrophobic and presumably form part of a signal sequence for the insertion of p61, the E2 precursor, into the host endoplasmic reticulum.

Three neutralization epitopes on E2, which together make up a significant neutralization site, have been mapped to residues 216 (epitope *a*), 232 and 234 (epitope *b1*) and 246, 248 and 251 (epitope *b2*). These epitopes are flanked in the primary amino acid sequence by asparagine-linked glycosylation sites at residues 200 and 262. This neutralization site is important in early virus–cell interactions, as judged by biological studies on RRV mutants selected during passage in cell culture or mice, or generated using infectious RNA derived from cDNA clones. The existence of a '6K' protein has not yet been reported in RRV-infected cells. The hemagglutinin, E1, is glycosylated at residue 141; it has an uncharged tract (residues 80–96) within an extended region (amino acids 72–130) which is highly conserved between alphaviruses. This region may be involved in fusion with cell membranes during entry of virus into cells. Comparative sequence data predict that the genomic RNA encodes two polyproteins (NSP1-2-3, the major species; NSP1-2-3-4, a minor species) which are processed to four NSPs of 533, 798, 531 and 611 residues respectively.

Replication and Virus Assembly

In cultured vertebrate cells RRV infection is cytopathic. The latent period is 3–5 h; maximum extracellular virus titers (10^8 PFU/ml) and levels of viral RNA synthesis are at 8–10 h, at which time the progressive shut-down of host cell protein synthesis is virtually complete. Virus-specific RNAs formed in BHK and *Aedes albopictus* cells include RF, RI, 45S, 26S and small amounts of 38S and 33S RNA (conformational variants of 45S and 26S RNAs respectively). Seven major virus-specific polypeptides are detected in vertebrate cells: p127, p95, p61 (E2 precursor), p52 (E1), p49 (E2), p32 (capsid protein) and E3. In Vero cells, RRV is commonly found in small cytoplasmic vesicles; 'type 1 cytopathic vacuoles' are observed. There is a pronounced accumulation of nucleocapsids, particularly late in infection.

In cultured *Ae. albopictus* cells at 28°C RRV generates a noncytopathic, persistent infection with peak titers (2×10^7 PFU/ml) at 2–3 days. At 12 and 48 h postinfection, 85 and 5% of cells respectively assay as 'infective centers'. Viral protein synthesis is sustained over the period 10–24 h postinfection but is quantitatively less than in vertebrate cells; no p95 is observed. There is no shut-down of host protein synthesis and cell division rate is unaffected by infection. Virus matures within large electron dense cytoplasmic inclusions and at the cell membrane. Free nucleocapsids are infrequent. When titers decrease during the later stages of infection, inclusions are transformed into microvesiculated vacuoles which may result from fusion with lysozomal vesicles.

Geographic and Seasonal Distribution

An extensive serological survey of neutralizing antibodies in human populations in southeast Asia and the Pacific islands established that, before 1979, RRV occurred only in Australia, the islands of New Guinea, New Britain and the Solomons, with activity in these islands decreasing from west to east. No alphavirus activity was detected in island groups further to the east. Chikungunya virus was the predominant alphavirus west and north of New Guinea through to southeast Asia.

RRV and epidemic polyarthritis occur in every state of Australia. Limited surveys in New Guinea indicate that activity is confined to the lowlands and deep valleys of the central mountain ranges. The South Pacific island epidemics which started in 1979 petered out early in 1981, and, although there have been occasional reports of isolated cases since then, there is no evidence that the virus has become enzootic.

Human infection occurs throughout the year in tropical and subtropical northern regions of Australia, with the highest incidence during the monsoon wet season, December to April. In temperate southern regions sporadic cases have been diagnosed in winter, but substantial outbreaks occur in late summer and autumn. In flood years, major epidemics involving thousands of cases occasionally occur from spring through to autumn in the irrigation towns in the Murray–Darling basin. Cases have occurred on the outer fringes of most of the large coastal cities, but there has not been a major urban epidemic. Localized outbreaks occur in coastal regions of eastern and southern Australia, including Tasmania. In Western Australia, outbreaks are more common on the southwest coast than in the tropical north, which probably reflects human population density.

Host Range and Virus Propagation

The vertebrate host range of RRV in the primary cycle is effectively limited to placental and marsupial mammals. There is a lack of specificity in mosquito vector species; the virus has been recovered from 11 species encompassing five genera. Horses are commonly infected in nature, sometimes resulting in lameness and constitutional or nervous disturbances of varying severity; RRV is suspected as a cause of equine death, but proof is lacking. No other domestic or native animal is known to show signs, but naturally acquired antibodies are found in most mammals, and virus has been recovered from marsupials and from a horse. Antibodies are rarely found in birds, and, in the laboratory, viremia has been demonstrated only in recently hatched chickens. In contrast, viremia has been readily produced in a range of small and large adult marsupial species and in a native rodent.

Newborn and weanling outbred mice have been the most commonly used experimental host, but many readily available cell lines are susceptible to RRV and are used to prepare stocks, and in the assay of virus by plaque forming units or cytopathic effect (cpe). RRV growth in chick embryo cells is generally poor. The C6/36 line of *Ae. albopictus* mosquito cells is the most sensitive available cell line, and is now the usual host for isolation of RRV from human and field material, but as infection cannot usually be visualized by cpe or plaques, they are used in conjunction with Vero or BHK cells. An even more sensitive means for primary isolation of RRV is the intrathoracic inoculation of intact mosquitoes; the most commonly used is the large, nonblood-feeding *Toxorhynchites amboinensis*.

Genetics and Evolution

Naturally occurring variants of RRV, some apparently associated with particular geographic regions, were first identified through observation of biological properties such as pathogenicity in infant mice and serologically determined antigenic variation. These observations were reinforced by the identification of RRV genetic types and subtypes on the basis of differences between *Hae*III restriction digest profiles of single-stranded cDNA to virion RNA, and by sequencing genomic RNA. The examination of 14 isolates of RRV led to the identification of three genetic types (I–III) with an estimated 1.5–5% nucleotide sequence divergence between each type. RRV is not a bird virus, and the relative immobility of mammal hosts may allow the emergence of geographic variants best adapted to available local hosts. However, no clear pattern of geographic distribution of RRV genetic types could be established in this extension of the earlier survey, nor was there an association between RRV genetic type and vector species.

Sequence studies on the genomes of the Townsville T48 prototype (genetic type I), and NB5092, a strain isolated at Nelson Bay on the central coast of NSW (genetic type III), showed 284 nucleotide differences over the viral genome, equivalent to 2.4% nucleotide sequence divergence. Transitions are over four times more frequent than transversions. In the coding region most of the nucleotide differences are silent. There are 36 amino acid differences in the nonstructural proteins (NSPs) and 12 in the structural proteins. The distribution of amino acid differences largely correlates with the location of nonconserved regions in the proteins of alphaviruses such as Sindbis (SIN) virus and Semliki Forest virus (SFV).

Despite abundant evidence of relatively minor variability in field isolates and the ease with which variation can be induced by laboratory manipulation, under conditions of natural selection the RRV genome appears to be remarkably stable. During the first 10 months of the epidemic in Pacific island communities, involving hundreds of thousands of human, domestic animal and mosquito infections, change in the envelope glycoprotein (E2) gene of the April 1979 Fijian RRV strain was confined to a single nucleotide which altered amino acid residue 219. This mutation was first detected in a strain from an American Samoan patient infected in August 1979, and there was no further change in a strain from a Cook Island patient infected in February 1980. Furthermore, in these two isolates no changes were observed in the 3′ noncoding region of the genome which can vary dramatically between RRV isolates (see earlier).

RRV exhibits the high mutation rate noted generally for RNA viruses. Evolution of the virus is presumably moderated by the various selection pressures imposed on all arboviruses: to be successful, a newly generated mutant must be present in high concentration during the period of viremia in the vertebrate host in order to be significantly represented in the small volume of blood imbibed by the feeding insect; it must also have a selective advantage during replication in, alternately, its vertebrate hosts and insect vectors.

Serological Relationships and Variability

RRV shares group- and genus-specific antigens with other alphaviruses. Based on antigenic relationships determined by hemagglutination inhibition (HAI), complement fixation and plaque reduction neutralization tests, RRV is in the SFV complex, and a subtype of Getah virus.

Differences between the surface antigens of geographic variants have been demonstrated with kinetic HAI, complement fixation and neutralization tests. Homologous and heterologous virus/polyclonal antibody kinetic tests with RRV strains collected over a period of 13 years in North Queensland indicated antigenic identity. Similarly, strains collected over 3 years at Nelson Bay, NSW, were antigenically identical by kinetic tests. However, heterologous kinetic tests between Queensland and Nelson Bay viruses and antibodies gave no cross-reaction in HAI, and significantly reduced cross-reaction in complement fixation and neutralization tests. The control cross-reaction tests using standard incubation times indicated that all virus strains were antigenically identical.

Epidemiology

The most important vectors in coastal regions are brackish-water anthropophilic mosquitoes breeding in mangrove or melaleuca swamps. In these habitats *Ae. vigilax* is the dominant vector in tropical and subtropical coastal regions, but in cooler southerly coastal regions, such habitats are shared with, or dominated by *Ae. camptorhynchus*. In the inland regions of Australia the major vector is the summer-breeding *Culex annulirostris*.

RRV persists in a wildlife–mosquito primary cycle, possibly augmented by a low-level transovarial cycle in mosquitoes. The introduction and rapid spread of the virus in the Pacific islands can be rationally explained only by a cycle involving humans and mosquitoes, a proposition corroborated by the demonstration of high viremias in patients. Viremias since detected in Australian patients have been relatively low, but this could be artefactual, and it is probable that viremias commonly occur in humans at levels sufficient to infect feeding mosquitoes. This would be consistent with the explosive epidemics in army camps and the occasional massive epidemics in irrigation districts during years of heightened mosquito activity. Year-round sporadic cases, and the initiation of epidemics, are due to mosquitoes that have been infected in the primary cycle.

RRV is probably enzootic throughout Australia and the lowlands of New Guinea, and incidence of human disease varies with seasonal and climatic factors and with population densities of humans, mosquitoes and native animals.

Pathogenicity

Based on pathogenicity for outbred infant mice there is a continuum of strains of RRV with degrees of mouse virulence ranging from those, such as the prototype T48 strain, which kill all infant mice in 5–6 days, through to those such as the Nelson Bay strains which kill 0–25% of infant mice in 10–12 days with the survivors either subclinically infected or recovering after suffering hindleg paralysis for 10–20 days. These observations led to the investigations outlined earlier.

There is no mortality associated with epidemic polyarthritis, and recognition of differences in the human pathogenicity of RRV strains must rely upon observations of variability in the duration of signs and symptoms and subjective assessment of their severity. In any localized outbreak of epidemic polyarthritis the individual response to infection might range from subclinical, through degrees of incapacity lasting 4–8 weeks, to a so-called chronic form where the patient suffers a series of relapses of reducing intensity at increasing intervals over a period of a year or more. In an outbreak lasting 2 or 3 months, it is far more likely that this is merely due to individual responses to infection with a single RRV genetic type than to a range of types circulating concurrently. However, in published reports of discrete epidemics there are differences in both the incidence and the duration of signs and symptoms. For instance, there are differences in the incidence, severity and persistence of rash; in the presence or absence of muscle pain as well as joint pain; in the average duration of incapacity; in the occurrence of relatively rare signs such as buccal and palatal enanthems; and in the correspondence of disease onset with viremia or antibody production. Disease expression in the 1956 epidemic in the Murray Valley seems to have been generally much milder than, say, during the 1984 epidemic at Griffith, in the same general region, or during the series of Pacific island epidemics, 1979 to 1981. An outbreak in the Riverland region of south Australia in 1976 was of intermediate severity. It is possible that this interepidemic variation is due to the involvement of different strains of RRV.

Clinical Features of Infection

There are three major manifestations of the disease: rheumatic, rash, and constitutional. Arthritis sometimes begins insidiously, but usually develops very rapidly; the most common combination of signs is pain on movement, tenderness and slight swelling. Wrists are most frequently involved, followed closely by knees, ankles and fingers. Elbows, toes and tarsal joints are also commonly affected. Pain is often more intense than indicated by observed signs. Rash occurs in about two-thirds of patients, but is rarely the sole manifestation of the disease. Most commonly it appears as erythematous macules and papules 1–5 mm across, distributed sparsely to thickly on trunk and limbs, and less frequently on face and scalp. Appearance on palms, soles and digital webs is characteristic, particularly when there is no rash elsewhere. Hyperesthesia of the palms frequently accompanies the rash. Scattered purpura may be found, usually on the feet and lower legs. There is usually no discomfort due to the rash, and it resolves within 7–10 days. Pyrexia, one of the most common of the constitutional effects of other virus infections, is usually absent or slight. Myalgia is common, and carpal tunnel syndrome can be induced or exacerbated. Fatigue is the most consistently apparent constitutional effect and seems to be independent of other manifestations. It has been proposed that the last manifestation might extend into a 'postviral syndrome' but this has not been clearly established.

Pathology and Histopathology

In the early stages, cells in the synovial fluid and joint effusions are predominantly mononuclear and remarkable for the proportion of mitotic figures and highly vacuolated and phagocytic macrophages. In later effusions, macrophages appear less activated and small lymphocytes predominate. RRV antigen has been detected by immunofluorescence on the surface of 20–30% of the larger cells in synovial fluid during the first few days after onset of symptoms, but attempts to isolate virus have failed. There is no erosion nor permanent derangement of joints. In a minority of cases, relapses occur over a year or more; these gradually decline in incidence and intensity until final resolution.

The histology of the rash is variable. The dermis shows a chiefly perivascular mononuclear cell infiltrate, vasodilation, and varying degrees of edema. Diffuse to dense erythrocyte extravasation is quite common. Histologically detectable changes in the epidermis are present in about half the cases, although rarely recognized macroscopically. It is not clear whether rash is due to the direct action of virus or is the result of immunological processes.

Signs and symptoms of epidemic polyarthritis can be confused with those of rubella infection. There is no evidence that RRV is teratogenic, so, if termination is being considered in first trimester pregnant women, it is important to differentially diagnose cases of polyarthritis with rash.

In comparison with other alphaviruses there are unusual aspects in the pathology and pathogenesis of RRV in laboratory mice, but the murine disease appears to be very different to that in humans and is thus of little relevance here.

Immune Response

The detection of RRV antibodies is routinely performed by ELISA, although their first appearance can usually be detected several days earlier by standard alphavirus HAI tests. ELISA can also be used in antibody class capture assays to detect IgM, but precautions must be taken to avoid false positives due to the presence of rheumatoid factor which is not causally present in epidemic polyarthritis. Specific IgM often persists for many months so is not a reliable indicator of recent infection; as with many other arbovirus infections, a rising titer of IgG antibodies from acute to convalescent stages is a more reliable diagnostic tool.

Signs and symptoms can persist in the presence of antibodies.

Although appropriate investigations have not been carried out in humans, it is likely that infection bestows immunity to all genetic types of RRV.

Prevention and Control

As with other zoonoses involving wildlife as vertebrate hosts, particularly those that are vectored by insects, there can be no prospect of eradicating RRV. A degree of control can be achieved by reducing the interaction of vectors and humans through education, and, at the community level, by carrying out mosquito abatement programs appropriate to the district. At the personal level, the avoidance of mosquito attack can be achieved by remaining indoors during periods of maximum vector mosquito activity, and by the use of efficient repellants.

Vaccination against a disease which is not life-threatening and is without permanent sequelae can only be administered on a request basis. Infected children rarely express signs and symptoms, which, superficially, augurs well for the development of a live virus childhood vaccine. However, little is known about the persistence or nature of immunity after subclinical or frank RRV infection, nor whether the wide range of individual responses to infection, including long-term relapses, is related in any way to the prior immune status of the patient. Before developing candidate vaccines, prospective studies should be carried out to assess the duration of effective immunity following natural infection.

See also: Diagnostic techniques; Pathogenesis.

Further Reading

Fraser JRE (1986) Epidemic polyarthritis and Ross River virus disease. *Clin. Rheum. Dis.* 12: 369.

Marshall ID and Miles JAR (1984) Ross River virus and epidemic polyarthritis. In: Harris KF (ed.) *Current Topics in Vector Research*, p. 31. New York: Praeger.

Peters CJ and Dalrymple JM (1990) Alphaviruses. In: Fields BN *et al.* (eds) *Virology*, 2nd edn, p. 713. New York: Raven Press.

ROTAVIRUSES

Contents

General Features
Molecular Biology

General Features

Robert D Shaw
State University of New York at Stony Brook
Stony Brook, New York, USA

and

Harry B Greenberg
Stanford University Medical School
Stanford, CA, USA

History

Diarrhea has long been recognized world-wide as a leading cause of morbidity and mortality, but the search for important etiologic agents of human disease was not fruitful until the early 1970s. Rotavirus was identified as a mouse diarrheal pathogen in the 1950s and as a simian and bovine pathogen in the 1960s, but it was not until 1972 when Kapikian and co-workers, using immune electron microscopy to identify Norwalk virus in diarrheal stools that a virus was positively implicated as a cause of human gastroenteritis. The following year, investigators in several locations identified human rotavirus in intestinal biopsies and diarrheal stools of children. During the past two decades rotaviruses have been identified as the leading cause of severe gastroenteritis in infants and children. Progress in the study of rotaviruses was enhanced with the advent of efficient *in vitro* cultivation of many human strains in the early 1980s. Within the past few years, the successful cloning of the rotavirus genome and expression of individual rotavirus proteins in recombinant systems has further enhanced detailed knowledge of rotavirus structure and function, as well as many aspects of rotavirus serology, immunity and pathogenesis.

Taxonomy and Classification

The *Rotavirus* genus is contained within the family *Reoviridae*, which also contains the genera *Reovirus*, *Orbivirus*, *Phytoreovirus* and *Fijivirus*. Rotaviruses share several common features that form the basis of the classification. Viral particles are approximately 70 nm in diameter and consist of two protein capsids (creatively identified as outer and inner). The extracellular icosahedral particles are not enveloped by a lipid membrane. The core of double-stranded RNA (dsRNA) is organized into ten mono- and one bi-cystronic genomic segments. The negatively stained electron microscopic appearance of the complete rotavirus particle is responsible for the name rotavirus, which is derived from the Latin word *rota*, meaning wheel. The outer capsid appears as a sharply defined rim, to which spokes appear to radiate from a large central hub. Recent advances in cryoelectron microscopy with computer-enhanced reconstructed images has provided a more detailed view of rotavirus structure (Fig. 1). Notable features include the presence of spikes on the outer capsid which extend over 10 nm and pores that penetrate from the virion surface into the viral genome.

Geographic and Seasonal Distribution

Rotaviruses are ubiquitous among humans and many animal species throughout the world, and are usually important causes of gastroenteritis wherever they occur. Infection in humans predominates in the cooler months in developed countries, usually peaking January and February, while they are unusual causes of gastroenteritis in the summer months. In the US, rotavirus infections are detected in a wave, starting in the Southwest in November and spreading on to New England and the Canadian Maritime provinces in March. This seasonality is not seen in tropical climates (10° latitude from the equator) where rotavirus infections occur year-round.

Host Range and Virus Propagation

Rotaviruses have been recovered from diarrheal stools shed by a multitude of animal species. As an indication of the breadth of the host range of Group A rotaviruses, some species other than human that are infected include simian, equine, porcine, canine, feline, lapine, murine, bovine, ovine and avian, although rotavirus is not an important cause of disease in all of these species. Among laboratory animals, Group A rotaviruses are not known to infect guinea pigs or rats, although the latter are infected by Group B rotaviruses. Animal strains, even those with serotypes indistinguishable from human strains, rarely infect humans in nature. Recently, a few human rotavirus isolates probably derived from feline, canine or bovine rotaviruses have been described. Also, remarkable antigenic similarities between some porcine and human strains have been reported. The potential of animal rotavirus reservoirs as a source of genetic diversity for the evolution of new human strains is unknown.

Rotaviruses were first propagated *in vitro* in 1963, when Mahlerbe and co-workers reported the isolation of simian SA11 from a vervet monkey kidney cell culture. Human rotaviruses were not successfully cultivated until the discovery that trypsin exposure, which cleaves a single site of the VP4 protein, dramatically enhances the ability of the virus to grow in tissue culture. Group B and C viruses have not generally been cultured successfully, with the exception of a single porcine and bovine Group C strain. Group A rotaviruses are usually cultivated in simian kidney cell lines in the presence of trypsin. The most efficiently cultured animal rotaviruses typically yield 10^7–10^8 plaque-forming units (PFU) per ml in tissue culture systems. Human viruses tend to produce one to two orders of magnitude fewer PFU in tissue culture, although they are often shed in diarrheic stools in quantities of 10^9 per ml.

Genetics

The dsRNA segments of the rotavirus genome have

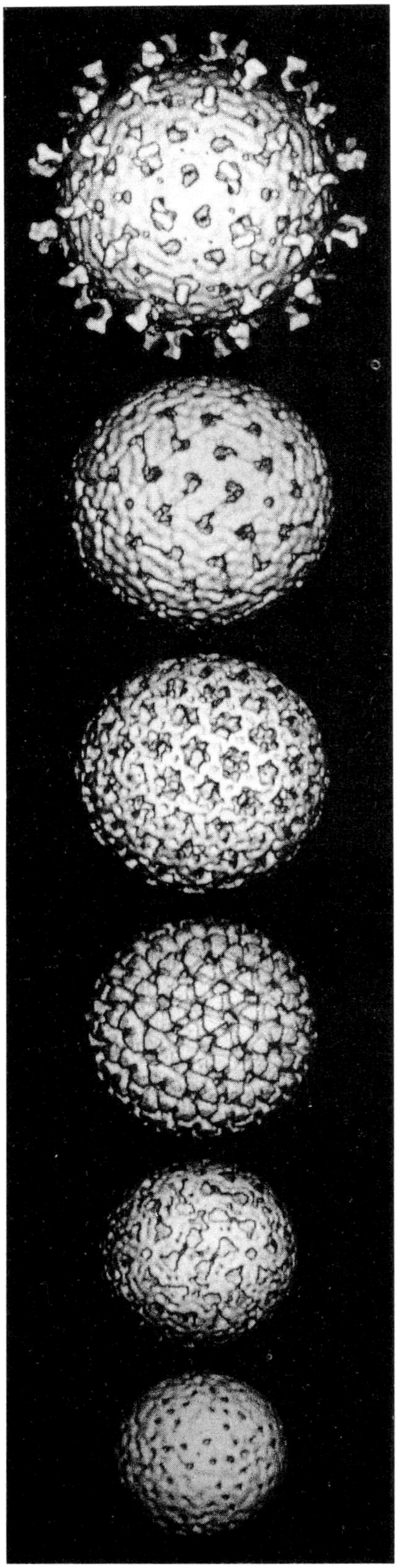

masses between 2×10^5 and 2.2×10^6 daltons. The genes are distributed by size into four classes that produce a characteristic pattern when the segments are separated by polyacrylamide gel electrophoresis (the 'electropherotype'). Electropherotype classification of rotaviruses was particularly important when human rotaviruses could not be cultivated *in vitro*. The wide variability of electropherotypes and the observation that serotypes and electropherotypes do not correlate well has led to a reduced importance of this technique in classification. However, detection and characterization of rotavirus electropherotypes remains useful for epidemiologic studies and for detection of nonGroup A rotaviruses.

Rotavirus RNA lacks 3′-terminal polyadenylated sequences and contains 5′-capped structures. Both ends of each segment contain short highly conserved regions of approximately eight nucleotides. It is speculated that these highly conserved sequences may be of importance in transcription, replication and assortment of the viral genome. These features are also characteristic of other *Reoviridae*.

The RNA itself is not infectious; rotaviruses contain within the single-shell particle an endogenous RNA-dependent RNA polymerase which transcribes the gene segments into mRNA. Transcripts are full-length positive-strands from which negative-strand synthesis occurs following the formation of replicase particles in the cytoplasm. RNA tran-

Fig. 1. Three-dimensional structure of rhesus rotavirus by cryoelectron microscopy and icosahedral image reconstruction (Yeager, *et al.* (1990) *J. Cell Biol.* 110: 2133). The surface-shaped representations were obtained by truncating the three-dimensional maps with spherical envelopes to reveal the internal structure. The 6 structures from top to bottom with their corresponding diameters (in parenthesis) are as follows: (**a**) The virion outer capsid surface displays 60 spikes attributed to the VP4 hemagglutinin (1020 Å) (**b**) The smoothly-rippled outer capsid surface, attributed primarily to VP7, is perforated by 132 aqueous holes (790 Å). (**c**) The space between the outer and inner capsids forms an open aqueous network that may provide pathways for the diffusion of ions and small regulatory molecules as well as the extrusion of RNA (720 Å). (**d**) The inner capsid has a 'bristled' outer surface composed of 260 trimeric columns, attributed to VP6 trimers (660 Å). (**e**) The VP6 trimers merge with a smooth, inner capsid shell which is perforated by holes in register with those in the outer capsid (580 Å). (**f**) A third protein shell (referred to as the 'core') is thought to be formed by VP1, VP2 and VP3 and encapsidates the dsRNA segmented genome (530 Å).

scripts can be identified 3 h postinfection. The proteins and structural requirements of RNA replication are not fully understood. Reassortment of gene segments occurs at high frequency during mixed infection with two or more rotavirus strains, although, unlike influenza, there is very little evidence that this is a mechanism for generation of serotypic diversity in nature.

There was initial confusion about the precise gene coding assignments for some of the rotavirus genes, but the assignments are now clear. Genes 5, 7, 8, 10, and 11 code for nonstructural proteins for which roles have not yet been conclusively determined. Gene 5, 7, and 8 products (NS53, NS35, and NS34 respectively) appear to be involved in RNA replication while the gene 10 product (NS28) appears to function as a receptor for single shell particles as they bud through the endoplasmic reticulum. Genes 1–4, 6, and 9 code for structural proteins VP1–4, 6, and 7, respectively (in most Group A viruses). VP4 and VP7 are the two surface proteins of the virion while VP6 is the major constituent of the single shell particle.

The electropherotypes of nonGroup A rotaviruses differ from Group A viruses primarily in gene segments 7, 8 and 9. The organization of the nonGroup A 7–9 genome segments lacks the tight triplet formation in polyacrylamide gel electrophoresis that characterizes the Group A electropherotype. NonGroup A rotaviruses are morphologically identical to Group A strains but they are antigenically distinct and they do not cross-hybridize in Northern blot analyses, even under conditions of low stringency. There appear to be at least five (B–F) nonGroup A rotavirus types. Groups B and C have been identified in man as well as in animals. Sequence data are now available for several nonGroup A rotaviruses and this has allowed some gene coding assignments to be made. The recent development of serologic reagents from expressed viral proteins will extend the understanding of the epidemiology and importance of nonGroup A viruses.

Evolution

Little is known about the evolution of rotaviruses. While strong similarities in genomic sequences exist among some human and animal strains, there remains little evidence as to the origin of any particular strain. High rates of mutation exist in RNA genomes, and the capacity of the segmented genome to reassort during mixed infections provides theoretical opportunities for rotaviruses to mutate and evolve rapidly. However, during multiple passages in cell culture without the presence of obvious selection pressure, rotaviruses do not appear to change appreciably.

Serologic Relationships and Variability

Six of the viral proteins are structural, but only three have played important roles in the classification of Group A rotaviruses by virtue of antigenic or functional properties. NonGroup A rotaviruses have not been widely cultivated *in vitro* and far less is presently known about the structure and serologic classification of these viruses. VP6, the major structural protein of the Group A inner capsid bears most of the common group antigens as well as the subgroup antigens. All Group A rotaviruses share antigenic determinants on VP6, and some shared epitopes appear to be present on group C rotaviruses as well. The glycoprotein VP7, the major neutralization protein of the outer capsid, induces serotype-specific neutralizing antibodies. VP4 is the other outer capsid protein. It is the viral hemagglutinin, an important determinant of virulence, and is involved in viral entry into cells. VP4 is also responsible for inducing neutralizing antibodies; but it is present in the outer capsid in smaller amounts than VP7. The contribution of VP4-specific neutralizing antibodies seems to be less important than VP7-specific neutralizing antibodies in hyperimmune sera used in determination of viral serotypes, but evidence delineating the relative roles of these antibodies in natural infections and following vaccination in many species is conflicting. A combined role of VP7- and VP4-specific immunity in the serologic classification of rotaviruses (such as the Influenza binary system using hemagglutinin and neuraminidase antigens) has been recently adopted. In this system the VP7 serotype is classified as a G (glycoprotein) type and the VP4 type as a P (protease) type. Reagents for classification of VP4 are not yet readily available.

Rotaviruses that cause most human disease are classified as Group A strains but occasionally Group B and C also infect man (see Epidemiology below). Animal viruses have been classified into Groups A–F. Human and animal Group A rotaviruses are subdivided into subgroups (I and II) on the basis on serologic reactivity with subgroup-specific monoclonal antibodies directed at VP6. At least 14 Group A G-types have been identified, eight of which have been isolated from man (1–4, 8–10, and 12) but four of which (only serotypes 1–4) cause most human disease. Initially, serotyping required tissue culture growth of the viruses in the presence of defined serotype-specific sera, but serotype-specific monoclonal antibodies as well as serotype-specific nucleic acid probes are now readily available and have dramatically reduced the time and expense of the serotyping procedure. At least four P types have been identified in human isolates and the number of P types found in animals has not been well-studied. For example, at least three bovine P types have been found. The number of serotypes is steadily increasing as larger numbers of isolates are tested. Although it is assumed that the serotype classification of rotavirus strains is important in determining immunity to these viruses, the relationship of serotype to protective immunity is not entirely clear.

Epidemiology

Group A rotaviruses are the principal cause of severe gastroenteritis in infants and young children throughout the world, accounting for 12–71% (median 34%) of all diarrheal episodes requiring hospitalization in children under the age of 2. In developing countries, the annual toll includes roughly 18 million cases of severe diarrhea and nearly a million deaths. In the US, over 70 000 children are hospitalized, and about 125 deaths are annually attributed to rotavirus infection.

Infants in the first 2–3 months of life seem to be relatively protected from severe rotavirus disease, probably because of residual maternal immunity mediated by transplacental antibodies. Beyond 3 years of age and into adult life, rotavirus infections occur but they are typically mild or asymptomatic. However, Group A rotavirus may occasionally cause severe diarrhea in immunologically competent adults. Epidemics are known to occur in institutional settings, especially among the elderly in nursing homes, and may rarely result in a fatal illness.

The epidemiology of rotavirus infections and viral shedding can be monitored by serologic assays or by electrophoretic patterns of viral RNA (electropherotype) obtained from fecal specimens. Electropherotype studies have demonstrated that several genomic patterns may co-exist within a community, but these patterns do not necessarily correlate with serotypic classifications. Several serotypes may also coexist within a community, but each season is usually dominated by a single serotype that may vary from year to year. Group A serotypes 1–4 cause most human disease and appear to be equally virulent. These serotypes have been identified in developed and underdeveloped settings and have been in circulation for at least thirty years.

Transmission and Tissue Tropism

Rotaviruses are transmitted by the fecal–oral route, as has been conclusively demonstrated by transmission of illness in human volunteers by oral inoculation of fecal filtrates. Transmission by the respiratory route has been considered but the evidence to support this route is weak. Rapid appearance of antibodies to rotaviruses is noted by 3 years of age in all areas of the world regardless of hygiene. Viral shedding is not always associated with symptoms, and asymptomatic infection occurs frequently in newborn nurseries and day care centers. The virus is quite stable on environmental surfaces for prolonged periods. These factors complicate efforts to control hospital outbreaks, which are not always successful even if patients are carefully monitored for virus excretion and appropriate control measures followed.

Rotaviruses replicate within and are primarily shed from mature small intestine epithelial cells located at the villous tips. Recent evidence of rotavirus infection of the liver in a small number of immunocompromised children has demonstrated that infection of other tissues is possible in some rare cases.

Pathogenicity

The genetic correlates of rotavirus pathogenicity have not yet been determined. Host range restrictions limit cross-infection between species in most cases. Animal strains have been used as human vaccine candidates as they possess antigenic similarity with human viruses but do not cause disease, except when given in very large doses. Animal rotavirus vaccine candidates appear to replicate in man at a low level and stimulate local and systemic immunity. All serotypes of rotavirus seem to be equally virulent, although some neonatal strains have been associated with asymptomatic infection. Other studies, however, have indicated that it is the new-

Table 1. Clinical features of rotavirus gastroenteritis

Symptom	Frequency (%)
Diarrhea	98
Diarrhea > 10 times daily	28
Vomiting	87
Vomiting > 5 times daily	51
Fever	84
Abdominal pain	18
Blood in stool	1
Hospitalization	39

Adapted from Uhnoo *et al.* (1986) *Arch. Dis. Child.* 61: 732.

born host rather than the rotavirus strain that determines the avirulent phenotype in the newborn.

Clinical Features of Infection

Rotavirus gastroenteritis is seen most commonly in children between the ages of 3 months and 2 years. Asymptomatic infection is common in infants of less than 2 months or individuals older than 2 years, although episodic severe disease in adults may result from Group A rotavirus infection. Group B rotavirus infection causes epidemics in older children and adults in China.

In some animal species, infection is strictly limited to the very young. Age-related restriction of rotavirus infection appears to be due, at least in part, to nonimmune mechanisms. The incubation period in humans and most animals appears to be 24–72 h. Malnutrition may increase the severity of the symptoms. In addition to the symptoms listed in Table 1, those related to severe volume depletion such as lethargy, irritability, confusion and eventually vascular collapse and death can be seen.

Pathology and Histopathology

The pathologic lesion resulting from rotavirus infection varies somewhat depending on the species and age in question. For instance, porcine rotavirus causes a particularly large amount of cellular damage while murine infection may be characterized by much more selective destruction. Infection of the very young of most species will characteristically produce more cell destruction than infection in adults. Blunting of intestinal villi and vacuolation of enterocytes may be seen within hours of infection, prior to the presence of detectable viral antigen. Also seen are mononuclear cell infiltration of the lamina propria, distended endoplasmic reticulum, mitochondrial swelling and denuded microvilli. Viral particles may be seen within columnar epithelial cells, goblet cells, phagocytic cells and M cells in the small intestinal mucosa (the colon is generally spared). Production of viral antigen in the intestine peaks around 48–72 h postinfection in most species. Large amounts of viral proteins accumulate in the cytoplasm (viroplasm), which may appear swollen and vacuolated. However, some damaged cells may be seen without detectable viral antigen present. Intestinal cellular morphology returns to normal in about 7 days, although much of the severe damage is repaired as quickly as 3 days after infection.

Adenylate cyclase is not activated by rotavirus infection. Water absorption of the small intestine is impaired, but can be corrected by the administration of glucose–salt solutions. Abnormal motility may contribute to rotavirus-induced diarrhea. Also, carbohydrate malabsorption and secondary osmotic diarrhea may occur. An integrated understanding of the roles of these many factors in the pathogenesis of diarrhea has not yet been achieved.

Group B and C rotaviruses also cause small intestinal lesions in several animal species as well as in humans. Villous blunting is seen in various small intestinal regions. Syncytia including up to 20 enterocytes are seen in Group B infection, a finding not observed in Group A infections.

Immune Response

The antibody-based immune response to rotavirus infection has been studied in many animals as well as in humans. Serum and mucosal antibodies are detected beginning several days after primary infection. Recently, cytotoxic T cells have been identified in the intestinal mucosa of mice undergoing experimental rotavirus infection. The rapid resolution of rotavirus diarrhea during an acute infection occurs somewhat before the immune response is fully developed, so at least some of the factors responsible for resolution of the illness are nonimmune. Immune factors are most likely to have substantial roles in the prevention of subsequent infections, although it is still unclear precisely which factors determine susceptibility to rotavirus infection.

Genetic studies using specific viral reassortants and passive transfer studies using monoclonal antibodies directed at specific rotavirus proteins have demonstrated that antibodies directed at either VP4 or VP7 (but not other rotavirus proteins) can neutralize virus and protect susceptible hosts. However, the bulk of antibodies elicited by rotavirus infection are directed at the major structural protein of the inner capsid VP6 (which constitutes 51% of the viral mass) and seem to have no functional role in the modulation, prevention or resolution of an infection.

The locations of the amino acid-defined regions of VP4 and VP7 that elicit neutralizing antibodies have been mapped. One large and complex conformationally determined neutralization domain exists on VP7 and at least two domains are found on VP4 (one on each of the two fragments resulting from trypsin cleavage of VP4, which are referred to as VP5* and VP8*). Most neutralizing antibodies elicited by VP7 are serotype-specific, although at least one epitope is heterotypic and binds antibodies that are broadly cross-reactive. VP4 serotypic diversity is still not well understood. There are at least two important neutralization regions on VP4, one on either side of the site of trypsin cleavage (cleavage of this site enhances growth in tissue culture; see under Host Range and Virus Propagation). The N-terminal fragment, VP8*, contains neutralization sites that are limited to particular strains. The C-terminal fragment VP5* has a domain that is cross-reactive among several human strains, and a similarly cross-reactive region is shared among several animal strains.

Primary rotavirus infection induces neutralizing antibodies to VP7 and VP4. There is conflicting evidence in various animal models and in humans concerning the relative importance of these two groups of antibodies in the establishment of protection against subsequent infections. Prevailing opinion at the present time could be simplified to state that a primary human infection results in predominantly serotype-specific immunity, although heterotypic immunity is frequently detected at lower levels. Individuals gradually establish broader immunity with re-infections, although whether this is due to accumulated diversity of homotypic responses to serial VP7 exposures or a gradual increase in the immune response to the major heterotypic regions on VP4 or VP7 remains unknown.

The complexities of the intestinal immune environment have impaired the development of a complete understanding of the immune response to rotavirus infection and vaccination. The discrete immunological environment of the intestinal mucosa is relatively difficult to monitor by either serum or intestinal fluid measurements. Animal studies have confirmed that most of the specific anti-rotavirus antibodies generated in response to infection are immunoglobulin A (IgA), which is secreted predominantly by lymphocytes in the small intestinal lamina propria. It has long been inferred from animal studies that replication of the virus in the intestinal tract is a prerequisite for the development of substantial local immunity. While infectious rotavirus administered directly into the intestinal tract has been demonstrated to be a powerful mucosal antigenic stimulus to specific antibody formation, it is not yet clear that replication is vital; or if it is, the mechanism by which replication enhances the response is not determined. Furthermore, rotavirus-specific cytotoxic lymphocytes have been identified in the intestinal mucosa following parenteral administration of killed rotavirus. In immunodeficient or normal murine model studies, passive transfer of immune cytotoxic T cells has been shown to prevent acute rotavirus infection and resolve ongoing rotavirus infection.

The mechanisms of rotavirus antigen processing and presentation in the mucosal immune compartment are largely unexplored. Rotaviruses, like many other particulate antigens including reoviruses, are known to bind to and be internalized by M cells overlying intestinal lymphoid aggregates, at least in a porcine model. The importance of this route of contact with immune cells, as opposed to penetration of virus directly into the lamina propria or presentation of viral antigens by major histocompatibility complex (MHC) Class I or II-bearing enterocytes, is unknown.

Prevention and Control of Rotavirus

Two avenues to prevention and control of rotavirus disease are vaccination and oral rehydration therapy. Treatment with oral rehydration solutions containing glucose and electrolytes is highly effective for ameliorating the consequences of rotavirus infection, but there are serious logistic, cultural and educational difficulties limiting the distribution of this treatment resource into underdeveloped areas. Recent evaluations of this approach in the US suggest that it is underutilized even in a developed setting.

Breast-feeding has been advocated as an inexpensive and effective means of rotavirus disease suppression, as breast milk is effective in the reduction of morbidity and mortality caused by bacterial gastroenteritis. However, despite the presence of anti-rotavirus antibodies in breast milk, breast-feeding has not consistently been shown to protect against rotavirus infection or serious rotavirus disease.

Vaccination strategies that have been tested have utilized the host range restrictions of animal rotaviruses in a 'Jennerian' approach to disease prevention. For example, simian and bovine rotaviruses have been used as naturally attenuated vaccine strains in children. Field trials demonstrated these vaccines to be safe and immunogenic, but efficacy seemed to be limited. Animal and human/animal reassortments or reassortments containing human VP7 genes on a simian or bovine background have induced both homotypic and heterotypic protection in some studies, but unfortunately in other circumstances, these vaccines have failed to protect in both developed and less developed settings. These findings have been encouraging, but further investigation will be required before an optimal vaccine strategy can be adopted that will routinely provide reproducible protective efficacy.

Future Perspectives

Vaccine strategies currently under consideration are varied. Both nonpathogenic neonatal rotavirus strains and quadrivalent collections of reassortant rotaviruses are presently undergoing intensive testing. Vaccines made from synthetic viral proteins or inactivated or recombinant virions administered systemically or enterally are planned or under investigation, but the ability of these constructs to stimulate protective immunity is in question.

A better understanding of the mechanisms of viral antigen processing and presentation, the determinants of the magnitude and specificity of the antibody and cytotoxic T cell responses, and a more precise determination of the mechanisms of naturally occurring protective immunity may permit a more efficient and effective design for synthetic vaccine products that will elicit sufficiently broad protective immunity.

See also: Enteric viruses; Norwalk and related viruses; Pathogenesis; Reoviruses; Vaccines and immune response.

Further Reading

Bellamy AR and Both GW (1990) Molecular biology of rotaviruses. *Adv Virus Res.* 38(1): 1.

Blacklow NR and Greenberg H (1991) Viral gastroenteritis. *N. Engl. J. Med.* 325: 252.

Estes M (1990) Rotaviruses and their replication. In: Fields BN *et al.* (eds) *Virology*, 2nd edn, p. 1329. New York: Raven Press.

Kapikian AZ and Chanock RM (1990) Rotaviruses. In: Fields BN *et al.* (eds) *Virology*, 2nd edn, p. 1353. New York: Raven Press.

Ramig R (ed.) (1991) Rotaviruses. *Curr. Top. Microbiol. Immunol.*

Theil K (1989) Group A rotaviruses. In: Saif LJ and Theil KW (eds) *Viral Diarrheas of Man and Animals*, p. 36. Boca Raton, FL: CRC Press.

Molecular Biology

Mary K Estes
Baylor College of Medicine
Houston, Texas, USA

Properties of the Virion

Rotaviruses are members of the *Reoviridae* and are characterized by a double-stranded RNA (dsRNA) genome and a nonenveloped icosahedral structure (Fig. 1). The rotaviruses were named (from the latin *rota*, meaning wheel) on the basis of the distinctive morphologic appearance of particles visualized by negative-stain electron microscopy (EM). Viral particles resemble a wheel, with short spokes and a well-defined rim. The virion contains six structural proteins (VP1, VP2, VP3, VP4, VP6 and VP7) which make up a triple-layered protein shell. In the interior of particles is a proposed, but still poorly characterized, structure called the 'subcore' which contains VP1, VP3 and the genomic dsRNA. Surrounding the subcore is the innermost shell which is composed of VP2. VP2 shells are also called core particles. The addition of VP6 to core particles results in structures called single-shelled particles. These particles possess an active transcriptase activity. The outer protein shell is composed of VP4, which forms spikes that emanate from the rippled surface of the particle which is composed of a glycoprotein, VP7.

Three-dimensional reconstructions of rotaviral particles using images of particles embedded in vitreous ice have provided the most detailed description of particle structure (see Fig. 1). The outer shell has a diameter of 76.5 nm, the second shell is 70.5 nm and the inner core shell is 50 nm in diameter. The two outer icosahedral layers have a $T = 131$ symmetry. The most distinctive feature of the outer shell is the presence of 60 spikes at least 10 nm in

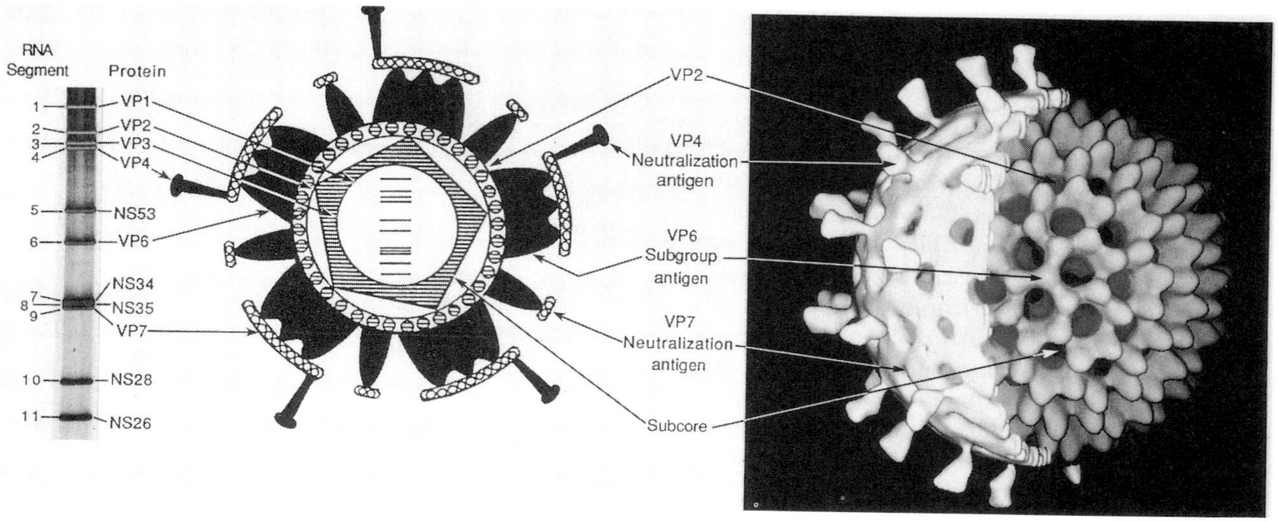

Fig. 1 Rotaviral genes, proteins and structure. The left panel shows the RNA segments and the gene coding assignments for the simian rotavirus SA11. See Table 1 for details on the genes and proteins. The location of the viral structural proteins in the different shells of the viral particles is shown in the schematic in the center. The major structural proteins that make up the outer shell of particles (VP4 and VP7), the inner shell (VP6) and the core shell (VP2) and the still poorly characterized subcore region are illustrated in the three-dimensional structure of particles shown in the right panel. This structure at 3.5 nm was kindly provided by BVV Prasad.

length, which extend from the particle surface. These spikes are made up of dimers of VP4, the protein product of genome segment 4. These spikes are situated at the edge of a subset of three types of 132 channels which lead from the viral surface to the center of the virion. The channels are likely involved in importing the metabolites for RNA transcription and exporting nascent RNA transcripts for subsequent viral replication processes. Cleavage of VP4 is associated with enhanced viral infectivity. This proteolytic cleavage results in the appearance of two products, VP5* and VP8*, which both remain associated with the virion. This effect is thought to be important in viral penetration into cells. The VP6 shell has a bristle-like structure composed of trimers of VP6. These trimers have a central indentation or hole and 132 channels lie in register with the channels in the outer capsid.

Properties of the Genome

The sequence of the genome of only one rotavirus strain (the simian rotavirus SA11) has been determined completely. The following summary of information about the genome is based on data obtained primarily for the Group A rotaviruses, of which SA11 is the prototype strain. The viral genome consists of 11 RNA segments which range in size from 667 (segment 11) to 3302 (segment 1) bp, with the total genome containing 18 556 bp (Table 1). The RNA segments are thought to be encapsidated in association with protein molecules. Each RNA segment encodes one protein with the exception of gene 11 which appears to encode two proteins.

Each genomic RNA segment contains a methylated cap 5′-sequence $m^7GpppG^{(m)}GPy$ followed by a 5′-nontranslated sequence, an open reading frame coding for the protein product, another set of noncoding sequences, and ending with a 3′-terminal cytidine. A poly(A) tract is not found in the genomic RNAs and viral RNA transcripts are not polyadenylated. The 5′ and 3′ ends of each genome segment contain terminal consensus sequences of 7–10 nucleotides. These consensus sequences are present in each RNA segment as part of the noncoding sequences. The 5′- and 3′-terminal consensus sequences are unrelated, implying that they have functional differences. The terminal consensus sequences are assumed to be important *cis*-acting signals, which presumably include, at the 3′ ends, the viral promoters. The termini are also thought to contain sequences important in packaging and in the regulation of rotavirus gene expression at the levels of transcription, replication and translation.

Properties of the Proteins

Descriptive properties of the six structural and five nonstructural proteins are quite well characterized (Table 1). Biochemical and antigenic information about the structural proteins is quite extensive, but molecular mechanisms of how these proteins function remain unknown. The two outer capsid proteins (VP4 and VP7) are the proteins studied in most detail, because of their expected important roles in viral replication and in the induction of protective immunity. Indeed, VP4 and VP7 have each been shown to induce antibodies with neutralizing activity, and they function to mediate early events (attachment and penetration) in the replication cycle. Viral serotypes are defined on the basis of antigenic properties of VP4 and VP7. VP7 serotypes (also called G for glycoprotein types) are now easily characterized and classified using monoclonal antibodies (MAbs). VP4 types (also called P for protease types) are still being characterized and VP4 typing MAbs are not yet available. VP4 and VP7 also interact with one another in still undefined ways, and these interactions can affect specific biologic and antigenic properties of virions. VP6 is the most immunogenic protein on the basis of ease of detection of antibodies to this protein following infection and the most sensitive diagnostic assays are based on detection of this protein or antibodies to it. The minor subcore proteins, VP1 and VP3, are part of the RNA-dependent RNA polymerase activity associated with single-shelled particles. VP1 may function as the polymerase in association with VP3, which is the guanylyltransferase responsible for capping the newly made transcripts. VP2 is an RNA-binding protein that can self-assemble into core-like particles. It is unclear whether VP2 has an active role in the transcription process. VP2 is needed for replication of genomic RNA, and it may be one of the key proteins responsible for the packaging of the nascent RNA segments into newly forming particles in infected cells. VP6 is needed for transcriptase activity. However, it is not known whether VP6 simply functions as a structural component required to keep VP1, VP2, VP3 and the RNA segments in the proper conformation to permit transcription or if it actively participates in the transcription process. On the basis of electron cryomicrograph studies, it is estimated that particles contain 180 molecules of VP2, 120 molecules of VP4, and 780 molecules each of VP6 and VP7.

The nonstructural proteins, found only in infected cells and not in mature virions, function in genome replication and virion assembly. However, the precise role of only one of the nonstructural proteins (NS28) in these processes is currently clear. Several of the nonstructural proteins involved in RNA replication may function as complexes. The nonstructural glycoprotein NS28 is a unique protein which functions in the usual rotavirus morphogenesis pathway which includes transport of subviral particles across the membrane of the endoplasmic reticulum (ER), and acquisition and subsequent loss of a transient envelope on particles. This process culminates in the assembly of the outer capsid glycoprotein onto particles in the lumen of the ER.

Physical Properties of the Virions

The three forms of particles (double-shelled, single-shelled and cores) are easily distinguished by EM and these also can be separated by centrifugation in gradients of cesium chloride (CsCl). The physical property of virions used most often for purification is virion density. In CsCl gradients, virions are relatively stable and viral infectivity is stable to extremes of pH (3.5–9.0). Virion stability is strain-dependent and some strains, particularly human virus strains, may be much less stable than viruses isolated from animals. Studies with reassortants suggest that virion stability during purification and storage at 4°C is determined by a particular gene 4 or its encoded protein VP4, and by specific interactions of VP4 with VP7. The outer capsid proteins can be removed by treatment of virions with chelating agents, such as 10 mM EDTA, or with ethanol, and this results in inactivation of infectivity. In some cases, viral infectivity is less stable when virions contain cleaved VP4. In addition, the VP4 spikes can be removed by treatment of virions at pH 11.2 with ammonium hydroxide; removal of VP4 may also occur during other physical or chemical manipulations such as treatment with organic solvents. The ease of removing the VP4 spikes suggests that most viral preparations must be quite heterogeneous and they probably contain many noninfectious particles (which lack a full complement of VP4 spikes). This heterogeneity of viral preparations may explain why the estimates of the approximate percentage of each protein in particles determined by biochemical methods do not agree with the calculations made from electron cryomicrographs.

Table I. Rotaviral genome RNA segments and protein products[a]

Segment	Number of base pairs	Number of noncoding sequences[b] 5′	3′	Protein product[c]	Nascent polypeptide mol. wt (no. of amino acids)[d]	Mature protein modified	No. of molecules calculated per virion (approx. % by weight of virion protein)[e]	*ts* mutant group[f]	Remarks[i]
1	3302	18	17	VP1	125 005 (1088)		(2)	C	Subcore protein, slightly basic
2	2690	16	28	VP2	102 431 (880)	Myristoylated, cleaved	180 (15)	F	Core protein, RNA binding, leucine-zipper (aa536 to 559; 665 to 686)
3	2591	49	34	VP3	98 120 (835)		(0.5)	B	Subcore protein, basic, guanlylytransferase
4	2362	9	22	VP4	86 782 (776)	Cleaved VP5* (529)[g] VP8* (247)[g]	120 (1.5)	A	Surface spike protein, dimer, hemagglutinin, protease-enhanced infectivity, neutralization antigen, cell-attachment protein, virulence, putative fusion region (aa384 to 401)
5	1611	30	93	NS53	58 654 (495)			NA	Slightly basic, zinc fingers (aa54 to 66; 314 to 327)
6	1356	23	139	VP6	44 816 (397)	Myristoylated	780 (51)	G	Inner capsid protein, trimer, hydrophobic subgroup antigen
7	1104	25	131	NS34	34 600 (315)			NA	Slightly acidic, RNA binding, oligomer
8	1059	46	59	NS35	36 700 (317)			E	Basic, role in RNA replication?
9	1062	48	33	VP7(1)	37 368 (326)	Cleaved signal sequence, N-linked high-mannose glycosylation trimming[h]	780 (30)	NA	Surface glycoprotein, RER integral membrane, glycoprotein, cell-attachment protein?, neutralization antigen, 2 hydrophobic N-terminal regions, bicistronic gene? putative Ca^{2+}-binding site (aa127 to 157)
		135	33	VP7(2)	33 919 (297)				

Table I. Continued

10	751	41	182	NS20	20 290 (175)	NS29 → NS28 uncleaved signal sequence, N-linked high-mannose glycosylation and trimming	NA	Nonstructural RER transmembrane glycoprotein, 2 hydrophobic N-terminal regions, role in morphogenesis, putative Ca^{2+}-binding site
11	667	20	49	NS26	21 725 (198)	NS26 →NS28, phosphorylated, O-linked glycosylation	NA	Nonstructural, slightly basic; serine-threonine-rich, RNA binding

[a]For the Group A/Si/SA11 strain. Modified from Estes MK and Cohen J (1989) *Microbiol. Rev.* 53: 410.

[b]Number of 5′ noncoding sequences is up to the first AUG; number of 3′noncoding sequences does not include the termination codon.

[c]Determined by biochemical and genetic approaches. The size (in thousands) of the primary translational product is given for the nonstructural (NS) proteins. The names of the nonstructural proteins will be changed to facilitate comparisons of cognate nonstructural proteins encoded by group A and non-group A rotaviruses (Mattion *et al.* 1993). The designations using the nomenclature (previous names are in parenthesis) are NSP1 (NS53), NSP2 (NS35), NSP3 (NS34), NSP4 (NS28), and NSP5 (NS26). This change is required because the mol.wt of these cognate proteins differ for rotaviruses in the different groups (Mattion *et al.*, 1993).

[d]Molecular weights are calculated from the deduced amino acid sequences from nucleotide sequence data. The molecular weights are calculated from the largest potential open reading frame.

[e]The calculated numbers of molecules of VP2, VP4, VP6 and VP7 are made from predicted structural analyses of individual particles using electron cryomicroscopy. These estimates are not totally consistent with the percentage of each protein determined by biochemical analyses of a population of purified virions.

[f]NA indicates none assigned.

[g]There are two trypsin cleavage sites in SA11 4fM VP4 at amino acid 241 and 247. The indicated mature products are those based on use of only the preferred second cleavage site.[h]Mature cleaved VP7 contains 276 amino acids.

[i]aa, amino acid; RER, rough endoplasmic reticulum.

Replication

General features of rotaviral replication are that infectious triple-shelled rotavirions bind to a host cell receptor and virions enter cells by still poorly characterized mechanisms. The outer shell of the virion is apparently disrupted or removed as part of the virus entry process and this allows activation activity of the RNA-dependent RNA polymerase (transcriptase) associated with the particles that contain VP1, VP2, VP3 and VP6. The virion-associated transcriptase synthesizes viral mRNAs which are end-to-end transcripts of each of the 11 genome segments. These mRNAs are also capped by the virion-associated capping enzyme. The mRNAs are subsequently translated giving rise to all the viral proteins. The accumulation of viral proteins in the cytoplasm results in the formation of large perinuclear electron dense inclusions termed viroplasms which are thought to be the sites of genome replication and assembly of progeny subviral particles. Subviral particles that contain a subset of newly made structural proteins (VP1, VP2, VP3, VP6), nonstructural proteins and mRNAs are apparently formed. Some subviral particles contain an associated RNA polymerase (replicase) activity which uses the mRNAs as templates for the synthesis of minus-strand RNA resulting in the formation of dsRNA. It has been proposed that the replicase particles mature into the double-layered (single-shelled) particles, and some of these nascent single-shelled particles will function as transcriptase particles and synthesize additional mRNAs, thus leading to an amplification of the level of RNA replication in the cell. Other newly formed single-shelled particles associate with regions of the ER containing the nonstructural protein NS28 and bud into the ER. During this process, the particles acquire a membrane that is subsequently lost, while VP4 and VP7 condense around the particles producing the outer shell of protein found on mature viral particles. Recent data indicate that VP4 associates with NS28 and VP6 prior to budding of particles into the ER.

Stages of the Replication Cycle

Adsorption, penetration and uncoating

The initial stages of viral replication have been examined by biochemical and morphologic (EM) procedures. Only double-shelled particles attach to cells when monitored by EM or by infectivity assays. It is unclear whether virus attachment occurs via VP4 or VP7, but increasing evidence favors attachment via VP4. Binding to cells does not require cleaved VP4 or glycosylated VP7.

The identity of the cellular receptor for rotaviruses is not known, but a study of the binding of radiolabeled SA11 to MA104 cells found approximately 13 000 receptor units per cell. Binding is sodium-dependent, pH-insensitive between 5.5 and 8 and dependent on sialic acid residues in the membrane. Virus binds but is not internalized at 4°C. The receptor has not yet been identified, but two proteins of mol. wt approximately 300 000 and 330 000 from brush border membranes of murine enterocytes have been reported to bind virus specifically.

Many rotavirus strains contain a hemagglutinin, as demonstrated by their ability to bind to erythrocytes. The hemagglutination activity has been mapped to VP8*, the N-terminal cleavage fragment of VP4. Studies of virus binding to erythrocytes were the first to show that neuraminidase treatment reduces virus binding, indicating a role for sialic acid in virus attachment. Sialic acid-containing compounds such as fetuin and mucin also inhibit virus binding to cells. These results add rotaviruses to an increasing number of viruses (such as reoviruses and influenza) that require sialic acid for binding to cells. However, these studies have not determined whether virus binds directly to sialic acid or whether sialic acid maintains the configuration of the binding site without directly interacting with the virus. Binding to sialic acid is not essential for rotavirus infectivity as most human rotaviruses infect cells in a sialic acid independent manner.

After binding to susceptible cells, virus is internalized. The enhancement of rotavirus infectivity by proteolysis is not due to increased efficiency of virus attachment to host cells, but to facilitation of the virus internalization (penetration) step. Internalization will not take place at 0–4°C, indicating that this step requires active cellular processes. All virus is internalized by 60–90 min after binding. The mechanism of internalization (penetration) into cells remains unclear and controversial.

Both morphologic and biochemical approaches have been used to investigate the mode of entry of rotaviruses into cells. Early EM studies suggested that virus entry (SA11 strain) occurs by endocytosis and that incoming particles are rapidly transported to lysosomes. Clear documentation by EM of uptake of trypsin-treated viral particles into coated pits, coated vesicles and secondary lysosomes confirms that rotavirus particles (porcine OSU strain) enter cells by receptor-mediated endocytosis and suggests that uncoating might occur by the effect of lysosomal enzymes. However, the calcium ionophore A23187 can increase the intracellular Ca^{2+} concentration during the early stages of replication and block uncoating. These results support the hypothesis that low Ca^{2+} concentrations in the intracellular microenvironment may be responsible for uncoating. This idea was originally proposed because it was known that removal of the outer capsid of particles and activation of the endogenous polymerase could be accomplished by calcium chelation.

Contrasting results have come from studies of the uptake of a human rotavirus, the infectivity of which is reported to be absolutely dependent on trypsin cleavage. Work with this virus suggests that the mode of rotavirus entry into cells differs depending on whether viral particles have been pretreated with trypsin. Infectious virus pretreated with trypsin has been observed to enter cells by direct penetration of particles through the cell membrane into the cell cytoplasm. In contrast, nontrypsin-treated particles were taken up by phagocytosis and such virions were sequestered into lysosomes 20 min after virus attachment to the cell membrane. Mere phagocytosis of particles into lysosomes was thought to be unrelated to rotaviral replication. Direct release of nucleic acid through spaces that form between the capsomeres and cell membrane pores after the attachment of trypsin-activated viral particles onto cells has been proposed; this mechanism is considered to be analogous to injection of nucleic acid into bacteria by phages. The inability to detect viral shells lacking nucleic acid attached to cells and dsRNA free in the cytoplasm of cells make this observation difficult to understand.

The concept that trypsin-treated and nontrypsin-treated virus enter cells by different mechanisms is supported by a recent study of the kinetics of entry of rhesus rotavirus into MA104. Trypsin activated rhesus rotavirus was internalized with a half-time of 3–5 min while nonactivated virus disappeared from the cell surface with a half-time of 30–50 min. Only trypsin-activated virus resulted in productive infection. Trypsin-treated virus was shown to mediate the rapid release of ^{51}Cr from infected cells and this activity did not occur at 4°C. The specificity of the permeability alterations was examined by showing that single-shelled particles did not mediate them and that neutralizing anti-VP4 MAbs (but not nonneutralizing antibodies to VP6 or VP7) inhibited the ^{51}Cr release. Unfortunately, permeability results for nontrypsin-treated virus were not reported, but these data suggest that virus enters by direct penetration of the cell membrane which results in its permeabilization.

Other viruses that initiate infection by mechanisms involving receptor-mediated endocytosis often depend on acidification of endosomes for partial uncoating or entry into the cell. The importance of acidification of endosomes for the initiation of infection of rotaviruses has been studied by several groups. In all cases, lysosomotropic agents (ammonium chloride, chloroquine, methylamine, amantadine) had little inhibitory effect on viral replication, as measured by RNA synthesis, polypeptide synthesis or virus yields. Thus, it seems clear that acidification of endosomes is not important for the entry of rotavirus into cells, unlike other viral systems, including reovirus. Energy inhibitors (sodium azide and dinitrophenol) have a minimal effect on rotavirus infection and this suggests that rotaviruses do not use endocytosis to enter cells. However, it is not known if these inhibitors (and the conditions tested) specifically affect the cell processes (if any) required for rotaviruses to enter cells.

It seems clear that the passage of rotaviruses from endocytic vesicles to the cytoplasm does not occur by a pH-dependent fusion mechanism, but this does not prove that rotaviruses are not taken up by endocytosis. The most direct support for the idea that rotaviruses enter the cell by direct penetration of the plasma membrane is the demonstration that trypsin-treated double-shelled rotavirus (but not nontrypsinized virus or single-shelled particles) causes release of a fluorophore encapsulated within liposomes. A putative fusion region in VP5* has been identified that has sequence homology with Sindbis virus, and it has been suggested that this region might mediate virus penetration into cells. This remains to be demonstrated. It also remains to be determined if the interaction of cleaved VP4 with lipids occurs only at the plasma membrane or if this might occur in the endosome, or during viral morphogenesis (see below). It is possible that more than one mechanism, including endocytosis and direct passage, is operative for rotaviruses, as has been proposed for poliovirus and reoviruses.

Transcription and replication

The synthesis of viral transcripts is mediated by a viral RNA-dependent RNA polymerase (transcriptase) which has a number of enzymatic activities. The transcriptase is a component of the virion and properties of this enzyme (or enzyme complex) have been inferred by studying the characteristics of products from *in vitro* transcription reactions. Rotavirus particles presumably contain the same enzymatic activities found in reoviruses including transcriptase, nucleotide phosphohydrolase, guanylyltransferase and two methylases. These activities are inferred because rotavirus transcripts made *in vitro* in the presence of *S*-adenosylmethionine possess a methylated 5′-terminal cap structure, m^7GpppGm, and transcription is inhibited by pyrophosphate. Particles also contain a poly(A) polymerase activity the precise function of which remains unknown; it has been postulated to be responsible for the synthesis of oligo(A) molecules.

The virus-associated transcriptase is latent in double-shelled particles and can be activated *in vitro* by treatment with a chelating agent or by heat shock treatment. Such treatments result in removal of the outer capsid proteins with conversion of double-shelled particles to single-shelled particles. In infected cells, double-shelled particles have been shown to be uncoated to single-shelled particles, so it is thought that transcription in cells occurs from such particles. Transcription is asymmetric and all transcripts are full-length plus-strands made off the minus dsRNA strand. The intracellular site of transcription is unknown.

Activation of transcriptase activity is a process that is not well understood. 'Activation' may be a misnomer since it has been suggested that, in reoviruses, this process does not actually modify the enzyme complex but instead releases the templates from structural constraints, allowing them to move past the transcriptase catalytic site. Rotavirus transcription requires a hydrolyzable form of ATP, and studies with analogs that inhibit transcription suggest that ATP is required in reactions other than polymerization. ATP may be used for initiation or elongation of RNA molecules as has been described for vesicular stomatitis virus or vaccinia virus RNA polymerases. It remains unclear whether distinct polypeptides in the transcriptively active particles perform distinct functions or the inner core polypeptides function as an enzymatic complex. VP1 has been cross-linked with a photoreactable nucleotide analog indicating that VP1 is a component of the transcriptase. VP3 in viral particles and expressed alone in insect cells can bind GTP, suggesting that this protein is the guanlylytransferase. Whether VP3 also possesses transcriptase activity alone or in association with VP1 remains unclear. Similarly, the role of VP2 in the transcription process is unclear.

The synthesis of plus- and minus-strand RNA has been studied in SA11-infected cells and in a cell-free system. These studies were facilitated by optimization of an electrophoretic system that allows separation of the plus- and minus-strands of rotavirus RNAs in acid–urea–agarose gels. In these gels, the complementary strands migrate at different rates. For SA11, the plus-strand migrates faster than its complementary minus-strand. Analysis of the kinetics of RNA synthesis in infected cells has shown that plus- and minus-strand RNAs are detected initially at 3 h postinfection, in agreement with other studies that looked at the time of incorporation of [^{3}H]uridine into rotavirus RNA. After 3 h, the level of transcription increases until 9–12 h at which time the levels of plus-strand RNAs are maximal. The ratio of plus- to minus-strand RNA synthesis changes during infection and the maximal level of minus-strand RNA synthesis is seen several hours prior to the peak of plus-strand RNA synthesis.

The delay in obtaining maximal plus-strand RNA synthesis has been hypothesized to be due to a requirement for the accumulation of stoichiometric amounts of a protein (e.g. VP6) necessary for the assembly of transcriptase particles. Both newly synthesized and preexisting plus-strand RNA can act as templates for minus-strand RNA synthesis throughout infection, an unexpected result based on earlier studies with reoviruses. The observation that the level of RNA replication does not increase continually in conjunction with the increasing levels of plus-strand RNA suggests that RNA replication is regulated by factors other than the level of plus-strand RNAs in the infected cell.

The synthesis of dsRNA has also been analyzed using a cell-free system to study the replication of rotavirus RNA. The components of this system include (1) subviral particles prepared from infected cells to template the synthesis of viral RNA and (2) an mRNA-dependent rabbit reticulocyte lysate to support protein synthesis. On the basis of nuclease sensitivity, approximately 20% of the RNA made *in vitro* is double-stranded and 80% is single-stranded. Although not examined directly, it is assumed that rotavirus RNA replication, like that of reovirus, takes place in a conservative fashion, i.e. both parental dsRNA strands remain within partially uncoated particles. The synthesis of dsRNA *in vitro* has been determined to be an asymmetrical process in which a nuclease-sensitive plus-strand RNA acts as template for the synthesis of minus-strand RNA. After its synthesis, dsRNA remains associated with subviral particles suggesting that free dsRNA is not found in cells.

This *in vitro* system also supports the initiation of minus-strand RNA using exogenous viral plus-strand RNA as template. The conversion of exogenous mRNA to dsRNA by subviral particles is potentially an exciting result as it suggests a method to study (1) the specificity of viral proteins in recognition and replication of rotavirus mRNAs and (2) the effect of adding exogenous synthetic RNAs containing specific mutations on replication. However, the efficiency of the system may not be adequate to allow such evaluations.

Finally, the possibility that subviral particles can be assembled from the viral proteins made in this *in vitro* replication system was investigated. Ribonucleoprotein complexes similar to subviral particles present in infected cells were assembled. Together, these results suggest that a cell-free system to support rotavirus RNA replication, transcription and the assembly of subviral particles can be established. This system should be useful to help define the defects in rotavirus mutants and to study the RNA sequences and proteins involved in viral replication and assembly.

Attempts have been made to characterize the subviral particles (complexes separable by sedimentation through sucrose gradients and by equilibrium centrifugation in CsCl gradients) in which dsRNA synthesis occurs in both infected cells and the cell-free system. The morphology of these particles is unknown. Replicase particles, like those with transcriptase activity, consist of core proteins, VP1, VP2 and VP3, small amounts of the protein VP6, large amounts of the nonstructural protein NS34 and lesser amounts of NS53 and NS35. Because of the inability to separate particles with transcriptase and replicase activities completely by these methods and the problems of contamination of some fractions with proteins from neighboring fractions, the complexes involved in RNA replication will require more analyses. The roles of individual proteins and specific protein complexes in RNA replication and viral morphogenesis will probably not be resolved until they are studied *in vitro* with pure species of native rotaviral proteins and viral RNAs.

The sites and precise details of RNA replication remain unclear. However, electron dense viroplasms are probably the sites of synthesis of the single-shelled particles that contain RNA. This conclusion is based on the localization of several of the viral proteins (VP2, NS35, gene 11 protein) to viroplasms and of VP4 and VP6 to the space between the periphery of the viroplasm and the outside of the ER, and on the observation that particles emerging from these viroplasms often seem to directly bud into the ER which contains VP7 and NS28.

Assembly

The distinctive feature of rotaviral morphogenesis is that subviral particles, which assemble in cytoplasmic viroplasms, bud through the membrane of the ER and maturing particles are transiently enveloped. This is one of the more interesting aspects of rotaviral replication, differing from members of other genera in the *Reoviridae* family. The envelope acquired in this process appears to be lost as particles move toward the interior of the ER, and it is replaced by a thin layer of protein which ultimately comprises the outer capsid of mature virions.

The sites of synthesis or localization of the viral proteins have been examined by ultrastructural immunocytochemistry using polyclonal monospecific or monoclonal antibodies and by studying the distribution of proteins by immunofluorescence or subcellular fractionation. Taken together, the morphologic and biochemical data are consistent with rapidly assembling single-shelled particles serving as an intermediate stage in the formation of double-shelled virions. Most of the rotaviral structural proteins and all of the nonstructural proteins are synthesized on the free ribosomes, although the nascent proteins on free ribosomes have not been analyzed. Instead, this conclusion has been drawn on the basis of the absence of signal sequences which would indicate targeting to the ER and lack of protection against digestion in *in vitro* protease protection studies. In contrast, the glycoproteins VP7 and NS28 are synthesized on ribosomes associated with the ER membrane and they are cotranslationally inserted into the ER membrane in response to signal sequences at their N termini. The glycoprotein NS28 is a homotetramer oriented with the C terminus on the cytoplasmic side of the ER membrane. This cytoplasmic domain of NS28 acts as a receptor to bind to nascent single-shelled particles. This binding is thought to be the first step which initiates the membrane budding event.

VP7 is detected in the ER of SA11-infected cells in two pools. One pool is found only in intact particles and is detected only by a neutralizing MAb. The second pool of VP7 is unassembled, it remains associated with the ER membrane and it is detected by a polyclonal antibody made to denatured VP7. A kinetic study of the assembly of VP7 and of other structural proteins into particles has shown that the incorporation of the inner capsid proteins into single-shelled particles occurs rapidly, while VP4 and VP7 appear in mature double-shelled particles with a lag time of 10–15 min. Kinetic analyses of the processing of the oligosaccharides on the two pools of VP7 have shown that the virus-associated VP7 oligosaccharides have a 15-min lag compared with that of the membrane-associated form, suggesting that the latter is the precursor to virion VP7. This lag appears to represent the time required for virus budding and outer capsid assembly. NS28, VP7 and VP4 can also form heterooligomers which are not associated with any known subviral particle; these heterooligomers are thought to be present at sites on the ER membrane where maturation to double-shelled particles begins. The proteins of the outer shell are apparently assembled on to the single-shelled particles either during the budding process or once the particles reach the ER lumen.

Rotavirus maturation is reported to be a calcium-dependent process, on the basis of the observation that virus yields are decreased when produced in cells maintained in calcium-depleted medium. Viruses produced in the absence of calcium were found to be exclusively single-shelled, and budding of virus particles into the ER was not observed. Among the viral proteins, reduced levels of VP7 were observed, and subsequent studies showed that such reduced levels were due to the preferential degradation, and not to the impaired synthesis, of VP7. An interesting finding of these studies is that unglycosylated (but not mature) VP7 made in the presence of tunicamycin is relatively stable in a calcium-free environment. It is possible that calcium stabilizes or modulates folding or compartmentalization of the newly glycosylated VP7 for subsequent assembly into particles. Alternatively, calcium deprivation may destabilize the ER or ER proteins required for the stable association of glycosylated VP7 with the membrane.

Virus release

EM studies of infected tissue culture cells have shown that the infectious cycle ends when progeny virus is released by host cell lysis. Extensive cytolysis during infection and drastic alterations in the permeability of the plasma membrane of infected cells resulting in the release of cellular and viral proteins have been demonstrated. In spite of cell lysis, most single-shelled and many double-shelled particles remain associated with the cellular debris, suggesting that these particles interact with structures within cells. Interactions with cell membranes and the cell cytoskeleton have been suggested to occur and these may play a role in movement of the viral proteins or particles within the cell. Whether the cytoskeleton provides a means of transport of viral proteins and particles to discrete sites in the cell for assembly or acts as a stabilizing element at the assembly site and in the newly budded virions or if particles are simply trapped by the cytoskeleton remains to be determined. It is also possible that virus may not be released from infected enterocytes because of cytopathic effects (CPE) and cell lysis. Instead, virus-infected enterocytes may merely be sloughed intact into the intestinal lumen. This possibility has been suggested by studies of rotaviral replication in polarized human intestinal epithelial cells. These cells are infected in a symmetric manner and cell functions are shut-off before the development of CPE and extensive virus release.

Future Perspectives

Future basic research is expected to define the functions of each of the nonstructural proteins in the replication cycle, leading to an understanding of the mechanisms of RNA replication and genome packaging. This may in turn lead to the ability to use reverse genetics to probe in great detail the functions of any gene and to construct virions with desired properties. Knowledge of the three-dimensional structure of these complex virions is awaited for further understanding of the interactions between the outer capsid proteins and between the proteins in each of the capsid shells.

See also: Reoviruses.

Further Reading

Bellamy AR and Both GW (1990) Molecular biology of rotaviruses. *Adv. Virus Res.* 38: 1.

Estes MK (1990) Rotaviruses and their replication. In: Fields BN *et al.* (eds), *Virology*, 2nd edn, New York: Raven Press.

Estes MK and Cohen J (1989) Rotavirus gene structure and function. *Microbiol. Rev.* 53: 410.

Mattion NM, Cohen J and Estes MK (1993) The rotavirus proteins. In: Kapikian AZ (ed.) *Virus Infections of the Gastrointestinal Tract*, 2nd edn. New York: Marcel Dekker Inc. (in press).

Prasad BVV and Chiu W (1992) Structure of rotavirus. *Curr. Top. Microbiol. Immunol.*, (in press).

RUBELLA VIRUS

Jerry S Wolinsky
The University of Texas–Houston
Health Science Center
Houston, Texas, USA

History

Rubella is predominantly a banal childhood infection. Endemic world-wide, rubella virus causes epidemics at irregular intervals. First described by German physicians in the eighteenth century as distinct from scarlet fever and rubeola, rubella only gained international acceptance as a unique disease in the late 1800s. Experimental transmission studies confirmed the viral nature of rubella in 1938. Three years later, Norman Gregg reported an epidemic of congenital cataracts often in association with cardiac malformations as a consequence of gestational rubella; this established rubella virus as a major teratogen. Rubella virus was not isolated in tissue culture until 1962. The last major pandemic of rubella to impact the US occurred 2 years later. During the American epidemic, 20 000 infants suffered permanent damage from *in utero* virus exposure. The development of a live-attenuated rubella vaccine and the introduction of an aggressive vaccination program in the US in 1969 proved successful in reducing the incidence of natural rubella and its consequences.

Rubella virus has retained its interest due to the unique properties of the virus that distinguish it from other togaviruses, the continued problems that congenital rubella poses world-wide, the intriguing but unexplained ability of the virus to persist in man for many years in the absence of new detectable signs or symptoms, and the potential for rubella to initiate immune-mediated human disease.

Taxonomy and Classification

Rubella virus is a member of the family *Togaviridae* and the only member of the genus *Rubivirus*. Like other members of this frequently revised family of viruses, rubella virus consists of an icosahedral core cloaked in a lipid envelope. The virion core contains a single copy of the single-stranded 40S RNA genome protected by a capsid (C) protein. Embedded in the envelope of the virion are two types of acylated glycoproteins, E1 and E2. Unlike other togaviruses, rubella virus has no invertebrate vector and the only known natural reservoir for rubella virus is man. Further, rubella virus is serologically distinct and certain features of the genomic organization of rubella virus also distinguish it from the alphaviruses.

Properties of the Virion

Rubella virions are 60–70 nm spherical particles composed of a 30 nm electron dense core separated by an electron lucent zone from the lipid envelope. The intact virion and nucleocapsid core have a density of 1.18–1.20. The central core contains a single copy of the single-stranded 40S RNA genome protected by phosphorylated C protein (~ 33 kD) arranged as paired, disulfide-linked homodimeric capsomeres. Embedded in the envelope are two types of acylated glycoproteins, E1 and E2, which project from the envelope as poorly resolved 6–8 nm spikes. The envelope is selectively released from the virus by detergents to produce rosettes that maintain the organization of the spikes. E1 has an monomeric molecular weight of ~ 58 kD and is present within mature virus in monomeric form, as an E1–E1 homodimer and as a heterodimer with E2. E2 is heterogeneous (~ 42–47 kD) due to variable glycosylation; most of E2 in intact virus forms heterodimers with E1. E1 contains monoclonal antibody-defined domains independently involved in virus attachment to red blood cells and the initiation of infection. Available structural and functional data suggest that E1 is the dominant surface molecule of the virus particle and the main target for the detection and subsequent elimination of virus by the host.

Properties of the Genome

The virion RNA is message-sense (positive polarity)

and naked viral RNA is infectious but inefficient. Genomic RNA is single-stranded, capped at the 5′ end, polyadenylated at the 3′ end and consists of 9757 nucleotides exclusive of the polyadenylated tail. The gene has an extraordinary guanine and cytosine content of nearly 70%. Two long and partially overlapping open reading frames are present; one from nucleotide 41–6506 encodes a predicted polyprotein precursor for the nonstructural proteins (nsPs) of 2205 amino acids, the other from nucleotide 6507–9696 encodes the polyprotein precursor of the three structural polypeptides. The rubella virus genome has several features that are highly conserved among togaviruses. The stem-loop structure at the 5′ end could provide the replicase binding site for the initiation of transcription of positive-strand genomic length RNA. A stretch of 46 nucleotides beginning at nucleotide 223 is analogous to a 51 base consensus region also involved in transcription. The final region of homology begins with nucleotide 6383 and extends to within 20 nucleotides of the subgenomic RNA start site. The analogous region in Sindbis virus appears to be critical for the synthesis of subgenomic RNA. A unique 22–30 nucleotide stem-loop structure has been noted at the 3′ end of the rubella genome centered between nucleotides 9685–9686. This structure seems to interact specifically with phosphorylated cytoplasmic host proteins that may co-participate with viral nsPs in RNA-dependent RNA replication to initiate negative-strand 40S RNA synthesis.

Properties of the Proteins

Little direct information is available on the nsPs of rubella virus. By analogy with other togaviruses and based on the presence of predicted regional motifs in common with known replicases and helicases, they must include a number of molecules derived from the predicted nsP precursor polypeptide that together are responsible for virus replication, transcription and perhaps protease activities. Ultimately these nsPs, in concert with several host cell proteins, interact with genomic RNA to produce the negative-strand templates needed for the generation of both subgenomic 24S RNA and full-length 40S progeny genomes.

The bulk of C protein found in the cell and virion appears to be composed of 300 amino acid residues and has a predicted mass of 32 967. It appears likely that the midportion of the E1 molecule ($E1_{202}$–$E1_{286}$) has reasonable surface exposure in the mature virion as several epitopes in this region have been defined by monoclonal antibodies capable of neutralizing viral infectivity

Physical Properties

Rubella virus is heat labile and rapidly inactivated at 56°C; infectivity is lost on storage at −20°C but stable for prolonged intervals at temperatures below −60°C. Hydrogen ion concentrations outside the broad physiologic normal range and physical agents that extract lipids, denature protein or alter nucleic acids all inactivate the virus.

Strategy of Replication of Nucleic Acid

Rubella virus is believed to adsorb to host cells by the interaction of its envelope glycoproteins with specific host cell surface receptors. It enters the cell by receptor-mediated endocytosis and is thought to require a low pH environment for fusion of its envelope and release of the genome into the cytosol. However, these events remain poorly defined and the host receptors are incompletely characterized.

Once within cells, the genomic RNA can serve as a direct template for translation to produce the nsPs essential for virus replication. Viral nucleic acids in infected cells are of three types, positive-polarity genomic and subgenomic length molecules, and negative-polarity genomic length replicative intermediates. The positive-polarity subgenomic 24S mRNA is capped, methylated and contains 3327 nucleotides exclusive of the polyadenylated tail which correspond to the 3′ one-third of the genomic RNA. This subgenomic message encodes the information for the three structural proteins of the virus in the order 5′-C-E2-E1-3′. Rubella viral RNA synthesis is first detectable after an 'eclipse' interval of 10 h.

In vitro translation of the 24S mRNA produces a precursor polyprotein (predicted mass 114 678) through which the processing of the structural proteins of rubella virus proceeds. *In vivo* the nascent polyprotein precursor is immediately cleaved following the translation of C by a signal peptidase (signalase) of the luminal surface of the endoplasmic reticulum. The C-terminal 23 amino acids of C provide a highly hydrophobic signal sequence that is immediately followed by a candidate signalase cleavage site. This signal sequence, required to effect translocation of E2 into the endoplasmic reticulum, is retained on C. The translation of E2 and E1 then proceeds on ribosomes anchored to the endoplasmic reticulum. A stretch of 20 uncharged amino acids immediately precedes the N terminus of E1. This region also has characteristics of a signal sequence and is required to effect the insertion and translocation of elongating E1 molecules into the lumen of the endoplasmic reticulum. A C-terminal hydrophobic region of 18 residues probably contributes to the anchoring of E2 into cytoplasmic membranes. Near the C-terminal end of the E1 protein is a sequence of 22 uncharged amino acids that forms a credible transmembrane region. The C-terminal 13 residues of E1 are probably free on the cytoplasmic side of the membrane.

Post-translational Processing

A number of co-translational and post-translational modifications of the rubella virus glycoproteins are required before they are incorporated into virions. Both E1 and E2 are altered by the addition of carbohydrates, predominantly mannose and glucosamine. Four possible N-linked glycosylation sites exist on the E2 of most rubella virus strains; all that are present appear to be used. The three possible N-linked glycosylation sites of E1 are used. Intracellular forms of the glycoproteins contain only high-mannose glycans. Additional modifications of these moieties take place during maturation to forms that contain both the high-mannose and complex glycans found in the virion. Both glycoproteins covalently incorporate palmitic acid in the rough endoplasmic reticulum. E1 and E2 contain numerous cysteines, some of which undoubtedly contribute to intrachain disulfide bond formation and the interchain bonds that stabilize both E1–E1 homodimers and E2–E1 heterodimers.

Assembly Site, Release, Cytopathology

The replication of rubella virus is seldom attended by reliable or distinctive cytopathic effects and intracellular inclusions are rarely seen. The infection proceeds with limited alteration of host cell protein synthesis. However, the virus can profoundly disturb the cytoskeletal organization of cells infected *in vitro*. Acute infection of cells is blocked by interferons alpha and gamma, and amantadine. When added early in the eclipse phase of infection, actinomycin D markedly impairs the synthesis of positive- and negative-polarity 40S RNA, positive-polarity 24S RNA, and viral structural proteins by blocking the synthesis of host cell factors required for the replication of viral RNA.

Maturation of rubella virus occurs when the RNA-containing nucleocapsid core buds from a virus-modified cellular membrane, thus acquiring an envelope consisting of host cell lipid and the viral glycoproteins E1 and E2. The budding process can start from either intracellular membranes or from the plasma membrane of cells grown in culture. The capacity of rubella virus to mature within intracellular vacuoles and thus avoid detection by the host immune system may be an important determinant of its propensity to establish persistent infections.

Geographic and Seasonal Distribution

While endemic world-wide, there is considerable geographic variation in rubella attack rates in different age groups. Attack rates throughout the year are low; in temperate zones seasonal peaks occur in the spring. Before widespread vaccine use, rubella epidemics occurred at 6–9 year intervals in the US and the UK, and slightly more frequently in other European countries. Peak infection in these countries occurred in 5–9-year-old school-age children. In much of Africa, highest attack rates occur in children under 5, and 80% of all children are immune by 10 years of age. In island and rural tropical populations the incidence of rubella infection is low, with high percentages of susceptible women of childbearing age.

Host Range and Virus Propagation

Rubella has no known natural host other than man. Nonhuman primates and a variety of laboratory animals can be infected with rubella virus, but no reliable animal model exists for the study of clinically symptomatic acquired or gestational rubella. Rubella virus grows routinely in many laboratory cell lines. However, the cytopathic effects of rubella virus are capricious. These are dependent on the state of adaptation of the virus isolate to the cell line and its passage history, and are even vulnerable to minor differences in the growth media used.

Primary African green monkey kidney cells

(AGMK) provide a sensitive culture system for the initial isolation of rubella virus. AGMK afford the advantage of being directly usable in interference assays with ECHO-11 virus within several days of the initial isolation attempt. The RK-13 rabbit kidney cell line, Vero cells, a continuous cell line derived from AGMK, and baby hamster kidney (BHK-21) cells are frequently used alternatives. Highest yields of cell-free virus are obtained in Vero cells.

Persistent infection of a wide variety of primary or continuous cell lines is readily established *in vitro* and the growth rate of chronically infected cells is slowed. Rubella virus has no ready mechanism for incorporating its genomic information into the gene pool of the host cell as it does not encode a reverse transcriptase. The persistent infection of cell cultures by rubella virus has been explained by mechanisms involving temperature-sensitive mutants and by the presence of defective-interfering particles. However, neither appears to be required to initiate persistence in culture.

Genetics

Comparison of the predicted amino acid composition of the structural proteins can be made only on a limited number of rubella virus strains for which reasonable sequence data are now available. Independent sequence data for E1, E2 and C show remarkable consistency despite the potential for divergence given different passage histories of the Therien virus strain used in the distant laboratories. Differences found at the nucleotide level result in only seven amino acid changes, three in E1, and two each in E2 and C; four of these seven changes are conservative. Differences between the predicted E1 molecules of the Judith strain and the E1 and C molecules of the Therien strain, and related M_{33} and HPV77 strains tend to follow a pattern of expected relatedness. A somewhat anomalous result obtains when the predicted E2 molecules are compared. These results suggest that changes on passage of the original M_{33} isolate resulted in a greater divergence of HPV77, including loss of a glycosylation site from HPV77 and later passaged versions of M_{33}.

Evolution

Rubella virus shows a pattern of genomic organization similar to other *Togaviridae*, but with few exceptions it lacks important nucleic acid or protein sequence homology with the alphaviruses. Three of four predicted RNA secondary structures common among alphaviruses and believed important in replication and transcription are present in rubella virus. However, the locations of predicted helicase and replicase motifs in the nonstructural region of rubella virus suggest that it may have arisen by genetic rearrangement of an alphavirus archetype.

Serologic Relationships and Variability

As might be anticipated by the lack of serologic cross-reaction with other togaviruses, there is no important homology between rubella virus and sequenced alphaviruses in the subgenomic region that specifies the structural polypeptides of the virion. It is also difficult to demonstrate important antigenic differences between strains or diverse isolates of rubella virus. Kinetic neutralization studies have teased out some relatively minor differences. Further, monoclonal antibodies to distinct domains on the E1 glycoprotein neutralize infectivity of a panel of diverse rubella virus clinical isolates as well as the RA27/3 vaccine strain. Thus, rubella virus is monotypic, with no evidence of important, naturally evolving antigenic variants.

Epidemiology

The development of pools of susceptible individuals generally explains the occurrence of epidemics at irregular, 3–9 year intervals in different geographic locations. Major world-wide pandemics occur every 10–30 years. Much of the regional variation in the age of onset, incidence, and appearance and spread of epidemics is explained by population densities, socioeconomic factors and levels of medical sophistication. Serologically defined attack rates are highest amongst closely housed susceptibles during an epidemic. Infection of adolescents and young adults is not uncommon, particularly in situations where a large number of susceptible individuals are housed closely together. Acquired rubella infection is usually mild so that morbidity is low and death attributed to acquired rubella is exceedingly uncommon. Only 0.05% of all cases of rubella reported to the Centers for Disease Control were fatal. The most devastating consequences of natural rubella are the abortions, miscarriages, stillbirths and particularly the fetal malformations that arise from maternal infection during the first trimester of pregnancy. Following the 1964 rubella epidemic in the US, an excess of 6250 cases of spontaneous abortion occurred and 5000 therapeutic abortions were performed with maternal rubella as the recorded indication.

Since the widespread introduction of vaccines in the US in 1969, the incidence of both acquired and congenital rubella infection has declined more than 99% to all-time low levels. As part of the 1990 health objectives, the United States Public Health Service sought to reduce the number of rubella cases to less than 1000 annually and the number of congenital rubella cases to less than 10 annually. Unfortunately, neither goal has been repetitively met. The incidence of rubella has declined least in those over 20 years of age.

Transmission and Tissue Tropism

Man is the only known natural host for rubella virus. Virus is transmitted between individuals by aerosolation. Congenitally infected infants shed virus at high levels and represent a potential virus reservoir and clear source of transmission to susceptibles. While vaccinees harbor recoverable virus, transmission of vaccine virus to susceptible individuals has not been observed. The epithelium of the buccal mucosa provides the primary site for rubella virus replication with the mucosa of the upper respiratory tract and nasopharyngeal lymphoid tissue serving as portals of virus entry. Systemic spread by lymphatics or a transient viremia then seeds regional lymph nodes where virus replicates further. After an incubation period of 7–9 days, virus appears in the serum and is shed into the nasopharynx where it can be further spread by aerosolation. The cell-free viremia ceases with the onset of detectable rubella-specific antibody shortly after the rash appears. A mononuclear cell borne viremia and virus in nasopharyngeal secretions can be detected for a week or more after the rash subsides.

Since the appearance of the rubella rash coincides with the detection of rubella-specific antibody, it has been postulated to be an immune complex-mediated phenomenon. However, circulating immune complexes have not been shown to cause the exanthem. Virus has been isolated from skin biopsies at the time of the exanthema from both involved and uninvolved skin. The development of the rash is not invariant, even following rubella induced experimentally with fresh virus isolated from active cases. When present, the rash can resolve within 12 h or persist for up to 5 days.

During viremia and especially during the first month of gestation, placental tissues are very susceptible to infection. Placental infection results in scattered foci of necrotic syncytiotrophoblast and cytotrophoblast cells and damage to vascular endothelium. Once placental infection is established virus can disseminate to the fetus, but this is not invariant. Rubella virus is more often recovered from placental than fetal products of conception. Rubella virus can spread throughout the developing fetus and almost any organ may be infected. A chronic and generally nonlytic infection is then established. Persistent shedding of virus by neonates may depend on early gestational infection. Rubella virus-infected products of conception obtained during the first trimester of gestation show a distinct lack of inflammation. However, consistent ultrastructural changes are found at this stage of the infection in the endothelial lining of blood vessels of brain and other organs. Precise mechanisms of rubella viral teratogenesis are unknown but are probably multifactorial.

Pathogenicity

The incidence of congenital rubella may vary following different pandemics and regional epidemics. This suggests the existence of rubella virus strains with distinct teratogenic potential; a yet unproven hypothesis. Similarly, certain vaccinees have been associated with different frequencies of complicating arthralgias. This has been associated with their distinctive capacity for tissue invasion in organized synovial explant cultures. The molecular correlates of teratogenic or arthrogenic capabilities of different rubella virus isolates and vaccinees are not established.

Clinical Features of Infection

Rubella acquired in early childhood or adult life is usually mild and often clinically inapparent. Symptomatic rubella encompasses combinations of maculopapular rash, lymphadenopathy, low-grade fever, conjunctivitis, sore throat and arthralgia. These follow an incubation period of 16–20 days. The rash is the most prominent and earliest feature. It begins as distinct pink maculopapules on the face that then spread over the trunk and distally on to the extremities. The maculopapules coalesce and the rash rapidly fades over several days. As associated posterior cervical and suboccipital adenopathy is characteristic. Fever is typically low grade. The entire clinical syndrome usually resolves in days, but thrombocytopenia and encephalopathy infrequently occur. Acute polyarthralgia and arthritis following natural rubella virus infections of adults are common; 52% of women and 9% of men develop objective evidence of arthritis with symptoms being more severe among women. Arthralgia usually resolves within several weeks. However, pain or arthritis may persist or recur over several years. The most common symptoms of rubella, lymphadenopathy, erythematous rash, and low grade fever, are nonspecific and easily confused with similar illnesses caused by other common viral and nonviral pathogens or drug-induced eruptions. Therefore, a definitive diagnosis of rubella requires isolation of rubella virus or demonstration of seroconversion.

Gestational rubella has dire consequences for fetal development and few fetuses escape infection during the first 2 months of gestation. The rate of serologically determined fetal infection falls after the eleventh week of gestation only to rise again near term. Fetuses infected before the sixteenth week of gestation rarely escape damage. By contrast, defects referable to congenital rubella are unlikely following confirmed infection after the seventeenth week of gestation. The most frequent clinical manifestations of symptomatic congenital rubella in those fetuses that survive early gestational infection include suspected or confirmed hearing loss, congenital heart disease most frequently manifest as patent ductus arteriosus or pulmonary artery or valvular stenosis, psychomotor retardation, cataract or glaucoma, retinopathy, neonatal thrombocytopenic purpura, hepatomegaly and/or splenomegaly, and intrauterine growth retardation. Less frequent features seen in 5–10% of affected infants include adenopathy, bony radiolucencies, hepatitis usually with jaundice and hemolytic anemia. Nearly 80% of congenital rubella children show some type of neural involvement, particularly neurosensory hearing loss.

Most clinical manifestations of congenital rubella are evident at or shortly following birth and some are transient. However, recognition of some cases of retinopathy, mental retardation and hearing loss may be delayed for several years. Very late consequences of congenital rubella are now increasingly appreciated as these children are carefully followed longitudinally. These predominantly involve endocrine dysfunction with diabetes mellitus and thyroid dysfunction being most common. A late-onset, chronic and progressive rubella panencephalitis has also been described that bears superficial resemblance to subacute sclerosing panencephalitis.

In the presence of naturally acquired rubella-specific antibody, re-infection with rubella virus is rare. When it occurs, it usually ensues without viremia, virus shedding, clinical illness or risk to the fetus. However, re-infection with clinical illness or with transmission of virus to the fetus has been reported. Anamnestic antibody responses are typical of re-infection, with rubella-specific IgG produced. IgM antibody is not produced or found only in low concentrations. Specific high-avidity IgG can distinguish rubella re-infection from subclinical primary infection. Failure to detect specific IgG3 in sequential sera collected after rubella exposure also suggests re-infection. Re-infection can be confidently diagnosed serologically if the initial sample is IgG positive and a subsequent sample demonstrates a rise in IgG titer and low-level or negative IgM response. In cases that present late after rubella exposure, diagnosis of re-infection must rely on evidence of pre-existing rubella antibody such as by two positive laboratory assays or history of vaccination followed by one positive laboratory assay. Vaccine-induced antibody levels are often lower than those acquired by natural infection, so that as many as 2–10 times more vaccinees experience re-infections than those with natural immunity. There are a small number of cases in which rubella re-infection in women with well-documented immunity induced by vaccination has been associated with the birth of malformed infants. Fortunately, fetal damage after maternal re-infection is rare.

Pathology and Histopathology

There is limited information on histopathological changes seen with uncomplicated rubella because of the benign nature of the illness. Complicated or fatal cases can show organ-specific inflammation, but virus-specific alterations and intracellular inclusions are conspicuously absent. In congenital rubella, affected organs are small for gestational age or show a limited number of well-recognized malformations. Histologically, a noninflammatory pathology predominates; notwithstanding, focal necrosis and mononuclear inflammatory change can be found in some infants.

Immune Response

Serologic responses are measurable from the onset of the rash and evolve over several weeks. Rubella E1-specific IgM antibodies predominate early and may persist for some time after the acute infection. Within a week, antirubella virus antibodies appear in all immunoglobulin classes. The dominant early and persistent IgG response is in the IgG1 subclass and IgG1 antirubella antibody titers correlate well with classically measured hemagglutination inhibition (HAI) titers. Antibodies of this class persist indefinitely after childhood infection in healthy individuals. Immunoprecipitation studies disclose that most of the immunoglobulin response is directed to the E1 glycoprotein, with proportionally less of the response directed at E2 or C. Immunoblot analyses using denatured viral protein confirm this pattern, but show relatively more reactivity to C. Rubella-specific immune complexes are frequent but transient after uncomplicated rubella infections.

Rubella virus-specific cellular immune responses are measurable within 1–2 weeks of onset of rubella. These decline over several years but persist at low levels indefinitely following natural rubella. Lymphocyte proliferative and cytotoxic responses stimulated by rubella virus are major histocompatibility antigen (MHC) restricted. Transient depression of nonspecific lymphocyte responsiveness to mitogenic stimulation and delayed-type hypersensitivity to unrelated antigens follows natural or vaccine infections of children and adults.

Infants infected during early gestation often demonstrate prolonged impairment of rubella virus-specific cytotoxicity, cell-mediated immune responses and lymphokine secretion *in vitro*, but do produce detectable serum rubella virus-specific IgM, IgA and IgG. The prolonged persistence of virus in fetal tissue suggests that neither transferred maternal IgG nor developing fetal immune responses are sufficient to eliminate virus introduced early *in utero*, even though maternal IgG and fetal IgM can neutralize the virus *in vitro*. Newborns actively secreting virus appear to have impaired cell-mediated immunity as measured by the response of their mononuclear cells *in vitro* to mitogens, perhaps reflecting the effects of rubella virus infection on lymphocytes. The infected fetus produces α-interferon that may contribute to impairment of cellular immunity without limiting the infection *in utero*. Rubella infants infected later in gestation have cellular responses quantitatively closer to those seen following postnatal rubella.

Prevention and Control of Rubella

The most devastating consequences of natural rubella are the abortions, miscarriages, stillbirths and particularly the fetal malformations that arise from maternal infection during the first trimester of pregnancy. Their prevention through prophylactic control of virus infection was the primary goal for the development of rubella vaccines. Attenuated live virus strains have been used as vaccines against rubella world-wide. The strain currently used in the US is RA27/3. RA27/3 was originally isolated in explant cultures from fetal human kidney and then attenuated by multiple passages in WI-38 human diploid cell cultures. Vaccine strains of rubella virus typically cause subclinical infection with transient viremia in susceptible recipients but without transmission of vaccine virus between susceptibles. The RA27/3 vaccine strain produces seroconversion and protective immunity in more than 90% of vaccinees, though seroconversion typically occurs in 99% or more of seronegative subjects by 4–5 weeks postimmunization. Seroconversion rates for rubella remain high when rubella vaccine is combined with measles and mumps virus components.

Initially low, antirubella IgG avidity increases markedly over the first 4 months following vaccination of seronegative subjects. Recent vaccinees have antibody to all three structural proteins of wild-type or vaccine strain rubella virus. Antibody to E1 persists for years postvaccination, and antibodies to E1 correlate with the presence of neutralizing antibody, underscoring the importance of E1 in immunity. Rarely there are individuals who, on repeated vaccination, are HAI seronegative but show rising antirubella virus antibodies by ELISA. Cancer and associated chemotherapy can interfere with antibody production, possibly causing loss of immunity in previously vaccinated children. Cell-mediated immune responses to rubella virus antigen after vaccination are similar to those seen after natural infection.

Immunization with live attenuated strains of rubella virus has been associated with acute, usually transient arthritis. The incidence of vaccine-associated joint reactions is higher in females and adults, but occurs at a reduced frequency compared to that after natural rubella. In a recent prospective study of 44 adult rubella seronegative women undergoing immunization, approximately half developed joint symptoms or signs within 2–4 weeks of vaccination. All individuals who developed frank post-vaccine arthritis had pre-existing IgG or IgA by ELISA even though seronegative by conventional HAI assay. Arguably, this suggests that rubella-associated arthritis is a consequence of re-infection rather than primary infection.

Of serious concern in the use of a live rubella virus vaccine is the risk to the fetus in mothers who are immunized during early pregnancy. Placenta and fetal tissue may become infected. However, none of 712 children born of mothers inadvertently immunized within 3 months of conception showed malformations compatible with the congenital rubella syndrome. These data include vaccination with the earlier Cendehill and HPV77 vaccines as well as RA27/3. Thus, vaccine strains of rubella virus appear to lack the teratogenic capacity of wild-type rubella virus. Since a potential risk to the fetus still exists, pregnancy remains a contraindication to rubella vaccination. However, inadvertent vaccination of a seronegative pregnant women is not generally considered an indication for termination of pregnancy.

Given the multiple problems associated with the use of live rubella vaccines and the unknown long-term effects of viral persistence, the development of a subunit vaccine may be desirable. The E1 glycoprotein appears to be the structural protein critical to virus–host interactions and induction of long-lasting immunity. Understanding of the molecular structure of T cell and B cell determinants of E1 should facilitate the development of synthetic peptide vaccines effective in an MHC-diverse population and avoid the risks associated with current rubella vaccines.

Future Perspectives

A remarkable aspect of rubella virus is its association with a diverse group of clinical diseases including natural rubella, rubella arthritis, postinfectious encephalopathy, congenital rubella and delayed polyendocrine autoimmune and neurologic diseases. The mechanism of viral involvement in each of these diseases, however, is not fully understood. Our present understanding of disease mechanisms in the rubella virus-related syndromes is hindered by the current lack of a suitable animal model system that fully mimics the infection seen in man. Several interesting features of rubella virus pathogenesis provide areas for future research. These include determining the mechanism of viral persistence in man, elucidating the details of the teratogenesis of wild-type rubella virus, achieving a complete analysis of protective and possible aberrant viral interactions with the immune system, and the development of a subunit vaccine.

See also: Defective-interfering viruses; Diagnostic techniques; Epidemiology of viral diseases; Immune response; Persistent viral infection; Vaccines and immune response.

Further Reading

Bowden DS and Westaway EG (1989) Rubella virus products and their distribution in infected cells. *Subcell. Biochem.* 15: 203.

Dominguez G, Wang C-Y and Frey TK (1990) Sequence of the genome RNA of rubella virus: evidence for genomic rearrangment during *Togavirus* evolution. *Virology* 177: 225.

Plotkin SA (1988) Rubella vaccine. In: Plotkin SA and Mortimer Jr EA (eds) *Vaccines*, p. 235. Philadelphia: WB Saunders.

Preblud SR and Alford Jr, CA (1990) Rubella. In: Remington JS and Klein JO (eds) *Infectious Diseases of the Fetus and Newborn Infant*, p. 196. Philadelphia: WB Saunders.

Wolinsky JS (1990) Rubella virus. In: Fields BN *et al.* (eds) *Virology*, p. 815. New York: Raven Press.

Wolinsky JS *et al.* (1993) An antibody- and synthetic peptide-defined rubella virus E1 glycoprotein neutralization domain. *J. Virol* 67: 961.

RUSSIAN SPRING SUMMER ENCEPHALITIS

See Encephalitis viruses

S

S13m BACTERIOPHAGE

See ϕ174 bacteriophage and related bacteriophages

SAN MIGUEL SEA LION VIRUS

See Vesicular exanthema virus and caliciviruses of pinnipeds, cats and rabbits

SCRAPIE

See Spongiform encephalopathies

SEMLIKI FOREST VIRUS

See Sindbis and Semliki Forest virus

SENDAI VIRUS

Morio Homma and Masato Tashiro
Kobe University School of Medicine
Kobe and Jichi Medical School, Tochigi, Japan

History

Sendai virus was first discovered in 1952 by Kuroya, Ishida and Shiratori at Tohoku University Hospital, Sendai, Japan, during an epidemic of fatal pneumonitis in newborn children. A new virus with hemagglutinating activity antigenically different from known influenza viruses was recovered from mice inoculated with the autopsied lung tissue. The virus was originally named newborn pneumonitis virus, type Sendai. However, the question soon arose as to the causative agent of human disease because of wide spread of this virus among laboratory rodents and pigs in Japan at that time. Since then, Sendai virus has been shown to be prevalent world-wide and to cause enzootic or epizootic infections in mice and rats. Sendai virus was designated hemagglutinating virus of Japan (HVJ) in 1955 by the Society of Japanese Virologists. Later, Sendai virus was shown to be related to human parainfluenza virus type 1, which was first isolated by Chanock in 1955 and named hemadsorption virus type 2 (HA2).

Since Sendai virus causes a respiratory infection in mice, this virus–animal system has been widely investigated in pathogenesis and immunology as a suitable model of respiratory viral infection. Okada's finding of cell fusion by Sendai virus in 1958 has developed a new field of cell biology, e.g. the production of hybrid cells such as heterokaryon and hybridoma. Proteolytic activation of the fusion glycoprotein was first described for Sendai virus by Homma and Ohuchi in 1973. Additionally, most of the information on the molecular biology of paramyxoviruses have been obtained from studies on Sendai virus, the prototype of the family *Paramyxoviridae*.

Taxonomy and Classification

Sendai virus belongs to the genus *Paramyxovirus* of the family *Paramyxoviridae* of RNA viruses. Based on serological relationships, Sendai virus is assigned to a murine subtype of human parainfluenza virus type 1, the classification being supported by phylogenical analyses of the genome nucleotide sequences among paramyxoviruses.

Properties of Virion and Genome

The virions are pleomorphic in size and shape. They are roughly spherical and 150 to 250 nm in diameter but larger particles are common. The particles consist of a nucleocapsid with helical symmetry (18 nm in width, 1 μm in length), enclosed by a host cell-derived lipid envelope. The nucleocapsid is composed of a nonsegmented, linear, negative-strand genomic RNA (15 384 nucleotides) containing six genes, covalently linked in tandem, and associated proteins NP, P and L. The envelope contains protein M beneath the inner layer, and two spike-like projections, composed of homotetramers of glycoproteins HN and F, respectively. They penetrate the lipid bilayer beyond the outer surface to make the fuzzy appearance of the envelope surface when viewed by electron microscopy. The NP protein (58 kD), bound directly to the RNA, is the major structural component of the nucleocapsid. Proteins P (72 kD) and L (255 kD) act in concert as RNA polymerase complexes with transcriptase and replicase activities. The M protein (40 kD) has a dual affinity for the NP protein and the cytoplasmic domain of the glycoproteins and plays an important role in virus assembly. The larger glycoprotein HN (68 kD) has receptor binding, hemagglutinating and neuraminidase activities, while the smaller glycoprotein F (63 kD) has fusion and hemolytic activities and mediates virus entry through envelope fusion.

Replication

Like other members of paramyxoviruses, virus replication takes place exclusively in the cytoplasm and does not require host DNA synthesis. A one-step replication cycle takes about 15 h.

Virus entry

The HN glycoprotein mediates adsorption of virus to the sialic acid-containing receptors on the glycoproteins or glycolipids of host cell surface. After adsorption, the F glycoprotein mediates an envelope fusion between the viral envelope and the cytoplasmic membrane, by which the nucleocapsid is introduced into the cytoplasm to initiate transcription.

Characterization of transcription

The genomic RNA consists of the leader sequence at the 3′ end and a related sequence at the 5′ end. There is a set of six genes ordered NP, P/C, M, F, HN and L, and bound by the consensus transcriptional regulatory sequences for the poly(A) initiation site (E), intergenic (I) and a gene start sequence (S). The polymerase complexes associated with the nucleocapsid catalyze the primary transcription to synthesize monocistronic mRNA species corresponding to each polypeptide. Several polycistronic mRNAs may also be synthesized by reading through the stop signals at the intergenic sequence. From the P/C gene, a second mRNA species, in addition to the P mRNA of the complementary copy of the genome RNA, is synthesized by RNA editing, with insertion of an additional G residue not coded by the genome, and thereby with a new reading frame downwards for a cysteine-rich protein, V. The secondary transcription is catalyzed by the polymerase complexes of the viral proteins newly synthesized in infected cells.

Characterization of translation

Each mRNA translates a corresponding polypeptide. With the P mRNA, however, four to five nonstructural proteins (C, C′, Y, Y′ and X) are synthesized from reading frames overlapping the P gene using different initiation codons. The glycoproteins HN and F are synthesized on membrane-bound polysomes and subjected to post-translational processing.

Post-translational processing

The F protein, a type I membrane protein, is translated and inserted into the membrane of the rough endoplasmic reticulum (ER) by the signal peptide at the N terminus, which is removed after insertion. On the other hand, the HN is a type II membrane protein, which is anchored to the lipid layer at a hydrophobic region in the N terminus. Both proteins are glycosylated and during transport to the smooth ER, the carbohydrate side-chains are processed to mature forms, and the HN and F polypeptides form homotetramers by disulfide or noncovalent linkage, respectively.

This precursor form of F protein is biologically inactive, and gains the fusion activity by a proteolytic cleavage into the disulfide-linked subunits, F_1 and F_2, by host proteases. Since the cleavage site of the F protein contains a single arginine residue, it is not cleavable by ubiquitous cellular proteases present in the trans-Golgi region that cleave preferentially dibasic or multibasic motif. Accordingly, Sendai virus produced in most tissue culture cells remains inactive; lacking cell fusion and hemolytic activities, and infectivity. These activities are restored by *in vitro* treatment with trypsin or a trypsin-like serine protease present in chorioallantoic fluid of chicken eggs. By contrast, progeny virus is produced in the activated form in several host cells, such as the chorioallantoic membrane of embryonated hen's eggs, primary monkey kidney cells and mouse lungs that have specific proteases required for the cleavage of the single arginine residue of the F protein.

Strategy of RNA replication

Replication of the genome RNA is mediated by polymerase complexes composed of gene products of the primary transcripts, P, L and probably some of the nonstructural proteins. Genome replication provides templates for secondary transcription and for further amplification of the genome, and finally supplies the mature progeny genome for incorporation into virus particles. The entire nucleotide sequence of the negative-strand genomic RNA is copied complementarily to the antigenomic positive-strand RNA that serves as an intermediate template for the synthesis of progeny genome RNA. Subgenomic RNA species are frequently synthesized and incorporated into viral particles producing defective-interfering (DI) particles.

Assembly and release

The progeny genome RNA is combined with the NP protein to form a helical structure to which the P and L proteins are incorporated to form the assembled nucleocapsid.

The glycoproteins F and HN are transported through the Golgi apparatus to the cytoplasmic membrane, where they replace host membrane proteins. In polarized epithelial cells, the glycoproteins are transported to the apical membrane domain, where viral assembly occurs. The M protein accumulates beneath the cytoplasmic domain of the glycoproteins and binds the glycoprotein tails and the nucleocapsid. Viral particles are then assembled by budding with incorporation of the lipid bilayer of the host membrane to form the viral envelope. The neuraminidase that resides on the HN protein facilitates the release of viral particles from infected cells by destroying cellular receptors.

Cytopathology

Sendai virus usually produces a lytic-type cytopathic effect. Since the F protein is not activated in ordinary tissue culture cells, syncytial formation mediated by 'fusion from within' is not observed, unless trypsin is added to culture medium. Inhibition of cellular RNA or protein synthesis is not remarkable. Sendai virus can set up a variety of persistent infections in tissue culture cells without killing infected cells.

Cell Fusion

Sendai virus causes cell fusion and produces multinuclear giant cells. The F protein is responsible for the fusion activity for which proteolytic activation is required. Cell fusion falls into two categories; 'fusion from without (FFWO)' and 'fusion from within (FFWI)'. FFWO occurs in cells within a few hours after infection at high multiplicity of virus infection and does not require virus replication. FFWO has been used for production of hybrid cells, such as heterokaryons and hybridomas, and for introducing macromolecules into cells. On the other hand, FFWI occurs late in the replication cycle and is mediated by the F protein synthesized in infected cells and expressed at the cell surface.

Geographic Distribution

Retrospective serological surveillance revealed that Sendai virus initially appeared in Japan in the early 1950s and became prevalent world-wide by 1970 as a most common contaminant in laboratory rodent colonies under conventional conditions.

Host Range and Virus Propagation

Mice and rats are natural hosts of Sendai virus while other rodents including hamsters, guinea pigs and rabbits and also pigs can be infected. Suckling and weanling mice are highly susceptible to Sendai virus infection. Susceptibility to experimental infections varies considerably among mouse strains, which is probably due to differences in immunological responses. Sendai virus can be grown efficiently in the chorioallantoic cavity of embryonated chicken eggs and replicates productively in a broad spectrum of established tissue culture cell lines as well. However, exogenous trypsin is usually needed for proteolytic activation of the F glycoprotein to support multiple cycles of replication. Primary monkey kidney cells support multiplication of the virus without exogenous proteases. Sendai virus replication is detected by hemagglutination of chicken erythrocytes and the virus is identified by the hemagglutination inhibition test.

Serological Relationship and Variability

The Sendai virus core soluble (S) antigen is mainly composed of the NP protein. It exhibits a serological relationship to, but distinguishable from, human parainfluenza virus type 1. Envelope components, HN, F and M proteins, share antigenicity with their human counterparts among the members of the *Paramyxoviridae*, as determined by analyses of the genome nucleotide sequences. Sendai virus is antigenically homogeneous, i.e. there are no subtypes.

Epidemiology

Sendai virus is a frequent contaminant maintained persistently in conventional colonies of laboratory mice and rats, while no evidence for the infection in wild rodents has been obtained. Both acute enzootic and epizootic infections occur in laboratory rodents.

Curiously, in the 1950s, Sendai virus was prevalent nationwide among pigs in Japan, but by 1961 the virus had disappeared from pigs. Infection of Sendai virus in pigs has not been reported in other countries.

Transmission and Tissue Tropism

Virus replication in the respiratory tract in mice reaches a peak 4–6 days after infection, declines gradually, and virus shedding terminates within 2 weeks. However, the virus may be recovered for up to 6 weeks after infection. In suckling mice, the virus persists longer. Virus transmission occurs either by direct contact with infected mice or by airborne virus. Vertical transmission is suggested to occur from pregnant mice to fetus, resulting in stillbirths or a variety of malformations in offspring.

Sendai virus is exclusively pneumotropic in weanling and adult mice, the target tissues being restricted to the epithelial cells of the upper respiratory tract, trachea, bronchi and bronchioles. Infection of alveolar and peritoneal macrophages results in abortive infections. Neither subepithelial invasion from the surface mucosa nor spread to systemic organs via viremia occurs. In newborn and suckling mice, the virus may spread from the nasal epithelium to the brain via the olfactory route. Intracranial inoculation of the mice causes infections in the ependyme and meninges.

Pathogenicity

Pneumotropism of Sendai virus in mice cannot be

explained by cellular receptors in the lungs, since receptors for the virus are also present in other organs that are not permissive for Sendai virus infection. Instead, pneumotropism has been proposed to be primarily determined by host protease-mediated activation of the F protein. Inactive Sendai virus containing an uncleaved F protein does not replicate in any organs of mice after intranasal inoculation. However, if the F protein has previously been cleaved in F1 and F2 subunits *in vitro* by trypsin, virus will infect the respiratory mucosa and replicate in multiple cycles, since the F protein of progeny virus is cleaved in the lungs by serine protease. As a result, infected cells are increased in number and extended lung lesions are induced. In contrast, infection by protease-activation mutants whose F protein is cleavable by chymotrypsin but not by trypsin or the lung protease terminates after a single cycle of replication in the lung, because the progeny virus cannot initiate infection. Various organs of mice other than lung, which lack the protease(s) required for activation of wild-type F protein, have a potential capacity to support replication of wild-type virus but only for a single cycle. On the other hand, a pantropic mutant whose F protein is changed to be cleavable by host proteases distributed ubiquitously in various organs of mice causes a systemic infection in mice. These results, together with similar observations on the virulence of Newcastle disease virus and avian influenza viruses, indicate that host range and organ tropism of Sendai virus are primarily attributed to the presence of activating proteases of the F protein in target tissues.

The mode of virus budding at the primary target of infection is considered an additional determinant for organ tropism. The budding site of wild-type Sendai virus in the bronchial epithelial cells is restricted to the apical membrane domain, whereas the pantropic mutant buds bidirectionally at the apical and basolateral domains. This may explain why infection by wild-type virus remains localized in the surface epithelium of the respiratory tract, whereas the pantropic mutant readily invades subepithelial tissues so that it spreads into systemic organs via viremia.

Clinical Features of Infection

With enzootic infections, the virus usually produces a subclinical infection in mice and rats. Experimental infections as well as epizootic infections cause symptoms characteristic of acute respiratory infection. Moderate fever, ruffled furs and nasal discharge are common signs of the upper respiratory infection which appear 2–3 days after exposure. Dyspnea, cough, anorexia and body weight loss, which usually appear on the sixth or seventh day after infection, will indicate progression to bronchopneumonia. With weanling mice, retardation of body weight gain reflects the severity of the disease. In the second week of infection, infected animals die of bronchopneumonia or begin to recover from the infection. In newborn or 1- to 2-day-old suckling mice, intracranial inoculation causes meningitis and ependymitis resulting in hydrocephalus.

Pathology and Histopathology

Major pathology of Sendai virus infection in mice is mild rhinitis, moderate tracheitis and severe bronchopneumonia. Immunohistological studies reveal that target tissue of the virus is confined to the epithelial mucosa of the upper respiratory tract, trachea, bronchi and bronchioles. Subepithelial tissues, alveolar epithelium and infiltrating cells are not usually involved.

Macroscopically, the lungs become swollen and hyperemic in a few days after intranasal infection, and on about the seventh day, lung consolidation that looks like the liver or spleen begins to appear. Light microscopic studies reveal infected epithelial cells to be swollen and pyknotic with destruction of cilia within 1 day. Submucosal edema and hyperemia occur with a peribronchial infiltration by neutrophils and mononuclear cells. For several days, the number of such infected cells increases progressively and meanwhile these cells become necrotic and desquamated. The cellular infiltration progresses for up to 2 weeks with massive edema and bleeding in the interstitium and alveolar spaces. Resolution of the lesion, if it occurs, begins on about the tenth day and proceeds rapidly, but complete resolution takes more than a month.

When 1- to 2-day-old mice are infected intracranially, meninges, ependyme, choroid plexus and labyrynth are involved with respective inflammations.

Immune Response

Recovery from Sendai virus infection involved both

humoral and cellular immunities. Envelope glycoproteins, HN and F, are mainly responsible for the humoral response and internal proteins, NP and M, are also of importance as target antigens for the cellular response. Virus is cleared mainly by CD8+, cytotoxic T lymphocytes in a class I-restricted manner, and therefore, in nude mice, virus infection persists for more than 2 months. Macrophages, natural killer cells and interferons contribute to the virus clearance in concert with the T lymphocytes. Lung consolidation is caused by CD4+ T cells primed by the internal proteins.

Mucosal IgA antibodies to the HN are primarily responsible for resistance to infection. Serum IgG antibodies are less effective for preventing replication by initial infecting virus, but can minimize further replications and lung lesions. The role of cellular immunity in protection is controversial.

The HN molecule contains at least four antigenic sites. Antibodies against HN inhibit hemagglutinating and neuraminidase activities, and neutralize infectivity. Antibodies to the F protein, with at least four antigenic sites, inhibit fusion and hemolytic activities, and can also prevent infection, specifically cell-to-cell infection. When mice are infected, considerable titers of serum antibodies with these activities become detectable. Since the HN, F, NP and M proteins share antigenic determinants with other members of paramyxoviruses, most closely to human parainfluenza virus type 1, heterotypic antibody responses may occur when Sendai virus-primed animals are immunized or boosted with a related virus, or vice versa.

Prevention and Control

Control of Sendai virus infection is of great practical importance for maintenance of laboratory animals and for the performance of animal experiments. Contamination with the virus will interfere with the experimental data, specifically in immune responses and lung histology, and often interrupts animal experimentation by causing epizootic acute infections, with or without devastating animal death. A drop in breeding efficiency as a result of infection can be critical for maintenance of animal strains.

Virus-free animals are produced by cesarean birth and can be maintained under specific pathogen-free conditions isolated from conventional colonies. Once Sendai virus-free colonies are established, frequent serological surveillance and quarantine of contaminated colonies are needed for maintenance of laboratory animals.

Inactive vaccines prepared from egg-grown virus are commercially available, but are presently far from being in general use. Experimental live vaccines of protease-activation mutants, temperature-sensitive mutants, or defective-interfering particles with considerable efficacy have been developed. Recombinant vaccinia viruses with genes encoding Sendai virus proteins or oligopeptides corresponding to the epitopes responsible for immune protection have been shown to induce protection in mice.

See also: Defective-interfering viruses; Genetics of animal viruses; Parainfluenza viruses.

Further Reading

Brownstein DG (1986) Sendai virus. In: Bhatt PN, Jacoby RO, Morse HC III and New A (eds) *Viral and Mycoplasmal Infections of Laboratory Rodents*, p. 37. New York: Academic Press.

Galinski MS (1991) *Paramyxoviridae:* transcription and replication. *Adv. Virus Res.* 39: 129.

Ishida N and Homma M (1978) Sendai virus. *Adv. Virus Res.* 23: 349.

Kingsbury DW (1991) *The Paramyxoviruses*. New York: Plenum Press.

Okada Y (1988) Sendai virus-mediated cell fusion. *Curr. Top. Memb. Trans.* 32: 297.

SHEEP POXVIRUS

See Poxviruses

SHOPE FIBROMA VIRUS

See Poxviruses

SHOPE PAPILLOMA AND BOVINE PAPILLOMAVIRUSES

Gary L Bream and William C Phelps
Burroughs Wellcome Company
Research Triangle Park, North Carolina, USA

History

Warts or papillomas have been recognized in animals for centuries; a stablemaster for the Caliph of Bagdad described equine warts in the ninth century AD. A viral involvement in papillomas was first demonstrated in 1907 when Ciuffo showed that sterile filtrates were capable of transmitting warts between humans. Bovine warts were also recognized to be contagious; however, demonstration of viral transmission did not occur until 1929 and it was not until 1978 when it was realized that bovine lesions could result from infection by more than one viral type. Warts of wild rabbits were first documented in the scientific literature in 1909 by the naturalist Ernest Seton. In 1933 Richard Shope succeeded in isolating the cottontail rabbit papillomavirus (CRPV) or Shope papillomavirus.

In 1935, Peyton Rous observed that benign rabbit papillomas occasionally progressed to carcinoma, which was one of the earliest indications of a virus's involvement in cancer. This model has been instrumental in the study of viral oncology and neoplastic progression, and has contributed to our recognition of cancer as a multistep process. The potential for malignant conversion of benign lesions is characteristic of the papillomaviruses (PV) and has contributed to the recognition of the role of the human papillomaviruses (HPV) in human genital malignancies such as cervical cancer.

Owing to a shortage of viral material and a lack of a permissive *in vitro* system for virus replication, progress in papillomavirus research has historically been relatively slow. The molecular cloning of the viral DNAs in the early 1980s greatly expanded the capacity for experimental manipulation of the viral genomes and brought renewed vigor to papillomavirus research.

Taxonomy and Classification

Papillomaviruses (PV) belong to genus A of the *Papovaviridae* family of DNA viruses. Viral types are distinguished on the basis of host range and DNA sequence relatedness. By convention a new virus type has less than 50% overall DNA sequence homology to other types of the same species, as determined by solution hybridization; anything greater than 50% but less than 100% is classified as a new subtype. Animal PVs are further subdivided on the basis of tissue tropism. Four major groups are currently recognized. Group I (CRPV and bovine papillomavirus (BPV) types 3 and 6) induce neoplasia of cutaneous epithelium. Group II (BPV-4) induce hyperplasia of nonstratified squamous epithelium. Group III (BPV types 1, 2 and 5) induce an underlying fibroma in addition to a cutaneous papilloma. Group IV (deer papillomavirus) induce primarily a fibroma with minimal hyperplasia of the cutaneous epithelium.

The virus contains a double-stranded, covalently closed, circular DNA of approximately 8000 bp enclosed within a 55 nm, lipid-free, icosahedral capsid. In comparison, SV40 and polyomavirus of the genus B polyomaviruses have a 45 nm virus particle containing an approximately 5000 bp genome. The PV virion consists of two structural proteins arranged in 72 capsomeres which represent 88% of the mass of the viral particle. The major structural polypeptide L1 is 55 kD; the minor capsid protein L2 is 72 kD. The viral DNA is complexed with cellular histones which are condensed into nucleosomes; viral DNA replication occurs within the host cell nucleus.

The virion is resistant to heat, desiccation and, because it lacks a lipid envelope, to ether and other lipid solvents. The particle is stable between pH 3 and 7 and tends to be more resistant to X rays than most viruses. The virus can be disrupted by alkali, phenol and detergents.

Geographic Distribution

PVs occur world-wide; however, virus types can display some geographic variability. For instance CRPV is endemic only in the cottontail rabbits of the midwestern US even though the host is widespread. Likewise, BPV-4 induced enzootic bovine hematuria is restricted to geographic regions where the cattle graze on bracken fern even though BPV-4 infection occurs over much larger regions. By contrast, BPV-induced cutaneous fibropapillomas have a world-wide distribution, yet in the US, lesions are normally found on the haired skin of the head, neck, chin, shoulder and dewlap while those in Germany are most common on the neck, ventral abdomen, legs, back and udder.

Factors which contribute to this diversity include multiple virus types, local differences in health and immunological status of the host, absence of vectors or other cofactors, and local variations in host management practices such as cattle breeding.

Host Range and Virus Propagation

PVs infect most higher vertebrates being especially prevalent in mammals, although they are also known to infect birds (Table 1). A PV or PV-specific antigen has been identified in the lesions of at least 30 different species. Papilloma-like lesions also occur in a number of other organisms including nonvertebrates such as clams, although these have yet to be definitively associated with a virus.

PVs show a pronounced host and tissue specificity. CRPV is restricted to the cutaneous epithelium of rabbits. The virus naturally infects wild cottontail rabbits but will also induce productive lesions in jackrabbits and snowshoe rabbits by experimental infection. Experimental infection also induces benign papillomas in domestic rabbits; however, because of genotypic differences between the wild and domestic rabbits, these lesions are strictly nonproductive. In addition, these lesions are three times more likely to progress to carcinoma.

Table 1. Characterized animal papillomaviruses

Host	Lesion
Cattle (6 virus types)	Fibropapilloma, papilloma
Deer (6 virus types)	Fibropapilloma, papilloma, fibroma
Horses (2 virus types)	Papilloma
Rabbits (2 virus types)	Papilloma
Dog (1 virus type)	Papilloma
Rodent (1 virus type)	Keratoacanthoma, papilloma
Birds (2 virus types)	Papilloma

BPV types 1 and 2 exhibit a slightly broader host range and tissue tropism, causing endemic fibropapillomas in cattle and nonproductive sarcoids in horses, and inducing nonproductive fibromas in hamsters, rabbits, pikas and mice after experimental infection. Experimental infection of cattle can induce benign tumors of the urinary bladder and genital mucosa, meningiomas after intracranial injection and moderate fibroblastic stimulation in diverse internal organs. The other BPV types (BPV types 3–6) have been shown by both natural and experimental infection to be restricted to cattle. Transmission from cow to man was suspected from the high incidence of cutaneous warts in butchers; however, the virus isolated from these lesions does not appear to be related to any known BPV type.

A transgenic mouse line has recently been developed with BPV-1 which, after 8–9 months, will develop nonproductive fibropapillomas on areas of skin which are prone to wounding. The warts occasionally become malignant and are locally invasive but have not been observed to metastasize. Interestingly, free episomal BPV-1 DNA is exclusively found in the wart tissue, while in all normal tissues including skin the BPV DNA is integrated and transcriptionally silent. This model may be useful for studying the secondary factors required for malignant progression.

As yet, no *in vitro* system exists for the propagation of virus and recovery of infectious particles. Limited supplies of CRPV are available since the disease is sporadic in the wild and nonproductive in laboratory animals. BPV can be recovered from experimental infections of cattle and is therefore in relatively higher supply. Recent efforts have focused on the development of surrogate animal systems in which cells from the host species are infected with virus and then transplanted to the renal capsule of athymic mice. This technique has very recently been successfully used for the limited propagation of CRPV and BPV in the laboratory.

Genetics

The emergence of techniques for molecular cloning in the late 1970s permitted a single viral genome to be isolated and propagated in the laboratory which has provided an unlimited source of homogenous viral DNA for study. In 1980, Doug Lowy and co-workers showed that BPV-1 virus and cloned viral DNA could morphologically transform established mouse cell lines. The viral DNA was shown to persist extrachromosomally and replicate as a stable episome. As a result of these features, BPV-1 has been the model system for the genetic analysis of the viral transformation and replication functions.

The genetic information of all PVs is expressed from a single strand of the double-stranded genome which, by convention with the polyomaviruses, is divided into three regions (Fig. 1). The long control region (LCR) is a 1 kb sequence containing many of the viral transcriptional regulatory signals as well as the origin of replication. The early region contains those opening reading frames (ORF) expressed primarily in nonproductively infected cells (labelled E 1–8 in descending order of size). The late region contains the ORFs for the capsid proteins L1 and L2, which are expressed only in productively infected cells and are dispensable for *in vitro* transformation and replication of BPV-1 in mouse cells. A polyadenylation site (poly(A)) for the early message is located between the early and late regions while the late transcripts use a second poly(A) site in the LCR. Transcriptional termination within the late region occurs in nonproductively infected cells and may prohibit expression of the late mRNAs in these cells.

Seven transcriptional promoters have been identified in the BPV-1 genome, of which one (the presumptive late promoter) is expressed only in productive warts. The activities of the six early promoters are regulated by host factors and the virally encoded E2 proteins. E2 exists in three forms within the cell, the result of expression from multiple promoters and alternative RNA splicing. Two have sequence-specific DNA binding and dimerization activity (E8/E2 and E2TR) while the third (E2TA) has, in addition, a transcriptional activating domain. E2TA stimulates viral promoters by binding to multiple 12 bp palindromic sequences in an E2-specific enhancer within the LCR. This activity is modulated by the E8/E2 and E2TR proteins which abrogate E2TA stimulation.

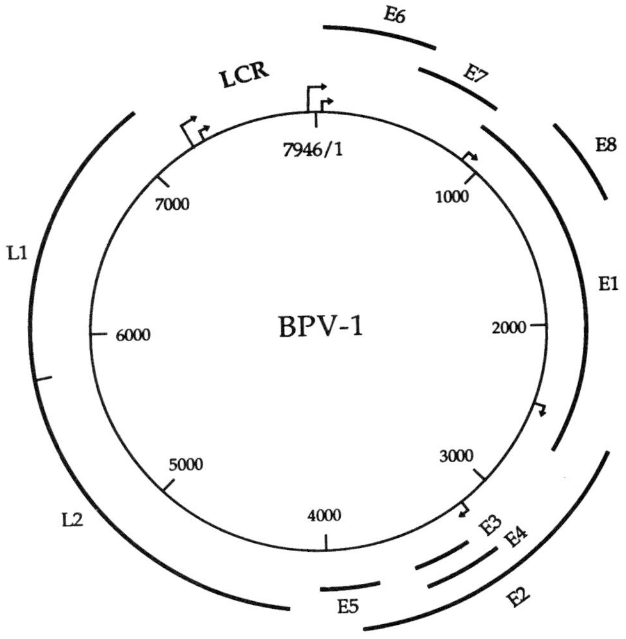

Fig. 1 The circular BPV-1 genome. Nucleotide numbers are noted within the map. E and L designate the early and late open reading frames respectively; promoters are indicated by arrows. LCR indicates the long control region.

E6 is a Zn^{2+} binding protein found in the nucleus and nonnuclear membranes. It is capable of transforming mouse C127 cells but not mouse NIH3T3 cells. The E7 protein has no effect on colony formation but does affect the extent of colony growth. Both E6 and E7 are thought to have a function in controlling the virus copy number in BPV transformed mouse cell lines. Recently, the E6 protein has been shown to have transcriptional activating activity in a heterologous assay system. The E6 and E7 proteins of the HPVs are known to bind to the cellular proteins p53 and pRB, respectively, both of which have been shown to have tumor suppressor activities. No analogous functions have yet been associated with the BPV or CRPV viral proteins.

E5 plays a central role in transformation of the established mouse cell lines C127 and NIH3T3 and, at 44 amino acids, is the smallest known transforming protein. It is a membrane protein localized to the Golgi apparatus and plasma membrane and is well conserved among the animal PVs which cause fibropapillomas. Acute expression of E5 results in rapid morphological transformation of C127 cells indicating that E5 is sufficient for transformation and that no secondary cellular event is required. Recent evidence suggests that E5 functions by activating the β type receptor for platelet-derived growth factor (PDGF).

E4 is a cytoplasmic phosphoprotein which is poorly conserved among the PVs. It is thought to be a late-associated protein since it is found in high concentration in the productively infected cells of the wart. Its function is unknown.

E1 is a nuclear phosphoprotein required for replication of the viral DNA. It requires for activity an ATP-binding sequence homologous to the ATP-binding domain of SV40 large T antigen. The viral E1 and E2TA proteins are both required for extrachromosomal replication of the viral DNA and a physical association of these two proteins will specifically localize the E1 protein to the E2 enhancer in the LCR. In BPV-1 transformed mouse cells, replication initiates within or near this enhancer. Viral replication also requires host replication factors which includes the host DNA polymerase.

A second protein translated from the E1 ORF is called the E1M protein. It contains only the N-terminal information of the E1 ORF and is thought specifically to prevent over-replication of the viral genome.

Evolution

All human and animal PVs sequenced to date have a similar genomic organization. Multiple, minimally related serological types occur within a single species implying that PVs evolve in response to pressures from the host immune system, especially since regions of the capsid proteins not exposed to the immune system have remained conserved. The PVs do not appear to be evolving at a fast rate; DNA sequence of two different isolates of BPV-1, one isolated in the US the other in Sweden, only differed by five nucleotides out of 8000.

Serologic Relationships and Variability

PVs exhibit both type-specific and genus-specific antigenic determinants. Antisera prepared against intact virions of one type are specific to that virus type. Antisera to detergent-disrupted virus particles reveal an internal, genus-specific antigenic determinant which is broadly cross-reactive against the spectrum of PVs. These sera have been used successfully to screen tissue from papilloma-like lesions for the presence of PV antigens.

Transmission and Tissue Tropism

Animal PV lesions are predominantly cutaneous, although some are specific for the mucosal epithelium. Most display a distinct cellular tropism for squamous epithelial cells with the exception of BPV types 1, 2 and 5 and certain PVs of other ruminants which also induce a dermal fibroblastic proliferation giving rise to a fibropapilloma or fibroma. BPV types 1 and 2 induced fibropapillomas are the most common skin tumor of cattle but can also occur on the less fully keratinized epithelium of the rumen, omasum, vagina, vulva, penis and anus. BPV-3 induces true papillomas in cattle and is transmissible to the skin of adult cows but not calves. BPV-4 also induces true epithelial papillomas commonly of the nonstratified epithelium of the alimentary tract. BPV-4 can be transmitted experimentally to the soft palate but not to the skin. BPV-5 is associated with 'rice-grain' lesions on the teat; BPV-6 infection causes a true papilloma which is also associated with the teat.

BPVs are readily transmitted in herd animals, principally through direct contact of the virus with abraded skin. Natural BPV infection of horses generally occurs after placing the animals in stalls which previously housed cattle. Sexual transmission of venereal warts in cattle is likely since such lesions are rare in animals that are artificially inseminated. PV virions have been detected in milk, suggesting that mammary epithelium may be susceptible to infection and could represent another route for transmission.

CRPV induces strictly cutaneous lesions around the head, shoulder, neck, back, abdomen and inner thigh. In addition to direct contact, CRPV infections can be spread experimentally by nematodes, mosquitoes, reduviid bugs and rabbit ticks but whether these are a source of natural transmission is not clear.

Infection requires direct access to the basal cells of the epithelium such as in a wound and may also require the activation of cell division which occurs during healing. The viral DNA replicates as a stable episome in the basal cells, as well as the dermal fibroblasts in BPV fibropapillomas; however, virus particle production occurs only in the terminally differentiating cells of the epithelia. A role for virus latency in PV infection is unclear; however, a reservoir of latent viral DNA may remain in asymptomatic epithelia and dermal fibroblasts.

Pathogenicity

PVs induce benign neoplasms which spontaneously regress and therefore present little danger to the host, although severe outbreaks of warts on immunocompromised animals could lead to death in the wild. Papillomas induced by CRPV and BPV-4 can undergo malignant transformation to squamous cell carcinoma. These tumors are highly invasive and will metastasize to the lungs and lymph resulting in the death of the host.

Progression of virus-induced benign tumors to carcinoma is a multifactorial process in which PVs are apparently necessary, but probably not sufficient, for malignant transformation. Progression to malignancy probably requires multiple secondary events, the occurence of these events being influenced by virus, host and external factors. Virus-specific effects are evident among the BPV viruses where, of the six known viral types, the benign lesions of only one (BPV-4) are observed to progress to carcinoma. Host-specific effects are evident in rabbits, where genotypic differences between wild and domestic varieties result in a threefold increase in the incidence of malignant progression of CRPV-induced carcinomas in the domestic host. External or environmental factors may also influence malignant conversion in both the rabbit and bovine systems. BPV-4 induced alimentary papillomas were observed to become malignant only in areas where the cattle grazed on bracken fern. Bracken fern contains radiomimetic and immunosuppressive agents but only when grown in specific geographic regions. It is only within these regions that bracken fern can act synergistically with BPV-4 to promote malignant transformation. In a similar fashion, tumorigenic substances such as methylcholanthrene increase the rate and frequency with which CRPV-induced lesions progress to carcinoma.

Clinical Features of Infection

PV infection is not associated with overt clinical symptoms. In cattle, exposure generally occurs in young calves and induces adequate immunity to account for the low incidence of disease in adults.

During natural outbreaks, BPV lesions arise in 101–126 days after exposure and regress spontaneously 18–173 days after formation of the wart. Little is understood about clinically inapparent primary and latent infections. For the domestic rabbit oral papillomavirus, a virus distinct from CRPV, viral DNA can be detected in clinically normal tissue and virus shedding is even known to occur in the absence of any clinical signs. These healthy animals will often develop papillomas later in life.

Pathology and Histopathology

Fibropapillomas in cattle occur either singly or as multiple nodules and may reach several centimeters in size. They may appear sessile or pedunculate and lobate, fungiform or verrucate. CRPV-induced papillomas occur as multiple, gray or black, well-keratinized masses. They range in size from 0.5 to 1 cm in diameter and can reach several centimeters in height resulting in cutaneous 'horns'. The carcinoma appears as a raised, fungoid or eroding lesion.

The PV life cycle is exquisitely atuned to the progressive vertical differentiation that occurs during maturation of the epidermis and it is presumably our inability to mimic precisely all phases of keratinocyte differentiation *in vitro* which has precluded the establishment of productive PV infections in cell culture. The actively dividing basal cells of the epithelium presumably are the target cells for initial viral infection and are believed subsequently to maintain the virus in a proviral, possibly latent, state. The suprabasal daughter cells of the stratum spinosum and stratum granulosum no longer divide and are commited to terminal differentiation which includes progressive keratinization and enucleation. Virus-induced hyperplasia, which is believed to be induced by the viral early gene products, results from increased division of the basal cells and delayed maturation of the commited keratinocytes of the spinous layer (acanthosis). The mass of cells coalesce into numerous papillary extensions and contain many koilocytic cells with clear cytoplasm. Late gene expression and virus particles are first detected in cells of the stratum spinosum. Vegetative viral replication begins in this layer and is likely to be dependent on the specific cellular environment of these more highly differentiated cells. Virus-specific cytopathic effects are most pronounced in the stratum granulosum; virus particles are found throughout the nucleus frequently organized into paracrystalline arrays.

Fibropapillomas show a marked proliferation of the underlying fibroblasts within a dense matrix of collagen. Fibroplasia occurs within the first week after infection while hyperplasia of the epithelia may not be visible before 4–6 weeks. This long delay in the appearance of epithelial changes is similar to that observed with the strictly epithelial PVs such as CRPV where it may take 3–8 weeks for a papilloma to appear.

Immune Response

The immune response to PV infection is not well understood. Warts persist for variable periods of time. They can then spontaneously regress, leaving the host immune to re-infection with the same PV type. Once regression of a single wart begins all papillomas begin to regress, indicating that the mechanism of immunity is systemic. The immune system is the major determinant of regression since animals which have had their immune system suppressed by methylprednisolone show almost a 20-fold reduction in regression frequency.

Regressing warts contain a mononuclear cell infiltrate. Leukocytes are present near the basement membrane but do not coincide with the areas of reduced cell proliferation, suggesting that soluble lymphokine-like substances are involved. Lymph node cells from regressor and nonregressor rabbits inhibit *in vitro* colony formation by papilloma- and carcinoma-derived cells. However, nonregressor sera can specifically block the inhibitory effect of lymph node cells, suggesting that immunosuppressive humoral factors may contribute to the persistence of tumors. These results also suggest that a cell-mediated response may be the primary factor for regression.

In rabbits, CRPV neutralizing antibodies (Ab) generated against the two structural peptides protect the animal from re-infection. However, these Ab are not responsible for regression since they are present in animals with both regressing and persistent warts. When Ab neutralization of innoculated virus is avoided by transplantation of infected skin or by transfection with viral DNA, papillomas occur in animals with persistent warts but not in those where the warts have regressed, again suggesting cell-mediated immunity is the primary factor in rejection.

While neutralizing Ab are not sufficient to bring about tumor rejection, vaccination with tumor cell preparations has been shown in some cases to increase the rate of regression. Regression and immunity to subsequent PV infections may result from a cell-mediated reaction against a tumor-associated antigen as opposed to a virus-specific antigen; however, such a cellular determinant has not been identified.

The information on the immune response of cattle to BPV infection is limited. Cattle demonstrate a typical primary response to antigen, with the development of a full IgG response by 6 weeks. Like CRPV, the presence of these Ab does not correlate with regression; the primary indicator for immunity remains the previous rejection of warts. Cell-mediated immunity is again thought to be the principle mechanism for rejection but unlike CRPV, vaccination with autologous tumor preparations has generally yielded poor results.

Prevention and Control

Treatment of papillomas is difficult to evaluate since the disease is self-limiting and the duration varies. There is little economic loss associated with animal PV outbreaks so most attempts at treating these lesions focus on an application toward human infections.

Inoculation with homogenized, autologous wart tissue has been successful in stimulating regression of CRPV-induced warts. Aromatic retinoids also successfully inhibit the induction and development of CRPV-induced lesions; up to 60% of the rabbits tested had complete and permanent regression of well-established tumors and all showed marked reduction of growth. In the BPV-1 system retinoids act primarily as a suppressive agent, failing to permanently cure transformed mouse cell lines of viral DNA. The BPV-1 system has responded more successfully to interferon. Mouse β-interferon treatment will cure a small fraction of C127 mouse cells harboring viral DNA and reduce the copy number in the others. Cured cells are no longer morphologically transformed and fail to raise tumors in nude mice. A photodynamic therapy using hematoporphyrin derivative (HPD) has recently been used to treat CRPV tumors. HPD is preferentially retained in warts over the surrounding normal tissue and is toxic to cells when activated by light. Exposure to white light results in loss of the warts and no regrowth for at least 18 months.

Prophylactic vaccination has been successful in cattle using a commercially available, formalinized homogenate of bovine fibropapillomas which is applied intradermally in two doses. Vaccinated herds show a lower wart incidence and cows are protected from experimental inoculations by intradermal injection or scarification. Vaccination with viral capsid proteins has also been encouraging: 12 out of 13 calves were successfully protected from an intradermal challenge of BPV-1 after an intramuscular inoculation with 5–10 mg of a bacterially expressed, BPV-1 L1 protein.

In a typical PV infection, virus and virus capsid antigens are present only in the more superficial regions of the skin which may not be readily accessible to circulating Ab. A more useful target for vaccine preparation may be the early proteins which might then target the proliferating cells of the lower epithelial strata. The current challenge is to determine which of these proteins are immunogenic and how these proteins can be successfully presented to the hosts immune system.

Future Perspectives

The lack of a productive *in vitro* system for passaging virus has severely hindered the study of PVs. *In vitro* transformation systems have partially circumvented the problem but a primary goal for PV research continues to be the development of a system which mimics all facets of the virus life cycle. Many questions remain to be answered, including what virus and host factors are responsible for malignant progression and what is the basis for the differences in tissue tropism exhibited by the many viral types. More also needs to be learned about the interaction of the virus with the host immune system, especially as it relates to regression. Is there a role for latency in the virus life cycle and what signals might activate latent virus? Finally, how can the animal models currently available be used or adapted to study drug efficacy in the treatment of PV infections in humans?

See also: Immune response; Latency; Papillomaviruses–human; Transformation; Virus–host cell interactions.

Further Reading

Lancaster WD and Olson C (1982) Animal papillomaviruses. *Microbiol. Rev.* 46: 191.

Pfister H (ed.) (1990) *Papillomaviruses and Human Cancer.* Boca Raton: CRC Press.

Salzman NP and Howley PM (eds) (1987) *The Papovaviridae; The Papillomaviruses.* Series *The Viruses.* New York: Plenum Press.

Sundberg JP (1987) Papillomavirus infections in animals. In: Syrjanen K, Gissman L and Koss LG (eds), *Papillomaviruses and Human Disease.* Berlin: Springer Verlag.

SIGMA RHABDOVIRUSES

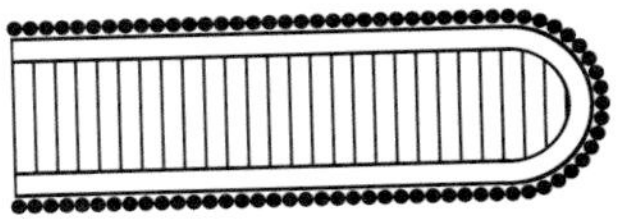

Danielle Teninges
Centre National de la Recherche Scientifique
Gif-Sur-Yvette, Cedex, France

History

The sigma virus of *Drosophila* is a harmless virus of a harmless insect and probably would never have attracted any attention were it not for the fact that geneticists use carbon dioxide (CO_2) as a mild narcotic for the flies they handle. This gas is generally not noxious and flies recover from narcosis within a few minutes after return to a normal atmosphere. Those infected by the sigma virus, in contrast, remain irreversibly paralyzed and die. CO_2 sensitivity in some *Drosophila* strains was first reported in 1937 by Ph. L'Heritier and G. Tessier as a hereditary trait which was not chromosome-linked. The viral etiology was not suspected at first due to the complete absence of horizontal transmission of CO_2 sensitivity in natural conditions. Later, L'Heritier observed that inoculation of acellular extracts from CO_2-sensitive flies into resistant flies produced the symptom after an incubation period which increased with the dilution of the inoculum. The size of the agent was deduced from the target size to X-ray inactivation. Studies by electron microscopy confirmed this information and identified sigma virus as a member of the rhabdovirus family. With its genetic background, *Drosophila melanogaster* was an ideal host to allow genetic analyses of the host factors involved in a virus–insect association and most efforts have been centered on researches of this kind.

Pathology

The CO_2 sensitivity symptom is the only sign of disease observed in infected flies. It affects larvae as well as adults and persists throughout their lives. The range of CO_2 concentrations that are lethal to the flies is not frequently met in nature and this cannot account for any counter-selection of infected flies. Among inoculated flies the expression of CO_2 sensitivity and the appearance of infectious material in the central nervous system are correlated. Immunocytochemistry reveals the presence of granular inclusions of viral material in the cortex of the cephalic and thoracic ganglia. In hereditarily infected flies, such inclusion bodies are present in all tissues, except muscle. Wild sigma virus-infected flies have been compared to uninfected flies for egg viability, male and female fertility, female longevity and sexual selection. Sigma virus infection only reduced egg viability whereas other parameters did not show any systematic or significant variations. Survival of adults under winter conditions in France was also studied and showed that the fitness of infected flies to overwintering was slightly lower. Wild sigma virus clones show a very low multiplication rate which may account for their innocuousness. Laboratory strains selected for high multiplication rates generally affect fertility and egg viability in such proportions that when they appear in natural populations they are probably severely counter-selected.

In *Drosophila* cells cultivated *in vitro*, sigma virus multiplies without any evidence of a cytopathic effect and a persistent infection is maintained throughout cell transfers.

Since a number of other rhabdoviruses and a bunyavirus (but no flaviviruses) were shown to induce CO_2 sensitivity after inoculation to *Drosophila* flies or to mosquitoes, exposure to CO_2 was proposed as a fast means of screening infected insects in nature.

Virion and Genome Structure

Sigma virus particles exhibit all the structural details of typical rhabdoviruses: a bullet shape, a coiled nucleocapsid, an envelope derived from the host cell and covered with surface projections. The diameter is 75 nm and the length 200 nm. The presence of shorter particles has also been recorded. The genome structure of a strain selected for high yields in *Drosophila* cell cultures *in vitro* has been studied. It consists of a single segment of negative-strand RNA of approximately 13 000 nucleotides. It encodes six proteins mapped in the following order on the genome strand: 3′ N-P-X-M-G-L 5′ where N is the nucleoprotein, P a phosphoprotein which is the equivalent of the polymerase-associated protein (also named NS) in VSV, X a protein of unknown function with many potential phosphorylation sites and which has no counterpart in other rhabdoviruses at the present time, a matrix protein M, a glycoprotein G and a polymerase L. The respective lengths predicted for the polypeptides are: 450 amino acids for N, 322 for P, 296 for X, 207 for M, 499 for G. The exact length of L is still unknown but its electrophoretic mobility corresponds to a protein of 250 kD. The consensus gene-start and gene-end sequences are, respectively, CAACANC and $CAUG(A)_7$ in the mRNA sense and thus conform to the consensus AACA and $AUG(A)_7$ observed in all the rhabdoviruses previously analyzed. Intergenic sequences are thought to intervene in transcriptional control. Some rhabdoviruses like VSV have constant intergenes, whereas others, as rabies virus, show variation. Sigma virus belongs to the second category and has 36 untranscribed nucleotides which separate N and P, six which separate P and X and four which separate X and M, then G overlaps M, starting 26 nucleotides upstream of the end of the M gene, so the polymerase has to read through the end sequence of the M gene to transcribe the G gene. In spite of this overlap, the transcription results in the regular synthesis of monocistronic M and G mRNAs. The start signal of the L gene is immediately adjacent to the end signal of the G gene. As deduced from UV target size analyses, the transcription of unsegmented negative-strand viruses proceeds sequentially from the 3′ extracistronic region of the genome to the 5′ end. A gene overlap implies that sites of initiation other than the most 3′ proximal one can be used by newly entered polymerases. Gene overlap could be a means of reducing the expression of the two most distal genes from the 3′ end.

Serology and Taxonomy

Fifty animal rhabdoviruses of diverse origins were

screened for antigenic cross-reactivity. Most of them could be classified in the two classical subgroups: the lyssaviruses and the vesiculoviruses. In this study, sigma virus was shown to cross-react with Tupaia, a vesiculovirus isolated from a shrew, and with Tibrogargan, Humpty doo and Parry creek, three lyssaviruses isolated from mosquitoes; it was thus proposed to bridge the gap between the two subgroups. Considering other criteria such as the scores of similarity between predicted protein sequences, sigma virus appears almost equally distant from the prototypes of both subgroups. The progress of molecular data on other rhabdoviruses may raise the necessity for the definition of new subgroups more closely related to sigma virus.

Virus–Host Interactions

Hereditary transmission is a very efficient means of propagation but its efficiency varies according to the virus genotype and to the sex and genotype of flies. 'Stabilized' females transmit the virus to all their descendants with few exceptions. They issue from oocytes that were infected very early, and in which the virus was able to multiply enough to invade all the germ line cells from the outset. Their male progeny may also transmit the virus but with a lower efficiency (it was shown that infection via a spermatozoid was an efficient means of cloning virus particles). The pattern of virus transmission from flies hatched from spermatozoid-infected eggs is different from that of stabilized females: males do not transmit the virus and females may transmit it but only to a small proportion of their descendants. This is an effect of the low virus genome concentration in the spermatozoid-infected eggs: cell differentiation and organogenesis probably result in barriers against the penetration of the virus into some tissues. There could be a race during embryogenesis between the virus invasion and the building up of these barriers. Male germ cells which are not infected at the very early stages of embryogenesis (before segmentation) are not invaded at a later stage of development. Female germ cells may be invaded at any stage, including adults, provided that the virus genotype is g^+ (g^- mutants are normally perpetuated in stabilized maternal lines and virions are infectious to somatic cells but they cannot invade the germinal cysts, once isolated by organogenesis). Some stabilized maternal lines (the ultra-ρ lines) perpetuate defective viruses which are not infectious and which do not induce CO_2 sensitivity, the only sign of their presence is an immunity to superinfection by homologous virus, immunity which shows exactly the same inheritance pattern as nondefective viruses. In other stabilized lines (the ρ lines) the same phenotype was observed but a still unexplained genetic instability of the viral genomes infecting these lines resulted in the occasional production of fully infectious virions. Some stabilized maternal lines, infected with temperature-sensitive mutants for maturation functions, express exactly the same phenotype as the ultra-ρ lines when bred for several generations at nonpermissive temperature, but the CO_2 sensitivity symptom is restored upon return to the permissive temperature. The persistent infection of germ cells through generations by viruses which are defective (or temperature restricted) proves that only the maturation functions are affected in these mutants and not the genome replication. It also proves that viral information may be transmitted by cellular continuity without a necessity for infectious virus production.

Natural populations of *Drosophila melanogaster* are often polymorphic for alleles of genes that confer resistance to sigma virus infection. These restrictive alleles map to six different loci: *ref(1)H, ref(2)M, ref(2)P, ref(3)O, ref(3)V* and *ref(3)D* (the number in parentheses represents the *Drosophila* chromosome carrying the gene, the capital letter is the particular name of the gene, and specific alleles of those genes are indicated by an exponent). The refractory loci (*ref*) do not represent a general antiviral system: not only do they not affect other insect viruses but none of them confers resistance to all strains of sigma virus (with a possible exception for *ref(3)V*; see later). Restrictive effects have been assessed for several parameters of viral infection such as the mean incubation time required for the expression of CO_2 sensitivity, the probability of initiating infection, the kinetics of virus production (either in flies or in cells cultivated *in vitro*) and the efficiency of virus transmission either by females or by males. A distinction can be made between restriction during either virus maturation or earlier stages of virus production (i.e. genome replication): since hereditary transmission in stabilized maternal lines does not require the production of infective virus, those alleles restricting this transmission necessarily affect the genome replication steps. The four alleles $ref(1)H^h$, $ref(2)M^m$, $ref(2)P^p$ and $ref(3)D^d$ reduce the probability of initiating infection and increase the incubation time for the manifestation of the CO_2 sensitivity symptom. The transmission through the maternal gametes is also reduced very strongly by $ref(1)H^h$ and $ref(2)P^p$, whereas $ref(2)M^m$ and $ref(3)D^d$

exert a weaker action. This implies that the products of these four genes intervene in the replication of the viral genomes. *Ref(3)O^e* does not affect the hereditary transmission but increases the incubation time of the viral clones sensitive to its action. The allele *ref(1)H^h* is the only fully dominant allele, flies heterozygous at all other loci express intermediate phenotypes. The *ref(3)V^p* restrictive allele does not affect the virus multiplication in somatic cells nor in female germ cells and its unique effect is to prevent transmission by spermatozoa. The specificity of this interaction with sigma virus is still an unanswered question since to date no viral strains have been found to be resistant. In contrast, the major effects described for the restrictive alleles at other loci apply only to sensitive viral strains.

Flies that are homozygous for either permissive or restrictive alleles are indistinguishable except for the difference in their capacity to permit the multiplication of sigma virus. The most extensively studied refractory gene is *ref(2)P*. Loss-of-function alleles (*ref(2)P^{null}*) were induced by mutagenesis. Homozygous flies for these alleles are all viable and display no phenotype other than male sterility. The mutation exclusively affects the structure of the spermatozoa in which the organization of the mitochondrion and axonema is perturbed. Nevertheless, the gene is expressed in a wide spectrum of tissues: it is expressed in female germ cells as seen from the strong inhibition of the virus transovarian transmission by restrictive alleles, but females are normally fertile. It is also expressed in somatic cells as seen, for instance, from the effects on virus yields and CO_2 sensitivity. This expression is autonomous: in organs transplanted into individuals bearing permissive or restrictive alleles, the action of *ref(2)P* on the virus conforms to the organ genotype. The effect of restrictive alleles is also observed at the single cell level in permanent *in vitro* cell lines. Homozygotes *ref(2)P^{null}* are permissive to sigma virus infection, thus the *ref(2)P* gene product serves no indispensable function in the virus cycle. The cloning and the sequencing of this gene did not throw much light on the role of the gene in the physiology of the fly: it encodes a single protein of 599 amino acids which shows no homology to known protein sequences. All the transcripts found in different tissues result from the same splicing pattern of three exons but different initiation sites are used. As a consequence, the 5′ untranslated region varies in length according to the cell type. The size distribution of the transcript varies according to the tissue: long transcripts are in a higher proportion in ovaries. This suggests a possible tissue-specific regulation of the gene expression at the transcriptional and possibly at the translational level.

Some alleles of a gene located on the third chromosome suppress the male sterility of *ref(2)P^{null}* homozygotes. These alleles are frequent in wild populations. The meaning of the distribution of restrictive alleles in nature is still unknown as it seems to be independent from any selective pressures exerted by sigma virus infection.

Temperature-sensitive mutants are frequent among viral mutants which escape the restrictive effect of *ref(2)P^p* (*hap* mutants). This high frequency suggests that both phenotypes result from the same mutation. Since temperature-sensitive gene products are generally proteins, the restrictive interaction between the *ref(2)P* gene and sigma virus most likely occurs at the protein–protein level. Another interesting observation is the simultaneous adaptation or disadaptation of a number of *hap* mutants to other *ref* genes such as *ref(1)H^h*, *ref(3)O^e* or *ref(2)M^m*. This co-variation suggests that the same viral protein interacts with the cellular proteins encoded by these *ref* genes.

Replication

The time between inoculation and the first detection of virus progeny is about 30 h at 20°C. All the present knowledge about the viral replication process results from the physiological study of temperature-sensitive and host-range mutants. These mutants identify four steps in the growth cycle of the sigma virus. At 20°C, after a rapid phase of adsorption–penetration (1 h), there is a phase lasting about 8 h which corresponds to the temperature-sensitive period of the mutant *hap*7. The next stage, from 4 to 15 h postinfection, is defined by the mutant *ts*4. The mutants *hap*7 and *ts*4 have been designated early mutants. *Ts*4 is defective in hereditary transmission at the restrictive temperature (28°C); *hap*7 is not, even at 30°C. The function affected in *hap*7 is indispensable before genome replication and facultative once genome replication has started. The next stage is defined by mutants such as *ts*9 in which germinal transmission is not temperature sensitive. In such mutants, designated late mutants, the genome replication functions are not affected and their temperature-sensitive period corresponds to a phase of virus assembly and budding. The functions necessary to initiate infection are not affected in *ts*9, as deduced from analyses of infectious center decay, but the virions are thermolabile. The *ts* mutations have not yet been assigned to any gene. In analogy to the molecular biology of other rhabdoviruses some predictions can be made. In VSV only the three proteins of the viral nucleocapsid (N, P and L) are required for the synthesis of monocistronic capped and polyadenylated mRNAs and, in the next step, of full genome length RNAs. The matrix protein M is required for the transport of newly synthesized nucleocapsids toward the membrane patches in which the glycoprotein G is inserted and for the budding or progeny virions.

The sigma virus counterparts of these proteins are likely to play the same roles. If so, the proteins modified in early mutants may be N, P or L, while late mutants may bear temperature-sensitive M or G proteins. We have no clues which could permit the prediction of the phenotypes of mutants affected in the additional protein X. Nevertheless, the absence of cytopathogenicity of the virus, even in the most permissive host genotype, and even in the most sensitive stages of the host development, strongly suggests that the sigma virus multiplication rate is self-restricted. It is tempting to speculate that the protein of yet unknown function, or the gene overlap, or both, could account for this feature. According to the first hypothesis, mutants with disfunctioning X protein would produce higher virus yields. Paradoxically, clones with *ts* mutations of X would be more invasive to the host and thus have enhanced temperature resistance. Viral strains with such a phenotype exist. High yield and high temperature resistance in sigma virus clones are always correlated with a pathogenic effect on the germ line and the loss of hereditary transmission. Such characteristics are shared by the vesiculoviruses (which do not have an X gene) when inoculated into *Drosophila*.

Ecology

The specific CO_2 sensitivity symptom makes the identification of infected flies very easy and has allowed significant exploration of *Drosophila* natural populations. CO_2 sensitive flies were found among several *Drosophila* species throughout the world. The viruses carried by the different fruitfly populations share the major characteristics of sigma virus (hereditary transmission, symptom, etc.) but they may be distinct. The data indicate that such viruses are endemic in all the populations of *Drosophila*, the infected flies being the minority (10–20% in French natural populations). The high proportion of uninfected flies does not correspond to a virus resistant fraction of the populations since they and their offspring may become infected experimentally in the laboratory.

The genetic approach to the study of an insect–virus relationship performed with sigma virus underlines the complexity of the interactions and the difficulties involved in the control of all the relevant parameters.

See also: Rabies virus; Rhabdoviruses; Vesicular stomatitis viruses.

Further Reading

Brun G and Plus N (1980) The viruses of *Drosophila*. In: Ashburner M and Wright TRF (eds) *The Genetics and Biology of Drosophila*, p. 625. New York: Academic Press.

Dezelee S *et al.* (1989) Molecular analysis of *ref(2)P*, a *Drosophila* gene implicated in sigma rhabdovirus multiplication and necessary for male fertility. *EMBO J.* 8: 3437.

Fleuriet A (1988) Maintenance of a hereditary virus: the sigma virus in populations of its host, *D. melanogaster*. In: Hecht M and Wallace B (eds) *Evolutionary Biology*, p. 2. New York: Plenum.

Teninges D *et al.* (1980) *Drosophila* sigma virus. In: Bishop DHI (ed.) *Rhabdoviruses*, p. 113. Boca Raton: CRC Press.

Teninges D *et al.* (1993) Genome organization of the sigma rhabdovirus: six genes and a gene overlap. *Virology* 193: 1018.

SIMIAN HEMORRHAGIC FEVER VIRUS

See Encephalitis viruses and Lactate dehydrogenase-elevating, equine arteritis and Lelystad viruses

SIMIAN HERPESVIRUS

See Herpesviruses saimiri and ateles

SIMIAN IMMUNODEFICIENCY VIRUSES

Hilary G Morrison and Ronald C Desrosiers
Harvard Medical School
New England Regional Primate Research Center,
Southborough, Massachusetts, USA

History

Simian immunodeficiency virus (SIV) was first isolated in 1984 from captive immunodeficient rhesus macaques (*Macaca mulatta*) at the New England Regional Primate Research Center. This virus, now called SIVmac, was originally called STVL-III because its morphology, growth characteristics and antigenic properties suggested a close relationship to the human immunodeficiency virus (HIV), then called HTLV-III, LAV or ARV. SIV was subsequently isolated from asymptomatic sooty mangabey monkeys at the Delta and Yerkes Regional Primate Research Centers. SIVs have also since been isolated from other captive macaques (*M. fascicularis, M. nemestrina, M. arctoides*) and from feral African primates, including African green monkeys (SIVagm), mandrills (SIVmnd) and chimpanzees (SIVcpz).

Taxonomy and Classification

Simian immunodeficiency viruses belong to the *Lentivirus* subfamily of *Retroviridae*. This subfamily includes the classic ungulate lentiviruses (visna virus of sheep, caprine arthritis encephalitis virus and equine infectious anemia virus) and the immunodeficiency viruses of humans (HIV), monkeys (SIV), cats (FIV) and cattle (BIV).

The retroviruses can be subclassified by morphologic and morphogenic criteria. Lentivirus particles are approximately 80–110 nm in size and consist of an RNA genome and viral enzymes enclosed in a core of viral proteins that is encased by a cell-derived membrane spiked with viral envelope glycoproteins. Lentiviruses can be distinguished from other subgroups of retroviruses by the presence of a cylindrical or rod-shaped nucleoid in mature particles and the absence of preformed particles in the cytoplasm. Lentiviruses also share a similarity in certain biological properties and the organization of their genomes. Unlike many of the other retroviruses, they are not oncogenic. Instead, they produce long-term, persistent infections which eventually lead to chronic debilitating disease. All lentiviruses studied to date replicate and persist in cells of the monocyte/macrophage lineage. In addition to the standard *gag, pol* and *env* genes that all retroviruses have, lentiviruses possess a number of additional genes not found in other retroviruses.

The SIVs are named according to the primate species of origin, e.g. SIVmac from macaques or SIVsmm from sooty mangabey monkeys. Based on genetic sequence analysis, four discrete groups of primate lentiviruses have been identified to date (Fig. 1, Table 1 and see later). These are HIV-1/SIVcpz, HIV-2/SIVsmm/SIVmac, SIVagm and SIVmnd.

Geographic and Seasonal Distribution

Thus far, only African green monkeys, sooty mangabey monkeys and mandrills have been shown to be infected with SIV in their natural habitats. Serological evidence suggests that 20–50% of green monkeys in Kenya, Ethiopia, South Africa and Senegal have SIV antibodies. However, green monkeys which became established in the Caribbean since the seventeenth century are sero-

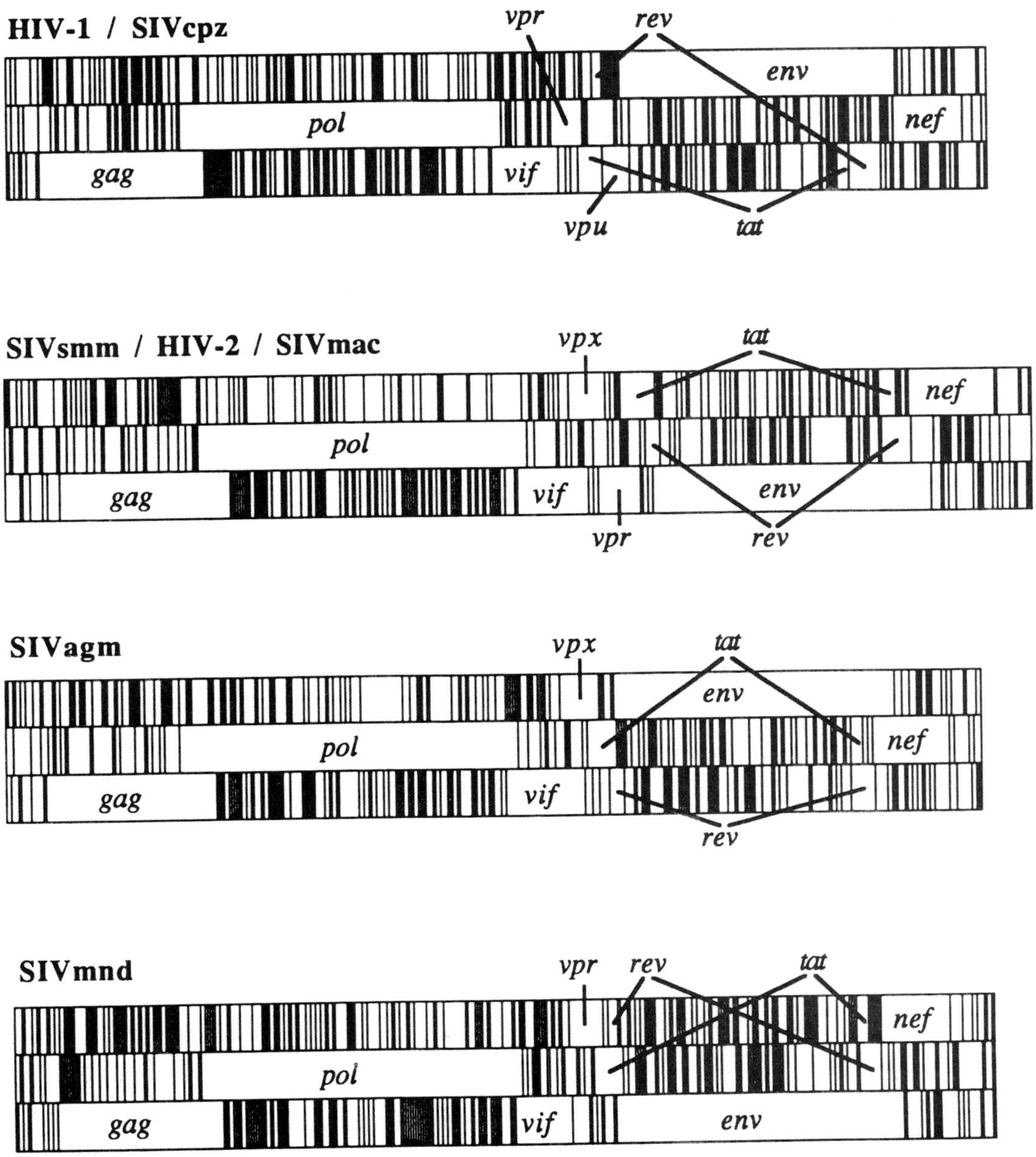

Fig. 1 Genomic organizations in the four groups of primate lentiviruses.

negative. Recent data also indicate that sooty mangabey monkeys in their native habitat in the coastal forests of western Africa are infected with their own SIV. The closely related HIV-2 is also endemic in this area, suggesting that HIV-2 may have originated from sooty mangabeys in this region and evolved into a human virus after cross-species transmission. Macaques do not appear to harbor SIV in their native Asian habitat. Apparently, some macaques became infected with SIV on rare occasion while in captivity at one or more US Regional Primate Research Centers.

Host Range and Virus Propagation

African green monkeys, sooty mangabey monkeys and mandrills appear to be natural hosts for their own discrete SIVs. Other species that are susceptible to infection and disease are limited to primates, possibly including humans. The natural routes of transmission for SIV are in general not known. Macaques have been infected experimentally via intravenous and intramuscular inoculation and by exposure of the genital mucosa. Some SIVs clearly have the capacity for crossing species barriers. For

Table 1. Primate lentivirus nonstructural genes

Gene	SIVagm	SIVsmm/SIVmac/HIV-2	SIVmnd	HIV-1/SIVcpz
vpu	–	–	–	+
vpx	+	+	–	–
vpr	–	+	+	+
tat	+	+	+	+
rev	+	+	+	+
vif	+	+	+	+
nef	+	+	+	+

example, SIVsmm is readily able to infect macaque monkeys. In contrast, numerous attempts to infect macaques and other Old World primates with HIV-1 have failed.

SIVs can be propagated in many tissue culture cell lines, primarily tumor-derived CD4+ T lymphocytes and monocytes. In many of the cell lines, viral infection results in the fusion of cellular membranes producing giant, multinucleated syncytial cells. Syncytium formation allows the virus to spread from cell to cell directly. Persistently infected virus-producing cell lines can be established from the cells surviving initial infection. Mitogen-stimulated primary peripheral blood lymphocytes from macaques and humans are often used for replication of virus *in vitro*. Some isolates also grow well in cultured macrophages derived from lung, blood or bone marrow. Virus can also be produced transiently in non-CD4 cells by transfection of the proviral DNA genome. It has not been clearly established whether SIV can infect some CD4– cells as has been reported for HIV-1. As with HIV-1, infection with SIV is predominantly a CD4-mediated event.

Genetics

Retroviruses contain an RNA genome that replicates via a DNA intermediate. The viral particle contains a diploid genome of single-stranded RNA, linked non-covalently near the 5′ ends of the molecules. The 5′ end of the viral RNA is capped, and the 3′ end is polyadenylated. DNA synthesis by the reverse transcriptase is primed by host tRNA that is base-paired to the viral RNA. The dsDNA provirus is integrated into the host cell chromosome by a viral-encoded integrase and further events in transcription, translation and assembly are host cell dependent. Particles assemble and bud through the plasma membrane.

All retroviruses possess certain basic features in their genomic organization. Regulatory sequences controlling DNA synthesis, integration, transcription and other functions are located in the long terminal repeat (LTR) at each end of the provirus. LTRs are common to the broader group of eucaryotic transposable elements (retrotransposons) that include the retroviruses. The open reading frames encoding the major structural proteins lie between the LTRs. Genes may be encoded in any of the three possible open reading frames, and overlaps between open reading frames are common. All retroviruses contain genes called *gag* (group-specific antigen) encoding the core proteins, *pol* (polymerase) encoding the viral reverse transcriptase, protease and integrase, and *env* encoding the envelope glycoproteins.

The SIVs similarly contain *gag, pol* and *env* genes and use a replication strategy similar to all retroviruses. However, SIVs, HIVs and other lentiviruses contain a number of genes not found in other subfamilies of retroviruses. The SIV genome is approximately 9.6 kb from the 5′ cap to the 3′ polyadenylation site. The sequences of several cloned SIVs have been reported (SIVmac251, SIVmac142, SIVmac239, SIVsmmH4, SIVsmmPBj14, SIVmndGB1, SIVagmTYO1 and SIVcpz). The envelope gene, encoding gp120 and gp41, often contains a premature translation termination signal, resulting in a truncated transmembrane protein. In addition to the major open reading frames *gag, pol* and *env*, the human and simian immunodeficiency viruses contain open reading frames for *tat* (transactivator protein), *rev* (regulator of gene expression), *vif* (viral infectivity protein) and *nef* (originally termed negative factor; may be a determinant of persistence or pathogenicity *in vivo*). The *vpu* open reading frame found in HIV-1 and SIVcpz is not present in HIV-2 nor in any of the other SIVs. HIV-1/SIVcpz, SIVsmm/HIV-2/SIVmac and SIVmnd contain an open reading frame called *vpr*, which is not found in SIVagm. SIVagm and SIVsmm/HIV-2/SIVmac contain a gene called *vpx* not found in SIVmnd and HIV-1/

SIVcpz. Genes *vpx* and *vpr* share sequence similarity and one probably arose from the other via a gene duplication event. The presence of these additional open reading frames in the four discrete groups of primate lentiviruses is summarized in Table 1. Several of these additional genes (specifically *vpx, vpr* and *nef*) can be deleted without abrogating the ability of the virus to replicate in tissue culture cells, but they are likely to have important functions *in vivo*. Recently, it has been reported that *nef* plays an important role in maintaining high virus loads *in vivo* and for disease development.

SIV proteins are translated from a complex population of unspliced, singly spliced, and multiply spliced mRNA molecules. The amount of each protein is regulated at least in part by the extent of splicing. The mechanisms that regulate splicing are currently under investigation, although it appears likely that the interaction of the Rev protein (encoded by fully spliced transcripts) with a region of RNA called the *rev*-responsive element (RRE) results in the accumulation of full-length transcripts that are translated into the major structural proteins.

SIVs, like other retroviruses, accumulate genetic changes rapidly *in vivo*, presumably because of errors introduced by the error-prone reverse transcriptase and because of the selective pressures of the host. Lentiviruses may differ from other retroviruses in being able to tolerate greater variation in their envelope glycoproteins and this may contribute to their ability to persist. In a recent report, the rate of fixation of mutations in the gp120 portion of the envelope gene of SIVmac239 was calculated to be 8.5×10^{-3} changes per site per year. Mutations in the envelope gene result in antigenic variations that may enable the virus to evade ongoing host immune responses and contribute to its ability to establish persistent infection. Other mutations in the envelope gene appear to affect cell and tissue tropism, for example, by altering the ability of the virus to replicate in macrophages vs lymphocytes. Another mechanism that may generate genetic diversity is the production of heterozygous dimer genomes or pseudotype particles in cells that are infected by more than one virus. Endogenous SIV sequences have not been detected in the germ line and gene conversion/recombination events have not been documented.

Evolution

Comparisons of genetic sequences among human and simian immunodeficiency viruses suggest that there are at least four discrete groups of primate

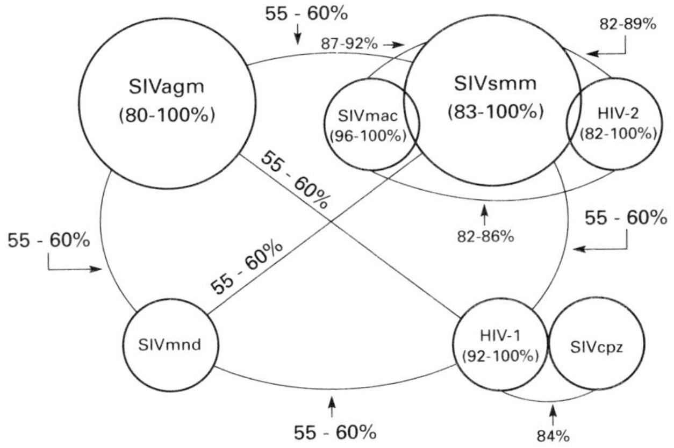

Fig. 2 Four groups of primate lentiviruses. The numbers refer to the approximate percentage amino acid identity in Pol.

lentiviruses currently in existence (Figs 1 and 2): HIV-1 and the closely related SIVcpz; SIVmnd; SIVagm; and SIVsmm/SIVmac/HIV-2. The evolutionary origin of each of these groups may never be known. It is possible that HIV-1 and HIV-2 evolved from simian viruses and entered the human population by cross-species transmission relatively recently in history. Cross-species transmission between primates in nature or in captivity may have resulted in the generation of new pathogenic variants, with SIVmac infection of macaques possibly being analogous to HIV-2 in humans. However, it is also possible that some of the primate immunodeficiency viruses may have always been present in the corresponding host population but not have been recognized until recently. The SIVs and HIVs are more closely related to one another than to any of the nonprimate lentiviruses. This suggests that the HIVs and SIVs are inherently primate viruses and that they were not derived from rodents, cats, ungulates or other nonprimates via cross-species transmission.

Serologic Relationships and Variability

The *gag* and *pol* genes are the most highly conserved among related primate lentiviruses; sequence comparison of these regions are often used to estimate relatedness rather than serologic tests. Antiserum to the Gag protein is generally cross-reactive among different isolates within a group, while antiserum to the envelope protein can be used to distinguish between them. Serologic cross-reactivity between a member of one group to one in another group is usually weak even to Gag proteins.

Epidemiology

SIV-induced disease has only been observed in macaques in captivity and upon experimental infection. SIV-associated disease has not been seen outside these settings.

Transmission and Tissue Tropism

The routes of transmission of SIV in the wild are not known but, as with HIV-1, transmission through blood and sexual contact seems likely. Transmission through bite and scratch wounds may also be significant in monkey populations. Macaques can be experimentally infected via intravenous or intramuscular inoculation, or by exposure of the genital mucosa. Like HIV, SIV is tropic for CD4-bearing cells; both viruses grow preferentially in CD4+ cells, and soluble CD4 or monoclonal antibodies specific to the CD4 antigen can block viral infection. The potential targets for infection in the macaque are helper/inducer T lymphocytes, mononuclear phagocytes, Langerhans cells and follicular dendritic cells of the lymphoid organs. The tissue macrophage appears to be a major cell type replicating virus in the infected host. Isolate SIVsmmPBj appears to be tropic for gut-associated lymphoid tissues.

Pathogenicity

Natural infection of mandrills, African green monkeys and sooty mangabeys with their own SIV appears to be nonpathogenic. SIVmac and SIVsmm can be pathogenic in rhesus, cynomologus, stump-tail and other macaques. Animals generally succumb within 1–3 years postinfection with a disease remarkably similar to AIDS in humans. The induction of AIDS-like disease in macaques following infection with SIVmac is currently used as a model for human AIDS, to investigate mechanisms of pathogenesis and to evaluate potential vaccines and therapies.

Clinical Features of Infection

Infection of rhesus macaques with SIVmac virus follows a disease course similar to that of AIDS in humans. Animals exhibit diarrhea and wasting, with losses of up to 60% of original body weight. They may also develop a cutaneous macular rash. Immune abnormalities include decreases in the peripheral CD4 lymphocyte population, decreases in the CD4/CD8 ratios and reduced responses to mitogens. The occurrence of lymphoma and lymphoproliferative disease has been reported in experimentally infected animals and occasionally occurs in captive macaques found to be infected with SIVmac. Opportunistic infections are common, and include many of the same agents that are found in HIV-1 infected humans: *Pneumocystis carinii*, cytomegalovirus, *Cryptosporidium, Candida*, adenovirus and mycobacterium. Central nervous system involvement is frequent, with the likelihood of brain lesions appearing to depend on the strain of SIV. Approximately 50% of SIV-infected macaques have died with a characteristic granulomatous encephalitis, very similar in its features to that often seen in humans dying from HIV-1 infection. The disease spectrum associated with SIVmac infection of macaques can be divided into four broad categories: diseases associated with opportunistic infections, neoplastic diseases, diseases associated with the infection of tissue macrophages, and diseases of unknown pathogenesis. The course of disease in pig-tailed macaques infected with SIVsmmPBj virus is extremely rapid in comparison. These animals develop a profuse, bloody diarrhea within one week of infection, and die from severe fluid and electrolyte loss.

Pathology and Histopathology

Pathologic findings in macaques are varied, depending on the clinical signs and course of disease. Approximately 50% of infected animals develop primary lentiviral meningoencephalitis, which is characterized by perivascular infiltrates of macrophages and multinucleated giant cells in the gray and white matter and in the leptomeninges. The infected macrophages contain mature lentiviral particles within cytoplasmic vacuoles. These findings are similar to reports on human patients with subacute AIDS encephalitis. Pathological findings in the CNS may also include SV40-associated progressive multifocal leukoencephalopathy, CMV disease and toxoplasma encephalitis. Another lesion associated with SIV infection in macaques is a dense, granulomatous infiltrate containing giant cells in the lymph nodes, spleen, lung, gastrointestinal tract and other organs.

Many animals develop interstitial pneumonia that is not associated with an opportunistic infection. This disease is characterized by a mononuclear interstitial infiltrate, syncytial cell formation, fibrin exudation and thickening of the alveolar septa. This lung inflammation appears to be the result of a cytotoxic immune response against SIV-infected alveolar macrophages.

In rhesus macaques, diarrhea is most frequently associated with pathogenic bacteria or protozoa; however, in some animals the diarrhea appears to be caused directly by SIV. In these cases, nonspecific enteropathy consisting of blunting of the small intestinal villi, shortening of the crypts of Lieberkuhn, a predominantly mononuclear inflammatory infiltrate within the lamina propria, and attenuation or immaturity of the epithelium is seen. A most prominent necropsy finding in pigtailed macaques infected with isolate SIVsmmPBj is a twofold to 10-fold increase in the mass of lymphoid organs (spleen, mesenteric lymph nodes and gut-associated lymphoid tissues). Pathological changes in the gastrointestinal tract appear to be caused by an immune reaction to the infected tissues.

Immune Response

Experimentally infected animals generally fall into two classes: those that develop a high, persistent humoral and cellular immune response, and those with little or no response. Animals in the first group remain persistently infected, develop a protracted disease course similar to AIDS in humans, and generally die one year and longer after infection. Animals in the second group die within a shorter period of time postinfection. These observations, and the results of vaccine studies, indicate that a potent immune response can be partially protective in delaying disease. Humoral immune responses can be conveniently measured by ELISA, Western blot and neutralization tests. As mentioned previously, infected macaques commonly develop immune abnormalities and many of the pathological findings may result from cytotoxic immune responses directed against infected tissues.

Prevention and Control

Extensive testing programs have essentially eliminated SIV from captive macaque colonies. However, continued vigilance is needed to maintain breeding colonies free from accidental exposure to the virus. Animals can be conveniently tested serologically for evidence of infection. As yet, no human cases of SIV infection have been documented, but SIV is handled with the same precautions as for work with HIV-1. Disposable surgical gloves and gowns are used, all work with live virus is carried out in a biosafety cabinet, procedures creating aerosols are avoided and use of glass and needles is minimized.

Note added in proof

A fifth group of primate lentiviruses, SIV from sykes monkeys has recently been identified.

Future Perspectives

The most important role of SIV will be its use in basic research relevant to AIDS. The induction of AIDS in macaques by infectious molecular clones of SIV represents the best existing animal model for AIDS in humans. SIVs are the closest known relatives of the HIVs and the disease induced in macaques is remarkably similar to AIDS in humans. Rhesus and other macaque species are not endangered, can be purchased at a reasonable cost and breed well in captivity. This system can be used to dissect the molecular determinants of AIDS pathogenesis, to define the role of the so-called nonessential genes, to map functional regions of the structural genes such as the envelope, and to evaluate the potential of new treatment and vaccine strategies. Finally, the ongoing investigation of 'new' isolates and their genetic relatedness to existing SIV and HIV isolates may shed light on the origins and evolution of the primate immunodeficiency viruses.

See also: Autoimmunity; Bovine immunodeficiency virus; Feline immunodeficiency virus; Human immunodeficiency viruses; Immune response; Persistent viral infection; Visna–Maedi viruses.

Further Reading

Desrosiers RC (1990) The simian immunodeficiency viruses. *Annu. Rev. Immunol.* 8: 557.

Desrosiers RC and Ringler DJ (1989) The use of simian immunodeficiency viruses for AIDS research. *Intervirology* 30: 301.

Eichberg JW (ed.) (1990) Nonhuman primate models for AIDS II. *J. Med. Primatol.* 19: 161.

SIMIAN VIRUS 40

Janet S Butel
Baylor College of Medicine
Houston, Texas, USA

History

Simian virus 40 (SV40) was discovered in 1960 as a contaminant of poliovaccines. Hundreds of millions of people world-wide were inadvertently exposed to SV40 in the late 1950s and early 1960s when they were administered contaminated virus vaccines prepared in rhesus macaque kidney cells. Infectious SV40 had unknowingly contaminated batches of both the inactivated and live attenuated forms of the poliovaccine, as well as preparations of some other viral vaccines. Although primary cultures of monkey cells were known to be commonly contaminated with indigenous viruses and safety testing was carried out, SV40 had escaped detection because it failed to induce cytopathic effects in rhesus cells. However, when it was inoculated into African green monkey kidney cells, a prominent cytoplasmic vacuolization developed. Originally christened as 'vacuolating virus,' the name was later changed to SV40 to conform with a numerical system of designating simian virus isolates.

Concern about the vaccine contaminations heightened considerably when it was found in 1962 that SV40 was tumorigenic in newborn hamsters and could transform many types of cells in culture. Because of the potential risk to public health posed by the previous distribution of contaminated poliovaccines, SV40 became the focus of intensive investigation. Fortunately, the individuals exposed to SV40-contaminated vaccines appear not to be at higher risk of developing cancer than those who received SV40-free vaccines. For the scientists, SV40 has turned out to be an invaluable tool for dissecting molecular details of eucaryotic cell processes. Numerous techniques now commonly used in molecular biology were pioneered in the SV40 system. It continues to serve as a leading model for basic studies of viral carcinogenesis.

Taxonomy and Classification

SV40 is classified as a member of the *Polyomavirus* genus in the *Papovaviridae* family (Table 1). The other well-studied member of the genus is polyoma virus of mice. The group also includes the human polyomaviruses, BKV and JCV, as well as isolates from other species, including hamsters, rabbits, birds and baboons. The papillomaviruses are classified in the other genus, *Papillomavirus*, in the family. The human and animal polyomaviruses are antigenically distinct, and there is only one recognized serotype for each virus.

The polyomaviruses are small and simple and share certain physical and chemical properties. These include an icosahedral capsid about 45 nm in diameter which contains three viral proteins, the lack of an envelope and a double-stranded (ds) circular covalently closed DNA genome about 5 kbp in size. The outstanding biological characteristics of the

Table 1. Properties of SV40

Classification:	Family *Papovaviridae*, genus *Polyomavirus*
Strain variation:	Genetically stable; one serotype
Virion:	Icosahedral, 45 nm in diameter, no envelope
Genome:	Circular covalently closed dsDNA, 5200 bp
Proteins:	Three structural proteins, VP1, VP2, VP3; cellular histones condense DNA in virion; nonstructural replication protein, T-antigen, is potent oncoprotein
Replication:	In certain primate kidney cells; nucleus; stimulate cell DNA synthesis; long growth cycle
Natural host:	Asian macaques, especially the rhesus monkey
Diseases:	Asymptomatic persistent infections in natural hosts; tumors in experimentally infected rodents
Historical note:	Contaminant in early poliovaccines administered to millions of people

polyomaviruses are that they stimulate cellular DNA synthesis in infected cells, and they are tumorigenic in the appropriate hosts.

Properties of the Virion

SV40 particles are small and spherical, with a diameter of approximately 45 nm. Infectious virions have a sedimentation coefficient of 240S and band at a density of $1.34\,g\,ml^{-1}$ in CsCl; empty capsids have a density of $1.29\,g\,ml^{-1}$. The molecular mass of the SV40 virion has been estimated at 270 kD. The DNA content is 12.5% (w/w). The major capsid protein (VP1) accounts for 75% of the total virion protein. VP2 and VP3 are minor capsid proteins. Cellular histones (H2A, H2B, H3, H4) are used to condense the viral DNA for packaging and are present in the core of the particle. There is no lipid envelope. SV40 does not agglutinate erythrocytes.

SV40 particles exhibit icosahedral symmetry. The virion is composed of 72 pentameric capsomeres composed of the VP1 protein arranged on a $T = 7d$ icosahedral surface lattice. This surprising structure (that the hexavalent capsomeres have pentameric substructure) demands nonequivalent contacts between pentamers. This seems to be accomplished by the C termini of the VP1 polypeptides, which extend as arms from one pentamer and fit into binding sites on adjacent pentamers. The arms can go in different directions, providing the necessary flexibility to build a capsid. The N-terminal arm of VP1 is completely internal in the viral particle. Minor capsid proteins VP2 and VP3 are predominantly internal as well and do not contribute to the basic structure of the viral outer shell.

The viral particles are very resistant to heat inactivation but are relatively labile when heated in the presence of divalent cations. Whereas SV40 is stable at 50°C for hours, incubation in the presence of 1 M $MgCl_2$ at 50°C for 1 h will inactivate the virus. At a higher temperature (60°C), ~ 99% of infectious virus is inactivated within 30 min in the absence of divalent cations. Purified virions can be disrupted by strong alkaline conditions (pH 10.5), by lower pH (9.2) plus a reducing agent, or by detergent treatment. Intact viral particles are not affected by nucleases, but in the presence of a reducing agent nuclease can enter the virion and cleave the viral DNA. SV40 is efficiently inactivated by UV light irradiation, following single-hit kinetics.

Properties of Viral Genome

The SV40 genome is a circular covalently closed dsDNA molecule. The native DNA assumes a superhelical configuration (form I) which sediments at 21S in a neutral sucrose gradient. A single-stranded (ss) nick generates relaxed circular dsDNA molecules (form II) which sediment at 16S, whereas a ds break produces linear dsDNA (form III, 14S). Alkaline denaturation of form I DNA produces dense cyclic coils which sediment at 53S. Form II DNA is converted to ss circular (18S) and ss linear (16S) molecules by denaturation. The supercoiled (form I) molecules can be separated from relaxed circular and linear forms by centrifugation of a DNA preparation in CsCl gradients with ethidium bromide. The form I molecules will band in a lower position in the gradient. The DNA forms also separate during electrophoresis in a neutral agarose gel; the supercoiled molecules migrate the fastest, the linear forms move at an intermediate speed, and the relaxed circles migrate the slowest.

The viral DNA both in virions and in infected cells is associated with cellular histones H2A, H2B, H3 and H4. The histones are assembled in 24–26 nucleosomes on the viral DNA. The nucleosome structure and histone composition of the viral minichromosome mimic the chromatin structure of cellular DNA.

SV40 DNA was the first eucaryotic viral genome to be physically mapped by restriction endonuclease analysis (1971) and to be completely sequenced (1978). It contains 5243 bp for a calculated molecular weight of 3.5×10^6. The genome is numbered in a clockwise direction from 1 to 5243, the center nucleotide of the unique *Bgl*I recognition site being assigned as 0/5243. Numbering continues through the late region in the sense orientation and the early region in the antisense orientation. The numbering system begins and ends (0/5243) in the middle of the functional origin of DNA replication. The unique *Eco*RI site at nucleotide 1782 was arbitrarily chosen as a point of reference and assigned a value of 0/1.0 on the circular map.

SV40 makes maximal use of a limited amount of genetic information, including having compact regulatory sequences and overlapping genes. The genome contains a single origin of replication (core *ori* = 64 bp in size) embedded within a nontranslated regulatory region. These elements control transcription and replication and span about 400 bp. The SV40 genetic map is divided into two halves, corresponding to regions that are expressed during the early and late stages of infection. These regions represent the 'early' nonstructural genes and the 'late' structural genes, respectively.

Properties of Virus Proteins

SV40 encodes six gene products: two 'early' nonstructural proteins [large T antigen (T-ag), small t antigen (t-ag)], three 'late' structural proteins (VP1, VP2, VP3) and a maturation protein (LP1 or agnoprotein).

The nonstructural proteins are expressed early in infection, before the onset of viral DNA synthesis. The coding regions of the two T-ags overlap; alternative splicing of viral transcripts determines each protein sequence. Large T-ag (Table 2) contains 708 amino acids (~ 90 kD), and small t-ag contains 174 residues (~ 20 kD). The large and small T-ags share 82 N-terminal amino acids, whereas the remainder of each protein is unique.

Large T-ag is an essential replication protein required for initiation of viral DNA synthesis. It stimulates host cells to undergo DNA synthesis and is the SV40 transforming protein. Large T-ag contains a nuclear transport signal (126-Pro-Lys-Lys-Lys-Arg-Lys-Val-132) that targets the protein into the nucleus. However, about 10% of the T-ag in the cell is found in the cytoplasm and the plasma membrane. The biology of small t-ag is enigmatic. It is a cytoplasmic protein that is not essential for viral replication in cultured cells. Perhaps it is required during natural infections by SV40 in host primates.

Large T-ag is a multifunctional protein which is chemically modified in several ways. Its functions in SV40 DNA replication are regulated by phosphorylation. The sites of phosphorylation are clustered near the ends of the molecule, one region lying between residues 106 and 124 and the other between residues 639 and 701. The majority of the phosphorylated residues are serines, although two threonine residues also become phosphorylated. Unlike many oncoproteins, T-ag is not phosphorylated at tyrosine residues.

T-ag is a DNA-binding protein that recognizes multiple copies of the sequence GAGGC in three T-ag-binding sites in the viral *ori*. The phosphorylation of one amino acid, Thr-124, is crucial for T-ag to be able to bind to site II and to initiate viral DNA replication. The minimal origin-specific DNA-binding domain of T-ag lies between residues 131 and 259. T-ag is predicted to have a zinc finger motif, typical of DNA-binding proteins, between amino acids 302 and 320. T-ag-specific ATPase and helicase activities are required in addition to DNA-

Table 2. Properties and functions of SV40 T-ag

A. Structural properties
 1. Size:
  708 amino acids
  82 N-terminal residues shared with t-ag
  81 632 Daltons
  M_r 90 000–100 000
 2. Modifications:
  Phosphorylation
  N-terminal acetylation
  O-glycosylation
  Poly-ADP-ribosylation
  Palmitylation
  Adenylation
 3. Supramolecular structure:
  Zinc finger
  Monomers, dimers, higher homooligomers
  Heterooligomers with cellular proteins

B. Subcellular distribution
 1. Nuclear:
  Nucleoplasmic
  Chromatin bound
  Nuclear matrix associated
 2. Plasma membrane:
  Nonidet P-40 soluble
  Nonidet P-40 insoluble (plasma membrane lamina)
  Butanol soluble

C. Functions
 1. Autoregulation of viral early transcription
 2. Activation of viral late transcription
 3. Specific DNA binding (viral origin of replication)
 4. Nonspecific DNA binding (cellular DNA)
 5. Initiation of cellular DNA replication
 6. Induction of cellular enzyme synthesis
 7. Re-activation of ribosomal RNA genes
 8. Initiation of viral DNA replication
 9. ATPase activity
 10. Helicase activity
 11. Complex formation with cellular protein p53
 12. Complex formation with DNA polymerase α
 13. Complex formation with p105 Rb and Rb-related p107
 14. Complex formation with heat-shock protein 70
 15. Complex formation with cdc2, cyclin and tubulin
 16. Determination of SV40 host range
 17. Adenovirus helper function
 18. Initiation and maintenance of cellular transformation
 19. Induction of immunity to SV40 tumor cells
 20. Target for cytotoxic T cells (TSTA)

binding activity in order for T-ag to function in initiation of DNA replication. The ATP-binding domain of T-ag is similar in structure to other ATP-binding proteins and is located between residues 418 and 528.

Large T-ag forms complexes with several cellular proteins. Such interactions are thought to be involved in T-ag functions in viral DNA replication and induction of cellular DNA synthesis. Target cellular proteins found in heterooligomeric structures with T-ag include p53 and p105 Rb (two cellular tumor suppressor gene products), DNA polymerase α, the molecular chaperone heat-shock protein (hsp)70, cell-cycle regulatory proteins cdc-2 and cyclin, and tubulin. The indicated cellular proteins are not all found in the same T-ag-associated complex; many subpopulations of T-ag exist in a cell.

The structural (capsid) proteins are expressed late in infection, after the onset of DNA replication. They are synthesized in much greater abundance than the early proteins. The major capsid protein, VP1, contains 362 amino acids ($\sim$ 45 kD). The minor structural proteins are VP2 (352 residues, $\sim$ 38 kD) and VP3 (234 residues, $\sim$ 27 kD). The coding regions for VP2 and VP3 overlap, and they are translated in the same reading frame, so the sequence of VP3 is identical to the C-terminal two-thirds of VP2. VP3 is synthesized by independent initiation of translation via a leaky scanning mechanism. It is not a proteolytic cleavage product of VP2. The N-terminal portion of VP1 is derived from sequences that encode the C termini of VP2 and VP3. However, VP1 is translated in a different reading frame from a different spliced transcript, so it shares no sequences with VP2 and VP3. VP1 is modified by phosphorylation and acetylation.

The late proteins are required only for the assembly of progeny virions during lytic infection. They are not involved in the early phases of viral replication. They are synthesized in the cytoplasm and move into the nucleus where particle morphogenesis occurs. The minor capsid proteins contain nuclear transport signals. The VP2/3 signal is Gly-Pro-Asn-Lys-Lys-Lys-Arg-Lys-Leu (VP2, residues 316–324; VP3, residues 198–206). The N-terminal eight residues of VP1 (Ala-Pro-Thr-Lys-Arg-Lys-Gly-Ser) are important in targeting it into the nucleus. Mutations in VP1 affect capsid assembly and/or virion stability. Mutations in VP2 and VP3 affect the uncoating process, when virions penetrate new host cells.

The agnoprotein LP1 is synthesized late in infection but is not found in viral particles. It is a small (62 residue, $\sim$ 8 kD) basic protein involved in particle assembly. It is believed that LP1 interacts with VP1 molecules to inhibit self-polymerization until they interact with viral minichromosomes in the nucleus to form virions.

Replication

Overview of SV40 replication cycle

The SV40 replication cycle is cleanly divided into early and late events, with the onset of viral DNA replication being the dividing landmark. SV40 virions attach to receptors on the cell surface, become internalized and are transported to the cell nucleus where the viral DNA is uncoated. After uncoating, the half of the genome that contains the early region is transcribed ('early' mRNAs). Viral early proteins (T-ags) are synthesized, cellular enzymes are induced and the cells enter S phase. Viral DNA replication then begins. 'Late' mRNAs are transcribed from the other half of the viral genome (the opposite strand), and viral structural proteins are synthesized. Virus particles are assembled. The majority of progeny virions stay associated with the cell until cell lysis occurs following cell death. The SV40 multiplication cycle is slow, taking 48–72 h.

Strategy of replication of nucleic acid

SV40 DNA is replicated in the cell nucleus as a free unintegrated minichromosome. The only viral components required are the viral origin of replication on the DNA and the T-ag protein; all other factors are provided by the host cell replication machinery. T-ag is required for the initiation of DNA replication. The specific T-ag functions required are its DNA-binding ability and its ATPase/helicase activities. The relative simplicity of the SV40 system has allowed the development of cell-free replication systems and the identification of factors involved in mammalian DNA replication.

T-ag binds to the viral *ori*, a 64 bp segment that contains binding site II for T-ag. In an ATP-dependent process, T-ag causes localized unwinding of the *ori* region; cellular single-stranded binding protein is required to stabilize the unwound single strands. The cellular DNA polymerase α-primase complex initiates DNA replication, and replication proceeds bidirectionally, with the two forks advancing at equal rates. The strand growing in the 5′ to 3′ direction is synthesized in a continuous fashion, whereas the strand growing in the 3′ to 5′ direction is synthesized as small pieces (Okazaki fragments) that are later ligated together. Elongation involves DNA polymerase α, DNA polymerase δ and proliferating cell nuclear antigen; presumably topoisomerase I or II releases torsional strain that is created as the DNA helix is unwound. Termination occurs 180° away from the viral *ori*; topoisomerase II segregates the newly synthesized daughter molecules. Cellular histones are added to the new strands during the process of DNA replication.

Only certain monkey and human cells support SV40 DNA replication. This cell permissiveness seems to depend on the nature of the DNA polymerase α–primase complex.

The onset of viral DNA replication is carefully coordinated with the host cell cycle by the phosphorylation of a specific site on T-ag. The T-ag synthesized soon after viral infection induces the cell to enter S phase. The binding of p53 and Rb cellular proteins is presumably important in this process. A cell-cycle-dependent kinase (cdc-2) becomes activated and phosphorylates T-ag at Thr-124. Only then can T-ag bind to site II in the viral *ori* and initiate viral DNA replication. This regulation assures that viral DNA does not become unwound for replication until the host cell has entered S phase and the necessary cell replication factors are in place.

Characterization of transcription

Transcription of the viral DNA is carried out by the cellular RNA polymerase II. In the noncoding region of SV40 DNA near the origin of replication are early and late promoter structures and enhancer elements. Early transcription begins at about 0.67 map units, proceeds in the counterclockwise direction, and ends at about 0.17 map units. The early promoter contains a TATA box about 30 bp upstream of the early RNA initiation site. (This start site is about 70 nucleotides upstream of the initiation codon shared by the early proteins.) There are three G+C-rich regions, the ‘21-bp repeats,’ located 40–103 nucleotides upstream which are binding sites for the Sp1 cellular factor. Even further upstream are the SV40 enhancer elements, the ‘72-bp tandem repeats,’ which contain binding sites for other cellular factors that regulate transcription. The primary early transcripts are differentially spliced to generate the mRNAs that code for large T-ag and small t-ag.

There is no requirement for virus-encoded proteins, but early transcription is regulated by T-ag. As the concentration of T-ag increases in the cell, it binds first to site I and then to sites II and III on the viral DNA. Early transcription is repressed when site II is occupied, because the presence of T-ag blocks the binding of RNA polymerase. Therefore, T-ag regulates its own synthesis.

Late transcription begins after viral DNA synthesis is underway. The abundance of late transcripts is much greater than the early transcripts because progeny DNA molecules are utilized as templates. Late transcription begins at multiple sites between 0.67 and 0.76 map units and proceeds in the clockwise direction, ending at about 0.16 map units. A heterogeneous collection of late mRNAs is made, as there is no TATA box upstream from the late RNA start sites. Both the 21 bp repeats and the 72 bp repeats have positive effects on late transcription. T-ag *trans*-activates late transcription by an unknown mechanism that does not involve DNA binding. The late transcripts are alternatively spliced into two size classes (19S, 16S). VP1 is synthesized from 16S RNA and both VP2 and VP3 are translated from the 19S species. The agnoprotein is synthesized predominantly from the most abundant species of 16S RNA.

Characterization of translation

Early gene products (T-ag, t-ag) are synthesized from differentially spliced early transcripts. Likewise, the structural proteins (VP1, VP2, VP3) are produced from differentially spliced late transcripts. VP3 is a truncated version of VP2, due to initiation of translation at an internal site on the same species of transcripts. The agnoprotein LP1 is translated from the leader region of late transcripts. The late gene products are produced in much greater abundance than the early proteins and seem to reflect the relative concentrations of the transcripts.

Post-translational processing

No post-translational cleavages are involved in the production of SV40 proteins. As noted above, T-ag and VP1 are modified in various ways, including phosphorylation.

Uptake and release of virions

SV40 particles attach to receptors on the cell surface. The receptors recognized by viral particles have not been identified, but they are expressed on the apical surface of polarized epithelial cells. There are an estimated 10^5 binding sites per cell.

Attached particles are internalized by endocytosis

and are transported to the nucleus by an unknown pathway. SV40 particles are karyophilic and probably enter the nucleus through nuclear pores. Uncoating of the viral genome occurs in the nucleus. Agents that affect lysosomes have no effect on uptake and uncoating of SV40.

Maturation of progeny virions occurs in the nucleus, where the viral nucleic acid is replicated. Viral proteins are synthesized in the cytoplasm off viral transcripts exported from the nucleus, and the proteins are then transported back into the nucleus. The structural proteins condense around the viral minichromosomes. During the maturation process, the agnoprotein is released and is not retained as a component of mature virions. Assembly intermediates that sediment more slowly than extracellular virions have been detected but details of particle morphogenesis are not known.

Some progeny virus is released from the cell by an unknown mechanism, but the majority stays associated with the cell until lysis caused by cell death. Host cells are killed as the result of a variety of effects, including the release of lysosomal enzymes into the cytoplasm and damage to the cell mitochondria, causing an impairment of aerobic metabolism. Late in infection, the monkey kidney cells develop a characteristic cytopathic effect, cytoplasmic vacuolization. From 10^4 to 10^5 viral particles are produced by each infected cell.

Geographic and Seasonal Distribution

The geographic distribution of SV40 can only be inferred, as no comprehensive surveys have been conducted. Its distribution in the wild presumably reflects its narrow host range. As far as is known, SV40 is found naturally in wild populations of certain Asian macaque species. Many captive primates can be infected if they have been in contact with an infected macaque. Infections in humans are probably more widespread geographically, as contaminated poliovaccines were broadly distributed. Nothing is known about seasonal effects on natural infections by SV40.

Host Range and Virus Propagation

Papovaviruses, in general, have a narrow host range, with each virus infecting only one or a few closely related species. Based on antibody surveys of wild populations of primates, the natural hosts for SV40 appear to be a few species of Asian macaque monkeys, especially the rhesus (*Macacca mulatta*). In captivity, several related species are easily infected, including the cynomolgus macaque (*M. fascicularis*) and the African green monkey which belongs to the same family as macaques (*Cercopithecidae*). The virus grows poorly in more distantly related primates. SV40 can infect humans, but does not replicate well.

SV40 is propagated in tissue culture in established cell lines derived from kidneys of African green monkeys. Characteristic vacuolated cells appear in response to viral replication. The virus grows in rhesus kidney cell lines in which it establishes a persistent infection but produces no cytopathic effects. The SV40 growth cycle is long, compared with those of other virus families.

SV40 does not cause tumors in its natural hosts. To demonstrate its tumorigenic potential, the virus must be inoculated into experimental animals (newborn hamsters are most susceptible). Many types of cells can be transformed in culture, especially those of rodent origin. Primate cells can be transformed, but only with difficulty and if experimental conditions are manipulated to prevent viral replication.

Genetics

SV40 is genetically stable, and there is no evidence that genetic variation plays any role in natural infections. However, many point mutations, deletions and substitutions have been introduced into the SV40 genome in the course of experimental studies designed to examine mechanisms of gene regulation, viral replication and cell transformation. Serial undiluted passage of the virus in cultured cells often results in the accumulation of defective-interfering particles containing DNA with extensive deletions and rearrangements. To produce high-titer stocks of virus, serial passage of undiluted preparations should be avoided.

Evolution

Different strains of SV40 do not evolve during natural infections or during serial passages in culture. The evolutionary origin of SV40 is obscure. Sequence comparisons have revealed short regions of similarity between portions of SV40-encoded gene products and cellular proteins. It may be that the viral proteins are composites of functional domains pirated from cellular progenitors. Because of size constraints imposed on the SV40 genome by capsid architecture, the bulk of the coding sequence for a cellular protein would have to be jettisoned, making identification of origins difficult. Among the polyomaviruses, SV40 is most closely related to BKV by base sequence homology. When all the polyomaviruses are compared, the lowest homologies are found in the noncoding regulatory sequences.

Serologic Relationships and Variability

Only one serotype of SV40 is known. The virus does not undergo noticeable antigenic variation. Perhaps restrictions imposed by the symmetry of the capsid permit only minimal deviation in amino acid sequence of the structural proteins, making most changes lethal for the virus.

There is a genus-specific antigenic determinant on the major capsid protein, VP1, which is shared by all animal and human polyomaviruses. It is internal in the virion, but antibodies are elicited against it by immunization with disrupted capsids or with purified VP1 protein. The determinant is expressed in infected cells. Antibodies against the shared determinant are not neutralizing, as the site is not exposed on the surface of viral particles. Although the structural proteins of SV40 and the two human polyomaviruses are antigenically distinct (with the exception of the genus-specific determinant on VP1), the T-ags of SV40, BKV and JCV show extensive antigenic cross-reactivity.

Epidemiology

Most adults of the Asian macaque species believed to be natural hosts for SV40 have neutralizing antibodies to the virus. Few of the juvenile animals of those species, in the wild, have antibodies. However, in captivity the young animals are readily infected if they have contact with a virus-positive animal.

Serologic surveys have detected low levels of SV40-reactive antibodies in humans not exposed to contaminated vaccines. Limited cross-reactivity with BKV and JCV does not explain the presence of such antibodies. This may indicate that SV40, or an unknown SV40-like agent, is circulating in the human population. However, it appears that SV40 infections in humans are rare, if they occur at all.

Transmission and Tissue Tropism

SV40 establishes persistent infections in the kidneys of susceptible hosts. The infections may become latent. The level of virus present may be very low. Modes of transmission are not known, but transmission probably occurs due to virus shed in the urine. Experiments have established that susceptible animals can be infected by the oral, respiratory or subcutaneous routes. Both viremia and viruria occur in infected animals.

The major known source of human exposure to SV40 was via the administration of contaminated viral vaccines before SV40 was recognized. That risk no longer exists. Human exposure could occur by contact with infected monkeys, a situation limited to small numbers of animal handlers. It is presumed that patterns of tissue tropism and transmission similar to those described in monkeys would be observed in humans infected by SV40.

Pathogenicity and Pathology

SV40 infections in monkeys appear to be asymptomatic and harmless. However, SV40 has been associated with a fatal case of pulmonary and renal disease, as well as with two cases of progressive multifocal leukoencephalopathy, in unhealthy rhesus. No tumors have ever been found in the natural hosts. Transgenic mice carrying wild-type SV40 DNA develop choroid plexus papillomas and die rapidly because of the physiological importance of the tumor site. When foreign tissue-specific regulatory sequences are substituted for the native promoter-enhancer of the virus, SV40 expression can be directed to other tissues in transgenic animals. Tumors usually appear in the targeted tissue and are lethal. In conventional animals, tumors induced by virus injection tend to stay localized and do not invade or metastasize, but rodents bearing such tumors usually succumb due to the tumor load. SV40-induced tumors are usually classified as undifferentiated carcinomas or sarcomas. Intravenous inoculation of SV40 into weanling hamsters induced leukemia, reticulum cell sarcoma and osteogenic sarcoma. SV40 DNA has been detected in a few human brain tumors, including those from children in the first decade of life. Any role SV40 may have played in the induction of those tumors is unknown.

Immune Response

SV40, like other members of the *Polyomavirus* genus, induces an asymptomatic persistent infection in natural hosts. An antibody response to capsid antigen is elicited that can be detected in neutralization assays. It is well documented with the human viruses BKV and JCV that impaired cell-mediated immunity is associated with virus re-activation, showing that viral replication is under the control of the immune system of the host; the same presumably applies to SV40.

Little is known about the immune response of humans to infection by SV40. Small numbers of individuals exposed to contaminated vaccines were analyzed for neutralizing antibody responses to SV40. Humoral responses were variable and dependent on the size of inoculum and route of inoculation.

Experimental studies have shown that animals with active infections by SV40 may produce humoral antibodies against the replication oncoprotein, T-ag. The responses were variable and probably reflected the extent of viral replication. It should be noted that a T-ag antibody response could not be used to monitor SV40 infections in humans because of the cross-reactivity among the T-ags of SV40 and the human viruses BKV and JCV.

SV40 tumor-bearing animals develop a strong immune response to T-ag. Both humoral and cell-mediated responses occur. No antibodies are produced against capsid antigens, as the structural proteins are not expressed in tumor cells. Cytotoxic T cells directed against T-ag determinants at the cell membrane help render the animals resistant to the growth of SV40 tumor cells. This system has been an important experimental model for helping to understand the immune response to neoplastic cells in humans.

Interferon is induced only weakly by the polyomaviruses and is not thought to be an important component of the host response to SV40.

Prevention and Control

No control measures are available to prevent SV40 infections.

Future Perspectives

The scattered reports of antibodies to SV40 in humans and the infrequent association of SV40 markers with human brain tumors suggest that SV40, or a closely related virus, may be circulating in the population. If its presence is substantiated, it will be important to determine the natural history of SV40 in humans, including modes of transmission and factors affecting susceptibility to infection. If tumors are produced in humans following exposure to SV40, it will be necessary to develop appropriate control measures to prevent such infections. Because of its small genetic content and dependence on host cell functions, SV40 will continue to be a useful model system for discerning mechanisms of cellular processes, such as mammalian cell DNA replication, cell cycle progression and growth control processes altered in neoplasia.

See also: Defective-interfering viruses; JC and BK viruses; Persistent viral infection; Polyomaviruses–murine; Virus structure.

Further Reading

Butel JS and Jarvis DL (1986) The plasma-membrane-associated form of SV40 large tumor antigen: biochemical and biological properties. *Biochim. Biophys. Acta* 865: 171.

Liddington RC *et al.* (1991) Structure of simian virus 40 at 3.8-Å resolution. *Nature* 354: 278.

Prives C (1990) The replication functions of SV40 T antigen are regulated by phosphorylation. *Cell* 61: 735.

Shah K and Nathanson N (1976) Human exposure to SV40: review and comment. *Am. J. Epidemiol.* 103: 1.

Tevethia SS (1990) Recognition of simian virus 40 T antigen by cytotoxic T lymphocytes. *Mol. Biol. Med.* 7: 83.

SINDBIS AND SEMLIKI FOREST VIRUSES

Milton J Schlesinger
Washington University School of Medicine
St Louis, Missouri, USA

History

Sinbis (SIN) virus was first isolated in 1952 from a group of *Culex univittatus* mosquitoes captured in a light trap in the Sindbis health district 30 km north of Cairo, Egypt. Isolation was by inoculation of triturated mosquitoes into 3-day-old mice. Size of the virus was estimated to be 40–48 nm and, based on its pathogenesis in neonatal but not adult mice, it was initially classified as a Coxsackie-type virus. Additional studies showed it to be distinct from these viruses and it was placed in a separate class.

Semliki Forest virus (SFV) was first isolated in Uganda in 1942 from *Aedes* (*Aedimorphus*) *abnormalis*. Its size was estimated to be 20–67 nm.

Taxonomy and Classification

SIN and SFV are members of the *Alphavirus* genus of the *Togaviridae* family. They are serologically related to Eastern equine, Western equine and Venezuelan equine encephalitis viruses. They have been classified also as members of the RNA virus superfamily I.

Geographic and Seasonal Distribution

SIN strains have been isolated world-wide. Two geographical subgroups exist: European–African and Asian–Australian. Virus has been isolated in the Near East, Africa, Southeast Asia, India, Borneo, Australia, former USSR and Czechoslovakia. A virus closely related to SIN has been isolated in Sweden.

SFV strains have been isolated from Africa, India and Southeast Asia.

Host Range and Propagation

The known or suspected natural vectors of SIN include *Culex univittatus*, *C. antennatus*, *C. annulirostris*, *C. pseudovishnue*, *C. tritaeniorhynchus*, *C. bitaeniorhhynchus* and *Mansonia fuscopennata*.

SFV is propagated by *Aedes abnormalis* and *Aedes* spp. Transovarial transmission occurs in *Aedes* spp.

Both viruses exist in an enzootic cycle involving small wild animals, birds, subhuman primates and mosquitoes.

Structure and Molecular Biology

The intact virion is spherical, 70–80 nm in diameter, with icosahedral symmetry and a $T = 4$ lattice. The internal nucleocapsid or core is also spherical, 40 nm in diameter, with icosahedral symmetry and a lattice, tentatively identified, as $T = 4$. The external spikes are arranged as 80 trimers, each consisting of heterodimers composed of two glycosylated transmembranal proteins. These spikes are embedded in a lipid bilayer which surrounds the core. An electron micrograph of negatively stained virions is shown in Fig. 1.

The genome consists of a nonsegmented, single-stranded RNA with a sedimentation coefficient of 49S. The RNA is capped at its 5′ end with m^7G and polyadenylated, average length of 70As, at its 3′ end. The genome RNA is of the positive orientation and isolated RNA is infectious. The complete sequence of approximately 11 700 nucleotides has been determined for SIN and SFV. Plasmids containing the complete genomic sequence as a cDNA have been constructed and these can be transcribed *in vitro* to yield infectious RNA.

The genome encodes four nonstructural and five structural proteins. The sizes and putative functions of these proteins are shown in Table 1.

The virions of SFV and SIN are stable to storage at 4°C and to repeated freeze-thawing. Infectious particles rapidly lose activity at 56°C (−1 log/10 min) and are labile to organic solvents and detergents.

Virus replication initiates by binding of virions to cell surface structures. Binding is sensitive to ionic conditions and host-range mutants of SIN exist that differ from their parental wild-type only in

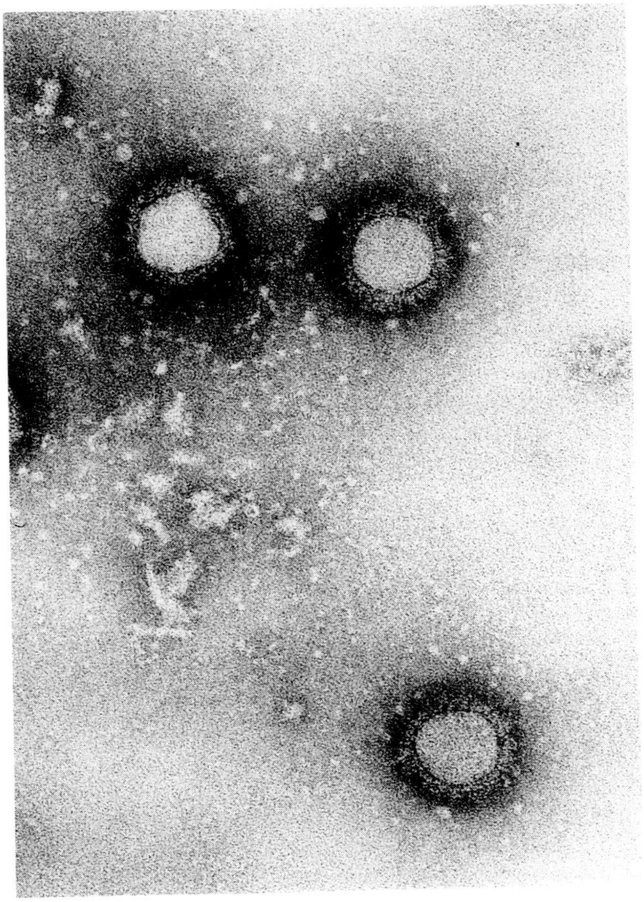

Fig. 1 Electron micrograph of a Sindbis virus particle negatively stained.

their net surface charge. A laminin receptor on chicken fibroblasts binds SIN but other receptors are also available. In most cells, the major route for uptake is via coated pits and acidified endosomes which are part of the host cell receptor-mediated endocytotic apparatus. Entry of virus cores into the cell cytoplasm occurs by fusion of the virus membrane with the cell's endosomal membrane. A pH <6 and presence of cholesterol in the host cell membrane are required for effective fusion. The E1 virus glycoprotein is the fusogenic component of the virus. The biochemical mechanism for uncoating of the nucleocapsid and release of the viral genome is unknown.

The biosynthetic events leading to new virus particles consist of the following steps (see Fig. 2). (1) The genomic mRNA is partially translated to form the four nonstructural proteins. Translation initiates at a single site near the 5′ end of the genomic mRNA and produces polyproteins that are cleaved to individual subunits (NSP1, 2, 3 and 4) by an autoprotease encoded in the virus NSP2 gene. For SIN, one polyprotein contains NSP1–3 and one poly-protein contains NSP1–4: the latter is made at one-fifth the amount of NSP1–3 and requires suppression of an Opal stop codon for NSP4 to be made. For SFV, one polyprotein that contains NSP1–4 is made. For both SIN and SFV, several stop codons located about two-thirds of the length of the genomic mRNA, terminate translation of the NSP1–4 polyprotein. (2) Components in a NSP1–4 complex transcribe the genomic RNA to form a negative-strand RNA template. A conserved sequence of 19 nucleotides adjacent to the 3′ polya-

Table 1. Sizes and putative functions of SIN and SFV proteins

	SIN	SFV	Function(s)
Nonstructural proteins			
NSP1	540	537	Methyl transferase (5′ capping activity)[a]
NSP2	807	798	Autoprotease (helicase, nucleotidase)[a]
NSP3	549	482	
NSP4	610	614	RNA polymerase
Structural proteins			
Capsid	264	267	Nucleocapsid; autoprotease
p62	487	482	Precursor to E2
E2	423	418	Spike component; major antigen for neut. antibodies
E3	64	64	N-terminal portion of p62
6K	55	60	Membrane-associated; has signal sequence for E1
E1	439	438	Spike component. Fusogen at pH <6

Figures are the number of amino acids in the protein. Note that the p62 does not appear as a final virion structural protein.

[a]Activities based on sequence motifs.

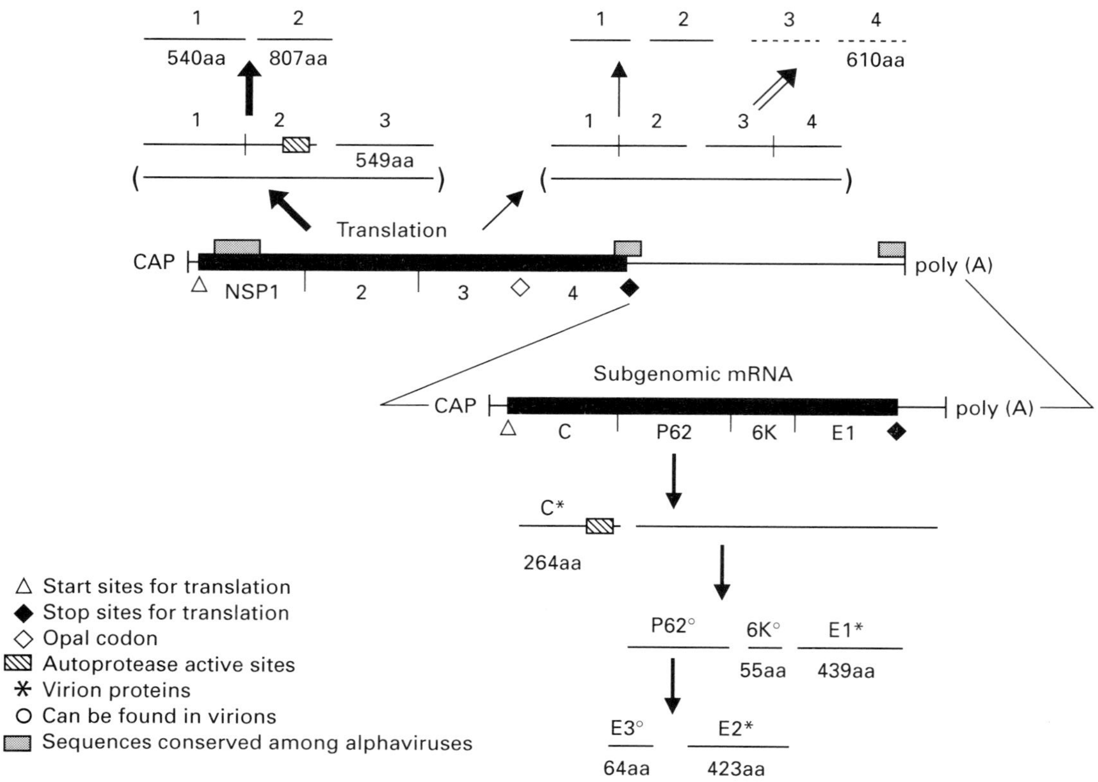

Fig. 2 Organization and expression of the Sindbis virus genome. nsP1–4 refer to the nonstructural proteins. The numbers under the bars indicate the number of amino acids in the protein. Co- and posttranslational proteolytic cleavages are shown by the arrows. Refer to text for details. Reproduced with permission from Schlesinger S and Schlesinger MJ (1991). In: Fields BN *et al.* (eds). *Virology*, 2nd edn, p 453. New York: Raven Press.

denylation site is postulated to be the site for initiation of negative-strand RNA transcription. (3) The negative-stranded RNA serves as a template for transcription, also by components of the NSP1–4 complex, of a subgenomic positive polarity mRNA that is about one-third the size of the genomic RNA and has a sedimentation coefficient of 26S. Subgenomic 26S mRNA contains sequences identical to those at the 3′ portion of the genomic RNA. The promoter for subgenomic mRNA transcription consists of 18 or 19 nucleotides upstream and five nucleotides downstream from the start site of 26S mRNA transcription. (4) The intact negative polarity RNA template is also transcribed by components of the NSP1–4 complex to form intact positive polarity RNA genomes. (5) The subgenomic 26S RNA is translated to form the virus structural proteins. Translation initiates at a single site near the 5′ end of the subgenomic mRNA and produces polyproteins that are cleaved to form the capsid, p62, 6K and E1. The first cleavage is by an autoprotease activity of the capsid which releases the capsid from the nascent polypeptide. The second and third cleavages are by the host cell signalase and form p62, 6K and E1. Transmembranal 'signal' sequences occur at the N terminus of the p62 and at the C terminus of the 6K protein and these allow for insertion and translocation of the p62 and E1 proteins into the lumen of the endoplasmic reticulum. (6) Several kinds of post-translational modifications occur during transport of the p62, 6K and E1 proteins from the endoplasmic reticulum to the plasma membrane of the cell. These include: (i) attachment of glycosyl groups to p62 and E1 in the endoplasmic reticulum and their modification in the Golgi vesicles; (ii) covalent attachment of fatty acyl groups (palmitate) to cysteinyl residues of p62, 6K and E1 in a transitional compartment between the endoplasmic reticulum and the *cis*-Golgi; and (iii) proteolytic processing of p62 in a *trans*-Golgi vesicle that produces E2 and E3. E3 of SIN is released but E3 of SFV is retained in the structure of the enveloped virus. A small amount of 6K is incorporated into the virus. (7) Assembly of new virus initiates by a self-association of capsid proteins to form a protein shell that incorporates the genomic RNA. There is specific binding between a domain near the 5′ end of the genomic RNA and sequences of the capsid protein. The final stages of assembly occur at the plasma membrane where intact nucleocapsids bind to cytoplasmic domains of the transmembranal E2 glycoproteins. Additional binding of trimer spikes to nucleocapsid leads to envelopment of the latter by the lipid bilayer. Fusion of the membrane releases the virus from the cell. In avian cells, 1000 to 10 000 new virus particles are released from each infected cell in 8 h.

Genetics

The frequency of mutation is about 10^{-4}, which is similar to other RNA viruses. Temperature-sensitive and host-range mutants have been isolated and placed into seven complementation groups. Site-directed mutations have been prepared using a cDNA clone encoding the entire virus genome.

Defective-interfering (DI) particles are generated within six to nine passages in tissue culture cells at high multiplicities of infections. DIs are about one-third the size of the wild-type genome and contain scrambled and repeated portions of the genome. However, they have three regions of the genome similar to wild-type virus genome: a 5′ domain, a packaging domain near the 5′ end of the RNA and 19 bases at the 3′ terminus.

SIN can undergo recombination. Western equine encephalitis virus has been shown to be a natural recombinant between SIN and eastern equine encephalitis viruses.

Evolution

SIN is the 'Old World' counterpart of western equine encephalitis virus. Persistence is believed to result from continuous cycling among birds via mosquitoes. The RNA sequences encoding the nonstructural proteins are related to the icosahedral plant viruses.

SFV is also an 'Old World' virus.

Pathology and Histopathology

Neither SIN or SFV are serious pathogens for adult vertebrates unless virus is injected directly into the brain. SIN is highly pathogenic to embryonated hens' eggs. Only in infected infant rodents is cytopathology observed when the virus is given intracerebrally or intraperitoneally. Lesions were confined to CNS and skeletal muscles. In CNS, focal cystic degeneration, vascular dilation and neurolysis occurs. In skeletal muscle, atrophy, focal necrosis and diffuse degeneration occurs. In rare infections in humans, SIN and SFV produced arthralgia, rash and fever. Genetic variations of SIN have been isolated that are more neurotropic and more virulent than the wild-type strains.

In vertebrate tissue culture cells, SIN and SFV have a broad host range and are highly cytopathic. Shortly after infection of cells in culture, host cell protein synthesis stops. In insect cells, virus grows somewhat slower to high titers but host cell protein synthesis is not inhibited and cells are not killed. After an initial burst of virus, the insect cell culture becomes persistently infected and lower levels of virus are continuously secreted.

Relationship

SIN and SFV are not serologically cross-reacting but antibodies raised against certain epitopes in several of the viral proteins are cross-reacting.

Immunological Response

SIN and SFV are highly antigenic and produce high titers of neutralizing antibodies. Humans with antibodies to SIN and SFV have been found worldwide.

See also: Defective-interfering viruses; Equine encephalitis viruses; Persistant of viral infection.

Further Reading

Schlesinger S and Schlesinger MJ (eds) (1986) The *Togaviridae* and *Flaviviridae*. Series *The Viruses*. New York: Plenum Press.

Schlesinger S and Schlesinger MJ (1990) In: Fields BN *et al.* (eds) *Virology* 2nd edn, p. 453. New York: Raven Press.

Strauss JH and Strauss EG (1986) In: Schlesinger S and Schlesinger MJ (eds) *Togaviridae and Flaviviridae*, p. 35. New York: Plenum Press.

SINGLE-STRANDED RNA BACTERIOPHAGES

Jan van Duin
Leiden University
Leiden, The Netherlands

Introduction

Since their discovery in 1961 by Loeb and Zinder, the RNA phages have served as a model system to explore a variety of problems in molecular biology. As a source of homogeneous and readily obtainable messenger RNA, they have been particularly helpful in solving questions on regulation of gene expression at the level of translation. The concepts of translational polarity and translational control by repressor proteins resulted from early studies on bacteriophage RNA.

Taxonomy and Classification

The RNA coliphages form the family *Leviviridae*. Within this family two genera can be distinguished, the *Leviviruses* and the *Alloleviviruses*. The leviviruses, also known as group A, are subdivided into groups I and II, whereas the alloleviviruses (group B) consist of the groups III and IV (Fig. 1).

The best characterized phage in group I is MS2. Close relatives are R17, f2, M12 and JP501 (>90% homology). Somewhat more distant but still in group I is fr, isolated by Hoffmann-Berling. The genomes of MS2 and fr are completely sequenced and from the others partial sequences are known. In group II, GA and KU1 have been fully sequenced and JP34 partly. Other members of this group are TH1 and BZ13. The prototype of group III is Q_β, which has been sequenced. Other members include VK and TW18. In group IV, SP is the best known representative, being the only one whose sequence has been determined. Other strains belonging to group IV are NL95, FI, TW19 and TW28. Most of the phages mentioned here are part of the Watanabe Collection at the Keio University, Japan.

Classification into the four groups is based on serological and physicochemical properties, such as the molecular weight of the phage particle or the sedimentation velocity of its RNA, and has been largely confirmed by the presently available sequence data. Serotyping by itself is sometimes ambiguous since a few amino acid substitutions in the major coat protein can change the immunological properties dramatically.

Virion Structure

In addition to one molecule of positive-strand RNA, each virion contains 180 copies of the coat protein and one copy of the maturation, or A, protein. The group III phages contain also 3–14 copies of the readthrough protein in their capsid, and this presumably also holds for group IV. For this reason, the protein shell of the single-stranded RNA phages is not isomeric like other icosahedral viruses such as poliovirus or satellite tobacco necrosis virus. The diameter of the phage is 26 nm, and the protein shell is 2.3 nm wide. The icosahedral shell has a $T = 3$ surface lattice. Crystals of phage MS2 which diffract X rays with high resolution have been obtained and the coat protein structure has been solved to 3.3 Å resolution. Unfortunately, the RNA and the A protein are not seen in the electron density map. These molecules are apparently not ordered in the crystal.

The Infection Process

The single-stranded RNA phages gain access to the interior of the cell by attachment to and transport via long, filamentous structures called pili. These can be of various origins and serotypes. In *Escherichia coli*, the sex (or F) pili are used as vehicles but in *Pseudomonas* and *Caulobacter* polar pili are used for this purpose. Transfer of RNA via pili is not essential for infection however. Cells that lack pili can be infected if they are converted to spheroplasts.

The attachment of the phage to the sides of the pili proceeds via the maturation (A) protein. The coat protein plays no role in this process, since binary complexes consisting of one copy of the A protein and one copy of (group I) RNA can infect a piliated

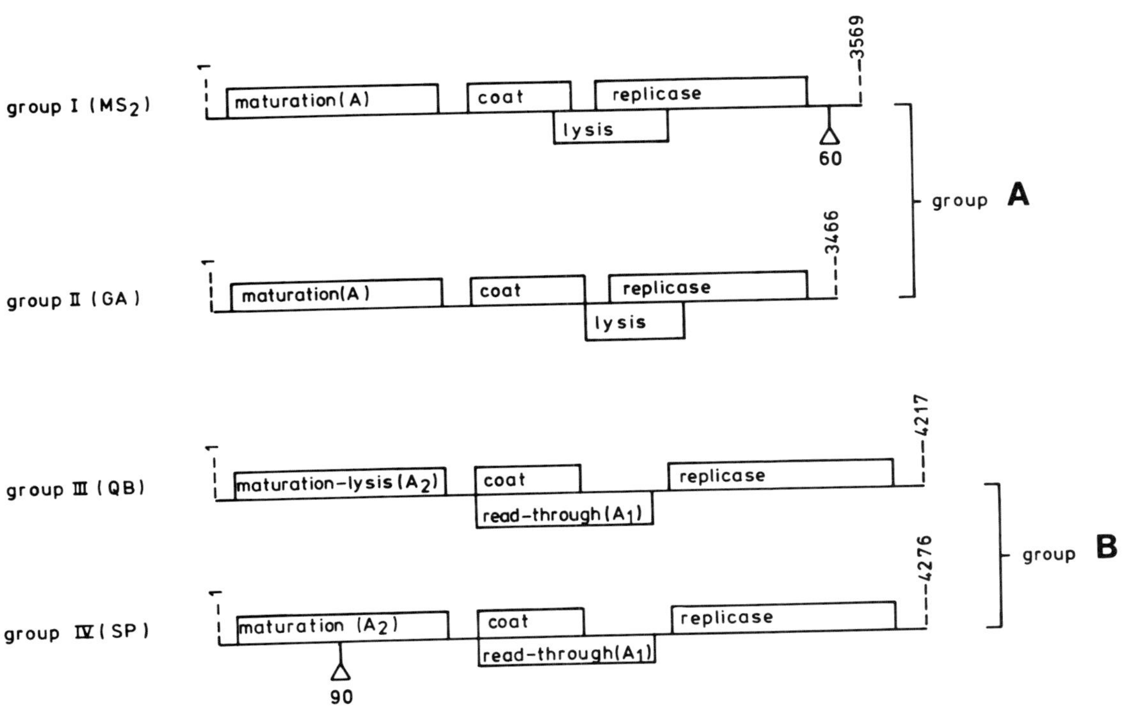

Fig. 1 Genetic map of group I–IV RNA coliphages. Inserts in group I with respect to group II and in group IV with respect to group III are given as triangles.

cell. In group III phages the minimal infection set requires the additional presence of the readthrough (A1) protein.

Contact of the phage with the pilus results in cleavage of the A protein into a 15 kD and a 24 kD fragment. This cleavage probably triggers the ordered ejection of the tightly packed RNA from the virus shell. The binding sites of the MS2 A protein on the RNA have been determined at nucleotide regions 3500 and 400, respectively. When cleavage of the A protein occurs between the two RNA binding domains of the protein this would potentially lead to the liberation of the two ends. Conceivably, the 5′ end of the RNA begins to move along the pilus towards the cell. This stage of infection corresponds to the RNase-sensitive step. The two A protein fragments remain associated with the RNA during penetration of the cell envelope. However, it is unlikely that these protein fragments play any further role, since the naked RNA is fully able to generate infectious progeny in the spheroplast infection assay.

Host Range and RNA Phages of Other Genera

RNA phages have also been found in several other gram-negative bacteria. In *Pseudomonas aeruginosa* PP7 and 7S have been characterized and ϕCb5, ϕCb12r, ϕCb8r and ϕCb23r were found in different *Caulobacter* strains. As judged by several criteria, these phages must be very similar to the coliphages. They have the same morphology, diameter and molecular weight range, and they probably also fall into four groups. A partial sequence from the *Pseudomonas* RNA phage PP7 is available. The absence of a coat readthrough sequence shows this phage to belong to group A. These noncoliphages also contain a major coat protein of the standard size (14 kD), together with a minor structural component that corresponds to the maturation (A) protein. In view of the established role of the A protein in absorption to the pili the presence of this sort of protein is expected.

RNA phages depend for infection on the presence of pili on the host. The coliphages infect via the F pili but *Pseudomonas* and *Caulobacter* phages enter the cell via polar pili. The dependence of RNA phages on pili as an entry to the bacterial cytoplasm is also reflected in the existence of the single-stranded RNA phage PRR1 that will infect many genera such as *Pseudomonas*, *E. coli* (Hfr and F+), *Salmonella typhimurium* and *Vibrio cholerae*, provided the host expresses the pili encoded in the drug resistance factor RP (P compatibility group, e.g. RP1,

RP4 or R1822). In this connection, we should mention that some members of the *Enterobacteriaceae* have been artificially converted to coliphage sensitivity. If the F factor of *E. coli* is introduced in *Shigella*, *Proteus* or *Salmonella*, these genera can be infected by the *E. coli* phages.

The host range of the single-stranded RNA phages is thus determined by the kind of pili that are exposed on the cell surface. Host specificity is basically an adsorption or pilus specificity. This phenomenon was exploited to screen sewage samples for the occurrence of RNA phages. The F factor was introduced into *Salmonella*. As *Salmonella* is not readily infected by DNA phages, the majority of plaques obtained represent the RNA coliphages.

Ecology

Furuse has examined and reviewed the geographical distribution of the single-stranded RNA phages as well as their present day habitat. They are most frequently encountered in sewage and feces of mammals, and their titers in sewage samples may be as high as 10^7 PFU/ml. RNA phages may constitute to up to 90% of total coliphages present in these samples, but the number can vary substantially. It is about 50% for Japan, Korea and the Philippines, but as low as 5% in, for example, India, Mexico or Brazil. An explanation for the difference is not available.

The geographical distribution of groups II and III shows a strong bias. In northern Japan, there is a relative abundance of group II over group III (6:1) per sampling site. Moving southward, this ratio drops dramatically until in Southeast Asia group II becomes rare. Furuse has suggested that the north–south gradient is related to differences in climate. Group III (and also I and IV) propagates well at 40°C but not at 20°C, whereas for group II the situation is reversed.

Although the natural host for the single-stranded RNA phages is not known with certainty, it is clear that they survive passing through the gastrointestinal tract of gnotobiotic mice and propagate stably in the intestines if *E. coli* is present as host. Thus *E. coli* can sustain the life cycle of the phage under 'natural' circumstances. In Japan and the Netherlands attempts have been made to determine whether certain phage groups are preferentially associated with certain animal species or with humans. So far, these studies have not been conclusive.

Index Organism

The RNA coliphages and enteroviruses share the same habitat. In addition, because of their structural similarity RNA phages show approximately the same inactivation characteristics as enteroviruses in sewage water treatment processes. For this reason RNA phages can be and are used as index organisms for the possible occurrence of pathogenic enteroviruses.

The Genetic Map

The map of the four groups is shown in Fig. 1. The conspicuous difference between groups A and B is the presence of a readthrough protein in the latter group. The major coat protein ends in a leaky UGA stop codon that is read as tryptophan with a probability of about 5%. This leads to a C-terminally extended coat protein that is incorporated in the capsid and required for infectivity. The other major divergence is the presence of a separate lysis gene in group A. This out-of-frame overlapping gene encodes a roughly 70 amino acids long hydrophobic peptide, not present in the virion. Upon synthesis it inserts in the cytoplasmic membrane of the bacterium which is de-energized as a result. Somehow short-circuiting the membrane distorts the balance between the components of the bacterial enzyme ensemble that extends the peptidoglycan network. This leads to cell lysis.

Differences within the genera are more subtle. In group A the most pronounced one is a 60 nucleotide insertion in the 3′ untranslated leader of group I. A 90 nucleotide insertion in the A2 protein gene of group IV represents the major physical difference with group III in supergroup B.

The Lysing Protein of Group B Phages

The strong similarity in gene arrangement and control of gene expression between the single-stranded RNA phages is fully lost in the way the phages organize their escape from the wasted host. Phenotypic analysis of bacteria harboring subgenomic Q_β cDNA fragments has shown that the overproduction of the maturation protein causes cell lysis. This finding is consistent with the fact that Q_β mutants carrying an amber mutation in the maturation protein do not lyse. The implication of this different lysis mechanism for the evolutionary relationship between the two major phage groups will be discussed later.

Gene Expression

The appearance of the phage-coded proteins is carefully controlled in timing and amount. Replicase is an early product, and the amount of coat protein exceeds by far that of the other products. Since no DNA intermediates occur in the life cycle of the phage, control is predominantly exerted by RNA secondary structure that restricts ribosome binding to translation initiation regions.

Upon entry in the cell, the RNA first serves as messenger, because multiplication requires the product of the replicase gene. After subunit II is made, the holoenzyme with the accessory host factors can be assembled and amplification can start.

The maturation protein is needed in small amounts, one copy per virion, and its translation is accordingly kept at a low level. A suggestion is that the A protein in MS2 is downregulated by base-pairing of its start region to an internal coding region. Accordingly, only nascent RNA chains on which the complementary sequence has not yet been produced would serve as messenger RNA. This view has not yet been confirmed using the presently available powerful recombinant DNA techniques. The proposed base-pairing is also not conserved in fr.

Replicase is needed in small amounts early in infection. Its synthesis is turned down 20 min after infection. This inhibition is effectuated by the coat protein which binds to a site in the start region of the replicase gene. In all phages this site can adopt a helical structure that, once formed, will be stabilized by the binding of a coat protein dimer. In addition, translation of the replicase (R) gene is coupled to that of the coat gene. This arises by base-pairing between part of the start region of the R gene to the coding region of the coat gene. Coat protein synthesizing ribosomes passing through this region will temporarily destroy the base-pairing and thus liberate the R start.

The coat protein is abundantly expressed during phage infection. Except for the competition between ribosome and replicase for the ribosome binding site of the coat gene (see later), the synthesis of this protein seems not to be negatively controlled in any way.

Like the replicase gene, the lysis gene is under translational control of the coat gene. Early nonsense mutations in the coat gene prevent the lysis protein from being expressed. The trigger to lysis gene expression is termination of coat gene translation at its natural stop codon. This conclusion rests mainly on the observation that turning the stop codon of the coat gene into a sense codon, leading to termination of eight codons further downstream, abolishes cell lysis. In the absence of coat gene translation the lysis cistron of MS2 is closed by RNA secondary structure. The terminating ribosome resolves this structure, hereby creating a chance for translation initiation at the previously closed lysis cistron.

Replication

Most of our knowledge of replication has been obtained in the Q_β system, but it is assumed that the principles also apply to the other groups. The replicase holoenzyme contains five different proteins called subunits I to IV, or subunits α, β, γ and δ; the fifth subunit is called host factor (HF). Subunit I was identified as ribosomal protein S1, and subunits III and IV are the translation elongation factors EF-Tu and EF-Ts. Thus, three proteins that normally function in the synthesis of proteins are recruited by the phage to assist in RNA synthesis. Subunit II is encoded in the viral genome, and all these four subunits occur in the enzyme complex in one copy. The fifth component (HF), whose function in the uninfected cell is not yet known, is present as a hexamer. There is a claim that group II uses a different host factor.

Replication proceeds via the synthesis of a free negative strand. The annealed positive and negative strands are not a substrate for the replicase enzyme. Thus one property of the enzyme must be to separate mother and daughter strands at the replication site.

The requirements for copying the positive and the negative strands are different. To copy the positive strand, only subunits II, III and IV are required, whereas positive strand replication needs also subunit I and the host factor. RNase protection studies have shown that in the absence of the initiator nucleotide GTP, the replicase binds to two internal RNA sequences called the M site and the S site. Probably, in this complex the 3′ terminus is in close vicinity of the polymerization site of the enzyme. The S site overlaps the start region of the coat gene. Therefore, the binding of replicase to M and

S sites clears phage RNA from ribosomes. This is necessary to permit its undisturbed replication.

Template specificity is not absolute. For instance, GA (group II) replicase works well with group I RNA and shows significant activity with RNA templates from group IV. Q_β replicase does not accept group A RNA as template, but SP RNA is replicated with limited efficiency. All replicases copy poly(C) under production of double-stranded poly(C)·poly(G).

6S RNA and Q_β RNA Variants

Infection of *E. coli* by Q_β leads, apart from phage multiplication, to the accumulation of what has been termed '6S' RNA. This is a nonhomogeneous collection of RNA molecules that vary in size from about 100 to 200 nucleotides and serve together with their negative strands as templates for Q_β replicase. They do not code for any protein nor do they contribute to the infection process. Three '6S' RNA representatives have been fully sequenced. Two of these show no homology with Q_β RNA. All are characterized by a high degree of secondary structure and like all phage RNA have at least three consecutive Cs at their 3′ ends. 6S RNA is also generated upon transformation of *E. coli* with the cDNA for Q_β replicase or by the *in vitro* incubation of Q_β replicase with the four nucleotide triphosphates in the absence of added template. The last experiments suggest that Q_β replicase can synthesize RNA *de novo*.

From an evolutionary point of view, it is interesting that these molecules that are not constrained by a coding sequence quickly respond to selection pressure. Mutants of 6S RNA that are adapted to *in vitro* replication under adverse conditions, such as the presence of ethidium bromide, T1 ribonuclease or limiting amounts of the building blocks can be easily obtained. Abbreviated Q_β RNA variants can be prepared *in vitro* by gradually reducing the time allowed for Q_β replication. After some 70 replication rounds the length of Q_β was reduced to about 12%. There seems no basic difference between such truncated Q_β RNA and the RNA present in the defective-interfering (DI) virus particles that accompany, for instance, influenza infection. Also here, the abridged molecules once created by replication errors can survive as long as they are templates for the replicase, and their multiplication does not endanger the survival of the virus population as a whole.

The rapid yield to selection pressure by 6S RNA and the Q_β RNA variants reflects the inaccuracy of the Q_β replicase, which has been estimated as between 10^{-3} and 10^{-4} per nucleotide per replication. The presumed absence of a 3′–5′ exonuclease editing activity in Q_β replicase would be consistent with its relatively low copying fidelity. At the same time the frequency with which deletions occur must also be unusually high.

Diversity

Considering the inaccuracy of phage RNA replication one might expect an endless scala of phage sequences, all fit to survive as self-replicating molecules. This turns out not to be true. Even considering the relatively small amount of sequence data available it is already clear that diversity is severely limited. It is smallest in group II and group I where it is difficult to isolate an RNA with less than 90% homology to the prototype. Greatest flexibility appears to be possible in group IV. This group also shows the greatest serological diversity.

Phylogeny of RNA Phages

An interesting but necessarily most difficult question is that of the origin and kinships of the RNA bacteriophages. It is generally assumed that the four groups derive from a common ancester. The basis for this assumption is the nearly identical genetic organization, the strong resemblance of the replicases, the use of the same host proteins as auxiliaries in the copying reaction, and the similarities of several control mechanisms, such as the translational coupling between coat and replicase genes and the repression of replicase synthesis by the coat protein. These properties are more easily explained by divergent than by convergent evolution.

There are two views on the relationship between supergroups A and B: the longer phages derive from the shorter by insertions, or the shorter derive from the longer ones by deletions. Furuse has proposed that group IV stands closest to the forebear, implying that 'recent' evolution was accompanied by deletions. Several arguments exist. Group IV contains the most diverse subgroups and has the broadest habitat. One of its members, ID2, shows serological cross-reactivity with all other groups. Some coat gene amber mutants of group IV phage SP produce viable progeny in intergroup complementation with MS2 and Q_{β}. Also within group IV there is more variability in the molecular weight of the coat protein and in several other physical parameters.

The tentative reconstruction would be that group IV has spawned group III by small deletions and group I by deleting the readthrough part of the coat protein in combination with several other small losses. The poor serological relationship between Q_{β} and group A makes it unlikely that group A is directly derived from group III. If indeed during younger times natural selection has favored shorter genomes, then of course group II derives from group I. Also, the 60% sequence similarity between GA RNA and MS2 RNA seems to exclude that these two groups arose independently from group IV.

Others have arrived at a different conclusion. Several coat gene sequences in Q_{β} are to some extent repeated in the same order in the coat gene readthrough part, which suggests that a gene duplication may have occurred. The appearance of a conditional stop codon between the two coat gene copies would then contribute to the genetic potential.

The position of the lysis gene in group A has interesting aspects against this background. In the gene contraction view, this gene must have evolved after group A emerged from group IV. This is consistent with the observation that codon usage in the lysis gene qualifies it as a late addition. Also in the gene expansion view the development of the lysis function must be a late event. If not, one would have to explain what possible advantage could be gained by inactivating the lysis gene through an insertion, whose usefulness still had to be crafted.

See also: Bacterial identification – use of bacteriophages; Bacteriophage ecology, evolution and speciation; Defective-interfering viruses.

Further Reading

Furuse K (1987) Distribution of coliphages in the environment: general considerations. In: Goyal SM (ed.) *Phage Ecology*, p. 87. New York: Wiley.

Van Duin J (1988) In: Fraenkel Conrat H and Wagner RR (eds) *The Bacteriophages*. Series *The Viruses*, p. 117. New York: Plenum Press.

SMALLPOX AND MONKEYPOX VIRUSES

Keith Dumbell
University of Cape Town
South Africa

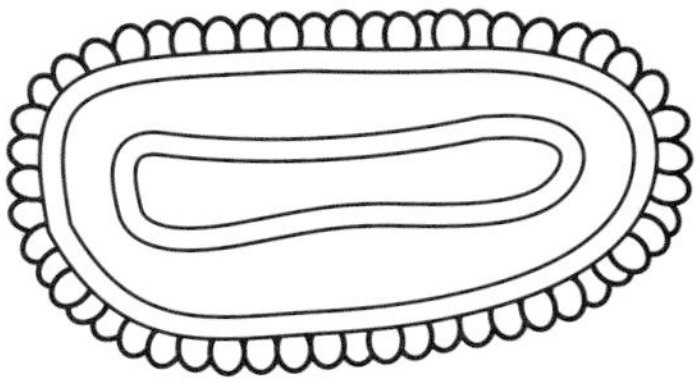

Smallpox (Variola) Virus

History

A Chinese account of a disease which was clearly smallpox was written in the 4th century AD. Over the next two centuries reliable accounts described smallpox in Japan, Korea, India and Egypt. Shortly after this the Islamic expansion carried smallpox into North Africa and Europe. From the 15th century smallpox was taken along with the spread of European colonists to America and Southern Africa. According to the Chinese account, smallpox had appeared in China during the first century, but it may have been current in India well before that. A few Egyptian mummies show evidence of a pustular rash which has some resemblance to smallpox, but there are no accounts of such a disease in contemporary Egyptian or Jewish writings.

Taxonomy and Classification

Variola (smallpox) viruses belong to the *Orthopoxvirus* genus of the *Chordopoxvirinae* subfamily of DNA viruses and have linear, double stranded DNA genomes of 180 kb. The virions are large (about 250 × 200 nm), brick-shaped particles. Like vaccinia they can be seen as naked, intracellular virions or as enveloped, extracellular virions. The particles have two lateral bodies and a central core.

Variola virus has a close antigenic relationship to other orthopoxviruses; it is distinguished as a species by its biological characters and, more recently, by the restriction patterns of its genome.

Virion Structure and Properties

What is known about the structure, physical properties and replication of variola virus conforms so closely to the much better studied vaccinia that the reader is referred to that article. One point of difference is that the variola genome has only a very short inverted terminal repeat sequence (ITR), unlike the 10 kb ITR of vaccinia virus.

Geographic and Seasonal Distribution

The last natural case of smallpox occurred in 1977; and in 1980 the disease was declared by WHO to have been eradicated. In its heyday smallpox had a world-wide distribution. The development of effective public health services and the routine practice of vaccination gradually eliminated the disease from North America and most of Europe, but outbreaks, apparently endemic in origin continued in these parts of the world through the first decades of the twentieth century. For a variety of reasons routine vaccination failed to eradicate smallpox from many tropical and sub-tropical areas. At the start of the global eradication campaign in 1967 there were still 31 countries in which the disease was endemic.

After smallpox eradication was complete, the number of laboratories maintaining stocks of variola virus was rapidly reduced. At present (1992) variola virus is maintained in just two high containment laboratories, situated in Atlanta and in Moscow. WHO is currently coordinating a programme to sequence the complete genome of representative strains of variola virus before the end of 1993, at which time it is planned to destroy all remaining stocks of variola virus.

Smallpox was maintained entirely by transmission from an active case to another human; latency did not occur. Consequently, active search would reveal cases throughout the year in countries where the disease was endemic. Nevertheless there were low seasons and high seasons. In tropical areas it seems likely that the seasonal changes in incidence were achieved more through the ease or difficulty of travelling about than from direct effects of climate on survival of virus or susceptibility to it.

Host Range and Virus Propagation

Variola virus had a narrow host range. Natural infection was confined to humans. Experimentally the virus produced mild disease in some species of monkeys and could be propagated, at least for a time, in rabbits and baby mice.

Under natural conditions, the host range of variola virus was limited to humans. There was no suitable experimental animal, though limited transmission has been reported in monkeys, rabbits and intracerebrally in mice. Variola virus was propagated in laboratories on the chick embryo chorioallantois or in cell cultures usually of human origin.

Genetics

DNA restriction maps of variola from different geographic areas were remarkably uniform. Consistent differences were found between variola major strains and variola minor (alastrim) strains from South America, but this did not apply to strains of variola minor from Africa. There is, at present, no evidence that the variola virus genome encodes genes which are unique to that virus. A project to obtain the complete DNA sequence of a few strains of variola is currently in progress and it is intended that these sequence data shall become the permanent record of this virus when all viable variola virus strains have been destroyed. Comparison of the entire variola sequence with that of vaccinia virus may reveal some of the secrets of this major pathogen.

Variola viruses readily recombine with other orthopoxviruses; a number of resulting recombinants were characterized biologically, but work with variola virus ceased before these results were correlated with exchange of particular segments of the parental genomes.

Evolution

The naturally occurring orthopoxviruses, cowpox, monkeypox and camelpox, currently occupy separate geographic and biologic niches in Europe,

Africa and the Middle East. The similarities between the genomes of these three viruses, variola and vaccinia are sufficient to postulate a common ancestor. Other orthopoxviruses exist but have not been so well studied. There is evidence that some orthopoxvirus genes have counterparts in other genera of poxviruses, such as leporipox, capripox and avipox. Cowpox and monkeypox viruses are able to infect a variety of host species, unlike the narrow host range of variola and camelpox. The availability of sufficient human hosts to support continued transmission of smallpox over some thousands of years may be the reason that the variola virus of modern times had lost any primitive ability to utilize other host species. Although monkeypox virus sporadically infects humans and may cause disease closely resembling clinical smallpox, there is evidence that monkeypox virus is not a direct ancestor of variola virus.

Serologic Relationships and Variability

There was never any evidence for antigenic variability in variola viruses. Indeed it was even difficult to separate variola from other orthopoxviruses by serologic tests. Any strain of vaccinia would protect against variola in any part of the world and in different years. Several different antigens of variola virus find their way to the surface of the infected cell before virus release. Because of this and also the length of the incubation period it is likely that minor variations in antigenic specificity would not confer any selective advantage.

Epidemiology and Transmission

The vast majority of smallpox was contracted by direct contact with an infected person. There were rare instances of indirect spread via fomites but animals played no part in sustaining or transmitting the variola virus.

Despite occasional startling incidents, smallpox required close contact for effective transmission and spread mainly within households or to other close contacts.

Those infected with smallpox remained well for the 10–12 days of the incubation period and, during this time, were not infectious to others. Severe cases were prostrated just before they became infectious and thus were likely to infect fewer secondary cases than people with milder disease who remained ambulatory. On the other hand funeral ceremonies for fatal cases were sometimes attended by relatives from distant communities and resulted in wider dissemination of the outbreak.

In natural smallpox (as opposed to variolation) virus was transmitted by droplet infection. The most usual source was the lesions which occurred in the pharynx or mouth (the enanthem). These ulcerated more quickly than the lesions developing in the skin (the exanthem) and were well situated for distribution to close contacts. Primary lesions were presumed to occur in the respiratory tract but were never seen or demonstrated post mortem. Throat swabs from contacts, who subsequently developed smallpox, were occasionally found to harbor virus during the incubation period but there was no evidence that transmission occurred before the end of the incubation period. Nor did those who were subclinically infected pass on the disease.

Cases were potentially infectious until all the lesions had completely healed but virus was not so readily dispersed from dried exudate or scabs. There was no evidence of latency and no spontaneous recurrence. Consequently once an outbreak had run its course or been controlled in a local community, smallpox would not recur unless re-introduced from outside that community.

Outbreaks varied in severity, but even in a severe outbreak there would be some milder cases. At times whole outbreaks of mild smallpox occurred with little or no fatality. This led to the distinction of two varieties of the disease: variola major and variola minor and to the presumption that the variola minor outbreaks were caused by a distinct strain of variola virus of lower virulence. In some areas the minor form became established as the endemic strain. The variety of variola minor known as alastrim apparently arose towards the end of the nineteenth century and produced consistently mild disease. This strain was first described in the South East USA and thence spread through North America, to Europe, to South America and to Australasia. Outbreaks of minor smallpox were still occurring in Europe in the 1950s and in Brazil through the 1960s. Isolates from outbreaks of alastrim in Europe and Brazil shared biologic characters and a DNA restriction pattern which distinguished them from variola major viruses. The position in Africa was less well defined. Outbreaks of both major and minor smallpox continued to occur. Although the endemic variola in some African countries was clinically and epidemiologically variola minor in recent years, laboratory tests could not distinguish the virus isolates from variola major virus.

Clinical Features

The incubation period of 10–12 days was terminated by the abrupt onset of fever and prostration, often with nausea or vomiting. On the third day of illness a macular rash appeared which became papular and then vesicular and eventually pustular. The rash tended to be more concentrated on the head and limbs than on the trunk. The lesions were firm and raised but also based deeply in the skin. Scabs formed as the ulcerated pustules dried and gradually separated over two to three weeks, leaving depigmented areas. Residual facial scarring, due to destruction of sebaceous glands, was a characteristic feature of the survivors of major smallpox. Blindness was another not uncommon sequel. The above description applies to classical cases of major smallpox, in which the fatality rate was from 16 to 30 per cent. Prognosis was worse if the rash was slower in development with the lesions remaining flat and soft or becoming hemorrhagic. Sometimes the initial prostration was severe, generalized hemorrhagic manifestations appeared on the second or third day and no distinct rash developed. These patients invariably died, usually between the 5th and 7th day of illness. A more favorable prognosis was signified if the rash evolved more quickly, and the lesions were more superficial. This was the usual course of events in people who had been vaccinated, but not recently enough to give complete protection.

When the causative virus was variola minor, prostration was notably less than in patients infected by variola major, who had a comparable extent of rash.

Pathology and Histopathology

The majority of smallpox lesions were in the skin and mucous membranes of the mouth and oro-pharynx. The few postmortem studies that were done showed few specific lesions in internal organs, except in cases of hemorrhagic smallpox. In the absence of obvious damage in the vital organs it was presumed that death had been due to the profound toxemia. Fuller accounts will be found in the texts cited at the end of this article.

Immune Response

Neutralizing antibody and hemagglutination inhibiting (HI) antibody appears by about the 6th day of the disease, corresponding to the 18th day after infection. This is slower than the antibody response to primary vaccination and may partly explain why vaccination soon after contact with smallpox often ameliorated the outcome. However, experience with the complications of vaccination showed that a failure to develop cell-mediated immunity was more important than failure to develop antibody. It is presumed that cell-mediated immunity also controlled the development and resolution of smallpox lesions, but this was not adequately investigated.

Recovery from smallpox left a lasting immunity; though second attacks of smallpox were not unknown. Neutralizing antibody persisted for many years; HI antibody was demonstrable for some months to a few years.

Prevention and Control

Successful vaccination gave complete protection against smallpox for at least three years and probably five or more. Immunity waned after longer intervals, but rarely, if ever, did anyone who had been vaccinated contract fatal smallpox. The International Certificate of vaccination against smallpox was given a validity of three years, but those at high risk were normally revaccinated at yearly intervals.

Compulsory routine vaccination should, theoretically have eradicated smallpox in any area where it was practised. However in many countries where the disease was endemic, the fraction of the population that escaped the vaccination program sufficed to maintain a chain of active cases and a reservoir of the virus.

The main factors in containing an outbreak of smallpox were the isolation of cases and the tracing of contacts before their incubation period was over. These measures were supplemented with vaccination or re-vaccination in the locality of the outbreak, and were effective in the control and elimination of smallpox introduced into a non-endemic area. The eradication of smallpox from endemic areas was achieved by detecting current outbreaks and containing them. This strategy effectively interrupted the local transmission of the virus and progressively reduced the number of chains of transmission which remained.

Future Perspectives

In 1980 the World Health Organization officially certified the eradication of smallpox, and no further cases have come to light in the 12 years that have elapsed since then. This major achievement seems to be firmly established. The only remaining viable variola viruses are the two reference collections of variola strains held in high containment laboratories at Atlanta and Moscow and the possibility that virus might have survived in any mummified bodies of smallpox victims. Extrapolation from the known rates of decay of variola virus in smallpox scabs indicates that this possibility is unrealistic. It has been agreed that the two collections of variola virus strains should be destroyed as soon as sufficient alternative archival records of this virus have been prepared. WHO first coordinated a program to clone fragments of the variola genome into bacterial plasmids, several sets of which were produced. More recently this program has been superseded by a program to obtain the complete DNA sequence of at least one strain of variola and to destroy remaining virus by the end of 1993. Of course much of the gene pool of variola virus will still remain, in the genomes of other, closely related orthopoxviruses; the complete DNA sequence will reveal the extent of this and any features unique to variola. The section on monkeypox includes a discussion of whether variola virus could emerge again from natural sources.

Monkeypox Virus

History

Monkeypox was first described as an apparently spontaneous outbreak of fever and rash in a colony of cynomolgus monkeys held at the State Serum Institute in Copenhagen in 1958. During the next few years similar outbreaks were reported in monkey colonies in the USA. In 1966 monkeypox infection was introduced into the Zoo at Rotterdam. Here the first animals affected were giant anteaters from South America, but the disease spread to various species of apes and monkeys. The viruses isolated from these animals were found to be similar and to represent a species of orthopoxvirus that had not been described before 1958. In two separate incidents in 1970 monkeypox virus was found to have caused a smallpox-like illness in humans in Zaire and in Liberia. Sporadic cases have continued to occur in the African rain forest belt, particularly in Zaire. The virology and epidemiology of monkeypox have been intensively investigated because of the possible threat to the success of smallpox eradication.

Taxonomy and Classification

Monkeypox virus is a separate species of the *Orthopoxvirus* genus. Biologically it can be differentiated from other species by the morphology and ceiling temperature of its pocks on the chick chorioallantois and by its pathogenicity to rabbits. Antigenically it crossreacts extensively with other orthopoxviruses but produces a few specific antigens which can be detected with suitably absorbed antisera or monoclonal antibodies. The genome, of approximately 190 kbp has a distinctive restriction map.

Virion Properties and Replication

There is no reason to believe that replication of monkeypox virus differs significantly from the more extensively studied vaccinia virus (see Vaccinia virus).

Geographic and Seasonal Distribution

The reservoir of monkeypox virus is confined to the tropical rain forest belt of Africa. On current information the reservoir hosts are arboreal rodents, from which sporadic transmissions occur to simians and to humans. Human infections are rare events; most of the reported cases have been in Zaire which harbors more than half of the present African rain forest. There is no marked seasonal incidence, but some correlation with times of maximum agricultural activity (see Epidemiology).

It is presumed that the outbreaks in Europe and America were initiated by subclinical infection in monkeys or other animals exported from West Africa. There have been no outbreaks since 1968 following a general tightening of regulations concerning exports of animals.

Host Range and Virus Propagation

Under experimental conditions monkeypox virus is capable of infecting most common laboratory animals and many simian species. The resulting disease was usually mild in African monkeys but Asian and South American monkeys suffered moderate to severe disease. Apes, with the exception of chimpanzees, are seriously affected and the wide range of susceptible exotic animals is illustrated by the two South American giant anteaters which became infected while in transit to the Rotterdam Zoo.

In its natural habitat, monkeypox virus has been recovered from many sporadically infected humans and from one sick squirrel (*Funisciurus anerythrus*) and a chimpanzee. Other information comes from serologic surveys; antibody specific to monkeypox has been detected in a significant proportion of squirrels belonging to the genera *Funisciurus* and *Heliosciurus*. Specific antibody was also detected in seven species of *Cercopithecus*, and also in *Cercocebus* and *Colobus* monkeys. Monkeypox virus was propagated in laboratories on the chick chorioallantois, where it produced characteristic small pocks with a central hemorrhage; cell cultures of many different species were susceptible.

Genetics

*Hin*dIII restriction maps of the monkeypox virus genome confirm the conservation of the central part of the orthopoxvirus genome throughout the genus. Although restriction site maps show more variation in the outer quarters of the genome, DNA fragments from these regions cross hybridize strongly with corresponding fragments of other orthopoxviruses and rule out any long region of sequence that could be unique to monkeypox. There must be short unique stretches to account for the monkeypox-specific antigens that have been demonstrated, but few studies of DNA sequence in monkeypox have yet been reported. There is some variation in restriction site maps of monkeypox isolates from different geographical areas but variants with more dramatic changes in genome structure can readily be isolated in the laboratory. These are the 'white pock' mutants, which also have been described among other orthopoxviruses with a hemorrhagic pock phenotype such as cowpox (q.v.). Restriction site maps have been constructed for several of the white-pock mutants of monkeypox virus. Some show the subterminal deletion, combined with copy of the opposite terminus which are typical of cowpox white pock mutants, but more complex genomic rearrangements also occur among the mutants of monkeypox. Nevertheless the restriction site maps of these mutants still sufficiently resemble those of parental monkeypox to cluster with them when compared to maps of other orthopoxviruses.

Evolution

Monkeypox virus, like cowpox virus, appears to be sustained by transmission in small rodents. The present distributions of these viruses do not overlap, but the antigenic and genetic similarities among all the orthopoxviruses are such that they must share a common ancestor. Each of these viruses is known to infect a variety of species, including humans under natural conditions. This could have led to the possibility of further speciation. It has been suggested that monkeypox might be the progenitor of variola. From a biological point of view this is plausible, because monkeypox virus can produce a human disease, clinically resembling smallpox. Also a white-pock mutant of monkeypox which had also lost its virulence for rabbits and become more thermosensitive would resemble the laboratory phenotype of variola. DNA analysis however reveals sufficient difference between their respective genomes that this derivation can be ruled out. Also it must be remembered that smallpox appeared in East Asia well before it was current in Africa (see the earlier section on variola virus).

Epidemiology and Transmission

In Zaire, 338 cases of human monkeypox were recorded in the six years from 1981 through 1986, more than half of them occurring in the Equator region. Nearly 90% of the cases lived in small rural villages of less than 1000 inhabitants and only 3 cases occurred in towns with a population of more than 5000. The age-specific incidence of primary monkeypox (i.e. excluding spread from person to person) is highest in young children between 1 and 8 years, peaking in the 3–4 year age group. Young children have access to the cleared areas between the village and the forest; in this area also are found squirrels of the genera *Funisciurus* and *Heliosciurus* in which monkeypox antibody is prevalent. The age-specific incidence of secondary cases is more evenly distributed. Infected animals are the source of most human monkeypox, and whether transmission is respiratory or by inoculation through small abrasions is not known. Variation in the reported incidence from area to area and from year to year may well depend on natural fluctuations in the wildlife reservoir but will also be affected by the efficiency of surveillance activities. Transmission from person to person is probably from the enanthem via the respiratory route, but the secondary attack rate among unvaccinated household contacts is much lower than that found in smallpox. Based on experience to date, the basic case reproduction rate does not exceed 1.0 and outbreaks would be self-limiting even in an unvaccinated community.

Clinical Features

The incubation period of human monkeypox appears to be about 12 days, the same as for smallpox, but there have been only a few instances where it could be accurately determined. The clinical picture of human monkeypox closely follows that of ordinary smallpox. Most cases in unvaccinated children have been severe, but equivalents of the flat or hemorrhagic types of smallpox have not been encountered. The illness begins with fever, followed in 1–3 days by the rash and an enanthem is usually present on the oral mucosa. Unlike smallpox, most patients develop a generalized lymphadenopathy or, less often, a regional lymphadenopathy. Mortality in the unvaccinated has been about 11%, though somewhat higher than this in children under two years of age. All deaths have been in children under the age of nine years.

Pathology and Histopathology

This has only been studied in experimental infection of monkeys. Histopathology of the lesions resembled that of smallpox. Following intramuscular inoculation, there was an intense local inflammatory response. Virus spread to the regional lymph nodes and thence to spleen, tonsil and bone marrow. Viremia occurred between the 3rd and 14th days and a generalized rash appeared about the 7th or 8th day.

Immune Response

The antibody response to infection with monkeypox virus can conveniently be detected by hemagglutination inhibition or by ELISA tests. High titers in straight antibody tests suggest a response to monkeypox rather than to vaccination, but a residual titer after absorption with vaccinia virus is required to demonstrate antibody specific to monkeypox. For this reason radioimmunoassay tests have been most useful, because of the high titers obtainable from unabsorbed sera by this technique.

Prevention and Control

Recent vaccination with vaccinia virus effectively protects against monkeypox; those who had been vaccinated some years previously were susceptible but were less likely to develop severe illness. Widespread routine vaccination would be necessary to protect all who might come into contact with monkeypox. But the incidence of monkeypox is so low that complications arising from the vaccination program might rival the morbidity to be expected from monkeypox itself.

Future Perspectives

The main incidence of monkeypox in humans has been in young children. Although this age group is currently unprotected by vaccination, there has not been any evidence of a rising incidence of monkeypox infections. At present monkeypox does not present a public health problem serious enough to require specific action. This situation may last as long as the monkeypox virus maintains its relatively low transmissibility from person to person. Stochastic modelling suggests that all outbreaks would be rapidly self-eliminating. However, some caution is in order, lest a variant arise with increased potential for person to person spread. Some continuing surveillance in areas where monkeypox has been most prevalent is justified to detect any such change in the epidemic pattern.

See also: Cowpox virus; Immune response; Vaccines and immune response; Vaccinia virus.

Further Reading

Fenner F, Wittek R and Dumbell KR (1989) *The Orthopoxviruses*. New York: Academic Press, Inc.

Fenner F, Henderson DA, Arita I, Jezek Z and Ladnyi ID (1988) *Smallpox and its Eradication*. World Health Organization.

Jezek Z and Fenner F (1988) Human Monkeypox. *Monographs in Virology* 17. Karger.

SOBEMOVIRUSES

OP Sehgal
University of Missouri
Columbia, Missouri, USA

Taxonomy and Classification

Southern bean mosaic virus (SBMV) is the prototype of the sobemovirus group. Sobemovirus virions are about 30 nm in diameter, sediment at about 115S, and require divalent cations for their structural stability. The viral capsid is constructed from identical 30 kD subunits according to $T = 3$ design. The genome is a linear, single-stranded, messenger-sense RNA of mol. wt 1.4×10^6. Virions also encapsidate subgenomic RNAs or viroid-like satellite RNAs. Sobemoviruses are not serologically cross-reactive nor are these related to other viruses. They are transmitted by insects, through seeds, and with mechanical inoculation. The host range of sobemoviruses is limited.

Besides SBMV, the following viruses are considered as definitive members of the sobemovirus group: blueberry shoestring (BSSV), cocksfoot mottle (CfMV), ginger chlorotic fleck (GCFV), lucerne transient streak (LTSV), olive latent-1 (OLV-1), rice yellow mottle (RYMV), *Solanum nodiflorum* (SNMV), sowbane mosaic (SoMV), subterranean clover mottle (SCMoV), turnip rosette (TRoSV) and velvet tobacco mottle (VTMoV) viruses. Three additional viruses, cocksfoot mild mosaic (CMMV), Cynosurus mottle (CyMV) and Panicum mosaic (PMV), are included as tentative members pending availability of additional data.

SBMV and TRoSV are the two most characterized sobemoviruses with respect to physicochemical properties and structural organization. In particular, SBMV has proved to be a model, *par excellence*, in elucidating the nature of virion stabilizing interactions and particle architecture. The viroid-like encapsidated RNAs of LTSV, SNMV, SCMoV and VTMoV have been the subjects of much biological and structural investigation.

Geographical Distribution and Natural Hosts

Sobemoviruses are distributed world-wide, although individually several occur in specific areas. Nine viruses affect dicotyledonous species, six affect monocotyledons, while none is transmitted to both.

SBMV is prevalent in tropical and subtropical parts of the world. In Africa, SBMV infection of cowpea (*Vigna unguiculata*) (caused by SBMV cowpea strain) is widespread, while in Central and South America, garden beans (*Phaseolus vulgaris*) are affected (SBMV, bean strain) primarily. The world-wide occurrence of SoMV is due to its seedborne nature and its identification in several *Chenopodium* spp. that are used widely as indicator hosts for virus identification. CfMV, CMMV, CyMV and PMV occur in grasses in the Western hemisphere, although CfMV and CyMV are reported also from New Zealand. RYMV, first reported from Kenya in 1970, is now widespread in several countries in West Africa; both cultivated and wild rice are the natural RYMV hosts. GCFV was found in ginger rhizomes imported in Australia from India, Malaysia, Mauritius and Thailand; it has not been detected in commercial plantings in Australia. BSSV is found in North America, OLV-1 occurs in Italy and TRoSV has been reported only from Scotland.

LTSV, SCMoV, SNMV and VToMV are confined to the Australian continent, except that LTSV occurs also in New Zealand and Canada. Lucerne (*Medicago sativa*) is the only natural host of LTSV. SCMoV is widespread on subterranean clover in Southwest Australia, SNMV is found in *S. nodiflorum* in the northeast coastal areas, while VTMoV is distributed in velvet tobacco (*Nicotiana velutina*) in the arid areas of South Australia.

Experimental Host Range and Symptomatology

Sobemoviruses have a restricted host range, most infect a few species in one or two families. LTSV, TRoSV and OLV-1, however, infect plants in three or four families, but the total number of susceptible species is small. SNMV and VToMV have several common solanaceous hosts. Interestingly, most other sobemoviruses have few or no common hosts.

For instance, SBMV, LTSV and SCMoV infect legumes, but of the many species tested, none is susceptible to all three viruses. Similarly, of the six viruses that affect *Gramineae*, only PMV is transmitted to maize (*Zea mays*).

Symptoms induced by sobemoviruses are persistent mosaic, mottle or chlorosis, often accompanied by leaf deformities and plant stunting. OLV-1 is symptomless in its primary host but causes localized infections in several herbaceous species.

Transmission

Sobemoviruses are transmitted easily by sap inoculation, which is a reflection of a high endogenous concentration and particle stability. Transmission through leaf abrasion during strong wind is possible, but actual proof is lacking. RYMV exuded with guttation fluid may contaminate irrigation water, which then serves as the inoculum source. SBMV survives in the soil and is transmitted via roots in the absence of a subterranean agent. Long-distance transport of diseased rhizomes probably contributes to GCFV dispersal.

SBMV, SCMoV, SoMV and PMV are seed-transmissible. SBMV is carried through embryo and seed coat. SBMV-like particles are present on the exines of pollen recovered from the infected plants although no direct evidence exists of its transmission via pollen. The extent of SCMoV transmission through seed is 3–10%, while SoMV is transmitted through seed with an unusually high (83%) efficiency. PMV is seed-borne in *Setaria italica* cv Red Siberian.

Insects are the principal vectors of sobemoviruses. The chrysomelid beetles transmit SBMV, CfMV, PMV, RYMV and TRoSV, while the coccinelled beetles transmit SNMV and VToMV. The mirid bug, *Cyrtopeltis nicotianae*, is also the vector of SNMV and VToMV. Aphids transmit BSSV, CMMV and CyMV. SoMV is transmitted by the leaf miner fly, *Liriomyza langei*, mechanically and is carried on the mouth parts as well as on the ovipositor.

The relationship of sobemoviruses with their beetle, aphid and mirid vectors is nonpropagative (semipersistent), and of a circulative type with the exception of CyMV, which is noncirculative in its aphid vector, *Rhopalosiphum padi*. SBMV is present in fairly high concentration in the intestinal tract, hemolymph and the regurgitant fluid of its beetle vector, *Ceratoma trifurcata*. It is most likely transmitted through contaminated mouth parts and during regurgitation. CfMV is excreted in fecal material by its vector, *Lema melanopa*, and it can initiate infection if deposited at the freshly damaged feeding sites. Preliminary evidence exists that CfMV increases the mortality rate of its vector.

Cytopathology

Sobemoviruses are distributed in most plant parts including epidermis, mesophyll, meristem, phloem and xylem. Virions occur in cytoplasm and vacuoles, and except for BSSV, RYMV and CyMV infections, also in the nuclei. Virions are not found in mitochondria or chloroplasts. In most cases, virions accumulate in large numbers without crystallizing, but occasionally these are arranged in crystalline arrays in cytoplasm and vacuole. Also, the cytoplasm of the infected cells contains fibrillar material, either enclosed in discrete membranous vesicles or distributed as diffused patches. These fibrillar structures resemble the double-stranded replicative form of RNA. In cells infected with BSSV, RYMV and SNMV, tubules aggregated as bundles are also present; their nature or function is not known.

Cellular organelles, such as chloroplasts and mitochondria in the infected cells, often exhibit persistent aberrations, like vesiculation and collapse of the internal membrane systems. Additionally, large osmiophilic or lipid globules develop in the chloroplasts. A large-scale disintegration of plasmalemma, tonoplast and nuclear membrane occurs in the chronically infected cells.

The Virions

Sobemovirus virions are isometric with diameters in the range 25–30 nm. When packed closely, virions appear distinctively hexagonal (Fig. 1). Negatively stained particles lack surface features, which is due to an unusually high surface protein density. Additionally, virions exclude the stain from penetrating to the core, which signifies a highly compact capsid organization.

Virions sediment as single components with $S_{20,w}$ values ranging from 109S (PMV) to 120S (BSSV). SBMV possesses a $S_{20,w}$ value of 115S. In gradients of CsCl, sobemoviruses develop single bands at the densities of 1.35 to 1.39 g/ml; some CMMV isolates yield two zones in CsCl however. In Cs_2SO_4 gradi-

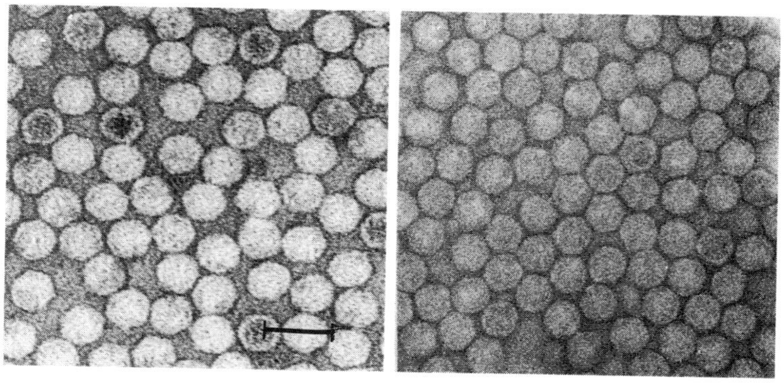

Fig. 1 Purified virions of Southern bean mosaic virus (SBMV) **(a)** and lucerne transient streak virus (LTSV) **(b)**. Ammonium molybdate was used as the negative stain. Scale bar=50 nm.

ents, most viruses exhibit a polydisperse banding pattern, but GCFV, OLV-1 and CyMV band homogeneously. The banding heterogeneity of SBMV in Cs_2SO_4 gradients is due to conformational variations among particles in a population rather than any real differences in their chemical composition.

Sobemoviruses comprise 21% RNA and 79% protein. SBMV, TRoSV and LTSV contain significant amounts of divalent cations (Ca, Mg), which are bound intimately to the capsid. The specific absorbance (1 mg/ml, at 260 nm) value for these viruses is approximately 5.5–6.0.

Capsid Organization and Stability

Sobemovirus capsid is constructed from a single protein species of about 30 kD. VTMoV coat protein is a polypeptide of 37 kD, but it is degraded into 33 kD and 31 kD products by leaf proteases during virion purification. In several sobemoviruses, a proportion of coat subunits exist as stable dimers. The complete amino acid sequence of the coat protein of SBMV cowpea strain is known; its N-terminal region is highly basic and also contains several proline and glutamine residues.

The basic capsid structure is $T = 3$ quasisymmetry of 180 subunits. High-resolution X ray diffraction has revealed three types of quasiequivalent SBMV units, A, B and C, which possess somewhat different conformations but are chemically identical. The A subunits cluster at the fivefold axes, while sets of B and C aggregate at the quasi sixfold vertices. Each subunit consists of a random domain, i.e. the N-terminal 'arm' located towards the virion interior and the surface (or shell) domain, which is organized into an eight-stranded antiparallel β-barrel and five α-helices. Considerable similarity exists in the overall structure of the coat proteins of SBMV and tobacco necrosis virus.

SBMV capsid stability is derived from intersubunit interactions, which are largely hydrophobic and involve tryptophan residues; ionic and hydrogen bonds further stabilize these interactions. Moreover, divalent cation-mediated bonds play an essential role in maintaining the proper conformation of the capsid. Additionally, there are pH-dependent interprotein bonds (effective at pH 5–6) and protein–RNA linkages that also contribute to particle stability. Perturbations in the pH-dependent and divalent cation-mediated bonds cause sobemovirus capsid to 'swell', and in this state, virions are rendered sensitive to proteases, salt or sodium dodecyl sulfate (SDS). Only CyMV among the sobemoviruses is sensitive to degradation by SDS in the native (unswollen) state.

Serology

Sobemoviruses are strongly immunogenic. In general, these viruses are not serologically cross-reactive. However, SNMV and VTMoV are closely related, while SCMoV and LTSV are distantly related. Isolates of CfMV and CyMV from New Zealand show serological relatedness, but not those from Europe.

Strains of SBMV or RYMV can be differentiated on the basis of spur formation in gel diffusion tests. The type and St Augustine decline strains of PMV show serological relatedness to phleum mottle and maize eye spot viruses; the last two viruses have not yet been assigned to any plant virus group.

Studies using monoclonal antibodies have indicated that new binding sites appear when SBMV virions 'swell'. These sites are unavailable on the native capsid and are probably located at the subunit : subunit interacting surfaces.

The Viral Genome

Sobemovirus genomes are linear, single-stranded RNAs of positive polarity (messenger-sense) and range in mol. wt from 1.3×10^6 (SoMV, OLV-1) to 1.5×10^6 (GCFV, SNMV, SCMoV, VTMoV). SBMV RNA has mol. wt 1.4×10^6. The sequence of the cowpea strain SBMV RNA is known. It consists of 4194 nucleotides, has a covalently linked genome protein (VpG, 12 kD) at the 5′ terminus and the 3′ terminus is ...UGG-OH. VpG is essential for infectivity since protease treatment inactivates SBMV RNA. TRoSV RNA also possesses VpG (12 kD), which is needed for infectivity. LTSV and RYMV genomic RNAs are rendered noninfectious when treated with proteinase K, indicating that these RNAs also possess VpG. Information is lacking regarding VpG presence in the other sobemoviruses.

Concerning sequence homologies between sobemovirus genomic RNAs, data is scanty. Approximately 17–20% homology exists between the RNAs of cowpea and bean strains of SBMV and each of these have 12–15% homology with TRoSV RNA. VTMoV and SNMV RNAs have 20–50% sequence homologies. RNAs of CfMV and CyMV have 5–8% common sequences.

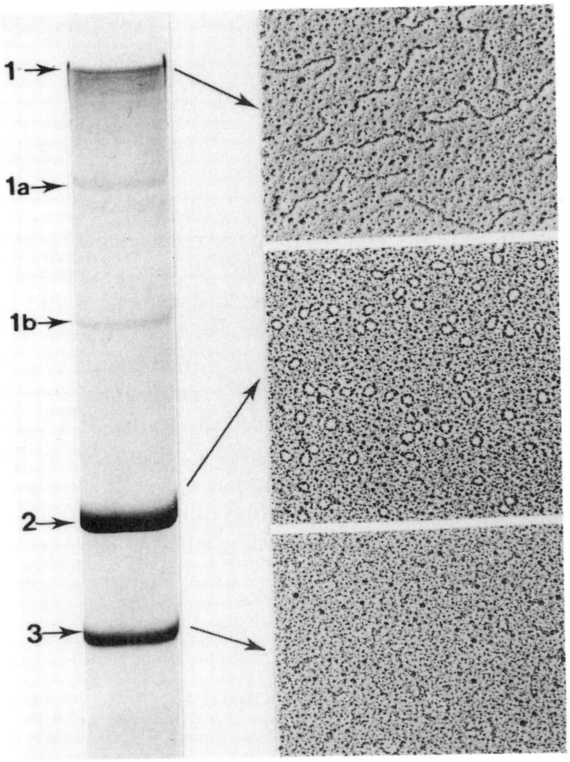

Fig. 2 Polyacrylamide gel electrophoresis and electron microscopy of the virion RNAs of velvet tobacco mottle virus (VTMoV). From Velvet tobacco mottle virus, *AAB Descriptions of Plant Viruses* No. 317, 1986, with permission; courtesy of Dr JW Randles.

Encapsidated Subgenomic and Satellite RNAs

Virions of SBMV, GCFV, LTSV, OLV-1, SNMV, SCMoV, TRoSV, VTMoV and CMMV encapsidate small amounts of heterogeneous subgenomic or putative subgenomic RNAs. Among the subgenomic SBMV RNAs is a mol. wt 0.38×10^6 component, which is the autonomized coat protein cistron. SBMV subgenomic RNAs contain VpG at their 5′ termini. CfMV virions encapsidate a discrete 0.5×10^6 RNA and several intermediate-sized putative subgenomic RNAs, while VTMoV contains two well-defined subgenomic RNAs, RNA 1a and RNA 1b (Fig. 2), of mol. wt 0.63×10^6 and 0.25×10^6 respectively.

LTSV, SCMoV, SNMV and VTMoV encapsidate discrete, low-molecular-weight, viroid-like satellite RNAs. No sequence homologies exist between the satellite RNAs and their respective viral genomes. The satellite RNAs exist as linear or circular molecules (Fig. 2) but circular molecules mostly predominate. SCMoV virions encapsidate two distinctive satellite RNAs; some virus isolates contain one RNA, while others contain both RNAs. Other SCMoV isolates lack satellite RNA(s).

Complete sequences are known of the satellite RNAs of LTSV (326 nucleotides), SCMoV (327 nt and 388 nt), SNMV (377 nt) and VTMoV (366 nt). Satellite RNAs of SNMV and VTMoV are quite similar in their sequences, but they differ from those of LTSV and SCMoV. The two satellite SCMoV RNAs have similar sequences in one half but divergent sequences in the other. The satellite RNA of a Canadian LTSV isolate (322 nt) shares homology with the Australian isolate. Another LTSV isolate from Australia possesses a satellite RNA (388 nt) which is similar in its primary structure with the 388 nt long satellite SCMoV RNA.

When co-inoculated with SBMV, SoMV and TRoSV in an appropriate host, satellite LTSV RNA replicates and is encapsidated by these viruses. None of the SBMV, SoMV or TRoSV isolates that have been investigated so far contains viroid-like satellite RNAs.

The Infection Process

Like most plant viruses, sobemoviruses cause infection following entry through wounds created by mechanical abrasion or insect damage. Some evidence exists, however, that SBMV virions enter uninjured cells by pinocytosis. Once inside the cell, the viral capsid swells and the translational and disassembly events proceed coincidentally. It is as yet unclear if any particular capsid element (monomer, dimer or a cluster of subunits) at a specific site is displaced or removed, which then allows an initial contact to be established between SBMV genome and ribosomes.

Genome Replication and Expression

Among the sobemoviruses, only for VTMoV are any substantive details available with respect to *in vivo* replication. RNA-dependent RNA polymerase activity in the inoculated leaves is first detected at 4 days after infection and increases rapidly during the next several days. The endogenous concentration of the genomic RNA remains low, but the subgenomic RNA 1b is easily detected 4–6 days following infection. About the same time, virions appear. The viroid-like RNAs can be detected as early as 2–3 days after infection. A double-stranded form of genomic RNA (mol. wt 2.8×10^6) and a putative replicative complex of the satellite RNA (mol. wt 3.6×10^6) are found 8–9 days after inoculation.

In vitro transcription studies using crude preparation of RNA-dependent RNA polymerase yielded, beside the double-stranded form of genomic RNA and the replicative complex of satellite RNA, a third double-stranded RNA (mol. wt 0.7×10^6); its significance is not known. The double-stranded RNAs and the polymerase activities are localized in the soluble cytoplasmic fraction.

Several sobemovirus RNAs have been translated in the *in vitro* systems. Figure 3 illustrates the expression strategy of SBMV RNA, cowpea strain. ORF-1 encodes a 21 kD protein (P4) of undefined function. ORF-2 overlaps the 3′ end of ORF-1 by 34 bases and encodes a 105 kD protein (P1). The 70 kD protein (P2) is probably derived from the

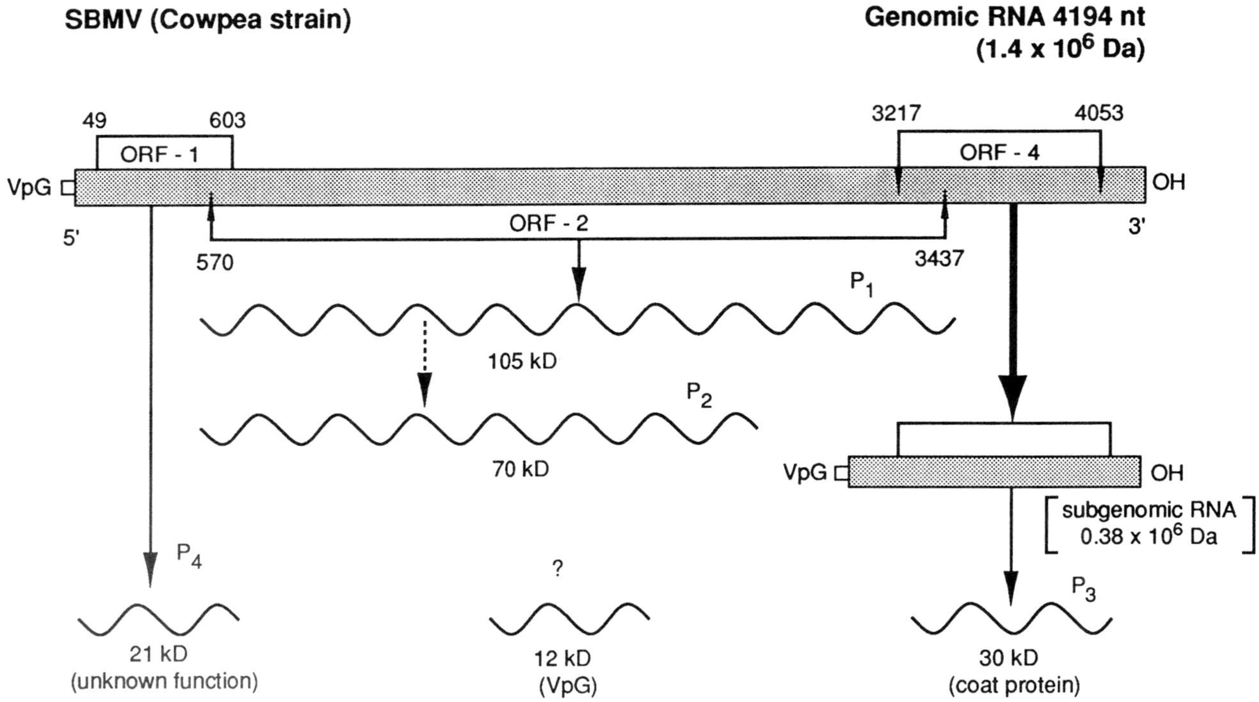

Fig. 3 A tentative scheme of the organization and expression of Southern bean mosaic virus (SBMV) genome. Based upon Wu S, Rinehart CA and Kaesberg P (1987) *Virology* 161 73.

105 kD protein by proteolytic processing. ORF-4 is expressed via a subgenomic RNA (mol. wt 0.38×10^6) and codes for the 30 kD coat protein (P3). Another open reading frame, ORF-3, extends from base 1895 to 2380 and is capable of encoding a 18.3 kD protein, but no such protein has been identified. The actual number of functional proteins derived from 105 kD protein and location of VpG are speculative. However, from information derived from the consensus sequences, it appears that RNA polymerase and VpG sequences are coded by ORF-2.

Translation of the unfractionated VTMoV virion RNA or of the genomic RNA yields 115, 60, 37, 30, 19, 15 and 14 kD proteins. The 115 kD protein is a readthrough protein comprising the 60, 30 and 19 kD proteins, while 15 and 14 kD proteins are degraded products. The subgenomic RNA 1b appears to be the messenger of the 37 kD coat protein.

The translation products of TRoSV RNA are 105, 57, 35 and 30 kD proteins. The three larger proteins are coded by the genomic RNA, while the 30 kD coat protein is expressed through the mol. wt 0.5×10^6 subgenomic RNA. The translational strategy of LTSV RNA is similar to that of TRoSV RNA.

The satellite RNAs of LTSV, SNMV and VToMV lack messenger activities.

Epidemology and Control

Field infections of SBMV and RYMV are often widespread and may reach epiphytotic proportions. The seed-borne inocula and weeds serve as the primary field sources of SBMV, which is then spread by beetles. RYMV incidence is significantly greater in areas of continuous rice cultivation than those with interrupted plantings. This suggests that RYMV inocula present locally contribute largely to disease initiation and spread. It has been well documented that RYMV survives during off-season in volunteer and wild rice (*Oryza longistaminata*) plants, regrowths of harvested crops, and in the ratoons. The beetle vectors (*Chaetocnema* spp.) move rapidly from plant to plant, causing RYMV dispersal. Sources of resistant germplasm for SBMV and RYMV have been identified for developing resistant cultivars. The infection of herbage grasses with CfMV, CMMV and CyMV is widespread but tolerant lines are available.

The potential of economic losses in forage legumes due to LTSV and SCMoV infections exists because these viruses induce severe symptoms. The extent of economic loss due to infection by BSSV, TRoSV, GCFV and OLV-1 is not known. SNMV and VTMoV naturally infect weed plants only, and presently do not pose any economic threat.

Conclusions

On the basis of physicochemical parameters and stability characteristics, sobemoviruses constitute a remarkably homogeneous and coherent group. Further, these viruses have a relatively simple structure and their genomes are the smallest among the plant viruses. The sobemoviruses are somewhat unique in certain other features as well. While serological relatedness is a commonly used criterion in grouping viruses, sobemoviruses, in general, are not serologically interrelated. Further, the markedly divergent and nonoverlapping host ranges of most sobemoviruses signify a high degree of biological specificity and host plant adaptation. Finally, sobemoviruses represent a class of genetically stable viruses because very few naturally occurring variants or strains have been recognized.

The number of proteins synthesized by SBMV genome far exceed its coding potential. A combination of different strategies, such as overlapping reading frames, internal sites of translation initiation and generation of subgenomic components, are involved in this genomic expression. Further, processes such as post-translational proteolysis and chemical modifications contribute to the protein diversity. The expression strategies of other sobemovirus genomes appear similar to that of SBMV. Furthermore, although data is fragmentary, the replication of the sobemovirus genome appears to follow the classical pattern: infection by positive-strand RNA leads to the formation of an intermediate double-stranded complex, and finally to production of the progeny positive-strand RNA.

The presence of an encapsidated viroid-like satellite RNA is a feature that distinguishes LTSV, SNMV, SCMoV and VTMoV from the other sobemoviruses. The fact that viruses such as SBMV, SoMV and TRoSV support replication of LTSV satellite RNA, which in turn is also encapsidated by these viruses, indicates that a satellite RNA can become associated with different sobemoviruses.

Although marked structural similarities exist between the coat proteins of SBMV and tobacco necrosis virus, there are only limited similarities in the polymerase sequences around the GDD motif. SBMV genome shows substantial sequence similarities in the GDD motif and polymerase environs with members of the carmovirus and luteovirus groups.

Further Reading

Brunt A, Crabtree A and Gibbs A (1990) *Viruses of Tropical Plants*. Wallingford, UK: CAB International.

Francki RIB, Milne RG and Hatta T (1985) Sobemovirus group. *Atlas of Plant Viruses*, vol. 1. Boca Raton, Florida: CRC Press.

Hull R (1988) The Sobemovirus Group. In: Koenig R (ed.) *The Plant Viruses*, vol. 3, *Polyhedral Virions with Monopartite RNA Genomes*. New York: Plenum Press.

Sehgal OP, and White JA (1989) Southern bean mosaic virus. In: Mandahar CL (ed.) *Plant Viruses*, vol. I, *Structure and Function*. Boca Raton, Florida: CRC Press.

SPO1 BACTERIOPHAGE

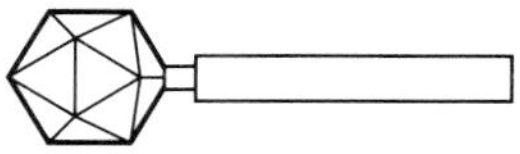

Charles R Stewart
Rice University
Houston, Texas, USA

History

SPO1 is a large virulent bacteriophage of the gram-positive bacterium *Bacillus subtilis*. Shunzo Okubo isolated it from soil in Osaka, Japan, during the early 1960s, and brought it with him to Chicago, where he collaborated with Bernard Strauss and Marvin Stodolsky on the first published studies of SPO1. Peter Geiduschek, Stodolsky's mentor, saw in SPO1 the opportunity to test the generality of the lessons then being learned about sequential gene action in the *Escherichia coli* phage T4, and began a series of experiments which illuminated the regulation of SPO1 gene action and stimulated the interest of many others.

Taxonomy and Evolution

SPO1 is a member of a family of *B. subtilis* phages, whose distinguishing feature is the presence of the unusual base, hydroxymethyluracil, instead of thymine in the DNA. All members of the family, which also includes SP82, Φe, 2C, SP8, H1 and SP5c, appear to be descended from a common ancestor, since they show many striking similarities in structure, restriction maps, genetic maps, gene products and gene regulation. However, they also show many differences in detail, and thus have clearly had the opportunity for significant divergence. Perhaps the most striking similarity is the presence, in each family member tested, of a group I self-splicing intron, discovered by David Shub and his colleagues by testing RNAs from phage-infected cells for their capacity to incorporate labelled GTP. This is the only intron known in gram-positive bacteria, and among the very few known in all procaryotes. The presence of introns both here and in T4 has been cited as an argument for the existence of introns in ancient evolutionary times, before the divergence of gram-positive from gram-negative bacteria. SPO1 is similar to T4 in many other ways as well, including size, structure, overall organization of the life cycle, and the presence of an unusual pyrimidine.

Virion Structure and Proteins

An SPO1 particle includes a single linear double-stranded DNA molecule and at least 53 different types of polypeptide, organized into an icosahedral head about 87 nm in diameter, a short neck and a contractile tail of 19 × 140 nm, ending in a complex base plate 60 nm in diameter. The average molecular weight of the 53 proteins, estimated by gel electrophoresis, is about 42 000, so the genes specifying structural proteins may be estimated to occupy at least 42% of the genome.

Genome Structure and Gene Function

Figure 1 shows a map of the SPO1 genome. Fifty-four genes have been identified, by conditional lethal mutations, DNA sequencing and/or *in vitro* expression; they have been mapped, by restriction and recombination mapping, on to a single linkage group that spans the entire genome. Judging from

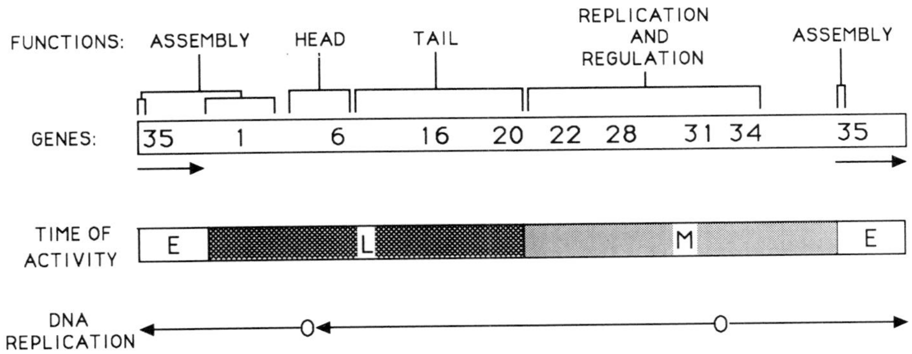

Fig. 1 Maps of the SPO1 genome. The 'Genes' box shows the approximate map location of selected genes representing various parts of the genome. The position of the terminal redundancy is shown by the arrows under the ends of that box. Gene *35*, being located in the terminal redundancy, is indicated twice. The clustering of genes according to function is indicated above the 'Genes' box. Most or all of the known genes in each region specify products involved in the function indicated. Genes for replication and regulation are interspersed among each other, and certain genes are required for both processes. Regions whose functions are not indicated have no genes with known functions. The 'Time of activity' box shows regions predominantly transcribed during early (E), middle (M), or late (L) times, as indicated by the unshaded, lightly shaded and heavily shaded areas respectively. A minority of each type of transcript is also scattered among the other areas of the genome; for instance, gene *28* is transcribed at early times. The 'DNA replication' line shows the simplest pattern of replication consistent with the data. The two Os represent origins of replication, and the arrows show the directions of replication proceeding from each origin.

the size of the genome and the close packing observed for those genes whose locations are known precisely, an additional 80 or 90 genes are expected to be present. Conditional lethal mutations have permitted analysis of the functions of 39 of the known genes, showing that genes involved in the same function, such as DNA replication, tail formation, head formation, or virion assembly, tend to be clustered together on the map.

The single DNA molecule is 145 kbp long, with a 12.4 kb terminal redundancy, and with hydroxymethyluracil (hmUra) completely replacing thymine as the base-pairing partner for adenine. The presence of hmUra gives the DNA a CsCl buoyant density substantially higher than that of other DNAs of similar GC content, a feature which has been useful in making experimental distinctions between phage DNA and host DNA. The two strands of the phage DNA are physically separable on CsCl, a characteristic that has been used extensively to identify the template strand for specific transcripts. For instance, that property of SP8 was used by Marmur and Greenspan to make the first demonstration that specific RNAs are complementary to one strand of the DNA from which they were transcribed.

Growth Cycle

Under optimal conditions, SPO1 has an eclipse period of 25–30 min, a latent period of 33–40 min, and produces a burst of 100–300 phage. Into this brief period is packed a remarkably complex sequence of events. Immediately after infection, some of the SPO1 early genes are turned on, with the others following during the next few minutes. Shortly thereafter occurs the shutoff of host DNA replication and gene expression. By 5 or 6 min after infection, SPO1 middle genes begin turning on and some of the early genes are turned off. The middle genes specify most of the enzymes necessary for SPO1 DNA synthesis, which begins about 10 min after infection. Shortly thereafter, SPO1 late genes begin to function, directing the synthesis of structural proteins, and some of the early and middle genes are turned off. By 25 min after infection, the first infectious phage particles have formed, and lysis begins as early as 33 min, resulting in the total destruction of the infected cells. The following sections describe each of these processes in detail.

Regulation of Gene Action

The transitions from early to middle to late gene activity are caused by a cascade of sigma factors. The major host RNA polymerase, with sigma factor A, transcribes from promoters whose −35 and −10 sequences approximate the consensus sequences

TTGACA . . . TATAAT. The promoters for the SPO1 early genes fit that consensus and thus are transcribed by the host polymerase. One of the early genes, gene *28*, specifies another sigma factor, which substitutes for sigma A on the RNA polymerase, changing the specificity of the polymerase so it then recognizes the promoters of the middle genes, whose consensus sequence is AGGAGA . . . TTT–TTT. Two of the middle genes, *33* and *34*, specify a third sigma, which causes the RNA polymerase to recognize the late gene promoters, with sequence CGTTAGA . . . GATATT.

This story was put together primarily from a series of elegant experiments done by Jan Pero, Peter Geiduschek and their colleagues. RNA pulse-labelled at early, middle or late times showed completely different patterns of hybridization competition and of hybridization to Southern blots of restriction digests of SPO1 DNA. Mutations in gene *28* prevented the transition from the early to the middle pattern, while mutations in genes *33* or *34* prevented the middle to late transition. Three different RNA polymerases, A, B and C, were extracted from SPO1-infected cells and used to transcribe SPO1 DNA *in vitro*, producing RNAs which showed the same hybridization patterns as the above pulse-labelled early, middle and late RNAs respectively. Analysis of the polypeptides associated with the three polymerases showed that A contains sigma A, B contains (instead) the gene *28* product, and C contains (instead) both the gene *33* and gene *34* products.

Thus, the sequential onset of early, middle and late transcription is explained by the sigma cascade. The mechanisms responsible for other regulatory events in the SPO1 life cycle are not yet understood, although, in some cases, gene products that play essential roles have been identified. One of the most interesting is TF1, a type II DNA-binding protein that is synthesized in large quantities during SPO1 infection, and that binds preferentially to specific sites on DNA containing hmUra, causing bending of the DNA. SPO1 mutants deficient in TF1 fail to shut off transcription of certain middle genes and fail to turn on transcription of certain late genes. The bending caused by TF1 may be, in itself, a cause of these regulatory changes, or may make those genes accessible to other regulatory factors.

Certain mutations in two other genes, *22* and *27*, also prevent activity of certain late genes. The effect of the gene *22* mutation may be an indirect result of its prevention of phage DNA synthesis, suggesting that some change in structure, associated with replication, may be necessary for the activation of certain late promoters. The effect of the gene *27* mutation, however, seems to be independent of its effect on replication.

A further complexity in late gene regulation is that certain of the late genes can be transcribed even in the absence of the gene *33* product. Thus, it appears that, while the sigma factor specified by gene *34* is essential for all late transcription, gene *33* specifies an accessory protein which is required for some, but not all, late transcription.

Other regulatory events whose mechanisms are not yet understood include: the delay in the onset of transcription of some early genes relative to others; the turn-on of translation of certain early mRNAs, requiring the activity of one or more SPO1 gene products; the shutoff of two different groups of early genes at two different times; the shutoff of those middle genes for which TF1 is not required; and probably others that are not yet so well defined.

By hybridization of pulse-labelled RNAs to Southern blots of restriction digests of the SPO1 genome, transcription maps have been prepared showing which regions of the genome are transcribed at which times. A summary of the overall trends revealed by such mapping is included in Fig. 1. Detailed analysis of individual transcription units, within certain regions of the genome, has been performed by S1 mapping and by *in vitro* transcription, and the results are summarized in the following paragraphs.

Most early transcription takes place in the terminal redundancy, each copy of which contains at least 13 promoters, most of which are very active and some of which are among the strongest promoters known. It has been suggested that this high density of active promoters in a duplicated region is for the purpose of competing effectively with host promoters for RNA polymerase. All 13 promoters direct transcription toward the middle of the redundant region, where two efficient transcription terminators halt transcription from the left and right sides respectively. Except for gene *28*, the activities of the early gene products are not known, although it seems likely that they include the shutoff of host biosyntheses, as discussed later.

Most middle transcription occurs in a 60 kb region that occupies most of the right half of the genome. A 28 kb subset of this region is the most intensely studied region of the SPO1 genome, including all five of the regulatory genes mentioned earlier, as well as all of the genes known to be required for SPO1 DNA replication. Thirteen active middle promoters and three relatively weak early promoters have been identified in this region; nine of the genes, including all of the regulatory genes, have been sequenced.

Each of the early promoters is in a tandem arrangement with a middle promoter, permitting some sequences to be transcribed by both the early-specific and middle-specific polymerases.

Most late transcription occurs in the left half of the unique region of the genome. This region includes nearly all of the genes known to be directly involved in head or tail morphogenesis, and it is assumed that most late genes do specify structural proteins.

DNA Replication

Conditional lethal mutations have identified 10 SPO1 genes as essential for SPO1 DNA replication. Except for gene *28*, most or all of these are middle genes, whose products begin to appear about 7 min after infection. Thus, the time at which replication is initiated, about 10 min after infection, may be determined simply by the time at which all necessary proteins and precursors have accumulated to a sufficient concentration. Genes *23* and *29* are required for the synthesis of hmUra, *32* and *21a* or *b* for initiation of replication, *22*, *30* and *31* for elongation, and the roles of the other two are unknown. Gene *31* specifies the DNA polymerase and is the site of the intron, which includes an open reading frame whose product may be involved in intron mobility. Other SPO1 gene products that play a role in replication, but whose genes have not been identified, include DNA gyrase, dCMP deaminase, dTTPase, dTMPase and an inhibitor of thymidylate synthetase.

There are at least two origins of replication, near the opposite ends of the unique region of the SPO1 genome. One growing point proceeds leftward from the right-hand origin, replicating most of the unique region. Another proceeds leftward from the left-hand origin, replicating from there to the end. There must be at least a third growing point to replicate the right end of the genome, but it is not clear whether that starts at the above right-hand origin and replicates rightward or whether there is a third origin, farther to the right, from which replication proceeds bidirectionally. Figure 1 includes a representation of the simplest interpretation of the data on directions of replication.

As the SPO1 DNA replicates, it forms concatemers of as many as 20 genomes, joined end to end by overlapping terminal redundancies. It has been proposed that the purpose of concatemer formation is to make up for the inability of DNA polymerase to synthesize the 5′ end of a linear DNA molecule. The two daughters of the replication of a linear molecule would each have protruding 3′ ends which, because of the terminal redundancy, would be complementary and would thus anneal to form an end-to-end dimer, a process which would be repeated again and again. When the concatemer is broken back down to unit genomes, cleavage would be staggered so as to produce protruding 5′ ends, whose complements could then be synthesized by DNA polymerase. This hypothesis predicts that, after formation of the first dimer, and before the second round of replication has begun, genetic markers in the terminal redundancy should have been replicated only to half of the extent of markers in the rest of the genome, a prediction that has been dramatically confirmed, for SPO1, by temperature-shift experiments with temperature-sensitive mutants affected in gene *32*.

Morphogenesis

Little is known about SPO1 morphogenesis. Although more than half of the known genes are required for head or tail formation or for virus assembly, and 53 proteins have been identified as part of the virus particle, only one gene has been identified with a specific viral protein (gene *6* specifies a particular head protein), and only one morphogenetic process has been studied. The proteolytic processing of a precursor polypeptide to produce the mature form of the major head protein requires the activity of both the gene *5* product and TF1. Nothing is known of the biochemical activity of the gene *5* product. TF1 may have any of several possible roles. Other type II DNA-binding proteins participate in the wrapping of DNA molecules into chromatin-like structures, suggesting that TF1 might play a similar role in folding SPO1 DNA for packaging into the head, and that processing of the head protein might be an integral part of the packaging process. Alternatively, TF1 might be necessary for expression of gene *5*, or for some other necessary gene.

Effect on the Host Cell

It is to the selective advantage of a virus to shut off the macromolecular syntheses of the host cells, so they will not compete with the comparable viral syntheses for energy and materials. Most synthesis of host DNA, RNA and proteins is shut off within a few minutes after SPO1 infection. The shutoff mechanisms are very selective, since not only do they have no effect on the synthesis of the comparable phage macromolecules, they also spare certain host syntheses. Host ribosomal RNA continues to be synthesized at nearly normal rates, which seems sensible since the phage has a use for the host ribosomes. The mechanism by which SPO1 distinguishes between host and phage DNAs is not clear. It must be more subtle than the presence or absence of hmUra, since some host genes are unaffected. Unlike T4, which causes the complete degradation of DNA without hydroxymethyl cytosine, SPO1 causes no detectable degradation of *B. subtilis* DNA.

No SPO1 genes have yet been shown to be responsible for host shutoff, but several restriction fragments have been identified as possible carriers of such genes. These fragments are unclonable in *B. subtilis*, as would be expected of a fragment expressing a host shutoff gene, and other explanations for the unclonability were ruled out. Several of these fragments are located in the terminal redundancy, which, as discussed earlier, includes a number of early genes whose functions have not yet been identified. At least one gene whose product is required for the shutoff of host DNA synthesis is among the early genes whose transcription is delayed relative to most other early genes.

Recombination and Mutagenesis

SPO1 has a very active recombination system, producing frequencies of recombination between nearby genetic markers of about 0.001% per base pair. Nothing is known about the mechanisms of recombination or the gene products involved, but the high frequency facilitates several types of experimentation. For instance, cloned SPO1 restriction fragments, as small as 200 bp, undergo significant recombination with the homologous region of the SPO1 genome, resulting in marker rescue of markers as little as 12 bp from the end of the fragment, permitting efficient fine-structure mapping.

This efficient recombination also permits new mutations to be constructed by *in vitro* mutagenesis and inserted into the SPO1 genome by marker rescue recombination. For instance, a phenotypically silent mutation of the TF1 gene, which differed from wild-type by 2 bp and no amino acids, was created *in vitro* by site-specific mutagenesis, introduced into *B. subtilis* on a 654 bp fragment of SPO1 DNA, and allowed to recombine with superinfecting wild-type SPO1. Of the progeny phage 5×10^{-4} carried the mutation, by the criterion of plaque filter hybridization.

Recombination is also an integral part of the process of transfection by SPO1 DNA (most studies of the phenomenon have been done with the close relative SP82). When a cell is infected with purified DNA, the DNA is damaged by a nuclease activity of the host cell. (This activity is inhibited, and therefore causes no problem, during normal infection.) Production of a single intact genome requires recombination between several of the damaged genomes.

Cloning Vehicle

SPO1 can also serve as a cloning vehicle. Entire plasmids, carrying a short region of homology to the SPO1 genome, can be inserted into the SPO1 genome by Campbell-mode integration. At least 5.6 kb of exogenous DNA can be added to the SPO1 genome in this way without apparent effect on the viability of the phage. Although too cumbersome to be used for routine cloning, this procedure offers a way to test the effect on SPO1 infection of adding specific genes to the genome, and to test the effect of the incorporation of hmUra on the functioning of any DNA of interest.

See also: *Bacillus subtilis* bacteriophages – groups 1–5; Bacteriophage recombination; Bacteriophage taxonomy and classification; Bacteriophages in soil; Bacteriophage transduction; Bacteriophages as cloning vehicles; T4 bacteriophage and related bacteriophages.

Further Reading

Geiduschek EP, Schneider GJ and Sayre MH (1990) TF1, a bacteriophage-specific DNA-binding and DNA-bending protein. *J. Struct. Biol.* 104: 84.

Goodrich-Blair H, Scarlato V, Gott JM, Xu M-Q and Shub DA (1990) A self-splicing group I intron in the DNA polymerase gene of *Bacillus subtilis* bacteriophage SPO1. *Cell* 63: 417.

Losick R and Pero J (1981) Cascades of sigma factors. *Cell* 25: 582.

Stewart CR (1993) SPO1 and related bacteriophages. In: Sonenshein AL, Hoch JA and Losick R. (eds) *Bacillus subtilis and other Gram-positive Bacteria*, p. 813. Washington: American Society for Microbiology.

SPONGIFORM ENCEPHALOPATHIES

Contents

Mule Deer, Elk and Bovine
Creutzfeldt–Jakob Disease, Scapie and Related Transmissible Encephalopathies

Mule Deer, Elk and Bovine

Maurizio Pocchiari
Istituto Superiore di Sanità,
Rome, Italy

History

Transmissible spongiform encephalopathy (TSE) of the mule deer (chronic wasting disease of deer) was recorded for the first time in 1980 in a wildlife facility in Colorado. Subsequently, the disease appeared also in Wyoming in animals born in captivity and in 1982 a similar disease was reported in Rocky Mountain elk held captive in Colorado.

The first case of the fatal neurodegenerative disease of cattle (bovine spongiform encephalopathy, BSE) was detected in November 1986 in the UK. Since then an increasing number of cases have been reported from farms all over Britain. BSE is a scrapie-like disease that originated from the combination of several factors, the most important of which is the change in the method of production of meat and bone meal which led to an increased level of scrapie agent contamination in the commercial foodstuffs and subsequently to infection by the oral route in cattle.

Taxonomy and Classification

BSE and TSE of mule deer and elk belong to the group of subacute spongiform encephalopathies or prion diseases of animals and man. The nature of these infectious agents is still unknown. It is still debated whether they have a small, still undetected, nucleic acid associated with one or more virus-encoded proteins (virus hypothesis), or whether they are composed of a self-replicating protein (prion hypothesis), or whether they are formed by an agent-specific nucleic acid associated with a host-encoded protein (virino hypothesis).

Geographic and Seasonal Distribution

BSE epidemic is restricted to Britain and Ireland. Sporadic cases have been reported in Switzerland, France, Oman and the Falkland Islands. BSE may theoretically appear in countries with a high incidence of natural scrapie and where the carcasses of infected sheep enter into the food-chain for domestic cattle. TSE of mule deer and elk has been mainly reported in animals held in wildlife facilities in the US. There is no apparent seasonal distribution of these diseases.

Host Range and Virus Propagation

The infectious agent of BSE commonly causes disease in bovine after 2–8 years of incubation. Other exotic ruminants kept in captivity have been accidentally infected by the same agent responsible for the epidemic of BSE in cattle. A similar event is likely to be responsible for a few cases of feline spongiform encephalopathy in cats.

BSE has been experimentally transmitted to bovine, pigs and mice. Mice are not naturally infected with BSE, but they develop a BSE-like disease about 1 year after the intracerebral injection of infected brain. Oral administration of infected brain also produces the disease in mice with an incubation period of 1.5 years. Only one in 10 pigs injected with the agent of BSE by multiple routes developed a BSE-like disease 16 months after infection.

It is not yet known if BSE can be infectious for man and if, as a consequence of that, it will result in an increased incidence of the human-related spongiform encephalopathy Creutzfeldt–Jakob disease. However, precautions have been taken by the British and by other European Governments to remove this theoretical and remote risk.

Intracerebral injection of infected brain of mule deer with TSE produces the disease in ferrets after 1.5–2 years incubation period. BSE and TSE of mule deer and elk agents have not been grown in any tissue culture system.

Genetics

The agents responsible for BSE and for TSE of mule deer and elk, as well as the other animal and human-related agents, have not yet been isolated and characterized. This, of course, causes difficulties in recognizing any eventual modification of their genomes. It is likely that there is only a single strain of the BSE agent.

Evolution

Epidemiological observations suggest that the BSE agent circulating in cattle originated approximately in the 1980s from scrapie agents in sheep. The available information suggests that a scrapie strain established itself in cattle and caused the first few cases of BSE in bovine. Subsequently, the agent adapted to cattle and gave rise to the current BSE epidemic in Britain. There is no information on the origin of the TSE of mule deer and elk.

Serologic Relationships and Variability

The BSE and TSE of mule deer and elk agents do not produce any detectable immune response in the host, and, therefore, there are no serologic markers for these diseases.

Transmission studies of different isolates of BSE into various strains of mice suggest that there is a single strain of the BSE agent which derives from one or more strains of the scrapie agent after subsequent passages from cattle to cattle, and that it does not undergo variation. In scrapie, the mouse model provided evidence of agent strains through variation in the distribution and intensity of spongiform pathology in the brain together with differences in the length of the incubation period.

Epidemiology

The available evidence indicates that the BSE epidemic in Britain resulted from the adaptation of a scrapie agent of sheep to cattle. Epidemiological data suggest that the exposure of cattle to the scrapie agent began in 1981/82 and that most of the affected animals were infected in calfhood. The earliest suspected cases occurred in April 1985, and since then only few cases were observed until September 1987 when the incidence constantly increased until the end of 1988. A constant incidence was then maintained until July 1989 when there was a further increase which levelled off in January 1990. The reason why BSE appeared only less than 10 years ago while scrapie decimated sheep and goats for more than 200 years probably comes from an abrupt change in the preparation of bone and meat meal included in commercial foodstuff. In the early 1980s, there was a reduction in the use of hydrocarbon solvent for the extraction of lipids during meat and bone meal production. This might have led to a decreased effect upon scrapie agent inactivation which was the first cause of sheep-to-cattle infection. In the following years, the recycling of carcasses of BSE-infected animals for the preparation of meat and bone meal was responsible for cattle-to-cattle infection which determined the sudden increase in incidence of BSE. The annual incidence of BSE is now 4.5 per 1000 adult cattle. The disease equally affects both sex and there is no breed predisposition. The incidence is greater in the south of England than in the rest of Britain and is higher in dairy than in beef herds. About 12% of the herds had at least one case of BSE. However, about 50% of affected herds experienced only a single case of BSE. The ban in July 1988 of ruminant-derived protein to ruminants will effect a drastic reduction in the incidence of BSE in 1992/93. The eradication of BSE can then be possible in a 10–15 year period unless maternal and lateral spread of the infection from cattle to cattle give rise to endemic BSE like scrapie in sheep and goats.

The epidemiology of TSE of mule deer and elk is unknown.

Transmission and Tissue Tropism

The infectious agent of BSE has been propagated from cattle to cattle by contaminated bone and meat meal included in the foodstuffs for ruminants which have been prepared from BSE-infected carcasses. It is not yet known whether the agent can spread from cattle to cattle by maternal or horizontal transmission. Nor is the mode of transmission of TSE of mule deer and elk known. The agents of BSE and of TSE of mule deer and elk are present at high concentration in the brain of infected animals as brain is the target organ of these agents. In BSE-affected cattle, extraneural tissues considered highly infected in sheep and goats with natural scrapie (i.e. lymph nodes, spleen, intestine), do not apparently contain any infectivity as shown by their failure to produce disease when injected into mice.

Pathogenicity

Natural infection of cattle with the BSE agent is established through the oral route. The agent then needs to enter and to replicate into the CNS in order to produce the disease. Very little is known of the pathogenesis of BSE. However, it is likely that the infectious agent behaves like scrapie in sheep and that early replication takes place in the lymphoreticular tissues (most likely in Peyer's patches) before reaching the CNS. Spread of infection to the CNS likely occurs via the enteric and sympathetic fibers, though viremia may also be responsible for neuroinvasion.

As for scrapie in sheep and goats and for the other spongiform encephalopathies of animals and man, the disease resulting from the BSE agent infection involves the formation and the accumulation of the disease-specific amyloid protein, named prion protein (PrP_{BSE}). PrP_{BSE} derives from a post-translational modification (not yet known) of a normal host cellular protein (PrPc). It is still debated whether PrP_{BSE} is a component of the infectious agent or a pathological response of the cell to infection.

The bovine PrP gene is similar to that of sheep but contains either five (as in sheep and in other species including man) or six copies of 24/27 nucleotides which encodes for a glycine-rich peptide located in the N-terminal region of the protein. Normal cattle have been found to be either homozygous for the gene with six copies of the peptide (6:6) or heterozygous (6:5), while two screened BSE-affected animals were homozygous (6:6). Since PrP gene in sheep and in mice is linked to the development of disease by controlling the incubation period of experimentally induced scrapie, it is under investigation whether homozygous (6:6) cattle are more susceptible to BSE infection than heterozygous (6:5) animals.

Clinical Features of Infection

The main clinical signs of BSE comprehend changes in behavior, apprehension, hyperesthesia to touch and sound, abnormal posture and hindlimb ataxia. Frequently, there are muscular tremors and teeth grinding. Signs consistent with pruritus are not as common as in natural scrapie in sheep. Loss of weight and of milk yield are usually observed. Clinical blood and CSF biochemistry does not reveal any significant abnormalities. The disease progresses and is inevitably fatal after a clinical duration from 2–3 weeks to 1 year. Most of the animals affected are 3–5 years of age, but older animals can be affected as well. The youngest recorded case was 22 months old.

TSE of mule deer and elk is characterized by abnormal behavior, loss of weight and teeth grinding. Polyuria and polydipsia are frequently observed. Death usually occurs in a few months from the appearance of clinical signs.

Pathology and Histopathology

BSE and TSE of mule deer and elk agents induce pathologic changes only in the brain. Both are characterized by spongiform change in the neuropil, neuronal loss and astrocytic reaction. In BSE the lesions are mostly found in the nuclei of midbrain, brainstem and cervical spinal cord with minimal changes in cerebral cortex, cerebellum hippocampus and basal nuclei. In contrast to natural scrapie in sheep and goats, the distribution and the severity of the lesions in a large series of BSE-affected brains remained constant. This suggested the use of a single standard coronary section of the most affected area, i.e. medulla oblongata cut at the obex, for routine diagnosis. Amyloid plaques are confined to the thalamus in a small number of cases and immunostain with prion protein (PrP) antisera.

Negative-staining electron microscopy of detergent-treated brain fractions from BSE-infected brains reveals the presence of infection-specific 4–6 nm diameter helically wound filaments, named scrapie-associated fibrils (SAF), in brain regions where the histological changes are more conspicuous (i.e. basal nuclei, thalamus, midbrain and medulla). Antibodies raised against PrP decorate SAF from BSE-infected brains suggesting that the major component of SAF is the amyloid protein PrP. In TSE of mule deer the lesions are mostly found in the brainstem, hypothalamus and in the olfactory regions of the frontal cortex. Cerebral amyloid plaques are frequently observed.

BSE and TSE of mule deer and elk do not show any pathological sign typical of viral infections such as inflammatory lesion, perivascular cuffing or mononuclear cell infiltration of the cerebral parenchyma.

Immune Response

The BSE and TME of mule deer and elk agents do not produce any cellular or humoral immune response in the host. The reason for this is presently unknown.

Prevention and Control of Bovine Spongiform Encephalopathy

A great number of measures have been taken by the British Governments and by the European Community to control the spread of BSE in cattle and to prevent the hypothetical risk to humans from BSE infection. They include the prohibition for human and animal consumption of brain, spinal cord, thymus, tonsil, spleen and intestine from any cattle over 6 months of age, the destroying of affected carcasses and of milk from suspect cases of BSE except for feeding the cow's own calf, the prohibition of feeding to ruminants of ruminant-derived protein, the restriction to import cattle above the age of 6 months from countries with a high incidence of BSE and progeny of affected females, and the monitoring of offspring from affected cattle. Moreover, BSE has been made legally notifiable in the European Community.

A combination of measures has also been specified for minimizing the risk of transmitting the BSE agent via bovine-derived medicinal products. The most important of them is the careful selection of bovine tissues which should be sourced from BSE-free countries which have an effective surveillance program of detecting BSE. It has also been suggested that cattle yielding source materials should not be older than 6 months. This comes from the observation that BSE, as well as natural scrapie in sheep and goats, has never been reported in animals under the age of 6 months. Moreover, on the basis of distribution of infectivity in sheep with natural scrapie, the various organs and tissues have been classified into four categories (high, medium, low and no detectable infectivity) bearing different potential risks. Finally, it has been suggested to validate the effectiveness of the extraction and purification procedures of medicinal products in removing the agent of BSE, using appropriate animal infection experiments (i.e. scrapie-adapted strains in rodents).

Future Perspectives

The BSE epidemic in Britain indicates that outbreaks of the disease are possible in countries with a high sheep/cattle population ratio which have a high incidence of natural scrapie in sheep and goats, and which feed ruminants with ruminant-derived protein.

Although the possibility that BSE infection may cause disease in humans is considered to be remote, epidemiological surveillance of the incidence of Creutzfeldt–Jakob disease in Britain and in other European countries in the following years has been established to monitor this theoretical risk.

See also: Kuru; Nervous system viruses.

Further Reading

Bradley R (1991) Bovine spongiform encephalopathy (BSE): the current situation and research. *Eur. J. Epidemiol.* 7: 532.

Hope J, *et al.* (1988) Fibrils from brains of cows with new cattle disease contain scrapie-associated protein. *Nature* 336: 390.

Wells GAH, Wilesmith JW and McGill IS (1991) Bovine spongiform encephalopathy: a neuropathological perspective. *Brain Pathol.* 1: 69.

Wilesmith JW, Ryan JBM and Atkinson MJ (1991) Bovine spongiform encephalopathy: epidemiological studies on the origin. *Vet. Rec.* 128: 199.

Williams ES and Young S (1980) Chronic wasting disease of captive mule deer: a spongiform encephalopathy. *J. Wildlife Dis.* 16: 89.

Williams ES and Young S (1982) Spongiform encephalopathy of Rocky Mountain elk. *J. Wildlife Dis.* 18: 465.

Creutzfeldt–Jakob Disease, Scrapie and Related Transmissible Encephalopathies

Laura Manuelidis and Elias E Manuelidis
Yale Medical School
New Haven, Connecticut, USA

History

Scrapie was probably first recognized as a disease in the English-speaking world shortly after the importation of Merino sheep from Spain in the fifteenth century. The name derives from the characteristic scraping syndrome, where afflicted animals denude large regions of wool by scratching against fences. The first experimental evidence that this was an infectious disease was published in 1936 by Cuillé and Chelle who transmitted scrapie to healthy sheep by intraocular inoculation. In the 1940s D.R. Wilson showed cell-free filtrates were infectious, and found that heat and formaldehyde treatments did not completely destroy infectivity. He also transmitted the disease in sheep for nine serial passages, and his experiments implicated a replicating infectious moiety. Sadly, his pioneering contributions were belittled. In 1963, Chandler transmitted scrapie from sheep to mice, providing a useful experimental model of the disease where logarithmic replication and other properties of the infectious agent could be more readily assayed. Eklund did experiments on heat inactivation and concluded that the agent was a virus, but in the 1970s Alpers raised questions about the existence of a nucleic acid genome from radiation studies. A 'self-replicating' membrane, or a proteinaceous agent devoid of nucleic acid was therefore first considered at this time. While biochemical and structural studies were inconclusive, A. Dickinson exploited inbred strains of mice and serial passage of different scrapie isolates to demonstrate at least six distinct scrapie agents by incubation time and lesion characteristics. He also identified a major host genetic determinant, designated *sinc* for *s*crapie *inc*ubation time, that controls susceptibility to disease. Kimberlin and Marsh isolated a scrapie strain (263K) that produced very high levels of brain infectivity in hamsters, and this strain was exploited by S. Prusiner and colleagues in the 1980s in attempts to purify the infectious agent. These workers first identified an ~27 kD protein in scrapie, but not control brain preparations that were subjected to limited proteolytic digestion. They baptized this protein the 'prion' protein (or PrP) and hypothesized this protein, devoid of nucleic acid, was the causative replicating agent. Simultaneously, Merz and others identified amyloid-like fibrils in infectious fractions (known as SAF or scrapie-associated fibrils). Subsequent characterizations have shown that PrP derives from a conserved, host-encoded membrane glycoprotein of 34 kD (Gp34) that is the major constituent of SAF amyloid. Although Gp34 can be sedimented only in afflicted animals, the mRNA for this host protein is not elevated during infection. Interestingly, the gene encoding this protein appears to reside within the *sinc* susceptibility locus.

The neutral term infectious agent is used most frequently by investigators in this field because of the continuing controversy about the interpretation and accuracy of some experimental data, and more importantly, the nature of the replicating causative molecule(s).

Readers should be aware however, that the terminology for these agents is not uniform, and descriptions used in experiments can be confusing. Thus when brain homogenates are inoculated, some authors may refer to this as inoculation of 'prions'. Other investigators can use viral-related terms to stress their belief in a nucleic acid scrapie genome, and to distinguish prion protein from assayed infectivity.

A number of rare human progressive neurological diseases, including Creutzfeldt–Jakob Disease (CJD), kuru and Gerstmann–Sträussler–Scheinker Disease (GSS) were initially classified with other degenerative and/or familial CNS diseases of unknown etiology. These diseases are now known to be caused by infectious agents as determined by transmission studies. Remarkably, in 1923 Jakob suggested CJD was caused by an infectious agent that might be experimentally transmitted to animals using brain inoculation. In 1963 Hadlow proposed that kuru was related to scrapie of sheep, but only the transmission of kuru in 1966, and of CJD in 1968 to chimpanzees by Gajdusek and Gibbs provided the compelling evidence for an infectious etiology. In the 1970s Manuelidis was able to serially transmit CJD to small rodents. These experimental models of CJD have been useful in unambiguously identifying human infections. They additionally have been used in controlled studies of pathogenesis showing viremia, elucidating maternal and other relevant routes of infection, and charting cellular changes that occur at late stages of infection.

Because CJD and related human agents readily infect rodents, direct comparisons can be made with rodent scrapie. Various sheep-derived scrapie agents and human-derived CJD agents are not identical, but do belong to the same general nosological class. Such differences are most apparent in host susceptibility, incubation (replication) time patterns and detailed neuropathology. On a molecular level, rodent CJD models continue to be exploited for delineation of agent characteristics. Such studies show most prion protein has little infectivity. Moreover physical and molecular studies in CJD suggest a conventional viral-like core structure (nucleic acid–protein complex) for these agents. From this perspective, abnormalities in host prion protein are interpreted as part of a pathological response, ultimately resulting in complex membrane changes, cell lysis and amyloid depositions that can be induced by a separable infectious agent.

Taxonomy and Classification

Up to the present time only host-encoded molecules have been detected in any infectious fractions. There are no agent-specific probes, and specific antigens or nucleic acid sequences have not yet been identified. Viral-like particles are inapparent, even in brain which contains the highest levels of infectivity. Thus the classification of these agents is still essentially descriptive, and often based on secondary morphological or clinical findings which are found only at later stages of disease. In humans, a number of dementing diseases may show similar or overlapping clinical or pathological findings. Metabolic, genetic and other etiological causes unrelated to an infectious agent of this class, as well as artifactual or agonal events may induce cellular and molecular alterations that resemble those produced by these infectious agents. In principal, this group of diseases should be restricted to *bona fide* transmissible examples.

Essentially all naturally occurring transmissible encephalopathies have been related to either a sheep or human source, and therefore these agents are most easily classified as either scrapie-like (including the mink and bovine infections) or CJD-like (including GSS and kuru). Characteristics of the transmissible disease in different species and inbred strains of animals have also allowed for the delineation of different strains within the CJD and scrapie agent categories. Nonetheless, the characteristic long incubation in animals, and the complexity of changes in the brain makes it difficult to further delineate these agent variants without more specific probes. Additionally, it is unlikely that all these agents are ultimately derived from exclusive sheep or human sources. The issue of natural disease in other species remains to be explored.

Agent Properties

The nature of the infectious agent is unknown, and the agent has never been purified to homogeneity. Biologically it behaves like a virus. Many of the 'unconventional' or nonviral physico-chemical properties ascribed to these agents are based on experiments with relatively crude brain preparations where recoveries of infectivity are not always well documented. Additionally, survival of a small amount of infectivity can be due to the presence of impenetrable tissue fragments or aggregates. Thus for complete inactivation, extended times of autoclaving are required for decontamination of tissue fragments, whereas careful studies with more dispersed preparations of scrapie have shown complete inactivation at 80°C, much like the profile of inactivation for the conventional virus hepatitis B. Early experiments in scrapie also showed no transmissibility from phenol-extracted nucleic acid, or from dispersed CJD homogenates digested with proteinase K. Such experiments indicate that at least one protein is required for effective infection. Because no agent-specific antibodies have been detected, affinity purification and other molecular cloning strategies based on an agent-specific protein are still unavailable. Additionally, more recent attempts to isolate an agent-specific nucleic acid, using total mRNA in a subtractive cloning approach, have failed to reveal nonhost sequences. This approach may be insufficiently sensitive to clone relatively rare viral messages, presumes a polyadenylated RNA, and excludes endogenous viral sequences that may be intimately associated with the infectious agent. Substantial purification of agent is probably necessary for isolating agent-specific nucleic acids, a difficult task given the complexity of brain, the lack of consistent ultrastructural particles or other agent markers, and the lengthy and labor-intensive animal bioassays necessary for accurate infectivity titrations.

There are essentially three hypotheses for the molecular structure of the agent: (1) The prion hypothesis considers the agent is constituted by prion protein with <50 bases of nucleic acid. More specifically, a post-translational change in this protein is thought to induce a similar (possibly conformational) change in other host Gp34 molecules. (2) The virino hypothesis assumes that a 'small informational' agent nucleic acid genome may be in a complex with PrP. (3) The viral hypothesis assumes a conventional viral structure, with a more sizeable genome protected by a nucleic acid binding or core-like protein.

The prion hypothesis has several difficulties. The established biological features of exponential agent replication and agent variation are most difficult to reconcile with this hypothesis. Protein conversion as well as 'replication' have both been proposed to occur via novel and unknown mechanisms of confirmational recruitment. Agent variations cannot be explained by these mechanisms. Despite intensive studies no evidence has been found for a conformational or other post-translational change in this protein that distinguishes its 'infectious' from uninfectious form. No recombinant or transgenic prion protein construct has yet yielded demonstrable and/or significant titers of infectivity. Finally, direct biochemical data has shown most sedimenting prion protein can be separated from infectivity. The last observation is solidified by animal experiments demonstrating prion protein levels do not correspond to infectious titers.

Physical studies as well have never shown reproducible evidence for an infectious monomeric or dimeric protein. All filtration and sizing experiments using membrane, chromatographic or sedimentation methods have reproducibly shown a minimum size of ~25 nm for infectivity. Other claims of small (protein-like) agent size by membrane filtration through 100 kD filters, by chromatography, or by gel electrophoresis are not reliable. Leaky filters, technical artifacts, and irreproducible data have compromised these claims which have not been formally retracted. These and other biochemical experiments below indicate the infectious agent may be larger and more complex than is generally assumed. Finally, all substantially purified scrapie and CJD preparations reported to date, including those treated with nucleases and Zn^{2+}, contain substantial amounts of nucleic acid, especially when assessed with ^{32}P-labeling or hybridization techniques that do not exclude larger nucleic acid species.

Both the virino and virus hypothesis can account for the biological data using established molecular mechanisms, but are still speculative because no agent-specific genome has yet been identified. Although most inactivation studies do not preclude a conventional viral core, the perplexing radiation data must be explained. An unusually small target size is therefore assumed by proponents of the virino hypothesis. However, relatively large viral genomes can be resistant to radiation and molecular mechanisms for this resistance have been established (*vide infra*).

Until recently, the physical characteristics of these agents could not be readily assessed because the majority of infectivity was aggregated with cellular components. In rodent CJD, gentle conditions for disaggregation, with no loss of infectious titer, have been discovered; the infectious agent behaved for the first time with homogeneous physical characteristics. Infectivity reproducibly separates as a single Gaussian peak at ~120S and has a homogeneous density of ~1.27g/ml in sucrose. These values indicate a protein–nucleic acid complex of significant viral-like size. Additional sedimentation field flow fractionation analyses indicate the CJD agent is ~30 nm. Moreover, exhaustive nuclease digestion does not alter these physical characteristics, results in higher purifications of agent than with any other methods, and yields protected nucleic acid species of up to 6 kb in length as detected by sensitive amplification protocols. They would seem to be favorable for subtractive and cloning procedures directed toward the identification of an agent specific genome. They may additionally be of value in identifying agent-specific proteins because the majority of uninfectious prion protein is removed.

Clearly, a viral hypothesis remains viable. In the light of the biological data from animal transmissions the potential germ line integration of these agents, and the observations on CJD tissue cultures showing transformation with the production of specific growth factors, we have favored a retroviral working hypothesis. The agent might have a truncated retroviral-like genome or an unrelated nucleic acid that is copackaged or transduced by an endogenous retrovirus. In fact, retroviral sequences of several thousand bases have been detected in highly purified CJD preparations. The noninflammatory retroviral infections of the nervous system that are slowly progressive, and produce spongiform changes in the CNS, provide an interesting parallel. In some of these infections viral particles are not readily apparent. Additionally, it is known that some endogenous retroviral particles are resistant to harsh inactivation treatments, including high levels of SDS and guanidinium. The perplexing ineffectiveness of radiation in destroying these transmissible agents can also be accounted for in a retroviral model. Retroviruses are highly resistant to UV and X irradiation. The mechanism for this resistance rests on the ability of reverse transcriptase to utilize template (strand) switching between two damaged RNA genomes within the core particle during cDNA synthesis. A similar mechanism could be operative for the CJD/scrapie agents.

Host Range and Agent Variations

Although transmission of CJD was originally considered to be restricted to primates, ~85% of typical human CJD cases positively transmit to rodent species. In American and European cases (Western CJD), Syrian hamsters give the highest percentage of positive takes, and mice are poor hosts for primary CJD transmission. Even after serial passage in hamsters or guinea pigs, the human CJD agent shows an incubation of >250 days in mice, and mice can die of old age before expressing any clear signs of infection. In contrast to Western CJD, several Asiatic cases of CJD are readily propagated in mice with short incubation times of ~90 days. Thus there are likely to be molecular differences between these Western and Asiatic CJD agents.

In susceptible species, primary transmissions are often very prolonged, but with serial passage the incubation time is reduced. In primary Western CJD transmissions to hamsters or guinea pigs, the incubation time generally ranges from 250 to 500 days, but rapidly declines with serial passage, reaching a minimum incubation time of ~130 days to terminal illness. The mechanisms underlying these changes are unclear. The 'adaptation' of the human CJD agent to a new host may be due to molecular changes (such as mutation) within the agent, to selection of more viable strains of the agent derived from a heterogeneous mixture, and/or to recombination with other host-derived molecules or endogenous helper viruses.

Both host genetic factors and agent strain characteristics are important in determining host range and the expression of neurological disease. Although various inbred strains of mice inoculated with CJD or GSS show very long incubation times, brain homogenates from old mice, with no clinical disease, readily transmit to hamsters with an incubation time of <150 days. Thus this CJD agent maintains its distinct characteristics despite serial passage through mice. Host genetic factors, rather than agent modifications, underlie these species differences in disease expression. In contrast, early experiments showed that after sheep scrapie is serially passaged in another species it can become incapable of re-infecting sheep. Similarly, the hamster-adapted 263K scrapie strain shows a nonexistent (or very prolonged) incubation time in mice, although many other strains of scrapie have relatively short incubation times in mice (~130 days). These types of observations emphasize differences between related scrapie agents and indicate that the infectious agent can exhibit selective and stable changes. Convention would suggest that in either case such variants must be encoded by an agent-specific nucleic acid genome.

Such agent differences however, can be modulated or obscured by host factors for disease expression. The importance of one specific genetic factor in determining host susceptibility and/or incubation time has been illuminated by recent transgenic studies with a prion protein recombinant. When multiple copies of the hamster Gp34 gene are inserted in transgenic mice, the mice become susceptible to the 263K agent, and rapidly develop disease. Thus this host-encoded gene product appears to be a major determinant of susceptibility to the 263K agent, and this determinant may map to one or more species-specific polymorphisms described for the hamster and mouse PrP genes. Host prion protein is probably essential for modulating at least one critical feature used in the life cycle of these agents. It may function as a receptor for cellular entry, or be linked to processes involved in agent replication or toxicity.

Recently, a large number of cows in England have been found to have a transmissible bovine encephalopathy (BSE). BSE presumably originated when cattle were fed scrapie-infected meal (after the rendering process was altered). An appreciation of agent variations is pertinent for the potential infection of humans from this new bovine source. Although humans eating large amounts of mutton in regions where sheep scrapie is endemic do not have an unusually high incidence of CJD-like diseases, and most people are probably resistant to sheep scrapie, it is clear that passage of scrapie in an alien species can result in an altered pattern of species-specific virulence. A scrapie agent could have an increased virulence in humans after passage in cows. Because the route of gastrointestinal entry is inefficient, and the dose of agent is likely to be low in cooked meat, an infection of humans with a bovine 'adapted' scrapie agent might not be recognized epidemiologically for 10–15 years or more after exposure.

At the present time only mammals are known to be susceptible to these infectious agents, but transmission to other species such as chickens has not been attempted. No similar natural disease has yet been described in nonmammalian species. Although the penetrance of these agents across the complete spectrum of mammals is not well studied, rabbits appear to be resistant to scrapie. Nevertheless, the bovine disease clearly demonstrates that many mammals may be susceptible to this group of agents.

Routes of Inoculation, Tissue Tropism and Replication

The natural route of infection in scrapie is not definitively known. An endemic infection has been considered although others contend that sheep acquire the disease from previously infected pastures and posts, via an oral or superficial wound point of entry. Several species appear to have been infected with scrapie via an oral–gastrointestinal route. Examples include transmissible mink encephalopathy, a rare spongiform encephalopathy of cats, and a similar encephalopathy of domestic mule deer. In human kuru, superficial wounds produced during ritual preparation of brains, or by eating infected brains would suggest the natural infection in this case occurs via the bloodstream or by an oral–gastrointestinal route to the bloodstream. Thirty years after the cessation of ritual cannibalism in New Guinea there have been no new cases of this disease.

Even less is known about the natural mode of infection, and the incubation period to develop clinical disease in humans with sporadic CJD or with familial GSS. The general low incidence of these diseases combined with their presumably very long incubation period precludes knowing how CJD is initially acquired in most instances. An endemic infection of low virulence may be considered. Peripheral routes of infection are presumably operative with dissemination of agent in the bloodstream, as in most conventional viral diseases. In both experimental CJD and rodent scrapie there is definitive evidence for cell-associated (buffy coat) infection. Similarly, viremia has been independently confirmed in natural human CJD cases. Such data raise the possibility of a blood-borne route of infection for some humans by transfusion. Although CJD has not shown a positive association with transfusion, the long incubation time does not rule out this route of infection in some instances. The CJD agent is infectious, but not highly contagious for humans, as spread from one individual to another appears to be extremely rare. Similarly, in experimental rodent CJD, the transmission of disease to healthy cage mates has not been detectable in animals housed together for 1200 or 400 days (guinea pigs or hamsters).

Unfortunate human iatrogenic transmissions indicate vast differences in incubation time to disease, with a potential incubation time of 40 or more years for clinical expression of neurological symptoms. In cases where CJD-contaminated electrodes or dura mater were used in neurosurgical procedures, the incubation time was quite short, and patients generally died within a year after exposure. On the other hand, people inoculated peripherally with growth hormone lots contaminated with low levels of CJD have shown clinical symptoms as long as 19 years after exposure. Several other anecdotal reports of transplant infections of humans have been verified in an experimental setting. For example, infection by corneal transplantation has been simulated by inoculating minced cornea into the anterior chamber of the guinea pig eye. Although the cornea is essentially acellular, the inoculation produced unambiguous CJD after a relatively prolonged incubation time.

In experimental CJD, as in scrapie, peripheral routes of inoculation (intraperitoneal, subcutaneous or intravenous) produce disease after a prolonged incubation time, whereas intracerebral inoculation yields the shortest incubation time. Additionally, as is the case with most viruses, the dose of infectious agent also determines the incubation time to clinical disease. In fact this dose response is often used for titering levels of the agent in a given specimen by bioassay. With peripheral and blood-borne routes of infection, tissue in the reticuloendothelial–lymphatic system, especially the spleen, develops appreciable titers of agent before replication in the brain. Moreover, in scrapie, pretreatment of animals with polyanions, known to be ingested by macrophages, lengthens the incubation time to develop neurological disease. It is likely that macrophages or other reticuloendothelial cells deliver the infectious agent to the brain, as is the case with many other neurotropic viral diseases. Once the nervous system is exposed to the infectious agent, neural spread of infection may also be operative, as has been shown by intraocular injection of scrapie, with radiation of infectivity along CNS visual pathways.

The levels of infectivity in the spleen plateaus, and never becomes as high as infectivity levels in brain. The reasons for limited agent replication in spleen as compared to brain are unknown, but presumably selected cell types, and possibly other tissue-associated and diffusible factors are essential for vigorous agent replication. Additionally, as yet unknown immunological responses may limit the replication in certain species. The ability of some scrapie variants (as compared to other scrapie isolates or CJD) to replicate to an ~100-fold higher level in the brain is not understood.

Maternal and Vertical Transmission

In experimental rodent CJD, extensive experiments have shown no evidence for maternal transmission, despite the presence of viremia in the parental animals. Similarly, in human kuru, there is no evidence for maternal transmission in large epidemiological studies. However, some experiments on scrapie-infected sheep suggest that maternal transmission may occur in this species. It is not known if maternal transmission can occur in human CJD. The positive transmission of an infantile form of human CJD ('Alpers' disease) to rodents raises the question of infection in very early life, possibly by a transplacental or germ line route of infection. In this transmissible infantile CJD case, neither parent showed signs of a neurological disease, but this does not preclude an inapparent infection in the parent's blood or germ cells. In the very rare familial cases of human GSS several investigators have postulated that infectivity may be transmitted via the germ line. This possibility cannot currently be distinguished from an altered susceptibility to a more widespread or inapparent human infection. No experiments have been done on sperm from infected humans or animals. If there is human maternal, or vertical transmission in very rare instances, the search for an endogenous virus becomes relevant.

Incidence and Distribution

Human CJD is a rare world-wide disease, estimated to occur with an incidence of 1–2 per million per year in most locations. This estimate is based on the prevalence of cases with typical clinical or neurological findings. Because there are still no molecular markers that are specific for the agent, lengthy animal transmission studies remain the only means of accurately assessing the incidence of infection, and these are not feasible for large-scale population studies. Nonetheless, in a pilot experiment, when buffy coat from healthy individuals was injected into rodents, several samples showed positive evidence of a transmissible agent. Moreover, repeat experiments using virgin instruments and new buffy coat samples from these same subjects (where available), showed the same pattern of positive and negative transmission for each of these individuals. Thus at least some people harbor these agents without clinically apparent disease. Additionally, transmission studies in our laboratory have shown that atypical cases, as well as some cases clinically diagnosed as Alzheimer's disease, or other dementing or familial diseases of unknown etiology, have proved to be transmissible. Some of these cases have provoked less widespread or minimal CJD-like neuropathological changes on primary passage. Other chronic neurological diseases or individual cases classified as 'degenerative' may also prove to be caused by infectious agents of this type.

The general low incidence of disease may represent only the tip of the iceberg, and considerably larger segments of the population may be infected without developing neurological disease. In several other conventional viral diseases, such as polio, only a small percentage of the infected population develops neurological sequelae. Other conventional viruses such as the 'JC' papova virus are carried by ~ 60% of the population, but cause neurological disease only in special circumstances. Because the incubation period in human CJD can be as long as 19 years (and possibly longer), some infected individuals may die of natural causes before developing neurological disease. The infection of mice with Western CJD is instructive in this context because some inbred strains of mice that clearly harbor the transmissible agent in the brain die of old age with no clear-cut clinical symptoms or obvious neuropathological changes. A similar situation exists for certain agent strain/mouse genotype combinations in scrapie. Moreover, the association of host Gp34 polymorphisms with susceptibility in certain populations is reminiscent of the situation in certain slow neurotropic viral infections such as the retroviral disease visna, where certain breeds of sheep express disease early, presumably due to the specificity of a host-encoded viral receptor. Other inbred breeds show increased resistance to the natural visna infection, with a significantly delayed development of disease.

An increased incidence of clinically expressed human disease has been correlated with host coding mutations within the host-encoded prion gene. At least five significant polymorphisms in the prion protein gene have been uncovered in different individuals, and these appear to correlate with incubation time and/or expression of particular pathological sequelae. In a few populations, including Libyan Jews and in a region of Slovakia, the incidence of clinical disease can be up to 100-fold more prevalent than in other populations. These populations have codon alterations in host prion protein gene. The overall data are most consistent with host-modulated susceptibility to these agents. Additionally in ~10% of CJD cases, there is an autosomal dominant pattern of infection, suggesting an important contribution of the host genome to disease expressions. GSS, a very rare disease with distinctive clinical and histological features (amyloid plaques and cerebellar changes) also has an autosomal dominant inheritance pattern in most cases. Familial GSS cases have been tested and are readily transmissible, as shown by independent experiments in several laboratories. In each of these families several changes in the Gp34 coding sequence have been identified, again linking this host sequence to disease expression. Similarly, in different breeds of sheep, polymorphisms in the prion sequence have also been linked to incubation time differences in scrapie expression, and several inbred strains of mice with distinct PrP sequences also show different patterns of neuropathological disease expression. In naturally occurring sheep scrapie one polymorphism is present in most sick animals but is relatively rare in healthy animals.

It should be noted that some cases considered to be atypical cases of CJD, such as 'amyotrophic forms' of CJD, have not been transmissible. Additionally, cases of fatal familial insomnia, which exhibit prion protein changes that some investigators consider to be diagnostic of CJD, have thus far shown no positive transmission in experiments in several laboratories. Similary, we have been unable to transmit CJD to rodents from a brain sample of the Indiana kindred with resistant prion protein. Such cases raise the issue of an infectious agent that is separable from prion protein. Indeed, some cases with prion protein changes may not have an infectious origin.

Clinical Manifestations and Treatment

Sporadic CJD manifests itself predominantly in middle-aged and older individuals. Sensory disturbances, confusion, inappropriate behavior and severe sleeping disorders are the most common presenting symptoms. Approximately 80% of patients develop myoclonic jerking movements, sometimes with a 'startle' myoclonus. There is typically a rapid progression of neurological deterioration leading to frank dementia and ultimately a comatose state within six months to one year. This rapid progression is usually helpful in distinguishing the disease from Alzheimer's disease in a clinical setting. However, the progression in CJD is sometimes less rapid, and clinical symptoms may evolve over several years. Recovery is extraordinarily rare, although one patient with a transmissible dementia showed an arrest in symptoms with partial recovery. Similarly, extremely rare recovery has been noted in sheep scrapie. In GSS, cerebellar symptoms usually predominate, and dementia usually occurs only late in the course of disease. The clinical course in familial GSS is typically longer than in sporadic CJD, and has been reported to last for ~5 years.

There are no effective treatments for transmissible dementias. In some rodent scrapie models, amphotericin B prolongs the incubation time to develop clinical disease. In this case the appearance of prion protein is also delayed. It is notable, however, that the titer of infectious agent is not similarly delayed, and the characteristic temporal replication of the infectious agent is the same as in untreated animals. This effect is only seen when amphotericin B treatment is instituted before challenge with the agent, and thus is unlikely to be of use for human patients. Polyanion prolongation of the incubation time is only observed when treatment is started before the brain becomes infected.

Cytological and Molecular Pathology

In transmissible encephalopathies only the brain exhibits pathological sequelae even though other organs such as spleen harbor the infectious agent. No virus-like particles have been consistently identified ultrastructurally although dense core particles of ~25 nm have been observed occasionally in a few synaptic boutons in some laboratories. The most characteristic lesions in human CJD are the spongiform (vacuolar) changes, with an absence of inflammatory lymphocytic infiltrates. Vacuoles in rodent scrapie are distinct from those in rodent or human CJD. Nonetheless, the general ultrastructural features are similar. Vacuolization is most prominent in neuritic processes, but may also be seen in some astrocytes. Membrane-bound vesicles are seen in swollen neural processes, and membrane disintegration and fusion are apparent ultrastructurally. Especially on a light microscopical level, the spongiform changes must be distinguished from nonspecific swelling, agonal and artifactual changes. Other histological changes, such as astrogliosis, an abundance of lysosomes, and an increase in microglial cells lack diagnostic specificity because such changes can be present in a variety of dementing illnesses of unknown etiology, as well as in brains from elderly individuals. Such changes, as well as amyloid depositions, are likely to represent epiphenomena, with common or similar final pathways for cell death. They are seen at end stages of disease in many types of dementia regardless of the etiological cause.

Although amyloid depositions can be secondary to an infection with transmissible agents, amyloid-associated changes may provoke additional pathological sequelae. In sporadic human CJD, amyloid-like plaques are generally inapparent. At the other end of the spectrum, kuru brains show large characteristic plaques in the cerebellum. Familial GSS also displays prominent cerebellar plaques, and as discussed above this may relate in part to human host genetic factors. Interestingly, rodents infected with GSS do not show extensive or large plaques comparable to those observed in the original human material. Moreover, only selected scrapie agent–host genotype combinations produce widespread prominent plaques. This feature is not necessarily a consequence of a prolonged incubation period in these particular models because several other long incubation time scrapie models do not show similar plaques. Infrequently, Alzheimer-associated stigmata of β-amyloid plaques and neurofibrillary tangles can be combined with spongiform changes in humans to present diagnostic problems. Whereas Alzheimer plaques are rich in the β-amyloid membrane glycoprotein, transmissible encephalopathy plaques contain the PrP host membrane glycoprotein.

Although most diagnostic neuropathologists look for spongiform changes, an inordinate emphasis on such changes is unwarranted. Despite the usefulness of spongiform changes in a diagnostic setting, CJD has been transmitted from material that is negative for such changes. Several experimental observations are pertinent for interpreting the significance of spongiform changes. First, the route of inoculation influences the degree of vacuolization. Whereas spongiform changes are prominent after intracerebral inoculation, these changes are considerably less severe with peripheral routes of inoculation. Second, the type of inoculum can also influence the degree of spongy change. Intracerebral inoculation of spleen or cornea produces minimal spongiform change in contrast to inocula of brain homogenates. This difference is likely, at least in part, to relate to differences in the infectious titer in the various inocula. Third and most importantly, temporal studies in experimental CJD demonstrate that florid spongiform changes occur at a relatively late stage of disease, well after the infectious agent has replicated by $\geq$4 logs. Finally, although spongiform changes have been reported in one transgenic model bearing a human recombinant prion protein construct, the transgenic mice do not reproducibly appear to harbor a transmissible agent. This study underscores the caveat that apparently similar histological and molecular changes may derive from very different etiological causes, i.e. infection-induced changes and unrelated genetic-membrane processing pathways.

Immunohistochemical or Western blotting for detection of prion protein can be of value in distinguishing CJD from Alzheimer's disease and other dementias of unknown etiology. In the absence of transmission, assay of the protein appears to be the most specific way to distinguish different dementias. However, cases that are negative for prion protein by Western blotting have been transmissible, so an absence of this protein does not reliably indicate a lack of infection. This type of observation is supported by animal studies where high titers of infectivity have been found without detectable prion protein. For example, the infectious agent replicates to significantly high levels before this host protein develops its abnormal sedimentation and limited protease resistance characteristics. Thus, this protein assay is of little help in sorting out most inapparent, or early infections with these agents. Additionally, the positive detection of prion protein is not invariably associated with transmissibility. Examples include tested cases of fatal familial (thalamic) insomnia or 'familial myoclonic dementia masquerading as CJD' as well as other rare dementia cases with prion protein alterations. Other molecular changes, including increased steady state levels of host mRNAs, including those encoding glial fibrillary protein (GFAP), membrane-associated apolipoprotein E, lysosomal cathepsin D, β-microglobulin and heat shock-related proteins have been identified in infected brains. Most of these are increased only in later stages of disease. Assays for these transcripts could have some diagnostic value, although many of these gene products can also be elevated in nontransmissible dementias. Interestingly, elevations of GFAP mRNA in experimental CJD can be found before detectable spongiform changes, suggesting that functional cellular alterations, or induced diffusible factors may have a role in the evolution of the pathology.

Tissue Culture Studies

Tissue culture studies have been hampered by the lack of pathological or other signs of infection. Thus long animal bioassays have been required to assess infectivity in tissue culture cells. There are, however, several salient findings from tissue culture studies. First, agent replication is relatively low in all cultures examined to date. The highest yields of scrapie infectivity have been reported in PC12 growth arrested cells and in some mouse neuroblastoma cultures. Nonetheless, both the CJD and scrapie agents clearly replicate in nonneuronal cell lines. Second, with extended serial passage cells can lose their infectivity. This observation may have relevance for incubation time differences as well as therapeutic ramifications. Third, there is no evidence that the infectious agent is released into the supernant, i.e. it remains cell-associated (as in blood). Fourth, metabolic and molecular labeling studies have shown PrP in infected culture can be resistant to release from the plasma membrane and can collect in lysosomes. Details of cellular processing of this protein in infected and control cells may be most easily approached in tissue culture. As with *in vivo* studies, some cultures have appreciable infectivity but no detectable resistant PrP. Fifth, both CJD brain-derived cultures and cells exposed to infectious preparations become transformed with a very high frequency. The time course of transformation resembles that seen with slowly oncogenic viruses, and many of these cultures also show an elevation of an α-TGF-like factor. Finally, although titers are low, tissue culture may still be useful for testing therapeutic strategies in a simplified setting.

Future Perspectives

It is essential to have an open mind about the molecular nature of these agents, and an overriding respect for the data and its most streamlined explanation (Occam's razor). Inexpensive, rapid and simplified assays for detecting these agents will be invaluable for exploring their molecular nature. Tissue culture models, which have not yet been useful in producing high titers of infectivity or diagnostic tests, deserve further attention. Clearly the identification of an agent-specific nucleic acid, or a specific antigen, such as a nucleic acid binding protein will be most useful. Progress in this area will also be invaluable for defining the role of these agents in dementias of unknown etiology, and in developing rational preventive and therapeutic strategies.

See also: Kuru; Nervous system viruses; Visna–Maedi viruses.

Further Reading

Bock G and Marsh J (ed.) (1988) *Novel Infectious Agents of the Nervous System*. Ciba Foundation Symposium 135. Chichester, UK: Wiley.

Chesebro BW (ed.) (1991) Transmissible Spongiform Encephalopathies. *Curr. Top. Microbiol. Immunol.* 172. Berlin: Springer Verlag.

SQUIRREL FIBROMA VIRUS

See Poxviruses

ST LOUIS ENCEPHALITIS VIRUS

See Encephalitis viruses

SWINE HERPESVIRUS-1

See Pseudorabies virus

SWINE VESICULAR EXANTHEMA VIRUS

See Vesicular exanthema virus and caliciviruses of pinnipeds, cats and rabbits

SWINE POXVIRUSES

See Poxviruses

T

T1 BACTERIOPHAGE

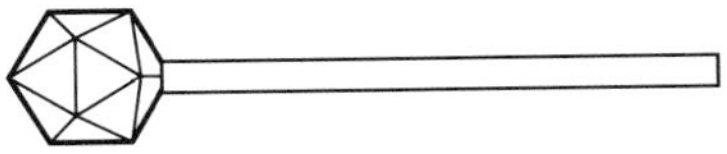

JR Christensen
University of Rochester Medical Center
Rochester, New York, USA

Introduction

T1 is one of the seven phages collected by Max Delbrück and renamed T1 through T7; they all make clear-centered plaques on *Escherichia coli* B. T1 is unrelated to any of the others. Its latent period at 37°C is 13 min, and the burst size is about 100. T1 also infects some other laboratory strains of *E. coli* (e.g. K-12 and C) and *Shigella dysenteriae*. Of 290 clinical isolates of *E. coli*, T1 could replicate in two, and kill a third without producing phage.

The best-known relative of T1 is the *Shigella* phage, D20, with which it readily hybridizes. T1-like phages can be isolated from sewage, but none has received much study. Among laboratory strains of T1, there are a few minor differences, revealed by restriction analysis.

The virion is highly stable in the dry state, so that careless technique may cause T1 (like other phages with this property) to become air-borne in the laboratory, and to unexpectedly lyse cultures contaminated accidentally.

The Virion

In the electron microscope, T1 closely resembles λ, with a polyhedral head about 55–60 nm across and a long flexible tail about 7 × 150 nm. Fifteen virion proteins have been recognized. One of these, P7, accounts for about 50% of the total protein, and two additional proteins for another 35%.

The genomic molecule is double-stranded DNA with approximately 48 500 bp. Only the four conventional bases are present. About 0.2% of the cytosine residues and 1.7% of the adenine residues are methylated, at the 5 and 6 positions respectively. The DNA molecule includes a terminal redundancy of about 2800 ± 530 bp (6%), so that the coding capacity is about 46 000 bp. Because of the way the genome is packaged during morphogenesis (see below), there is a limited set of cyclic permutations of the nucleotide sequence within a population of virions.

Early Steps in Infection

Adsorption of T1 is a two-step process: an initial reversible interaction with an outer membrane protein coded by *fhuA* (*tonA*), followed by an irreversible interaction involving the *tonB*-encoded protein. Curiously, sequentially these proteins are involved in the transport of iron into the cell. Irreversible interaction of T1 with the host requires that the cell membranes be in an energized state; there is no such requirement for the other T phages.

During the first 1–2 min after infection, there are marked changes in the membrane properties and in the energy state of the bacterium: there is a large efflux of K^+, the proton motive force (PMF) decreases, but does not vanish, and intracellular ATP levels fall, as a result of activity of the proton-translocating ATPase. Those transport systems driven by ATP or PMF are inhibited, but the activity of the sugar phosphotransferase systems is stimulated. None of these changes requires phage gene activity.

During this same period of time, entry of phage DNA into the cell and shutdown of host protein synthesis occur. A model that puts these observations together has been proposed (although there is some dispute about it). Before infection, at least two cation gradients contribute strongly to the energized state of the membrane: K^+ is higher inside the cell, and H^+ is higher outside. The entry of T1 DNA is effected by a proton symport that involves *tonB*. This process tends to deplete the PMF, but partial activity is maintained by H^+ efflux, driven partially by ATP hydrolysis and partially by K^+ symport. The resulting fall in intracellular ATP leads to a fall in GTP. This, and perhaps other changes in the intracellular ionic environment, produces an inhibition of translation of host proteins at the initiation level. (Presumably, the translation of phage proteins is resistant to these changes.)

Within a few minutes, a 'resealing' process occurs, although it is not clear that the membrane is restored exactly to its original condition. Resealing and the continued maintenance of the remaining level of the PMF are necessary for the infectious process to continue.

Another early event is the appearance of a phage-encoded DNA methyltransferase. Although the specificity of this enzyme is identical to the host's *dam* enzyme, it is quite distinct in other enzymological properties, and T1 DNA shows no hybridization to that of the cloned host gene. The phage-encoded enzyme almost totally methylates the adenines occurring within 5′-GATC sequences, even when the phage is grown in *dam* mutants. The biological role of this methylation is unclear.

Transcription

T1 depends on the host RNA polymerase for transcription throughout its cycle. Both early and late, all or nearly all transcripts are read from the same strand of DNA. All major regions of the genome are transcribed early, but there is a relative shift at later times towards the regions coding for virion proteins. The basis for the shift remains unknown. Considerable transcription (but not translation) of host genes continues after infection.

Protein Synthesis

The synthesis of 31 phage-encoded proteins has been documented in infected cells. From their combined molecular weights, these account for about 80% of the coding capacity of the genome. Five temporal patterns of synthesis have been noted: proteins synthesized only early, only late, continuously but at a declining rate, continuously and steadily, continuously and at an increasing rate. As with the synthesis of host proteins, much of this regulation is probably at the translational level.

DNA Synthesis

Among virulent double-stranded DNA phages, T1 is unusual in that it depends on the *polC*-encoded α subunit of Pol III for DNA synthesis. This is a property typical of temperate phages. It also depends on most of the other host-encoded proteins involved in the elongation phase of DNA synthesis, except for *dnaB*, but does not require host proteins involved in the initiation of DNA synthesis.

Host DNA synthesis is stopped very early in infection. This inhibition requires protein synthesis, presumably for the expression of a phage gene, but the gene responsible has not been identified.

Two sets of phage genes are required for DNA synthesis. Mutants in genes *1* and *2* are totally defective in DNA synthesis. Presumably, the products of these genes function in initiation of synthesis on T1 templates. The continuing function of both these genes is required throughout the growth cycle. Lysis of the host is delayed after infection by these mutants, but most phage-encoded proteins are produced normally.

Two other genes, *3.5* and *4*, encode a recombination function called Grn (for **g**eneral **r**ecombi**n**ation–pronounced 'green'). This function is obligatory for phage production (see below), but once expressed, it becomes progressively dispensible as the infection proceeds. Under some conditions, the host's RecE or λ's Red recombination system can substitute, at least partially, for Grn. Evidence suggests that gene *4* encodes an exonuclease, but no such enzyme has been isolated.

Early in infection, the products of DNA synthesis are linear monomeric molecules. Later, under the influence of Grn, linear concatemeric molecules are produced; these have a broad distribution of sizes up to about 8- to 10-mers, presumably produced mainly by head-to-tail recombination between homologous redundant ends on two molecules. In the total absence of Grn function, no concatemers are found, and DNA synthesis ceases prematurely, about 6 min into the infection. Why synthesis stops under these conditions is not clear, particularly since, once Grn has been expressed, synthesis of both monomers and concatemers continues even though further Grn activity is blocked. In the total absence of Grn, the cells lyse at the normal time, but no phages are released.

As with other virulent phages, T1 infection leads to degradation of the host's DNA. The liberated material provides about two-thirds of the precursors for the synthesis of T1 DNA. Mutants in gene *2.5* are deficient in host DNA breakdown, but phage DNA synthesis proceeds normally.

However, while phage DNA synthesis can proceed in the absence of host breakdown, the converse is not true. In T1 there is an unusual functional dependence of the degradation of host DNA on ongoing synthesis of phage DNA. If phage DNA synthesis is prevented, whether by use of mutants or naladixic acid, degradation of host DNA does not occur. If degradation has already begun and a synthesis block is imposed, degradation stops. Thus, no conditions have been found under which free degradation products can be detected. With T4 and other virulent phages that have been studied, if phage DNA synthesis is blocked, host DNA degradation nevertheless occurs, with released materials leaking into the medium.

Morphogenesis

Relatively little is known about the pathways of T1 capsid assembly. Some virion proteins, including P7, are cleaved from larger precursors, as occurs during capsid assembly with several other phages. Mature DNA-filled heads can join to tails and form infectious particles *in vitro*.

More is known about the manner in which DNA becomes encapsidated. Linear genomic concatemers are the substrate for packaging. The process is initiated at a site called *pac*, located between gene *1* and gene *2* on the map, and processive head-filling proceeds toward gene *1* (leftward as the map is conventionally represented). As mentioned above, a 'headful' is about 1.06 genomes worth of DNA. Only two or three particles are produced from a single initiation event, so that a limited set of cyclic permutations is produced. Thus, about 18% of the genome will, in some of the particles, be represented twice, once at each end of the packaged DNA molecule.

A mutation, *pip*, located between markers in genes *2.5* and *3* (and which likely represents a separate gene), has a marked influence on DNA packaging. If the host happens to be a λ lysogen, T1*pip* has an enhanced frequency of initiating packaging 'mistakenly' at a site, *esp*-λ, located on the λ prophage (see under Transduction below). The *pip* mutant is also deficient in processive packaging, so that almost all of the packaged molecules are initiated at *pac*, even though, judging by the small burst size, the efficiency of initiation at *pac* is probably reduced by the mutation.

Mutants in most of the genes involved in head production are grossly defective in processing concatemeric DNA to monomeric DNA, suggesting that nearly complete head structures are required for the maturation of concatemers into monomeric 'headfuls'. However, mutants in gene *12* process concatemers normally, although they do not produce heads; presumably they fill DNA normally into head precursors that are unstable because of the lack of the gene *12* product. Finally, *am*383, the sole mutant in gene *13.3*, which maps in the head region and encodes a virion protein, is only partially defective in processing concatemers. It is not clear whether this mutation is phenotypically 'leaky', or whether this observation points to some special role for this gene in DNA packaging.

Extracts, prepared from cells infected with T1 bearing amber mutations in both gene *1* and gene *2* are capable of packaging either homologous or heterologous DNA added *in vitro*. Two packaging pathways have been identified.

If the extract is given homologous DNA extracted from virions of T1 (or T1-like phages), it produces phage. Presumably, the pathway involves two steps: the production of concatemers, via recombination, followed by packaging, initiated at *pac*.

Given heterologous DNA extracted from T3, T7 or λ *nin*, all of which are about 80% the length of T1 DNA and lack any known *pac*-like sites (*esp*-λ is in the *nin* region), the extract produces the corresponding phage by a *pac*-independent pathway. DNA from wild-type λ is packaged less efficiently than that from λ *nin*, suggesting that the second pathway prefers shorter molecules. (The first pathway does not act on wild-type λ DNA, despite the presence of *esp*-λ, since there are no redundant ends to facilitate concatemer formation.)

Mutants and Maps

The current genetic map contains 23 essential genes, identified by complementation tests between conditional-lethal mutants, and the nonessential gene *2.5*, identified by *tar* mutants (which enhance transduction frequency). These are in numerical order, with fractional gene numbers for those genes identified since the first 18 genes were mapped.

At the left end are the genes discussed above: *1*, *2*, *2.5*, *3.5* and *4*, as well as *pac* and *pip*, all of which have roles in DNA metabolism. Curiously, also within this cluster is gene *3*, which has no role in DNA metabolism – rather it is essential for tail formation. Perhaps it has a regulatory role, rather than coding for a virion protein.

Next come eight more genes, *5* through *11.5*, which are required for the production of phage

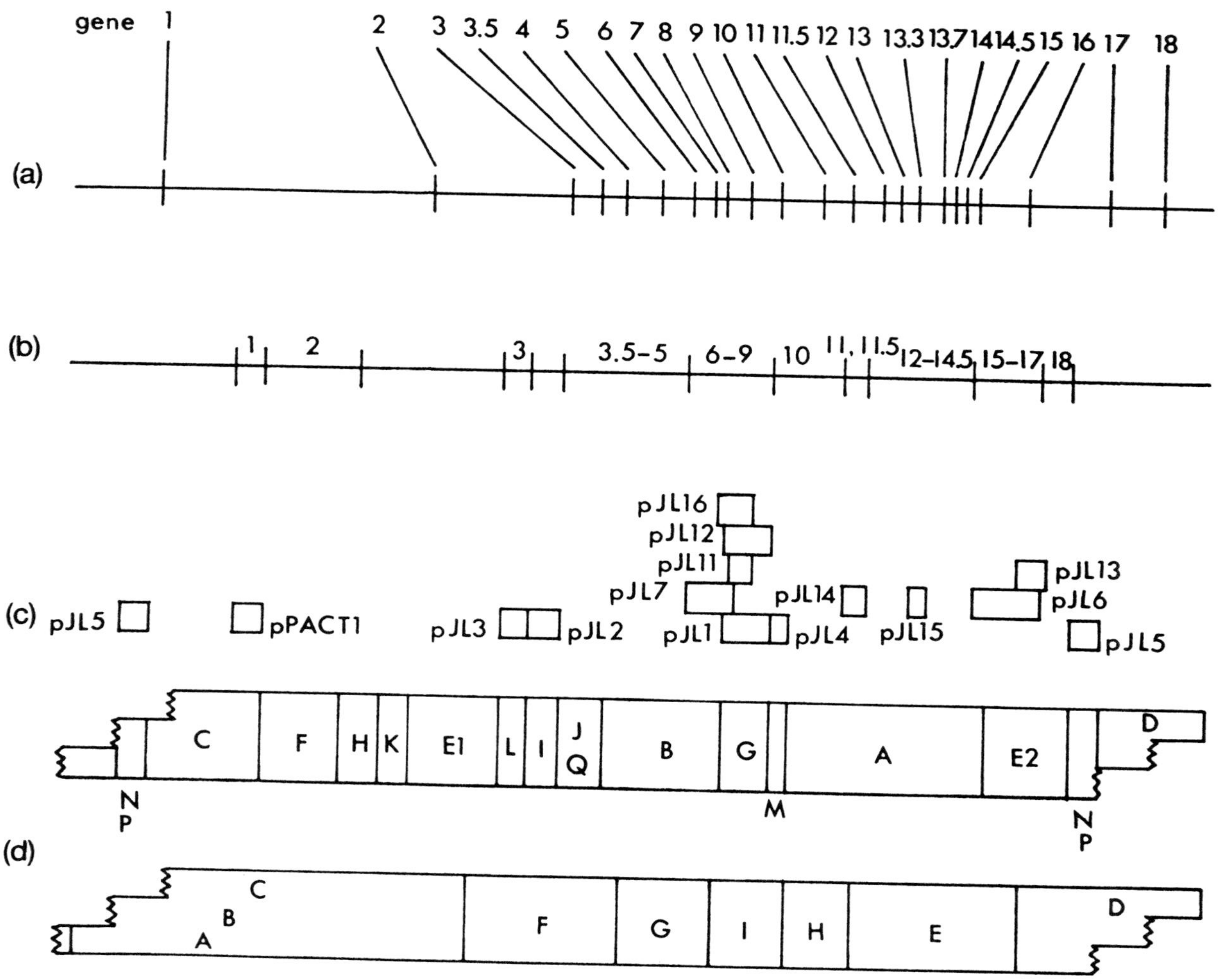

Fig. 1 Genetic and physical maps of T1. (**a**) The genetic map, with each gene identified by the *am* or *ts* mutant used to map the gene. (**b**) Positions of the T1 genes on the DNA molecule. Each interval represents the outer limits of the regions occupied by the *am* mutant tested, relative to the physical map below. (**c**) Locations of the T1 cloned fragments relative to the physical map. (**d**) Physical map of T1 DNA with the *Bgl* II (upper) and *Bgl* I (lower) cleavage sites. [Reprinted, with permission from Liebeschuetz J, Harris RD and Ritchie DA (1987) *J. Gen. Virol.* 68: 2049.]

tails. The *hr* mutation, which allows T1 to infect *tonB* (but not *fhuA*) mutant bacteria, maps just to the left of the available gene *5* markers, and may be within that gene. Finally come ten genes, *12* through *18*, required for the production of phage heads. There are probably at most a very few undiscovered head or tail genes, judging from the number present in the morphologically similar phage, λ.

Several tail and head genes, plus the *pac* region, have been cloned in the positive selection vector, pLV59. These clones, which together include about one-third of the total genome, have allowed a comparison of the genetic and physical maps (see Fig. 1). Genetic markers are relatively far apart at both ends, but especially so at the left end. The occurrence of head-to-tail recombination during concatemer formation would be expected to increase genetic distances between markers at the map ends (this does not mean that the genetic map is 'wrong', merely that it represents a different kind of information). But this effect will not be found to the right of *pac*, and so the wide genetic distances between markers in the region from gene *2* to gene *5* cannot be due to this effect. It is almost certain, however, that more genes remain to be discovered, particularly genes for nonessential functions, since these will not be discovered in collections of conditional-lethal mutants. Unless these nonessential genes are scattered among the head and tail genes, which seems highly implausible, their likely location is in the leftward portion of the map.

Restriction and Modification

T1 passes freely among such *E. coli* strains as B, K-12 and C, so it is not subject to either B or K restriction. At one time, it was felt that the high level of methylation of T1 DNA might account for this, but since 5′-GATC sequences are the main sequence methylated (at least, the total level of adenine methylation is consistent with the number of such sequences that might be expected on a genome of T1's size), and since this sequence is not part of the recognition sequence for either the B or K restriction enzyme, this seems unlikely.

However, in P1 lysogens, T1 grown in a nonlysogen is subject to restriction and modification; P1-modified T1 plates with full efficiency on both lysogens and nonlysogens. When unmodified T1 infects a lysogen, about 80% of phage DNA is degraded and excreted into the medium within 5 min. After this, degradation stops, and several observations indicate that biologically active fragments of the restricted genome persist in the lysogenic cells for a considerable period of time.

Complementation occurs in 10–25% of the cells when unmodified T1*am*$^+$ *hr* and modified T1*am hr*$^+$ phage co-infect lysogens. This is true for most of the genes tested, but not all. Most of the progeny phase are *am hr*$^+$, so this is not primarily due to marker rescue, although that also occurs (see below). Complementing activity is but little diminished when infection by the modified phage comes 4, or even 8, min after infection by the unmodified one.

In co-infection experiments that use plaque-morphology markers to distinguish modified from unmodified phage (no complementation required), recombinational rescue of the alleles of the unmodified parent occurs in a few percent of the cells. Certain markers are rescued more frequently than others. In three-factor crosses, usually only a single marker is rescued in a given cell. The alleles of the nonmodified phage remain available for rescue for at least 10 min.

In addition, under special conditions, unmodified T1 alone can successfully infect lysogens. The required conditions are: high multiplicity of infection (about 10), strong aeration in nutrient medium and occurrence of protein synthesis during the first few minutes after infection. (These special conditions are not required for the phenomena described above.) Up to 10% of the cells produce phages; this is called cooperative infection. The progeny phages are mostly modified, and they have undergone extensive recombination. Again, dividing the infecting phage into two portions, with up to 6 min intervening, does not interfere with cooperation.

Finally, among the rare (~ 1in10^4) lysogens that do yield phages after infection by a single unmodified T1, individual cells lyse and produce their progeny (most of which are unmodified) over a period of 3–5 h.

Transduction

Like many other phages that package DNA by a headful mechanism, T1 is a generalized transducing phage. To demonstrate this, it is necessary to use T1*am* phage (typically a double *am* stock), permissive donors and nonpermissive recipients, otherwise, potential transductants are killed on the assay plate by the large excess of viable virulent phages in the transducing lysate. The fact that, for a given marker, transduction frequencies are quite reproducible from experiment to experiment makes T1 a good subject for studying various aspects of the transduction process.

Although all markers tested can be transduced, the frequency of transduction varies widely among different markers. It appears likely that packaging of bacterial DNA can be initiated at many sites in the chromosome that mimic, to varying degrees, the *pac* site of T1. This idea is strengthened by the discovery of two specific sites that have a special property: markers to one side of the site, but not the other, are transduced at high frequency; this is what would be expected of transductional 'pick-up' initiated at a *pac*-like site. One of these, *esp*, is located between *att*λ and *gal* on the bacterial chromosome. From this site, *bio* (but not *gal*) markers in nonlysogens, or λ plaque-forming units (PFU) in lysogens, are transduced at relatively high frequency. The second site, *esp*-λ, is within λ, between genes P and Q, and from this site, PFU are readily transduced from polylysogens. Experiments with tandem heteroimmune dilysogens show that packaging proceeds leftward from this site. While initiation of packaging at *pac*-like sites probably contributes greatly to T1 transduction, *pac*-independent packaging of heterologous DNA can occur *in vitro* (see above), and a similar process may be involved in the transduction of low-frequency markers.

Small plasmids can also be transduced. The frequency is markedly enhanced by cloning *pac* or *esp*-λ into the plasmid. Transducing particles carry head-to-tail multimers of plasmid DNA; perhaps Grn can stimulate circle-into-circle recombination.

The phage-induced degradation of host DNA would be expected to compete with transduction. It does: most of the transducing particles are formed early in the infectious cycle, and *tar* mutations (gene *2.5*), which block degradation, enhance the formation of transducing particles.

See also: Bacteriophage transduction; Lambda bacteriophage.

Further Reading

Drexler H (1988) Bacteriophage T1. In: Calendar R (ed.) *The Bacteriophages*, vol. 1, p. 235. New York: Plenum Press.
Figurski DH and Christensen JR (1974) Functional characteristics of the genes of bacteriophage T1. *Virology* 59: 397.
Liebescheutz J, Harris RD and Ritchie DA (1987) Further characterization of phage T1 DNA clones. *J. Gen. Virol.* 68: 2049.
Wagner EF, Auer B and Schweiger M (1983) *Escherichia coli* virus T1: genetic control during viral infection. *Curr. Top. Microbiol. Immunol.* 102: 131.

T4 BACTERIOPHAGE AND RELATED BACTERIOPHAGES

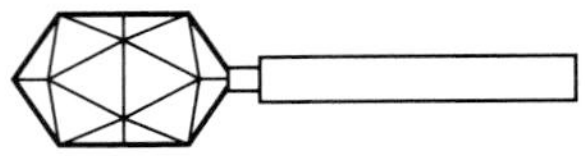

Gisela Mosig
Vanderbilt University
Nashville, Tennessee, USA

History and Overview

Bacteriophages T2, T4 and T6, the first members of the T-even phages, were among the seven *Escherichia coli* phages ('Snow White and the Seven Dwarfs') selected by Max Delbrück to study fundamentals of viral replication. The concerted efforts of several phage groups led early on to the recognition of DNA as the genetic material and to the first formulations of several fundamental biological concepts: (1) the operational differences in defining the gene by mutational, recombinational or functional analyses (the concepts of muton, recon and cistron), (2) the demonstration of mRNA, (3) the triplet code and nonsense codons as termination signals for protein synthesis, (4) homologous genetic recombination as exchange between DNA molecules, (6) DNA repair mechanisms occurring in the light or in the dark, (5) restriction and modification of DNA as an important aspect of host–virus interactions, (7) pathways of macromolecular assembly, i.e. protein machines, and the more recent demonstrations (8) of introns in procaryotes and (9) that ribosomes can skip introns in mRNA during translation.

The T-even phages are serologically and morphologically related. They share more than 80% DNA sequence homology, and their genomes exhibit similar organization and regulatory patterns. T-even DNA contains hydroxymethylcytosine (HMC) instead of cytosine. These residues are glycosylated to different extents. The homology between these phages allows considerable interspecies recombination. However, a few regions of complete heterology are interspersed between the homologous segments. They are postulated to be evolutionary remnants of recombination with foreign elements. Some of these heterologies correspond to introns and at least two are mobile. Some of the introns are self-splicing. Movement is mediated by intron-encoded site-specific nucleases and by homologous recombination.

Most of the recent work on T-even phages has been concentrated on T4, mainly because a large collection of conditional-lethal mutants has provided a powerful impetus for molecular analyses of phage development. The work produced by the combined efforts of all groups working on T4 is summarized as of 1988 and a revision of the T4 book will be published in 1994.

T-even phages are some of the most successful molecular parasites. Like all viruses, they depend for their propagation on many vital functions of their host – energy metabolism, membranes, transcriptional and translational machines – and they manage to subvert host functions gradually to their own purpose in an exquisitely timed choreography. This subversion is accomplished at several interconnected levels by the following phage-encoded proteins: most, if not all, proteins required for DNA replication and recombination, proteins that degrade the host DNA and modify the T4 DNA, and proteins that modify host RNA polymerase, ribosomes and membranes.

1. A cascade of phage-induced enzymes and other proteins modifies the host core RNA polymerase covalently and noncovalently. These modifications ultimately turn off host transcription, and lead to timed initiation from different classes of phage promoters and termination–antitermination control of certain transcripts.
2. RNA processing by phage and host enzymes, translational repression and poorly understood modifications of ribosomes modulate translation from the primary and processed phage transcripts. The emphasis on translational controls may facilitate the rapid development of these phages: one growth cycle lasts about 28 min at 37°C.
3. The host genome and host mRNA, present at the time of infection, are rapidly degraded, and the breakdown products are reused to synthesize phage RNA and DNA.
4. The onset of the first phage DNA replication from specific origins requires host RNA polymerase. Most subsequent initiations of replication forks depend on phage-encoded recombination proteins. This switch to recombination-dependent mode(s) of DNA replication is one of the critical components of the T4 strategy to escape the host's controls.
5. DNA packaging proteins eventually compete with recombination–replication proteins for the same DNA structures.
6. These processes are interconnected at several levels. For example, temporal control of transcription is dependent on DNA replication and, in turn, influences the pattern of and competition between recombination, replication and packaging. The multiple interconnections between these processes, together with multiple pathways for some of them, allows a flexibility of development which is the recipe for the success of these viruses.

Genome Structure and Map

The genome of T4 contains about 168 000 bp, the cytosine residues of which are hydroxymethylated and glycosylated. This modification makes T4 DNA refractory to T4 enzymes designed to destroy the host's DNA and to host restriction enzymes. Using genetic tricks, T4 DNA without these modifications can be generated. The use of such DNA has been instrumental in cloning and sequencing most of the T4 genome (>99%).

Mature DNA molecules, packaged into virions, are linear and contain 3–5% terminal redundancies. During packaging, they are cut at nearly random positions from highly branched 'concatemeric' intracellular DNA and used to fill preformed heads. Consequently, the ends of individual chromosomes are almost randomly permuted over the circular genetic map.

Mutations, their assignments to open reading frames (ORFs) and complementation analyses have allowed definition of approximately 130 genes with known functions. Approximately 100 additional unidentified ORFs, revealed by sequencing, are in search of functions. This information is continuously being updated. The maps constructed from (1) recombinational analyses, (2) probabilities of cutting during packaging and (3) restriction enzyme analysis and sequencing of DNA are nearly congruent, except for a major distortion due to a strong hot spot of recombination in or near the tail fiber gene *34*. This hot spot is not apparent in phage T2.

Genes with functionally related products are clustered on the genetic map to only a limited extent. In contrast to the lambdoid phages, early and late gene clusters and transcription units are interdigitated, and interacting proteins can be encoded by distant genes. Most remarkably, in spite of the large genome size and the presence of many redundant functions, T4 genes and ORFs are tightly packed. There are many overlapping coding regions; occasionally they overlap on the two complementary DNA strands. Only a few short regions are devoid of coding capacity. These observations suggest that the apparent redundancies confer selective advantage.

Particle Structure and its Relationship to Assembly, Infection and Host Range

T-even phages build some of the most complex virus particles, which resemble lunar landing modules (Fig. 1). They devote more than 40% of their total genetic information to synthesis and assembly of the protein components of this particle. The complex assembly pathways appear to be designed to allow packaging of the DNA and, at the start of a

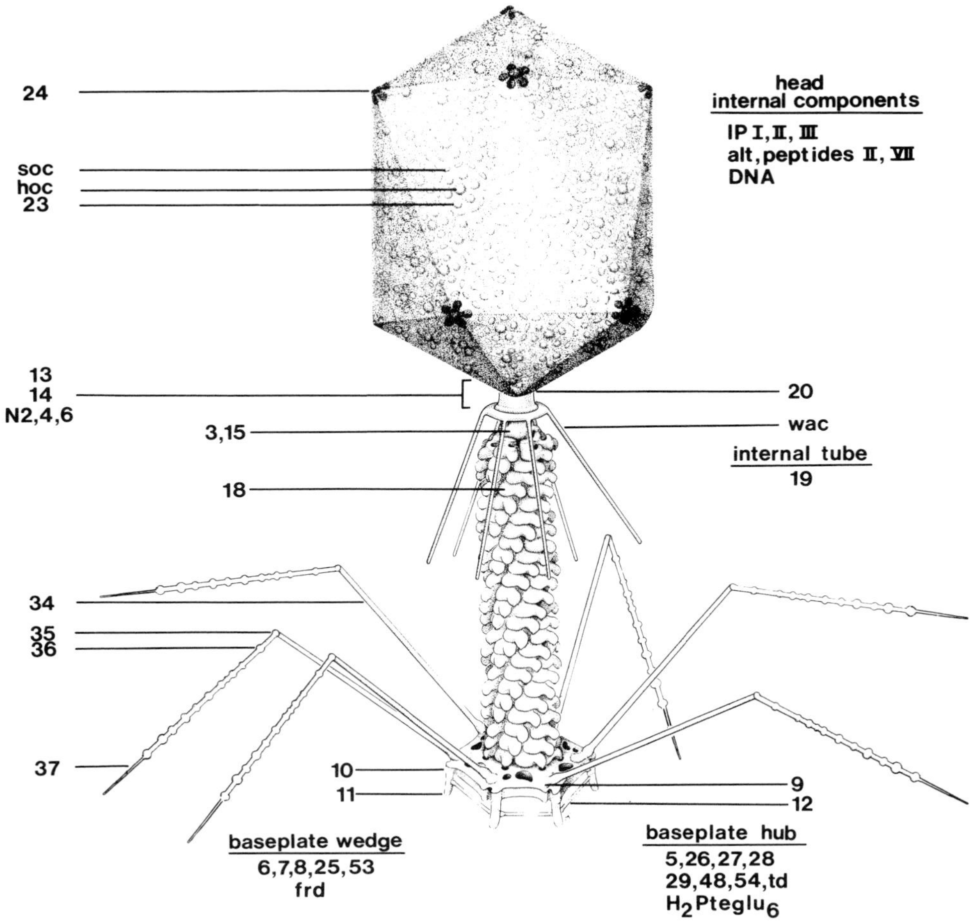

Fig. 1 Structure of the T4 virion, based on electron microscopy at 2–3 nm resolution. The locations of protein components are indicated by gene number except for several unknown connector proteins, called N2, 4 and 6, and the baseplate component dihydropteroylhexaglutamate, called $H_2PteGlu_6$. The portal vertex composed of gp20 is attached to the upper ring of the neck structure inside the head itself. The internal tail tube is inside the sheath and itself contains a structural component in its central channel. The baseplate contains short tail fibers made of gp12; these are shown in a stored, folded conformation. [From Mosig G and Eiserling F (1988) Phage T4 structure and metabolism. In: Calendar R (ed.) *The Bacteriophages*, vol 2. New York: Plenum Press. With permission of Plenum Press.]

subsequent infection, to facilitate DNA release into its host bacterium with utmost efficiency. Thus, the particle is built like a complex machine cocked to inject the packaged DNA. Twenty-four genes are involved in head morphogenesis, more than 25 encode structural proteins of tails and tail fibers and five others are needed for assembly. Heads, tails and tail fibers are assembled in independent pathways and put together after the heads are filled with DNA. Assembly of head subunits requires a host chaperonin, GroEL.

The components of heads, tails and baseplates are similar among the T-even phages, except that two nonessential T4 proteins that decorate the head (Hoc and Soc) are absent from some T2 strains.

Heads

The complex head assembly requires sequentially T4-encoded scaffolding proteins, extensive controlled proteolysis by a specific T4-encoded protease, and cooperative conformational changes which alter the shape and size of the head precursors while they are being assembled. Together, these factors render the process irreversible.

The normal T4 head, containing a full genome length, can be described as an elongated icosahedron. The size of the head is, however, not uniquely determined. Heads of different sizes (mainly isometric icosahedra) and anomalous shapes are made in low proportions in wild-type and in higher proportions in certain mutants. Head lengths and shapes are determined at a very early stage, prior to the formation of the unprocessed prohead. The anomalous sizes and structures are thought to reflect chaotic events during assembly which depends on numerous interactions of proteins that can exist in multiple conformational states and on the relative concentrations of these proteins. Initially, it was assumed that assembly would follow the same geometric pattern, consistent with a unique triangulation number (13). It now appears that different mutant proteins may preferentially assemble in atypical geometric patterns.

Consistent with the head-filling mechanism of DNA packaging, heads of different lengths contain different sizes of DNA; the smaller heads (petites) contain incomplete genomes which represent nearly random permutations of the genetic map. On the other hand, the larger 'giants' contain multiple genomes as one continuous linear DNA concatemer.

Tails

The baseplate of the tail which serves as a perfect injection valve is perhaps the most remarkable structure of this machine. It consists of a central hub, six outer wedges and six tail spikes, each of these structures being assembled from several different subunits.

The multifunctional baseplate contains assembly information for building the tail. During infection, it actively punctures the cell wall from the outside by built-in lysozymes and it can undergo conformational transitions (from a hexagon to a star configuration), thereby opening the tail of the particle for passage of the DNA. Several other baseplate proteins are thought to contain enzymatic activities, but their functions are obscure. Baseplate formation combines aspects of catalyzed assembly, proteolysis and self-assembly.

The long tails are rigid. They consist of a tubular inner core surrounded by a contractile sheath. In contrast to heads, tail sizes are remarkably constant. Assembly as well as contraction of the sheath during infection are accomplished by conformational changes of the individual proteins.

Tail fibers

Superficially, the six tail fibers of different T-even phages also appear similar. Each fiber consists of two half-fibers which are joined at an angle. The inner (proximal) half-fibers join the baseplate. Most significantly, the outer (distal) half-fibers, which recognize phage-specific receptors on the surface of the bacterial cell, are different in the different T-even phages, and their genes are heterologous. In fact, they are the basis for distinguishing the different T-even phages by their host range. The tips of the T4 fiber (gp37) interact with diglucosyl residues of a lipopolysaccharide of *E. coli* B or with the OmpC protein of *E. coli* K-12. In contrast, the tips of T2 tail fibers (gp38) interact with OmpF, and those of phage K3, another member of the group, with OmpA protein. The corresponding genes provide most remarkable examples of viral–host coevolution.

The modular design of particle assembly allows rapid evolutionary adaptation to different hosts. It is also partly responsible for 'phenotypic mixing', the disguise of T2 genomes in particles that adsorb like T4 (and vice versa).

Temporally Controlled Gene Expression During the Life Cycle

T-even phage infection inactivates host translation and transcription by as yet poorly understood mechanisms. Host translation is thought to be turned off by viral particles that lack DNA ('ghosts'). Injection of the phage chromosomes initiates temporally and spatially controlled expression of different sets of phage genes. In terms of timing, gene classes are distinguished as early (immediate early, IE), middle (delayed early, DE) or late. Operationally, IE genes are distinguished from the other classes, in that they can be transcribed when protein synthesis is inhibited after infection. Expression of all other genes requires protein synthesis for various reasons.

The temporal regulation of gene expression is exerted at many levels: transcript initiation, termination–antitermination, stability, translation and combinations thereof. The distinction between the classes is blurred because most T4 genes are under dual or multiple controls which affect transcription initiation and several subsequent processes. Furthermore, T4 codes for many more proteins that modify RNA polymerase than would appear necessary to accomplish transitions between different classes of genes. The corresponding genes are 'nonessential' under laboratory conditions. The fact that they have been maintained in the phage genome suggests that they confer selective advantage by modulating the basic transitions. The temporal regulation of T4 gene expression would be better described by neural network analogies than by simple classification schemes. Because of overlapping and interdigitated transcription, it is often impossible as well as misleading to assign a specific gene to a specific promoter or a class of transcription units.

Transcription

Changes in the timing of transcription initiation from different promoters are accomplished by a cascade of covalent and noncovalent RNA polymerase modifications, by certain DNA-binding proteins and by the process of DNA replication. The first set of promoters ('early promoters') are recognized by host RNA polymerase containing the major sigma factor σ^{70}. These promoters resemble the consensus sequence of *E. coli* promoters but have a higher information content (Fig. 2a). It is not yet established why these promoters are preferred over those of the host's DNA which is still intact at that time. It is plausible to postulate that DNA-binding proteins play roles in transitions of transcription from bacterial to viral genes. At the time of infection, the host DNA is associated with abundant nonspecific (e.g. HU, NS) and less abundant but more specific DNA-binding proteins (e.g. IHF, FI5). In contrast, the infecting phage DNA is at first not covered by these or phage-encoded proteins and may therefore be much more readily accessible to the host's RNA polymerase.

Initiation from the set of promoters that are expressed next ('middle') requires the prior expression of the T4 MotA protein, a DNA-binding protein that recognizes *motA* boxes of middle promoters (Fig. 2b). Initiation from early or middle promoters can occur concomitantly, since initiation from middle promoters does not depend on a different σ factor. Middle promoters may be favored because of one or more of the following modifications of the host RNA polymerase after T4 infection: adenosylribosylation of an arginine residue of the α subunits (by either one of two T4 gene products: Alt and Mod) or association with several small (10–15 kD) T4-coded peptides. None of these modifications, singly or in combination, are, however, essential. The last class of promoters ('late') (Fig. 2c) functions only after DNA replication has begun (or after recombination enzymes have generated a competent template). These promoters require three phage proteins to associate with the host's RNA polymerase: a new σ factor (σ^{gp55}), a helper protein (gp33) and gp45, which is also the sliding clamp that holds the replisome to the DNA. Elegant recent experiments have established that DNA replication serves several functions in activating these promoters: it provides entry sites on the DNA *in cis* with the late promoter for gp45 and perhaps other accessory proteins, and the joining of these proteins requires that they track along the DNA.

Post-initiation effects on gene expression

Genes downstream of any of the different T4 promoters may be poorly expressed because of *rho*-dependent transcription termination or because of RNA processing or degradation. In these cases, expression is thought to require putative phage-encoded antitermination functions to transcribe through the terminators or mRNA stabilizers which would prolong the life of the few complete transcripts. T4 mutations that compensate for certain *rho* (terminator) host mutations have been found in three (nonessential) T4 genes. In addition, a T4-encoded protein (gpAlc), which binds to DNA and RNA polymerase and inhibits transcription from cytosine (instead of HMC)-containing DNA, is now known to affect transcript elongation, i.e. premature termination.

Specific nucleases are important in processing or degradation of primary transcripts. The nonessential T4 *reg-B*-encoded nuclease destroys transcripts mainly in the ribosome-binding site of a few specific T4 genes. Several host nucleases and autocatalytic cleavages are important in processing the precursor RNAs for eight T4-encoded tRNAs and two tRNA-like structures of unknown function. These tRNAs supplement host tRNAs for those codons that are rare in *E. coli* but frequent in T4.

The self-splicing introns in at least three T4 pre-mRNAs (for thymidylate synthetase, nucleotide reductase B and putative unaerobic nucleotide reductase SunY) and appropriate mutations serve to define the active centers of 'ribozymes' in exquisite detail.

Translational controls

The 'sloppiness' of the interdigitated T4 transcription implies, and it is experimentally found, that many early transcripts are extended into late genes situated downstream of early or middle promoters. Nevertheless, few or no protein products of these late genes are made early. It is remarkable that expression of all ten such late genes investigated so far follows the same pattern: in the long early transcripts, a hairpin sequesters the translation initiation region. There is, however, an additional late promoter immediately upstream of the late gene,

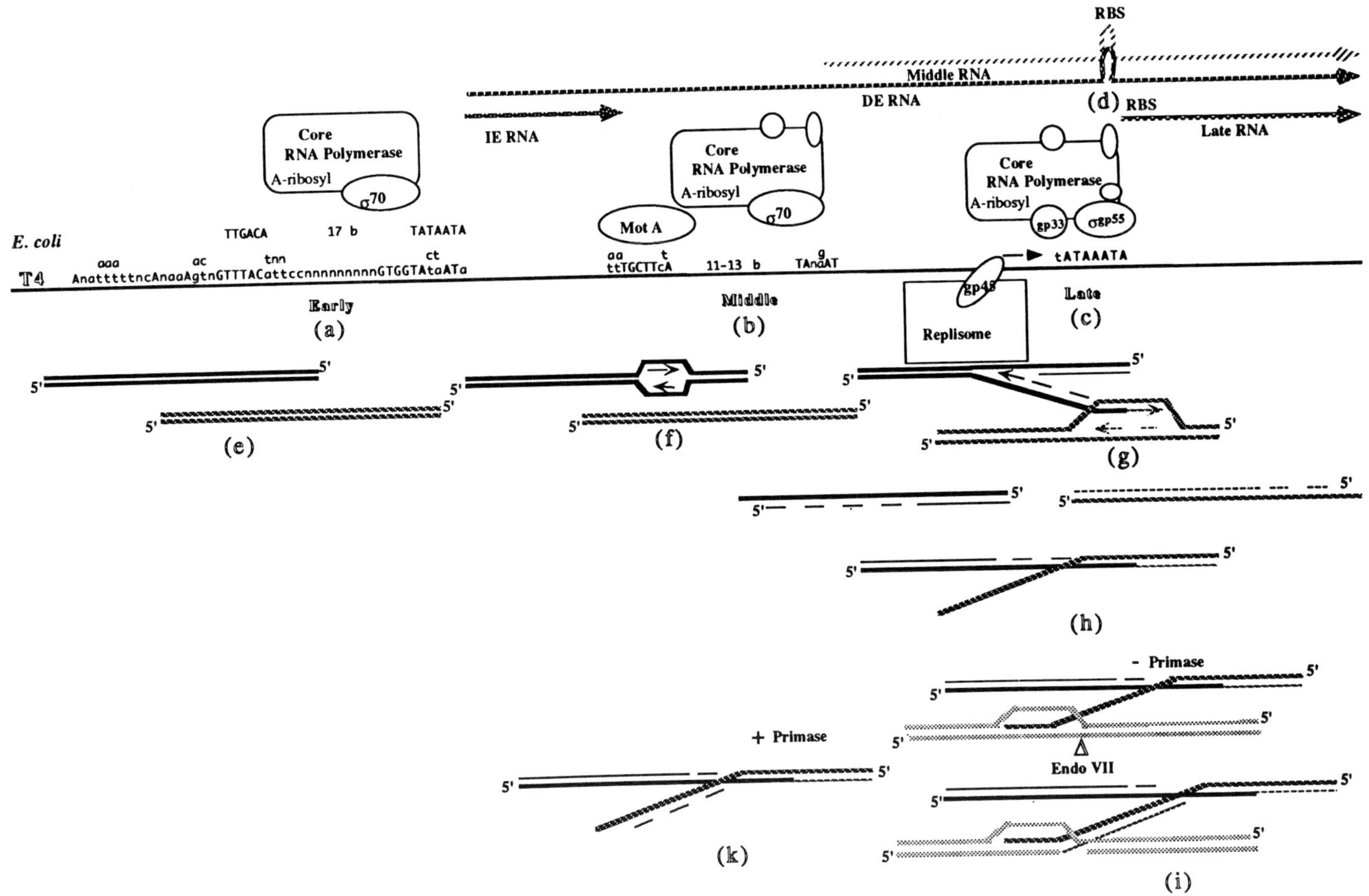

Fig. 2 Integration of gene expression and different replication modes in phage T4. The upper panel, (**a**) through (**d**), shows the early, middle and late promoters, the concomitant modifications of the RNA polymerase and the sequestering of ribosome-binding sites (RBS) described in the text. There are multiple interdigitated transcription units and the spatial order of promoters (early vs middle vs late) is different in these units. The lower panel shows the different stages of DNA replication and recombination. (**e**) illustrates only two of the many possible permuted chromosomes infecting a single bacterium. Once sufficient replication proteins have been synthesized, bidirectional origin initiation (**f**) is initiated. As soon as the first growing point reaches an end (only one is shown), the 3′ end of the initial template for lagging-strand synthesis can invade a homologous region of another chromosome (**g**) or its terminally redundant region (not shown). If DNA replication is initiated at the junction, structures shown in (**h**) are generated. Some of their ends can reiterate the process. The displaced stand in (**h**) can be copied by primase-initiated synthesis (**k**) or, if primase is deficient (**i**), after invasion of another molecule, from an endo VII-generated cut in the invaded molecule. Only 5′ ends are marked. Parental molecules are drawn as bold lines, filled with different patterns. Newly synthesized DNA is drawn as thin lines in the same pattern as the template. Discontinuous synthesis of Okazaki pieces is indicated by dashed lines, continuous synthesis by solid lines. Arrowheads mark the direction of transcription or DNA synthesis.

and transcripts initiated from the late promoters cannot form the hairpin (Fig. 2d). Remarkably, phage evolution has conserved the early transcription of these late genes while incorporating other means of preventing their translation.

In addition three regulatory systems of T4 depend on translational components: the genes for the major single-stranded DNA-binding protein involved in DNA replication, recombination and transcription (*32*) and for DNA polymerase (*43*) are autogenously regulated by translational repression. A more general (nonessential) translational repressor (RegA) affects synthesis of several T4 replication proteins as well as of some host proteins.

Collectively, prereplicative genes encode (1) nucleases that degrade the host DNA, (2) enzymes of the deoxyribonucleotide biosynthesis complex, (3) proteins of the replication and recombination machines, (4) proteins that modify the host membrane, (5) proteins that modify the T4 DNA (hydroxymethylase and glycosylases) to protect it from restriction, (6) several tRNAs, (7) proteins that modify the RNA polymerase, among them at least one σ factor (gp55) and (8) at least one RNase (gpRegB) that selectively destroys certain transcripts and proteins that affect translation (gpRegA and probably others). In addition, some prereplicative transcripts are required to initiate the first round(s) of DNA replication at specific origins. The association of σ^{gp55} with the RNA polymerase, by default, inactivates transcription from early and middle promoters, including origin promoters.

The late genes code for components of the viral capsids and for enzymes that resolve recombinational junctions (endonuclease VII) and cut and package the complex vegetative DNA into capsids (heads). A soluble lysozyme, different from but evolutionarily related to the baseplate lysozyme, ultimately destroys the cell walls of the host bacteria from the inside to release the progeny phage.

DNA Replication and Recombination *In Vivo*

Early studies established that DNA replication and recombination are tightly interwoven in T-even phages. It has become evident that this interconnection has evolved, at least in part, from a response of the replication strategy to the changing transcription pattern. Furthermore, the existence of alternative modes of replication and recombination ensures that both processes work under many different conditions and during different stages of development. The known interrelationships are shown diagrammatically in Fig. 2.

The first round(s) of DNA replication can be initiated from one of several origins. Because of the permutation of the chromosomes (indicated in Fig. 2e), the origins are at different distances from the ends. Initiation from only one origin is shown in Fig. 2f. Different origins share the requirement for transcription and the presence of early or middle promoters. However, each origin that has been closely investigated has a different sequence and overall structure, presumably related to preferred usage under different conditions. Three of the origins (A, F and G) are *motA*-dependent but *oriE* is not. The transition from early and middle to late transcription inhibits initiation of DNA replication from these origins, either by default (because the origin promoters are no longer recognized by RNA polymerase) or because a late protein actively inhibits the use of the origin, or for both reasons. Subsequent DNA replication is initiated from recombinational intermediates (Fig. 2g and i), generating a complex branched network of so-called vegetative DNA. Recently, it became evident that there are at least two modes of recombination-dependent initiation of DNA replication. An early mode, which can start as soon as the first growing point has reached an end (Fig. 2g), is independent of the late T4 recombination endonuclease VII (which cuts Holliday junctions and helps packaging of DNA). A late mode (Fig. 2i) uses endonuclease VII-generated cuts in recombination Y junctions to prime DNA replication. This mode can bypass the requirement for T4 primase or topoisomerase in DNA replication and can be used when these enzymes might become limiting late after infection. Ultimately, reiteration of recombination-dependent initiation of replication generates a highly branched network in which no individual chromosomes can be distinguished.

Of course, not all recombination junctions need to be converted to replication forks. T4 recombination can occur when DNA replication is inhibited. Electron micrographs of such recombining T4 DNA intermediates provided compelling evidence for the importance of 'branch migration' of recombination junctions in homologous recombination. Under these conditions, no viable progeny is produced, because no packageable concatemers are formed and late gene expression is severely inhibited because the DNA is largely incompetent for late transcription and there are fewer copies of the late genes.

DNA Packaging

T4 has no sequence-specific *pac* sites. Instead, the packaging T4 terminase (gp16 and gp17) is thought to recognize and cut the concatemeric DNA at junctions of single- and double-stranded DNA generated by replication–recombination. Like terminases of other phages, the T4 terminase also recognizes the vertex of a preformed head to fill it with linear uninterrupted DNA. Endonuclease VII is required to cut and trim the recombinational branches. DNA ligase, endonuclease V, topoisomerase and a degradation product of the major head subunit are also involved in the complete process.

Because the T4 packaging system does not require *pac* sites, it is being developed as a promising vehicle for packaging large pieces of foreign DNA.

DNA Replication *In Vitro*

Virtuoso biochemical characterization of replication proteins, singly and in combination, supported by results of genetic experimentation, has led to an understanding of the precise functions and interactions of the basic replication proteins and their interactions in the machine (replisome) that moves the replication fork. Seven proteins, corresponding to genes *43*, *44*, *62*, *45*, *32*, *41* and *61*, form an active complex which moves with *in vivo* speed on model templates. Leading- and lagging-strand DNA synthesis are coupled in this model system by the interactions of the primase–helicase (gp41 and gp61) with the DNA polymerase (gp43) and its accessory proteins (gp44, gp62 and gp45). A most remarkable result is that these basic reactions and protein functions are universal among procaryotes and eucaryotes.

Origin initiation has not yet been reconstituted *in vitro*, but recombination-dependent initiation has been achieved in elegant experiments. Consistent with the genetic analyses, this *in vitro* reaction requires several recombination proteins in addition to the basic fork proteins: the T4 RecA analogue gpUvsX, which facilitates invasion of single strands into a duplex, the gpUvsY helper protein and gp59, which helps to position the gene *41* helicase on to the displaced strands. Distinctions between the multiple recombination pathways that operate *in vivo* remain to be demonstrated *in vitro*.

Restriction–Modification

The complex modification and restriction of T4 DNA can best be rationalized as the result of an ongoing evolutionary battle between the phage and its host. T-even phage DNA contains HMC instead of cytosine. This modification protects the T4 phage DNA against T4-encoded restriction endonuclease(s) (II and IV) designed to attack the host DNA. However, HMC renders DNA susceptible to the Mcr restriction systems of the host. To escape this restriction, the HMC residues are glycosylated (to different extents) by glycosyltransferases encoded by the different phages. In T4 DNA, all HMC residues are modified; 70% with α- and 30% with β-glycosyl linkages. In T2 and T6 DNA, there are no α-glycosyltransferases, and 25% of the HMC residues remain unglycosylated. T6 DNA contains many β-diglycosylated HMC residues. In HMC-containing DNA, glycosylation provides protection against host restriction mechanisms (initially called Rgl) which are now known to restrict DNA containing methylcytosine or HMC in specific sequences. Therefore, these restriction systems are now called McrA and McrB. A T4-encoded early antirestriction endonuclease (Arn) protects nonglycosylated T4 DNA against one but not all of these host restriction enzymes.

T2 and T4 DNAs, but not T6 DNA, are further modified by N^6-adenine methylation of 0.5–1.5% of the adenine residues, mostly, but not exclusively, at GATC sequences. The corresponding enzyme is encoded by T2 and T4 but not by T6. The T4 enzyme exhibits patches of similarity, at the protein level only, to the *E. coli* Dam methylase and the *Dpn* II methylase of *Diplococcus pneumonia*. The only proven physiological role of adenine methylation is protection against the phage P1 restriction system, when the HMC glycosylation systems are inactive.

Further Reading

Black LW (1989) DNA packaging in ds DNA bacteriophages. *Annu Rev. Microbiol.* 43: 267.

Geiduschek EP (1991) Regulation of expression of the late genes of bacteriophage T4. *Annu. Rev. Genet.* 25: 437.

Gold L (1988) Posttranscriptional regulatory mechanisms in *Escherichia coli*. *Annu. Rev. Biochem.* 57: 199.

Keppel F, Fayet O and Georgopoulos C (1988) Strategies of bacteriophage DNA replication. In: Calendar R (ed.) *The Bacteriophages*, vol. 2, p. 145. New York: Plenum Press.

Kutter E *et al.* (1993) Genomic map of bacteriophage T4. In: O'Brien SJ (ed.) *Genetic Maps*, 6th edn. Cold Spring Harbor, NY: Cold Spring Harbor Press in press.

Mosig G and Eiserling F (1988) Phage T4 structure and metabolism. In: Calendar R (ed.) *The Bacteriophages*, vol. 2, p. 521. New York: Plenum Press.

T5 BACTERIOPHAGE AND RELATED BACTERIOPHAGES

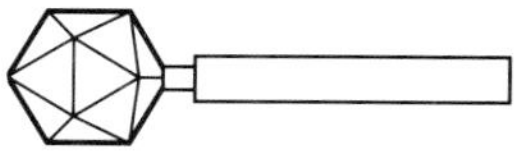

D James McCorquodale
Medical College of Ohio
Toledo, Ohio, USA

Morphology

Bacteriophage T5 and its relatives BF23, PB, BG3 and 29-α have a general morphology that consists of an icosahedral head and a long non-contractile flexible tail. The head of T5 has an average diameter of 90 nm. The tail is attached to one of the head apices via a head–tail linker protein and has three L-shaped tail fibers attached near its distal end. A ring-like structure is formed at this site as a result of the attachment of these tail fibers. The tubular tail undergoes a transition at the tail fiber attachment site to a conical form and tapers into a single straight tail fiber. The tail has a diameter of 12 nm with a length of 190 nm. The cone (12 nm) plus the single straight tail fiber (50 nm) bring the total length of the tail to about 250 nm. The phage protein (Oad) that binds irreversibly to the host receptor is located in this conical region. The tail is hollow and provides a route for transfer of phage DNA from the head of the phage particle into the host cell. Fifteen different polypeptides have been detected in mature T5 phage particles and the number of copies of each polypeptide per particle depends upon the part of the phage structure that it forms. For example, the major head polypeptide is present at 730 copies per particle whereas the straight tail fiber polypeptide is present at only five copies per particle.

Structure of the Genome

The DNA within mature T5 or BF23 particles is linear, double-stranded and about 121 300 bp long. It contains only the four common bases, adenine, guanine, cytosine and thymine, none of which are methylated. Two features of the DNA stand out. First, it has unusually long direct terminal repetitions of about 10 100 bp and second, it is nicked at precise sites, and all nicks are in one strand only. The nicks consist of a missing phosphoester bond between the 5′-phosphate group of one nucleotide residue and the 3′-OH group of its adjacent nucleotide residue. Thus, these nicks can be ligated with DNA ligase. The nicks are introduced by a site-specific nicking enzyme coded by genes *sci*A and *sci*B, which map at the right end of the 'late' region of the genetic map (see below). Nicks are divided between 'major' and 'minor', with the major nicks occurring in virtually all phage DNA molecules and the minor nicks occurring in only a fraction of these molecules. The frequency of the major nicks (four per 121 300 bp) predicts a recognition sequence of seven or eight nucleotides. A prominent sequence on the 5′ side of nicks in T5 DNA is 5′-GCGCGGTG-3′, and the sequence in the unnicked strand around the major nick at 64.8% of the length from the left end of both T5 and BF23 DNA is 5′-CCCGCGCCC-3′. Thus the sequence most efficiently recognized by the nicking enzyme appears to be

```
intact strand → 5′    (C) | C C G C G C C | C 3′
nicked strand → 3′ G  (T) | G G C G C G G | G 5′
                                      ↑
                                     nick
```

The boxed in sequence would generate major nicks whereas minor variations in this sequence would presumably generate minor nicks.

The DNA can be deleted in three regions without affecting viability. The major deletable region is between positions 20.0 and 35.7% of the length of the DNA measured from the left end. The left end is defined as the end that always enters the host cell first during normal infections. Although this deletable region spans about 19 bp, only about 13.3 bp can be deleted from the DNA and still be packaged. Larger deletions yield DNA too small to be packaged. This region contains genes that code for at least one tRNA for each of the 20 amino acids found in proteins. Another deletable region is between positions 4.1 and 7.1% and is therefore in the terminal repetition. It follows that the genes in the right half of the terminal repetition are mostly unessential. The third deletable region is between positions 67.8 and 69.5% and is within a nonessential gene that codes for the L-shaped tail fibers of the mature phage. Because regions of the T5 and BF23 genomes can be deleted, these phages could be used as cloning vehicles, but such use has not yet been pursued.

About 17.3% (or ~ 21 000 bp) of the nucleotides in T5 or BF23 DNA have been sequenced. A stretch from position 58.3 to 68.5% is the largest sequenced region (about 12 370 bp). This region includes gene *D9* (the phage DNA polymerase), *D10* (a putative helicase), *D11*, *D12* and *D13* (nucleoside triphosphate-binding proteins, probably involved with DNA replication, recombination and/or repair), *D14*, *D15*(a 5′-exonuclease) and the portion of the *ltf* gene coding for the C-terminal portion of the L-shaped tail fibers. Three other interesting regions of T5 or BF23 DNA have been sequenced. About 1000 bp have been sequenced in the region where the first step of DNA transfer stops, which is very close to the right end of the left terminal repetition (positions 7.4–8.3%). Present in this sequence are several direct repeats and palindromes, Dna A-binding sites, and simple repeats with a periodicity suggestive of DNA bending. Another sequence of about 1000 bp (positions 2.4–3.3%) defines gene *A2–A3*, an unidentified open reading frame (ORF) on its right, and the beginning of gene *A1* on its left. The final region (positions 22.4–27.8%) that has been sequenced codes for tRNAs, but contains short ORFs interspersed between the tRNA genes.

The genetic maps for both T5 and BF23 have been correlated with their physical genomes such that pre-early genes are located from 0 to 8.3%, and are repeated between 91.7 and 100%. These regions are the terminal repetitions in their DNAs. Early genes are located between positions 8.3 and 67.7%, and late genes between 67.7 and 91.7%.

Restriction maps for T5 DNA have been developed for the restriction endonucleases *Bal*I, *Bam*HI, *Bgl*I, *Bst*EII, *Eco*RI, *Hin*dIII, *Hpa*I, *Kpn*I, *Pst*I, *Sac*I, *Sal*I, *Sma*I, *Sst*I and *Xho*I, and for BF23 DNA for *Bal*I, *Bam*HI, *Eco*RI, *Hpa*I and *Sal*I.

The Infection Process

Attachment of T5 to host cells is facilitated by the L-shaped tail fibers which bind reversibly to O-antigens on the surface of the host cell and allow the phage to move across the surface until a receptor is located. The host cell receptor for T5 is *fhuA*, the receptor for ferrichrome, and for BF23 is *btuB*, the receptor for vitamin B12. Irreversible binding occurs between the phage Oad protein and these host receptors, and is accompanied by a covalent cross-linking of three copies of a minor tail protein, pb4. The single straight tail fiber, pb2, rearranges to form a channel that spans the outer membrane, the periplasm and the inner membrane, and through which the DNA enters the host cell in a unique two-step manner. The formation of this channel depolarizes the membrane, and DNA transfer proceeds through the depolarized membrane.

Transfer of phage DNA is unidirectional such that the left terminal repetition is always transferred first. DNA transfer stops when the left terminal repetition has been transferred and resumes only after the genes in this terminal repetition have been expressed. The remaining 92% of the phage DNA is then quickly transferred to the host cell such that early and late genes enter the cell at essentially the same time.

Pre-early Genes

Since the genes in the terminal repetition are the first phage genes to be expressed in the infected host cell, they have been termed 'pre-early' genes. Pre-early genes that have been identified include *dmp* (coding for a deoxyribonucleoside-5′-monophosphatase), *A1* (coding for a protein required for completion of DNA transfer, for shutdown of expression of pre-early genes, and for degradation of host DNA), and *A2–A3* [coding for a protein that is also required for completion of DNA transfer and that binds to DNA, lipopolysaccharide (LPS) and host RNA polymerase]. Other functions induced by genes in the terminal repetition include the inactivation of (1) host restriction endonucleases, (2) the host cell re-activation system, (3) DNA methylases and (4) the total inhibition of host DNA, RNA and protein synthesis. The product of gene *A2–A3* (gp*A2–A3*) is also crucially involved in the abortive response that ensues when either T5 or BF23 infects host cells that harbor a ColIb plasmid or, in some cases, a ColIa plasmid.

Effect on Host Cell Metabolism

In addition to the inactivation of certain host enzyme

systems, infection by T5 or BF23 results in a rapid and complete degradation of host DNA to individual deoxyribonucleotides. In turn, these deoxyribonucleotides are partially degraded further to ribonucleosides free bases and deoxyribose l-phosphate. The mixture of free bases and deoxyribonucleosides is excreted by the infected cell so that all deoxyribonucleoside triphosphates used in the synthesis of phage DNA are synthesized via *de novo* pathways of nucleotide anabolism. A possible reason for the clearance of all nucleotides derived from host DNA is that the phage-induced nuclease that degrades host DNA may only attack methylated DNA. Phage DNA is not methylated and so is protected from attack, but if any methylated bases derived from host DNA were incorporated into phage DNA, it would be attacked. Thus, the elimination of host-derived bases would be a requirement for a successful infection.

Another requirement for a successful infection is the inactivation of host cell restriction endonucleases. Neither T5 nor BF23 DNA contain *Eco*R1 restriction sites in their terminal repetitions but do have them in the central nonredundant portion of their genomes. Thus, inactivation of host restriction endonucleases by the product of one or more pre-early genes, and therefore prior to entry of the portion of the genome containing susceptible restriction sites, allows the susceptible portion of the phage genome to escape the action of host restriction endonucleases. If, on the other hand, the terminal repetition contains even a single restriction site that is cleaved by a host restriction endonuclease, the infection is unsuccessful.

Early Genes

After pre-early genes are expressed, phage DNA transfer resumes and early but not late genes begin their expression as soon as the rest of the phage DNA enters the host cell. Early genes code mostly for enzymes and proteins required for biosynthesis of deoxyribonucleotides, replication of phage DNA and regulation of transcription. Early gene expression begins about 5 min after infection at 37°C, and continues in the case of some early genes until about 20 min after infection, but in the case of other early genes until lysis. Thus, early genes can be divided into two subclasses on the basis of their period of expression.

Products of early T5 genes that have been identified include DNA polymerase (gene *D9*), deoxynucleoside monophosphokinase (gene *dnk*), dihydrofolate reductase (gene *B3*), 5′-exonuclease (gene *D15*), ribonucleotide reductase (possibly *B1* or *B2*), thioredoxin, thymidylate synthase (gene *thy*), t-RNAs (genes within the major deletable region) and RNA polymerase modifying proteins (genes *C2*, *D5* and *14* and *10* in the case of BF23).

Late Genes

Expression of late genes begins at 10–12 min after infection and continues until lysis. Phage DNA replication begins 8–9 min after infection, shortly after synthesis of early proteins begins, and continues until lysis. Thus, there is a well-regulated temporal sequence for the synthesis of pre-early, early and late proteins which corresponds to the same temporal sequence of the synthesis of mRNAs. Lysis of the infected cell presumably depends upon a late gene which codes for a lysis protein, but this has yet to be demonstrated for the T5 system. Most late genes code for structural proteins of the mature phage particle, but two late genes code for the protein that introduces nicks into the phage DNA.

Regulation of Transcription

The orderly appearance of phage-specified proteins in T5- or BF23-infected cells is regulated at the level of transcription. However, since all classes of T5 or BF23 genes (pre-early, early and late) are efficiently transcribed *in vitro* by unmodified host RNA polymerase (with σ^{70}), and this pattern of transcription is the same whether nicked or ligated phage DNA is used as a template, the temporal expression of phage genes *in vivo* must be regulated by mechanisms that prevent the simultaneous expression of all classes of genes if the phage DNA entered the cell in one step. This regulation appears to be accomplished in part by sequential modifications of the host RNA polymerase. Expression of pre-early genes on the other hand is temporally separated from early and late gene expression because of the two-step mechanism of phage DNA transfer, whereby pre-early genes enter the host cell first and must be expressed before early and late genes enter the cell. Pre-early genes are therefore transcribed *in vivo* by the pre-existing unmodified host RNA polymerase. The first modification to host RNA polymerase is the binding of the pre-early proteins coded by gene *A2–A3* (gpA2–A3) and gene *A1* (gpA1). The modification by gpA1 causes shutdown of pre-early gene expression, whereas the modification by gpA2–A3 prevents the premature expression of late genes when the phage DNA carrying early and late genes enters the host cell after pre-early genes are expressed. The RNA polymerase is further modified by the products of two early genes, gpC2 and a 15 kD protein in the case of T5 and gp14 and gp10 in the case of BF23. Gp10 displaces gpA2–A3, which then allows this further modified form of RNA polymerase to transcribe late genes.

DNA Replication

T5 DNA contains multiple origins of initiation, which suggests that the DNA is replicated linearly. However, T5 DNA can be found in a circular form in infected cells, and the length of such circles is equal to a genome length minus one terminal repetition. Formation of these circles could therefore arise from a recombinational event between the terminal repetitions of incoming parental DNA. The occurrence of circles suggests a rolling circle model of DNA replication. Sedimentation studies of replicating T5 DNA from infected cells shows a fast-sedimenting fraction, which could be linear concatemers or the rolling circle intermediate, and a slow-sedimenting fraction which corresponds to genome-length T5 DNA. These findings indicate that packaging of phage DNA into immature heads is not coupled to excision of genome-length DNA molecules from larger precursors.

Morphogenesis

Morphogenesis follows two separate pathways, head formation and tail formation. The immature head is filled with a precise length of phage DNA which is cut from a linear concatemer of phage genomes. Two possibilities for this precise cutting are (1) the linear concatemers consist of genome-length segments that have a palindromic recognition sequence at each junction and both strands are cut bluntly at a site within this palindrome, or (2) the linear concatemers consist of genomes minus the length of one terminal repetition. A staggered single-strand cut at each end of an internal terminal repetition followed by fill-in by a DNA polymerase, starting from the 3′-OH at the single-strand cuts, would thereby generate two double-stranded terminal repetitions one of which would be at the end of one genome-length DNA and the other at the end of the adjacent genome-length DNA. Thus complete genomes with double-stranded terminal repetition at each end can be formed.

Both head and tail morphogenesis involve cleavage of polypeptides that form these structures. Tails can be connected to heads *in vitro*, but packaging of T5 or BF23 DNA *in vitro* has not yet been accomplished.

Abortive Infection in ColIb Hosts

If T5 or BF23 infects a host cell harboring the colicinogenic plasmid, ColIb (or some ColIa plasmids), the infection is abortive. In such infections, the phage adsorbs to the host cell normally, and the phage DNA is transferred into the host cell in the usual two-step manner without being degraded. Pre-early genes are expressed and shutdown normally, but early genes barely begin expression before the host cell prematurely lyses, resulting in death of both the host cell and the infecting phage. Gene products from the phage, the host cell and the plasmid are necessary for this abortive response. The phage gene is pre-early gene *A2–A3*, which binds to both DNA and host RNA polymerase. The host cell genes involved are *cmrA* and *cmrB*, which map suspiciously close to *trkA* and *trkB* which code for potassium-transport proteins located in the cell membrane. The plasmid gene is *abi* (*ab*ortive *i*nfection), which codes for a polypeptide of 89 amino acids which is strongly hydrophobic and may therefore interact with cell membranes. How these gene products from three sources interact to cause the abortive response has yet to be elucidated.

Transfection

DNA isolated from mature phage particles can transfect *E. coli* protoplasts. Successful transfection, however, is a two-hit process in wild-type protoplasts. This process is understandable in terms of hostile host functions that are normally inactivated by products from pre-early genes before the bulk of the DNA enters the host cell in the usual two-step DNA-transfer mechanism. Presumably, the terminal repetition of the first phage DNA molecule that enters a protoplast successfully carries out the functions of the pre-early genes, but the remainder of this DNA molecule is degraded by one or more of the hostile host functions. However, such a transfected protoplast can now accept a second DNA molecule, and the infective cycle can be completed. If the protoplast is *recB*$^{-}$, however, the transfection is a one-hit process. The RecB protein must therefore be lethal to incoming T5 DNA.

Cloning Genes of T5 or BF23

Most restriction fragments from T5 or BF23 DNA are not directly clonable because they either code for lethal products or contain such strong promoters that the cell harboring them cannot survive. Fragments that have been cloned from T5 include a fragment from position 2.1 to 3.4% which contains gene *A2*, *Hin*dIII fragment 15 (24–25%) which codes for some tRNAs of T5, *Bal*I fragments 11 and 12 (58.2–61.3%) which together code for T5 DNA polymerase, *Eco*R1 fragment 6 (60.5–63.5%), *Pst*I fragments 10, 9 and 7 (63.0–76.7%), *Bam*HI fragment 4 (67.2–71.3%) and *Hin*dIII fragments 16 and 7 (69.3–75.7%) and 12 (83.7–87.0%).

Fragments that have been cloned from BF23 include *Hpa*I fragment 16 (63.3–65.3%) and a *Bam*HI–*Hpa*I fragment spanning positions 89.9–92.3%. Thus, many of the late genes have been cloned, but only a few of the early genes.

The T5 DNA polymerase gene (*D9*) has been successfully placed in an expression vector to overproduce this highly processive polymerase. Similarly the T5 5′-exonuclease (*D15*) has been expressed and purified from an overexpression system. This 5′-exonuclease can be viewed as a complement to *E. coli* exonuclease III, which is a 3′ – 5′ exonuclease.

Future Perspectives

Future contributions from the T5 and BF23 systems include the identification and use of several of their gene products. Identification and purification of the pre-early gene products that inactivate several host functions should prove very informative and useful. Similarly, the availability of the 'nicking' enzyme would add to our battery of enzymes for manipulation of DNA. The nucleotide sequence of all T5 and BF23 promoters and their strength of binding to unmodified and modified forms of host RNA polymerase should sharpen our understanding of promoter function. The elucidation of the mechanisms by which the two-step transfer of phage DNA to host cells and the ColIb-directed abortive response is accomplished will probably reveal heretofore unknown cellular interactions. Finally, the complete nucleotide sequence of the T5 and BF23 genomes would greatly help our understanding of this system.

See also: Bacteriophage taxonomy and classification; Bacteriophages as cloning vehicles; Host-controlled modification and restriction; Replication of viruses.

Further Reading

Bujard H *et al.* (1982) The interaction of *E. coli* RNA polymerase with promoters of high signal strength. In: Rodriguez RL and Chamberlin MJ (eds) *Promoters Structure and Function*, p. 121. New York: Praeger Publishers.

Duckworth DH, Glenn J and McCorquodale DJ (1981) Inhibition of bacteriophage replication by extrachromosomal genetic elements. *Microbiol. Rev.* 45: 52.

McCorquodale DJ and Warner HR (1988) Bacteriophage T5 and related phages. In: Calendar R (ed.) *The Bacteriophages*, p. 439. New York: Plenum Press.

T7 BACTERIOPHAGE

Ian J Molineux
University of Texas
Austin, Texas, USA

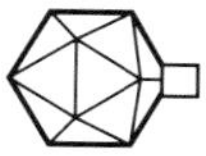

General Properties, Ecology and Evolution

Bacteriophage T7 is the prototype of a group of phages having a $T = 7$ icosahedral head, approximately 60.2 nm in diameter, and a stubby noncontractile tail (about 20 nm in length and 10 nm wide) plus six thin tail fibers. A distinguishing characteristic is the synthesis, early after infection, of an approximately 100 kD single-polypeptide RNA polymerase that is resistant to the antibiotic rifampicin and is specific for promoters present only in phage DNA. T7-like phages that infect one or more of a variety of Gram-negative bacteria have been described, but none are yet known that infect Gram-positive hosts.

Most studies on this group of phages have been performed on the coliphage T7 and, to a lesser extent, T3; properties of other phages are usually described relative to those of T7. T7 does not form plaques on most *Escherichia coli* strains newly isolated from nature because it cannot adsorb specifically to smooth or capsulated bacteria. The receptor for T7 is the R-core portion of the lipopolysaccharide (LPS) of the outer membrane (at least in *E. coli* B), and in smooth strains this is inaccessible to the phage tail fibers specified by gene *17*, the latter specifying the adsorption host range. It is not clear whether LPS is the sole receptor for the T7 family of phages; T3 probably also utilizes an outer membrane protein as a receptor in K-12 strains. T7-like phages are known that grow on smooth bacteria; these possess virion-associated hydrolases that degrade some of the polysaccharide on the bacterial cell surface to allow specific adsorption to the receptor.

Bacteriophages T7 and T3 were isolated in 1945 as phages that grew on *E. coli* B. Similar coliphages were subsequently isolated from different parts of the world and about 60 representatives are now known; these can be placed into three groups based on the promoter specificity of the phage-encoded RNA polymerase. The largest group contains T7 itself; three phages, BA14, BA127 and BA156, comprise a second group, and T3 is the sole member of the third. Recombination between phages in a group is very efficient; however, recombination between groups occurs only rarely, but is of significance in the evolution of various individual phages. Electron microscopic analyses of heteroduplexes of T3 and T7 DNA have shown that they exhibit varying degrees of homology, containing regions of extremely high sequence conservation (>90%) and others with no apparent similarity. Furthermore, one promoter in T3 DNA has the sequence of a T7 promoter; this promoter is not utilized during T3 infection, although in T7 it is the major late promoter $\phi 17$. These observations are highly suggestive of multiple recombination events involving various T7-like phages.

The T7-like phages are virulent and they also exhibit superinfection exclusion, raising the question of how and where they recombine in nature. Recombination between organisms requires coinfection, and the concentrations of these phages in the environment are not particularly high. Perhaps pseudolysogeny, known to occur in starved *E. coli* infected by T3 (but not T7), is not infrequent in nature, or perhaps reversible adsorption to inert particulate matter increases the local concentrations of both phages and susceptible hosts.

Genetic Map

The genetic map of T7 is based on the nucleotide sequence of 39 937 bp. Numbers define genes, and these are ordered sequentially from the genetic left end of the DNA; this is also the end that first enters the cell (Fig. 1). Three classes of genes have been identified: Class I, or early, genes are expressed until about 8 min after infection (at 30°C) and these prepare the cell for the exclusive development of the phage. Class II genes, responsible for DNA metabolism, are expressed from about 6 to 15 min after infection, and Class III genes, containing maturation and morphogenetic functions, are expressed from about 8 min until lysis (about 25 min at 30°C). Fifty-six known or potential T7 genes have been described – about 20 are known to be essential for phage growth on usual laboratory strains; mutant hosts have allowed the functions of several other genes to be elucidated. Coding sequences occupy 91.9% of the T7 genome, and most of the remaining DNA contains recognizable genetic signals. There is little overlap of coding sequences and the genetic organization is best described as close-packed; only two genes predominantly overlap the sequences of a larger gene but are read in different reading frames, and, by means of an inframe internal initiation, gene *4* codes for two essential proteins, gp4A and gp4B. In addition, ribosomal frameshifting to the +1 frame during translation of gene *0.6*, and to the −1 frame during translation of gene *5.5* and gene *10*, yields gp0.6B, a *5.5–5.7* fusion protein, and gp10B respectively. No biological role for these frameshifted products is known, but it is significant not only that gp10B is assembled into phage particles but also that a comparable T3 gp10B is made, even though the sequences thought to cause frameshifting have diverged in the two phages.

Genetic signals include promoters for the host and phage RNA polymerases, RNase III recognition sequences, the primary origin of DNA replication,

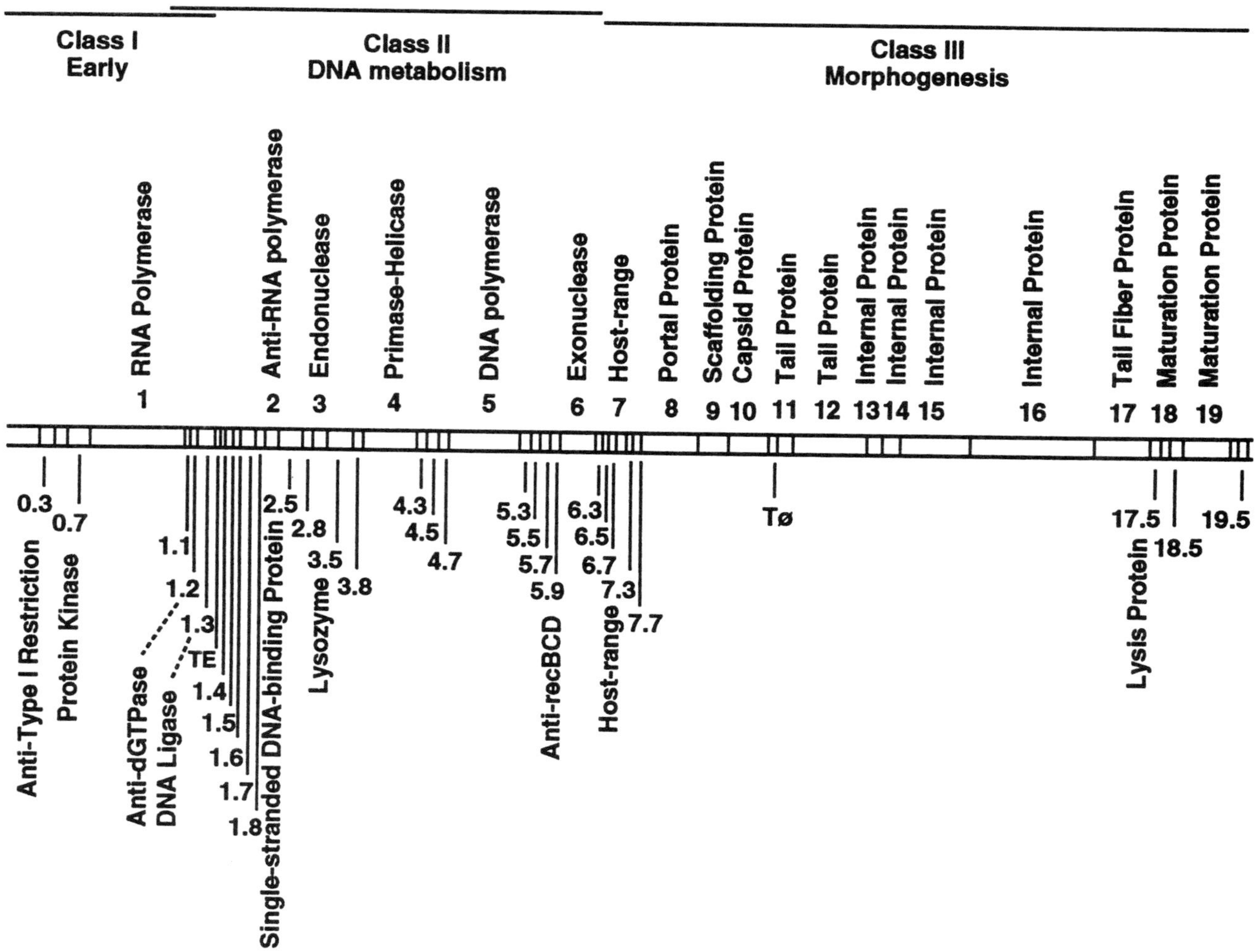

Fig. 1 Genetic map of bacteriophage T7.

a terminal repetition of 160 bp used in forming concatemers, and arrays of 12 imperfect copies of a heptamer sequence adjacent to each terminal repeat. These arrays are thought to be involved in forming the mature ends of T7 DNA during packaging. Although there is no obvious bias against di- or trinucleotide sequences in the T7 genome, the sequence GATC (a major regulatory sequence in *E. coli*) is distinctly underrepresented. This sequence, statistically expected to occur 156 times in a 40 kb phage genome, occurs only six times in T7, ten times in T3 and not at all in BA14. T7 inactivates Type I restriction enzymes but it remains susceptible to cells harboring Type II and III enzymes. However, the 6-base recognition sequences of many common restriction enzymes are absent in T7 DNA, suggesting that the phage may have utilized hosts containing enzymes with these specificities during its evolution.

The Phage Particle

The phage particle is primarily composed of gp10A, plus some of the frameshifted product gp10B, which together constitute the phage head. At one vertex, 12 molecules of gp8 form a portal through which the DNA molecule passes during both packaging and ejection. This portal also serves as the connector for the phage tail consisting of gp11 and gp12 plus six fibers, each comprised of gp17 trimers, which are attached to the tail just below its junction with the capsid. The particle also contains a number of internal proteins that form a cylindrical core structure, attached to the inner surface of the capsid and lying coaxial with the tail. These core proteins are essential for infectivity, although their precise functions are not known.

The T7 Infection Cycle

Class I genes

Upon infection, the left end of T7 DNA enters the cell first. The host RNA polymerase recognizes three strong promoters located within the first 750 bp; transcription from these promoters leads to the expression of the nine early genes that occupy 19% of the genome. Transcription terminates at a site that is not completely efficient *in vivo* or *in vitro*. It is not clear where in the cell this transcription occurs; for the first 6 or 7 min of infection while early genes are being expressed, the phage DNA cannot be cleaved by resident restriction enzymes. Some data suggest that only a small portion of the genome is ejected from the phage head directly into the cell, and that the majority is brought in by transcription; however, it is also possible that the genome is sequestered in some specialized cellular compartment that is not accessible to all host enzymes. There is no evidence for preexisting phage proteins entering the cell along with the DNA and thus this hypothetical compartment might also serve to protect the ends of the genome from host nucleases. By whatever route the DNA may enter the cell, it is clear that the classic syringe mechanism of phage infection does not apply to T7.

The biochemical functions of five early proteins are known; gp0.3 inactivates the host Type I restriction enzyme and gp0.7 is a serine-threonine protein kinase that phosphorylates several host proteins, and also itself. Gp0.7 is also responsible for shutdown of host-catalyzed transcription, although this function is separable from that of kinase action. Gene *1* specifies RNA polymerase, gp1.2 inhibits deoxyguanosine triphosphate triphosphohydrolase and gp1.3 is a DNA ligase. Of these gene products, only gp1 is essential for growth in normal laboratory strains and the remainder are required only in mutant hosts or in cells growing in nutrient-poor medium.

In the first few minutes of infection, T7 early gene products thus inactivate some potentially deleterious host enzymes and subsume the bacterial transcription machinery for the exclusive benefit of phage development. Since host-catalyzed transcription is inactivated by the action of gp0.7, early phage genes are also shutdown. Furthermore, even though phage mRNAs appear to be stable in the infected cell, synthesis of most early proteins (and residual host proteins) also abruptly ceases about 8 min after infection.

Class II genes

The three rightmost early genes, and the remaining 81% of the phage DNA are sequentially transcribed by T7 RNA polymerase in a process that not only provides mRNAs but may also assist in the entry of phage DNA into the cell. Sequential transcription of the phage genes achieves temporal regulation and ensures expression of the 24 Class II genes; these are preceded by weaker promoters than those in the Class III region of the genome. Class II gene products primarily function in T7 DNA metabolism, although the gp3 endonuclease and the gp6 exonuclease also degrade host DNA to mononucleotides. These are reutilized in T7 DNA synthesis, and more than 80% of the nucleotides in progeny phage are derived from the host chromosome. *In vitro*, these two nucleases can degrade T7 DNA; how their activities are controlled *in vivo* is not known.

DNA Replication

Replication *in vivo* requires the T7 proteins RNA polymerase, DNA polymerase, helicase–primase, single-stranded DNA-binding protein (SSB), endonuclease and exonuclease. Two other phage gene products are required indirectly in T7 DNA synthesis: gp2 inactivates host RNA polymerase, which, although unable to transcribe DNA by the time DNA synthesis begins, can interfere with the maturation and packaging of T7 DNA. Secondly, gene *3.5* (lysozyme) mutants have a reduced replication capacity; replication *in vivo* may require the reduced activity of RNA polymerase that results from inhibition by lysozyme (see under Transcription); alternatively, replication may require a lysozyme–RNA polymerase complex. Other than thioredoxin, which complexes with gp5 to form DNA polymerase, no host proteins are known to be required for T7 DNA replication.

Replication is initiated at an AT-rich region located 15% from the left end, although deletion mutants lacking this region grow well in most laboratory hosts (secondary origins are used by these mutant phages). Since the origin is not centrally located on the genome, bidirectional replication enlarges the replication 'bubble' until the molecule becomes Y-shaped, and the arms of the 'Y' then elongate to give two linear molecules. Both 'bubbles' and 'Y's, together with more complex structures (the primary origin can be used more than once), have been observed by electron microscopy

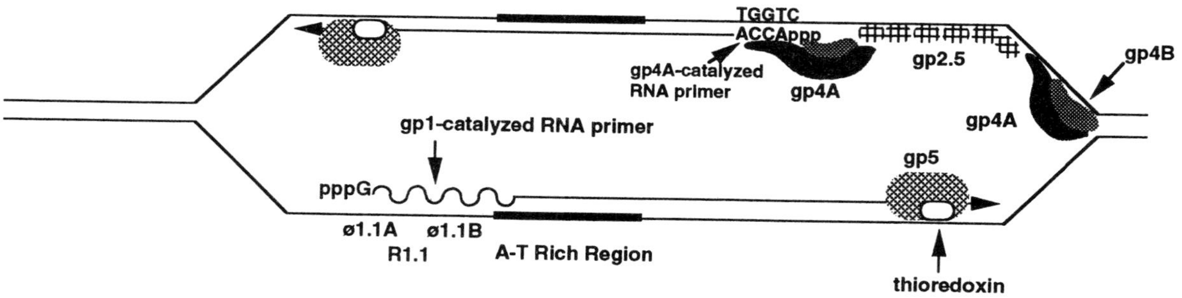

Fig. 2 Schematic diagram of the initial events at the primary origin of DNA replication.

of replicating DNA. Replicated linear molecules necessarily contain unreplicated 3′ ends but electron microscopy studies have not revealed rolling-circle molecules such as those found in phage λ-infected cells. T7 circumvents the problem of completely replicating linear DNA by forming concatemers via the terminal repeats.

Plasmid-based assays have also revealed secondary origins of replication active *in vivo*; these are thought to function in later stages of replication. Plasmids containing sequences including the promoters *ϕ6.5* and *ϕ13* are replicated following T7 infection as efficiently as plasmids containing the primary origin, and those containing *ϕOR* are replicated even more efficiently. Plasmids containing other promoters are, at best, poorly replicated after infection, but the sequences that are important for a promoter to serve as part of a replication origin are not yet understood.

As replication proceeds, fast-sedimenting DNA molecules appear that consist of a mixture of linear concatemers, complex branched molecules, and a small percentage of circular molecules. The fast-sedimenting DNA complex is ultimately converted by gp3 into linear concatemeric molecules, which then can be packaged into phage particles.

In vitro, bidirectional replication from the primary origin has been achieved using purified T7 RNA polymerase, DNA polymerase, SSB, and primase–helicase (Fig. 2). RNA polymerase synthesizes primers of 10–60 nucleotides from both promoters that lie immediately upstream of the origin; primers are then processively elongated by DNA polymerase. The gp4B helicase binds the displaced strand and, in a reaction requiring the energy of ribo- or deoxy-NTP (preferably dTTP) hydrolysis, translocates 5′ to 3′ to unwind duplex DNA ahead of the polymerase. The gp4A protein serves as both a helicase and a primase; the N-terminal 63 residues not present in gp4B likely contain a 'zinc finger' that interacts with the sequences 3′-CTGG(G/T)-5′ or 3′-CTGTG-5′ in single-stranded DNA and synthesizes tetraribonucleotide primers, 5′-ACC(A/C) or 5′-ACAC. These primers serve to initiate DNA synthesis on the lagging strand. The spectrum of primers synthesized by gp4A is affected by the presence of gp4B the two may exist as a mixed oligomer *in vivo*. Although *E. coli* SSB stimulates T7 DNA polymerase in simple *in vitro* reactions, it cannot substitute for T7 SSB for bidirectional synthesis from the primary replication origin of the phage DNA, and does not substitute for T7 SSB *in vivo*. *In vitro* T7 SSB interacts with both DNA polymerase and primase to stimulate lagging strand synthesis and to promote bidirectional replication from the primary origin.

Class III genes

Expression of Class III genes appears to be completely independent of DNA replication; the times of appearance, rates of synthesis, and final accumulation of Class III proteins are unaffected by preventing phage replication. Class III genes are also expressed temporally by sequential transcription of the genome although there are more mRNAs containing gene *10* in the infected cell than other Class III genes. This apparent dichotomy is due to the synthesis of many overlapping polycistronic RNAs, many of which contain gene *10* (Fig. 3). Essentially all of the Class III genes are involved in morphogenesis of the phage particle or in its release from the cell.

Capsid Assembly and DNA Packaging

Detailed information on morphogenesis has been obtained with both T7 and T3. It is assumed here that the mechanism of particle assembly is common to both phages.

Packaging of T7 DNA starts with the assembly of a DNA-free procapsid. The major capsid protein

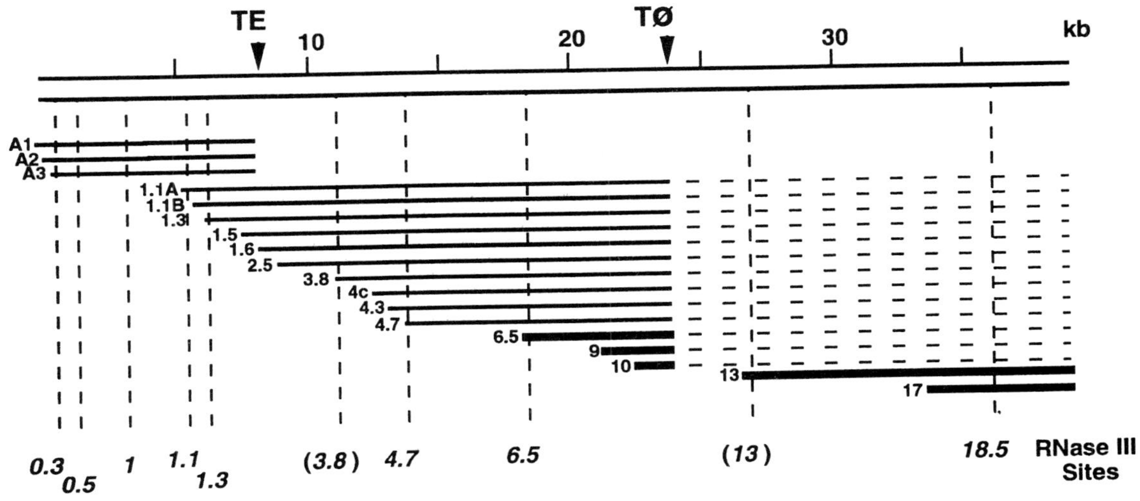

Fig. 3 Schematic of T7 RNAs. Promoters are indicated at the 5′ ends of transcripts. Horizontal dashed lines represent the major readthrough RNAs. Vertical dashed lines indicate the positions of cleavage by RNase III; numbers in parentheses are sites where cleavage is inefficient.

gp10 assembles around a scaffold of gp9 in a reaction that has been accomplished *in vitro* using purified proteins. One vertex of the icosahedral procapsid is modified by the addition of gp8 to form the portal. In addition, the procapsid contains the internal protein core and also the maturation protein gp19. The latter is likely on the outside of the procapsid since it recognizes the gp8 portal protein and, perhaps by association with gp18, interacts with phage DNA. Both gp18 and gp19 are required for packaging although neither are found in the mature phage particle.

After association of the procapsid with concatemeric DNA, packaging proceeds from a genomic right end leftwards in a process, the details of which are not understood. A complex series of reactions is necessary for duplication of the terminal repeat sequence of T7 DNA, which in concatemers is present in only one copy between genomes. The DNA is nicked by an unknown nuclease at a palindromic sequence located to the left of the right terminal repeat, thereby creating a primer for DNA polymerase that synthesizes back on itself forming a hairpin structure (Fig. 4). Synthesis extends the hairpin through the terminal repeat and into the genome being packaged providing duplex DNA that can be converted into the mature left end. On the remaining concatemeric DNA, primase-initiated synthesis on the strand displaced during elongation of the hairpin proceeds through the terminal repeat, thereby providing sequences that can be converted into the mature right end of the next genome to be packaged. The nuclease that recognizes the termini of mature T7 DNA in a concatemer is not yet known, although both gp18 and gp19 are required to create both the right end and, after packaging, the mature left end. Gp19 is also known to possess a nonspecific endonucleolytic activity that is suppressed by gp18. T7 RNA polymerase is also required for DNA packaging; although its biochemical role is not yet established, it is known to interact with gene *19*.

During packaging of DNA into the procapsid, the latter undergoes a conformational change characterized by an increase in size and the conversion from a rounded to an icosahedral morphology. The gp9 scaffolding protein likely exits the procapsid as these changes occur. Packaging of a single genome from concatemeric DNA is estimated to take about 90 s, a rate that has been achieved *in vitro*.

The final stage of phage development is lysis of the host cell, a process poorly understood despite the fact that lysis of T7-infected cells is more abrupt and complete than lysis of other phage-infected cells. Genes *3.5* and *17.5* are known to be required, perhaps together with another component (gp18.5 and/or nucleic acid?). Gp17.5 may disrupt the membrane to form nonrefractile cell ghosts; the lysozyme activity of gp3.5 can then release T7 particles from these ghosts, or from cell debris. The potential role of DNA in lysis is unclear, but it is interesting that mutants defective in replication are as lysis-defective as gene *3.5* mutants (even though replication defects do not affect late gene expression). Perhaps a DNA structure is the signal to initiate lysis.

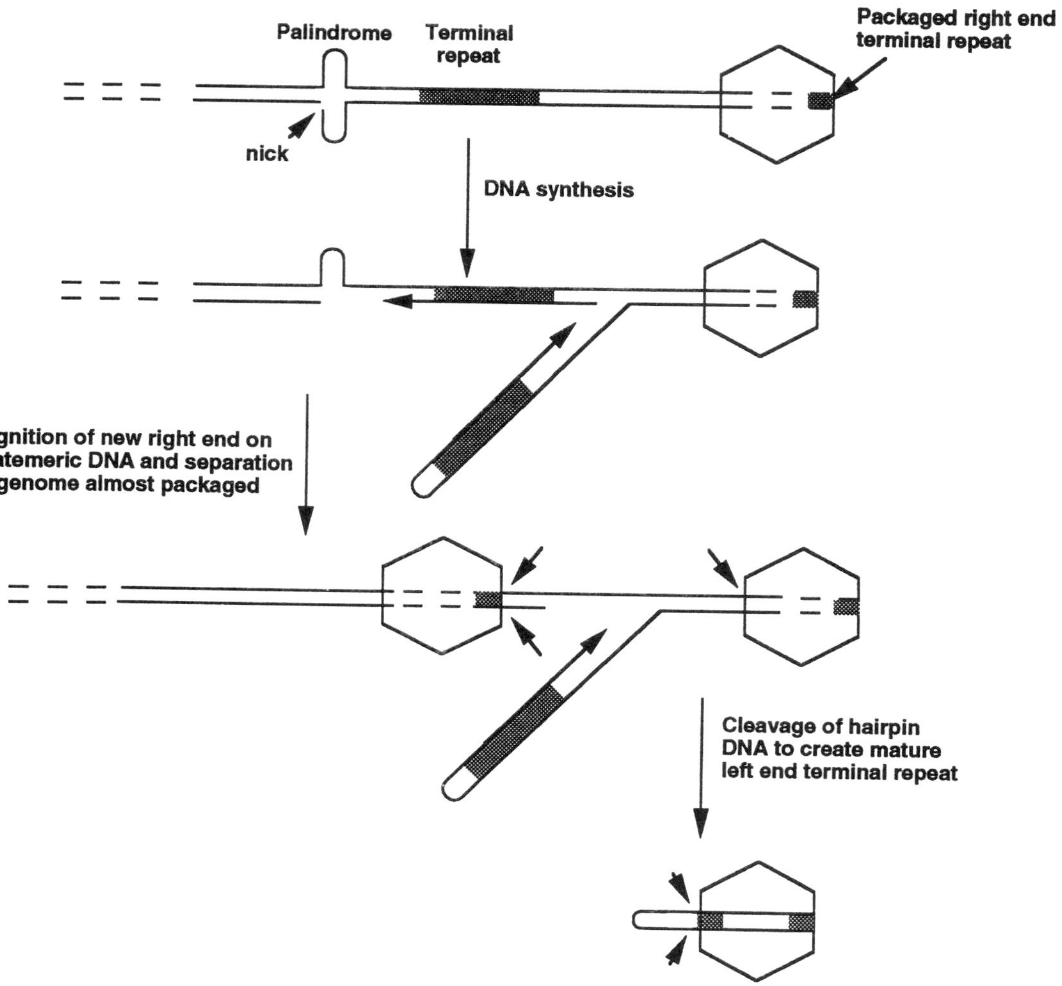

Fig. 4 Schematic pathway for duplication of the terminal repeat during packaging from concatemeric DNA.

Transcription of T7 DNA

All transcription of T7 DNA *in vivo* goes from left to right on the genetic map (Fig. 3). The three promoters for *E. coli* RNA polymerase are very efficiently utilized by the enzyme, the A1 promoter being among the strongest promoters known. Additional minor promoters have been identified by *in vitro* transcription studies or predicted from the nucleotide sequence; however, none of these are known to have any biological significance. In the absence of the host enzyme RNase III, transcription from the major promoters usually terminates at the early terminator TE producing three early RNAs, all containing identical coding information. Transcription that reads through TE terminates inefficiently either near the 3′ end of gene *3.5* or distal to gene *10* at the T7 RNA polymerase terminator.

Processing of primary transcripts by the host enzyme RNase III is a major feature of T7-infected cells even though it is not essential for phage development. The stability of T7 RNAs during infection is likely to be due, at least in part, to the potential formation of base-paired structures at the 3′ ends of the processed mRNAs. The early RNAs are processed into an initiator RNA, which contains no coding information, and five mRNAs, only one of which is monocistronic.

Promoters for T7 RNA polymerase consist of a highly conserved 23 bp segment that runs from −17 to +6, relative to the transcription start site. There are 17 such promoters in the T7 genome, ten initially expressing Class II genes, five expressing Class III genes, ϕOR, thought to be primarily involved in replication, and ϕOL, the function of which is unclear. Transcription from Class II promoters, and that from the first three Class III promoters, terminates at Tϕ and results in a nested set of RNAs that differ only in their 5′ ends. The significance of synthesizing such a complex array of transcripts, especially of Class II genes, is not known; however, it does provide a means of producing large quantities of gene *9* and especially gene *10* RNAs. mRNA dosage is one reason for the relative abundance of gp9 and gp10, the latter being the protein made at highest rates in the infected cell.

Termination of T7 polymerase-catalyzed transcription at Tϕ is about 90% efficient, and genes *11* and *12* are expressed only from readthrough RNAs. These RNAs, together with transcripts from the *ϕ13* and *ϕ17* promoters, terminate at the end of the genome, although it is not known whether RNAs transcribed from concatemeric DNA (resulting from replication) also terminate at this site.

The sequences of the five Class III promoters and the replication promoter *ϕOR* are identical; these promoters are stronger than the Class II promoters, the sequences of which differ from the consensus at two to seven positions. This difference in promoter strength allows the selective shutdown of Class II transcription; gp3.5 complexes with T7 RNA polymerase and inhibits its transcriptional activity. As polymerase becomes limiting, the stronger Class III promoters are preferentially utilized. Since gp3.5 is itself a Class II gene product, its synthesis is self-regulating, and premature high-level expression of gene *3.5* is inhibitory for phage growth. The specificity of T7 RNA polymerase for its cognate promoter and the regulation of its activity by gp3.5 have made it extremely useful as an expression system in gene cloning experiments.

Promoter Specificities of Other Phage RNA Polymerases

The RNA polymerases encoded by other coliphages are similar to that of T7; a single amino acid change allows T7 RNA polymerase to specifically recognize T3 promoters. Similarly, a single amino acid change allows T3 RNA polymerase to specifically recognize T7 promoters. RNA polymerases encoded by the *Salmonella* phage SP6 and the *Klebsiella* phage K11 have diverged more extensively from the T7 enzyme, although their respective promoters retain substantial homology. However, there is little or no recognition of any of these enzymes for noncognate promoters. Of possible evolutionary importance is the significant levels of homology between T7-like phage RNA polymerases and the *Saccharomyces cerevisiae* mitochondrial RNA polymerase.

Host Functions in T7 Development

The roles of *E. coli* RNA polymerase and thioredoxin in T7 development have already been described. Inhibitors of DNA gyrase affect replication and transcription of T7 DNA, but an actual requirement for the enzyme is less clear since T7 grows normally in some *gyr* mutants under conditions nonpermissive for growth of the host. DNA gyrase may simply bind to T7 DNA, and, when inhibited, poison further phage development.

A number of natural *E. coli* strains, and other enterobacterial hosts, are nonpermissive for T7 and, in some cases, for other members of the T7 group of phages. The host genes involved have not been fully identified, but may prove to be mutant (relative to *E. coli* B) genes normally required for T7 development. Several prophages or resident plasmids are also known to have the potential to inhibit growth of T7 by a process(es) distinct from adsorption or DNA restriction; however, the incoming phage often contains a gene that prevents inhibition. The λ *rex* genes exclude certain mutants of T7, perhaps by a comparable (but unknown) mechanism to that of exclusion of T4 *r* II mutants. The Col Ib plasmid inhibits the growth of *0.7* mutants; the basis of exclusion is not understood but may involve the failure of gene *2* to inactivate *E. coli* RNA polymerase. T7 and many of its coliphage relatives are also excluded from growth in F plasmid-containing cells, although T3 is an exception. The *pif* gene of F interferes with the normal functions of T7 genes *1.2* and *10*; interaction of either of these genes with *pif* causes inhibition of all macromolecular synthesis and of membrane functions. The rapid loss of metabolic potential of the abortively infected, F-containing *E. coli* suggests that some key cellular component(s) is inactivated by the interaction of phage and *pif* genes. The identity of this component is not known but is likely to be important for T7 development in normal hosts.

Expression Systems Based on T7 RNA Polymerase

The specificity of T7 RNA polymerase for its promoter has allowed the development of high-level regulated expression systems for cloned DNA that can achieve >50% of total cellular proteins in the form of a desired product. These systems have been adapted for eucaryotic cells as well for bacteria. The T7 promoter sequence is very rare, even in mammalian cells, and with the appropriate constructs highly selective cloned gene expression can be attained. Typically, the Class III gene *10* promoter is employed, with or without gp10 translational start sequences, and T7 RNA polymerase is supplied from a resident plasmid or prophage, or by phage infection.

See also: Bacteriophage recombination; Host-controlled modification and restriction.

Further Reading

Beck PJ, Gonzalez S, Ward CL and Molineux IJ (1989) Sequence of bacteriophage T3 DNA from gene *2.5* through gene *9*. *J. Mol. Biol.*, 210: 687.

Chung Y-B, Nardone C and Hinkle DC (1990) Bacteriophage T7 DNA packaging. *J. Mol. Biol.*, 216: 939.

Dunn IJ and Studier FW (1983) Complete nucleotide sequence of bacteriophage T7 DNA and the locations of T7 genetic elements. *J. Mol. Biol.*, 166: 477.

Molineux IJ (1991) Host-parasite interactions: recent developments in the genetics of abortive phage infections. *The New Biologist*, 3: 230.

Serwer P (1990) In: Adolph KW (ed.) Double-stranded DNA packaged in bacteriophages; conformation; energetics and packaging pathway. *Chromosomes: Eukaryotic, Prokaryotic, and Viral*, vol. 3, p. 203. Boca Raton: CRC Press.

Studier FW (1991) Use of bacteriophage T7 lysozyme to improve an inducible T7 expression system. *J. Mol. Biol.* 219: 37.

TANAPOX VIRUS

See Yabapox and Tanapox viruses

TAXONOMY AND CLASSIFICATION–GENERAL

Claude M Fauquet
The Scripps Research Institute
La Jolla, California, USA

History

Humans have a tendency to classify and name everything and viruses are no exception. Classifications are extremely useful for showing similar characteristics and properties. Thus, appropriately chosen classification criteria become extremely informative in the case of newly discovered viruses. Unfortunately for virus taxonomy there are no fossils, so evolutionary relationships are very speculative. Only a virus classification would be able to provide indications of the evolution of viruses. In theory, nomenclature and classification are totally independent, but for viruses both issues are often considered at the same time. As a result, virus taxonomic names have always been the subject of passionate discussions.

Classification of viruses is a fairly new exercise considering the first evidence of the existence of a virus was made at the end of the last century. Johnson, a plant virologist, drew attention to the need for virus nomenclature and classification as early as 1927. The first efforts to classify viruses employed a range of ecological and biological properties including pathogenic properties for human and animal viruses and symptoms for plant viruses. For example, viruses that share the pathogenic property of causing hepatitis (e.g. hepatitis A virus, hepatitis B virus, yellow fever virus, Rift Valley fever virus) were grouped together as 'the hepatitis viruses'. Virology developed substantially in the 1930s and the first classifications of viruses reflected this development. Holmes published in 1939 a classification of plant viruses based on host reactions and differential hosts using a binomial–trinomial nomenclature based on the name of the infected plant, but only 89 viruses were classified. In the 1950s, with the development of electron microscopy and biochemical studies, the first groupings of viruses based on common virion properties emerged: the Herpesvirus group, the Myxovirus group, and the Poxvirus group. During this period, there was an explosion of newly discovered viruses. In response, several individuals and committees independently proposed virus classification systems but none was widely used. It became obvious that only an international association of virologists would be able to propose a comprehensive and universal system of virus classification.

At the International Congress for Microbiology held in Moscow in 1966, the International Committee on Nomenclature of Viruses (ICNV) was established by an international group of 43 virologists. An international organization was set up with the aim of developing a unique world-wide recognized taxonomy and nomenclature system for all viruses. The name of the ICNV was changed in 1974 to a more appropriate one: the International Committee on Taxonomy of Viruses (ICTV), which is active today. The ICTV is now considered the official body for all matters related to taxonomy and nomenclature of viruses.

Since the founding of the ICTV, all virologists agreed that the hundreds of viruses isolated from different organisms should be classified together in a unique system, but separate from other microorganisms such as bacteria and mycoplasma. However, there was much controversy on the way to do it. Lwoff, Horne and Tournier argued for the adoption of a system for the classifying of viruses into subphyla, classes, orders, suborders and families. Descending hierarchical divisions would have been based on nucleic acid type (DNA or RNA), strandedness (single or double), presence or absence of an envelope, capsid symmetry and so on. This hierarchical system has never been recognized by the ICTV; nevertheless, the rest of the proposal became the basis of the universal taxonomy system now in place and all ICTV reports reflect this scheme. Until 1990, the scheme did not utilize any hierarchical classification level higher than the family, but the system has recently begun to move in this direction. A first order, *Mononegavirales*, has been accepted in 1990, and a second one, *Caudovirales*, has been proposed for consideration in 1993. In its non-Linnean structure, the scheme is quite different from that used for the taxonomy of bacteria and other organisms. The usefulness of the scheme is being demonstrated by its wide application. It has replaced all competing classification schemes for all viruses.

At the first meeting of the ICNV in Mexico City (1970), two families with a corresponding two genera and 24 floating genera were accepted to begin the grouping of vertebrate, invertebrate and bacterial viruses. In addition, 16 plant virus groups were designated. The Fifth ICTV Report describes one order, 40 families, nine subfamilies, 102 genera, two floating genera and two subgenera for vertebrate, invertebrate, bacterial and fungal viruses and 32 groups and seven subgroups for plant viruses (Table 1). While most virologists shifted to the grouping of viruses in families and genera, plant virologists have persisted in clustering plant viruses in 'groups' until very recently. It is only in 1993 that the ICTV will propose a uniform system for all viruses with two orders, 50 families, 9 subfamilies, 126 genera, 23 floating genera and 4 subgenera encompassing 2644 assigned virus species.

The descriptions of virus families can provide valuable information for new 'unknown' members. Therefore, the ICTV work is not only a taxonomic exercise for evolutionists but a valuable source of information for virologists, teachers, medical doctors and epidemiologists. Since the establishment of the ICTV, five virus taxonomic reports have been published and new reports will appear every three years.

How Does the ICTV Operate?

The ICTV is a Committee of the Virology Division of the International Union of Microbiological Societies. The ICTV operates through a number of committees, subcommittees and study groups of more than 372 eminent virologists with expertise in human, animal, insect, protozoal, bacterial, mycoplasmal, fungal, algal and plant viruses. Taxonomic proposals are initiated and formulated by the study groups. These proposals are revised and accepted by the subcommittees and presented for Executive Committee approval. All decisions are finally affirmed at a plenary session held at each virology congress where all members of ICTV and more than 50 representatives of national microbiological societies are represented. Presently, there are 45 study groups working in concert with six subcommittees, namely, the vertebrate, invertebrate, plant, bacteria, fungus and virus data subcommittees. The ICTV is a non-profit association composed of

Table 1. List of orders, families and groups of viruses[a]

Nature of the presentation criteria		Order	Family or group	Morphology	Genome configuration	Genome size (kbp)	Virus host	Number of species: Members	Tentative	Total
dsDNA	Enveloped		*Baculoviridae*	Bacilliform	1 circular supercoiled	90–230	Invertebrate	14		14
			Hepadnaviridae	Isometric	1 circular	3	Vertebrate	5		5
			Herpesviridae	Isometric	1 linear	120–220	Vertebrate	19	4	23
			Lipothrixviridae	Rod	1 linear	16	Bacteria	2		2
			Plasmaviridae	Pleomorphic	1 circular	12	Bacteria	2	5	7
			Polydnaviridae	Rod, fusiform	1 circular supercoiled	2–28	Invertebrate	2		2
			Poxviridae	Ovioid	1 linear	130–375	Vertebrate, invertebrate	61	16	77
			SSV-1 group	Lemon-shape	1 circular supercoiled	15	Bacteria	3		3
dsDNA	Nonenveloped	(*Caudovirales*)	*Myoviridae*	Tailed phage	1 linear	336	Bacteria	83		83
		(*Caudovirales*)	*Podoviridae*	Tailed phage	1 linear	40	Bacteria	51		51
		(*Caudovirales*)	*Siphoviridae*	Tailed phage	1 linear	53	Bacteria	111		111
			Adenoviridae	Isometric	1 linear	32–48	Vertebrate	111		111
			Caulimovirus	Isometric	1 circular	8	Plant	11	6	17
			Commelina yellow mottle virus group	Bacilliform	1 circular	8	Plant	4	11	15
			Corticoviridae	Isometric	1 circular supercoiled	10	Bacteria	1	1	2
			Iridoviridae	Isometric	1 linear	160–400	Vertebrate, invertebrate	70	2	72
			Papovaviridae	Isometric	1 circular	5–8	Vertebrate	28		28
			Phycodnaviridae	Isometric	1 linear	250–350	Algae	47		47
			Rhizidiovirus	Isometric	1 linear	27	Fungus	1		1
			Tectiviridae	Isometric	1 linear	16	Bacteria	8		8
ssDNA	Nonenveloped		*Geminivirus*	Isometric	1 or 2 circular	3–6	Plant	35	13	48
			Inoviridae	Rod	1 circular	7–20	Bacteria, mycoplasmas	32		32
			Microviridae	Isometric	1 circular	6	Bacteria	28		28
			Parvoviridae	Isometric	1 – strand	6–8	Vertebrate, invertebrate	4	11	15
dsRNA	Enveloped		*Cystoviridae*	Isometric	3 segments	17	Bacteria	1		1
dsRNA	Nonenveloped		*Birnaviridae*	Isometric	2 segments	6	Vertebrate, invertebrate	5		5
			Cryptovirus	Isometric	2 segments	3–5	Plant	20	10	30
			Partitiviridae	Isometric	2 segments	4–10	Fungus	9	5	14
			Reoviridae	Isometric	10–12 segments	19–62	Vertebrate, invertebrate, plant	136	33	169
			Totiviridae	Isometric	1 segment	5–7	Fungus	4	8	12

Nature of the presentation criteria		Order	Family or group	Morphology	Genome configuration	Genome size (kb)	Virus host	Number of species		
								Members	Tentative	Total
ssRNA	Enveloped; no DNA step; positive sense genome		*Coronaviridae*	Pleomorphic	1 + segment	28–33	Vertebrate	11	3	14
			Flaviviridae	Isometric	1 + segment	10–22	Vertebrate, invertebrate	35	19	54
			Togaviridae	Isometric	1 + segment	10–13	Vertebrate, invertebrate	29	2	31
ssRNA	Enveloped; no DNA step; negative non-segmented genome	*Mononegavirales*	*Filoviridae*	Bacilliform	1 − segment	13	Vertebrate	2		2
		Mononegavirales	*Paramyxoviridae*	Helical	1 − segment	15–16	Vertebrate	32	4	36
		Mononegavirales	*Rhabdoviridae*	Bacilliform	1 − segment	10–13	Vertebrate, invertebrate, plant	75	100	175
ssRNA	Enveloped; no DNA step; negative segmented genome		*Arenaviridae*	Spherical	2 − segments	11	Vertebrate	15		15
			Bunyaviridae	Spherical	3 − segments	12–23	Vertebrate invertebrate, plant	253	45	298
			Orthomyxoviridae	Helical	8 − segments	13–14	Vertebrate	3	2	5
ssRNA	Enveloped; DNA step		*Retroviridae*	Spherical	dimer 1 + segment	7–10	Vertebrate	32		32
ssRNA	Nonenveloped; monopartite genome; Isometric particles		*Caliciviridae*	Isometric	1 + segment	8	Vertebrate	4	1	5
			Carmovirus	Isometric	1 + segment	4	Plant	8	9	17
			Leviviridae	Isometric	1 + segment	3–4	Bacteria	43		43
			Luteovirus	Isometric	1 + segment	6	Plant	14	12	26
			Maize chlorotic dwarf virus group	Isometric	1 + segment	9	Plant	1	2	3
			Marafivirus	Isometric	1 + segment	6–7	Plant	3		3
			Necrovirus	Isometric	1 + segment	4–5	Plant	2	2	4
			Parsnip yellow fleck virus group	Isometric	1 + segment	10	Plant	2	1	3
			Picornaviridae	Isometric	1 + segment	7–8	Vertebrate, invertebrate	215	13	228
			Sobemovirus	Isometric	1 + segment	4	Plant	10	6	16
			Tetraviridae	Isometric	1 + segment	5	Invertebrate	1	14	15
			Tombusvirus	Isometric	1 + segment	5	Plant	12		12
			Tymovirus	Isometric	1 + segment	6	Plant	18	1	19
ssRNA	Nonenveloped; monopartite genome; rod-shaped particles		*Capillovirus*	Rod	1 + segment	7	Plant	2	2	4
			Carlavirus	Rod	1 + segment	7–8	Plant	27	29	56
			Closterovirus	Rod	1 + segment	7–18	Plant	10	12	22
			Potexvirus	Rod	1 + segment	6	Plant	18	21	39
			Potyvirus	Rod	1 + segment	8–10	Plant	73	84	157
			Tobamovirus	Rod	1 + segment	6	Plant	12	2	14

Continued

Table I. Continued

Nature of the presentation criteria	Order	Family or group	Morphology	Genome configuration	Genome size (kb)	Virus host	Number of species		
							Members	Tentative	Total
ssRNA Nonenveloped; bipartite genome; isometric particles		*Comovirus*	Isometric	2+segments	9	Plant	14		14
		Dianthovirus	Isometric	2+segments	4	Plant	3		3
		Fabavirus	Isometric	2+segments	10	Plant	3		3
		Nepovirus	Isometric	2+segments	12	Plant	28	8	36
		Nodaviridae	Isometric	2+segments	5	Invertebrate	6		6
		Pea enation mosaic virus group	Isometric	2+segments	9	Plant	1		1
ssRNA Nonenveloped; bipartite genome; rod-shaped particles		*Furovirus*	Rod	2+segments	9–11	Plant	5	6	11
		Tobravirus	Rod	2+segments	9–11	Plant	3		3
ssRNA Nonenveloped; tripartite genome; bacilliform particles		Alfalfa mosaic virus group	Bacilliform	3+segments	8	Plant	1		1
ssRNA Nonenveloped; tripartite genome; isometric particles		*Bromovirus*	Isometric	3+segments	8	Plant	6		6
		Cucumovirus	Isometric	3+segments	9	Plant	3	1	4
		Ilarvirus	Isometric	3+segments	8	Plant	20		20
ssRNA Nonenveloped; tripartite genome; rod-shaped particles		*Hordeivirus*	Rod	3+segments	10	Plant	4		4
ssRNA Nonenveloped; tetrapartite genome		*Tenuivirus*	Rod	4 – ? segments	19	Plant	3	4	7
Total no. species							1970	530	2500

[a] The taxa are listed according the Fifth ICTV Report with the following criteria: nature and strandedness of the nucleic acid, presence or absence of a lipoprotein envelope, the single-stranded (ss)RNA enveloped viruses are arranged on the basis of genome strategy and the ssRNA nonenveloped viruses are arranged on the basis of the number of segments of their genome and their particle morphology. For each family or group of viruses, also indicated are the morphology of the virions, the genome configuration, the genome size in kb, the virus host, the number of species and tentative members in the taxa, and the total number of species listed in 1990.

prominent virologists representing countries from throughout the world and names and taxa are accepted following a democratic process. ICTV does not impose any taxonomic word or taxa but ensures that the propositions are compatible with ICTV rules for homogeneity and consistency. The ICTV regularly publishes reports that describe all the virus taxa with a list of classified viruses as well as compilations of virus families and genera. A last report was published in 1991 and the next will be published in 1994. With the increasing number of viruses and virus strains and the explosion of data on many descriptive aspects of viruses and viral diseases, ICTV decided to launch an international virus database project. This project, termed ICTVdB®, is scheduled to be fully operational and accessible to the scientific community around the year 2000.

System for Virus Classification

There are two systems for classifying organisms: the Linnean and the Adansonian systems. The Linnean system is the monothetic hierarchical classification applied by Linnaeus to plants and animals while the Adansonian system is a polythetic hierarchical system proposed by Adanson in 1763. Maurin and collaborators proposed to apply the Linnean classification system to viruses in 1984. Although the system is very convenient to use, there are shortcomings when it is applied to the classification of viruses. First, it is difficult to appreciate the validity of a particular criterion. For example, it may not be appropriate to use the number of genomic components as a hierarchical criterion. Second, there are no reasons for privileging a particular criterion from another so it is difficult to rank all the available criteria. For example, is the nature of the genome (DNA/RNA) more important than the presence of an envelope or the shape of the virus particles?

The Adansonian system considers all available criteria at once and makes several classifications considering the criteria successively. The criteria leading to the same classifications are considered as correlated and are therefore not discriminatory. Subsequently, a subset of criteria are considered. The process is repeated until all criteria can be ranked to provide the best discrimination of the species. This system was not frequently used due to its labor-intensive nature, but with present-day computers it can be easily implemented. Furthermore, qualitative and quantitative data can be simultaneously considered to generate such a classification. In the case of viruses, it has been determined that at least 60 characters would be needed for a complete virus description. Thus, the limiting factor for applying the Adansonian system is the lack of data in many instances. The increasing number of viral nucleic acid sequences allows the comparison of viruses to generate different phylogenetic trees according to the gene or set of genes used. To date, none of them has satisfactorily provided a clear classification of all viruses. A multidimensional classification, taking into account all the criteria necessary to describe viruses, would probably be the most appropriate way of representing the virus classification but it would not be very easy to use.

For nearly the last 20 years, ICTV has been classifying viruses essentially at the family and genus levels using a nonsystematic polythetic approach. This has clustered viruses first in genera and then in families. A subset of characters including physicochemical, structural, genomic and biological criteria has then been used to compare and group viruses. This subset of characters may change from one family to another according to the availability of the data and the importance of a particular character. It is obvious that there is no homogeneity in this respect throughout the virus classification and that virologists weigh differently the criteria in this subjective process. Nevertheless, we can see a rather good stability of the current ICTV classification. When sequence, genomic organization and replicative cycle data are used for taxonomic purposes they usually confirm the actual classification. It is also obvious that hierarchical classifications above the family level will encounter conflicts between phenotypic and genotypic criteria and that virologists will have to consider the entire classification process to progress in this direction.

Currently, and for practical reasons only, virus classification is structured according to the presentation indicated in Tables 1 and 2. This order of presentation of virus families and groups does not reflect any hierarchical or phylogenetic classification but only a convenient order of presentation. Since a taxonomic structure above the level of family or group (with the exception of the order *Mononegavirales* and the pending order *Caudovirales*) has not been developed extensively, any listing must be arbitrary. The order of presentation is generally the same as in the Fifth ICTV Report. The order of presentation of virus families and groups follows three criteria: (1) the nature of the viral nucleic acid, (2) the strandedness of the nucleic acid and (3) the presence or absence of a lipoprotein

Table 2. Order of presentation of virus classification in the Fifth ICTV Report

A. DNA/RNA
B. Double-stranded/single-stranded
C. Enveloped/nonenveloped
    for the ssRNA enveloped viruses only:
        DNA/no DNA step in the replication cycle
    Positive/negative sense genome
  Monopartite/multipartite genome
    for the ssRNA nonenveloped viruses only:
        Mono/bi/tri/tetrapartite genome
        Isometric/bacilliform/rod-shaped particles

envelope. There are no known single-stranded (ss)DNA viruses with envelopes, so these three criteria give rise to seven clusters comprising the 73 families and groups of viruses (comprising one floating genus). Within two of these clusters, the ssRNA enveloped and nonenveloped viruses, the families have been arranged as follows: the ssRNA enveloped viruses are arranged on the basis of genome strategy, i.e. DNA/no DNA step in the replication cycle, positive/negative sense genome and monopartite/multipartite genome. The ssRNA nonenveloped viruses are arranged on the basis of number of segments of RNA of their genome, i.e. mono-/bi-/tri-/tetrapartite genome and their virion morphology: isometric/bacilliform/rod-shaped particles.

Nomenclature of Virus Taxa

The debate over virus nomenclature has generated significant controversy and discussion over the years and was the primary reason for virologists to establish the ICNV. In the earliest examples of virus taxonomy, Gibbs proposed to adopt a cryptogram to add precision to the vernacular names of the viruses. The cryptograms used a combination of letters and numbers to describe the structure, the biochemical composition of the genome, the host type and the transmission properties of the virus. This system of virus identification was set up in the first ICTV report but was never used and therefore abandoned.

When a family, genus or virus group is approved by ICTV, a type species or type member is designated. However, none of these type species has received an official name and only English vernacular names are indicated. Use of latinized binomial names for virus names was supported by animal and human virologists of ICTV for many years, but has never been implemented. This suggestion was in fact withdrawn from ICTV nomenclature rules in 1990 and consequently such names as *Herpesvirus varicella* or *Polyomavirus hominis* should not be used. For several years, plant virologists have set up a different nomenclature, using the vernacular name of a virus but replacing the word 'virus' by the group (genus) name; for example, cucumber mosaic cucumovirus and tobacco mosaic tobamovirus. Though this usage is favored by many scientists and examples of such practice can be found for human, animal and insect viruses (e.g. human rhinovirus, canine calicivirus, *Acheta* densovirus), it has not been adopted by the ICTV.

The ICTV has a set of rules for virus nomenclature and orthography of taxonomic names. The international genus names universally end in '-virus', the international subfamily names end in '-virinae', the international family names end in '-viridae' and the international order names end in '-virales'. In formal taxonomic usage, the virus order, family, subfamily and genus names are printed in italics (or underlined) and the first letter is capitalized. Species names, which are used in English vernacular form, are not capitalized or italicized (or underlined). In formal usage, the name of the taxon precedes the name of the taxonomic unit; for example, 'the family *Picornaviridae*' or 'the genus *Rhinovirus*'. In informal vernacular usage, virus order, family, subfamily, genus and species names are written in lower case Roman script; they are not capitalized or italicized (or underlined). Additionally, in informal usage, the name of the taxon should not include the formal suffix, and it should follow the term for the taxonomic unit; for example, 'the mononegavirales order', 'the adenovirus family', 'the avihepadnavirus genus' or 'the tobamovirus group'.

To avoid ambiguous identifications, it has been recommended to journal editors to follow ICTV guidelines for proper virus identification and nomenclature, and to cite viruses with their full taxonomic terminology when they are first cited in an article, as in the following examples.

- Order *Mononegavirales*, Family *Paramyxoviridae*, Subfamily *Paramyxovirinae*, genus *Paramyxovirus*, avian paramyxovirus 1.
- Order *Mononegavirales*, Family *Rhabdoviridae*, Plant rhabdovirus group, Plant rhabdovirus subgroup A, lettuce necrotic yellows virus.
- Family *Iridoviridae*, genus *Iridovirus*, *Chilo* iridescent virus.

- Family *Podoviridae*, genus T7 phage group, coliphage T7.

A Universal Classification System

The present universal system of virus taxonomy is set arbitrarily at hierarchical levels of order, family (in some cases subfamily), genus and species. Lower hierarchical levels, such as subspecies, strain, variant, pathotype and isolate, are established by international specialty groups or/and by culture collections, but not by the ICTV.

Virus species

The species taxon is always regarded as the most important taxonomic level in classification but it has proved to be the most difficult to apply for viruses. The ICTV definition of a virus species was long considered to be 'a concept that will normally be represented by a cluster of strains from a variety of sources, or a population of strains from a particular source, which have in common a set or pattern of correlating stable properties that separates the cluster from other clusters of strains'. This was a general definition which was in fact not very precise for delineating species in a particular family or in all families. Furthermore, this definition directly addressed the definition of a virus strain, which had never been attempted in the history of virus taxonomy. In 1990, Van Regenmortel proposed another species definition which has been accepted by the ICTV Executive Committee in 1991. This definition states: 'A virus species is a polythetic class of viruses that constitutes a replicating lineage and occupies a particular ecological niche.' The major advantage in this definition is that it can accommodate the inherent variability of viruses and it does not depend on the existence of a unique characteristic. Members of a polythetic class are defined by more than one property and no single property is absolutely essential and necessary. Thus in each family it might be possible to determine the set of properties of the taxonomic level 'species' and to check if the family members are species of this family or if they belong to a lower taxonomic level. The ICTV is currently conducting this exercise throughout all virus families. This should ultimately result in an excellent evaluation of a precise definition of each virus species in the entire classification.

Several practical matters are related to the definition of a virus species with the goal of a better usage of a virus classification. These include: (1) homogeneity of the different taxa; (2) diagnostic related matters; (3) virus collections; (4) evolution studies; (5) biotechnology; (6) sequence database projects; and (7) virus database projects.

Virus families, genera and groups

There is no formal definition for a genus, but it is commonly considered as: 'a population of virus species that share common characteristics and are different from other populations of species'. Although this definition is somewhat elusive, this level of classification seems stable and useful; some genera have been moved from one family to another but the composition and description of these genera have remained stable over the years. The characteristics defining a genus are different from one family to another and there is a tendency to create genera with fewer differences between them. Upon examination, there is more and more evidence that the members of a genus have a common evolutionary origin. The use of subgenera is very limited in current virus classification (see Table 1); only one subgenus classification exists in the entire family *Baculoviridae* and there are three other examples in plant virus groups. However, these may disappear when plant virus groups are reorganized into families and genera (see below). Since the creation of the ICTV, plant virologists have always kept the classification of plant viruses in 'groups' and strongly refused to place them in genera and families. However, due to obvious similitude, plant reoviruses and rhabdoviruses have been integrated into the families *Reoviridae* and *Rhabdoviridae* (Table 1). This position was mostly due to the refusal of plant virologists to accept binomial nomenclature. Since this form of nomenclature has been withdrawn from the ICTV rules, they have subsequently accepted classification of plant viruses into genera and families. The current classification still presents plant viruses in groups but the next report will only have families and genera for all 'virus kingdoms'. Five plant virus families and 39 genera have been proposed for the next ICTV Report.

Virus orders

As mentioned previously, the upper hierarchical levels of the virus classification are extremely difficult to establish. Despite several general propositions in the past, none of them have been accepted.

Table 3. List of descriptive characters used in virus taxonomy at the family level

I. Virion properties

A. Morphology properties of virions

1. Virion size
2. Virion shape
3. Presence or absence of an envelope and peplomers
4. Capsomeric symmetry and structure

B. Physical properties of virions

1. Molecular mass of virions
2. Buoyant density of virions
3. Sedimentation coefficient
4. pH stability
5. Thermal stability
6. Cation (Mg^{2+}, Mn^{2+}) stability
7. Solvent stability
8. Detergent stability
9. Radiation stability

C. Properties of genome

1. Type of nucleic acid – DNA or RNA
2. Strandedness – single stranded or double stranded
3. Linear or circular
4. Sense – positive, negative or ambisense
5. Number of segments
6. Size of genome or genome segments
7. Presence or absence and type of 5′-terminal cap
8. Presence or absence of 5′-terminal covalently linked polypeptide
9. Presence or absence of 3′-terminal poly(A) tract (or other specific tract)
10. Nucleotide sequence comparisons

D. Properties of proteins

1. Number of proteins
2. Size of proteins
3. Functional activities of proteins (especially virion transcriptase, virion reverse transcriptase, virion hemagglutinin, virion neuraminidase, virion fusion protein)

E. Lipids

1. Presence or absence of lipids
2. Nature of lipids

F. Carbohydrates

1. Presence or absence of carbohydrates
2. Nature of carbohydrates

II. Genome organization and replication

1. Genome organization
2. Strategy of replication of nucleic acid
3. Characteristics of transcription
4. Characteristics of translation and post-translational processing
5. Site of accumulation of virion proteins, site of assembly, site of maturation and release
6. Cytopathology, inclusion body formation

III. Antigenic properties

1. Serological relationships
2. Mapping epitopes

Table 3. Continued

IV. Biological properties
1. Host range, natural and experimental
2. Pathogenicity, association with disease
3. Tissue tropisms, pathology, histopathology
4. Mode of transmission in nature
5. Vector relationships
6. Geographic distribution

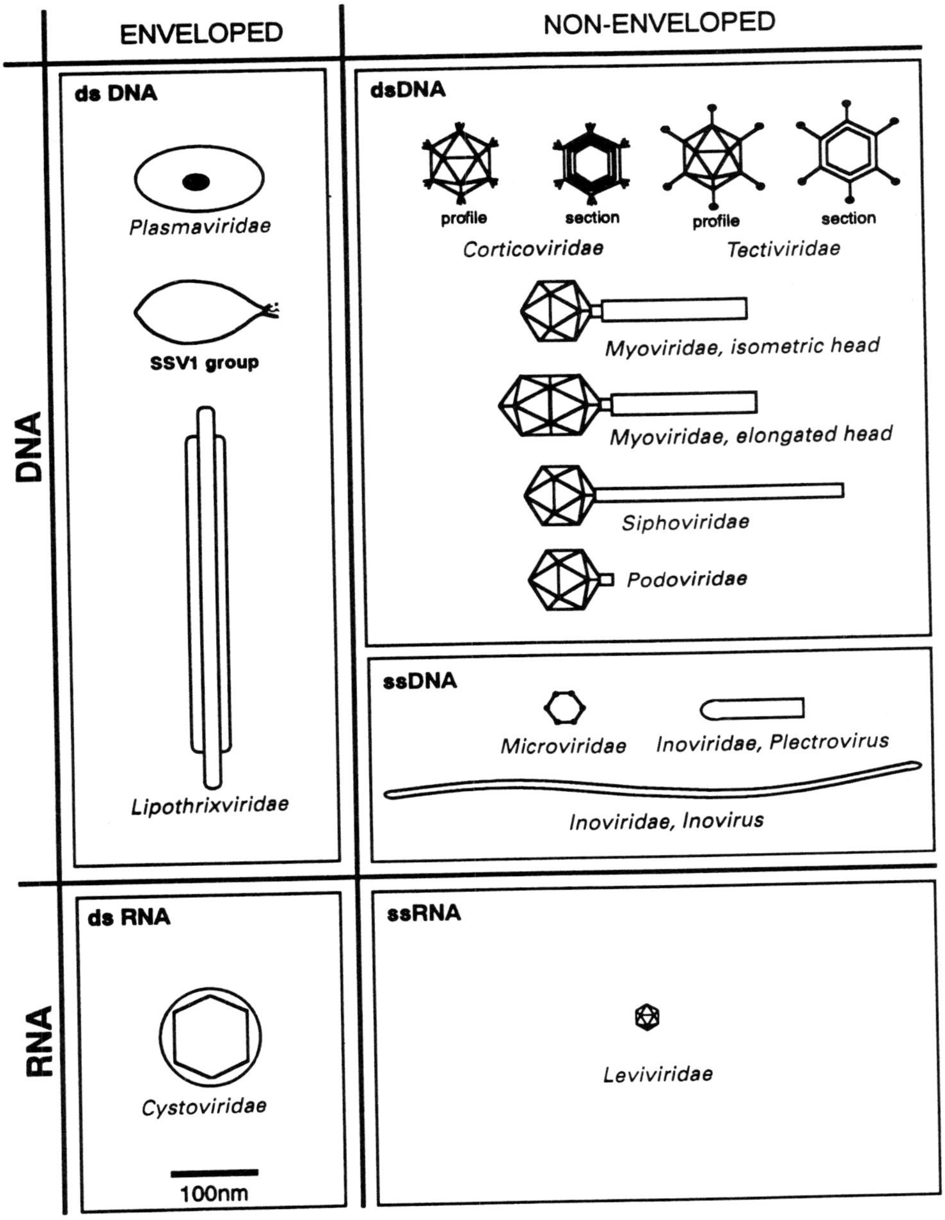

Fig. 1 Diagrammatic representation of the families of viruses infecting bacteria, grouped according to the nature and strandedness of their genome and the presence or absence of an envelope. Reproduced with permission from Springer-Verlag.

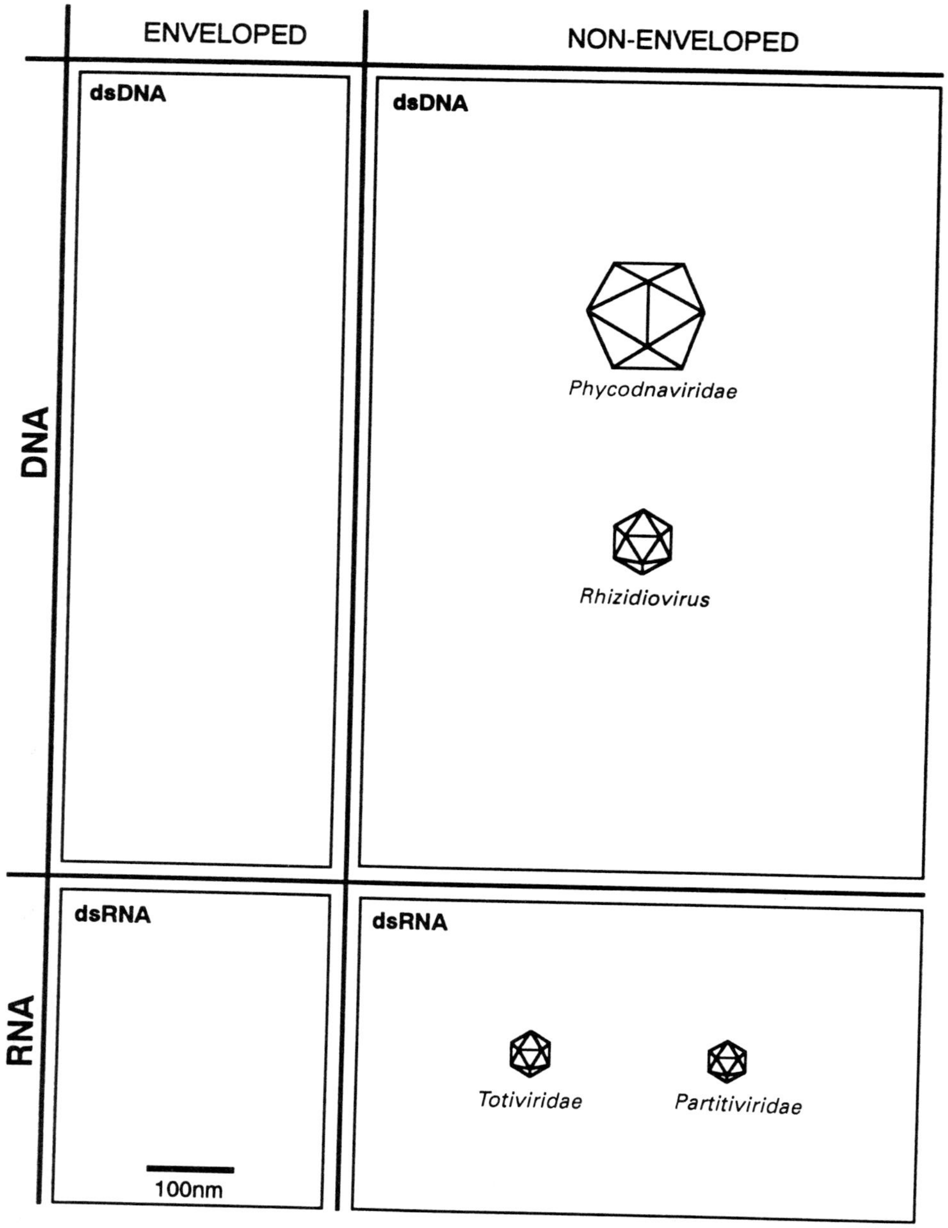

Fig. 2 Diagrammatic representation of the families of viruses infecting algae, fungi and protozoa, grouped according to the nature and strandedness of their genome and the presence or absence of an envelope. Reproduced with permission from Springer-Verlag.

Nevertheless, it has been stated several times that the creation of orders could be considered on a case-by-case basis. The first virus order *Mononegavirales* was established in 1990. This order comprises the non-segmented ssRNA negative-sense viruses, namely the families *Filoviridae*, *Paramyxoviridae* and *Rhabdoviridae*. This decision has been taken because of the great similitude between these families at many points of view including the replication strategy of these viruses. A second order is under consideration; it is named *Caudovirales*, and it includes all the families of dsDNA phages having a tail, including *Myoviridae*, *Podoviridae* and *Siphoviridae*. Many members of the ICTV advocate the creation of many more orders, but it has been decided to proceed cautiously to avoid creation of short-life orders. The creation of formal taxa higher than orders, for example, kingdoms, classes and subclasses, has not been considered by ICTV.

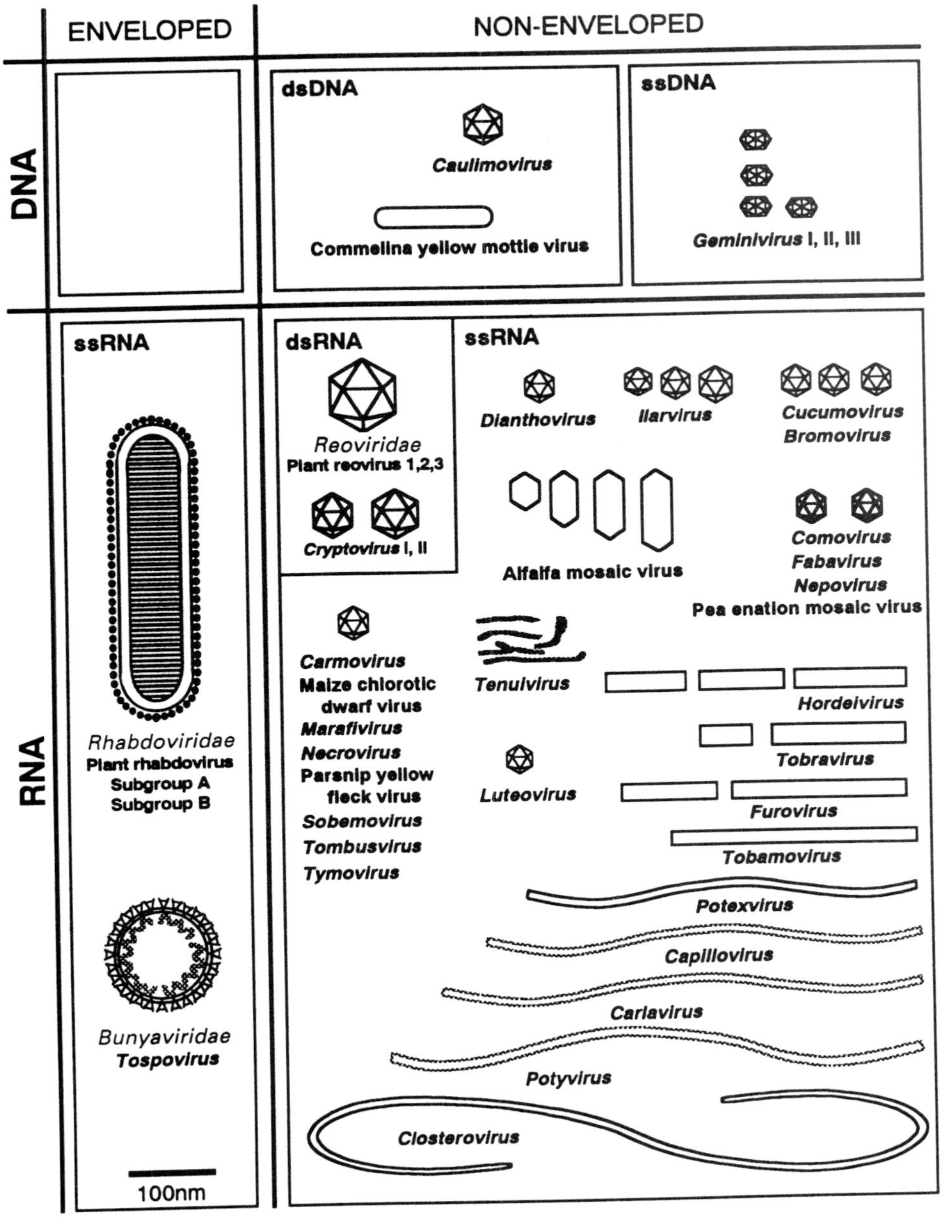

Fig. 3 Diagrammatic representation of the families of viruses infecting plants, grouped according to the nature and strandedness of their genome and the presence or absence of an envelope. Reproduced with permission from Springer-Verlag.

Virus Taxa Descriptions

Virus classification continues to evolve with the technologies available for describing viruses. The first wave of descriptions, before 1940, mostly took into account the visual symptoms of the diseases caused by viruses and their modes of transmission. A second wave, between 1940 and 1970, brought an enormous amount of information from studies of virion morphology (electron microscopy, structural data), biology (serology and virus properties) and physicochemical properties of viruses (nature and size of genome, number and size of viral proteins). Since 1970, the third wave of virus descriptions has included genome and replicative information (sequence of genes, sequence of proteins), as well as molecular relationships with virus hosts. There has been a correlative modification of the list of virus descriptors and Table 3 lists the family and genera

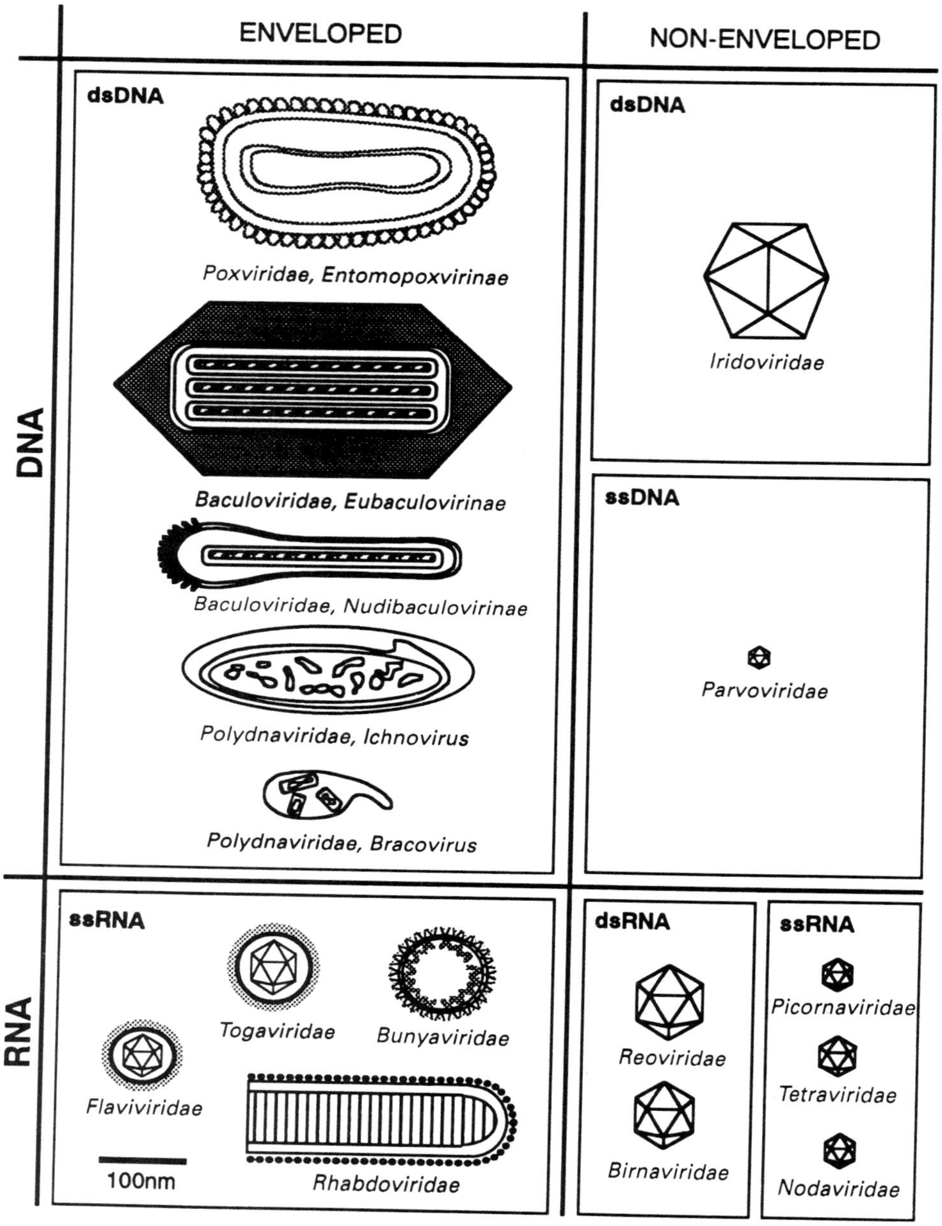

Fig. 4 Diagrammatic representation of the families of viruses infecting invertebrates, grouped according to the nature and strandedness of their genome and the presence or absence of an envelope. Reproduced with permission from Springer-Verlag.

descriptors which are used in the ICTV Fifth Report.

The impact of descriptions on virus classification has been particularly influenced by electron microscopy and the negative staining technique for virions. This technique had an immediate effect on diagnostics and classification of viruses. With negative staining, viruses could be identified from poorly purified preparations of all types of tissues and information about size, shape, structure and symmetry could quickly be provided. As a result, virology progressed simultaneously for all viruses infecting animals, insects, plants and bacteria. Thin sections of infected tissues brought a new dimension to virus classification by providing information about virion morphogenesis and cytopathogenic effects. These techniques in conjunction with the determination of the nature of the genome provided a major source of information for the system of virus classification established in the 1980s (Figs 1–5).

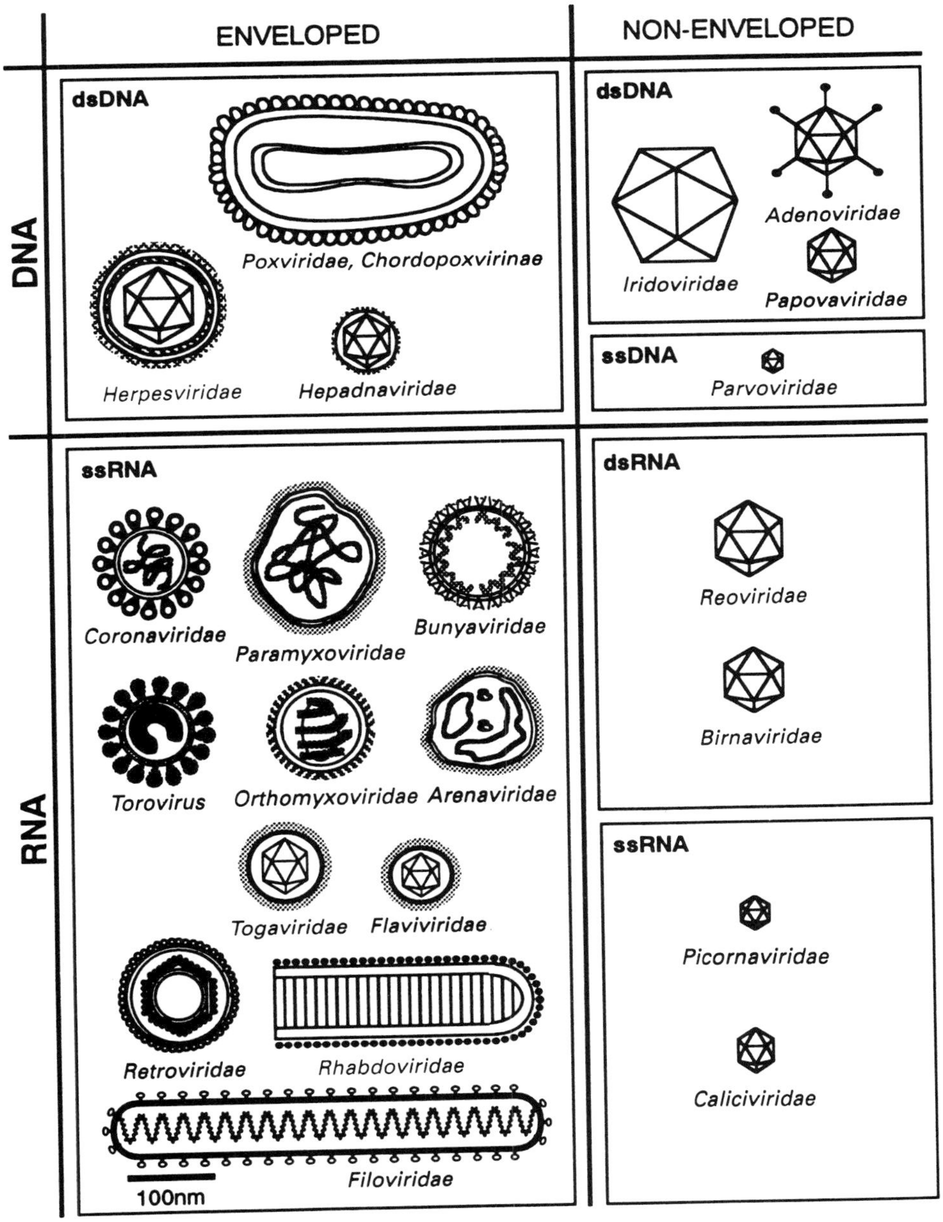

Fig. 5 Diagrammatic representation of the families of viruses infecting vertebrates, grouped according to the nature and strandedness of their genome and the presence or absence of an envelope. Reproduced wtih permission from Springer-Verlag.

In many instances the properties of viruses belonging to the same genus are correlated. Thus, the classification of a few of them will likely be sufficient to allow the classification of a new virus into an established genus. For example, a plant virus with filamentous particles of 700 to 850 nm and transmitted by aphids is likely to be a potyvirus. Establishment of new genera in the future will require more information. Most of the properties listed in Table 3 will have to be rigorously analyzed to warrant the formation of a new genus.

Table 3 lists 45 different categories of properties but each category includes many items. Lists of virus descriptors usually comprise between 500 and 1000 descriptors. The establishment of a universal list of virus descriptors is under way and should be adopted by ICTV in 1993. It will contain a common set of descriptors for all viruses and discrete subsets for specific viruses in relation to their specific hosts (human, animal, insect, plant and bacterial).

See also: Bacteriophage taxonomy and classification.

Further Reading

Francki RIB, Fauquet CM, Knudson DL and Brown F (1991) *Classification and Nomenclature of Viruses.* Fifth Report of the International Committee on Taxonomy of Viruses. *Arch. Virol.* suppl. 2 (whole issue).

Francki RIB, Milne RG and Hatta T (1985) *Atlas of Plant Viruses.* Boca Raton, Florida: CRC Press.

Lwoff A, Horne R and Tournier P (1962) A system of viruses. *Cold Spring Harbor Symp. Quant. Biol.*

Matthews REF (1983) *The History of Viral Taxonomy. A Critical Appraisal of Viral Taxonomy*, p. 1. Boca Raton, Florida: CRC Press.

Murphy FA and Kingsbury DW (1990) Virus taxonomy. In: Fields BN *et al.* (eds) *Virology*, 2nd edn, p. 9. New York: Raven Press.

Van Regenmortel MHV (1990) Virus species, a much overlooked but essential concept in virus classification. *Intervirology* 31: 241.

TENUIVIRUSES

Bryce Falk
University of California, Davis
Davis, California, USA

Taxonomy and Classification

The tenuiviruses are a relatively newly recognized group of plant viruses, however, the diseases they cause have been known since the early 1900s. There are currently five recognized tenuiviruses including: rice stripe virus (RStV); maize stripe virus (MStV); rice hoja blanca virus (RHBV); rice grassy stunt virus (RGSV); and European wheat striate mosaic virus (EWSMV) (see also Table 1). Based upon their similar biological properties these viruses have been loosely grouped together for several years. However, only since 1981 have some of the unique and interesting molecular properties of the tenuiviruses become known.

Biological Properties

All tenuiviruses are transmitted to plants by specific delphacid planthoppers (Homoptera: Delphacidae, see Table 1). They are not mechanically transmissible even experimentally. The plant host ranges of all tenuiviruses are limited to monocotyledonous species within the family *Poaceae*. The symptoms induced in infected plants are generally similar for the different tenuiviruses and includes general leaf striping, a distinct white coloring of the leaf stripes and stunting. The similarity of their biological properties and symptomatology led to an early artificial grouping of these viruses. Recent work on the physical, chemical and molecular properties of the tenuiviruses has confirmed their relationships to each other, and shown them to be distinctly different from most other plant viruses.

Virus Structure and Composition

The name for the tenuivirus group is derived from the slender (tenuous), filamentous ribonucleoprotein particles associated with these viruses. Such particles have been identified in cells of tenuivirus-infected plant and insect hosts. Electron microscopic analysis has shown the particles to be threadlike, very thin in diameter (8–10 nm) and often without defined lengths. These particles have often been referred to as virions or virus particles. However, recent evidence obtained by examining the filamentous particles using high-resolution electron microscopy, and by *in vitro* characterization of the particles suggests that they are likely to be ribonucleoprotein components of a yet to be identified larger, more complex virion.

Filamentous ribonucleoprotein particles (RNPs) have been purified from plants infected by MStV, RStV, RGSV and RHBV. Electron microscopic

Table 1. Tenuiviruses, their planthopper vectors and plant hosts, and geographic incidence.

Virus	Vector[a]	Plant hosts[b]	Range[c]
European wheat striate mosaic virus (EWSMV)	*Javasella pellucida*	Wheat (*Triticum aestivum*)	Europe
Maize stripe virus (MStV)	*Peregrinus maidis*	Maize (*Zea mays*)	World-wide tropics and subtropics
Rice grassy stunt virus (RGSV)	*Nilaparvata lugens*	Rice (*Oryza sativa*)	Asia, Far East
Rice hoja blanca virus (RHBV)	*Tagosodes oryzicola*	Rice (*O. sativa*)	American tropics and subtropics
Rice stripe virus (RStV)	*Laodelphax striatellus*	Rice (*O. sativa*)	Asia, Far East

[a] All tenuivirus insect vectors are planthoppers in the Delphacidae. The name indicates the main vector.
[b] Indicates the primary economically important host. All tenuiviruses have more extensive experimental host ranges (see Gingery, 1988).
[c] Indicates the main natural geographic distribution.

analyses have shown that the purified RNPs resemble the particles seen in cells of tenuivirus-infected plants and planthoppers. However, various studies have shown slightly different and indistinct morphologies for different tenuivirus RNPs. The initial reports for RNP morphologies ranged from very fine-stranded (3 nm diameter) filamentous structures for MStV, to larger diameter (9 nm) spiral filaments for RHBV. Recent electron microscopic studies of purified tenuivirus RNPs have shown a more uniform morphology at least for the RNPs of RGSV, RStV, MStV and RHBV. The RNPs appear as circular filamentous strands (Fig. 1). The strands composing the circular RNPs have a diameter of about 8 nm, and under various conditions the RNPs can exhibit varying degrees of supercoiling.

Four to five RNPs have been resolved for different tenuiviruses by rate-zonal sucrose density gradient centrifugation analysis. The sedimentation coefficients of individual MStV RNPs range from about 70S to 190S for the slowest to fastest sedimenting RNPs, respectively. The RNPs are composed of viral RNA encapsidated by a single protein species, the nucleocapsid (N) protein. The N proteins for all tenuiviruses are very similar in size, mol. wt around 32 000–35 000.

Five RNAs can be isolated from the RNPs of RStV and MStV. The overall pattern and sizes of the RStV and MStV RNAs, as determined by denaturing agarose gel electrophoresis, are very similar. For MStV, sizes are estimated to be about 8.3, 3.5, 2.4, 2.2 and 1.3 kb for RNAs 1–5, respectively. Only four RNAs have been found for RGSV, and these compare in size to the four larger RNAs of RStV and MStV. Analysis of individually purified MStV and RStV RNPs has shown that the different-sized RNA molecules are separately encapsidated in different RNPs. The smallest RNA (RNA 5) can be isolated from the slowest sedimenting RNP, with larger RNAs being found in faster sedimenting RNPs, respectively, based on the size of the RNA molecule. Careful studies with RStV have shown that the circumference of the filamentous, circular RNPs is proportional to the size of the encapsidated RNA. The five RStV RNPs are estimated to be about 290,

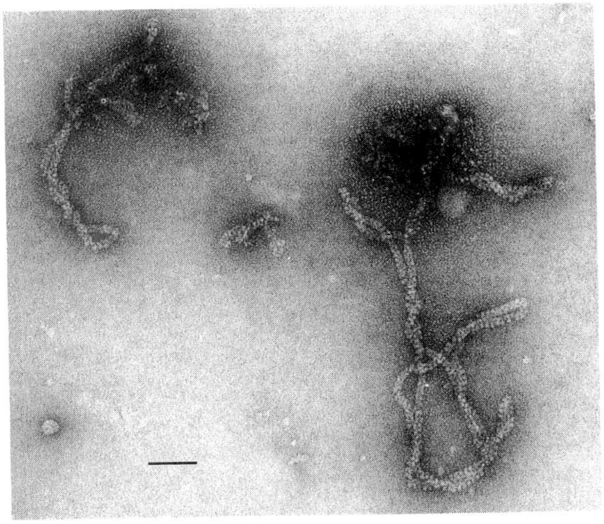

Fig. 1 Electron micrograph showing purified rice hoja blanca virus (RHBV) RNPs. Note circular filamentous nature of the particles. Scale bar = 50 nm. Courtesy of Dra. A. M. Espinoza Esquivel, University of Costa Rica.

510, 610, 840 and 2110 nm in circumference for RNPs 5–1, respectively.

The RNP RNAs appear to be largely single-stranded. However, under appropriate conditions, both single- and double-stranded RNAs have been detected in RNAs extracted from purified RNPs of MStV, RGSV and RStV. This is somewhat surprising, and it seems unlikely that both single- and double-stranded RNAs would be contained within the RNPs. An alternative possibility which is supported by limited data, suggests that only single-stranded RNAs are encapsidated, but that for each size RNA (e.g. RNA1), complementary molecules are separately encapsidated in different RNPs. This is supported by data which show that mostly single-stranded RNAs are detected when MStV RNPs are disrupted with detergent and the RNAs immediately analyzed by gel electrophoresis. However, when RNP RNAs for MStV and RStV are isolated using SDS and phenol, concentrated by ethanol precipitation, and subsequently analyzed by nondenaturing gel electrophoresis, both single- and double-stranded RNAs are detected. The double-stranded RNAs presumably arise by hybridization of complementary strands for each RNA.

Nucleic acid hybridization results have definitively shown that opposite polarity, complementary molecules for each of the five RNAs are present in the RNPs. However, unequal amounts of the complementary molecules for all five MStV RNAs are present. Northern hybridization studies of denatured and nondenatured MStV RNP RNAs suggest that each of the five RNP RNAs is represented primarily as one polarity; however, smaller amounts of the complementary polarity molecules for each RNA also are present.

The above characteristics are consistent and characteristic features of the tenuiviruses. However, the properties of the RNPs suggest that the circular particles may not be virions, but components of a more complex virion. At this time the exact nature of tenuivirus virions remains in question.

Genome Composition and Structure

The different-sized tenuivirus RNAs are most likely distinct genome components. Infectivity experiments demonstrating whether or not all five of the tenuivirus RNAs are necessary for competent infections have not been done. This is largely because neither the tenuivirus RNAs nor RNPs are mechanically transmissible to plants. The purified RNPs are infectious upon injection of the planthopper vector; however, the more highly purified the RNP preparation, the less infectious it is. Also at least for MStV, no infections have resulted when individuals of the planthopper vector, *Peregrinus maidis*, were injected with MStV RNAs. Thus, classical mixing and matching of individual tenuivirus RNAs to establish the necessity of each for a competent infection remains to be done.

However, nucleic acid hybridization analyses have shown that the five MStV RNAs do not contain a high degree of nucleotide sequence homology with each other, and analyses with the four largest RStV RNAs have yielded similar results. Thus, it seems likely that the tenuivirus genome is composed of five separate RNA components. The additive size of the RNAs is in the range of 18 kb (estimates for MStV) to 20 kb (estimates for RStV). Such a genome size is relatively large for plant viruses, and is second in size only to the approx. 23 kb genome of phytoreoviruses.

The terminal structures of the RNAs have not been definitively identified; however, it is very likely that the 5′ termini do not contain a VPg, nor are they capped. RNAs of both RStV and MStV can be readily radiolabeled *in vitro* by using polynucleotide kinase. Similarly, the 3′ ends of the RStV RNAs have been labeled using T4 RNA ligase, suggesting that the 3′ terminus is an OH group.

Nucleotide sequence analyses of RStV and MStV RNAs have shown that the various genomic RNAs have conserved terminal nucleotide sequences. There are 11 nucleotides at the 5′ termini and 10 nucleotides at the 3′ termini which are largely conserved not only among the five RNAs of a given tenuivirus, but also between them (at least for the two tenuiviruses so far examined, MStV and RStV, Table 2). The minor exceptions are that RStV RNA 1 and MStV RNA 5 have only nine of the 10 conserved nucleotides at their 3′ termini. It is also of interest that the terminal eight nucleotides at both the 5′ and 3′ termini are identical to those found in the vertebrate-infecting phleboviruses and uukuviruses (*Bunyaviridae*).

Another interesting and characteristic feature of tenuiviruses is that the 3′ and 5′ termini of the genomic RNAs are complementary. The 3′ 17 terminal nucleotides are complementary to 17 of the 18 5′-terminal nucleotides. A model has been proposed for RStV which would allow the eleventh nucleotide from the 5′ end to bulge out so that the 17 complementary nucleotides at each terminus could base-pair and form a hairpin structure. Such complementarity of RNA terminal nucleotide sequences is largely unique among the plant viruses, with the exception that tomato spotted wilt virus (Tospovirus) recently has been shown also to exhibit this feature. However, this is a feature which is shared among

Table 2. Terminal nucleotide sequences of tenuivirus RNAs

RNA 1	5′-**ACACAAAGUCC**
	3′-**UGUGU**A**UCAG***
RNA 2	5′-**ACACAAAGUCC**
	3′-**UGUGUUUCAG***
RNA 3	5′-**ACACAAAGUCC**UGGGUAA
	3′-**UGUGUUUCAG***ACCCAUU
RNA 4	5′-**ACACAAAGUCC**AGGGCAU
	3′-**UGUGUUUCAG***UCCCGUA
RNA 5	5′-**ACACAAAGUCC**UUGGCAC
	3′-**UGUGU**A**UCAG***AACCGUG

The nucleotide sequences shown were determined by direct RNA sequencing RStV RNAs 1 and 2, and from cloned cDNAs representing MStV and RStV RNAs 3 and 4, and MStV RNA 5. The nucleotides shown in bold are conserved among the RNA sequences determined so far. The asterisk (*) in the 3′-terminal sequences represents a space inserted so as to align 5′ and 3′ nucleotides for complementary base pairing.

animal viruses with segmented, negative-sense RNA genomes in the *Bunyaviridae, Arenaviridae* and *Orthomyxidae*.

For La Crosse virus, a member of the vertebrate-infecting *Bunyaviridae*, the complementary RNA termini have been shown to exist in cells and in RNPs as base-paired, stable panhandle structures. It is believed that because the La Crosse virus RNA termini are base-paired within the RNPs, this could play a role in determining the circularity of the RNPs. Interestingly, the RStV and RGSV circular filamentous nucleoproteins are morphologically very similar to those of La Crosse virus, and it is tempting to speculate that the tenuivirus RNA 5′ and 3′ termini base pair, and that this feature at least in part gives rise to the circular nature of the characteristic tenuivirus RNPs.

Gene Expression Strategies

So far, only RNAs 3 and 4 of RStV and RNAs 3, 4 and 5 of MStV have been analyzed in detail. Nucleotide sequence and Northern hybridization analyses suggest that the tenuivirus RNAs 3 and 4 are very similar, not only in their sizes, but also in their organization. RStV RNAs 3 and 4 are about 2.4 and 2.1 kb, respectively, while MStV RNAs 3 and 4 are 2.4 and 2.2 kb, respectively. Two separate open reading frames (ORFs) have been identified within cDNA clones representing both RNAs 3 and 4, suggesting that RNAs 3 and 4 each encode two proteins.

For RNA 3, one ORF is located in the 5′ region of the RNA. Nucleotide sequence and *in vitro* translation analysis suggest that for MStV, this ORF encodes for a nonstructural protein (NS_3) of mol. wt 22 741. The second MStV RNA 3 ORF encodes the nucleocapsid protein (N protein, mol. wt 34 574). However, the N protein ORF is located in the opposite end of RNA 3, and is in the complementary sense, or in 5′ region of the viral complementary RNA (vcRNA) (Fig. 2). As opposite and complementary RNA 3 molecules are both encapsidated, the majority of RNA 3 which is encapsidated is of 'sense' polarity for the NS_3 ORF. Lesser amounts of the vcRNA 3 are detected in the RNPs. This also is reflected when RNP RNAs are analyzed by *in vitro* translation analysis. Translation products for each ORF are detected but only minor amounts of the N protein are detected relative to the NS_3 protein.

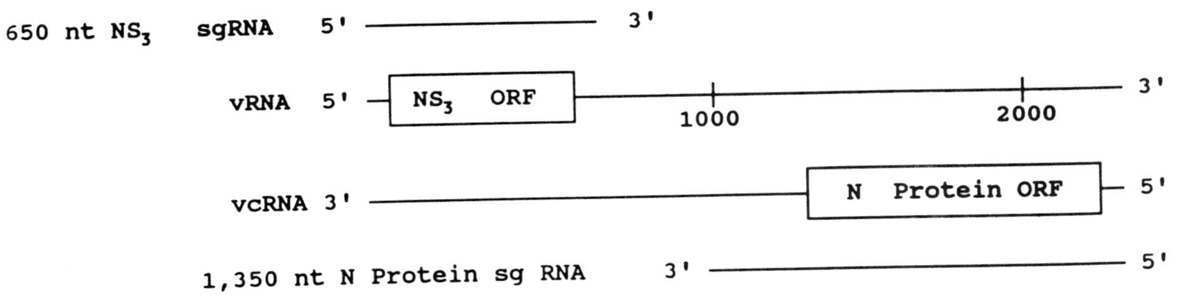

Fig. 2 MStV RNA 3 gene organization and expression strategy. The figure shows the ambisense open reading frame (ORF) organization and subgenomic RNAs for MStV RNA 3. A similar organization has been identified for RStV RNA 3, and for both MStV RNA 4 and RStV RNA 4.

RNA 4 exhibits a similar overall organization to RNA 3. One ORF is located in the RNA 4 5′ region and encodes the major noncapsid protein (NCP, mol. wt 19 815 for MStV). The majority of RNP RNA 4 is of 'sense' polarity with respect to the NCP ORF, and when MStV RNP RNAs are translated *in vitro*, the NCP is a prominent product. The second RNA 4 ORF is located in the 5′ region of the vcRNA 4, and most likely encodes a nonstructural protein (NS_4, mol. wt 31 903 for MStV). The opposing ORFs of both RNAs 3 and 4 are separated by relatively large A-U rich intergenic regions of about 660 nucleotides for RNA 3 and 720 nucleotides for RNA 4.

The proteins encoded by these ORFs are most likely translated *in vivo* from subgenomic mRNAs. Four subgenomic RNAs representing the RNA 3 and RNA 4 ORFs have been detected in tissues from MStV-infected *Zea mays* plants, but not in purified RNPs. The RNA 3 subgenomic RNAs are about 650 and 1350 nucleotides in size for the NS_3 and N protein ORFs respectively. The RNA 4-related subgenomic RNAs for the NCP and NS_4 protein ORFs are about 950 and 1000 nucleotides, in size respectively. In agreement with the polarities of the two ORFs contained in each of these RNAs, the subgenomic mRNAs for RNA 3 are of opposite polarity to each other, as are the two subgenomic mRNAs for RNA4.

The organizational and gene expression strategies exhibited by MStV and RStV RNAs 3 and 4 are very similar to those exhibited by the S RNAs of the phleboviruses, uukuviruses (*Bunyaviridae*), arenaviruses (*Arenaviridae*), and the plant virus, tomato spotted wilt virus. This organization and means of gene expression is termed ambisense, referring to the opposing nature of the polarities for ORFs contained within each RNA, and the use of opposite polarity subgenomic mRNAs for translation of each ORF.

Of the other tenuivirus RNAs, only MStV RNA 5 has been analyzed. MStV RNA 5 is 1317 nucleotides in length. Nucleotide sequence analysis of MStV RNA 5 cDNA clones suggests that RNA 5 is probably a monocistronic RNA, and that vcRNA 5 contains a 1125 nucleotide ORF capable of encoding a protein of mol. wt 44 237. No subgenomic RNAs corresponding to this ORF have been detected, and it is probably translated directly from the vcRNA. The protein encoded by RNA 5 has not yet been detected in MStV-infected plants or planthoppers. Analysis of the potential ORF 5 protein show it to be extremely hydrophilic and to contain 21% arginine and lysine.

Geographic Distribution and Economic Importance

The tenuiviruses are not widespread, but generally are somewhat limited in their geographical distribution, probably due to limits for the natural geographical ranges of their planthopper vectors. RStV and RGSV appear to be limited to rice-growing areas of Asia. RHBV is limited to tropical and subtropical rice-growing regions of the Americas, and EWSMV is limited to wheat-producing regions of Europe. Only MStV is found in both the New and Old Worlds. It has been reported from North, Central and South America, Australia and parts of Africa. However, MStV is limited only to subtropical and tropical maize producing regions in these areas.

Even though they are somewhat limited in their natural incidence, the tenuiviruses are still of considerable economic importance. They cause damaging diseases in important food crops including rice (*Oryza sativa*) and maize (*Zea mays*). As such they have received considerable attention by plant virologists, plant pathologists, agronomists and plant breeders. For example, RHBV, which is the only virus disease of rice in the Western hemisphere, has sporadically affected rice production in many tropical and subtropical rice production areas of the Americas. When RHBV was identified in Florida, USA, in the 1950s, a quarantine on Florida rice production was implemented in the hope of preventing RHBV spread into rice-producing areas of nearby states. Rice production in southern Florida did not resume until the late 1970s. RHBV is still of considerable importance in the rice-growing regions of Latin America. In their respective geographic regions, RStV and RGSV cause similarly important diseases in rice, and MStV causes annual losses in maize.

Virus–Vector Interactions and Transmission

Tenuiviruses are transmitted by specific planthopper vectors in a persistent-propagative manner. The time between when the planthopper acquires the virus, by either feeding on an infected plant or by intrathoracic injection of inoculum, and when the insect can then subsequently transmit the virus to plants is fairly long, measured in days. This relatively long period of time is suggestive of viral multiplication in the insect which most likely must occur before the insect can subsequently transmit the virus to plant hosts. Further evidence strongly suggesting replication of tenuiviruses in their insect vectors is that RHBV, RSV and MStV have been shown to be transovarially transmitted from viruliferous female planthoppers to their progeny. Recently, direct serological analysis has shown that MStV antigens increase over time in *Peregrinus maidis* after MStV acquisition (either by plant feeding or intrathoracic injection), providing conclusive evidence that MStV replicates in *P. maidis*, its planthopper vector.

Even though transmission of a given tenuivirus is very specific, often with insects of only one or a few species being able to serve as vectors, it appears that vector specificity may be even further affected by the genetic composition within the species population. It has been demonstrated that there are 'transmitters' and 'nontransmitters' for RHBV within *Tagosodes oryzicola* populations. The 'transmitters' are competent vectors of RHBV, while the 'nontransmitters' are not. Genetic studies have shown that 'transmitters' are capable of supporting RHBV replication, while 'nontransmitters' are not. By making crosses between individuals of these two types, the ability to support RHBV replication, and thus transmit RHBV to plants, was shown to be due to a single recessive gene.

Virus–Host Relationships

One of the most diagnostic features of tenuivirus infections of plants is the presence of large amounts of a tenuivirus-encoded protein in plant tissues. This protein, the major noncapsid protein (NCP), can easily be detected by serological methods, or directly by light and/or electron microscopy. The NCP is the most abundant protein in MStV-infected plant tissues at levels exceeeding 10 mg/g. Interestingly, the NCP has not been detected in MStV-infected *P. maidis*, the MStV planthopper vector. The reason for the differential abundance of this viral-encoded protein in plant as opposed to insect hosts is not understood, but the subgenomic RNA which corresponds to the NCP ORF is also extremely abundant in MStV-infected plants, and not in MStV-infected *P. maidis*. In contrast, the nucleoprotein (N protein) can easily be detected both in MStV-infected plant and insect hosts.

Electron microscopic analyses have shown the characteristic filamentous nucleoprotein particles within tissues of both plants and planthoppers infected by tenuiviruses. In addition, abundant inclusion bodies are evident in tenuivirus-infected plants. The inclusion bodies are detectable by light and/or electron microscopy and are of two morphological types. Amorphous, semielectron opaque inclusion bodies are by far the most abundant, sometimes appearing to fill the entire cytoplasm. For RHBV, these inclusion bodies appear to be composed of NCP because they specifically react with NCP antisera. However, for MStV, the second type of tenuivirus inclusion body, a filamentous, electron opaque body, was found to react with antiserum to the MStV NCP.

Serological and Nucleic Acid-based Relationships

Antibodies to two tenuivirus proteins, the NCP and N proteins, have been generated for most tenuiviruses, and limited serological comparisons have been done. Antibodies to each of these proteins have generally only given positive reactions with corresponding proteins of their respective viruses. However, by using less stringent serological tests, slight serological cross-reactions have been obtained in some instances between various tenuiviruses. The N proteins of RStV and MStV are distantly serologically related, as are those of RStV and RGSV. However, the MStV N protein does not show even a distant serological relationship to those of RGSV or RHBV. Only very slight serological cross reactions were detected when the MStV and RHBV NCPs were compared. Thus, current data suggest that tenuiviruses are largely serologically distinct from each other.

So far, no nucleic acid hybridization studies comparing the RNAs of different tenuiviruses have been reported. However, computer-based comparisons of nucleotide sequences from cloned cDNAs, and similar comparisons of deduced amino acid sequences have shown that the tenuiviruses are very closely related. Also, when the amino acid sequences for the MStV N protein and NS_3 proteins were compared with those of the corresponding proteins encoded by the other viral ambisense RNAs, the nonstructural (NS) proteins showed no homology. However, the MStV N protein shows significant homology at the amino acid sequence level to the corresponding N proteins of the vertebrate-infecting phleboviruses and uukuviruses. This suggests that not only do these viruses exhibit similar gene arrangements and expression strategies, but that some of their gene products are related as well.

Epidemiology and Control

Tenuivirus epidemiology is largely determined by the biology and behavior of their planthopper vectors. Planthopper vector populations fluctuate seasonally, and disease also fluctuates seasonally and from year to year. However, when crop incidence corresponds with high planthopper populations or activity, diseases caused by tenuiviruses often result. For example, in Florida MStV disease in *Z. mays* occurs in fall-planted maize crops. *P. maidis* populations build up on alternate grasses during the summer months. As the fall-planted maize plants emerge, planthoppers move into the young succulent plants and severe disease can result. Later fall, or winter or spring plantings are not as severely affected. A similar situation has been reported from Japan for RStV, and when high planthopper populations coincide with young plants, resulting disease can be severe.

Disease control strategies based upon epidemiological data (i.e. late planting to avoid planthopper activity) may offer some help in controlling diseases caused by tenuiviruses. However, at present, forecasting tenuivirus epidemics has not proved reliable, and this has hampered using tactics such as delayed planting. Genetic resistance seems to be the most useful and effective strategy for attempting to control RHBV, RGSV and RStV in rice. Active efforts to develop cultivars resistant to these tenuiviruses have been at least partly successful. Continued plant breeding efforts will probably be necessary to provide long-term tenuivirus disease control.

See also: Bunyaviruses.

Further Reading

Espinoza AM *et al.* (1992) *in situ* immunogold labeling analysis of the rice hoja blanca virus nucleoprotein and major noncapsid protein. *Virology* 191: 619.

Falk BW, Klaassen VA and Tsai JH (1989) Complementary DNA cloning and hybridization analysis of maize stripe virus RNAs. *Virology* 173: 338.

Gingery R (1988) In: Milne RG (ed.) The rice stripe virus group. *The Plant Viruses*, vol. 4. New York: Plenum Press.

Huiet L, Klaassen VL, Tsai JH and Falk BW (1991) Nucleotide sequence analysis and RNA hybridization analyses reveal an ambisense coding strategy for maize stripe virus RNA3. *Virology* 182: 47.

Ishikawa K, Omura T and Hibino H (1990) Morphological characteristics of rice stripe virus. *J. Gen. Virol.* 70: 3465.

Takahashi M *et al.* (1990) Complementarity between the 5′- and 3′- terminal sequences of rice stripe virus RNAs. *J. Gen. Virol.* 71: 2817.

TETRAVIRUSES

Donald Hendry
Rhodes University,
Grahamstown, South Africa

and

Deepak Agrawal
Purdue University,
West Lafayette, Indiana, USA

Introduction

The *Tetraviridae* (formerly known as the *Nudaurelia* beta virus group) are a family of small insect-pathogenic viruses containing single-stranded RNA of positive polarity as their genetic material. The family name reflects the unique $T = 4$ icosahedral surface symmetry demonstrated for certain of its members. These viruses occur world-wide and infect species of Lepidoptera, mainly Saturniid, Noctuid and Limacodid moths. They are of economic importance, serving as natural control agents of various insect pests.

Taxonomy and Classification

The type member of this family is *Nudaurelia* beta virus (NβV), which is one of at least five small isometric viruses recurrently infecting the pine emperor moth, *Nudaurelia (= Imbrasia) cytherea capensis* (Lepidoptera: Saturniidae) in South Africa. Viruses are classified as tetraviruses if they are serologically related to, or have capsids and protein subunits of similar sizes to, recognized members of this family. Table 1 lists the recognized as well as potential members of the *Tetraviridae*.

Table 1. Members of the family *Tetraviridae*

Virus name	Insect host	Origin
Recognized members		
Nudaurelia beta virus (NβV)	*Nudaurelia cytherea capensis*	South Africa
Antheraea eucalypti virus (AeV)	*Antheraea eucalypti*	Australia
Darna trima virus (DtV)	*Darna trima*	Malaysia
Thosea asigna virus (TaV)	*Thosea asigna*	Malaysia
Trichoplusia ni virus (TnV)	*Trichoplusia ni*	USA
Dasychira pudibunda virus (DpV)	*Dasychira pudibunda*	UK
Philosamia × virus (P×V)	*Philosamia cynthia × ricini*	UK
Nudaurelia omega virus (NωV)	*Nudaurelia cytherea capensis*	South Africa
Pseuduoplusia includens virus (PiV)	*Pseudoplusia includens*	USA
Potential members	*Saturnia pavonia*	UK
	Acherontia atropas	Canary Islands
	Setora nitens	Pakistan
	Eucocytis meeki	New Guinea
	Hypocrita jacobeae	UK
	Agraulis vanillae	Argentina
	Limantria ninayi	New Guinea
	Euploea corea	Australia
	Hyalophora cecropia	USA
	Heliothis armigera	Australia

Virion Composition

The tetravirus particle has a molecular mass of about 16 MD, a sedimentation coefficient ($s_{20,w}$) of 194–210, and is stable at pH 3.

The tetraviral genome consists of single-stranded messenger-sense RNA that is not polyadenylated. Electrophoretic analyses and physicochemical techniques indicate a genome size of 1.8–1.9 MD, corresponding to about 5500 bases and representing about 11% of the particle mass. The base composition (expressed as molar percentages of A:U:G:C) of NβV RNA is 24:23:28:25 and of *Pseudoplusia includens* virus (PiV) RNA is 26:21:31:22. The RNAs of two members of this family, *Trichoplusia ni* virus (TnV) and *Dasychira pudibunda* virus (DpV), have been shown by hybridization analysis to share little homology. The tetraviral genome is monopartite, except in the case of *Nudaurelia* omega virus (NωV) which has two RNAs of 0.8 and 1.8 MD. A recently isolated virus of *Heliothis armigera, H. armigera* stunt virus (HaSV), that is structurally similar but serologically unrelated to NωV, also contains a bipartite genome. These two viruses may herald the need for subgroups within the *Tetraviridae*.

A unique feature of the tetraviruses, and one that first alerted virologists to the existence of this family, is the presence of an unusually large capsid protein for a small RNA virus (see Table 2). Among members, this protein varies in size from 61 to 68 kD, although PiV has a smaller capsid protein of only 55 kD. The large tetraviral protein appears to be the sole capsid component, although the exceptions again are NωV and HaSV which possess in addition a second small capsid protein of about 6–8 kD.

The tetraviruses are characterized by a relatively low particle buoyant density of about 1.29 g ml^{-1}. This is consistent with the low average packing density of the RNA inside the capsid, deduced from three-dimensional image reconstruction from electron micrographs of NβV. PiV has a higher density of 1.33 g ml^{-1}, probably reflecting the reduced influence of its smaller capsid protein.

The biophysical properties of the most extensively characterized tetraviruses are listed in Table 2, and partial amino acid compositions of NβV, TnV and *Darna trima* virus (DtV) capsid proteins are presented in Table 3.

Virion and Genome Structure

The particle diameters determined for the various tetraviruses vary from 35 to 40 nm (Table 2). The

Table 2. Biophysical properties of the *Tetraviridae*

Virus*	RNA ($\times 10^6$)	Capsid protein ($\times 10^3$)	Buoyant density in CsCl ($g\,ml^{-1}$)	Sedimentation coefficient (s)	Diameter (nm)
NβV	1.8	61	1.295	210	39.7
AeV	+	–	–	215	50
TnV	1.9	67–68	1.3	200	35–38
DtV	+	62–66	1.289	199	35–38
TaV	+	60.8	1.275	194	35
PxV	+	62.4	1.275	206	35
DpV	1.8	66	1.31	–	38
NωV	0.8, 1.8	8, 62	1.285	–	40
PiV	1.9	55	1.33	190	40

Key: +, contains RNA; –, not determined.
* See Table 1 for full virus names.

most accurate figure, determined for NβV using cryoelectron microscopy which minimizes particle distortion, is that of 39.7 nm. The figure of 50 nm reported in 1965 for AeV needs to be re-examined using modern techniques.

Detailed capsid structures have been elucidated for both NβV and NωV, using image reconstruction from electron micrographs of negatively stained or frozen specimens (Figs 1 and 2). Both virions have a distinct icosahedral shape and contain 240 copies of the large capsid polypeptide arranged with $T = 4$ symmetry. Each triangular face of both the NβV and NωV capsids is composed of 12 protein subunits clustered in four Y-shaped trimeric aggregates. The subunits are packed such that each face is nearly planar, contains three deep pits (absent in NωV) and is separated from neighboring faces by deep grooves (Figs 1 and 2). The computed density distribution of the NβV particle indicates that each of the subunits consist of two domains: small domains that associate at radii between 13 and 16.5 nm to form the contiguous capsid shell (Fig. 2), and large cylindrical domains that each associate with two similar neighboring domains to form the trimeric aggregates in the outer capsid surface.

Table 3. Amino acid compositions (residues per 1000) of tetraviral coat proteins

Amino acid	TnV	NβV	DtV
Asp	106.15	117.90	103.65
Thr	62.56	96.97	83.33
Ser	61.77	46.25	69.42
Glu	121.00	84.27	87.92
Pro	48.60	84.16	68.81
Gly	96.14	95.08	107.40
Ala	87.55	100.37	88.99
Val	75.72	86.38	87.75
Met	10.42	25.58	16.48
Ile	53.14	41.72	52.00
Leu	74.32	66.36	69.75
Tyr	21.77	15.15	26.07
Phe	39.02	38.05	54.16
His	15.87	20.10	16.88
Lys	78.29	24.98	20.26
Arg	47.81	56.68	47.40

See Table 1 for full virus names.

The *Tetraviridae* are apparently the only viruses whose capsid structure is solely organized with $T = 4$ icosahedral symmetry. However, both Sindbis and Semliki Forest viruses, which have inner nucleocapsids with $T = 3$ symmetry, are surrounded by a lipid bilayer containing glycoproteinaceous spikes arranged on a $T = 4$ lattice. Also, the $T = 16$ outer capsid of herpes simplex virus appears to surround a $T = 4$ intermediate layer.

Nucleotide sequence analysis of the small RNA (RNA2) of NωV has revealed that it consists of 2450 nucleotides and encodes the capsid polypeptide (Fig. 3). RNA2 contains a single long open reading frame (ORF) consisting of 644 codons initiating at the second $A^{367}UG$ codon from the 5′-end, terminating at UAA^{2301}, and representing a translation product of 70 kD. The 5′-proximal AUG is followed by a short ORF (30 codons) of unknown function. The deduced capsid protein

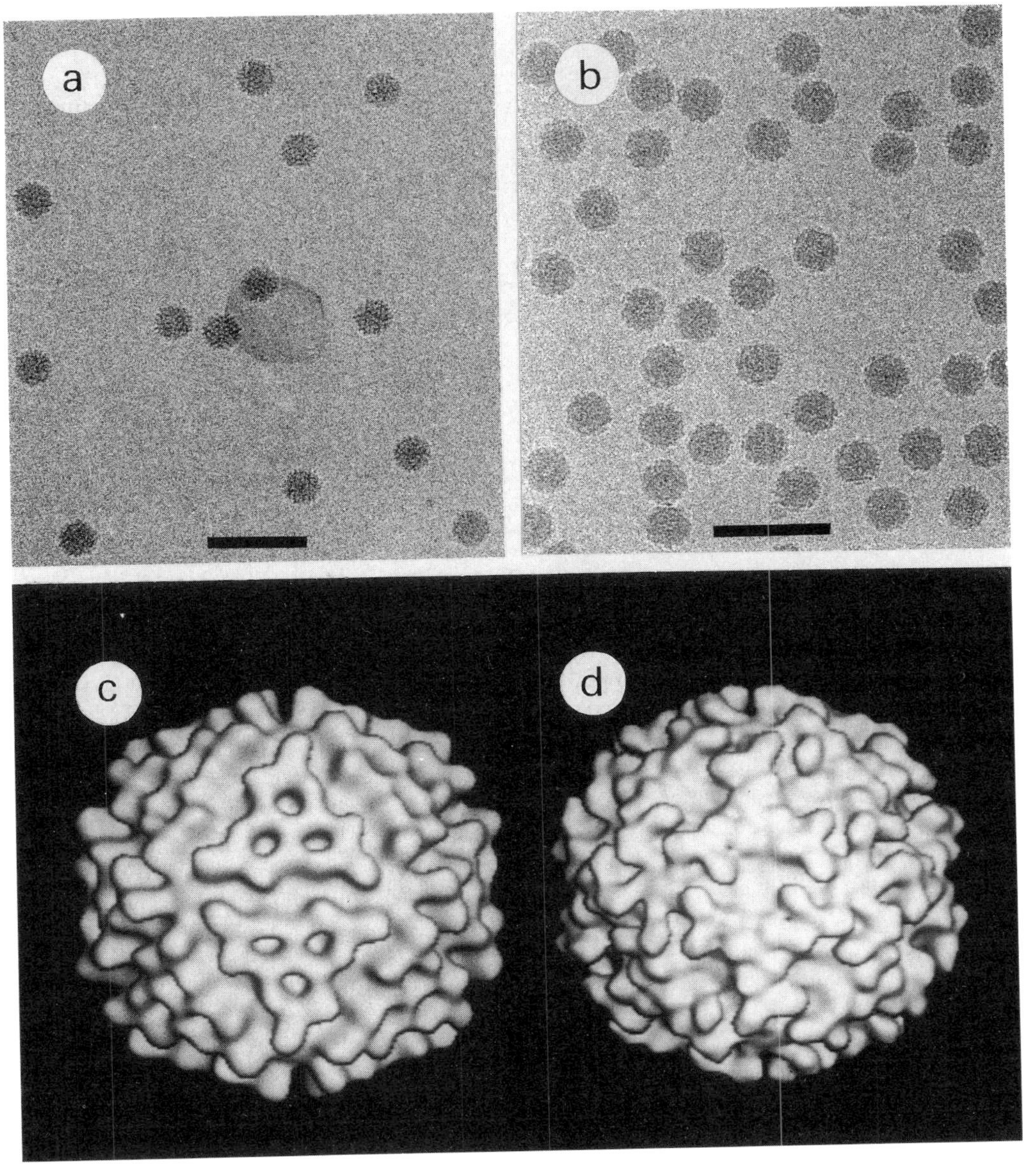

Fig. 1 Electron micrographs and image reconstructions of frozen-hydrated *Nudaurelia* viruses NβV (**a,c**) and NωV (**b,d**). Scale bars = 100 nm (note the different magnifications in **a** and **c**). (Kindly provided by Norman Olson and Tim Baker.)

sequence has characteristics also present in other virus structural proteins. In addition to a basic N terminus (net charge +11 over the first 40 residues), it contains many regions that are predicted to fold in β-sheet structures. The chemically determined N-terminal sequence of the small 8 kD capsid protein corresponds to a portion of the long ORF beginning at residue 571. The lack of a nearby methionine codon precludes the possibility that this small component is independently translated. It is more likely that it is produced as a result of post-translational cleavage from the 70 kD precursor (the product of the long ORF), thereby also giving rise to the 62 kD species observed by SDS–PAGE of purified virions. This idea is also supported by the nearly 1:1 stoichiometry of the large and small polypeptides. In this respect, NωV resembles the nodaviruses, the capsids of which also contain a large and a small protein that are generated by the cleavage of a precursor translated off the smaller of two genomic RNAs. However, definite evidence for such a cleavage in the case of the tetraviruses awaits the ability to synthesize the putative capsid precursor.

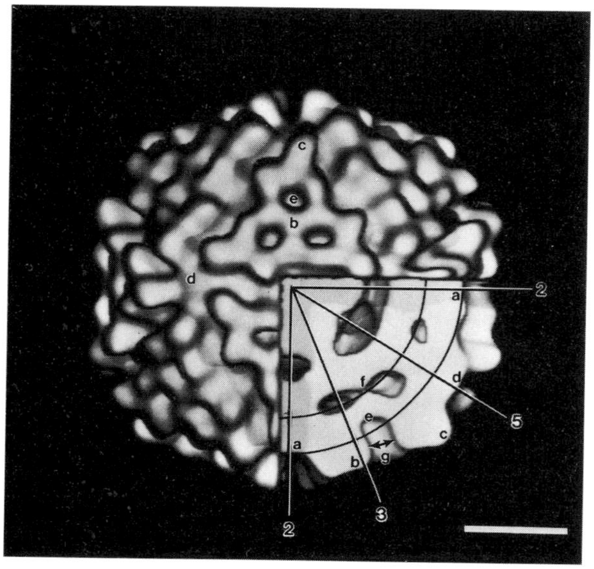

Fig. 2 Shaded surface representation of NβV reconstruction viewed along a two-fold axis with a lower right octant removed to show internal features. Approximate boundaries of the contiguous shell of protein are defined by circular arcs. Icosahedral twofold, threefold and fivefold axes in the equatorial plane are identified. The radial dimensions (nm) of the labeled features are as follows: **a** twofold axis, 16.6; **b** threefold axis, 19.4; **c** highest radial feature, 21.3; **d** fivefold axis, 18.8; **e** base of pit, 15.2; **f–d** maximum contiguous shell thickness, 6.1; **f–e** minimum contiguous shell thickness, 2.5; **g** diameter of pit, 2.4. Scale bar = 10 nm. (Reprinted with permission from Academic Press Inc.)

Serology

The *Tetraviridae* are good immunogens, readily yielding hyperimmune sera. Immunodiffusion, enzyme-linked immunosorbent assay (ELISA), crossed-over immunoelectrophoresis (COIE) and complement fixation reactions indicate that most members of this family are related in varying degrees to NβV, the type member (Table 4). The most closely related pairs, as judged by comparative ELISA, are NβV and DpV, followed by NβV and DtV, DtV and DpV, DpV and TnV, NβV and TnV and DtV and TnV. TnV shares the fewest antigenic determinants with the tetraviruses listed, while NωV is apparently totally unrelated serologically to NβV.

Virus–Host Interaction and Pathology

Insects appear to acquire viruses of this family by ingestion, as larvae reared in the laboratory readily become infected if fed with virus suspensions. Although it is not clear how this occurs in the wild, emerging larvae could become infected as they bite their way out of eggs superficially contaminated with virus. Virus is present in insect frass, but whether this could play any role in transmission is unclear. Transovarial transmission from infected adults is another possibility. Injection is an effective means of artificial infection in the laboratory.

The members of this group of viruses exhibit a wide range of pathogenic effects, mainly on the larvae of their hosts, although NβV can be isolated from

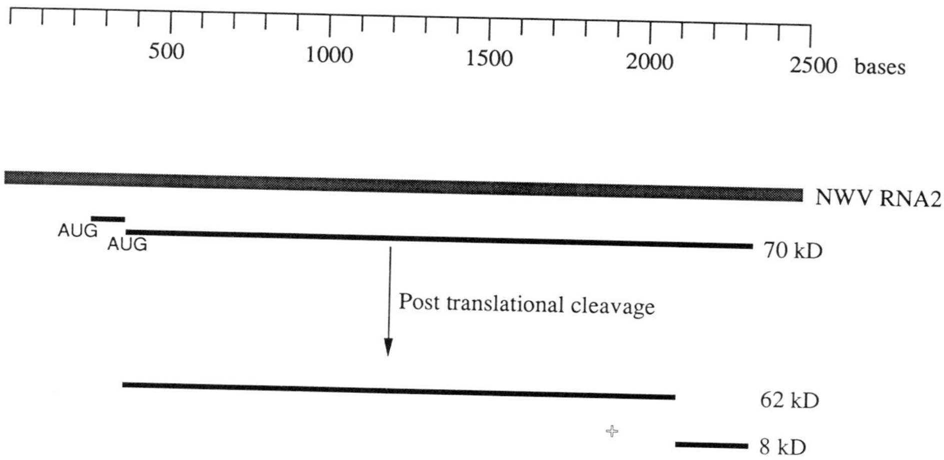

Fig. 3 Organization of NωV RNA2. The RNA2 segment of NωV (gray bar) encodes the capsid polypeptides. Two start codons are present, with the second one initiating the translation of a single long open reading frame coding for a 70 kD precursor that is probably post-translationally cleaved to yield the 62 kD and 8 kD capsid proteins of NωV. The function of the small upstream open reading frame is unclear.

Table 4. Serological interactions among the *Tetraviridae*

Virus	Antiserum							
	NβV	DpV	DtV	TnV	TaV	P×V	AeV	PiV
NβV	++	++	+	−	+	+	+	*
DpV	++	++	+	(+)	*	*	*	*
DtV	+	+	++	−	+	+	*	*
TnV	(+)	−	−	++	*	*	*	−
TaV	+	*	+	*	+	+	*	*
P×V	+	*	+	*	+	+	*	*
NωV	−	−	*	*	*	*	−	*
AeV	+	*	*	*	*	*	+	−
PiV	*	*	*	−	*	*	−	+

Key: ++, strong positive reaction (obtained using comparative ELISA); +, positive reaction; (+), weak reaction (obtained using either ELISA, high virus concentrations or COIE); −, negative reaction; *, not tested. For full virus names see Table 1.

infected adults and pupae. Larvae of *T. ni* dosed with as much as 10 μg of TnV show no appreciable mortality apart from a marked weight reduction, which is observed even with as little as 0.1 ng of virus. Pupation is often delayed in infected larvae and the pupae are smaller. Except for size, these larvae are morphologically similar to their healthy counterparts and show no abnormal body proportions. At dosages lower than 0.1 ng, no weight loss occurs, although virus can be detected by ELISA. TnV is detectable in the frass of infected larvae.

Most other tetraviruses cause severe symptoms, usually resulting in death, in host larvae of all instars. Insects infected with *Antheraea eucalypti* virus (AeV), NβV, NωV, DpV, *Philosamia* × virus (P×V), DtV or *Thosea asigna* virus (TaV) become moribund, usually accompanied by discoloration and flaccidity. Vomiting can occur accompanied by the voiding of loose or discolored feces. Infected larvae of *A. eucalypti* and *N. c. capensis* often hang from branches by the last pair of prolegs or fall to the ground, *N. c. capensis* larvae often littering the forest floor during heavy insect outbreaks. The final stages of the disease are often characterized by complete liquefaction of the internal organs. Infected larvae which manage to pupate often die before reaching the adult form, or emerge as underdeveloped or malformed adults.

The multiplication site of TnV, NβV, P×V and DpV has been shown by electron microscopy of tissue sections and by fluorescent antibody staining to be the cytoplasm of foregut and midgut cells. Virus particles are visible as crystalline arrays within cytoplasmic vesicles. Cells infected with TnV also show a mitochondrial structural alteration, the matrix granules becoming prominent and of increased size and number. If organs of infected insects are examined by ELISA, virus antigens are detected only in gut cells until very late in the infection cycle. Thereafter, antigens are detected in the immature sex organs and fat-bodies. No inclusion bodies have been observed in infected cells.

Replication

Investigation of the replication of tetraviruses has been hampered by the lack of an insect cell line supporting virus multiplication. This has resulted in our understanding of tetraviral replication and molecular biology lagging far behind our understanding of tetraviral structure. However, NβV does replicate in primary ovarian cell cultures obtained from both *N. c. capensis* and the silkworm, *Bombyx mori*, without causing any cytopathic effects.

Tissue infected with NβV contains only one major double-stranded RNA (dsRNA) that hybridizes with a radiolabeled NβV RNA probe and is the correct size to be the genomic replicative form. This is further evidence for NβV having a monopartite genome, and suggests that replication does not involve subgenomic RNAs. In contrast, NωV-infected tissue contains two dsRNAs of the correct size to be genomic replicative forms, confirming the bipartite nature of this virus. Tetraviral RNAs are, in general, poor messengers in cell-free translation systems, seldom stimulating an incorporation of radioactive amino acids that is more than 10 times background. A common feature is that the genomic RNA does not act as messenger for the viral coat protein in these systems. Translation of NβV, NωV, TnV and DpV RNAs in rabbit reticulocyte lysates results in the production of a range of polypeptides, with the largest (possibly the polymerase) being in excess of 100 kD and with no significant polypeptide comigrating with authentic coat protein in polyacrylamide gels. Sequencing of the smaller RNA or NωV indicates that the coat proteins probably arise by cleavage of a precursor; in fact, products slightly larger than coat protein are often observed when tetraviral RNAs are translated. However, attempts to precipitate these with virus antiserum have proved inconclusive.

In the absence of a cell line supporting tetraviral replication, elucidation of their genome structure and replication strategy will have to await the complete sequencing of a tetraviral genome.

Ecology

Little is known of the basic ecology of these viruses, mainly due to their sporadic occurrence in the wild. Thus, aspects such as the existence of natural virus reservoirs, their methods of spread, and host specificities are all poorly understood. With some of these viruses, it would appear that low levels could remain dormant in the insects, their emergence possibly being triggered by stress conditions such as high population density or unseasonably high temperatures.

Tetraviruses can occur in mixed infections. NβV was originally detected among a complex of at least five other viruses infecting *N. c. capensis*, although whether they all occurred simultaneously within individual insects is unclear. Likewise, NωV was first isolated from an insect population co-infected with NβV. Virus succession occurs, at least in *N. c. capensis*, as NωV was seen over several years to gradually emerge in regions previously endemic for NβV, where NβV has subsequently disappeared and NωV has become the dominant pathogen of the insect. Mixed infections also occur with viruses from other families. TnV was first detected in *T. ni* larvae that had been infected with contaminated inocula of the nuclear polyhedrosis virus (AcMNPV) of the alfalfa looper, *Autographa californica*. Bioassays indicate that the addition of TnV to AcMNPV inoculum only slightly increases its potency against *T. ni* larvae. TnV and AcMNPV can be demonstrated in the cytoplasm of the same *A. californica* or *T. ni* midgut cells, and TnV also becomes occluded within AcMNPV polyhedra.

Although tetraviruses are regarded as being specific for insects, antibodies reactive against certain of these viruses have been isolated from mammals. Antibodies to DtV have been detected in sera from several species of domestic animals (including cows, sheep and pigs) in the UK and from some wild animals from East Africa. Precipitating antibodies to DtV have also been observed in a small percentage of human sera obtained from both Malaysia and the UK. These antibodies were present in adolescents and adults of both sexes in whom no associated illness was identified. The Malaysian results could be ascribed to exposure following the deliberate release of DtV in an attempt to control *D. trima* in oil palm plantations. However, the presence of these antibodies in human sera obtained in the UK suggests a prior infection with a virus or viruses serologically related to DtV. The presence of nonviral heterophile antigens in the environment cannot be excluded. Human antibodies to NβV have also been detected in southern Africa.

The phenomenon of vertebrate antibodies to an invertebrate virus is not unique to the *Tetraviridae*, as mammalian antibodies have also been detected to an insect picornavirus infecting *Gonometa podocarpi* (Lepidoptera: Lasiocampidae).

Economic Importance

A number of tetraviruses serve as natural control agents of pest insects. *Darna trima* and *Thosea asigna* (both Lepidoptera: Limacodiidae) are pests of coconut and oil palms in South East Asia. Large populations of particularly the former insect can produce almost complete defoliation, and DtV has been used successfully as an applied control agent of *D. trima* on oil palms in Sabah and Sarawak. Similarly, *N. c. capensis* is a severe defoliator of commercial plantations of the pine, *Pinus radiata*, in certain regions of South Africa. Both NβV and NωV periodically cause widespread decimation of the insect, resulting in drastically reduced numbers for the following few seasons until the populations build up and the cycle repeats itself. Foresters use insect counts conducted early in the insect growing season to predict whether costly control measures will be necessary later. An ability simultaneously to estimate the extent of virus infection and thereby also the probable mortality would undoubtedly assist them in their predictions. Viruses have not been used to control *N. c. capensis* artificially.

Apart from any logistical and economic considerations, the use of tetraviruses as applied control agents of insects should not be attempted until the presence of mammalian antibodies to these viruses has been explained.

See also: Honey bee viruses; Nodaviruses; Picornaviruses; Sindbis and Semliki Forest viruses.

Further Reading

Moore NF (1990) In: Kurstak E (ed.) *Viruses of Invertebrates* p. 277. New York: Marcel Dekker.

Moore NF, Reavy B and King LA (1985) General characteristics, gene organization and expression of small RNA viruses of insects. *J. Gen. Virol.* 66: 647.

Olson NH, Baker TS, Johnson JE and Hendry DA (1990) The three-dimensional structure of frozen-hydrated *N. c. capensis* β virus, a $T = 4$ insect virus. *J. Struct. Biol.* 105: 111.

THEILER'S VIRUSES

Howard L Lipton
Evanston Hospital, Evanston, IL USA

History, Geographic Distribution and Host Range

The mouse encephalomyelitis viruses are naturally occurring enteric pathogens of mice. Discovered by Max Theiler in the early 1930s and originally called murine polioviruses, these agents are frequently referred to as Theiler's murine encephalomyelitis viruses (TMEV). Theiler initially recovered isolates from several mice with spontaneous paralysis that were being housed in a research colony; subsequently the TMEV have been found to be present in virtually all nonbarrier mouse colonies, where they cause asymptomatic intestinal infections. While the TMEV are widely distributed throughout the world, their host range appears to be quite narrow. Serological evidence indicates that *Mus musculus* (the house mouse) is the natural host but that other species of mice, voles and rats may also serve as hosts. As is the case for other picornaviruses, TMEV can spread to the central nervous system (CNS) producing encephalitis, or more commonly, spontaneous paralysis, i.e. poliomyelitis. The incidence of spontaneous paralysis is low, around one paralyzed animal per 1000 to 5000 mice in a colony. Since the TMEV may go undetected unless appropriate serological testing for the virus is performed, these agents are a potential hazard for investigators using mice in biomedical research.

In recent years, this group of viruses has assumed additional importance because TMEV infection in mice provides one of the few available experimental animal models for multiple sclerosis. TMEV-induced demyelinating disease is perhaps the most relevant animal model for multiple sclerosis because: (1) chronic pathological involvement is limited to CNS white matter; (2) myelin breakdown is accompanied by mononuclear cell inflammation; (3) demyelination results in clinical disease from involvement of upper motor neurons, e.g. spasticity; (4) myelin breakdown is immune-mediated; and (5) the disease is under multigenic control with a strong linkage to certain major histocompatibility complex genes.

Classification and Serologic Relationships

Based on the complete nucleotide sequence and genome organization, TMEV have recently been classified as cardioviruses in the family *Picornaviridae* along with encephalomyocarditis virus (EMCV) and Mengo virus (Table 1). The TMEV constitute a separate serological group of cardioviruses since polyclonal antisera show no cross-neutralization between TMEV and EMCV or Mengo virus. Because the coat proteins share a high level of amino acid sequence identity with the other cardioviruses, cross-reactions are seen on ELISA when disrupted virions are used as antigens and on complement fixation tests.

Three-dimensional Virion Structure

Picornavirions have a relative molecular mass of $\sim 8.5 \times 10^6$ D, of which $\sim 30\%$ is RNA. The spherical virus particles have an external diameter of about 300 Å. The capsid proteins are arranged in icosahedral symmetry with 60 protomers each composed of a single copy of VP1, VP2, VP3 and VP4.

Recently, the three-dimensional structures of the BeAn and DA strains have been determined at ~ 3 Å resolution by X ray crystallography. The overall architecture is very similar to that of other picor-

Table 1. Classification of TMEV in the family *Picornaviridae*

Human enteroviruses: Polioviruses, Coxsackieviruses, Echoviruses
Human rhinoviruses
Hepatitis A viruses
Aphthoviruses: Foot and mouth disease viruses
Cardioviruses
Group A: EMCV, Mengo, MM, Columbia-SK, Maus-Elberfeld
Group B: TMEV (GDVII, BeAn8386, DA, FA, WW, TO4, etc strains)

naviruses whose structures have been determined, and closely resembles that of Mengo virus. Each of the three major capsid proteins VP1, VP2 and VP3 consist of a wedge-shaped eight-stranded antiparallel β barrel as demonstrated for other picornaviruses. The N termini of the capsid proteins form an extensive, intertwined network on the inner surface of the protein shell. The loops connecting the β strands form the outer surface features of the protein shell and provide the surface differences with Mengo virus. The pit area which has been proposed as the viral receptor is a broad depression over the junction between VP1 and VP2 along the twofold axis. The BeAn and DA surface structures can be differentiated from that of Mengo virus in having: (1) a larger VP1 CD double loop with loop I containing an extra five residues and shifted more toward the VP2 EF puff at the twofold axis while loop II points more toward the fivefold axis; (2) an 11 residue insertion in the VP2 EF puff forming a double loop in which the inserted loop interacts with the VP1 FMDV GH loop; and (3) the tip of the VP3 knob (a loop inserted in βE) points straight outward on the rim of the pit.

RNA Genome

The genetic component of the virion is a positive-sense, single-strand RNA molecule that has a sedimentation coefficient of 35S and 8100 nucleotides long. Virion RNA has a poly(A) tract on the 3′ end and a small basic protein, VPg, covalently linked to the 5′ end. The complete genomes of the GDVII, DA and BeAn strains have been cloned and sequenced. With the notable absence of a poly(C) tract in the 5′ noncoding region, the organization and sequence of the TMEV genome is remarkably similar to that of EMCV. The polyprotein of the BeAn strain, a typical TMEV, initiates at the AUG codon at nucleotide 1065 and extends for 6909 (or 2303 codons) ending at the single UGA termination triplet at base 7972. The polyprotein coding region is flanked by 5′ and 3′ noncoding sequences of 1064 and 125 nucleotides, respectively. In BeAn the 5′ noncoding region contains a stretch of 11 pyrimidines interrupted by a single purine before the AUG at nucleotide 1065. In picornaviruses, the 5′ noncoding region is thought to mediate cap-independent translation and also serves as an internal ribosome entry site (IRES) when present in the intercistronic region in a bicistronic mRNA. The TMEV 5′ noncoding sequences have been predicted to form stable secondary structures which in the 500 nucleotides upstream of the authentic AUG (at 1065) are nearly identical to those predicted for EMCV and the foot and mouth disease viruses. In BeAn, eight AUGs precede the initiator AUG, but none of them has an optimum Kozak context sequence. Hence, selection of the authentic initiator AUG after binding of ribosomes to TMEV RNA could be argued to not involve internal ribosome binding. However, BeAn nucleotides ~ 500 to 1065 determine a structure that serves as an IRES in bicistronic mRNAs both *in vitro* (rabbit reticulocyte lysate) and *in vivo* (BHK-21 cells). On the 3′ end of the viral genome, a poly(A) tail of indeterminate length is present.

Polyprotein and Post-translational Processing

As is the case for other picornaviruses, the final TMEV gene products are the result of post-translational cleavages of the polyprotein. The 2303 amino acid polyprotein has a calculated molecular weight of 255 990. (This applies to BeAn virus; the polyprotein of GDVII virus contains no insertions or deletions; however, two VP1 amino acids are deleted in that of DA virus). The processing scheme follows the standard L4-3-4 picornavirus polypeptide arrangement, i.e. the leader peptide (L), four capsid polypeptides in part one (P1) of the genome, three polypeptides in P2 and four polypeptides in P3 (Fig. 1 and see below). The coding limits of individual polypeptides have been predicted by analogy with those of EMCV, since the only confirmation to date of the deduced sequence is that of the N terminus of 1D. The eight amino acids flanking the putative cleavage sites are highly conserved for the two viruses. All of the cleavage sites in the polyprotein except for two, 1A/1B and 2A/2B, are believed to be processed by the viral protease 3C. The TMEV 3C protease therefore processes the Q-C, as well as the Q-S and Q-A, dipeptides and, in addition, the E-N dipeptide at the 1D/2A cleavage. However, only 6 of 8 Q-G, 2 of 13 Q-S and 1 of 7 Q-A dipeptides in the polyprotein are cleaved by 3C, indicating that involvement of secondary, tertiary, or both types of structure is also important for recognition of these particular dipeptides. The 2A/2B site may be cleaved autocatalytically as in EMCV.

Cleavage of the polyprotein gives rise to three primary products, the first of which (116 530 mol. wt) contains the leader protein (8593 mol. wt), the P1 capsid proteins, and the first P2 polypeptide 2A

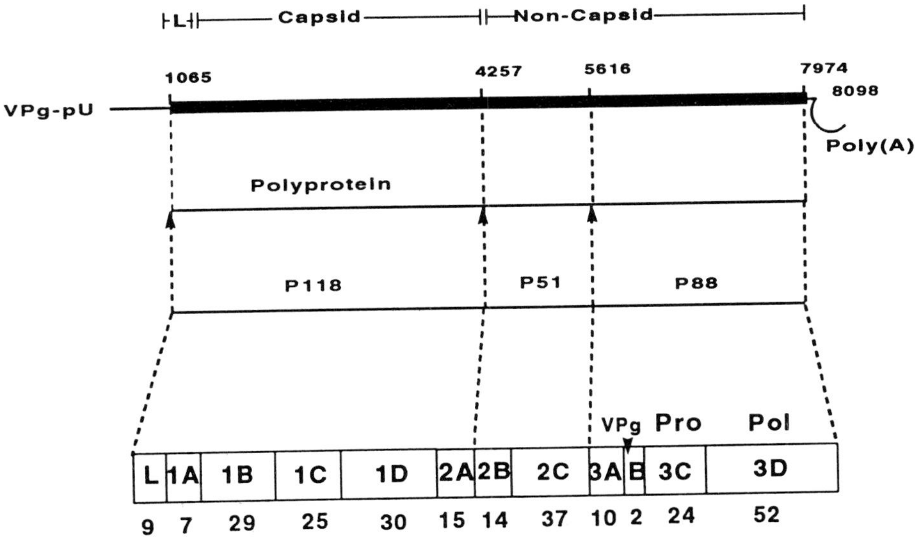

Fig. 1 TMEV-specific protein cleavage scheme. The numbers at the base are the molecular weights (in thousands) of each of the proteins as calculated from their predicted amino acid sequences.

(15 353 mol. wt). Thus, the initial precursor released from the polyprotein is like that of the other cardioviruses and differs from that of other groups of picornaviruses. The capsid proteins are arranged in the following order: 1A (VP4; 7102 mol. wt), 1B (VP2; 29 433 mol. wt), 1C (VP3; 25 463 mol. wt), and 1D (VP1; 30 457 mol. wt). The second processing precursor (2BC) is 51 708 mol. wt and gives rise to 2B (14 863 mol. wt) and 2C (36 845 mol. wt). The P2 proteins 2A, 2B and 2C have not been assigned functions as yet for the cardioviruses. The third or C-terminal precursor protein is 87 950 mol. wt and is processed into the four mature proteins 3A (9934 mol. wt), 3B (2169 mol. wt), 3C (23 612 mol. wt), and 3D (52 235 mol. wt). Protein 3B, also designated VPg, is a small basic protein which is 20 amino acids in size and is found covalently linked to the 5′ end of viral RNAs. This peptide may be important in viral replication. By analogy with other viruses, the 3C polypeptide is a viral protease and 3D is the viral polymerase.

Physical Properties

Since the TMEV do not have an envelope, they are more resistant than lipid-containing viruses to chemicals and physical agents. TMEV are insensitive to chloroform, ether, nonionic detergents (such as deoxycholate, NP40 and Tween-80) and the ionic detergent sodium dodecyl sulfate, but are inactivated by 0.3% formaldehyde and 0.1 N HCl. TMEV are rapidly destroyed at temperatures over 50°C and lose some infectivity upon lyophilization. Purified virions can be stored for long periods of time at −70°C without loss of infectivity, but slowly lose infectivity on storage at −20°C.

As enteroviruses the TMEV require stability at low pH to pass through the acidic conditions of the stomach. The TMEV are stable over the entire pH range from 3 to 9.5. However, in contrast to the other cardioviruses, such as Mengo virus and EMCV, they are not highly thermolabile in the presence of 0.1 M chloride or bromide in the pH range 5 to 7.

The virions have a sedimentation coefficient of 150S by velocity centrifugation in sucrose and a buoyant density of 1.34 g/ml by isopycnic centrifugation in cesium salts.

Replication

The reader is referred to chapters on human poliovirus and cardioviruses for the strategy of picornavirus RNA replication since no information is available on this topic for the TMEV.

Mapping Genomic Determinants Important in Pathogenesis

The existence of two distinct TMEV neurovirulence

groups makes the TMEV particularly useful for molecular pathogenesis studies. The difference in virulence between the highly virulent and less virulent TMEV groups is on the order of $\geq 10^5$ plaque-forming units per LD_{50}. Further, full-length cDNA clones of the highly virulent GDVII virus and the less virulent DA and BeAn viruses have been constructed in different laboratories, and RNA transcribed from these cDNAs has been shown to be infectious upon transfection of BHK-21 cells. To identify the determinants important in pathogenesis, e.g. virulence and persistence, recombinant chimeras between parental cDNAs have been assembled and analyzed *in vitro* and after inoculation of mice.

Neurovirulence has been mapped primarily to the P1 region encoding the leader and the coat proteins, while virus persistence appears to have been narrowed even further to the sequences encoding VP1. These results suggest that the mechanism for the pathogenetic properties of virulence and persistence are likely to involve the exterior surface of the virus, and that immunological or receptor-mediated events may be involved. Chimeric virus constructs in the coat protein region resulting in the interaction of potentially disparate protomeric subunits of two parental viruses have been found to be prone to assembly defects. Such chimeric viruses may exhibit compromised growth *in vitro*. Thus, the pathogenetic phenotypes of chimeric viruses need to be interpreted with caution and require analysis of their growth properties *in vitro*.

Transmission and Tissue Tropism

TMEV are transmitted by the fecal–oral route but can be separated into two biological groups based on neurovirulence (Table 2). The first group consists of three isolates, GDVII, FA and Ask-1 viruses, which are highly virulent and cause a rapidly fatal encephalitis in mice. All of the other TMEV (some 10–20 isolates), including viruses recovered from the CNS of spontaneously paralyzed mice and from the feces of asymptomatic mice, form a second less virulent group. Experimentally, the less virulent viruses produce poliomyelitis (early disease) followed by demyelinating disease (late disease); however, the poliomyelitis phase is subclinical when cell culture adapted viruses are used for inoculation. Brain-derived stocks produce both less virulent disease phases.

Pathogenesis

Little information is available about the pathogenesis of TMEV infection following peripheral routes of infection, including feeding of virus. In general, isolates from either of the TMEV neurovirulence groups do not readily produce CNS disease following peripheral routes of inoculation with the exception of one strain, TO(B). TO(B) is a mutant selected for its invasiveness from the intestinal tract. When mice are inoculated intracerebrally with the highly virulent strains, the virus replicates throughout the brain and spinal cord causing encephalitis or encephalomyelitis. Thus, neurons as well as glial cells (astrocytes and oligodendrocytes) are infected in the cerebral cortex, hippocampus, basal ganglia, thalamus, brainstem and spinal cord. Affected mice develop a hunched posture and have hindlimb paralysis. The rapid demise of these animals is the result of widespread lytic infection. The

Table 2. Two TMEV neurovirulence groups

	Highly virulent	Less virulent
TMEV isolated	GDVII, FA, ASK1 VIE 415_{HTR}	DA, BeAn 8386, Yale, WW TOB, VL, TO4, Vilyuisk
Disease	Encephalitis	Polio/demyelination
Incubation period[a]	1–10 days	7–20 days/>30 days
CNS target cell[b]	Neurons	Motor neurons/macrophages
Mean LD_{50}	< 10 PFU	$>10^6$ PFU
Persistent infection	No	Yes
Temperature sensitive[c]	No	Yes

[a]Incubation period following experimental infection by the intracerebral route of inoculation.
[b]Preferential site of virus replication in the central nervous system (CNS).
[c]Inability to replicate at 39.8°C compared to 33 or 37°C.

following sections focus on the pathogenesis of the biphasic disease produced by the less virulent TMEV which provides a model system for multiple sclerosis.

Clinical Features of Infection

Although TMEV are enterically transmitted, the pathogenesis of the infection has been primarily studied using the intracerebral route of inoculation which maximizes the incidence of neurological disease. Following intracerebral inoculation, the less virulent strains produce a distinct biphasic CNS disease in susceptible strains of mice, characterized by poliomyelitis during the first few weeks postinfection (p.i.), followed by a chronic, inflammatory, demyelinating process that begins during the second or third week p.i. and becomes manifest clinically between 1 and 3 months p.i. Mice with poliomyelitis develop flaccid paralysis, usually of the hindlimbs; only one limb may be affected, or paralysis may spread to involve all limbs and lead to death. In contrast to the fatal outcome of paralysis produced by the Lansing strain of human poliovirus type 2, complete recovery from TMEV-induced poliomyelitis is unusual. Occasionally, residual limb deformities are seen as the result of extensive anterior horn cell infection and severe paralysis (early disease).

Gait spasticity is the clinical hallmark of the demyelinating or late disease. Late disease is first manifest by slightly unkempt fur and decreased activity, followed by an unstable, waddling gait. Subsequently, generalized tremulousness and ataxia develop, and the waddling gait evolves into overt paralysis. Incontinence of urine and priapism are commonly seen. As the disease advances, prolonged extensor spasms of the limbs followed by difficulty in righting can be induced. The clinical manifestations of late disease are progressive and lead to the animal's demise in several to 14 months.

Pathogenesis and Histopathology

Motor neurons in the brainstem and spinal cord are the main targets of infection during poliomyelitis (early disease), but sensory neurons and astrocytes are also infected. TMEV does not replicate in endothelial and ependymal cells or initially in oligodendrocytes. A brisk microglial reaction is elicited with the appearance of numerous microglial nodules, particularly in the anterior gray matter of the spinal cord. Examples of neuronophagia are quite frequent at this time, but very little lymphocytic response is seen. The poliomyelitis phase lasts 1–4 weeks, after which time little residual gray matter involvement is apparent other than resolving astrocytosis.

As early as 2 weeks p.i., inflammation of the spinal leptomeninges begins to appear, followed by involvement of the white matter. Initially, the inflammatory infiltrates are almost exclusively composed of lymphocytes, but at later times plasma cells and macrophages are numerous. The influx of macrophages is in close temporal and anatomic relationship with myelin breakdown. Both light and ultrastructural studies of the demyelinating process show that myelin breakdown is strictly related to the presence of mononuclear cells, which either actively strip myelin lamellae from otherwise normal-appearing axons or are found in contact with myelin sheaths undergoing vesicular disruption. Foci of inflammation and myelin destruction extend from the perivascular spaces into the surrounding white matter, leading to sharply demarcated plaques of demyelination. The ultrastructure of oligodendrocytes during the initial phase of myelin breakdown has not shown alterations in oligodendroglial loops, which are in close apposition with naked but otherwise normal axons, suggesting that myelin injury is not directly related to oligodendrocytopathology.

Sites of TMEV persistence

TMEV persistence involves active or ongoing virus replication, since infectious virus can be readily isolated from the CNS. TMEV replication has been shown to be restricted by a block at the level of minus strand viral RNA synthesis. Highly restricted virus production has also been demonstrated in macrophages isolated from the CNS of diseased mice. But greater details of the kinetics of replication remain to be elucidated, such as the length of the replicative cycle in infected cells and whether infected cells are lysed or continue to produce infectious virus. It is also not known whether the oligodendrocyte or macrophage is the primary target of persisting virus. Some oligodendrocytes are infected and an ultrastructural study reported finding an isolated oligodendrocyte containing crystalline arrays of Theiler's virions. However, increasing evidence favors the macrophage as the target cell for persistence. A conservative estimate suggests that

90% of chronic TMEV infection takes place in these phagocytic cells.

Immune Response

During the first week, TMEV-infected mice mount a virus-specific humoral immune response that reaches a peak by 1 to 2 months p.i. and may be sustained for the life of the host. Neutralizing and other virus-specific antibodies have been measured. The majority of the antiviral IgG response in persistently infected, susceptible mice is of the IgG2a subclass, with little antiviral IgM detected by day 14 p.i., whereas IgG1 antiviral antibodies appear to predominate in resistant and immunized mice. Murine CD4+ T cells of the Th1 subset mediate delayed-type hypersensitivity (DTH) and regulate Ig2a production via IFN-γ production, whereas CD4+ Th2 cells regulate IgG1 and IgE production via IL-4. Thus, the predominant IgG2a antiviral response in susceptible mice may be an *in vivo* measure of preferential stimulation of a Th1-like pattern of cytokine synthesis.

When infected, susceptible mouse strains also produce substantial levels of virus-specific cellular immune responses. T cell proliferation (prlf) and DTH appear by 2 weeks p.i. and remain elevated for at least 6 months. Both DTH and T $cell_{prlf}$ have been shown to be specific for TMEV and mediated by CD4+ class II restricted T cells. A temporal correlation has also been found between the onset of demyelinating lesions and development of these virus-specific T cell responses, as well as between the high levels of virus-specific DTH and the susceptibility of mice of different genetic backgrounds and mixes. Recently, DTH and T $cell_{prlf}$ responses in infected and immunized SJL mice (a susceptible strain) have been found to be primarily directed toward the VP2 virus coat protein, and specifically to an epitope contained in VP2 amino acids 74–86, VP2 74–86 is believed to be the immunodominant epitope in SJL mice, and responsible for the immunopathologically directed host responses (see later).

Although mice mount virus-specific humoral and cellular immune responses on early virus exposure and peak virus titers fall by 100–1000-fold, TMEV somehow evades immune clearance to persist at low levels indefinitely in the CNS of the host as described above. Extraneural persistence has not been observed. Current dogma holds that humoral immunity is more important than cellular immunity in clearing infections by nonenveloped viruses, such as picornaviruses, but this has not been established for TMEV. The precise mechanism by which TMEV evades immune surveillance is not known but does not involve antigenic variation. Although complement and virus–antibody deposition in the CNS parenchyma have not been detected, extracellular transport of virus as infectious virus–antibody complexes, in aggregates, or contained or enveloped within cell membranes are means whereby virus could be protected from TMEV-specific immune responses and enable continued virus persistence. This is an area for further study to enable a better understanding of how TMEV evades immune surveillance.

Immune-mediated Mechanism of Demyelination

Appropriately timed immunosuppression can prevent and reverse the clinical signs and pathological changes of TMEV-induced demyelinating disease, indicating that myelin breakdown is immune-mediated. A number of different immunosuppresive modalities have proven to be effective, including cyclophosphamide, antilymphocyte serum, antitumor necrosis factor antibodies, and monoclonal anti-IA, CD4+ and CD8+ antibodies. If given too early in the course of the infection, potentiation of early neuronal infection and encephalitis ensues with high mortality. Thus, immunosuppression may be most effective when administered after the first week of infection. Recently, SJL mice infected with a dose of virus that normally produces a low incidence of disease were adoptively immunized with TMEV VP2-specific T cell line, resulting in an increased incidence of demyelinating disease. This observation further supports a role for CD4+ T cells in mediating TMEV-induced demyelinating disease.

The effector mechanism by which a nonbudding virus, such as TMEV, might lead to immune-mediated tissue injury is unknown. Because TMEV antigens have been primarily found in macrophages and not oligodendrocytes, it has been proposed that myelin breakdown may result from an interaction between virus-specific T cells trafficking into infected areas of the CNS and the virus. Thus, myelinated axons may be nonspecifically damaged as a consequence of a virus-specific immune response, i.e. an 'innocent bystander' response. In this system, cytokines produced by MHC class II-restricted, TMEV-specific T_{DTH}

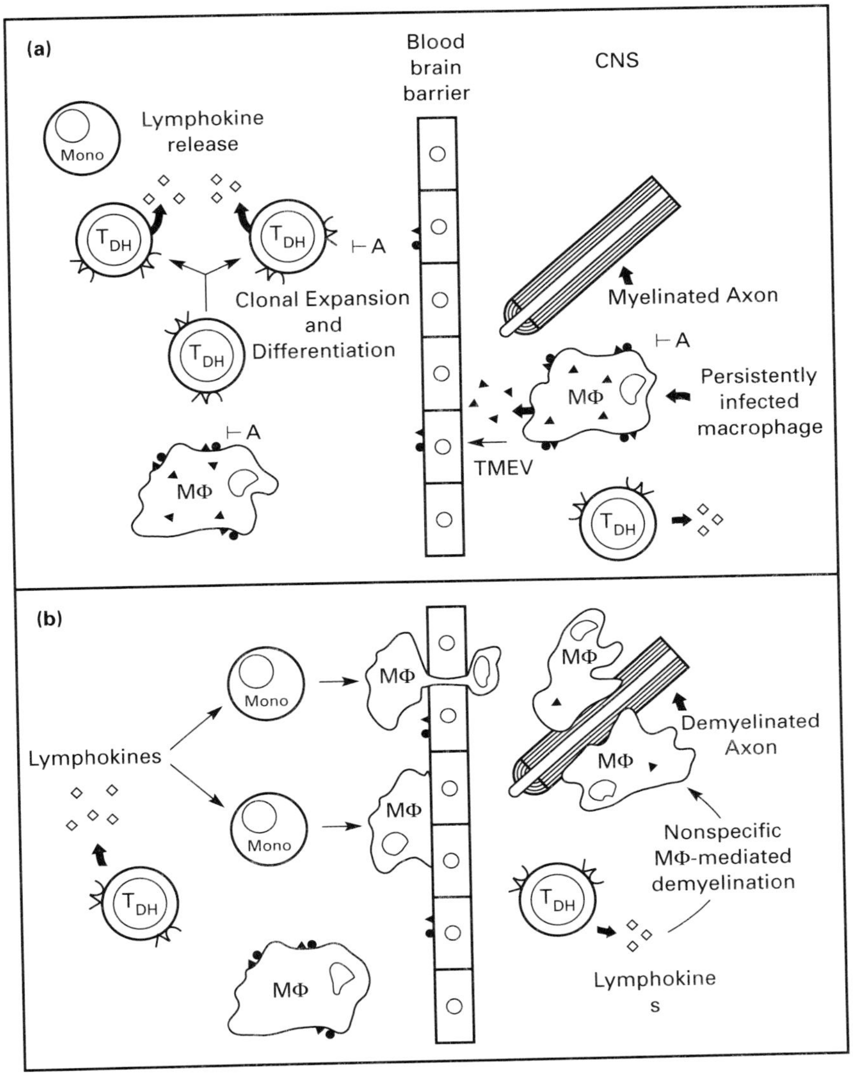

Fig. 2 Proposed DTH-mediated mechanism of TMEV-induced demyelination. (**a**) Virus antigen presentation to Th1 lymphocytes, here designated T_{DH} cells, by antigen presenting cells in either systemic lymphoid organs (left) or inside the blood brain barrier (right). (**b**) Th1 lymphocyte response resulting in the recruitment of monocytes into the CNS and their differentiation into macrophages which then mediate demyelination.

cells primed by interaction with infected macrophages would lead to the recruitment and activation of additional macrophages in the CNS, resulting in nonspecific macrophage-mediated demyelination (Fig. 2). This hypothesis is consistent with the CNS pathological changes observed in mice exhibiting TMEV-induced demyelinating disease and the fact that antigen-specific T cells and T cell lines have been shown to cause bystander CNS damage via macrophage activation in several other model systems. Alternatively, in the case of extensive infection of oligodendrocytes, demyelination might result from immune injury to these myelin maintaining cells expressing TMEV antigens in conjunction with H-2 class I determinants. CD8+ T cells would then be the likely T cell to kill infected oligodendrocytes; however, widespread degeneration of oligodendrocytes has not been observed and no one as yet has adequately studied CD8+ T cell killing in TMEV-induced demyelinating disease.

See also: Cardioviruses; Immune response; Persistent viral infection; Polioviruses; Virus structure.

Further Reading

Borrow P and Nash AA (1992) Susceptibility to Theiler's

virus-induced demyelinating disease correlates with astrocyte class II induction and antigen presentation. *Immunology* 76: 133.
Gerety SJ *et al.* (1991) Class II-restricted responses in Theiler's murine encephalomyelitis virus-induced demyelinating disease. IV. Identification of an immunodominant T cell determinant on the N-terminal end of the VP2 capsid protein in susceptible SJL/J mice. *J. Immunol.* 146: 2401.
Grant RA, Filman DJ, Fujinami RS, Icenogle JP and Hogle JM (1992) Three-dimensional structure of Theiler virus. *Proc. Natl Acad. Sci. USA* 89: 2061.
Luo M, He C, Toth KS, Zhang CX and Lipton HL (1992) Three-dimensional structure of Theiler's murine encephalomyelitis virus (BeAn strain). *Proc. Natl Acad. Sci. USA* 89: 2409.
Miller SD and Gerety SJ (1990) Immunologic aspects of Theiler's murine encephalomyelitis virus (TMEV)-induced demyelinating disease. *Sem. Virol.* 1: 263.
Peterson JD, Waltenbaugh C and Miller SD (1992) IgG subclass responses to Theiler's murine encephalomyelitis virus infection and immunization suggest a dominant role for TH1 cells in susceptible mouse strains. *Immunology* 75: 652.
Rodriguez M, Patick AK, Pease LR and David CS (1992) Role of T cell receptor Vβ genes in Theiler's virus-induced demyelination in mice. *J. Immunol.* 148: 921.

TICK-BORNE ENCEPHALITIS VIRUS

See Encephalitis viruses

TIPULA IRIDESCENT VIRUS

James Kalmakoff
University of Otago
Dunedin, New Zealand

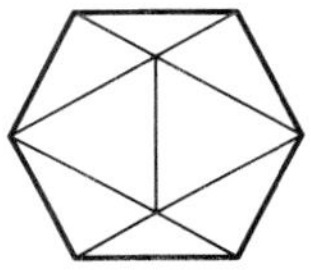

Introduction

Iridoviruses are icosahedral, double-stranded (ds)DNA viruses belonging to the family *Iridoviridae*. Members of this family are able to infect both invertebrate and vertebrate hosts. The viruses are ubiquitous in nature and have been found on all continents of the world and have been isolated from a variety of insect species. The genus *Iridovirus* consists of the small insect viruses (120–140 nm) with blue to purple iridescence, whereas the genus *Chloriridovirus* consists of the large insect virus group (>180 nm) with a yellow–green iridescence. The type species for the *Iridovirus* group is *Chilo* iridescent virus (CIV) (type 6).

Taxonomy and Classification

The first iridovirus to be described was the iridescent insect virus from the larvae of the crane fly *Tipula paludosa* discovered by Claude Rivers in 1954. The systemic infection conferred a blue coloration to the larvae and the term iridescent virus was coined. Since then similar viruses have been isolated from different parts of the world and different invertebrate species. Iridoviruses have been isolated from invertebrates indigenous to all the major continents. The viruses are sometimes referred to by their historical designation such as *Tipula* iridescent virus (TIV) and CIV or by the proposed types (type 1 and type 6).

The current taxonomic scheme is an interim measure based on the host of origin and the date of isolation. Since TIV was the first to be discovered it has been designated iridovirus type 1, while CIV, being the sixth to be reported, is iridovirus type 6. This scheme provides no real criterion for classification and has been made unworkable by the number of iridoviruses isolated. To date, 33 insect iridescent viruses have been recognized and typed (Table 1). Iridescent viruses have also been reported in invertebrates other than insects, including terrestrial isopods, nematodes and brachiopods.

Table 1. Recognized insect iridescent viruses

Iridescent virus type	Host	Iridescent virus type	Host
1	*Tipula paludosa*	17	*Pterostrictus madidus*
2	*Sericesthis pruinosa*	18	*Opogonia* sp.
3*	*Aedes taeniorhynchus*	19	*Odontria striata*
4*	*Aedes cantans*	20	*Simocephalus expinosus*
5*	*Aedes annulipes*	21	*Heliothis armigera*
6	*Chilo suppressalis*	22	*Simulium* sp.
7	*Simulium ornatum*	23	*Heteronychus arator*
8*	*Culicoides* sp.	24	*Apis cerana*
9	*Wiseana cervinata*	25	*Tipula* sp.
10	*Witlesia sabulosella*	26	'Mayfly'
11*	*Aedes stimulans*	27	*Nereis diversicolor*
12*	*Aedes cantans*	28	*Lethocerus columbiae*
13	*Corethralla brakeleyi*	29	*Tenebrio molitor*
14*	*Aedes detritus*	30	*Heliothis zea*
15*	*Aedes detritus*	31	*Armadillidium vulgare*
16	*Costelytra zealandica*	32	*Porcellio diatatus*
		33	*Chironomus plumosa*

*Members of the *Chloriridoviridae*.

Not all of the different types mentioned above are distinct viruses. For instance, type 9 was 100% homologous to type 18 by both DNA and serological comparisons and these two viruses are probably the same. Type 19, which was isolated from the coleopteran host *Odontria striata*, is very likely to be the same virus as type 16, which was isolated from a similar coleopteran host *Costelytra zealandica*. These two insect hosts are difficult to distinguish from each other and often occur as mixed populations in the same locality. In view of these and other observations, it is unlikely that the scheme proposed in Table 1 is of any taxonomic value. Now that techniques are available for sequence analysis, it should be possible to determine the exact relationships among all the isolates.

Virion and Genome Structure

The first 'spherical' virus shown by shadowing and electron microscopy to have icosahedral symmetry was TIV (Fig. 1) in 1958. An icosahedron consists of 20 equilateral triangular faces each of which can consist of smaller subunits called capsomeres. Disruption of TIV gave triangular and pentagonal subunit aggregates. Each pentasymmetron contained 31 subunits and each trisymmetron contained 55 subunits. This suggested a possible triangulation number of $T = 147$ (1472 subunits). Negative staining shows the particle size to be 120 nm.

The genome is a single-copy linear dsDNA, 150–280 kbp in size. It is circularly permuted with ter-

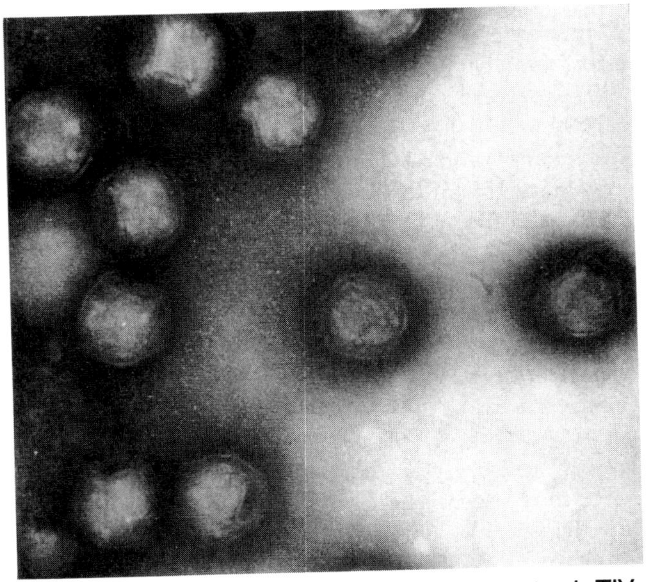

Fig. 1 Electron micrographs of negative-stained TIV particles.

minal redundancy and has internal repetitive DNA sequences. CIV has a genome size of 210 kbp with a terminal redundancy (the amount of DNA duplicated at each end of the genome) of 12%, and the circular permutation (the range of fragments that formed the ends) was 100%, i.e. no unique ends. These genome structures are common to some procaryotic viruses such as bacteriophages T4 and λ, but have not been demonstrated in any other eucaryotic viruses.

Virion Composition

The virions contain from 11 to 18% DNA packed within an electron dense core in association with six or more DNA-binding proteins. The GC content of the DNA is about 30%. This DNA core is surrounded by an internal lipid bilayer (about 5–9% of the virus) which has an abundance of phosphatidylinositol and diglycerides. This differs from the lipids present in host cell membranes which are predominantly phosphatidylcholine, phosphatidylethanolamine and triglycerides. Since the lipid composition does not change with the host in which the virus replicates, it has been concluded that the membranes are not host-derived, but synthesized *de novo* during replication. This feature has been demonstrated in only a small number of viruses such as frog virus-3 (a member of the *Iridoviridae*), poxvirus and bacteriophage PM2. The third outer concentric domain is a proteinaceous icosahedral capsid which makes up 50% of the viral mass. Depending on the method used, iridoviruses have been shown to have between 25 and 30 structural polypeptides ranging in molecular mass from 12 000 to 150 000 kD. They all have a major protein between 50 and 55 kD which has been estimated to comprise 45% of the total proteins. Various enzymes have been found to be associated with viral particles. These include a protein kinase, an alkaline protease, a deoxyribonuclease, an RNA polymerase, a nucleotide phosphorylase and an ATPase. None of these enzymes have been shown to be virally encoded. There are reports that the protein kinase phosphorylates the structural proteins. Studies on TIV have shown that none of the virion structural proteins appear to be glycosylated. The DNA of the iridoviruses is not methylated, unlike other members of the *Iridoviridae* which infect vertebrates. Some isolates of TIV contain two DNA components L (176–247 kbp) and S1 (10.8 kbp) within the DNA core.

Virus–Host Interaction

The host range of some iridoviruses such as TIV appears to be wide, since they are able to infect dipteran, lepidopteran and coleopteran hosts, while others, e.g. *C. zealandica* iridescent virus (CzIV), are restricted to their original coleopteran host. The mode of transmission for these viruses has not been elucidated. Many routes of infection have been suggested, including oral ingestion, infection through the spiracles, transovarial transmission and the contamination of eggs. The oral ingestion of virus appears to be one of the most obvious routes but generally it is thought that these viruses are not very infectious by oral ingestion, the virus being degraded shortly after entering the midgut. In the foregut, the peritrophic membrane is a barrier to infection. Transovarial infection appears to be the method of transmission for the *Chloriridovirus* group. Other possible routes include cannibalism, wounding or infection via other parasites (such as nematodes). Cannibalism of infected cadavers would result in a massive uptake of concentrated virus. Wounding, which is common among aggressive species, would mimic the intrahemoceolic injection that is used to infect insects in the laboratory.

Serology

Serological methods include serum neutralization, complement fixation, enzyme-linked immunosorbent assay (ELISA), latex agglutination, immunoperoxidase staining and radioimmunoassay. In general, serum neutralization is impractical for screening because of bacterial contamination of triturated larvae. Complement fixation is relatively insensitive when larval extracts are used and is not quantitative. Latex agglutination was found to be 50 times less sensitive than ELISA, while radioimmunoassay is technically demanding. ELISA is probably the best serological technique available for viral detection, but has the disadvantage of being only semiquantitative. Moreover, many iridescent viruses were propagated in experimental laboratory hosts such as *Galleria mellonella*, and it is possible that there has been cross-contamination between iridescent viruses. For example, there were antigenic differences in TIV propagated in *Lymantria dispar* or *T. paludosa* when compared with TIV propagated in *G. mellonella*. Despite these problems, serology has been used to compare different isolates of iridoviruses and some general conclusions can be made. The members of the *Chloriridovirus* genus (iridescent viruses types 3, 4, 5, 8, 11, 12, 14 and 15) appear to be serologically related to each other, but not to any members of the *Iridovirus* genus. Within the *Iridovirus* genus, TIV and *Sericesthis* iridescent virus (SIV) have been shown to be virtually identical by most serological tests, including immunodiffusion, immunoprecipitation and immunoelectron microscopy, although complement fixation and neutralization tests successfully discriminated between them. Other isolates tested showed relatedness to varying degrees with the exception of CIV and iridovirus type 24. Iridoviruses types 1, 2, 9, 10, 16, 18, 21, 22, 23, 24, 25 and 28 all showed common antigens and formed a broad serological group. Three groups of closely related viruses could be distinguished: (1) types 1, 22 and 25, (2) types 21 and 28, and (3) types 9 and 18. Analysis by Western blotting and the purification of individual polypeptides has shown that the antigenic relationships between the iridoviruses is very complex, and some of the earlier findings have to be revised. For example, CIV was shown not to be as antigenically distinct from the other iridoviruses (type 9 and type 16) as previously reported.

Morphogenesis

The assembly of a virus with icosahedral symmetry is complex, particularly one consisting of 25–30 structural proteins. There are three main models for iridovirus assembly: (1) formation of empty viral shells without the DNA core, (2) development of the core followed by encapsidation by the proteins, and (3) concurrent formation of viral shells and DNA fibrils. Results of studies with ferritin-labeled antibodies to capsid proteins suggested that the viral shell was assembled prior to the entry of the DNA, similar to that for herpesvirus and adenovirus. This suggests that the DNA is inserted into preformed empty viral shells.

Inclusion Bodies

The major common feature of all iridescent viruses is the iridescent, or opalescent, blue–green coloration of heavily infected larvae or purified pellets of the virus. This coloration is caused by closely packed paracrystalline arrays of viral particles within heavily infected tissues causing Bragg reflections of light. The color of the iridescence is determined by the interparticle spacing. For TIV, the viral particles are packed in a face-centered array with an interparticle distance of 250 nm, or about twice the size of virions observed by negative staining. The production of iridescence in infected larvae does not always occur, probably because paracrystalline virus arrays do not always form. This has been shown both *in vivo* and *in vitro* and may be due to the virus budding continuously from infected cells Massive amounts of virus can result from the infection of larvae with the highest concentration in the fat body and epidermis. It has been estimated that the virus can constitute as much as 25% of the body weight of a dead larva. Since the fat body acts as the storage organ for the larvae, infection usually results in flaccid larvae 5–10 days after infection. It has been suggested that iridescence may be a trivial characteristic of these viruses as it bears no relationship to any functional feature of the virus and is due only to physical structure.

By electron microscopy, viroplasmic centers can be seen as membrane-free pockets of cytoplasm 18–24 h after infection (Fig. 2). These contain 'dark matrix' where DNA synthesis is occurring and 'light matrix' where ribosomes are synthesizing viral proteins and where virion assembly occurs.

Pathology

The pathogenesis of these viruses is poorly under-

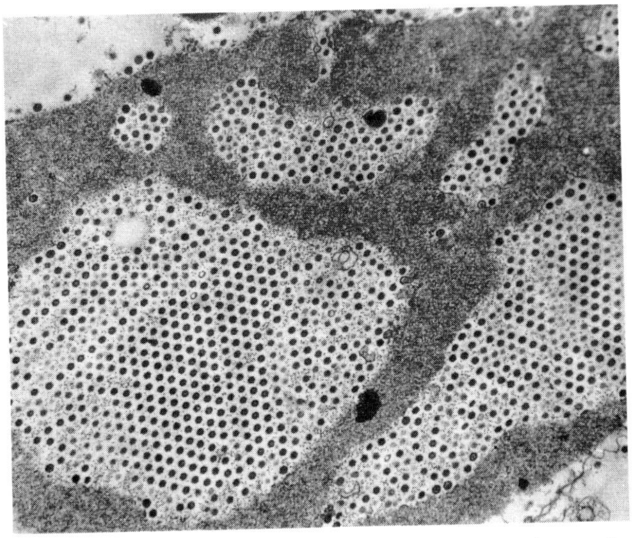

Fig. 2. Viral crystalline array of TIV in the cytoplasm of a virus-infected cell.

stood since little is known about how the virus gains access to the initial site of replication. When administered orally these viruses have low infectivity and under natural conditions they are probably transmitted by cannibalism and wounds. The exception may be an iridovirus from *Scapteriscus vicinus* (Southern mole cricket) which has been shown to cause infections *per os*. The primary sites of replication are probably the hemocytes and fat bodies. The rate of pathogenesis is temperature-dependent, and at optimum temperatures the virus spreads through the insect to produce a systemic infection. Larvae become blue and flaccid about 5–10 days after infection. The cytopathology has mainly been studied in hemocytes or cells in culture. Viral particles enter cells by pinocytosis or by phagocytic engulfment into membrane-bound vesicles within 1.5 h of inoculation. A general breakdown of the virus occurs within the vesicles where presumably they are uncoated, and viroplasmic centers appear in the cytoplasm after 4 h. Release of virus occurs by cell lysis or exocytosis and the virus is dispersed into the hemolymph. The cytoplasm has been described as the site of viral assembly, morphogenesis and DNA synthesis. However, the necessity for a functional cell nucleus has been proposed for frog virus-3. Infection with CIV rapidly induces the formation of syncytia which is independent of viral replication and due to components of the invading CIV particle.

Replication

With CIV infection host cell macromolecular synthesis is rapidly shut down and this appears to be caused by unknown viral structural protein(s). The viral inner membrane proteins, which can be solubilized by octylglucoside, contain all the factors important in the initial stages of infection, including a cell-fusion activity (independent of capsid proteins), an inhibitory protein which switches off host cell synthesis and an activating factor which stimulates the transcription of immediate early viral genes. Cellular DNA polymerase, thymidine kinase and RNA polymerase levels increase during infection. If frog virus-3 can be taken as the model for iridoviruses, the focus of infection at this stage moves to the nucleus where 30% of the total viral DNA replication takes place, producing molecules that are one to two genomes in length. The DNA is then transported to the cytoplasm where it ligates to produce long concatamers which are used as templates for further DNA replication. Early mRNA synthesis also occurs in the nucleus and involves a virally modified host RNA polymerase. At the macromolecular level of infection, proteins are produced in two phases; early and late. Early protein synthesis can be further divided into the immediate early (IE) and early/late (EL) phases (other terms used for these phases are α, β and γ). The IE proteins require no prior synthesis of viral proteins and appear to be involved in host cell shutdown. The expression of EL proteins requires the presence of functional IE proteins and the synthesis of viral mRNA. Late proteins appear after DNA replication and require functional EL proteins. The involvement of both the nucleus and cytoplasm as sites of DNA replication in frog virus-3 has no other counterpart and is unique among eucaryotic viruses.

The genome of these viruses has been shown to be circularly permuted and terminally redundant. The terminal repetition (the amount of DNA duplicated at each end of the genome) varies, being 12% for CIV. This genome arrangement is thought to be due to an imprecise mechanism of headful packaging, where concatemeric DNA is cleaved in such a way that the amount of DNA placed into a virion head is larger than one set of genes. Repetitive DNA sequences have also been found in a number of different iridoviruses including *Wiseana* iridescent virus (WIV), CIV and TIV. Repeat sequences in other viruses have been found to have various functions such as the origin of replication in adenovirus and herpesvirus, early mRNA production in vaccinia virus and host range determinants in herpesviruses. There has only been a limited number of genes sequenced, but the three that have been published (two from TIV, and one from IV type 22) appear to have eucaryotic-like transcription signals (TATA and CAAT box-like structures and GC motifs).

Economic Importance

Conventional wisdom has it that the iridoviruses are not good biological control agents since they do not cause a lethal infection and are difficult to transmit in the field. However, the iridovirus isolated from southern mole crickets, *Scapteriscus vininus*, operates in an epizootic fashion, killing greater than 70% of the collected mole crickets. When the virus is added to the diet of laboratory-reared nymphs, at least 65% became infected. This virus also infects

Trichoplusia ni larvae and insect cell cultures and may be a pathogenic strain of iridovirus.

CIV might be of agricultural importance since it has been shown to infect the green rice leafhopper, *Nephtholettix cincticeps*, and is lethal for the leafhopper, *Colladonus montanus*, the vector of a mycoplasma agent of stone fruits. CIV could be used to eliminate insects which act as vectors in transmitting plant diseases.

WIV infects *Wiseana* spp. a major soil dwelling pasture pest in New Zealand and infection rates can be as high as 30% based on the iridescence of the infected larvae. The actual infection rate would have been much higher since iridescence is a poor indicator of infection. Infection of the larvae results in cessation of feeding and death within 14 days. A similar situation applies to CzIV, a potential pathogen of coleopteran larvae. However, naturally occurring epizootics have not been reported since 1969.

There is an exciting potential in the field of genetic modification of these viruses. The terminal redundancy and circular permutation can be used to introduce toxins or to modify the virus easily. There is no necessity to find nonessential regions to insert genes since the DNA contains at least 112% of the genome and the duplication is at random. Furthermore, the promoter for the coat protein gene would seem to be a good candidate site since this gene is highly expressed during viral infection, up to 25% of the weight of an insect being virus.

Ecology

Evans and Entwistle (1987) have discussed the division of the *Iridoviridae* into the large (chloriridovirus) and small (iridovirus) insect iridescent viruses, and their evolution. They have suggested that, based on ecologically different routes of evolution, the *Iridovirus* with its wider host range is the older genus and the *Chloriridovirus*, which has evolved exclusively in association with the aquatic diptera, is a more specialized derivative. The iridoviruses are thought to have only a horizontal route of transmission, while the chloriridovirus have a vertical route of transmission in addition to larval cannibalism. It has been suggested that the internal membrane structure gives these viruses greater stability in aquatic environments. This idea has been extrapolated to suggest that insect viruses were the first viruses and that they later adapted to life in vertebrates which fed on the insects. WIV and CzIV, isolated from indigenous sympatric hosts in New Zealand, are genetically closely related and probably share a common ancestor. The WIV genome is 24 kbp larger than the CzIV genome and the region of nonhomology between the WIV and CzIV genomes is also 24 kbp in size. It is possible that in the course of evolution WIV gained that extra piece of DNA or CzIV lost it. In considering insect viruses in general, it has been postulated that the evolution of the virus group is probably not older than the host taxon itself. Using this assumption one could postulate that, since the coleoptera (Late Permian) are more ancient than the lepidoptera (Middle Cretaceous), then CzIV is probably older than WIV, in which case WIV acquired the extra piece of DNA as it diverged from CzIV. Since CzIV has a limited host range, one could speculate that the piece of DNA which was gained by WIV coded for the host range determinants. This idea can also cover CIV which has a genome of 209 kbp, 17 kbp larger than the WIV, and CIV has a larger host range than WIV. It would be worthwhile to return to the sites of the original WIV and CzIV isolations to see whether there have been any changes in these viruses in the years since the original isolation. One iridovirus isolate, TIV, appears to have diverged into two evolutionary lines. The TIV virion comprises two DNA components, L and Sl. Recently an isolate from Ireland was found to contain only the L component. Since this isolate was from the same host as the original isolate, one might speculate that time and differences in climate and environmental conditions have enabled a variant TIV strain to evolve. Of interest would be the natural occurrence of the iridoviruses from isopods and nematodes and their relationship with soil insects. Equally interesting is the ecology of the isolates from aquatic environments. For instance, *Corethrella brakeleyi* is a small chaoborid found in Louisiana, a predator of first-instar mosquito larvae which can be infected with an iridovirus, but no transmission occurs to mosquitoes or vice versa. In New Zealand, a copepod was found to be infected with an iridovirus while in the same pool an epizootic of an iridovirus, was occurring among mosquitoes, yet the two viruses appeared to be unrelated.

Now that specific and sensitive DNA probes are available to detect iridoviruses from field samples, the ecology offers many Cinderella-like opportunities.

See also: Frog virus-3; Lymphocystis disease virus.

Further Reading

Darai G (ed.) (1990) *Molecular Biology of Iridoviruses*. Boston, Dordrecht and London: Kluwer Academic Publishers.

Evans HF and Entwistle PF (1987) Viral diseases. In: Fuxa JR and Tanada Y (eds) *Epizootiology of Insect Disease*, p. 257. New York: John Wiley & Sons.

Ward VK and Kalmakoff J (1991) Invertebrate *Iridoviridae*. In: Kurstak E (ed.) *Viruses of Invertebrates*, p. 197. New York: Marcel Dekker.

Willis DB (ed.) (1985) *Iridoviridae*. *Curr. Top. Microbiol. Immunol.* 116: 173.

TOBAMOVIRUSES

Dennis J Lewandowski and William O Dawson
University of Florida Citrus Research and Education Center, Lake Alfred, Florida, USA

History

The mosaic disease of tobacco has been recognized since the 1870s and early work with the disease agent demonstrated its filterability and led to the discovery of the phenomenon of viruses. Since that time, tobacco mosaic virus (TMV), the type member of the tobamovirus group, has had a central role in many fundamental discoveries in virology. The first quantitative biological assay for plant viruses was the use of *Nicotiana glutinosa* which produces local lesions when inoculated with TMV. TMV was the first virus to be purified and crystallized, which led to the discovery of the nucleoprotein nature of viruses and determination of the atomic structure of the coat protein and the virion. TMV was the first virus to be visualized in the electron microscope, confirming the predicted rod shape. The genetic material of TMV was shown to be RNA, a property previously thought to be restricted to DNA. TMV was the first virus to be mutagenized. The first viral protein for which an amino acid sequence was determined was the coat protein of TMV. Subsequent determination of coat protein sequences from a number of strains and mutants helped to establish the universality of the genetic code. Methods to infect plant protoplasts with viruses were developed with the tobacco–TMV combination, creating a synchronous system to study events in the infection cycle.

Taxonomy and Classification

Tobamoviruses have been classified into the 'Sindbis virus supergroup', which includes alpha viruses of animals, and several plant virus groups including tripartite, tobra-, potex- and hordeiviruses. There are 12 members within the tobamovirus group (Table 1). Although tobamoviruses are one of the most intensively studied groups of viruses, the taxonomy is often confused. Historically, plant viruses with rigid virions of approximately 18×300 nm were usually classified as a strain of TMV. Most viruses referred to as strains of TMV are actually different tobamoviruses. For example, the tobamovirus that has been referred to as the tomato strain of TMV, which is approximately 80% similar to TMV at the nucleotide level, is actually tomato mosaic tobamovirus. Different tobamoviruses for which the entire nucleotide sequence is known are 60–80% similar to TMV at the nucleotide level.

Host Range

Tobamoviruses are exceptional in that they have extraordinarily narrow natural host ranges but wide

Table 1. Definitive tobamoviruses

TMV	tobacco mosaic tobamovirus
ToMV	tomato mosaic tobamovirus
TMGMV	tobacco mild green mosaic tobamovirus
ORSV	Odontoglossum ringspot tobamovirus
PMMV	pepper mild mottle tobamovirus
CGMMV	cucumber green mottle mosaic tobamovirus
SHMV	sunn-hemp mosaic tobamovirus
RMV	ribgrass mosaic tobamovirus
FrMV	frangipani mosaic tobamovirus
MaMV	maracuja mosaic tobamovirus
SOV	Sammon's Opuntia tobamovirus
UMMV	Ullucus mild mottle tobamovirus

experimental host ranges. Tobamoviruses tend to adapt to specific plant species. Even where more than one host species occurs in the same community, different plant species are likely to contain different tobamoviruses. In contrast, several tobamoviruses are able to infect a large number of plant species in laboratory experiments. For example, TMV was shown to infect 199 of 310 species tested from 29 families.

Epidemiology

Tobamoviruses cause significant losses in tobacco, tomatoes and peppers throughout the world. The age at which the plants become infected greatly affects disease progression and yield reduction. In general, reductions in yield are greater for fruit-bearing crops. For example, yield losses due to tobamovirus infection have been estimated to be about 1% in tobacco and as high as 20% for tomato.

Tobamoviruses are transmitted mechanically in nature through human handling, contaminated machinery and contact between plants. Vegetative propagation plays a role in long-range dispersal of some tobamoviruses, such as Odontoglossum ringspot tobamovirus in cultivated orchids. Another significant means of spread is through soil transmission due to the presence of infected plant debris, particularly in fields grown without rotation of a resistant crop. There have been some reports of seed transmission, usually by the virus being carried on the seed coat. The primary source of inoculum in a field is usually a few plants that become infected from virus-contaminated debris in the soil. Many of the crops that are infected by tobamoviruses are cultivated throughout the growing season resulting in polycyclic disease progression.

The high titer of virion accumulation, which increases the likelihood of transmission from minor contact, and the unusual stability of the virions, which allows survival for long periods of time in dried tissue or soil, compensate for the absence of efficient vectors and allow effective transmission of tobamoviruses between plants.

Generally, the only successful means of control is through the planting of resistant varieties. Cross-protection, in which plants are intentionally inoculated with a mild strain to prevent infection by a more virulent related strain, has been used successfully in some commercial crops, such as tomato. Although not as practical in field-grown crops, sanitation practices can be important for reducing viral transmission in vegetatively propagated plants. Virus-resistant transgenic plants expressing part of a viral genome have been produced and provide new sources of resistance.

Virus Structure and Composition

Tobamovirus virions are straight tubes approximately 18×300 nm with a central hollow core 4 nm in diameter. Virions are 95% protein and 5% RNA. Approximately 2100 individual subunits of a single coat protein are arranged in a right-hand helix around a single RNA molecule, with each subunit associated with three nucleotides.

Purified coat protein and viral RNA assemble into infectious particles *in vitro*. Protein–protein associations are the essential first event of virion assembly. Coat protein subunits assemble into several types of aggregates. Coat protein monomers and small heterogeneous aggregates of a few subunits are collectively referred to as 'A-protein'. The equilibrium between A-protein and larger aggregates is primarily dependent upon pH and ionic strength. The larger aggregates that have been characterized are disks, which are composed of two individual stacked rings of coat protein subunits, and protohelices. Protohelices contain approximately 40 coat protein subunits arranged in a spiral around a central hollow core, similar to the arrangement within the virion.

A sequence-specific stem-loop structure in the RNA, the 'origin of assembly', initiates encapsidation and prevents defective packaging which could result from multiple initiation events on a single RNA. The origin of assembly is located within the open reading frame (ORF) for the 30 kD protein of most tobamoviruses and within the coat protein ORF of sunn-hemp mosaic (SHMV) and cucumber green mottle mosaic (CGMMV) tobamoviruses. Subgenomic mRNAs containing the origin of assembly are encapsidated into shorter virions that are not required for infectivity. The level of expression of a particular subgenomic mRNA containing the origin of assembly determines the relative proportion of that particular virion species. Consequently, in most tobamoviruses, in which the origin of assembly is located in the ORF for the 30 kD protein, subgenomic mRNAs account for only a minor fraction of the total virion population. In contrast, where the origin of assembly is located within the highly expressed coat protein subgenomic mRNA, as in SHMV and CGMMV, a significant proportion of smaller virions are produced. Hybrid nonviral RNAs containing an origin of assembly will assemble with coat protein into virus-like particles of length proportional to that of the RNA.

Virion assembly initiates as the primary loop of the origin of assembly is threaded through a coat protein disk or protohelix with both ends of the RNA trailing from one side. The conformation of the coat protein protohelix changes as the RNA becomes embedded within the groove between the two layers of subunits. Elongation is bidirectional, proceeding rapidly towards the 5′ end of the RNA as the RNA loop is extruded through the elongating virion and additional coat protein disks are added. There is disagreement about the mechanism of elongation towards the 3′ end of the RNA, but it appears that this slower process involves the addition of smaller protein aggregates.

Genome Structure and Molecular Biology

The tobamovirus genome consists of one single-stranded (ss) messenger-sense RNA of approximately 6400 nucleotides (Fig. 1a). There is a methylguanosine cap at the 5′ terminus, followed by an AU-rich leader. The 3′ nontranslated end of the RNA consists of sequences that can be folded into pseudoknot structures, followed by a tRNA-like terminus. The tRNA-like terminus can be aminoacylated *in vitro*, and in most cases specifically accepts histidine. The exception is SHMV, which accepts valine and appears to have arisen by a recombination event between a tobamovirus and a tymovirus.

Four ORFs that are contained within the tobamovirus genome correspond to the proteins found in infected tissue (Fig. 1a). Two overlapping ORFs begin at a common start codon near the 5′ terminus. Termination at the first inframe stop codon produces a 125–130 kD protein. A 180–190 kD protein is produced by readthrough of this termination codon approximately 10% of the time. Both proteins are necessary for replication.

The remaining proteins are expressed from individual 3′ coterminal subgenomic mRNAs, from which only the 5′ proximal ORF is expressed (Fig. 1c). The next ORF encodes the 30 kD protein, which is involved in cell-to-cell movement of the virus. The 3′ most ORF encodes the coat protein (17–18 kD). A subgenomic mRNA containing an ORF for a 54 kD protein that encompasses the readthrough region of the 180–190 kD ORF has been isolated from infected tissue, although no protein has been detected.

Within the protein-coding regions of the genome, there are nucleotide sequences that have additional functions as *cis*-acting elements, such as subgenomic RNA promoters and the origin of assembly. The promoter elements for subgenomic mRNA synthesis are located on the genomic length complimentary RNA, presumably upstream from the respective initiation sites. Gene expression from subgenomic mRNAs is regulated both temporally and quantitatively. The 30 kD protein is produced

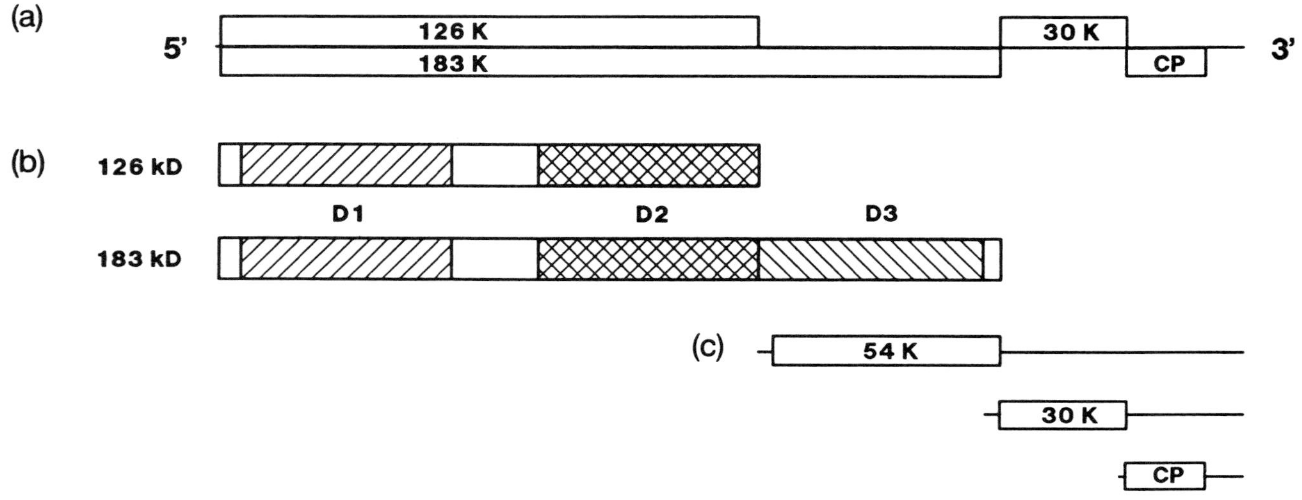

Fig. 1 (**a**) Tobamovirus genome organization. ORFs designated as narrow open boxes. (**b**) 126/183 kD replicase proteins. Domains of amino acid sequence similarity to viruses within the 'Sindbis virus supergroup' designated as hatched boxes. (**c**) Subgenomic mRNAs with expressed ORF labeled. CP, coat protein.

early and accumulates to low levels, whereas the coat protein is produced late and in massive quantities. Several factors are probably involved in regulation of gene expression from the various subgenomic mRNAs. The subgenomic mRNA for the 30 kD protein is not capped and has a long 5′ leader, whereas the coat protein subgenomic mRNA is capped, has a short AU-rich leader, and is an efficient mRNA. Also, the levels of 30 kD and coat protein production in TMV mutants are directly related to distance from the 3′ terminus. There are no obvious sequence similarities between the promoters for the 30 kD and coat protein subgenomic mRNAs, suggesting that secondary structure may be important for recognition.

Virus Genetics

The TMV 126/183 kD proteins are involved in viral replication. Both are contained in crude replicase preparations, and temperature-sensitive replication-deficient mutants map to these ORFs. There are three domains of amino acid sequence similarity shared with replicase proteins from other ssRNA plant and animal viruses (Fig. 1b). The first domain, located at the N termini of the 126/183 kD proteins, shares sequence similarity with a domain having methyltransferase activity, thought to be associated with viral RNA capping. The second domain, located at the C terminus of the 126 kD protein (internally in 183 kD), is proposed to have helicase activity, based upon conserved sequence motifs. The third domain, which has a signature sequence for RNA polymerase function, is located within the readthrough region of the 183 kD protein. Additionally, the 126/183 kD proteins are symptom determinants, as mutations in mild strains map to these ORFs.

The 30 kD protein has a plasmodesmatal binding function associated with the C terminus and a single-stranded nucleic acid-binding domain associated with the N terminus. The 30 kD protein–host interaction determines whether the virus can systemically infect some plant species.

Although principally a structural protein, the coat protein is also involved in other host interactions. Coat protein is required for efficient long-distance movement of the virus. Coat protein is also a symptom determinant in some susceptible plant species and an elicitor of plant defense mechanisms in other plant species.

Cytology

Most cells of all tissues become infected with tobamoviruses. The exceptions are the embryo of seeds and dark green islands of uninfected cells in leaves that develop after infection. Cells in dark green islands are resistant to infection and develop normally, providing photosynthetically active tissue. Within infected cells, virions accumulate primarily in the cytoplasm. Several types of inclusion bodies can be found in tobamovirus-infected tissue. Virions of different tobamoviruses aggregate to form characteristic elongated (paracrystalline) or plate-shaped (crystalline) inclusion bodies. Additionally, amorphous inclusions and X-bodies consisting of viral proteins and cell membranes often occur in the cytoplasm.

Multiplication and Spread

Virions or free viral RNA will infect plants. Since tobamoviruses have a genome consisting of messenger-sense RNA that is infectious, one of the first events is translation of the RNA to produce proteins required for replication of the genomic RNA and production of subgenomic mRNAs. When virions are the infecting agent, the first event is thought to be cotranslational disassembly, in which the coat protein subunits at the end of the virion surrounding the 5′ end of the RNA loosen, making the RNA available for translation. Ribosomes then associate with the RNA, and translation of the 126 kD and 183 kD replicase proteins occurs as the coat protein is displaced from the RNA.

After the formation of an active replicase complex, complementary minus-strand RNA is synthesized from the genomic plus-strand RNA template. Minus-strand RNA serves as template for both genomic and subgenomic RNAs. Asymmetric replication produces substantially more plus-strand RNA relative to minus-strand RNA. Early in infection, genomic RNA functions as template for minus-strand RNA synthesis and as mRNA for production of the 126/183 kD replicase proteins. Later in the infection cycle, most of the newly synthesized genomic RNA is encapsidated into virions. Subgenomic mRNAs express the 3′ ORFs not translated from the genomic RNA.

Within an infected cell, replication proceeds rapidly between approximately 16 and 96 h post-infection, after which replication ceases. Even though the infected cells are packed with virions, they metabolize normally for long periods of time. During the early stages of infection of an individual cell, the infection spreads through plasmodesmatal connections to adjacent cells. This event requires the 30 kD protein which modifies plasmodesmata to accommodate larger molecules. Movement through plasmodesmata does not require the coat protein. A second function of the 30 kD protein appears to be binding to the viral RNA to assist its movement through the small plasmodesmatal openings. As the virus spreads from cell to cell throughout a leaf, it enters the phloem for rapid long-distance movement to other leaves and organs of the plant. This process requires the coat protein.

Available evidence suggests that the interaction of viral proteins with host factors are important determinants of viral movement and host ranges. Amino acid substitutions in the 30 kD protein can alter the movement function in different hosts. Some viruses, including tobamoviruses, can assist movement of other viruses that are incapable of movement in a particular plant species. These interactions suggest that there are more precise associations of viral proteins with host factor(s) than with viral RNA. Additionally, precise coat protein–plant interactions are required for movement to distal positions within the plant.

Symptoms

Tobamoviruses usually cause symptoms when they move throughout, and systemically infect, a plant. Symptoms typically are a mosaic of normal (green) and chlorotic areas in developing leaves. This type of symptom occurs in leaves that develop in the presence of the virus infection and are due to a lack of normal chloroplast development. Lack of normal chloroplasts in a large percentage of leaf tissues causes a photosynthetic deficiency that reduces growth, and results in stunted plants. Some tobamoviruses also induce yellowing of infected areas of expanded leaves and a yellow–dark green mosaic in developing leaves. Other tobamovirus–host combinations result in leaf deformation or systemic necrosis.

The inducation of light green–dark green mosaic symptoms in developing leaves is not correlated with viral coat protein because mutants without coat protein induce the same symptoms as wild-type virus. However, the coat protein can induce or modify disease symptoms. The yellowing of mature leaves, which is due to disorganization of mature chloroplasts, and the appearance of bright-yellow mosaics in developing leaves of some tobamovirus infections are correlated with the accumulation of nonassembled coat protein aggregates within the cytoplasm.

Resistance

One important type of resistance mechanism that is present in several plant species against tobamoviruses is the hypersensitive response (HR). The resistance is often conferred by a single dominant gene. Once the plant recognizes a specific viral infection, multiple defense mechanisms against pathogens are induced, some of which are antiviral. This response usually restricts the viral infection to a small necrotic area surrounding the infection site. Current thought is that initiation of the defense response occurs after the recognition of a specific viral gene product (elicitor) by the resistance gene product.

The best understood system is resistance to tobamoviruses in *Nicotiana sylvestris* which is controlled by the N' gene. Coat protein is the elicitor of the hypersensitive response in this system. The N' gene provides resistance against most tobamoviruses that are able to replicate in *N. sylvestris*, except TMV, which does not elicit the HR. However, specific single amino acid substitutions in the TMV coat protein result in mutants that induce the HR. Most of the HR-inducing amino acid substitutions are localized at the interface between two adjacent coat protein subunits (as found in disks or protohelices). The location and nature of the amino acid substitution can affect the timing and degree of the HR. These HR-inducing amino acid substitutions may perturb subunit association, which could shift the equilibrium between monomers and disks, or alternatively, the affinity of a receptor might be affected. Somehow the plant is able to recognize these very small alterations in the coat protein structure.

In contrast, the *N* gene that controls resistance in *N. glutinosa* against tobamoviruses recognizes some viral product other than the coat protein. TMV mutants with deletion of parts or all of the coat protein continue to elicit the HR. Frangipani mosaic tobamovirus (FrMV) and some tobamoviruses isolated from pepper do not elicit the HR and systemically infect *N. glutinosa*.

Other types of resistance have been characterized. The *Tm-1* resistance gene in tomato prevents tomato mosaic tobamovirus (ToMV) from replicating in inoculated cells. Resistance-breaking strains of

ToMV have been isolated that are able to replicate and spread in tomato cultivars carrying the *Tm-1* gene. The amino acid substitutions for these strains are located within the 130/180 kD replicase proteins. The *Tm-2* and *Tm-2*^{2} resistance genes in tomato prevent ToMV from moving cell to cell within infected plants. Resistance-breaking strains that overcome *Tm-2* resistance have been mapped to amino acid substitutions within the 30 kD protein.

The mechanism of resistance and the viral component directly or indirectly involved varies between different plant genotypes. Resistance controlled by the *N* and *N'* genes in *Nicotiana* species is an active plant response to a threshold level of a particular viral gene product. Conversely, *Tm-1* and *Tm-2* resistance genes in tomato apparently fail to provide host components required for viral replication and movement, respectively.

The production of virus-resistant transgenic plants expressing portions of a viral genome was first demonstrated with TMV, and the technology has been applied to other systems. Plants expressing high levels of TMV coat protein are resistant to infection by TMV and some tobamoviruses. Coat-protein-expressing plants are only resistant to low levels of inoculum, but this may be sufficient to protect against levels of inoculum encountered naturally in the field. Plants expressing the ORF for the TMV 54 kD protein are immune to infection by higher levels of inoculum, but only from the homologous virus. The higher level of resistance from plants expressing portions of viral replicase genes may have practical value because often only one virus is found to be naturally infecting a particular host.

Satellite Tobacco Mosaic Virus

Satellite tobacco mosaic virus (STMV), a tobamovirus-dependent satellite virus, has been isolated from *Nicotiana glauca* infected with the tobamovirus (TMV-U5) in California. The genome consists of one plus-sense ssRNA of 1059 nucleotides. The 240 3'-nucleotides share approximately 65% sequence similarity with TMV-U5 and TMV, contain two pseudoknot structures and have a tRNA-like terminus. No sequence similarity to tobamoviruses exists over the remainder of the genome. Two overlapping ORFs that are expressed in *in vitro* translation reactions are present in the genomic RNA of most STMV isolates. The 5' proximal ORF encodes a 6.8 kD protein that has not been detected *in vivo*. The second ORF encodes the 17.5 kD coat protein, which is not serologically related to any tobamovirus coat protein. The 17 nm icosahedral virions are composed of a single STMV genomic RNA encapsidated within 60 STMV coat protein subunits. Replication of natural populations of STMV is supported by other tobamoviruses, but at lower levels than with the natural helper virus, TMV-U5. The host range of STMV parallels that of the helper virus. No effects on symptom expression by the helper have been reported.

See also: Plant virus disease – economic aspects.

Further Reading

Culver JN, Lindbeck AGC and Dawson WO (1991) Virus–host interactions: Induction of chlorotic and necrotic responses in plants by tobamoviruses. *Annu. Rev. Phytopathol.* 29: 193.

Dawson WO and Lehto KM (1990) Regulation of tobamovirus gene regulation. *Adv. Virus Res.* 38: 307.

Gibbs AJ (1977) Tobamovirus group. *CMI/AAB Descriptions of Plant Viruses*, No. 184.

Matthews REF (1991) *Plant Virology*, 3rd edn. New York: Academic Press.

van Regenmortal MHV and Fraenkel-Conrat H (eds) (1986) *The Plant Viruses*, vol. 2. *The Rod-Shaped Plant Viruses*. New York: Plenum Press.

TOBRAVIRUSES

Alexander Mathis and Huub JM Linthorst
Institute of Molecular Plant Sciences
Gorlaeus Laboratories
Leiden, The Netherlands

Taxonomy and Classification

The tobraviruses (type virus: *tob*acco *ra*ttle virus) are bipartite plant viruses which are divided into three subgroups. These are (1) tobacco rattle viruses (TRV), (2) pea early browning viruses (PEBV) and (3) pepper ringspot virus (PRV), which was formerly known as the CAM strain of TRV. The first two subgroups contain several to numerous distinguishable strains or isolates, whereas only one PRV has yet been identified. Their classification into one group of tobraviruses was originally based on common properties such as particle morphology, transmission vector and the unusual ability to cause a systemic infection in plants with only part of the genome. Recently, this grouping has been supported by molecular studies. Comparisons based on molecular characteristics have also indicated that the tobraviruses are members of the supercluster of Sindbis-like viruses. Another plant virus member of this superfamily is the tobamovirus tobacco mosaic virus (TMV), which, both in morphology and genome organization, closely resembles the tobraviruses.

Virus Structure

The tobravirus genome is composed of two single-stranded RNA molecules, which are capped at the 5′ termini and which fold into a tRNA-like tertiary structure at the 3′ termini. However, this tRNA-like structure cannot be aminoacylated as is the case in other viruses showing such a structure at the 3′ terminus.

The RNAs, designated RNA1 and RNA2, are separately encapsidated by a large number of identical polypeptide subunits. The genomic RNA is infectious on its own, indicating that it is of positive polarity (or messenger sense). The virions are rigid rod-shaped particles with similar diameter (20–23 nm) but with a different length. The L (large) particles of TRV, PEBV and PRV have a similar size (180–210 nm); they encapsidate RNA1. The length of the S (small) particles varies considerably (50–110 nm), even between different strains in one subgroup. This difference in size of the S particles reflects the difference in length of the encapsidated RNA2. The structure of the particles is more or less comparable to that of TMV, with a helical arrangement of coat protein subunits around a central axis.

During a normal infection cycle, both L particles and S particles are generated, and homogenates of infected plants are highly infectious (M-type infections, where 'M' stands for 'multiplying'). However, sometimes a second type of natural infection is apparent in which no particles are formed (NM-type infections, 'non-multiplying'). While homogenates of plants with NM-type infections are not or hardly infectious, infectious RNA is produced, as is evident after phenol extraction. The recent advancement of our knowledge of plant viruses has made it possible to explain this unique property of the tobraviruses in molecular genetic terms and to relate it to the specific way in which the genes are separated over the two genome segments.

Genome Organization and Molecular Biology

RNA1 molecules are highly homologous between isolates within the same subgroup, but little homology was identified between TRV, PEBV and PRV. The homology between RNA2 molecules varies from high to undetectable, reflecting also the variation in serological relationship between these isolates.

For each of the three tobraviral subgroups, cDNA clones corresponding to the two genomic segments have been characterized. Both RNA1 of TRV strain SYM and RNA1 of PEBV strain SP5 were completely sequenced and shown to encode four open reading frames (ORFs). The first ORF of TRV strain SYM terminates at a UGA stop codon and encodes a protein of 134 kD. Suppression of this UGA termination codon results in a readthrough protein of 194 kD. Also, the first two ORFs of

PEBV strain SP5 are fused by a UGA termination codon, which, upon suppression, allows the translation of two partially overlapping polypeptides. *In vitro* translation studies have indeed shown that both TRV RNA1 and PEBV RNA1 code for two polypeptides with sizes corresponding to the ones deduced from the nucleotide sequences. Similar mechanisms of genome expression via readthrough translation have been found with viruses belonging to several groups, among which are plant viruses from the tobamo-, luteo-, tymo- and carmovirus groups. The 3′-terminal one-third of tobraviral RNA1 contains two ORFs encoding smaller proteins (29 kD and 16 kD for TRV, and 30 kD and 12 kD for PEBV), which are probably expressed via subgenomic mRNAs.

These data, supplemented with partial sequence data from other tobraviruses, have indicated that extended similarities exist between the various proteins encoded by the different members of the tobravirus group and even with viruses from other plant virus groups. For instance, the 134 and 194 kD TRV-SYM proteins contain regions with high homology with similar regions in the TMV 126 and 183 kD proteins and in the proteins encoded by RNA1 and RNA2 of the tricornaviruses. Based on these similarities, functions have been inferred for the different encoded proteins. The second half of the 134 kD TRV RNA1-encoded peptide contains a nucleotide-binding motif present in helicases, whereas the readthrough portion of the 194 kD protein accommodates a so-called ‘GDD box’, which is present in the catalytic subunit of RNA-dependent RNA polymerases. The presence of these motifs is a strong argument in favor of the involvement in genome replication of the large tobravirus RNA1-encoded proteins.

The 29 kD protein encoded by TRV RNA1 has homology with the TMV 30 kD movement protein. The functional homology of these two proteins was proven by mutation analysis. A TRV mutant with a defect in the gene for the 29 kD protein could not infect whole tobacco plants completely. This defectiveness can be complemented by transgenic expression of the TMV 30 kD protein. The TRV 29 kD protein may also play a role in symptom induction as this same mutant complemented with TMV 30 kD protein did not produce necrotic spots on test plants as does the wild-type TRV.

The small protein encoded by the 3′-terminal ORF in tobraviral RNA1 is conserved between all different tobraviruses characterized so far. However, the lack of homology with proteins encoded by other plant viruses hampers the elucidation of its function. Infected protoplasts accumulate relatively large amounts of this protein. It has been suggested that the cysteine and histidine residues form a zinc-finger domain, which would point to a nucleic acid-binding capability. In this respect it is of interest to note that in plants the 16 kD protein is associated with nuclei. Infectious transcripts containing mutated 16 kD ORFs indicated that the protein is not required for replication or cell-to-cell spread of RNA1.

Thus, RNA1 encodes the genes for viral replication and movement, and this combination allows RNA1 to replicate and spread on its own to give rise to the NM-type infections.

In contrast to similarity in size of RNA1, the RNA2 molecules of the tobraviruses differ considerably. PRV RNA2 is 1799 nucleotides long, whereas RNA2 of PEBV-SP5 has a size of 3374 nucleotides; even between different strains of TRV the length of RNA2 has been found to vary between 1905 and 3389 nucleotides (TRV-PSG and TRV-TCM respectively). This variation in size is a reflection of two distinctive properties of tobraviral RNA2 molecules: (1) RNA2 may encode a variable number of genes, and (2) it may contain a variable-sized 3′-terminal region homologous with the 3′ terminus of RNA1. The 5′-proximal gene on all tobraviral RNA2 molecules is the coat protein gene, which for TRV isolate PSG and for PRV is the only RNA2-encoded ORF. Coat protein is translated from a subgenomic messenger (depicted as RNA2a). RNA2 of PEBV-SP5 encodes another three genes with expected protein sizes of 9, 29 and 23 kD, all of which have unknown functions. The 29 kD protein is predominantly hydrophilic and was shown to be expressed in infected plants.

TRV isolate TCM has, besides the coat protein cistron, two more ORFs encoding proteins of 29 kD (which is homologous with the SP5-encoded 29 kD protein) and 16 kD. The latter is identical to the 3′-proximal ORF of RNA1, as a consequence of the long 3′-terminal region that is identical to the 3′ terminus of RNA1.

Virus Genetics

In all tobraviruses sequenced up till now, very high homology at the 3′ terminus between both genomic RNAs of each isolate was found. The homologous sequence varies between about 500 and 1100 nucleotides (TRV isolates) and was found to be 266 nucleotides in PEBV-SP5 and 459 in PRV. Whereas these sequences are identical in the RNAs of the TRV strains and in PRV, there are nine differences in the PEBV-SP5 homologous stretch. The nature of the selection pressure that leads to perfect 3′-terminal sequences between the RNAs of TRV and PRV isolates is not known. Maintenance or acquisition of identical 3′ termini is not a necessity for a stable genotype as is evident from the field isolate SP5. Earlier work with pseudo-recombinants, which are the experimental combination of RNA1 (or RNA1-containing L particles) from one strain with RNA2 (or S particles) from another to create an M-type isolate, showed that they are stable genotypes. In the greenhouse, a pseudo-recombinant was inoculated mechanically on to tobacco, and it was shown that, after 25 serial passages, each RNA had retained its specific 3′ sequences. The ability to form pseudo-recombinants is independent of serological relationship between the strains used, but is only possible within one subgroup, e.g. between different TRV strains but not between a TRV strain and a PEBV strain. Thus, bearing in mind that RNA1 molecules within one subgroup are highly homologous, there seems to be a necessity for RNA2 to have a highly homologous or even identical 3′ terminus with RNA1 in order to establish a general compatibility with RNA1 and the replication functions it specifies. The importance of homologous termini between the genome parts of multipartite viruses was also shown with other such viruses (e.g. bromoviruses, alfalfa mosaic virus, nepoviruses). In tobraviruses, the acquisition of the 3′ terminus of RNA1 by RNA2 is thought to occur by a copy-choice mechanism of the viral replicase. The junction between RNA2- and RNA1-specific sequences shows a sequence that closely resembles the one that occurs at 5′ termini of subgenomic TRV RNAs.

The need for homologous termini in tobraviruses was further demonstrated by analyzing the so-called 'anomalous' isolates. These isolates were called 'anomalous' because they combined properties of TRV (symptom expression, ability to create pseudo-recombinants with other TRV isolates) and PEBV (serological relationship). These 'anomalous' isolates were shown to consist of an RNA1 homologous with that of other TRV isolates, whereas RNA2 (including the coat protein cistron) was PEBV-derived with both 3′- and 5′-terminal sequences homologous with TRV. Whereas the 3′-terminal sequence is probably acquired from TRV RNA1 by the mechanism mentioned above, the 5′-terminal sequence is presumably derived from TRV RNA2 by recombination within a conserved sequence.

Thus, the 'anomalous' isolates are members of the TRV subgroup that have captured genes of the PEBV gene pool. Because of the existence of such 'anomalous' isolates, serological tests to distinguish tobraviruses are not reliable. Nucleic acid hybridization with sequences homologous with RNA1 is the method of choice to classify a new tobraviral isolate.

Diseases Caused

Tobraviruses have wide host ranges, infecting herbaceous and a few woody plant species. They cause several types of disease. One of the diseases led to the name of this virus group: when the wind blows through a heavily infected tobacco field a rattle-like sound is produced by the TRV-infected dried-out leaves.

Among tobraviruses, TRV has by far the widest host range, probably the widest of any plant virus. TRV infects more than a hundred species in nature, including several common weed species (e.g. *Stellaria media*, *Capsella bursa-pastoris*, *Senecio vulgaris*). Experimentally, it was shown that more than 400 plant species in over 50 families can be infected with TRV by inoculation with sap. In about half of these infected plant species, the virus can spread systemically.

In many naturally infected species, the virus remains localized at the initial site of infection. Other species are invaded systemically but remain symptomless (as *Stellaria media*) or may develop a wide variety of symptoms. On leaves, symptoms range from necrosis to all kinds of yellow markings (blotching, mottling, mosaic, ringspot), often accompanied by a variable degree of distortion. Of economic importance is the damage caused in bulbous ornamental crops such as tulip, narcissus, lily and crocus, in which leaves become mottled, and gladiolus, which develops notched leaves. Symptoms on underground plant parts include corky arcs in potato tubers (spraing), which lowers the value of the crops, and necrotic spots (malaria) in hyacinth bulbs.

Furthermore, vigor and yield are decreased in tomato, sugarbeet, spinach, tobacco, artichoke, celery, pepper and lettuce.

The PEBV subgroup includes the tobravirus isolates that systemically infect leguminous plants. Only four crop species (pea, bean, broad bean and lucerne) are reported to become naturally infected. Symptoms caused by PEBV include large necrotic spots on pea leaflets, leaf distortion with chlorotic

V-shaped markings in lucerne and mosaic in bean leaves. Only early-browning of pea is known to be a widespread disease. PEBV probably also has hosts among weed species, but these have not been well studied.

No extensive data are available for PRV. This virus is of local concern in Brazil and was reported to cause leaf markings in artichoke and tomato. In addition, wild plants are known to be hosts.

Tobraviruses, especially TRV and PEBV, exist as many serological variants and are known to be very variable in type and severity of symptoms produced. This variability in symptoms may be the result of the specific transmission of tobraviruses by trichodorid nematodes (see below), with different viral strains being transmitted by different nematode species. Furthermore, variation in symptom expression on one host species is well known to occur independently of antigenic variation of the viruses. Experimentally, it was shown that 14 TRV isolates of the same serotype which are naturally transmitted by the same nematode species produced symptoms ranging from severe systemic malformation to mild chlorotic spots on the mechanically inoculated leaves only.

Detection and identification of tobraviruses is further complicated by the existence of NM-type isolates (in which only RNA1 replicates without RNA2). Such infections rapidly spread from cell to cell but only slowly systemically; the symptoms caused are usually more severe than those induced by M-type isolates (in which both RNA1 and RNA2 are present).

Geographic Distribution

TRV is the most widespread of the tobraviruses and has been recorded throughout Europe, in North America, New Zealand and in Japan. PEBV is known to occur in Europe and North Africa, whereas PRV has only been described in South America (Brazil).

Virus Transmission

Nematode transmission

Tobraviruses are naturally transmitted by root-feeding nematodes of the genera *Trichodorus* and *Paratrichodorus*. It has also been suggested that transmission can occur in the absence of trichodorid nematodes, e.g. through soil water and root contacts.

At present, eight species within the genus *Paratrichodorus* and five within *Trichodorus* are known to be vectors of tobraviruses. In Europe, 50% of (*Para*)*Trichodorus* species are tobravirus vectors.

A highly specific relationship has been found between TRV strains and the nematode species involved in their transmission: individual species of (*Para*)*Trichodorus* transmit serologically distinct isolates of tobraviruses. In addition, certain species are able to transmit more than one virus within the tobravirus group.

The factors that determine the specificity of viral transmission have hardly been elucidated. With electron microscopy, TRV particles were shown to be associated with the lining of the pharynx and the esophagus. As this lining is stained with carbohydrate stains, it was suggested that virus retention involves the interaction of the carbohydrates with viral particles. The release of the viral particles is thought to occur by a pH change caused by the saliva flow produced by the nematode when it begins to feed on a new plant. Successful transmission of a viral strain only takes place if there is a balance between retention and subsequent release of viral particles. How delicate this balance may be was shown by the difference in transmission efficiency of several serologically indistinguishable viral isolates by a nematode population: whereas one isolate was transmitted very frequently, two isolates were not transmitted at all, and the others ranked intermediately.

There is no latent period between acquisition of the virus by the nematode and inoculation to new plants. The viruses do not replicate in the nematodes but can persist there for months or years. Neither are they retained through the molt (as the lining of the esophagus is also shed during molting), nor is there evidence that they are passed to progeny nematodes.

NM-type isolates cannot be acquired and inoculated by nematodes. However, naturally infected plants may contain only NM-type isolates as a result of the chance separation of long and short particles during transmission.

Seed transmission

Seed transmission of TRV is not known to occur in crop plants but is possible in several weed species. PEBV and PRV on the other hand are reported to be seed-borne in pea and tomato respectively.

Virus Epidemiology and Control

The occurrence of tobraviruses depends on the distribution of their nematode vectors, which tend to be prevalent on lighter, sandy or loamy soils. Tobraviruses can survive at sites in three main ways: they can persist in the nematode vector, in perennial plants and in infected seeds.

Viral spread at a site depends on the number, activity and transmitting efficiency of vector nematodes. The number of trichodorids depends mainly on the type of previous crops and on the degree and type of weed infestation (as nematodes multiply differently on different host plants). Weed infestation is also of importance in determining the proportion of nematodes that carry viral particles. A wide range of weed species (e.g. *Stellaria media*) can harbor TRV with a high incidence of systemic infection. Such plants are a constant source of viral particles to transmitting nematodes. Tests on naturally occurring *S. media* plants are a reliable indication of whether or not TRV occurs at a site.

The activity of nematodes is mainly determined by soil water content as nematodes need a water film on soil particles to be able to move through the soil. Therefore, incidence of tobravirus-caused diseases is increased after wet periods or in irrigated crops. Optimum temperature for transmission was found to be 15–20°C with little transmission occurring at 4°C.

In Scotland it was shown that 80% of the trichodorid populations in arable land were carrying TRV. The rate of viral transmission by populations of viruliferous nematodes was found to be rather low. This probably reflects the small proportion of virus-carrying individuals. Experiments with single nematodes, however, revealed that once a nematode has acquired viruses subsequent transmission to healthy plants can be very efficient.

Another factor affecting the spread of tobraviruses is the vertical and horizontal distribution of vector nematodes. Infected plants typically are patchily distributed in crops. These patches do not necessarily represent the horizontal distribution: as trichodorid nematodes seem to occur in considerable numbers in somewhat deeper soil layers (below the depth of cultivation) but above hard layers, such patches may occur where the topsoil is shallow.

As nematodes move only small distances laterally (probably less than 50 cm per year), spread of virus-carrying nematodes to new sites occurs by agricultural activities. Transport of vector nematodes in wind-blown soil is probably inefficient, as the nematodes are very susceptible to desiccation. Virus-infected seed and vegetative plant material can also be carried for long distances to sites with previously virus-free populations of vector nematodes. Thus it could be observed that trichodorids appeared soon after colonization of a sand dune by grasses but TRV was not recorded until the flora included *Viola tricolor*, in which the virus is seed-borne.

The control of tobraviruses depends largely on the use of tolerant or resistant cultivars (potato, pea) and on the application of expensive nematode-controlling chemicals to vector-infested land. Weeds are both a virus source and a vehicle for virus spread (via seeds). Rigorous control of weeds in order to eliminate virus sources, however, can actually increase TRV infections because virus-carrying trichodorids that may prefer to feed on weed roots are then obliged to feed on crop plants.

Virus dissemination can be minimized by using virus-free planting material (pea, flower bulbs).

See also: Sindbis and Semliki Forest viruses; Tobamoviruses.

Further Reading

Harrison BD and Robinson DJ (1986) Tobraviruses. In: Van Regenmortel MHV and Fraenkel-Conrat H (eds) *The Plant Viruses*. Vol. 2. *The Rod-Shaped Plant Viruses*, p. 339. London and New York: Plenum Press.

Goulden MG *et al.* (1990) The complete nucleotide sequence of PEBV RNA-2 reveals the presence of a novel open reading frame and provides insights into the structure of tobraviral subgenomic promotors. *Nucleic Acid Res.* 18(15): 4507 (and references therein).

Ploeg AT, Brown DJF and Robinson DJ (1992) The association between species of *Trichodorus* and *Paratrichodorus* vector nematodes and serotypes of tobacco rattle tobra virus. *Ann. Appl. Biol.* 121: 619.

Ziegler-Graff V, Guilford PJ and Baulcombe DC (1991) Tobacco rattle virus RNA-1 29k gene product potentiates viral movement and also affects symptom induction in tobacco. *Virology* 182: 145 (and references therein).

TOMBUSVIRUSES

DM Rochon
Agriculture Canada Research Station
Vancouver, British Columbia, Canada

Taxonomy and Classification

Tombusviruses have nonenveloped 30 nm isometric particles containing 180 copies of a single type of coat protein subunit of approximately 41 kD. The genome is a single copy of a linear monopartite positive-polarity RNA approximately 4.7 kb in length. Some tombusviruses are associated with satellite RNAs and at least three members are associated with symptom-attenuating defective-interfering (DI) RNAs. The tombusvirus group presently consists of the type member, tomato bushy stunt virus (TBSV), nine additional serologically interrelated members [artichoke mottled crinkle virus (AMCV), carnation Italian ringspot virus (CIRV), cymbidium ringspot virus (CyRSV), eggplant mottled crinkle virus, grapevine Algerian latent virus, Moroccan pepper virus, Neckar river virus, pelargonium leaf curl virus (PLCV), petunia asteroid mosaic virus (also referred to as the cherry strain of TBSV; TBSV-Ch)] and the serologically unrelated member cucumber necrosis virus (CNV). Nucleic acid hybridization studies using seven members of the tombusvirus group have demonstrated sequence similarity between each of the members examined. In addition, the complete nucleotide sequences of three distantly related tombusviruses (CNV, CyRSV and TBSV-Ch) and a partial sequence of AMCV, which is serologically closely related to TBSV-Ch and TBSV-type, have revealed that the genomic RNAs and protein products of these viruses share a high degree of relatedness (64.4%–69% overall sequence similarity in the protein products). The greatest sequence diversity is in the capsid proteins (37–44.5%) with the protruding domain of the coat protein exhibiting the most extensive variation. The hypervariable nature of the P domain may explain the serological differences observed among tombusviruses, since the P domain is the most externally exposed portion and therefore might be the dominant epitope of the capsid. The high degree of nucleic acid sequence similarity among members of the tombusvirus group suggests that tombusviruses should be considered related strains with distinctive biological properties (see below) rather than separate viruses.

Evolutionary relationships

The putative polymerases of tombusviruses and those of carmoviruses (represented by carnation mottle virus), the dianthovirus, red clover necrotic mosaic, the unclassified maize chlorotic mottle virus, the necrovirus, tobacco necrosis, the luteovirus, barley yellow dwarf, and the animal virus, hepatitis C, have been shown to share a high level of amino acid sequence similarity. In addition, these viral polymerases share only minimal amino acid sequence similarity with the polymerases of members of the proposed alphavirus and picornavirus-like 'supergroups'. Interestingly, each of the six plant viruses lack the helicase motif found in most positive-strand RNA viruses with larger genomes, suggesting that a host component may serve the helicase function for tombusviruses and these small RNA plant viruses.

Significant amino acid sequence similarity also exists among the shell domains of the capsid proteins of tombusviruses and other small RNA plant viruses with spherical capsids in the carmovirus, dianthovirus, sobemovirus, luteovirus and necrovirus groups and the unclassified maize chlorotic mottle virus. Less striking but significant homology is also observed between these coat proteins and the coat proteins of the picornavirus-like comoviruses and nepoviruses as well as those of spherical satellites associated with tobacco necrosis, panicum mosaic, maize white line mosaic and the satellite of the alphavirus tobacco mosaic virus. The tombusvirus coat protein gene appears to have evolved independently from the polymerase gene which is consistent with the modular theory proposed for positive-strand RNA virus evolution.

Particle Structure

The structure of the TBSV particle has been determined at 0.29 nm resolution. The particle is a $T = 3$ icosahedron approximately 30 nm in diameter which consists of 180 identical coat protein subunits 387 amino acids in length. Each of the subunits folds into three major domains: an N-terminal inward-facing R domain (66 amino acids); a tightly bonded globular shell domain (S; 167 amino acids); and a C-terminal outward-facing protruding domain (P; 114 amino acids). The R domain is flexibly tethered to the rest of the subunit by the arm (a; 35 amino acids) and the S and P domains are connected by a short hinge (h; 5 amino acids). The 180 TBSV coat protein subunits can adopt three conformations denoted A, B and C; the A and B conformations are very similar with disordered N-terminal arms, whereas the C conformation has an ordered arm that interdigitates with two other C subunit arms around the icosahedral threefold axis. The P domain projects from the surface of the particle in 90 pairwise clusters. The hinge which connects the S and P domains exists in two states which differ by 2 nm.

The S and P domains consist largely of β-sheet with a jellyroll conformation. The S domain consists of two four-stranded antiparallel β-sheets, and the P domain is composed of two antiparallel β-sheets, one of six strands and the other of four strands. The S domain contains loops and helices which appear to contain residues important in variable intersubunit or interdomain contacts. The tombusvirus protruding domain was previously suggested to have evolved from the shell domain as a result of a gene duplication event. This belief was based partly on the observation that the S and P domains shared similar jellyroll topologies. It has, however, recently been demonstrated that the two jellyroll conformations of the S and P domains are distinct and that the jellyroll conformation of P is unique among proteins so far examined. This unique jellyroll topology may therefore have been acquired from an as yet unidentified cellular protein.

The R domain and the inward-facing parts of the S domain contain a preponderance of basic amino acid residues. The positive charges of these residues neutralize most of the phosphates of encapsidated viral RNA allowing dense packaging of the RNA.

Calcium ions play an important role in capsid stability. The calcium ions link two neighboring subunits together via carboxylate groups from the side chains of glutamate and aspartate amino acid residues in the shell domain. In the absence of calcium ions, TBSV swells when the pH is raised above neutrality. Swelling is due to the repulsive forces of negatively charged carboxylate groups previously neutralized by divalent cations. Such observations suggest that calcium bridges in the TBSV capsid are important in the release of viral nucleic acid in an infected cell. Since the calcium concentration in plant cells is low, calcium may be sequestered from viral capsids causing the particles to swell and subsequently release viral nucleic acid. The actual uncoating mechanism following swelling may occur cotranslationally in the presence of ribosome-associated factors. In addition to the stabilizing forces contributed by calcium-mediated shell domain contacts, it has been suggested that P-domain dimer contacts may also contribute to TBSV particle stability.

Structural studies of the particle and coat protein subunits of the turnip crinkle carmovirus and southern bean sobemovirus have shown that tombusviruses share similarities with these viruses in both the three-dimensional structure of their particles and the tertiary structure of their coat proteins. Comparisons of the coat protein amino acid sequences of tombusviruses, carmoviruses, southern bean mosaic and several other spherical RNA plant viruses suggest that the capsid proteins of these viruses share a common ancestor (see under Taxonomy and Classification).

Genome Structure and Molecular Biology

The complete nucleotide sequences of CNV, CyRSV and TBSV-Ch and the sequence of the 3′-terminal region of AMCV have been determined. The structure of the genomes deduced from these sequences are highly similar to each other and all of the corresponding nucleotide sequences and predicted protein amino acid sequences show extensive similarity. The structure of the tombusvirus genome as exemplified by CNV is shown in Fig. 1. The tombusvirus genome consists of five long open reading frames (ORFs) encoding proteins with approximate molecular masses of 33, 92, 41, 21 and 20 kD. The 33 kD protein terminates with an amber codon. Readthrough of this amber terminator would produce a protein of 92 kD. The 41 kD coat protein ORF has an internal location and the 20 kD ORF is completely nested within a different reading frame of the 21 kD ORF.

The tombusvirus 92 kD protein contains the canonical Gly-Asp-Asp tripeptide and surrounding hydrophobic amino acids characteristic of putative polymerase domains of most RNA viruses sequenced to date and thus likely forms at least part of the viral RNA replicase. The tombusvirus genome lacks coding information for the helicase domain found in most RNA viral genomes, leading

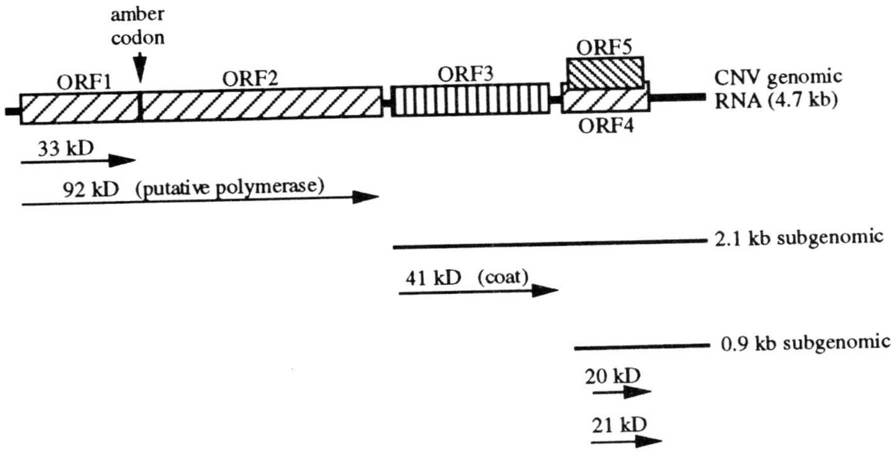

Fig. 1 Organization and expression of the tombusvirus genome as exemplified by CNV. The shaded boxes correspond to the five open reading frames (ORFs) encoded by tombusviruses (the different shading indicates the different reading frames utilized). The arrows indicate the origins and sizes of encoded proteins. The horizontal lines correspond to the two subgenomics which are generated during tombusvirus infection.

to speculation that such a function may be acquired from a host protein during infection. The 41 kD protein has been demonstrated to be the coat protein based on *in vitro* translation and immunoprecipitation studies as well as amino acid sequence similarities with the coat protein of the BS-3 strain of TBSV, which was sequenced directly. The function of the 33 kD protein is presently not known, but *in vitro* mutagenesis studies have suggested possible functions for the 21 and 20 kD proteins (see below).

Plant tissue infected with tombusviruses contains at least two viral-specific RNA species in addition to genomic RNA. The RNA species have sizes of approximately 2.1 and 0.9 kb, are 3′ coterminal and colinear (see Fig. 1) and serve as the templates for the 41 and 21/20 kD proteins respectively. Three prominent viral-specific double-stranded RNAs are found in tombusvirus-infected leaves. The largest band presumably corresponds to genomic RNA and the two smaller ones to the two subgenomic RNAs. It is not known if the smaller RNAs correspond to independently replicating subgenomic RNAs or if they are an artifact of the dsRNA extraction procedure.

Infectious synthetic transcripts have been derived from full-length cDNA clones of CNV, CyRSV and TBSV-Ch. These clones are highly infectious even in the absence of added cap analog in the transcription reaction. Site-directed *in vitro* mutagenesis of CNV infectious clones has been used to study the expression and function of CNV genes and gene products.

The 21 and 20 kD proteins

Point substitutions introduced into the AUG codons for the 21 and 20 kD proteins have demonstrated that both proteins are likely produced *in vivo* since full-length or subgenomic-length transcripts which lack either of the AUG codons alter the infectivity, symptomatology and *in vitro* translation profiles of CNV transcripts. The 21 kD protein would appear to be essential for viral RNA accumulation since inoculation of plants with full-length transcripts containing a mutated 21 kD AUG codon does not result in symptoms, and viral RNA cannot be detected by Northern blot analysis. It is possible that the 21 kD protein has a role in cell-to-cell movement but this has yet to be substantiated. Plants inoculated with mutant transcripts which lack the AUG codon for the 20 kD protein or which have an in-phase terminator shortly following the 20 kD AUG codon show symptoms of viral infection, but the symptoms are very mild in comparison to the symptoms produced by wild-type transcripts. Examination of infected leaf extracts several days postinoculation or a few days following sap transfer of transcript-inoculated plants has revealed that 20 kD mutant plants accummulate high levels of symptom-attenuating CNV DI RNAs which are generated *de novo* from the mutant transcript. Since such a rapid accumulation of DI RNA is not observed in wild-type inoculated plants, it would appear that the CNV 20 kD protein plays a role in viral RNA replication preventing the formation or accumulation of aberrant replication products such as DI RNAs. It is also likely that the DI RNAs formed in 20 kD mutant infected plants are responsible for the observed symptom attenuation. However, even early in infection when DI RNAs are not detected, 20 kD mutant plants still show only mild symptoms.

In vitro and *in vivo* studies have demonstrated that a single subgenomic mRNA acts as the template for both the 21 and 20 kD proteins. Expression of both proteins from this bifunctional subgenomic mRNA may be regulated by a 'leaky scanning' mechanism, as the upstream AUG codon (21 kD protein) has been found to be in a suboptimal context for efficient expression in animal or plant cells whereas the downstream AUG codon (20 kD protein) is in a more favorable context.

The coat protein

Mutant CNV transcripts in which the coding sequence of the CNV coat protein protruding domain has been deleted produce infections in plants. However, such infections can be distinguished from wild-type infections due to smaller necrotic lesions and a delay in the formation of systemic symptoms. Examination of extracts of transcript-inoculated plants for CNV particles or coat protein by electron microscopy, immunoelectron microscopy and enzyme-linked immunosorbent assay (ELISA) using antisera raised to CNV particles has failed to detect particles or coat protein, suggesting that coat protein subunit lacking the protruding domain is not produced or is not stable in infected plants. Analysis of viral-specific RNA by Northern blotting reveals that genomic RNA evolves from a size in transcript-inoculated leaves that would lack the coding sequence of the protruding domain to a size in systemically infected leaves that would lack the coding sequence for the entire coat protein. It would appear therefore that deletion mutants arising during replication of protruding-domain-less mutants are selected for and become dominant. In addition, the replication and spread of a CNV mutant which lacks all or part of the coat protein coding sequence indicates that the CNV coat protein is not essential for cell-to-cell and systemic movement, although its presence may facilitate both types of movement.

Symptomatology, Host Range and Geographic Distribution

Tombusviruses as a whole share a wide geographic distribution having been reported in North and South America, several locations in Europe and the Mediterranean and also in Algeria. Individual members of the tombusvirus group, however, generally have a restricted geographic distribution with the exception of TBSV and PLCV which were likely spread internationally through infected geraniums. Most tombusviruses naturally infect only a single or a few hosts. This is in contrast to their experimental host range which is wide and diversified comprising as a whole over 120 plant species in 20 different monocotyledonous and dicotyledonous plant families. Tombusviruses systemically invade naturally infected plants but remain localized in the majority of experimentally infected plants; common exceptions to this rule are the experimental hosts *Nicotiana clevelandii* and *Nicotiana benthamiana*. Tombusvirus-infected crops generally show stunting and diffuse mottling and malformation of leaves. Characteristic lesions associated with the indicator plants basil, globe amaranth and *Chenopodium quinoa* Willd. can be used to identify tombusviruses at the group level. However, identification of individual tombusviruses using host responses alone cannot be relied upon. Finally, many tombusviruses can be inactivated *in vivo* by growing plants at elevated temperatures (*c.* 36° C) for 1–2 months. The diseases, host ranges and geographical distribution of tombusviruses have recently been reviewed by Martelli *et al.* (see Further Reading section).

Transmission

Tombusvirus particles are very stable and reach high concentrations in infected tissues. In addition, tombusviruses are efficiently spread and can become established in diverse environments. Different tombusviruses have been shown to be spread in nature by a number of means including seed and pollen transmission, transmission through propagation material and possibly mechanical transmission. Experimentally, tombusviruses are readily transmitted via sap, and infected leaf extracts retain infectivity after freezing for several years. There is little doubt that several tombusviruses are spread through the soil but attempts to transmit them (except CNV; see below) by soil-borne vectors have not been successful. In addition, attempts to transmit different tombusviruses using aphids, whiteflies or mites have also been unsuccessful.

Several reports have identified tombusviruses in association with rivers and lakes throughout the world, and there is evidence suggesting that tombusviruses can enter a field through irrigation water. TBSV is infectious after passage through the human alimentary tract, suggesting that man may be an important carrier as tombusviruses may possibly enter rivers through sewage treatment of infective feces.

In spite of the lack of evidence for a soil-borne vector in the transmission of most tombusviruses, CNV has been demonstrated to be transmitted in a non-persistent fashion by the root-inhabiting Chytrid fungus *Olpidium radicale*. Transmission of CNV by this fungus is mechanistically similar to that demonstrated for the transmission of tobacco necrosis necrovirus by *Olpidium brassicae*. In both cases, viral particles and *Olpidium* zoospores are released independently into the soil from roots of infected plants, and viral particles adsorb to the axonemal sheath of the zoospore flagellum. Virus gains entry into plant roots when the flagellum, along with its axonemal sheath and adsorbed virus, is withdrawn prior to encystment of the zoospore. Different *O. radicale* isolates have varying abilities to transmit CNV and other *O. radicale*-transmitted plant viruses (melon necrotic spot carmovirus and the unclassified cucumber leaf spot virus), suggesting that there is a specific mechanism by which *O. radicale* zoospores and the viruses transmitted by them recognize each other. In addition, it has been suggested that other soil-borne tombusviruses may also have highly specific fungal vectors that have yet to be identified.

Cytopathology

Cytopathic inclusions known as multivesicular bodies and virus-containing bleb-like evaginations of the tonoplast are associated with infection by tombusviruses. Multivesicular bodies are composed of vesicles intermingled with or surrounding granular electron-dense material and are believed to originate from either mitochondria, peroxisomes and/or chloroplasts depending on the virus/host combination examined. Clumps of densely staining amorphous material found scattered or loosely aggregated in the cytoplasm of systemically infected cells (dense granules) are also found in tombusvirus-infected cells but not in all virus–host combinations tested. Multivesicular bodies are believed to be sites of RNA replication and dense granules to be accumulations of excess coat protein subunit.

Satellites and DI RNAs

Tombusvirus infections have been found associated with both satellite and DI RNAs. Virion RNA preparations of and plants infected with five tombusviruses (AMCV, CyRSV, CIRV, TBSV-Ch, PLCV) and two isolates of TBSV (TBSV-type, TBSV-BS-3) have been found to contain satellite RNAs of approximately 700 nucleotides. These satellite RNAs were shown by hybridization analyses to have sequences in common with each other but not with genomic RNA. In addition, the presence of satellite RNAs was associated with a decrease in symptom severity. The nucleotide sequence of one such satellite RNA associated with CyRSV infection has been determined and found to contain limited sequence similarity with genomic RNA. The sequence similarity might suggest that tombusvirus satellites have arisen by mutational drift of DI RNAs.

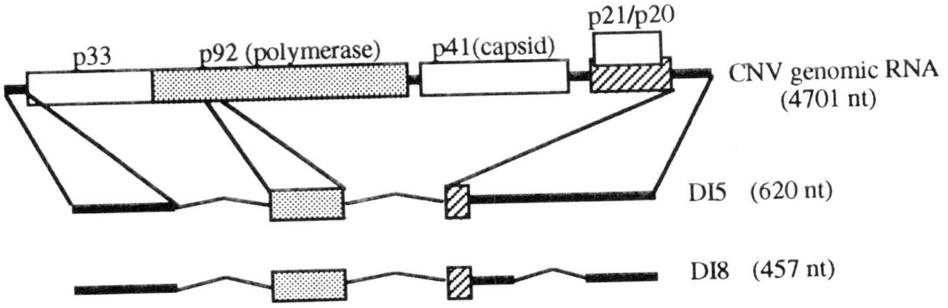

Fig. 2 Location of CNV DI RNA sequences relative to the CNV genome. The approximate genomic origins of the sequences of two DI RNAs (DI5 and DI8) generated *de novo* from infections initiated with a CNV 20 kD mutant are shown below the CNV genome (not in scale). Bent lines between the blocks of DI RNA sequence correspond to deletions in DI RNAs relative to CNV RNA. The general structures of these CNV DIs typify those of DI RNAs associated with the previously characterized TBSV-Ch and CyRSV DI RNAs.

Three tombusviruses (CNV, CyRSV, TBSV-Ch) have been demonstrated to be associated with DI RNAs. TBSV-Ch DI RNAs were the first tombusvirus DI RNAs to be characterized and moreover provided the first definitive demonstration of DI RNAs associated with plant viral infection. Tombusvirus DI RNAs are small (*c.* 400–700 nucleotides) colinear deletion mutants of the helper virus genome which are dependent on the helper virus for replication and encapsidation and are associated with a dramatic attenuation of the severe necrosis typical of tombusvirus infection. TBSV DI RNAs interfere with replication of genomic RNA in protoplasts, and such interference in plants is presumed to be responsible for the reduction in symptom severity associated with the presence of DI RNAs. Figure 2 provides a diagrammatic representation of the location of CNV DI RNA sequences relative to genomic RNAs. Previously characterized DI RNAs associated with TBSV-Ch and CyRSV share similar sequence relationships with the genomic RNAs of their respective helper viruses.

It has been shown that tombusvirus DI RNA, like the DI RNAs associated with animal virus infections, can be generated *de novo* upon serial high-multiplicity passage of synthetic genomic length RNA, probably as a result of selection for particular deletion mutants the replication of which is favored. Interestingly, synthetic CNV mutants, which do not express the 20 kD nonstructural protein, rapidly generate DI RNA without serial passage and even at low multiplicity of infection. This observation suggests a role for the CNV 20 kD nonstructural protein in preventing the generation or accumulation of errors during RNA replication and may provide insight into the means by which DI RNAs naturally arise in wild-type virus infections.

See also: Carmoviruses; Defective-interfering viruses; Dianthoviruses; Necroviruses; Virus structures.

Further Reading

Harrison SC (1983) Virus structure: High-resolution perspectives. *Adv. Virus Res.* 28: 175.

Martelli GP, Gallitelli D and Russo M (1988) Tombusviruses. In: Koenig R (ed.) *The Plant Viruses*, vol. 3, p. 13. New York: Plenum Publishing Corporation.

Morris TJ and Knorr DA (1990) Defective interfering viruses associated with plant virus infections. In: Brinton MA and Heinz FX (eds) *New Aspects of Positive-Strand RNA Viruses*, p. 123. Washington, D.C.: American Society of Microbiologists.

Rochon DM, Johnston JC and Riviere CJ (1991) Molecular analysis of the cucumber necrosis virus genome. *Can. J. Plant Pathol.* 13: 142.

TOROVIRUSES

Marian C Horzinek
University of Utrecht
Utrecht, The Netherlands

History

Over the past ten years, toroviruses have been defined as a result of a trilateral partly collaborative study between groups led by the late F. Steck in Berne, Switzerland (isolation in cell culture), G.N. Woode in Ames/Iowa, USA (discovery of Breda virus and propagation in calves) and M. Horzinek in Utrecht, The Netherlands (characterization).

Berne virus (BEV) was accidentally isolated in equine kidney cells in 1972 from a rectal swab taken from a horse with diarrhea. Upon postmortem examination, pseudomembranous enteritis and miliary granulomas and necrosis in the liver were diagnosed; *Salmonella* Lille (0, 6, 7, Z_{38}) was initially considered to be the causative agent. BEV can be propagated in lines of equine dermis or embryonic mule skin cells, where it causes a cytopathic effect that results in cell lysis. The virus was not neutralized by antisera against known equine viruses. Serological cross-reactions were observed in neutralization and enzyme-liked immunosorbent assays (ELISA) with sera from calves that had been experimentally infected with morphologically similar particles, the Breda viruses.

Breda virus (BRV) was discovered in 1979 during

investigations in a dairy herd in Breda (Iowa), in which severe neonatal calf diarrhea had been a problem for three subsequent years. Despite repeated attempts, BRV has not been adapted to growth in cell or tissue culture which has hampered its biochemical, biophysical and molecular characterization. The pathogenesis and pathology of BRV infections have been studied in gnotobiotic calves.

Torovirus-like particles have been observed by electron microscopy in fecal preparations from pigs, cats, children and adults. Results of cDNA hybridization studies indeed proved the presence of toroviral sequences in stools from humans with diarrhea. They were confirmed by successful polymerase chain reaction (PCR) amplification using RNA extracted from hybridization-positive feces.

Taxonomy and Classification

Toroviruses possess an RNA genome of positive (messenger) polarity. On the basis of their morphological and physicochemical characteristics, the toroviruses were originally proposed to constitute a new family of enveloped RNA viruses. In the light of recent genetic data, toroviruses and coronaviruses have been classified as two separate genera within the family *Coronaviridae*. The name is derived from *torus* (Latin) which denominates the lowest convex molding in the base of a column or pilaster; it indicates the doughnut (biconcave disk) shape of the virion which is determined by a tubular capsid of helicoidal symmetry, surrounded by a peplomer-bearing envelope.

Virion Properties

Extracellular BEV particles have a helical nucleocapsid coiled into a hollow tube (diameter 23 nm, average length 104 nm, periodicity 4.5 nm) which is either straight or bent into an open torus. A tightly fitting envelope 11 nm thick surrounds this structure. Consequently, the virion assumes an erythrocyte-like or a kidney shape, depending on whether the envelope follows the small curvature of the nucleocapsid or not. The largest diameter of BEV is estimated at 120–140 nm. Club-shaped projections (average length 20 nm) are present on the virion surface. Intracellularly, the particles are predominantly rod-like. In cross-sections through tubular virions, they appear as three concentric circles of high electron density with an electron-lucent center.

Negatively stained BRV virions appear to be either kidney shaped measuring 30–120 nm or approximately circular and measuring 75–90 nm. They have an envelope, which bears drumstick-shaped peplomers 8–10 nm long. Longer peplomers are seen occasionally (17–24 nm), especially on the viruses of the Ohio and second Iowa strain. In virus-infected intestinal cells of calves killed 48–96 h after infection, tubules of 21 nm diameter and indeterminate length are found in both the cytoplasm and nuclei.

In thin-sectioned preparations, intracellular torovirions show a bacilliform morphology (rod-like, both ends rounded, in contrast to the circular outline of coronavirions); extracellular particles may reveal twin circular structures resulting from cross-sections through both limbs of the C-shaped tubular nucleocapsid.

A schematic model of a torovirion is given in Fig. 1.

Properties of the Genome

BEV virions (sedimentation coefficient 380S) contain one species of polyadenylated RNA; in agarose gel electrophoresis its length appears to be ≥ 20 kb. When assayed under hypertonic transfection conditions, genomic RNA is found to be infectious and RNase-sensitive. The BEV genome contains six open reading frames.

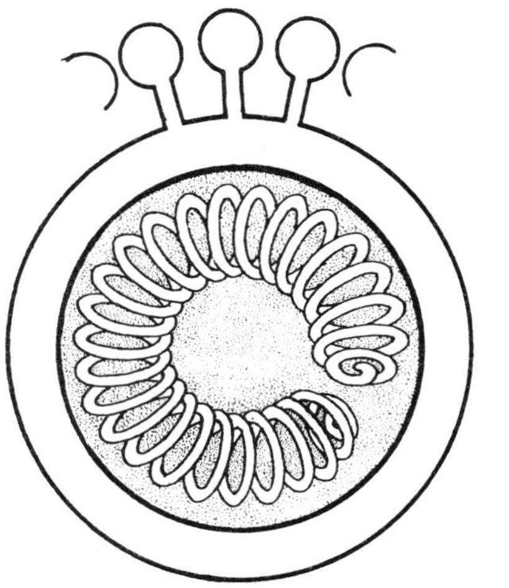

Fig. 1 A schematic model of a torovirion. Illustrated by A. Veer.

Proteins of the Virus

Proteins with molecular masses of 19, 22, 37 and 80–100 kD are identified in labeled BEV virions. Detergent treatment releases the 22, 37 and 80–100 kD species from the virion, which indicates their association with the envelope. Only the 19 kD protein is present in purified BEV nucleocapsids and was accordingly named nucleocapsid (N) protein. The heterogeneous, *N*-glycosylated 80–100 kD protein is recognized by both neutralizing and hemagglutination-inhibiting monoclonal antibodies and is therefore identified as the peplomer (P) protein. Another membrane-associated polypeptide is the nonglycosylated envelope protein (E; 22 kD); the 37 kD molecule also occurs in close association with the viral membrane, but its virus specificity could not be established.

From the deduced amino acid sequence of the nucleocapside (N) protein gene, a basic protein of 18.3 kD is predicted. *In vitro* transcription and translation, followed by immunoprecipitation, were used to identify the gene. Identification was confirmed by metabolic labeling, using the knowledge that cysteine residues are absent from the amino acid sequence of the N protein. Smaller N-related polypeptides encountered in BEV-infected cell lysates are products of aberrant translation, due to initiation on AUG codons further downstream in the N protein gene.

The 26.5 kD product of the BEV envelope (E) protein gene was identified by *in vitro* transcription and translation. Computer analysis revealed the characteristics of a class III membrane protein lacking a cleaved signal sequence but containing three successive transmembrane α-helices in the N-terminal half. Only small portions of either end of the polypeptide are exposed on opposite sides of the vesicle membranes; the C terminus protrudes at the cytoplasmic side of the membrane. The E protein accumulates in intracellular membranes, predominantly those of the endoplasmic reticulum.

The nucleotide sequence of the peplomer (P) protein gene encodes an apoprotein of 1581 amino acids with a molecular mass of about 178 kD. The deduced amino acid sequence contains domains typical for type I membrane glycoproteins: an N-terminal signal sequence, a putative C-terminal transmembrane anchor and a cytoplasmic tail.

Since BRV has not been labeled in cultured cells, its protein composition was studied by means of surface radioiodination of purified virus. Polypeptide species of 105, 85, 37 and 20 kD were identified, of which the former two probably represent the BRV peplomeric surface structures. Rabbit antisera raised against purified BRV recognize the BEV P protein in immunoprecipitation experiments.

Physical Properties

Buoyant densities of 1.16, 1.18 and 1.14 g ml^{-1} are reported for BEV, BRV serotype 2 and human toroviruses respectively.

Under experimental conditions BEV is remarkably stable, even to the action of phospholipase C or deoxycholate; Triton X-100 and organic solvents destroy its infectivity. BRV1 appears to be less stable than BEV, as changes in its sedimentation behavior and density have been observed after prolonged storage at −70°C. The infectivity of a fecal preparation containing BRV1 was lost completely after 3 weeks at 4°C whereas BEV in cell-free supernatant remained stable for 92 days. Two cycles of freeze–thawing of purified BRV2 resulted in loss of peplomers.

Replication

BEV replication occurs in the cytoplasm via a 3′-coterminal nested set of five subgenomic mRNAs. Preformed tubular capsids bud through membranes of the Golgi stack and endoplasmic reticulum. A host cell nuclear function seems to be required since UV preirradiation of cells, actinomycin D and α-amanitin reduce viral yields.

Transcription

In BEV-infected cells, five virus-specific polyadenylated RNA species are found with molecular sizes > 20.0, 7.5, 2.1, 1.4 and 0.8 kb. Northern (RNA) blot hybridizations with restriction fragments from cDNA clones showed that the five BEV mRNAs form a 3′-coterminal nested set. Sequence analysis revealed the presence of four complete open reading frames of 4743, 699, 426 and 480 nucleotides, with initiation codons coinciding with the 5′ ends of BEV RNA2 through 5 respectively; RNA5 is contiguous on the consensus sequence. BEV RNA 1, 2 and 3 are transcribed independently, as shown by UV transcription mapping. Upstream of the AUG codon of each open reading frame, a conserved sequence pattern is encountered, probably a core promoter sequence in subgenomic RNA transcription. In the area surrounding the core promoter region of the two most abundant subgenomic BEV RNAs, a number of homologous sequence motifs occurs.

Translation

The 7.5, 2.1 and 0.8 kb RNAs encode a 151 kD product (possibly the precursor to the peplomer protein), the envelope protein and the nucleocapsid protein, respectively, as shown by *in vitro* translation.

The 3′ part (8 kb) of the polymerase gene of BEV contains at least two open reading frames (designated ORF 1a and ORF 1b) which overlap by 12 nucleotides. The complete sequence of ORF 1b (6873 nucleotides) is known. Like corona- and arteriviruses, BEV expresses its ORF 1b by ribosomal frameshifting during translation of the genomic RNA; also the predicted tertiary RNA structure (a pseudoknot) in the frameshift-directing region is similar. The amino acid sequence of the predicted BEV ORF 1b translation product contains homologies with the ORF 1b product of coronaviruses. Four conserved domains are present: the putative polymerase domain, an area containing conserved cysteine and histidine residues, a putative helicase motif, and a domain apparently unique for toro- and coronaviruses.

Post-translational Processing

The *N*-glycosylated peplomer protein is derived from processing of a 200 kD precursor present in infected cells but not in virions. Eighteen potential *N*-glycosylation sites, two heptad repeat domains and a possible 'trypsin-like' cleavage site exist in the peplomer gene. The mature P protein consists of two subunits, and their electrophoretic mobility upon endoglycosidase F treatment suggests that the predicted cleavage site is functional *in vivo*. The heptad repeat domains are probably involved in the generation of an intrachain coiled-coil secondary structure; similar interchain interactions can play a role in the formation of the observed P protein dimers. The intra- and inter-chain coiled-coil interactions may stabilize the elongated BEV peplomers.

Assembly

About 10 h after infection, BEV particles are seen within parts of the unaltered Golgi apparatus and extracellularly. At that time tubular structures of variable length, diameter and electron density appear in the cytoplasm and in the nucleus of infected cells, probably representing preformed nucleocapsids. It is unknown whether the accumulation of nucleocapsids in the nucleus reflects a nuclear phase in the replication of BEV or some sort of defective assembly.

Viruses predominantly bud into the lumen of Golgi cisternae. The preformed nucleocapsid tubules approach the Golgi membrane with one of both rounded ends and attach to it along one side. During budding the nucleocapsid is apparently stabilized, leading to a higher electron density and a constant diameter (23 nm).

Intracellular BRV virions are rod-shaped with rounded ends; they measure 35–40 nm in diameter and are 80–100 nm long.

Defective-interfering Virus

Defective-interfering (DI) genomes of BEV can be generated by serial undiluted passages. Isokinetic sucrose gradient analysis showed that they are packaged into virus-like particles with smaller S values than standard virions. DI RNAs contain sequences from the presumed 5′ end and the proven 3′ end of the BEV genome. Using probes from the 5′ end, a consensus nucleotide sequence of about 800 nucleotides and the 5′ end of the putative BEV polymerase gene were identified. A conserved sequence motif, probably involved in subgenomic RNA transcription, is situated immediately downstream of the 5′ end of the DI RNAs. There is no evidence for the presence of a common leader sequence in BEV RNAs.

In the gut of a BRV-infected calf, a wave of simultaneous infections progresses through a population of susceptible cells. In view of the immense particle numbers encountered in the feces, enteroabsorptive epithelial cells are probably infected at high multiplicities; DI particles may also be generated *in vivo* and may play a role in modulating the pathogenesis of torovirus infections.

Geographic and Seasonal Distribution

Toroviruses in cattle have been demonstrated by

ELISA serology in Europe, North America and Asia. Seasonal patterns of infection have been described in calves in relation to herd management (pasture/stable). Most adult horses in Switzerland possess neutralizing antibodies to BEV. Possible human toroviruses have been found in France, UK, USA and The Netherlands.

Host Range and Virus Propagation

Using the neutralization assay, antibodies to BEV have been found in sera from horses, cattle, goats, sheep, pigs, rabbits and feral mice, but not in humans or carnivores. Torovirus-like particles have been seen in fecal samples of cats with a transmissible diarrhea, but serological identification has not been obtained.

By electron microscopy, pleomorphic virions have been observed in the feces of children and adults with diarrhea; the particles were coated and aggregated after the addition of anti-BRV calf sera. The stool specimen reacted in an ELISA for the detection of BRV antigen in calves, and possessed a low titer of hemagglutinin (HA) for rat erythrocytes, which was blocked by antisera to BRV.

With the aid of cDNA hybridization tests using a broad range of BEV probes, the presence of toroviral sequences in stools from humans with diarrhea has been demonstrated and confirmed by PCR amplification. The use of fresh material (without freeze–thawing) is essential to obtain unequivocal results. No antibody to either BRV or BEV was detected by ELISA or neutralization assay respectively in sera collected from veterinarians and farm workers, indicating that human toroviruses may be only distantly related to BRV and BEV.

Cultured cells of equid origin (horse, mule) are infectable with BEV; no other cell species tested supported growth. BRV could not be propagated in any culture of primary cells or permanent lines and has been passaged in gnotobiotic calves. Recently, torovirus isolations in culture have been reported from young calves with respiratory symptoms (pneumonia), the strains could be propagated, though with difficulty, in MDBK cells and passaged in other lines (e.g. Vero cells).

Genetics

No information is available.

Evolution

Toroviruses and coronaviruses are ancestrally related by divergence of their polymerase and envelope proteins from common ancestors. In addition, their genomic organization and expression strategy, which involves the synthesis of a 3′-coterminal nested set of mRNAs, are comparable. Nucleotide sequence analysis of the BEV genome has revealed the results of two independent nonhomologous RNA recombinations: Berne virus ORF 4 encodes a protein with significant sequence similarity (30–35% identical residues) to a part of the HA esterase proteins of coronaviruses and influenza virus C. The sequence of the C-terminal part of the predicted BEV polymerase ORF 1a product contains 31–36% identical amino acids when compared with the sequence of a nonstructural 30/32 kD coronavirus protein. The cluster of coronaviruses that contains this nonstructural gene does not express it as a part of their polymerase, but by synthesizing an additional subgenomic mRNA.

Although there are several important differences between the coronaviruses, toroviruses and arteriviruses, striking similarities on the level of genomic organization and expression have been observed. This implies that these viruses may be ancestrally related. The existence of such a relationship is underlined by the presence of four domains of amino acid sequence homology in the product of ORF 1b of the *POL* gene. On the basis of these similarities and homologies they may be classified into a third superfamily of positive-stranded RNA viruses, the 'coronavirus-like' superfamily.

Serological Relationships and Variability

One strain of BEV has been isolated (and re-isolated from the same material), but all attempts to obtain a second equine isolate were fruitless.

Two strains of BRV have been reported in addition to the original isolate by Woode and colleagues: one had been detected in feces from a 5-month-old diarrheal calf in Ohio, and a second Iowa strain was recovered from a 2-day-old experimental animal. On the basis of their reactivity in ELISA, immunoelectron microscopy and HA/HA inhibition assays using rat erythrocytes, the three isolates were assigned to two serotypes: BRV1, represented by the Iowa 1 isolate, and BRV2 comprising the Ohio and the second Iowa isolate.

The occurrence of antigenically different toroviruses is not unlikely. Two serotypes of BRV have been described, and more probably exist. It is anticipated that serologically unrelated toroviruses will be identified with the aid of nucleic acid probes.

Epidemiology

The high prevalence of BRV antibodies in cows cannot be explained by the few BRV infections that were found in calves and adult cows with diarrhea. The viruses may circulate in herds through subclinically or chronically infected animals as described for rota- and coronaviruses. The level of maternal BRV-specific antibodies influences the clinical outcome of the infection, as differences in the severity of diarrhea were observed between colostrum-fed and colostrum-deprived animals.

With the aid of solid-phase immunoelectron microscopy, torovirions can be identified in fecal material; in direct preparations, virion pleomorphism makes diagnosis by electron microscopy ambiguous.

Transmission and Tissue Tropism

Toroviruses spread through direct and indirect feco–oral contact. In fecal samples from experimentally BRV-infected calves, HA titers in excess of 10^7 units per ml are measured which would correspond to particle concentrations of $> 10^{11}$. Therefore, once an outbreak is underway, the infection spreads rapidly, especially when highly susceptible hosts are on the premises (e.g. in the calving season).

Very young calves are extremely susceptible to enteric BRV infection. However, in view of recent information, attention should be focused on respiratory diseases, which generally occur at an older age. Aerogenic infections are likely to occur: experimental infection of nasal epithelial cells has been reported; high numbers of seroconversions to toroviruses have been found in association with respiratory disease in calves at 3 months of age; cytopathogenic isolates have recently been made from calves with severe respiratory symptoms.

Pathogenicity

Torovirus infections play a role in diarrhea in breeding calves up to 2 months of age and in winter dysentery of adult cattle. Torovirus was detected four times more often in diarrheal calves than in healthy animals. Torovirus-associated diarrhea of calves starts later (average 12.7 days of age) than enteritis due to rota- or coronaviruses (average 7.7 and 8.3 days respectively). Seroconversion was found significantly more often after winter dysentery outbreaks than on farms without a disease history; coronavirus seroconversion was less common.

Recent evidence suggests a role of toroviruses in respiratory disease of calves. Even generalization by viraema and intrauterine infection cannot be excluded since antibodies have been found in precolostral sera.

Clinical Features of Infection

Seroconversions to BEV occurred in all horses between 10 and 12 months of age, but without symptoms. Experimentally infected animals (intravenous route) seroconverted without clinical signs. To the author's knowledge, oral infection experiments in horses have not so far been reported.

All BRV strains are pathogenic, although with varying virulence, for newborn gnotobiotic and non-immune conventional calves after oral infection. Most calves develop anorexia and a watery yellow–green diarrhea which lasts 4–6 days, and shed virus for 3–4 days. In some calves the diarrhea is preceded by a mild temperature reaction (40°C). In the calves with a normal intestinal flora, the diarrhea is generally more severe than in gnotobiotic calves. Reduction of D-xylose resorption may reach 65% in severely affected calves. In some animals, depression and dehydration is observed, occasionally with shivering, hyperpnea and watery eye discharge. Mortality rate in experimental infections approaches 25%.

When fed to specified pathogen-free kittens, the feline torovirus-like agent produces diarrhea and pyrexia.

Pathology and Histopathology

Target organs of BRV in calves are the lower half or two-thirds of the small intestine and the entire large intestine, particularly the spiral colon. There is little macroscopic evidence of the infection. Histological examination shows villus atrophy and epithelial desquamation from the mid jejunum to the lower small intestine, and areas of necrosis in the large intestine. Both crypt and villus epithelial cells contain antigen as shown by immunofluorescence. The watery diarrhea is probably a result of loss of resorptive capacity of the colonic mucosa, combined with malabsorption in the small intestine. Infection of crypt epithelium may affect the duration of diarrhea, as regeneration of villus epithelium starts in the crypts. The germinal centers of the Peyer's patches are depleted of lymphocytes and may occasionally show fresh hemorrhage. The dome epithelial cells, including the M cells, display the same cytopathic changes as seen in the absorptive cells of villi. Virions are found in cells of both the small and large intestine. Extracellular virus is closely associated with microvilli of absorptive cells and in the coated pits between microvilli, indicating receptor-mediated endocytosis. In addition, virions are found between enterocytes at the basal and lateral plasma membranes. Virions in various stages of degradation are found within macrophages in the lamina propria.

Antigen is detected as early as 48 h after infection in epithelial cells of the lower half of the villus and the crypts of the affected areas, as well as in dome epithelium. Fluorescence is cytoplasmic (although a few nuclei may be faintly stained) and generally most pronounced in the intestines with the least tissue damage. The mid-jejunum is infected first, the infection eventually reaching the large intestine. Diagnosis by immunofluorescence techniques should be performed preferentially on sections of the large intestine from calves killed after the onset of diarrhea (i.e. several days after the infection of epithelium).

Immune Response

Up to the age of 4 months, all calves in a sentinel experiment regularly excreted BRV in the feces. They showed early serum IgM responses despite the presence of IgG1 isotype maternal antibodies, but no IgA seroconversion. Antibody titers then decreased below detection, and persistent IgG1 titers developed in only a few animals. After introduction into the dairy herd at 10 months of age, all calves developed diarrhea and shed virus. Seroconversion for all antibody isotypes was observed at this stage, indicating lack of mucosal memory. In contrast, coronavirus infection in the presence of maternal antibodies leads to isotype switch and a memory response.

Prevention and Control

No control strategies have been implemented.

Future Perspectives

Conclusions on an evolutionary relationship between toro-, corona- and arteriviruses can be drawn on the basis of sequence analysis of the arteriviral polymerase gene. The pathogenic significance of toroviruses for animals and man, also as agents of nonenteric infections needs to be established. Diagnostic procedures for the discovery of more distantly related viruses of this cluster will have to include procedures for the recognition of nucleotide sequence motives.

See also: Coronaviruses; Defective-interfering viruses; Lactate dehydrogenase-elevating, equine arteritis virus and Lelystad viruses; Simian hemorrhagic fever virus.

Further Reading

Horzinek MC and Weiss M (1984) *Toroviridae*: a taxonomic proposal. *Zbl. Vet. Med.* B31: 649.

Koopmans M and Horzinek MC (1993) Toroviruses of animals and humans: a review. *Adv. Virus Res.* in press.

Snijder EJ and Horzinek MC (1993) Toroviruses: replication, evolution and comparison with other members of the coronavirus-like superfamily. *J. Gen. Virol.* in press.

Woode GN *et al.* (1982). Studies with an unclassified virus isolated from diarrheic calves. *Vet. Microbiol.* 7: 221.

TOSPOVIRUSES

Peter de Haan
Zaadunie Biotechnology,
Enkhuizen, The Netherlands

History and Taxonomy

A disease in tomato, called spotted wilt, was first observed in Australia in 1915. The causal agent of this plant disease turned out to be a virus, from then on designated tomato spotted wilt virus (TSWV). Subsequently, this virus became known under various different names, such as kromnek virus, vira cabeça, tip blight virus etc. The unique properties of TSWV among the plant viruses led to its classification into a monotypic plant virus group: the tomato spotted wilt virus group. In 1990 a serologically distinct TSWV-like virus was isolated from *Impatiens*, which differed from TSWV in symptomatology and host range. On the basis of its induced symptoms in *Impatiens*, this new virus is called *Impatiens* necrotic spot virus (INSV). Recently, several other TSWV-like viruses have been isolated and characterized, for instance, tomato chlorotic spot virus (TCSV), groundnut ringspot virus (GRSV), groundnut bud necrosis virus (GBNV) and peanut yellow spot virus (PYSV). Extensive nucleotide sequence information on the genomic RNAs of TSWV and INSV is now available. Both serological and molecular studies have considerably extended our knowledge of these viruses and ultimately enabled their proper classification. Taking all data into account, it is clear that they represent a distinct genus within the family *Bunyaviridae* (Table 1), for which the genus name *Tospovirus* has been accepted by the ICTV. The type species of this newly created genus is TSWV, and the other TSWV-like viruses should be considered as distinct viral species within this genus (Table 2).

Virus Structure

Over the past 30 years, extensive electron microscopic studies have been performed on tospoviruses in infected plant cells. These analyses have shown that viral particles are spherically shaped (80–110 nm in diameter) and consist of a granular core of nucleocapsids, bounded by a lipid membrane, which is covered with surface projections. In contrast to the animal members of the *Bunyaviridae*, tospoviral particles are found clustered within dilated cysternae of the rough endoplasmic reticulum and most likely mature by budding of nucleocapsids through the endoplasmic reticulum membrane. So far, there is no direct evidence that the Golgi complex is involved in maturation or transport of the virus. Besides enveloped viral particles, other structures associated with tospoviral infection can be observed by electron microscopy. The cytoplasm of infected plant and insect cells also contains clusters of free nonenveloped nucleocapsids (also described as diffuse, electron dense masses or viroplasm) and fibers, fibrous structures or tubuli.

Only for TSWV and INSV are sufficient data available in terms of protein composition and particle morphology. *In vitro*, viral particles are highly unstable, as can be concluded from the short half-life in plant sap (30–60 min) and the thermal inactivation point of 45°C. Purified TSWV preparations contain three major structural proteins, with molecular masses of 29, 58 and 78 kD, and one minor large protein (> 200 kD). The 29 kD protein is tightly associated with the genomic RNA to stable circular nucleocapsid structures. This protein is therefore called the nucleocapsid (N) protein. Viral preparations contain three distinct nucleocapsids, each bearing a copy of genomic RNA. The nucleocapsids, together with a few copies of the large (L) protein, form the interior of the viral particle. The 78 and 58 kD proteins are glycosylated and therefore called glycoproteins G1 and G2 respectively. Since the G proteins are located at the surface, it can be assumed that they both represent the observed surface projections (spikes). So far, the structure of these spikes has not been elucidated.

Genome Structure and Coding Capacity

Tospoviruses have a genome consisting of three linear single-stranded RNA molecules, with pyro-

Table 1. Some properties of TSWV compared with those of other members of the *Bunyaviridae*

Property	TSWV	*Bunyaviridae*
Morphology		
Shape	Spherical	Spherical
Diameter (nm)	80–110	90–120
Envelope	+	+
Surface projections	+	+
Circular nucleocapsids	+	+
$s^{\circ}_{20,w}$	520–530S	350–470S
Bouyant density in CsCl ($g\,cm^{-3}$)	1.21	1.20
Morphogenesis		
Maturation	Budding into RER	Budding into RER
Localization	Cysternae of ER	Golgi complex
Structural proteins ($M_r \times 10^{-3}$)		
N	29	19–54
G1	78	55–120
G2	58	29–70
L	331.5	230–250
Genome		
Type	ssRNA	ssRNA
Number of segments	3	3
Polarity	Negative/ambisense	Negative/ambisense
Length of the segments (kb) and coding properties		
S RNA	2.9 (N,NSs)	0.8–2.0 (N,NSs)
M RNA	4.8 (G1,G2,NSm)	3.2–4.9 (G1,G2,NSm)
L RNA	8.9 (L)	6.9–13.0 (L)
Transmission		
Vector	Thrips	Ticks, mosquitoes, sandflies and other arthropods
Vertical transmission	?	+
Replication in the vector	+	+

Abbreviations: RER, rough endoplasmic reticulum; ER, endoplasmic reticulum; ss, single-stranded; S RNA, M RNA and L RNA, small, medium and large RNA respectively.

Table 2. Classification of the tospoviruses

Taxonomic status		N protein homology (%)
Family	*Bunyaviridae*	
Genus	*Tospovirus*	
Type species	Tomato spotted wilt virus (TSWV)	100
Other members	*Impatiens* necrotic spot virus (INSV)	55
	Tomato chlorotic spot virus (TCSV)	76
	Groundnut ringspot virus (GRSV)	78
Possible members	Groundnut bud necrosis virus (GBNV)	N.D.
	Peanut yellow spot virus (PYSV)	N.D.
	Watermelon silver mottle virus (WSMV)	N.D.

N protein homology is given relative to TSWV.
N.D., not determined.

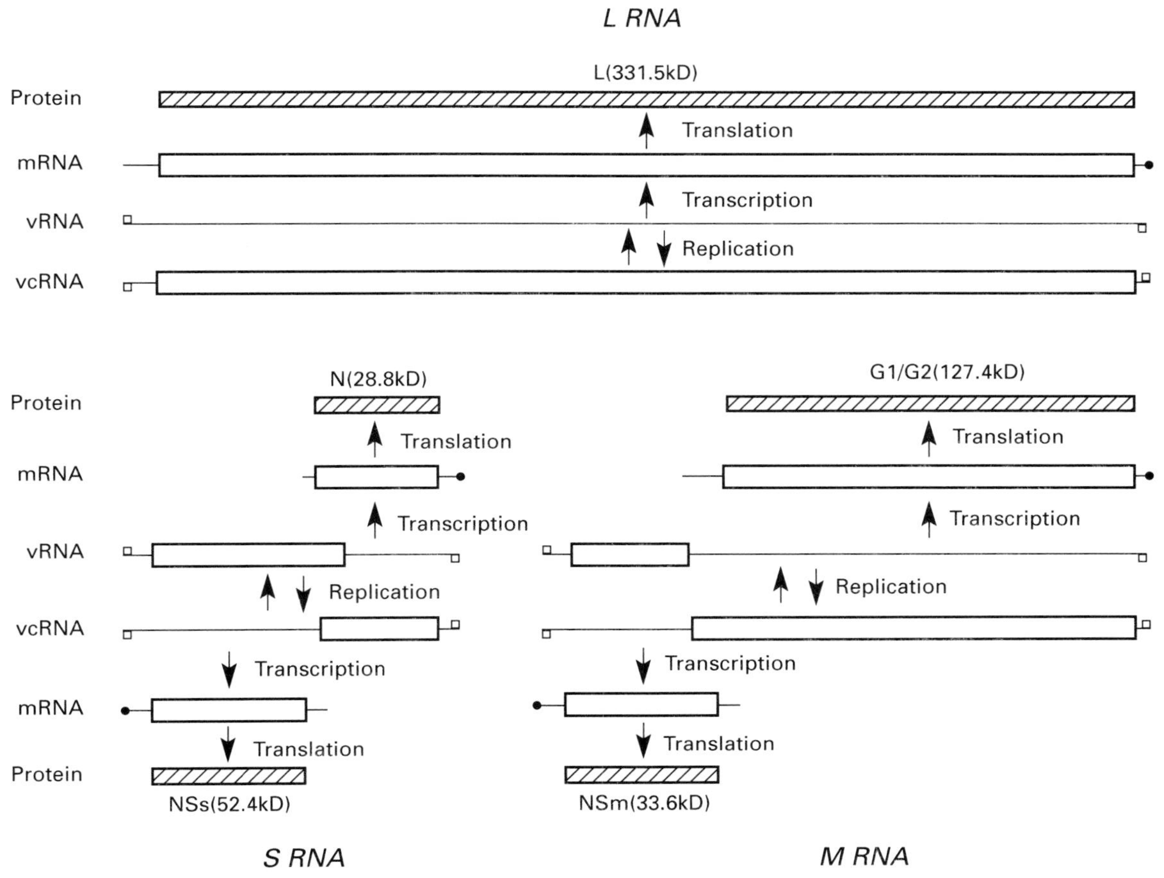

Fig. 1 Schematic representation of the ambisense S and M RNA segments and the negative-stranded L RNA of tospoviruses.

phosphate groups at their 5′ ends and free hydroxyl groups at their 3′ ends. As for other bunyaviruses, the RNAs are designated small (S) RNA, medium (M) RNA and large (L) RNA. The complete nucleotide sequence of the genomic RNA of TSWV is now available. For INSV, the sequences of the S and M RNA have been determined. In addition, nucleotide sequences corresponding to the N protein genes have recently been determined for several other tospoviruses. From these molecular data, it can be anticipated that the standard tospovirus genome consists of two ambisense RNA molecules (e.g. S and M RNA) and one RNA segment of negative polarity (L RNA) (Fig. 1).

The S RNA segment of TSWV is 2.9 kb in length and has an ambisense coding strategy. The viral N protein is encoded by a viral complementary-sense subgenomic mRNA species and the nonstructural protein (NSs of 52.4 kD) by a viral-sense subgenomic mRNA. Comparison of the N proteins from a number of tospoviruses showed 55–78% amino acid sequence homology, which is consistent with the divergence levels of other members of the *Bunyaviridae* (Table 2). Antisera have been raised against S RNA-encoded proteins of both TSWV and INSV, and immunogold decoration experiments revealed that NSs forms, or is associated with, the cytoplasmic fibrous structures found in infected plant cells. In viruliferous thrips, both N and NSs can be found in muscle cells associated with the midgut epithelium and, more prominently, in the salivary glands.

The M RNA is 4.8 kb long and also has an ambisense gene arrangement, which is a unique property among the *Bunyaviridae*. The gene encoding the 127.4 kD precursor protein of the G1 and G2 membrane glycoproteins is located on the viral complementary RNA strand. The mature membrane proteins are formed by cleavage of this large precursor protein. The amino acid sequence of the putative G2 protein revealed the presence of an RGD ‘cell attachment’ site, which might be involved in binding of viral particles to receptor molecules located on the cellular membranes of midgut epithelium cells of the thrips. The viral M RNA strand contains a smaller open reading frame (ORF), capable of coding for a 33.6 kD protein, designated NSm. The function of this putative nonstructural protein remains unknown. The genes on the ambisense RNA molecules are expressed by the formation of subgenomic mRNAs.

TSWV L RNA has a length of 8.9 kb and is completely of negative polarity. There is one large ORF on the viral complementary strand, which corresponds to a primary translation product of 331.5 kD. Comparison with the polymerase proteins of other negative-strand viruses revealed the presence of conserved amino acid signature sequences, which are specific for all polymerases using RNA as a template. It can be deduced that the predicted 331.5 kD protein represents the viral polymerase (L). Comparative studies among polymerase signature sequences of the L proteins from members of the *Bunyaviridae* demonstrated that TSWV is more closely related to *Bunyamwera* virus than members of the other genera analyzed so far are to the prototype of this large virus family.

The genomic RNA molecules have complementary terminal sequences, which is a property characteristic of negative-strand viruses. The RNAs can be folded into panhandle structures, which explains the appearance of circular nucleocapsids in the viral particle. Moreover, it can be assumed that the termini contain the recognition signals for the viral polymerase and hence play an important role in transcription and replication of the genomic RNAs.

Hybridization experiments have revealed that both the genomic and the antigenomic RNAs are wrapped with N proteins to yield nucleocapsids. Viral particles contain the viral-sense L RNA, but in case of the ambisense S and M RNA minor amounts of the antigenomic sense molecules are also included in the viral particles. The mRNAs are not encapsidated and contain 12–20 nontemplated nucleotides at their 5′ ends, which are most likely obtained from host cellular mRNAs in a process referred to as cap-snatching. Based on studies of animal negative-strand viruses, a plausible model can be proposed to explain the molecular events during the tospoviral infection cycle. The RNA polymerase, present in the viral particle, is in the transcriptive mode. Early in the infection cycle, capped mRNA molecules are synthesized and immediately scanned by host ribosomes, thereby preventing their encapsidation. Transcription termination occurs at signals located on the encapsidated template RNA or, more likely, on the 3′ ends of the nonencapsidated mRNAs. Later in the infection cycle, when free unassembled N protein accumulates in the cytoplasm, the polymerase switches to the replicative mode, and antigenomic RNAs are synthesized. Since these RNAs are not recognized and covered by ribosomes, they are immediately and cooperatively wrapped with N protein. Encapsidation of the progeny RNA molecules prevents premature termination, yielding genome-length RNA molecules.

Geographic Distribution and Host Range

The economical impact of tospoviruses is enormous, not only because of their wide geographic distribution, but also because of their extremely broad host range. Up to 360 species of plant from 49 different families, both monocotyledons and dicotyledons, have been reported to be susceptible to TSWV infection. Infections with tospoviruses cause a wide variety of different symptoms such as necrosis, chlorosis, enation, stunting and local lesions, depending on the plant species. Both genetic and environmental factors seem to affect the susceptibility of the host and the severity of the induced symptoms.

The most important host plant species of TSWV for which considerable yield losses have been reported include tomato, potato, tobacco, groundnut, (sweet) pepper, lettuce and papaya. Although TSWV mainly

Table 3. Thrips species reported as vectors of tospoviruses

Thrips species	Tospovirus transmitted
Thrips tabaci Lind.	TSWV (TCSV, GRSV ?)
Frankliniella schultzei Tryb.	TSWV, GBNV (TCSV, GRSV ?)
Frankliniella occidentalis Perg.	TSWV, INSV (TCSV, GRSV ?)
Frankliniella fusca Hinds.	TSWV
Thrips setosus Moult.	TSWV
Thrips palmi Karny.	GBNV, WSMV
Scirtothrips dorsalis Hood.	GBNV

occurs in (sub)tropical regions or in areas with relatively warm summers, severe outbreaks have recently also been reported in many countries within the temperate climate zones (e.g. North America and Europe). In these areas, TSWV has become an actual threat in greenhouses, in the cultivation of tomato, sweet pepper, egg-plant and a still growing number of ornamental plants like *Chrysanthemum, Ageratum* and *Impatiens*. The incidence of INSV is less frequent. Its occurrence is restricted to ornamental plant species, where it causes severe outbreaks. Recently a new tospovirus in cucurbit hosts, causing silver mottle symptoms and hence designated watermelon silver mottle virus (WSMV), has been found in Taiwan and Japan. GBNV covers all major groundnut-growing areas in Asia, while PYSV has only been found in India. Not much is known about the occurrence of the other species. So far, they have been occasionally found in (sub)tropical areas.

Transmission

Tospoviruses are transmitted by thrips (*Thysanoptera*). So far, seven species have been reported as possible vectors, of which the first four listed in Table 3 seem to be the most important. The recently observed revival of tospoviruses in the temperate climate zones of both the Old and the New World is most likely due to the introduction of the 'Western Flower thrips', *Frankliniella occidentalis (Perg)* into these areas.

Recently, it has been demonstrated that TSWV and INSV multiply in the vector. As a consequence, tospoviruses are propagatively transmitted by thrips and hence they are among the few viruses that can replicate in both plant and animal cells. Acquisition of virus by its vector takes place only during the first and second larval stages with an acquisition threshold of 5–45 min. The latency period is about 5–10 days, and virus can be retained in the vector for its whole lifespan.

In the laboratory, tospoviruses are easily transmitted from plant to plant by sap inoculation. In the field, neither mechanical transmission nor transmission by seed or pollen plays a significant role in the spread of the virus. In general, naturally occurring weeds serve as viral reservoirs and hence play a crucial role in the survival and distribution of vector and virus. The epidemiology of thrips and tospoviruses is still poorly understood and more detailed studies are needed to develop feasible management procedures.

Disease Management

In the field it is very difficult to limit the incidence of tospoviral infections, since it is almost impossible to control thrips properly, and weed host plants are abundantly present in the neighborhood of the threatened crops. In the greenhouse, the situation is less complicated. Several sanitary measures can be taken to minimize losses due to tospoviral infections, such as early destruction of infected plants and biological or chemical control of thrips. In the case of vegetative crop propagation, it is important to maintain virus-free stock material. To accomplish this, sensitive and early detection of tospoviruses is a prerequisite. Polyclonal antisera and recently monoclonal antibodies also have been raised against viral preparations, or against purified nucleocapsids, in a number of laboratories. With these antibodies, serological detection assays have been developed for diagnostic purposes and for fundamental research. Recently, several sensitive molecular techniques, including dot-blot hybridization and polymerase chain reaction (PCR), have also been described.

Another important line of research is dedicated to gain crops with increased resistance or tolerance to tospoviral infections. However, limited progress has been made in breeding tomato, tobacco and lettuce with increased resistance to TSWV, partly because of the lack of suitable forms of resistance. In most cases, naturally occurring TSWV resistance or tolerance is polygenic, based on complex interactions between virus, vector and plant.

In addition to these traditional breeding approaches, genetically engineered resistance to TSWV has been described recently. Transgenic tobacco plants, containing TSWV N gene sequences, show high levels of resistance to mechanical inoculation with this virus. Besides being protected against mechanical inoculation, transgenic tobacco plants are also resistant to inoculation via viruliferous thrips, e.g. *Frankliniella occidentalis* (Perg.), the most important natural vector species. This new approach may be an important contribution to the control of tospoviruses in the near future.

See also: Bunyaviruses; Plant virus disease – economic aspects.

Further Reading

de Avila AC *et al.* (1992) Distinct levels of relationships between tospovirus isolates. *Arch. Virol.* 128: 211.

De Haan P *et al.* (1991) Tomato spotted wilt virus L RNA encodes a putative RNA polymerase. *J. Gen. Virol.* 71: 2207.

De Haan P *et al.* (1992) Characterization of RNA-mediated resistance to tomato spotted wilt virus in transgenic tobacco plants. *Bio/Technology* 10: 1133.

German *et al.* (1992) Tospoviruses: diagnosis, molecualr biology, phylogeny and vector relationships. *Annu. Rev. Phytopathol.* 30: 315.

Hsu MDH-T and Lawson RH (eds) (1991) Virus–thrips–plant interactions of TSWV. Proceedings of USDA Workshop, Beltsville, Agric. Res. Serv. ARS–87.

Kormelink R *et al.* (1992) The nucleotide sequence of the M RNA segment of tomato spotted wilt virus: A bunyavirus with two ambisense RNA segments. *J. Gen. Virol.* 73: 2795.

TOTIVIRUSES

Contents

General Features
Ustilago maydis viruses

General Features

Said A Ghabrial
University of Kentucky College of Agriculture
Lexington, Kentucky, USA

Introduction

The discovery of the killer phenomenon, in the sixties, in the yeast (*Saccharomyces cerevisiae*) and in the smut fungus (*Ustilago maydis*) eventually led to the discovery of the isometric double-stranded (ds)RNA mycoviruses with undivided genomes, presently classified in the family *Totiviridae*. Yeast or smut killer strains secrete a protein toxin to which they are immune, but which is lethal to sensitive cells. Genetic and biochemical studies have conclusively shown that toxin production and immunity are cytoplasmically inherited, and that dsRNAs of viral origin comprise the cytoplasmic determinants. The killer toxin is encoded by a satellite dsRNA, denoted M-dsRNA, which is dependent on a helper virus with undivided dsRNA genome for encapsidation. The helper viruses [*Saccharomyces cerevisiae* virus-L-A (ScV-L-A) and *Ustilago maydis* virus-H1 (UmV-H1)], which are autonomously replicating viruses that do not require M-dsRNA for replication, also encode the RNA polymerase protein required for the replication of both the M-dsRNA and their own genomic monopartite dsRNAs. The helper virus dsRNA and M-dsRNA are separately encapsidated in identical capsids coded for by the helper virus. Unlike the helper viruses associated with the yeast and smut killer systems, the member viruses in the family *Totiviridae* that infect filamentous fungi are not known to be associated with killer phenotypes. However, purified preparations of these viruses are often associated with dsRNA species suspected of being satellite or defective dsRNAs. The yeast virus ScV-L-A and associated killer system as well as the smut viruses and associated killer system are considered in detail in separate entries in this volume.

Taxonomy and Classification

The family *Totiviridae* includes a single genus, *Totivirus* (monopartite dsRNA mycovirus group). *Giardiavirus* (the protozoal *Giardia* virus group) is proposed as a possible genus within the *Totiviridae* family. The protozoal viruses are considered in a separate entry (see Giardiaviruses). The yeast virus ScV-L-A (synonym ScV-L1) is the type species of the family. Other members of the genus *Totivirus* are: the smut virus UmV-H1 and *Helminthosporium victoriae* 190S virus (HvV-190S). The list of possible members include: *Aspergillus foetidus* virus S (AfV-S), *Aspergillus niger* virus S (AnV-S), *Gaeumannomyces graminis* virus 87-1-H (GgV-87-1-H), *Mycogone perniciosa* virus (MpV), *Saccharo-*

myces cerevisiae virus L-BC (ScV-L-BC; synonym ScV-La) and *Yarrowia lipolytica* virus (YLV). The possession of an undivided dsRNA genome, as in the case of the fungal and protozoal viruses of the family *Totiviridae*, is an unusual property for viruses with dsRNA genomes. All dsRNA viruses of bacteria, higher plants and animals have divided genomes.

Particle Properties

The sedimentation coefficients $S_{20,w}$ (in Svedberg units) for members of the *Totivirus* genus are in the range 160S–190S. Particles lacking nucleic acid sediment at a rate of $S_{20,w} = 98-113S$. Buoyant density in CsCl $[\rho CsCl\ (g\,cm^{-3})] = 1.40-1.43$. Isolates of ScV-L-A and UmV-H1 may have additional components, containing satellite or defective dsRNAs, with different sedimentation coefficients and buoyant densities. Purified viral preparations from killer yeast strains contain, in addition to the ScV-L-A virions, two density components: ScV-M-light ($\rho = 1.3513$) and ScV-M-heavy ($\rho = 1.3834$), with one and two molecules of M-dsRNA (1.8 kbp) respectively. These two ScV-M components can be separated by CsCl equilibrium density gradient centrifugation. Yeast viral preparations may also contain small quantities of one or more dsRNA species with sizes between 0.7 and 1.6 kbp, denoted as (suppressive) S-dsRNA. All S-dsRNAs are derived from M-dsRNA by internal deletion, in a manner similar to the evolution of defective-interfering (DI) animal viral RNAs. Suppressive sensitive strains of yeast, in which the killer functions have been lost, are known to contain S-dsRNA in place of M-dsRNA, but still have helper virus dsRNA (L-dsRNA).

Particle Structure and Composition

The totiviruses have isometric particles, approximately 40 nm in diameter, with icosahedral symmetry. The capsids are single-shelled and comprised of a single major polypeptide. In some viruses, e.g. HvV-190S, phosphorylated forms of the capsid polypeptide occur in the virions (see next section). The capsid consists of 120 subunits of molecular mass in the range 69–88 kD, arranged in $T = 1$ lattices.

The virions of the totiviruses encapsidate a single molecule of dsRNA 4.7–6.7 kbp in size. Some isolates of totiviruses may contain additionally or alternatively to satellite dsRNAs subgenomic or defective dsRNAs which are encapsidated separately in capsids encoded by the totivirus genome. For example, preparations of AfV-S contain in addition to the 6.0 kbp genomic dsRNA two other dsRNA species, 3.8 and 0.4 kbp in size, that are separately encapsidated.

The complete nucleotide sequence (4579 bp) of ScV-L-A (L1) dsRNA is deposited as EMBL accession number J04692 (X13426). The plus-strand (4580 bases; contains unpaired A at the 3′ terminus) has two large open reading frames (ORFs) that overlap by 130 bases (129 coding bases). The 5′-proximal ORF (ORF 1) encodes the viral major capsid protein with a predicted size of 76 kD. Estimates of the size of the major capsid protein, based on sodium dodecyl sulfate–polyacrylamide gel electrophoretic (SDS–PAGE) analyses, are generally higher than 76 kD and in the range 80–88 kD. This may be accounted for by post-translational modification of a primary translation product of the coat protein gene. ORF 2, which is in the −1 frame with respect to ORF 1, codes for an RNA-dependent RNA polymerase (RDRP) with features similar to the RDRPs of plus-strand single-stranded (ss)RNA viruses and dsRNA viruses. The two ORFs together encode, via a −1 ribosomal frameshifting event, a Gag–Pol-like fusion protein of a predicted size 171 kD. There are a number of *cis*-acting sites in the dsRNA molecule including: a viral binding site (VBS) believed to be involved in particle assembly, an internal replication enhancer (IRE) and a site controlling viral interference (INS).

Capsid Protein Heterogeneity

Although the capsids of totiviruses are encoded by a single gene, and typically are composed of a single major polypeptide, capsid protein heterogeneity of an unknown nature has been observed in purified preparations of several totiviruses. The origin of capsid protein heterogeneity in the totivirus HvV-190S was the subject of a recent comprehensive study. The capsids of HvV-190S contain three polypeptides with molecular mass 78, 83 and 88 kD (denoted p78, p83 and p88 respectively). Two of these polypeptides occur as major capsid proteins whereas the third polypeptide occurs as a minor component; p88 is always a major capsid protein, while the relative abundance of p83 and p78 varies with the viral preparation. The three capsid proteins show similar peptide profiles when subjected to selective chemical cleavage at tryptophan residues or to limited proteolysis using V8 protease. No evidence could be obtained to support the idea that the smaller polypeptides are generated from p88 as a result of proteolysis during viral purification or storage.

Recently, evidence was presented that capsid protein heterogeneity of HvV-190S represents post-translational phosphorylation of a primary translation product. Thus, HvV-190S p83 and p88 represent respectively partially and highly phosphorylated forms of p78. Moreover, purified preparations of the HvV-190S virus have been shown to separate into two closely spaced sedimenting components (190S-1 and 190S-2) distinguishable by capsid protein composition, state of phosphorylation, and transcriptional efficiency. The slower sedimenting component (190S-1), contains p88 and p83 as the major capsid proteins, and the faster component (190S-2) contains p88 and p78. Evidence has also been recently presented for the presence of protein kinase activity in purified virions of HvV-190S, which utilizes ATP as a phosphoryl donor. Capsid protein heterogeneity has also been reported for several other totiviruses; GgV-87-1-H, ScV-L-BC and YlV are reported to possess two or more related capsid polypeptides of comparable size to those of the HvV-190S. Although the origin of capsid protein heterogeneity in these viruses is not known, post-translation phosphorylation of a primary translation product, as in the case of HvV-190S, could account for it. Phosphorylation of the capsid protein may be a common feature of viruses in the family *Totiviridae* and may play a regulatory role in viral dsRNA transcription/replication.

dsRNA Replication

Because of the lack of suitable infectivity assays for the totiviruses, there is little information on whether viral dsRNA replication is synchronized with cell division. Although synchronized cultures of yeast can readily be obtained, there is disagreement on whether dsRNA synthesis occurs throughout the cell cycle or occurs through the G1 phase and ceases in the S phase when DNA synthesis takes place. *In vivo* studies, involving density-transfer experiments, indicate that the replication of the yeast ScV-L-A dsRNA (as well as the M-dsRNA) occurs conservatively in a sequential manner. This mode of replication is analogous to that of the animal reoviruses. Thus, the parental dsRNA duplex remains intact, with the two strands of the progeny dsRNA molecules being synthesized asynchronously. The plus-strand RNA is synthesized first (on the parental dsRNA template), followed by minus-strand synthesis on the plus-strand which is presumably released from the first particle and now packaged in a separate particle.

Information on the replication cycle of totivirus dsRNA has mainly been derived from *in vitro* studies of virion-associated RNA polymerase activity and the isolation of particles representing various stages in the replication cycle. In *in vitro* reactions, the RNA polymerase activity associated with virions of ScV-L-A, UmV-H1 or HvV-190S, isolated from lag phase cultures, catalyzes end-to-end transcription of dsRNA by a conservative mechanism to produce mRNA for capsid polypeptide, which is released from the particles. Purified viral preparations of the yeast ScV-L-A, isolated from log phase cells, contain a less dense class of particles which package only plus-strand RNA. In *in vitro* reactions, these particles exhibit a replicase activity that catalyzes the synthesis of minus-strand RNA to form dsRNA. The now mature particles, which attain the same density as that of the dsRNA-containing virions isolated from the cells, are capable of synthesizing and releasing plus-strand RNA. The plus-strand RNA is the species that is packaged to form progeny particles, and serves as the template for minus-strand synthesis. This completes the replication cycle. It is not known whether the plus-strand RNA associates with coat protein subunits or a preformed capsid.

Replication of the satellite M-dsRNA associated with the yeast virus ScV-L-A also occurs conservatively. However, the plus-strand RNA transcripts from M-dsRNA are often retained within the particle. The retained plus-strand serves as a template for minus-strand RNA synthesis to form a second molecule of dsRNA in the same particle (a headful replication mechanism). Particles containing one and two molecules of M-dsRNA have been isolated from cells, as indicated earlier. The rationale for the headful mechanism is that the satellite dsRNAs are packaged in helper virus-encoded capsids which are designed to accommodate a much larger dsRNA molecule. Therefore, the particles may contain one or more molecules of satellite, defective or mutant dsRNAs, depending on when full capacity is attained.

Transmission

There are no known natural vectors for the transmission of totiviruses. They are transmitted intracellularly during cell division, sporogenesis and cell fusion (see the entry on Partitiviruses for a more comprehensive discussion of intracellular transmission of fungal viruses). Although the yeast viruses are effectively transmitted via ascospores, the totiviruses infecting the ascomycetous filamentous fungi, e.g. GgV-87-1-H, are essentially eliminated during ascospore formation.

Conventional infectivity assays for totiviruses using purified virus are not presently available. Extracellular transmission of the yeast viruses may be possible during natural mating of yeast mating pairs. Successful transfection of the protozoan *Giardia lamblia* has been accomplished via electroporation with plus-strand RNA transcribed *in vitro* from the *Giardia lamblia* virus (GlV) dsRNA, an isometric dsRNA virus with affiliation to the *Totiviridae* family. Because of the monopartite nature of their genomes, the totiviruses present ideal systems for studies on developing infectivity assays for fungal viruses based on the use of purified virions, full-length *in vitro* transcripts of genomic dsRNA or cloned cDNA to dsRNAs.

Host Range

There are no known experimental host ranges for the totiviruses because of the lack of suitable infectivity assays. As a consequence of their intracellular modes of transmission, the natural host ranges of totiviruses are limited to individuals within the same or closely related vegetative compatibility groups. Furthermore, mixed infections with two or more unrelated viruses are common, probably as a consequence of the ways by which fungal viruses are transmitted in nature. Examples of mixed infections involving totiviruses include: the totiviruses ScV-L-A and ScV-L-BC; the totivirus AfV-S and the unclassified AfV-F; the totivirus HvV-190S and the possible partitivirus HvV-145S. There are apparently no structural interactions between these pairs of viruses as evinced by the lack of heterologous encapsidation in mixed infections.

Virus–Host Relationships

The yeast killer system, comprised of a helper totivirus and associated satellite dsRNA, is one of a very few known examples where viral infection is beneficial to the host. The ability to produce killer toxins by immune yeast strains confers an ecological advantage over sensitive strains. The use of killer strains in the brewing industry provides protection against contamination with adventitious sensitive strains.

Because of the extensive knowledge of yeast genetics and molecular biology, and because the killer phenotype can be readily scored, the yeast killer/virus system provides an ideal system to study the effect of chromosomal genes on viral replication. There are over 30 chromosomal genes, termed *MAK* genes for maintenance of killer, the products of which are necessary for replication of M-dsRNA, and which are important or essential for cell growth. Genes of known function that are required for M-dsRNA maintenance include: *PET18*, which is required for maintenance of mitochondrial DNA; *SPE2*, which encodes adenosylmethionine decarboxylase required for polyamine synthesis; *MAK8* which is identical to *TCM1*, the gene for trichodermin resistance, and encodes ribosomal protein L3; and *MAK1*, which is the gene encoding DNA topoisomerase I. Only three host genes (*MAK3*, *MAK10* and *PET18*) are needed for the maintenance of L-dsRNA. It is of interest that, like M-dsRNA, a 530 bp deletion mutant (X-dsRNA) that is derived entirely from L-dsRNA requires all the *MAK* genes tested. Both X- and M-dsRNAs replicate by the headful mechanism which entails the presence of multiple copies of dsRNA copies per viral particle (see section on dsRNA replication).

Unlike the *MAK* genes which regulate dsRNA replication by controlling the supply of host factors required for the replication function, the gene products of a set of six chromosomal superkiller (*SKI*) genes act as negative regulators by lowering the copy number of M- and L-dsRNAs. Mutations in any of these *SKI* genes lead to the development of the superkiller phenotype as a result of the increased copy number of M-dsRNA.

See also: Furoviruses; Giardiaviruses; Partitiviruses; Yeast RNA viruses.

Further Reading

Buck KW and Ghabrial SA (1991) *Totiviridae*. In: Francki RIB *et al.* (eds) *Classification and Nomenclature of*

Viruses. Fifth Report of the International Committee on Taxonomy of Viruses. New York: Springer-Verlag.
Ghabrial SA and Havens WM (1992) *Helminthosporium victoriae* 190S mycovirus has two forms distinguishable by capsid protein composition and phosphorylation state. *Virology* 188: 657–665.
Wickner RB (1989) Yeast virology. *FASEB J.* 3: 2257.

Ustilago maydis viruses

Aliza Finkler and Yigal Koltin
Tel Aviv University
Ramat Aviv, Israel

Taxonomy and Classification

Ustilago maydis virus (UmV) belongs to the *Totiviridae* family. The viruses in this family include the monopartite double-stranded (ds)RNA viruses of which the L-A virus of *Saccharomyces cerevisiae* (scV-L-A) is the type species. These viruses consist of a single molecule of dsRNA, and some viruses contain additional satellite dsRNA molecules some of which encode the polypeptides, known as 'killer factors', that are toxic to sensitive strains of the host and to other species related to the host. The function of other dsRNA molecules is unknown. All of these molecules are encapsidated in capsids encoded by the helper virus, and the satellite dsRNA is encapsidated separately from the viral dsRNA. Three viral subtypes, P1, P4 and P6, are distinguished on the basis of the satellite dsRNAs associated with the helper, and by the killer proteins encoded by these satellite molecules. The killer toxins are referred to as KP1, KP4 and KP6, respectively.

Virus Structure and Composition

UmV is isometric and *c*. 41 nm in diameter. The capsid consists of a major polypeptide of 75 kD and a few minor proteins. Based on its mass, each capsid consists of 120 units of the major polypeptide and contains at most 30% dsRNA. Based on the size of the viral dsRNA segment, each capsid can accommodate one molecule of the viral dsRNA or multiples of the satellite molecules. These data are derived from the fact that, on purification of the virus, multiple viral fractions are obtained with distinct sedimentation coefficients and distinct densities, which is indicative of various masses of nucleic acids being encapsidated in different fractions. Characterization of the dsRNA from each fraction reveals clearly that the satellite molecules are encapsidated separately from the viral dsRNA and that each capsid contains a single viral dsRNA or multiples of the satellite molecules.

The viruses encapsidating the helper viral dsRNA segment sediment at 172S and have a buoyant density in CsCl of 1.418 g cm^{-3}. The empty capsids sediment at 98S and have a density of 1.278 g cm^{-3}. The major capsid polypeptides of P1, P4 and P6 are immunologically interrelated and cross-react, and the viral dsRNA cross-hybridizes. A dsRNA-dependent RNA polymerase is associated with the capsids and functions *in vitro*, requiring no host factors for efficient transcription.

Genome Structure

The viral dsRNA consists of a single molecule with a mol. wt of 4.2×10^6 and 6.7 kbp. The viral subtypes, classified according to the 'killer protein' expressed, are typified by a series of satellite molecules, as detected in the natural isolates that are the type forms. To simplify reference to the dsRNA segments, the entire series of molecules found in the three subtypes is divided into three classes, heavy (H), medium (M) and light (L). With the exception of L, all satellite molecules are unique. L is homologous to one 3′ end of one M segment in each viral subtype but it is not known how this segment is generated. P1 is typified by five satellite segments: H2-1 (4.5 kb), M1-1 (1.4 kb), M2-1 (1.0 kb), M3-1 (0.92 kb), L1-1 (0.36 kb). P4 is typified by six satellite segments: H2-4 (4.5 kb), H3-4 (3.2 kb), H4-4 (2.8 kb), M2-4 (0.98 kb), M3-4 (0.92 kb), L1-4 (0.36 kb). P6 is typified by four satellite segments: H2-6 (4.5 kb), M2-6 (1.2 kb), M3-6 (1.06 kb) and L1-6 (0.36 kb). The H1 segments are similar among the three subtypes. Similarly, all the L segments share the same 3′ termini. The H3 and H4 segments have no detectable homology with H1. Both the hybridization and sequencing results show that M1-1 shares homology with M(2 + 3)-4 and M2-6. The 3′ ends of UmV, as are those of ScV-L-A, fall into the consensus sequence A/U A/U A/U A/U A/U A/U A/U U/G C A/G_{OH}, which may reflect some functional similarity. Even though the UmV and ScV dsRNAs show no homology by hybridization, they do have recognizable sequence homology at the 3′ termini. Some of the ends of UmV dsRNA terminate with a sequence UUUUUCA/G_{OH}^{-}, which is the transcriptase initiation consensus sequence of the ScV. Many other dsRNA species that appear to be unrelated to the viral system have been recovered recently from strains of *U. maydis*. Some of these species are cell-cycle dependent and fluctuate during the cell cycle, reaching values of a few thousand copies per cell at one point in the cell cycle and as low as a few copies per cell at other points. dsRNA hybridizations suggest that these molecules comprise an independent system of dsRNA plasmids. One species of the L segments found in association with P1, referred to as L minor (0.3 kb) since it is difficult to resolve, is thought to confer resistance to KP1. Similar relations are not known for P4–KP4 and P6–KP6.

Virus Multiplication

Details of viral replication are not known. However, the replication is coordinated with cell division and, very rarely, cells devoid of the virus are detected from infected cultures. In fact, the segments that are lost are occasionally the satellites, and a range of variants is detected in the natural population. The viral dsRNA segment is found in all natural isolates of the host. Viral replication appears to be at its peak in the mid-log phase, and the maximal level of transcription occurs at the late-log phase. Translation of the viral information continues in the stationary phase, leading to massive accumulation of assembled capsids devoid of dsRNA.

Virus Genetics

Genetic analysis, including tests with partial genomes, exchange of viral segments and *in vitro* translation experiments, served to localize viral functions to specific dsRNA segments. These tests indicate that the viral H1 segment encodes the coat proteins. This segment contains sufficient information to also encode the transcriptase and polymerase. The dsRNA sequence of this segment is unavailable as yet, and it is not known whether the transcriptase and polymerase are encoded by this segment. Genetic tests indicate that the toxin KP4 is encoded by M2-4, and M2-6 encodes KP6. Similar tests suggest that segments M1-1, M2-1 or both encode KP1. Sequence data are available only for M2-6, confirming the genetic data. The sequence of one L segment is also available, and it localizes to the 3′ end of the M1-1 plus-strand. It contains no significant open reading frames. This segment bears a high degree of homology to the 3′ terminus of the M2-6 (77%). The function of these segments and the way in which they are generated is not known. However, they are encapsidated like all other satellite dsRNAs.

By immunological tests all the viral capsids from P1, P4 and P6 cross-react. The infected cells normally contain one type of satellite molecule which encodes the toxin. The different viral subtypes can be mixed by cytoplasmic mixing which occurs during crosses of different strains of the host, in which each is infected with a different viral subtype. Most often the progeny of such crosses segregate so that either all the progeny contain only one of the two subtypes or a majority of one type, and, very rarely, the two types are found together in an unstable association. The relations between the different subtypes are constant and repeatable. The pattern of segregation follows the difference in size of the satellite dsRNA that confers the phenotype, namely the dsRNA that encodes the toxin. This phenomenon has been referred to as the exclusion phenomenon, but it differs from typical exclusion via cleavage of molecules, as known for T-even phages, and is more similar to the phenomenon of exclusion displayed by defective-interfering particles. In some instances, even dsRNA segments of equal size appear to exclude segments of the same size, a phenomenon that can be followed by the expression of the toxins. Strains carrying such viruses are referred to as suppressive strains. The suppressive strains contain the excluding dsRNA segment that is mutated in the sequence encoding the toxin. It is unclear how such mutations affect the rate of replication which appears to be the major factor in the exclusion phenomenon of UmV.

Virus–Host Relationship

The first indications for the occurrence of viruses in strains of *U. maydis* were obtained in plate assays designed to identify the mating types of wild-type strains of *U. maydis*. Mating tests of some strains failed as a result of interstrain inhibition by a substance secreted by one of the two mates. The secretion of this substance was correlated with the occurrence of viral particles in the secreting strains and the transfer of these particles by cytoplasmic exchange between mates. As in most other fungi, the viral particles were not infective but were of uniform size and contained dsRNA. Later studies indicated that every strain in nature contains viral information. Not all strains contain the same pattern of satellite molecules, and only 1% of natural strains contain the satellite molecule that encodes the toxin. Many strains contain similar molecules in an inactive form. Nuclear-encoded functions related to maintenance or replication of the UmV have not been so far reported. Beyond toxin secretion, no other phenotype can be attributed to the dsRNA. *U. maydis* is a pathogen of corn, but no correlation has ever been established between virulence or growth rate and the dsRNA content of the viral subtype harbored by the host.

Virus Transmission

The transmission of UmV in nature appears to be via cytoduction. *U. maydis* is a heterothallic fungus with a very efficient outbreeding mechanism based on a multiple allelic mating system. Effective infection occurs only by heterocaryotic mycelia or diploid mycelia consisting of different mates. The system regulating this stage consists of one gene with two alleles. The sexual stage of the fungus, which develops on the infected plants following the formation of the zygotes, as a prelude to meiosis, occurs only between different mates that are distinguished by a multiple allelic system at one locus. This process allows easy transmission of the viral particles. Infection of intact cells with viral particles, or naked dsRNA, is not effective even when spheroplasts are used. The frequency of cells that are competent and can be infected is very low, and it is unclear whether intact cells can be infected at all. The development of transformation vectors for *U. maydis* opened the way to identifying competent cells that occur at a low frequency. As a part of a contransformation, cells can be infected with the helper virus and with the viruses containing satellite molecules, which allow detection of infected cells based on the secretion of the toxin. The frequency of infected cells among the transformants is low, no more than 1–8% of the transformed cells. This procedure is highly relevant for experimental purposes. In nature, since the virus does not induce cell lysis at any point, the viral particles are not found in abundance extracellularly, unless the fungal cells lyse as a part of senescence. Therefore, cytoduction appears to be the major or sole mode of transmission in nature.

Biological Properties

The helper virus supports the satellite viruses that encode the toxin that affects sensitive strains of *U. maydis* and related species, such as the pathogens of other Gramineous plants (wheat, oats, barley and other species). The toxins encoded by the viruses, KP1, KP4 and KP6, are all proteinaceous. Each toxin displays a unique specificity, and resistance to each toxin is conferred by a nuclear gene, one gene for each toxin. Resistance is a recessive trait, and each of the natural isolates carrying the satellite that expresses the toxin is also resistant to the toxin. Among the natural isolates, some 30% of the strains are resistant to one of the three toxins. Resistance determined by the dsRNA was only detected for KP1. Resistance to KP6 is not determined by the viral information and appears to be conferred only by a nuclear gene.

The three toxins differ in some of their physical characteristics. KP1 is the most unstable of the toxins and is effective over a narrow range of pH. KP4 and KP6 are very stable and function over a wide range of pH from 4 to 8. Also, even after denaturation, the activity of KP4 and KP6 can be reconstituted following renaturation. KP6 is not inactivated by sodium dodecyl sulfate (SDS). The β-polypeptide of KP6 is highly sensitive to mercaptoethanol, even at room temperature. The α-polypeptide of KP6 is sensitive to mercaptoethanol after heat denaturation of the polypeptide. Both α and β are rich in cysteines, and the intramolecular S–S bonds are affected by the reducing agent. The toxins bind with different affinities to various cellulose matrices.

Of the three toxins, KP6 is the best characterized and is a binary toxin consisting of two noncovalently bound polypeptides of 78 and 81 amino acids, referred to as α and β, which are encoded as a single preprotoxin of 219 amino acids. The activity of the toxin can be reconstituted *in vitro* by combining the purified polypeptides. The effect of the toxin on the cell appears to be at the level of the membrane, with some initial interaction at the level of the cell walls. The β-polypeptide is intolerable to sensitive cells and is suspected as being the toxic component of the binary toxin. However, even when spheroplasts are exposed to the toxin, both polypeptides are essential. The information related to the structure, organization, expression and processing of KP6 is fully supported by sequence data of the M2-6 cDNA, total purification of the two polypeptides, partial amino acid sequence of each polypeptide and *in vitro* transcription and translation of M2-6 dsRNA. KP4 has been fully characterized and appears to consist of one polypeptide. It is somewhat larger than KP6, and it is not known whether this is a fused form similar to the KP6 binary toxin.

The specificity of the toxins is narrow and limited to species of *Ustilago* that are pathogens of grain crops. The toxins do not affect other yeast forms of various classes of the fungi and mycelial forms. Also, bacteria are not sensitive to the toxins.

Economic Significance

The host of UmV is a plant pathogen of corn which causes significant damage throughout the world in regions in which corn is grown. Virulence is not known to be affected by UmV. However, since the toxin encoded by UmV affects this pathogen, current efforts are directed towards the generation of transgenic plants expressing the viral information. The same approach is also proposed for other grain crops, as limited surveys of resistance among Ustilaginales of grain crops have not detected any resistant isolates.

Geographic Distribution

The distribution of UmV is not restricted to a specific geographical location. Surveys conducted reveal that the same viruses have been isolated from Mexico, US, UK, Poland and most recently from the Peoples Republic of China. The incidence of isolates secreting the virus-encoded toxin is *c.* 1%, and thus far only three specificities have been detected. Among the recent isolates in the Peoples Republic of China, no new toxin specificities were detected.

Molecular Biology

Transcription

The RNA-dependent RNA polymerase copurifies with the viral particles, as in all other fungal viruses. Maximum specific activity of the enzyme is obtained by virions purified from late log–early stationary phase cells. The enzyme has an optimum pH of 8.9 and salt requirement, with maximum activity at 20 mM Na^+ and 5 mM Mg^{2+}. The reaction products of the polymerase activity are single-stranded (ss)RNA molecules, with molecular weights half the size of the corresponding dsRNA segments, suggesting that the polymerase is a transcriptase. The transcripts are homologous to the three classes of viral dsRNA. The molar ratio of the transcripts is 1:4:4 for the H, M and L transcripts, suggesting a relationship between M and L. The H transcript is the first to be synthesized, followed by M and L. The transcripts are released from the particles as soon as transcription is completed.

Translation

UmV encodes killer proteins, of which the best characterized is KP6, composed of two noncovalently linked polypeptides, α and β. The polypeptides are synthesized as one preprotoxin of 219 amino acids with one glycosylation site and a typical signal sequence with a Pro-Arg signal sequence processing site and two KEX1 and KEX2 processing sites. The resulting polypeptides, α (78 amino acids) and β (81 amino acids), are not glycosylated and are secreted independently. Both polypeptides are rich in cysteine. The calculated isoelectric points are both near neutrality: 7.21 for α and 6.94 for β. Neither of the polypeptides are toxic to sensitive cells, but activity can be restored by *in vitro* and *in vivo* complementation.

See also: Furoviruses; Partitiviruses; Plant virus disease – economic aspects; Yeast RNA viruses.

Further Reading

Koltin Y (1986) The killer systems of *Ustilago maydis*. In: Buck KW (ed.) *Fungal Virology*, p. 109. Boca Raton, FL: CRC Press.

Koltin Y (1988) The killer system of *Ustilago maydis*: secreted polypeptides encoded by viruses. In: Koltin Y and Leibowitz M (eds) *Viruses of Fungi and Simple Eukaryotes*. p. 209. New York: Marcel Dekker.

Peery T *et al.* (1987) Virus encoded toxin of *Ustilago maydis*: Two polypeptides are essential for activity. *Mol. Cell. Biol.* 7: 470.

Tao J *et al.* (1990) *Ustilago maydis* KP6 killer toxin: structure, expression in *Saccharomyces cerevisiae*, and relationship to other cellular toxins. *Mol. Cell. Biol.* 10: 1373.

TRANSFORMATION

Ron Wisdom and Inder M Verma
The Salk Institute
San Diego, California, USA

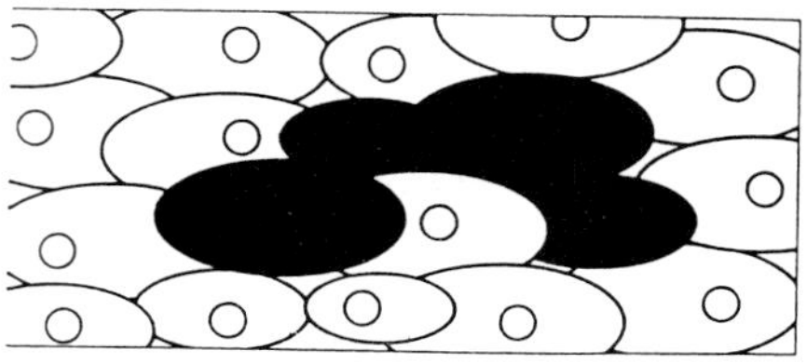

Introduction

The observation that certain viruses rapidly and reproducibly induce tumors in the host was first made nearly a century ago. The ability to accurately duplicate this process in tissue culture has made possible the investigation of neoplastic transformation *in vitro*. This, in turn, has allowed the techniques of molecular biology and genetics to be applied to the problem of cancer, especially cancer induced by viruses. Much of our current knowledge of the molecular mechanisms of cancer today is derived from the study of virally induced tumors. Virus-induced tumors are not only important as laboratory tools for investigation of the fundamental mechanisms of tumor formation; some human cancers, including hepatocellular carcinoma, cervical carcinoma, Burkitt's lymphoma and acute T-cell leukemia, appear to have a viral etiology.

The types of viruses that are capable of inducing tumors are remarkably diverse, and include viruses with RNA and DNA genomes. Nonetheless, certain common features have emerged from the study of neoplastic transformation by different viruses. The first is that tumor formation induced by viruses is the result of the acquisition by the host cell of the viral genome. In all cases of virally induced tumors so far described, at least a part of the viral genome is present. Furthermore, the continued expression of at least a part of the viral genome is required for the transformed phenotype to be present. In the case of the transforming RNA viruses, the viral genome is present in the form of a DNA provirus integrated into the host genetic material. An important consequence of the presence of the viral genetic material is the fact that the transformed phenotype induced by the virus is heritable.

Cancer cells show many properties that distinguish them from normal cells. The hallmark of malignant cells is their ability to grow in the absence of appropriate signals from the extracellular environment. While the extracellular signals that are either not sensed or not required may vary from tumor to tumor, the property of malignancy is cell autonomous, that is, it is a property of the transformed cell and not the surrounding normal tissue. Genetic events that give rise to cancer, then, alter the cellular response to external stimuli. The investigation of cancer over the last decade has served to outline a cell signalling pathway that allows cells to respond to changes in the external environment. Support for the idea that this signalling pathway is the direct target for cancer-inducing mutations has been found in a wide range of both viral and nonviral tumors.

Retroviruses

The study of retroviruses has contributed more to our understanding of cancer than the study of any other group of viruses. The relevant features of the retroviral life cycle involve reverse transcription of the RNA genome into double-stranded proviral DNA, integration of the proviral DNA in the host genome, and expression of the viral genome from the chromosomal site of integration. The retroviruses have been of particular importance for several reasons. First, the small size of the retroviral genome and the small number of proteins expressed from the retroviral genome (a transforming retrovirus frequently encodes only one protein) has made them experimentally favorable for molecular analysis. Second, the proteins that produce neoplastic transformation are not required for the virus to complete its life cycle. Third, the retroviral transforming genes are mutated forms of normal cellular genes, giving their study direct relevance to nonviral forms of tumor formation.

Retroviruses can be divided into those that induce transformation acutely (in days to weeks) and form polyclonal tumors, and those that induce transformation over long periods of time (months) and form monoclonal or oligoclonal tumors. The distinction is important because the mechanism of tumorigenesis is fundamentally different in the two cases. Acutely transforming retroviruses consist of a mixture of replication-defective viruses that contain the transforming activity, and replication-competent nontransforming (helper) viruses that serve to propagate the transforming virus. The transforming virus in these cases often encodes only a single protein, and it is this protein that has the oncogenic activity. These transforming viruses are potent carcinogens, in the sense that the infection of a cell is sufficient to induce transformation with a very short latency. The transforming viral genes are transduced mutated copies of cellular genes, and the transforming proteins correspond to mutant versions of the corresponding cellular proteins. Structural and functional analyses of the viral and cellular forms of transforming proteins allow elucidation of the features that convert a normal cellular protein into an oncogenic one.

In contrast to the acutely transforming retroviruses, the retroviruses that induce tumors only after long latency have several important distinctions. They are replication-competent viruses that have not transduced cellular genes and so do not contain a transforming gene. Tumor formation by these viruses is often the result of the mutation of cellular genes during the integration process. Expression of cellular proto-oncogenes may be transcriptionally activated by the occasional nearby integration of retroviral promoter and enhancer elements, as in the case with c-*myc* activation by avian leukosis virus (ALV) and murine leukemia virus (MLV). As retroviruses integrate into a wide range of sites in the host DNA, only rare integration events will give rise to the activation of a protooncogene, explaining the long latency and the monoclonal nature of the tumors. In addition to being potentially tumorigenic, the juxtaposition of the retroviral transcriptional control signals with cellular proto-oncogenes may be of importance for a second reason. It is likely that production of a fusion virus–oncogene mRNA is the first step in the viral transduction event that results in the generation of an acutely transforming retrovirus through an aberrant viral recombination event.

Cell signaling

The ability of cells to respond to their external environment is mediated by the interaction of factors outside the cell (peptide and nonpeptide hormones and growth factors, cell matrix and other cells) with specific receptors located on the cell surface. These cell surface receptors contain at least two distinct functions. First, they contain a ligand-binding function which allows sensing of the presence or absence of the appropriate factor in the external environment, and second, they contain or control some type of enzymatic activity that generates an intracellular change in response to the presence of the ligand. Often, this signal is in the form of reversible phosphorylation in response to ligand. The initial signal frequently generates a cascade of intracellular events, which include the generation of small molecule second messengers, such as cyclic nucleotides, calcium ions, and phospholipids, the activation of signals generated by the small GTP-binding proteins, and the activation of other protein kinases. Any long-lasting change in cell behavior, such as growth, requires new gene expression, and an important consequence of the signaling cascade is the activation of a set of transcription factors that directly influence the expression of genes required to alter cell behavior. Theoretically, it should be possible to generate mutant cells defective in the response to extracellular changes by alteration or mutation in any part of the signaling pathway, and the last ten years has witnessed the experimental confirmation of this idea. Thus, neoplastic transformation can be due to mutations in proteins that function as autocrine growth factors, growth factor receptors, membrane-bound and cytoplasmic components of the cell signaling apparatus, and transcription factors, all the normal activities of which are subject to control by changes in the external environment.

Growth factors and receptors

The simian sarcoma virus (SSV) contains a single open reading frame and encodes a single protein (v-Sis). This protein is a fusion protein between the retroviral *env* gene product and the product of a gene derived from the c-*sis* gene of the SSV host, the woolly monkey. The *env* sequences comprise little besides the signal sequence of the Env protein, and the remainder of the protein is the product of the transduced c-*sis* gene. Nucleotide sequence analysis has shown that c-*sis* is the platelet-derived growth factor (PDGF) B chain gene. The v-Sis protein is therefore a form of PDGF, a peptide mitogen for a number of cell types. That the v-*sis* gene product functions as an autocrine growth factor to induce transformation is supported by several lines of evidence. Transformation by v-Sis is limited to cells that express the PDGF receptor. The PDGF receptor is a protein tyrosine kinase the activity of which is dependent on the presence of ligand. In v-Sis-transformed cells, there is increased activity of the PDGF receptor and an increase in total cellular phosphotyrosine-containing proteins when compared to nontransformed cells. As both the v-Sis protein and the PDGF receptor are present in the endoplasmic reticulum, binding and activation of the receptor may occur at sites other than the cell surface. That this is the case is suggested by the observation that activated intracellular forms of the receptor can be found in v-Sis-transformed cells, but not in cells exposed to PDGF. Therefore, the v-*sis* gene product functions as an autocrine growth factor in susceptible cells.

A second class of oncogenic mutations predicted from the cell signaling model outlined above are mutations that separate activation of growth factor receptors from the binding of ligand. The first described example of such a mutation was the v-*erbB* gene of avian erythroblastosis virus (AEV). The v-*erbB* gene is a virally transduced mutant form of the cellular epidermal growth factor (EGF) receptor. AEV infection of chickens results in the rapid induction of erythroblastosis, and although AEV carries a second gene (v-*erbA*), expression of v-*erbB* is necessary and sufficient for the induction of erythroblastosis. The EGF receptor is typical of many growth factor receptors in its molecular architecture. It contains an extracellular domain involved in ligand (EGF) binding, a hydrophobic membrane-spanning domain, a cytoplasmic domain that contains protein tyrosine kinase activity and a C-terminal region involved in regulation of the kinase activity. Several observations suggest that the v-ErbB protein is a constitutively active form of the EGF receptor. In comparison to the EGF receptor, v-ErbB has suffered several structural changes. These include deletion of the extracellular ligand-binding domain and alterations of the C-terminal regulatory region. Molecular analysis has shown that both of these alterations are important for the transforming activity of the protein. Consistent with its proposed function, v-ErbB shows protein tyrosine kinase activity in the absence of EGF, and AEV-transformed cells show increased phosphotyrosine-containing proteins when compared to nontransformed cells. All mutations that impair the kinase activity of v-ErbB impair the transforming activity of the virus in a coordinate manner. Thus, the v-*erbB* oncogene product is a ligand-independent form of the EGF receptor.

Intracellular signal transduction molecules

A number of retroviral oncogenes have been identified that function as intracellular effectors of changes in the external environment. The v-Src protein of Rous sarcoma virus (RSV) is a cytoplasmic tyrosine kinase anchored to the inner surface of the plasma membrane by an N-terminal modification by myristic acid addition. The v-Src protein contains several mutations when compared to c-Src. Important among these is mutation of a C-terminal tyrosine in c-Src that serves as a negative regulator of the c-Src tyrosine kinase activity. Although it is not known what the physiologic mechanisms are for activating c-Src tyrosine kinase activity, v-Src functions as a constitutively activated enzyme, and RSV-transformed cells contain increased amounts of phosphotyrosine-containing proteins. An additional important feature in the Src protein is the presence of regulatory sequences known as SH2 and SH3 domains. These domains were initially recognized as conserved motifs in other proteins, first in tyrosine kinases and then in other molecules, including the v-*crk* oncogene and a number of substrates for tyrosine kinase enzymes, such as phospholipase C, phosphatidylinositol 3′-kinase (PI3 kinase), and the *ras* GTPase-activating protein (GAP). The investigation of the v-*crk* oncogene has been particularly instructive. The transforming protein, v-Crk, is composed only of SH2 and SH3 sequences, without any tyrosine kinase catalytic domain. However, v-Crk-transformed cells contain elevated levels of phosphotyrosine-containing proteins. Investigation of this phenomenon has led to the realization that SH2 domains function as ligands for proteins containing phosphotyrosine. As all active tyrosine kinase enzymes are themselves phosphorylated on tyrosine, potential substrates may bind through their SH2 domains to the active phosphotyrosine-containing forms of the enzymes. This may be a way of controlling availability of substrates for these enzymes, and the presence of SH2 domains on the substrates of tyrosine kinases is likely to accelerate the identification of substrates, a process that until now has been frustratingly slow.

A second class of cytoplasmic signaling molecule that can be rendered oncogenic by mutation is the Ras family of proteins. Initially identified as v-Ras, the transforming proteins of Harvey murine sarcoma virus (Ha MuSV), the Ras proteins have been conserved as essential functions from yeast to man, with very little sequence variation. The Ras proteins are small GTP-binding proteins with GTP-hydrolyzing activity. The GTPase activity is essentially a way to control the signaling activity, as only the GTP-bound form is active. Recent genetic and biochemical evidence suggests that at least one function of Ras proteins is to mediate the signals generated by protein tyrosine kinases. In contrast to the normal cellular form of Ras, the oncogenic v-Ras contains two point mutations that impair the GTPase activity and so trap the molecule in an 'on' or signal-generating state. Unfortunately, the nature of the signals generated by active Ras remains elusive.

Studies of the regulation of Ras GTPase activity have revealed a second mechanism of oncogenic activation through the Ras signaling pathway. The intrinsic GTPase activity of Ras proteins is augmented by GTPase-activating proteins, or GAPs. The GAP molecules increase the hydrolysis of GTP, and so convert active Ras to inactive Ras. An important observation has been the realization that the gene that confers predisposition to neurofibromatosis (NF1), an inherited form of cancer, encodes a GAP. Inactivating mutations in NF1 GAP are associated with a decrease in GAP activity, a consequent increase in the amount of Ras protein in the active (GTP-bound) state and the formation of tumors.

Cell signaling through transcription factors

The generation of second messengers and cytoplasmic signals by external factors eventually results in changes in gene expression. Several retroviral oncogenes are transduced forms of cellular transcription factors involved in regulating the response to external factors. The transcription factor AP1 is a heterodimeric transcription factor with sequence-specific DNA-binding activity composed of peptides derived from the Fos and Jun families of proteins. The *fos* family of genes was identified as the cellular homologs of v-*fos*, the transforming gene of FBJ MuSV, while the *jun* family was identified by homology to v-*jun*, the transforming gene of avian sarcoma virus (ASV) 17. AP1 activity is tightly regulated, and is rapidly and transiently induced by a wide variety of mitogens, including the peptide mitogens PDGF and EGF. The induction of AP1 activity requires *de novo* protein synthesis, and is a consequence of the early induction of the *fos* and *jun* genes in response to mitogens. Several lines of evidence suggest that transformation by v-Fos is due to overexpression of AP1 activity. Overexpression of either c-Fos or v-Fos protein is sufficient to result in neoplastic transformation, and mutant proteins that are unable to bind AP1 target DNA sequences are nontransforming. Cells transformed by Fos protein show increased levels of AP1 activity compared to nontransformed cells, consistent with the idea that transformation is the result of activation of a specific set of genes that contain AP1 regulatory sequences in their upstream regions. The identification of the relevant genes the activation of which is important for transformation has proven to be difficult.

A somewhat different situation is represented by the transforming gene, v-*rel*, of the reticuloendotheliosis virus (REV-T). The Rel family of proteins constitute a large family of proteins with NF-KB activity. NF-KB is a transcription factor that is involved in the response of lymphoid cells to a variety of external stimuli. NF-KB activity is regulated by a number of complex mechanisms, but an important form of regulation appears to be retention in the cytoplasm through its interaction with specific inhibitor proteins. External stimuli then result in the dissociation of the inhibitor and allows translocation of NF-KB to the cytoplasm. An usual feature of transformation by v-Rel is that mutant v-Rel proteins that are localized in the cytoplasm can induce transformation. This suggests that direct transcriptional activation by v-Rel is not the mechanism of neoplastic transformation, and suggests that v-*rel* may function as a dominant negative allele of c-*rel*, and this idea has received some experimental support. According to this model, NF-KB activity would function to promote cellular differentiation and inhibit cell growth, and oncogenesis would be the consequence of decreasing NF-KB activity.

Oncogene cooperation

Spontaneously occurring tumors are recognized to be the result of a multistage process in which multiple mutations arising at different times conspire to produce a tumor. On the other hand, neoplastic transformation due to infection with the acutely transforming retroviruses usually occurs as the result of the introduction of a single gene. Yet even the acutely transforming retroviruses contain examples of cooperativity among oncogenes. AEV contains two genes, v-*erbA*, a transduced mutant form of the thyroid hormone receptor gene, and v-*erbB*, a constitutively active form of the EGF receptor gene. Although v-*erbB* is sufficient to direct neoplastic growth of infected erythroid cells, the expression of v-*erbA* is required for growth of the erythroid cells under some culture conditions. Thus, the fully transformed phenotype is the product of a cooperative interaction between the two genes. This theme of cooperation between various oncogenic proteins has been extensively documented in the process of neoplastic transformation of primary cells.

Transformation by human retroviruses

Retroviruses may acquire transforming activity by transduction of cellular proto-oncogenes, they may activate cellular proto-oncogenes by insertional mutagenesis, or they may transform through the action of regulatory genes required for the viral life cycle to proceed. The causative agent of acute T-cell leukemia, human T-cell leukemia virus (HTLV-1), appears to operate through the last type of mechanism. The viral gene *tax* is a regulator of transcription from the viral long terminal repeat, as well as from certain cellular promoters. Interestingly, Tax activates both the interleukin 2 (IL-2) promoter and the IL-2 receptor promoter, generating a loop with potential autocrine activity in T cells. The mechanism of Tax activation of transcription remains unclear. It is able to activate transcription directed by NF-KB and cyclic AMP-responsive sites, although there is no evidence that Tax itself binds DNA. The mechanism of transformation induced by HTLV-1 is likewise unclear; the tumors arise only after a long latency, sometimes 20 years, and are monoclonal.

DNA Tumor Viruses

Several classes of DNA viruses are able to induce neoplastic transformation in culture. In contrast to the situation with transforming retroviruses, in which transformation is the result of the activation of a gene that is not required for the completion of the viral life cycle, transformation by DNA viruses is the result of expression of proteins intimately involved in the replication of the viral genome.

Simian virus 40 (SV40)

Transformation by SV40 is associated with nonproductive infection and is the result of the expression of a single viral protein, the SV40 large T antigen (T Ag). In contrast to productively infected cells in which the SV40 genome is episomal, in transformed cells the viral genome is integrated into the host DNA in a manner which allows for expression of T Ag. T Ag is a multifunctional protein that forms noncovalent interactions with a number of cellular proteins. Key among these are the products of the cellular tumor suppressor genes p53 and Rb. Oncogenic activation of the p53 and Rb genes is associated with loss, rather than gain of function. p53 was initially identified as a protein stably associated with T Ag in transformed cells. The normally rapid turnover of p53 in uninfected cells is greatly reduced by virtue of complex-formation with T Ag, giving rise to increased levels of the protein in transformed cells. The net result of this interaction is not to increase p53 function, however, but rather to sequester it in an inactive complex. The protein product of the retinoblastoma susceptibility gene, pRb, is also noncovalently associated with T Ag, and this complex results in loss of pRb function. Transformation by T Ag has therefore served as the mechanism of identification of one tumor suppressor gene (p53), and has been the entry point for biochemical investigation of another (pRb).

Adenovirus

Transformation by the human adenoviruses has been observed only *in vitro*. As is the case with transformation by SV40, transformation by adenovirus is the result of a nonproductive infection. DNA transfection experiments reveal that the expression of two different viral genes, E1a and E1b, is required for full transformation. E1a is a multifunctional protein with a bewildering array of activities. The important feature for transformation appears to be the ability to interact with pRB. Expression of the E1a gene alone is sufficient for immortalization of primary cells, but not for full transformation, which requires expression of the E1b gene. The product of the E1b gene is complexed physically with p53. Thus, the combined activities of E1a and E1b are reminiscent of the activities of SV40 T Ag.

Human papilloma virus

Investigation of the transforming activities of the human papillomaviruses (HPV) is of interest for reasons of medical importance as well as scientific interest. Nearly all human cervical carcinomas are associated with the presence of DNA from either HPV 16 or HPV 18. Completion of the life cycle of the HPVs is unusual in that the early and late stages of the life cycle appear to take place in different stages of epithelial cell differentiation, and complete viral replication takes place only in benign papillomas, or warts. This feature has so far made it impossible to grow the HPVs in culture, a problem that has slowed analysis. However, molecular analysis has been informative. Cervical carcinomas are associated with the presence of the viral genome integrated into the host DNA in a manner that allows for expression of the early region of the viral genome. This region is quite complex, with at least eight open reading frames; there is evidence that at least five are involved in generating the transformed phenotype. In a scenario reminiscent of adenovirus, the products of the E6 and E7 open reading frames are physically associated with p53 and pRB respectively. These interactions result in the functional inactivation of p53 by accelerated proteolysis, and pRb by sequestration.

Hepatitis B virus

The hepadnaviruses should be mentioned because of their potential role in human cancers. The hepatitis B virus, especially when acquired by vertical transmission, is associated with the development of hepatocellular carcinoma. At this point, there is no decisive evidence that this association is causal, but strong circumstantial data are present. Experimental evidence suggests two potential mechanisms: activation of cellular oncogenes by integration in a manner similar to insertional mutagenesis by retroviruses, and repetitive mitogenic stimulation due to the toxic effects of chronic exposure to the surface antigen proteins.

Epstein–Barr virus

Nonproductive infection with Epstein–Barr virus (EBV) is associated with Burkitt's lymphoma, a malignancy of B lymphocytes. Once again, there is no decisive evidence for a causal role, but the ability of the virus to immortalize human B lymphocytes suggests a potential mechanism. Infection with EBV may result in the generation of a large pool of preneoplastic cells that are predisposed to further oncogenic mutations.

Future Perspectives

Much of what we understand about the molecular mechanisms of cancer is the result of investigation of virus-induced tumors. This area of investigation has demonstrated, clearly and convincingly, that cancer is a disease caused by mutations in specific genes. We have learned the identity of many of these genes, whether they are carried by the virus, activated by viral insertion, or inactivated by the physical association of their products with viral proteins. The products of these genes form a network of signaling elements that allows cells to respond to the external environment. Where understood, oncogenic mutations result in the production of proteins that propagate mitogenic signals in the absence of appropriate monitoring of the extracellular environment, a result consistent with the proposed normal functions of these proteins. Although many oncogenes were initially identified by virtue of their presence in viral genomes, it is now clear that these genes are also the targets of oncogenic mutations in tumors of nonviral origin. It seems certain that future investigations of virus-induced transformation will further contribute to our understanding of cancer, an understanding that does not appear to be as elusive as a decade ago.

See also: Adenoviruses; Avian leukosis viruses; Epstein–Barr virus; Hepatitis B viruses; Human T-cell leukemia viruses; Murine leukemia viruses; Oncogenes; Papillomaviruses; Reticuloendotheliosis viruses; Simian virus 40; Tumor viruses–human.

TRANSPLANTATION

Helen E Heslop, Robert A Krance and Malcolm K Brenner
St Jude Children's Research Hospital,
Memphis, Tennessee, USA

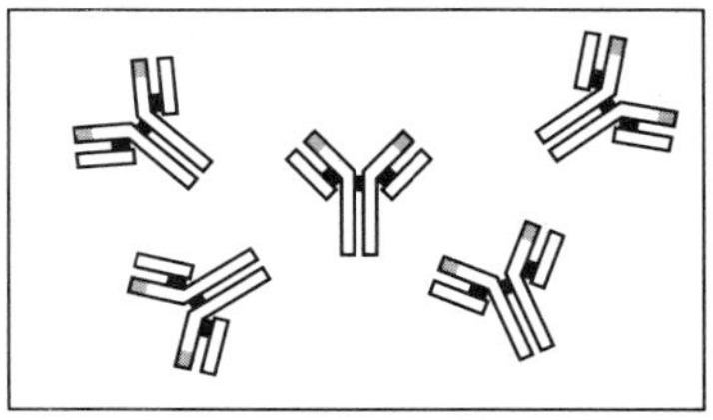

Introduction

Organ transplantation is being used to correct an ever broader range of medical abnormalities. In large part the expanding application and success of transplantation has been based upon improved methods of producing immunosuppression in the recipient, which allow acceptance of a genetically disparate graft. This is particularly true following bone marrow transplantation (BMT), where immunosuppression must be especially profound, since the aim is not only to prevent rejection of the graft by the host but to prevent the incoming immune system from 'rejecting' the recipient to produce graft versus host disease (GVHD).

Unfortunately, the immunosuppression required

by transplant recipients has a number of adverse consequences. Amongst the most important of these is the suppression of host immune responses directed against endogenous and exogenous micro-organisms. Suppression of immune surveillance allows invasion by opportunist fungi and protozoa and permits recrudescence of those endogenous viruses that are usually maintained in the latent state. Immunosuppression may also increase the severity of infection after exposure to exogenous viruses. All transplant recipients are vulnerable to viral infections, but BMT recipients, who are the most intensely immunosuppressed, are the most vulnerable of all. Notwithstanding the introduction of a number of different therapeutic agents, viral disease remains one of the leading causes of morbidity and mortality after transplantation.

Herpesviruses

The herpesviruses, cytomegalovirus, Epstein–Barr virus, herpes simplex virus and varicella–zoster virus, represent the greatest potential threats to transplant recipients. All four are present in the latent state in a high proportion of individuals, and all four can cause serious or fatal disease in the immunocompromised host.

Cytomegalovirus (CMV)

Background

In a normal individual, defense against CMV is mediated by specific cytotoxic T lymphocytes (CTL), production of specific antibody and perhaps also by major histocompatibility complex (MHC)-unrestricted cellular effector mechanisms. In the transplant recipient, immunosuppression-mediated inpairment of these defense mechanisms places the recipient at risk of CMV disease. Depending on location and socioeconomic status, 40–90% of transplant recipients have a past history of primary CMV infection and are at risk of viral re-activation. The remaining seronegative recipients are at risk of acquiring primary infection, either from the donor tissue or from blood products required during the post-transplant period. CMV re-activation or infection does not always lead to CMV disease in the transplant recipient and it is important to distinguish between these phenomena.

While detection of CMV is not always predictive of disease, there may be a high probability that disease will subsequently occur in patients receiving intensive immunosuppression. Following solid organ grafts, the risk of CMV disease rises with the intensity of the immunosuppression administered and is especially increased by the use of monoclonal anti-T-cell antibodies, such as OKT3. BMT recipients who have had their own immune system ablated as part of the conditioning regimen and who have not yet regenerated a donor-derived system are particularly likely to develop CMV disease.

Incidence of CMV infection/re-activation and CMV disease

From blood product support. The risk of a seronegative recipient of seronegative donor tissue acquiring CMV infection from blood product support was previously significant, but the incidence of this mode of transmission has been substantially reduced by the use of CMV seronegative or leukocyte-poor blood and is now less than 5%.

From solid organ grafts. Seropositive recipients have a 50–80% chance of developing evidence of viral re-activation. Between 10 and 25% of these individuals will develop evidence of CMV disease. The donor organ is also a source of infection and 50–100% of seronegative recipients will develop infection if they receive a transplant from a seropositive donor. Disease occurs in up to one-half of such patients.

From BMT. Some 50–60% of both allograft and autograft recipients who are seropositive or have a seropositive donor will have evidence of either primary infection or re-activation, with a peak incidence in the first 3 months post-transplant. Patients receiving autologous marrow have a low incidence of disease with less than 5% of seropositive recipients developing pneumonitis. By contrast, half of the allograft recipients who develop a primary infection or re-activation will develop pneumonitis which is fatal in 80–90%. The presence of moderate to severe GVHD markedly increases the risk of CMV disease, especially pneumonitis, and some studies have shown a reduced incidence of CMV in allograft recipients who receive more effective GVHD prophylaxis.

Pathogenesis

CMV initially infects a variety of cells including fibroblasts, marrow stromal cells and hemopoietic progenitors. The mechanism of latency is not well understood, but re-activation results in viral shedding. In some situations, such as in acquired immunodeficiency syndrome (AIDS) patients, viral replication alone results in disease, but in the transplant patient an immunopathological component is implicated, and both host versus graft and graft versus host reactions correlate with CMV disease. For example, a host versus graft relationship is seen in recipients of liver transplants, who are particularly likely to develop CMV hepatitis. In allogeneic BMT recipients, there is evidence that damage to the lungs in CMV pneumonitis (Fig. 1a and b) results from an abnormal response to antigen, and that this is exacerbated by GVHD. The activated immune system may also damage or destroy uninfected 'bystander cells', since antiviral agents that eliminate viral shedding and prevent antigen expression have limited effect on disease progression (see below under Treatment).

Diagnosis

CMV infection may be diagnosed by either serological means or direct isolation of the virus. The virus is generally cultured from blood, throat, urine or tissues. CMV is cultured on human fibroblasts, and it usually takes 1–4 weeks for the characteristic cytopathic effect to become evident. The use of monoclonal antibodies directed at early or intermediate antigens in conjunction with centrifugation of the specimen on to monolayers of fibroblasts – the shell vial technique – allows diagnosis within 24–48 h. The presence of the virus can be diagnosed even more rapidly by the use of the polymerase chain reaction (PCR) to amplify CMV DNA and RNA, but how accurately detection of viral infection/reactivation by such sensitive methodology predicts subsequent CMV disease is unknown.

Serological evidence of infection is usually obtained using enzyme-linked immunosorbent assay (ELISA) or indirect fluorescence. Diagnosis requires a fourfold rise in titer and may be problematic after BMT when patients are so severely immunocompromised that antibody production is impaired.

Clinical aspects

CMV infection in the immunosuppressed host ranges from an asymptomatic illness – similar to that in normal individuals – to disseminated disease. However, there is a higher incidence of diffuse organ involvement which may present with a variety of clinical manifestations including hepatitis, enteritis, pneumonitis, fever and bone marrow suppression. Disease presentation may also vary according to the organ transplanted. In recipients of renal transplants, pneumonitis is rare and patients are more likely to present with mononucleosis-like symptoms or hepatitis. Recipients of lung, heart or liver transplant have a higher risk of more severe CMV infections such as pneumonitis. Liver transplant recipients are particularly at risk of CMV hepatitis which is more common after primary infection and may be difficult to distinguish from rejection. Finally, following allogeneic BMT, pneumonitis is the most common presentation of CMV disease and will occur in 40–50% of recipients with evidence of infection on blood culture. Pneumonitis presents with fever and cough dyspnea, and chest X ray shows an interstitial pattern. Some 80–90% of BMT recipients who develop CMV pneumonitis will die of this complication.

Treatment

A number of measures have been employed to treat established CMV disease. All are of most benefit following solid organ grafting, and of least value after BMT.

Interferon has antiviral activity and may reduce morbidity in renal transplant recipients, but has little effect in BMT patients. Similarly, acyclovir and gancyclovir may reduce viral shedding and may be effective in CMV disease in solid organ recipients, but given alone they do not significantly modify morbidity and mortality from CMV disease in allogeneic BMT recipients.

Intravenous immunoglobulin (IVIG) as a single agent is also ineffective in most studies. However, the combined use of gancyclovir and IVIG in allogeneic BMT recipients with CMV pneumonitis results in disease resolution of about 70%. This response rate appears significantly better than the 10% response rate of historical controls, although no randomized trials have been performed and the improved survival may in part reflect changes in diagnostic criteria and techniques. Indeed, subsequent studies where the drug combination was administered to patients who were ventilated yielded lower response rates.

Prophylaxis (Table 1)

The most effective preventative measure is to use CMV-negative blood products for seronegative recipients who receive seronegative organ grafts. Several studies have convincingly shown that this policy can almost completely eliminate CMV disease. Seronegative recipients of seropositive allogeneic marrow also benefit from seronegative blood

(a)

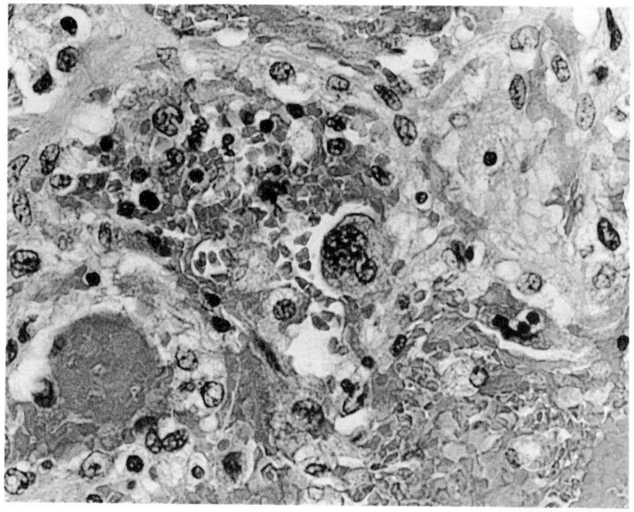

(b)

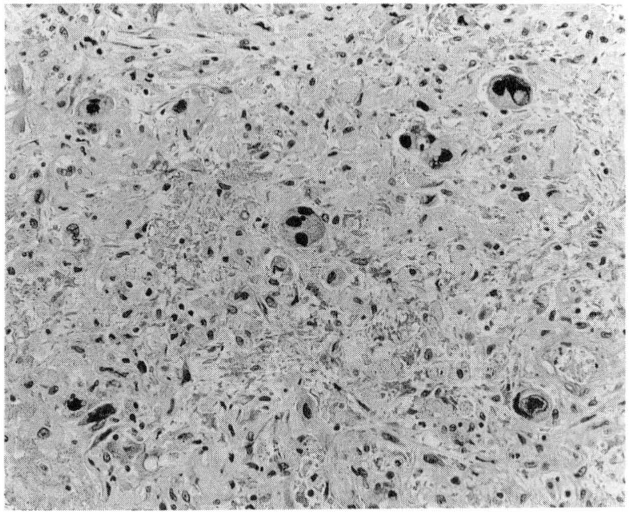

(c)

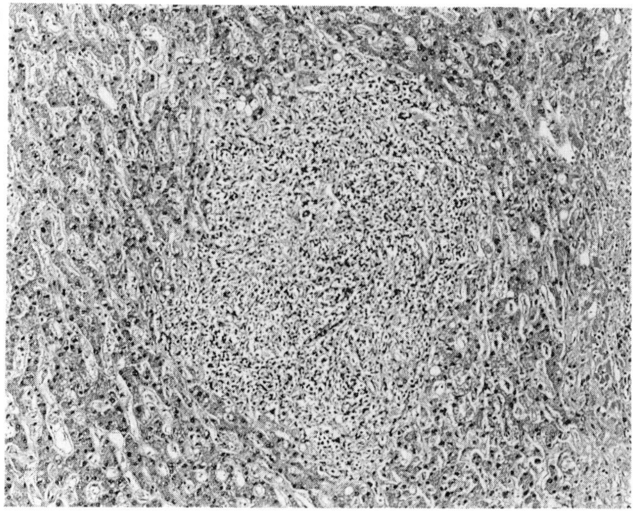

(d)

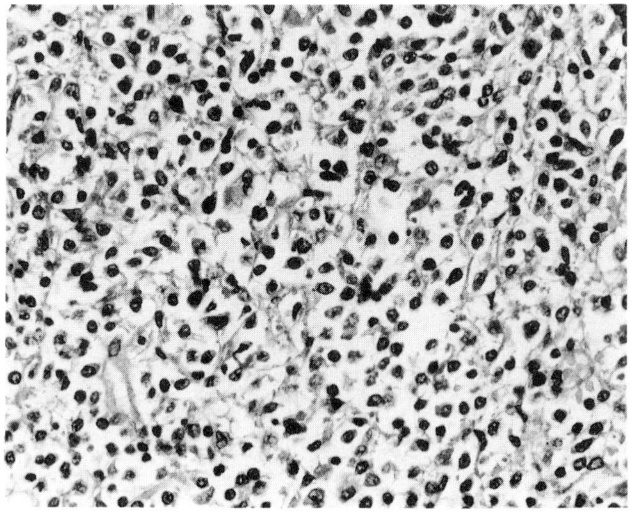

Fig. 1 **(a)** Section of lung, infected with CMV, containing intraalveolar hemorrhage and multinucleated giant cell. This is an histologically atypical infection producing only scattered cells with relatively small intranuclear and intracytoplasmic inclusions and no large cells with 'owls-eye' nuclei (hematoxylin and eosin stain; ×315). **(b)** Same infection as in (a) demonstrating numerous cells positive for CMV antigen using monoclonal antibody and immunoperoxidase technique (avidin–biotin complex technique; ×315). **(c)** (EBV low power) Section of liver with lymphoid infiltrate (EBV infection) causing portal expansion (hematoxylin and eosin stain; ×80). **(d)** Same infection as in (c). The portal infiltrate is polymorphous and contains cells ranging from small mature lymphocytes to larger immunoblasts (hematoxylin and eosin stain; ×315).

Table 1. Effect of prophylaxis in preventing CMV re-activation/disease

Prophylactic regimen	Effect on		
	Re-activation/infection	Disease	Mortality
CMV-negative blood products	+	+	+
Acyclovir (low dose)	+/−	+/−	−
Acyclovir (high dose)	+	+	+/−
Gancyclovir	− (see text)	+	+
IVIG	−	+	+/−

products. It has been suggested that seropositive recipients should also receive CMV-negative products to reduce exposure to new strains. In practice most blood banks do not have enough CMV-negative products to supply all these populations, and leukocyte-poor products, prepared by filtration of blood, are used instead. It appears that leukocyte-poor products may be as effective, at least in low-risk patients.

Antiviral agents have also been explored as prophylactic agents. Acyclovir in renal transplant recipients reduces the rate of CMV infection and disease but did not improve survival. Neither α-interferon nor low-dose acyclovir have any convincing effect in BMT recipients. High-dose acyclovir 500 mg/m^2 administered three times daily from day −5 to day +30 was shown to reduce the incidence of both CMV infection and invasive disease, but the control group in this series had a higher incidence than the historical rate, and further studies are needed to confirm these observations. Gancyclovir appears to be effective prophylaxis for reducing the incidence of pneumonitis in allograft recipients with CMV infection detected by the shell viral technique on broncheoalveolar lavage performed at day +35.

IVIG administered following BMT, resulting in passive transfer of specific antibody, does not reduce the rate of viral re-activation or infection but does decrease the incidence of pneumonia and death. Such treatment may therefore be justified in seropositive recipients. The expense of IVIG does not appear to be justified in seronegative recipients receiving seronegative blood products, in whom risk of CMV infection and disease is negligible. Similarly, in seronegative renal recipients receiving a seropositve graft, IVIG reduces the incidence of CMV disease though not the rate of viral isolation or seroconversion. Neutralizing human monoclonal antibodies are currently being evaluated in Phase-I–II trials.

Immunomodulation

Several studies have correlated the pattern of CMV infection in allogeneic BMT recipients with the ability to generate CMV-specific CD8+ CTLs, which may allow the host to control CMV infection. These observations suggested that adoptive transfer of specific CTLs (expanded *in vitro*) to high-risk patients may be therapeutically beneficial and trials of such an approach are currently underway. Similarly, the use of recombinant cytokines, such as interleukin 2 (IL-2), which promote growth of the immune system may accelerate immune reconstitution in these patients.

Epstein–Barr virus (EBV)

Background

Even though more than 90% of adults have serological evidence of EBV infection, disease produced by re-activation of this herpesvirus has, until recently, been less prevalent in transplant recipients than disease produced by CMV. However, the development of more aggressive and effective techniques for immunosuppression has not only increased the success of tissue transplantation, but has also substantially increased the incidence of the lymphoproliferative syndrome associated with uncontrolled EBV re-activation in the immunocompromised host. Nowhere has this complication become more frequent than in patients receiving mismatched or unrelated donor BMT.

Incidence

Following solid organ grafting, up to 60% of patients may show evidence of EBV re-activation or infection, assessed on the basis of rising antibody titers to viral proteins. Less than one-third of these patients will develop the clinical features of a viral illness. The incidence of lymphoproliferative syndrome is between 1 and 13%, and is lowest in patients receiving renal allografts and highest in those receiving heart–lung double transplants. Within each group, patients receiving the most intensive immunosuppression – with cyclosporin A (CSA), antilymphocyte globulin or anti-T-cell monoclonal antibody (e.g. OKT3) – have the highest incidence.

After BMT, the incidence is even more variable. At one extreme, adult patients receiving MHC-identical sibling allografts as treatment for leukemia have an incidence of 0.25%. At the other extreme, children transplanted for congenital immunodeficiency syndromes who receive marrow from an MHC-nonidentical or unrelated donor have a 100-fold greater incidence of lymphoproliferative disease, which occurs in up to 25% of such recipients. This wide variation is in part attributable to the increased immunosuppression required by recipients of mismatched/unrelated marrow, but probably also reflects the long delay in immune reconstitution exhibited by patients receiving a genetically disparate immune system. Because the numbers of mismatched/unrelated donor BMTs are steadily rising, some recent BMT series show that death from EBV-induced lymphoproliferation now exceeds that from CMV disease.

Pathogenesis and diagnosis

The primary pathologic process is uncontrolled proliferation of mature B lymphocytes. Analysis of proliferating lymphoblasts following solid organ transplant using restriction fragment length polymorphisms (RFLPs) or minisatellite probes, has shown that EBV lymphoproliferation in these patients almost always arises from recipient B cells. In contrast, lymphoproliferation after BMT commonly arises from cells of donor origin. In all cases, the transformed B cells closely resemble, morphologically and phenotypically, the lymphoblastoid cell lines generated *in vitro* when human B cells are exposed to EBV in the absence of T cells. The infected lymphocytes may have an immunoblastic or plasmacytoid appearance and are usually CD19+, CD20+, CD21+ and CD24+ (see under Treatment). Like Burkitt lymphoma cells, the lymphoblasts are positive for EBV nuclear antigens 1 (EBNA1), but, unlike Burkitt lymphoma cells, they usually express all of the five other virus-encoded latent cycle nuclear antigens and are positive for most latent membrane proteins. Again, in contrast to Burkitt lymphoma cells, lymphoblasts in transplant patients express a number of cell adhesion molecules and ligands. All these phenotypic features would make them intensely vulnerable to cytotoxic T-cell killing in a normal individual, and it is only the profound suppression of the immune system after transplantation which permits their outgrowth (see under Future Perspectives).

The B-cell proliferation may be oligoclonal or monoclonal, and may be associated with production of oligoclonal or monoclonal peaks of immunoglobulin which can be detected by serum electrophoresis. Within the B cells, the virus may be present in linear (replicative) or circular (latent) form. Analysis of these attributes may predict the response to therapy.

Clinical manifestations

Early in the illness, fever, malaise and circulating atypical lymphocytes are usually seen. The proliferating lymphoblasts may then produce a number of disease patterns. They may diffusely infiltrate different organ systems, including lungs, liver, kidney, gut, bone marrow and central nervous system (CNS). Infiltration may be so extensive that organ failure results. Alternatively, a classical 'lymphoma' pattern is observed, with lymphadenopathy, hepatosplenomegaly and a biopsy appearance of a diffuse immunoblastic lymphoma. Rarely, a predominantly leukemic picture occurs. Although these distinctive clinical patterns may be seen in isolation, a combination of features often evolves (Fig. 1c and d).

Clinical course

Initial re-activation of EBV, associated with an infectious mononucleosis-like illness, may resolve spontaneously. But once viral-induced lymphoproliferative syndrome has occurred, the course is generally rapidly progressive. Death may result from renal, hepatic or pulmonary failure, from hemorrhage due to ulceration of bowel tumor or from CNS involvement.

Treatment

As yet there is no wholly satisfactory treatment for EBV-induced lymphoproliferation. Withdrawal of immunosuppression is associated with spontaneous remission of the tumors in up to 50% of recipients of solid organ grafts; regression is more likely to occur in oligoclonal than in monoclonal tumors. Withdrawal of immunosuppression alone is rarely effective when the lymphoproliferation has occurred after BMT.

Acyclovir may disrupt the replicative lytic cycle of the (linear) virus, and interferon may prevent infection of fresh lymphocytes, but neither approach can significantly modify the growth of already transformed B cells which contain nonreplicating (circular) virus. Nonetheless, remissions have been reported in response to both drugs, although these are more likely in recipients of solid organ grafts and if the tumor is oligoclonal. More recent reports suggest that infusion of monoclonal antibodies to CD21 and CD24 – molecules present on lymphoblasts – may rid the patient of infected B lymphocytes. As with other treatments, this approach seems most effective in solid organ recipients with oligoclonal lymphoproliferative syndrome.

Future perspectives

It seems likely that effective treatment for lymphoproliferative syndrome will be dependent on devising either drugs that can disrupt the function of EBV latent-cycle genes when the virus is in its nonreplicating form in transformed cells or approaches such as adoptive transfer that will restore effective cytotoxic T-cell function to transplant recipients, without inducing graft rejection. Another possibility is the administration of IL2 which induces MHC-unrestricted activated killer cells capable of destroying EBV-infected lymphoblastoid cell lines *in vitro*. If this cytotoxic action can be safely reproduced *in vivo*, eradication of the proliferating lymphoblasts might occur.

Finally, the development of mouse models in which proliferating human lymphoblasts can be studied may help in the development and assessment of novel therapeutic strategies.

Varicella–zoster virus (VZV)

Incidence

Re-activation of VZV infection occurs in 5–25% of solid organ transplant recipients and in as many as 50% of BMT patients. Primary infection develops only rarely, and usually in children.

Epidemiology and pathogenesis

Like other herpesviruses, the probability of VZV re-activation is increased when cell-mediated immunity is compromised. In solid organ transplant recipients, the eruption of clinical shingles often follows upon the initiation or increase of medications to prevent graft rejection. The risk of VZV reaction in BMT patients is even greater because the attendant immune deficiency is more severe and prolonged than that of solid organ transplant patients. Unlike herpes simplex virus (HSV) re-activation where patients with high HSV antibody titer are more likely to develop re-activation, pretransplant VZV antibody titer does not predict post-transplant infection.

Although VZV infection may develop in the first weeks following transplantation, the median time to re-activation is 4–5 months for patients not receiving prophylactic acyclovir. Late graft rejection requiring aggressive immunosuppressive measures is frequently accompanied by VZV re-activation in solid organ transplant patients. In allogeneic BMT, VZV infection may be more common in patients with GVHD, whereas for autologous transplantation, VZV infection is more common in patients with Hodgkin's disease or nonHodgkin's lymphoma.

Manifestations and treatment

Classical cutaneous herpes zoster (shingles) is the most common manifestation of VZV infection among transplant patients. Depending upon the dermatome affected, VZV re-activation may cause post-herpetic neuralgia, cutaneous scarring or corneal opacification. Solid organ transplant recipients rarely develop disseminated or invasive VZV infection.

In BMT patients, VZV infection may present as typical dermatomal herpes zoster, as herpes zoster with cutaneous dissemination, or even as varicella. Serology in these patients indicates past VZV infection. Prior to the availability of acyclovir, cutaneous zoster infection progressed to invasive disease in up 50% of patients, and 10% died. The consequences of visceral dissemination included hepatitis, pancreatitis and encephalitis; however, VZV pneumonia was the usual cause of death.

Intravenous acyclovir is the most effective and least toxic therapy for VZV infection in transplant patients. In BMT patients, prophylactic administration of acyclovir can prevent VZV re-activation, but re-activation is common upon discontinuing the drug. Consequently, prophylactic administration of acyclovir to prevent VZV infection is not recommended. Vidarabine and α-interferon are effective against VZV infections but they are secondary treatments since they are more toxic and less efficacious. Their use is limited to situations when VZV resistance to acyclovir is suspected.

In those patients without prior VZV infection, varicella–zoster immune globulin administered within 72 h of VZV exposure may prevent or modify illness. Varicella vaccine is a promising therapy for children who may require transplantation but have not yet acquired their primary immunity.

Herpes simplex virus (HSV)

Incidence

Re-activation of latent HSV infection has occurred in between 30 and 90% of seropositive transplant patients. The majority have suffered a mild, usually self-limited, illness with fever accompanied by lesions of the oral or genital mucosa. However, in patients undergoing BMT, HSV disease has been more severe.

Epidemiology and pathogenesis

Less than 5% of HSV illness following transplantation has been due to primary HSV infection. There are two possible sources of primary HSV exposure: intimate contact with an individual shedding HSV or transfer of HSV through the transplanted organ. Transfer of HSV by liver, pancreas, heart and kidney transplantation has been documented. Presumably, the virus, latent in neural tissue within the organ, became re-activated and initiated primary HSV illness in the host.

Between 30 and 100% of adults are HSV seropositive, and 15–30% of these experience recurrent vesicular eruptions. Although the events leading to HSV re-activation are incompletely understood, the combination of high prevalence of HSV seropositivity and iatrogenic immunosuppression have undoubtedly promoted HSV illness among transplant patients. Interestingly, a high antibody titer to HSV pretransplant has been linked to a high risk of viral re-activation.

Clinical manifestations and treatment

The incidence of HSV infection peaks during the first weeks following transplantation when cell-mediated immunity is most suppressed. The classical perioral vesicular eruption is the typical manifestation of HSV infection in solid organ transplant recipients. Genital HSV infection is also common. In BMT patients, the diagnosis of HSV infection may be problematic because the classic lesions are often superimposed on mucositis caused by the pretransplant chemotherapy and/or radiotherapy.

Visceral dissemination of HSV has been rare following kidney and liver transplantation, more frequent following heart, lung and heart–lung transplantation, and most common following BMT. Manifestations of visceral dissemination include esophagitis, pneumonia, hepatitis and encephalitis; HSV esophagitis and pneumonia are thought to develop from contiguous spread of oral/perioral disease. HSV has been isolated from 5% of BMT patients with pneumonia. Often a co-isolate with cytomegalovirus (CMV), the role of HSV in the pathogenesis of pneumonia is unclear.

The incidence of hepatitis or encephalitis due to HSV has been less than 0.5%. Unlike HSV pneumonia or esophagitis, the onset of hepatitis has not always followed oral/perioral lesions. In the first weeks following liver transplantation, immediate liver biopsy is recommended for patients who develop hepatic dysfunction, since prompt initiation of acyclovir has arrested the progression of HSV hepatitis to fulminant and usually fatal liver failure.

Treatment

Acyclovir is the principal treatment for HSV infection. It is effective for prophylaxis as well as treatment of established infection. Because the likelihood of disseminated HSV disease is enhanced by immunosuppression, acyclovir prophylaxis is routine for seropositive recipients of BMT. In solid organ transplantation, where immunosuppression is neither as intense nor as prolonged, opinion is divided on the necessity for prophylaxis. If prophylaxis is used, HSV re-activation often occurs once acyclovir is discontinued. In addition, approximately 10% of patients develop HSV re-activation even while receiving prophylaxis.

Disconcerting reports of acyclovir-resistant HSV, particularly among human immunodeficiency virus (HIV)-infected patients, have been confirmed. Vidarabine and foscarnet, an investigational agent, may be effective alternatives.

Hepatitis Viruses

Hepatitis A virus or hepatitis delta virus have been uncommon sources of hepatitis in transplant patients. In contrast, infections with hepatitis B virus (HBV) or hepatitis C virus (HCV) have been frequent. Transplant patients may be more susceptible to the complications of HBV and HCV infection, but the uncertainty regarding epidemiology and pathogenesis of these infections complicates decisions regarding the suitability of patients with hepatitis for transplant.

More than a few impediments have obscured understanding the consequence of hepatitis. Until the recent development of the anti-HCV assay, the course of patients with HCV infection could not be distinguished from the course of patients with other hepatic abnormalities. Even though patients with HCV can now be identified, the current assay is insensitive. Meanwhile, assays for HBV infection, long considered sensitive and specific, do not adequately detail the potential for HBV re-activation and infectivity. Thus, the outcome of patients previously infected with HBV or HCV and reexposed to one or both of these viruses during transplantation is presently unclear. Transplant patients may also have serologic evidence of infection by both viruses, and distinguishing the active agent by serologic assays is difficult. Finally, liver dysfunction in the post-transplant patient is a common event and other sources of hepatic injury are often present concurrently.

Incidence

The incidence of HBsAg positivity varies widely. In Taiwan where 15–20% of the population are HBsAg-positive, 93% of BMT recipients were positive prior to transplant. In the USA and Italy, 0.1% and 6% of BMT patients respectively were antigen-positive. Among renal dialysis patients, 3% in the USA and 46% in France were chronically infected. The prevalence of HBsAg seropositivity among liver transplant recipients is skewed by the particular admission requirements to transplant centers. Because of the high rate of HBV re-activation and subsequent hepatic failure, HBsAg-positive patients are not accepted for liver transplantation in some programs. Nonetheless, HBV chronic active hepatitis with hepatic failure is among the most common indications for liver transplantation.

Non-A, non-B chronic active hepatitis with cirrhosis is the most commonly diagnosed condition in patients undergoing liver transplantation. Anti-HCV has been identified in half of these patients. In addition, 37% and 27% of patients undergoing transplant for Laennec's cirrhosis and chronic HBV infection respectively were seropositive for HCV. Among kidney transplant patients, up to 26% tested positive for anti-HCV before transplantation. The incidence of HCV infection in BMT recipients is uncertain. The widespread use of IVIG preparations has been blamed for the prevalence (up to 100% of BMT patients) of anti-HCV seropositivity; however, immunoglobulin concentrates have not been implicated as a source of HCV transmission.

Epidemiology and pathogenesis

HBV or HCV infection during transplantation is either due to transmission through infected blood or transplanted organ in a previously uninfected patient or more commonly to re-activation or progression of illness in a previously infected patient. There is little evidence that prognosis differs following either primary or secondary infection. Virus re-activation is fomented by the immunosuppression required to establish the allograft. Re-activation does not immediately lead to changes in liver function, and indeed laboratory evidence of hepatocellular injury may not emerge until withdrawal or reduction of immunosuppression.

Within the HBV-infected group, patients at greatest risk for re-activation are asymptomatic HBsAg-positive carriers. Anti-HBs-positive but HBsAg-negative patients are considered immune. However, the intensive immunosuppression administered for bone marrow transplantation, has promoted HBV re-activation in immune patients.

Transmission of HCV by allograft has also been documented, but the frequency of this event has not yet been established. Transfusion of contaminated blood has been assumed to be a major source of HCV infection in transplant patients. Infection by other routes is possible.

Details of HCV pathogenesis are largely unknown. The assay for anti-HCV antibody has not proved enlightening in that regard, since anti-HCV does not appear for weeks to months after infection, and seropositivity may persist for years without evidence of clinical disease. Alternative assays may provide the missing details. For example, serum HCV RNA can now be measured and its presence correlates with infectivity. Although prolonged viremia is common in patients with chronic hepatitis, it remains to be determined whether clearance of HCV viremia indicates permanent immunity or merely a subclinical state capable of re-activation.

Clinical manifestations and treatment

Hepatitis B

The reported incidence of HBV re-activation following bone marrow transplantation has been between 5 and 12%. However, these figures undoubtedly underestimate the true incidence of re-activation.

Most HBsAg-positive BMT recipients develop transient abnormalities in liver function tests, but clinically significant hepatic dysfunction is uncommon. Patients who were HBsAg-positive pretransplant rarely cleared antigenemia post-transplant, whereas antigenemia appearing post-transplant persisted for months but almost always resolved. Regardless of whether HBsAg was cleared, intermittent elevation of liver enzymes persisted in over half these patients. The long-term prognosis for these patients is presently unknown. For now, HBsAg seropositivity is not an absolute contraindication to bone marrow transplantation.

By 8 years after kidney transplant, 25–40% of HbsAg patients have developed chronic active hepatitis, cirrhosis, or hepatocellular carcinoma. Data from one study in which HBV re-activation was documented by PCR showed that HBV DNA was detected post-transplant in the 20% of HBV-immune patients. Although these patients remained HBsAg negative, their incidence of chronic hepatitis was equivalent to HBsAg seropositive patients.

Following orthotopic liver tranplantation, more than 80% of HBsAg-positive patients will develop re-infection. Overall survival in these patients was significantly reduced compared to other liver-transplant patients. HBV-immune patients have not developed re-infection, and this has led to therapies designed to eradicate HBsAg antigenemia: passive immunization with anti-HBs immunoglobulin, active immunization with HBV vaccine, combined active and passive immunization, and α-interferon. It appears that HBV re-infection can be delayed by passive and combined immunization, but late re-infection has occurred. Nonhepatic reservoirs of HBV have been implicated as the cause of re-infection.

Hepatitis C

There are few reports that document the course of HCV infection in transplant patients. In one, 22% of renal transplant recipients tested repeatedly positive for anti-HCV; 50% of these developed chronic hepatitis. Cirrhosis would be the expected outcome in half of the patients developing chronic hepatitis. Most HCV seropositive patients undergoing liver transplant remained seropositive afterwards. Again hepatitis and cirrhosis were noted, but the impact on survival has not been established.

Transplantation of solid organs from HCV seropositive donors has infected over half the recipients. Typically, liver disease progressed to subfulminant hepatic failure or became chronic. At present, transplantation of organs from anti-HCV-positive donors is contraindicated.

Therapy

Some patients with chronic hepatitis secondary to HBV or HCV infection have responded to α-interferon therapy, and this response was occasionally sustained once therapy was completed. Whether α-interferon will improve survival is unknown. Vidarabine and acyclovir have also been effective in a proportion of patients with chronic HBV although their activity was less predictable and vidarabine was more toxic than α-interferon. α-Interferon may enhance graft-versus-host mechanisms in BMT recipients. Limited use of α-interferon in patients undergoing orthotopic liver transplantation has failed to prevent HBV re-infection.

Prevention

Prevention of HBV and HCV infections is the single most effective means of controlling post-transplant hepatitis. In France, immunization against HBV infection has not been the common practice; 46% of chronic dialysis patients are HBsAg positive and 71% of the HBsAg-negative patients are anti-HBs positive. In the USA, measures to prevent HBV infection including isolation of infected patients and HBV immunization have reduced the prevalence to 2.7% and 12% for HBsAg and anti-HBs respectively. The ability to test blood for anti-HCV antibodies will further reduce the incidence of transmitted hepatitis.

Polyomavirus

Primary and usually clinically undetected infections with BK virus (BKV) and JC virus (JCV) occur during childhood, since measurable antibodies to JCV and BKV are present in 70% and 90% of adults respectively. Such persistence of seropositivity suggests equal persistence of infection, and autopsy studies have identified the viral genone, presumably in a latent state, in various tissues including the kidney. Thus most infections with JCV or BKV in transplant patients are attributed to viral re-activation. However, seronegative renal transplant recipients have acquired primary infection from the transplanted kidney.

Situations in which host immunity, especially cell-mediated immunity, are compromised favor the re-activation of BKV and JCV. Some 4–8 weeks following kidney transplantation, 26–44% of patients develop viuria. The incidence is higher in patients treated with antilymphocyte serum to prevent graft rejection. Up to half of BMT patients also develop viuria with BKV or JCV. Viuria onset has been noted between weeks 2 and 8 and typically persisted for 3–4 weeks. JCV was isolated from 6.7% of seropositive patients, while 55% of BKV-seropositive patients excreted the virus. This disparity between the frequency of JCV viuria and BKV viuria is unexplained and contrasts with the experience in renal transplant patients where the incidence of viuria is similar.

BKV or JCV re-activation has been associated with stenosis of the ureteral anastomosis in kidney transplant patients, since virus has been identified in the tissue at the stenosis site. In BMT patients, BKV has been linked to hemorrhagic cystitis, with more than 80% of viuric patients developing this complication, an incidence four times that of patients who were not viuric. Compared to autologous or syngeneic transplants, the likelihood of hemorrhagic cystitis was markedly increased among viuric patients undergoing allogeneic BMT. The explanation for this is not apparent. Presumably the more profound immunosuppression that occurs with allogeneic transplantation contributes to the virulence of BKV re-activation.

BKV/JCV have also been identified in the lesions of multifocal leucoencephalopathy, a progressive encephalitis which occurs in a minority of patients on long-term immunosuppression. The contribution of these viruses to the disease state is unclear, but responses to intrathecal cytosine arabinoside have been reported. In general, however, there is no effective therapy identified for BKV or JCV.

Respiratory Viruses

Respiratory syncytial virus (RSV)

Infection with parainfluenza and RSV can lead to severe, even fatal, pulmonary disease in transplant patients. Although RSV is a common pathogen among children, infection confers only transient resistance; most adults will be susceptible at the time of transplant. Some 20–50% of patients exposed to RSV in the hospital will be nosocomially infected, and RSV infection in transplant patients has coincided with community outbreaks of RSV disease. The communicability of RSV emphasizes the need for stringent isolation measures.

Pneumonia is the most serious consequence of RSV infection. RSV pneumonia has seldom been fatal in solid organ transplant recipients, but the mortality rate among BMT patients has been 50%. The importance of RSV as a pathogen appears to be increasing. In one transplant center, it was identified in 27% of patients with pulmonary disease.

After BMT, the onset of pneumonia is early in the post-transplant course, usually before bone marrow engraftment. Sinusitis and otitis media often precede the pulmonary symptoms, an important diagnostic clue, since other viral pneumonias are not accompanied by these upper respiratory tract findings. Diagnosis of infection before the onset of pneumonia has been made by finding RSV antigen in upper airway secretions. RSV infections that occur 100 days after transplant have not been fatal.

Identification of RSV prior to the onset of pulmonary symptoms and treatment with nebulized ribavirin has arrested disease progression. Once the lower respiratory tract is clinically involved, ribavirin treatment has been less effective. Improved methods of drug delivery, including intravenous ribavirin, may improve the outcome in this latter group.

Adenovirus

Incidence

Although adenovirus is not recognized as a major cause of post-transplant infections, it has been isolated in 5–10% of patients following solid organ transplantation and between 5% and 25% of patients undergoing allogeneic BMT. Less than half of the patients in whom adenovirus was isolated had symptoms of viral infection and in some that were ill, CMV or HSV was isolated concurrently, confusing the assignment of cause. Nonetheless there is undeniable evidence that adenovirus has caused invasive and fatal disease in transplant recipients.

Pathogenesis and epidemiology

Most patients are seropositive for adenovirus prior to transplant. Tonsils, adenoids, other lymphoid tissue and kidneys are recognized as latent virus reservoirs. This, plus the absence of a contagious source, suggest that re-activation of latent virus has been the usual cause of adenovirus infection following transplantation. Rarely, latent virus residing in the transplanted tissue has been implicated as an infectious source.

Although many adenovirus serotypes have been isolated post-transplant, not all isolates have been associated with invasive disease. Several serotypes, notably types 11, 12, 34 and 35, which rarely cause community outbreaks, have been isolated in immunosuppressed patients with pneumonia. Adenovirus type 5, implicated as a cause of intussusception and sporadic cases of hepatitis, has commonly been recovered from liver transplant recipients with adenovirus-associated hepatitis. The disproportionate representation of certain serotypes in transplant patients with invasive disease is unexplained.

Immunosuppression promotes adenovirus infection. In liver transplant patients, invasive disease has been most common when additional therapy was needed to prevent graft rejection. Severe GVHD, which is immunosuppressive in itself and is managed by immunosuppressive agents, has been noted to promote adenovirus infection in BMT patients.

Clinical manifestations and treatment

Adenovirus infections have occurred as early as the second week or as late as several months following transplantation. The virus has been isolated from the throat, stool, urine, blood and tissue parenchyma. Manifestations include gastroenteritis, hemorrhagic cystitis, hepatitis, pneumonia, meningoencephalitis and hemophagocytic syndrome. In some circumstances, infections have been mild and recovery complete. Renal dysfunction suggestive of kidney allograft rejection has followed adenovirus viuria and hemorrhagic cystitis. Renal failure requiring dialysis has complicated the course of BMT patients, and in these patients adenovirus was isolated from the renal parenchyma. Adenovirus hepatitis following liver transplantation and pneumonia following BMT have been almost uniformly fatal.

At present there is no specific therapy for adenovirus infection. Treatment consists of aggressive support. There are anecdotal reports of successful management of disseminated adenovirus infection with immunoglobulin and other nonspecific humoral factors. One patient, after undergoing liver transplant, developed adenovirus hepatitis and liver failure and was successfully retransplanted. Undoubtedly measures such as these will prevail until adenovirus-specific treatment is available.

Immunization

Vaccination

Although viral illnesses occur with high frequency and often devastating effect in transplant recipients, vaccination presently has only a limited role in disease prevention. There are a number of reasons why this should be so. Transplant patients are heavily immunosuppressed, and make poor antibody or T-cell responses to killed/subunit vaccines. Administration of live attenuated vaccines, which are potentially more immunogenic, is fraught with peril, since the immune response may be so feeble that even attenuated viruses may produce fatal disease. Finally, effective vaccines to herpesviruses in man are simply not available. Since this is precisely the virus group that is responsible for most viral-related morbidity and mortality after transplantation, vaccination can at best have a limited impact on outcome. For example, several studies have examined the effect of attenuated CMV (Towne) strain in renal transplant recipients. Unfortunately, the vaccine induces minimal specific CTLs in seronegative renal transplant recipients, many of whom also fail to make antibody. In clinical trials there was no benefit to seropositive recipients, and no change in the incidence of infection in seronegative recipients. Morbidity was, however, decreased in seronegative recipients.

Despite these limitations, vaccination should still be considered for all transplant patients. In particular, killed or subunit vaccines to poliomyelitis and hepatitis B are readily justified and often produce protective levels of antibody. After BMT, antibody responses to vaccines may be greatly enhanced by immunizing both donor and recipient before BMT to allow adoptive transfer of high-titer responses. In contrast, live vaccines should be withheld whilst patients are receiving immunosuppression. After BMT especially, such vaccines should not be given for at least 2 years; in the presence of chronic GVHD, this time period may need to be extended indefinitely.

Passive immunization

While vaccination has a limited role after transplantation, it has been repeatedly suggested that passive immunization with pooled immunoglobulins reduces the incidence or severity of disease associated with viral infection or re-activation after transplantation in general and after BMT in particular (see section on CMV above). These claims have yet to be fully substantiated.

See also: Adenoviruses; Antivirals; Cytomegaloviruses; Epstein–Barr virus; Hepatitis A virus; Hepatitis B viruses; Hepatitis C virus; Herpes simplex viruses; Immune response; Interferons; JC and BK viruses; Latency; Parainfluenza viruses; Respiratory syncytial virus; Varicella–zoster virus.

Further Reading

Degos F *et al.* (1988) Hepatitis B virus and hepatitis B-related viral infection in renal transplant recipients. *Gastroenterology* 94: 151.

Fischer A *et al.* (1991) Anti-B-cell monoclonal antibodies in the treatment of severe B-cell lymphoproliferative syn-

drome following bone marrow and organ transplantation. *N. Engl. J. Med.* 324: 1451.
Grundy JE, Shanley JD and Griffiths PD (1987) Is cytomegalovirus interstitial pneumonitis in transplant recipients an immunopathological condition? *Lancet* ii: 996.
Winston DJ and Gale RP (1991) Prevention and treatment of cytomegalovirus infection and disease after bone marrow transplantation in the 1990s. *Bone Marrow Transplantation* 8: 7.

TRANSPOSABLE BACTERIOPHAGES

See Mu and related bacteriophages

TREE SHREW HERPESVIRUSES

Gholamreza Darai and Angela Rösen-Wolff
Institute of Medical Virology
University of Heidelberg
Heidelberg, Germany

History

The tree shrew (*Tupaia*), a member of the family *Tupaiidae*, is regarded as one of the earliest prosimians bridging the gap between insectivores and primates. The first discovery of herpesvirus-like particles of the tree shrew was reported by Mirkovic in 1970. This *Tupaia* herpesvirus (THV-1), isolated from a degenerating lung tissue culture from an apparently healthy animal, was characterized electron microscopically by McCombs in 1971. Between 1979 and 1985, six additional THVs (THV-2–7), were isolated and characterized in our laboratory.

The second THV was isolated in 1979 by us from a degenerating cell culture of a high-grade malignant lymphoma of a moribund 8-year-old tree shrew and termed THV-2. The third (THV-3) was isolated from a degenerating cell culture of a Hodgkin's sarcoma (Hodgkin's disease, lymphocytic depletion type) from a moribund 9-year-old tree shrew. The other four THVs (THV-4–7) were isolated from spleen tissues of moribund animals aged 4–11 years.

Taxonomy and Classification

THV is a still unclassified member of the family *Herpesviridae*. According to pertinent genetic and phenotypic properties, it differs significantly from the alpha, beta and gammaherpesviruses. Therefore, it should be proposed as a separate genus of the family *Herpesviridae*.

Properties of the Virion

Naked viral capsids have a diameter of about 100 nm. Extracellular herpesvirus particles have an envelope studded with small surface projections. Beneath this envelope, an electron dense zone of high density is detectable. Some of the enveloped particles contain several viral capsids. The diameter of the envelope ranges from 200 to 350 nm.

Properties of the Virus Genome

THV genomes have attracted attention because they consist of a unique linear double-stranded DNA sequence without any detectable long inverted repeat sequences longer than 40 bp within, or at either end of, the viral genome. A precise analysis of the DNAs of THV strains 1–4 by analytical ultracentrifugation gave a value of $\rho = 1.724\,\mathrm{g\,ml^{-1}}$. The UV-absorbance–temperature profile of the DNA in 0.1 × SSC gives a T_m value of 81.2 ± 0.8°C corresponding to a G+C content of 64.5 ± 1.9% (where 1 × SSC = 0.15 M NaCl/0.015 M sodium citrate).

Electron microscopic measurement of the contour length of viral DNA molecules reveals molecular sizes of about 194.8 ± 5, 200.8 ± 3, 196.3 ± 3 and 196.3 ± 5 kbp for THV strains 1–4 respectively. Physical maps of THV-2 genome have been constructed.

THV-2 DNA synthesis reaches a maximum between 24 and 36 h postinfection, preceded by a transient stimulation of host cell DNA synthesis. Linear concatemeric and/or circular viral DNA molecules are found during DNA synthesis. The nucleotide sequence of terminal DNA regions of the molecularly cloned THV genome is characterized by a relatively high G+C content of 75% and furthermore contains numerous short repeat elements. A sequence (A3C8AAAGGCACC6G5) postulated to be a consensus signal for site-specific endonucleolytic cleavage in terminal regions of the genomes of herpes simplex virus 1 and 2 (HSV-1, HSV-2), Epstein–Barr virus (EBV), and varicella–zoster virus occurs at the terminal region of the THV-2 genome. It is intriguing that the sequences for HSV and THV are located at similar distances from the genomic termini (at nucleotide positions 432 and 470 bp). This consensus sequence, in combination with several short repeats, which are asymmetrically bracketed by GC-rich arrays, may play a role in forming the mature ends of THV DNA. Alternatively, this consensus sequence could be the site of binding of the cleavage enzyme or the site for the processing of concatemeric viral DNA molecules into unit-length molecules and for the packaging of the processed unit-length molecules into viral capsids.

THV-1–4 DNAs cause infections in tree shrew embryonic fibroblasts.

Virus Proteins

The polypeptide patterns of the three purified THVs (THV-1–3) are remarkably similar, each consisting of at least 35 polypeptides ranging in molecular mass from 120 to 230 kD. While the majority of analogous polypeptides of the three viruses are of indistinguishable electrophoretic mobility, some (e.g. 82–86 kD polypeptides) showed small differences in apparent molecular mass which were characteristic of the viral strain. By comparative sodium dodecyl sulphate (SDS)–polyacrylamide gel electrophoresis, it is possible to distinguish the THV isolates from each other. At least five glycoproteins are found in purified THV virions. Two-dimensional electropherograms reveal at least 47 discernible protein spots, some of which are specific for a given THV isolate and which are detectable in lysates of THV-infected cells.

Virus Enzymes

A protein kinase activity is associated with THV. A divalent cation such as Mg^{2+} or Mn^{2+} is necessary as well as ATP as the phosphate donor. The predominant sites of phosphorylation are the β-OH groups of the serine and threonine residues of these THV proteins. Distinct THV polypeptides (molecular masses of 100, 82 and 53 kD) were found to be phosphate acceptor proteins when 5 mM Mg^{2+} was used. At a higher Mg^{2+} concentration (20 mM), additional viral proteins (220, 71, 31 and 20 kD) were phosphorylated.

Geographic and Seasonal Distribution

THV infection of tree shrews is endemic only to South and Southeast Asia corresponding to the geographic distribution of the animal. Seasonal dependence for viral isolation has not been documented.

Host Range and Virus Propagation

The host range of THVs *in vitro* clearly shows that tree shrew embryonic fibroblasts are the most susceptible cells for viral replication, indicating that this animal is the natural host. Primary rabbit kidney cells, human foreskin fibroblasts and marmoset skin fibroblasts are less susceptible than tree shrew cells, whereas cells from rodents are not susceptible. Tree shrew embryonic fibroblasts are the cells of choice for propagation and plaque assays of THVs.

Genetics

A high degree of DNA sequence homology between

the different THV isolates has been detected using DNA/DNA hybridization, heteroduplex mapping of the viral DNA molecules and generation of intratypic recombinant viruses. This indicates the close genetic relationship between the individual THVs. The seven isolates of THV are genetically grouped into five strains according to DNA fragmentation patterns. THV strains 1, 2, 3 and 5 comprise isolates 1, 2, 3 and 7 respectively, whereas THV strain 4 comprises isolates 4, 5 and 6. However, minor differences between isolates of THV strain 4 with respect to the mobility of the individual DNA fragments of different restriction enzymes are detectable.

Evolution

As the original host (family *Tupaiidae*) is placed at a low stage of the evolutionary tree of mammals, it seems rational to grade the THVs on the same evolutionary level beyond the mammalian herpesviruses.

Serological Relations and Variability

Serological cross-reaction between different THV strains can usually be demonstrated. Cross-reaction is sufficiently strong and can be detected with antisera against individual viral strains. However, it is possible to distinguish THV strains from each other in neutralization tests by significant differences in the resulting titers.

Epidemiology

THV epidemiology has not been the subject of intensive studies. However, analysis of the sera of a limited number of apparently healthy animals in capture reveals that less than 1% of tree shrews have neutralizing antibodies against THV.

Transmission and Tissue Tropism

The natural route of transmission of THV is probably the same as for other herpesviruses. Most infections occur as a result of contact between open lesions and/or moist surfaces. However, spontaneous re-activation of latent virus should also be considered. The target organ for latent THVs is the spleen, since this is the only organ from which infectious viruses can be recovered even after a long period of chronic infection of tree shrews (24 months) and rabbits (14 months). Analysis of the genomes of recovered viruses by different restriction enzymes shows the same DNA fragment patterns when compared to the DNA of the originally inoculated viruses.

Pathogenicity

The pathogenicity of THV-1–4 for tree shrews as their indigenous host was studied using juvenile, young adult and adult animals inoculated intravenously or intraperitoneally. *In vivo* pathogenicity studies show that the tree shrew is most susceptible to these viruses. Intravenous inoculation leads to death of infected animals (lethality 100%). In contrast, the majority of intraperitoneally inoculated animals survive the infection (lethality 25%). The major pathology is inflammatory hemorrhagic necrosis of the lungs. High virus titers are detectable in the tissues and whole blood of the infected animals.

It is remarkable that two of the four known THV strains were directly isolated from metastasizing tumors, in one case from a malignant lymphoma and in the other case from a Hodgkin's-like disease. However, only one malignant lymphoma has been induced experimentally in tree shrew using THV-2. The tumor developed 3.5 years after intraperitoneal administration of the virus. Infectious virus that was genetically identical to the original inoculum was recovered from cultured tumor cells. The failure to routinely induce malignant lymphomas in the tree shrew after experimental inoculation with THV-2 and -3 is probably due to the short period of observation, since tree shrews can reach an age of 14 years.

The response of rabbits to THV-1, -2, -3 and -4 infections is well documented. THV-2 and -3 induce in newborn New Zealand rabbit hyperplasia of the thymus (80%), which in 8% of cases develop malignant thymomas. Infectious virus can be recovered only from established spleen cultures of the infected rabbits. The spontaneous degeneration of the rabbit spleen cultures always develops by the first or second tissue culture passage. The genome of the recovered viruses is identical when compared to the genome of the originally inoculated viruses.

Clinical Features of Infection

Clinical illness usually appears in juvenile animals infected with THV-1–4 on the second day after inoculation. The general symptoms grow continuously worse. Intravenous inoculation of these viruses leads to death of the infected animals. Death occurs from 5 to 18 days postinfection as a result of inflammatory hemorrhagic necrosis of the lungs. High virus titers are detectable in the tissues and whole blood of infected animals.

Pathology and Histopathology

It is of particular interest that THV-2 and -3 were isolated from degenerating cultured tumor cells from a high-grade malignant lymphoma and a Hodgkin's-like sarcoma respectively. The lymphoma was detected in an adult (7–8-year-old) female tree shrew. Histopathological investigation revealed a discordant hyperplasia of lymphoreticular tissues with formation of nodal follicle-like proliferates. Two types of cell were observed. The first was small with polymorphic nuclei of clear appearance and small nucleoli; the second was of medium to large size with large rounded nuclei. The chromatin showed marginal aggregations with marginal nucleoli. Mitosis was relatively frequent. Several thick reticulin fibers were sparsely distributed after silver-staining. Tumor cell infiltration was observed in the parenchyma of the pancreas, the peribronchial lymph follicles, the connective tissue of the pelvis of the kidney, the subserous membrane of duodenum and the cerebral meninges.

A generalized lymphoproliferative disorder resembling human Hodgkin's disease was observed in a female tree shrew of 9 years of age. Histological examination revealed hyperplasia of the lymphoreticular tissue with a polymorphic cellular population consisting of immunoblasts, numerous Hodgkin-like cells and Reed–Sternberg-like cells, eosinophilic granulocytes and a marked lymphocytic depletion. The spleen showed a diffuse infiltration pattern; liver and peripelvin tissue of kidneys were likewise infiltrated.

Future Perspectives

Many other interesting aspects of THV DNA replication strategy, such as transcription, posttranscriptional and post-translational processing, viral enzyme activity, etc., are still not understood and deserve intensive study in the future. Furthermore, genetic analysis of other tree shrew viruses, such as tree shrew retro-, rhabdo-, paramyxo-, and adenovirus, which have all been isolated in our laboratory, is necessary for the final determination of their evolutionary roles.

See also: Epstein–Barr virus; Herpes simplex viruses.

Further Reading

Darai G *et al.* (1981) DNA of tupaia herpesviruses. In: Becker Y (ed.) *Herpesvirus DNA*, p. 345. The Hague: Nijhoff.
Darai G *et al.* (1982) Tupaia herpesviruses: characterization and biological properties. *Microbiologica* 5: 185.
Koch H-G (1985) Molecular cloning and physical mapping of the tupaia herpesvirus genome. *J. Virol.* 55: 86.

TUMOR VIRUSES – HUMAN

Herbert Pfister and Bernhard Fleckenstein
Institute for Clinical and Molecular Virology
Friedrich-Alexander University
Erlangen-Nürnberg, Germany

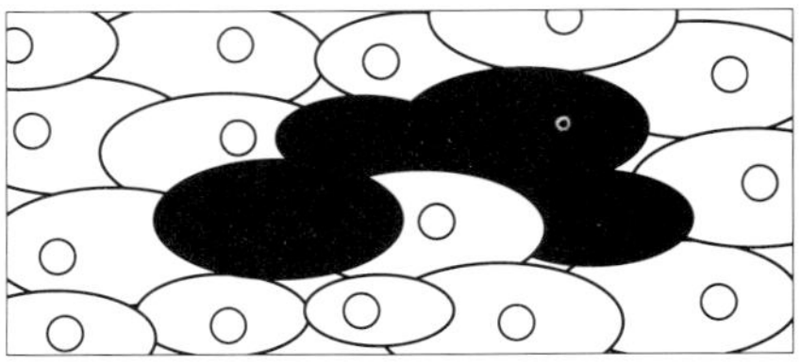

Criteria for Causal Relationship between Virus and Cancer

Although there are numerous animal models for viral oncogenesis, only in rare cases is the virus sufficient by itself for tumor induction. In those systems, experimental infection may lead to sizeable tumors within a few weeks, thus leaving no doubt about the etiologic role of the particular pathogen. In humans, none of the viruses incriminated as being oncogenic appear to be independent of other factors. Many years, and often several decades, pass by before tumors develop from primary infection in a small number of infected individuals. Monoclonal tumors arise from the pool of virus-infected cells. This suggests that the virus is, at best, one of several factors which together increase the probability of a cell undergoing malignant transformation. Possible cofactors are chemical or physical carcinogens. Furthermore, the oncogenic activity of the virus may be restricted to a specific genetic background of the patient or to a susceptible stage of cell development. These complex interrelations and the ethical ban on experiments in humans render it extremely difficult to prove the causative role for the virus.

Viruses may contribute to tumor development at various stages of the multistep carcinogenesis process and by different mechanisms. If viral activity is necessary, the virus should be found at some stage in every case in which the tumor occurs. Viral footprints do not have to appear in cancer cells, however. Human immunodeficiency virus (HIV), for example, significantly increases the risk of development of Kaposi sarcomas and B-cell lymphomas, probably because of its suppressive effect on the immune system. For some herpesviruses and papillomaviruses, a 'hit and run' mechanism has been proposed, which implies virally induced irreversible damage at some point early in tumorigenesis, but no role later on in the maintenance of the malignant phenotype; the virus may therefore completely disappear from the cancer. Viruses might act as mutagens or, even more indirectly, by increasing the risk of mutations via induction of cell proliferation in the course of chronic inflammatory reactions. The etiologic significance of a virus is very difficult to prove in this case. Large-scale and long-term prospective epidemiological studies are the only way to define the risk of a given infection. As the virus may be present for short periods only, the infection is best monitored by the detection of specific antibodies. Previous studies focused on anti-capsid antibodies, but with many DNA viruses the production of particles and oncogenic transformation appear to be mutually exclusive so that it might be more promising to screen for antibodies against early non-structural viral proteins.

Following the classical concepts of virus-induced cell transformation, a tumor virus introduces new genetic material into host cells by establishing a persistent infection. The expression of one or more viral oncogenes may then affect cell proliferation and/or differentiation. This can be tested *in vitro* by transformation assays which may result in cells that are tumorigenic in nude mice. If a virus acts in this way, one has to find viral genomes as well as virus-specific transcripts and oncogenic proteins in transformed cells and also in cancer cells if the viral functions are essential for maintaining the malignant phenotype. Tumor viruses may alternatively contribute to carcinogenesis by inserting viral transcription control elements in the vicinity of cellular proto-oncogenes. Viral DNA has to be detectable in these cases adjacent to relevant proto-oncogenes of cancer cells. A harmful integration is typically a very rare event among many random insertions, which is selected for by tumor growth *in vivo*. This mechanism cannot be mimicked by *in vitro* transformation assays.

The physical association of viral genes with tumor cells, in combination with data on oncogenic activity from cell culture or animal experiments, is usually regarded as most convincing evidence for a role for a virus in human carcinogenesis. However, even better proof of the effective contribution of the virus is obtained from longitudinal epidemiological studies. Finally, the most convincing argument in favor of a necessary role for a tumor virus will be cancer prevention by intervention directed at the virus such as specific vaccination.

Burkitt's Lymphoma

This malignant lymphoma which was recognized by Denis Burkitt in 1958 as a distinct disease entity in Africa was the first human neoplasm to be linked with a viral etiology. Burkitt's lymphoma (BL) is the most frequent type of childhood tumor found in the hot and humid lowland areas of Central and Eastern Africa, but sporadic cases of the disease, defined by the peculiar histopathology and chromosome translocations, occur throughout the world. Although, in the majority of cases, BL presents in younger children around 8 years of age as a unilateral swelling of the jaw, the disease is usually multifocal at diagnosis due to early metastasis; it characteristically involves liver, kidneys, ovaries or testes, lymphoid tissues in the gut, and endocrine or exocrine glands. In older children and young adults, mostly in the sporadic cases outside Africa, abdominal masses are the first sign of disease.

Lymphoid tumor cell lines derived from BL biopsies led to the discovery in 1964 by Epstein and colleagues of a new herpesvirus, termed Epstein–Barr virus (EBV). The cell lines usually contain EBV genomes as nonintegrated covalently closed circular double-stranded DNA (dsDNA) genomes of about 172 kb in high multiplicity; some lymphoblastoid cell lines carry the EBV genomes integrated into the cellular genomes. EBV is capable of immortalizing human B lymphocytes. The transformed cells express a viral nuclear antigen complex, termed EBNA, which consists of at least six proteins, and in addition latent membrane proteins (LMP). Genetic experiments identified EBNA-2 and LMP as the growth-transforming factors. BL-derived lymphoblastoid cell lines typically express EBNA-1, while, remarkably, the other EBNA proteins and LMP are not detectable. Some lymphoblastoid lines are in a semipermissive state; they produce mature herpesvirus particles spontaneously or can be induced to do so by substances such as the tumor-promoting phorbol ester, phorbol 12-myristate 13-acetate (PMA). BL lymphoma cells and lymphoblastoid tumor cell lines have the surface antigen markers of B lymphocytes; this correlates with a narrow host range of the virus; the virus selectively adsorbs and penetrates by binding the virion envelope glycoprotein gp350/220 to the complement receptor 2 (CR2; CD21) of B lymphocytes or their precursors.

EBV-producing lymphoblastoid cell lines provided the basis for seroepidemiology. It showed that EBV is a ubiquitous virus that occurs in the majority of populations on a worldwide scale. Primary infection can be accompanied by developing infectious mononucleosis or may remain inapparent without overt disease. BL typically follows early seroconversion during infancy; tumor-bearing children have high antibody titers against structural and early EBV proteins.

In spite of a general association between development of BL and the viral markers, the exact role of EBV in tumor pathogenesis has remained controversial. More than 90% of the tumors in areas of endemic BL contain EBV DNA and express EBNA, and the children have generally high titers of antibodies. Sporadic forms of BL, however, that occur in other parts of the world contain the viral DNA and protein markers only to about 15–20%, and elevated antibody titers are not observed in the case of tumors devoid of viral markers. Thus, the presence of EBV persisting in certain compartments of B cells or their precursors may be a risk factor for the development of BL, but it cannot be the sole driving force for oncogenesis. So far, it is also difficult to exclude strictly the possibility that EBV is only a passenger of the tumor without contributing to its generation. A common denominator of all BL forms, endemic and sporadic, are certain chromosome translocations. Typically, the gene locus for the proto-oncogene c-*myc* (chromosome 8q29) is juxtaposed to the immunoglobulin heavy chain locus on the long arm of chromosome 14. The break points on the c-*myc* gene are located upstream of the promoter, in the first exon, or between first and second exon of the proto-oncogene. Less frequently, c-*myc* is translocated to the immunoglobulin κ (chromosome 2p11) or λ (chromosome 22q11) light chain genes. It is generally believed that the translocations result in some form of c-*myc* deregulation, maybe at the level of transcriptional control, mRNA processing or protein synthesis. A unifying concept to explain the role of EBV in BL proposes that the virus initiates the process of tumorigenesis through growth stimulation of a certain B-cell population. Polyclonal B-cell expansion, possibly favored by malaria-induced impairment of T-cell functions, may enlarge the number of target cells for the necessary key event, chromosome translocation and deregulation of the proto-oncogene c-*myc*. An alternative may be the immortalization by EBV of a malaria-amplified pool of pre-existing c-*myc*-translocated B-cells. In both cases, early EBV infection and viral persistence would be an important risk factor for BL, primarily in its endemic forms.

EBV-associated Lymphoproliferative Syndromes other than BL

While primary infection with EBV usually remains without overt disease or manifests as benign infectious mononucleosis, rare cases develop into fatal lymphoproliferations with features of malignant B-cell lymphoma. About half of the cases occur in boys or male adolescents with an inherited defect in the immune system termed X-linked lymphoproliferative syndrome (XLPS) or Duncan syndrome. The other cases are sporadic without a sex preponderance. Proliferating cells were shown to contain persisting EBV DNA and to express the viral latency genes EBNA and LMP. The infiltrating B-cell derivatives have been shown to be polyclonal in some cases; other processes appeared oligoclonal. The defects of the immune system leading to XLPS or fatal mononucleosis are not precisely known, as the children are not generally immunodeficient. Possibly, it relates to functional impairment of natural killer (NK) cells, cytotoxic T cells or changes in the B cells that are the targets of EBV-induced growth stimulation. Notably, some of the patients do not have the normal antibody response to EBV antigens, particularly EBNA-1.

EBV sometimes induces fatal lymphoproliferative disease or lymphoma in globally immunodeficient individuals, either in cases of severe congenital immunodeficiencies or acquired immunosuppression in transplant recipients. Diffuse B-cell lymphomas, often localized in the central nervous system or other extranodular sites, are a frequent cause of morbidity and mortality in acquired immunodeficiency disease (AIDS). The tumors are often polyclonal initially, as they arise from several independent transformation events; they gradually evolve into an oligoclonal or even monoclonal proliferation. The tissues usually have the markers of persisting and actively transforming EBV, including EBNA and LMP expression. Similarly, the lymphomas of owl monkeys (*Aotus trivirgatus*) and cotton top marmosets (*Saguinus oedipus*) that are experimentally inoculated with EBV are considered to be immediate outgrowth of EBV-transformed B lymphoblasts. Presumably, deficiency in certain immune effector mechanisms, genetically determined or acquired, that are necessary to eliminate EBV-transformed B lymphoblasts from the body leads to progressive lymphoproliferative diseases. On occasion, it has been reported that human lymphomas with T-cell markers, such as nasal T-cell lymphomas, contain persisting EBV. A viral etiology of those tumors remains difficult to establish; the cell surface receptor for EBV on T cells is not yet known.

EBV genomes have been detected more recently in a large proportion (up to 60%) of cases of Hodgkin's disease. Distinct LMP-specific membrane and cytoplasmic staining has been found exclusively in Hodgkin and Reed–Sternberg cells of virus DNA-positive specimens, while EBNA-2 is not detected. This suggests a pattern of EBV gene expression different from that of B lymphoblastoid cells and BL. In view of the transforming potential of the LMP gene, it suggests a causal role of EBV in the case of Hodgkin's disease. Viral persistence and preferential expression of LMP has also been seen in some cases of CD30-positive anaplastic large cell lymphoma, a heterogeneous group of high-grade malignant lymphomas at the borderline between Hodgkin's and non-Hodgkin's lymphomas. The role of EBV in angioimmunoblastic lymphoadenopathy remains to be determined.

Nasopharyngeal Carcinoma

Nasopharyngeal carcinomas (NPC) are highly malignant neoplasms which mostly occur in adults between the ages of 20 and 50 years. Though the incidence is generally low in Europe and North America, NPC is a frequent tumor in southern parts of China, Tunisia, Central Africa and in the native population of Alaska. The prognosis is poor; most frequently, the presenting symptoms of NPC are signs of early metastasis into cervical lymph nodes and the skull. The association of EBV with NPC has been clearly established. Viral DNA is regularly found in all cases of poorly differentiated or lymphoepithelial nasopharyngeal carcinoma. Studies on the structure of the persisting episomal EBV genomes have indicated that the neoplastic epithelial cells contain a single clone of EBV in each case. Carcinoma cells consistently express EBNA-1 and, frequently, latent membrane proteins, while the EBNA-2 gene is not transcribed. There is a strong serologic association between EBV and NPC. The sera of patients with NPC typically have high titers of immunoglobulin A (IgA) antibodies to structural viral components, latency proteins and replicative antigens. A similar association was found between EBV and several rare tumor forms that are also assumed to be derived from embryonic branchial cleft remnants. High IgA titers against EBV proteins and persisting viral DNA in the tumors have also been found in some forms of neoplasms from the parotid, thymus and lymphatic tissue of the oropharynx.

T-Cell Leukemia

Adult T-cell leukemia (ATL) was the first human disease to be causally linked to a retrovirus. ATL, first described as a frequent tumor form in Southern regions of Japan, is a distinct disease entity that occurs in many other parts of the world including Central Africa and the Caribbean basin. Epidemiology clearly demonstrated the association with human T-cell leukemia virus type 1 (HTLV-1). Serological surveys showed that antibodies against HTLV-1 were regularly found in leukemia patients, and frequently in their close contacts. Tumor cells mostly contain integrated proviral DNA; clonality of the tumor correlates with the monoclonal or oligoclonal integration pattern of the viral genomes. HTLV-1 can be detected by polymerase chain reaction (PCR) in peripheral lymphocytes of latently infected persons and ATL patients. If peripheral mononuclear blood cells are infected with HTLV-1 in cell culture, $CD4^+$ T cells are immortalized, resulting in continuous growth in suspension culture. The cell surface marker phenotype largely resembles the tumor cells. HTLV-1 has also been shown to be oncogenic in animal model systems.

ATL usually develops 30–50 years after perinatal infection with HTLV-1. The virus is most frequently transmitted by breast feeding. The leukemia, arising after the long carrier phase, may be manifested as an acute leukemia with high white blood cell count and a poor prognosis, usually being fatal within 6–8 months. Smoldering and chronic forms of ATL can convert into an acute course. Approximately 3% of the perinatally infected will develop ATL during their lifetime. Unlike retroviruses of the other subgroups, HTLV-1 contains a 1.6 kb proviral genomic region coding for at least two regulatory polypeptides, the 42 kD transcriptional transactivator Tax, and the 27 kD phosphoprotein Rex which is required for the cytoplasmic targeting for structural protein mRNA. Tax, a promiscuous transactivator of numerous cellular genes, appears to mediate the oncogenic effects of the virus; it induces polyclonal expansion of $CD4^+$ T lymphocytes. The early virus-induced helper T-cell proliferation may precede a long period of tumor progression through several decades as growth becomes autonomous, turning independent from Tax expression and interleukin 2 (IL-2) receptor function.

Human T-cell leukemia virus type 2, which frequently occurs in some tribes of native Americans, has been isolated repeatedly from the T-cell variant of hairy cell leukemia, but the etiology of T-cell malignancies by HTLV-2 has not been substantiated further until now. Like type 1, HTLV-2 is increasingly spread among intravenous drug abusers in European and American cities.

Genital Cancer

Clinical studies and especially data from molecular biology suggest that certain types of human papillomaviruses (HPV) are of etiological importance for genital cancer. In developing countries, carcinoma of the cervix uteri is the most frequent type of female cancer. HPV DNA can be found in 80–90% of the tumors, and the early genes E6 and E7 are usually expressed. The most prevalent type in epidermoid carcinomas is HPV 16. HPV 18 may be preferentially associated with adenocarcinomas. Other types like HPV 31, 33, 35, 39, 45, 51, 52 or 56 have been detected in a few cases of squamous cell carcinoma each. HPV DNA was also demonstrated in the less prevalent carcinomas of the vulva, the vagina, the penis and the anus; HPV 16 is again the most frequent type, followed by HPV 18, HPV 6 and HPV 11. The significantly elevated prevalence of individual HPV types in cancers compared with the normal population led to the concept of HPVs with higher (HPV 16, 18, 45) and lower (HPV 6, 11) carcinogenic potential. This grouping is supported by first follow-up studies of the natural history of precancerous lesions associated with different HPVs and by differences in the *in vitro* transformation of keratinocytes.

HPV 16 and other members of the high-risk papillomavirus group immortalize primary human keratinocytes and induce resistance to differentiation stimuli. Histological abnormalities can be observed in stratifying keratinocyte cultures that resemble those in precancerous, intraepithelial lesions *in vivo*. The cells are not tumorigenic in nude mice initially, but quickly change to an aneuploid karyotype, which is in keeping with frequently occurring abnormal mitoses in HPV 16-positive lesions. At higher passage level, malignant clones reproducibly arise, which indicates that HPV infection is sufficient to induce cancer cells in combination with additional spontaneous or virus-induced modifications.

The viral genes E6 and E7 are required to trigger these effects. They encode proteins that are transactivators of transcription and are able to interact with the cellular proteins p53 and p105-RB (the retinoblastoma protein) respectively. The known cell-cycle regulating functions of these proteins are likely to be disturbed by this complex formation with the viral proteins. The E6 and E7 proteins of low-risk viruses display much lower affinities to the cellular proteins in parallel with a lower or not detectable transforming potential *in vitro*.

Much attention has been paid to the possible role of viral DNA integration in tumor progression. HPV 18 DNA appears integrated into the cellular genome in almost all cervical cancers and HPV 16 DNA in about two-thirds of the cases. This is in contrast with benign and premalignant lesions, where the viral DNA usually persists extrachromosomally. There is no evidence for a specific integration site, but HPV DNA has been repeatedly detected in the vicinity of the *myc* proto-oncogene in combination with an overexpression of the cellular gene.

The opening of the circular viral genome during integration frequently disrupts the regulator genes E1 and/or E2. Engineered mutants in these genes revealed increased transformation efficiency *in vitro*, so that naturally occurring inactivation may quantitatively enhance cell transformation.

In addition to the disruption of viral control mechanisms, there seems to be a failure of a cellular regulation of viral gene expression in malignant cells. The analysis of hybrids between HPV DNA-positive cervical cancer-derived cells and primary keratinocytes suggested that an inducible control system, possibly encoded by chromosome 11, can normally suppress viral transcription.

The persistence of viral DNA and the continual expression of transforming genes in advanced cancers suggest that HPV functions are also involved in the maintenance of the malignant phenotype. An experimental suppression of E6 and E7 expression inhibited the proliferation of HPV-positive cervical cancer cell lines and reduced the cloning efficiency in semisolid medium, thus indicating that the viral proteins are still modulating the growth of malignant cells.

The genital tract HPVs are also responsible for many HPV infections at extragenital mucosal sites such as the oral cavity and most notably the larynx. However, cancers arising in this field only rarely harbor HPV DNA. Case reports describe the presence of HPV 2, 6, 11, 16, or 30 in carcinomas of the tongue, the oral and nasal cavity, the larynx, hypopharynx and the lung. The reason for the striking difference in the association of HPV and cancer between genital and aerodigestive tract is not known. Either the etiology of oral and laryngeal cancers is unrelated to HPV or the relevant HPV types are not yet characterized or the viral DNA is no longer necessary for cancer cells and finally lost.

Skin Cancer

HPVs induce various proliferative skin lesions like plantar, common and flat warts that are benign. An association between HPV and skin cancer becomes obvious in epidermodysplasia verruciformis (EV). EV patients are infected with a subgroup of HPVs, which induce characteristic persisting macular lesions disseminated over the body. Many EV patients develop squamous cell skin carcinomas mainly at sun-exposed sites, which suggests a co-carcinogenic effect of UV light. The DNA of HPV 5 or 8 persists extrachromosomally in high copy number in more than 90% of the cancers. HPV 14, 17, 20 or 47 were occasionally detected. The prevalence of specific HPVs is in striking contrast with the plurality of HPV in benign lesions and has been interpreted as reflecting a higher oncogenic potential of these types.

Oncogenes of EV-HPVs were mainly identified by their effects on rodent fibroblasts. The E6 gene induces altered morphology, reduced serum requirement and anchorage-independent growth. In contrast with genital HPV, no complex formation could be detected between HPV 8 E6 and the cellular p53 protein which suggests different strategies of transformation. The HPV 8 E2 gene, which encodes a transactivator of transcription, leads to reduced serum requirement and growth in soft agar. There is some indication of an increased transforming activity of E6 from HPV 5, 8 and 47 when compared with E6 from related HPVs, which have not yet been detected in carcinomas.

Clinically apparent infections with EV-HPVs are largely restricted to EV patients, although the viruses appear to be widespread in the general population. EV-specific skin lesions occasionally occur in immunodepressed patients, and HPV 5 or 8 persist in some of their skin cancers. Most attempts to detect the DNAs of these HPV types in skin carcinomas of the general population have failed. Among immunocompetent patients it is difficult to detect any HPV genomes in skin carcinomas. HPV 41 was identified in a few cases. Remarkable is a strong association between genital HPV 16 and squamous cell neoplasms from the finger.

In summary, the role of HPV in skin cancer is not definitely settled although the carcinogenic potential of papillomaviruses was first noted in the context of skin cancer arising from Shope papillomavirus-induced warts of rabbits. The possibilities discussed above for carcinomas of the aerodigestive tract are also valid for skin carcinomas.

Hepatocellular Carcinoma

Primary liver cancer is among the most common fatal malignancies of man worldwide. An association with hepatitis B virus (HBV) from the hepadnavirus family was suggested by the geographical coincidence of a high incidence of hepatocellular carcinoma in southeast Asia and equatorial Africa with high rates of chronic HBV infections generally contracted congenitally. Prospective studies demonstrated about a hundredfold increased risk of hepatoma among carriers of the HBV surface antigen (HBsAg). Integrated HBV DNA can be detected in a large proportion of the tumors from high-risk areas and in hepatoma-derived cell lines.

Liver cancer usually develops only after several decades of chronic HBV-induced hepatitis and may thus be triggered by accumulating genetic damage due to inflammation and continuous cell regeneration. A specific contribution of HBV might be expected from cis effects following integration of viral DNA, but except for a few case reports no consistent evidence has been obtained for the activation of particular proto-oncogenes. A transactivation of transcription may be more relevant, which can be achieved by the viral X protein and by a truncated $preS_2/S$ protein. The viral $preS_2/S$ gene, which normally encodes a surface protein, appears frequently disrupted in cancers as a consequence of DNA integration and then gives rise to the transactivator. Both X and truncated $preS_2/S$ protein exert pleiotropic effects via the protein kinase C-controlled signal pathway and a second, not yet clearly defined, pathway. The analysis of viral integration patterns and functional assays suggest that at least one transactivator may function in most hepatomas. Multifocal nodular hyperplastic liver disease developed in mice transgenic for the HBV surface protein genes, and liver cancer arose in mice transgenic for the X gene. The notion of a specific role of HBV in carcinogenesis is supported by the wood-chuck model system where another member of the hepadnavirus family induces liver cancer in a background of minimal hepatitis and no cirrhosis. HBV is the first human tumor virus against which vaccination programs have been initiated on a broad basis. Although installed first to fight hepatitis they will also clarify the viral role in cancer development.

More recently, seroepidemiological evidence was obtained for a correlation between hepatitis C virus (HCV) infections and hepatoma. Antibodies against HCV were detected in 50–60% of liver cancer patients in Japan and Taiwan in contrast with less than 5% of controls. The prevalence of anti-HCV was significantly higher in patients with HBsAg-negative hepatomas, i.e. in patients without evidence for chronic HBV infection, than in those with HBsAg-positive liver cancer. Finally, hepatomas appeared significantly more frequently in anti-HCV-positive patients than in negative patients after adjustment for other known risk factors such as alcoholism or chronic HBV infection. HCV is related to flavi- and pestiviruses and is the first human tumor virus with a RNA genome and no DNA intermediate during replication. A possible oncogene has not yet been identified. HCV leads to chronic persistent infections, and liver injury may be responsible for malignant conversion.

Kaposi's Sarcoma

Kaposi's sarcoma arises from malignant proliferations of endothelial cells of small blood vessels in skin and mucous membranes of the oral cavity and gastrointestinal tract. The tumor typically arises from multiple foci. Kaposi's sarcoma had been a rare, rather benign tumor of elderly males, mostly of Mediterranean or Jewish origin. Aggressive forms have been known to be endemic in Central Africa for many decades; the rapidly progressive forms of the tumor have now become frequent by spreading with the AIDS epidemic. There is an unexplained preponderance of Kaposi's sarcoma as symptoms of advanced AIDS in the major risk group of male homosexuals. Kaposi's sarcoma has also been observed in immunosuppressed patients following organ transplantation. An early hypothesis tried to link Kaposi's sarcoma with human cytomegalovirus infections, but sensitive hybridization studies confirmed that cytomegalovirus was not found in tumor biopsies. HPV DNA was detected by the polymerase chain reaction. The significance of this finding remains to be established. It has been hypothesized that the transactivator protein Tat of human immunodeficiency virus may favor the proliferation of endothelia, thus triggering Kaposi's sarcoma; further studies will be required to substantiate this.

Future Perspectives

About 15 years ago, a single human virus, EBV was accepted as a candidate human tumor virus. Since then, the oncogenic potential of human papilloma-, hepadna- and retrovirus has been clearly established. Now it has become clear that virus infections that may result in tumor formation are quite frequent. On a worldwide scale, more than 20% of all malignant tumor forms in females and about 8% of the tumors in males may be the late consequence of a previous virus infection. In most cases, long latency periods of many years or several decades

Table 1. Latency periods for virus-associated tumors

Virus	Tumor	Latency periods (years)
Hepatitis B	Hepatocellular carcinoma	30–50
HTLV-1	T-cell leukemia	20–50
Epstein–Barr	Burkitt lymphoma	3–12
Epstein–Barr	Nasopharyngeal carcinoma	30–40
HPV 5, 8	Squamous cell skin carcinoma of EV patients	5–15
HPV 16, 18	Cervical cancer	5–25
HPV 16, 18	Penile and vulval carcinoma	20–50

Source: zur Hausen H (1986) Intracellular surveillance of persisting viral infections. *Lancet* ii: 489.

elapse between primary infection by tumor viruses and first symptoms of cancer (Table 1). All human tumor viruses are widespread in the world population. They contribute to tumor disease, mainly by initiation of oncogenesis, but not sufficiently by themselves to completely cause a tumor. Thus, all human tumor viruses are important risk factors for cancer, but require additional events. This implies that mere proof of an infection with a tumor virus is of limited value for the management of patients and cancer prevention. Specific diagnostic tests have to be designed, which evaluate parameters of the viral infection more closely related to malignant conversion. Often the viruses cannot be traced unambiguously in the tumor; conventional virus isolation procedures usually remain unsuccessful; tumor tissues are usually not infectious. The neoplastic phenotype of HPV-positive genital carcinoma cells seems to be affected by viral functions, which raises the prospect of virus-specific pharmacological interference for adjuvant cancer therapy. In many cases, however, continuous viral expression is not detectable in malignant tumors. Primary infection and initial growth transformation apparently lead through tumor progression to a constitutive form of proliferation where viral gene products are not necessary for growth and the receptor molecules may not be functional any longer. Thus, even in the longer range, antiviral gene therapy may not be successful for the treatment of the tumors, as they have become autonomous. Also in the distant future, the most efficient method for prophylaxis of virus-induced human tumors will probably be vaccination trials to prevent primary infection with the viruses.

See also: Epstein–Barr virus; Hepatitis B viruses; Hepatitis C virus; Human immunodeficiency viruses; Human T-cell leukemia viruses; Oncogenes; Papillomaviruses; Transformation.

Further Reading

Lorincz AT, Reid R, Jenson AB, Greenberg MD, Lancaster W and Kurman RJ (1992) Human papillomavirus infection of the cervix: relative risk associations of 15 common anogenital types. *Obstet. Gynecol.* 79, 328.

Meyer M *et al.* (1992) Hepatitis B virus transactivator MHBs[t]: activation of NF-KB, selective inhibition by antioxidants and integral membrane localization. *EMBO J.* 11, 2991.

Miller G (1990) Epstein–Barr Virus: Fields BN, Knipe DM *et al.* (eds.). In: *Virology*, 2nd edn, p. 1921. New York: Raven Press.

Pfister H (ed.) (1990) *Papillomaviruses and Human Cancer.* Boca Raton: CRC Press.

Rickinson AB (ed.) (1992) Viruses and human cancer. *Sem. Cancer Biol.* 3: 2049.

Sherker AH and Marion PL (1991) Hepadnaviruses and hepatocellular carcinoma. *Annu. Rev. Microbiol.* 45, 475.

Smith MR and Greene WC (1991) Molecular biology of the type 1 human T-cell leukemia virus (HTLV-1) and adult T-cell leukemia. *J. Clin. Invest.* 87, 761.

zur Hausen H (1986) Intracellular surveillance of persisting viral infections. Lancet ii: 489.

zur Hausen H (1991) Human papillomaviruses in the pathogenesis of anogenital cancer. *Virology* 184, 9.

TURKEY HERPESVIRUS

See Marek's disease virus

TYMOVIRUSES

Adrian Gibbs
Australian National University
Canberra, Australia

Distinguishing Features

At least 20 tymoviruses are known, and they are named after turnip yellow mosaic virus (TYMV), the first species to be described. They infect dicotyledonous angiosperms and are transmitted by beetles. Their virions are isometric and 25–30 nm in diameter. Their genomes are single molecules of translatable single-stranded (ss)RNA of about 6300 nucleotides of which 34–42% are cytidylic acid and only 15–17% guanylic acid. Their 5′ termini are capped, and most have a tRNA-like 3′ terminus. The genomes replicate in small vesicles which form as invaginations of the outer chloroplast membrane.

Taxonomy and Classification

Tymoviruses have only been isolated from dicotyledonous plants, most of them wild. Tymovirus species have restricted host and geographical ranges. The species (and their normal geographical ranges) include Andean potato latent (the Andes), belladonna mottle (Europe), cacao yellow mosaic (Sierra Leone), clitoria yellow vein (Kenya), desmodium yellow mottle (USA), dulcamara mottle (UK), eggplant mosaic (Antilles), erysimum latent (Europe), kennedya yellow mosaic (Australia), okra mosaic (Ivory Coast), ononis yellow mosaic (UK), passiflora yellow mosaic (Brazil), physalis mosaic (USA) (confusingly called belladonna mottle virus-Iowa strain in some publications), plantago mottle (USA), scrophularia mottle (Germany), turnip yellow mosaic (Australia and Europe), voandezia necrotic mosaic (Ivory Coast and Upper Volta) and wild cucumber mosaic (USA) viruses, and also, probably, poinsetta mosaic virus (world-wide). These species have distinctive host ranges, and their virions differ serologically to a greater or lesser extent. The serological relationships correlate with known genomic sequence differences, especially of the virion protein gene, and place them in four groups; the turnip yellow mosaic viruses, the legume-infecting tymoviruses plus cacao yellow mosaic, the solanaceous tymoviruses plus ononis yellow mosaic and wild cucumber mosaic, and erysimum latent, which is the most distinctive of the tymoviruses. Some of the species include isolates that are biologically indistinguishable yet have genomic sequences that differ by amounts that distinguish other species; the TYMVs fall into two strain groups and the Kennedya yellow mosaic viruses (KYMV)s into three, and there is clearly a great complex of tymovirus strains infecting the *Solanaceae*, especially in South and Central America.

The sequence motifs of the replicase protein (RP) of tymoviruses show clear similarities to those of the RPs of viruses of the 'alphavirus/tricornavirus' supergroup; all three enzyme motifs of the tymovirus RP are most closely related to those of the potexviruses, carlaviruses and closteroviruses, implying that the RPs of all these viruses have a common ancestor. However, the virion protein (VP) of tymoviruses comes from different ancestors, as the potex-, carla- and closteroviruses have filamentous helically constructed virions, whereas those of the tymoviruses are isometric and have a shell constructed from 180 subunits of the β-barrel VP. The third tymovirus gene, the overlapping (OP) gene, does not occur in other viruses of the 'alphavirus/tricornavirus' supergroup or in any other known viruses, and hence it probably arose *de novo* in the ancestral tymovirus.

Virions

Tymoviral virions are isometric and *c.* 28 nm in diameter. Their shells are regular icosahedra of 180 subunits of a single protein species which cluster in fives or sixes in the surface of the virion shell to form the 32 morphological subunits, which can be seen in negatively stained virions. Purified virions sediment as two components. Those that sediment at 110–120S are nucleoprotein, and each contains a single copy of the viral genome, which constitutes 35% of their mass. Some also contain a variable number of tRNA or VP mRNA molecules. The other virion component sediments at 50–55S and consists of the protein shell of the virion, and those of the solanaceous tymoviruses and ononis yellow mosaic virus also contain variable amounts of nongenomic RNA molecules (VP mRNA and host tRNAs).

Genome and Encoded Proteins

Tymoviral genomes are single molecules of about 6300 nucleotides of ssRNA, and, when separated chemically from the virions, they are infectious. The genomes have a characteristic nucleotide composition; G 15–20%, A 17–24%, C 31–42% and U 20–29%.

There are untranslated regions of the genome at both termini and between some open reading frames (ORFs), but these constitute only about 3% of the genome. They have a 5′-terminal m^7GpppGp cap structure. Most tymovirus genomes also have a 3′-terminal sequence that can form a tRNA-like structure and that can be specifically valylated, but three tymoviruses, belladonna mottle, dulcamara mottle, erysimum latent, have a 3′-terminal sequence that can only form part of a tRNA-like structure, and dulcamara mottle has a 3′-terminal poly(A) sequence. All tymovirus genomes have three ORFs which are of similar size and arrangement in the genome. The largest is also the most conserved and spans most of the genome. It encodes a large replicase protein (RP; 1747–1874 residues, mol. wt 194–210K) which has sequence motifs (NH_2- to COOH-) characteristic of *N*-methyltransferase, protease, helicase/nucleotide-binding fold and replicase enzymes. Overlapping the 5′-terminal third of the RP, and always starting seven nucleotides to the 5′ side of its start codon, is an ORF that encodes an overlapping protein (OP) of variable length (440–750 residues, mol. wt 49–82K) and the least conserved sequence. The OPs are very basic (pI 10.9–11.9), and may aid systemic spread of the virus in the plant. The third ORF, at the 3′ end of the genome, is of the virion protein (VP; 188–202 residues, mol. wt 19.6–21.5K), which is probably, like most VPs of isometric virions of *c.* 25 nm diameter, an eight-stranded antiparallel *β*-barrel. The VP ORF is in different reading frames, relative to the RP/OP gene doublet, in different tymoviruses, and, in some, overlaps by a few nucleotides the 3′-terminal region of the RP ORF.

A region of about 50 nucleotides to the 5′ side of the start of the VP ORF, and hence in the 3′-terminal part of the RP ORF, is similar in all tymoviruses, and two blocks of it are particularly conserved. One, which has been named the 'tymobox', is 28–44 nucleotides from the start of the VP ORF and is 16 nucleotides in length; 11 tymoviruses have tymoboxes with the same sequence (-GAGUCUGAAUUGCUUC-) there is a single nucleotide difference in three, and there are four differences in the tymobox of wild cucumber mosaic virus, hence the tymobox is useful as a target for a nucleic acid probe as it will identify most tymoviruses. Part of the tymobox sequence encodes the near-terminal sequence, -ELL-, found in most tymoviral RPs. The second conserved region occurs between the start of the VP ORF and the tymobox, 7 or 8 nucleotides from the latter. This is the translation initiation box -CAAU- (-CAAG- in one), and includes the 5′-terminal AAU found in the VP subgenomic mRNA of three tymoviruses.

Replication Strategy

Tymovirus genomes probably replicate in the characteristic chloroplast vesicles. It is probable that the OP and RP ORFs are translated directly from genomic RNA, but that of the VP is translated from a subgenomic mRNA transcribed from the negative genomic strand using the tymobox sequence as a promoter. Tymovirus proteins are produced by cytoplasmic ribosomes. The RP polyprotein is probably *cis*-hydrolysed into its component parts by a viral protease in the C-terminal half of the methyltransferase region, the RP being cut first between the helicase and replicase regions, and then between the helicase and protease regions.

Symptoms

Tymoviruses cause yellow mosaics, vein-clearing and mottles. All plants infected with tymoviruses develop small characteristic vesicles within their chloroplasts attached to their outer membranes. These vesicles form as invaginations of the outer chloroplast membranes, and the bilayer membranes of these membranes and the vesicles are confluent.

Host Ranges

Individual tymoviruses have restricted natural host ranges and usually infect a few species from one dicotyledonous plant family. Most have been isolated from wild plants, whereas species of all other groups of plant viruses have mostly been isolated from crop plants. Tymoviruses cause bright yellow mosaic symptoms, and depress the growth of their wild natural hosts, but may protect these hosts against herbivores, whereas most other viruses that naturally infect wild plants cause few or no symptoms and do not affect growth. They have broader, but still restricted, experimental host ranges, and there is only one unconfirmed report of a tymovirus experimentally infecting a monocotyledonous plant. They infect more species from the family of their natural hosts, than from other families, and also a greater proportion of plant species from the same major division of the dicotyledons as their natural hosts, than of species from the other division. These divisions, crassinucelli or tenuinucelli, were defined phenetically by Young and Watson and have recently been confirmed by analysis of the sequences of the Rubisco large subunit gene. Only a few C_4 plants are susceptible to tymoviruses, which mostly infect plants that use the C_3 photosynthetic pathway.

Transmission

The known natural vectors of tymoviruses are all chrysomelid beetles (Halticidae and Galerucidae) except that of TYMV in Australia, which is transmitted by a pill beetle (*Byrrhidae*). The vector beetles prefer feeding on plants that are tymovirus-infected rather than virus-free. A few tymoviruses are transmitted by seed, but not pollen. All may be transmitted by sap inoculation, but none by plant contact.

Host Interactions

The chloroplast vesicles found in TYMV-infected plants have been shown to contain the viral replicase complex and double-stranded RNA 'replicative intermediate' of the virus. Virions of the virus seem to assemble at or near where these vesicles connect with the cytoplasm. Virions and genome-free virion shells are found throughout the cytoplasm and all parts of infected plants. Virion shells also accumulate in large numbers in nuclei.

Further Reading

Gibbs AJ (1980) A plant virus that partially protects its wild legume host against herbivores. *Intervirology* 13: 42.

Guy PL *et al.* (1984) A taxonomic study of the host ranges of tymoviruses. *Plant Pathol.* 33: 337.

Koenig R (1979) Tymovirus group. *CMI/AAB Descr. Plant Viruses*. No.214.

Matthews REF (1991) *Plant Virology*, 3rd edn. New York: Academic Press.

Young DJ and Watson L (1970) The classification of dicotyledons: a study of the upper levels of the hierarchy. *Aust. J. Bot.* 18: 387.

TY ELEMENTS

See Retrotransposons of fungi

U/V

USTILAGO MAYDIS VIRUSES

See Totiviruses

VACCINES AND IMMUNE RESPONSE

Gordon L Ada
Australian National University
Canberra, Australia

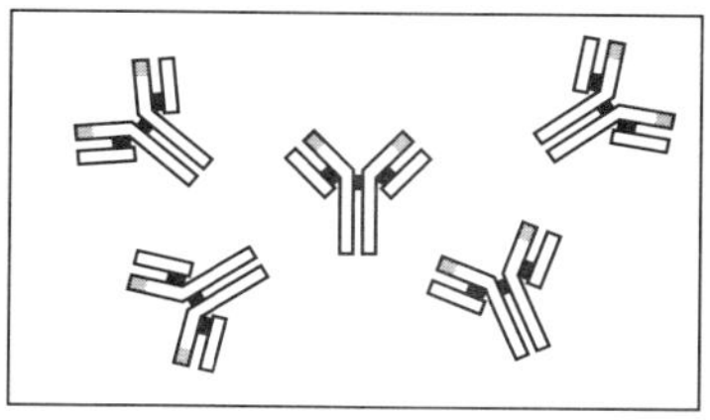

History

Any history of vaccination to control viral infections, however brief, begins with the story of smallpox and its final eradication. There will be bicentenary celebrations in 1996 to recall the famous experiment of Edward Jenner, in which he inoculated a boy, James Phipps, with cowpox material and demonstrated his subsequent resistance to challenge with the smallpox preparation. Almost 100 years later, Pasteur developed the rabies virus vaccine, and, in 1935, another highly successful attenuated viral vaccine was made, this time to combat yellow fever. The first inactivated influenza viral vaccine became available in 1936. Production of the yellow fever and influenza viral vaccines was possible because of the newly developed technique of growing virus in the embryonated eggs of chickens.

The major event that accelerated viral vaccine development was the ability to grow many viruses in tissue culture. This began in the 1920s, but it was the work of John Enders and colleagues in successfully growing polioviruses in cell culture which initiated an explosion of activity in this area. Jonas Salk produced the trivalent formalin-inactivated poliovirus vaccine and this was followed shortly afterwards by the oral live attenuated polio vaccines of Albert Sabin. Most viral vaccines in use today and those under clinical trial and close to registration (Table 1) are attenuated preparations.

A subunit preparation, consisting of the glycoproteins of the influenza virus, was made and is administered to children because of its low reactogenicity compared to intact virus. A new era of viral vaccine development was initiated with the preparation of the surface antigen of the hepatitis B viral vaccine, first from the plasma of infected people and in the late 1980s from DNA-transfected yeast cells.

Smallpox was declared to be eradicated in 1980, 3 years after the last case of endemic smallpox was cured, and nearly 14 years after the World Health Organization (WHO) initiated an intensified campaign to eradicate this disease. In 1989, WHO initiated a similar campaign to eradicate wild-type poliomyelitis by the year 2000.

Table 1. Viral vaccines

Type	Agents	
	Current	Under trial
Live attenuated	Vaccinia Measles Yellow fever Mumps Polio (OPV)* Adeno* Rubella Varicella–zoster	Cytomegalo Hepatitis A Influenza Dengue Rota* Parainfluenza Japanese encephalitis Polio (OPV)*
Inactivated	Polio (IPV) Influenza Rabies Japanese encephalitis	Hepatitis A
Subunit	Hepatitis B Influenza	

* Administered orally.
OPV, live polio vaccine; IPV, inactivated polio vaccine.

Two Requirements of a Vaccine

There are two essential requirements of a vaccine. The first is safety. Current vaccines vary in this respect. Many vaccines, such as the yellow fever and the adenovirus vaccines, have remarkable safety records. In contrast, the smallpox vaccine had a high level of side effects, about 1 death per 10^5 recipients, and a higher level of other effects. Attenuated live vaccines may pose a risk to immunodeficient/compromised people.

The second requirement is efficacy, which is assessed as the ability to protect recipients from disease following a subsequent exposure to/infection by the wild-type agent. Vaccine efficacy depends almost entirely on the nature and persistence of the induced immune response. Most vaccines are given prophylactically and, as the time interval between vaccination and the subsequent exposure may be short (weeks) or long (years), efficacy depends mainly upon stimulation of the adaptive components of the immune response, i.e. the lymphocytes, which are responsible for immunological memory. If a vaccine is used therapeutically (which may be deliberate or occur naturally if the vaccine is used in a disease-endemic area), nonadaptive components may contribute to a protective effect. For a number of years, the efficacy of some vaccines has been predicted by measuring the increase in protective antibody levels after vaccination. The neutralization of viral infectivity for cultured cells by antibody in an *in vitro* test is the standard test. It is only quite recently that interest in measuring different parameters of the cell-mediated immune (CMI) response is seen to be important.

Roles of Antibodies

Specific antibody has three main roles: to neutralize the infectivity of the virus, to lyse infected cells which express viral antigens at the cell surface, and to complex with viral debris and help in its removal. The continuing presence after vaccination of antibody which complexes with virus and prevents infection is the first line of defense and regarded by vaccinologists as a critical requirement of a vaccine. Other than those that are vector-borne, nearly all viruses naturally infect via a mucosal surface so that S.Iga would fulfill this role (see below). Many vaccines are given parenterally, in which case the IgM or IgG so generated will not prevent the initial infection but should prevent systemic spread. Under optimum conditions, vaccine-induced antibody might give 'sterilizing immunity' but it is more likely to reduce the infecting virus load to such an extent (say >99%) that the host's own immune system comfortably copes with the remainder and prevents disease.

The mechanism of infectivity neutralization by antibody varies between viruses, and even different immunoglobulin isotypes may have different effects. Whether the antibody prevents adsorption to, penetration or expression of the viral genome within the susceptible cell, the net result is prevention of viral replication.

Antibody to those antigens that may be expressed on the infected cell surface may contribute to destruction of the cell by antibody-dependent cell cytotoxicity (ADCC). An alternative mechanism is complement-dependent lysis by certain immunoglobulin isotypes. There is little reason to doubt that these mechanisms help to clear a viral infection. Experiments to see whether they can completely clear an infection have given variable results; they need to be carried out in a model system in which the contributions of effector T cells have been eliminated.

Epitopes Recognized by Protective Antibodies

Generally, only one or a few antigens of a virus – usually the surface antigens – contain epitopes recognized by antibodies which neutralize infectivity or lyse infected cells. A major role of antibody to internal antigens is to help dispose of viral debris. The epitopes recognized by such antibodies may be continuous or discontinuous.

Roles of Effector T Cells

Generally, the two major classes of effector T cells, CD4+ and CD8+, have distinct effector functions. The former may mediate help (Th cells) and/or delayed-type hypersensitivity (Td cells); there may be two classes of Th cells, Th1 and Th2, distinguished by their pattern of interleukin production. CD8+ cells may be cytotoxic (Tc cells) or have suppressor activity (Ts cells). Most T cells have α, β receptors and are major histocompatibility complex (MHC) restricted; some T cells, usually a small proportion which often recognize heat-shock proteins, have γ, β receptors and are not MHC

Table 2. Examples of clearance of infectious virus in animal models following transfer of specific CD8+ cytotoxic T cells

Orthomyxovirus (influenza virus)
Paramyxovirus (Sendai virus)
Poxvirus (vaccinia virus)
Togavirus (Sindbis)
Pneumovirus (respiratory syncytial virus)
Arenavirus (lymphocytic choriomeningitis virus)
Alphaherpesvirus (herpesvirus)
Betaherpesvirus (murine cytomegalovirus)

restricted. In viral infections in both humans and mice, and at the time after infection when MHC-restricted Tc cell activity is detected *in vivo* (without culture of the cell *in vitro*), only Class I MHC-restricted Tc cell activity has so far been found. Table 2 lists some examples of murine viral infections where transfer of viral-specific CD8+ Tc cells has greatly reduced or cleared the viral infection. Such cells may be primary (direct from the host), secondary (memory cells activated in culture) or cloned T cells. In mice, transfer of virus-specific CD4+ cells may exacerbate a viral infection; transfer of individual clones of such cells has either helped to clear, has had little effect or has exacerbated the infection. Individual clones of CD4+ human cells specific for a variety of viruses are frequently cytotoxic, as are short-term cultured cells. It is not clear whether the precursors of these cells *in vivo* are cytotoxic but if so, they would represent only a small proportion of the total population of Tc cells.

Sources and Production of T-Cell Epitopes

In contrast to the restricted number of antigens important as sources of epitopes recognized by protective antibodies, potentially many viral antigens may be the source of T-cell epitopes. These epitopes have been detected in three ways:

1. Individual antigens are screened, perhaps using cells infected with a chimeric live vector, as a source of the epitopes.
2. If the amino acid sequence of the antigen is known, algorithms have been described which predict active sequences.
3. A series of overlapping peptides is synthesized and each tested for its ability to sensitize a target cell for lysis. This approach should indicate the largest number of epitopes in a given antigen; however, it needs to be established that infected people produce effector T cells recognizing all such epitopes.

There were some general findings.

1. With the few viruses for which comprehensive studies have been made, e.g., the influenza virus, the majority of antigens may be sources of these peptides.
2. There are a number of examples where the same peptide may serve as an epitope for Class I and II MHC-restricted responses/cell lysis.
3. As internal antigens of a virus usually show less antigenic variation compared to surface antigens of viruses displaying antigenic variation, T-cell epitopes from the former are likely to be more conserved sequences, and hence more attractive to include in a vaccine.
4. Although there are a few known peptide epitopes which bind to MHC antigens of a number of different specificities, many peptides form a complex which may be recognized by only one or a few MHC antigenic specificities.

Protection Following Immunization by Different Routes

The area of mucosal surface in mammals is far greater than the area of skin. Other than vector-borne viruses, nearly all other viruses infect via a mucosal surface. The main sites of entry are oral, rectal, respiratory, urogenital, the mammary glands and ocular. Some of the attenuated live virus vaccines in Table 1 are administered parenterally even though the virus naturally infects mucosally. The spleen and draining lymph nodes are involved after parenteral administration, and IgM, IgG and sometimes IgE are the main immunoglobulin isotypes formed. The major immunoglobulin isotype induced after mucosal administration is IgA; on passage, for example from the Peyers Patches to the gut lumen, the molecule dimerizes and binds to a secretory component to form a complex (S.IgA) which is protease resistant and is the first line of defense against an organism invading via a mucosal surface.

A comparison between the live polio vaccine (OPV) and the inactivated polio virus (IPV) is instructive. OPV induces both serum immunoglobulin responses and a marked local resistance, which is most likely S.IgA; this protects against a wild-type poliovirus challenge. Type 3 virus in the vaccine may revert to a virulent state; paralysis occurs at a frequency of about 1 in 3×10^6 vaccine doses. Use of this vaccine has almost completely eradicated infection by wild-type virus in North and South America, a remarkable achievement. In contrast, IPV is administered parenterally, and induces IgM, IgG and low levels of IgA. Substantial doses are required to induce the level of antibody needed to prevent viremia. It is a safe vaccine and has been used very successfully in some European countries.

The adenovirus vaccine exemplifies a special feature of the mucosal system. It is administered in an enteric-coated capsule to protect from gastric acidity and to prevent respiratory tract infection if the tablet is regurgitated. The virus replicates in the gut-associated lymphoid tissue and sensitizes cells which circulate to other mucosal sites, so preventing a disease-causing infection in the respiratory tract by wild-type virus. The measles vaccine is currently administered parenterally, but the possibility of using a mucosal route, ocular, intranasal or oral, is being explored.

Immunological Requirements of a Vaccine

There are four immunological requirements of a vaccine:

1. Activation of antigen-presenting cells to initiate antigen processing and production of interleukins;
2. Activation of both T and B cells to give a high yield of memory cells;
3. Generation of Th and Tc cells to several epitopes to overcome the variation in the immune response in the population due to MHC polymorphism;
4. Persistence of antigen, probably on dendritic cells in lymphoid tissues, where B memory cells are formed and recruited to form antibody-secreting cells so that antibody is continually present.

Factors Affecting the Feasibility of Development of Viral (and other) Vaccines

Several factors are important. They include the following.

Animal models

The availability of animal models that mimic the human infection, immune responses and disease processes greatly aids development of a vaccine. Two examples illustrate this. Human immunodeficiency virus (HIV) infects the chimpanzee – an endangered species and very expensive to maintain – but disease does not occur. The ready availability of a smaller, much less expensive, animal model would greatly facilitate vaccine development. Secondly, vaccine development to respiratory syncytial virus, in progress since the 1960s, would have proceeded more rapidly if an animal model had been available which mimicked the early experience in infants given a formalin-inactivated whole viral preparation. Infants so vaccinated developed an acute respiratory illness following exposure to the wild-type virus.

Two general developments offer promise. One is the ability to reconstitute severe combined immunodeficiency (SCID) mice with cells/serum from an immunized person and so check the effect of the immune components on growth of a virus, such as HIV, in the transferred human cells. The second is the ability to produce transgenic mice expressing a critical component such as a human cellular receptor for a virus, such as poliovirus, thus making the mouse susceptible to infection.

Complexity of the virus

With simple viruses, such as myxo- or picornaviruses, it has been straightforward to determine the contribution of each component to different responses, such as epitopes recognized by neutralizing antibody and effector T cells. This is a more formidable task with complex viruses such as the poxviruses.

The disease

Most current vaccines are against agents that cause acute infections and which the induced immune response rapidly clears in most people. It will in general be more difficult to develop vaccines to viruses that cause chronic persistent infections, such as HIV. There are several reasons for this, such as infection by infected cells as well as free virus, the presence of suppressor sequences and integration of the viral DNA/cDNA into the host cell genome during viral replication.

Protective immunity

The occurrence of protective immunity following a

natural infection (e.g. smallpox) is a good indicator of the feasibility of vaccine development. Its absence does *not* mean an effective vaccine cannot be developed (e.g. hepatitis B), but indicates that there may be some difficulties (e.g. HIV).

Antigenic diversity

This may vary from none, a little (a few serotypes), a large amount (many serotypes) to antigenic drift/shift. Generally, the more such diversity, the more difficult it is to develop a completely protective vaccine. Thus, new influenza vaccines with changed antigenic specificities are produced each year to cope with antigenically different emerging viruses. Because of this property, a vaccine against rhinoviruses, which cause much morbidity in young children, has yet to be developed.

See also: Immune response; Replication of viruses.

Further Reading

Ada GL (1989) Vaccines. In: Paul WE (ed.) *Fundamental Immunology*, 2nd edn, p. 985. New York: Raven Press.

Ada GL (1990) Modern vaccines: The immunological principles of vaccination. *Lancet* 335: 523.

Plotkin SA and Mortimer EA (eds) (1988) *Vaccines*. Philadelphia: WB Saunders.

Woodrow GC and Levine MM (1990) *New Generation Vaccines*. New York and Basel: Marcel Dekker.

VACCINIA VIRUS

Riccardo Wittek
Université de Lausanne
Lausanne, Switzerland

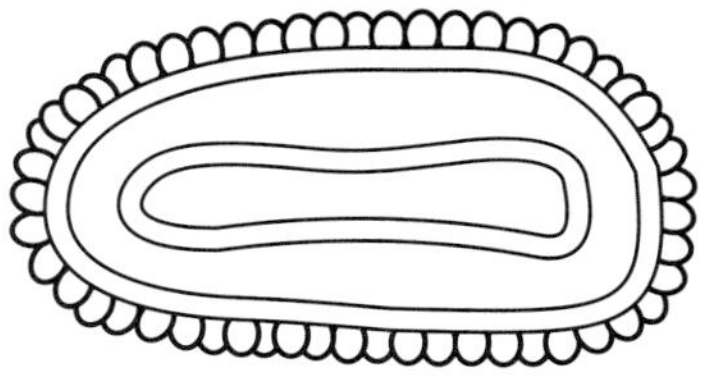

History

In 1980, the World Health Assembly announced that smallpox, once the most serious infectious disease of mankind, had been eradicated from all the countries of the world. Several factors were important in making the eradication campaign a unique success in the history of medicine. Among these was the active role played by WHO which organized the campaign and provided expertise in all relevant areas. Furthermore, variola virus, the agent of smallpox, had no animal reservoir and, consequently, interrupting the chain of human-to-human transmission eliminated the virus. Finally, with vaccinia virus, an excellent vaccine for smallpox was available. The vaccine had a low production cost, could be freeze-dried and in this form was very heat stable, obviating the need of a cold chain. In addition, the vaccine was easy to administer even by relatively untrained field workers and left a characteristic scar providing permanent evidence of vaccination which was important for WHO in assessing vaccination coverage.

It is not known when vaccinia virus was introduced as a smallpox vaccine. Edward Jenner, who was the first to demonstrate, at the end of the 18th century, that inoculation of a related poxvirus provided protection against challenge with variola virus, isolated his vaccine virus from a milkmaid infected with 'cowpox'. It is unlikely that this virus was vaccinia virus. Although, in modern times, vaccinia virus has occasionally been isolated from skin lesions of infected cows, these cases were usually associated with contact with vaccinated humans, and therefore the cow does not seem to be the natural host of the virus. The origin of vaccinia virus thus remains a matter for speculation.

Taxonomy and Classification

Vaccinia virus is a member of the family *Poxviridae* which comprises a group of complex animal DNA viruses. The family is further subdivided into the two subfamilies, *Chordopoxvirinae* and *Entomopoxvirinae*, the representatives of which infect vertebrate and insect hosts respectively. Within the subfamily, *Chordopoxvirinae*, vaccinia virus belongs to the genus *Orthopoxvirus*.

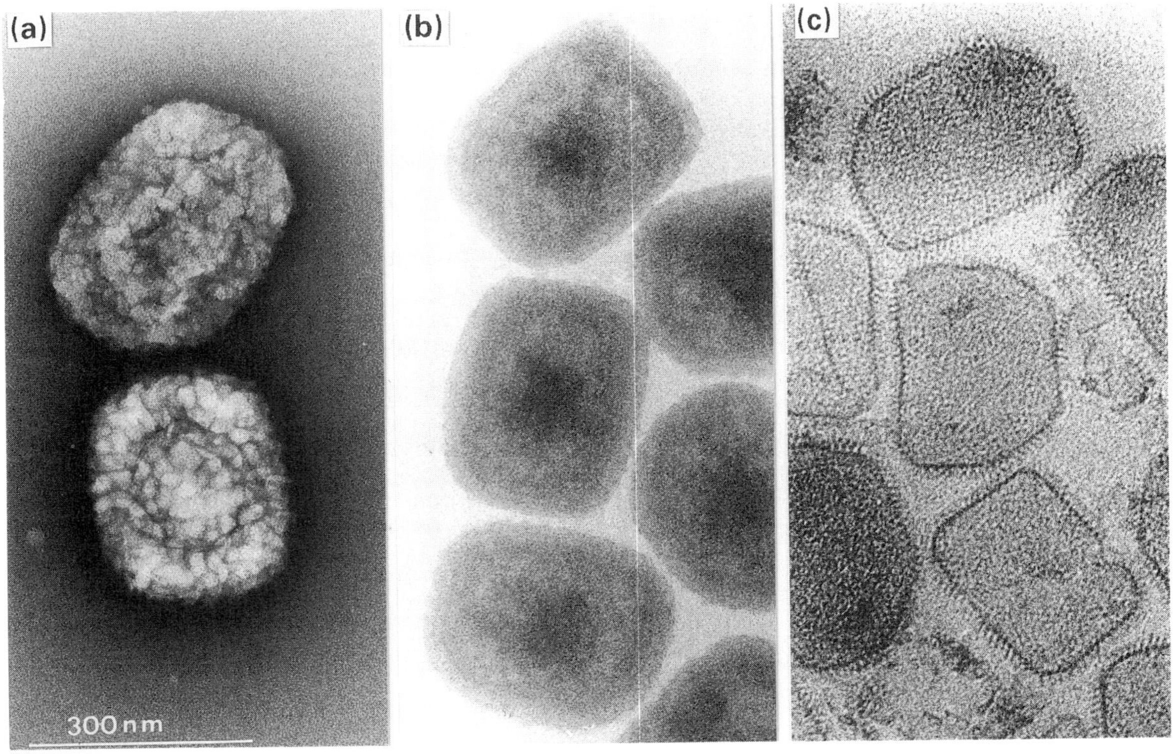

Fig. 1 Structure of vaccinia virus including the viral core. (**a**) Conventional preparation of the virus by negative staining. (**b**) Vitrified sample observed by cryoelectron microscopy. The virus particles are in the native state and are floating in a thin layer of vitrified solution kept for observation at *ca.* − 160°C and are neither stained nor chemically fixed. (**c**) Core particles prepared as in (**b**). (Courtesy M. Adrian and J. Dubochet.)

Properties of the Virion

Poxviruses are among the largest and most complex viruses known. The particles have a typical brickshape and measure about 300 × 250 nm. Two types of particles can be distinguished. Virions that are naturally released from the infected cells are surrounded by a Golgi-derived envelope. Virions that are released by experimental cell lysis lack this envelope, but are also infectious. Electron microscopy of negatively stained specimens reveals that the surface of nonenveloped particles is composed of tubular structures ('surface tubules'), which give the particles a characteristic appearance (Fig. 1a). However, more recent cryoelectron microscopy (Fig. 1b), which allows examination of hydrated viral particles, provided no evidence of surface tubules and it is conceivable that these structures represent a shrinkage artifact resulting from dehydration of particles prepared for electron microscopy by conventional procedures.

In thin sections of virions (Fig. 1c), the internal core structure can be visualized, which consists of an oval biconcave disk. The concavities of the core accommodate the two lateral bodies.

Properties of the Genome

Vaccinia virions contain a large double-stranded DNA genome which has been entirely sequenced in the Copenhagen strain and shown to consist of 191 636 bp. A characteristic feature of the molecules are the cross-links that join the two DNA strands at both ends. The terminal 100 nucleotides or so consist of single-stranded loops. The genome is further characterized by the presence of long inverted terminal repeats of about 10 kbp. The terminal 3.5 kbp of these are mainly composed of a tandemly repeated 70 bp sequence.

Properties of the Protein

The DNA sequence of the viral genome provides evidence for the presence of about 200 potential protein-coding sequences, but only relatively few proteins have been assigned a function. The majority of polypeptides with known or suspected functions are enzymes involved in nucleic acid metabolism. Examples are the subunits of RNA polymerase, enzymes involved in capping and polyadenylation of mRNA, DNA polymerase, thymidine and thymidylate kinase, DNA ligase and several more. Other proteins are structural components of the virion. Together with some enzymes, which are also packaged in the viral particle, the total number of virion polypeptides may be as high as 100. The envelope of extracellular virions contains at least eight proteins, seven of which are glycosylated. The 37 kD major envelope antigen is acylated.

Physical Properties

The development of a stable smallpox vaccine was greatly facilitated by the fact that the infectivity of vaccinia virus is relatively unaffected by environmental conditions which inactivate most other viruses. This property, together with the large particle size, was also the reason why vaccinia virus was the first animal virus to be purified extensively. Heating at 55°C for 60 min, or at 50°C for 90 min, completely destroys infectivity. Other methods for destroying infectivity were tested in attempts to produce an inactivated smallpox vaccine. UV irradiation, formaldehyde treatment, photodynamic inactivation with methylene blue and γ irradiation were all found to inactivate vaccinia virus, but some of these procedures also resulted in a loss of antigenicity.

Complete solubilization of vaccinia virus is achieved by heating at 100°C in the presence of sodium dodecyl sulfate and reducing agents.

Strategy of Replication of Nucleic Acid

Poxviruses are unusual among DNA viruses in that their replication cycle takes place in the cytoplasm of the infected host cell. Upon penetration, the virus sheds its outer protein layers resulting in the release of the viral core containing the DNA and enzymes involved in the synthesis and modification of mRNA. A first burst of RNA synthesis occurs. Following expression of the early genes, the DNA is released from the core and replicated. This allows expression of the intermediate genes, the transcription of which requires a naked DNA template. After transcription of intermediate genes, late genes, many of which encode structural proteins, are expressed.

Replication of the genome occurs through synthesis of long concatemeric intermediates which are subsequently resolved into unitlength genomes. Concatemer resolution is a highly specific process and depends on a 20 bp element located adjacent to the hairpin loop in the mature DNA molecule. DNA replication itself does not appear to require specific origins of replication since any DNA transfected into vaccinia virus-infected cells is replicated.

Characterization of Transcription

Transcription of early genes depends on early promoter elements, which are about 30 bp long. An early transcription factor binds to these regulatory sequences. Termination of early transcription occurs about 50 bp downstream of the termination signal TTTTTNT. Early mRNAs are capped and polyadenylated and typically contain short 5′ untranslated leader sequences.

Intermediate gene transcription requires the presence of two intermediate gene transcription factors which are synthesized early in infection but which are able to activate intermediate gene transcription only after the DNA has been replicated.

Three intermediate gene products are required for late gene transcription. Late gene promoters also consist of about 30 bp and contain the highly conserved TAAAT motif in which transcription initiation occurs. Late mRNAs are heterogeneous in length, are polyadenylated and have a capped poly(A) leader sequence of about 35 A residues. The leader is not encoded in the genome but is produced by stuttering of RNA polymerase in the TAAAT motif.

Transcription in vaccinia virus is characterized by a cascade in which transcription of each temporal class of genes requires the presence of specific transcription factors which are made by the preceding temporal class of genes. Thus, early gene transcription factors are made late in infection, packaged in virions and used in the subsequent round of infection. Intermediate gene transcription factors are encoded by early genes, and late transcription factors by intermediate genes.

Characterization of Translation

Infection of cells by vaccinia virus results in rapid shut-off of host cell DNA, RNA and protein synthesis. Inhibition of protein synthesis is mediated by a virion structural protein and consequently can occur in the absence of viral gene expression. Other mechanisms leading to inhibition of cellular protein synthesis have also been proposed.

The preferential translation of temporal classes of mRNAs is probably a direct consequence of the abundance of mRNAs at a given time of infection, although it is tempting to speculate that the poly(A) leader of late mRNAs confers on these mRNAs selective translatability late in infection.

Post-translational Processing

Several types of post-translational processing events occur in the maturation of viral proteins. The mature forms of the two major core proteins and of the viral growth factor are generated by cleavage of higher-molecular-mass precursors. Several proteins are glycosylated. Examples are the envelope proteins with the exception of the 37 kD protein which is acylated. Finally, some viral proteins are phosphorylated.

Assembly Site, Uptake, Release and Cytopathology

The penetration of cells by naked virions differs from that of enveloped particles. Some naked virions enter cells by fusion of the outer membrane with the cell membrane. The strong temperature dependence of penetration of these virions and its sensitivity to sodium fluoride and cytochalasin B indicate that the majority of particles enter cells by endocytosis. In contrast, uptake of enveloped particles is relatively efficient at low temperature and insensitive to the above compounds suggesting that such particles penetrate by fusion of the viral envelope with the cell membrane.

Since vaccinia virions are composed of a very large number of proteins, it is not surprising that viral assembly is a complex process which is still poorly understood and which requires several hours to be completed.

Morphogenesis results in the production of infectious virions which are surrounded by a lipid-containing outer membrane. A small fraction of these particles are further enveloped by two Golgi-derived membranes containing several viral antigens. Wrapped virions migrate along actin-containing microfilaments to the cell surface where the outer of the two membranes fuses with the plasma membrane. This results in the release of enveloped virions which are composed of the original naked particle enclosed in the inner Golgi membrane. This process of envelopment and release is relatively inefficient. Depending on viral strains and host cell, between 1 and 30% of the progeny is released as enveloped particles. The majority of naked virions remain cell-associated even after cell death.

Infection of cells in monolayer cultures induces several changes. Within a few hours after infection, the cells become rounded and retract from each other. Basophilic areas, which are the sites of viral replication, then appear in the cytoplasm. These areas develop into B-type inclusion bodies ('Guarneri bodies') which are distinct from the more prominent A-type inclusion bodies seen in cells infected with other orthopoxviruses.

Vaccinia Virus as a Eucaryotic Expression Vector

Foreign genetic information has been inserted into the vaccinia virus genome by homologous *in vivo* recombination. The standard procedure consists of fusing the foreign gene of interest to a vaccinia virus promoter and inserting the chimeric gene into the nonessential viral thymidine kinase gene contained in a recombinant plasmid. The resulting DNA is amplified and transfected into cells that have also been infected with wild-type virus. Upon viral DNA replication, the thymidine kinase DNA sequences in the viral genome and recombinant plasmid undergo homologous recombination resulting in insertion of the foreign gene into the viral genome. Since this inactivates the viral thymidine kinase gene, recombinant virus can conveniently be selected on the basis of the thymidine kinase-negative phenotype.

Recombinant vaccinia virus has become an invaluable tool for overexpressing proteins that are normally made only in trace amounts in the cell, such as transcription factors, receptors or other proteins of particular biological interest. Advantages of vaccinia virus compared to procaryotic expression systems are the correct post-translational processing and solubility of the proteins.

An exciting application of recombinant virus expressing antigens of other infectious agents is their use as live virus vaccines in human and veterinary medicine. Experimental animals immunized with recombinants expressing antigens of other pathogens produce neutralizing antibodies and a cell-mediated immune response and are protected against infection with the live pathogen. Advantages of such vaccines are the low production cost, stability and ease of administration, features that were major determinants for the success of the smallpox eradication program. The main argument against the introduction of recombinants is the residual virulence of vaccinia virus. Considerable effort is therefore currently being made to produce a safer vaccinia virus vector.

Geographic and Seasonal Distribution

Vaccinia virus is not associated with a naturally occurring disease in humans. The smallpox vaccine was extensively used world-wide but, after eradication of smallpox, vaccination was abandoned.

Rabbitpox virus, a highly virulent vaccinia virus strain has caused outbreaks of a lethal pox disease in colonies of laboratory rabbits in Utrecht and New York. Several outbreaks of buffalopox occurred in Maharashtra state in India, where buffaloes are used for milk production. The disease is caused by an orthopoxvirus which is closely related to vaccinia virus, but different enough to justify classification as a subspecies of vaccinia virus.

Host Range and Virus Propagation

Vaccinia virus has a very broad host range and can infect most vertebrate animals. Mice and rabbits are the most commonly used experimental animals for studying the pathogenicity of viral mutants. The chorioallantoic membrane of developing chick embryos was widely used for virus propagation and for differentiating vaccinia virus from other poxviruses, but has been replaced by cultured cells. High virus titers can be obtained in most cell lines. Some of the most frequently used cells are HeLa cells, either grown in suspension or as monolayer cultures, mouse L cells and CV-1 African green monkey kidney cells. Chinese hamster ovary cells do not support virus multiplication.

Genetics

The first vaccinia virus mutants were isolated on the chorioallantoic membrane of chick embryos where most strains normally produce red hemorrhagic lesions (pocks). White pock lesions arise with a relatively high frequency and the viruses isolated from such pocks have been designated white pock mutants. Analysis of the genome of these mutants revealed large deletions near the termini of the DNA or more complex rearrangements involving both ends. In some white pock mutants, the deletions were also associated with an altered host range. A large number of temperature-sensitive and drug-resistant mutants have also been isolated. For many of these, the mutation was mapped in the genome by marker rescue experiments.

Evolution

Several hypotheses for the origin of vaccinia virus have been advanced, but none is entirely convincing. According to some of these, vaccinia virus was either derived from variola or cowpox virus, or is a hybrid between the two. Other authors propose that vaccinia virus is a 'fossil' and represents the maintenance in the laboratory of a virus of a domestic or wild animal that has otherwise become extinct. With the exception of buffalopox virus, all vaccinia virus strains form a homogeneous group of viruses both with respect to biological properties and genome structure and are not more closely related to cowpox virus or variola virus than to other orthopoxvirus species.

At the genome level, differences between vaccinia virus strains are mainly due to variability in the terminal regions of the molecule, whereas the central part is highly conserved. Comparison of protein-coding regions between different strains typically show greater than 99% identity in amino acid sequences.

Buffalopox virus is also considered to be a vaccinia virus strain and was probably derived from smallpox vaccine. The virus appears to have evolved more rapidly and has become a separate subspecies which is maintained in buffalo herds in certain parts of India.

Serologic Relationships and Variability

All orthopoxviruses show extensive serological cross-reactivity which is even more marked between individual vaccinia virus strains. Strains cannot, therefore, be distinguished by serological means. Restriction enzyme analysis of the viral genome has become the most reliable method for differentiating between vaccinia virus strains, and has largely replaced methods based on biological criteria such as pock morphology on the chorioallantoic membrane or ceiling temperature.

Epidemiology

Vaccinia virus is occasionally spread from newly vaccinated humans to nonvaccinated individuals. This could lead to serious complications in cases where contraindications for vaccination of the unvaccinated contact exist. Transmission also occurred to domestic animals, such as cows and pigs, and these animals in turn, represented a source of infection for humans. Buffalopox virus, which probably also originated from vaccinia virus, was spread between animals by farmers. Several severe infections of persons handling the buffaloes were reported from India. With the cessation of smallpox vaccination, it was expected that buffalopox would disappear, but severe outbreaks still occurred in 1985. Most outbreaks of rabbitpox in colonies of laboratory rabbits were caused by a vaccinia virus strain that was accidentally transmitted to the animals by laboratory workers. In some outbreaks, the source of infection was not traced.

Transmission and Tissue Tropism

The smallpox vaccine was usually introduced into the epidermis over the deltoid muscle. Jenner inoculated human subjects by a light scratch of the skin through a drop of vaccine and variations of this technique were used for a very long time. During the smallpox eradication campaign, other techniques for delivering vaccine were developed. Jet injectors were widely used in West Africa and Brazil. The major advance was the development of the bifurcated needle which could easily be used even by inexperienced vaccinators and which required no special maintenance.

Accidental transmission of vaccinia virus between humans or between humans and animals occurred by direct contact through minute skin lesions. In most outbreaks of rabbitpox, spread appeared to occur by the respiratory route; direct contact between animals was not necessary.

After smallpox vaccination of humans, virus replicated in the epidermis and remained confined to the site of inoculation. Spread to other parts of the body was only observed in the rare cases of complications. After infection of rabbits with rabbitpox virus by the respiratory route, virus spread through the body in a stepwise manner and replication occurred in various tissues.

Pathogenicity

The frequency of the rare, but serious, complications of smallpox vaccination was related to the use of particular vaccinia virus strains. For instance, after the Bern strain, once used in Austria, Switzerland and West Germany, had been replaced by the Lister strain in 1971, the occurrence of complications of the central nervous system declined. Other strains with a good safety record are the New York City Board of Health strain and the Japanese LC16m8 strain. Since the residual virulence of vaccinia virus is the major obstacle for the introduction of recombinant virus as live vaccines against other pathogens, considerable effort is being made to make the virus a safer vector. Several nonessential genes have been deleted from the viral genome and, in each case, the resulting mutant virus has shown an attenuated phenotype in experimental animal models.

Rabbitpox virus is a highly virulent vaccinia virus strain for rabbits causing a rapidly lethal disease. Frequently death occurs before skin lesions have time to develop.

Clinical Features of Infection

After primary smallpox vaccination, a papule developed in 3–5 days, rapidly became a vesicle and later became pustular, reaching its maximum size after 8–10 days. A scab then formed, which separated at 14–21 days leaving a typical vaccination scar. Vaccination produced a generalized infection with swelling and tenderness of the draining lymph nodes and mild fever. The vaccinees frequently felt miserable for a few days.

Apart from the 'normal' vaccination reactions, which were more intense than those seen with other human vaccines, a number of rare but serious complications occurred. These were of two kinds; those affecting the skin and those affecting the central nervous system. In the first group, progressive vaccinia was the least frequent, but most serious complication and was observed only in vaccinees with a deficient cell-mediated immune system. The clinical features of progressive vaccinia were a failure of the primary vaccination lesion to heal, appearance of lesions elsewhere on the body and progression of all lesions until the patient died. The fatality rate was extremely high. Eczema vaccination was the second most serious skin affection and occurred among persons with eczema. This complication was characterized by the appearance of lesions on areas of the skin that were eczematous at the time or had previously been so. The fatality rate was about 30%. Generalized vaccinia followed virus spread via the blood stream. Lesions similar to the vaccination lesions appeared on many parts of the body. Generalized vaccinia had a good prognosis. Accidental infection occurred among laboratory workers in research facilities or vaccine production centers or in contacts of newly vaccinated persons and were most serious when they affected the eye. Complications affecting the central nervous system were encephalopathy, usually in children less than 2 years of age, and encephalomyelitis in older children and adults. These complications were of particular concern since they occurred in persons with no obvious contraindications for vaccination and had a high case-fatality rate.

Pathology and Histopathology

The pathology and histopathology of 'normal' skin lesions at various stages after smallpox vaccination were studied in biopsies. For more information, the reader is referred to the specialized literature which also deals with the pathology of the rare complications of vaccination.

Immune Response

Smallpox vaccination with vaccinia virus provided complete protection against smallpox for at least 5 years and some protection for over 30 years. Both humoral and cell-mediated immune responses were observed. Studies in mice indicated that cytotoxic T cells are more important for protection against mousepox than circulating antibodies. In humans, progressive vaccinia developed in patients with an impaired cell-mediated immune system but normal humoral antibody responses, underlining the importance of cytotoxic T cells for limiting vaccinia virus spread and presumably also for providing protection against smallpox.

Prevention and Control

Since the eradication of smallpox, all countries have abandoned general compulsory smallpox vaccination. Some countries, however, have continued to vaccinate their armed forces, at least until very recently. Vaccination is still recommended for the staff in research institutes working with vaccinia virus.

Future Perspectives

The future of vaccinia virus as a vaccine will depend on whether recombinant virus expressing antigens of other pathogens will be used as live virus vaccines. The main argument against the introduction of such recombinants are the rare complications associated with the use of vaccinia virus as the smallpox vaccine. Considering the great potential of recombinant virus, the decision of whether or not to introduce such vaccines should only be made after a thorough risk–benefit evaluation, and the development of safer vectors might influence the decision favorably.

See also: Cell structure and function in virus infections; Epidemiology of viral diseases; Genetics of animal viruses; Mousepox and rabbitpox viruses; Poxviruses; Smallpox and monkeypox viruses; Vaccines and immune response; Vectors.

Further Reading

Fenner F, Wittek R and Dumbell KR (1989) *The Orthopoxviruses*, New York: Academic Press.

Moss B (1990) *Poxviridae* and their replication. In: Fields BN *et al.* (eds) *Virology*, 2nd edn, p. 2079. New York: Raven Press.

VARICELLA-ZOSTER VIRUS

Contents

General Features
Molecular Biology

General Features

Jeffrey I Cohen and Stephen E Straus
National Institutes of Health
Bethesda, Maryland, USA

History

Descriptions of vesicular rashes characteristic of chickenpox (varicella) date back to the ninth century. In 1875 Steiner showed that chickenpox was an infectious agent by transmitting the disease from chickenpox vesicle fluid to previously uninfected people. Shingles (zoster) has been recognized since ancient times. In 1909 Von Bokay suggested that chickenpox and shingles were related infections, an idea that was experimentally confirmed in the 1920s and 1930s when children inoculated with fluid from zoster vesicles were shown to contract chickenpox. By virtue of the remarkable dermatomal confinement of most zoster lesions, Garland suggested in 1943 that zoster was due to reactivation of varicella virus that had remained dormant or latent in sensory nerve ganglia.

The viral agents of varicella and zoster were first cultivated by Weller in 1952 and shown on morphologic, cytopathic and serologic criteria to be identical. In 1984 Straus and colleagues showed that viruses isolated during sequential episodes of chickenpox and zoster from the same patient had identical restriction endonuclease patterns, proving the concept of prolonged latent carriage of the virus. However, demonstration of latent virus within dorsal root ganglia awaited subsequent molecular studies using *in situ* hybridization and polymerase chain reaction techniques.

Taxonomy and Classification

Varicella–zoster virus (VZV) is a member of the *Alphaherpesvirus* subfamily of the family *Herpesviridae*. Other alphaherpesviruses that infect humans include herpes simplex viruses 1 and 2, and cercopithicine herpes virus (B virus). All of these agents exhibit relatively short replicative cycles, destroy the infected cell, and establish latent infection in sensory ganglia.

Three subgroups of simian viruses cause severe varicella-like illnesses in nonhuman primates. These include the deltaherpesvirus, Medical Lake macaque virus and chimpanzee herpesvirus. While these simian viruses are antigenically related to VZV, they are more homologous to each other than to VZV.

Geographic and Seasonal Distribution

Varicella and zoster infections occur world-wide. Over 90% of varicella occurs during childhood in industrialized countries located in the temperate zone, but infection is commonly delayed until adulthood in tropical regions. Zoster may occur less frequently in tropical areas, because of later acquisition of primary infection.

Varicella infection is epidemic each winter and spring, while zoster occurs throughout the year, without a seasonal preference.

Host Range and Virus Propagation

The reservoir for VZV is limited to humans. The virus inherently grows poorly in nonhuman animals or cell lines. Meyers and colleagues, however, developed a guinea pig animal model of VZV infection by adaptation of virus for growth in guinea pig embryo cells and subsequent inoculation of animals with a resulting self-limited viremic infection and the emergence of both humoral and cellular immunity. Virus replicates initially in the nasopharynx and can be transmitted to other guinea pigs. Recently, the same investigators extended their animal model to congentially hairless Hartley guinea pigs with the advantage that their inoculation with VZV results in viremia and an evident papular erythematous rash.

An alternative, but less ideal, animal model involves the common marmoset (*Callithrix jacchus*). VZV replicates in the lungs with a mild viral pneumonia and a subsequent humoral immune response. None of these animal models have, as yet, reproduced the disease pattern seen in humans, namely a vesicular rash and spontaneous re-activation from latency.

VZV is usually cultured in human fetal diploid lung cells in clinical laboratories. The virus has been cultivated in numerous other human cells including melanoma cells, primary human thyroid cells, astrocytes, schwann cells and neurons, and can be grown in simian cells including African green monkey kidney cells and Vero cells, and in guinea pig embryo fibroblasts.

VZV is extremely cell associated. The titer of virus released into the cell culture supernatant is usually very low, and preparation of cell-free virus, by sonication or freeze–thawing cells, usually results in a marked drop in the viral titer. Therefore, virus propagation is usually performed by passage of infected cells on to uninfected cell monolayers. VZV is detected by its cytopathic effect with refractile rounded cells that gradually detach from the monolayer, or by staining with fluorescein-labeled antibody.

Genetics

Several markers have been described to distinguish different strains of the virus. These markers include temperature sensitivity, plaque size, different effects on host cell lipid metabolism, antiviral sensitivity and restriction endonuclease cleavage patterns. The molecular basis for most of these strain differences are unknown, but viruses that are resistant to acyclovir are usually found to have mutations in their pyrimidine deoxyribonucleoside kinase gene. Other resistant strains have mutations in the DNA polymerase gene.

Little formal genetic analysis of VZV has been carried out. The entire genome consists of 124 884 bp. The identifications of some viral genes were made by analogy to herpes simplex virus type 1 genes with similar sequences and by genetic complementation studies in which cell lines expressing selected VZV proteins were used to support the growth of temperature-sensitive mutants of herpes simplex type 1 viruses. More recently, a VZV gene was shown to substitute for a herpes simplex type 1 gene when recombined into the genome of the latter virus. Recombinant VZVs containing an Epstein–Barr virus gene or the gene encoding hepatitis B virus surface antigen have been constructed.

Evolution

Comparison of the nucleotide and predicted amino acid sequence of VZV with herpes simplex virus type 1 indicates that these two viruses originated from a common ancestor. The two viruses have very similar arrangements of their genes and only five genes of VZV do not appear to have a herpes simplex virus type 1 counterpart.

VZV is more distantly related to the other human herpesviruses, but many of the nonstructural proteins involved in viral replication have conserved elements. Comparison of VZV, for example, with Epstein–Barr virus shows that the majority of VZV genes are homologous with Epstein–Barr virus. Three large blocks of genes are conserved, although rearranged within the two genomes.

Serologic Relationships and Variability

There is only one serotype of VZV. Antibodies detected by the complement fixation test and virus-specific IgM antibodies decline rapidly after convalescence from varicella. Other, more sensitive, serologic tests recognize antibodies that persist including immune adherence hemagglutination (IAHA), fluorescent antibody to membrane antigen (FAMA) and enzyme-linked immunosorbent assay (ELISA). VZV-specific antibodies are boosted by recrudescent infection (zoster).

Variability of VZV strains has been shown primarily by differences in restriction endonuclease patterns. Passage of individual strains *in vitro* eventually results in minor changes in restriction endonuclease patterns, predominantly through deletion or reiteration of small repeated elements scattered throughout the genome. Other than these sites, the genome sequence is remarkably stable. For example, the sequence of the pyrimidine deoxyribonucleoside kinase gene has been determined for several wild-type and acyclovir-resistant strains and found to possess >99% nucleotide and amino acid identity.

Epidemiology

Varicella may occur after exposure to chickenpox or to herpes zoster. Over 95% of infections with VZV result in symptomatic infection known as chickenpox. Over 90% of individuals in temperate countries are infected with VZV before the age of 15.

Zoster is due to re-activation of VZV in patients who have had chickenpox in the past: some of these patients may not recall a primary infection. Zoster is not clearly related to exposure to chickenpox or to other cases of zoster. About 10–20% of individuals ultimately develop herpes zoster; the majority of cases occur in the elderly. Severely immunocompromised patients, such as those with the acquired immunodeficiency syndrome (AIDS), have a particularly high incidence of zoster. Recurrent zoster is uncommon; less than 4% of patients experience a second episode.

Transmission and Tissue Tropism

Transmission of VZV occurs by the respiratory route. Intimate, rather than casual, contact is important for transmission. Chickenpox is highly contagious; about 60–90% of susceptible household contacts become infected. Herpes zoster is less contagious than chickenpox. Only 20–30% of susceptible contacts become infected. Patients are infectious from 2 days before the onset of the rash until all the lesions have crusted.

Primary infection with VZV results in viral replication in the upper respiratory tract and oropharynx with lesions present on the respiratory mucosa. The infection subsequently spreads to the lymphatics and a mild viremia develops. Mononuclear cells are thought to support viral replication and convey virus throughout the body. Further viral replication occurs in the reticuloendothelial system. A brisk secondary viremia results in spread to the periphery. Skin lesions begin with infection of the vascular endothelial cells with spread to epithelial cells. At some time in the course of the infection the virus becomes latent in sensory dorsal root ganglia. Recent studies using *in situ* hybridization indicate that the site of latency in the ganglia is the satellite cells surrounding the neurons. Conflicting data argue for neuronal latency. In either case, viral DNA can be detected in thoracic and trigeminal ganglia by the polymerase chain reaction.

Zoster is due to re-activation of virus in the sensory ganglia. The factors leading to re-activation are not known, but are associated with neural injury and cellular immune impairment. Re-activated virus spreads down the sensory neuron to the skin where the resulting vesicles are typically confined to a single dermatome. Viremia and subsequent cutaneous or visceral dissemination of lesions may occur with zoster, especially in immunocompromised patients.

Pathogenicity

Passage of wild-type VZV in cell culture to produce the Oka vaccine strain resulted in attenuation of the virus and changes in temperature sensitivity and infectivity for certain cell lines. The Oka vaccine strain has differences in restriction endonuclease patterns from its parental wild-type strain, but it is not known which of these differences, if any, contribute to attenuation. For example, the gene for glycoprotein gpV has four 42 bp repeats in the Oka strain, while a wild-type clinical isolate has seven of the repeats.

Clinical Features of Infection

The incubation period for chickenpox is 2 weeks, with a range of 10-21 days. The disease begins with fever and malaise, followed by a generalized vesicular rash. Lesions tend to present first on the head and trunk and then spread to the extremities. New lesions usually follow viremic waves for 3–5 days and, in the normal host, most lesions are crusted and healed by 2 weeks. Lesions are often present in different stages at the same time in an individual. The disease is usually self-limited in the normal host.

Complications of varicella are more common in neonates, in children with malnutrition, in immunocompromised patients (e.g. malignancy or immunosuppressive therapy), in pregnant women and in older adults. These complications include bacterial superinfection of the skin, pneumonia, hepatitis, encephalitis, thrombocytopenia and purpura fulminans. Reye syndrome occurs in rare children who take aspirin to treat varicella fevers. There are about

3 million cases of varicella with about 100 deaths each year in the US.

Zoster usually presents with pain and dysesthesias 1–4 days before the onset of the vesicular rash. The rash is usually painful and confined to a single dermatome, but may involve several adjacent dermatomes. Fever and malaise often accompany the rash. Vesicles are often pustular by day 4 and become crusted by day 10 in the normal host.

Postherpetic neuralgia, manifested by pain lasting for weeks to several years in the area of the initial rash, is the most common and disconcerting complication of zoster. Less common complications include encephalitis, myelitis, the Ramsay–Hunt syndrome (lesions in the ear canal, with auditory and facial nerve involvement), ophthalmoplegia, facial weakness and pneumonitis. Immunocompromised patients with zoster are more likely to develop disseminated disease with neurologic, ocular or visceral involvement. Patients with AIDS have a high frequency of zoster and may develop recurrent or chronic disease with verrucous hyperkeratotic skin lesions.

Pathology and Histology

Varicella lesions are most often recognized in the skin and mucous membranes. However, lesions also occur in the mucosa of the respiratory and gastrointestinal tracts, liver, spleen and any tissue. With severe disease there is inflammatory infiltration of the small vessels of most organs. Zoster causes inflammation and necrosis of the sensory ganglia and its nerves, and skin lesions which are histopathologically identical to those seen with varicella.

Cutaneous lesions due to VZV begin with infection of capillary endothelial cells followed by direct spread to epidermal epithelial cells. The epidermis becomes edematous with acantholysis and vesicle formation. Mononuclear cells infiltrate the small vessels of the dermis. Initially, vesicles contain clear fluid with cell-free virus, but later the vesicles become cloudy and contain neutrophils, macrophages, interferon and other cellular and humoral components of the inflammatory response pathways. Subsequently, the vesicles dry leaving a crust that heals usually without scarring.

Cells infected with VZV show eosinophilic intranuclear inclusions with multinucleate giant cell formation. These changes are not specific for VZV, but are seen with other herpesvirus infections.

Immune Response

Infection with VZV elicits both a humoral and cellular immune response. The ability of VZV immunoglobulin (VZIG) to attenuate or prevent infection in exposed children (see under Prevention and Control below) indicates that virus-specific antibody is important in protection from primary infection. The presence of VZV-specific IgG does not correlate, however, with protection from zoster. Antibody to VZV is often present at the time the rash of varicella first appears. Virus-specific IgM, IgG and IgA are present within 5 days of symptomatic disease; however, only IgG persists for life. Antibodies to viral glycoproteins, gpI, gpII, gpIII and the immediate-early gene product ORF 62 have been detected during acute infection, and the titers to these proteins increase during recurrent infection. The presence of antibody to the three glycoproteins in children with leukemia who had received live varicella vaccine was not adequate to prevent breakthrough varicella or zoster.

The cellular immune responses are thought to be more important in recovery from acute varicella infection and for prevention of and recovery from zoster. The level of cellular immunity correlates with disease severity during acute varicella. Cytotoxic T cells that lyse virus-infected cells are present 2–3 days after the onset of the rash of varicella. Cell-mediated immunity, as measured by lymphocyte proliferative response, is directed to glycoproteins gpI–gpV, the immediate-early gene product ORF 62, and presumably other gene products as well. Interferon is present in VZV vesicles.

Most varicella infections result in lifelong immunity. Second episodes of varicella are rare; these individuals tend to have reduced humoral and cellular immunity to VZV at the time of the second infection. Zoster is associated with a reduction in cellular immunity to VZV that, in the normal host, is partially restored in response to this recurrent infection. Recurrent zoster is very uncommon, except in severely immunodeficient patients, such as those with AIDS.

Prevention and Control

Prevention of varicella can be achieved by restricting exposure or by resorting to either immunoglobulin prophylaxis or vaccination with live attenuated virus. VZIG prevents or attenuates varicella in seronegative persons, if given within 4 days of exposure to the virus; the preparation has no effect in modifying zoster. VZIG is recommended for individuals (1) with recent close contact to patients with varicella or zoster, (2) who are susceptible to varicella and (3) who fall in a high-risk category. The latter include premature or certain newborn infants, pregnant women and patients with congenital or acquired immunodeficiencies.

A live attenuated varicella vaccine (Oka strain) protects normal children and adults, as well as children with malignancies, from infection with varicella. Most vaccine recipients develop adequate humoral and cellular immunity to varicella after a single dose of vaccine; additional doses enhance the degree of immunity. A rash may follow vaccination. It is usually mild, but can be severe if the vaccine is given to patients experiencing periods of profound cellular immune impairment. The live vaccine virus establishes neural latency and can re-activate. Thus zoster has been reported in vaccinees, especially those who are immunocompromised, but the rate appears to be no higher than that following natural infection. Vaccination may be combined with VZIG for postexposure prophylaxis.

Patients with varicella or zoster should be isolated from susceptible persons until all lesions have crusted. This is particularly important for hospital workers and immunodeficient patients.

Acyclovir, vidarabine and leukocyte interferon have been used in the treatment of varicella and zoster in immunocompromised patients. Interferon proved to be an inadequate and impractical therapy. Vidarabine must be administered intravenously and is also less effective than acyclovir. Acyclovir, then, is the current treatment of choice for selected infections. It results in a shorter duration of symptoms and decreased visceral dissemination of varicella or zoster in the immunocompromised host. Acyclovir also prevents spread of trigeminal zoster to the eye and modestly shortens the duration of varicella and zoster symptoms in the normal host. Acyclovir-resistant strains of VZV have been reported in patients with AIDS; these infections are best treated with the experimental drug, foscarnet.

Future Perspectives

Vaccination of children (both normal and immunocompromised) with the attenuated live varicella vaccine should reduce the incidence and severity of varicella. The increase in humoral and cellular immunity after vaccination of elderly patients suggests that the vaccine might also reduce the frequency and severity of zoster; however, this has not yet been confirmed. Because of the potential for the live vaccine to cause zoster, subunit vaccines (containing viral glycoproteins) may prove preferable.

See also: Antivirals; Diagnostic techniques; Herpes simplex viruses; Immune response; Interferons; Latency; Persistent viral infection.

Further Reading

First International Conference on the varicella–zoster virus (1992) *J. Infect. Dis.* 166(Suppl. 1): S1.

Hyman RW (1987) *Natural History of Varicella–Zoster Virus*. Boca Raton, Florida, CRC Press.

Straus SE *et al.* (1988) Varicella–zoster virus infections: biology, natural history, treatment and prevention. *Ann. Intern. Med.* 108: 221.

Takahashi M (1988) Varicella vaccine. In: Plotkin SA and Mortimer EA (eds) *Vaccines*, p. 526. Philadelphia: W.B. Saunders.

Molecular Biology

William T Ruyechan and John Hay
State University of New York at Buffalo
Buffalo, New York, USA

Properties of the Virion

The morphology of the varicella–zoster virus (VZV) virion is very similar to that of other herpesviruses. The virion is 180–200 nm in diameter and is made up of four structurally distinct elements. The linear duplex viral DNA genome is packaged in a central 75 nm core, and the core is contained within an icosahedral nucleocapsid 100 nm in diameter, with 5:3:2 axial symmetry. The nucleocapsid is composed of 162 hexameric and pentameric capsomeres with central hollows wider on the outside than the inside. A proteinaceous tegument surrounds the nucleocapsid and is, in turn, surrounded by a lipid bilayer derived from the nuclear membrane and Golgi transport vesicles. This lipid coat contains five major virus-encoded glycoprotein species which are visualized by electron microscopy as spikes or studs approximately 8 nm in length.

A minimum of 30 proteins ranging in size from about 250 to 17 kD have been identified as components of the VZV virion. The capsid is comprised primarily of the 155 kD VZV major capsid protein which makes up the capsomeres. In addition, some eight to ten other polypeptides are also present in purified nucleocapsids based on sodium dodecyl sulfate–polyacrylamide gel electrophoretic (SDS–PAGE) analysis. Thus the capsid proteins and the glycoproteins account for somewhat less than half of the virion proteins. The remainder are, by a process of elimination, contained in the tegument of the virus. Little is currently known about the VZV tegument proteins. Recent work, however, has shown that two of the components are the VZV homologues of the herpes simplex virus (HSV) α-trans inducing factor (α-TIF) and the immediate-early transcriptional regulatory protein ICP4 (IE175). In addition, the product of VZV gene 47, the predicted amino acid sequence of which contains a highly conserved protein kinase motif, also appears to be present in the virion. The significance of the presence of these proteins will be discussed in subsequent sections.

Properties of the Genome

The genome of VZV is a linear duplex DNA molecule with a buoyant density of 1.706 g cm^{-3} and an overall G+C content of 46%. The genome of VZV strain Dumas has been sequenced in its entirety and contains a total of 124 884 bp. The size and organization of VZV DNA indicated by this sequence are in good agreement with experimental data derived from a number of different strains over the last decade and it has been taken as the paradigm by workers in the field. The VZV genome is composed of two covalently joined segments designated L (long segment) and S (short segment) and is subdivided into four distinct domains based upon sequence redundancy. These include a long unique region, U_L, about 105 kbp in length and a short unique region, U_S, about 5.2 kbp in length, both of which are composed primarily, although not exclusively, of single-copy sequences. The U_L region is bounded by a short 88.5 bp element (R_L) which is present in inverted orientation at the external (TR_L) and internal (IR_L) termini of the L segment. The S segment is also bounded by a set of inverted repeats termed TR_S and IR_S. These repeat elements, however, are much larger, being about 7.3 kbp in size. Both repeat elements are considerably higher in GC content than the U_L and U_S segments with average values of 58% and 70% for R_S and R_L respectively as compared to about 43% for the single copy sequences.

The VZV genome exists as two predominant isomeric forms which result from inversion of the S segment. These two isomers which have been designated P (prototype) and I_S (inverted S) are present in equimolar amounts and represent 90–95% of packaged VZV DNA. The remaining two isomeric possibilities, involving inversion of the L segment (I_L) and of both L and S (I_{LS}) with respect to the prototype, are also present but represent only 5–10% of the viral DNA. The four isomeric forms of VZV DNA are believed to result from general recombination during viral DNA synthesis. Why only two isomers predominate as opposed to the HSV case of four equimolar populations is not clear, although this finding could indicate a preferential recognition of specific L–S joints by the VZV cleavage apparatus. Finally, a small and variable percentage of molecules (0.1–5.0%) can be isolated from virions as full-sized circular genomes. The origin of these molecules, the modes and temporal processes of their packaging and circularization, and the role they may play in the infectious cycle of the virus are unknown.

Properties of Virus Proteins

Computer analysis of the sequence of the VZV genome has revealed that the viral DNA is capable of encoding a minimum of 68 proteins ranging in size from 8079 to 306 325 daltons. Currently, 20 of these proteins and/or their associated activities have been identified in VZV-infected cells. The functions of another 23 have been inferred based upon comparison of their genome location and predicted primary amino acid sequences with those of known HSV functions. Five of the open reading frames (ORFs) appear to be unique to VZV but the function of only one of these unique genes, thymidylate synthetase, has been characterized. This information is summarized in Table 1.

The most extensively studied VZV protein species are the viral glycoproteins designated gpI, gpII, gpIII, gpIV and gpV which are encoded by ORFs 67, 31, 37, 68 and 14 respectively and which show varying degrees of homology to the HSV glycoproteins gE, gB, gH, gI, and gC. These glycoproteins are incorporated into the lipid coat of the virion during maturation and also appear on the plasma membrane of infected cells. All of the VZV glycoproteins appear to be heavily modified post-translationally.

Table 1. Functions of VZV open reading frames

Gene	HSV	Mol. size	Properties	Function
1		12 103	Phobic/C	
2		25 983		
3	U_L55	19 149		
4	U_L54	51 540	Philic/N	(IE) transcriptional regulator
5	U_L53	38 575	Phobic	Membrane protein/*syn*? essential*
6	U_L52	122 541		DNA helicase–primase complex*
7	U_L51	28 245		
8	U_L50	44 816		dUTPase
9	U_L49	32 845	Philic	
10	U_L48	46 573		Structural (tegument)
11	U_L47	91 825	Philic/acidN	Late/tegument*
12	U_L46	74 269		
13		34 531		Thymidylate synthetase
14	U_L44	61 350		Glycoprotein V (=HSV gC)
15	U_L43	44 522	Phobic	Membrane?
16	U_L42	46 087		DNA polymerase accessory*
17	U_L41	51 365		Virion*
18	U_L40	35 395	Acid	Ribonucleotide reductase (small)
19	U_L39	86 823		Ribonucleotide reductase (large)
20	U_L38	53 969		Virion?*
21	U_L37	115 774		Late/transcription regulation?*
22	U_L36	306 325		Tegument*
23	U_L35	24 416	Phil/STQrich	
24	U_L34	30 451	Phobic/C	Phosphoprotein*
25	U_L33	17 460	Philic/acidN	Virion*
26	U_L32	65 692		Virion?*
27	U_L31	38 234	Philic/baseN	
28	U_L30	134 041		DNA polymerase
29	U_L29	132 133		Major (ss)DNA-binding protein
30	U_L28	86 968		Virion?*
31	U_L27	98 026		Glycoprotein II (=HSV gB)
32		15 980	Philic/acid	
33	U_L26	66 043		Protease*
34	U_L25	65 182		Virion*
35	U_L24	28 973	Basic	
36	U_L23	37 815		Pyrimidine deoxynucleoside kinase
37	U_L22	93 646		Glycoprotein III (=HSV gH)
38	U_L21	60 395		
39	U_L20	27 078	Phobic	Membrane?*
40	U_L19	154 971		Major capsid
41	U_L18	34 387		
42/45	U_L15 ex1/2	82 752 (spliced)		
43	U_L17	73 905		
44	U_L16	40 243		
46	U_L14	22 544		
47	U_L13	54 347		Predicted protein kinase
48	U_L12	61 268		Deoxyribonuclease
49	U_L11	8 907	Philic	Virion*
50	U_L10	48 669	Phobic	
51	U_L9	94 370		*ori*-binding protein

Table 1. Continued

Gene	HSV	Mol. size	Properties	Function
52	U_L8	86 343		DNA helicase–primase complex*
53	U_L7	37 417		
54	U_L6	86 776		Virion*
55	U_L5	98 844		DNA helicase–primase complex*
56	U_L4	27 166	S,T rich	
57		8 079	Philic/basic	
58	U_L3	25 093	Philic/basic	
59	U_L2	34 375		Uracil-DNA glycosylase*
60	U_L1	17 616	Acidic	
61	IE110	50 913	Philic	(IE) transcriptional regulator
62	IE175	139 989		(IE) transcriptional regulator
63	US1	30 494	Philic/acid	(IE?) (=HSV IE68)
64	US10	19 868		Virion*
65	US9	11 436	Phobic	Tegument*
66	US3	43 677		Protein kinase*
67	US7	39 362		Glycoprotein IV (=HSV gI)
68	US8	69 953		Glycoprotein I (=HSV gE)

*Inferred from data with HSV.

GpI and gpIV are believed to form a VZV-specific Fc receptor and thus may have a role similar to that of their HSV counterparts. GpII and gpIII have not as yet been extensively studied; however, based on reasonable homology to gB and gH, it is currently assumed that they are involved in viral attachment and fusion. GpV shows a weak C-terminal homology to HSV gC and, like that glycoprotein, is apparently dispensable for growth in tissue culture. The expression of gpV has been tentatively linked to increased severity of response to inoculation with a live attenuated vaccine strain and hence this glycoprotein may have a role in the pathogenesis of VZV in the human host.

Homologues of all of the seven genes required for origin-dependent DNA synthesis in the HSV-1 system are present in the VZV genome. Of these seven predicted VZV gene products, only two have been identified in infected cells and characterized to any significant extent. These are the viral DNA polymerase and the VZV gene *29* product which is the homologue of the HSV major DNA-binding protein, ICP8. The VZV DNA polymerase has not, as yet, been purified to homogeneity owing to the extreme lability of its activity. Thus all of the data gathered concerning this protein have been obtained with partially purified extracts. The DNA polymerase is activated by monovalent cations and has a peak activity at approximately 80 mM KCl. The pH optimum of the enzyme is 8.0 and it shows a preference for oligopurine-primed deoxypyrimidine templates. The enzyme is sensitive to thiol-blocking reagents and 2,3-butadione suggesting that both thiols and arginine residues are required for activity. Finally, the VZV DNA polymerase is sensitive to phosphonoacetate and acyclovir although to a considerably lesser extent than the HSV-1 DNA polymerase.

The VZV gene *29* product is a major protein species in VZV-infected cells. This protein has been shown to interact with single-stranded DNA in a manner similar to the HSV-1 U_L*29* gene product, with which it shares over 50% homology at the amino acid level. The VZV gene *29* product contains a potential divalent metal-binding site (zinc finger) although the metal content of the protein has yet to be determined. Immunofluorescence staining of VZV-infected cells has shown that the gene *29* product is present primarily in the nuclei of infected cells. The staining pattern observed is similar to the punctate and localized patchy staining seen with the HSV-1 major DNA-binding protein. These results suggest that this VZV protein may have a role in organization of proteins involved in a DNA replication complex analogous to that of the HSV ICP8.

The remainder of the putative VZV origin-dependent replication genes (*6, 16, 51, 52* and *55*) have not been examined in any detail. Owing to their relatively low abundance and the poor growth characteristics of VZV in tissue culture, extensive characterization of these proteins must await their purification from recombinant expression systems.

VZV encodes homologues of four of the five HSV immediate-early genes. All of these gene products are known to be involved in transactivation and transrepression of VZV transcription. The transactivators are the products of genes *4* and *62*. VZV gene is homologous in genome location and, to a limited extent, in predicted amino acid sequence with the HSV U_L*54* gene which encodes IE63 or ICP27. Based on transient expression assays, the VZV gene *4* product is a potent transactivator of both homologous and heterologous promoters and can act synergistically with the VZV IE62 gene product to enhance transcription. No evidence of a transrepressor activity has as yet been identified and this property appears to differentiate the VZV gene *4* product from its HSV analogue which can act both as a transactivator and a transrepressor.

The VZV IE62 gene product is the homologue of the HSV ICP4 or IE175. The gene for this protein is diploid and maps in the inverted repeats bounding the U_S region of the VZV genome. The expected size of the protein based both on its predicted amino acid sequence and the size of its transcript is approximately 140 kD whereas the estimated molecular mass obtained by SDS–PAGE is approximately 175 kD. Both of these pieces of data suggest that the VZV IE62 is highly modified like its HSV homologue. This protein is synthesized very early during infection and appears to be continuously expressed throughout the replication cycle. The IE62 gene product is incorporated into viral particles as a component of the tegument at late times during infection, and levels comparable to that of the major capsid protein are found in mature virions. Transient expression assays have shown that the IE62 protein is a potent transactivator of transcription from VZV genes encoding both structural and nonstructural proteins. The VZV IE62 gene product is capable of rescuing HSV *ts* mutants with lesions in the ICP4 gene indicating that it is a true functional homologue of ICP4.

The third VZV IE protein that will be considered here is the product of gene *61*. The predicted amino acid sequence of this protein indicates the presence of a cysteine-rich region which may be involved in metal binding. A similar region is found in the N-terminal region of the HSV-1 IE110 or ICP0 protein which is known to be a promiscuous transactivator of transcription and to which VZV gene *61* has weak homology. The VZV ORF *61* gene product, in contrast, has been shown to be a potent transrepressor of VZV promoters in nonpermissive Vero cells, whereas it acts as a transactivator of heterologous promoters. The reasons for this apparently VZV-specific transrepression and its consequences in the life cycle of the virus are currently the objects of extensive investigation. More recent data suggest that the transrepression may be cell type-specific since the VZV ORF 61 gene acts synergistically to increase transactivation by the IE4 and IE62 proteins in permissive or semipermissive continuous human T-cell lines.

The fourth IE-type gene product in VZV infected cells is the product of gene 63. This protein can act as both a transactivator and a transrepressor. The transactivation domain is present in the carboxy-terminal half of the protein. The transrepression appears to be cell type dependent as was the case with the gene 61 protein.

Physical Properties and Sensitivity to Environment

Owing to the highly cell-associated nature of VZV and the relatively low yields of virus obtained from infected cells, very little work has been done on the physical properties and chemical composition of the virion. It is assumed that, based on its overall structural similiarity to other herpes virions, that the relative proportions of protein, nucleic acid and lipid are similar, as are the buoyant density and overall virion mass.

The VZV virion is highly sensitive to physical and chemical agents. This sensitivity is manifested in the extremely labile nature of the infectious particle resulting in a particle/infectivity ratio of approximately 10^7 for cell free virus. Vesicle-fluid virus, as a rule, appears to be somewhat more stable than cell-free virions to the agents and manipulations discussed below. VZV is temperature-sensitive and is inactivated rapidly and completely at 60°C and somewhat more slowly at lower temperatures. VZV is also sensitive to freezing, a property that contributes to difficulties in preparation and storage of high-titer stocks. Quick freezing of the virus or infected cells and subsequent storage at −70°C, however, can result in less than 10% loss in titer. In contrast, the titer is rapidly lost upon storage at −10°C. Recent reports indicate that lyophilization of both infected cells and cell-free virus in the presence of sugar followed by storage at temperatures as high as −20°C results in complete preservation of titer.

VZV is sensitive to pH, based on studies with

infected cells, with loss of infectivity occurring below pH 6.2 and above pH 7.8. It is also sensitive to UV irradiation to a similar extent to that of HSV-1 and HSV-2. VZV is also sensitive to mechanical disruption with 80–99% of infectivity lost following 2 min of sonication. This property also appears to be manifested in the loss of infectivity seen upon purification of the virus by ultracentrifugation when a large number of damaged cell-free virions are observed in the electron microscope. The specific reason for the fragility of the VZV virion is not as yet known. However, HSV virions treated in the same manner remain largely intact. One possibility is that VZV virions derived from infections in tissue culture co-purify with vacuoles containing lysosomal enzymes. Thus, the low infectivity seen in tissue culture may be caused, at least in part, by digestion of virion components by glycosidases and peptidases.

Replication

Figure 1 depicts the VZV lytic replication cycle. Various aspects of the cycle and gene expression are described in the following sections.

Nucleic acid

The replication of VZV DNA has not been subject to extensive analysis primarily because of the difficulty in obtaining sufficient numbers of synchronously infected cells. Consideration of the relative proportions of isomeric forms of the viral DNA and the frequency of novel junctions has led to the following model. It has been proposed that the viral genome circularizes upon entry into the infected cell nucleus forming a novel L–S joint. These circular molecules then undergo a limited number of rounds of bidirectional replication. During this phase, segment inversion could take place by intramolecular recombination between the inverted repeats. Replication, generating head-to-tail concatemers via a rolling circle mechanism, then accounts for the bulk of viral DNA synthesis. Finally, the newly synthesized DNA is cleaved into unit length molecules and packaged into preassembled capsids. The relative proportions of the isomeric forms are postulated to result from a differential recognition of the normal and novel L–S joints by the viral cleavage system. Experimental evidence addressing several aspects of this model is currently being gathered.

The full complement of viral and cellular proteins required for the complete replication of VZV DNA

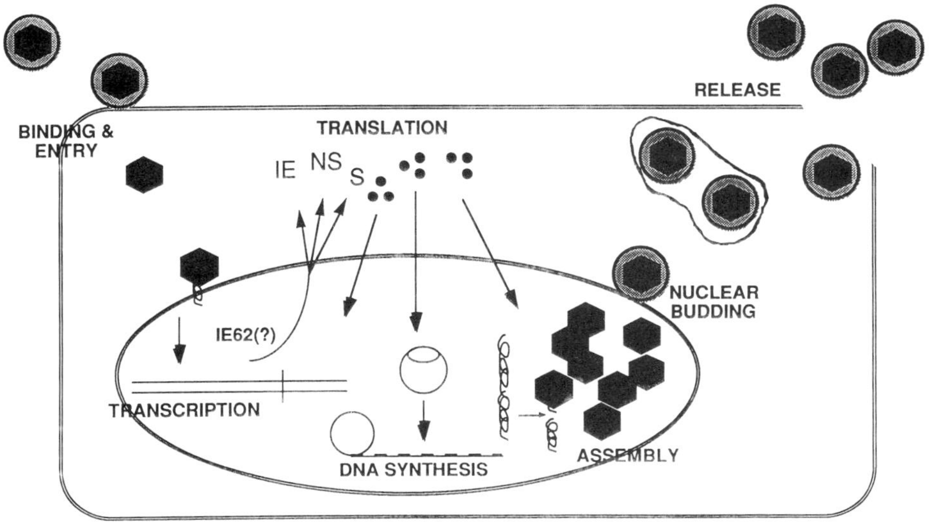

Fig. 1 Schematic diagram of the proposed infectious cycle of VZV during lytic infection. Infection is initiated by binding of the virus to the plasma membrane of susceptible cells followed by fusion of the viral envelope and entry of the virus. Following uncoating, the viral DNA enters the nucleus where three temporal classes of proteins, IE, E and L are produced. Transcription of IE mRNA may be enhanced by the IE62 protein which is carried into the cell as a major virion component. DNA synthesis takes place during the E phase, possibly utilizing both bidirectional and rolling circle mechanisms. During the L phase, progeny DNA are packaged into preformed nuclear capsids which, in turn, are enveloped by budding through the nuclear membrane. Enveloped particles acquire a second envelope by budding into cytoplasmic vesicles derived from the Golgi. The membranes surrounding these vesicles ultimately fuse with the cell membrane resulting in release of the virus.

has not, as yet, been identified. However, as indicated above and in Table 1, analogues of the seven 'Challberg' genes required for origin-dependent replication of HSV have been identified in the VZV system and these genes represent a starting point for future investigations.

Characterization of Transcription

The enzyme responsible for transcription of VZV mRNAs is, most likely, the RNA polymerase II encoded by the host cell. As with other herpesviruses, no viral-encoded RNA polymerase activity has been identified. Thus far, some 78 relatively abundant transcripts have been mapped to the VZV genome. Forty-two are transcribed right to left based on the prototype configuration of the genome, 35 are transcribed left to right and one is of indeterminate directionality.

Two additional classes of transcript have been identified. The first includes 33 transcripts of lesser abundance ranging in size from 0.7 to 4.6 kb in length. The second class, made up of large transcripts varying in size from 6 to 11 kb is comprised of at least 29 separate species and spans the majority of the viral genome. What function these low abundance transcripts play in the infectious cycle of the virus and which, if any, virus-specific polypeptides are translated from them are unknown.

No evidence of splicing has, as yet, been observed in VZV transcripts. This finding is consistent with computer analysis of the VZV genome, which predicted the existence of only one spliced message, derived from ORFs 42 and 45. It should, however, be noted that these data are preliminary and that only a very few VZV transcripts have been analyzed in detail. Thus the potential for splicing in specific VZV messages remains a definite possibility.

Based on the extensive colinearity of the VZV and HSV genomes and the relatively high degree of functional homology predicted between VZV and HSV gene products, it is reasonable to assume that VZV transcription is regulated in the coordinated cascade scheme seen with other alphaherpesviruses. Direct experimental evidence for this scheme has been difficult to obtain because of the low titers of VZV obtained in tissue culture. However, time course studies of protein synthesis have been carried out, which indicate that under conditions of nearly synchronous infection three major temporal classes of VZV-specific proteins are synthesized in infected cells. They have been named immediate-early (IE), early (E) and late (L) by analogy with HSV; each temporal class contains proteins that are representative of such classes in other herpesvirus systems. For example, the ORF 62 gene product is synthesized in the IE phase, the gene *29* major DNA-binding protein is synthesized during the E phase and the major capsid protein is synthesized at peak levels during the L phase of protein synthesis. Further delineation of VZV protein synthesis into temporal subphases and an evaluation of relative amounts of a given protein present within a specific class have thus far proved intractable.

Despite these similarities, VZV transcription presents questions quite distinct from those encountered with HSV-1. Foremost among these relates to the fact that the VZV gene product corresponding to α-TIF lacks the acidic C-terminal region which is responsible for the strong transinducing activity of that protein. The VZV protein continues to fulfill the structural role played by α-TIF, a requirement for virion maturation, but VZV has apparently evolved a separate strategy for the efficient transcription of its immediate-early genes.

As of this writing, detailed transcript mapping has been carried out on only four VZV genes: the viral deoxypyrimidine kinase (dPyK), the ORF 62 product and the glycoproteins, gpV and gpIV. The most striking finding from these studies is the fact that VZV appears to use atypical *cis*-acting transcriptional control elements. In each of these genes, several consensus TATA box elements are present upstream of the transcriptional start site but are not used, whereas nonconventional AT-rich regions are present within putative promoter regions. Potential ATG-start sites are found in the 5′-untranslated regions of the dPyK and gpV genes but are not used. Both glycoprotein genes exhibit heterogeneous start sites, raising the possibility that initiation of transcription may occur at a pyrimidine nucleotide as well as the more conventional purine nucleotide.

Analysis of the 3′ ends of the dPyK, and the gpV and gpIV genes also indicates differences in the utilization of *cis*-acting polyadenylation sites in VZV. The true polyadenylation signals of both the dPyK and gpV transcripts are atypical and are utilized preferentially to consensus polyadenylation sites also present in the sequence. In contrast, the 3′ terminus of the primary gpIV transcript is 20 bp downstream of two concensus polyadenylation signals, AATAAA and ATTAAA. It is not clear why VZV uses atypical control sequences for some genes and conventional ones for others. We will have to await more data in this regard in order to come to any reasonable conclusions. It is clear, however, that prediction of the sites of potential regulatory sequences in VZV genes from nucleotide sequence data is not reliable and that the location of such regulatory elements must be experimentally determined.

Characterization of Translation

Little is known concerning translation of viral message in VZV-infected cells. It is assumed that mRNA is translated in a fashion similar to that seen in HSV-infected cells and that translation occurs both on free and membrane-bound polyribosomes. Following translation, the majority of viral proteins are efficiently transported to the nucleus where they carry out their roles in viral replication.

Post-translational Processing

Thus far, the most extensively studied VZV proteins with regard to post translational processing are the five major viral glycoprotein species. All five are N-linked glycosylated and the majority show additional modifications including O-linked glycosylation, sulfation and phosphorylation. VZV gpI contains both O-linked and N-linked glycans. Its fully glycosylated form has an apparent molecular mass of 98 kD on SDS–polyacrylamide gels. This glycoprotein is also heavily sulfated and phosphorylated, with phosphorylation occurring at the level of both the polypeptide chain and glycosyl residues. The amino acid side chains that are phosphorylated are those of serine and threonine. Phosphorylation of tyrosine is not observed. The enzymes responsible for phosphorylation are most likely cellular casein kinases I and II. The enzyme responsible for the phosphorylation of the complex oligosaccharides has not, as yet, been identified, nor has a specific role for the phosphorylated or sulfated residues been established.

The mature form of gpII is a disulfide-linked heterodimer, composed of 66 kD and 68 kD subunits which are derived via proteolytic cleavage of the nascent 100 kD polypeptide backbone. Like gpI, gpII contains both N- and O-linked oligosaccharides with the N-linked saccharides including both sialic acid and fucose. This glycoprotein is also sulfated, although to a much lesser extent than is gpI. The sulfation occurs at the level of O-linked glycans. In contrast, gpIII contains only N-linked oligosaccharides, predominantly of complex type.

The smallest of the VZV glycoproteins is gpIV. The mature form of this protein is 55–60 kD and it has been less characterized than several of the other glycoproteins because of a lack of immunological reagents. It is assumed to contain N-linked oligosaccharides and is known to be phosphorylated; this modification occurs at the level of the polypeptide chain.

The final VZV glycoprotein which has been characterized as to post-translational modification is gpV. The predicted primary sequence contains five potential N-linked glycosylation sites and the mature form of this protein contains both N- and O-linked oligosaccharides. It is not yet known if gpV is sulfated or phosphorylated. The apparent molecular mass of the protein is strain dependent and has been shown to vary between 80 and 170 kD. This variation is primarily due to the presence of variable numbers of a 14-amino acid repeat element located near the N terminus of the protein. The extent to which the presence or absence of specific oligosaccharides or other modifications also contributes to this variability is unknown.

The IE62 transactivating protein is also subject to post-translational modification. This protein, which has a predicted molecular mass of 140 kD, is heavily phosphorylated, resulting in a species which migrates at approximately 175 kD on SDS–PAGE. The kinases that modify the IE62 protein are unknown. However, it has been suggested that the gene *47* virion kinase is capable of phosphorylating this protein, as it is found in viral particles. The functional significance of the phosphorylation of the IE62 protein is currently under investigation.

Assembly Site, Uptake, Release and Cytopathology

The current model for assembly and release of VZV virions is derived from extensive,well-established electron microscopic studies and more recent analysis of VZV glycoprotein biosynthesis and trafficking in infected cells. Fully assembled capsids containing unit length genomes comprising all four possible isomeric forms are released from the nucleus into the perinuclear space and, in the process, acquire a lipid envelope derived from the nuclear lamella. These particles migrate into the cytoplasm and acquire a second envelope by budding into large cytoplasmic vesicles. These cytoplasmic vesicles are derived from small transport vesicles released at the *trans*-

Golgi region and which contain fully processed viral glycoproteins. This second envelopment completes the maturation of the virion. The mature virions are then transported to the cell membrane within the cytoplasmic vesicles, and release of the virus presumably takes place by fusion of the membrane surrounding the vacuole with the cell membrane. Why so little of the virus is actually released and why so many damaged virions remain closely associated with the cell membrane in tissue culture is apparently because gpI in mature virions contains mannose 6-phosphate residues. As a result, virions are shunted to lysosomal vacuoles via mannose 6-phosphate receptors and are exposed to lysosomal enzymes. This does not occur in the human host since the cells in the superficial layers of the skin lack lysosomal vesicles. This could also explain why vesicle fluid virus tends to have a higher and more stable titer.

In tissue culture, cells infected with VZV initially show increased refractility followed by rounding and swelling. Multinucleated cells are common although the extent of syncytium formation is considerably less than that seen in HSV strains which exhibit a syncytial plaque morphology. Eventually the infected cells detach from the support surface producing readily discernable plaques. The nuclei of infected cells are larger than those of uninfected cells with marginally located chromatin and peripherally located nucleoli. Eosinophilic proteinaceous intranuclear inclusion bodies are present which occupy the sites of viral DNA replication. The nature and function of these inclusion bodies are unknown. Chromosomal aberrations and chromosome and chromatid breakage have also been observed.

Future Perspectives

A variety of interesting and fundamental questions remain to be answered concerning this important human herpesvirus. First and foremost is the specific cellular location and nature of latent VZV infection. Several early reports indicated that the sites of latency were the neuronal cells within the dorsal root ganglia. More recent studies indicate that the actual site is more likely satellite cells surrounding the neuron. This has led to a model for re-activation of latency wherein re-activated virus in the satellite cells lytically infects the neuron resulting in the destruction of this cell as the virus propagates. This model is attractive in that it is consistent both with the usually singular recrudescence of the virus and the neuronal damage and post-zoster neuralgia associated with re-activation of VZV.

A second important consideration concerning VZV latency is the state of expression of the genome. Unlike HSV, which appears to produce only a single transcript in the latent state known as the latency associated transcript (LAT), several messages are expressed in cells latently infected with VZV. Recent work has shown that two of these messages are derived from the gene for the major transcriptional activator (the IE62 protein) and from the gene *29* major DNA-binding protein. Other probable messages appear to be derived from genes *4*, *61* and *63*, all of which are capable of regulating VZV gene expression *in vitro*. Since the HSV gene *29* analogue has been suggested to exhibit properties consistent with a transcriptional regulatory gene, it would appear that all five of the functions expressed during VZV latency are involved in gene expression and that one possibility for maintenance of and re-activation from the latent state may involve changes in the relative ratios of these *trans*-acting functions.

Alternatively, the gene *29* product may be capable of interacting with a specific sequence and/or structure such as a stem-loop with partial single-stranded character. Such structures could exist at or near the palindromic sequences forming the replication origin of the DNA, suggesting the possibility that the gene *29* product may somehow maintain the DNA in a quiescent state.

A second question involves the factors delineating the host range of VZV and the specific cell types which the virus can infect in a productive manner. Currently there are two obvious areas for investigation. The first involves the identification and understanding of both the *cis*- and *trans*-acting factors involved in the regulation of expression of VZV genes. Why, for example, has the virus evolved to use noncanonical transcription start and stop sites? Does VZV use specific viral polypeptides for recognition of such sequences or are cellular factors also involved? What role does the IE62 polypeptide play in overall regulation of the viral infection? Can the amounts of IE62 found in the virion be sufficient to offset the lack of a functional α-TIF protein or is there another protein that can substitute for this activity? The roles of the known transcriptional modulatory proteins (62, 4, 63 and 61) will have to be evaluated in permissive, semipermissive and nonpermissive cell lines in order to identify their true roles during productive infection in the human host. Such an approach should also lead to the identification of host cell factors which influence the tissue tropism of VZV.

The second area with regard to the question of tissue tropism is why VZV encodes its specific set of major glycoproteins. First of all, only five major species are present as compared to at least eight for HSV. Second, the HSV homologues of three of these five proteins (gpI=gE, gpIV=gI, gpV=gC) have been shown to be dispensable for growth in tissue culture whereas VZV lacks a homologue of gD, which is required for penetration of HSV. Thus far, the data indicate that gpV is also largely dispensable for growth of VZV in tissue culture. Some tantalizing hints regarding function are coming from new understanding of the roles played by the 'non-essential' HSV glycoproteins. Specifically, gC is required for infection of the apical surface of polarized cells whereas gE and gI are required for infection of the basal surface. Whether or not this will be true for the homologous VZV proteins remains to be established, as does the mechanism for VZV attachment in the absence of a gD homologue.

A final consideration to be addressed lies in the fact that a live attenuated vaccine strain (Oka) has been developed against VZV and is in relatively widespread use in the industrialized world. This vaccine has been licensed for general use and utilized to vaccinate a large population in Japan, and is currently completing clinical trials for licensure in the US. The specific molecular nature of attenuation is unknown, although the initial characterization of the Oka strain following passage in guinea pig cells indicated a slight thermolability; subsequent studies show that the initial isolate of the vaccine strain is deficient in its production of gpV. The vaccine has been shown to be particularly efficacious as a prophylactic against VZV infection in immunocompromised children in Japan, Europe and the US.

Thus far, no incidence of zoster significantly greater than that observed in the general population has been seen in normal vaccinees, and less than 5% of vaccinees have exhibited clinically significant varicella-like symptoms following inoculation. These data not withstanding, concern remains regarding the potential for re-activation particularly in the light of our lack of understanding of the nature of the attenuation of the Oka strain. These questions as well as more generic issues regarding the use of live virus vaccines in the US and reports indicating that VZV is capable of transforming primary hamster embryo cells clearly point to the need for a subunit vaccine in the form of either a mixture of VZV antigens or possibly recombinant vaccinia virus vectors. Data are currently being gathered on a variety of individual VZV proteins in order to assess their roles as targets of both the humoral and cellular immune responses.

See also: Herpes simplex viruses; Latency; Persistent viral infection.

Further Reading

Croen K and Straus SE (1991) Varicella zoster virus latency. *Annu. Rev. Microbiol.* 45: 265.

Ellis RW, Kuter BJ and Zajac BA (1991) Varicella zoster vaccine. In: Cryz S (ed.) *Vaccines and Immunotherapy*, p. 325. New York, Pergamon Press.

Gelb L (1990) Varicella zoster virus. In: Fields BN *et al.* (eds) *Virology*, p. 2011. New York: Raven Press.

Grose C (1990) Glycoproteins encoded by varicella–zoster virus: biosynthesis, phosphorylation and intracellular trafficking. *Annu. Rev. Microbiol.* 44: 59.

Ostrove JM (1991) Molecular biology of varicella zoster virus. *Adv. Virus Res.* 38: 46.

Ruyechan WT, Ling P, Kinchington PR and Hay J (1991) The correlation between varicella–zoster virus transcription and the sequence of the viral genome. In: Wagner EK (ed.) *Herpesvirus Transcription and its Regulation*, p.301. Boca Raton, FL: CRC Press.

VARIOLA VIRUS

See Smallpox and monkeypox viruses

VECTORS

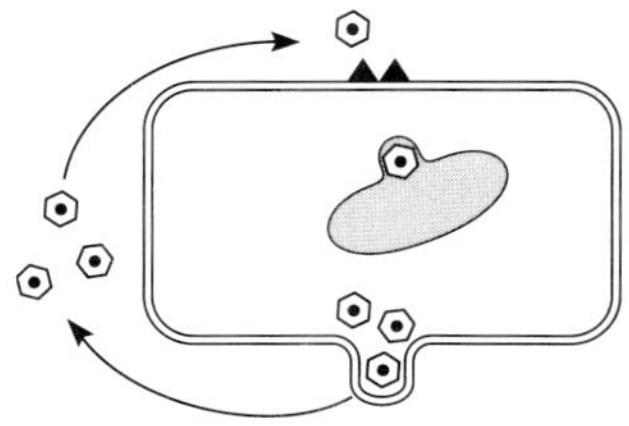

Contents

Animal Viruses
Plant Viruses

Animal Viruses

James Tartaglia, Russell Gettig and Enzo Paoletti
Virogenetics Corporation
Troy, New York, USA

Introduction

The ability to manipulate viruses for the expression of genes from heterologous sources was realized upon the advent of molecular cloning techniques in the 1970s. Initially, papovaviruses [i.e. simian virus 40 (SV40)] and papillomaviruses (i.e. bovine papillomavirus) were utilized to vector foreign gene-containing expression cassettes into mammalian cells. The early exploitation of these viral systems as vectors was made possible by prior knowledge of the viral genomic organization, the small size of their double-stranded DNA genomes and the infectious nature of purified genomic DNA upon calcium phosphate-mediated transfection on to mammalian cell systems.

Advances in the knowledge of the biology of other DNA-containing viruses (baculovirus, adenovirus, herpesviruses and poxvirus) and RNA viruses have since enabled their development as vector systems. Among the notable advances leading to the incorporation of foreign genetic material into larger DNA genomic structures were (1) increased knowledge of their genomic organization and viral-specific gene expression, and (2) the demonstration of recombination between intact viral genomes and plasmids containing DNA sequences homologous to the particular virus in infected cells. The ability to isolate infectious cDNA clones from RNA-containing viruses and to manipulate these cDNA clones has brought forth the potential development of such viruses as vectors. Manipulation of RNA viruses in this fashion, a strategy referred to as reverse genetics, has been illustrated with poliovirus, togaviruses and retroviruses.

The development and refinement of viral vectors over the past two decades has provided the scientific community with valuable reagents. During this period, scientists have realized the diverse utility of these vector systems to study gene regulation and mRNA processing, to achieve high level expression of biologically active proteins, to address questions of the structure–function relationship of specific polypeptides, to investigate the immunobiology of specific pathogens, to develop recombinant vaccine candidates and serodiagnostic reagents and to evaluate the potential for gene replacement therapy.

Adenoviruses as Eucaryotic Vectors

General features

Properties of adenoviruses that have provided the rationale for vector development include (1) the close association between viral-induced macromolecular synthesis and host cell functions, (2) the ability to introduce genes efficiently into a wide variety of nonproliferating cell types, (3) the ability to synthesize high levels of viral-specific proteins, and (4) the ability to transform a variety of rodent and human cell lines for establishing integrants. Over the past decade a number of foreign genes have been expressed by adenovirus recombinants. In general, authentic expression of foreign gene products has been achieved with efficient post-translational modifications.

The generation of adenovirus recombinants is facilitated by the infectious nature of isolated viral DNA. Conventional techniques to introduce isolated exogenous genetic material into mammalian cells have allowed the generation of novel adenovirus recombinants by two basic schemes. The first approach employs adenovirus genomic fragments derived using specific restriction endonucleases and plasmids containing the foreign gene flanked by appropriate adenovirus sequences in co-transfection experiments. The second approach utilizes infec-

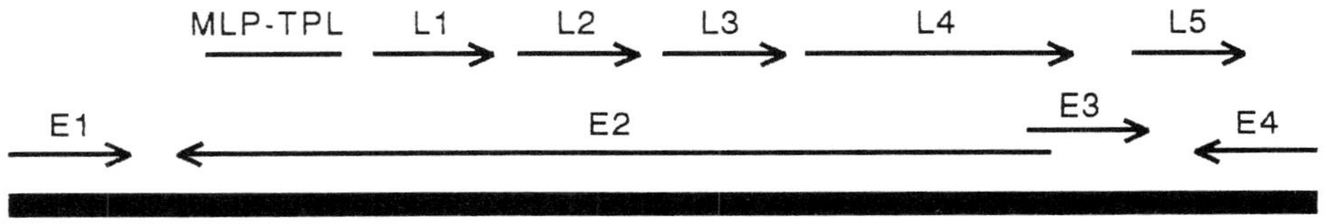

Fig. 1 Schematic representation of the adenovirus transcription map. Early (E) and late (L) transcription units are represented by solid lines relative to the 36 kb adenovirus genome. Arrowheads indicate direction of transcription. Each transcription unit gives rise to families of mRNAs derived by differential splicing events. The late mRNAs are generated by differential splicing and polyadenylation site utilization and all are initiated from the major late promoter (MLP) and contain the tripartite leader (TPL).

tious plasmids containing the entire adenovirus genome and an appropriately engineered gene containing plasmid in co-transfection experiments. Resultant adenovirus recombinants have been generated which are helper-independent and have either a conditional or nonconditional replication phenotype or are helper-virus dependent. The type of adenovirus recombinant generated is dependent on the insertion locus and the adenovirus sequences retained in the recombinant virus. For instance, insertion of exogenous sequences into the nonessential E3 region (Fig. 1) results in the generation of a helper-independent nonconditional recombinant virus. Insertion into the essential E1 region (Fig. 1), however, results in the generation of helper-independent conditional viruses which are propagated on 293 cells to provide the E1 gene functions. Recombinant viruses that are helper-virus dependent have also been derived by the deletion of large amounts of adenovirus sequences, while retaining the inverted terminal repetitions and packaging signals.

Considerations which need to be recognized in generating adenovirus recombinants that express the foreign gene of interest are as follows. Approximately 105% of the total genomic size can be efficiently packaged to produce infectious progeny. The restriction imposed upon packaging coupled to the essential nature of most of the viral DNA permits the incorporation of 5.3 kb of foreign DNA. Further manipulation of the adenovirus genome has provided the potential for substituting up to 7.5 kb. Secondly, successful expression of the foreign gene product requires careful consideration of the complex RNA-splicing patterns which occur during viral gene expression. The potential for altering these RNA-splicing patterns must be investigated when engineering the viral genome. Lastly appropriate regulatory sequences must be used to express the foreign gene of interest during either the early phase of the viral replicative cycle (E1 or E3 promoters) or expressed at late times postinfection (major late promoter–tripartite leader (MLP–TPL); see Fig. 1).

Applications of adenovirus recombinants

A major impetus for generating adenovirus recombinants was to express high levels of a heterologous gene product. This was fueled by the large amounts of adenovirus-specific proteins that are produced, particularly at late times postinfection. Recombinants have therefore been engineered containing the gene of interest under the regulation of the MLP–TPL (Fig. 1) sequences. Adenovirus recombinants have also been employed to investigate tissue-specific gene expression, establish stable cell lines, generate vector-based vaccine candidates and to investigate gene therapy strategies. These latter two applications have generated the greatest enthusiasm for adenovirus vectors over the past several years.

The use of adenovirus vectors expressing extrinsic immunogens as live vaccines is based, in part, on the successful application over the past 20–30 years of live adenovirus types 4 and 7 vaccines for the prevention of acute respiratory disease in military recruits. Adenovirus recombinants as live viral vaccine candidates were initially described using human adenovirus/hepatitis B sAg recombinants. These recombinants elicited a humoral response to both the vector and the foreign immunogen in rabbits and hamsters. Upon administration of these recombinants into chimpanzees via the oral route significant humoral responses specific to the hepatitis B sAg were induced, and partial protection against acute hepatitis following hepatitis B virus challenge was afforded. Following the initial demonstration of the potential for adenovirus recombinants as vaccine candidates, a number of other extrinsic immunogens have been inserted into the adenovirus genome including the genes encoding the vesicular stomatitis virus G protein, the human immunodeficiency virus (HIV) envelope glycoprotein, the herpes simplex virus gB, and the pseudorabies virus gp50. All of these recombinant adenoviruses were capable of eliciting immunological responses specific to the foreign gene product.

The broad host range of adenoviruses, which includes both cells of the hemopoietic cell lineage and epithelial cells, the ability to accommodate somewhat large pieces of foreign DNA (7.5 kb), the ability to introduce genes into nonproliferating cells, and the ability to generate stable integrants, at least with some human cell types, has brought forth the potential for adenoviruses, or derivatives thereof, as vectors for human gene therapy. Such an approach has been tested successfully in animal models in transferring the human α-1 antitrypsin gene and the human cystic fibrosis transmembrane conductance regulator (CFTR) gene to the respiratory epithelium. In the case of the CFTR gene, expression in rat lung cells was evident for up to 6 weeks. Although these initial studies using adenovirus vectors for gene therapy appear encouraging, certain issues pertaining to safety need to be addressed. Furthermore, although gene transfer efficiencies are greater with adenovirus vectors than standard DNA-mediated transfer techniques, the efficiencies are still below that obtained with retrovirus vectors. Optimization of the adenovirus system for gene therapy is therefore an active area of research.

Poxvirus-based Vector Systems

History and development

The demonstration of marker rescue by Sam and Dumbell and Paoletti and colleagues in 1981 and 1982 respectively in the poxvirus system were the critical events leading to the development of vaccinia virus as a eucaryotic expression system. An extension of the marker rescue technique then led to the insertion of exogenous genetic elements into nonessential regions of the vaccinia virus genome in 1982 by Paoletti and Panicali.

Since this initial report describing the use of vaccinia as a eucaryotic expression vector, virtually hundreds of heterologous proteins of biological, as well as medical, significance have been expressed by vaccinia virus. With proper DNA engineering, the proteins expressed by vaccinia-based vectors are of the predicted molecular mass and undergo appropriate post-translational modifications including N- and O-linked glycosylation, proteolytic cleavage, phosphorylation, myristylation and proper cellular transport and subcellular localization. Furthermore, owing to the broad host range of vaccinia virus, cell-type specific limitations imposed upon authentic expression of foreign gene products have, in general, been overcome.

The general methodology used to generate vaccinia virus recombinants is well documented in the literature. In short, the foreign gene of interest by virtue of being flanked by vaccinia DNA sequences becomes incorporated into intact viral genomes via homologous recombination events. Chimeric viral genomes generated by homologous recombination are then packaged into infectious progeny. An important feature of the vaccinia methodology in contrast to procedures used to generate recombinants of other DNA-containing viruses (i.e. adenovirus, baculovirus, herpesviruses) is that transfected DNA can only be rescued by infectious vaccinia particles. This stems from the noninfectious nature of isolated poxvirus DNA due to the cytoplasmic site of viral replication and the obligatory requirement for virion-associated enzymatic functions during the initial stages of the viral replicative cycle. Another important aspect of poxvirus recombinants relative to recombinants derived from other DNA-containing viruses is the apparent lack of packaging constraints wherein large amounts of DNA can be inserted into the genome. This feature is advantageous with respect to allowing the simultaneous expression of multiple heterologous proteins for biochemical analyses and the generation of multivalent vaccine candidates.

The recovery of infectious vaccinia virus recombinant progeny expressing the foreign gene of interest is dependent on several factors. These include (1) insertion of foreign DNA into nonessential sites, (2) appropriate construction of expression cassettes with poxvirus-specific regulatory elements and (3) the insertion of cDNA sequences encoding nonspliced mRNAs. Furthermore, to facilitate the recovery of recombinant vaccinia viruses, screening and selection schemes have been developed which are based on *in situ* plaque hybridization and immunoscreen, plaque size, β-galactosidase expression, expression of drug-resistance markers (i.e. neomycin resistance, *Escherichia coli* guanyl phosphoribosyl transferase), bromodeoxyuridine resistance and host range functions.

Vaccinia-based vector systems fall into two basic categories: (1) recombinant viruses and (2) recombinant plasmids that express the foreign gene product upon transfection into vaccinia virus-infected cells. The latter provides for transient expression assay systems. Further manipulation of such vector systems has also provided inducible expression systems based upon the bacteriophage T7 and T3 RNA polymerase promoter and the *Escherichia coli lac* operator–repressor systems.

More recently, the vaccinia-vector technology has been extended to other members of the poxvirus family. The rationale for developing such vector systems was primarily to provide poxvirus vectors with more limited host range to be used for specific applications. Other poxviruses that have been developed as vectors include avipoxviruses (i.e. fowlpox, canarypox, pigeonpox), capripoxvirus, raccoonpoxvirus and suipoxvirus. All of these viruses are members of the chordopoxvirus subfamily, as is vaccinia virus. Aside from raccoonpox, however, all these viruses are classified in separate genera from vaccinia virus.

Practical applications of poxvirus expression vectors

Since the initial description in 1982 of the development of vaccinia virus as a eucaryotic expression vector, the contribution of this relatively new technology to the study of biology has been invaluable. Vaccinia vector systems based on recombinant viruses or transient assay systems using recombinant plasmids have been used to express foreign genes for purposes of large-scale protein purification, analysis of protein structure–function relationships, the investigation of the immunobiology of bacterial, parasitic and viral pathogens, epitopic mapping and the generation of vector-based vaccine candidates.

Over the past decade, the development of vaccinia virus as a live vector-based vaccine candidate has received the most attention. This is primarily due to the successful history of vaccinia virus in the eradication of smallpox. Furthermore, a large number of vaccinia virus recombinants have been described expressing immunogens from human and veterinary pathogens. Many studies have shown that such recombinants elicit immunological responses, both humoral and cell-mediated, to the extrinsic immunogens. A large number of examples have now been gathered showing that animals immunized with vaccinia virus recombinants are protected against subsequent challenge with the corresponding pathogen. Specifically engineered vaccinia virus recombinants have also demonstrated the feasibility of using vaccine candidates for the immunoprophylaxis and immunotherapy of certain cancers.

Recently, the practical application of vaccinia virus vaccine candidates has been demonstrated in both the veterinary and human fields. In Europe and North America, extensive field testing has been carried out with a vaccinia-based rabies G recombinant. Both the safety and efficacy of this recombinant have been ascertained from these studies. Clinical trials with a vaccinia-based HIV-1(IIIB) *env* recombinant has also been initiated in the USA. Results from these studies clearly demonstrated the ability of this recombinant to prime an immunological response in the recipients.

Although promising, safety concerns that emerged from the smallpox eradication program have slowed the development of genetically engineered vaccinia virus strains as vaccine candidates. Several approaches have been taken to address the safety issues. One approach is to use alternative poxvirus vector systems which do not have as wide a vertebrate host range as vaccinia virus. For instance, the extension of the vector technology to fowlpoxvirus has provided a vector appropriate for the poultry industry.

Major advancements in the field of vaccinia vectored vaccine candidates for human and veterinary application have been in the development of two highly attenuated poxvirus vectors which have pro-perties consistent with providing safer alternatives to existing vaccinia vaccine strains. One vector, designated as NYVAC, was derived from the Copenhagen strain of vaccinia virus by the deletion of 18 endogenous open reading frames implicated in the virulence and host range of the virus (Fig. 2). The other vector was derived from a vaccine strain of canarypoxvirus and designated ALVAC. ALVAC is restricted to avian species for productive replication; nevertheless, when ALVAC recombinants are inoculated into nonavian species, the foreign gene is expressed and protective immunity is induced. Both NYVAC and ALVAC have characteristics which suggest the potential safety of these viruses as general laboratory virus strains and as immunization vehicles. NYVAC and ALVAC have also been shown to have highly attenuated *in vivo* characteristics compared to several established vaccinia virus vaccine strains in immunocompetent and immunocompromised animal model systems. Despite these attenuated characteristics, the NYVAC and ALVAC vectors have the ability to mount significant immunological responses against extrinsic immunogens in mammalian species, including primates. An ALVAC-based recombinant expressing the rabies G glycoprotein has been assessed in phase I clinical trials in Europe and the USA with extremely encouraging results.

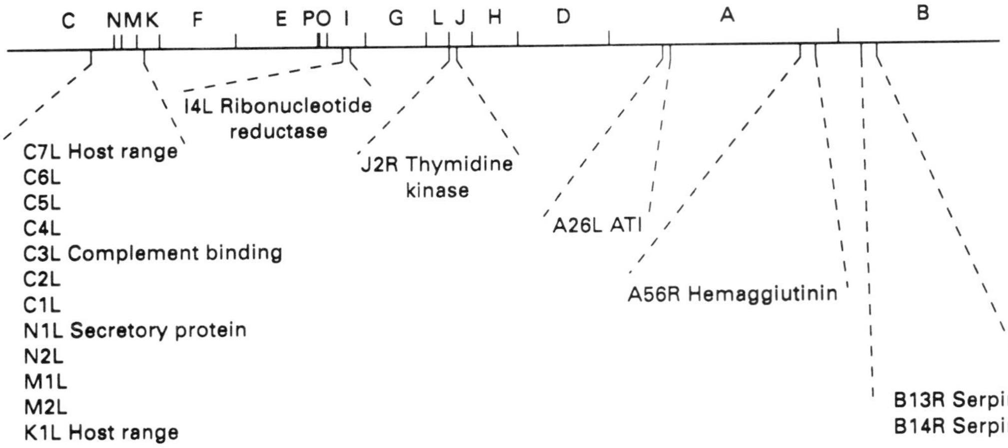

Fig. 2 Derivation of NYVAC. At the top is depicted the *Hin*dIII restriction of the vaccinia virus genome. Expanded are the regions that were sequentially deleted in the generation of NYVAC. Below each deletion locus are listed the open reading frames which were deleted along with functions or homologies of their gene products.

Retrovirus Vectors

General description

Retroviruses with their high transduction efficiency in a wide array of host cells and their inherent ability to stably integrate proviral sequences into the host cell genome have provided an extremely useful eucaryotic vector. Initially, retrovirus vectors were derived from prototypic murine and avian retroviruses. More recently, however, other retroviruses, including HIV, have been engineered as vectors for specific applications. For instance, HIV-based vectors have been used to efficiently introduce foreign genes into CD4+ T lymphocytes.

Since their initial description in the early 1980s as a eucaryotic vector system, three basic retrovirus vector types have emerged. These three archetypes are termed (1) double expression (DE) vectors, (2)

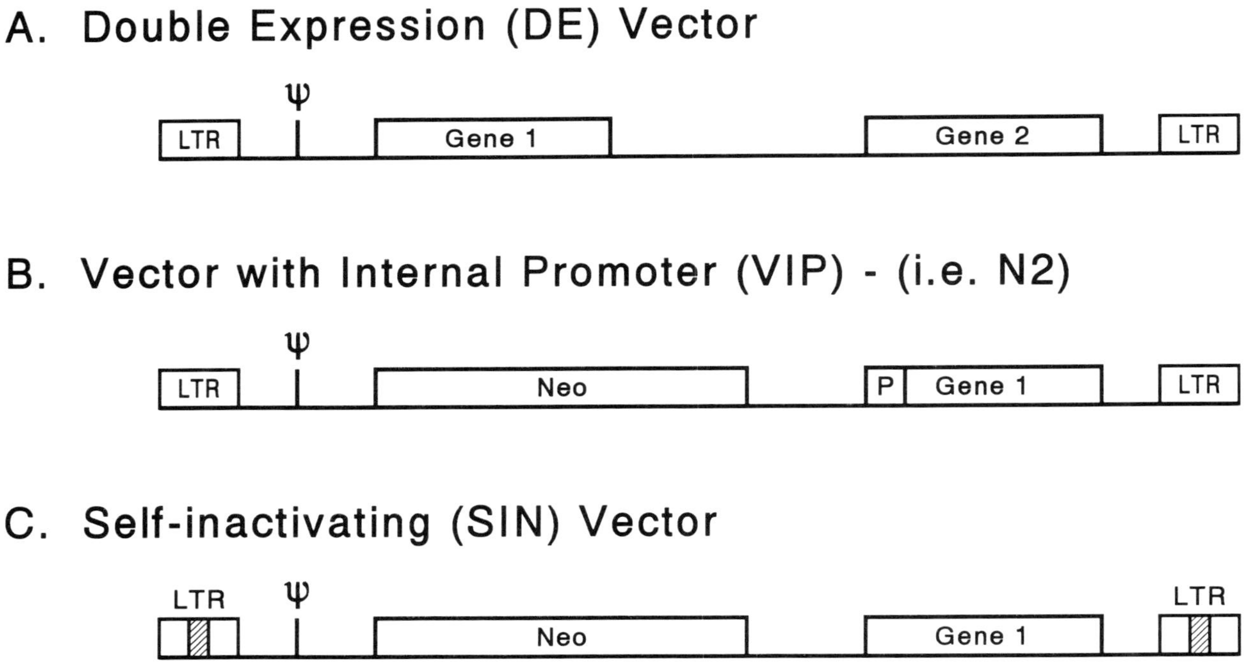

Fig. 3 Schematic representation of the proviral structures of the three basic retroviral vector types. The deletion of the long terminal repeat (LTR) sequences engineered to derive the SIN vectors is represented by the hatched area. Neo refers to the neomycin-resistance gene and Ψ refers to the psi packaging sequences.

Table 1. Advantages and disadvantages of the three basic retrovirus vector types

Vector type	Advantages	Disadvantages
Double expression (DE) vectors	Expression of two foreign genes	Splicing required for downstream gene expression Cell-type specific expression from retrovirus promoter Activation of downstream cellular genes (i.e. oncogenes)
Vectors using internal promoters (VIP)	Selectable gene (i.e. Neo gene in N2 vector) Downstream gene under separate promoter Splicing not required for downstream gene expression N2 vector engineered for production of higher-titered virus stocks N2 vector successful as vector for cells in suspension (bone marrow cells; lymphoid cells)	Internal promoter insertion affects vector functions (i.e. vector production) Activation of downstream cellular genes (i.e. oncogenes) Suppression of foreign gene expression by epigenetic mechanisms or promoter occlusion
Self-inactivating (SIN) vectors	Transcriptionally disabled No occlusion of downstream promoter Inability to activate cellular genes Provide safe vectors for *in vivo* gene therapy	Low-production titers

vectors with internal promoters (VIP), which include the commonly employed N2 vector and (3) self-inactivating (SIN) vectors. Provided in Fig. 3 are schematic representations of the proviral forms of these vector types.

Although each retrovirus vector type has unique structural and functional properties, utilization of each follows the same general principles of retrovectorology. These principles involve (1) replacement of retrovirus proviral sequences encoding products that can be provided in *trans* (*env*, *gag*, *pol*) with the foreign gene of interest, (2) retaining in the vector *cis*-acting sequences necessary for encapsidating (psi sequences) RNA transcripts into virions as well as the integration and expression of proviral sequences, (3) encapsidation of the defective hybrid RNA using packaging cell systems derived, in most cases, from NIH 3T3 cells infected with helper virus lacking the psi sequences and (4) infection of actively dividing target cells for integration and eventual expression of proviral vector sequences. Each of these retrovirus vector types has certain advantages and limitations. These are outlined in Table 1.

Certain general limitations of the retrovirus vectors, including potential recombination and oncogene activation, have been addressed over the years leading to specific modifications. With regard to the potential emergence of replication-competent virus by recombination events between vector sequences and sequences within the packaging cell systems or endogenous retroviral sequences, certain additional genetic manipulations have been introduced into the basic retrovirus vectors described in Fig. 3. Manipulations have been made to delete specific long terminal repeat (LTR) sequences and to introduce a translation termination codon at the initiation codon of the *gag* coding sequence. The vector used in initial gene therapy studies in humans (see below) was, in fact, derived from the N2 vector (Fig. 3) by positioning a termination codon at the beginning of the *gag* coding sequence (LNL6 vector). The concern for recombination has also been addressed by designing hybrid packaging cell lines that express structural genes from retroviruses derived from multiple animal species. The concern regarding the activation of nearby quiescent cellular genes resulting from the insertion of a transcriptionally active provirus sequence provided the interest leading to the development of the SIN vectors and its derivatives (Fig. 3, Table 1). These vectors are transcriptionally disabled and can therefore not activate other cellular genes, including proto-oncogenes.

Applications of retrovirus vectors and future directions

Despite the limitations cited above, retroviral vectors have represented the most efficient means for integrating DNA sequences into large numbers of target cells. Retroviral vectors have been used to investigate biological events in the fields of virology and cell biology. One of the major impacts of these vectors was their ability to introduce genetic material into lymphoid and myeloid cell lines. This has enabled the investigation of gene expression in these cell systems which was previously impossible using other DNA transfection techniques. Initial examples demonstrating the utility of these vectors has now been extended to hemopoietic lineage analyses and somatic cell gene therapy. In the development of gene therapeutic approaches involving retroviral-mediated gene transfer, examples have been reported implicating their utility in the therapy for hemophilia B, Maroteaux-Lamy syndrome (mucopolysaccharidosis type VI) and severe combined immunodeficiency disease.

The involvement of retrovirus vectors in a therapeutic approach for the treatment of cancer involves the introduction of genes into tumor cells or other cell types that alter the behavior of neoplastic cells in cancer patients. The four basic modes of retroviral-based cancer therapy that have been introduced are (1) the expression of antisense RNA molecules, (2) expression of cytokines in tumor cells or tumor-infiltrating lymphocytes, (3) virus-directed enzyme/prodrug therapy (VDEPT) to exploit transcriptional differences between normal and neoplastic cells for the selective elimination of neoplastic cells, and (4) restoration of functional tumor suppressor genes (i.e. retinoblastoma gene, p.53). The potential of certain somatic gene therapeutic approaches using retroviral-mediated gene transfer are presently being evaluated in the clinic. Initial phase I trials are assessing the ability to correct adenosine deaminase deficiency in severe combined immunodeficiency disease and to modulate tumor-infiltrating lymphocytes in cancer patients.

Current areas of active research to optimize retrovirus vectors for use particularly *in vivo* for somatic cell gene therapy and cancer therapy include (1) consistent production of high titered virus stocks, (2) development of vectors that efficiently express multiple genes and (3) development of vectors that can be used *in vivo* to infect selected cells and to integrate in a site-specific manner into the host cell genome.

Baculovirus Expression Vectors

Background

Baculoviruses are pathogenic insect viruses which have been studied for use as biological control agents in controlling insect pests. Recently, these viruses have proven useful in the development of a helper-independent vector for the overexpression of foreign genes in eucaryotic cells. The primary rationale behind the development of baculovirus vectors involves utilization of a strong viral promoter which expresses a very late gene product (polyhedrin) at high levels. This protein surrounds the mature virus particles forming an occlusion body (1–5 μM) which functions to stabilize the virus and provide for the horizontal spread of the disease. Polyhedrin expression in infected tissue culture cells can continue for days following initial infection and often amounts to 50% of the total cell protein. Maximum yields of polyhedrin can reach 1 mg/ml per 2×10^6 cells. Other considerations for developing baculovirus vectors are (1) the polyhedrin gene is nonessential for replication of the virus in tissue culture, (2) the viral genome is infectious and can easily be cotransfected with plasmid DNA to generate recombinants, (3) the rod-shaped capsid has considerable capacity for packaging foreign DNA and (4) the virus does not infect vertebrates.

Although very little was known about the molecular biology or genomic organization of baculovirus, several groups in the early 1980s were able to generate recombinants in *Autographa californica* nuclear polyhedrosis virus (AcMNPV) which expressed foreign genes under the control of the polyhedrin promoter. Since then, the AcMNPV system has been used successfully to express a wide range of viral, bacterial and human genes in insect cells. The first human clinical trials for an AIDS vaccine used an HIV-1 (IIIB) *env* gene product purified from the baculovirus expression system.

General description and applications

Generation of baculovirus recombinants involves homologous recombination between the infectious viral DNA genome and a donor plasmid containing the coding sequences of the foreign gene coupled to the polyhedrin promoter all of which is flanked by viral sequences which target the recombination to the polyhedrin locus. Using standard transfection procedures, recombinants can be generated in SF21AE insect cells. Frequencies are generally high enough to allow screening by differences in plaque morphology between occlusion-positive (wild-type) and occlusion-negative (recombinant) viruses. Once purified, expression of the foreign gene is accomplished in SF21AE (or equivalent) cells by harvesting the cells and/or media at 48–96 h postinfection. Large-scale insect cell culture techniques have been developed and used to produce larger quantities of the product. An alternative approach involves infection of susceptible insect larvae, although purification of the product has proven more difficult.

Certainly much needs to be learned concerning the processing and stability of foreign genes expressed in insect cells. Some general observations on post-translational modification of baculovirus-expressed proteins include (1) signal sequences and other specific proteolytic sites are recognized and cleaved, (2) cellular localization signals are recognized and proteins accurately transported to the proper cell compartment, (3) proteins can be phosphorylated and acylated and (4) both N-linked and O-linked glycosylation can occur.

Early reports indicated that insect cells, although capable of N-linked glycosylation, were unable to process the high-mannose glycan to the more complex types because of a lack of specific glycosidase and transferase enzymes in these cells. However, recent reports on human plasminogen expressed in SF cells show these enzymes can be activated with glycan processing identical to that found in the human-expressed protein. These are important considerations if baculovirus-expressed products are to have the same biological and immunological properties as that of the native protein.

Additional Viral Vector Systems

Researchers are investigating the potential of RNA- and DNA-containing viral vectors other than poxviruses, baculoviruses, adenoviruses and retroviruses. Genetic and molecular analysis of herpesviruses, such as the human cytomegalovirus, herpes simplex virus, pseudorabies virus and varicella–zoster virus have provided a means for engineering these viruses as eucaryotic vectors. Herpesviruses have been manipulated in much the same way as described above for adenoviruses using isolated viral DNA and molecularly cloned subgenomic fragments in transfection experiments. Only limited information, however, exists on the use of these viruses as vectors. Researchers have suggested a role for such vectors in the development of live recombinant vaccine candidates and gene therapy.

Another viral system gaining attention as an alternative vector system, especially for gene therapy, is the adeno-associated virus (AAV). AAV is a human parvovirus that propagates as both a lytic virus and as a provirus. Potential advantages of the AAV system are as follows: (1) safety, in that AAV does not replicate in the absence of a helper virus (such as adenoviruses and herpesviruses), (2) integration of the AAV genome to form the proviral state has been shown to occur at a specific site on chromosome 19 and (3) much of the AAV genetic material can be substituted with a substantial amount of foreign genetic material for introduction into eucaryotic cells. These characteristics of AAV have provided the impetus for investigating the value of this viral system as a vector for gene therapy. The inability to obtain helper-free AAV stocks and a lack of knowledge relating to any long-term consequence of integration of these sequences in cellular genomes, however, has limited their utilization.

Reverse genetics has been extended to the engineering of other RNA viruses as eucaryotic expression vectors. Poliovirus antigen chimeras have been introduced as epitope-presentation systems for the purpose of developing vaccine candidates and reagents for diagnostic or therapeutic purposes. This approach takes advantage of the tremendous knowledge base pertaining to the three-dimensional structure of the picornavirus, poliovirus and its relationship to antigenicity. Such chimeras have been developed by replacing antigenic site 1 with known immunologically pertinent epitopes from other pathogens. To date, poliovirus chimeras have been generated containing epitopes from hepatitis A virus, rhinovirus-14, human papillomavirus-16, foot and mouth disease virus and HIV. The poliovirus–HIV (amino acids 735–752 of gp41) chimera has been shown to elicit anti-HIV antibodies in rabbits. Furthermore, this antiserum neutralized a wide range of American and African HIV–1 isolates of 1.

Alphavirus (Sindbis virus and Semliki Forest virus)-based vector systems have also been introduced. The value of such systems is derived from the ability to utilize these vectors in a wide range of mammalian cells, as a plasmid, or a recombinant progeny virus. Limited examples for the use of such vectors have, thus far, been reported in the literature.

Future Perspectives

There is no doubt that viral vectors will continue to provide valuable reagents for future investigations of biological systems. Further, these vectors have the potential to provide novel approaches for addressing urgent needs in the human and veterinary fields, particularly for diagnostics, vaccines and gene therapy. Future investigation of these viruses will hopefully lead to their optimization as vectors and perhaps alleviate safety concerns surrounding their use in human and veterinary medicine.

See also: Adenoviruses; Baculoviruses; Fowlpox virus; Polioviruses; Retroviruses – Type D; Sindbis and Semliki Forest viruses; Vaccinia virus.

Further Reading

Baxby D and Paoletti E (1992) Potential use of non-replicating vectors as recombinant vaccines. *Vaccine* 10: 8.
Berkner KL (1988) Development of adenovirus vectors for the expression of heterologous genes. *BioTechniques* 6: 616.
Gilboa E (1986) Retrovirus vectors and their uses in molecular biology. *BioEssays* 5: 252.
Gutierrez AA, Lemoine NR and Sikora K (1992) Gene therapy for cancer. *Lancet* 339: 715.
Luckow VA and Summers MD (1988) Trends in the development of baculovirus expression vectors. *BioTechnology* 6: 47.
Maeda S (1989) Expression of foreign genes in insects using baculovirus vectors. *Annu. Rev. Entomol.* 34: 351.
Moss B (1991) Vaccinia virus: A tool for research and vaccine development. *Science* 252: 1662.
Mulligan RC (1993) The basic science of gene therapy. *Science* 260: 926.
Rose CSP and Evans DJ (1991) Poliovirus antigen chimeras. *Trends BioTechnol.* 9: 415.
Tartaglia J and Paoletti E (1990) Live recombinant viral vaccines. In: van Regenmortel MHV and Neurath AR (eds) *Immunochemistry of Viruses*, II. *The Basis for Serodiagnosis and Vaccines*, p. 125. Amsterdam: Elsevier Science Publishers.

Plant Viruses

Thomas Hohn
Friedrich Miescher Institute
Basel, Switzerland

and

Rob Goldbach
Agricultural University
Wageningen, The Netherlands

Introduction

Virus vectors have become useful tools for introducing genes into a variety of organisms. The first successful virus vectors were derivatives of the bacteriophage λ, with the help of which the first homologous and heterologous gene libraries were established in *Escherichia coli*. Likewise, a number of animal viruses were used to introduce foreign genes into animal cells where they replicated and were expressed. The two most important types are based on the double-stranded (ds)DNA papovaviruses, e.g. bovine papillomavirus, and those based on retroviruses. While papovavirus vectors replicate as circular dsDNA, retroviral vectors mediate the incorporation of the passenger gene into the host chromatin. All of the cases mentioned above lead to relative stability of the passenger gene in the transfected cell, since their replication as dsDNA involves a proofreading mechanism.

Neither true dsDNA viruses nor viruses mediating DNA integration have yet been found amongst the large variety of plant viruses. Accordingly, experiments with plant virus vectors are based on single-stranded (ss)DNA viruses, pararetroviruses and RNA viruses; replication in all these cases can lead to genome rearrangements, because single-strand intermediates are involved on which proofreading mechanisms do not apply and where template switching could occur. If the design and use of the virus vector interferes with viral replication, expression or other vital functions, the passenger gene will be removed from the vector population within a few generations. The passenger gene will in any case be finally eliminated unless it provides some selective advantage. This instability can be also looked upon as a benefit, since it provides biological containment: hybrid plant virus derivatives released purposely or accidentally into the environment are unlikely to maintain a nonselective gene. Stable transformation of plants, on the other hand, can be achieved either with the help of *Agrobacterium*-derived vectors, which mediate integration of passenger DNA into the plant chromatin, or by direct gene transfer.

Another peculiarity of plant viruses is their mechanism of spreading within the host tissue. Spreading usually makes use of the plasmodesmata (cytoplasmic connections between individual cells) which are thought to become modified to allow passage of viral nucleic acid or particles. Since plant cells in culture lack these connections, their infection with viruses or viral genomes usually depends upon artificial methods, such as electroporation or polyethylene glycol (PEG) treatment, and further spread will not occur. In contrast, animal viruses have developed methods to both leave and enter cells, e.g. by exo- and endocytosis, and are, therefore, able to spread through host cell cultures after the initial infection.

Inoculation of Plants with Virus and Vectors

Many plant viruses, irrespective of whether their genome consists of DNA or RNA, can be inoculated mechanically on to plant leaves using an abrasive. In these cases, naked nucleic acid or manipulated derivatives can also usually be used as an inoculum. A further method is agroinoculation, which was first developed with CaMV (see Table 1 for definitions of virus abbreviations) and potato spindle tuber viroid and makes use of the DNA-transfer machinery of *Agrobacterium tumefaciens,* the cause of crown gall disease in a number of dicotyledons. Using this method, multimers, or at least 'one-and-a-bit-mers', of the virus genome are cloned between the T-DNA borders of the *Agrobacterium tumefaciens* Ti-plasmid and this manipulated T-DNA is introduced into the plant upon bacterial infection. Once in the plant cell, probably in the nucleus, the viral genome can escape either by recombination or by the direct production of monomeric replicative intermediates, such as the CaMV 35S pregenomic RNA.

Certain viruses and their nucleic acids are not infectious on mechanical inoculation. This group contains many viruses that are restricted to the phloem tissue, being directly introduced there by their insect vectors. Among these are many viruses that have members of the *Gramineae* as hosts, plants that are also not susceptible to crown gall disease. It was thus surprising that maize plants could be infected with MSV by agroinoculation. This success indicated that *Gramineae* are also susceptible to T-DNA transfer, and opened a route to the introduction of a number of different viruses that could not enter by other artificial means, e.g. WDV into wheat and RTBV into rice.

Pararetrovirus Vectors

Pararetroviruses, like retroviruses, use both the host nucleus and the cytoplasm for replication. In the nucleus, viral DNA is transcribed into terminally redundant RNA, in the cytoplasm the RNA is reverse transcribed into DNA, which is transported either directly or via new infections into the nucleus. Pararetrovirus DNA accumulates (but does not replicate) in the form of supercoiled circles, while retrovirus DNA is inserted in a terminally redundant form into the host chromatin. Plant pararetroviruses exist as two groups, the icosahedral caulimoviruses and the bacilliform badnaviruses. The genomes of these viruses are all around 8000 bp in length.

So far, only CaMV, a caulimovirus from this group, has been used as a vector. The seven open reading frames (ORFs) of this virus are arranged densely packed on the pregenomic 35S RNA, and it is thought that all but one of the ORFs are translated from this polycistronic messenger. Infection can be achieved in the laboratory by mechanical inoculation with virus particles or virus DNA and thus ORF 2, which codes for an aphid transmission factor, becomes dispensable. In view of this, an ORF-2-replacement vector was constructed and used to clone, express and spread systemically payloads through turnips. By this means, the ORFs for bacterial dihydrofolate reductase (DHFR; 240 bp), the Chinese hamster metallothionein (200 bp), human αD interferon (500 bp), and BNYVV ORF N (171 bp) were cloned, spread systemically and expressed in plants.

From these successful and other unsuccessful experiments, it was learned that sequences introduced into the CaMV genome are frequently lost, probably as a result of illegal template switching of the nascent minus or plus DNA strands as well as negative selection against the inserted DNA sequence. These recombination events could be minimized by avoiding homologies at the insertion site, and selection for the deletions could be minimized by avoiding whatever might interfere with

Table 1. Plant virus names and their abbreviations

Virus	Abbreviation	Group
African cassava mosaic[a]	ACMV }	Bipartite gemini
Cassava latent[a]	CLV }	
Alfalfa mosaic	AlMV	—
Beet curly top	BCTV	Monopartite gemini
Brome mosaic	BMV	Bromo
Beet necrotic yellow vein	BNYVV	Furo
Barley stripe mosaic	BSMV	Hordei
Barley yellow dwarf	BYDV	Luteo
Beet western yellows	BWYV	Luteo
Cauliflower mosaic	CaMV	Caulimo
Cowpea chlorotic mottle	CCMV	Bromo
Cucumber mosaic	CMV	Cucumo
Cucumber necrosis	CNV	Tombus
Cowpea mosaic	CPMV	Como
Cymbidium ringspot	CyRSV	Tombus
Digitaria streak	DSV	Monopartite gemini
Maize streak	MSV	Monopartite gemini
Odontoglossum ringspot	ORSV	Tobamo
Pea early browning	PEBV	Tobra
Plum pox	PPV	Poty
Potato X	PVX	Potex
Red clover necrotic mosaic	RCNMV	Diantho
Rice tungro bacilliform	RTBV	Badna
Tomato bushy stunt	TBSV	Tombus
Turnip crinkle	TCV	Carmo
Tomato golden mosaic	TGMV	Bipartite gemini
Tobacco mosaic	TMV	Tobamo
Tobacco rattle	TRV	Tobra
Tobacco vein mottling	TVMV	Poty
Turnip yellow mosaic	TYMV	Tymo
White clover mosaic	WClMV	Potex
Wheat dwarf	WDV	Monopartite gemini

[a]The same virus.

the virus life cycle as follows:

1. The total size of the genome should not, or should only slightly, exceed the original viral genome length of 8000 bp. About 700 bp can be capitalized by deletion of ORF 2, as mentioned above, and ORF 7, which is also dispensable for virus growth. Allowing for overpackaging of 300 bp, a payload of about 1000 bp seems to be the upper limit. CaMV vectors to clone a larger payload might be developed from artificial bipartite CaMV genomes, which together contain all essential functions and lack recombination targets.
2. The tightly packed ORF organization should be maintained, probably because of the unusual CaMV translation mechanism. Experience has shown that noncoding sequences between CaMV and passenger ORFs diminish vector stability.
3. Inserted sequences should not interfere with CaMV gene expression and replication. For instance, the insertion of an RNA polyadenylation signal would interfere with the production of pregenomic RNA, insertion of a new promoter might occlude the original CaMV promoters and an intron would not survive the transcription/reverse transcription cycle for very long.
4. Sequences leading to gene products that interfere with virus replication either directly or by affecting the host cell must be avoided.

CaMV and its derived vectors do not proliferate efficiently in cell culture; while DNA replicative intermediates can be found in infected protoplasts, mature virus particles are absent or rare. This might be explained by developmental control of genome replication versus translation. Both processes depend on the same 35S RNA, and one can assume a control mechanism that switches between these two functions. In protoplasts and the derived cell cultures, the switch is apparently set to genome replication and particle formation is inhibited.

It is also likely that other plant pararetroviruses can be developed as gene vectors, collectively covering a large host range. This would also include the *Gramineae* if, e.g. RTBV is considered. Since this virus has a bacilliform capsid, overpackaging might be less of a problem.

Geminivirus Vectors

Geminiviruses (Table 2) contain 2.5–3 kb of circular ssDNA. They are characterized by twinned (geminate) icosahedral particles. Geminiviruses of dicotyledonous plants are usually, but not always, bipartite and transmitted by white flies; geminiviruses of monocotyledonous plants are monopartite and transmitted by leafhoppers. Usually, component 'A' of the bipartite geminiviruses is sufficient for replication in single cells, while component B is required for systemic spread and needs component A for replication. With respect to their bidirectional expression strategy, geminiviruses resemble the animal polyomaviruses and simian virus 40 (SV40). Their intranuclear dsDNA forms contain two noncoding regions, one harboring the clockwise and counterclockwise promoters and the other containing the two polyadenylators. Transcription in one direction (in consensus clockwise) produces RNAs coding for structural proteins, and transcription in the other direction gives RNAs required for DNA replication.

The coat proteins of geminiviruses are dispensable for genome replication but are required for insect transmission, and in the case of the monopartite viruses are also required for virus spread within the plant. Accordingly, mutants in the coat protein gene of bipartite geminiviruses can still replicate in single cells and spread through the host plant. Coat protein mutants of monopartite geminiviruses do not spread.

Geminivirus coat protein replacement vectors have been introduced into plants by either mechanical means or agroinoculation. Vectors based on component A DNA of bipartite geminiviruses can spread if component B is provided either as a transgene or as a co-inoculum. However, when transgenes of these vectors are used or transient expression in protoplasts or leaf disks is studied, component B is not required. Monopartite geminivirus vectors are so far restricted to transient expression experiments in localized patches of plant tissue disks, seed-derived embryos or cell cultures.

The following additional observations have been made.

1. In addition to the coat protein gene, the small ORF of unknown function located further upstream can also be replaced by the payload. Deletion or replacement mutants of the counterclockwise coding regions, however, are defective in replication.
2. Systemically spreading vectors are unstable if they are either smaller or larger than wild-type virus; rearrangements occur which restore the original size. The reason for this size dependence in the absence of packaging is not known. Non-spreading vectors, on the other hand, can exceed normal genome size by up to 4 kb, for instance when they harbor the long β-glucuronidase (GUS) or neomycin phosphotransferase (NPT) II coding regions.
3. Geminivirus vectors also tolerate the insertion of additional promoters, e.g. the CaMV 35S promoter, which leads to a large increase in payload expression without affecting replication.
4. An *E. coli* replicon can also be incorporated into

Table 2. Geminivirus vectors

Virus	Partite	Target	Host	Payload
TGMV	Bi-	Plants	*Nicotiana tabacum*	CAT,GUS,NPT-II
ACMV	Bi-	Plants	*Nicotiana benthamiana*	CAT
MSV	Mono-	Protoplasts	*Zea mays*	NPT-II,GUS,DS1
WDV	Mono-	Protoplasts	*Triticum monococcum*	NPT-II,CAT,βGAL

monopartite geminivirus genomes, creating shuttle vectors that replicate and express a payload gene in both bacteria and protoplasts of *Gramineae*.

RNA Viruses as Gene Vectors

Infectivity of cloned cDNA

To develop RNA viruses as gene vectors, it is essential that infectious RNA can be generated from cloned cDNA copies. This has been achieved for a growing number of plus-strand RNA viruses (but not minus-strand viruses) using four different approaches.

1. With a cDNA clone directly. Only limited results have been reported so far (excluding the viroids), namely for AlMV RNA3 and TMV RNA; in both cases, the cDNAs gave rise to virus at a very low level. The infectivity found here is probably due to transcription from unidentified promoters.
2. As cDNA fused to the CaMV 35S promoter. This results in more efficient production of progeny virus, as reported for BMV and BNYVV RNA 3 and RNA4.
3. By agroinoculation. This has been achieved for BWYV, a luteovirus, using a cDNA clone provided with a ribozyme sequence to enable *in vivo* run-off transcription.
4. As *in vitro* run-off transcripts, from linearized full-length cDNA clones, using phage SP6, T7, T3 or *E. coli* RNA polymerase, in combination with their respective promoters. This is by far the most frequent approach, and infectious transcripts have been synthesized for more than 20 different RNA viruses belonging to 15 distinct taxonomic groups (Table 3). It is essential for infectivity that the transcripts are not only of full length but also (almost) identical to the natural viral RNAs. Some problems may arise here since most promoters extend into the transcribed region, resulting in transcripts that start with extra nonviral nucleotides. In general, it has been found that nonviral residues at the 3′ end are better tolerated than extra nucleotides at the 5′ end, which mostly lead to a significant drop in infectivity.

The best results have been obtained with *in vitro* transcription systems using SP6, T3 or T7 promoters and the corresponding specific polymerases.

Table 3. Plant viral cDNAs reported to generate infectious *in vitro* transcripts

Group	Virus
Alfalfa mosaic	AlMV (RNA3)
Bromo	BMV, CCMV (RNA1–3)
Carmo	TCV
Como	CPMV (RNA1–2)
Cucumo	CMV (RNA1–3)
Diantho	RCNMV (RNA1–2)
Furo	BNYVV (RNA1–4)
Hordei	BSMV (RNA α,β,γ)
Luteo	BYDV, BWYV
Potex	PVX, WClMV
Poty	PPV, TVMV
Tobamo	TMV
Tobra	PEBV, TRV (RNA1–2)
Tombus	CNV, CYRSV, TBSV
Tymo	TYMV

These systems are highly efficient (up to 20 μg of transcript from 1 μg of template) and can be designed to yield almost authentic termini. However, if the first viral nucleotide is not *a priori* a G residue, one extra G residue must be tolerated at the 5′ end. It should be noted that, in most cases studied, the extra nonviral nucleotides were deleted in progeny RNAs. In cases of virus with capped genomic RNAs, the addition of a cap structure at the 5′ end of the transcripts often enhances infectivity. For BSMV RNAs capping seems to be an absolute requirement. There is an additional complication with viruses belonging to the 'picorna-like superfamily', e.g. the como-, nepo- and potyviruses. The genomic RNAs of these viruses possess a protein (denoted VPg) covalently linked to the 5′ end. Nevertheless, both capped and uncapped *in vitro* transcripts have been shown to be infectious.

Insertion and expression of foreign genes

Cloned full-length copies from which infectivity can be recovered by following one of the approaches discussed above provide a good starting point for constructing replicons that can be used as gene vectors (Table 3). Of course, insertion of foreign sequences should not interfere with any viral function or process (e.g. RNA replication, subgenomic mRNA synthesis, translation, encapsidation) required for (helper-independent) infectivity.

With respect to noncoding sequences, the terminal sequences should at least be retained for replication, while for some viral RNAs (e.g. BMV RNA3) internal noncoding sequences also appear to contain obligatory *cis*-acting replication signals.

With respect to the introduction of payloads into coding sequences ('replacement' vectors), the following points can be made.

Genes encoding proteins involved in RNA replication

All plant RNA viruses so far studied possess one or more cistrons encoding replication proteins (e.g. putative helicase, viral polymerase). It is obvious that any manipulation of these genes will lead to nonreplicating transcripts unless the functions can be provided in *trans*. This is feasible, e.g. by transformation of host plants with functional copies of the affected genes (transgenic tobacco plants expressing the RNA1 and RNA2 products of AlMV are able to support RNA3 replication) or by co-inoculation with helper virus (smaller TMV RNA-derived replicons co-replicate and spread systemically with TMV helper virus). In both cases, however, the viral vector has lost its truly independent character.

The movement protein gene(s)

A considerable number of plant RNA viruses encode a so-called 'movement protein' which is actively involved in cell-to-cell spread of the viral genome (e.g. tobamoviruses) or viral particle (e.g. comoviruses) through plasmodesmata. Its coding sequence should be retained if systemic spread of the viral vector is desired. This is true for TMV for the 30 kD protein, for BMV for the RNA3-encoded 32 kD protein and for CPMV for the M-RNA-encoded 58kD/48kD protein pair. For expression in protoplast systems or cultured cell suspensions, the movement protein cistron can be omitted and the site exploited.

The coat protein gene(s)

The viral coat protein gene can, in some instances, be manipulated in order to express foreign sequences, though it always leads to lower fitness in terms of systemic spread. Although for some viruses the coat protein is dispensable for cell-to-cell movement (e.g. TMV), this structural protein cannot be omitted for long-distance transport of the inoculated vector (via the vascular system) throughout the whole plant. Therefore, replacement of coat protein genes by reporter genes (e.g. CAT, Table 4) always leads to naked, infectious RNA which is confined to the inoculated leaf. Such replacement vectors can, however, be efficiently expressed in protoplast systems (Table 4). The problem of limited spread has recently been circumvented for TMV by creating a viral vector (TB2) which contained a duplicated subgenomic promoter of the coat protein gene, enabling the expression of both the coat protein and the payload gene (NPT II, DHFR). It was found that the duplicated promoter should be cloned from a different tobamovirus with enough sequence divergency (i.e. ORSV) to avoid removal by homologous recombination.

Table 4. Heterologous gene expression from plant RNA viral genomes

Group	Virus	Insertion site	Gene	Expression system
Bromo	BMV	Coat protein gene	CAT	Protoplasts
Furo	BNYVV	25 kD gene (RNA3)	GUS	Inoculated leaf
Hordei	BSMV	βb gene (RNAβ)	LUC	Protoplasts
Tobamo	TMV	Coat protein gene	CAT	Inoculated leaf
		Coat protein gene	ENK	Protoplasts
		Extra site[a]	NPT-II	Whole plant
		Extra site[a]	DHFR	Whole plant

[a]From inserted ORSV coat protein promoter.

Miscellaneous genes

Further genes with potential for exploitation as vectors exist in some viruses. Their function has usually not been resolved, although some of them have been shown to be dispensable for both viral RNA replication and systemic spread. For example, in the quadripartite RNA virus BNYVV, both RNA3 (specifying a 25 kD protein) and RNA-4 (specifying a 31 kD protein) have been shown to be dispensable and are in fact lost after serial mechanical inoculation. Large internal parts of both these RNAs can be deleted without affecting replication. Insertions of the bacterial GUS gene in a deletion-containing RNA3 cDNA has been shown to give rise to detectable GUS activity in the inoculated leaf (Table 3) and amplification of the transcript. A further example of successful replacement, this time with BSMV, is insertion of the firefly luciferase (LUC) coding sequence into the βb gene (located on β-RNA and encoding a 58 kD protein with a helicase motif), which resulted in infectious virus expressing high-level luciferase activity in both tobacco and maize protoplasts. However, luciferase activity was not detected in extracts of whole plants inoculated with BSMV RNAs α, β or the

LUC-gene-containing RNA, which is consistent with the replaced βb protein being essential for multiplication in whole plants.

Independent versus 'disarmed' vectors

From the previous paragraph, it is clear that viral RNA vectors able to replicate and spread independently through whole plants should contain all *cis*- and *trans*-acting factors involved in RNA replication, encapsidation and spread. Two approaches have been shown to be successful for the development of such helper-independent vectors, i.e. the insertion of a ('recombination-proof') extra promoter, as shown for TMV, and the replacement of non- or less essential genes (e.g. BNYVV 25 kD gene).

As an alternative, 'disarmed' viral vectors could be constructed for foreign gene expression in plants, which lack one or more essential functions and are able to multiply only in target plants transformed with the required gene(s). The properly constructed transgenic plant which is required provides a bonus in the form of control over the spread of field-released genetically engineered virus. This approach is indeed feasible, as shown by the replication of engineered AlMV RNA3 molecules in tobacco plants transformed with the AlMV replicase genes, and the complementation of spread-deficient TMV strains in transgenic plants expressing the TMV 30 kD movement protein.

Plant Viruses as Vectors for Studying DNA and RNA Rearrangements

Virus vectors have also been used for purposes other than expression of a payload ORF.

Splicing

Cloning of introns in virus vectors allows precise measurement of the efficiency and accuracy of splicing in monocotyledons and dicotyledons. The precise excision of an intron introduced into a CaMV vector verified the intron and was a proof for reverse transcription of CaMV.

Recombination

Experiments can be designed which make use of essentially nonviable virus hybrids which become viable upon specific genome rearrangements. The most obvious examples, already mentioned in the course of discussing agroinoculation, are given by 'one-and-a-bit-mer' of manipulated caulimo- and geminivirus genomes which release viable virus upon recombination. If the redundant portions of these constructs are derived from different viral strains, questions can be answered concerning the recombination mechanism, the presence of recombination hot spots, the degree of homology required, etc. Escape of pregenomic virus RNA by transcription was observed if the promoter/polyadenylator region of CaMV constituted the terminal redundancy. If transgenic *Brassica napus* plants were produced containing CaMV one-and-a-bit-mers arranged such that CaMV could not escape by simple transcription, true recombination events could be scored by appearance of viral symptoms. Sequence analysis of recombinant molecules suggested that mismatch repair was linked to the recombination process. Intermolecular recombination could also be studied, i.e. between transgenic CaMV ORF 4 and supertransfected complementary virus sequences.

Infectious transcripts from cloned RNA viruses have been very useful for studying RNA recombination, e.g. in bromoviruses, and for analyzing and understanding the *cis*- and *trans*-acting factors in RNA replication.

Agrobacterium T-DNA transfer

A CaMV-based system was used to analyze independent *Agrobacterium* transfer DNA (T-DNA) transfer events. The complete T-DNA without the border sequences was replaced by the virus genome such that a viable replicon could be produced by circularization upon transfection of plants. Analysis of this replicon revealed rather conserved right border remnants, while sequences remaining at the left border were more variable. The presence of small direct repeats between some of the joined ends showed that linear T-DNA had been transported to the plant.

Transposition

Using WDV and MSV vectors that contained the maize transposable element *Ac* or its defective *Ds* derivatives, excision of the transposable element could be studied in protoplasts and whole maize plants. Excision of the *Ds* element was dependent on the presence of *Ac*. The junction sequences left on the viral genomes after excision revealed the typical footprints.

Conclusions

Based on the limited and (sometimes) rather preliminary results obtained so far, it may be concluded that plant viruses can be engineered into gene vectors able to express desired genes. Systems are described that allow expression in single plant cells; others allow replication, expression and spread through the plant, either independently or as 'disarmed vectors' in the presence of a helper virus, or within a transgenic host carrying the missing functions.

The advantages of the viral vector transfection systems are ease of handling, the short time periods required to obtain results and the replication of the vectors to high copy numbers which can result in expression levels much higher than in transgenic plants.

Because of the lack of any proofreading mechanism and the large number of replication cycles (even in a single cell), one might imagine that payload genes in RNA virus and pararetrovirus vectors would rapidly accumulate point mutations, leading to inactivation of these nonessential inserts. The (limited) experimental data obtained so far, however, indicate that at least for the duration of a single protoplast batch or plant infection, functional proteins can be obtained in desirable amounts. It is not excluded that viral vectors eventually lose their capacity to encode a functional protein upon serial passage, although this would merely provide an advantage in view of biological containment.

A second issue of concern is the risk of recombination involving viral vectors. Indeed, it has now been well documented that viral genomes, including those of RNA viruses, are frequently the subject of recombinational events. One obvious mechanism for recombination in all classes of plant viral vectors relies on template switches of the nascent DNA and RNA strands. On the one hand, nonviral inserts in viral vectors might become lost by recombination because of the lack of selective pressure; the consequences seem only to be beneficial with respect to biological containment, since viral vectors will not survive for long in nature. On the other hand, new functional viral genomes may arise from recombinational events involving (disarmed) RNA vectors, coinfecting (helper-) viruses and transgenes, which could lead to undesired spread of pathogens. Although so far no evidence has been obtained for such events, it is clear that critical risk-assessment analyses should be performed prior to possible release of viral vectors in agricultural practice.

See also: Badnaviruses; Bromoviruses; Caulimoviruses; Geminiviruses; Lambda bacteriophage.

Further Reading

Ahlquist P and Pacha RF (1991) Gene amplification and expression by RNA viruses and potential for further application to plant gene transfer. *Physiol. Plant.* 79: 163.

Bakkeren G *et al.* (1989). Recovery of *Agrobacterium tumefaciens* T-DNA molecules in whole plants early after transfection. *Cell* 57: 847.

Brisson N and Hohn T (1988) Plant virus vectors: CaMV. In: Weissbach A and Weissbach H (eds) *Methods for Plant Molecular Biology*, p.437 San Diego: Academic Press.

Donson J *et al.* (1991) Systemic expression of a bacterial gene by a tobacco mosaic virus-based vector. *Proc. Natl Acad. Sci. U.S.A.* 88: 7204.

Hohn B *et al.* (1987) Plant DNA viruses as gene vectors. *CIBA Found.* 133: 185.

Joshi L, Joshi V and Ow DW (1991) BSMV genome mediated expression of a foreign gene in dicot and monocot plant cells. *EMBO J.* 9: 2663.

Matzeit V *et al.* (1991) Wheat dwarf virus vectors replicate and express foreign genes in cells of monocotyledonous plants. *Plant Cell* 3: 247.

Takamatsu N *et al*. (1990) Production of enkephalin in tobacco protoplasts using tobacco mosaic virus RNA vector. *FEBS Lett.* 269: 73.

VENEZUELAN EQUINE ENCEPHALITIS VIRUS

See Equine encephalitis viruses

VESICULAR EXANTHEMA VIRUS AND CALICIVIRUSES OF PINNIPEDS, CATS AND RABBITS

Michael J Studdert
University of Melbourne
Parkville, Victoria, Australia

History

In 1932 there occurred in southern California, in pigs fed uncooked garbage, a vesicular disease which was provisionally diagnosed as foot and mouth disease. The disease was eradicated by slaughter and quarantine. However, in 1933 it reoccurred at which time it was realized that it was not foot and mouth disease but a new disease which was called vesicular exanthema of swine (VES). Between 1932 and 1951, VES continued to occur in southern California and more than 2.5 million pigs or 21% of the total pig population of southern California were involved in outbreaks. There was a clear link between the disease and the feeding of raw, i.e. uncooked, garbage containing, it was assumed, pork scraps. Outbreaks of the disease continued to occur despite the introduction of laws requiring that all garbage fed to pigs be cooked. In late 1951, a train left San Francisco and off-loaded in Wyoming garbage containing pork scraps which were fed to pigs. When a few of the pigs in the herd became lame and developed vesicular lesions they were shipped for sale following which, between 1952 and 1956, outbreaks of VES occurred in 40 of the 48 states. The last outbreak occurred in New Jersey in 1956. Later that year, as a result of rigid enforcement of laws requiring that all garbage fed to pigs be cooked or a total ban on the feeding of garbage, the disease was declared eradicated from the national swine herd in the USA, the only country in which the disease has ever occurred. No further outbreaks of the disease have been recorded since.

In 1972, a calicivirus was isolated from sea lions during an investigation of an outbreak of abortion on San Miguel Island off the southern Californian coast. This virus closely resembles VESV and is transmissible to pigs. It is believed that the origin of VESV was the feeding of uncooked sea lion carcasses, which were washed up on the beaches of southern California, to pigs. Subsequently, caliciviruses have been isolated from other pinniped species inhabiting the western seaboard of North America, including Northern fur seals inhabiting the Pribilof Islands of Alaska.

Acute upper respiratory disease of cats is common. In 1957 the first feline calicivirus, originally identified as a picornavirus, was isolated and shown to be one of the two major viral causes of respiratory disease in cats (the other virus being feline herpesvirus 1).

In 1984, a major new epidemic disease of rabbits (*Oryctolagus cuniculus*) emerged in Nanjing Province, China. The virus, called rabbit hemorrhagic disease (RHD) virus, spread rapidly through China and westward through Europe and North Africa causing the deaths of uncounted millions of rabbits, both wild and domestic.

Viruses with typical calicivirus morphology have been frequently identified, but not cultivated, from enteric infections of other species including man, cattle, swine (new type), dog, rabbit, chicken, reptile, amphibian and insects. Other than this brief mention, these viruses will not be considered further in this entry.

Classification and Properties

In negatively stained electron micrographs, caliciviruses have a distinctive morphology. Virions are 35–40 nm in diameter and nonenveloped. The icosahedral capsid has 32 cup-shaped depressions, hence the name calici (= cup). The capsid is composed predominantly of a single protein of 67 kD. The genome is a single-stranded plus-sense RNA molecule of about 7.9 kb which has a poly(A) tail at the 3′ terminus and a cap protein at the 5′ terminus. Although they superficially resemble picornaviruses, with which they were originally classified, they have some distinctive properties and this led to the creation of a new family, *Caliciviridae*.

Geographic Distribution

As already noted, between 1932 and 1951, VES occurred only in southern California and subsequently in 40 of the 48 states of the USA before it was eradicated in 1956. Caliciviruses, including San Miguel sea lion virus, remain endemic in pinniped species notably along the western seaboard of North America. Feline calicivirus is the cause of a common everyday disease of cats throughout the world. RHD has occurred in China, eastern and western Europe, North Africa, Korea and probably Mexico and has the undoubted potential to spread throughout the rest of the world.

Host Range and Virus Propagation

Pigs are a significant alternative host for at least some of the pinniped caliciviruses. The extent to which pinniped and perhaps other marine species share as hosts for caliciviruses has not been fully defined. Feline calicivirus is restricted to members of the family Felidae although infections are most commonly recognized in domestic cats. In addition to rabbits, RHDV is known to infect and cause disease in the European brown hare (*Lepus europaeus*) and some other *Lepus* spp.

The pig, pinniped and feline viruses grow readily and rapidly in monolayer cell cultures derived from pig and feline tissues respectively; most strains of VEV grow in Vero cells. RHD calicivirus has not been cultivated in cell culture.

Serological Relationships and Variability

Based on serum neutralization assays and cross-protection studies in pigs, VESV has a very large number of antigenic types. It was not uncommon for more than one virus type to be isolated during a single outbreak of disease or indeed for more than one virus type to be isolated from a single pig. The exact number of types cannot be recorded with certainty since some of the early collections of viruses were lost. However, at least 13 distinct antigenic types exist in one collection of viruses.

A similar pattern of antigenic variation (types) was recognized among feline caliciviruses when rabbit antisera were used in serum neutralization assays, i.e. a large number of different antigenic types was identified in collections examined by the few individual laboratories that attempted these studies. (All feline caliciviruses have never been examined by a single set of typing criteria.) Curiously, when antisera raised in specific pathogen-free cats were used in virus neutralization assays to examine reasonably large collections of feline caliciviruses, extensive cross-reactions were identified essentially between all viruses, although considerable antigenic variation was recognized. It was concluded that feline caliciviruses were related as a single antigenic type and these findings paved the way for the development of monotypic vaccines.

Epidemiology

It now seems certain that vesicular exanthema was initially transmitted on an on-going basis to pigs by feeding uncooked pinniped carcass meat. Additional transmission occurred by feeding uncooked garbage containing pork scraps. Within a herd, pig-to-pig transmission occurred because ruptured vesicles shed large quantities of virus into the environment causing ready transmission by contact or via fomites.

Feline caliciviruses are transmitted by contact and particularly by sneezing when cats are closely confined as in multiple-cat households, breeding and boarding establishments, cat shelters and veterinary hospitals. Recovered cats remain persistently infected with virus for many months or years presumably as a consequence of low-grade infection in pharyngeal tonsillar tissues.

RHDV is very readily transmitted between rabbits via the fecal–oral route and over longer distances apparently by fomites including contaminated vesicles. Infected rabbit carcass meat could be carried over considerable distances by predatory species as when the fox goes out on a chilly night and returns to feed the little ones.

Pathogenesis

VESV gains entry via abrasions usually around the snout and mouth or on the feet. Secondary vesicles may occur as a result of direct local spread or following viremia. Abortions as well as death of baby pigs from agalactia in their dams also occur.

Feline calicivirus produces vesicular lesions on the muzzle, and within the oral cavity and the respiratory tract. These tend to be ruptured quickly because of the incessant licking propensity of the affected cat.

RHD is a generalized infection in which viremic spread results in lesions in a wide range of tissues. At postmortem, these are particularly evident as hemorrhagic necrosis of the liver and lung. There is debate as to whether the hemorrhages result from decreased synthesis of blood-clotting factors, because of massive liver necrosis, or from disseminated intravascular coagulopathy mediated by immune complexes; the latter seems most likely.

Clinical Signs

VES is clinically indistinguishable from the other three so-called vesicular diseases that affect pigs, i.e. foot and mouth disease, vesicular stomatitis and swine vesicular disease. Following an incubation period of 12–48 h, there is a marked febrile response, anorexia and listlessness. Primary vesicles are blanched raised areas of epithelium up to 3 cm in diameter and 1 cm high filled with a serous virus-rich fluid. They easily rupture leaving raw bleeding exceedingly painful ulcers which subsequently become covered with a fibrinous pseudomembrane. Secondary vesicles appear 48–72 h postinfection. Notably these appear on the soles of the feet, in the interdigital space and at the coronary band. There is severe four-footed lameness. Secondary bacterial infection of lesions, particularly of the feet, occurs and prolongs recovery where slaughter and eradication are not immediately pursued.

Feline calicivirus infection may produce a subclinical, or an acute or subacute disease, usually characterized by conjunctivitis, rhinitis, tracheitis, pneumonia (usually in young kittens) and by vesiculation/ulceration of the epithelium of the oral cavity and muzzle. There is fever, anorexia, lethargy, stiff gait and usually a profuse ocular and nasal discharge. Morbidity is high and mortality in untreated cases may reach 30%.

RHD (and the disease in hares) curiously affects only rabbits that are older than 3 months; younger rabbits, even where there is no maternal antibody, are not susceptible. The disease is often peracute, characterized by sudden death following a 6–24 h period of depression and fever. In acute and subacute forms of the disease, there is a serosanguinous nasal discharge and several central nervous signs including excitement, incoordination, opisthotonus and paddling. Morbidity rates approach 100%, and mortality rates of 90% occur in virgin soil epidemics.

Immunity

Recovered pigs are immune to the particular antigenic type of VESV with which they were infected but not to other types. Since slaughter and eradication policies are pursued, the questions of long-term immunity and vaccine development are not at issue.

Cats recovered from feline calicivirus infection or immunized with feline calicivirus vaccine appear to remain relatively free of disease when exposed further. This appears to be the case despite the considerable degree of antigenic variability recognized among feline caliciviruses. It is the practice to recommend annual boosting of immunity.

Vaccines have been developed for RHDV and are widely used particularly in those countries, such as China and Italy, where rabbit farming for meat and pelt production is a major industry.

The basis for immunity has been best studied for the feline calicivirus in which antibody and cell-mediated immunity, including the generation of cytotoxic T-lymphocyte responses able to lyse autologous cells, occur.

Prevention and Control

VES is effectively controlled by slaughter and is now considered an extinct virus disease.

Feline calicivirus infection is most difficult to control in large open cat populations. Vaccination is an important means of control. Clinically ill cats should be isolated, and incoming cats of uncertain status should be held in isolation for at least a week before being introduced into the general colony. Although recovered cats remain persistently infected, the amount of virus shed is not usually large, therefore they do not pose the same threat to contact cats as cats with obvious clinical disease.

For rabbits raised in captivity, strict attention to hygiene and the maintenance of closed colonies is probably the best most inexpensive practical means of controlling RHD. Where feed and other supplies are delivered by a common supplier to multiple farms, special care is required to prevent movement of virus between farms and over long distances. Feeding of pellets that are sterilized should minimize transmission from the feed itself. Vaccines are available and appear to be effective.

Future Perspectives

Vigilance will be required to avoid the re-emergence of VES from either marine sources or the laboratory. A clearer understanding of the molecular basis for antigenic variation among porcine and feline caliciviruses should be a research objective, although research budgets for this work are difficult to justify because the disease in pigs is extinct and currently used vaccines for feline caliciviruses appear to be reasonably effective. The nature of the carrier state for feline caliciviruses, however, requires further understanding.

The molecular details of the replication cycle of caliciviruses in general, including a definition of the total number of mRNA transcripts and their regulation, should emerge in the coming few years. The sequencing of the genome of at least one VESV and one feline calicivirus should be undertaken. For RHDV, in some ways the future has arrived; although it is the most recently identified calicivirus, the entire nucleotide sequence, comprising 7437 nucleotides, has been determined. Comparative sequence analysis among the caliciviruses will be as interesting and fascinating as similar studies of other viruses.

At a more prosaic level, the need to cultivate RHDV is a high priority, as is the need to understand why rabbit kittens less than 3 months of age are not susceptible to the disease, a phenomenon contrary to and without precedent in any other virus disease. Finally, for those countries, such as Australia, where wild rabbits are a plague upon the nation, reducing profits from farming and degrading the land, the possibility of introducing RHDV to control rabbits should be explored. This, however, will cause conflict among farming and animal welfare groups and those other 'greenies' among us concerned with the preservation of the environment and its return to a more natural state as was found before the arrival of the rabbit.

See also: Foot and mouth disease viruses; Persistent viral infection; Polioviruses; Vesicular stomatitis viruses.

Further Reading

Meyers G, Wirblich C and Thiel H-J (1991) Rabbit hemorrhagic disease virus – molecular cloning and nucleotide sequencing of a calicivirus genome. *Virology* 184: 664.

Morisse J-P (1991) Viral hemorrhagic disease of rabbits and European brown hare syndrome. Scientific and Technical Review. *Office International des Epizooties*, No. 10.

Schaffer FL (1979) Caliciviruses. *Comprehensive Virol.* 14: 249.

Smith AW and Madin SH (1986) Vesicular exanthema. In: eds *Diseases of Swine*, p. 358. Ames: Iowa State University Press.

Studdert MJ (1978) Caliciviruses *Arch. Virol.* 58: 187.

VESICULAR STOMATITIS VIRUSES

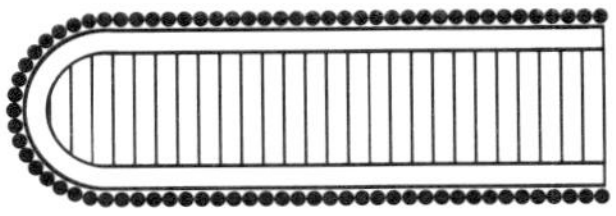

Stuart T Nichol
University of Nevada
Reno, Nevada, USA

History

Vesicular stomatitis virus (VSV) primarily infects cattle, horses, pigs and humans. Affected animals exhibit vesicular lesions on the mouth, udder teats and hoof coronary bands. In addition, insects can become infected. Descriptions of apparent disease outbreaks can be found dating back to the nineteenth century. One early record from 1862 described what appeared to be a large VS disease outbreak in army horses in the USA during the Civil War. Suspected sporadic cases of VS in horses and cattle in the USA were reported over the next several decades. The first large epizootic of VS in the USA to be described in detail occurred in 1916. The contagious febrile disease spread rapidly from Colorado to the East Coast, affecting large numbers of horses and mules and to a lesser extent cattle. Large epizootics have continued to occur in the USA on an approximately 10–15-year cycle with the last major outbreak occurring in 1982–1983. As more sophisticated means of differential diagnosis became available, different forms of VSV were recognized. In 1925 the prototype strain of the Indiana serotype classical type 1 viruses was isolated from an outbreak in cattle in Richmond, Indiana. A year later the prototype virus strain of the VS New Jersey serotype was isolated from cattle in New Jersey. In 1961, Cocal virus was isolated from mites in Trinidad and became the prototype strain of the VS Indiana serotype, type 2 viruses. In 1964 another Indiana-related virus was isolated from a mule in Alagoas, Brazil. This became the prototype strain of VS Indiana type 3 viruses. In addition, in the 1980s two other VS Indiana-related viruses Maraba and Carajas were identified, although they are yet to be studied in detail.

Taxonomy and Classification

VSVs belong to the *Vesiculovirus* genus of the *Rhabdoviridae* family of RNA viruses. These viruses all share a characteristic bullet- or rod-shaped morphology and basic biochemical properties. The VSVs are grouped together on the basis of antigenic cross-reactivity and genetic and biochemical relatedness. Two main serotypes, New Jersey and Indiana, have been defined based on cross-neutralization properties (see the Serologic Relationships section). The Indiana serotype is further divided into at least three serologically related but distinct virus types, type 1 (classical Indiana), type 2 (Cocal-like) and type 3 (Alagoas-like) viruses. Recent genome nucleotide sequence comparisons have defined three separate subtypes of VS New Jersey viruses and four for Indiana type 1 subtypes.

Properties of the Virion

VSV particles are rod-shaped (hence the name rhabdo, from the Greek for rod) viruses of approximate dimensions 70 × 180 nm (Fig. 1). The virion core consists of a helical ribonucleoprotein (RNP) structure containing the RNA genome. The protein components of RNP cores consist of nucleocapsid (N) protein, the polymerase-associated phosphoprotein (P or NS) and the large (L) polymerase protein. There are estimated to be 1,258, 466 and 50 molecules of N, P and L respectively in the virion cores of VS Indiana type 1 viruses. The matrix (M) protein forms a shell surrounding the core structure. This is then enveloped by a bilipid membrane derived from the host cell. Embedded in the membrane is the surface glycoprotein (G). The short cytoplasmic tail of G contacts the underlying M protein. The majority of the G molecule protrudes to the exterior of the viral particle. VS Indiana type 1 virions contain 1826 and 1205 molecules of M and G proteins respectively.

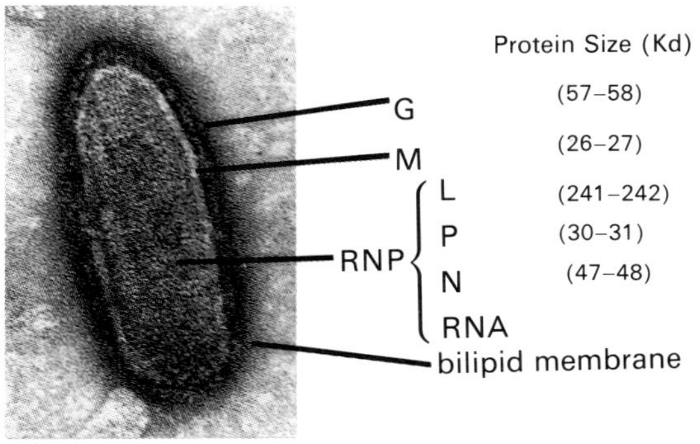

Fig. 1 Electron micrograph of purified virion. Sizes and locations of VS virion components are indicated.

Properties of the Genome

The single-stranded non-segmented negative-sense RNA genome of VSVs are approximately 11 kb in length. These linear RNA molecules possess 5′-phosphate and 3′-hydroxyl groups, and lack 5′ cap or 3′ poly(A) tail structures. The termini exhibit limited self-complementarity which may be involved in the circularization of the genome. The nucleotide sequences of the complete genome of both VS New Jersey and Indiana type 1 viruses have been determined. The naked RNA is non-infectious.

Properties of Proteins

Each of the different VSV genomes code for five proteins of apparently similar size and equivalent function. The following details relate to the VSV New Jersey proteins unless otherwise stated. The N protein has a pI of 5.8–6.3 and is 422 amino acids in length. The primary function of the protein appears to be to encapsidate the virion RNA. It is an essential component of actively transcribing or replicating viral cores and functions in close association with the P protein.

The P protein is an acidic (pI 4.2–4.3), highly phosphorylated protein, 274 amino acids in length, associated with viral polymerase activity. The primary function of the protein appears to be to mediate the binding of the L protein to the nucleocapsid cores and to facilitate the displacement of the tightly bound N protein to allow active polymerase to access the RNA template during transcription and replication. At least three functional domains have been defined for the P protein. The N-terminal half, domain I, is negatively charged and has been thought to be involved in N protein displacement. This domain contains several constitutively phosphorylated serine and threonine residues. Domain II located at amino acids 213–247 appears to be involved in the binding of the P protein to the L protein. This domain contains two serine residues whose phosphorylation may be essential for transcription. Finally, the basic C-terminal 21 amino acids, termed domain III, are probably involved in the tight binding of the P protein to N–RNA complex.

The large L protein is a basic protein (pI 8.6–8.8), 2109 amino acids in length. It appears to be a multifunctional enzyme required in only catalytic amounts and probably performs most of the polymerase-associated functions of the virus such as RNA synthesis, capping, methylation and poly(A) addition. The L protein has also been shown to possess protein kinase activity which preferentially phosphorylates serine residues on the P protein. Inhibition of P protein phosphorylation prevents transcriptase activity.

The matrix protein is highly basic (particularly the N-terminal domain) with a pI of 9.9. This is the most abundant protein of virions. The protein is 229 amino acids in length and appears to have two main functions. First, the protein plays a critical role during the assembly of virion particles prior to budding from the cell surface membrane. The M protein specifically binds to G protein monomers and promotes their trimerization. In addition, M is thought to associate with the cellular lipid bilayer through the C-terminal domain and promotes the condensation of viral RNP cores into tightly coiled helical structures subsequent to release within virion particles. Second, the M protein in tight association with RNPs inhibits genome transcription and may play an important role in the correct regulation of viral genome transcription.

The G protein is a surface glycoprotein 517 amino acids in length. It is a typical class I membrane-associated glycoprotein, with the N-terminal 90% of the molecule projecting from the surface of the virion or infected cell, a hydrophobic transmembrane domain anchoring the protein in the membrane, and a C-terminal 28-amino acid cytoplasmic domain projecting to the interior of the virion or infected cell. The protein contains two asparagine-linked complex oligosaccharide structures linked at amino acids 179 and 340.

Physical Properties

VS virions consist of approximately 74% protein, 20% lipid, 3% RNA and 3% carbohydrate. Infectivity is unstable at pH 3, but relatively stable in the pH range 5–10. It is rapidly inactivated at 56°C and by UV and X ray irradiation. Infectivity is also sensitive to lipid solvents and various detergents.

Replication

RNA synthesis

Unless otherwise stated, the following description of viral replication will be limited to VS Indiana type 1 virus, although it is likely to be representative of events occurring with the other VSVs. Viral replication is initiated after virion uptake by the infected cell and release of the viral ribonucleoprotein (RNP) cores into the cytoplasm (see Fig. 2, step 1). The negative-strand genome RNA is transcribed by the virion-associated polymerase (transcriptase) to produce five functional mRNAs (step 2). The genes are located 3′-NPMGL-5′ and are 1333, 822, 838, 1672 and 6380 nucleotides in length respectively. Active transcribing complexes consist of the template, made up of the RNA genome in tight association with the N protein, together with polymerase consisting of the phosphoprotein (P) and large (L) protein. The majority of experimental data are consistent with the virus transcriptase entering the RNP core at the 3′ end of the genome. The P protein may contact the genome at bases 16–30 from the 3′ terminus and mediate the binding of the L protein to the RNP. The complex initiates transcription by completing a 47-nucleotide transcript termed the leader RNA which is an exact copy of the genome 3′ end. This is followed by an untranscribed junctional sequence AAA (AAAA in VSV New Jersey).

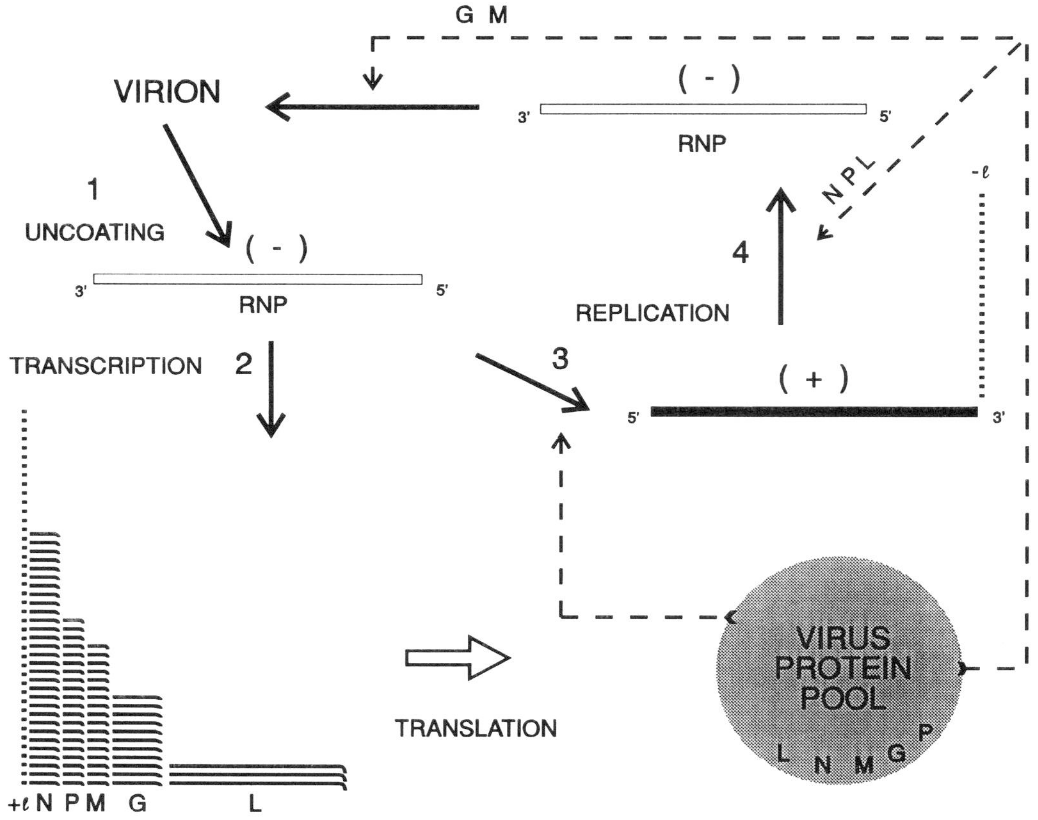

Fig. 2 Model of VSV replication cycle. The major steps in the replication cycle of VSV are shown in diagram form. Open bars represent viral negative-sense RNA and closed bars represent viral positive-sense RNA. Refer to the text (Replication section) for discussion of steps 1 through 4.

The leader transcript is neither capped nor polyadenylated. Attenuation occurs at this junction whereby only a proportion of polymerase molecules transcend the junction and initiate the next transcript which is the N mRNA. During synthesis of the nascent N mRNA, the 5′ end of the transcript is capped by the virion polymerase complex (probably the L protein). At the end of the N gene, and indeed all five viral genes, is the sequence 5′-AGUUUUUUUCAUA-3′ which signals the termination and polyadenylation of the mRNA. The mRNA poly(A) tract is probably synthesized by the viral polymerase slipping or chattering on the 7 U stretch. Again attenuation occurs at the junction between the N and P genes (a simple dinucleotide this time). An estimated 30% of transcribing polymerase molecules exit the RNP complex at each gene junction. This leads to differential levels of expression of each gene with transcript abundance directly reflecting proximity to the 3′ end of the genome, i.e. leader $> N > P > M > G > L$ (represented in Fig. 2). This attenuated transcription mechanism appears to be the primary means of regulation of viral expression. In an essentially identical manner, the polymerase completes the sequential synthesis of the remaining four mRNAs. Following the L gene transcription termination and polyadenylation signal, there remains a stretch of 57 untranscribed bases constituting the genome 5′ terminus. Transcription of input virion RNPs is resistant to inhibitors of protein synthesis and is termed primary transcription. Subsequently, a switch is made to synthesis of full-length positive-sense copies of the genome (see step 3). This switch mechanism is not fully understood. The critical step is the readthrough of the first junction encountered by the polymerase, i.e. the leader–N gene junction. The polymerase must be altered so as to no longer recognize the termination signal at this junction but continue down the full length of the genome. It appears that once the polymerase is committed past the first junction none of the other gene junctions are recognized and the full-length positive-strand is made. The simplest model fitting most of the experimental data proposes that rising levels of N protein encapsidation of the nascent positive-strand leader RNAs reaches a threshold whereby it pushes the polymerase complex through the termination signal and initiates replication of the entire template. However, analysis of mutants suggest that readthrough of the leader–N gene junction is possible in the absence of elevated levels of N protein, suggesting that altered N protein interaction may be involved. At least two different populations of N protein exist in the infected cell cytoplasm, one RNP bound and the other free. In addition, numerous species of P protein with varying degrees of phosphorylation exist. These may also play a crucial role in the transcription–replication mode switch mechanism.

Once full-length positive-strand synthesis occurs, production of the virus negative-sense RNA genomes necessary for packaging into released virions commences (step 4). A negative-sense leader RNA which represents a 46-nucleotide copy of the 3′ end of the positive-sense template can be found in infected cells. The simplest explanation for the synthesis of these molecules is analogous to that proposed for the positive-sense leader RNAs, i.e. in the presence of limiting levels of N protein the nascent negative-sense RNA synthesis terminates after the leader RNA transcript. In the presence of optimal levels of N protein the efficient encapsidation of the nascent strand pushes the polymerase through the termination signal and allows the polymerase complex to complete the synthesis of the entire negative-sense RNA genome. Such a proposed mechanism is attractive as it closely couples the production of virion RNA with the availability of viral proteins necessary for virion production.

Protein synthesis

The five viral monocistronic mRNAs are transcribed in the cytoplasm of infected cells, and their translation is directly coupled to the transcription process. The ratio of individual viral proteins synthesized is similar to that of individual viral mRNAs. Four of the viral mRNAs, N, P, M and L, are translated on free polysomes in a manner essentially analogous to cellular mRNAs. The G mRNA is translated on membrane-bound polysomes, the N terminus of the newly emerging G protein possessing a 16-amino acid hydrophobic signal peptide which targets the protein to the rough endoplasmic reticulum (ER). The signal peptide is cleaved in the lumen of the ER. The protein is glycosylated as it is transported through the Golgi complex and is eventually expressed on the cell surface membrane in a trimeric form.

Assembly and release

The glycoprotein forms patches on the surface of infected cells. The M protein appears to form a bridge between these G protein membrane patches and virion RNP cores and promotes the tight condensation of the cores and their assembly into mature virion particles. These are then released by budding from the surface of the cell. Considerable virus release occurs prior to the eventual disruption and death of the host cell. VSVs are generally highly cytopathic causing rapid cell death (in less than 12 h in some cases) in susceptible cells.

Geographic and Seasonal Distribution

VSVs have been isolated almost exclusively from New World mammals and insects. Disease epizootics occur throughout Canada, USA, Mexico, regions of the Caribbean, Central America, Panama, Venezuela, Colombia, Ecuador, Peru, Brazil and Argentina. VS New Jersey viruses are the most common, and have the widest distribution, with isolations as far north as Canada and as far south as Peru. VS Indiana type 1 viruses have a similar wide geographical distribution but are less frequently encountered. Disease outbreaks often follow natural features of the land (e.g. spreading throughout river valley systems rather than along road systems or trucking routes) and tend to have a marked seasonality. In temperate regions, they begin in the late spring, peak in the late summer and cease with the first frosts of late fall/early winter. In more tropical regions, the epizootics frequently appear at the cessation of rainy seasons. In both regions, the seasons of high VSV activity tend to correlate with high insect population levels. The major regions of enzootic disease activity include areas of the southeastern USA, Mexico, Central America and northern South America. VS Indiana type 2 (Cocal) viral infections have been detected in Trinidad, Brazil and Argentina. VS Indiana type 3 (Alagoas) viral infections have been detected in Brazil and Colombia. Owing to the close antigenic similarity of type 2 and 3 viruses with the type 1 Indiana viruses, they may be frequently misreported as type 1 outbreaks and thus their geographical distribution may be greater than is currently known.

There are only two reports of VS epizootics outside of the New World. The first was an apparent VS outbreak in horses and mules in the Transvaal, South Africa in the late 19th century. The second was a VS outbreak in France during the First World War which was initiated by a shipment of infected US army horses. There is no evidence of VSV activity persisting for prolonged periods in either of these continents.

Host Range and Virus Propagation

VSVs appear to infect a number of natural mammalian and insect hosts. Clear evidence of clinical disease has been reported for cattle, horses, swine and humans. In addition, there have been numerous reports of serological evidence for various VSV infections in white-tailed deer, mule deer, pronghorn antelope, wild turkeys, goats, sheep, ducks and dogs. Isolations of VSVs from insects have included: *Lutzomyia* sandflies, *Simulidae* (black flies), *Culicoides* (midges), *Culex nigripalpus, Hippilates* (eye gnats), *Musca domestica* (houseflies) and *Gigantolaelaps* sp. (mites). No disease symptoms are observed in these insects and their importance in the maintenance and transmission of VS disease is unclear. However, the ability of VSVs to replicate and be transovarially transmitted in *Lutzomyia* sandflies suggests a significant role for these insects in the VS disease process.

Numerous experimental hosts have been utilized for the study of VSVs. Viral dose and site of injection have been shown to influence the outcome of infection. For instance, intracerebral or intranasal inoculation of VSV in many rodents including mice, hamsters, guinea pigs, rats and chinchillas invariably results in a fatal encephalitis. However, intralingual inoculation of virus into cattle, horses, swine and guinea pigs or intradermal inoculation in foot pads of guinea pigs or rats can cause vesicular lesions at the injection sites but animals quickly recover. Newborn mice or hamsters are more sensitive to VSV infection, the virus being rapidly lethal following peripheral inoculation. The growth of VSVs is supported by a large number of vertebrate and insect tissue culture cell lines. Commonly utilized vertebrate lines include BHK-21, Vero and L cells, and common insect lines include *Aedes albopictus* cells.

Genetics

The single-strand, negative-sense RNA genomes of VSVs are approximately 11 kb in length. They contain five nonoverlapping genes coding for: a nucleocapsid protein (N), polymerase-associated phosphoprotein (P or NS), matrix protein (M), glycoprotein (G) and a large polymerase protein (L). Extensive analysis of temperature-sensitive mutants has identified five or six nonoverlapping complementation groups for these viruses. Viral genetic variability involves the accumulation of point mutations generated by the error-prone viral polymerase. The error rate is on the order of 1×10^{-4} to 4×10^{-4} base substitutions per base incorporated. There is no evidence of recombination playing a major role in the genetics of these viruses.

Evolution

The ancestral origin of the VSVs is currently unclear. A relatively conserved gene order and limited sequence similarity with other nonsegmented negative-strand RNA animal viruses suggest a common, albeit very ancient, ancestor. Phylogenetic analysis of VSV genome nucleotide sequences reveals a complex evolutionary pattern with multiple coexisting lineages and sublineages present in several regions of the Americas. There is good correlation between viral phylogeny and the location of virus isolation, indicating that many lineages and sublineages are relatively stably maintained in infection foci of limited geographical distribution (Fig. 3). There is no correlation between lineages and host species, i.e. there is not a distinct VS of any particular host species. Prediction of ancestral nodes on VSV evolutionary trees of New World viruses analyzed to date suggests that they may have originated in insects in Central America/northern South America. The rate of evolution of VSVs can be rapid, as seen during tissue culture passage [high multiplicity of infection (m.o.i.) or persistent infection] and by considerable genetic diversity in naturally occurring disease epizootics. However, genetic stability can also be observed during both tissue culture passage (at low m.o.i.) and natural maintenance of the virus in an enzootic focus. The relative rates of evolution of the N, P and G genes appears to be fairly constant. However, the rates of evolution of their encoded proteins is quite different. Comparison of different natural isolates have shown the order of amino acid sequence variability to be $P > G > N$. This is in contrast to several other RNA viruses where genes encoding surface antigens are most variable.

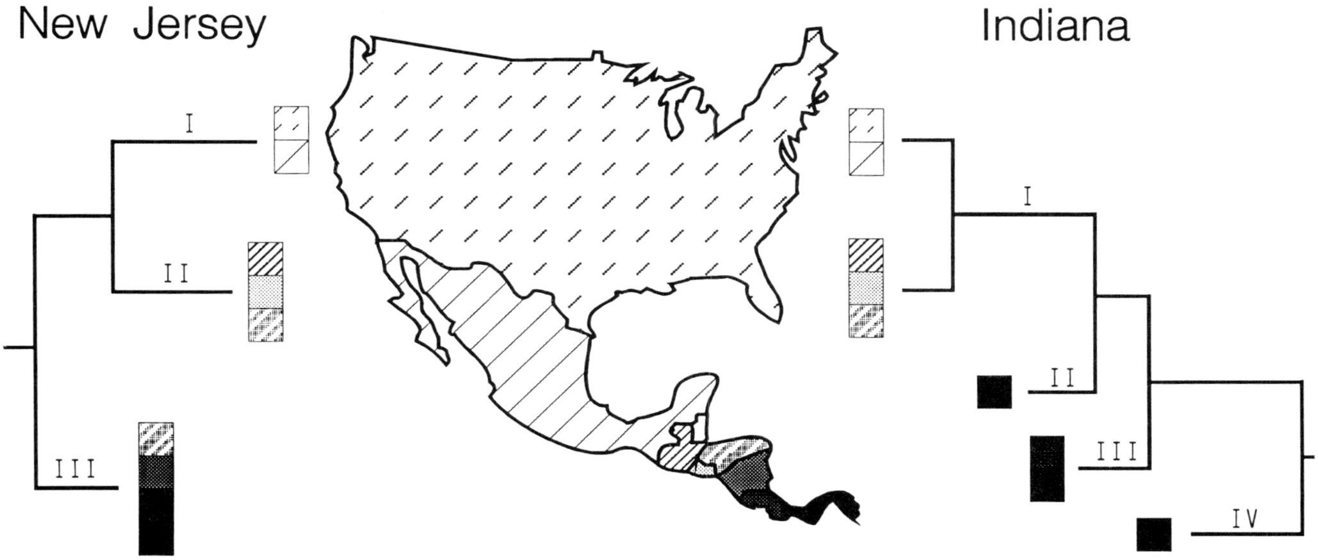

Fig. 3 Geographic and evolutionary relationships of the major subtypes of VSV New Jersey and Indiana serotype viruses. A summary of the VSV New Jersey and Indiana serotype evolutionary trees is shown on the left and right of the map respectively. These are based on maximum parsimony analysis of viral glycoprotein gene sequence differences among the major subtypes of these viruses. Horizontal branch lengths are proportional to the extent of gene sequence differences. The figure shows how virus isolates from more southern regions are located closer to the ancestral node of the trees than those further north.

Serological Relationships and Variability

All VSVs share common N protein epitopes which are group specific and usually detected by complement fixation assay. Neutralizing antibodies are elicited by the surface glycoprotein (G). Two major VSV serotypes, New Jersey and Indiana, have been defined on the basis of significant cross-neutralization among viruses of the same serotype. Limited antigenic variation has been observed among natural virus isolates of the VSV New Jersey serotype. Greater antigenic diversity is seen among Indiana serotype viruses with three distinct types identified. Antisera raised to virus of one Indiana type will neutralize virus of another type but to a lesser degree. Although some antigenic variation can be found, there is no evidence of distinct antigenic drift of VSVs in nature.

Epidemiology

VS New Jersey disease activity is the most common and has been studied in the greatest detail. The virus continues to cause periodic large-scale disease epizootics of cattle, horses and swine in the USA with peaks of activity approximately every 10–15 years. In addition, the disease has also persisted in the southeastern US in wild pigs on Ossabaw Island, Georgia over at least the last decade. Mexico, Central America, and northern South America generally exhibit more frequent disease epizootics, and various regions appear heavily enzootic for the disease. The last major US outbreak occurred in 1982–1983, affecting primarily cattle throughout 14 western states. Genetic analysis of viral isolates demonstrated no link between these viruses and those persisting in Georgia, or those found in Central America. However, a close genetic link was found with viruses persisting in Mexico, suggesting a possible source of virus initiating US epizootics. A similar epizootiological pattern is evident for VS Indiana type 1 viruses although outbreaks are generally less frequent and widespread. VS Indiana type 2 virus activity has been detected in Trinidad, Brazil and Argentina, including isolation from cattle, horses and insects. VS Indiana type 3 has been detected in Columbia and Brazil in numerous hosts including cattle, horses and insects. Little is known regarding the epizootiology of these two virus types although both can cause disease epizootics.

Transmission and Tissue Tropism

Little is known regarding the reservoir of VSVs, their transmission or their tissue tropism. Several lines of evidence suggest that VSVs are insect viruses and the infection of animals may be incidental. These include: (1) transovarial transmission of VSVs in experimentally infected *Lutzymia* sp. sandflies; (2) bite transmission of VSVs from experimentally infected mosquitoes to laboratory animal hosts; (3) seroconversion of caged sentinel animals placed in areas of enzootic VSV activity; (4) numerous reports of isolation of VSVs from blood-sucking insects in enzootic and epizootic disease areas; and (5) seasonality and geographic pattern of disease outbreaks (see Geographic and Seasonal Distribution section). Tissue tropism of VSVs in insects has not been studied extensively, although the midgut and fat body have been shown to be the major site of VS New Jersey viral replication in sandflies.

Viral activity is most obvious in cattle, horses and pigs because of the clinical illness induced. However, it is likely these animals play little role in the maintenance and transmission of the virus in nature. Infected animals exhibit low and transient viremias, with little or no virus shedding in milk, feces or urine. Direct transmission, although infrequent, is possible, as demonstrated by continuation of the 1982–1983 VS epizootic during the winter months in the USA. Most direct transmission is thought to involve teat to teat transfer in cattle dairy herds due to contaminated milking equipment. The saliva of animals exhibiting mouth lesions can contain high titers of virus and may provide a means of mouth to mouth transmission. Teat to mouth transmission has been reported between infected cows and their calves. Experimental data would suggest that the epidermal layer probably needs to be broken or abraded for successful direct viral transmission. Attempts to reproduce direct viral transmission have met with only occasional success. Transmission of VSVs to humans is not uncommon. Serological survey of humans in areas enzootic for VSVs suggests that human infection can be relatively common. Numerous cases of infection of laboratory workers, veterinarians and animal handlers have been reported. These are usually associated with viral entry through breaks in the skin or exposure to high virus titer aerosols such as infected animals sneezing in the faces of susceptible individuals. Tissue tropism of the virus in naturally infected animals is poorly understood. It seems likely that the majority of viral replication is found close to the site of entry of the virus and remains relatively localized. However, low levels of virus can be found elsewhere in the body including the blood.

Pathogenicity

The majority of VS infections of animals appear to be asymptomatic, as seroconversion of animals in the absence of obvious clinical symptoms is frequently observed. However, when clinical symptoms do appear it is unclear whether this is due to differences in the virulence of the virus, route or dose of virus inoculum, or host-specific factors. Tissue-culture-grown virus appears to quickly lose the ability to produce clinical disease when reintroduced into animals. However, viruses isolated from one host species can cause infection in another.

Clinical Features of Infection

VS disease in cattle, horses and swine is characterized by the appearance of vesicular lesions on the mouth (frequently the tongue), teats (of dairy cattle), or, relatively rarely, hoof coronary band epithelium layer approximately 2–4 days after introduction of the virus. Animals rarely exhibit lesions at more than one site. Depression, lameness, fever and excessive salivation are often seen before vesicles are detected. Alterations in hepatic enzyme levels suggest that VSV infection may also affect the liver in cattle and man. In dairy cattle, milk production can virtually cease and weight loss can be as much as 300 pounds. This is especially true if mouth lesions become too sore for animals to eat, or they become too lame to get to food and water or mastitis or other secondary infections become severe on udder teats preventing complete milking. In severe cases the udder can be lost to gangrenous mastitis. In most cases healing of lesions occurs within 7–10 days.

Human VSV infections can be abundant in rural areas with high VS disease activity. Reported cases of infections in laboratory workers or veterinarians are characterized by an influenza-like illness after 2–6 days incubation. The majority of cases develop a biphasic fever accompanied by malaise, myalgia, headache and chills. Fewer cases exhibit signs of pharyngitis, coryza and vomiting. Vesicles appear on the tongue, buccal and pharyngeal mucosa, lips or nose in a minority of cases. There is one report of VSV causing severe encephalitis in a child in Panama.

Pathology and Histopathology

Pathological changes appear to be limited to the epithelium of affected areas. Vesicles are easily ruptured, leaving raw reddened erosions surrounded by torn epithelium. In severe tongue lesions, much of the tongue epithelium may slough off.

Immune Response

VSV infection elicits both an humoral and cell-mediated immune response. Virus-specific serum antibody levels can be detected approximately 4 days p.i. and peak around 2–5 weeks p.i. Serum neutralizing antibodies are directed to the G protein, four major neutralization epitopes having been defined for both VS New Jersey and Indiana type 1 G proteins. Cattle with high levels of VSV antibodies are susceptible to re-infection, suggesting that such antibodies have limited protective effect. This may be explained by the majority of viral replication being localized in the epithelium. Serum neutralizing antibodies can persist for many years after infection and frequently fluctuate considerably. This has been suggested to be indicative of viral persistent infection although this has not been clearly demonstrated in naturally infected animals.

Experiments in cattle and swine have demonstrated proliferative responses of peripheral blood mononuclear (PBM) cells to VSV antigens. VSV-specific PBM proliferative responses could be detected at 3 weeks p.i. of swine or postvaccination of cattle. In both cases, these responses could still be detected 6 months later. Experiments with mice have shown that Class I major histocompatibility complex (MHC)-restricted T lymphocytes are primarily specific for VSV nucleocapsid protein. In contrast, Class II MHC-restricted T lymphocytes show specificity for the viral G protein.

Prevention and Control of VSVs

Strategies for the prevention and control of VSVs are poorly developed. They are complicated by the lack of understanding of the mechanisms of maintenance and transmission of these viruses in nature, although evidence is mounting for insect involvement in these processes. In the face of explosive VS epizootics, attempts to link viral spread center around quarantine of animals expressing clinical symptoms and insect vector abatement programs. Another complicating factor is the lack of understanding of the role of the host immune system in the regulation of VSV infections. Natural VSV infection of cattle does not appear to provide protection against subsequent infection by antigenically indistinguishable virus, suggesting that successful vaccination of animals may be difficult. However, several experimental vaccines have been designed including inactivated, live attenuated and vaccinia recombinant VSV New Jersey vaccines. None are currently utilized in the field and they all lack efficacy trials in a natural host system. International control of VS is by means of import/export control on the movement of animals with significant neutral-izing antibody titers to VSVs.

Future Perspectives

The current limited ability to control VS disease outbreaks is likely to be improved in the near future by studies leading to: (1) identification of VSV reservoirs in nature; (2) determination of the role of insects in virus maintenance and transmission; (3) increased understanding of the localized and cell-mediated immune response to VSV infections in natural hosts; (4) the increased use of viral cDNA clones to produce biologically active virus or viral components.

See also: Epidemiology of viral diseases; Immune response; Pathogenesis; Rhabdoviruses; Sigma rhabdoviruses.

Further Reading

Baer GM, Bellini WJ and Fishbein DB (1990) Rhabdoviruses. In: Fields BN *et al.* (eds) *Virology*, 2nd edn, p. 911. New York: Raven Press.

Banerjee AK and Barik S (1992) Gene expression of vesicular stomatitis virus genome RNA. *Virology* 188: 417.

Wagner RR (1990) *Rhabdoviridae* and their replication. In: Fields BN *et al.* (eds) *Virology*, 2nd edn, p. 867. New York: Raven Press.

VIRAL MEMBRANES

John Lenard
University of Medicine and Dentistry of New Jersey
Piscataway, New Jersey, USA

Introduction

Viruses of many kinds possess lipids as integral components of their structure. Lipid-containing, or enveloped, viruses include the *Corona-, Orthomyxo-, Paramyxo-, Bunya-, Rhabdo-, Toga-, Retro-, Herpes-* and *Poxviridae*. Despite the great diversity of these viruses in regard to structure, replicative strategy and pathogenicity, the function of the lipid is the same in all of them: to form a membrane surrounding the encapsulated viral genome. In all the lipid-containing viruses that have been studied, the lipids form a continuous bilayer that functions as a permeability barrier protecting the viral nucleocapsid from the external milieu. The bilayer is studded with virally encoded transmembrane proteins that serve specific functions facilitating virus entry, notably attachment to cell surfaces and fusion with cell membranes.

The membrane is acquired by the virus during assembly. In all cases (except perhaps one), membrane acquisition occurs by budding of the viral nucleocapsid through a particular cellular membrane, which is characteristic for each enveloped virus. Thus, during a normal infection most of the viruses mentioned above bud through the plasma membrane, but bunyaviruses bud through the Golgi apparatus, coronaviruses take their membranes chiefly from the endoplasmic reticulum and herpesviruses bud from the nuclear membrane. The single exception may be the poxviruses, which appear to acquire their membranes during *de novo* assembly in the cytoplasm.

Viruses that lack lipids possess capsids or shells consisting only of viral protein. These structures perform the same functions as viral membranes; i.e. protection of the genome and facilitation of its entry into the host cell.

The Viral Bilayer

Knowledge of the structure of viral bilayers comes chiefly from the study of a few viruses that are easily grown in large quantities in the laboratory, namely the orthomyxo-, paramyxo-, rhabdo- and togaviruses. In these, the bilayer arrangement of the lipids has been directly demonstrated using physical methods, and the lipid composition of various viruses grown under different conditions has been described in detail. Since all of these viruses acquire their bilayer by budding through the host cell's plasma membrane, the viral membrane generally contains the lipids present there. Wide variations of lipid composition are tolerated, and most of the lipids present display the properties characteristic of lipids in a bilayer, not those of protein-bound lipids. The precise content of each individual phospholipid present in a viral membrane may reflect some interaction with viral proteins, however. Intact virions are impermeant to proteases and other enzymes; indeed, the virions can swell and shrink in response to changes in osmolarity, showing that the viral membrane is impermeant to small ions as well, and forms an effective permeability barrier. It is generally assumed that these generalizations extend to all enveloped viruses, and not just to those that have been studied.

Viral Membrane Proteins

The proteins of enveloped viral membranes, like those of other membranes, may be classified as either integral or peripheral. Integral proteins are those that span the membrane one or more times, and thus cannot be removed without disrupting the bilayer, e.g. with detergents. Peripheral proteins do not cross the membrane, and can be removed from it by treatment with aqueous salts or chaotropic agents, which do not destroy the bilayer.

Integral membrane proteins of enveloped viruses generally span the membrane only once; an exception is the E1 protein of coronavirus, which spans the membrane three times. As much as 90% of the polypeptide chain of a viral membrane protein may be external to the bilayer, where it is accessible to degradation by added proteases. The remaining 'tail', consisting of the protected transmembrane and internal portions, can often be recovered after proteolytic digestion. These proteins possess carbohydrate side chains identical in attachment position and structure with those of cellular proteins. This normal glycosylation pattern reflects the viral proteins' synthesis at, translocation through, and assembly within, the cell's rough endoplasmic reticulum.

Peripheral proteins are attached to the viral membrane by a combination of electrostatic and hydrophobic interactions; although they may penetrate the bilayer, they do not cross it. Viral peripheral proteins include the M proteins of paramyxo-, orthomyxo- and rhabdoviruses.

Attachment of Viruses to Host Cells

The first step in infection, attachment of the virus to the outer surface of the host cell, is performed by the membranes of enveloped viruses. The feature of the host cell membrane recognized by the envelope of each virus is unique. This specificity is conferred by the binding properties of the membrane glycoproteins. Thus, the nearly total specificity of the retrovirus human immunodeficiency virus (HIV) for cells expressing CD4 protein is conferred by the affinity of the viral envelope protein gp120 for this cell surface 'receptor'. Orthomyxoviruses and paramyxoviruses have broader specificity; their hemagglutinin (HA and HN) glycoproteins respectively bind to sialic acid residues on the surface of the host cell. Although sialic acids in different glycosidic linkages bind to these proteins, some preference for specific linkages is shown by different influenza viral strains. The most nonspecific of the enveloped viruses may be the rhabdoviruses, represented by vesicular stomatitis virus (VSV), which are apparently able to bind indiscriminately to clusters of negative charges, whether created by lipids, proteins or saccharides.

The tissue-specific tropism of particular viruses during infection is determined in many cases by the presence of a specific 'receptor' on the surface of the susceptible cells, which is recognized by the relevant viral membrane protein. Receptor recognition provides a basis for viral tropism of both enveloped and nonenveloped viruses.

Viral Fusion

Before viral transcription and replication can commence, the viral genome must cross the barriers presented by both the viral envelope and the cell membrane; fusion of the viral and cellular membranes accomplishes this. Fusion, like cell attachment, is a property conferred upon each virus by a specific envelope glycoprotein. In contrast to cell attachment, however, fusion does not require cell proteins, since virions fuse readily with protein-free liposomes. Similarly, 'virosomes' reconstituted from pure viral fusion protein and lipid also have full fusion activity when tested with liposomes.

The fusion protein of paramyxoviruses (named F) differs from those of orthomyxo-, rhabdo- and togaviruses in regard to the pH at which it acts; while the former is active at neutral pH, the latter require a more acidic pH, usually 5–6, for fusion. This difference has profound consequences, since it reflects two distinctly different modes of viral entry (Fig. 1). Those viruses capable of fusing at neutral pH (only paramyxoviruses of the four classes mentioned above) can fuse with the cell's plasma membrane under normal conditions, i.e. at the neutral pH of extracellular fluid or culture medium. Those viruses that fuse only at acidic pH, on the other hand, must first be internalized from the plasma membrane by endocytosis; fusion, and consequent uncoating of the viral genome can only occur in the acidic endosomes. The pH dependence of viral fusion (conveniently measured as virus-induced hemolysis or by a variety of more direct techniques) serves to distinguish between viruses that enter the cell by these two distinct routes (Fig. 1). Another difference is that cell entry by acid-requiring viruses can be inhibited by any of a variety of membrane-permeant amines (notably chloroquine, ammonia and methylamine). Such weakly basic compounds are referred to as lysosomotropic, or acidotropic, in that they neutralize intracellular acidic compartments, thus preventing the activation of the acid-requiring viral fusion proteins.

Despite intensive study, the mechanism of viral fusion remains largely unknown, chiefly because it has proved impossible to isolate fusion intermediates for characterization. The influenza binding and fusion protein, named HA, has been the most thoroughly studied, since the three-dimensional structure of its extracellular portion (which contains the active sites for both fusion and binding) has been determined by X ray crystallography. Activation of HA occurs by proteolytic cleavage at a single position on the polypeptide chain; a hydrophobic sequence, often called the 'fusion peptide', becomes the newly created N terminus, and this participates in the fusion reaction. The paramyxovirus fusion

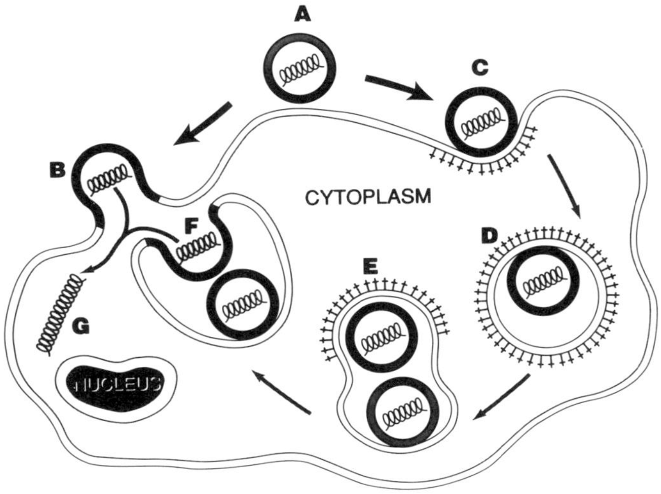

Fig. 1 A diagrammatic representation of the two major pathways for cellular penetration and uncoating of enveloped viruses. Uncoating begins with attachment of the virion to the cell surface (A), through the binding of an integral viral membrane protein to a 'receptor' on the cell surface, which may be a specific cell protein, an oligosaccharide, or a patch of charged lipids. Attachment is followed by fusion, which is mediated by a viral protein which may or may not be of the same type as the attachment protein. If the viral fusion protein is active at neutral pH, fusion can occur directly with the plasma membrane (B). Alternatively, if fusion requires an acidic pH, the virion must first be endocytosed via the coated pit–coated vesicle pathway (C,D). The viral fusion protein is activated in the acidic endosomes (F). Both pathways result in the introduction of the viral nucleocapsid, containing the viral genome, into the cytoplasm (G). For viruses that undergo transcription and replication in the nucleus (such as myxoviruses and DNA viruses), uncoating is followed by transport of the nucleocapsid across the nuclear membrane by an unrelated process.

protein, called F, shows considerable sequence homology with the orthomyxovirus HA, and the sequence of the fusion peptide is especially conserved. Togavirus fusion is unique in requiring the presence of cholesterol in the target membrane; one togavirus fusion protein has been shown to be a cholesterol-binding protein. Other viruses have less exacting lipid requirements for their target membranes, although fusion rates and efficiencies can vary markedly with alterations in lipid composition of the target membrane. A more detailed understanding of the mechanisms of viral fusion would be of both fundamental and therapeutic importance.

Membrane Synthesis

Viruses make maximal use of mechanisms already in place in the infected cell to perform their functions. Hence, viral protein synthesis is carried out on host cell ribosomes. Specifically, synthesis of viral membrane proteins occurs on membrane-bound ribosomes, from which they are inserted in the correct orientation into the endoplasmic reticulum membrane. There they are glycosylated and assembled into multimeric form. The influenza HA protein, for example, is assembled into the trimers that form the 'spikes' on the surface of the fully assembled virion. The viral glycoproteins may then be further processed through the Golgi and on to the plasma membrane (Fig. 2). In fact, the membrane proteins of VSV and influenza have provided valuable tools for the study of this cellular transport process. Each of these viruses possesses only one major membrane protein, which is expressed in infected cells at high levels. Further, host-cell protein synthesis is inhibited by both VSV and influenza infection, so large amounts of a single membrane protein are produced and correctly processed in infected cells.

Viral proteins are targeted to specific cellular locations by the same mechanisms that the cell uses for its own membrane proteins. Viral proteins, in fact, are widely used in the study of these targeting mechanisms, which are still incompletely understood. The VSV G protein, for example, is located exclusively in the basolateral plasma membranes of polarized cells. This is thought to occur by a 'default' mechanism, i.e. as a result of the lack of

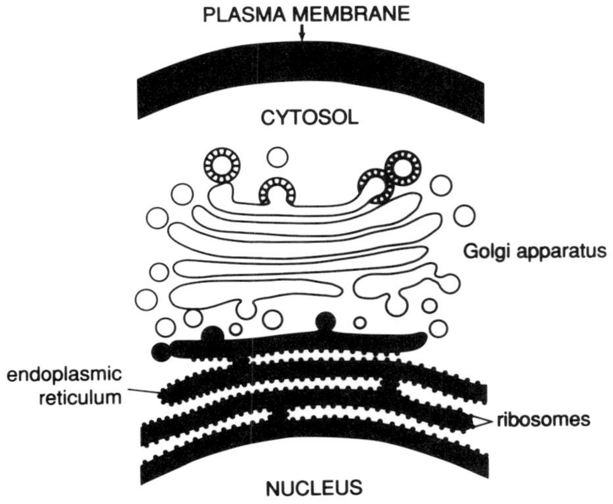

Fig. 2 A diagrammatic representation of the endoplasmic reticulum–Golgi–plasma membrane system of a cell. All viral and cellular membrane proteins are synthesized by ribosomes bound to the endoplasmic reticulum membrane. Proteins destined for the plasma membrane then undergo vesicular transport to the nearest lamella of the Golgi apparatus (called the 'cis face'). A series of sequential vesicular transport steps then carries these proteins through the Golgi to the 'trans face' and out to the plasma membrane. Assembly and budding of different viruses occurs at characteristic points in this membrane system.

any specific retention or targeting signal in G protein. The influenza HA protein, on the other hand, is delivered to the apical plasma membranes of the same polarized cells, presumably through the operation of a specific targeting sequence. Similarly, the retention of coronavirus glycoproteins by the endoplasmic reticulum, and of bunyavirus proteins by the Golgi are thought to reflect the operation of the same cellular mechanisms that retain normal resident proteins in these organelles in uninfected cells. The localization of viral membrane proteins is of particular importance, since it determines the location of viral assembly and budding.

As described above, the lipids of the viral membrane are taken from the host cell membrane during budding. No new lipids are specifically synthesized in response to viral infection, and viruses seem to tolerate wide variations in their lipid composition. Alterations in cellular lipid metabolism have been reported to result from some viral infections *in vitro*, but these are most likely secondary to other cytopathic effects; there is no indication that they play an important role in the progress of the infection.

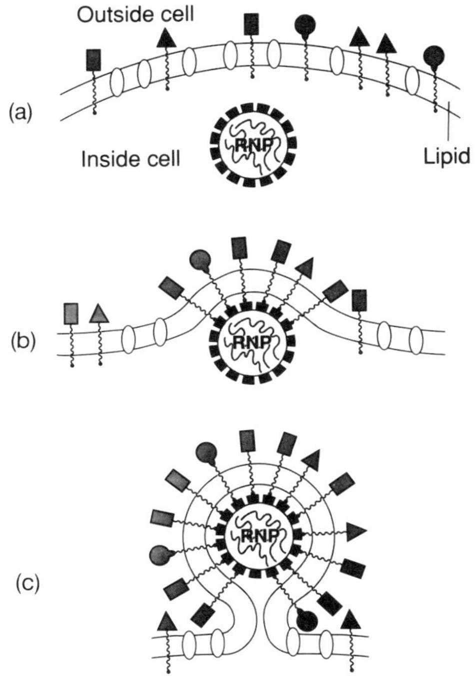

Fig. 3 A diagrammatic representation of virus budding. Viral glycoproteins, inserted into the cellular membrane at the endoplasmic reticulum (Fig. 2), associate with the assembled viral nucleocapsid. The direct association pictured here is characteristic of togaviruses; in other viruses, possessing helical nucleocapsids, the association is mediated by a peripheral membrane protein, called M. Cellular membrane proteins are excluded from the envelope of the mature virion. This may occur during assembly, as pictured, or by formation of a viral membrane 'patch' before the nucleocapsid binds the membrane.

Virus Assembly

The budding process consists of the wrapping of a specific piece of membrane around the previously assembled nucleocapsid, which contains the viral genome; it is shown diagrammatically in Fig. 3. The specificity of the process is remarkable in that the completed viral envelope contains only viral glycoproteins, while host cell membrane proteins are almost completely excluded.

Viruses can bud anywhere in the endoplasmic reticulum–Golgi–plasma membrane pathway shown in Fig. 2. While the paramyxo-, orthomyxo-, rhabdo- and togaviruses (and many others) generally bud from the plasma membrane, they have also been shown to bud intracellularly under certain conditions. Other viruses normally bud intracellularly, from the endoplasmic reticulum or Golgi apparatus, e.g. coronaviruses and bunyaviruses. In these cases the nucleocapsid, assembled in the cytoplasm, buds into the lumen of the appropriate organelle. The assembled virus is often seen inside vesicles, giving it a double-shelled appearance. Eventually, the newly formed virion may be secreted out of the cell through the normal secretory pathway, although this does not always occur efficiently.

Very little is known about the mechanism of viral budding, since budding has not yet been successfully accomplished in a cell-free system. In orthomyxo-, paramyxo- and rhabdoviruses, budding is mediated by a peripheral membrane protein (called M), which is thought to act as an interface between the viral nucleocapsid and the appropriate patch of membrane; however, specific interactions between M and other viral proteins have not been demonstrated. Togaviruses, which lack an M protein, possess an icosahedral nucleocapsid which interacts directly with the viral membrane protein. Completed virions contain an equal number of nucleocapsid and membrane protein molecules, both in a similar geometric arrangement, so in this case specific interaction between the two proteins appears likely.

This entry has been concerned with the general principles governing viral membrane structure and function. The reader is referred to the entries covering the individual enveloped viruses for more detailed descriptions.

See also: Human immunodeficiency viruses; Influenza viruses; Replication of viruses; Sendai virus; Sindbis and Semliki Forest virus; Vesicular stomatitis viruses.

Further Reading

Alberts B *et al.* (1989) Intracellular sorting and the maintenance of cellular compartments. In: *Molecular Biology of the Cell*, 2nd edn, p. 341. New York: Garland Publishing.

Wiley DC and Skehal J (1990) Viral membranes. In: Fields BN *et al.* (eds) *Virology*, 2nd edn, p. 63. New York: Raven Press.

Wilschut J and Hoekstra D (1991) *Membrane Fusion. Part III, Fusogenic Properties of Viruses*, p. 275. New York: Marcel Dekker.

VIRAL RECEPTORS

Horacio U Saragovi, Gordon J Sauvé and Mark I Greene
University of Pennsylvania School of Medicine
Philadelphia, Pennsylvania, USA

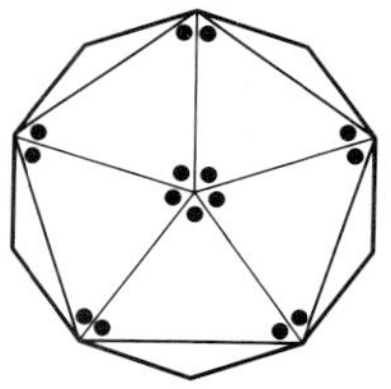

Introduction

Virus–host interactions are complex processes that span binding to cells, entry, dissemination and finally lytic or persistent infection. The first step, binding to host molecules that serve as viral receptors, is critical in all subsequent events. However, host cell alterations can occur in the virtual absence of entry and infection. These effects are due to viral binding to the cellular receptor and the activation of second messenger systems linked to the receptor.

Several viruses reportedly exploit cellular molecules for binding (see Table 1). Two interesting examples are the human immunodeficiency virus (HIV) and the vaccinia virus. HIV gp120 appears to bind to two distinct molecules, CD4 in T cells and galactosylceramide in brain. No homologies have been reported between gp120 and Class II major histocompatibility complex (MHC) molecules which are the biological ligand of CD4. Furthermore, it seems that the sites of CD4–Class II interaction are distinct from sites of CD4–gp120 interaction. This example emphasizes the evolution of the gp120 molecule as an effective binding agent for two completely different receptors. In contrast, the vaccinia VGF protein appears to bind to the epidermal growth factor (EGF) receptor. Vaccinia virus VGF is partially homologous to EGF which may explain its binding to EGF receptors. Therefore, some viruses may mimic the biological ligand of the cellular receptor.

Viruses have been proven to be invaluable tools in the molecular dissection of cellular processes and immune regulation. We have been studying the cellular receptor for reovirus type 3 and the role of this receptor in growth and development. Reoviruses

Table 1. Cell surface molecules that serve as viral receptors

Virus	Putative receptor
Vaccinia	EGF receptor
Sendai	Gangliosides
Epstein–Barr	C3d receptor
Lactic dehydrogenase	Ia and Fc receptors
Rabies	Acetylcholine receptors
Rhinovirus	ICAM-1 adhesion molecules
HIV (in T cells)	CD4
HIV (in brain)	Galactosylceramide

Abbreviations: EGF, epidermal growth factor; HIV, human immunodeficiency virus.

provide an excellent model for the study of viral binding which includes biological effects. Furthermore, the as yet unknown structure of the reovirus receptor and its putative biological ligand provides an interesting system in which receptor biological function can be analyzed. In the following section we describe the role of reovirus type 3 cellular receptor in growth and differentiation of a variety of tissue types. New findings also point to the development of novel vaccines and drugs to treat viral infections.

The Reovirus Type 3 Cellular Receptor (Reo3R)

Reo3R was initially analyzed by Fields and colleagues when mammalian lymphocytes and neurons were recognized to express a molecule utilized by the reovirus as its attachment site. Initially, cellular tropism by reovirus type 3 was linked to the expression of the viral hemagglutinin (HA) encoded by the $\sigma 1$ viral gene. Later, receptor-ligand assays using murine cells and labeled virus demonstrated that reovirus type 3 binds to lymphoid cells and fibroblasts with similar affinity. In contrast, reovirus type 1 or a recombinant construct of reovirus type 3 expressing type 1 HA was found to bind to fibroblast L cells but did not bind to lymphocytes (Table 2). The viral HA determines both the serotype specificity and the ability of reovirus to bind to cells. Thus, the HA type 1 (HA1) and HA type 3 (HA3) appear to recognize and bind to different and nonoverlapping structures on cell surfaces. These cell surface molecules bound by viral HA are exploited as cellular receptors for reovirus.

To analyze Reo3R in detail, we obtained a murine monoclonal antibody (MAb) directed against the receptor. The anti-Reo3R MAb 87.92.6 was raised against MAb 9BG5 which binds and neutralizes HA3. Thus, the interactions of MAb 9BG5 with MAb 87.92.6 or HA3, and the interactions of HA3 and MAb 87.92.6 with the cellular receptor, can be studied. An important feature of this model is that binding to Reo3R is accomplished by a structure that is shared by MAb 87.92.6, the HA3 and their analogs. It has been demonstrated that the light chain variable region (V_L) of MAb 87.92.6 (specifically the complementarity determining region 2; CDR2) bears the internal image of HA3 (Table 3), emphasizing the relevance of structural domains in binding to Reo3R. Conversely, it has been hypothesized that Reo3R may bear the internal image of MAb 9BG5. Work in progress towards the molecular cloning of Reo3R cDNA will answer this question.

Table 2. Binding affinity of *Reoviridae* to murine cell lines

Cell type	K_d for reovirus type 3 (nM)	K_d for reovirus type 1 (nM)
R1.1 thymoma	0.6	Not detected
Fibroblast cells	0.8	0.4

The R1.1 murine thymoma cells express ~ 50 000 Reo3R molecules per cell and the murine L fibroblasts express ~ 100 000 Reo3R molecules per cell.

Given the primary sequence and structural similarity between HA3 and MAb 87.92.6, it is not surprising that the MAb prevents reovirus type 3 attachment to cells and primes mice to develop humoral or cellular immunity to the virus. Most importantly, binding of MAb 87.92.6 or its peptide analogs to Reo3R elicits receptor-mediated biological effects identical to those elicited by binding of inactivated virus to cells. Functional effects include the inhibition of mitogen-induced proliferation of T cells, maturation of oligodendrocytes and demyelination of neurons. Peptide analogs derived from V_L amino acids 45–55, as well as a synthetic nonamino-acid-based β-loop structure which mimics this region, also bind to Reo3R and elicit functional responses, indicating that receptor ligation generates the biological effects. Thus the technology developed using the Reo3R system might lead to the development of new drugs to modulate immune responses and affect neural development.

Effects mediated upon Reo3R binding are so profound and significant that an important role for the cellular receptor is suggested, resulting in an increased interest in understanding the cognate

Table 3. Sequence comparison of MAb 87.92.6 variable heavy (V_H) and light (V_L) chains with HA3

87.92.6 V_H	43	**Q**	**G**	**L**	**E**	**W**	**I**	**G**	R	I	D	P	A	N	G				56
Reovirus HA3	317	**Q**	**S**	**M**	**-**	**W**	**I**	**G**	I	V	S	**Y**	**S**	**G**	**S**	**G**	**L**	N	332
87.92.6 V_L	39	K	P	G	K	T	N	K	L	L	I	**Y**	**S**	**G**	**S**	**T**	**L**	Q	55
									| Reo3R binding |									

Amino acid sequence similarity of the reovirus type 3 $\sigma 1$ protein and MAb 87.92.6 light chain CDR2 is shown using the single-letter code. Amino acid numbers of the proteins are shown, and the regions within HA3 and V_L required for Reo3R binding are indicated. Bold letters indicate amino acid homology between viral protein and mAb.

function of this surface molecule. Unlike other viral receptors the biological functions of which are known (e.g. CD4 is the HIV receptor and its biological ligands are Class II MHC gene products), it is not known whether Reo3R is also a true receptor with a true ligand. We will discuss evidence addressing this question in the next section.

Functional Studies of Reo3R

Reoviral infection takes place primarily in the gut epithelium, and can be spread to neural tissue by an as yet unclear mechanism. The uncoupling of viral entry and replication processes is emphasized by the fact that, while productive infection occurs in the epithelium, latent chronic infection can take place in lymphoid cells. Obviously Reo3R must be expressed on epithelial, neural and lymphoid tissues for viral binding, and studies with MAb 87.92.6 have determined that the viral receptor is the same for these three cell types. However, the varied effects manifested upon receptor ligation suggest that expression of Reo3R has different repercussions for different tissues. Therefore, in order better to understand the function of Reo3R, we have studied its expression and the functional consequences of receptor engagement with the use of anti-Reo3R MAb 87.92.6.

Lymphoid cells

Most (80%) freshly isolated murine splenic B cells express Reo3R. In contrast, only a small proportion (5–15%) of resting splenic T cells and thymocytes express Reo3R. Interestingly, activation of T cells with concanavalinA (ConA) mitogen induces expression of Reo3R to detectable levels after 24 h and optimally after 36 h. Therefore, Reo3R is constitutively expressed in splenic B cells, but is inducible in T cells, suggesting that differential regulation of transcription or translation occurs. The kinetics of expression in T cells parallel those of other inducible T-cell activation markers such as interleukin 2 (IL-2) receptor p55 (α) subunit and transferrin receptor, suggesting that Reo3R plays an important role in cellular immunity. After expression of Reo3R has been induced in T cells, binding by inactivated virus, MAb 87.92.6, peptides derived from the MAb V_L domain or synthetic compounds that mimic these structures to the cell surface leads to a dramatic inhibition of proliferation even in the presence of the powerful stimulator ConA.

Little is known about the mechanism of inhibition, but it does not involve cell death. Alternative possibilities include (i) blocking of an as yet unidentified biological ligand to the receptor or (ii) induction of receptor-mediated inhibitory signals. An exciting recent development has been the demonstration that inhibitory effects require ligation of Reo3R at the cell surface. In contrast to the negative growth regulation induced by Reo3R extracellular ligation, intracellular ligation has no effect on T cell proliferation. The data, however, do not clarify which of the possible mechanisms for inhibition may be operational.

Central nervous system (CNS)

The CNS is an important target of reovirus type 3 infection. Mature oligodendrocytes and both type 1 and 2 astrocytes, but not glial progenitor cells, express Reo3R. Surface Reo3R appears at an early stage of development prior to expression of myelin basic protein (MBP).

The effects of MAb 87.92.6 and its peptides in neural tissue have been tested *in vitro* and *in vivo*. In both systems neural tissue biology is altered and demyelination is induced, suggesting that Reo3R normally plays an important role in oligodendrocyte differentiation. Thus, it may be possible to create analogs of MAb 87.92.6 that stimulate myelin synthesis *in vivo* leading to the development of new drugs to modulate immune responses and affect neural development. A recent advance in this area has shown that synthetic analogs of MAb 87.92.6 bind Reo3R and induce functional effects, emphasizing the usefulness of viruses in analyzing receptor biology.

Future studies will reveal the structure and function of Reo3R, define the true physiological ligand(s) and provide insights into the signal transducing pathways by which this receptor exerts its important effects.

See also: Cell structure and function in virus infections; Human immunodeficiency viruses; Pathogenesis; Reoviruses; Replication of viruses; Virus–host cell interactions.

Further Reading

Cohen JA *et al.* (1989) In: Sercarz E (ed.) *Antigenic Determinants and Immune Regulation*, vol 46, p. 126. Basel: Karger.
Williams WV *et al.* (1990) *Trends Biotechnol.* 8: 256.

VIROIDS

Robert A Owens
United States Department of Agriculture
Beltsville, Maryland, USA

Viroids are the smallest known agents of infectious disease; they are small (246–375 nucleotides) highly structured single-stranded RNA molecules which lack both a protein capsid and detectable mRNA activity. Whereas viruses have been described as 'obligate parasites of the cell's translational system' and supply some or most of the genetic information required for their replication, viroids can be regarded as 'obligate parasites of the cell's transcriptional machinery.' Thus far, viroids are known to infect only plants.

History

While scientific investigation of plant diseases now known to be caused by plant viruses did not begin until the late 19th century, the earliest written records of plant viral disease date to the 8th century AD. The first viroid disease to be studied by plant pathologists was potato spindle tuber disease. In 1923, its infectious nature and ability to spread in the field led Schultz and Folsom to group potato spindle tuber disease with several other 'degeneration diseases' of potatoes. These maladies, long attributed to senility, 'reversion' or loss of vigor caused by prolonged asexual reproduction, are now known to be caused by infection with conventional RNA viruses. Nearly 50 years were to elapse between the first published descriptions of potato spindle tuber disease and Diener's 1971 demonstration of the fundamental differences between the structure and properties of its causal agent, potato spindle tuber viroid (PSTVd), and those of conventional plant viruses.

Growers and plant pathologists are unlikely to have simply overlooked diseases with symptoms as severe as those of chrysanthemum stunt or cucumber pale fruit, two other viroid diseases first reported after World War II. Thus, some aspect of modern agricultural practice seems to have favored the appearance of viroid diseases in the 20th century. Large-scale monoculture of genetically identical crops and the commercial propagation/distribution of many cultivars are two comparatively modern developments which can facilitate the development of serious disease problems following the chance transfer of viroids from their wild hosts to cultivated plants.

Genome Structure

Efforts to understand how viroids replicate and cause disease without the assistance of any viroid-encoded polypeptides have prompted detailed analysis of their structure. Viroids possess rather unusual properties for single-stranded RNAs (e.g. a pronounced

Table 1. Viroid species of known nucleotide sequence (1993)

Group[a]	Subgroup[a]	Viroid	Abbreviation	Length (nucleotides)
Self-cleaving viroids	–	Avocado sunblotch	ASBVd	246–250
		Peach latent mosaic	PLMVd	337
		Carnation stunt associated[e]	CarSAVd	275
PSTVd	PSTVd	Chrysanthemum stunt	CSVd	354, 356
		Citrus exocortis	CEVd	369–375
		Coconut cadang-cadang	CCCVd	246, 247
		Coconut tinangaja	CTiVd	254
		Columnea latent	CLVd	370
		Cucumber pale fruit[b]	CPFVd	303
		Hop latent	HLVd	256
		Hop stunt	HSVd	297–303
		Indian bunchy top[c]	CEVd-t	372
		Potato spindle tuber	PSTVd	356–360
		Tomato apical stunt	TASVd	360
		Tomato planta macho	TPMVd	360
	ASSVd	Apple scar skin	ASSVd	330
		Australian grapevine	AGVd	369
		Citrus bent leaf	CBLVd	318
		Grapevine yellow speckle	GYSVd	367
		Grapevine yellow speckle-2[d]	GYSVd-2	363
		Pear blister canker	PBCVd	315
	CbVd	*Coleus blumei*	CbVd	248

[a]Classification scheme essentially that of Koltunow AM and Rezaian MA (1989) (*Intervirology* 30: 194).
[b]Actually an HSVd sequence variant.
[c]Actually a CEVd sequence variant.
[d]Formerly designated 'Grapevine viroid 1B'.
[e]Proof autonomous replication required.

resistance to digestion by ribonuclease and a highly cooperative thermal denaturation profile), leading to an early realization that they might have an unusual higher-order structure. Their small size also made viroids tempting objects for detailed structural investigation. As a result of a series of subsequent studies probing their *in vitro* secondary and tertiary structures, only the structure of tRNA is better understood.

To date, nearly 20 different viroid species plus a number of sequence variants have been completely sequenced (see Table 1). All known viroids are single-stranded circular RNAs which contain 246–375 unmodified nucleotides and lack any unusual 2′,5′-phosphodiester bonds or 2′-phosphate moieties. Electron microscopy, optical melting and other physicochemical studies, and theoretical calculations of their lowest free energy secondary structure all indicate that viroids assume a highly base-paired rod-like conformation *in vitro* (i.e. the so-called 'native' structure shown in Fig. 1). Pairwise sequence comparisons of PSTVd with several related viroids suggest that the series of short double helices and small internal loops which comprise this structure are organized into five domains. Domain boundaries are defined by sharp changes in sequence similarity.

As implied by its name, the 'conserved central domain' is the most highly conserved viroid domain and is believed to contain the site where multimeric viroid RNAs are cleaved and ligated to form circular progeny. The 'pathogenicity domain' contains one or more structural elements which modulate symptom expression, and the relatively small 'variable domain' exhibits the greatest sequence variability between otherwise closely related viroids. The two 'terminal domains' appear to play an important role in viroid replication and evolution.

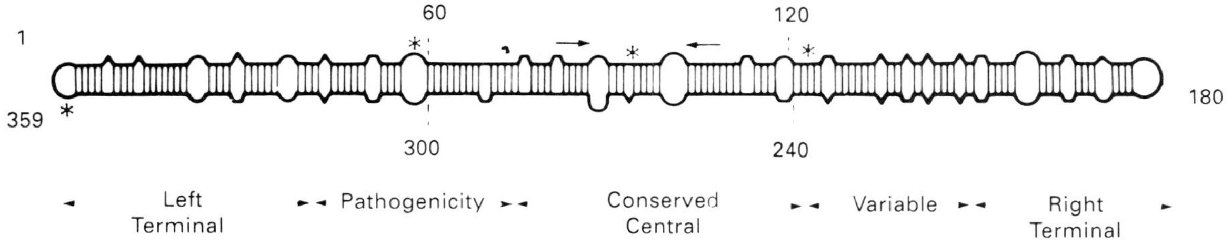

Fig. 1 Structural organization of PSTVd and related viroids. The native structure of PSTVd contains a series of short helices and internal loops organized into five structural domains. Boundaries of each domain were determined by pairwise sequence comparison, and inverted repeats within the conserved central domain (arrows) bracket a strictly conserved U-bulged helix (asterisk). Locations of several conserved sequence motifs in the other domains are also indicated by asterisks: CCUC in the left terminal loop, a polypurine region in the upper portion of the pathogenicity domain and a short oligopurine:oligopyrimidine helix in the variable domain.

Although these five domains were first identified in members of the PSTVd viroid group, ASSVd and related viroids also contain a similar domain arrangement. Certain viroids such as CLVd appear to be 'mosaic' molecules formed by exchange of domains between two or more viroids infecting the same cell. RNA rearrangement/recombination can also occur within individual domains, leading, in the case of CCCVd, to duplication of the left terminal domain plus part of the variable domain, as disease progresses. The presence of domains in ASBVd and CbVd remains uncertain pending identification of additional group members.

Much less is currently known about viroid tertiary structure, especially *in vivo* where these molecules accumulate as ribonucleoprotein particles within the cell nucleolus. Although the extended rod-like nature of the 'native' structure might suggest that viroids lack significant tertiary structure, the ability of UV irradiation to cross-link certain nucleotides within the conserved central domain of PSTVd provided the first definitive evidence for such tertiary interactions. Similar UV-sensitive structural elements have also been discovered in a number of other RNAs including the viroid-like domain of the hepatitis delta virus genome. Studies of the spontaneous self-cleavage of various ASBVd-related RNAs as well as the nuclease-dependent conversion of PSTVd multimers into monomers have provided additional evidence for the functional importance of viroid tertiary structure.

Taxonomy and Classification

Based upon marked differences in the structural and functional properties of their genomes, viroids and conventional viruses have long been assigned to separate taxa. To highlight these differences further, a change in viroid nomenclature has recently been proposed, i.e. the designation of potato spindle tuber viroid as PSTVd rather than PSTV, etc. Although this change has yet to be universally accepted, the new nomenclature will be used throughout this entry.

Comparative sequence analysis has defined a viroid species as one or more independently replicating sequence variants showing $\geq$90% sequence homology in pairwise comparisons. Based upon a combination of overall sequence similarity and the presence/absence of certain structural features within the conserved central domain (see Fig. 1), three viroid classification schemes have been proposed. All three schemes separate ASBVd, the only viroid known to undergo spontaneous self-cleavage, from three or more additional groupings whose type members include ASSVd and CbVd. As shown in Table 1, Koltunow and Rezaian consider the remaining viroids to form one large grouping the type member of which is PSTVd. The consensus phylogenetic tree of Elena *et al.*, on the other hand, distributes these same species among three smaller groupings (i.e. the PSTVd-like viroids, the CCCVd-like viroids and HSVd; see Fig. 2). Although the recently reported CbVd was not considered in this analysis, the inclusion of several viroid-like satellite RNAs and the viroid-like domain of hepatitis delta virus (HDV) has provided evidence for an evolutionary link between viroids and satellite RNAs (see below).

Host Range and Transmission

All viroids are mechanically transmissible, and most are naturally transmitted from plant to plant by man and his tools. Nevertheless, individual viroids vary

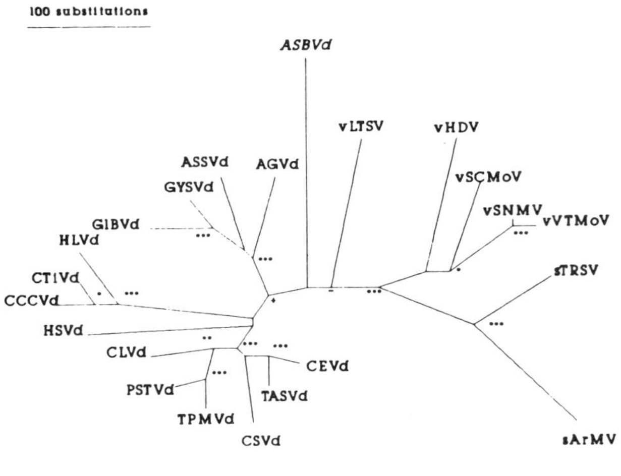

Fig. 2 Consensus phylogenetic tree containing 22 viroids, viroid-like satellite RNAs and the viroid-like domain of hepatitis delta virus (HDV) RNA. ASBVd has been taken as outgroup. *** Group monophyletic in all 1000 bootstrap replicates; ** monophyletic in more than 99%; * monophyletic in more than 95%; +, monophyletic in more than 90%; –, monophyletic in more than 80% of all replicates. From ASBVd to the left of the Figure, groups are considered as being within the viroid family, and from ASBVd to the right (including the viroid-like domain of HDV RNA) as within the satellite family. For example, satellite tobacco ringspot virus (sTRSV) and satellite *Anabis* mosaic virus (sARMV) (satellite family) or CCCVd, CTiVd and HLVd (viroid family) conformed to two well-defined monophyletic groups in all bootstrap replicates. LTSV, Lucerne transient streak virus; ScMoV, subterranean clover mottle virus (SCMoV); SNMV, *Solanum modiflorum* mosaic virus; VTMoV, Velvet tobacco mottle virus. [From Elena, SF *et al.* (1991) *Proc. Natl Acad. Sci. USA* 88: 5631, with permission.]

greatly in their ability to infect and replicate in different plant species. PSTVd can replicate in about 160 primarily solanaceous hosts, while only two members of the Lauraceae are known to support ASBVd replication. The particularly wide host range of HSVd includes several herbaceous species as well as woody perennials (e.g. grapes, citrus and various *Prunus* spp.). Many natural hosts are either vegetatively propagated crops such as potato and chrysanthemum or those subjected to repeated grafting or dressing operations (e.g. citrus, coconuts and hops). Both PSTVd and ASBVd can be vertically transmitted through pollen and/or true seed, but this mode of transmission is unlikely to be significant in the natural spread of disease. Insect transmission of viroids is also believed to be unimportant.

Commonly used techniques for the experimental transmission of viroids include the standard leaf abrasion methods developed for use with conventional viruses, various 'razor slashing' methods in which phloem tissue in the stem or petiole is inoculated via cuts made with a razor blade previously dipped into the inoculum, and, in the case of CCCVd, high-pressure injection into folded apical leaves. PSTVd and HSVd have also been experimentally transmitted by 'agroinoculation,' a technique in which a modified *Agrobacterium tumefaciens* Ti plasmid is used to introduce a greater-than-full-length viroid cDNA full-length viroid cDNA into the potential host cell. In both cases, agroinoculation was able to overcome a marked host resistance to mechanical inoculation. Identification of the molecular mechanisms that determine host range remains an important research goal.

Symptomatology

In general, viroids and conventional plant viruses induce a very similar range of macroscopic symptoms (and presumably metabolic changes) in their hosts. Prominent among such metabolic changes are dramatic alterations in growth regulator metabolism. Stunting and leaf epinasty (a downward curling of the leaf lamina resulting from unbalanced growth within the various cell layers) are often considered the classic symptoms of viroid infection, and other commonly observed symptoms include vein clearing, veinal discoloration or necrosis, and localized chlorotic/necrotic spots or mottling in the foliage. Symptom expression is usually optimal at the relatively high temperatures (30–33°C) that promote viroid replication. Only rarely does a viroid infection actually kill its host.

Viroid infections are also accompanied by a number of cytopathic effects, for instance various chloroplast and cell wall abnormalities, the formation of membranous structures (so-called 'plasmalemmasomes' or 'paramural bodies') in the cytoplasm and the accumulation of electron dense deposits in both chloroplasts and cytoplasm. A combination of subcellular fractionation and *in situ* hybridization experiments has shown that both viroid plus- and minus-strands accumulate in the nucleolus of infected cells. Precise partitioning between the nucleoplasm and nucleolus remains to be established, but this nucleolar localization may have important implications for viroid replication and pathogenicity. Unfortunately, only hypotheses exist as yet concerning the mechanism(s) of viroid pathogenicity.

Although viroid infection does not appear to cause gross disturbances in the synthesis or degradation of host nucleic acids, quantitative changes in a variety of host-encoded proteins have been described. Certain of these proteins, including a 14 kD polypeptide of unknown function and a 68 kD double-stranded (ds)RNA-dependent protein kinase present in PSTVd-infected tomato leaf tissue, may be 'pathogenesis related' or stress proteins, the synthesis and/or activation of which is part of a general reaction of plants to both biotic and abiotic stresses. Other proteins, such as a 140 kD protein accumulation which requires the presence of a replicating low-molecular-weight RNA, do not readily fit into this category, however. Precise cause and effect relationships remain to be established, especially at the molecular level.

Geographic Distribution

Several viroids (e.g. PSTVd, HSVd, CEVd and ASBVd) are widely distributed throughout the world, while others have never been detected outside the areas where they were first reported. Several factors may contribute to this variable distribution. Among the crops most affected by viroid diseases are a number of valuable woody perennials such as grapes, citrus, various pome and stone fruits, and hops. The propagation and distribution of improved selections is becoming increasingly commercialized, with the result that many cultivars are now grown world-wide. The international exchange of plant germplasm has also continued to increase at a rapid rate. In both instances, the large number of latent (i.e. asymptomatic) hosts facilitates viroid spread. Finally, several newly discovered viroids and viroid diseases affect either tropical or subtropical crops. A combination of the generally high temperature optimum for viroid replication and an increased interest in disease affecting tropical crops is likely to cause a continued shift in the geographic distribution of viroid diseases.

Epidemiology and Control

While many viroids were first detected in ornamental or crop plants, most viroid diseases are thought to be the result of their chance transfer from an endemically infected wild species to susceptible cultivars. Several lines of circumstantial evidence are consistent with this hypothesis:

1. The experimental host ranges of several viroids include many wild species, and these wild species often tolerate viroid replication without the appearance of recognizable disease symptoms.
2. Although coevolution of host and pathogen is often accompanied by the appearance of gene-for-gene vertical resistance, no useful sources of resistance to PSTVd has been identified in the cultivated potato. Furthermore, PSTVd has not been detected in wild potato species growing in the Peruvian Andes, its center of origin.
3. Closely related viroids and/or viroid-related RNAs have occasionally been detected in weeds and other wild vegetation growing near fields containing viroid-infected plants.

Viroid diseases may also arise by transfer between cultivated crop species. Studies conducted in the Peoples Republic of China have shown that pears provide a latent reservoir for ASSVd. Likewise, while there is no obvious correlation between disease occurrence and the presence of HSVd in grapes, this viroid is known to cause severe disease in hops. In both instances, the two crops are often grown in close proximity.

All viroid diseases pose a potential threat to agriculture, and several are of considerable economic importance. Coconut cadang-cadang has killed over 30 million palms since it was first recognized in the early 1930s, and estimates of the resulting loss in copra production range between \$80 and \$100 US per tree. Ready transmission of PSTVd by vegetative propagation, foliar contact and true seed or pollen poses a potentially serious threat to potato production, germplasm collections and breeding programs. For many plant viruses, the preferred method of prevention involves incorporation of genetic resistance into the genomes of commercially desirable cultivars. Unfortunately, no useful sources of resistance to viroid disease are known, and various thermotherapy and/or meristem culture protocols have not been widely adopted. Thus, suitable diagnostic tests for the rapid, specific and reliable detection of viroids play a prominant role in disease control efforts.

Tests based upon their unique physical or chemical properties have largely supplanted biological assays for viroid detection. Problems associated with viroid bioassays include the often extended period of time required for completion (weeks to months) and difficulties in detecting mild or latent strains of the pathogen. Several rapid (1–2 day) protocols involving two-dimensional or bidirectional polyacrylamide gel electrophoresis have been developed which take advantage of the circular nature of viroids. Using these protocols, subnanogram amounts of viroid can be unambiguously detected without the use of radioactive isotopes, but neither bioassay nor gel electrophoretic assays are well-suited for the routine analysis of large numbers of samples. Because viroids lack a protein capsid, antibody-based diagnostic techniques are not applicable.

In recent years, diagnostic procedures based upon nucleic acid hybridization have become widely used. The simplest methods involve the hybridization of a highly radioactive viroid–complementary DNA or RNA probe to viroid samples that have been bound to a solid support, followed by autoradiographic detection of the resulting DNA–RNA or RNA–RNA hybrids. Such conventional 'dot blot' assays can detect picogram amounts of viroids using clarified plant sap rather than purified nucleic acid as the viroid source, but sample preparation is often a significant stumbling block. Protocols based upon the polymerase chain reaction are finding increasing acceptance in those cases where either this level of sensitivity is inadequate or a number of closely related viroids are present in the same sample.

Molecular Biology

Although apparently devoid of mRNA activity, viroids replicate autonomously and induce disease in a wide variety of plant species. There are many gaps in our present understanding of the biological properties of these unusual molecules, and these have been aptly summarized by Diener as a series of questions:

1. What molecular signals do viroids possess (and cellular RNAs evidently lack) that induce certain host enzyme(s) to accept them as templates for the synthesis of complementary RNA molecules?
2. What are the molecular mechanisms responsible for viroid replication? Are these mechanisms also operative in uninfected cells? If so, what are their functions?
3. How do viroids induce disease? In the absence of viroid-specific proteins, disease must arise from direct interaction(s) of viroids (or their complementary RNA molecules) with as yet unidentified host cell constituents. Viroid accumulation in the nucleus/nucleolus suggests that viroid pathogenesis may provide a useful model for eucaryotic gene regulation.
4. What are the molecular determinants of viroid host range? Are viroids restricted to higher plants, or do they have counterparts in animals?
5. How did viroids originate?

Over the past several years, considerable information has accumulated concerning the molecular biology of viroid replication, pathogenesis and host range determination. Nevertheless, the precise nature of the molecular signals involved remains elusive.

Replication

Viroid replication is believed to proceed via a 'rolling circle' mechanism involving the synthesis of a minus-strand RNA template. A variety of multimeric plus- and minus-strand RNAs can be detected by nucleic acid hybridization. ASBVd replication appears to utilize a symmetric replication cycle in which the multimeric minus-strand is first cleaved to unit-length molecules and circularized before serving as template for the synthesis of multimeric ASBVd plus-strands. PSTVd and related viroids appear to utilize an asymmetric cycle in which the multimeric minus-strand is directly copied into a multimeric plus-strand precursor. As yet, neither the symmetric nor the asymmetric model for viroid replication can be taken as correct in all details.

A variety of host-encoded enzymes have been implicated in different aspects of viroid replication. Low concentrations of α-amanitin specifically inhibit the synthesis of both PSTVd plus- and minus-strands in nuclei isolated from infected tomato, strongly suggesting the involvement of RNA-dependent RNA polymerase II in the replication of PSTVd and related viroids. Localization of both mature viroids and their replicative intermediates within the nucleolus would seem to contradict the involvement of RNA polymerase II in their synthesis, but, until more information becomes available, this contradiction may be more apparent than real. One or more host-encoded nuclease activities appear to be required for the specific cleavage of multimeric PSTVd plus-strands, while both plus- and minus-strand ASBVd RNAs transcribed *in vitro* undergo a spontaneous self-cleavage to form linear monomers. The final step in viroid replication is the ligation of linear monomers to form mature circular progeny. Plant cells are known to contain RNA ligase activities which can act upon

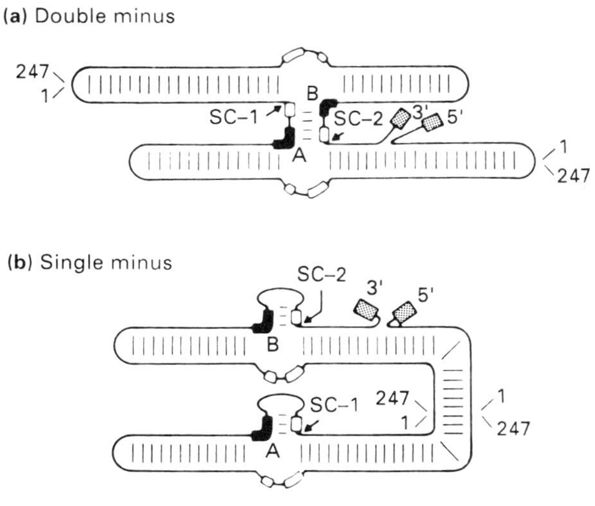

Fig. 3 Cleavage of multimeric viroid RNAs requires rearrangement of the native structure. Schematic representation of a dimeric minus-sense ASBVd RNA transcribed from a *Bst*NI dimeric cDNA clone *in vitro* and folded to contain double- (**a**) and single- (**b**) hammerhead structures. Self-cleavage sites, labeled SC-1 and SC-2, are indicated by arrows; stippled boxes, vector sequences at 5′ and 3′ ends; closed boxes, conserved GAAAC sequences labeled A and B; open boxes, remaining conserved nucleotides. Base-pairing is represented by lines between RNA strands. [From Davies C *et al.* (1991) *Nucleic Acids Res.* 19: 1893, with permission.]

the 5′-hydroxyl and 2′,3′-cyclic phosphate termini formed during either cleavage pathway, and ribonuclease T1 is able to generate circular RNA molecules from PSTVd-specific RNA transcripts by cleavage and intramolecular ligation *in vitro*.

During replication, the rod-like native structure of viroids must rearrange to assume one or more as yet undefined alternative conformations. There is increasing evidence to suggest that the thermodynamic domains responsible for the highly cooperative thermal denaturation of viroids *in vitro* also have important biological functions. How viroids 'switch' between these different conformations remains to be determined. As shown in Fig. 3, there is compelling biochemical and molecular genetic evidence for the ability of dimeric ASBVd minus-strands to undergo spontaneous *in vitro* cleavage via two different (but related) structures. The preferred pathway for enzymatic processing of multimeric PSTVd plus-strands seems to involve a cleavage site formed by rearrangement within the upper portion of the conserved central domain (i.e. a structure somewhat similar to the 'double minus' hammerhead shown in Fig. 3), but other less efficient sites can also be used *in vivo*. It is also possible that PSTVd and related viroids will yet be shown to undergo self-cleavage *in vitro*.

Pathogenicity

Both viroid cDNAs and their RNA transcripts are infectious when inoculated on to susceptible plants, a fact that provides a unique opportunity to relate pathogenicity to sequence (and hence structural) variation. Analysis of naturally occurring PSTVd and CEVd strains has shown that almost all sequence variation is restricted to the pathogenicity and variable domains. Infectivity studies with novel CEVd chimeras constructed by exchanging the pathogenicity domains from naturally occurring mild and severe strains have clearly shown that the pathogenicity domain contains important symptom expression determinants. Sequence variation within the variable domain of CEVd may influence viroid titer in the infected plants. Application of a similar experimental strategy to TASVd has shown that the left terminal loop also contains important determinants of symptom expression.

The ability of such viroid chimeras to replicate and move normally from cell to cell implies certain basic similarities between their *in vitro* and *in vivo* structures but provides no information about the nature of the molecular interactions responsible for symptom development. Sänger and co-workers have attempted to explain the pathogenicity of PSTVd and related viroids in terms of the ability of nucleotides within a portion of the pathogenicity domain known as the 'virulence modulating' region to interact with certain unidentified host components. This model is based upon an apparent correlation between increased symptom severity and decreased stability of this region for several naturally occurring PSTVd variants. One possible cellular target for such viroid–host interactions is the 5′ terminus of a host 7S RNA species that is thought to be involved in protein translocation across membranes via the so-called 'signal recognition particles'. PSTVd and small nuclear RNA (snRNA) U3 also share certain sequence similarities, leading to suggestions that viroid replication may interfere with ribosomal RNA processing.

Unfortunately, such models have yet to receive experimental support and are probably overly simplistic. Theoretical analyses involving several naturally occurring CEVd variants failed to yield a similar correlation between symptom severity and stability of the virulence-modulating region. More recently, mutagenesis within the PSTVd pathogenicity domain has produced several variants for which symptom severity did not fit the proposed correlation. Thus, viroid pathogenicity seems to be modulated by subtle differences in secondary or tertiary structure rather than structural stability *per se*, but the molecular mechanism(s) remains unknown.

Host range

While a variety of evidence suggests an important role for the conserved central domain in viroid replication, the biological properties of CLVd suggest that this domain may also contain one or more host range determinants. As described above, CLVd appears to be a natural mosaic of sequences present in other viroids; phylogenetic analysis (see Fig. 2) suggests that it can be considered to be a PSTVd-related viroid whose conserved central domain has been replaced by that of HSVd. Like HSVd (but unlike other PSTVd-related viroids), CLVd is able to replicate and cause disease in cucumber. It should soon be possible, using the polymerase chain reaction to exchange the conserved central domains among several infectious viroid cDNAs, to assess its role in determining viroid host range.

Origin and Evolution

As discussed by Matthews, available evidence suggests three possible origins for viruses: descent from molecules present in a prebiotic RNA world; descent from normal host RNAs; or degeneration from simple obligate cellular parasites. Present-day plant viruses form four clusters [i.e. single-stranded positive-sense RNA viruses (three superfamilies); plant reoviruses, rhabdoviruses and bunyaviruses; caulimoviruses; and geminiviruses], and the existence of these clusters may reflect their different origins. Much of the early speculation about viroid origin involved their possible origin as 'escaped introns' (i.e. descent from normal host RNAs). More recently, however, viroids have been proposed to represent 'living fossils' of a precellular RNA world which assumed an intracellular mode of existence sometime after the evolution of cellular organisms.

According to Diener, the chief proponent of this view, the inherent stability of viroids and viroid-like RNAs which arises from their small size and circularity would have enhanced the probability of their survival in primitive error-prone RNA self-replicating systems and assured their complete replication without the need for initiation or termination signals. Most viroids (but not satellite RNAs or random sequences of the same base composition) also display structural periodicities with repeat units of 12, 60 or 80 nucleotides. The high error rate of prebiotic replication systems would be expected to have favored the evolution of polyploid genomes, and the mechanism of viroid replication (i.e. rolling-circle transcription of a circular template) provides an effective means of genome duplication.

Viroids and viroid-like satellite RNAs all possess efficient mechanisms for the precise cleavage of their oligomeric replication intermediates to form monomeric progeny. As discussed earlier, most viroids appear to require proteinaceous host factor(s) for cleavage, but others (the viroid-like satellites and ASBVd) function as self-cleaving RNA enzymes far smaller and simpler than those derived from introns. Thus, ASBVd may represent an evolutionary link between viroids and satellite RNAs. No viroid is known to code for protein, a fact that is consistent with the possibility that viroids are phylogenetically older than introns. It is conceivable that introns may be 'captured' viroids, rather than viroids being 'escaped' introns.

See also: Hepatitis delta virus; Plant resistance to viruses.

Further Reading

Diener TO (1979) *Viroids and Viroid Diseases.* New York: Wiley Interscience.

Diener TO (ed.) (1987) *The Viroids.* New York: Plenum Press.

Matthews REF (1991) *Plant Virology,* 3rd edn. New York: Academic Press.

Semancik JS (ed.) (1987) *Viroids and Viroid-Like Pathogens.* Boca Raton: CRC Press.

Symons RH (ed.) (1990) Viroids and related pathogenic RNAs. *Semin. Virol.* 1: 75.

VIRUS STRUCTURE

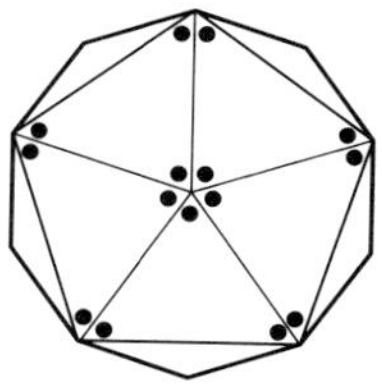

Contents

Atomic Structure
Principles of Virus Structure

Atomic Structure

Ming Luo
University of Alabama
Birmingham, Alabama, USA

Architecture of Viruses

Viruses have two essential components: protein and nucleic acid. A closed capsid may be formed by one type or a few types of proteins to encapsidate the nucleic acid genome. The protein capsid can have a helical (filamentous virus) or icosahedral (spherical virus) symmetry. The symmetry allows a small protein unit to assemble into a large particle. The helical symmetry is described by the diameter *d*, the pitch **P**, and the number of subunits per turn. There are as many capsid proteins as is necessary to completely cover the nucleic acid genome. The icosahedral symmetry is defined by six fivefold axes, ten threefold axes and 15 twofold axes. A number *T*, called the triangulation number, indicates how many quasisymmetrical subunit interactions there are within one asymmetrical region of the icosahedron. There are a total of 60 *T* copies of proteins in one icosahedral capsid. In some viruses, there is a membrane envelope wrapped around the protein–nucleic acid core which contains proteins on its surface.

Methods of Structure Determination

X ray diffraction is the common technique used for studying the atomic structure of proteins and nucleic acids. When X rays strike on electrons of the atoms in a stationary specimen, a diffraction pattern of different intensities is generated and recorded. By analysis of the diffraction pattern and the intensities, a three-dimensional electron density map (EDM) can be calculated by Fourier transformation. A three-dimensional chemical structure is built based on the interpretation of the EDM. Two types of X ray diffraction experiments are useful for virus structure studies; fiber diffraction (for filamentous viruses) and crystallography (for spherical viruses and globular viral proteins).

Atomic Structure of Helical Viruses

The disc of the tobacco mosaic virus (TMV) coat protein has been crystallized and its atomic structure resolved by X ray crystallography. The intact TMV structure containing the nucleic acid could only be determined by X ray fiber diffraction experiments, as also that of Pf2 phage. The coat proteins of TMV and Pf2 contain mainly α-helices and the nucleic acid interacts with the coat protein with one base (Pf2) or three bases (TMV) per protein unit. The axis of the coat protein helix coincides with that of the nucleic acid. The coat proteins of TMV have many aggregation forms, depending on Ph or ionic strength. The TMV RNA is inserted into the coat protein helix in the growing virus particle. The coat proteins of pf2 are added one by one to the DNA helix emerging from the membrane.

Atomic Structure of Spherical Viruses

Spherical viruses without a membrane envelope form large single crystals under appropriate conditions. Their atomic structure can be determined by X ray crystallography with the aid of supercomputers and synchrotron X ray sources. Since 1978, numerous atomic structures of viruses have been determined. These include plant RNA viruses (tomato bushy stunt virus (TBSV), southern bean mosaic virus (SBMV), satellite tobacco necrosis virus (STNV), bean pot mottle virus (BPMV), cow pea mast virus (CPMV)), animal RNA viruses (human rhinovirus 14 (HRV14) and 1A (HRV1A), poliovirus, black beetle virus (BBV), Mengo virus, foot and mouth disease virus (FMDV) and Theiler's murine encephalomyelitis virus (TMEV)) and animal DNA viruses (canine parvovirus (CPV), simian virus 40 (SV40) and adenovirus hexon), as well as bacteriophages (MS2 and ϕX174).

Most capsid proteins of these viruses contain an antiparallel eightstranded β-barrel motif. The motif has a wedge-shaped block with four β-strands (BIDG) on one side and another four (CHEF) on the other. There are also two conserved α-helices (A and B), one between βC and βD, the other between βE and βF. In animal viruses, there are large loops inserted in between the β-strands. These loops form the surface features of individual viruses. This shape is best suited for making a concealed icosahedral shell.

A viral capsid may contain multiple copies of the β-barrel fold with the same amino acid sequence (such as $T = 3$ (TBSV) or $T = 1$ (CPV)) or different amino acid sequences (such as pseudo $T = 3$ (HRV14)). In some cases, there are two β-barrel folds in a single polypeptide (such as CPMV and the adenovirus hexon).

Atomic Structure of Viral Proteins

There are many functional viral proteins that do not have any symmetrical quaternary structure in virus particles. Therefore, their atomic structure has to be analyzed by crystallizing isolated proteins. Crystal structures have been determined for the hemagglutinin (HA) and neuraminidase (NA) of enveloped influenza virus and the protease of human immunodeficiency virus (HIV).

The HA has two domains in the subunit, and the functional molecule is a trimer. The domain extending from the membrane contains α-helices and β-sheets. This domain forms the base interacting with the membrane envelope. The distal domain has an eight-stranded β-barrel fold similar to that seen in the spherical viruses. This domain bears the binding site for sialic acid, the receptor for influenza virus on the cell surface. The membrane fusion peptide at the N terminus of HA2 is located in the membrane-interacting domain.

NA is a tetrameric molecule and its subunit contains six β-sheets of four β-strands each. The six β-sheets are arranged like the blades of a propeller. The enzymatic site is at a hydrophobic depression in the center of the β-sheets. The antibody-recognition site has been identified on the external surface near the enzymatic site by the atomic structure of the NA complexed with Fab fragments.

The HIV protease is an aspartic acid protease with two β-sheets in each subunit. The enzyme has to dimerize before it becomes active. This activation mechanism has an important role in HIV assembly. The viral assembly complex attached to the membrane will not proceed during maturation until all the necessary components are present to initiate dimerization of the protease.

Nucleic Acid–Protein Interaction

The viral nucleic acid genome is always packaged inside the protein capsid. Usually the structure of the nucleic acid cannot be observed in single-crystal X ray diffraction experiment because of the random orientation of the icosahedral particles in the crystal. However, in rare cases, the nucleic acid might assume icosahedral symmetry by interacting with the protein capsid. Fragments of the complete genome make the same conformation, although with different nucleotide sequences, at locations related by icosahedral symmetry. Such structures have been seen in BPMV (RNA virus) and CPV (DNA virus). The bases are stacked either as in A-type RNA helix (BPMV) or to form a coiled conformation to fit the interactions with the protein capsid. These viruses readily form empty viral particles and have a hydrophobic pocket on the interior surface of the capsid. The nucleic acid generally interacts nonspecifically with the protein.

Evolution

The highly conserved β-barrel motif of the viral capsid protein indicates that many viruses must have evolved from a single origin. The unique three-dimensional structure of this motif is required for capsid assembly and it is generally conserved over a longer period of time than the amino acid sequence. The superposition of the capsid proteins from different viruses can be used to estimate the branch point in the evolutionary tree for each viral group. The structure alignment not only relates plant viruses to oenomel viruses and RNA viruses to DNA viruses, but also viruses to other proteins such as concanavalin A which has a similar fold and competes with poliovirus for its cell receptor. The evolutionary relationship of these viruses is supported by amino acid sequence alignment of more conserved viral proteins such as the viral RNA polymerase.

Assembly

The icosahedral capsid is assembled from smaller units made of several protein subunits. In small animal RNA viruses, a protomeric unit is first formed with one copy of each polypeptide after translation, The termini of the subunits are intertwined with each other to hold the subunits together in the protomer. The protomers are then associated as pentamers which in turn form the complete icosahedral virion while encapsidating the viral RNA. In $T = 3$ $T = 1$ plant RNA viruses, the pentamers are formed by dimers of the capsid proteins. In adenovirus and SV40, the capsid proteins form hexon units (three polypeptides, each with two β-barrels) or pentamers before they assemble into an icosahedral shell.

Host Receptor Recognition Site

Animal viruses have to recognize a specific host cell receptor for entry during infection. Host receptor binding is the initial step of the viral life cycle and could be an effective target for preventing viral infection. Based on the atomic structure of animal viruses, it was found that the receptor recognition site is located in an area surrounded by hypervariable regions of the antigenic sites. Usually the area is in a depression (called the 'canyon') on the viral surface which cannot be reached by antibodies. This structural feature is present in HRV14 and the active site of influenza virus NA. The receptor-binding site on influenza virus HA does not have a deep depression, but is surrounded by antigenic sites.

Antigenic Sites

Antibodies are the first line of defense by the immune system against a viral infection. The epitopes combined with the neutralizing antibodies are mapped on a few isolated locations on the surface of viral proteins. The structure of the influenza virus NA complexed with Fab fragments showed that the antibody makes contact with an area about 6 nm_2 and the epitope spans four discontinuous polypeptides. Therefore, an effective vaccine needs to include a complete viral protein or a large fragment. The binding of the antibodies does not significantly change the structure of the antigen. The exact mechanism by which antibodies neutralize antigens is still unclear.

Antiviral Agents

Viral infectious diseases can be cured if an agent can be administered to stop viral infection. Such agents have been synthesized and shown to bind to the capsid of HRV14 in the crystal structure. The compounds were inserted into the hydrophobic pocket within the β-barrel of the major capsid protein VP1. Binding of the compounds prevented uncoating of the virion, which prevented the release of viral RNA into the cytoplasm. These compounds inhibit infections of several other RNA viruses and may be effective against other viruses after modification, as the β-barrel structure exists in many viruses.

See also: Antivirals; Human immunodeficiency viruses; Influenza viruses; Rhinoviruses.

Further Reading

Rossman MG and Johnson JE (1989) Icosahedral RNA virus structure. *Annu. Rev. Biochem.* 58: 533.

Fields BN *et al.* (eds) (1990) *Virology*, 2nd edn. New York: Raven Press.

Principles of Virus Structure

John E Johnson and Andrew J Fisher
Purdue University
West Lafayette, Indiana, USA

Introduction

The virion is a nucleoprotein particle designed to move the viral genome between susceptible cells of

Table 1. Functions of the viral capsid proteins

Function	Comments
Assembly	Subunits must assemble to form a protective shell for the nucleic acid
Nucleic acid recognition and packaging	Subunits must specifically package the viral nucleic acid
Interactions with the host	The capsid may actively participate in the viral infection process by binding to receptors and mediating cell entry (nonplant viruses)
	Interactions between the virion and host cellular components may assist viral transport (plant viruses)
Mutations	Regions on some animal viruses are readily mutable producing changes in capsid structure that inhibit recognition by antibodies produced by previous infection or vaccination by another serotype

a host and between hosts. An important limitation on the size of the viral genome is its container, the protein capsid. The virion has a variety of functions during the virus life cycle (Table 1); however, the principles dictating its architecture result from the need to provide a container of maximum size with a minimum amount of genetic information. The universal strategy evolved for the packaging of viral nucleic acid requires multiple copies of one or more protein subunit types arranged symmetrically or quasisymmetrically about the genome. The assembly of these subunits into nucleoprotein particles is, in many cases, a spontaneous process that results in a minimum free energy structure under intracellular conditions. The two broad classes of symmetric virions are helical rods and spherical particles.

The nucleoprotein helix can, in principle, package a genome of any size. Extensive studies of tobacco mosaic virus (TMV) show that the protein subunits will continue to add to the extending rod as long as there is exposed RNA. Protein transitions required to form the TMV helix from various aggregates of subunits are now understood at the atomic level. It is clear that the subunit oligomers display significant polymorphism in the course of virus assembly; however, excluding the two ends of the rod, all subunits are in identical environments in the mature helical virion. This is the ideal protein context for a minimum free energy structure. In spite of these packaging and structural attributes, the simple helical viron must be deficient in functional requirements that are common for animal viruses because they are found only among plant and bacterial viruses. Even among plant viruses only 10 of the 34 recognized groups are helical. The large majority of all viruses are roughly spherical in shape.

Icosahedral Capsids

The architectural principles for constructing a 'spherical' virus were first described in 1956 by Crick and Watson. The relatively large protein shell produced with limited genomic information suggested that identical protein subunits were probably distributed with the symmetry of Platonic polyhedra (the tetrahedron, 12 equivalent positions; the octahedron, 24 equivalent positions; or the icosahedron, 60 equivalent positions). Protein subunits distributed with the symmetry of the icosahedron (Fig. 1a) generate the maximum sized particle in which all copies lie in identical positions. The repeated interaction of chemically complimentary surfaces at the subunit interfaces leads naturally to such a symmetric particle. The 'instructions' required for assembly are contained in the tertiary structure of the subunit (Fig. 1b). The assembly of the capsids is a remarkably accurate process. The use of subunits for the construction of organized complexes places strict control on the process and will naturally eliminate defective units. The reversible formation of noncovalent bonds between properly folded subunits leads naturally to error-free assembly and a minimum free energy structure.

Early ideas explaining spherical viral architecture were extended on the basis of physical studies of small isometric RNA plant viruses. The large yields and ease of preparation made them ideal subjects for investigations requiring substantial quantities of material. Protein subunits forming viral capsids of this type are usually 20–40 kD. An example of a virus consistent with the Crick and Watson hypothesis is satellite tobacco necrosis virus (STNV) which is formed from 60 identical 25 kD subunits. The particle outer radius is 8 nm and the radius of the internal cavity is 6 nm providing a volume of 9×10^2 nm^3

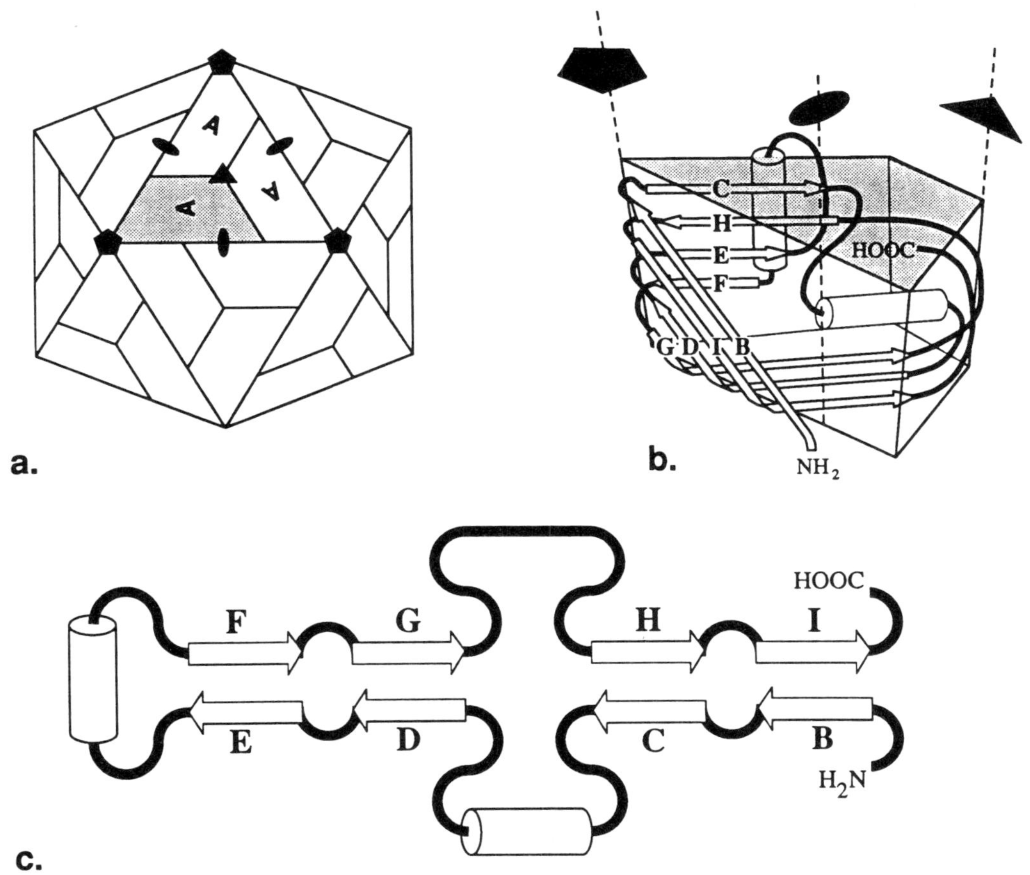

Fig. 1 (**a**) The icosahedral capsid contains 60 identical copies of a protein subunit labeled A. These are related by fivefold (vertices), threefold (faces) and twofold (edges) symmetry elements. For a given sized subunit this point group symmetry generates the largest possible assembly in which every protein lies in an identical environment. (**b**) A schematic representation of the subunit building block found in many RNA and some DNA viral structures. Such subunits have complimentary interfacial surfaces which, when they repeatedly interact, lead to the symmetry of the icosahedron. The tertiary structure of the subunit is an eight-stranded β-barrel with the topology of the jelly roll (see **c**). Subunit sizes generally range between 20 and 40 kD with variation among different viruses occurring at the N and C termini and in the size of insertions between strands of the β-sheet. These insertions generally do *not* occur at the narrow end of the wedge (B–C, H–I, D–E and F–G turns). (**c**) The topology of viral β-barrel showing the connections between strands of the sheets (represented by arrows) and positions of insertions between strands. The cylinders represent helices that are usually conserved. The C–D, E–F and G–H loops often contain large insertions.

for packaging RNA. A single hydrated ribonucleotide in a virion will occupy roughly 60–70 nm^3. The STNV volume is adequate to package a genome of only 1200–1300 nucleotides. STNV is a satellite virus and the packaged genome codes for only the coat protein. Proteins required for RNA replication are supplied by the 'helper' tobacco necrosis virus. Most simple ribovirus genomes contain coding capacity for at least two proteins, roughly 1200 nucleotides for the capsid protein and 2500 nucleotides for an RNA-directed RNA polymerase. A virion must have an internal cavity with a radius of at least 9 nm to package a minimal genome. Consistent with this requirement were experimental studies showing that the majority of simple spherical viruses had outer radii of at least 12.5 nm which corresponds to inner radii of roughly 10 nm. Such particles had to be formed by more than 60 subunits, yet X ray diffraction patterns of crystalline tomato bushy stunt virus (TBSV) were consistent with icosahedral symmetry. A number of investigators developed hypotheses explaining the apparent inconsistent observations; however, Caspar and Klug derived a general method for the construction of icosahedral capsids that contained

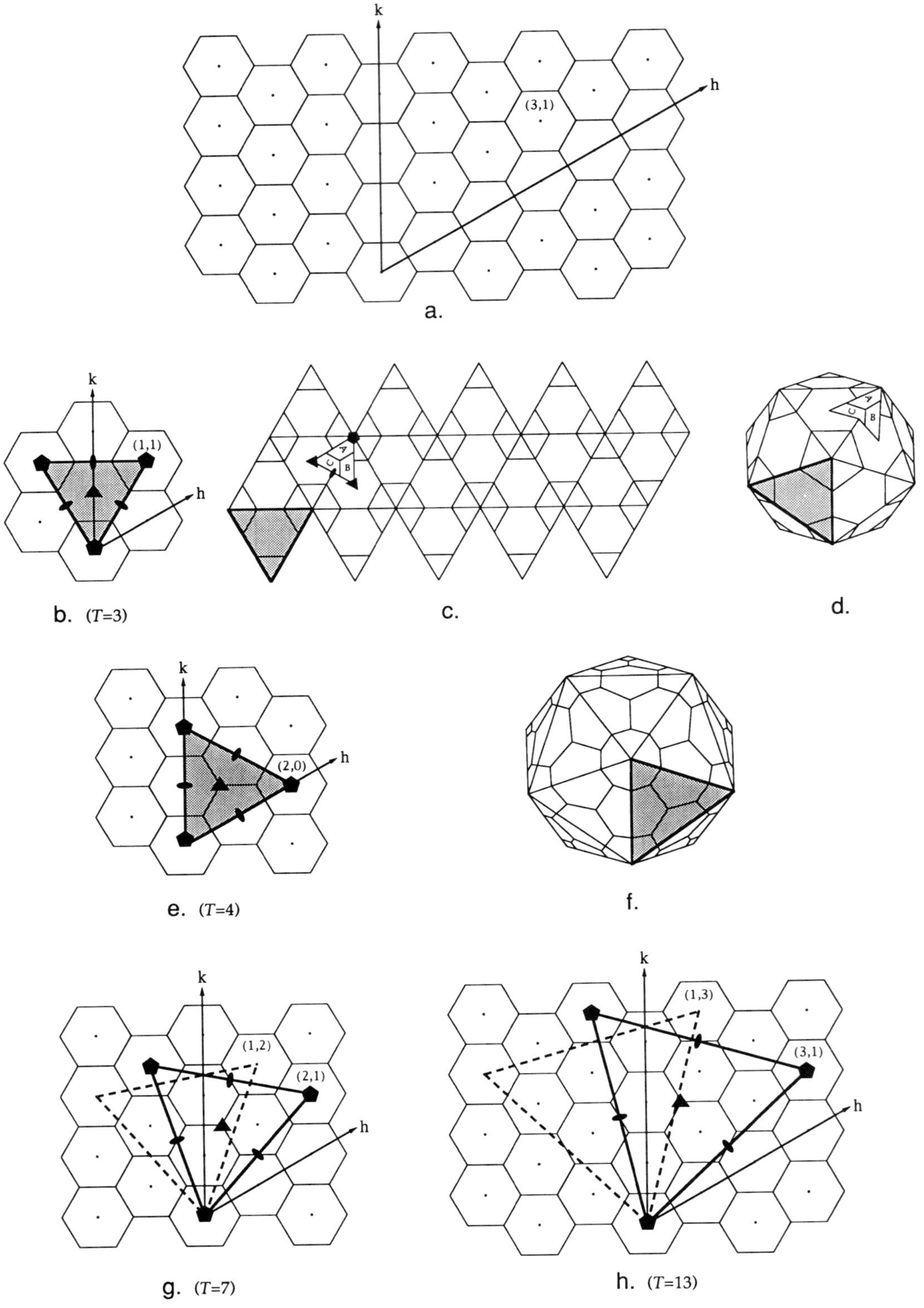

Fig. 2 The geometric principle for generating icosahedral quasiequivalent surface lattices and four examples of its application are illustrated. These constructions show the relation between icosahedral symmetry axes and quasi-equivalent symmetry axes. The latter are symmetry elements that hold only in a local environment. (**a**) It is assumed in quasiequivalence theory that hexamers and pentamers can be interchanged at a particular position in the surface lattice. Hexamers are initially considered planar (an array of hexamers forms a flat sheet as shown) and pentamers

multiples of 60 protein subunits. The method for systematically enumerating all possible quasiequivalent structures was similar to that used by Buckminister Fuller in constructing geodesic domes. The quasiequivalent theory of Caspar and Klug has explained the distribution of morphological units (features identifiable at low resolution by electron microscopy often corresponding to hexamer, pentamer, trimer or dimer aggregates of the subunits) on all structures observed to date, but the results from high-resolution crystallographic studies have shown some remarkable inconsistencies with the microscopic principles upon which the theory is based.

Quasiequivalent Icosahedral Shells

Quasiequivalence is best visualised graphically. Formally, subunits forming quasiequivalent structures must be capable of assembling into both hexamers (which are conceptually viewed as planar) and pentamers (which are convex because one subunit has been removed from the planar hexamer and yet similar, quasiequivalent, contacts are maintained). If subunits assembled as all hexamers, the result would be a flat sheet and a closed shell could not form (Fig. 2a). The rules of quasiequivalence described a systematic procedure for inserting

are convex, introducing curvature in the sheet of hexamers where they are inserted. The closed icosahedral shell composed of hexamers and pentamers is generated by inserting 12 pentamers at appropriate positions in the hexamer net. The positions at which hexamers are replaced by pentamers are identified by the indices **h** and **k** along the labeled axes in the drawing. Once an origin is defined, every hexamer in the lattice can be uniquely identified by the number of steps along each axial direction required to reach that lattice point (**hk**). (The lattice point $\mathbf{h} = 3$ $\mathbf{k} = 1$ (3,1) is shown as an example.) To construct a model of a particular quasiequivalent lattice, one face of an icosahedron is generated in the hexagonal net. The origin is replaced with a pentamer and the (**hk**) hexamer is replaced by a pentamer. The third replaced hexamer is identified by threefold symmetry (i.e. complete the equilateral triangle of the face). Each quasiequivalent lattice is identified by a number $T = \mathbf{h}^2 + \mathbf{hk} + \mathbf{k}^2$ where **h** and **k** are the indices described above. T indicates the number of quasiequivalent units in the icosahedral asymmetric unit (a hexamer contains six units and a pentamer contains five units). For the purpose of these constructions it is convenient to choose the icosahedral asymmetric unit as 1/3 of an icosahedral face defined by the triangle connecting a threefold axis to two adjacent fivefold axes. Other asymmetric units can be chosen such as the triangle connecting two adjacent threefold axes and an adjacent fivefold axis (Fig. 3a). The total number of units in the particle is $60T$, given the symmetry of the icosahedron. The number of pentamers must be 12 and the number of hexamers is $(60T - 60)/6 = 10(T - 1)$. (**b**) One face of the icosahedron for a $T = 3$ surface lattice is identified (this corresponds to a face of the icosahedron in Fig. 1a). The hexamer replaced has coordinates $\mathbf{h} = 1$, $\mathbf{k} = 1$. The icosahedral asymmetric unit is 1/3 of this face and it contains three quasiequivalent units (two units from the hexamer coincident with the threefold axis and one unit from the pentamer). (**c**) The three-dimensional model of the quasiequivalent lattice can be generated by arranging 20 identical faces of the icosahedron as shown. Three quasiequivalent units labeled A, B and C are shown. These correspond to the three quasiequivalent asymmetric units shown in Fig. 3. (**d**) The folded icosahedron is shown with hexamers and pentamers outlined. The shaded face represents the triangle originally generated from the hexagonal net. The $T = 3$ surface lattice represented in this construction has the appearance of a soccer ball. The trapezoids labeled A, B and C identify quasiequivalent units in one icosahedral asymmetric unit of the rhombic triicontahedron discussed in Fig. 3. (**e**) An example of a $T = 4$ icosahedral face ($\mathbf{h} = 2$, $\mathbf{k} = 0$). In this case the hexamers are coincident with icosahedral twofold axes. (**f**) A folded $T = 4$ icosahedron with the shaded face corresponding to the face outlined in the hexagonal net. Note that folding the lattice has required that the hexamers have the curvature of the icosahedral edges. (**g**) A single icosahedral face generated from the hexagonal net for a $T = 7$ lattice. Note that there are two different $T = 7$ lattices ($\mathbf{h} = 2$, $\mathbf{k} = 1$ (bold outline) and $\mathbf{h} = 1$, $\mathbf{k} = 2$ (dashed outline). These lattices are the mirror images of each other. To fully define such a lattice, the arrangement of hexamers and pentamers must be established as well as the enantiomorph of the lattice. (**h**) A single icosahedral face for a $T = 13$ lattice is shown. The two enantiomorphs of the quasiequivalent lattice ($\mathbf{h} = 3$, $\mathbf{k} = 1$ (bold) and $\mathbf{h} = 1$, $\mathbf{k} = 3$ (dashed)) are outlined. The procedure for generating quasiequivalent models described here does not exactly correspond to the one described by Caspar and Klug. Caspar and Klug distinguish between different icosadeltahedra by a number $P = \mathbf{h}^2 + \mathbf{hk} + \mathbf{k}^2$ where **h** and **k** are integers that contain no common factors but 1. The deltahedra are triangulated to different degrees described by an integer f which can take on any value. In their definition $T = Pf^2$. The description in this figure has no restrictions on common factors between **h** and **k**, thus $T = \mathbf{h}^2 + \mathbf{hk} + \mathbf{k}^2$ for all positive integers. The final models are identical to those described by Caspar and Klug.

pentamers into the hexagonal net in a way that forms a closed shell with exact icosahedral symmetry. Figure 2 illustrates this principle and the selection rules for inserting pentamers.

The quasiequivalence theory has been universally successful in describing surface morphology of spherical viruses observed in the electron microscope and, prior to the first high-resolution crystallographic structure of a virus, it was assumed that the underlying assumptions were essentially correct. The structure of TBSV determined at 0.29 nm resolution revealed an unexpected variation from the concept envisioned by Caspar and Klug. Quasiequivalence was defined as 'any small nonrandom variation in a regular bonding pattern that leads to a more stable structure than does strictly equivalent bonding'. The structure of TBSV showed that differences occurring between pentamer interactions and hexamer interactions were not small variations in bonding patterns, but almost totally different bonding patterns. Figure 3 shows diagrammatically the subunit interactions found in the shells of TBSV, southern bean mosaic virus and black beetle virus. Bonding contacts between quasi threefold related subunits are maintained with little deviation from exact symmetry while quasi twofold contacts and icosahedral twofold contacts (which are predicted to be very similar) are quite different. The hexamer quasisymmetry is better described as a trimer of dimers. Unlike the conceptual model, particle curvature results from nonplanar interactions in both pentamers and hexamers.

The high-resolution viral structures with $T = 3$ symmetry (see Fig. 2 for explanation of T numbers) showed that the lattice predictions of the quasiequivalent theory were correct, but that the underlying concepts of quasiequivalent bonding had to be revised. The first low (2.25 nm)-resolution structure of a $T = 7$ virus required an even greater adjustment to the underlying principles of quasiequivalence. Rayment *et al.* reported in 1982 that the polyomavirus capsid contained 72 capsomeres as previously determined from electron microscopy studies, but that all the capsomeres were pentamers of protein subunits even though they were located at hexamer lattice points. The $T = 7$ surface lattice predicts 12 pentamers and 60 hexamers, thus the prediction of the number and position of the morphological units was correct, but the fine structure of the morphological units was not as predicted by quasiequivalence theory. Although the result was highly controversial when first reported, additional studies have confirmed the initial analysis. In contrast to polyomavirus, a recent study of the $T = 7$ cauliflower mosaic virus clearly shows the presence of hexameric and pentameric subunit oligomers in positions predicted by quasiequivalent theory. The high-resolution X ray and electron microscopic studies show the limits of theory in predicting viral structure and indicate that further understanding of capsid architecture must come from experimental studies. Toward this end, a substantial number of complex viral structures have been determined by cryo-electron microscopy (Table 2). The surface lattices agree well with the predictions of quasi-equivalence, but the detailed capsomere structure and therefore number of subunits must be confirmed in each case.

A number of viral capsids are constructed with pseudo $T = 3$ symmetry. These structures contain β-barrel subunits (Fig. 1b) in the quasiequivalent environments observed in $T = 3$ structures, but each of the three β-barrels in the asymmetric unit has a unique amino acid sequence (Fig. 4). Such structures are referred to as $P = 3$ lattices because they contain 60 copies each of three different subunits rather than 180 identical subunits. These structures do not require quasiequivalent bonding because each unique interface will have different amino acids interacting, rather than the same subunits forming different contacts. They do, however, require three times the genetic information of that for $T = 3$ particles. The animal picornaviruses have capsids of this type. Animal virus capsids undergo rapid mutation to avoid recognition by the circulating immune system. Capsids composed of three unique subunit types can mutate in one subunit without affecting the other two. This would be less likely to affect assembly or other functions of the particle in $P = 3$ shells than it would in $T = 3$ shells. At least one plant viral group displays $P = 3$ shells, the comoviruses. An interesting variation occurs in these capsids when compared with the picornaviruses. Two of the β-barrel domains forming the shell are contained in a single polypeptide chain. This phenomenon is readily understood in the context of the synthesis of the subunits in picorna- and comoviruses. In both cases the proteins are synthesized as a polyprotein which is subsequently cleaved into functional proteins by a virally encoded protease. One of the cleavage sites in picornaviruses is missing in the comoviruses, resulting in two domains forming a 'polyprotein'.

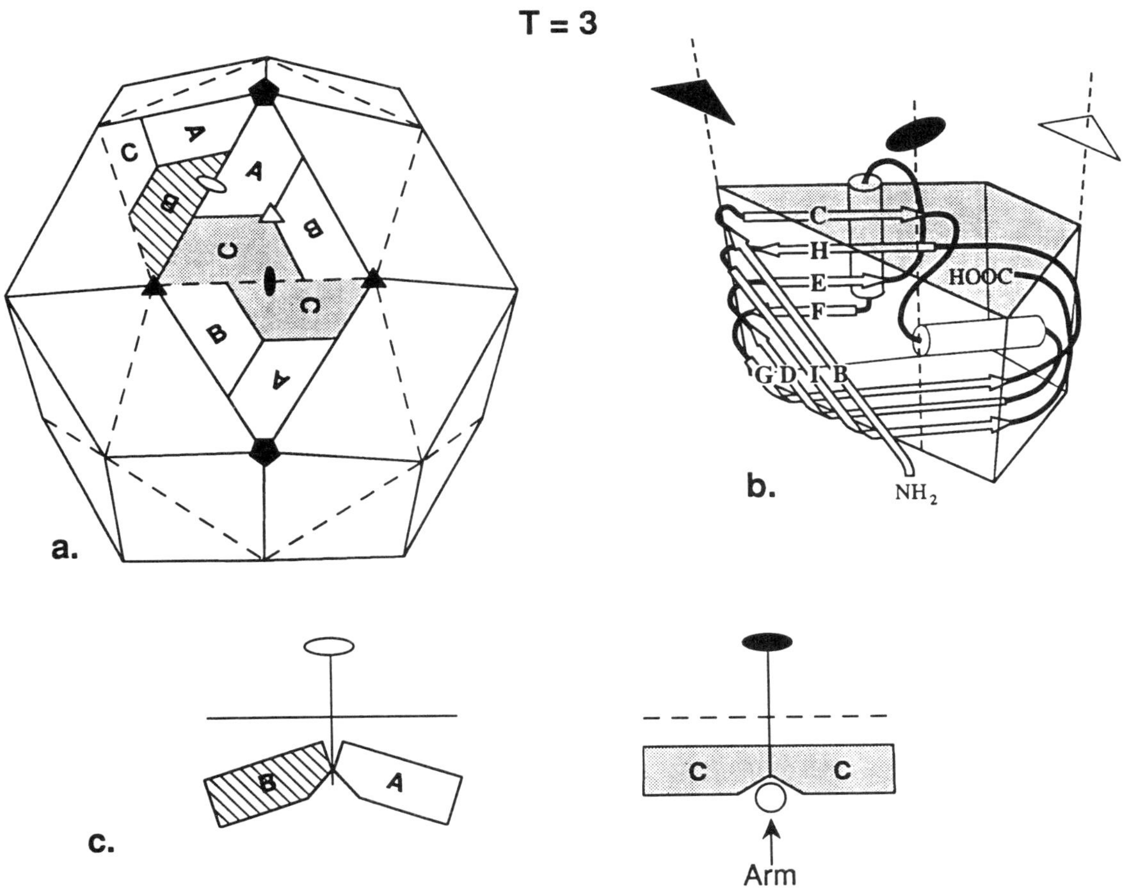

Fig. 3 Although quasiequivalence theory can predict, on geometrical principles, the organization of hexamers and pentamers in a viral capsid, the detailed arrangement of subunits can only be established empirically. High-resolution X ray structures of $T = 3$ plant and insect viruses show that the particles are organized like the rhombic triicontahedron. (**a**) The rhombic triicontahedron has icosahedral symmetry and is constructed by placing rhombic faces perpendicular to icosahedral twofold symmetry axes (solid ellipse). A convenient definition of the icosahedral asymmetric unit is the wedge defined by icosahedral threefold axes (solid triangles) left and right and an icosahedral fivefold axis (solid pentamer) at the top. Quasi threefold and twofold axes are represented by open symbols. The quasi sixfold axes are coincident with the icosahedral threefold axes as shown in Fig. 2b, c and d. The shape of the $T = 3$ soccer ball model in Fig. 2d is different from the shape of the rhombic triicontahedron. The quasisymmetry axes are, however, in the same positions relative to the icosahedral symmetry axes for both models. The icosahedral asymmetric unit contains three subunits labeled A, B and C (see Fig. 2c and d). (**b**) The shape of the subunit in $T = 3$ plant and insect viruses is nearly identical to the shape of the subunit in the $T = 1$ virus and they pack in a very similar fashion. The $T = 1$ subunits in one face (Fig. 1a) are related by an icosahedral threefold axis while the $T = 3$ subunits in one face are related by a quasi threefold axis. (**c**) The dihedral angle between subunits A and B (related by quasi twofold axes) is 144° and is referred to as a bent contact (left side), while the dihedral angle between two C subunits (related by icosahedral twofold axes) is 180° and is referred to as a flat contact (right side). Two dramatically different contacts between subunits with identical amino acid sequences are generated by the insertion of an extra polypeptide from the N-terminal portion of the C subunit into the groove formed at the flat contact. This polypeptide is called an 'arm'. This region of the A and B subunits is disordered and A and B are in direct contact.

Subunit Tertiary Structure

The dominant tertiary fold observed in high-resolution X ray structures determined to date is the eight-stranded β-barrel illustrated in Fig. 1(b). This fold exists in a wide range of viruses as given in Table 2. The wedge shape is ideally suited to form pentamers and hexamers as it does in $T = 3$ virus (Fig. 3) and $P = 3$ viruses (Fig. 4). In other viruses, however, the wedge is not found in this geometric environment.

Table 2. Description of virus structures

Family/Group	Size(nm)[a]	Host	Genome no.-type	Typical member	Nucleocapsid proteins[b]	Capsid symmetry[c]
Nonenveloped spheroidal						
Adenoviridae	80	Animal	dsDNA 1-linear	Adenovirus	720 copies 109 kD 70 copies 85 kD	$T = 25$
Birnaviridae	60	Animal	dsRNA 2-linear	Infectious bursal disease virus	4 proteins 29–105 kD	$T = 9$
Bromovirus	26	Plant	ssRNA (+) 3-linear	Brome mosaic virus	180 copies 20 kD	$T = 3$
Caliciviridae	23–38	Animal	ssRNA(+) 1-linear	Norwalk virus	180 copies 67 kD	$T = 3$
Caulimovirus	50	Plant	dsDNA 1-circular	Cauliflower mosaic virus	420 copies 42 kD	$T = 7$
Comovirus	28	Plant	ssRNA(+) 2-linear	Cowpea mosaic virus	60 copies 22 and 42 kD	$P = 3$
Cryptovirus	30–38	Plant	dsRNA 2–3-linear	White clover cryptic virus	53–63 kD	?
Cucumovirus	28	Plant	ssRNA(+) 3-linear	Cucumber mosaic virus	180 copies 24 kD	$T = 3$
Dianthovirus	34	Plant	ssRNA(+) 2-linear	Carnation ringspot virus	40 kD	?
Fabavirus	30	Plant	ssRNA(+) 2-linear	Broad bean wilt virus	27 and 43 kD	?
Ilarvirus	26–35	Plant	ssRNA(+) 3-linear	Tobacco streak virus	25 kD	?
Leviviridae	27	Bacteria	ssRNA(+) 1-linear	MS2	180 copies 13–17 kD	$T = 3$
Luteovirus	28	Plant	ssRNA(+) 1-linear	Barley yellow dwarf virus	24 kD	?
Machlovirus	30	Plant	ssRNA(+) 1-linear	Maize chlorotic dwarf virus	18 and 30 kD	?
Marafivirus	31	Plant	ssRNA(+) 1-linear	Maize rayado fino virus	29 and 22 kD (1:3 ratio)	?
Microviridae	27	Bacteria	ssDNA(+) 1-circular	ϕX174	60 copies 48, 19, 5 kD 12 copies 36 kD	$T = 1$
Necrovirus	28	Plant	ssRNA(+) 1-linear	Tobacco necrosis virus	180 copies 22 kD	$T = 3$

Table 2. Continued

Nepovirus	28	Plant	ssRNA(+) 2-linear	Tobacco ringspot virus	60 copies 58 kD	$P = 3$
Nodaviridae	29–31	Animal	ssRNA(+) 2-linear	Nodamura virus	180 copies 46 kD	$T = 3$
Papovaviridae	45–55	Animal	dsDNA 1-circular	Human papillomavirus	360 copies 42–55 kD	Pseudo $T = 7$ (72 pentamers)
Parsnip yellow fleck	30	Plant	ssRNA(+) 1-linear	Parsnip yellow fleck virus	31, 26 and 22.5 kD	?
Parvoviridae	18–26	Animal	ssDNA(-) 1-linear	Canine parvovirus	60 copies 64 kD	$T = 1$
Pea Enation Mosaic	28	Plant	ssRNA(+) 2-linear	Pea enation mosaic virus	180 copies 22 kD	$T = 3$
Phytoreovirus	75	Plant	dsRNA 12-linear	Wound tumor virus	outer: 130, 96, 36, 35 kD inner: 160, 118, 58 kD	$T = 13$
Picornaviridae	27–30	Animal	ssRNA(+) 1-linear	Human rhinovirus	60 copies, 7, 26, 29, 32 kD	$P = 3$
Reoviridae	60–80	Animal	dsRNA 10–12-linear	Reovirus	outer: 72, 49 and 41 kD inner: 144, 142, 137, 83 and 47 kD	$T = 13$
Satellite viruses	17	Plant	ssRNA(+) 1-linear	Satellite tobacco necrosis virus	60 copies 23 kD	$T = 1$
Sobemovirus	27–30	Plant	ssRNA(+) 1-linear	Southern bean mosaic virus	180 copies 30 kD	$T = 3$
Tectiviridae	65	Bacteria	dsDNA 1-linear	PRD 1	20 proteins	?
Tetraviridae	35–40	Animal	ssRNA(+) 2-linear	*Nudaurelia capensis* ω virus	240 copies 61–65 kD	$T = 4$
Tombusvirus	33	Plant	ssRNA(+) 1-linear	Tomato bushy stunt virus	180 copies 40 kD	$T = 3$
Totiviridae	30–48	Fungi	dsDNA 1–3-linear	*Saccharomyces cerevisiae* virus L1	73–88 kD	?
Tymovirus	29	Plant	ssRNA(+) 1-linear	Turnip yellow mosaic virus	180 copies 20 kD	$T = 3$

Table 2. Continued

Family/Group	Size(nm)[a]	Host	Genome no.-type	Typical member	Nucleocapsid proteins[b]	Capsid symmetry[c]
Nonenveloped nonspheroidal						
Alfalfa mosaic	18 × 58, 48, 36, 28	Plant	ssRNA(+) 3-linear	Alfalfa mosaic virus	24.3 kD	Bacilliform particles 
Capillovirus	12 × 600–700	Plant	ssRNA(+) 1-linear	Apple stem grooving virus	27 kD	Flexuous filaments 
Carlavirus	12–15 × 600–700	Plant	ssRNA(+) 1-linear	Carnation latent virus	32–36 kD	Flexuous filaments 
Closterovirus	12 × 1250–1800	Plant	ssRNA(+) 1-linear	Beet yellows virus	24 kD	Flexuous filaments 
Furovirus	20 × 130–300	Plant	ssRNA(+) 2-linear	Soil-borne wheat mosaic virus	19.7 kD	Rigid rods 
Geminivirus	17 × 33	Plant	ssDNA 1-circular	Maize streak virus	31 kD	Two incomplete $T = 1$ particles 
Hordeivirus	22 × 100–150	Plant	ssRNA(+) 3-linear	Barley stripe mosaic virus	25 kD	Rigid rods 
Inoviridae	6 × 750–2000	Bacteria	ssDNA(±) 1-circular	M13	5 kD	Flexuous threads 
Myoviridae	Head: 80 × 110 Tail: 16 × 80–455	Bacteria	dsDNA 1-linear	T2, T4	Head: ~ 1600–2000 copies 43 kD	Elongated or isometric head and contractile tails 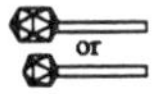
Podoviridae	Head: 65 Tail: 17	Bacteria	dsDNA 1-linear	T7, P22, ϕ29	Head: ~ 450 copies 38 kD	Head and short noncontractile tail 
Potexvirus	13 × 470–580	Plant	ssRNA(+) 1-linear	Potato virus X	18–28 kD	Flexuous filaments 
Potyvirus	11 × 720–770	Plant	ssRNA(+) 1-linear	Potato virus Y	33 kD	Flexuous filaments 
Tenuivirus	8 × varying length	Plant	ssRNA(+) 4–5-linear	Rice stripe virus	32 kD	Filaments (branched) 
Siphoviridae	Head: 80 Tail: 64–539	Bacteria	dsDNA 1-linear	λ	Head: ~ 420 copies 38 kD (gpE)	Isometric head and long noncontractile tail 

Table 2. Continued

Tobamovirus	18 × 300	Plant	ssRNA(+) 1-linear	Tobacco mosaic virus	17.6 kD	Rigid rods 
Tobravirus	22 × 200 (L) 46–114 (S)	Plant	ssRNA(+) 1-linear	Tobacco rattle virus	22 kD	Rigid rods 
Enveloped spheroidal						
Arenaviridae	100–300	Animal	ssRNA(-) 2-linear	Lymphocytic choriomeningitis virus	63 kD	
Bunyaviridae	95–120	Animal	ssRNA(-) 3-linear	Bunyamwera virus	22–45 kD	
Coronaviridae	60–220	Animal	ssRNA(+) 1-linear	Human coronavirus 229-E	50–60 kD	
Corticoviridae	60	Bacteria	dsDNA 1-circular	PM 2	43, 32, 12 and 5 kD	
Cystoviridae	60	Bacteria	dsRNA 3-linear	$\phi 6$	10.5 kD	
Flaviviridae	40–50	Animal	ssRNA(+) 1-linear	Yellow fever virus	14 kD	
Hepadnaviridae	42–47	Animal	dsDNA 1-circular	Hepatitis B virus	22 and 16 kD	
Herpesviridae	120–300	Animal	dsDNA 1-linear	Herpes simplex virus	5, 19, 23, 24, 12 kD	Icosahedral core $T = 16$ 
Iridoviridae	130–300	Animal	dsDNA 1–2-linear	*Tipula* iridescent virus	10–25 proteins, 10–250 kD	Icosahedral core 
Orthomyxoviridae	80–120	Animal	ssRNA(-) 8-linear	Influenza virus	50–60 kD	
Paramyxoviridae	150–300	Animal	ssRNA(-) 1-linear	Parainfluenza virus	44–62 kD	
Retroviridae	80–130	Animal	ssRNA(+) duplicate linear copies	Human T-lymphotropic virus	24–30 kD	
Togaviridae	40–70	Animal	ssRNA(+) 1-linear	Sindbis virus	14–32 kD	$T = 4$ core 

Table 2. Continued

Family/Group	Size(nm)[a]	Host	Genome no.-type	Typical member	Nucleocapsid proteins[b]	Capsid symmetry[c]
Tomato Spotted Wilt	85	Plant	ssRNA(?) 3-linear	Tomato spotted wilt virus	27 kD	
Enveloped Nonspheroidal						
Baculoviridae	50 × 300	Animal	dsDNA 1-circular	Baculovirus	6.9 and 39 kD	
Filoviridae	80× up to 14 000	Animal	ssRNA(-) 1-linear	Marburg virus	96–104 and 28–30 kD	
Phytorhabdovirus	45–95 × 135–380	Plant	ssRNA(-) 1-linear	Potato yellow dwarf virus	56 kD	
Poxviridae	170–260 × 300–450	Animal	dsDNA 1-linear	Vaccinia virus	74, 62, 25 and 11 kD	
Rhabdoviridae	50–95 × 130–380	Animal	ssRNA(-) 1-linear	Rabies virus	50 kD	

[a] Size is diameter of spherical viruses, or dimensions of nonspherical viruses.
[b] Molecular mass of major capsid proteins.
[c] Triangulation number is given when known, question mark indicates icosahedral virus whose T number is not determined. Virion structure is schematically illustrated for nonspherical and enveloped viruses (adapted from Hull *et al.* (1989) In: *Virology: Directory and Dictionary of Animal, Bacterial and Plant Viruses*. New York: Stockton Press).

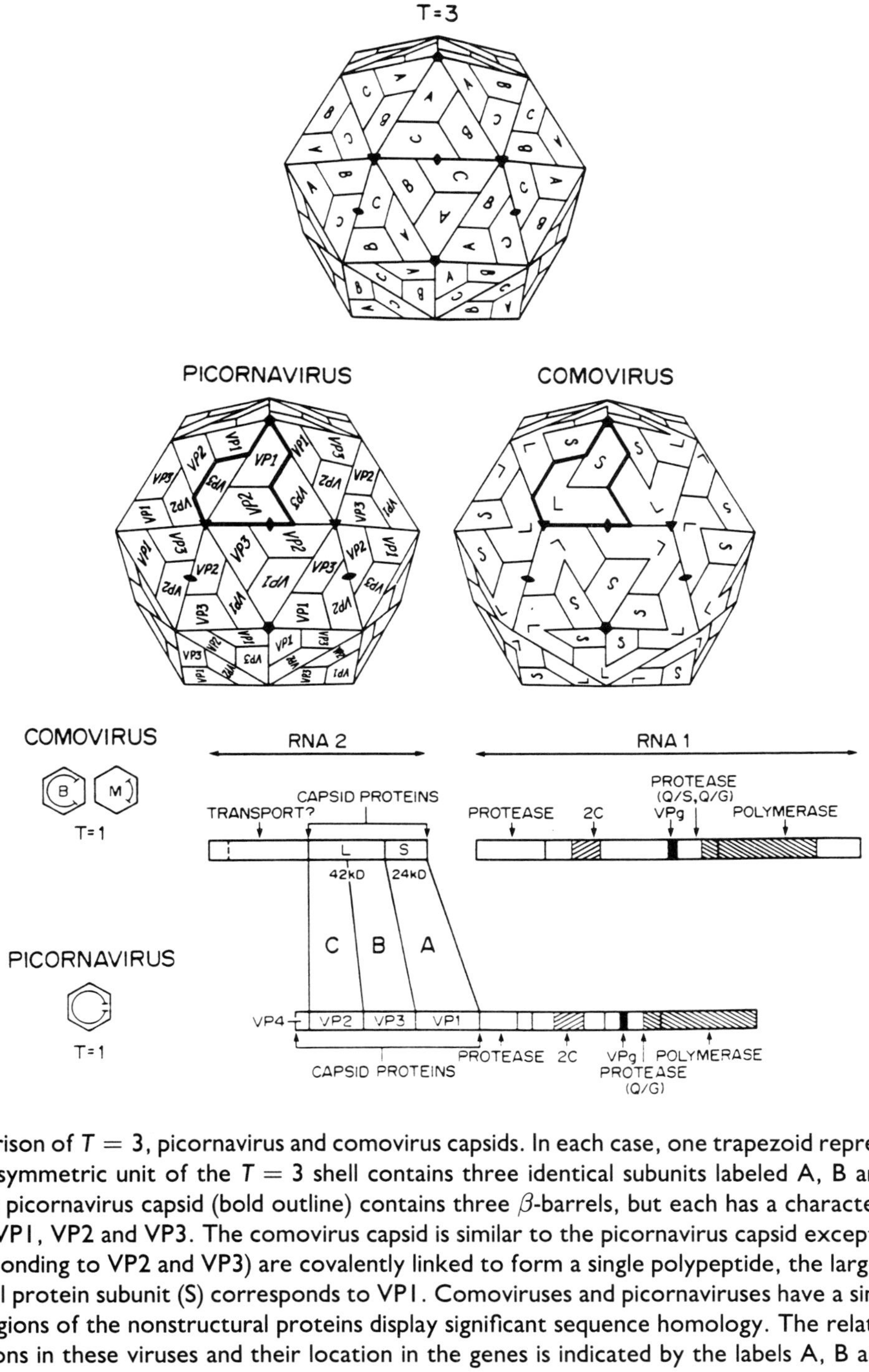

Fig. 4 A comparison of $T = 3$, picornavirus and comovirus capsids. In each case, one trapezoid represents a β-barrel. The icosahedral asymmetric unit of the $T = 3$ shell contains three identical subunits labeled A, B and C. The asymmetric unit of the picornavirus capsid (bold outline) contains three β-barrels, but each has a characteristic amino acid sequence labeled VP1, VP2 and VP3. The comovirus capsid is similar to the picornavirus capsid except that two of the β-barrels (corresponding to VP2 and VP3) are covalently linked to form a single polypeptide, the large protein subunit (L), while the small protein subunit (S) corresponds to VP1. Comoviruses and picornaviruses have a similar gene order, and the shaded regions of the nonstructural proteins display significant sequence homology. The relationship between the subunit positions in these viruses and their location in the genes is indicated by the labels A, B and C in the gene diagram.

Canine parvovirus, ϕX174, SV40 and adenovirus have the β-barrel fold, but its location relative to symmetry elements is different from that observed in the $T = 3$ and $P = 3$ viruses. In all cases, the residues in the β-barrel form most of the contiguous shell, while insertions between strands of the barrel project outward. In some cases these insertions can be 200 residues or more, creating additional domains of the protein. Extensions at the N terminus are generally on the interior of the shell, and in many plant and some animal viruses this portion is extremely basic. These regions are not visible in electron density maps because they are assuming different structures and do not obey icosahedral symmetry. Viruses that do not contain basic regions in the capsid protein often package polyamines to neutralize the charge of the RNA. Extensions at the C terminus are generally external. In the case of TBSV, an entire protruding domain is created by the 114 residues following the polypeptide that forms the contiguous shell.

Two other tertiary structures have been found in viral capsid subunits. The RNA phage MS2 has an entirely different fold from the β-barrels observed in other $T = 3$ structures. Furthermore the quaternary structure of MS2 is more closely related to the quasiequivalence of subunit interactions originally envisioned by Caspar and Klug. The differences between quasi and icosahedral twofold axes are extremely small, and the difference in contacts at the pentamers and hexamers is regulated by a loop between two strands that is extended at hexamer axes and folded down at pentamer contacts.

The other tertiary structure found is in the core protein of Sindbis virus. In a study by Choi *et al.*, subunits were purified and crystallized because the intact nucleoprotein core was not stable enough to form crystals that diffract to high resolution. The fold found was that of chymotrypsin! It was previously known that this subunit functions as an enzyme because it cuts itself out of a polyprotein after synthesis. Generally the structure has shorter loops than found in chymotrypsin but the topology and active site are conserved. The C-terminal tryptophan, where the cleavage occurs, is still in the active site of the subunit. The subunits form $T = 4$ shells with the viral RNA. In this case a protein fold with a totally different function has been adapted to form shells. The N-terminal region of this protein (residues 1–113) is composed predominantly of basic residues and is disordered in the crystal structure.

The principles of viral architecture discussed reflect the level of current understanding for a few systems. To date a total of 17 unique structures have been determined to 0.35 nm resolution or higher by X ray crystallography. These include ten different virus groups. Approximately the same number have been determined by high-resolution cryoelectron microscopy and image-processing methods (Table 2).

Complex Virus Structures

Many viruses are composed of complex particles with specific functions associated with different structural elements. Complex bacteriophages, for example, have been a subject of study for decades, and the details of their morphogenesis and low-resolution structure have been determined. A variety of viruses contain multiple copies of different subunits in their capsids, and the structural roles of each subunit type are still being determined. In most cases these particles are too large to be analyzed by crystallography, but some have been successfully examined by high-resolution electron microscopy. Many viruses of medical importance are enveloped by a membrane which contains functionally important proteins. Usually a quasisymmetric nucleoprotein particle assembles in the cytoplasm and the membrane and associated proteins are acquired when the particle buds through the plasma membrane of the host cell. In the paramyxoviruses, two proteins in the membrane (neuraminidase and hemagglutinin) have been purified, crystallized and analyzed at high resolution (Table 2). There are a number of examples in which envelopes have been removed and nucleoprotein cores have been analyzed by cryoelectron microscopy and image analysis. Their structures display a variety of T numbers, and they are not significantly different from nonenveloped protein capsids. Table 2 is a comprehensive listing of virus structures which is divided into four structural categories: nonenveloped icosahedral viruses, nonenveloped viruses of nonicosahedral shape, enveloped spherical viruses and enveloped nonspherical viruses. In the latter three categories, representative sketches of the virus particle are provided.

See also: Comoviruses; Influenza viruses; Polioviruses; Taxonomy and classification–general.

Further Reading

Caspar DLD (1980) Movement and self-control in protein assemblies. *Biophys. J.* 10: 103.

Caspar DLD and Klug A (1962): *Cold Spring Harbor Symp. Quant. Biol.* 27: 1.

Crick FHC and Watson JD (1956) Structure of small viruses. *Nature (London)* 177: 473.

Murialdo H (1991) Bacteriophage lambda DNA maturation and packaging. *Annu. Rev. Biochem* 60: 125.

Rosen R (1972) Morphogenesis. In: *Foundations of Mathematical Biology II*, p.1. New York: Academic Press.

Valegård K *et al.* (1990). The three-dimensional structure of the bacterial virus MS2. *Nature (London)* 345: 36.

VIRUS–HOST CELL INTERACTIONS

Patricia Whitaker-Dowling and Julius S Youngner
University of Pittsburgh School of Medicine
Pittsburgh, Pennsylvania, USA

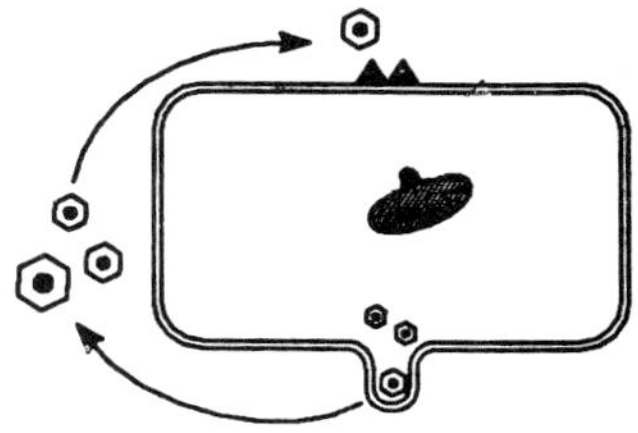

Introduction

The topic of virus–host cell interactions spans all of virology and provides some of the most important insights into this field. Since viruses are intracellular parasites, they rely on their host cells for the energy, macromolecular synthesis machinery and the work benches for genome replication and particle assembly. Because of this dependence, viruses have evolved a myriad of mechanisms for exploiting normal host cell functions. Often this exploitation is associated with damage to the host cell which may be one of the major factors in the pathology and disease caused by viruses. The material in this entry is confined to model systems of virus–host cell interactions that involve the infection by animal viruses of cells in culture.

The past few decades have witnessed a dramatic expansion of our knowledge of animal viruses. These advances have provided a detailed understanding of the structure and composition of the viral genome and the virus particle as well as insight into the replication strategies used by viruses and the regulation of viral gene expression during infection. Development of an understanding of the virus growth cycle has proved easier than a clear comprehension of the interaction of the virus with the host cell. Owing to the complexity of the cell, many of the effects of virus infection on the host occur by mechanisms yet to be determined.

Types of Virus Infections

When a virus infects a cell, the outcomes that may occur can be grouped into several general categories which are determined by the particular virus involved, as well as by the type of cell and its functional state. Productive infections result in the formation of progeny virus and usually cause the destruction of the host cell. In some cases the host cells are not all destroyed, leading to persistent infections in which the surviving cells multiply and continue to produce progeny viruses. When persistent infections occur in which the viral genome is present but no infectious virus is produced, these infections are referred to as latent infections. In such infections some level of viral gene expression is usually detectable although virions are not produced. When genetic information of the virus is integrated as DNA into the host cell genome or is carried as episomal DNA, transforming infections may take place. Such infections can cause an oncogenic alteration of the growth properties of the cell. Abortive infections occur when viruses infect cells that are nonpermissive or only partially permissive. In this instance, the virus is able to enter the cell but because some step essential for viral replication is absent, the replication cycle does not go to completion and no progeny are produced. Such abortive infections may or may not cause cell death.

A few examples follow which demonstrate that different outcomes of infection are dependent on the particular virus and host cell involved, as well as on the state of the host cell. For example, influenza A virus causes a productive, cytolytic infection of a line of canine kidney cells (MDCK). However, when the same virus infects the L cell line of mouse fibroblasts an abortive infection occurs because of a block at the level of virion RNA replication. The same mouse L cell line supports a productive, cytolytic infection by vesicular stomatitis virus (VSV). However, if the L cells are pretreated with interferon, the functional state of the cells is altered; the VSV replication cycle is blocked at the level of protein synthesis and an abortive infection results. When VSV infects insect cell lines derived from *Aedes* or *Drosophila*, productive noncytolytic infections occur. Continuous passage of the insect cell lines reveals that they have become persistently infected and continuously produce infective virus without any signs of cytopathology. Adeno-associated virus (AAV), a parvovirus, is capable of a productive or a latent infection depending upon whether or not the host cells are co-infected with a helper virus such as adenovirus. In some host cells, AAV can cause a latent infection by integrating its DNA into the host cell genome.

There are instances in which infection with a second unrelated virus can dramatically alter the type of infection produced by certain viruses. As mentioned in the preceding paragraph, adeno-associated viruses are capable of productive, cytolytic infections only in cells co-infected with adenovirus. Human adenoviruses can multiply in monkey cells only in the presence of SV40, a simian papovavirus that supplies a helper function that permits translation of adenovirus mRNAs. In rabbit corneal cells, co-infection with vaccinia or some other poxviruses can convert nonproductive infections with VSV into cytolytic infections.

Effects of Virus Infection on the Host Cell

Effects on host cell morphology and viability

The most readily recognized effects of viruses on host cells are those that involve morphologic changes or cell death. Enders defined viral cytopathogenicity as 'the capacity to induce any demonstrable departure from the normal either in the morphological or functional properties of cells'. The space available for this entry precludes a comprehensive survey of all the cytopathic effects induced by infection with the various families of animal viruses. However, one of the most striking observations that emerges from an overview of the effects of viruses on host cells is how little is known of the mechanisms by which viruses induce cytopathology. The production of cytopathic effects has been observed with most families of viruses and in many cases the viral gene(s) involved or implicated in these morphological changes has been defined. However, in most cases the mechanisms responsible for cell destruction have not been identified. It is fair to say that one of the most fundamental questions of virology, namely, how viruses kill cells, remains for the most part unanswered.

Effects on host cell macromolecular synthesis

Many of the investigations of the effect of virus infection on the host cell have centered on virus-induced alterations of host cell macromolecular synthesis. While these studies are important to an understanding of the viral growth cycle and have yielded significant insights into the control of host cell gene expression, there is no direct evidence that inhibition at this level is the direct cause of visible cytopathology or cell death. In fact, treatment of host cells with drugs such as actinomycin D and cycloheximide, which inhibit nucleic acid and protein synthesis, does not mimic the morphological changes produced by virus infection. Nevertheless, viruses do employ a variety of strategies to affect the host cell at the level of gene expression.

Effects on host cell DNA and RNA synthesis

A variety of DNA and RNA viruses are capable of affecting gene expression by directly altering the host cell genome. For example, the host cell DNA is degraded after infection with poxviruses. Herpes-, picorna- and reoviruses cause a displacement of the cellular chromatin, while an inhibition of host DNA synthesis has been reported following infection with herpes-, pox-, adeno-, picorna-, reo-, alpha- and rhabdoviruses. This inhibition of host DNA synthesis may be a direct effect of a virus factor in the nucleus or a secondary consequence of the inhibition of host cell protein synthesis.

Viral products can directly affect the activity of cellular RNA polymerases and cause an inhibition of host RNA synthesis. Such an inhibition has been seen with VSV and polioviruses. In the case of VSV, both a small viral-encoded RNA molecule (leader RNA) and a viral protein (the matrix M protein) have been implicated in the inhibition of the cellular polymerases at the level of RNA synthesis initiation. Another mechanism to inhibit host RNA synthesis is employed by polioviruses. These agents encode a protease that is capable of cleaving transcription factors required by host RNA polymerases II and III. Reo- and alphaviruses also block host RNA synthesis but the mechanism of this inhibition is not known. Synthesis is not the only level at which viruses can affect host mRNA. Infection with herpes- and poxviruses increases the rate of host mRNA degradation. A unique effect on host cell mRNA is produced by influenza viruses. These agents cleave the cap structure and the first 10–13 nucleotides from the 5′ ends of newly synthesized host mRNAs, and utilize this oligomer as a primer for viral mRNA synthesis. Another mechanism that affects host RNA is seen with adenoviruses; in this instance, infection inhibits the transport of host mRNA out of the nucleus.

Effects on host cell protein synthesis

Although much effort has been directed at understanding the effect of virus infection on host protein synthesis, it is unlikely that an inhibition of host protein synthesis is required for successful virus replication. Many viruses, such as paramyxo-, papova- and retroviruses do not normally inhibit host protein synthesis during their replication. Furthermore, mutant viruses that are defective in their ability to shut down host protein synthesis are not necessarily defective for virus growth. In fact, VSV mutants selected during a persistent infection have a reduced ability to inhibit the host's translational machinery; nevertheless, these mutants grow to higher titer during a lytic growth cycle than the parental wild-type virus. With several virus families, infection causes a selective inhibition of the translation of host cell mRNA. Such viruses include picorna-, pox-, herpes-, adeno-, rhabdo-, reo- and orthomyxoviruses. In many cases this inhibition is accompanied by a decrease in the overall rate of protein synthesis in the infected cell. It is likely that this overall inhibition occurs at the level of initiation of protein synthesis since, where it has been examined, the average size of the polysomes in the infected cells is reduced.

The most clearly defined case of virus-induced damage to the translational machinery of the host cell is the effect of poliovirus on one of the translation initiation factors. Following infection with poliovirus, the cap binding complex responsible for recognition of the capped 5′ end of cellular mRNA is inactivated by a proteolytic cleavage of the p220 component of the complex. It has been speculated that the destruction of the p220 protein confers a selective advantage on the translation of poliovirus messages which are uncapped. Infection with poliovirus also causes the release of host mRNA from the cytoskeleton.

Virus-mediated inactivation of other initiation factors for protein synthesis has also been reported. Translational extracts prepared from VSV-infected cells are deficient in eucaryotic initiation factor 2 (eIF-2) activity in one report and eIF-3 in another, while infection with reoviruses impairs the function of eIF-2. It has recently been shown that vaccinia virus, a poxvirus, encodes a small protein which has significant homology to the α subunit of eIF-2. This protein may function as a replacement initiation factor since there is evidence that it may be important in making the virus resistant to inhibition by interferon.

Another viral strategy to inhibit host protein synthesis involves a direct competition of viral and host RNAs. VSV and reoviruses compete successfully with the host for the translational machinery through sheer abundance of viral transcripts. Mengovirus, a picornavirus, produces mRNA which initiates translation more efficiently than the bulk of the host message and, in addition, synthesizes a factor that causes an overall inhibition of protein synthesis in infected cells.

It has been suggested that selective translation of viral mRNA may also occur following changes in intracellular ion concentrations during infection. Increased plasma membrane permeability is a common cytopathic effect of virus infection which can alter the intracellular levels of sodium and potassium ions. Under conditions that cause increased intracellular sodium ion concentrations, the translation of viral mRNAs may be unimpaired while the translation of host mRNAs is severely reduced. Such a differential effect on virus and host protein synthesis has been reported for cells infected with poliovirus, encephalomyocarditis virus, VSV, reovirus and Sindbis virus. In the case of Sindbis virus, the shutdown of host protein synthesis following infection has been correlated temporally with an increase in permeability of the plasma membrane.

Effects on host cell membranes and cytoskeleton

In addition to altering membrane permeability, virus infection can cause other changes in the membranes of the host cell. Insertion of viral proteins into the plasma membrane can induce syncytia formation by fusing infected cells with neighboring uninfected cells. This fusion can be induced either from without by input virions or from within by newly synthesized viral fusion protein made during infection. The ability to fuse cells, which is characteristic of the paramyxovirus family, is also seen with herpes-, flavi-, lenti-, pox- and coronaviruses. Flaviviruses can also affect internal membranes by causing the proliferation of the rough endoplasmic reticulum, a site associated with the assembly of viral particles. Reo-, picorna- and alphavirus infections frequently produce a significant increase in vesicle formation in the cytoplasm.

Cytolytic virus infections generally cause a progressive loss of integrity of the lysosomal membranes. Two phases of damage are recognized. In the first phase, which in some cases is reversible, the lysosomes become permeable to small molecules and are able to concentrate dyes such as neutral red. Visible evidence of this phenomenon is seen with a mutant strain of Newcastle disease virus, a paramyxovirus, which produces red plaques when assayed using an agar overlay containing neutral red. Concentration of this vital stain in lysosomes can also be detected in cells infected with certain strains of influenza A virus. In this instance a ring of darkly staining cells surrounds a clear area of unstained dead cells. In the second phase of lysosomal damage the membrane becomes so permeable that lysosomal enzymes are released into the cytoplasm. As a rule, this release occurs late in the replicative cycle. The release of lysosomal enzymes into the cytoplasm has been described for a wide variety of viruses such as picorna-, pox-, herpes-, orthomyxo-, paramyxo-, corona-, adeno- and papovaviruses. The mechanism responsible for this type of virus-induced cytopathology and the role it plays in cell death have not been clearly defined.

One of the most common signs of virus-induced cytopathology is cell rounding, a morphological change which has been correlated with alterations in the cytoskeleton. Disruption of one or more of the elements of the cytoskeleton has been described after infection with several viruses. Early gene products of herpes, vaccinia and SV40 viruses produce a disassembly of the actin-containing microfilaments, while infection with polio- and reoviruses causes an alteration and reorganization of the virimentin-containing intermediate filaments of the cytoskeleton. Microtubules, another element of the cytoskeleton, are depolymerized following infection with herpes simplex virus 1 (HSV-1), canine distemper virus and frog virus-3. It has been reported that infection with VSV causes a sequential disassembly of all three filament components of the cytoskeleton. The mechanisms by which virus infections disrupt the cytoskeleton are not known and it is not clear whether these morphologic changes are a direct effect of some virus product or a secondary consequence of some other aspect of virus-induced cytopathology. It is interesting to note that in normal cells polyribosomes are closely associated with the cytoskeleton, and on the basis of this association it is possible to speculate that some of the effects of virus infection on the host translational apparatus may be caused by virus-induced changes in the integrity of the cytoskeleton.

Viruses also use the structural elements of the cytoskeleton as the work benches for virion assembly and for transport of viral products within the cell. Examples of this function of the cytoskeleton include adenoviruses which appear to use the microtubules for movement within the infected cell; Newcastle disease virus, the viral products of which are associated with actin filaments; and reoviruses which produce inclusion bodies found in association with microtubules and are the site of viral RNA synthesis and virion assembly. It has also been suggested that, in VSV infections, assembly of nucleocapsids occurs in close association with the cytoskeleton.

Inclusion bodies

Another commonly recognized form of virus-induced alteration of the infected host cell is the formation of intracellular masses called inclusion bodies. It should be noted that at the beginning of this century the discovery of a characteristic cytoplasmic inclusion, the Negri body, in cells infected with rabies virus provided an effective diagnostic test for this disease. Depending upon the virus, these intracellular masses may consist of either virions or unassembled viral products. Inclusion bodies may occur in the cytoplasm, as in cells infected with pox-, paramyxo-, orthomyo-, reo-, rubella or rabies viruses, or may be found in the nucleus in cells infected with adeno- and herpesviruses.

Transformation of host cells

In addition to producing various forms of cell destruction, some families of animal viruses are capable of inducing cell transformation. In most cases, transformation is associated with integration of the viral genome into the host cell DNA or maintenance of viral DNA in an episomal state. Only one family of RNA viruses, the retroviruses, is capable of transforming cells. This family of viruses induces transformation through the action of a variety of oncogenes that are cellular in origin and that are not part of or necessary to the virus replicative cycle. There are several families of DNA viruses that are the cause of or are associated with tumor induction in animals and cell transformation in cultured cells. These include polyoma-, adeno-, herpes-, papilloma-, hepadna- and poxviruses. In contrast to the RNA viruses, the genes of DNA viruses responsible for transformation are viral in origin and required for virus replication.

Is It Murder or Suicide?

It is clear from the information reviewed above that the mechanisms responsible for cell death following virus infection have not been clearly defined. Perhaps the reason it has been so difficult to explain how viruses kill cells is that they do not do this directly. An alternative to a direct cell killing is the induction by viruses of a suicide function in infected cells. It would be advantageous for a cell, as part of a metazoan, to induce an apoptosis-like function in response to viral infection rather than to continue on as a factory producing a constant stream of progeny virus. Some recent evidence has appeared that lends support to this possibility. The cytopathic effect of human immunodeficiency virus (HIV) infection has been associated with apoptosis; and in another report, a noncytopathic latent infection of B cells with Epstein–Barr virus (EBV) has been associated with an inhibition of the apoptosis function. In this connection, it is interesting to note that latent infection with EBV blocks the killing of B cells by VSV with little or no effect on the replicative ability of this RNA virus. These observations provide some basis for suggesting that virus-associated cell killing may involve the induction of apoptosis or some other suicide function in the infected cell.

Resistance of Cells to Virus Infection

The major determining factor of the susceptibility of a cell to a particular virus is the ability of the viral attachment proteins to recognize and interact with specific receptors on the cell surface. In many cases, cells are resistant to infection by a particular virus simply because of the lack of appropriate surface receptors. A dramatic example of this type of resistance is seen when chicken fibroblast cells, which lack specific receptor molecules on their plasma membranes, are exposed to poliovirus. Infection does not take place because the viruses cannot adsorb to the cell membrane. However, the avian cells are fully able to support the growth of poliovirus if transfected with the virion RNA rather than infected with intact virions. In addition to cell surface viral receptors, the host range of some viruses can be determined by other factors such as host cell transcriptional regulators. There is evidence that suggests that viruses from the herpes-, polyoma-, retro- and hepadnavirus families can replicate only in cells that express the appropriate factors that permit recognition of the viral enhancers.

Although viruses have an adaptive advantage in terms of genetic plasticity, cells are not totally powerless to mount a defensive response to viral infection. The best characterized defense that cells have evolved for protection against viral infection is the interferon system. The interferon family of proteins that is produced in response to viral infection promotes the development of an antiviral state through the induction of a second group of proteins. Two of these proteins, the 2′-5′A synthetase and the protein kinase, have been well characterized and evidence has accumulated that demonstrates their role in the development of the interferon-mediated antiviral state. Perhaps the best evidence to suggest that these proteins are actually involved in the interferon-induced antiviral state comes from the fact that several families of viruses have evolved factors that are capable of blocking the activity of the 2′-5′A synthetase (herpes- and poxviruses) and the protein kinase (herpes-, pox-, adeno-, reo- and orthomyxoviruses).

Viruses as Tools for Probing the Host Cell

Many of the crucial discoveries concerning cellular processes were offshoots of investigations into the replication cycle of viruses or derived from the use of viruses as model systems. This is particularly true for understanding the mechanisms involved in gene expression. It is apparent that all viruses must use the host cell translational apparatus for the synthesis of viral proteins and that many DNA viruses depend on the host transcriptional and DNA replication machinery as well. Consequently, investigation of the intricacies of viral gene expression has led to the discovery of nearly all identified host factors involved in host genome replication, RNA splicing, enhancer sequences, the scanning model for the initiation of protein synthesis, the use of translational frameshifting for gene expression, and the manner in which proteins are targeted within the cell. This list, which is far from exhaustive, will surely be expanded in the future.

It would be difficult to overestimate the impact of the study of tumor viruses on our understanding of the mechanisms involved in transformation and the nature of the cancer cell. In spite of the fact that most naturally occurring cancers of humans and animals are not caused by viruses, investigation of transforming viruses, and retroviruses in particular, has led to an understanding of the major mechanisms and cellular genes responsible for transformation. A detailed review of this subject can be found elsewhere in this volume.

See also: Cell structure and function in virus infections; Enteroviruses; Host genetic resistance; Influenza viruses; Interferons; Pathogenesis; Persistent viral infection; Polioviruses; Replication of viruses.

Further Reading

Frankel-Conrat H and Wagner RR (eds) (1984) *Comprehensive Virology*, Vol. 19, *Viral Cytopathology*. New York: Plenum Press.

Knipe DM (1990) Virus–host cell interactions. In Fields BN *et al.* (eds) *Virology*, 2nd edn, p. 1091. New York: Raven Press.

VISNA–MAEDI VIRUSES

Opendra Narayan
Johns Hopkins University School of Medicine
Baltimore, Maryland, USA

History

Visna and maedi are Icelandic terms for two sheep diseases characterized by wasting paralysis and progressive labored breathing respectively. These diseases broke out in epizootic proportions among Icelandic sheep following introduction of European sheep into the local flocks. The newly introduced Karakul rams had gone through a long preimportation quarantine period and had come from flocks with no history of the type of disease that broke out in Iceland. These animals were intended to provide a new gene pool for native Icelandic animals which had been maintained in isolation on the island for several centuries. The new diseases spread rapidly among the Icelandic sheep, involving as many as 50% of the animals in some flocks. Maedi was the predominant disease with visna occurring mainly as a complication. The disease complex was finally eradicated from the islands during the 1960s by slaughter of all sheep on the farms that had affected animals. The farms were then restocked with sheep from other parts of the island that had had no contact with the foreign sheep or local sheep with the disease syndromes. Virus obtained from sick animals was the origin of the prototype visna–maedi virus. In 1974, a virus genetically and serologically related to visna–maedi virus was obtained from goats affected with arthritis and encephalitis in Washington State. This virus was named caprine arthritis encephalitis virus (CAEV).

Clinical and Pathological Criteria

Clinical signs of visna are gradual weight loss and gradual weakening of the hind legs followed by ataxia. These signs progress slowly during a period of months to cachexia and paralysis of the hindlimbs. Rarely, forelimbs are also involved in the paralytic syndrome. Signs of maedi are characterized by weight loss, labored breathing, a dry cough and inability to keep up with the rest of the flock. Both of these diseases occur in adult animals, 2–3 years old, and the disease syndromes last 6–8 months. Arthritis synovitis also occurs in some sheep and is the hallmark of the infection in goats. Similar to visna–maedi, this disease appears in adult life of the animals. Clinically, signs of swelling of the carpal joints ('knees') appear first and this progresses to arthritis and eventual lameness. The encephalitis in goats is confined to kids and becomes apparent a few weeks after birth. This progresses to paralysis and eventually death. In all cases, the animals have normal blood counts. They do not develop fever and all have normal appetites. They do not succumb to opportunistic infections.

Pathology

Histologically, all of the lesions in sheep and goats are characterized by infiltration and proliferation of mononuclear cells consisting of macrophages, B lymphocytes and T lymphocytes in the affected tissues. In the brain, this inflammation is acccompanied by demyelination and necrotizing changes in the neural parenchyma. In the lungs, these mononuclear cells infiltrate into the interalveolar interstitial areas of the lung causing consolidation of the organ and physical obliteration of the alveoli. The arthritis consists of thickening and infiltration of mononuclear cells into the synovial lining of the joint and accumulation of large amounts of synovial fluid. This is followed later by degeneration and calcification of the articular facets of the joint and eventually degeneration in the bone. Massive lymphadenopathy accompanies these organ-specific diseases.

There are only a few reports of visna from other parts of the world but pulmonary lesions in sheep similar to maedi had been described previously in the USA and subsequently in several other countries except Australia and New Zealand. In the Netherlands the disease was called Zwoegerziekte, in South Africa, Graaf Reinet, in France, La Bouhite, in the USA, Montana lung disease and/or progressive pneumonia, etc. This provides a clear indication at the pathological level that visna–maedi

viruses are present in sheep throughout the world. Modern diagnostic procedures have confirmed this, but unlike the experience in Iceland, the disease in most sheep populations occurs mainly sporadically and usually only in adult animals. The caprine disease complex has also been reported in most of the industrialized countries of the world, including New Zealand and Australia. The infection is rare in most underdeveloped countries where herds are small and isolated.

Host Range and Epizootiology

Visna–maedi virus replicates best in sheep and CAEV in goats. However, reciprocal infection has been documented. Further, some strains of ovine virus such as those found in the USA are biologically more similar to CAEV than Icelandic visna–maedi virus. These viruses are infectious for other small ruminant animal species. Such cross-over infections have been documented in zoos. The viruses are spread mainly by colostrum, milk and respiratory exudates. Under field conditions, viral transmission from mother to offspring is very efficient but localized unless the milk is used for feeding other animals. Virus is also disseminated by poor management practices and this has led to local epizootics. A high rate of transmission occurs in poorly ventilated barns during winter months and this is exacerbated further during intercurrent respiratory infections (caused by other agents) that cause increased production of respiratory exudates. Minimal spread occurs during summer months while animals are on pasture. Infected macrophages in colostrum milk and respiratory exudates are a source of the infection. Intercurrent mastitis causes increased numbers of infected macrophages in the milk and this increases the efficiency of transmission. The major mechanism of dissemination of CAEV has been the dairy husbandry practice by which all kids were fed pooled milk. This guaranteed infection in all kids when only a single infected lactating female may have been present in the herd.

Taxonomy and Classification

By the early 1960s, Icelandic investigators had established that the visna–maedi syndrome was caused by a new virus which replicated productively in stationary ovine cell cultures and caused acute cytopathic effects characterized by fusion of the cells into multinucleated giant cells. Paradoxically, sheep inoculated with the virus at that time remained clinically well for several months. Onset of clinical disease occurred insidiously and progressed slowly but inexorably, leading to death. These findings led to the definition of the viruses as 'slow viruses' , the forerunner of the present term, lentiviruses. CAEV had similar biological properties. Some 10 years after the discovery of reverse transcriptase in retroviruses by Temin and Baltimore, the ovine and caprine viruses were both shown to be retroviruses. As lenti-retroviruses, these new agents were distinct from the oncogenic retroviruses which require dividing cells for replication and the oncogenic and transforming properties of which are associated with defective helper virus-dependent replication. The viruses also differ from spuma retroviruses which are not associated with disease. This new ungulate lentivirus, visna, was shown to share similar genetic and biological properties with the lentiviruses of horses, cattle, cats, macaques and humans. Its genome has approximately 9.5 kb of nucleotide sequences and its reverse transcriptase enzyme requires Mg^{2+} instead of Mn^{2+} to catalyze transcription of viral RNA to DNA. In addition to the long terminal repeat (LTR)-gag-pol-env-LTR arrangement of the proviral DNA of retroviruses, visna virus and other lentiviruses also have open reading frames that encode regulatory genes important for replication of the virus. *Tat* and *rev* are examples of these genes.

Variability

The lentiviruses of sheep are genetically and biologically highly heterogeneous. The high mutation rate of the virus has been attributed in part to mistake-prone reverse transcription of viral RNA to DNA. The high mutation rate may be more apparent than real, however, because many mutations in the *env* gene of the virus are viable and viruses with distinct biological properties can be selected by various host systems. Mutant viruses with slow and fast replication rates have been selected by various tissues. Virus-neutralizing antibodies select for neutralization-escape variants (antigenic drift viruses) and macrophages in various tissues such as the brain, lung, synovium and mammary glands select for specific viral phenotypes that replicate optimally in these tissues.

Immune Responses

The envelope proteins of the lentivirus elicit biological responses of greatest importance in the infection. Neutralization epitopes in the envelope consist of either linear peptides or conformational structures and induce antibodies that inhibit replication of the virus. Other epitopes in the viral envelope induce antibodies that bind viral particles but do not neutralize infectivity. These antibodies thus bind to viral particles and enhance entry into target macrophages by Fc receptor-mediated endocytosis. Such antibodies comprise the so-called enhancing antibodies. Cytotoxic CD8 T lymphocytes are also induced by the virus. While these cells, along with neutralizing antibodies, probably lower the virus load *in vivo*, they are incapable of curing the infection, because the virus usually persists for the lifetime of the animal. Part of the failure of the immune responses to eliminate the virus may be due to the integrated unexpressed viral genome in precursor cells in the animal.

Cell Biology and Pathogenesis

The main cell type that supports visna replication *in vivo* is the macrophage-lineage cell, and virus-laden macrophages can be found in all tissues with lesions such as the encephalitic brain, the pneumonic lung, the arthritic joint and mastitic mammary gland. Intense viral replication in local macrophage populations along with local overproduction of cytokines lead to worsening of the inflammatory lesion and more viral replication. Hence, the progressive nature of the lesion and the disease. In animals with organ-specific disease, there is also a high level of infection in precursor cells in bone marrow and in dendritic cells in the blood. In subclinically infected animals that do not have organ-specific lesions, the virus is found at a low rate in the bone marrow and in rare cells in blood but not in tissues. Factors causing the upregulation of viral replication in specific macrophages in tissues have not yet been identified. Studies on viral replication in monocytes (immature macrophages) and in mature macrophages have shown that the viral life cycle is incomplete in the monocytes and mature viral particle formation occurs only in the mature macrophages. The linkage between the life cycle of the virus to that of the cell is probably due in part to the requirement for DNA binding proteins c-Fos and c-Jun (present only in mature cells) to activate viral transcription in the virus LTR. The LTR of the proviral DNA has AP-1 and AP-4 sites which specifically interact with these proteins in mature cells and result in increased transcription of viral RNA. Linkage of the viral life cycle to the physiology of the macrophage as the cell begins its life cycle in the bone marrow and ends in the tissues has relevance in both the disease state in the animals and the mechanism of transmission of the virus in body fluids to other animals.

Prevention and Control

The most effective mechanism for control of lentiviruses in sheep and goats has been the prevention of infection. This has been achieved by removal of infected animals and their young from flocks, prevention of consumption of colostrum and correction of management practices that facilitate enhancement of the spread of virus in respiratory exudates and milk. The sexual route is not a major mechanism for transmission of these viruses. Lentivirus proteins are poor at inducing protective immunity, and such mechanisms have not been investigated thoroughly.

Futhur Perspectives

These viruses are agricultural pathogens with effects that are in general too subtle to grasp the attention of regulatory agencies. They provide excellent models of human immunodeficiency virus (HIV) infection since they represent an example of natural lentivirus infection in which the host usually wins the battle. More studies in the future may be directed to investigations on mechanisms by which these hosts keep their viruses in check and mechanisms by which virulent strains of these viruses cause disease.

See also: Bovine immunodeficiency virus; Feline immunodeficiency virus; Human immunodeficiency viruses; Pathogenesis; Simian immunodeficiency viruses.

Further Reading

Haase AT (1988) Pathogenesis of lentivirus infections. *Nature (London)* 322: 130.

Narayan O and Clements JE (1989) Biology and pathogenesis of lentiviruses. *J. Gen. Virol.* 70: 1617.

Narayan O and Cork LC (1985) Lentiviral diseases of sheep and goats: chronic pneumonia, leukoencephalomyelitis and arthritis. *Rev. Infect. Dis.* 7: 89.

W

WESSELSBRON VIRUS

See Encephalitis viruses

WEST NILE ENCEPHALITIS VIRUS

See Encephalitis viruses

WESTERN EQUINE ENCEPHALITIS VIRUS

See Equine encephalitis viruses

YABAPOX AND TANAPOX VIRUSES

HA Rouhandeh
Southern Illinois University,
Carbondale, Illinois, USA

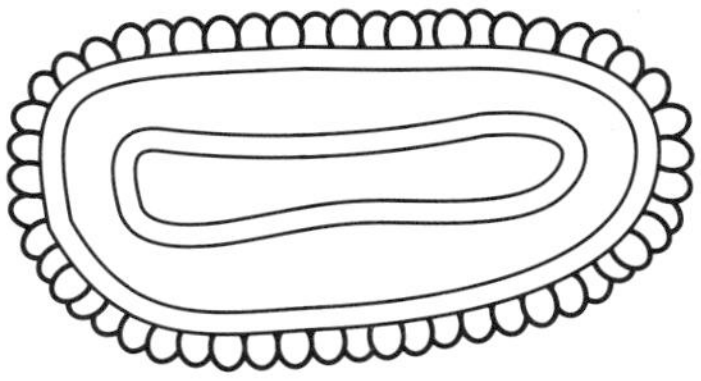

History

The disease produced by Yabapox virus is characterized by formation of tumors that regress slowly. The disease was first observed in 1958 in captive Rhesus monkeys housed in open pens in Yaba, Nigeria. At first the disease was confined to Asian monkeys and one baboon. Later one species of African monkey (*Cercopithecus aethiops tantalus*) became infected with the virus. In 1969 a spontaneous outbreak of the disease occurred in monkeys housed in Roswell Park Memorial Institute in Buffalo, New York. Infection of a laboratory worker reported in 1963 showed the virus to be pathogenic for humans.

The disease produced by Tanapox virus is characterized by a mild febrile illness in humans with one or two pock-like lesions developing on the upper part of the body. Two outbreaks occurred in 1957 and 1962 among the indigenous population living in the Tana River Valley in Kenya. In 1966 outbreaks occurred in monkey colonies in the US and in humans having contact with the animals. Although Tanapox virus was first isolated from humans during an outbreak of the disease, it, like Yabapox virus, primarily causes a disease of macaque monkeys.

Taxonomy and Classification

The International Committee for Taxonomy of Viruses (ICTV) has proposed a classification of *Yatapoxviruses* as the genus name for Yabapox virus and Tanapox virus (as well as Yabavirus-like disease) within the *Poxviridae*.

Evolution

Both Tanapox and Yabapox viruses have been identified within the last 40 years. As far as is known, both viruses originated in monkeys. Human infections of Tanapox are believed to have originated from monkeys through the mosquito as a vector.

Properties of the Virion

Yabapox virus measures 250–280 μm in the long axis. The surface consists of thread-like structures which are randomly formed around the viral particle. Both Yabapox and Tanapox viruses are symmetrical and dumbbell-shaped; both contain a core with lateral bodies in the concavities and a double lipoprotein bilayer envelope comparable to other poxviruses.

Properties of the Genome

Yabapox and Tanapox viruses contain linear double-stranded DNA; the number of segments of 3′ terminal poly(A) tracts present or absent is presently unknown for these viruses. The size of both DNAs is estimated to be approximately 145 kbp with a molecular mass of 119×10^3 kD. The G+C content is $32.5 \pm 0.5\%$.

Properties of Virus Proteins

Tanapox virus contains approximately 55 proteins ranging in size from about 200 to 11 kDa. Yabapox virus contains approximately 44 proteins of which 21 are core-associated, ranging in molecular mass from 220 to 10 kDa. Four enzymatic activities have been identified with purified Yaba virions: DNase with pH optimum at 5.0; DNase with pH optimum at 7.8; RNA polymerase; and nucleotide phosphohydrolase.

Physical Properties

Yabapox viral DNA has a density of 1.6905 g ml^{-1} in CsCl and its T_m value in 0.015 M citrate in saline is 82.3°C.

Replication of Yabapox virus

Nucleic acid

Viral DNA is detected 3 h after infection. At 6 h postinfection 7–10S RNA is detected which increases in amount after 12 h. At 24 h after infection, 14–15S RNA as well as 7–10S RNA is detected. The first and largest peak of mRNA synthesis occurs between 11 and 12 h postinfection and a second slightly smaller peak occurs between 21 and 23 h after infection. Late in the infection cycle, viral DNA is present in the host cell nucleus.

Protein synthesis

Yabapox viral proteins are synthesized at different times after infection and can be grouped into two classes, early and late. Early proteins are synthesized before the onset of viral DNA replication (3 h postinfection). Some of the proteins in this group are structural and continue to be synthesized in the presence of DNA inhibitor. Late viral proteins are detected at 6 h postinfection and continue to increase in number during the infection period. Viral infection does not inhibit host protein synthesis.

Virus synthesis

Yabapox virus is synthesized in the cytoplasm. At 35°C, the minimum length of replicative cycle is 35 h; however, maximum virus yields are not obtained until 75 h postinfection. Synthesis of at least two viral structural antigens occurs in the presence of the DNA inhibitor, cytosine arabinofuranoside, indicating potential transcription and translation of these antigens from parental DNA. The first progeny DNA is completed after 20 h postinfection, but is not detected in infectious form until 35 h postinfection. The maximum rate of progeny DNA synthesis occurs between 20 and 30 h postinfection. Viral DNA synthesis continues until 45–50 h after infection.

Morphogenesis

Within 3 h postinfection, the adsorption and phagocytosis of Yabapox virus particles by the cells can be seen by electron microscopy. This is followed by the disruption of the phagocytic vacuole membrane, with the release of viral DNA into the cytoplasm. At 24 h postinfection, large cytoplasmic inclusions termed 'factories' are observed. A typical factory contains a large number of viral particles, particulate glycogen, DNA-containing electron dense material, and small membranous spherical structures (40 nm in diameter) designated as 'micelles'.

Transformation

Various cell lines of monkey kidney cells have been transformed with UV-irradiated Yabapox virus. Two different cloned restriction enzyme fragments can also transform CV-1 monkey kidney cells. Transformed cells do not produce virus but exhibit biological characteristics typical of transformed cells. These include increased saturation density, reduced serum requirements for growth, and ability to grow in soft agar. The morphological alterations of transformed cells are similar to Yabapox virus-induced tumor cells and are characterized by loss of contact inhibition, multinucleated cells, and cytoplasmic lipid droplets. Southern blot hybridization showed that sequences homologous to low molecular weight viral DNA (5.1, 4.8 and 3.9 kbp) are present in transformed cells. Virus-specific antigens detected by immunofluorescence assays are found in the cytoplasm of transformed cells. Four virus-specific proteins, with molecular masses of 160, 140, 108 and 74 kD, are contained in transformed cells immunoprecipitated with sera from tumor-bearing monkeys.

Serological Relationships and Variability

Serologically Tanapox and Yabapox viruses show minimal to moderate cross-reactivity. The presence of type-common and type-specific antigens has been demonstrated. Yabapox viral infection fully protects primates against Tanapox virus challenges. In monkeys infected with Tanapox virus then challenged with Yabapox virus, symptoms were delayed in developing, and histiocytomas were smaller.

Complement fixing (CF) and complement fixing inhibiting (CFI) antibodies were demonstrated in the clinical and convalescent stages respectively of

Rhesus monkeys infected with either Tanapox or Yabapox viruses. The persistence of CF antibody in monkeys infected with Tanapox virus is from 10 to 12 weeks; in monkeys infected with Yabapox virus CF antibody persists up to 35 weeks postinfection.

Epidemiology

The epidemiology of Tanapox virus is believed to be a reservoir of monkeys in the wild in Kenya from which the natives of the Tana River Valley are occasionally infected as a result of mosquito transmission of the virus. An outbreak occurred in monkey colonies in California, Oregon and Texas in 1965 and 1966 and spread to men in contact with the housed diseased animals. Tanapox of man is essentially a zoonosis. Yabapox is believed to be epidemic in scope in African and Asian monkeys. Infection in man occurs only through injection of the virus.

Transmission and Tissue Tropism

Yabapox and Tanapox viruses are both believed to be transmitted by insect vectors. Yabapox virus transforms fibrocytes of the dermis and subcutaneous cells to pleomorphic polygonal cells. The histiocyte gives rise to the tumor. In monkeys inoculated with Yabapox virus, histiocytes migrate into the infected area by 48 h postinoculation. After 3–5 days, the histiocytes undergo striking morphologic alterations and proliferate rapidly, leading to tumor formation. Intravenous inoculation results in many tumors in the heart, lungs, muscles and subcutaneous tissues of susceptible monkeys.

Tanapox viral infection is histiologically distinct from Yabapox. It affects the epidermis almost exclusively, resulting in hypertrophy and thickening of the epithelial layers of the skin with swelling and ballooning of the deeper epithelial cells; there is little cellular infiltration into the underlying dermis.

Pathogenicity

Because of the lack of studies on the prevalence of either Yabapox or Tanapox disease in the population of monkeys in the wild, little is know about pathogenicity. However, both viruses are of low pathogenicity in housed monkey populations. Pathogenicity from monkeys to humans is probably nonexistent in the case of both viruses in that each depends upon a vector or artificial means of transmission.

Clinical Features of Infection

In humans, Tanapox starts with a short febrile illness lasting 3–4 days and sometimes a severe headache, backache and pronounced prostration also occur. During the course of the febrile illness, one or two pock-like lesions appear on the upper part of the body, usually on the upper arm, face, neck and trunk. The lesion resembles a modified smallpox lesion in a vaccinated individual, except there is no pustulation in the Tanapox lesion. The illness is nonfatal and of short duration.

In monkeys, Yabapox is characterized by tumors on the hairless areas of the face, on the palms and interdigital areas and on the mucosal surfaces of the nostrils, sinuses, lips and palate. The benign tumors develop 5 days after inoculation, grow to 25–45 mm in diameter and project up to 25 mm in diameter. Tumor growth proceeds steadily, reaching a maximum in 6 weeks, after which regression occurs and is completed by 12 weeks postinoculation.

Pathology and Histopathology

In Yabapox infection tumor cells which develop are characterized by the appearance of multinucleated cells, cytoplasmic granulation, nuclear enlargement, nucleolar hypertrophy and the formulation of numerous lipid vacuoles in the cytoplasm. Granular inclusions in the cytoplasm stain positively for DNA with acridine orange.

Tanapox virus infection is characterized by hypertrophy and thickening of the epithelial layers with swelling and ballooning of the deeper epithelial cells, which show vacuolation of cell nuclei and eosinophilic cytoplasmic inclusions.

Immune Response

In Yabapox, circulating neutralizing antibody is ineffective in preventing growth of established tumors. Immunity to superinfection is present when tumors are present or regressing, but after total regression of tumors, re-infection results in new tumor formation.

In Tanapox infection immunity persists in monkeys for at least nine months following healing of the skin lesions.

Prevention and Control

Prevention of Tanapox in the human population is thought to be controllable by controlling the mosquito population and avoidance of mosquito-infested areas where the disease is epidemic in monkeys. Control of both diseases in housed monkeys has been accomplished by control of insects and by strict isolation of infected animals.

Future Perspectives

The control of Tanapox in the human population does to present a serious problem because of the highly isolated and rare epidemiology of the disease and its non-serious nature. More studies of wild monkey populations where the disease is epidemic are needed in order to determine for certain how the disease is spread. If both diseases are spread by mosquitoes as postulated, complete control is unlikely to be achieved.

See also: Smallpox and monkeypox viruses; Zoonoses.

Further Reading

Knight C *et al.* (1989) Studies on Tanapox virus. *Virology* 172: 116.
Rouhandeh H (1988) Yaba Virus. In: Darai G (ed.) *Virus Diseases in Laboratory and Captive Animals.* Boston: Martinus Nijhoff Publishers.

YEAST RNA VIRUSES

Reed B Wickner
National Institute of Diabetes, Digestive and Kidney Diseases
Bethesda, Maryland, USA

History

Some strains of the yeast *Saccharomyces cerevisiae* secrete a protein toxin that kills other strains, but to which the secreting strain is itself immune. Toxin production ability is inherited as a non-Mendelian trait. This finding (in 1963) by Makower and Bevan led to the discovery of viral particles containing various double-stranded RNAs (dsRNA), one of which, called M, was correlated with the killer phenomenon. Studies of this area have taken two directions, one concerned with the virology of the system, and the other with the processing, secretion and action of the toxin and the mechanism of immunity.

Yeast RNA Replicons

Four completely distinct RNA to RNA replication systems have been found in various strains of *S. cerevisiae*: L-A and its satellites (including M), L-BC, 20S RNA and T dsRNA. The L-A and L-BC systems are viral, while the 20S RNA and T dsRNAs apparently replicate naked in the cells. None of the viruses are known to have a natural extracellular infectious cycle, although they can be introduced into spheroplasts along with transforming DNA plasmids. All are efficiently transmitted from cell-to-cell by the cytoplasmic mixing that occurs on mating. Perhaps because mating is a very frequent event for yeasts in nature, these viruses are widespread and most strains have more than one. The best studied is the L-A virus and its satellite viruses, M_1, M_2, M_3, and M_{28}, encoding different killer toxins and immunity functions. Defective interfering deletion mutants of L-A and M_1 have been described. Yeast strains also contain at least four families of retroviruses. Yeast RNA replicons are summarized in Table 1.

Table 1. RNA replicons in *Saccharomyces cerevisiae*

dsRNA viruses	kb	Proteins encoded	Comments
L-A	4.6	76 kD major coat protein = *gag* 170 kD Gag-Pol fusion protein	Single segment
M_1, M_2, M_{28} ...	1.0–1.8	preprotoxin-immunity protein	Satellites of L-A
L-BC	4.6	?	Single segment
ssRNA Replicons			
20S RNA	2.8	90 kD protein-homologous to RNA-dependent RNA polymerases	W dsRNA is RF; circular form seen
23S RNA	3.2	protein-homologous to RNA-dependent RNA polymerases	T dsRNA is RF

L-A Replication Cycle

L-A is a dsRNA virus that has a single 4.6 kb segment, and viral particles contain a single molecule of L-A dsRNA. The viral particles have a 76 kD major coat protein (Gag) and a minor 170 kD protein (Gag-Pol) whose N-terminal portion is a Gag monomer and whose C-terminal portion binds single-stranded RNA and is the RNA polymerase. A conservative transcriptase activity of the particles produces viral plus-strands and extrudes them from the particles. The plus ssRNA is translated to form viral particle proteins (see below) and is also the species packaged to form new viral particles. These particles then make the minus-strand on the plus-strand template to form L-A dsRNA-containing particles and complete the cycle (Fig. 1).

Satellite viruses of L-A, such as M_1, use the L-A-encoded coat proteins. Again, a single viral plus-strand is packaged to form new particles, but after replication and transcription, the new transcripts often remain inside the viral particle where a second (or more) dsRNA molecule is formed by a second replication event. This is called 'headful replication' to distinguish it from the 'headful packaging' phenomena seen in many bacteriophages. In the case of L-A and its satellites, a single viral plus-strand is packaged and then it replicates inside the particle until the head is full. Then all new transcripts are extruded from the particle.

The *cis* signals responsible for packaging and replication have been defined on L-A and M_1 (Fig. 2). The packaging proteins recognize a stem–loop structure located about 400 nucleotides from the 3′ end of either L-A or M_1. The sequence of the stem is not important, only that it is a stem, but an A residue that protrudes on the 5′ side of the stem is critical, as is the loop sequence. The replication reaction requires that the template has both an internal site that largely overlaps with the packaging site (and may be identical to it) and a specific 3′-terminal sequence and structure.

L-A Expression: Ribosomal Frameshifting in a dsRNA Virus

L-A plus-strands encode the 76 kD viral major coat protein (called Gag in analogy with retroviruses) and a 170 kD Gag-Pol fusion protein, formed, as in the case of retroviruses, by a -1 ribosomal frameshift (Fig. 3). The structure responsible for the frameshifting is a slippery site, GGGUUUA, followed by an RNA pseudoknot, both located in the region of overlap of the *gag* and *pol* open reading frames. The efficiency of frameshifting appears to be critical for viral propagation, suggesting that the balance of Gag and Gag-Pol fusion proteins may be important in the assembly process. Since no eucaryotic cellular gene is known to use ribosomal frameshifting as part of its expression strategy, these results suggest that drugs affecting this process may be useful for antiviral therapy.

Why do L-A, retroviruses and a number of plus ssRNA viruses use ribosomal frameshifting and readthrough of termination codons to make Gag-Pol and other fusion proteins? dsRNA viruses, plus ssRNA viruses and retroviruses all use their plus-strands as mRNA, as the form of the genome that is packaged to make new particles and as a template for replication. If they were to use splicing or RNA editing for the purpose of fusing open reading frames, they would be generating mutants – unless

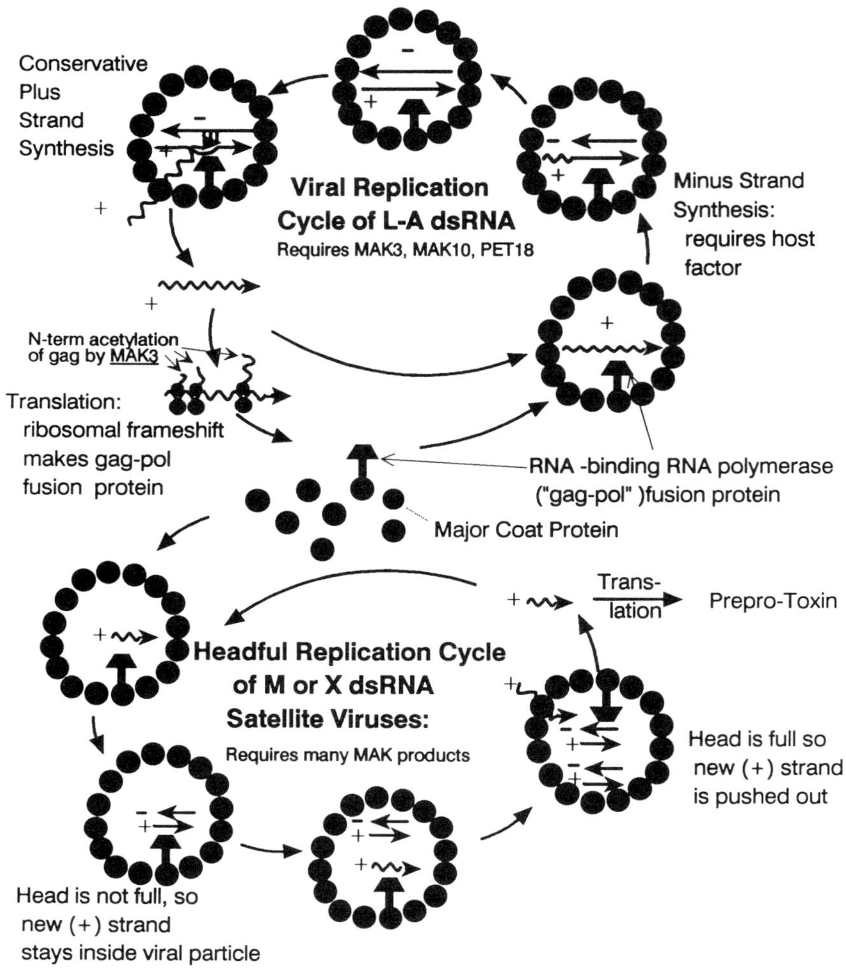

Fig. 1 Replication cycles of L-A and its satellites.

the spliced out region contained a site essential for packaging or replicating the genome. Retroviruses often splice, but they remove the packaging site in so doing so that the spliced mRNAs will not be packaged. Insofar as is known, plus ssRNA viruses and dsRNA viruses do not splice, perhaps for this reason. Minus ssRNA viruses do splice and edit their RNAs.

Host Functions

Over 40 chromosomal genes affecting the replication and expression of the L-A or M dsRNA genomes in yeast have been defined and a single gene is needed for L-BC (Table 2). Chromosomal (host) mutants that lose the M_1 genome are generally called *mak* mutants (for maintenance of killer). Among the 30 genes of this type are three, *MAK3, MAK10*, and *PET18* which are necessary for the propagation of L-A. *MAK3* is now known to encode an *N* acetyltransferase responsible for acetylating the N terminus of the major coat protein (Gag) encoded by L-A. This acetylation appears to be necessary for the assembly of viral particles. This parallels the N terminal myristylation necessary for proper assembly of poliovirus and proper localization of retrovirus Gag protein assembly.

MAK1 encodes DNA topoisomerase I, *MAK8* is the gene for ribosomal protein L3, *MAK11* encodes an essential membrane-associated protein, and *MAK16* is necessary for progression through the G1 phase of the cell cycle. But in none of these cases is the role of these genes in viral propagation understood.

A set of 6 genes, called SKI genes (for superkiller, the phenotype of the mutants), constitute an intracellular antiviral system. Deletion of any of the three of these genes that have been cloned produces

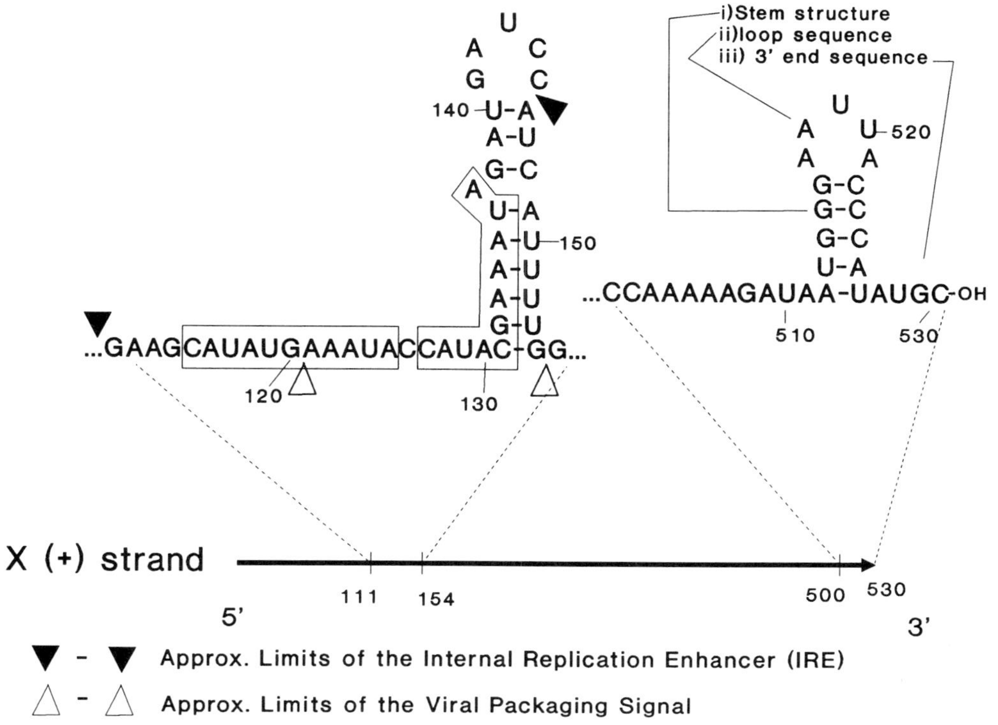

Fig. 2 Packaging and replication signals on L-A plus-single stranded RNAs. X is a deletion mutant of L-A that was used to analyze replication template requirements and the packaging signal. All but X's 5′ 25 nucleotides come from the 3′ end of L-A. The 3′ end site necessary for replication template activity has three known components, shown in the upper right. The internal replication enhancer is necessary as well for optimal template activity.

no discernible phenotype unless the M satellite virus is present. Then, however, the cells are cold-sensitive, temperature-sensitive or unable to grow at any temperature, depending on the presence of other non-chromosomal factors. The *SKI* system acts on L-A, M, L-BC and 20S RNA (see below), to lower their copy numbers in the cell. The *SKI3* protein is a 163 kD nuclear protein, but its function, and that of the other *SKI* proteins remains unknown.

A distinct system, affecting only L-A, is defined by mutations in the mitochondrial outer membrane porin and the major mitochondrial nuclease (*NUC1*). Growth on non-fermentable carbon sources is also known to elevate the copy number of L-A, indicating that a complex relationship exists between these viruses and mitochondrial functions.

Killer Toxin Processing and Mammalian Prohormone Processing

The killer toxin is encoded by M_1 as a precursor protein which apparently gives immunity to cells carrying M_1. The toxin has two subunits which, in close analogy to insulin, are processed from the preproprotein by removal of a signal sequence and a peptide between the subunits (Fig. 4). This analogy with mammalian prohormone processing is more than superficial. The *kex1* and *kex2* mutations were originally isolated based on their inability to secrete K_1 killer toxin and the yeast α mating pheromone. The *KEX2* product was then found to be a protease cleaving specifically C-terminal pairs of basic residues, and *KEX1* is another protease that removes the two basic residues. This is just like the processing of insulin, pro-opiomelanocortin and other mammalian hormones. Indeed, using sequence information from *KEX2* or functional complementation of *kex2* mutants, several mammalian protease genes have been isolated which are candidates for the physiological prohormone processing enzymes. Moreover, the yeast *KEX2* gene can substitute for the mammalian proteases in processing prohormones.

Toxin Action

The K_1 killer toxin must bind first to $(1 \rightarrow 6)$ β-D-

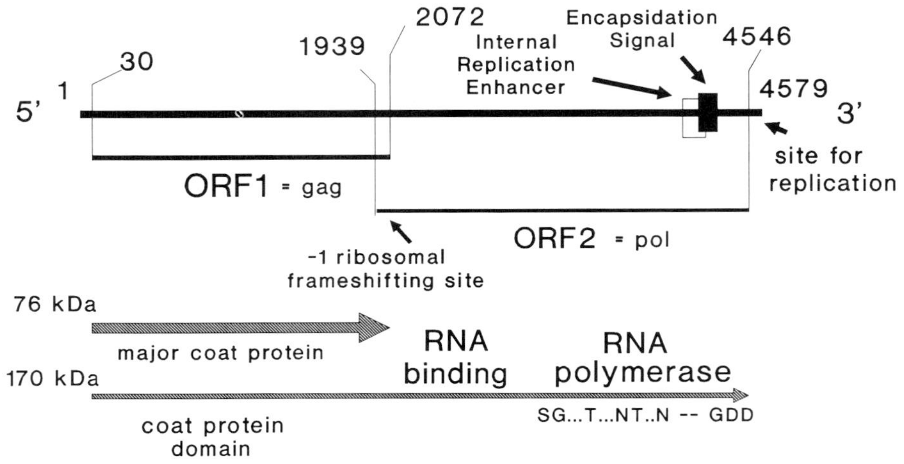

Fig. 3 Expression of L-A-encoded information. The two L–A ORFs encode a chimeric RNA polymerase–RNA binding protein with a major coat protein domain. ORF1 (Gag) is the major coat protein, while ORF2 (Pol) encodes a region with the consensus sequence patterns for the RNA-dependent RNA polymerases of plus ssRNA viruses and dsRNA viruses. Pol also has a ssRNA binding activity that is thought to play a role in the assembly process.

glucan, a major structural component of yeast cell walls. The K_{28} toxin binds first to cell wall mannans. However, since spheroplasts remain sensitive to these toxins, there must be a receptor further downstream in the process. The K_1 toxin acts by creating proton pores in the membrane.

20S RNA Replicon and T and W dsRNA

20S RNA was discovered in 1971 as a species whose synthesis is induced when cells are transferred to the media that are used to induce meiosis and sporulation in yeast (potassium acetate medium). Subsequent studies showed that some strains were unable to produce 20S RNA under these conditions, and that there was no role for 20S RNA in sporulation. Thus, some strains that could sporulate made no 20S RNA and some that could not sporulate could make 20S RNA. It was also shown that the ability to induce 20S RNA synthesis was inherited as a non-Mendelian factor.

Recently, most of 20S RNA has been cloned and sequenced. 20S RNA is not encoded by any cellular DNA (chromosomal or otherwise). It replicates via an RNA to RNA mechanism. Most remarkably, electron microscopy of 20S RNA coated with T4 gene 32 protein and cross-linked with glutaraldehyde showed mostly circles! That the molecule was circular was also suggested by the inability to label either the 3′ or 5′ ends of isolated molecules and the aberrant migration of 20S RNA on gel electrophoresis. Dimer length molecules of both plus and minus polarity have also been detected *in vivo*, consistent with a rolling circle mode of replication. However, the inability to clone all of the way around the 20S RNA circle has suggested that there may be an unusual structure at the uncloneable site, or a fundamentally linear molecule whose ends are held together by some structure.

Two low copy number dsRNAs, called T and W, were described in 1984. Neither was homologous to

Table 2. Chromosomal genes affecting dsRNA viral replication

Name	#	M_1 replication	L-A replication	L-BC replication	20S RNA
MAK3, 10, PET18	3	needed	needed	–	–
Other *MAK* genes	> 20	needed	–	–	–
SKI	6	represses	represses	represses	represses
porin, NUC1	2	–	represses	??	??
CLO1	1	–	–	needed	??

– = no effect

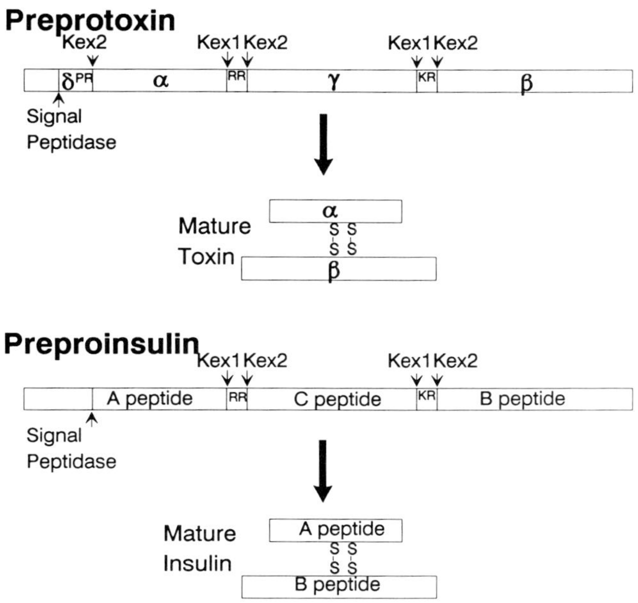

Fig. 4 Preprotoxin processing resembles mammalian hormone processing. The two subunits of both insulin and killer toxin are cleaved out of their precursor hormones by Kex-like proteases. These enzymes were first discovered in the killer system and this led to their discovery in mammalian cells.

cellular DNAs and the copy number of both was induced by growth at elevated temperatures. W was recently cloned and sequenced and was found to be identical to 20S RNA. Indeed, 20S RNA copy number is also induced at high temperature. Since the copy number of 20S RNA is always 10-fold or more that of W, W may be viewed as a replicative form of 20S RNA. Their sequence shows a single long open reading frame encoding a protein of about 90 kD with some homology to the RNA-dependent RNA polymerases of plus-strand RNA viruses.

Applications

Since killer strains kill nonkiller strains when co-cultivated, brewing strains have been modified to include a killer virus so that strains present in the grapes (barley, hops, etc.) will not spoil the brewing process. This application is in actual use today. Vectors based on the toxin secretion signals have also been developed.

Future Perspectives

The ease with which yeast is manipulated genetically and the high yields of L-A virus that can be easily obtained from engineered strains have made possible extensive characterization of the genetics of this virus and its satellites and their relationship with their host. The development of *in vitro* replication, packaging and transcription systems is unique among the dsRNA viruses of eucaryotes, and few systems have been developed for any RNA viruses. It is likely that the mechanisms found for L-A will not be unique to this system. Among the important problems to be solved are (1) the mechanisms by which the antiviral system acts and the virus uses host genes to oppose it, (2) development of an RNA transfection system so that the L-A virus may be used as a vector, (3) further pursuit of the mechanisms of transcription, replication and packaging and application of this information to viruses of higher systems, and (4) study of the role of various host components in ribosomal frameshifting and search for pharmacologic agents affecting this process – a new approach to antivirals.

The 20S RNA system (and the related T dsRNA system) have only begun to be studied. The question of the nature of the circularity of 20S RNA and the mechanism by which its propagation is controlled, including the massive 10 000-fold amplification on acetate medium, are of great interest. If it is a true circle of RNA, it is likely to self-cleave and self-ligate as do related viroid systems.

See also: Furoviruses; Partitiviruses; Retroviruses – Type D; Totiviruses; Viroids.

Further Reading

Bussey H (1988) Proteases and the processing of precursors to secreted proteins in yeast. *Yeast* 4: 17.

Fuller RS, Stearne RE and Thorner J (1988) Enzymes required for yeast prohormone processing. *Annu. Rev. Physiol.* 50: 345.

Wickner RB (1989) Yeast virology. *FASEB J.* 3: 2257.

Wickner RB (1991) Yeast RNA virology: the killer systems. In: Broach J, Jones E and Pringle J (eds) *The Molecular and Cellular Biology of the Yeast Saccharomyces* p. 263.

YELLOW FEVER VIRUS

Thomas P Monath
Oravax, Inc.
Cambridge, Massachusetts, USA

History

Yellow fever was first described as a disease entity in 1648 in Mexico. The origins of the disease are in doubt, but the susceptibility of New World – but not African – monkey species to lethal infection (suggesting contact with the virus in relatively recent times) indicates an African origin of the virus. Whether or not yellow fever in its enzootic form (transmitted between monkeys and sylvan mosquito species) predated the Spanish Conquest in tropical America, it was the slave trade that led to the introduction from Africa of the domestic mosquito vector, *Aedes aegypti* and the emergence of epidemic (urban) yellow fever. Yellow fever was one of the great scourges of mankind during the 18th and 19th centuries, with epidemics affecting coastal cities in the Americas, Europe, and West Africa. The mode of spread of the disease was the subject of great debate. Mosquito transmission was suggested as early as 1848, was emphasized by the Cuban physician, Carlos Finlay in 1881, and finally was proven by Walter Reed and his colleagues in 1900. Reed *et al.* also demonstrated that yellow fever was a filterable 'virus', although the etiology of the disease remained in dispute until isolation of the virus in 1927. In the decade that followed, quantitative virological and serological methods were established, allowing precise diagnostic, epidemiological and pathogenesis studies as well as the development and evaluation of live, attenuated vaccines.

Taxonomy and Classification

Yellow fever virus is the prototype member of the family *Flaviviridae*, genus *Flavivirus*, and the virus after which the family and genus were named (*flavus*, Lat. yellow). The genus consists of at least 68 viruses grouped on the basis of shared antigenic determinants and physicochemical properties (see below). The relatively specific neutralization test using polyclonal antisera has been used to distinguish at least eight antigenic complexes to which closely-related flaviviruses are assigned. Neutralization epitopes of yellow fever virus are sufficiently different from other flaviviruses to preclude assignment to a complex.

Properties of the Virion

Virions are spherical, 40–50 nm in diameter, and have short surface projections. The nucleocapsid has icosahedral symmetry, contains the RNA genome and a single core protein, and is surrounded by a lipid bilayer envelope. Approximately 66% of the virion is composed of protein, 12% carbohydrate, and 17% lipid. Viral infectivity is rapidly inactivated by heat (56°C for 30 min), ultraviolet radiation, and lipid detergents.

Morphogenesis of yellow fever virus occurs at intracellular (endoplasmic reticular, ER) membranes. Mature virions accumulate in these membranes, are transported to the cell surface, and released by exocytosis. Budding of virions into the lumen of the ER is rarely observed and may be an extremely rapid event. In secretory cells (such as the mosquito salivary gland and a variety of exocrine and endocrine glands infected in mammals), virion assembly occurs in concert with host cell secretory components and virions are released in secretory granules. Flavivirus infection often proceeds without markedly disturbing host cell function or macromolecular synthesis.

Properties of the Genome

The viral genome (17D vaccine strain) is composed of a single linear strand of infectious (positive polarity) RNA, 10 862 nucleotides in length, with a mass of 3750 kD. The 5′ terminus has a type 1 cap, and the 3′ terminus is not polyadenylated. The GC content is 49.7%. The viral genome contains a single long open reading frame 10 233 nucleotides in length, encoding 10 viral proteins (Fig. 1). The remainder of the genome comprises short 5′ (118 nucleotides) and 3′ (511 nucleotides) noncoding regions. Nucleotide conservation is seen between the 5′ terminus of the plus strand and the 3′ terminus of the minus strand, serving as common recognition sequences for the viral polymerase. The 3′ terminus forms a stable hairpin loop involved in binding to the capsid protein.

Viral Proteins

The three virion structural proteins are encoded by the 5′ one-fourth of the genome, in the sequence C (capsid), prM [precursor to the mature membrane (M) protein], and E (envelope) (Fig. 1). The remainder of the open reading frame encodes seven nonstructural proteins, in the order illustrated. Translation begins at the 5′ terminus of the viral genome, and the individual proteins are produced after translation by a series of enzymatic cleavages. The C protein (mol. wt 12–14 kD) interacts with RNA in the virion nucleocapsid; its hydrophobic C terminus anchors nucleocapsids to ER membranes, and provides signal sequence for prM. The prM glycoprotein (mol. wt 18–19 kD), present intracellularly, is cleaved at the time of virus maturation to form M in the extracellular virion. The M protein spans the viral membrane, and has exposed antigenic domains that may play a minor role in the induction of protective immunity. However, the E glycoprotein (mol. wt 53–54 kD) forms the structural spike protruding from the virion surface and subserves many biological functions, including cell attachment, hemagglutination, and neutralization. Important to these functions is the three-dimensional configuration of E protein, determined by disulfide bonding. Epitopes with strain-specific, type-specific and flavivirus group-reactive specificities are present in the E glycoprotein. As is the case for similar proteins, the C terminus contains hydrophobic sequences forming the protein anchor in the lipid membrane. The nonstructural NS1 glycoprotein is found both within infected cells and extracellularly (as circulating antigen with complement-fixing properties). Expressed on cell surface, NS1 is a target for antibodies involved in clearance of virus infection. The functions of other nonstructural proteins are incompletely understood. The large NS3 protein (mol. wt 68–70 kD) is believed to have both helicase and protease functions. NS5 (mol. wt 103–105) is the RNA polymerase. Because of their replicase functions, the amino acid sequence structures of NS3 and NS5 are highly conserved among flaviviruses.

Replication

After gaining entry to the cell, genomic, plus-strand RNA is translated and RNA replication proceeds by synthesis of complementary minus strands. New plus strands are transcribed from genome-length minus-strand templates. Duplex RNA molecules (replicative forms) are thus formed in the infected cell. New genome-length plus strands serve as mRNAs for translation of structural and nonstructural proteins, as templates for transcription of new minus strand for replication, and as genomes for inclusion into nucleocapsids for production of mature virus particles.

Variation and Evolution

Flaviviruses are unusual in having a high proportion of purine nucleotides and a low frequency of CG and UA dinucleotides. Although yellow fever and many flaviviruses are adapted to replicate in mosquitoes and ticks, a low CG content appears to favor translation of mRNA in mammalian hosts. It is interesting to note that a number of flaviviruses are highly

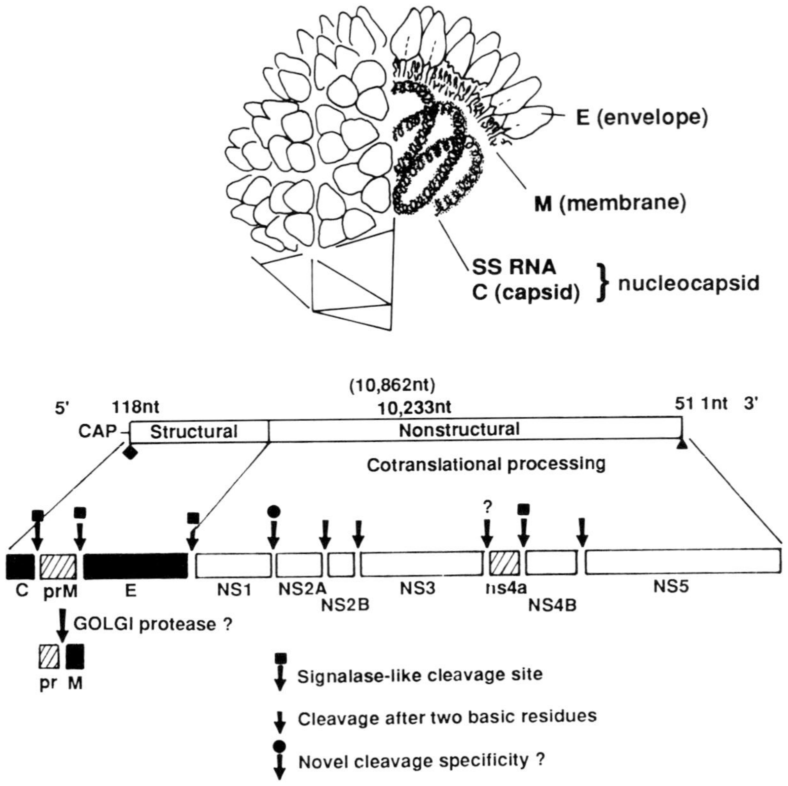

Fig. 1 Structure of the yellow fever (flavivirus) virion and genome, showing the structural and nonstructural protein coding regions. The genome is characterized by type I 5′ cap structure, followed by a short noncoding region and a long open reading frame of over 10 000 nucleotides encoding the structural (C-prM-E) and nonstructural (NS1–NS5) proteins. The genome lacks a 3′ poly-A tail.

adapted for replication in mammals [bats (e.g. Rio Bravo, Montana Myotis leukoencephalitis, Dakar bat viruses) or rodents (e.g. Modoc, Jutiapa, Cowbone Ridge viruses)] and are transmitted directly between these hosts without intervening arthropod vectors. It is possible that flaviviruses arose in mammals and subverted blood-feeding arthropods for transmission at a later stage in their evolution. It is interesting to note that the overall genome organization of the flaviviruses resembles that of the picornaviruses, which are principally viruses of mammals.

Yellow fever virus is distantly related to other flaviviruses by serologic tests and in terms of homology in RNA sequences. Strains of yellow fever virus from different geographic regions can be distinguished by RNA fingerprinting and are presently classified into 4 groups ('topotypes'), 2 from West Africa, one from Central and East Africa, and one from South America. A high degree of homology has been noted between strains belonging to one topotype.

The 17D vaccine strain of yellow fever and its virulent parent have been compared at the sequence level. These viruses are separated by over 240 passages and were found to differ at 68 nucleotides (32 amino acids). The specific changes responsible for attenuation have not yet been determined.

Host Range and Virus Propagation

Viral determinants involved in cell attachment have not been defined, but are presumably conserved sequences located on the envelope (E) glycoprotein spike. Cell membrane receptors also remain to be elucidated. Because of the broad host range of the flaviviruses, cell receptors may be molecules with conserved structure across the chordate and arthropod phyla. Yellow fever and other flaviviruses enter cells by typical receptor-mediated endocytosis.

Yellow fever virus replicates and produces cytopathic effects and plaque formation in a wide variety of cell cultures, including primary chick and duck embryo cells; continuous porcine, hamster, rabbit, and monkey kidney cell lines; and cells of human origin (e.g., HeLa, KB, Chang liver, SW-13 cells). The virus replicates in Fc-receptor bearing macrophages and macrophage cell lines, and replication is enhanced by antibody. Mosquito cell lines, especially *Ae. pseudoscutellaris* (AP-61) cells, are highly susceptible and are often used for primary isolation or efficient laboratory propagation of virus.

The most sensitive method for virus isolation and assay is the intrathoracic inoculation of mosquitoes, such as *Toxorhynchites* spp. or *Aedes aegypti*. Infected mosquitoes show no signs of illness, and the presence of virus must be demonstrated by immunofluorescence or subpassage to a susceptible host. Infant mice succumb within 6–8 days to encephalitis; at about 8 days of age, mice become resistant to lethal infection by the peripheral route, while remaining susceptible to intracerebral challenge. After parenteral virus infection, a number of sub-human primate species develop fatal hepatitis resembling the human disease. The only non-primate species that develops lethal hepatitis in response to yellow fever infection is the European hedgehog, *Erinaceus europaeus*. Antibodies have been found in a wide variety of wild vertebrates collected in the field. Wild animals have also been experimentally infected with yellow fever virus, including rodents, bats, and marsupials. With the possible exception of opossums in South America, the data do not support a role for non-primate species in transmission cycles.

Geographic and Seasonal Distribution

Yellow fever virus presently occurs in tropical areas of South America and Africa. *Aedes aegypti*-infested regions of North and Central America, the Caribbean, and southern Europe were intermittently invaded by the disease until the early part of this century and are still considered at risk, should the virus be introduced. Despite the prevalence of *Ae. aegypti* in tropical Asia, yellow fever has never reached that continent, possibly because transmission in Africa and South America occurs in relatively inaccessible areas and because cross-protection is afforded by immunity to dengue viruses; dengue immunity is nearly universal in many parts of Asia. Asian populations of *Ae. aegypti* are relatively inefficient vectors; thus, vector competence may also provide a partial barrier to the spread of yellow fever in Asia.

The breeding of many mosquito species engaged

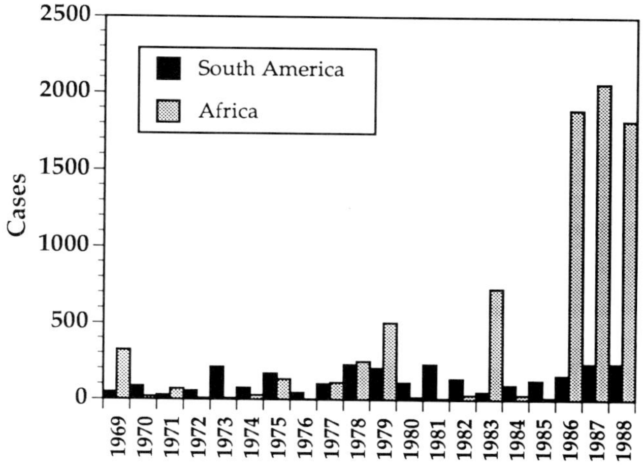

Fig. 2 Officially-reported incidence of yellow fever cases in South America and Africa. The recent upsurge in incidence is due to a series of severe epidemics in Nigeria. Official reports greatly underestimate the true morbidity due to this disease.

in yellow fever transmission occurs in tree-holes and is highly dependent on rainfall. Transmission waxes during the tropical rainy season and wanes or ceases during the dry season. Breeding of the domestic mosquito, *Ae. aegypti*, occurs in man-made containers used for water storage and is thus less influenced by seasonal rainfall; where this species becomes involved in virus transmission, outbreaks may occur during the dry season.

Disease Incidence

Official notifications underestimate the true incidence of the disease by 10- to 250-fold. The number of cases officially reported annually from South America and Africa is shown in Fig. 2, and their geographic distribution in Fig. 3. The incidence in Africa has increased dramatically in recent years, due principally to a series of epidemics in Nigeria. Some epidemics in Nigeria and elsewhere in Africa have been very large, involving 100 000 or more cases, with case-fatality rates of over 20%.

Epidemiology and Transmission

Yellow fever virus is present in the blood of infected, susceptible hosts (humans and sub-human primates) for several days, during which mosquito vectors

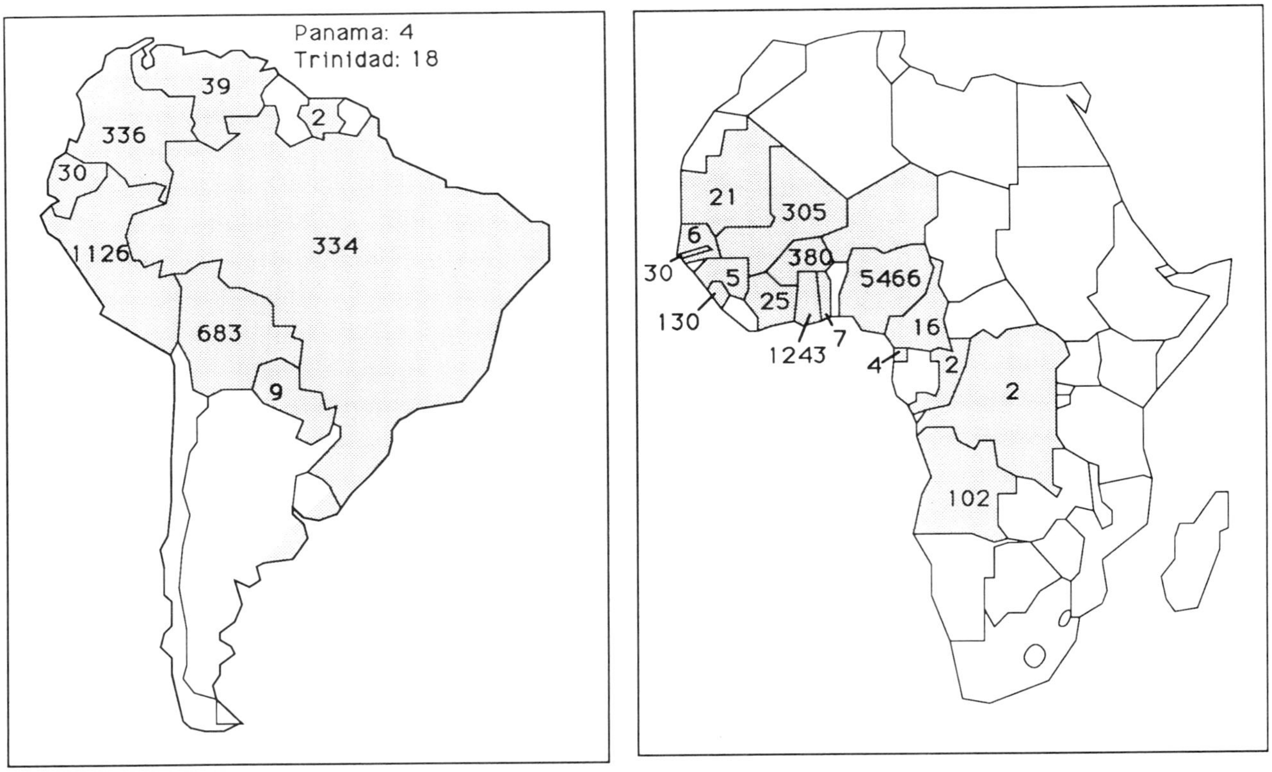

Fig. 3 Geographic distribution of yellow fever cases, 1970–1988.

taking a blood-meal may become infected. The virus undergoes sequential replication in the midgut epithelium, body, and salivary glands of the mosquito, a temperature-dependent process that takes a week or more to complete – the so-called extrinsic incubation period – before the vector is capable of transmitting virus by refeeding on a second host.

Two cycles of transmission are distinguished by the hosts and vectors involved. The urban cycle involves humans as the viremic host and *Aedes aegypti* mosquitoes, breeding in man-made containers. The forest cycle involves virus transmission between sub-human primates and tree-hole breeding mosquito vectors. The rate of transmission of yellow fever virus in the forest cycle can vary greatly, depending on the density of vectors and the available population of immunologically-susceptible monkeys. In South America, some primate species succumb to fatal disease, and die-offs of these animals may provide a clue to yellow fever activity; in Africa, monkeys generally develop subclinical infections. Humans exposed in the forest to infected tree-hole mosquitoes may acquire yellow fever. In South America, this form (so-called jungle yellow fever) accounts for all cases in the last 50 years. Jungle yellow fever strikes mainly young adult males engaged in timbering and agricultural pursuits in the Amazon and Orinoco river basins. The principal mosquito vectors are species of the genus *Haemagogus*. The disappearance in the early 1940's of urban yellow fever, once a major endemic and epidemic disease in South American towns and cities, was attributed to successful programs to control the domestic vector, *Ae. aegypti*. However, in the last 10–15 years, *Ae. aegypti* has reinvaded much of the territory in South America from which it had formerly been eliminated, raising the spectre of urban epidemics in the future. Although vaccine-induced immunity provides a degree of protection against such outbreaks, the level of coverage is presently incomplete.

The epidemiology of yellow fever in Africa is more complex than in the New World. The virus is present in a vast area of tropical Africa. Sporadic cases of jungle yellow fever occur in the rainforest zone, where the virus is maintained in a cycle involving monkeys and *Ae. africanus* mosquitoes, but such cases are rarely recognized because of inadequate surveillance. Unlike South America, tree-hole breeding *Aedes* spp., are also involved in epidemic transmission. Amplification of the virus transmission cycle occurs in moist savannah and forest-savannah transition zones, where the density of tree-hole *Aedes* reaches high levels during the rainy season. In these circumstances, humans are frequently infected. In urban areas or areas of low rainfall, where water storage practices favor breeding of domestic *Ae. aegypti*, this vector may be involved in epidemic spread, with humans serving as the viremic host.

Clinical Features

Subclinical or abortive infections occur more frequently than full-blown yellow fever. In its classical form, the disease begins suddenly, 3–6 days after the bite of an infected mosquito, with fever, headache, low back pain, muscle aches, loss of appetite and nausea. This period of infection is clinically nonspecific, and corresponds to the viremic phase when the patient is infectious for feeding mosquitoes. It may be followed by a brief period of remission, during which signs and symptoms abate. The patient then becomes increasingly ill, entering the period of intoxication, with reappearance of fever, vomiting, dehydration, abdominal pain, and the appearance of jaundice, protein in the urine, signs of renal failure, and hemorrhages, most notably the vomiting of blood. Virus is no longer present in the blood. Approximately 50% of patients who progress to this stage die between the 7th and 10th day after onset, with deepening renal and hepatic failure, shock, delirium, and convulsions.

Pathology and Pathogenesis

In humans, pathologic changes in the liver include swelling and necrosis of hepatocytes in the midzone of the liver lobule, with sparing of cells in the portal area and surrounding the central veins. Viral antigen and RNA are demonstrable by immunocytochemistry and nucleic acid hybridization in cells undergoing these pathologic changes, and cytopathology is mediated by direct viral injury. Inflammatory changes are absent or minimal, and patients with hepatitis who recover do not develop residual scarring or cirrhosis. The kidneys show acute tubular necrosis, probably the result of reduced perfusion of blood rather than direct viral injury. Focal degener-ation of muscle cells may be present in the heart. Spleen and lymph nodes show necrosis of B cell areas. The brain shows edema and petechial hemorrhages, but viral invasion and encephalitis are very rare events. Hemorrhage results principally from decreased synthesis of clotting factors by the liver. The mediators of hypotension and shock remain to be elucidated.

The pathogenesis of yellow fever has been studied in sub-human primates and mice, but our level of understanding is at the descriptive rather than the mechanistic level. Susceptible monkeys develop an illness similar to humans. Initial sites of virus replication have not been clearly defined, but probably include lymphatic tissue draining the site of virus inoculation. Viremia follows, with infection of the liver. Early virus replication occurs in fixed macrophages (Kuppfer cells) in the liver, which undergo necrosis. The virus then invades hepatocytes, which develop accelerated cytopathologic changes during the 24 hours before death. Yellow fever virus demonstrates both viscerotropism (infection of liver, lymphoid and other visceral tissues) and neurotropism (infection and inflammation of brain). In primates, viscerotropic infection is the rule; animals inoculated intracerebrally replicate virus in brain tissue and develop histopathologic evidence of encephalitis, but die of hepatitis. In contrast, mice develop brain infection without evidence of hepatitis.

Immune Response

Antibodies of both IgM and IgG subclass measurable by binding assays (immunofluorescence, ELISA), hemagglutination inhibition, and neutralization appear 5–7 days after onset of illness (8–14 days after infection). IgM antibodies tend to wane to low or undetectable levels between 6 and 12 months after infection, although, in one study, IgM neutralizing antibodies were still detectable several years after yellow fever immunization. Complement-fixing antibodies appear in the second week after infection, wane between 3–6 months later, provide a reliable indicator of recent wild-type yellow fever infection, and are rarely demonstrable after yellow fever vaccination. IgG neutralizing antibodies persist for at least 35 years, probably for life, and provide solid protection against reinfection. Antibodies to heterologous flaviviruses, such as Zika, dengue, and Wesselsbron viruses, provide partial cross-protection against yellow fever. Complement-dependent antibody-mediated cytolysis of cells with exposed NS1 protein sequences may play a role in virus clearance and recovery, as well as in protection in the previously-immunized host. Little is known about cell-mediated responses in yellow fever. Limited studies in mice suggest that nonspecific resistance is mediated by NK cells and macrophages, and that cytotoxic T cells are important in clearance of yellow fever infection.

Prevention and Control

The control of domestic *Ae. aegypti* mosquitoes is an important measure, but has proven difficult to sustain. The most effective approach to prevention is immunization of the human population in endemic areas. Yellow fever 17D is a live, attenuated vaccine produced in embryonated eggs. Over 200 million people have been immunized with this inexpensive product, which has proven safe and highly effective. Recent efforts have focused on the production of a new vaccine derived from a full-length cDNA clone, that generates infectious transcripts. By this approach, it may ultimately be possible to produce chimeric vaccines with heterologous cassette genes incorporated into the 17D yellow fever vaccine backbone.

See also: Dengue viruses; Pathogenesis; Replication of viruses; Zoonoses.

Further Reading

Chambers TJ, Hahn CS, Galler R and Rice CM (1990) Flavivirus genome organization, expression, and replication. *Annu. Rev. Microbiol.* 44: 649.

Monath TP (1986) Pathobiology of the flaviviruses. In: Schlesinger S and Schlesinger MJ (eds) *The Togaviridae and Flaviviridae*, p. 375. New York: Plenum Press.

Monath TP (1987) Yellow fever: a medically neglected disease. *Rev. Infect. Dis.* 9: 165.

Monath TP (ed.) (1989) Yellow fever. In: *The Arboviruses: Epidemiology and Ecology*. Vol. V., p. 139. Boca Raton: CRC Press.

Monath TP (1991) Yellow fever: Victor, Victoria? *Am. J. Trop. Med. Hyg.* 45: 1.

Rice CM *et al.* (1985) Nucleotide sequence of yellow fever virus: implications for flavivirus gene expression and evolution. *Science* 229: 726.

Z

ZOONOSES

Thomas M Yuill
University of Wisconsin
Madison, Wisconsin USA

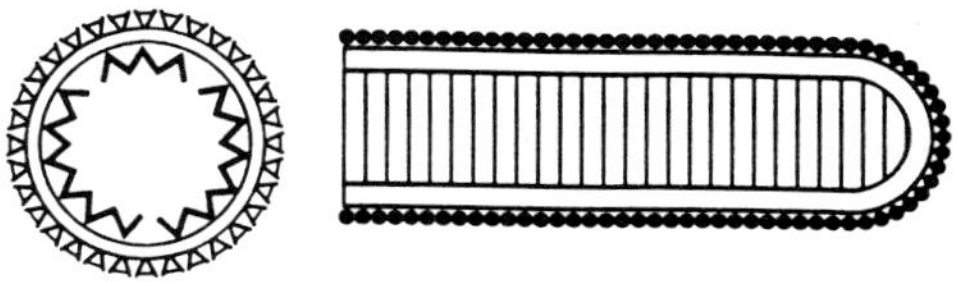

Introduction

Zoonoses are diseases transmissible from vertebrates, other than man, to humans. Hundreds of viruses in a wide range of families are zoonotic. Mammals, birds, reptiles and probably amphibians are reservoir or amplifier hosts for these viral zoonoses. Frequently, these viruses cause little or no overt disease in their nonhuman vertebrate hosts. Some zoonotic viruses have very limited host ranges, others may infect a wide range of vertebrates. Human infection may vary from inapparent to fatal disease. Some viral zoonoses have been recognized since ancient times, others have become public health problems recently. Rabies, because of its characteristic and violent symptoms, was recognized as a zoonotic disease in antiquity, and was described by the ancient Greeks. The hantaviruses, the causative agents of Korean hemorrhagic fever (HF) and other HF with renal syndrome, are still being isolated, characterized, and the nature and distribution of human disease is being defined. It is certain that many of the old viral zoonoses will continue to cause problems and new ones will be discovered in the coming years.

Transmission

Transmission of zoonotic viruses may occur by a variety of routes. Rabies is transmitted by *direct contact* through the bite of an infected animal. *Indirect contact* transmission of other zoonoses occurs by ingestion of contaminated food and water, as is the case with the hantaviral and arenaviral hemorrhagic fevers. *Nosocomial* arenavirus infection has been a problem in hospitals caring for patients where isolation has not been adequate. These hospital infections have been especially severe, probably due to exposure to a large dose of infecting virus. *Aerosol transmission* of Venezuelan equine encephalitis (VEE) virus has occasionally been a problem in the laboratory, where virus-bearing microdroplets have been widely circulated, with many resulting human cases. *Vertical* (*in utero*) transmission of arenaviruses results in persistent virus infection in the offspring. A large number of viral zoonoses throughout the world are *vector- or arthropod-borne*. Some of these viruses, such as yellow fever, may have maintenance cycles in wildlife, but cause massive epidemics, with human–vector–human transmission, in urban settings where peridomestic or domestic vector species are abundant.

Some Common Viral Zoonoses Around the World

Viral zoonotic diseases occur on every continent except, perhaps, Antarctica. Some are found around the world, in a variety of ecological settings. Others are found only in very limited ecologic and geographic foci. The panorama of viral zoonotic diseases is constantly changing. Although hundreds of viruses have been shown to infect both humans and animals, the importance of many of these viruses, as agents of either human or animal disease has not yet been established. New zoonotic viruses are being reported regularly from around the world. The distribution of well recognized zoonoses is changing, either because of actual shifts in their geography, or because of improved surveillance. Some of the more important viral zoonoses will be discussed briefly.

Widely Distributed Zoonoses

Rabies virus

Rabies is one of the oldest reported zoonoses. Rabies virus infection causes nervous system disease that ends in death. Animals can become infected without nervous system involvement or disease, develop antibodies, and survive, but play no role in transmission. Rabies is found all around the world except in Antarctica, Britain, the Hawaiian Islands, Australia and New Zealand. Transmission occurs by the bite of an infected mammal that has virus in its saliva. Aerosol (droplet) transmission to humans has been reported, although rarely. Dogs and cats are the main reservoirs, especially in the developing countries. In the industrialized countries, wild mammals are the main reservoir species; the species involved vary from region to region. The principal species are: in North America, skunks, raccoons and foxes; in Europe, foxes; and in the Caribbean, mongooses. Bats in all enzootic regions harbor rabies, with vampire bats especially important in the Neotropics. Rabies virus is classified in the *Lyssavirus* genus of the family *Rhabdoviridae*. Other rabies-related lyssaviruses, some zoonotic, are found in Africa. Diagnosis is based on characteristic altered behavior of infected mammals, confirmed by isolation of virus or demonstration of intracellular antigen by immunofluorescence. Postexposure treatment is accomplished by thorough washing of the bite wound, administration of hyperimmune serum or globulin, and administration of antirabies vaccine. Dogs and cats in enzootic areas should be vaccinated or, if strays, eliminated. Other domestic animals and humans at high risk should be also vaccinated. Some free-ranging wildlife species (especially foxes and raccoons) can be vaccinated by oral administration of new recombinant vaccinia vectored vaccines.

Korean hemorrhagic fever and hantaviruses

Hantaviruses are a newly-recognized complex of public health importance. Hantaan virus is the etiologic agent of Korean hemorrhagic fever, and occurs in east Asia, especially in Korea and China. Closely related rodent associated hantaviruses cause milder hemorrhagic fever with renal syndrome in humans and have been reported from Scandinavia (Puumala virus) and Asia as well as port areas around the world (Seoul virus). Seoul viruses have also been found in laboratory rats, with transmission to personnel there. Other hantaviruses are present in the US but have not been associated with human disease. Hantaviruses are harbored by wild rodents which often live in close association with humans. The similarity of many *Rattus*-associated viruses around the world suggests that movement of rats through commercial shipping has spread the virus as well. Virus is shed in urine and other excreta. These viruses belong to the *Hantavirus* genus of the *Bunyaviridae* family. Diagnosis has been complicated by the lack of efficient and sensitive isolation and serological methods, which explains why these 'new' viruses have only recently been isolated and characterized. Cell cultures and immunofluorescence are used to isolate or detect virus, and to test for antibody. Because some of these agents are highly pathogenic to humans, work with them must be carried out in facilities that provide a high degree of biocontainment. Rodent control is currently the only method available for prevention of transmission to humans.

Arenavirus hemorrhagic fever viruses

Arenavirus hemorrhagic fevers (HF), like the hantaviruses, are rodent associated, and can also cause severe disease in humans. Junin virus is the cause of Argentine HF, Machupo virus is the etiologic agent of Bolivian HF, and Lassa fever, caused by a virus of the same name, occurs in West Africa. A complex of related arenaviruses that do not cause human disease occur in South America. Like lymphocytic choriomeningitis virus, these pathogenic arenaviruses establish persistent infection in their rodent hosts, and virus is shed in excreta, infecting humans that live in close contact with these reservoirs. Lassa fever is also transmitted in rural hospitals to other people in contact with blood from viremic patients. Diagnosis is accomplished by virus isolation in cell culture, but detection of virus may require use of specific antibody and immunofluorescent or enzyme labeling. These techniques, as well as cDNA hybridization probes, are also used to detect viral products in tissues of infected humans or rodents. Diagnosis may also be established by a diagnostic rise of specific IgG, or detection of IgM antibodies. Virus isolation requires the use of biosafety level 4 facilities. Serological tests can be done with inactivated antigen at a lower level of biosafety. Control of these diseases is attempted mainly by rodent control, as was done successfully with Bolivian HF. Prevention of human disease by vaccination is possible. A live attenuated vaccine has been developed for Argentine HF, and a vaccinia-vectored vaccine has been developed for Lassa fever.

Yellow fever virus

Yellow fever (YF), first recognized clinically in the Neotropics in the sixteenth century, ravaged human populations in the Americas, Europe and Africa periodically until as recently as the beginning of this century. Infection causes hemorrhagic disease with severe liver damage and death in up to half of the most acute cases. Presently, the virus is maintained enzootically in Africa and the tropical Americas in primates and arboreal mosquitoes. Humans or primates transport the virus from its sylvan cycle in forested areas to rural or more urban areas, where other vector mosquitoes transmit it. YF may be a public health time bomb. There has been an alarming increase in human yellow fever cases in Africa in recent years. Although the number of 'jungle' yellow fever cases in the American tropics has not increased as dramatically, the recent establishment and spread of a potential YF mosquito vector, the Asian tiger mosquito (*Aedes albopictus*), in the hemisphere is a cause for real concern. YF virus is a flavivirus in the *Togaviridae* family. Diagnosis is by virus isolation in cell culture or in inoculated mosquitoes, or by demonstration of viral inclusion bodies or antigen in tissues taken at postmortem. Insecticide spraying can be used for vector control in epidemic situations. Disease can be prevented in humans by vaccination. Unfortunately, many countries have not been able to maintain an adequate level of immunity in populations in areas of risk. There is serious question about the ability of many national health services to provide adequate vaccination coverage quickly, in the face of a massive YF outbreak.

Encephalitis viruses

The tick-borne encephalitis (TBE) complex has broad geographic distribution. The complex contains 14 viruses, eight of which are zoonotic. In eastern Russia, Russian Spring–summer encephalitis has a case fatality rate of up to 20%. Central European encephalitis has a lower fatility rate. Nervous system sequelae commonly occur following both types. The virus is transmitted by *Ixodes* supp. ticks. Several species of rodents are the amplifying hosts, but infection of large animals occurs, also. Humans have been infected from drinking unpasteurized milk from infected cows. Human infection is most common in adults who frequent foci of transmission. TBE group agents are flaviviruses in the *Togaviridae*. Diganosis is by virus isolation or serological means. Prevention is through vaccination and avoidance of tick bites or of drinking raw milk. Tick control has not proven feasible.

Japanese encephalitis (JE) is found in a broad area from far eastern Russia, northeastern Asia through Southeast Asia and westward into India. Of the world's arthropod-borne encephalitides, JE causes the most number of clinical human cases, predominantly in children. It produces encephalitis in humans and horses, and acute febrile disease with abortion in swine, an amplifying host. Herons and egrets are wildlife amplifying hosts. The virus is transmitted by *Culex* spp. mosquitoes. The overwintering mechanism in temperate Asia is unknown. JE virus is a member of a complex of four related flaviviruses in the *Togaviridae* family. RNA oligonucleotide analysis indicates that there are differences in JE virus isolates from different vertebrate species, and between strains from different geographic areas. Diagnosis is by means of virus isolation or demonstration of IgG antibody rise, or of specific IgM in acute sera. Prevention of disease is mainly through vaccination of humans, horses and swine. A bivalent vaccine is used that incorporates the two recognized antigenic variants. Insecticides and integrated pest control measures that include natural compounds (*Bacillus thurengiensis* toxins) and larvacidal fish and larval habitat modification have been successfully used in China.

West Nile (WN) virus, a close relative of JE virus, occurs from India and Pakistan westward through the Middle East and into Africa, and northward into Europe and the republics of the former USSR. Mild infection to acute febrile disease with rash, and occasional encephalitis (mainly in the elderly) is produced in humans. *Culex* spp. are the main vectors, and birds are the vertebrate hosts. The virus has been isolated from ticks, also. Diagnosis is by virus isolation and serological means. Unlike the related flavivirus encephalitides, WN virus can be isolated from the blood of human patients relatively commonly.

Chikungunya virus

Chikungunya (CHIK) virus has been responsible for acute febrile disease with rash and severe arthralgia in people in Africa and Asia. The epidemiology in Africa is similar to yellow fever; CHIK virus is maintained in sylvan or savanna cycles involving wild primates and arboreal *Aedes* mosquitoes. In both Africa and Asia, the virus also has an urban cycle involving humans and *Aedes aegypti* mosquitoes. Virus can be recovered frequently from the blood of acute phase patients. The virus is an alphavirus of the *Togaviridae* family. The virus can be isolated in mammalian or mosquito cell cultures, although suckling mouse and mosquito inoculation are also sensitive systems. The virus has been transmitted to laboratory workers in the laboratory via aerosols created during virus isolation and passage. Serologic tests (neutralizing antibody or IgM capture ELISA test) are useful for diagnosis. Control of the sylvan and savanna enzootic cycles is not feasible. The urban cycle can be prevented or controlled by *Ae. aegypti* control measures. No vaccine is available.

Sindbis virus

Sindbis (SIN) is one of the most widely distributed arthropod-borne viruses in the world, being found in Africa, Europe, Asia and Australia. Disease in humans is usually mild, and is characterized by acute fever, with arthralgia and rash. SIN virus is maintained in wild bird populations, with transmission by *Culex* mosquitoes; in Africa and the Middle East, often in the same ecosystems where WN virus is being transmitted. The virus is an alphavirus of the *Togaviridae*. Oligonucleotide mapping and RNA hybridization studies indicate that the African and European strains are similar and form one subgroup, with Australian and Asian strains forming another. Diagnosis is accomplished by virus isolation or serological means. There is no vaccine available. Since many of the mosquito vectors breed in extensive rice fields, large-scale control would be expensive.

Crimean-Congo hemorrhagic fever virus

Crimean-Congo hemorrhagic fever (CCHF) virus is also very widely distributed, and is found from eastern Europe and the Crimean, eastward through the Middle East to western China, and southward through Africa to South Africa. CCHF is characterized by severe hemorrhagic fever with hepatitis, with case mortality of 10–50%. Asymptomatic infections occur, also. Maintenance of CCHF virus involves horizontal transmission from *Hyalomma* ticks to mammals, and vertical transmission in ticks, through the eggs. Immature ticks feed on small mammals, and the adult ticks feed on cattle and other large mammals. *Hyalomma* ticks have also been found on birds migrating between Europe and Africa – a mechanism for long-distance dispersal of the virus. Diagnosis is by virus isolation or by serological techniques. There are no vaccines or tick control measures available.

Sandfly fever viruses

Sandfly fever (Sicilian, Naples and Toscana) viruses are endemic in the Mediterranean area (southern Europe, North Africa, and Southwest Asia). They cause acute febrile disease in humans, with occasional aseptic meningitis. The viruses are members of the *Phlebovirus* genus of the *Bunyaviridae*. They are transovarially and horizontally transmitted by phlebotomine sandflies. Wild mammals are presumed reservoirs.

Viruses Occurring in North America

Encephalitis viruses

Eastern (EEE) and western (WEE) equine encephalitis viruses occur in epidemic form in North America, but have also been found in Central and South America. Generally, EEE is maintained east of the Mississippi River and WEE west of it in the prairie states and provinces and along the West Coast. Both involve wild birds and mosquito vectors, with spill-over into equine populations and humans, causing clinical encephalitis and death. Central nervous sequelae may occur among survivors. Diagnosis is by isolation of the virus in mice or cell cultures, or by antigen capture ELISA. Isolation from brain tissue of clinical encephalitis cases is difficult. Demonstration of IgG antibody by a variety of tests, or IgM antibody by capture ELISA are of diagnostic value. Effective vaccines are available commercially for equine animals, and experimental vaccines are used for laboratory personnel. Effective mosquito abatement to control vector populations has been carried out in the West for many years. Insecticide application is used for vector control in epidemic situations.

St Louis encephalitis (SLE) is widely distributed in the US, with most epidemics occurring in the West, down the Ohio and Mississippi Valleys into Texas, and in Florida. In humans, SLE is characterized by febrile disease with subsequent encephalitis or aseptic meningitis. Wild passerine birds are amplifying hosts, and SLE virus is transmitted by *Culex* mosquitoes. SLE virus is a flavivirus of the *Togaviridae*, and is closely related to Japanese encephalitis virus. RNA oligonucleotide fingerprint patterns indicate that the virus varies by geographic region. Since no vaccine is available, SLE prevention and control relies on surveillance and vector control.

Powassan (POW) virus is a North American member of the flavivirus tick-borne encephalitis complex. Although POW virus is widely distributed across the US and Canada, and westward into far eastern Russia, disease (febrile, with encephalitis) has only been detected in the eastern states and provinces. The transmission cycle involves small mammal and *Ixodes* ticks.

La Crosse (LAC) and other California encephalitides are human pathogens in the US. LAC is the most common arthropod-borne virus disease in North America, affecting mainly preschool-aged children. It is endemic in the Upper Midwest, but occasional cases occur elsewhere. Although fatality is uncommon, the disease is severe enough to cause prolonged hospitalization. LAC virus is maintained tranovarily in treehole breeding *Aedes triseriatus* mosquitoes with horizontal transmission to small forest mammal reservoirs and to humans. The other California group viruses affecting man have similar epidemiologies, but do not cause disease as commonly. California encephalitis virus was isolated in California, and has occasionally caused human disease there. Snowshoe hare (SSH) virus occurs in the Northern US and across Canada, and has caused human encephalitis in the eastern provinces. Jamestown Canyon (JC) virus is widely distributed across the US, has been shown to cause human disease, mainly in adults in the Midwest, and to infect deer. Like LAC virus, these other viruses have the same close epidemiological relationship with their *Aedes* vectors. These viruses are members of the California serogroup of the *Bunyavirus* genus of the *Bunyaviridae* family. SSH virus is an antigenic variant of LAC virus. Oligonucleotide analysis has shown differences in LAC viruses from the eastern and the western parts of its geographic range in the Midwest. Diagnosis is nearly always by serological means, with demonstration of an antibody rise during illness. There are no vaccines for the CAL group viruses. LAC has been controlled by eliminating vector breeding sites through filling treeholes and eliminating discarded rubber tires.

Colorado tick fever virus

Colorado tick fever (CTF) occurs at the higher elevations in the mountains of the western states and provinces. Although seldom fatal, CTF can cause serious disease with prolonged convalescence. CTF may present as a hemorrhagic or a central nervous system disease, and is most severe in preadolescent children. The virus is transmitted by and overwinter in *Dermacentor andersoni*. Wild rodents are the vertebrate hosts, and develop a prolonged viremia. CTF virus is classified in a serogroup of the same name (along with Eyach virus from Europe) in the *Orbivirus* genus of the family *Reoviridae*. Diagnosis is by virus isolation from erythrocytes, or demonstration of CTF viral antigen by immunofluorescence, during the long viremia. Serologic diagnosis may be problematic, because antibodies develop late in the course of infection. Avoidance of tick bites is the main preventive measure available, but control of rodents and the ticks that inhabit their burrows can be applied in foci of virus maintenance in the field.

Viruses Occurring in Central and South America

Encephalitis viruses

Venezuelan equine encephalitis (VEE) has been a major public and animal health problem in Central and South America. The VEE viruses form a closely related complex, whose members have differing epidemiologies, geographic distributions and disease importance. The epizootic/epidemic (VEE IAB) virus variants are of greatest concern. In equine animals, the virus causes acute encephalitis and case fatality may approach 80%, and survivors may have serious neurological deficits. Although the case fatality rate in human is low (less than 1%), the large numbers of acutely infected people that occur during an epidemic may completely overwhelm the local health care system. VEE IAB virus is maintained in northern South America, where it has periodically swept through Venezuela and Colombia in epidemic waves, with occasional extension into Ecuador and massively through Central America into Mexico and South Texas. Epidemic spread depends on the availability of susceptible equine populations (the amplifying host) and abundant mosquito vectors of several species. Although the inter-epidemic maintenance systems remain undefined, there is some recent evidence that the epidemic form may have become enzootic in Central America. The enzootic strains are maintained in limited foci involving rodents and *Culex* (*Melaconion*) mosquitoes from Florida to Argentina. These virus strains do not cause disease in equine animals, but can cause acute febrile illness in humans. The VEE complex viruses are in the *Alphavirus* genus of the *Togaviridae*. Diagnosis is usually done by antibody detection (IgG or IgM ELISA test), because virus isolation from clinical cases is difficult (except from the blood of early febrile cases in infected herds of equine animals). There is an effective live, attenuated vaccine for both human and equine use. Because the maintenance of equine herd immunity is costly, most animal health agencies do not carry out ongoing, intensive vaccination campaigns. Thus, the risk of reoccurrence of explosive outbreaks remains.

Other viruses

Oropouche virus, a Simbu serogroup bunyavirus, causes acute febrile disease with arthralgia in humans in Brazil. During rainy season epidemics, the virus is transmitted by culicoid biting midges. Enzootic maintenance cycles are believed to involve forest mammals and arboreal mosquitoes.

Mayaro (MAY) virus occurs epidemically in the Amazon Basin, and has also been associated with human disease in Trinidad. In humans, the febrile disease with rash is clinically similar to CHIK, an alphavirus to which is it antigenically and taxonomically related. MAY virus appears to be maintained in nature in a cycle similar to that of yellow fever, with arboreal mosquito vectors and primate hosts.

Rocio virus was first isolated from fatal human encephalitis cases during an explosive outbreak in coastal Brazil in 1975, after which time sporadic outbreaks have continued. This virus is an ungrouped flavivirus in the *Togaviridae*. The epidemiology is unclear, but probably involves wild birds and several mosquito species are suspected vectors.

Vesicular stomatitis virus is of major economic concern as a cause of acute, febrile vesicular disease in cattle, mainly in Central and northern South America. Both of the major serotypes, VS-Indiana and VS-New Jersey, cause influenza-like illness in human, as is an occupational hazard to people handling cattle. The VS viruses comprise a complex of related serotypes and subtypes in the Americas, with related vesiculoviruses (*Rhabdoviridae* family) viruses in Africa and Asia. Many of these viruses are transmitted horizontally and transovarially by phlebotomine sandflies, with evidence for infection of wild rodents and other small mammals. However, the role of these mammals in the epidemiology of VS viruses is unclear, since they do not develop viremia. Thus, the host and vector transmission cycles of VS viruses in the Americas are not well understood. The experimental vaccines developed for use in domestic animals have not yet been commercialized.

Viruses Occurring in Europe

Tahyna (TAH) virus is widely distributed in Europe, and has been reported in Africa. TAH virus produces an influenza-like febrile disease, with occasional central nervous system involvement. The virus is a bunyavirus of the California serogroup, in the *Bunyaviridae*. Like La Crosse virus, small forest mammals are TAH virus reservoirs, and the virus is horizontally and transovarially transmitted by *Aedes* mosquitoes. There are no effective control measures.

Omsk hemorrhagic fever occurs in a localized area of western Siberia. Disease can be severe, with up to 3% case fatality, and sequelae are common. This virus is a member of the tick-borne encephalitis (TBE) complex of the flaviviruses. The virus is epizootic in wild muskrats, which had been introduced into the area, and is associated with ixodid ticks. Muskrat handlers are at highest risk of infection. Water voles and other rodents are also vertebrate hosts of the virus. TBE virus vaccine is used in high risk individuals to provide protection.

Cowpox virus has a wide host range. In addition to cattle, this virus has produced severe, generalized infections in a variety of animals in zoos and circuses, including elephants and large cats, which may die. Humans develop typical poxvirus lesions (vesicle and pustule formation), usually on the hands. Milkers are at high risk of infection. Wild rodents are the reservoirs in nature. Laboratory diagnosis (characterization of isolated virus) is required to differentiate cowpox from other poxviruses.

Viruses Occurring in Africa

Rift Valley fever virus

Rift Valley fever (RVF) is among the most serious arbovirus infection in Africa today. Repeated RVF epidemics in sub-Saharan East Africa caused serious disease in small ruminant animals and humans. RVF has expanded its range, causing a massive epidemic in Egypt in 1977–1978, and appearing in epidemic form along the Mauritania–Senegal border ten years later. Cattle, sheep and humans are affected. Abortion storms with febrile disease and bloody diarrhea occur in ruminant animals, and mortality may be heavy in young stock. Most infected humans develop febrile disease, with prolonged convalescence. A few individuals develop more severe disease, with liver necrosis, hemorrhagic pneumonia, meningoencephalitis and retinitis with vision loss. The human case fatality rate is less than 1%. RVF virus is in the *Phlebovirus* genus of the *Bunyaviridae*. In sub-Saharan Africa, RVF virus is closely tied to its *Aedes* mosquito vector. Like La Crosse virus, RVF virus is transovarially and horizontally transmitted. The virus persists in mosquito eggs laid around seasonal pools called 'dambos'. When the rains come and the pools flood, the eggs hatch and infected mosquitoes emerge and begin transmission. Agricultural development projects in Africa must take into account that creation of larval habitats (artificial dambos) may lead to epidemics of RVF, as was apparently the case in West Africa. Diagnosis depends on virus isolation and serologic testing. Field and laboratory worker needs to exercise caution to not become infected by exposure to the virus during postmortem examination of animals, or processing materials in the laboratory. A high level of biosecurity is required. Several effective vaccines, both live attenuated and inactivated, have become available for animals, and there is an inactivated vaccine available for laboratory and field workers at high risk to infection.

Marburg and Ebola viruses

Marburg and Ebola viruses have sporadically caused severe hemorrhagic fever in humans. Marburg virus, although of African origin, first appeared in laboratory workers in Germany who had handled cell cultures originating from African primates. Later, epidemics of severe hemorrhagic fever occurred in the Sudan and in Zaire, and Ebola virus was isolated. These viruses produce hemorrhagic shock syndrome and visceral organ necrosis, and have the highest case fatality rate (30–90%) of the hemorrhagic fevers. These viruses, with their bizarre filamentous, pleomorphic morphology, belong to the *Filoviridae* family. These viruses are presumed to be zoonotic, but their hosts in nature and mechanisms of transmission in the field have not been determined. An outbreak of simian hemorrhagic disease, caused by a Marburg/Ebola-like filovirus, occurred in a primate holding facility, but without evidence of related human disease among animal care personnel. Nosocomial transmission of Marburg and Ebola viruses has occurred frequently; a high level of patient isolation and biosafety containment are essential to avoid hospital and laboratory-acquired infection. The virus can be detected by electron microscopic examination of tissue, or isolation of the virus in cell culture. Serologic diagnosis is accomplished by means of indirect immunofluorescence or ELISA test, with antigen specificity confirmed by western blot. No vaccines or control measures are available.

O'nyong-nyong virus

O'nyong-nyong virus is considered a subtype of Chikungunya virus, and is widespread in eastern sub-Saharan Africa. The disease it produces and its epidemiology are similar to CHIK, with *Anopheles* spp. as the main mosquito vector. Igbo Ora virus is also antigenically related to CHIK virus, and has been isolated from humans in West Africa.

Wesselsbron virus

Wesselsbron (WES) virus is mainly a problem in domestic animals in Africa, although human febrile illness, occasionally with central nervous system signs, has also been described. Infection causes abortion storms in sheep, with death of the ewe frequently, similar to Rift Valley fever. The virus has also been isolated in Thailand, but without significant disease in humans or animals. The vertebrate reservoirs have not been defined, but domestic livestock are capable of virus amplification. A variety of *Aedes* spp. are implicated as vector mosquitoes.

Viruses Occurring in Asia

Kyasanur Forest disease (KFD) was first recognized in India in 1957, when a hemorrhagic disease appeared in wild monkeys and people frequenting forested areas. KFD has been slowly spreading in India. KFD virus is a member of the tick-borne encephalitis complex of flaviviruses. The basic virus maintenance cycle involves forest mammals (rodents and insectivores) and ixoded ticks, mainly *Haemaphysalis spinigera*. The virus can be isolated in mice and cell cultures, including tick cells. An inactivated vaccine provides some protection to people at risk of infection.

Viruses Occurring in Australia

Murray Valley encephalitis occurs primarily in southeastern Australia, with cases occasionally occurring in other parts of the country and in New Guinea. Febrile disease leads to encephalitis and, in severe cases, death. Neurologic sequelae are common in survivors. Children are predominantly affected. Large water birds are the main vertebrate amplifying hosts, but mammals are reservoirs, also. The virus is transmitted mainly by *Culex* mosquitoes. MVE virus is a flavivirus closely related to Japanese encephalitis. RNA sequencing indicates that the Australian strains of MVE virus are similar, but different from New Guinea isolates. Like JE, MVE diagnosis depends on virus isolation and serological testing. No vaccine is available. Control is achieved though application of larvacides.

Ross River (RR) virus has caused annual epidemics of febrile disease with polyarthritis and rash in tropical and temperate Australia. Within the past 15 years, RR virus has spread through several Pacific islands in epidemic form. Convalescence can be long. RR virus is an alphavirus of the *Togaviridae* family. The enzootic maintenance cycles of RR virus in Australia are not well defined, but wild and domestic mammals appear to be the reservoir hosts, and the principal mosquito vectors are saltmarsh *Aedes* spp. and freshwater *Culex* sp. In the Pacific Islands, the virus was probably transmitted from person to person by *Aedes* mosquitoes.

Diagnosis

Diagnosis of viral zoonoses depends on the demonstration of the presence of the virus or its components in the host or vector, or the development of specific antibodies against the virus by the infected vertebrate. Individuals or populations should be sampled based on objective clinical or epidemiological criteria. It is essential that the appropriate blood, tissue or vector samples be taken, maintained under conditions (usually chilled or frozen) to prevent degradation of virus or antibody, and promptly delivered to the diagnostic laboratory. Virus is isolated in living laboratory hosts, such as mice, embryonated eggs or in cultured cells from vertebrates or arthropods. In some cases, xenodiagnosis – the inoculation of whole arthropods (especially mosquitoes or ticks) – is the most sensitive system for virus isolation. Viruses or their components also may be detected directly in tissue of the diseased host (such as rabies antigen demonstration in brain tissue by immunofluorescence) or infected vector. Although many techniques developed 20 or more years ago are still used routinely for diagnosis, modern biotechnology and molecular biology have brought powerful new tools to the diagnosis of viral zoonoses. Recent techniques have made viruses much easier and quicker to detect, and include radioimmunoassays, nucleic acid hybridization probes, antigen capture enzyme-linked immunosorbent assay (ELISA), and, most recently, polymerase chain reaction (PCR) that can multiply minuscule amounts of specific viral nucleic acid sequences by a millionfold or more, to readily detectable concentrations. Antigen and nucleic acid analysis techniques have been valuable for epidemiological studies. Monoclonal antibodies (MAB) have been used to detect virus variants having species- or geographically-limited epitopes. In rabies, for example, epitope differences have been used to track virus movement from one host population to another. Differences in viral RNA, demonstrated by oligonucleotide fingerprinting, have been used to establish the distribution of California group bunyavirus variants within a broad range, providing insights into the degree of geographic isolation that occurs in nature. Electrophoretic analysis (PAGE) of polypeptides of closely related viruses, such as some hantaviruses, has also been useful for establishing taxonomic and geographic relationships among these emerging zoonotic agents.

Techniques to more quickly and accurately detect specific viral antibodies in infected or convalescent hosts have also made significant advances in recent years. Although the traditional serological tests such as virus neutralization, hemagglutination, passive agglutination, complement-fixation and immunoprecipitation, are still commonly used to diagnose viral zoonoses, they are rapidly being complimented by other techniques to measure the presence of specific virus antibody. ELISA tests have become extremely useful for measuring antiviral antibody. IgM capture ELISAs have been developed to detect immunological response and make early serological diagnosis of zoonotic infections such as Japanese encephalitis and VEE. Antibody binding to specific viral polypeptides can be directly measured by western blotting, and determine which of several antigenically similar, cross-reacting viruses is the infecting agent.

Control

Control of zoonotic virus diseases is accomplished by breaking the cycle of transmission. This is usually accomplished by eliminating or immunizing vertebrate hosts, and reducing vector populations. Reduction of reservoir host populations is usually not accomplished because it is too expensive, not environmentally safe, and not technically or logistically feasible. However, there have been some notable exceptions. Bolivian hemorrhagic fever, caused by Machupo virus, was controlled through reduction of its rodent hosts through intensive rodenticiding. The principal vampire bat reservoir of rabies, *Desmodus rotundus*, is being controlled by the application of warfarin-type anticoagulants either directly to the bats themselves, to the cattle upon which they feed (with no harm to the cattle). Control programs like these have to be continuous to be effective. Their reduction or discontinuation results in host population recovery through reproduction and immigration, which may result in reemergence of disease sweeping through the increasing, susceptible cohort.

Immunization of hosts is both promising and discouraging. After decades of research, safe and effective rabies vaccines are being used for immunization of humans, domesticated animals and some wildlife species. The human diploid cell vaccine is extremely effective, free of adverse effects, and widely available but at a cost too high for use in many developing countries. Safe, effective animal vaccines of cell culture origin are on the market. After some initial public resistance, raccoon populations in the eastern United States and wild foxes in Europe are being successfully immunized by means of an oral, vaccinia-vectored recombinant vaccine. This experience illustrates the need for public understanding, in order to counteract fear of the unknown – in this case, field use of a genetically engineered virus. Other vectored rabies vaccines are under development for control of mongoose-transmitted rabies in the Caribbean. Vaccines will not be developed for many zoonotic viral diseases that affect relatively few people, and may be of very limited concern geographically. With limited markets for new vaccines, there is no economic incentive to justify the several millions of dollars for the research and testing required for licensure and commercialization of these products.

Vector control is another promising, but difficult area of zoonoses reduction or elimination. Insecticide application has become more problematic because both vectors themselves, as well as public opinion, have become more resistant to their use. Integrated pest management techniques, well developed for the control of many crop insects, along with the use of natural pesticides such as BT toxin, offers promise for the effective, environmentally safe control of dipterous vectors. Control of tick vectors is likely to remain a problem for some time to come.

Problems With Control of Zoonoses

Ecological change

Human disturbance has become a feature of nearly every part of the planet. All too often, these disturbances create habitats that favor increases in populations of key hosts and vectors, with subsequent increased transmission of viral zoonoses. Nowhere are ecological changes happening more rapidly and profoundly than in the world's tropics. Conversion of tropical forests to agricultural ecosystems simplifies diverse ecosystems and provides either native or introduced host or vector species the conditions necessary to become more abundant, and sustain intensified virus transmission in areas where people live and work. In Africa, recent yellow fever epidemics have been increasing dramatically in agricultural areas. Some agricultural irrigation development projects have created extensive vector breeding habitats, with an increase in mosquito-transmitted disease. The extensive dams constructed in Senegal were followed by epidemics of Rift Valley fever, with numerous cases of disease in humans and small ruminant animals. The public health consequences of development projects must never be overlooked.

Global climate change will also bring ecological changes and shifts of human populations that will affect the occurrence of viral zoonoses. Although experts debate the geography, severity and rapidity of the oncoming greenhouse effect, there is general consensus that changes in the global climate will happen with unprecedented speed. With those changes will come alterations in the geography of natural and agricultural ecosystems, with corresponding changes in the distribution of zoonotic diseases and the intensity of their transmission. While it is not possible to accurately predict what the world will be like in 100 years, nor what zoonotic diseases are likely to be most troublesome, it is certain that things will be different, and constant surveillance will be essential to avoid serious problems or deal promptly and effectively with the ones that arise.

Movement of zoonotic viruses can result from the displacement of infected animals, contaminated animal products and virus-carrying arthropod vectors. It is clear that illegal translocation of wild raccoons for hunting purposes, with inadvertent movement of rabies-infected animals, was responsible for initiating the recent rabies epizootic in the Middle Atlantic States. Pets, sport, laboratory and agricultural animals are moving around the world as never before. Although international and national regulations have been established to prevent movement of infected individuals, it is not possible to test for all possible zoonotic viruses, and prevent them from crossing international boundaries. Moreover, significant numbers of animals of high commercial value move illegally. Psittacine and other exotic birds worth hundreds to thousands of dollars each, cross illegally into the US, despite intensive government efforts to halt this smuggling. The importation of highly virulent Newcastle disease (ND) virus has been occasionally linked to smuggled birds. The costly ND outbreak (more than $50 million to control) in Southern California in the early 1970s was linked to the importation of an infected pet bird from Asia. Subsequently, highly lethal ND virus has been found in smuggled parrots. Although not a serious infection in humans, ND is highly transmissible and can be extremely lethal in poultry and costly to that industry.

Zoonotic viruses may be transported by movement of arthropod vectors, too. Just as the yellow fever mosquito, *Aedes aegypti*, moved around the world in water casks aboard sailing vessels, mosquitoes are transported around the world in international commerce. Ships still transport mosquito vectors. The Asian tiger mosquito, *Aedes albopictus*, has become established in the Western Hemisphere after multiple introductions in eggs deposited in used tires. This mosquito is capable to transmitting yellow fever and La Crosse encephalitis. It remains to be seen if this highly competitive exotic vector, with its peridomestic habits, will carry yellow fever virus from its jungle cycle to the urban, human amplified cycle in the American tropics, or increase the numbers of La Crosse encephalitis cases in south temperate or subtropical areas of the United States. Perhaps of greater concern, modern transport aircraft have been shown to move vector mosquitoes internationally. Nature can move vectors as well. For example, windblown biting insects, such as culicoid midges, have been shown to account for the spread of some arthropod-borne virus diseases of animals, in some cases over long distances. There is no reason to believe that vectors of human diseases cannot move similarly.

Social change

Increasing human populations place great demands on the public health and other government services, especially in developing countries where needs for zoonosis diagnosis, control and prevention are greatest and resources are most limited. Some preventive measures can be implemented by the people who live in the affected areas themselves, and at minimal cost if they knew why and how they needed to do it. Public education and information is essential for control and prevention of zoonotic diseases. However, it takes more than civic action to deal with zoonotic disease. Delivery of public education, disease surveillance and diagnosis and the technical materials and logistical support for control or preventive programs depend on national or international scientific and financial support. Because serious zoonotic viral diseases such as rabies, yellow fever, venezuelan equine encephalitis and the hemorrhagic fevers, know nothing of international boundaries, international technical cooperation and financial support are imperative.

See also: Bunyaviruses; Diagnostic techniques; Encephalitis viruses; Hantaviruses; Lassa, Junin, Machupo and Guanarito viruses; Rhabdoviruses.

Further Reading

Fenner F *et al.* (eds) (1987) *Veterinary Virology*. Orlando: Academic Press.

Fields BN *et al.* (eds) (1990) *Virology*, 2nd edn. New York: Raven Press.

Hubbert WT, McCulloch WF and Schnurrenberg PR (eds) (1975) *Diseases Transmitted from Animals to Man*. 6th edn. Springfield Il: Charles C. Thomas.

Steele JH (ed.) (1981) *Handbook Series in Zoonoses*, Boca Raton: CRC Press.

Appendix
Viruses Included in the Encyclopedia

ANIMAL VIRUSES INCLUDED IN THE ENCYCLOPEDIA

Virus	Family/Group	Synonym/(Abbreviation)
Abelson murine leukemia virus	Retroviridae	(Ab-MuLV)
Adeno-associated viruses	Parvoviridae	(AAV)
Adenovirus	Adenoviridae	
Adult T-cell leukemia virus	Retroviridae	(ATLV)
African horse sickness virus	Reoviridiae, Orbivirus	(AHSV)
African swine fever virus	Floating Genus	
Akabane virus	Bunyaviridae, Bunyavirus	
AKR murine leukemia virus	Retroviridae	(AKR-MuLV)
Alcelaphine herpesvirus 1, 2	Herpesviridae	Hartebeest herpesvirus
Aleutian mink disease virus	Parvoviridae	(ADV)
Amapari virus	Arenaviridae	
Arteriviruses	Togaviridae	
Astroviruses	Astroviridae	
Aura virus	Togaviridae, Alphavirus	
Australian wood duck hepatitis B virus	Hepadnaviridae	
Aviadenoviruses	Adenoviridae	
Avian acute leukemia virus	Retroviridae	
Avian encephalomyelitis virus	Picornaviridae, Enterovirus	
Avian erythroblastosis virus	Retroviridae	(AEV)
Avian hepatitis B viruses	Hepadnaviridae	
Avian infectious bronchitis virus	Coronaviridae	(IBV)
Avian influenza viruses	Orthomyxoviridae	
Avian leukemia sarcoma viruses	Retroviridae	(ALSV)
Avian leukemia virus	Retroviridae	(ALV)
Avian leukosis virus	Retroviridae	(ALVs)
Avian myeloblastosis virus	Retroviridae	(AMV)
Avian nephritis virus	Picornaviridae, Enterovirus	
Avian paramyxoviruses	Paramyxoviridae	(PMV)
Avian reoviruses	Reoviridae	
Avian sarcoma virus	Retroviridae	(ASV)
Avian sarcoma-leukemia viruses	Retroviridae	(ASLV)
Avipoxvirus	Poxviridae	

Baboon endogenous retroviruses	Retroviridae	(BaEV)
Baboon herpesvirus	Herpesviridae, Gammaherpesvirinae	
Barmah Forest virus	Togaviridae, Alphavirus	(BFV)
Bebaru virus	Togaviridae, Alphavirus	
Berne virus	Coronaviridae	(BEV)
Birnaviruses	Birnaviridae	
BK virus	Papovaviridae, Polyomavirus	Polyomavirus hominis 1
Bluegill hepatic necrosis reovirus	Reoviridae	
Bluetongue virus	Reoviridae, Orbivirus	(BTV)
Bluetongue virus 10	Reoviridae, Orbivirus	(BTV-10)
Bolivian hemorrhagic virus fever	Arenaviridae	(BHF) Machupo virus
Border disease virus	Flaviviridae, Pestivirus	(BDV)
Borna disease virus	Unclassified	(BDV)
Bovine adeno-associated viruses	Parvoviridae	(BAAVs)
Bovine coronavirus	Coronaviridae	(BCV)
Bovine enterovirus	Picornaviridae, Enterovirus	(BEV)
Bovine herpesvirus 1	Herpesviridae	(BHV1); Infectious bovine rhinotracheitis virus
Bovine herpesvirus 2	Herpesviridae	(BHV2); Bovine mammilitis virus: Allerton virus; Pseudolumpy skin disease virus
Bovine herpesvirus 3	Herpesviridae	(BHV3)
Bovine herpesvirus 4	Herpesviridae	(BHV4); Movar herpesvirus
Bovine herpesvirus 5	Herpesviridae	(BHV5); Bovine encephalitis herpesvirus
Bovine immunodeficiency virus	Retroviridae, Lentivirinae	(BIV)
Bovine leukemia virus	Retroviridae, Oncovirinae	(BLV)
Bovine papillomavirus	Papovaviridae	(BPV)
Bovine papillomavirus type 1	Papovaviridae	(BPV-1)
Bovine papillomavirus type 4	Parovaviridae	(BPV-4)
Bovine papular stomatitis virus	Poxviridae, Chordopoxviridae, Parapoxvirus	
Bovine parainfluenza virus type 3	Paramyxoviridae	(PIV-3)

Bovine parvovirus	Parvoviridae, Parvovirus	(BPV) HADEN virus
Bovine polyomavirus	Papovaviridae	
Bovine respiratory syncytial virus	Paramyxoviridae, Pneumovirus	(BRSV)
Bovine syncytial virus	Retroviridae	(BSV)
Bovine syncytium-forming virus	Retroviridae, Spumavirinae	(BSFV)
Bovine viral diarrhea virus	Flaviviridae, Pestivirus	(BVDV)
Breda virus	Coronaviridae	(BRV)
Budgerigar fledgling disease virus	Papovaviridae	(BFDV)
Buffalopox virus	Poxviridae, Orthopoxvirus	
Bunyamwera virus	Bunyaviridae	
Bunyavirus	Bunyaviridae	
Caliciviruses	Caliciviridae	
California encephalitis virus	Bunyaviridae, Bunyavirus	
Californian volepox virus	Poxviridae, Orthopoxvirus	
Camelpox virus	Poxviridae	
Canarypoxvirus	Poxviridae	
Canine adeno-associated viruses	Parvoviridae	(CAAVs)
Canine adenovirus	Adenoviridae	(CAV)
Canine coronavirus	Cornonaviridae	(CCV)
Canine distemper virus	Paramyxoviridae, Morbillivirus	(CDV)
Canine parainfluenza virus 5	Paramyxoviridae	(PIV-5)
Canine parvovirus	Parvoviridae	(CPV)
Caprine arthritis encephalitis virus	Retroviridae, Lentivirinae	(CAEV)
Caprine herpesvirus 1	Herpesviridae	Goat herpesvirus
Caprine respiratory syncytial virus	Paramyxoviridae, Pneumovirus	(CRSV)
Cardioviruses	Picornaviridae	
Carp pox herpesvirus	Herpesviridae, Gammaherpesvirinae	
CELO virus	Adenoviridae	
Central European encephalitis virus	Flaviviridae, Flavivirus	(CEEV)
Cercopithecine herpesvirus-12	Herpesviridae, Gammaherpesvirinae	
Cervid herpesviruses	Herpesviridae	Red deer herpesvirus; Reindeer herpesvirus
Chandipura virus	Rhabdoviridae, Vesiculovirus	(CHP)
Channel catfish virus	Herpesviridae, Gammaherpesvirinae	(CCV)
Chick syncytial virus	Retroviridae, Oncovirinae	(CSV)
Chicken herpesvirus	Herpesviridae	
Chicken influenza virus	Orthomyxoviridae	

Chicken syncytial virus	Retroviridae	(CSV)
Chickenpox virus	Herpesviridae, Alphaherpesvirinae	Varicellazoster virus
Chikungunya virus	Togaviridae, Alphavirus	(CHIK)
Chimpanzee herpesvirus	Herpesviridae	
Chordopoxvirinae	Poxviridae	
Cichlid virus	Incertae	
Colorado tick fever virus	Reoviridae	(CTFV)
Coltiviruses	Reoviridae	
Columbia-SK virus	Picornaviridae, Cardiovirus	
Coronavirus-like particles	Unclassified	
Coronaviruses	Coronaviridae	
Cottontail rabbit papillomavirus (Shope papillomavirus)	Papovaviridae	(CRPV)
Cowpox virus	Poxviridae, Orthopoxvirus	
Coxsackievirus	Picornaviridae, Enterovirus	
Coxsackievirus A	Picornaviridae, Enterovirus	
Coxsackievirus A21	Picornaviridae, Enterovirus	
Coxsackievirus A23	Picornaviridae, Enterovirus	
Coxsackievirus A24 variant	Picornaviridae, Enterovirus	(CA24v)
Coxsackievirus B	Picornaviridae, Enterovirus	
Coxsackievirus B3	Picornaviridae, Enterovirus	(CVB3)
Crimean-Congo hemorrhagic fever virus	Bunyaviridae, Nairovirus	(CCHF)
Cytomegaloviruses	Herpesviridae, Betaherpesvirinae	(CMV)
Deltaherpesvirus	Herpesviridae	
Dengue 1 virus	Flaviviridae, Flavivirus	(DEN-1)
Dengue 2 virus	Flaviviridae, Flavivirus	(DEN-2)
Dengue 3 virus	Flaviviridae, Flavivirus	(DEN-3)
Dengue 4 virus	Flaviviridae, Flavivirus	(DEN-4)
Dengue viruses	Flaviviridae, Flavivirus	
Dependovirus	Parvoviridae	
Distemper viruses	Paramyxoviridae, Morbillivirus	
Duck adenovirus	Adenoviridae	
Duck hepatitis B virus	Hepadnaviridae	(DHBV)

Duck infectious anemia virus	Retroviridae, Oncovirinae	(DIAV)
Duvenhage virus	Rhabdoviridae, Lyssavirus	
Eastern equine encephalitis virus	Togaviridae, Alphavirus	(EEE)
Ebola virus	Filoviridae	
Ebola-Zaire virus	Filoviridae	
Echovirus	Picornaviridae, Enterovirus	
Echovirus type 9	Picornaviridae, Enterovirus	
Ectromelia virus	Poxviridae, Orthopoxvirus	Mousepoxvirus
Edge Hill virus	Flaviviridae, Flavivirus	
Eel virus European	Birnaviridae	(EVE)
Egtved virus	Rhabdoviridae	Viral hemorrhagic septicemia virus
Encephalitis viruses	Flaviviridae	
Encephalomyocarditis virus	Picornaviridae	(EMCV)
Enterovirus (EV) serotype 68	Picornaviridae	(EV68); Fermon virus
Enterovirus (EV) serotype 69	Picornaviridae	(EV69); Toluca virus
Enterovirus (EV) serotype 70	Picornaviridae	(EV70)
Enterovirus (EV) serotype 71	Picornaviridae	(EV71)
Enterovirus (EV) serotype 72	Picornaviridae	(EV72)
Enteroviruses	Picornaviridae	
Epizootic hemorrhagic disease (of deer) virus	Reoviridae, Orbivirus	(EHDV)
Epstein–Barr virus	Herpesviridae, Gammaherpesvirinae, Lymphocryptovirus	(EBV)
Equine abortion virus	Herpesviridae, Alphavirinae	(EHV-1); Equine rhinopneumontis virus
Equine arteritis virus	Togaviridae, Arterivirus	(EAV)
Equine coital exanthema virus	Herpesviridae, Alphavirinae	EHV-3
Equine cytomegalovirus	Herpesviridae, Betaherpesvirus	(ECMV) EHV-2
Equine encephalitis viruses	Togaviridae, Alphavirus	
Equine encephalosis virus	Reoviridae, Orbivirus	(EEV)
Equine herpesviruses	Herpesviridae	
Equine infectious anemia virus	Retroviridae, Lentivirus	(EIAV)
Equine rhino-pneumonitis virus	Herpesviridae, Alphavirinae	EHU-1; Equine abortion virus
European eel virus	Birnaviridae	(EV)

Eyach virus	Reoviridae, Coltivirus	
FBJ murine osteogenic murine sarcoma virus	Retroviridae	(FBJ-MSV)
FBJ murine sarcoma virus	Retroviridae	(FBJ-MSV)
Feline calicivirus	Caliciviridae	(FCV)
Feline enteric coronavirus	Coronaviridae	(FECV)
Feline immunodeficiency virus	Retroviridae, Lentivirinae	(FIV)
Feline infectious peritonitis virus	Coronaviridae	(FIPV)
Feline leukemia virus	Retroviridae, Oncornavirinae	(FeLV)
Feline panleucopenia virus	Parvoviridae, Parvovirus	(FPV)
Feline parvovirus	Parvoviridae	(FPV)
Feline sarcoma viruses	Retroviridae, Oncornavirinae	(FeSVs)
Feline syncytium-forming virus	Retroviridae, Spumavirinae	(FeSFV)
Fermon enterovirus	Picornaviridae	Enterovirus serotype 68
Filoviruses	Filoviridae	
Fish herpesviruses	Herpesviridae, Gammaherpesvirinae	
Fish lymphocystis disease virus	Iridoviridae	(FLDV)
Fish pox virus	Herpesviridae	
Fish rhabdoviruses	Rhabdoviridae	
Flaviviruses	Flaviviridae	
Flexal virus	Arenaviridae	
Foamy viruses	Retroviridae, Spumavirinae	
Foot and mouth disease viruses	Picornaviridae, Aphthovirus	(FMDVs)
Fort Morgan virus	Togaviridae, Alphavirus	
Fowl plague orthomyxovirus	Poxviridae	
Fowlpox virus	Poxviridae, Chordopoxviridae, Avipoxvirus	(FPV)
Friend murine leukemia virus	Retroviridae	(MuLV)
Friend spleen focus-forming virus	Retroviridae	(SFFV)
Frog virus 3	Iridoviridae	(FV3)
Frog virus 4	Gammaherpes-viridae	(FV4)
Fujinami sarcoma virus	Retroviridae	(FuSV)
Gardner–Rasheed feline sarcoma virus	Retroviridae	(GR-FeSV)
Gibbon ape leukemia virus	Retroviridae	(GaLV)
Goat poxviruses	Poxviridae, Capripoxvirus	
Golden shiner virus	Reoviridae	
Goose hepatitis virus	Parvoviridae	(GPV)
Goose parvovirus	Parvoviridae	(GPV)
Gossas virus	Rhabdoviridae	
Grass carp reovirus	Reoviridae	(GCRV)

Gross virus	Retroviridae	
Ground squirrel hepatitis virus	Hepadnaviridae	(GSHV)
Guanarito virus	Arenaviridae	Venezuelan hemorrhagic fever
H-1 virus	Parvoviridae	
HADEN (hemadsorbing enteric virus)	Parvoviridae	Bovine parvovirus
Hamster polyomavirus	Papovaviridae	(HaPV)
Hamster syncytium-forming virus	Retroviridae, Spumavirinae	(HaSFV)
Hantaan virus	Bunyaviridae, Hantavirus	Korean hemorrhagic fever virus
Hantaviruses	Bunyaviridae	
Hardy–Zucherman feline sarcoma virus	Retroviridae	(HZ-FeSV)
Hare fibroma virus	Poxviridae, Leporipoxvirus	(HFV)
Harvey murine sarcoma virus	Retroviridae	(Ha-MSV)
Hawaii agent	Caliciviridae	
Hemadsorption virus type 2	Paramyxoviridae, Paramyxovirus	(HA2)
Hemagglutinating virus of Japan	Paramyxoviridae, Paramyxovirus	(HVJ); Sendai virus
Hemorrhagic fever viruses	Flaviviridae	
Hepadnaviruses	Hepadnaviridae	
Hepatitis A virus	Picornaviridae, Hepatovirus	(HAV)
Hepatitis B virus	Hepadnaviridae	(HBV)
Hepatitis C virus	Flaviviridae, Flavivirus	(HCV)
Hepatitis delta virus	Delta Virus	(HDV)
Hepatitis E virus	Caliciviridae	(HEV)
Hepatoencephalomyelitis virus	Reoviridae	(HEV)
Heron hepatitis B virus	Hepadnaviridae	
Herpes simplex viruses	Herpesviridae, Alphaherpesvirinae	(HSV)
Herpesvirus 6	Herpesviridae, Gammaherpesvirinae	Human B lymphotropic virus
Herpesvirus ateles	Herpesviridae Gammaherpesvirinae	
Herpesvirus pan	Herpesviridae, Gammaherpesvirinae	Pongine herpesvirus-1
Herpesvirus papio	Herpesviridae, Gammaherpesvirinae	Cercopithecine herpesvirus-12
Herpesvirus saimiri	Herpesviridae, Gammaherpesvirinae	Simian hepatitis A virus
Herpesvirus salmonis	Herpesviridae, Gammaherpesvirinae	
Herpesvirus sylvilagus	Herpesviridae, Gammaherpes-virinae	(HVS)
Herpesvirus of turkeys	Herpesviridae	(HVT); Turkey herpesvirus
Herpesviruses	Herpesviridae	
Highlands J virus	Togaviridae, Alphavirus	

Hirame rhabdovirus	Rhabdoviridae	
Hog cholera virus	Flaviviridae, Pestivirus	(HCV)
Human B lymphotropic virus	Herpesviridae, Gammaherpesvirinae	Human herpesvirus-6
Human caliciviruses	Caliciviridae	(HCV)
Human cytomegalovirus	Herpesviridae, Betaherpesvirinae	(HCMV) Human herpesvirus-5
Human enteric coronaviruses	Coronaviridae	(HECV)
Human foamy virus	Retroviridae, Spumavirinae	(HFV)
Human herpesvirus 4	Herpesviridae, Lymphocryptovirus	(HHV-4); Epstein-Barr virus
Human herpesvirus 5	Herpesviridae, Betaherpesvirus	(HHV-5); Human cytomegalovirus
Human herpesvirus 6	Herpesviridae, Betaherpesvirinae	(HHV-6); Human B lymphotropic virus
Human herpesvirus 7	Herpesviridae, Gammaherpesvirinae	(HHV-7)
Human immunodeficiency virus 1	Retroviridae, Lentivirinae	(HIV-1)
Human immunodeficiency virus 2	Retroviridae, Lentivirinae	(HIV-2)
Human immunodeficiency virus	Retroviridae, Lentivirinae	(HIV); Human T-cell leukaemia virus type III (HTLV-III)
Human monkeypox virus	Poxviridae	
Human papillomavirus type 1	Papovaviridae	(HPV1)
Human papillomavirus type 2	Papovaviridae	(HPV2)
Human papillomavirus type 3	Papovaviridae	(HPV3)
Human papillomavirus type 5	Papovaviridae	(HPV5)
Human papillomavirus type 6	Papovaviridae	(HPV6)
Human papillomavirus type 8	Papovaviridae	(HPV8)
Human papillomavirus type 11	Pavovaviridae	(HPV11)
Human papillomavirus type 16	Papovaviridae	(HPV16)
Human papillomavirus type 18	Papovaviridae	(HPV18)
Human papillomaviruses	Papovaviridae	(HPV)
Human parainfluenza viruses	Paramyxoviridae, Paramyxovirus	
Human parvovirus B14	Parvoviridae	
Human parvovirus B19	Parvoviridae	
Human respiratory coronaviruses	Coronaviridae	(HCV)
Human respiratory syncytial virus	Paramyxoviridae, Pneumovirus	(HRSV)
Human rhinovirus type 14	Picornaviridae	(HRV14)
Human rhinovirus type 39	Picornaviridae	(HRV39)
Human rhinovirus type 87	Picornaviridae	(HRV87)
Human rhinoviruses	Picornaviridae	(HRV)
Human spumavirus	Retroviridae, Spumavirinae	(HS_RV)
Human T-cell leukemia virus	Retroviridae	(HTLV)
Human T-cell leukemia virus type 1	Retroviridae	(HTLV-1)

Human T-cell leukemia virus type 2	Retroviridae	(HTLV-2)
Human T-cell leukemia virus type III	Retroviridae, Lentivirinae	(HTLV-III) Human immunodeficiency virus
Ibaraki virus	Reoviridae, Orbivirus	
Icosahedral cytoplasmic deoxyriboviruses	Iridoviridae	(ICDV)
Infectious bursal disease of chickens virus	Birnaviridae	(IBDV)
Infectious hematopoietic necrosis virus	Rhabdoviridae	(IHNV)
Infectious pancreatic necrosis virus of fish	Birnaviridae	(IPNV)
Influenza A virus	Orthomyxoviridae	
Influenza B virus	Orthomyxoviridae	
Influenza C virus	Orthomyxoviridae	
Influenza viruses	Orthomyxoviridae	
Ippy virus	Arenaviridae	
Isfahan virus	Rhabdoviridae, Vesiculovirus	(ISF)
Japanese eel iridovirus	Iridoviridae	
Japanese encephalitis virus	Flaviviridae	(JEV); Polyomavirus hominis 2; JE virus
JC virus	Papovaviridae, Polyomavirus	
JE virus	Flaviviridae	Japanese encephalitis virus
Juncopox virus	Poxviridae, Avipoxvirus	(JPV)
Junin virus	Arenaviridae	Argentine hemorrhagic fever virus
Kirsten murine sarcoma virus	Retroviridae	(Ki-MSV)
Klamath virus	Rhabdoviridae	
Korean hemorrhagic fever	Bunyaviridae, Hantavirus	(KHF) Hantaan virus
Kotonkan virus	Rhabdoviridae, Lyssavirus	
KV (polyomavirus)	Papovaviridae	
Kyasanur Forest disease virus	Flaviviridae, Flavivirus	(KFDV)
La Crosse virus	Bunyaviridae, Bunyavirus	
Lactate dehydrogenase-elevating virus	Togaviridae, Arterivirus	(LDV)
Lagos bat virus	Rhabdoviridae, Lyssavirus	
Lapine parvovirus	Parvoviridae	(LPV)
Lassa virus	Arenaviridae	
Latino virus	Arenaviridae	
Lelystad virus	Togaviridae	(LV)
Lentiviruses	Retroviridae	
Leporipoxvirus	Poxviridae	
LPV (polyomavirus)	Papovaviridae	
Lucké tumor herpesvirus	Gammaherpesviridae	(LTHV)

Lumpy skin disease virus of cattle	Poxviridae	(LSD)
Lymphocryptovirus	Herpesviridae	
Lymphocystis virus	Iridoviridae	
Lymphocytic choriomeningitis virus	Arenaviridae	(LCMV)
Lymphoproliferative disease virus of turkeys	Reotroviridae	(LPDV)
Lyssavirus	Rhabdoviridae	
Machupo virus	Arenaviridae	Bolivian hemorrhagic fever virus
Malignant rabbit fibroma virus	Poxviridae, Leporipoxvirus	(MRV)
Marburg virus	Filoviridae	
Marek's disease virus	Herpesviridae	(MDV)
Marin county agent	Astroviruses	
Mason-Pfizer monkey virus	Retroviridae	(M-PMV)
Mastadenovirus	Adenoviridae	
Maus-Elberfeld virus	Picornaviridae	(ME); mouse encephalomyelitis virus
Mayaro virus	Togaviridae, Alphavirus	(MAY)
Measles virus	Paramyxoviridae, Morbillivirus	
Medical Lake macaque virus	Herpesviridae	
Mengo virus	Picornaviridae	
Mill Hill-2 virus	Orthomyxoviridae	(MH2)
Mink cell focus-forming murine leukemia virus	Retroviridae	(MCF-MLF)
Mink enteritis virus	Parvoviridae, Parvovirus	(MEV)
Minute virus of canines	Parvoviridae, Parvovirus	(MVC)
Minute virus of mice	Parvoviridae, Parvovirus	(MVM)
Mo T-cell virus	Retroviridae	
Mobala virus	Arenaviridae	
Mokola virus	Rhabdoviridae, Lyssavirus	
Molluscipoxvirus	Poxviridae	
Molluscum contagiosum virus	Poxviridae, Chordopoxviridae, Molluscipoxvirus	(MCV)
Moloney murine leukemia virus	Retroviridae	(Mo-MuLV)
Moloney murine sarcoma virus	Retroviridae	(Mo-MSV)
Monkeypox virus	Poxviridae, Orthopoxvirus	
Montgomery County virus	Caliciviridae	
Mopeia virus	Arenaviridae	
Morbillivirus	Paramyxoviridae	
Mouse cytomegalovirus	Herpesviridae, Betaherpesvirinae	(MCMV)
Mouse encephalomyelitis virus	Picornaviridae	Maus-Elberfeld (ME) virus
Mouse hepatitis virus	Coronaviridae	(MHV)
Mouse hepatitis virus type 4	Coronaviridae	(MHV4)

Mouse mammary tumor virus	Retroviridae, Oncornavirinae	(MMTV)
Mousepox virus	Poxviridae, Orthopoxvirus	Ectromelia virus
Muchupo virus	Arenaviridae	
Mucosal disease virus	Flaviviridae, Pestivirus	(MD)
Mumps virus	Paramyxoviridae, Rubulavirus	(MuV)
Murine cytomegalovirus	Herpesviridae, Betaherpesvirus	(MCMV)
Murine leukemia viruses	Retroviridae	(MuLVs)
Murine parvovirus	Parvoviridae	
Murine poliovirus	Picornaviridae	Theiler's murie encephalomyelitis virus
Murray Valley encephalitis virus	Flaviviridae, Flavivirus	(MVEV)
Myeloblastosis associated virus	Retroviridae	(MAV)
Myelocytomatosis virus 29	Retroviridae	(MC29)
Myeloproliferative sarcoma virus	Retroviridae	(MPSV)
Myxoma leporipoxvirus	Poxviridae	
Myxoma virus	Poxviridae, Chordopoxvirinae, Leporipoxvirus	(MYX)
Nairovirus	Bunyaviridae	
Navarro virus	Rhabdoviridae	
NeVTA virus	Herpesviridae, Gammaherpesvirinae	
Newbury agent	Caliciviridae	
Newcastle disease virus	Paramyxoviridae, Paramyxovirus	(NDV)
Norwalk virus	Caliciviridae	
Obodhiang virus	Rhabdoviridae, Lyssavirus	
Omsk hemorrhagic fever virus	Flaviviridae, Flavivirus	(OHFV)
Oncorhynchus masou virus	Herpesviridae, Gammaherpesvirinae	(OMV)
Oncoviruses	Retroviridae	
O'Nyong Nyong virus	Togaviridae, Alphavirus	(ONN)
Orbiviruses	Reoviridae	
Oregon sockeye salmon disease virus	Retroviridae	(OSDV)
Orf viruses	Poxviridae, Chordopoxviridae, Parapoxvirus	
Oropouche virus	Bunyaviridae, Bunyavirus	
Orthohepadnaviruses	Hepadnaviridae	
Orthomyxoviruses	Orthomyxoviridae	
Orthopoxvirus	Poxviridae	

Orthoreovirus	Reoviridae	
Ovine herpesviruses	Herpesviridae	Sheep pulmonary adenomatosis associated herpesvirus; Sheep associated malignant catarrhal fever of cattle herpesvirus
Ovine respiratory syncytial virus	Paramyxoviridae, Pneumovirus	(ORSV)
Oyster virus	Birnaviridae	(OV)
Papillomaviruses	Papovaviridae	
Paraflumorbillivirinae	Paramyxoviridae	
Parainfluenza viruses	Paramyxoviridae, Paramyxovirus	(PIV)
Paramyxoviruses	Paramyxoviridae	
Parana virus	Arenaviridae	
Parapoxviruses	Poxviridae, Chordopoxviridae, Parapoxvirus	
Pararotaviruses	Reoviridae	Rotaviruses
Parotid virus	Papovaviridae	
Parvoviruses	Parvoviridae	
Perch iridovirus	Iridoviridae	
Perch rhabdovirus	Rhabdoviridae	
Peste des petits ruminants virus	Paramyxoviridae, Morbillivirus	(PPR)
Pestivirus	Flaviviridae	
Phlebotomus fever virus	Bunyaviridae, Phebovirus	
Phlebovirus	Bunyaviridae	
Phocid distemper virus	Paramyxoviridae, Morbillivirus	(PDV)
Pichinde virus	Arenaviridae	
'Picobirnaviruses'	Birnaviridae	
Picornaviruses	Picornaviridae	
Pike fry rhabdovirus	Rhabdoviridae	(PFV)
Piry virus	Rhabdoviridae, Vesiculovirus	
Pneumonia virus of mice	Paramyxoviridae, Pneumovirus	(PVM)
Pneumovirus	Paramyxoviridae	
PO-1-Lu virus	Retroviridae	
Polioviruses	Picornaviridae	
Polyomaviruses	Papovaviridae	
Pongine herpesvirus	Herpesviridae, Gammaherpesvirinae	
Porcine hemagglutinating encephalomyelitis virus	Coronaviridae	(HEV)
Porcine parvovirus	Parvoviridae	(PPV); SMEDI virus
Porcine respiratory coronavirus	Coronaviridae	(PRCV)
Powassan virus	Flaviviridae	(POW)
Poxviruses	Poxviridae	

Prospect Hill virus	Bunyaviridae, Hantavirus	
Pseudocowpox virus	Poxviridae, Chordopoxviridae, Parapoxvirus	
Pseudorabies virus	Herpesviridae, Alphaherpesvirinae	(PRV); Swine herpesvirus-1
Puumala virus	Bunyaviridae, Hantavirus	
Quail adeno-associated viruses	Parvoviridae	(Quail aa, AVs)
Quailpox virus	Poxviridae, Avipoxvirus	
Rabbit coronavirus	Coronaviridae	(RbCV)
Rabbit fibroma virus	Poxviridae, Leporipoxvirus	
Rabbit hemorrhagic disease virus	Caliciviridae	(RHD)
Rabbit papilloma virus	Retroviridae	
Rabbitpox virus	Poxviridae, Orthopoxvirus	
Rabies virus	Rhabdoviridae	
Rabies-like viruses	Rhabdoviridae, Lyssavirus	
Raccoon parvovirus	Parvoviridae, Parvovirus	(RPV)
Rat coronavirus	Coronaviridae	(RCV)
Rat virus	Parvoviridae	(RV)
Reovirus type 1 Lang	Reoviridae	(T1L)
Reoviruses	Reoviridae	
Respiratory syncytial virus	Paramyxoviridae, Pneumovirus	(RSV)
Reston virus	Filoviridae	
Reticuloendotheliosis virus	Retroviridae, Oncovirinae	(REV)
Reticuloendotheliosis virus strain T	Retroviridae, Oncovirinae	(REV-T)
Reticuloendotheliosis-associated virus	Retroviridae Oncovirinae	(REV-A)
Retroviruses	Retroviridae	
Rhabdoviruses	Rhabdoviridae	
Rhinoviruses	Picornaviridae	
Rift Valley fever virus	Bunyaviridae, Phlebovirus	
Rinderpest virus	Paramyxoviridae, Morbillivirus	(RV)
Rio Grande cichlid rhabdovirus	Rhabdoviridae	
Rocio virus	Flaviviridae	
Rodent parvoviruses	Parvoviridae	
Ross River virus	Togaviridae, Alphavirus	(RRV)
Rotaviruses	Reoviridae	Pararotaviruses
Rous sarcoma virus	Retroviridae	(RSV)
Rubella virus	Togaviridae, Rubivirus	

Rubivirus	Togaviridae	
Russian spring summer encephalitis virus	Flaviviridae, Flavivirus	(RSSEV)
SA11 (simian rotavirus)	Reoviridae	
SA12 polyomavirus	Papovaviridae	
Sacramento River Chinook disease virus	Retroviridae	(SRCDV)
St Louis Encephalitis virus	Flaviviridae, Flavivirus	(SLEV)
Salmonid herpesviruses	Herpesviridae, Gammaherpesvirinae	
San Miguel sea lion virus	Caliciviridae	
Sandfly fever viruses	Bunyaviridae, Phleboviruses	
Sea lion foamy virus	Retroviridae, Spumavirinae	
Semliki Forest virus	Togaviridae, Alphavirus	(SFV)
Sendai virus	Paramyxoviridae, Paramyxovirus	Hemagglutinating virus of Japan
Seoul virus	Bunyaviridae, Hantavirus	
Sepik virus	Flaviviridae, Flavivirus	
Sheeppox viruses	Poxviridae, Chordopoxvirinae, Capripoxvirus	
Shope fibroma virus	Poxviridae, Leporipoxvirus	(SFV)
Shope papillomavirus	Papovaviridae	
Sialodacryadenitis virus of rats	Coronaviridae	(SDAV)
Simian cytomegalovirus	Herpesviridae, Betaherpesvirus	(SCMV)
Simian foamy virus	Retroviridae, Spumavirinae	(SFV)
Simian hemorrhagic fever virus	Flaviviridae, Flavivirus	(SHFV)
Simian hepatitis A virus	Herpesviridae, Gammaherpesvirinae	(HAV) Herpesvirus saimiri
Simian herpesvirus	Herpesviridae, Gammaherpesvirinae	Herpesvirus ateles
Simian immunodeficiency virus	Retroviridae	(SIV)
Simian retrovirus type 1	Retroviridae, Lentivirus	(SRV-1)
Simian retrovirus type 2	Retroviridae, Lentivirus	(SRV-2)
Simian retrovirus type 3	Retroviridae, Lentivirus	(SRV-3)
Simian retrovirus type 4	Retroviridae, Lentivirus	(SRV-4)
Simian retrovirus type 5	Retroviridae, Lentivirus	(SRV-5)
Simian rotavirus	Reoviridae	(SA11)
Simian sarcoma associated virus	Retroviridae	(SSAV)

Simian sarcoma virus	Retroviridae	(SSV)
Simian T-cell leukemia virus	Retroviridae	(STLV)
Simian T-cell leukemia virus type III	Retroviridae	(STLV-III)
Simian T-lymphoma virus-1	Retroviridae	(STLV-1)
Simian virus 5	Paramyxoviridae, Paramyxovirus	(SV5)
Simian virus 40	Papovaviridae, Polyomavirus	(SV40) Vacuolating virus
Simian virus 41	Paramyxoviridae	(SV41)
Simplexvirus	Herpesviridae	
Sindbis virus	Togaviridae, Alphavirus	(SIN)
Smallpox virus	Poxviridae, Orthopoxvirus	Variola virus
SMEDI enteroviruses	Picornaviridae	
Snakehead rhabdovirus	Rhabdoviridae	
Snow Mountain agent	Caliciviridae	
Snowshoe hare virus	Flaviviridae	(SSH)
Snyder-Theilin feline sarcoma virus	Retroviridae	(ST-FeSV)
South African ovine maedi-visna virus	Retroviridae, Lentivirinae	(SA-OMVV)
Southampton virus	Caliciviridae	
Spleen focus-forming virus	Retroviridae	(SFFV)
Spring viremia of carp virus	Rhabdoviridae	(SVCV)
Spuma retroviruses	Retroviridae	
Spumaviruses	Retroviridae, Spumavirinae	
Squirrel fibroma virus	Poxviridae, Leporipoxvirus	(SqFV)
Squirrel monkey retrovirus	Retroviridae	S(MRV)
Stomatitis papulosa virus	Poxviridae, Chordopoxviridae, Parapoxvirus	
Suipoxvirus	Poxviridae	
Susan McDonough feline sarcoma virus	Retroviridae	(SM-FeSV)
Swine herpesvirus-1	Herpesviridae	Pseudorabies virus
Swine vesicular disease virus	Picornaviridae, Enterovirus	(SVDV)
Swinepox virus	Poxviridae, Chordopoxvirinae, Suipoxvirus	(SPV)
Tacaribe virus	Arenaviridae	
Tahyna virus	Bunyaviridae	(TAH)
Tamiami virus	Arenaviridae	
Tanapox virus	Poxviridae, Yatapoxvirus	
Taunton agent	Caliciviridae	
Teschen disease virus	Picornaviridae	(TMEV)
Theiler's murine encephalomyelitis virus	Picornaviridae	Murine poliovirus
Theiler's viruses	Picornaviridae	
Thottapalayam viruses	Bunyaviridae, Hantavirus	

Tick-borne encephalitis virus	Flaviviridae, Flavivirus	(TBEV)
Togaviruses	Togaviridae	
Toluca-1 virus	Picornaviridae, Enterovirus	Enterovirus serotype 69
Toroviruses	Coronaviridae	
Toscana virus	Bunyaviridae, Phlebovirus	
Trager duck spleen necrosis virus	Retroviridae, Oncovirinae	(SNV)
Transmissible gastroenteritis virus of swine	Coronaviridae	(TGEV)
Transmissible spongiform encephalopathy of mule deer and elk virus	Unclassified	(TSE)
Tree shrew herpesviruses	Herpesviridae	
Tupaia herpesvirus	Herpesviridae	(THV-1)
Tupaia herpesvirus	Herpesviridae	(THV-2)
Tupaia herpesvirus	Herpesviridae	(THV-3)
Turbot herpesvirus	Herpesviridae, Gammaherpesvirinae	
Turkey bluecomb coronavirus	Coronaviridae	(TCV)
Turkey herpesvirus	Herpesviridae	Herpesvirus of turkeys (HVT)
Turkey rhinotracheitis virus	Paramyxoviridae, Pneumovirus	(TRTV)
Uukuviruses	Bunyaviridae	
Vaccinia virus	Poxviridae, Chordopoxvirinae, Orthopoxvirus	
Vacuolating virus	Papovaviridae, Polyomavirus	Simian virus 40
Varicellazoster virus	Herpesviridae, Alphaherpesvirinae	(VZV); Chickenpox virus
Variola virus	Poxviridae, Orthopoxvirus	Smallpox virus
Venezuelan equine encephalitis virus	Togoviridae, Alphavirus	
Venezuelan hemorrhagic fever	Arenaviridae	(VHF) Guanarito virus
Vesicular exanthema virus of swine	Caliciviridae	(VES)
Vesicular stomatitis virus	Rhabdoviridae, Vesiculovirus	(VSV)
Vesiculovirus	Rhabdoviridae	
Viral hemorrhagic septicemia virus	Rhabdoviridae	(VHSV) Egtved virus
Virus diarrhea virus	Flaviviridae, Pestivirus	(VD)
Visna virus	Retroviridae, Lentivirus	
Visna–maedi viruses	Retroviridae, Lentivirus	
Walleye herpesvirus	Herpesviridae	

Wesselsbron virus	Flaviviridae, Flavivirus	(WSLV)
West Nile virus	Flaviviridae, Flavivirus	(WNV)
Western equine encephalitis virus	Togaviridae, Alphavirus	(WEE)
Woodchuck hepatitis virus	Hepadnaviridae	(WHV)
Y62-33 virus	Togaviridae	
Yaba monkeypox virus	Poxviridae, Chordopoxvirinae, Yatapoxvirus	
Yamagushi 73 sarcoma virus	Retroviridae	
Yamame tumor virus	Herpesviridae, Gammaherpesvirinae	
Yatapoxviruses	Poxviridae	
Yellow fever virus	Flaviviridae, Flavivirus	

PLANT VIRUSES INCLUDED IN THE ENCYCLOPEDIA

Virus	Family/Group	Synonym/(Abbreviation)
Abutilon mosaic virus	Geminivirus	(AbMV)
African cassava mosaic virus	Geminivirus	(ACMV)
Agaricus bisporus virus 4	Partitiviruses	(AbV-4)
Alfalfa cryptic virus	Cryptoviruses	(ACV1)
Alfalfa latent virus	Carlavirus	(ALV)
Alfalfa mosaic virus	Alfamovirus	(AlMV)
Alligatorweed stunting virus	Closteroviruses	(AwSV)
American hop latent virus	Carlavirus	(AHLV)
Andean potato latent virus	Tymoviruses	
Andean potato mottle virus	Comoviruses	(APMV)
Anthoxanthum latent blanching virus	Hordeiviruses	(ALBV)
Anthriscus yellows virus	Sesquiviruses	(AYV)
Apple chlorotic leaf spot virus	Trichoviruses	(ACLSV)
Apple latent virus type, 2	Capilloviruses	
Apple mosaic virus	Ilarvirus	(ApMV)
Apple scar skin viroid	Viroid	
Apple stem pitting virus	Closteroviruses	
Apple stem grooving virus	Capilloviruses	(ASGV)
Arabis mosaic virus	Nepoviruses	(ArMV)
Arracacha A virus	Nepoviruses	
Arracacha B virus	Nepoviruses	
Arrhenatherum blue dwarf virus	Reoviridae, Phytoreoviruses	(ABDV)
Artichoke Italian latent virus	Nepoviruses	
Artichoke latent virus	Carlavirus	(ArLV)
Artichoke mottled crinkle virus	Tombusviruses	(AMCV)
Artichoke vein banding virus	Nepoviruses	
Artichoke yellow ringspot virus	Nepoviruses	

Aspergillus foetidus virus S	Totiviridae	(AfV-S)
Aspergillus niger virus S	Totiviridae	(AnV-S)
Aspergillus ochraceous virus	Partitiviruses	(AoV)
Australian grapevine viroid	Viroid	
Avocado sunblotch viroid	Viroid	
Banana streak virus	Badnaviruses	(BSV)
Barley stripe mosaic virus	Hordeiviruses	(BSMV)
Barley yellow dwarf virus	Luteoviruses	(BYDV)
Barley yellow striated mosaic virus	Rhabdoviridae	(BYSMV)
Bean golden mosaic virus	Geminivirus	(BGMV)
Bean leaf roll virus	Luteoviruses	
Bean mild mottle virus	Carmoviruses	(BMMV)
Bean pod mottle virus	Comoviruses	(BPMV)
Bean rugose mosaic virus	Comoviruses	(BRMV)
Bean yellow vein banding complex virus	Luteoviruses	(BYVBV)
Beet cryptic virus	Cryptoviruses	(BCV)
Beet curly top virus	Geminiviridae	(BCTV)
Beet mild yellowing virus	Luteoviruses	
Beet necrotic yellow vein virus	Furoviruses	(BNYVV)
Beet pseudo-yellows virus	Closteroviruses	(BPYV)
Beet soil-borne virus	Furoviruses	(BSBV)
Beet western yellows virus	Luteoviruses	(BWYV)
Beet yellow stunt virus	Closteroviruses	(BYSV)
Beet yellows virus	Closteroviruses	(BYV)
Belladonna mottle virus	Tymoviruses	
Belladonna mottle virus - Iowa strain	Tymoviruses	
Blackgram mottle virus	Carmoviruses	(BGMV)
Blueberry leaf mottle virus	Nepoviruses	(BLMV)
Blueberry scorch virus	Carlavirus	(BCV)
Blueberry shoestring virus	Sobemoviruses	(BSSV)
Broad bean mottle virus	Bromoviruses	(BBMV)
Broad bean necrosis virus	Furoviruses	(BBNV)
Broad bean stain virus	Comoviruses	(BBSV)
Broad bean true mosaic virus	Comoviruses	(BBTMV)
Broad bean wilt disease virus	Fabaviruses	(BBWV)
Broccoli necrotic yellows virus	Rhabdoviridae	
Brome mosaic virus	Bromoviruses	(BMV)
Bromovirus	Ilarvirus	
Burdock yellows virus	Closteroviruses	(BuYV)
Cacao swollen shoot virus	Badnaviruses	(CSSV)
Cacao yellow mosaic virus	Tymoviruses	
Cactus virus 2	Carlavirus	(CaV2)
Cactus virus X	Potexviruses	
Canna yellow mottle virus	Badnaviruses	(CaYMV)
Caper latent virus	Carlavirus	(CapLV)
Caper vein binding virus	Carlavirus	(CapVBV)
Carnation cryptic virus	Cryptoviruses	(CCV)
Carnation Italian ringspot virus	Tombusviruses	(CIRV)
Carnation latent virus	Carlaviruses	(CLV)
Carnation mottle virus	Carmoviruses	(CarMV)
Carnation necrotic fleck virus	Closteroviruses	(CNFV)
Carnation ringspot virus	Dianthoviruses	(CRSV)

Carnation stunt associated viroid	Viroid	
Carrot red leaf luteovirus	Luteoviruses	
Carrot yellow leaf virus	Closteroviruses	(CYLV)
Cassava American latent virus	Nepoviruses	
Cassava green mottle virus	Nepoviruses	
Cassava latent virus	Geminiviruses	(CLV)
Cassia mild mosaic virus	Carlavirus	(CasMMV)
Cauliflower mosaic virus	Caulimoviruses	(CaMV)
Cereal tillering disease virus	Reoviridae, Phytoreoviruses	(CTDV)
Cherry leafroll virus	Nepoviruses	(CLRV)
Cherry rasp leaf virus	Nepoviruses	
Chicory blotch virus	Carlavirus	(ChiBV)
Chicory yellow mottle virus	Nepoviruses	(CYMV)
Chlorella NC64A viruses	Phycodnaviridae	
Chlorella Pbi viruses	Phycodnaviridae	
Chlorella viruses	Phycodnaviridae	
Chloris striae mosaic virus	Geminiviruses	(CSMV)
Chrysanthemum stunt viroid	Viroid	
Chrysanthemum virus B	Carlavirus	(CVB)
Citrus bent leaf virus	Ilarviruses	
Citrus exocortis viroid	Viroid	(CEVd)
Citrus leaf rugose virus	Ilarviruses	(CLRV)
Citrus tatter leaf virus	Capilloviruses	(CTLV)
Citrus tristeza virus	Closteroviruses	(CTV)
Citrus variegation virus	Ilarviruses	(CVV)
Clitoria yellow virus	Tymoviruses	
Clover yellow mosaic virus	Potexviruses	(ClYMV)
Clover yellows virus	Closteroviruses	(CYV)
Cocksfoot mild mosaic virus	Sobemoviruses	(CMMV)
Cocksfoot mottle virus	Sobemoviruses	(CfMV)
Cocoa necrosis virus	Nepoviruses	
Coconut cadang-cadang viroid	Viroid	
Coconut tinangaja viroid	Viroid	
Coffee ring spot virus	Rhabdoviridae	
Cole latent virus	Carlavirus	(CoLV)
Coleus blumei viroid	Viroid	
Columnea latent viroid	Viroid	(CLVd)
Commelina yellow mottle virus	Badnaviruses	(CoYMV)
Cowpea chlorotic mottle virus	Bromoviruses	(CCMV)
Cowpea mild mottle virus	Carlavirus	(CpMMV)
Cowpea mosaic virus	Comoviruses	(CPMV)
Cowpea mottle virus	Carmoviruses	(CMeV)
Cowpea severe mosaic virus	Comoviruses	(CPSMV)
Crimson clover latent virus	Nepoviruses	
Cucumber green mottle mosaic tobamovirus	Tobamoviruses	(CGMMV)
Cucumber mosaic virus	Cucumoviruses	(CMV)
Cucumber necrosis virus	Tombusviruses	(CNV)
Cucumber pale fruit viroid	Viroid	
Cucumber soil-borne virus	Carmoviruses	(CSBV)
Cucumber virus 1	Cucumoviruses	(CV1)
Cucumovirus	Ilarvirus	
Cycas necrotic stunt virus	Nepoviruses	
Cymbidium ringspot virus	Tombusviruses	(CyRSV)

Cynodon mosaic virus	Carlavirus	(CynMV)
Cynosurus mottle virus	Sobemoviruses	(CyMV)
Dandelion latent virus	Carlavirus	(DLV)
Dandelion yellow mosaic virus	Sesquiviruses	(DYMV)
Daphne virus S	Carlavirus	(DaVS)
Dark green epinasty virus	Capilloviruses	
Dasheen mosaic virus	Potyvirus	
Dendrobium vein necrosis	Closteroviruses	(DVNV)
Desmodium yellow mottle virus	Tymoviruses	
Digitaria streak virus	Geminivirus	(DSV)
Diodia vein chlorosis viruses	Closteroviruses	(DVCV)
Dioscorea bulbifera bacilliform virus	Badnaviruses	(DbBV)
Dulcamara carlavirus A	Carlavirus	(DuCVA)
Dulcamara carlavirus B	Carlavirus	(DuCVB)
Dulcamara mottle virus	Tymoviruses	
E-36 virus	Capilloviruses	
Echinochola ragged stunt virus	Reoviridae, Phytoreoviruses	(ERSV)
Eggplant mild mottle virus	Carlavirus	(EMMV)
Eggplant mosaic virus	Tymoviruses	
Eggplant mottled crinkle virus	Tombusviruses	
Eggplant mottled dwarf virus	Rhabdoviridae	(EMDV)
Elderberry carlavirus	Carlavirus	(ECV)
Elderberry latent virus	Carmoviruses	(ELV)
Elm mottle virus	Ilarvirus	(EMtV)
Erysimum latent virus	Tymoviruses	
European wheat striate mosaic virus	Tenuviruses	(EWSMV)
Festuca leaf streak virus	Rhabdoviridae	(FLSV)
Festuca necrosis virus	Closterovirus	(FNV)
Fig virus S	Carlavirus	(FVS)
Figwort mosaic virus	Caulimoviruses	(FMV)
Fiji disease virus	Reoviridae, Phytoreoviruses	(FDV)
Fijivirus	Reoviridae, Phytoreoviruses	
Foxtail mosaic virus	Potexviruses	(FoMV)
Frangipani mosaic tobamovirus	Tobamoviruses	(FrMV)
Fuchsia latent virus	Carlavirus	(FuLV)
Furcraea necrotic streak virus	Dianthoviruses	(FNSV)
Gaeumannomyces graminis virus 019/6-A	Partitiviruses	
Gaeumannomyces graminis virus 87-1-H	Totiviridae, Totiviruses	(GgV-87-1-H)
Gaeumannomyces graminis virus T1-A	Partitiviruses	
Galinsoga mosaic virus	Carmoviruses	(GMV)
Garlic mosaic virus	Carlavirus	(GaMV)
Gentiana carlavirus	Carlavirus	(GeCV)
Ginger chlorotic fleck virus	Sobemoviruses	(GCFV)
Gladiolus carlavirus	Carlavirus	(GCV)
Glycine mosaic virus	Comoviruses	(GMV)
Glycine mottle virus	Carmoviruses	(GMeV)

Grapevine Algerian latent virus	Tombusviruses	
Grapevine Bulgarian latent virus	Nepoviruses	
Grapevine chrome mosaic virus	Nepoviruses	(GCMV)
Grapevine corky bark associated virus	Closteroviruses	(GCBAV)
Grapevine fanleaf virus	Nepoviruses	(GFLV)
Grapevine leafroll disease virus	Closteroviruses	(GLRAVs)
Grapevine Tunisian ringspot virus	Nepoviruses	
Grapevine virus A	Closteroviruses	(GVA)
Grapevine yellow speckle viroid	Viroid	
Groundnut bud necrosis virus	Bunyaviridae, Tospoviruses	(GBNV)
Groundnut crinkle virus	Carlavirus	(GrCV)
Groundnut ringspot virus	Bunyaviridae, Tospoviruses	(GRSV)
Groundnut rosette assistor luteovirus	Luteoviruses	(GRAV)
Gynura latent virus	Carlavirus	(GyLV)
Helenium virus S	Carlavirus	(HelVS)
Helleborus carlavirus	Carlavirus	(HeCV)
Helminthosporium victoriae 145S virus	Partitiviruses	(HvV-145S)
Helminthosporium victoriae 190S virus	Totiviridae, Totiviruses	(HvV-190S)
Heracleum latent virus	Closteroviruses	(HLV)
Heracleum virus 6	Closteroviruses	(HV6)
Hibiscus chlorotic ringspot virus	Carmoviruses	(HCRSV)
Hibiscus latent ringspot virus	Nepoviruses	
Hop latent viroid	Viroid	
Hop latent virus	Carlavirus	(HopLV)
Hop mosaic virus	Carlavirus	(HopMV)
Hop stunt viroid	Viroid	(HSVd)
Hop trefoil cryptic virus 1	Cryptoviruses	(HTCVi)
Hpochoeris mosaic virus	Furoviruses	(HMV)
Impatiens necrotic spot virus	Bunyaviridae, Tospoviruses	(INSV)
Indian bunchy top viroid	Viroid	
Indian cassava mosaic virus	Geminivirus	
Indonesian soybean dwarf virus	Luteoviruses	
Kalanchoe carlavirus	Carlavirus	(KaCV)
Kalanchoe latent virus	Carlavirus	(KLV)
Kalanchoe top-spotting virus	Badnaviruses	(KTSV)
Kennedya yellow mosaic virus	Tymoviruses	(KYMV)
Lamium mild mosaic virus	Fabaviruses	(LMMV)
Lettuce big vein virus	Furoviruses	
Lettuce infectious yellows virus	Closteroviruses	(LIYV)
Lettuce necrotic yellow vein virus	Rhabdoviridae	(LNYV)
Lilac chlorotic leafspot virus	Capillovirus	(LCLSV)
Lilac mottle virus	Carlavirus	(LMV)
Lily symptomless virus	Carlavirus	(LSV)
Lily virus X	Potexviruses	(LVX)
Lolium enation viruses	Reoviridae, Phytoreoviruses	(LEV)

Lonicera latent virus	Carlavirus	(LonLV)
Lucerne Australian latent virus	Nepoviruses	
Lucerne Australian symptomless virus	Nepoviruses	
Lucerne transient streak virus	Sobemovirus	(LTSV)
Lychnis ringspot virus	Hordeiviruses	(LRSV)
Maize chlorotic dwarf virus	Machlomoviruses	(MCDV)
Maize chlorotic mottle virus	Machlomoviruses	(MCMV)
Maize mosaic virus	Rhabdoviridae	(MMV)
Maize rough dwarf virus	Reoviridae, Phytoreoviruses	(MRDV)
Maize streak virus	Geminiviridae	(MSV)
Maize stripe virus	Tenuiviruses	(MstV)
Maracuja mosaic tobamovirus	Tobamoviruses	(MaMV)
Melon necrotic spot virus	Carmoviruses	(MNSV)
Miscanthus streak virus	Geminivirus	
Moroccan pepper virus	Tombusviruses	
Mulberry latent virus	Carlavirus	(MuLV)
Mulberry ringspot virus	Nepoviruses	
Mung bean yellow mosaic virus	Geminivirus	
Muskmelon vein necrosis virus	Carlavirus	(MmVNV)
Mycogone perniciosa virus	Totiviridae, Totiviruses	(MpV)
Myrobalan latent ringspot virus	Nepoviruses	
Nandina stem pitting virus	Closteroviruses	(NSPV)
Narcissus mosaic virus	Potexviruses	(NaMV)
Narcissus tip necrosis virus	Carmoviruses	(NTNV)
Nasturtium mosaic virus	Carlavirus	(NasMV)
Nasturtium ringspot virus	Fabaviruses	(NRSV)
Neckar river virus	Tombusviruses	
Nerine latent virus	Carlavirus	(NeLV)
Nicotiana velutina mosaic virus	Furoviruses	(NVMV)
Northern cereal mosaic virus	Rhabdoviridae	
Oat golden stripe virus	Furoviruses	(OGSV)
Oat sterile dwarf virus	Reoviridae, Phytoreoviruses	(OSDV)
Odontoglossum ringspot tobamovirus	Tobamoviruses	(ORSV)
Okra mosaic virus	Tymoviruses	
Olive latent-1 virus	Sobemovirus	(OLV-1)
Olive latent ringspot virus	Nepoviruses	
Ononis yellow mosaic virus	Tymoviruses	
Pangola stunt virus	Reoviridae, Phytoreoviruses	(PSV)
Panicum mosaic virus	Sobemovirus	(PMV)
Papaya mosaic virus	Potexviruses	(PaPMV)
Paramecium bursaria Chlorella virus 1	Phycodnaviridae	(PBCV-1)
Parsnip yellow fleck virus	Sesquiviridae	(PYFV)
Paspalum striate mosaic virus	Geminivirus	
Passiflora latent virus	Carlavirus	(PaLV)
Passiflora yellow mosaic virus	Tymoviruses	
Pbi viruses (*Chlorella*)	Phycodnaviruses	

Pea early browning virus	Tobraviruses	(PEBV)
Pea enation mosaic virus	Enamovirus	(PEMV)
Pea leaf roll virus	Luteoviruses	
Pea mild mosaic virus	Comoviruses	(PMiMV)
Pea streak virus	Carlavirus	(PeSV)
Peach latent mosaic viroid	Nepoviruses	
Peach rosette mosaic virus	Nepoviruses	
Peanut chlorotic streak virus	Caulimoviruses	
Peanut clump virus	Furoviruses	(PCV)
Peanut stunt virus	Cucumoviruses	(PSV)
Peanut yellow spot virus	Bunyaviridae, Tospoviruses	(PYSV)
Pear blister canker viroid	Viroid	
Pelargonium flower break virus	Carmoviruses	(PFBV)
Pelargonium leaf curl virus	Tombusviruses	(PLCV)
Pelargonium line pattern virus	Carmoviruses	(PLPV)
Penicillium brevicompactum virus	Partitiviruses	(PbV)
Penicillium chrysogenum virus	Partitiviruses	(PcV)
Penicillium cyaneo-fulvum virus	Partitiviruses	(Pc-fV)
Penicillium stoloniferum virus F	Partitiviruses	(PsV-F)
Penicillium stoloniferum virus S	Partitiviruses	(PsV-S)
Pepino latent virus	Carlavirus	(PepLV)
Pepper mild mottle tobamovirus	Tobamoviruses	(PMMV)
Pepper ringspot virus	Tobraviruses	(PRV)
Petunia asteroid mosaic virus	Tombusviruses	
Petunia ringspot virus	Fabaviruses	
Phialophora radicicola virus 2-2-A	Partitiviruses	(PrV-2-2-A)
Physalis mosaic virus	Tymoviruses	
Pineapple virus	Closteroviruses	(PV)
Piper yellow mottle virus	Badnaviruses	(PYMV)
Plantago mottle virus	Tymoviruses	
Plantain virus 6	Carmoviruses	(PlV6)
Plantain virus 8	Carlavirus	(PlV8)
Plum pox virus	Potyviruses	(ppV)
Poa semilatent virus	Hordeiviruses	(PSLV)
Poinsettia mosaic virus	Tymoviruses	
Poplar mosaic virus	Carlavirus	(PMV)
Potato aucuba mosaic virus	Potexviruses	(PoAMV)
Potato black ringspot virus	Nepoviruses	
Potato leaf roll virus	Luteoviruses	(PLRV)
Potato mop-top virus	Furoviruses	(PMTV)
Potato spindle tuber viroid	Viroid	(PSTVd)
Potato U virus	Nepoviruses	
Potato virus M	Carlavirus	(PVM)
Potato virus S	Carlavirus	(PVS)
Potato virus T	Capilloviruses	(PVT)
Potato virus X	Potexviruses	(PVX)
Potato virus Y	Potyviridae	(PVY)
Potato yellow dwarf virus	Rhabdoviridae	(PYDV)
Potato yellow mosaic virus	Geminivirus	(PYMV)
Prune dwarf virus	Ilarvirus	(PDV)
Prunus necrotic ringspot virus	Ilarvirus	(PNRV)
Quail pea mosaic virus	Comoviruses	(QPMV)

Radish mosaic virus	Comoviruses	(RaMV)
Raspberry ringspot virus	Nepoviruses	(RRSV)
Red clover mottle virus	Comoviruses	(RCMV)
Red clover necrotic mosaic virus	Dianthoviruses	(RCNMV)
Red clover vein mosaic virus	Carlavirus	(RCVMV)
Rhizoctonia solani virus 717	Partitiviruses	(RsV-717)
Ribgrass mosaic tobamovirus	Tobamoviruses	(RMV)
Rice black streaked dwarf virus	Reoviridae, Phytoreoviruses	(RBSDV)
Rice dwarf virus	Reoviridae, Phytoreoviruses	(RDV)
Rice gall dwarf virus	Reoviridae, Phytoreoviruses	(RGDV)
Rice grassy stunt virus	Tenuiviruses	(RGSV)
Rice hoja blanca virus	Tenuiviruses	(RHBV)
Rice ragged stunt virus	Reoviridae, Phytoreoviruses	(RRSV)
Rice stripe necrosis virus	Furoviruses	(RSNV)
Rice stripe virus	Tenuiviruses	(RStV)
Rice transitory yellowing virus	Rhabdoviridae	
Rice tungro bacilliform virus	Badnaviruses	(RTBV)
Rice tungro spherical virus	Sequiviridae	(RTSV)
Rice yellow mottle virus	Sobemoviruses	(RYMV)
Rubus chinese seedborne virus	Nepoviruses	
Rygrass cryptic virus	Cryptoviruses	(RGCV)
Rygrass spherical virus	Cryptoviruses	(RGSV)
Saccharomyces cerevisiae virus L-A	Totiviridae, Totiviruses	(ScV-L-A)
Saccharomyces cerevisiae virus L-BC	Totiviridae, Totiviruses	(ScV-L-BC)
Saguaro cactus virus	Carmoviruses	(SCV)
Sammon's Opuntia tobamovirus	Tobamoviruses	(SOV)
Satellite tobacco mosaic virus	Tobamoviruses	(STMV)
Satellite tobacco necrosis viruses	Necroviruses	(STNV)
Satsuma dwarf virus	Nepoviruses	
Schefflera ringspot virus	Badnaviruses	(SRV)
Scrophularia mottle virus	Tymoviruses	
Shallot latent virus	Carlavirus	(SLV)
Soil-borne wheat mosaic virus	Furoviruses	(SBWMV)
Solanaceous tymoviruses	Tymoviruses	
Solanum nodiflorum virus	Sobemoviruses	(SNMV)
Solanum yellows virus	Luteoviruses	
Sonchus yellow net virus	Rhabdoviridae	(SYNV)
Sorghum chlorotic spot virus	Furoviruses	(SCSV)
Southern bean mosaic virus	Sobemoviruses	(SBMV)
Southern bean sobemovirus	Sobemoviruses	
Southern potato virus	Carlavirus	(SPV)
Sowbane mosaic virus	Sobemoviruses	(SoMV)
Sowthistle yellow vein virus	Rhabdoviridae	(SYVV)
Soybean dwarf virus	Luteoviruses	
Squash leaf curl virus	Geminivirus	(SqLCV)
Squash mosaic virus	Comoviruses	(SqMV)
Strawberry latent ringspot virus	Nepoviruses	(SLRSV)

Strawberry mild yellow edge-associated virus	Potexviruses	(SMYEAV)
Strawberry pseudomild yellow edge virus	Carlavirus	(SPMYEV)
Subterranean clover mottle virus	Sobemoviruses	(SCMoV)
Subterranean clover redleaf virus	Luteoviruses	
Sugarcane bacilliform virus	Badnaviruses	(ScBV)
Sunn-hemp mosaic tobamovirus	Tobamoviruses	(SHMV)
Sweet clover necrotic mosaic virus	Dianthoviruses	(SCNMV)
Tephrosia symptomless virus	Carmoviruses	(TSV)
Tobacco black ring virus	Nepoviruses	(TBRV)
Tobacco etch virus	Potyviruses	(TEV)
Tobacco mild green mosaic tobamovirus	Tobamoviruses	(TMGMV)
Tobacco mosaic virus	Tobamoviruses	(TMV)
Tobacco necrosis virus	Necroviruses	(TNV)
Tobacco necrotic dwarf virus	Luteoviruses	(TNDV)
Tobacco rattle virus	Tobraviruses	(TRV)
Tobacco ringspot virus	Nepoviruses	(TRSV)
Tobacco streak virus	Ilarvirus	(TSV)
Tobacco stunt virus	Furoviruses	
Tobacco vein mottling virus	Potyviruses	(TVMV)
Tom-P	Rhabdoviridae	
Tomato apical stunt viroid	Viroid	
Tomato aspermy virus	Cucumoviruses	(TAV)
Tomato black ring virus	Nepoviruses	(TBRV)
Tomato bushy stunt virus	Tombusviruses	(TBSV)
Tomato chlorotic spot virus	Bunyaviridae, Tospoviruses	(TCSV)
Tomato golden mosaic virus	Geminiviridae	(TGMV)
Tomato mosaic virus	Tobamoviruses	(ToMV)
Tomato planta macho viroid	Viroid	
Tomato ringspot virus	Nepoviruses	(ToRSV)
Tomato spotted wilt virus	Bunyaviridae, Tospoviruses	(TSWV)
Tomato top necrosis virus	Nepoviruses	
Tomato vein yellowing virus	Rhabdoviridae	
Tomato yellow leaf curl virus	Geminivirus	(TYLCV)
Tomato yellow top virus	Luteoviruses	
Tulare apple mosaic virus	Ilarvirus	(TAMV)
Tulip breaking virus	Potyvirus	
Tulip virus X	Potexviruses	
Turnip crinkle carmovirus	Carmoviruses	(TCV)
Turnip rosette virus	Sobemoviruses	(TRoSV)
Turnip yellow mosaic virus	Tymoviruses	(TYMV)
Ullucus C virus	Comoviruses	(UCV)
Ullucus mild mottle tobamovirus	Tobamoviruses	(UMMV)
Ustilago maydis virus-H1	Totiviridae, Totiviruses	(UmV-H1)
Ustilago maydis virus	Totiviridae, Totiviruses	(UmV)
Ustilago maydis virus	Totiviridae, Totiviruses	(UmV-P1)
Velvet tobacco mottle virus	Sobemoviruses	(VTMoV)
Vicia cryptic virus	Cryptoviruses	(VCV)

Voandezia necrotic mosaic virus	Tymoviruses	
Voandzeia mosaic virus	Carlavirus	(VoMV)
Watermelon silver mottle virus	Bunyaviridae, Tospoviruses	(WSMV)
Wheat dwarf virus	Geminivirus	(WDV)
Wheat yellow leaf virus	Closteroviruses	(WYLV)
White bryony mosaic virus	Carlavirus	(WBMV)
White clover cryptic virus I	Partitiviridae	(WCCVI)
White clover cryptic virus II	Partitiviridae	(WCCVII)
White clover mosaic virus	Potexvirus	(WClMV)
Wild cucumber mosaic virus	Tymoviruses	
Wineberry latent virus	Potexvirus	
Wound tumor virus	Reoviridae, Phytoreoviruses	(WTV)
Yarrowia lipolytica virus	Totiviridae, Totiviruses	(YLV)
Yeast dsRNA virus	Unclassified	(ScV-L)

INSECT VIRUSES INCLUDED IN THE ENCYCLOPEDIA

Virus	Family/Group	Synonym/(Abbreviation)
Acute paralysis virus	Honey bee viruses	(APV)
Aedes aegypti densonucleosis virus	Parvoviridae	(AA: DNV)
Amsacta moorei entomopoxvirus	Poxviridae, Entomopoxvirinae	(AmEPV)
Antheraea eucalypti virus	Tetraviridae	(AeV)
Apis iridescent virus	Honey bee viruses	
Arkansas bee virus	Nodaviridae	(ABV)
Ascoviruses	Unclassified	
Autographa ascovirus	Ascoviruses	
Autographa californica nuclear polyhedrosis virus	Baculoviridae	(NPV)
Baculoviruses	Baculoviridae	
Bee virus X	Honey bee viruses	(BVX)
Bee virus Y	Honey bee viruses	(BVY)
Berkeley bee virus	Honey bee viruses	
Black beetle virus	Nodaviridae	(BBV)
Black queen cell virus	Honey bee viruses	(BQCV)
Bombyx mori densonucleosis virus	Parvoviridae	(BmDNV)
Boolarra virus	Nodaviridae	(BoV)
Bracoviruses	Polydnaviridae	
Cydia pomonella granulosis virus	Baculoviridae	
Chilo iridescent virus	Iridoviridae	(CIV)
Chironomus luridus entomopoxvirus	Poxviridae, Entomopoxivirinae	
Chloriridovirus	Iridoviridae	
Choristoneura biennis entomopoxvirus	Poxviridae, Entomopoxvirinae	(CbEPV)

Choristoneura fumiferana entomopoxvirus	Poxviridae, Entomopoxvirinae	(CfEPB)
Chronic paralysis virus associate	Honey bee viruses	(CVPA)
Chronic paralysis virus	Honey bee viruses	(CPV)
Cloudy wing virus	Honey bee viruses	(CWV)
Costelytra zealandica iridescent virus	Iridoviridae	(CzIV)
Cricket paralysis virus	Picornaviridae	(CrPV)
Cricket paralysis virus	Picornaviridae	$(CrPV)_{ark}$
Cricket paralysis virus	Picornaviridae	$(CrPV)_{brk}$
Cypoviruses	Reoviridae	
Cytoplasmic polyhedrosis viruses	Reoviridae	(CPV)
Darna trima virus	Tetraviridae	(DtV)
Dasychira pudibunda virus	Tetraviridae	(DpV)
Deformed wing virus	Honey bee viruses	(DWV)
Densonucleosis viruses	Parvoviridae	
Drosophila C virus	Picornaviridae	(DCV)
Drosophila line virus	Nodaviridae	(DLIV)
Drosophila X virus	Birnaviridae	(DXV)
Egypt bee virus	Honey bee viruses	
Entomopoxviruses	Poxviridae, Entomopoxvirinae	(EPVs)
Filamentous virus	Honey bee viruses	(FV)
Flock house virus	Nodaviridae	(FHV)
Galleria mellonella densonucleosis virus	Parvoviridae	(GmDNV)
Gonometa virus	Picornaviridae	(GoV)
Granulosis viruses	Baculoviridae	(GVs)
Gypsy moth virus	Nodaviridae	(GMV)
Harrisina brillians granulosis virus	Baculoviridae	
Helicoverpa ascovirus	Ascoviruses	
Heliothis armigera stunt virus	Tetraviridae	(HaSV)
Honey bee viruses	Unclassified	
Hz-1 baculovirus	Baculoviridae	
Ichnoviruses	Polydnaviridae	
Iridoviruses	Iridoviridae	
Junonia coenia densonucleosis virus	Parvoviridae	(JcDNV)
Kashmir bee virus	Honey bee viruses	(KBV)
Lymantria dispar nuclear polyhedrosis virus	Baculoviridae	
Manawatu virus	Nodaviridae	(MwV)
Melolontha entomopoxvirus	Poxviridae, Entomopoxvirinae	
New Zealand black beetles virus	Nodaviridae	(BBV)
Nodamura virus	Nodaviridae	(NOV)
Nodaviruses	Nodaviridae	

Nuclear polyhedrosis viruses	Baculoviridae	(NPVs)
Nudaurelia beta virus	Tetraviridae	(NβV)
Nudaurelia omega virus	Tetraviridae	(NωV)
Orgyria pseudotsugata multiple nuclear polyhedrosis virus	Baculoviridae	
Oryctes baculovirus	Baculoviridae	
Philosamia X virus	Tetraviridae	(PXV)
Picornaviruses	Picornaviridae	
Plodia interpunctella granulosis virus	Baculoviridae	
Polydnaviruses	Polydnaviridae	
Pseudaletia separata virus	Poxviridae, Entomopoxivirinae	
Pseudoletia unipuncta granulosis virus	Baculoviridae	
Pseuduoplusia includens virus	Tetraviridae	(PiV)
Rhabdoviruses	Rhabdoviridae	
Sacbrood virus	Honey bee viruses	(SBV)
Sericesthis iridescent virus	Iridoviridae	(SIV)
Sibine fusca densonucleosis virus	Parvoviridae	
Sigma rhabdoviruses	Rhabdoviridae	
Slow paralysis virus	Honey bee viruses	
Spodoptera ascovirus	Ascoviruses	
Spodoptera litura granulosis virus	Baculoviridae	
Sporoptera frugiperda virus	Ascoviruses	
Striped jack nervous necrosis virus	Nodaviridae	(SJNNV)
Tellina virus	Birnaviridae	(TV)
Tetraviruses	Tetraviridae	
Thailand sacbrood virus	Honey bee viruses	(TSBV)
Thosea asigna virus	Tetraviridae	(TaV)
Tipula iridescent virus	Iridoviridae	(TIV)
Trichoplusia ascovirus	Ascoviruses	
Trichoplusia ni granulosis virus	Baculoviridae	
Trichoplusia ni virus	Tetraviridae	(TnV)
Wiseana iridescent virus	Iridoviridae	(WIV)

BACTERIAL VIRUSES INCLUDED IN THE ENCYCLOPEDIA

Virus	Family/Group
2C bacteriophage	*B subtilis* group
7S bacteriophage	Leviviridae
186 bacteriophage	Myoviridae
434 bacteriophage	Siphoviridae
933J bacteriophage	Podoviridae
933W bacteriophage	Podoviridae
Alpha 3 bacteriophage	Myoviridae
Archaebacterial bacteriophage	Unclassified

B278 bacteriophage	Myoviridae
β bacteriophage	Siphoviridae
BA14 bacteriophage	Podoviridae
BA127 bacteriophage	Podoviridae
BA156 bacteriophage	Podoviridae
Bacillus subtilis bacteriophages	Unclassified
Bacillusphage	Tectiviridae
BZ13 bacteriophage	Leviviridae
CWP bacteriophage	Archaebacterial
Cyanobacteria bacteriophage	Unclassified
D108 bacteriophage	Myoviridae
D3112 bacteriophage	Myoviridae
DAV1 bacteriophage	Archaebacterial
f1 bacteriophage	Inoviridae
f2 bacteriophage	Leviviridae
ϕ6 bacteriophage	Cystoviridae
ϕ15 bacteriophage	Siphoviridae
ϕ29 bacteriophage	*B subtilis* group
ϕ105 bacteriophage	Siphoviridae
ϕ105J119 bacteriophage	*B subtilis* group
ϕCb12r bacteriophage	Leviviridae
ϕCb2r bacteriophage	Leviviridae
ϕCb8r bacteriophage	Leviviridae
ϕCb23r bacteriophage	Leviviridae
fd bacteriophage	Inoviridae
Φe bacteriophage	*B subtilis* group
Ff bacteriophage	Inoviridae
ϕH bacteriophage	Archaebacterial
Fl bacteriophage	Siphoviridae
ϕN bacteriophage	Archaebacterial
fr bacteriophage	Leviviridae
ϕX174 bacteriophage	Microviridae
G4 bacteriophage	Microviridae
GA bacteriophage	Leviviridae
GA-1 bacteriophage	Podoviridae
H1 bacteriophage	*B subtilis* group
H19A bacteriophage	Siphoviridae
H19B bacteriophage	Siphoviridae
Halobacterial phage	Archaebacterial
Hh1 bacteriophage	Archaebacterial
Hh3 bacteriophage	Archaebacterial
Hs1 bacteriophage	Archaebacterial
Inovirus virions	Inoviridae
Ja1 bacteriophage	Archaebacterial
JP34 bacteriophage	Leviviridae
JP501 bacteriophage	Leviviridae

KU1 bacteriophage	Leviviridae
L2 bacteriophage	Plasmaviridae
L17 bacteriophage	Tectiviridae
Lambda bacteriophage	Siphoviridae
Leviviruses	Leviviridae
M12 bacteriophage	Leviviridae
M13 bacteriophage	Inoviridae
Microvirus virions	Microviridae
MS2 bacteriophage	Leviviridae
Mu bacteriophage	Myoviridae
N4 bacteriophage	Podoviridae
NL95 bacteriophage	Leviviridae
P1 bacteriophage	Myoviridae
P2 bacteriophage	Myoviridae
P4 bacteriophage	Tectiviridae
P7 bacteriophage	Tectiviridae
P22 bacteriophage	Podoviridae
PB bacteriophage	Myoviridae
PBS1 bacteriophage	Myoviridae
PBSX bacteriophage	Myoviridae
Pf2 bacteriophage	Inoviridae
PG bacteriophage	Archaebacterial
PM2 bacteriophage	Corticoviridae
PMS1 bacteriophage	Archaebacterial
PP7 bacteriophage	Leviviridae
PR3 bacteriophage	Tectiviridae
PR4 bacteriophage	Tectiviridae
PR5 bacteriophage	Tectiviridae
PR772 bacteriophage	Tectiviridae
PRD1 bacteriophage	Tectiviridae
PS42-D bacteriophage	Myoviridae
ΨM phage	Archaebacterial
PZA bacteriophage	*B subtilis* group
Q_β bacteriophage	Leviviridae
Qβ levivirus	Leviviridae
R17 bacteriophage	Leviviridae
RNA bacteriophage	Leviviridae
S13 bacteriophage	Microviridae
S45 bacteriophage	Archaebacterial
S5100 bacteriophage	Archaebacterial
SP bacteriophage	Leviviridae
SP5c bacteriophage	*B subtilis* group
SP6 bacteriophage	Leviviridae
SP8 bacteriophage	Myoviridae
SP50 bacteriophage	Myoviridae
SP82 bacteriophage	*B subtilis* group

SP82G bacteriophage	Leviviridae
SPβ bacteriophage	Siphoviridae
SPO1 bacteriophage	*B subtilis* group
SPO2 bacteriophage	*B subtilis* group
SPP1 bacteriophage	Siphoviridae
SSV1 bacteriophage	Fuselloviridae
T1 bacteriophage	Siphoviridae
T2 bacteriophage	Myoviridae
T3 bacteriophage	Podoviridae
T4 bacteriophage	Myoviridae
T5 bacteriophage	Siphoviridae
T6 bacteriophage	Myoviridae
T7 bacteriophage	Podoviridae
T12 bacteriophage	Myoviridae
TH1 bacteriophage	Leviviridae
TP-13	*B thuringiensis* group
TTV1 bacteriophage	Lipothrixviridae
TTV4 bacteriophage	Lipothrixviridae
TW18 bacteriophage	Leviviridae
TW19 bacteriophage	Leviviridae
TW28 bacteriophage	Leviviridae
VcA1 bacteriophage	Myoviridae
Vi bacteriophage	Myoviridae
Vill bacteriophage	Siphoviridae
VK bacteriophage	Leviviridae
VLP J2	Archaebacterial

Index

Guidance to use of the Index

Page numbers

Page numbers in **bold** refer to individual articles.

Order

Entries are in *word-by-word* alphabetical order (in which a group of letters followed by a space is filed before the same group of letters followed by further letters). Hyphens are given in an 'intermediate' value, and en-rules are placed after these e.g. in the following order 'Cell, Cell cultures, Cell-mediated, Cell–cell fusion'. Characters within brackets e.g. (s) are excluded from the alphabetical order. In this way 'Gene(s)' has been placed ahead of 'Gene cloning, Gene therapy'.

Entries

1. Bacteriophages are listed after the main index entry 'Bacteriophage', unless they are the subjects of entire articles, in which case they are located in the correct alphabetical section e.g. Bacteriophage T4 is indexed as 'T4 bacteriophage'. Cross-references have, however, been included to avoid confusion. ϕ is alphabetized as 'phi'. ψ is alphabetized as 'psi'.
2. Characteristics of virus families are listed in Table 2 (pages 1580–1584).
3. US spellings are used throughout the index.

Abbreviations

Abbreviations used in index subentries are as follows:

CMV	Cytomegalovirus
EAV	Equine arteritis virus
EBV	Epstein–Barr virus
HBV	Hepatitis B virus
HSV	Herpes simplex virus
ORF	Open-reading frame
LCMV	Lymphocytic choriomeningitis virus
LDV	Lactate dehydrogenase-elevating virus

Abbreviations of other important viruses appear within the index. All abbreviations appear within the Appendix.

For the viral entries the superscripts (A), (P), (B), (I) indicate which are animal, plant, bacteriophage and insect viruses, respectively.

Volume page numbers

Volume 1	1–516
Volume 2	517–1046
Volume 3	1047–1622

A
Abdominal pain, coxsackievirus infection 270
Abelson murine leukemia virus (Ab-MuLV)$^{(A)}$ 884, 887
  oncogenes 936
abl oncogene 936
Abortions, equine 423, 427, 763, 769
Abortive infections 1170, 1587
Abutilon mosaic virus (AbMV)$^{(P)}$
  DNA replication 521
  genome organization/expression 519–520
  genome structure 518
Ac element 1232, 1542
Acariviruses$^{(I)}$, glycoprotein homology with baculoviruses 440
Acetylcholine receptors, as virus receptors 229
Acid conditions, viral fusion 1557
Acid stability, of viruses 1077
Acquired immunodeficiency syndrome, *see* AIDS
Acquired resistance, *see* Plant resistance (to viruses)
Acridine orange stain 342
Actin
  changes due to frog virus 3 (FV3) 507
  measles virus assembly 842
  microfilaments 231, 1590
  virus-induced changes 1590
Actinomyces, phage in 124, 125
Actinomycin D 1293
  birnavirus sensitivity 146
Acute erythroblastosis virus (AEV)$^{(A)}$ 941
Acute hemorrhagic conjunctivitis (AHC), *see* Conjunctivitis
Acute infections 1076
Acute infectious dropsy, fish 478
Acute paralysis virus (APV)$^{(I)}$ 656, 657–658
Acycloguanosine, *see* Acyclovir
Acyclovir 43, 44
  cytomegalovirus infection 298, 1481
  Epstein–Barr virus (EBV) infections 409, 1482
  herpes simplex virus (HSV) infections 43, 592
    after transplantation 1484
    resistance to 1484
  oral
    bioavailability problem 46
    in herpes simplex virus infections 592–593
  phosphorylation by HSV 592
  varicella–zoster virus (VZV) infections 1483, 1518
Acyclovir triphosphate 43, 592
Adenine, methylation 1383
Adeno-associated viruses (AAV)$^{(A)}$ 1052
  bovine 1052
  canine 1052
  clinical features of infection 1055
  densonucleosis virus genome similarity 332
  DNA integration into host 1060
  gene expression 1058–1059
  genome and DNA replication 1058
  helper viruses 1052, 1060, 1587
  latent infection 1060, 1587
  mRNA 1058, 1059
  proteins 1059
  rep gene 1058
  Rep proteins 1059
  replication strategy 1060, 1587
  structure and properties 1057
  as vector in gene therapy 1060, 1535
  virus types 1052
Adenocarcinoma
  mammary in mice, *see* Mouse mammary tumor virus (MMTV)$^{(A)}$
  renal 503, 744
Adenoid-degenerating (AD) agent 1
Adenosine arabinoside, *see* Vidarabine
Adenosine triphosphate, *see* ATP
S-Adenosyl-methionine 670
Adenoviridae 973, 1398, 1409
  characteristics 442
  eye infections 444
  structure 1580
Adenovirus$^{(A)}$ **1–23**
  Ad2
    genetic map 8, 9
    replication in hamster cells 13, 14
  Ad12
    abortive infection of hamster cells 13, 14
    integration into host genome 13, 14
    major late promoter 13
    see also Adenovirus$^{(A)}$, *malignant transformation and oncology (below)*
  adsorption and penetration 11, 1204

Adenovirus$^{(A)}$ *continued*
animal **14–17**
subgenera 1, 8, **14–17**, 973
virion properties 15
see also Mastadenovirus$^{(A)}$; *specific subentries above/below*
antigens, type-specific 16
avian, *see* Aviadenoviruses$^{(A)}$
canine 17
cytopathic effects 337
defective particles 1171
DNA 2, 8–10
animal adenoviruses 15
integration into host genome 13–14
methylation 12–13
packaging 1529
downregulation of MHC expression 701
duck 16–17
E1A gene
conserved (CR) regions 18, 19–20
E1B cooperation 19
exogenous sequence insertion into 1529
functions 18, 19–20, 1476
interferon transcription prevention 737
structure 1, 11, 18
E1A proteins (289R and 243R) 18
cellular protein association 20–21
functions 19–20
transcription factor activation 21
transcription repression 18, 19, 20
E1A-induced tumors 7, 10, 19, 20, 1476
E1B gene
functions 21–22
loss of 176R, phenotypes 22
structure 18–19
E1B proteins (496R and 176R) 18, 19, 21–22, 1476
cellular protein association 22
E3 gene
functions 22–23
map 22
MHC complex 19, 23
tumor necrosis factor and 23, 1094
E3 protein (19K) 23, 701
protection from lysis by TNF 1094
enteric 1, 2, 5, 374–375
replication 375
subgenera (A-F) 375
epitopes 2, 4
evolution 2
animal adenoviruses 16
variants 2
in feces 373
future perspectives 7
in gene therapy 1530
general features **1–7**
genetics 2
animal adenoviruses 16
genome structure 1, 8–10, 18
animal adenoviruses 15
E1 region 10
E1A 1, 11, 18
E1B 18–19
E2A and E2B 12
E3 12, 19
E4 12
early genes 8, 10, 11, 12, 18
late genes (L1) 8, 12, 18
map and mapping 8, 9, 10, 1529
terminal protein (tp) 8, 11
VA RNAs 10, 15, 413
geographic and seasonal distribution 2
history 1
animal adenoviruses 14
host range 2, 1530
animal adenoviruses 16
hostvirus interactions 10, 1588, 1589
human subgenera 1, 3, 4–5, 8, 15, 18
life-cycle 18
see also subgenera (below)
identification 341
immune evasion 701, 1094
importance of 8
infections, *see* Adenovirus infections *(below)*
inside nuclei 233
intramolecular recombination with SV40 527
latency 788
malignant transformation and oncology 10, 14, **17–23**, 631, 1476
E1A, E1B and E3 genes 7, 10, 18–19, 20, 1476
history 1, 8
mechanism 19, 21–23
see also Adenovirus$^{(A)}$, E1A gene, E3 gene, E1B gene *(above)*
MHC class I gene transcription 21
molecular biology **8–14**, 17–23
pathogenicity 5, 17
as persistent infection 1090
physical properties, animal adenoviruses 15
propagation 2
animal adenoviruses 16
proteins 11, 12
animal adenoviruses 15
functions 19–20
interfering with cytokine function 7, 1094
properties 11, 12, 15
recombinants 1528, 1529
applications 1529–1530
replication 11, 18, 973
animal adenoviruses 16
early and late events 18
initiation 11
serological relationships and variability 2, 4
animal adenoviruses 16
structure 8–10
animal adenoviruses 15
polypeptides 1, 3, 4, 8
subgenera 2, 4–5
DNA homology 2, 3, 15
properties 3, 15
serotypes 3, 4, 16
subgenus A 1, 3, 4, 18
subgenus B:1 3, 4, 18
subgenus B:2 3, 4
subgenus C 3, 4–5
subgenus D 3, 5
subgenus E 3, 5
subgenus F 1, 3, 5
taxonomy and classification 1–2
tissue tropism 5, 10
animal adenoviruses 16–17
transcription 11–12, 17, 1529
control 12
RNA splicing 12, 1529
transcription factors 20, 21
transformation, *see* Adenovirus, malignant transformation
transmission 5, 449
animal adenoviruses 16
iatrogenic 449
tumors from, *see* Adenovirus$^{(A)}$, *malignant transformation and oncology (above)*
type 2, inclusions 339
type 5 1487
VA RNA 10, 15, 413, 1094
vaccines 7, 17, 1506, 1529
as vectors 7, 1528–1530
applications 1529–1530
co-transfection by plasmids 1529
considerations in 1529
rationale and properties 1528
techniques/approaches 1528–1529
Adenovirus infections
after transplantation 1487–1488
clinical features and treatment 1488
epidemiology and pathogenesis 1487–1488
incidence 1487
clinical features 6, 1488
animal adenoviruses 17
diarrhea 6
epidemic keratoconjunctivitis 1, 2, 6
epidemiology 4–5
animal adenoviruses 16
eye 449
frequency and respiratory disease types 1223
heart 965
immune response in 6–7, 701
animal adenoviruses 17
kidney 973
microtubules role in 232
pathogenicity 5, 17
pathology and histopathology 6
animal adenoviruses 17
pharyngitis 1223
pneumonia 1224
prevention and control 7
animal adenoviruses 17
respiratory disease 4, 6, 1223
Adenylate cyclase, in rotavirus infection 1279
β-Adrenergic receptors, as virus receptors 229, 1200
Adsorption of viruses 1169
see also Attachment of viruses; Receptors for viruses
Adult T-cell leukemia (ATL) 682, 684, 686, 1496
acute 684
$CD4^+$ leukemic cells 685
chronic 684
HTLV-1 association 682, 685, 1496
onset and features 684, 1496
smoldering type 684
without HTLV-1 685
Adult T-cell leukemia virus (ATLV)$^{(A)}$, *see* Human T-cell leukemia virus type 1 (HTLV-1)$^{(A)}$
Aedes aegypti
Chikungunya (CHIK) virus transmission 239, 1615
control 330
dengue virus transmission 324, 325, 328
factors increasing 327
larval habitats 327, 330
yellow fever virus transmission 1606, 1609, 1610
see also Mosquito transmission
Aedes aegypti densonucleosis virus (AA: DNV)$^{(I)}$ 331, 332, 333
histopathology 334
Africa, zoonoses in 1618–1619
African cassava mosaic virus (ACMV)$^{(P)}$ 520, 521, 524, 1538
as vector 1539
African horse sickness 942, 945, 953
African horse sickness virus (AHSV)$^{(A)}$ 941, 943
epidemiology 950
genetics 946

host range and propagation 945
pathogenicity 952
vaccines 954, 955
African swine fever virus(A) **23–29**
attachment protein (p12) 25
evolution 27–28
genetic variability 27
genome 24–25
genes 24
hairpin loops 24, 27
history and geographic distribution 23–24
host range 27
infection
clinical features 28
immune response 29
neutralizing antibodies absent 29
pathogenicity 28
pathology and histopathology 29
prevention and control 29
Paramecium bursaria Chlorella virus 1 (PBCV-1) difference 36
penetration 25
propagation 27
proteins 24
induced in infected cells 26
p12 25
p72 28, 29
post-translational processing 26–27
replication 25
RNAs 26
serologic relationships and variability 28
structure 24
taxonomy and classification 24
tissue tropism 28
transcription 25–26
transmission 28
Agaricus bisporus virus 4 (AbV-4)(P) 1047, 1051
Agarose gel electrophoresis (AGE), orbiviruses 946
Age
neural cell vulnerability changes 910
outcome of infection and 1082–1083
Agnoprotein 1138
Agrobacterium tumefaciens 1537
geminivirus infections 522
T-DNA 1537, 1542
Ti-plasmid 522, 1537, 1566
transposable phage (Psi) 875
Agroinoculation 1537, 1566
geminiviruses 522
AIDS 674, 707
animal model, feline immunodeficiency virus (FIV) 459
antiviral treatment 681
autoimmune features 65
clinical features 680–681
CNS dementia 65–66
coronavirus-like particles from patients 258
cytomegalovirus (CMV) infections 296, 702, 1479
EBV infections and lymphomas due to 1495
human herpesvirus 6 (HHV-6) infection 624, 626
human papillomavirus (HPV) infections 1018
interferon clinical use, *see* Interferon-α
interferon-α production in 734
JC and BK infections in 755
Kaposi's sarcoma and 1499
modeling role 403
molluscum contagiosum virus (MCV) infection 852
myocardial disease in 966
non-Hodgkin's lymphomas in 407, 408, 410
oligodendrocyte infection 910
pathology and histopathology 681
pericarditis in 966, 968
as persistent infection 1089
SIV used in research 1321
thyroid function 976
type D retrovirus from patient 1242
varicella–zoster virus (VZV) infections 1516, 1517
see also Human immunodeficiency virus (HIV)(A)
AIDS-like disease, simian immunodeficiency virus (SIV) causing 1320
Air pollution
respiratory virus infections 1219
virus-infected plants tolerating 1112
Akabane virus(A) 190
AKR murine leukemia virus (AKR-MuLV)(A) 887, 888
Alanine aminotransferase (ALT)
in hepatitis A 552
in Lassa fever 782
Alastrim 1341
Albuminuria 544
Alcelaphine herpesvirus 1(A) 155
Alcelaphine herpesvirus 2(A) 156
Aleutian mink disease 1055, 1061, 1066
clinical features 1055, 1066
hypergammaglobulinemia in 1056, 1066
immune response 1067
pathogenesis 1054
pathology and histopathology 1056, 1066
prevention and control 1067
Aleutian mink disease virus (ADV)(A) 1052, **1061–1067**
epidemiology 1053, 1065
gene expression 1058
genetics and genome 1064
geographic and seasonal distribution 1061–1062
history 1061
host range and viral propagation 1062
pathogenicity 1065
serologic relationships and variability 1065
structure 1062
taxonomy and classification 1061
see also Parvoviruses(A)
Alfalfa cryptic virus (ACV1)(P) 277
Alfalfa latent virus (ALV)(P) 215
Alfalfa mosaic virus (AlMV)(P) **30–35**, 139, 1538
abbreviations (AMV, AlMV) 30
coat protein 30, 31, 35
deletion analysis 35
economic significance 34
epidemiology and control 34
genetics 32–33
genome structure (RNA) 30, 31–32
RNAs1, 2, 3, 4 31–32, 32
history 30
host range and geographic distribution 34–35
host relationship 33
infectivity of cDNA 1540
molecular biology 35
protein P3 31, 35
pseudorecombinants 33
replication 32
resistance of transgenic plants 34
structure and composition 30–31, 1582
particles (4) 30
taxonomy and classification 30
temperature-sensitive (*ts*) mutants 32
transmission 34
as vectors 1540, 1542
YSMV strain 31, 33
Alfalfa mosaic virus (AlMV) group 30, 1400
Algal viruses(P) **35–40**
families 1406
history 35–36
NYs-1 39
XZ-6E 39
see also Chlorella viruses(P); *Paramecium bursaria Chlorella* virus 1 (PBCV-1)(P)
Alimentary tumors, feline leukemia virus causing 462
Alkaline phosphatase, serum, elevated in hepatitis A 552
Alleles 664
Alligatorweed stunting virus (AwSV)(P) 243
Allograft recipients
lymphomas in 407
see also Transplantation
Alloleviviruses (Group B) 1334, 1338
Alpers disease 1366
Alpha 3 bacteriophage(B), *see* Bacteriophage ϕX174(B)
Alphaherpesvirinae 155, 156, 157
bovine herpesviruses 155, 156, 157
equine herpesviruses 424
herpes simplex viruses 587
HSV relationship to other members 600
pseudorabies virus 1174
Simplexvirus, herpes simplex viruses 587
see also Herpes simplex viruses (HSV)(A); Varicella–zoster virus (VZV)(A); *other specific viruses*
Alphaherpesvirus(A) 1514
simian viruses 1514
Alphavirus (alphaviruses)(A) 237, 1269, 1330, 1615
antigenic complexes 238
classification 417, 445
encephalitic, *see* Equine encephalitis viruses(A)
evolution 238, 419
plant viruses relation 238
genetics and mutations 238, 419
genome 1269
isolation 338
replication 238
serologic relationship and variability 238–239
structure and characteristics 443, 445
as vectors 1535–1536
see also individual viruses (pages 238, 445)
Alzheimer's disease 1367, 1368
Amantadine 44, 47
influenza 714
Amapari virus(A) 777
Ambisense RNA genome
arenaviruses 776, 802, 807
bunyaviruses 1207, 1461
tenuiviruses 1413, 1414
tospoviruses 1461
American hop latent virus (AHLV)(P) 215
Amphibian herpesviruses(A) **40–42**
frog virus-4 (FV4) 40, 41–42
history 40–41
Lucké tumor herpesvirus (LTHV) 40–41
Amphotericin B, in transmissible spongiform encephalopathies 1367
Ampligen 738
Amsacta moorei entomopoxvirus (AmEPV)(I) 392, 394
cell culture 396
DNA and genome structure 394, 395
gene regulation 397
infection characteristics/pathology 395

Amsacta moorei entomopoxvirus (AmEPV)(I) *continued*
replication 396–397
spheroidin 394
Amyloid plaques 1359, 1368
in kuru 761
Amyloid protein 1359
Anabaena, cyanophage 289, 290
Anal lesions, HPV causing 1019
Anaplastic large cell lymphoma 1495
Andean potato latent virus(P) 1500
Andean potato mottle virus (APMV)(P) 249
Anemia
aplastic, feline leukemia virus causing 462
equine 434
see also Equine infectious anemia virus (EIAV)(A)
in human parvovirus B19 infection 1054, 1056
Animal(s), virus propagation in 1171–1172
Animal bites, rabies 1183, 1184
Animal cells, structure, *see* Cell structure/function
Animal models, vaccine development 1506
Animal viruses 1409
atomic structures 1572
history, *see* History of virology
host cell receptor recognition 1573
isolation methods 335, 336
see also Diagnostic techniques
see also individual families/viruses (page 1409)
Anogenital carcinogenesis
HPV causing 1019
see also Human papillomaviruses (HPV)(A)
Anterograde axon transport 908
Antheraea eucalypti 1417, 1421
Antheraea eucalypti virus (AeV)(I) 1417
Anthoxanthum latent blanching virus (ALBV)(P) 661
Anthriscus yellows virus (AYV)(P) 1042
parsnip yellow fleck virus (PYFV) relationship 1042, 1044
Anti-idiotypic responses, development by viral infections 64
Anti-virals, *see* Antiviral agents
Antibodies 1092
anti-idiotypic 64
antitumor in cats (FOCMA) 462
Borna disease virus (BDV) infection 153, 154
bovine herpesviruses 157
bovine immunodeficiency virus (BIV) 165
in CNS 908
complement-fixing, coxsackieviruses 639
cross-reacting, autoimmunity development 64, 65, 66
detection, *see under* Diagnostic techniques
enhancing 702
epitopes recognized, in vaccination 1504
functions 1504
influenza virus 699
neutralizing, *see* Antibodies, neutralizing *(below)*
neutralizing antibodies to 64
to peptides/carbohydrates, in viral infections 699
primary response 345, 1092
see also Humoral response; Immune response
role in vaccination 1504
'solid phase' 345
synthesis 698
continued production 699
varicella–zoster virus (VZV) 1517
vertebrate, to invertebrate virus 1422
viral, detection, *see under* Diagnostic techniques
see also Immunoglobulin(s)
Antibodies, neutralizing 699
absence in African swine fever virus infection 29
avian hepatitis B virus infections 568
birnaviruses 147
caprine arthritis encephalitis virus 204
cardioviruses 213
cottontail rabbit papillomavirus (CRPV) 1310
cowpox virus 266
coxsackievirus infections 272, 639
development with interferon therapy 744
echovirus infection 359
equine arteritis virus (EAV) 771
equine herpesviruses 428
to hemagglutinin 699
in hepatitis A 553
hog cholera virus (HCV) 653
human cytomegalovirus infections 297
human respiratory syncytial virus (HRSV) 1214, 1217, 1218
Japanese encephalitis virus 750
Junin virus 785
Marek's disease virus 836
murine cytomegalovirus (MCMV) 311
orbiviruses 954
rabies virus 1185
respiratory viruses 1222
rotaviruses 1277, 1280, 1283
SV40 1329
vaccination 1504
vesicular stomatitis virus (VSV) 1555
visna–maedi viruses 1594
see also Humoral response; Immune response
Antibody secreting cells (ASCs) 699
Antibody-dependent cell-mediated cytotoxicity (ADCC) 700, 1092, 1222, 1504
'Antibody-mediated enhancement' (AME) 1079
Antidiuretic hormone (ADH) 975
Antigen
assays, *see under* Diagnostic techniques
binding to class I antigens 726
internal 699, 701
persistence 699
presentation 697, 698, 704, 725
to B cells 698, 705, 725, 1092
to T cells 698, 704, 705, 725, 1092
processing 697, 726
in cell-mediated immune response 705
class I-associated 726, 1092
class II-associated 727
influenza virus proteins 726–727, 727
pathways 705
rotavirus 1280
source, for serological diagnoses 345
Antigen-presenting cells (APCs) 697
antigen processing 697
viral infection 64
Antigenic determinants (epitopes) 724
B-cell 724, 725–726
T-cell 1505
vaccination and 1504, 1505
in viral infections 727
Antigenic diversity, vaccine development 1507
Antigenic drift 402, 700–701, 1221
influenza viruses 711
polioviruses and 1116
Antigenic shift 402, 701, 1221
influenza viruses 711
see also Genetic reassortment
Antigenic sites, viruses 1573
Antigenic variation 402, 700–701
alphaviruses 239
caliciviruses 1545
immune evasion by 1093
internal antigens for vaccines and 1505
Antigen–antibody complexes, *see* Immune complexes
Antigen–antibody reactions, historical aspects 696
Antioncogenes
retinoblastoma (RB) gene 20, 1024, 1476, 1477, 1496
see also p53 gene; Tumor suppressor genes
Antiphage chemicals 119–120
Antisense mRNA, in control of phage 121
Antisera
monoclonal, *see* Monoclonal antibodies
polyclonal 344
Antitermination 819
Antiviral agents **42–49**
atomic structures 1573
avian hepatitis B virus infections as model for 568
cytomegalovirus (CMV) infections 46, 298, 1479, 1481
dominant-negative mutants as 731–732
equine herpesviruses 429
experimental and approved, characteristics 44–45
future prospects 49
hepatitis viruses 45
herpesviruses 43, 44, 47
myxoviruses and paramyxoviruses 44, 47
picornaviruses 44–45, 47–48
progressive multifocal leukoencephalopathy (PML) 757
respiratory tract infections 1226
rhinoviruses 1258, 1259
strategies 43, 46
varicella–zoster virus (VZV) infections 1483
see also Acyclovir; Vidarabine
AP1 transcription factor 1475
Aphid transmission 633
alfalfa mosaic virus (AlMV) 34
carlaviruses 216
caulimoviruses 225
closteroviruses 246
cucumoviruses 283–284
fabaviruses 453
luteoviruses 795, 796
parsnip yellow fleck virus (PYFV) group 1044
pea enation mosaic virus (PEMV) 1086
potyviruses 1148, 1152
rhabdoviruses 1248
Aphthovirus(A) 489
properties 489
Apis cerana, sacbrood virus 656, 657
Apis iridescent virus(I) 655, 656, 660
Apis mellifera
Kashmir bee virus (KBV) 658
sacbrood virus 656, 657
Aplastic anemia, feline leukemia virus causing 462
Aplastic crisis, in human parvovirus B19 infection 1054, 1056
Apodemus mice 538, 542
Apoptosis 1590
inhibitor, gene in granulosis virus (*iap*) 128–129
Apple chlorotic leaf spot virus (ACLSV)(P) 244
disease and economic significance 247
genome structure/organization and

  expression 245
Apple latent virus type 2(P) 198
Apple mosaic virus (ApMV)(P) 33
Apple scar skin viroid(P) 1564
Apple stem grooving virus (ASGV)(P) 197, 198, 242, 244
  genome structure/organization and expression 245
Apple stem pitting virus(P) 243
Ara A, *see* Vidarabine
1-β-[SC]D[sc]-Arabinofuranosyl-E-5-(2-bromovinyl)uracil 44
Arabis mosaic virus (ArMV)(P) 904, 906
Arboviruses(A) 361, 403, 444
  bunyaviruses 186
  Chandipura, Piry and Isfahan viruses as 234–235
  dengue viruses 324
  discovery 628
  encephalitis viruses 361, 363, 911
  Group A 237, 445
    serologic relationship and variability 238
    see also Alphavirus (alphaviruses)(A); Equine encephalitis viruses(A)
  Group B 445, 746
    see also Flaviviridae
  heart infections 965
  host factors determining infection outcome 1082
Archaea 50
Archaebacteria 50
Archaebacterial bacteriophage(B) **50–58**
Arenaviridae 776, 1399, 1409
  evolution 779
  eye infections 447
  lymphocytic choriomeningitis virus (LCMV) 802, 806
  Old and New World 776
  replication/transcription strategy 1207
  structure and characteristics 443, 447, 1583
  see also Arenaviruses(A); *specific viruses (page 776)*
Arenaviruses(A) 776, 777, 1614
  characteristics 776, 777
  epidemiology 779–781
  evolution 803
  granules in 776, 806
  kidney infections 975
  pathogenesis of infection 781–782
  transmission 779, 781, 1614
  see also Guanarito virus(A); Junin virus(A); Lassa virus(A); Lymphocytic choriomeningitis virus (LCMV)(A); Machupo virus(A)
Argentine hemorrhagic fever (AHF) 776, 975, 1614
  clinical features 783–784
  epidemiology 780–781
  mortality 784
  passive antibody therapy 786
  pathogenesis 782
  pathology and histopathology 784
  therapy and prevention 785, 786
  see also Junin virus(A)
Arkansas bee virus (ABV)(I) 656, 660, 919
Arracacha A virus(P) 904, 905
Arracacha B virus(P) 905
Arrhenatherum blue dwarf virus (ABDV)(P) 1096
Arterivirus 763
Arteriviruses(A) 763
  toroviruses relationship 1456
Arthralgia
  Chikungunya (CHIK) virus infection 240
  rubella 1296
Arthritis
  caprine arthritis encephalitis syndrome 203
  Ross River virus (RRV) 1272, 1273
  rubella 1296
  rubella vaccine causing 1297–1298
  visnamaedi virus infections 1592
  see also 'Epidemic polyarthritis'
Arthritis with rash, epidemics 1268
Arthropod transmission 401, 403
  bunyaviruses 189
  history 632
  see also Arboviruses(A); *specific vectors*; Zoonoses
Artichoke Italian latent virus(P) 904
Artichoke latent virus (ArLV)(P) 215
Artichoke mottled crinkle virus (AMCV)(P) 1447
  genome 1448
Artichoke vein banding virus(P) 905
Artichoke yellow ringspot virus(P) 904
Ascoviruses(I) **58–63**
  composition 59
  cytopathology 61
  economic importance 63
  future perspectives 63
  genome structure 59
  histopathology 62–63
  history 58–59
  host interactions and ecology 61–62
  pathology of disease 62
  replication and morphogenesis 61
  structure 59, 60
  taxonomy and classification 59
  transmission 61–62
  variants 59
Asia, zoonoses 1619
Asialoglycosphingolipids, as virus receptors 229
Asparagine-linked sugars, proteins of hantaviruses 539, 541
Aspartate aminotransferase (AST)
  in hepatitis A 552
  in Lassa fever 782
Aspartyl protease 167
Aspergillus foetidus virus S (AfV-S)(P) 1464
Aspergillus niger virus S (AnV-S)(P) 1464
Aspergillus ochraceous virus (AoV)(P) 1047
Assembly of viruses 1169, 1209–1210, 1559–1560
  host cell cytoskeleton role 1590
  icosahedral 1574
  process and structures 1573
  see also Budding; Replication of viruses
Asthma 1222, 1258
Astrocytes, HIV gp41 protein cross-reaction 66
Astrocytoma cell lines, human 294
Astrogliosis, reactive 153
Astroviruses(A) 376
  gastroenteritis due to 376
  serotypes 376
Ataxia, feline parvovirus (FPV) infection 1066
Atherosclerosis 966–967
  role of viruses in pathogenesis 967
ATP
  rotavirus transcription 1288
  T1 phage infection 1371
ATPase, fish lymphocystis disease virus (FLDV) 799
Attachment of viruses 228, 229, 1081, 1169, 1203–1204, 1557
  see also Entry of viruses; Receptors for viruses; Replication of viruses
Aujeszky's disease (AD) 1173
  see also Pseudorabies virus (PRV)(A)
Aura virus(A) 417, 418
Australia, zoonoses in 1619–1620
Australia antigen 554
  see also Hepatitis B virus (HBV)(A), HBsAg
Australian grapevine viroid(P) 1564
Australian wood duck hepatitis B virus(A) 564
Australian X disease 361
Autoantibodies
  Epstein–Barr virus (EBV) infection 64, 409
  lactate dehydrogenase-elevating virus (LDV) infection 770
Autographa ascovirus(I) 59
Autographa californica nuclear polyhedrosis virus (NPV)(I) 128, 129, 131, 1534
  genome 132
  life cycle and pathogenesis 133, 134
  polyhedrin gene inserted in *Oryctes* baculovirus 137
  tetravirus detection in contaminated inocula 1422
  as vector 1534
Autoimmune diseases 703
  interferon-α production in 734
  viral infections associated 702, 703
Autoimmunity **63–66**, 703
  in AIDS 65
  diabetes and 971
    pathogenesis 65
  mechanisms 63–64
  models 64–66
  tissue injury mechanisms 65
Aviadenovirus(A) 1, 14, 16
Aviadenoviruses(A)
  genome 15
  host range and propagation 16
  infections 17
  physical properties 15
  serologic relationships 16
  structure and properties 15
  see also Adenovirus(A)
Avian acute leukemia virus(A), oncogenes 935
Avian encephalomyelitis 387, 389
  clinical features 390
  pathology and histopathology 390
  vaccines 391
Avian encephalomyelitis virus(A), transmission and tissue tropism 389
Avian erythroblastosis virus (AEV)(A), oncogenes 936, 1474, 1475
Avian infectious bronchitis virus (IBV)(A) 255, 258
Avian influenza viruses(A) 710
Avian leukemia sarcoma viruses (ALSV)(A), *see* Avian sarcoma-leukemia viruses (ASLV)(A),
Avian leukemia virus (ALV)(A), as insertional mutagens 937
Avian leukosis viruses (ALVs)(A) 66, **66–71**, 1473
  distribution and host range 69
  DNA integration 68
  as endogenous proviruses 69, 71
  evolution 70
  future perspectives 71
  genetics 70
  genome structure/properties 67
  history 66–67
  infection
    immune response 70–71
    pathogenicity 70
    prevention and control 71
  malignancies due to 70
  oncogene-containing 69, 70

- Avian leukosis viruses (ALVs)[A] *continued*
  - physical properties 68
  - propagation 69
  - proteins, properties 67–68
  - receptors 68, 71
  - replication 68–69
  - resistance, receptor alleles 68, 71
  - structure and properties 67
  - taxonomy and classification 67
  - tissue tropism 70
  - transcription and translation 67, 68–69
  - transformation 69
  - transmission 67, 70
- Avian myeloblastosis virus (AMV)[A], oncogenes 936
- Avian nephritis virus[A] 390
- Avian paramyxoviruses (PMV)[A] 1036
- Avian reoviruses[A] 1195
- Avian sarcoma virus (ASV)[A], oncogenes 936
- Avian sarcoma-leukemia viruses (ASLV)[A]
  - lymphoproliferative disease virus (LPDV) of turkeys genome similarity 812
  - reticuloendotheliosis viruses differences 1227, 1229
- Avidity assays 346
- *Avihepadnaviruses* 564
  - *see also* Duck hepatitis B virus (DHBV)[A]; Hepatitis B viruses (HBV), avian[A]
- *Avipoxvirus*[A] 497
- Avocado sunblotch viroid[P] 1564, 1565
  - host range and transmission 1566
  - replication and cleavage 1568, 1570
  - *see also* Viroids
- Axons, anterograde/retrograde virus transport 908, 1079
- Axoplasmic transport systems 908, 909
- Azidothymidine (AZT), *see* Retrovir
- Aziridines 496
- AZT, *see* Retrovir

B

- B cells 697
  - antigen presentation 698, 705, 725, 726
  - baboon and chimpanzee herpesviruses specificity 610, 612–613
  - bovine leukemia virus in 171, 172
  - epitopes 724
    - difference from T-cell epitopes 725–726
  - Epstein–Barr virus (EBV) infection 407, 411, 790, 1477, 1591
    - immortalization of 404, 411, 412, 1494
    - polyclonal activation 63, 408
    - proliferation of 1482, 1494
  - infectious bursal disease of chickens virus (IBDV) tropism 145, 148
  - memory cells 699
  - mumps virus in 882
  - polyclonal activation by viruses 63, 1494
    - EBV infection 63, 408
    - LDV infection 770
  - reovirus type 3 receptor (Reo3R) 1562
  - response to Lassa virus 784–785
  - surface receptor (CD21) on 407
  - viral infection, as immune evasion mechanism 702
- B-cell lymphoma
  - in bursa of Fabricius 70
  - retroviruses (avian leukemia virus) causing 937, 938
- B-cell malignancy
  - HTLV-2 infection and 693
  - tartrate-resistant acid phosphatase-positive (TRAP+) 693
- *Babesia bovis* virus[A] (BBV) 534
- Baboon endogenous retroviruses (BaEV)[A] 440
- Baboon herpesvirus[A], *see* Herpesvirus papio[A]; *under* Herpesviruses[A]
- Baby hamster kidney (BHK) cells, LCMV infection 802
- *Bacillus*, strains, economic aspects 78
- *Bacillus amyloliquefaciens* 981
- *Bacillus liqueniformis* 981
- *Bacillus pumilus* 981
- *Bacillus subtilis* 72
  - $\sigma^{(A)}$, RNA polymerase 986, 987
  - soil culture, phage in 126
  - strain 168 72
    - protoplasts 77
    - transfection 76–77
    - transformation 77
  - strain W23 bacteriophage PMB12 74
- *Bacillus subtilis* bacteriophages[B] **72–78**, 980
  - as cloning vectors 77
    - strategies 77
  - future perspectives 77–78
  - phagebacterial DNA hybrid 77
  - prophages 75
  - pseudotemperate 73, 74–75
    - clear-plaque mutants 74
    - PBS1 73, 74–75
  - temperate 75–76
    - defective 75, 76
    - nondefective 75
  - transduction
    - general 76
    - specialized 76
  - virulent 72–74
    - ϕ29 73–74
    - SPO1 72–73, 1352
  - *see also* Bacteriophage ϕ29[B]; SPO1 bacteriophage[B]
- *Bacillus thuringiensis* bacteriophage TP-13[B] 74
- Bacillusphage[B], hydroxymethylcytosine (HMC) in DNA 671
- Bacteria
  - characteristics of strains 78
  - fermentations, *see* Industrial fermentations
  - identification by bacteriophage, *see* Phage typing
  - lysis, bacteriophage discovery 644, 645, 646
  - phage-resistant strains 119, 120–121
  - survival, prophage genes promoting 82
- Bacterial viruses, *see* Bacteriophage
- Bacteriocin, betacin 75
- Bacteriophage
  - 'adapted phage' 80
  - adsorption and receptors 80
  - *Bacillus subtilis*, *see Bacillus subtilis* bacteriophages[B]
  - bacterial identification by, *see* Phage typing
  - bacterial protection against, M/R systems 671
  - biological concepts formulated from 1376
  - as cloning vehicles, *see* Gene cloning
  - conversion 101
  - cyanobacteria, *see* Cyanobacteria bacteriophage[B]
  - defective 109
  - densities, quantitation 81
  - detection
    - direct counts in soil samples 122–123
    - enrichment procedures 122
    - in industrial fermentations 118
    - plaques 80, 81
    - in soil 122–123
  - discovery 627, 628, **643–648**
  - distribution 82
  - DNA 1405
    - *pac* site 108
    - unusual bases in 671
    - *see also* DNA bacteriophage
  - DNA packaging into 108–109, 999
    - cellular DNA 108, 109
    - *see also specific phage*; Transduction
  - DNA phage, *see* DNA Bacteriophage
  - double-strand DNA breaks (DSB), tolerance 92
  - ecology 81
  - evolution 82, 437, 440
  - families (diagrammatic presentation) 1405
  - filamentous, *see* Bacteriophage f1[B]; Filamentous bacteriophage
  - frog virus 3 (FV3) shared features 505, 507, 508
  - head 108
  - helper 109
  - history 629–630, **642–648**
    - d'Herelle's discovery 628, 629, 644–646, 646–647
    - genetics and molecular biology 629
    - Hershey and Chase experiment 629, 642
    - importance of 642
    - lysogeny 629–630, 647
    - RNA phage 630
    - Twort *vs* d'Herelle 645–646, 646–648
    - Twort's discovery 628, 643–644, 645, 647
    - as 'vaccine' 629, 645, 646, 648
  - host interactions, in soil 124–126
  - host ranges 81
  - incompatibility of prophage 1000
  - in industrial fermentations, *see* Industrial fermentations
  - lysogenic, *see* Bacteriophage, temperate; Lambda bacteriophage[B]; Lysogeny; Prophage
  - lytic 80
    - *Bacillus subtilis* phage 72–74, 981
    - phage in soil 124
    - T1 phage 1372
  - mutation, in industrial fermentations 118
  - nonconverting 101
  - plaque assay method, development 630
  - population biology 82
  - recombination, *see* Genetic recombination, bacteriophage
  - RNA, *see* RNA bacteriophage
  - selective advantage on host 82
  - single-stranded DNA, *see* DNA bacteriophage
  - in soil **121–127**
    - abundance and host abundance 124
    - burst size and death rates 125
    - community dynamics 126
    - detection 122–123
    - ecology 122
    - extraction efficiency and variables 123
    - filamentous hosts 125
    - numbers 122
    - pH/temperature effect 123–124
    - phagehost interactions 124–126
    - single species dynamics 125–126
    - soil type influence 123, 125
    - species isolated by enrichment method 122, 123
    - SPO1 phage 1352
    - sporulation effect 123, 125
    - stability of free phage 123–124
  - speciation 82–83
  - species 82–83
  - stability, determination in soil 123
  - T-even, *see individual bacteriophage*; T-

even phage
tailed 95–96
taxonomy and classification **93–100**
description of families 95–100
families and orders 94–95
genera 94
history 93
ICTV universal system 93–95
new phages, description 95
properties used in 93, 94
species 94
temperate 80, 81, 772, 814
Bacillus subtilis phage 75–76
coliphage P2 and coliphage 186 1003
coliphage P4 1003
P22 bacteriophage 1009
toxinogenic converting, *see toxins and phage conversion (below)*
toxins-encoded 101
see also Lambda bacteriophage$^{(B)}$; Lysogeny; Prophage
toxinogenic converting 101
toxins and phage conversion **101–106**
botulinum 105–106
diphtheria 101–102
evolution of phage 106
gene expression 106
pyrogenic of *S. aureus* 103–105
pyrogenic of *S. pyogenes* 103–105
Shiga-like toxins 102–103
transduction, *see* Transduction
transposable, *see* Mu bacteriophage$^{(B)}$; Transposable bacteriophage
treatment of bacterial disease and 629, 645, 646, 648
typing, *see* Phage typing
virulence-associated genes 106
virulent, *see* Bacteriophage, lytic
see also specific coliphage/bacteriophage/bacterial species
Bacteriophage 2C$^{(B)}$ 1352
Bacteriophage 7S$^{(B)}$ 1335
Bacteriophage 29-$\alpha^{(B)}$ 1384
Bacteriophage 186$^{(B)}$, *see* Coliphage 186$^{(B)}$
Bacteriophage 434$^{(B)}$, λ phage recombination 817
Bacteriophage 933J$^{(B)}$ 103
Bacteriophage 933W$^{(B)}$ 103
Bacteriophage B278$^{(B)}$ 875
Bacteriophage $\beta^{(B)}$
diphtheria toxin (*tox*) 101–102
genome and integration into host 102
prophage map 102
Bacteriophage BA14$^{(B)}$ 1389
Bacteriophage BA127$^{(B)}$ 1389
Bacteriophage BA156$^{(B)}$ 1389
Bacteriophage BF23$^{(B)}$ 1384
cloning genes 1388
see also T5 bacteriophage$^{(B)}$
Bacteriophage BG3$^{(B)}$ 1384
Bacteriophage BZ13$^{(B)}$ 1334
Bacteriophage CS112$^{(B)}$, pyrogenic toxins of *S. pyogenes* 104
Bacteriophage CWP$^{(B)}$ 51
Bacteriophage D20$^{(B)}$ 1371
Bacteriophage D108$^{(B)}$ 868
DNA modification 873–874
DNA transposition mechanism 873
genetic map 868, 869, 871
invertible G-loop 874
left end regulatory regions 871
life cycle 868–871
mod gene 873–874
see also Mu bacteriophage$^{(B)}$
Bacteriophage D3112$^{(B)}$ 868, 874–875
DNA transposition mechanism 873
genetic map 869, 871
left end regulatory regions 871
repressor protein 872
transposase gene 872
Bacteriophage DAV1$^{(B)}$ 51
Bacteriophage f1$^{(B)}$ 464
genome structure 465, 466
structure 464, 465
Bacteriophage f2$^{(B)}$ 1334
Bacteriophage fd$^{(B)}$ 464
genome structure 465
Bacteriophage Ff$^{(B)}$ 464
Bacteriophage FI$^{(B)}$ 1334
Bacteriophage fr$^{(B)}$ 1334
Bacteriophage G4$^{(B)}$, *see also* Bacteriophage ϕX174
Bacteriophage GA$^{(B)}$ 1334, 1335
Bacteriophage GA-1$^{(B)}$, genome 981
Bacteriophage H1$^{(B)}$ 1352
Bacteriophage H19A$^{(B)}$ 103
Bacteriophage H19B$^{(B)}$ 103
Bacteriophage Hh1$^{(B)}$ 51
Bacteriophage Hh3$^{(B)}$ 50, 51
Bacteriophage Hs1$^{(B)}$ 50, 51
Bacteriophage Ja1$^{(B)}$ 51
Bacteriophage JP34$^{(B)}$ 1334
Bacteriophage K11$^{(B)}$ 1395
Bacteriophage KU1$^{(B)}$ 1334
Bacteriophage L2$^{(B)}$, characteristics and taxonomy 96
Bacteriophage L17$^{(B)}$ 1165
Bacteriophage lambda$^{(B)}$, *see* Lambda bacteriophage$^{(B)}$
Bacteriophage M2Y$^{(B)}$ 981
Bacteriophage M12$^{(B)}$ 1334
Bacteriophage M13$^{(B)}$ 115, 464
as cloning vehicle 115
genome structure 465
in λZAP vector 116
packaging of genome and length of 110
'phagemids' 115, 116
Bacteriophage MP501$^{(B)}$ 1334
Bacteriophage MS2$^{(B)}$ 1334, 1335
A protein 1335, 1337
genetic map 1335
structure 1586
Bacteriophage Mu$^{(B)}$, *see* Mu bacteriophage$^{(B)}$
Bacteriophage N4$^{(B)}$, *see* N4 bacteriophage$^{(B)}$
Bacteriophage NL95$^{(B)}$ 1334
Bacteriophage P1$^{(B)}$, *see also* P1 bacteriophage$^{(B)}$
Bacteriophage P2$^{(B)}$, *see* P2 coliphage$^{(B)}$
Bacteriophage P4$^{(B)}$, *see* P4 bacteriophage$^{(B)}$
Bacteriophage P7$^{(B)}$ 997
Bacteriophage P22$^{(B)}$, *see* P22 bacteriophage$^{(B)}$
Bacteriophage PB$^{(B)}$ 1384
Bacteriophage PBS1$^{(B)}$ 73, 74–75
transduction 76
Bacteriophage PBSX$^{(B)}$ 73, 76
Bacteriophage Pf2$^{(B)}$, atomic structure 1571
Bacteriophage PG$^{(B)}$ 51, 54
Bacteriophage $\phi6^{(B)}$ **978–980**
classification 100
discovery and classification 978
genome and genetics 978–979
in vitro assembly 980
life cycle 979
structure and properties 100, 978
temperature-sensitive mutants 978
Bacteriophage $\phi15^{(B)}$, genome 981
Bacteriophage $\phi29^{(B)}$ 73–74, **980–989**
deletion mutants 981, 986
DNA
encapsidation 988
structure/sequences 981
DNA polymerase 983, 984–985
DNA-linked terminal protein (P3) 74
genes 73, 982
genetics 981
genome 981
map 981
isolation 980
lytic phage 981
morphogenesis 987–988
proheads 988
promoters 986, 987
protein p7 987
replication 74, 983–986
DNA polymerase 984
gene 5 product 986
gene 6 product 985
nontemplate strand 984
origins 983
proteins in 983
replicative intermediates 983
TP and DNA polymerase roles 983
TP domains 984
RNA 988
single-stranded DNA binding protein (SSB) 983, 986
structure and proteins 982–983
sus mutants 981
temperature-sensitive (*ts*) mutants 981, 985
terminal protein (TP) 981, 983
TP-DNA packaging 988
transcription 73, 986–987
gene 4 product 987
map 986
regulation 987
Bacteriophage $\phi105^{(B)}$ 73, 75
repressor gene 75
Bacteriophage ϕ105J119$^{(B)}$, 'direct transfection' for cloning DNA 77
Bacteriophage ϕAmp$^{(B)}$, *see* Bacteriophage P7$^{(B)}$
Bacteriophage ϕCb2r$^{(B)}$ 1335
Bacteriophage ϕCb8r$^{(B)}$ 1335
Bacteriophage ϕCb23r$^{(B)}$ 1335
Bacteriophage Φe$^{(B)}$ 1352
Bacteriophage ϕH$^{(B)}$ 50, 51, 52, 53–54
genetic map 53
genetic structure 53–54
lysogeny 52, 53
transcription 53–54
variant ϕH1 53
variant ϕHL 54
see also Halobacterial phage
Bacteriophage ϕN$^{(B)}$ 50, 51, 52
Bacteriophage ϕX174$^{(B)}$ **989–996**
adsorption 990
cell lysis 994, 995
cis-acting proteins 995
classification and ecology 989
DNA (circular single-stranded) 989
'reduction sequence' 990
gene A protein 990, 995
gene E protein 990, 994
gene expression 995
gene K protein 990
genetic map 989
genome structure 85, 989–990, 991
in vitro replication 996
mRNA 995
primosome 991
recombination 996
RecA-dependent 84
RecA-independent 84–85
related phage 989
replication 990–995
importance to DNA replication understanding 996

Bacteriophage ϕX174(B) *continued*
stage I 991–993
stage II 992, 993–994
stage III 992, 994
replicative forms (RFI, RFII, RFIII) 993
structure and proteins 989
taxonomy 989
transcription and translation 995
Bacteriophage PM2(B) 1432
characteristics and taxonomy 97
Bacteriophage PMS1(B) 51, 54
Bacteriophage PP7(B) 1335
Bacteriophage PR3(B) 1165
Bacteriophage PR4(B) 1165
Bacteriophage PR5(B) 1165
Bacteriophage PR772(B) 1165
Bacteriophage PRD1(B), *see* PRD1 bacteriophage(B)
Bacteriophage PS42-D(B), pyrogenic toxins of *S. aureus* 104
Bacteriophage psi (ψ)(B) 875
Bacteriophage psiM (ΨM)(B) 54–55
Bacteriophage PZA(B), genome 981, 982
Bacteriophage Q_β(B) 1334
6S RNA 1338
genetic map 1335
lysing protein 1336–1337
replication 1337–1338
RNA replication, discovery 630
RNA variants 1338
Bacteriophage Qβ levivirus
evolution 437, 440 §see query
genomic sequence 437
protein structure 440
Bacteriophage R17(B) 1334
Bacteriophage S13(B) 989
Bacteriophage S45(B) 51, 52
Bacteriophage S5100(B) 52
Bacteriophage SP5c(B) 1352
Bacteriophage SP6(B) 1395, 1540
Bacteriophage SP8(B) 1352, 1353
Bacteriophage SP50(B) 73
Bacteriophage SP82(B) 72, 1352
Bacteriophage SP82G(B) 72
Bacteriophage SP(B) 1334, 1335
Bacteriophage SPβ(B) 73, 75–76
specialized transduction 76
Bacteriophage SPO1(B), *see* SPO1 bacteriophage(B)
Bacteriophage SPO2(B) 73, 75
Bacteriophage SPP1(B) 73
Bacteriophage SSV1(B), characteristics and taxonomy 96
Bacteriophage T1(B), *see* T1 bacteriophage(B)
Bacteriophage T2(B), *see* T2 bacteriophage(B)
Bacteriophage T3(B), *see* T3 bacteriophage(B)
Bacteriophage T4(B), *see* T4 bacteriophage(B)
Bacteriophage T5(B), *see* T5 bacteriophage(B)
Bacteriophage T6(B), *see* T6 bacteriophage(B)
Bacteriophage T7(B), *see* T7 bacteriophage(B)
Bacteriophage T12(B)
see T12 bacteriophage(B)
pyrogenic toxins of *S. pyogenes* 104
Bacteriophage TH1(B) 1334
Bacteriophage TTV1(B) 51, 52, 55–56
characteristics and taxonomy 96
variants 56
Bacteriophage TTV4(B) 51, 56
Bacteriophage TW18(B) 1334
Bacteriophage TW19(B) 1334
Bacteriophage TW28(B) 1334
Bacteriophage VcA1(B) 875
Bacteriophage Vi(B) 79
Bacteriophage ViII(B) 79
Bacteriophage VK(B) 1334
Baculoviridae 1398, 1408
structure 1584
Baculovirus expression system 1534–1535
bluetongue virus 963
Norwalk virus 929, 932
Baculoviruses(I) **127–139**
acarivirus envelope glycoprotein homology 440
granulosis viruses (GVs) **127–130**
nonoccluded 131–132, **136–139**
biological control using 138–139
defective interfering (DI) particles 137
enveloped nucleocapsids 138
genome structure and organization 137
host range 136–137, 138
infection route 138
protein synthesis 138, 1534
replication in insects 138
replication in tissue culture 137–138
structure 137
Norwalk virus cloning 929, 932
nuclear polyhedrosis viruses (NPVs) **130–136**
recombinants 1535
as vectors 1534–1535
see also Granulosis viruses (GVs)(I); Hz-1 baculovirus(I); Nuclear polyhedrosis viruses (NPVs)(I); *Oryctes* baculovirus(I)
Badnaviruses(P) **139–143**, 223
control 142–143
economic importance 141
epidemiology and detection 142
genome structure 139, 140
geographic distribution 141
host range and symptomatology 141
molecular biology 140–141
multiplication 140
origin 140
serology 141
structure and composition 140
taxonomy and classification 139–140
transmission 142
virushost relationship 141–142
BALB/c mice, LCMV infection 810
BALB/cfC3H mice, mouse mammary tumor virus (MMTV) infection 858
Banana streak virus (BSV)(P) 139, 141, 142
Barley stripe mosaic virus (BSMV)(P) 512, 661, 1538
antisera 663
coat protein mutants 663
cytopathology 661
economic importance 663
genome structure and organization 661–662
RNAs (α, β, γ) 661, 662
host range 663
infectivity of cDNA 1540
molecular biology and mutations 662
pathogenicity 663
proteins 662
recombinants 662, 663
structure and properties 661
translational control in 663
as vector 663–664, 1540, 1541
Barley yellow dwarf virus (BYDV)(P) 792, 793, 1538
control 796
cytopathology 795
economic significance 797, 1112
effect on fungal infections 1112
host range 796
pea enation mosaic virus (PEMV) relationship 1085
resistance to 797
symptomatology 794
tobacco necrosis virus relationship 900
transmission 795
Barley yellow striated mosaic virus (BYSMV)(P) 1246
Barmah Forest virus (BFV)(A) 1269
Basidiospores, partitivirus transmission 1050
Bats
Machupo virus transmission 1621
rabies-like viruses from 1186, 1187, 1189
BB rats, lymphocytic choriomeningitis virus (LCMV) infection 65
bcl-2 expression, EBV LMP-1 role 413
Bean golden mosaic virus (BGMV)(P) 517, 522
Bean leaf roll virus(P) 792
helper virus for pea enation mosaic virus 1087
Bean mild mottle virus (BMMV)(P) 218, 219
Bean pod mottle virus (BPMV)(P) 249, 250, 254
Bean rugose mosaic virus (BRMV)(P) 249
Bean yellow vein banding complex virus (BYVBV)(P) 1087
Bebaru virus(A) 1268
Bee virus X (BVX)(I) 656, 659
Bee virus Y (BVY)(I) 656, 659
Bees, viruses, *see* Honey bee viruses
Beet cryptic virus (BCV)(P) 274, 276, 277
Beet curly top virus (BCTV)(P) 517, 1538
DNA replication 521
evolution 524
genome structure and expression 518, 520
host range 522
Beet mild yellowing virus(P)
economic significance 1112
effect on fungal infections 1112
Beet necrotic yellow vein virus (BNYVV)(P) 509, 1538
deletion mutants 514
epidemiology and control 516
genetics and replication 513
genome structure and expression 512–513
as vector 1541
Beet pseudo-yellows virus (BPYV)(P) 247
Beet soil-borne virus (BSBV)(P) 509
Beet western yellows virus (BWYV)(P) 792, 793, 1538
cytopathology 795
economic significance 797
host range 796
infectivity of cDNA 1540
management 797
pea enation mosaic virus (PEMV) relationship 1085
symptomatology 794
Beet yellow stunt virus (BYSV)(P) 243
Beet yellows virus (BYV)(P) 242, 244
disease and economic significance 247
genome structure/organization and expression 245
transmission 246
Beetle transmission
bromoviruses 184
carmoviruses 219
comoviruses 254
tymoviruses 1502
bel gene 485
Belladonna mottle virus(P) 1500
Belladonna mottle virus - Iowa strain(P) 1500
Berkeley bee virus(I) 656, 660
Berne virus (BEV)(A) 256, 377, 1452
clinical features of infection 1457
envelope 1454
strains 1456
see also Toroviruses(A)
Berry-Dedrick transformation 1154

β-adrenergic receptors 229
  reovirus receptor 1200
β-barrel motif 250, 1575
  comoviruses 250, 251
  evolutionary significance 1572
  foot and mouth disease viruses 489
  nodaviruses 920
  picornaviruses 250, 1578
  poliovirus 642, 1119
  rhinoviruses 642, 1254
  structure 1579
    atomic 642, 1572
  tertiary fold 1579
β-cells, *see* Islet cells
Betacin 75
Betaherpesvirus, *see Herpesviridae*
BHK cells, orbivirus culture 951
BHK-21 cell line
  foot and mouth disease viruses (FMDVs) culture 488, 494, 495
  nodavirus culture 924
Bile ducts, reovirus infection 1193
Bile salts, virus entry prevention 1077
Biliary atresia, neonatal, reovirus infection 1194
Bilirubin, elevated in hepatitis A 552
Biological control
  cytoplasmic polyhedrosis viruses (CPV) 318
  densonucleosis viruses 334
  disadvantages 397
  entomopoxviruses 397–398
  history 644–645
  insect picornaviruses 1103
  luteoviruses, by parasitoid wasps 796–797
  Oryctes baculovirus 138–139
  tetraviruses 1422
  see also Insecticides
Biological response modifiers, with interferon-α 741
Biopesticides, *see* Biological control; Insecticides
Birds
  encephalitis 421
  equine encephalitis virus hosts 418–419, 420, 421
  smuggled and zoonoses spread 1622
Birnaviridae 143, 475, 1398, 1408, 1409
  fish 475, 476
  structure 1580
  transcription/replication strategy 1207–1208
Birnaviruses(A) **143–149**
  antigenic structure 147
  cell tropism 145
  future perspectives 148
  gene products, processing and functional significance 146–147
  genome structure 146–147
  history 143–144
  host range and epidemiology 144
  immune response 147–148
  in vitro replication 145–146
  infection
    clinical signs and pathology 145
    laboratory diagnosis 148
  pathogenic properties 145
  prevention and control 148
  stability 144–145
  taxonomy and classification 144
  transcription and replication 146, 147, 1207–1208
  see also Infectious bursal disease of chickens virus (IBDV)(A)
Bismuth subsalicylate 932
BK virus(A) **752–757**
  antibodies to 754, 756
  antigenic determinant 754
  antigenic variants 754
  archetype 753
  classification 752
  DNA replication 752
  epidemiology 754–755
  evolution 754
  genetics 754
  genome
    hypervariability 753, 754
    promoterenhancer regions 754
    properties 752–753
  geographic and seasonal distribution 753
  history 752
  host range and viral propagation 753
  infection
    after transplantation 1486–1487
    immune response 754, 756, 1486
    kidney 974
    pathogenicity 756
    persistent 755, 1090
    prevention and control 756–757
  reactivation 1486, 1487
  serologic relationships and variability 754
  structure and properties 752–753
  T antigens 752, 753
  transmission and tissue tropism 755
  see also JC virus(A)
Black beetle virus (BBV)(I) 919
Black queen cell virus (BQCV)(I) 656, 658–659
Blackgram mottle virus (BGMV)(P) 218, 219
Bleeding
  Argentine/Bolivian hemorrhagic fevers 782, 783
  Lassa fever 782, 783
  Marburg and Ebola viruses 831
  see also Hemorrhages
Blindness, Borna disease virus (BDV) infection 153
Blood
  spread of viruses through 909, 910, 1078–1079
  virus transport from 1079
Blood products
  hepatitis B transmission 557
  hepatitis C transmission 572
  HIV transmission 681
Blood transfusions
  Creutzfeldt-Jakob disease (CJD) transmission 1365
  cytomegalovirus transmission 293, 295, 296
  HTLV-1 infection 684, 686
Bloodbrain barrier 907, 907–908, 1093
  virus invasion 908, 909
Blueberry leaf mottle virus (BLMV)(P) 904
Blueberry scorch virus (BCV)(P) 215
Blueberry shoestring virus (BSSV)(P) 1346
  see also Sobemovirus(P)
Bluegill hepatic necrosis reovirus(A) 479
Bluetongue 942, 950, 952
Bluetongue virus (BTV)(A) 941, 942, 943, 956
  adsorption and infection 962, 963
  age and neural cell vulnerability 910
  assembly and budding 963
  genetics 946
  genome sequence 956
  genome structure/organisation 957
  geographic and seasonal distribution 944
  host range and propagation 945
  immune response to 954
  mRNA 962, 963
  outbreaks and prevention 942
  pathogenicity 952
  pathology 953
  phylogenetic associations of serotypes 949
  proteins 958–962
    capsid proteins (VP2 and VP5) 948, 952, 959, 963
    core proteins (VP3 and VP7) 948, 959
    minor core proteins (VP1, VP4 and VP6) 960
    nonstructural (NS1, NS2 and NS3) 961–962
    NS3/NS3a protein 951
    synthesis 959, 960, 963
  replication 953, 962–963
  RNA 960
  RNA polymerase 962
  serologic relationships and variability 948
  structure and topography of proteins 961
  tissue tropism 951
  transmission 951, 953
  'vaccine' 954, 955
  virulence 952
  see also Orbiviruses(A)
Bluetongue virus 10 (BTV-10)(A), genome sequence 956, 957–958
Bluegreen algae 285–286
  see also Cyanobacteria
Bolivian hemorrhagic fever (BHF) 776, 1614
  clinical features 783–784
  epidemiology 781
  mortality 784
  pathogenesis 782
  pathology and histopathology 784
  prevention and control 785, 1621
  see also Muchupo virus(A)
Bollinger bodies 496, 502
Bombyx mori, cytoplasmic polyhedrosis viruses (CPV) in 312
Bombyx mori densonucleosis virus (BmDNV)(I) 331, 332, 333
  economic significance 334
Bone marrow, in human parvovirus B19 infection 1056
Bone marrow transplantation (BMT) 1477
  adenovirus infections after 7, 1488
  cytomegalovirus infection
    risk/incidence 1478
    treatment and prophylaxis 1479, 1481
  echovirus infection after 359
  Epstein–Barr virus (EBV) infections 1482
  hepatitis virus infections 1485, 1486
  immunization before 1488
  polyomavirus infections (JC virus and BK virus) 1486, 1487
  varicellazoster virus (VZV) infections 1483
  see also Transplantation
Boolarra virus (BoV)(I) 919
Border disease, *see* Bovine viral diarrhea virus (BVDV)(A), infection
Border disease virus (BDV)(A) **175–181**, 649
  hog cholera virus (HCV) relationship 649, 651
  see also Bovine viral diarrhea virus (BVDV)(A)
Bordet, J. 646, 648
Borna disease virus (BDV)(A) **149–154**
  cDNA clones 150
  epidemiology 151–152
  experimental hosts and model system 151
  future perspectives 154
  genome structure 150
  geographic distribution 150–151
  history 149
  host range and virus propagation 151
  infection
    clinical features 152–153

Borna disease virus (BDV)[A] *continued*
human infections 151, 154
immune response 153–154
pathology and histopathology 153
prevention and control 154
routes 152
pathogenicity 152
proteins 150
sequence 150
replication 150
serologic relationship and variability 151
structure and properties 150
taxonomy and classification 149
tissue tropism 152
transmission 152
Bornholm disease 271, 639
see also Coxsackieviruses[A]
Borrelia, linear plasmids, African swine fever virus similarity 27–28
Boston exanthema disease 358
Botulinum toxins 105–106
mechanism of action 106
phage conversion 105
structural genes 105
synthesis 106
types 105
Botulism 105
forms of 105
'Boules hyalines' 129
Bovine adeno-associated viruses (BAAVs)[A] 1052
Bovine coronavirus (BCV)[A] 255, 258
Bovine enterovirus (BEV)[A] 385
genome 385
host range and propagation 387
proteins 385
serotypes 387, 388
transmission and tissue tropism 389
see also Enteroviruses[A], animal
Bovine herpesvirus 1 (BHV1)[A] 155
diseases caused by 155, 157
prevention and control 158
Bovine herpesvirus 2 (BHV2)[A] 155, 157, 158
Bovine herpesvirus 3 (BHV3)[A] 155
Bovine herpesvirus 4 (BHV4)[A] 155, 157, 304
Bovine herpesvirus 5 (BHV5)[A] 155
Bovine herpesviruses[A] **155–158**
antigenic relationships 156
epidemiology 156–157
future perspectives 158
genome structure 155–156
geographic distribution 156
history 155
infection
clinical features/diseases 157
immune response 157
pathogenesis 157
prevention and control 158
nucleocapsid, tegument and glycoproteins 155
replication 156
structure 155–156
taxonomy and classification 155
vaccines 157
Bovine immunodeficiency virus (BIV)[A] **158–166**
capsid (CA) protein 161, 164
cell lines infected 162–163
evolution 163–164
future perspectives 165
gag gene 160, 161, 164
genome structure 160, 161
geographic distribution 162
history 158–159
host range and virus propagation 162–163
immune response 165
morphogenesis 159–160
nucleocapsid (NC) protein 161, 164
pathogenicity 165
pol genes 160, 161, 164
prevention and control 165
serologic relationships and variability 164–165
structure and properties 159–160
taxonomy and classification 159
tissue tropism 165
transcription and translation 160–162
transmission 165
vaccination 165
Bovine leukemia virus (BLV)[A] 159, **166–174**
cell tropism 171, 172
diagnostic techniques 173
dsDNA proviral genome 167, 169
envelope 168
evolution 170
future perspectives 174
genetic and serologic variability 171
genome structure 167, 169
mutations 171
tax/rex genes 167, 169, 170
geographic distribution and epidemiology 170
history 166–167
host range and virus propagation 170–171
HTLV-1 relationship 167, 170, 171, 174, 682
immune response 173
antibodies 171
laboratory source 171
pathogenesis 171–173
prevention and control 173
proteins 167–169
capsid (CA;p24) 167
matrix (MA;p15) 168
nucleocapsid (NC;p12) 167
Orf-1 169
regulatory (Tax/Rex), *see below*
surface glycoprotein (SU;gp60/51) 168
transmembrane glycoprotein (TM;gp30) 168
replication 169–170
sites 172
reverse transcriptase 167
Rex protein 169, 170
antibodies to 173
post-transcriptional regulation of mRNA 169, 170
structure 167
Tax protein 167, 168–169, 170
expression affecting host transcription 172
trans-activation of transcription 168, 169, 170
taxonomy and classification 167, 682, 687
transmission 171
tumor development 172, 173
mechanism 167, 172–173
vaccination 173
Bovine malignant catarrhal fever 155, 157
prevention and control 158
Bovine papillomavirus type 1 (BPV-1)[A] 1306
cDNA 1307
genetics 1307
genome 1307
long control region (LCR) 1307
proteins 1307–1308
transcription and promoters 1307
Bovine papillomavirus type 4 (BPV-4)[A] 1306
malignant transformation of papillomas 1309
Bovine papillomaviruses (BPV)[A] **1305–1311**
DNA sequences in transgenic mouse 531, 1306
epidermal development and 1309
evolution 1308
future perspectives 1311
genetics 1307–1308
geographic and seasonal distribution 1306
history 1305
host range and viral propagation 1306–1307
surrogate animal systems 1307
infections
clinical features 1309
fibropapillomas 1306, 1308, 1309, 1310
immune response 1310
lesions 1306
pathology and histopathology 1309–1310
prevention and control 1026, 1310–1311
malignant transformation 1308, 1309
pathogenicity 1309
proteins (E1-E7) 1307–1308
serologic relationships and variability 1308
taxonomy and classification 1305–1306
tissue tropism 1308
transgenic mouse line 1306
transmission 1306, 1308
vaccination 1311
see also Cottontail rabbit papillomavirus (CRPV)[A]
Bovine papular stomatitis 1037
Bovine papular stomatitis virus[A] 1039
Bovine parainfluenza virus type 3 (PIV-3)[A] 1031–1035
classification and properties 1031
cytopathic effects 1032
epidemiology 1033
genetics and evolution 1032–1033
genomic sequence 1031
geographic and seasonal distribution 1031
hemagglutinin 1031, 1032–1033
amino acid sequence comparison 1034
history 1031
host range and viral propagation 1032
infections
clinical features and pathology 1033
immune response 1034
prevention and control 1034
M-YN substrain 1032, 1034, 1034–1035
monoclonal antibodies 1034
MR-YN substrain 1032, 1034–1035
neuraminidase 1031, 1032–1033
proteins 1031, 1032
replication 1032
serologic relationships and variability 1033
substrains 1032
syncytium-inducing ability 1034–1035
transmission and tissue tropism 1033
Bovine parvovirus (BPV)[A] 1052, **1067–1075**
gene expression 1058
genome and replication 1069
geographic and seasonal distribution 1071
history 1068
host range and viral propagation 1071, 1072
infection
clinical features 1055, 1074
prevention and control 1075
properties and structure 1068–1069
serologic relationships and variability 1072, 1073
taxonomy and classification 1068
tissue tropism and pathogenicity 1074
transcription and translation 1070

  transmission and epidemiology 1073
  see also Parvoviruses(A)
Bovine polyomavirus(A) 1135
Bovine respiratory syncytial virus (BRSV)(A) 1210
  human RSV (HRSV) antigenic cross-reactivity 1214–1215
  see also Respiratory syncytial virus(A)
Bovine spongiform encephalopathy (BSE) 1357
  clinical features 1359
  Creutzfeldt-Jakob disease (CJD) relationship 1358, 1364
  epidemics 1357, 1358
  future perspectives 1360
  host range and viral propagation 1357
  incidence 1358
  infectious agent 1357
  intracerebral injection 1357
  pathogenicity 1359
  pathology and histopathology 1359–1360
  prevention and control 1360
  serologic relationships and variability 1358
  transmission 1359
  see also Spongiform encephalopathies
Bovine syncytial virus (BSV)(A) 159
Bovine syncytium-forming virus (BSFV)(A) 481
  geographic and seasonal distribution 486
Bovine viral diarrhea virus (BVDV)(A) **175–181**, 649
  antigenic variability 178
  assembly 177
  epidemiology 178
  evolution 178
  future perspectives 180
  genetics 178
  genome structure 176
  geographic and seasonal distribution 177
  history 175
  hog cholera virus (HCV) relationship 649, 651
  host range and virus propagation 177–178
  infection
    clinical features 179–180
    immune response 180
    pathology and histopathology 180
    prevention and control 180
  monoclonal antibodies 178
  noncytopathic (NCP) and cytopathic (CP) 178
  pathogenicity 179
  persistent infections (PI) 178, 179
  properties 176
    physical 176
  proteins 176
    glycoprotein gp116 177
    p14 176
    p20 (autoprotease) 176, 177
    p125 177, 178
  replication 176–177
  serologic relationships 178
  taxonomy and classification 175–176
  transcription, translation and processing 177
  transmission and tissue tropism 179
  see also Border disease virus (BDV)(A)
Bovine viral diarrhea-mucosal disease (BVD-MD)(A) 175
Bowenoid papules 1019
Bowen's disease 1019
Bracken fern 1309
Bracoviruses(I) 1133
  see also Polydnaviruses(I)
Brain
  Borna disease virus (BDV) infection 151, 152
  Creutzfeldt-Jakob disease (CJD) pathology 1367–1368
  JC virus infection 755, 756, 757
  kuru 759, 761
  spongiform encephalopathy agent tropism 1365
  viral infections rarity 909
Breast cancer, interferon therapy 744
Breast-feeding
  HTLV-1 transmission prevention 686
  rotavirus infections prevention 1280
Breda virus (BRV)(A) 377, 1452–1453
  antibodies 1457, 1458
  clinical features of infection 1457
  strains 1456
  see also Toroviruses(A)
Brefeldin A 1130
BRL39123 44, 46
BRL42810 44, 46
Broad bean mottle virus (BBMV)(P) 181, 185
Broad bean necrosis virus (BBNV)(P) 509
Broad bean stain virus (BBSV)(P) 249
Broad bean true mosaic virus (BBTMV)(P) 249
Broad bean wilt disease virus (BBWV)(P) 253, 451
  host range 452
  serology 453
  serotype I 453
  serotype II 453
  transmission 453
Broccoli necrotic yellows virus(P) 1246
Brome mosaic virus (BMV)(P) 181, 1538
  barley stripe mosaic virus (BSMV) differences 661
  cucumber mosaic virus (CMV) similarity 278
  infectivity of cDNA 1540
Bromelain treatment, influenza virus 722
(E)-5-(2-Bromovinyl)-2'-deoxyuracil 44
Bromovinyluracil nucleosides 44, 46
Bromovirus(P) 1400
Bromoviruses(P) 30, **181–185**
  3'-termini 183
  cDNA in plasmids, infectivity 182
  cis-acting sequences 183–184
  coat protein transcription/translation 183–184
  cucumoviruses relationship 181
  genetics 184
  genome structure 181–182
    RNA1, RNA2, RNA3, RNA4 181, 182, 183
  host range 184–185
  pathology 185
  proteins 1a and 1b 182, 183
  replication 182–183
  RNA amplification 183
  structure 181–182, 1580
  transmission 184
Bronchiolitis 1221, 1223, 1224
  causative agents 1224
  human parainfluenza virus (PIV-3) 1027, 1030
  respiratory syncytial virus (RSV) 1222
Bronchitis 1221, 1223, 1224
Bronchoalveolar carcinoma, interferon therapy 744
Bronchopneumonia 1221, 1223
Bronchus-associated lymphoid tissue (BALT) 1222
BSE, *see* Bovine spongiform encephalopathy (BSE)
Budding (of viruses) 227, 1209, 1559, 1586
  bovine immunodeficiency virus (BIV) 160
  bovine viral diarrhea virus (BVDV) 177
  bunyaviruses 196, 1556
  cell surface role 231
  coronaviruses 257
  equine arteritis virus (EAV) 766
  frog virus 3 (FV3) 506
  lactate dehydrogenase-elevating virus (LDV) 766
  membrane acquisition 1556
  mumps virus (MuV) 880
  nonoccluded baculoviruses 138
  process 1559–1560
  see also Release of viruses
Budgerigar fledgling disease virus (BFDV)(A) 1135
Buffalopox 1511, 1512
Buffalopox virus(A) 267
  epidemiology 1512
  evolution 1511
Bulb growers, economic significance of viruses 1111, 1112
Bunyamwera virus(A) 193, 446
  tospoviruses relationship 1462
Bunyaviridae 186, 1399, 1407, 1408, 1409
  ambisense RNA 1207, 1461
  Bunyavirus, see Bunyavirus(A)
  in Europe 1618
  eye infections 446–447
  genera 186
  Hantavirus 186, 538
  Nairovirus 186
  Phlebovirus 186, 187, 190–191
  replication/transcription strategy 1207
  structure and characteristics 186, 443, 446, 1459, 1460, 1583
  Tospovirus 186, 1459
  see also Bunyaviruses; *Hantavirus*(A); *Nairovirus*(A); *Phlebovirus*(A); *Tospovirus*(P)
Bunyavirus(A) 187, 190, 446
  California serogroup 188, 189, 190
  Simbu complex 190
  structure and genome 188
  viruses included 190
  see also Bunyaviruses(A); La Crosse virus(A); Oropouche virus(A)
Bunyaviruses(A,P) **185–196**
  cell culture 188
  defective interfering 188
  ecology and epidemiology 189–190
  epidemics and epizootics 190
  genera and serogroups 186, 187
  genome
    coding strategies 193, 194
    in persistent infections 196
    sequences 193, 194
  genome structure 186, 188, 192–193, 1207
    consensus terminal sequences 193
    open reading frame 188, 193
    reassortment of segments 188
    RNA segments (L,M,S) 186, 192, 193
    terminal complementarity 188, 192, 1412
  host range 186
  host susceptibility 189
  human infections/viruses 187, 190, 191
  invertebrate host infection 189, 196
  maturation and budding 196, 1556
  nucleocapsid 192
  persistent infections 196
  plant infection 189, 191–192
  replication 186, **192–196**, 1207
    assembly and release 196, 1556
    attachment, entry and uncoating 193, 195
    genome 196
    transcription and translation 195

Bunyaviruses(A,P) *continued*
  structure 186, 188, 192–193
    glycoproteins 188, 192, 193
    proteins 188, 192, 193
  taxonomy and classification 186
  transmission 189
  see also *Bunyavirus*(A); *Hantavirus*(A); *Nairovirus*(A); *Phlebovirus*(A); Tospoviruses(P)
Burdock yellows virus (BuYV)(P) 243
Burkitt's lymphoma 404, 1477, 1482, 1494–1495
  EBV role theories 408, 1494–1495
  epidemiology and clinical features 406, 1494
  see also Epstein–Barr virus (EBV)(A)
BV-araU 44
BVdU 44

C
C57BL strain, *see under* Mice
c-*erb* 937
c-*erbA* 941
c-*erbB* 938
c-Ha-*ras* proto-oncogene 937
c-*mos* 940
c-*myc* 937, 1494
c-*onc*, *see* Oncogenes; Proto-oncogenes
c-*raf* 940
c-*ras* 940
c-*rel* 1230
CAAT box 169
  lymphoproliferative disease virus (LPDV) of turkeys 812
Cacao swollen shoot virus (CSSV)(P) 139, 140, 141
  control 1113
  economic significance 1110, 1111
Cacao yellow mosaic virus(P) 1500
Cactus virus 2 (CaV2)(P) 215
Cactus virus X(P) 1146
Cadang-cadang 1111
  viroid 1564, 1567
Calcium
  rotavirus uncoating and 1286–1287
  tombusviruses capsid stability 1448
Caliciviridae 375, 1399, 1409, 1544
  hepatitis E virus in 581
  replication/transcription strategy 1206–1207
  structure 1580
Caliciviruses(A) 375–376, **1544–1547**
  classification and properties 1544
  diarrhea due to 375
  feline, *see* Feline calicivirus
  genomic organization 929, 1544, 1547
  hepatitis E virus relationship 581
  history 1544
  human 375
  Norwalk virus relationship 375, 928, 929, 933
  pinnipeds 1544–1547
  rabbit, *see* Rabbit hemorrhagic disease (RHD) virus(A)
  see also Vesicular exanthema virus of swine (VES)(A)
California encephalitis virus(A) 446
Californian volepox virus(A) 267
Calomys callosus 778, 781
Calomys musculinus 780
Camelpox virus(A), smallpox virus relationship 1340
Campbell model, λ prophage excision 109
Canarypoxvirus(A), ALVAC vector 1531
Cancer 1472
  interferon clinical use 738, 741–745
  model systems, feline leukemia virus 463
  properties of cells 1472
  treatment, retrovirus vector application 1534
  see also Carcinogenesis; Transformation; Tumor formation
Canine adeno-associated viruses (CAAVs)(A) 1052
Canine adenoviruses (CAV)(A) 17
Canine coronavirus (CCV)(A) 255
Canine distemper
  clinical features 1265
  geographic and seasonal distribution 1261
  incubation period 1265
  pathology and histopathology 1266
Canine distemper virus (CDV)(A) 838, 1260, **1260–1268**
  'biotypes' 1265
  cytopathic effects 1261
  envelope and proteins 1262–1263
  epizootiology 1264
  evolution 1263
  genetics and genome structure 1261–1263
  host range and viral propagation 1261
  infection, *see* Canine distemper
  pathogenicity 1265
  serologic relationships and variability 1263–1264
  taxonomy and classification 1211
  translation 1262
  vaccines 1266, 1267
  see also *Morbillivirus*(A); Rinderpest virus(A)
Canine parainfluenza virus 5 ((PIV-5)(A) 1035
Canine parvovirus (CPV)(A) 1052, 1055, **1061–1067**
  epidemiology 1053, 1065
  evolution 1064
  future perspectives 1067
  genetics 1053, 1064
  genome 1053, 1064
  geographic and seasonal distribution 1061–1062
  history 1061
  host range and viral propagation 1053, 1062
  infection
    clinical features 1065–1066
    immune response 1067
    pathology and histopathology 1056, 1066
    prevention and control 1057, 1067
  pathogenicity 1065
  serologic relationships and variability 1064
  structure 1062, 1063
  taxonomy and classification 1061
  vaccine and prevention 1057
  variants 1064
  see also Parvoviruses(A)
Canna yellow mottle virus (CaYMV)(P) 139, 141, 142
Cannibalism 759, 760, 761
'Canyon hypothesis' 207, 1121, 1123, 1255, 1573
Cap-snatching 1462
Caper latent virus (CapLV)(P) 215
Caper vein binding virus (CapVBV)(P) 215
Capillary endothelium, viruses entering 909
Capillary leak syndrome 328, 329
Capillary leakage
  Argentine/Bolivian hemorrhagic fevers 782
  Lassa fever 782
Capillovirus 1399
Capilloviruses(P) **197–198**, 242
  detection and control 198
  geographic distribution 198
  'ghosts' 197
  host range and symptoms 198
  relationships and variability 198
  structure and composition 197–198, 244–245, 1582
  taxonomy and classification 197, 244
  transmission 198
Caprine arthritis encephalitis (CAE) syndrome 199, 203
  clinical features and histopathology 203–204
  immune response 204
  prevention and control 204
Caprine arthritis encephalitis virus (CAEV)(A) **199–205**, 1592
  epidemiology 202
  future perspectives 204–205
  genetics and evolution 202
  genome structure 199–201
  history 199
  host range 201–202, 1593
  pathogenicity 203
  propagation 201–202
  regulatory proteins 199, 200–201, 203
  replication 201
    in macrophage 203
    polyprotein processing 201
  serologic relationships and variability 199, 202
  structure 199–200
    $p25^{gag}$ 201, 202
    proteins and glycoproteins 199
    Q, Tat and Rev proteins 199, 200–201
  taxonomy and classification 199
  tissue tropism 202–203
  transcription and translation 201
  transmission 202–203
    horizontal 201, 202
Caprine herpesvirus 1(A) 157
Caprine respiratory syncytial virus (CRSV)(A) 1210
  antigenic cross-reactivity 1214
Capripoxvirus 954, 1160
  see also Poxviruses(A), sheep and goats
Capsid
  maturation 1169
  proteins, functions 1574
  see also Helical symmetry; Icosahedral symmetry; Structure of viruses
Carbohydrates, glycocalyx 228
Carbon dioxide, *Drosophila* sensitivity 1311, 1312, 1313, 1315
Carbonyl cyanide *m*-chlorophenylhydrazone (CCCP) 290, 291
Carcinoembryonic antigen (CEA), family, coronavirus receptor 256
Carcinogenesis
  human papillomaviruses and cervical cancer 1024–1025
  multistep 1493
  see also Oncogenesis; Tumor formation
Carcinoid tumors, interferon therapy 744
Cardiac disease, *see* Heart, virus infections
Cardiomyopathy, dilated (DCM), *see* Dilated cardiomyopathy (DCM)
Cardiovascular infections, *see* Heart, virus infections
Cardioviruses(A) **205–213**, 1423
  antigenic determinants 207
  capsid
    depressions ('pits') 207
    proteins 206, 207
    three-dimensional structure 206, 207
  cell cultures 211
  cytopathology 212–213

genetics and evolution 211, 212
genome structure 207–209
sequences 211, 212
geographic distribution 211
groups and members 1423
history 205–206
host range 211
host RNA/protein synthesis shutdown 210, 491
immune response 213
initiation of infection 209
pathogenicity in mice 212
prevention and control 213
propagation 211
properties 206
replication 209
assembly 210–211
RNA 209–210
synthesis of viral components 209–210
RNA virus core (RVC) motif 207
serotype stability 211
taxonomy and classification 206
transmission and tissue tropism 212
vaccine production 213
see also Theiler's murine encephalomyelitis virus (TMEV)$^{(A)}$
Carlavirus 1399
structure 1582
Carlaviruses$^{(P)}$ **214–218**, 1399
cytopathology 216–217
economic significance and ecology 217–218
epidemiology and control 217
evolution 218
genome expression and translation 216
genome structure 214–216
triple gene block proteins 216
geographic distribution 217
history 214
host range and symptomatology 217
physicochemical properties 214
potexviruses similarity 1145
replication 216
serology 217
structure and composition 214, 1582
taxonomy and classification 214, 215
transmission 216
Carmovirus 1399
Carmoviruses$^{(P)}$ **218–223**
coat-protein-binding sites 220, 222
cytopathology 219
distribution and host range 218–219
economic significance 219
genetics and gene function 220–222
genome structure 220
open reading frames (ORFs) 220, 221
RNAs (C,D,F,G) 222–223
hepatitis C virus relationship 570
in vitro assembly 220
mutagenesis studies 222
replication and regulation 222
satellites and defective-interfering RNAs 222–223
serology 219
structure and assembly 219–220
amino acid sequences 221, 222
coat protein 219–220, 221–222
taxonomy and classification 218
tobacco necrosis virus classification 900
tombusviruses relationship 1447
transmission 219
Carnation cryptic virus (CCV)$^{(P)}$ 275
Carnation Italian ringspot virus (CIRV)$^{(P)}$ 1447
Carnation latent virus (CLV)$^{(P)}$ 214, 215
Carnation mottle virus (CarMV)$^{(P)}$ 218
genome structure 220, 221
tobacco necrosis virus similarity 899
transmission 219
Carnation necrotic fleck virus (CNFV)$^{(P)}$ 244
Carnation ringspot virus (CRSV)$^{(P)}$ 349
control and transmission 352–353
host range and economic significance 352
Nodavirus relationship 919
Carnation stunt associated viroid$^{(P)}$ 1564
Carp pox herpesvirus$^{(A)}$ 470
'Carrier culture' 1170
Carrier state
halobacterial phage 52
see also individual viruses
Carrot red leaf luteovirus$^{(P)}$ 792
host range 796
transmission 795
Carrot yellow leaf virus (CYLV)$^{(P)}$ 243
Case control studies 399
Case-fatality rate, definition 399
Cassava American latent virus$^{(P)}$ 904
Cassava green mottle virus$^{(P)}$ 904
Cassava latent virus (CLV)$^{(P)}$ 1538
Cassia mild mosaic virus (CasMMV)$^{(P)}$ 215
CAT box 595
Cat-bite associated diseases 456
Cats
AIDS-like disease 454–455
see also Feline immunodeficiency virus (FIV)
cowpox virus infection 263, 264, 265
lymphomas in 458, 459
see also viruses beginning Feline
Cattle
Breda virus (BRV) infections 1453
foot and mouth disease 494
virus infections, *see specific viruses beginning* Bovine
Cattle plague 1260
see also Rinderpest virus$^{(A)}$
Caudovirales 95, 96, 1397, 1406
characteristics 96
Cauliflower mosaic virus (CaMV)$^{(P)}$ 223, 224, 225, 1538
cell culture 1539
genome 1537, 1538
Hepadnaviridae relationship 556
inoculation of plants 1537
as vector 1537–1539
applications 1542
see also Caulimoviruses$^{(P)}$
Caulimovirus 1398, 1407
structure 1580
Caulimoviruses$^{(P)}$ **223–226**, 1398, 1407
epidemiology and control 226
genetics 224
genome structure 224, 1537
conserved genes 224, 225
history 223
host relationship 225
multiplication 224–225, 1537
proteinaceous inclusion bodies 141
structure and composition 223–224, 1580
proteins 224, 225
taxonomy and classification 223
transmission 225–226
vector relationship 225, 1537–1539
see also Cauliflower mosaic virus (CaMV)$^{(A)}$
Caulobacter 1335
CD4 molecules
HIV adsorption 48, 678, 707, 1557, 1560
as virus receptors 229, 1557, 1560
see also Human immunodeficiency virus (HIV)$^{(A)}$
$CD4^+$ cells, *see* T cells, $CD4^+$ helper (Th)
$CD8^+$ cells, *see* Cytotoxic T lymphocytes (CTL); T cells, $CD8^+$
CD21 407, 413
gp350/220 glycoproteins of EBV 415
monoclonal antibodies 1483
CD23 413
expression, activation by LMP-1 (EBV) 413
CD24, monoclonal antibodies 1483
CD45RO marker 699
cdc2 protein 20
cDNA 530
infectivity, RNA viruses as vectors 1540
see also under specific viruses
Cell
homeostasis 229
polarity 226–228
Cell attachment proteins, *see* Attachment of viruses; Receptors for viruses
Cell cultures 336, 1171
Amsacta moorei entomopoxvirus (AmEPV)$^{(I)}$ 392
Chikungunya (CHIK) virus, O'Nyong Nyong (ONN) virus, Mayaro virus (MAY) 238
coronaviruses 255, 256
coxsackieviruses 268
cytomegaloviruses, human (HCMV) 293, 305
cytoplasmic polyhedrosis viruses (CPV)$^{(I)}$ 316, 318
defective-interfering particles 1171
dengue viruses 325–326
echoviruses$^{(A)}$ 355
enterovirus (EV) serotype 70 (EV70)$^{(A)}$ 381
equine infectious anemia virus (EIAV)$^{(A)}$ 432
foamy viruses 484–485
foot and mouth disease viruses (FMDVs) 488, 494
history 630, 639
infection process 1171
for isolation of viruses 336–338
centrifugation culture in shell vial 34, 338
conventional methods 336–338
cytopathic effects in tube culture 336–337
differential cell susceptibility 337–338
plaque formation 338
Marek's disease virus isolation/propagation 834
multiplicity of infection (MOI) 1171
optimal conditions 1171
zoonoses diagnosis 1620
see also under other specific viruses
Cell cycle
checkpoints 940
cyclins role 20
oncogene action 939, 940
Cell division 232
Cell fusion 1170
biological significance 608
'fusion from within' (FFWI) 1302
'fusion from without' (FFWO) 1302
HSV glycoproteins role 599, 607–608
in vitro, retroviruses, type D 1239
Sendai virus 1300, 1302
see also Syncytium formation
Cell growth, genes involved 938
Cell killing, host cells 1590–1591
Cell lines 1171
animal enterovirus culture 387
continuous 336
fish herpesviruses 471

Cell lysis 698, 704
  by animal viruses, bovine herpesviruses 156
  in autoimmunity 65
  of bacteria by phage 79
    see also Bacteriophage, lytic
  host cells 1169–1170, 1589, 1590–1591
  T cells, *see* Cytotoxic T lymphocytes (CTL)
Cell membrane, *see* Plasma membrane
Cell proliferation, oncogenes controlling 938, 940
Cell rounding 1590
Cell signaling 1473, 1474, 1475
  see also Signal transduction; Transformation
Cell structure/function in virus infections (animal cells) **226–233**
  cytosol 231–232
  diagrammatic representation 227
  membrane network 230–231
  nuclear–cytoplasmic exchanges 232–233
  nucleus 232
  surface 226–230
    extracellular coats 228
    plasma membrane 228–230
    tissue organization and cell polarity 226–228
  see also Cytopathic effects
Cell-cell fusion, *see* Cell fusion
Cell-mediated immunity **703–709**, 1091
  antigen processing 705
    endogenous pathway 704, 705
    exogenous pathway 704, 705
  compromised, in leporipoxvirus infections 1159
  cytotoxic 698, 703–704, 754, 756
    staphylococcal enterotoxin-dependent 105
    see also Cytotoxic T lymphocytes (CTL)
  effector mechanism 703
  evasion of 701–702
  general features 703–705
  history 708
  stimulation (in viral infections) 705–706
  to vaccines 1504
  in viral infections 706
    adenovirus 7
    African swine fever virus 29
    bovine herpesviruses 157
    caprine arthritis encephalitis virus 204
    cardioviruses 213
    cowpox virus 267
    coxsackievirus 273
    dengue virus 330
    ectromelia (mousepox) virus 865
    Epstein–Barr virus (EBV) 409
    equine encephalitis virus 422
    hepatitis A virus 553
    herpes simplex virus 591
    herpesviruses saimiri and ateles 620
    human cytomegalovirus 297
    human immunodeficiency virus (HIV) 680, 707–708
    human papillomaviruses 1020, 1026
    human respiratory syncytial virus (HRSV) 1217
    influenza viruses 714
    lactate dehydrogenase-elevating virus (LDV) 769
    Lassa fever virus 785
    Marek's disease virus 836
    measles virus 846
    molluscum contagiosum virus (MCV) 852
    mumps virus (MuV) 882
    murine cytomegalovirus (MCMV) 311
    polioviruses 1118
    pseudorabies virus 1177
    rubella virus 1297
    Theiler's viruses 1428
    varicella–zoster virus (VZV) 1517
    vesicular stomatitis virus (VSV) 1555
  see also T cells
Cell-mediated lysis (CML), *see* Cell lysis
Cell-surface adhesion molecules
  down regulation and immune evasion strategy 1094
  see also ICAM-1
Cell-to-cell spread, of viruses in nervous system 909
CELO virus(A) 15
Central America, zoonoses in 1617–1618
Central European encephalitis 366
Central European encephalitis virus (CEEV)(A) 362, 363
  vaccine 367
  see also Encephalitis viruses(A); Tick-borne encephalitis virus (TBEV)(A)
Central nervous system (CNS) 1117
  antibodies in 908
  in Creutzfeldt-Jakob disease (CJD) 1367
  equine encephalitis virus replication 420
  pathways for virus invasion 908–909, 1079–1080
  reovirus type 3 receptors (Reo3R) 1562–1563
  site for persistent infections 1093
  spongiform encephalopathies, bovine 1359
  Theiler's viruses
    'bystander' damage 1429
    features 1423, 1429
  virus infection
    animal enteroviruses 390
    Borna disease virus (BDV) 153
    echoviruses 358–359
    encephalitis viruses 366
    enterovirus (EV) serotype 71 (EV71) 383, 384
    equine encephalitis virus 420, 421, 422
    herpes simplex virus 589, 590
    Japanese encephalitis virus 749
    mumps virus (MuV) 882
    pseudorabies virus 1176
    rabies virus 1183
    rabies-like virus 1189
    reovirus 1193, 1194
  see also Nervous system, viruses; Neurological disease
Centrifugation culture, in shell vial 338, 348
Cercopithecine herpesvirus-12(A), *see* Herpesvirus papio
Cereal tillering disease virus (CTDV)(P) 1096
Cerebellum, in kuru 761
Cerebral infections, murine cytomegalovirus 310
Cervical carcinoma
  causative agents (HPV types) 1017, 1018
  development 1024–1025
  human papillomaviruses (HPV) and 1014, 1017, 1021, 1024, 1476, 1496–1497
  squamous, herpes simplex virus 2 (HSV-2) in 592
  see also Human papillomaviruses (HPV)(A)
Cervical intraepithelial neoplasia (CIN) 1017, 1020, 1024
  human papillomavirus (HPV) types 1019
Cervid herpesviruses(A) 156
Chamberland filter 627
Chandipura (CHP) virus(A) **233–236**
  cross-reactions 234
  genome structure 234, 235
  host range 234–235
  infection features 235
  pathogenicity and genetics 235
  proteins 234
  temperature-sensitive (*ts*) mutants 235
  transmission and distribution 234–235
Channel catfish virus (CCV)(A) 470, 475
  attenuated 473
  clinical features and pathology 473, 475
  evolution 471, 473
  genome structure 471–472
  geographic and seasonal distribution 471, 475
  history and classification 470–471, 473
  host range and virus propagation 471, 475
  immune response 473, 475
  pathogenesis and transmission 473
  prevention, control and vaccine 473–474
  proteins and expression of 471
  reference strain (Auburn: VR-665) 471
  see also Fish herpesviruses(A)
Chaperonin 773
  GroEL 1378
Cherry leafroll virus (CLRV)(P) 904, 906
Cherry rasp leaf virus(P) 905
Chick embryo cells
  cowpox virus, white pocks 262, 266
  cultivation of viruses, history 630
  infectious bursal disease of chickens virus (IBDV) infection 146
Chick inoculation, Marek's disease virus isolation/propagation 834
Chick syncytial virus (CSV)(A) 1227
  DNA integration 1231
  as helper virus to REV-T 1231
Chicken hepatocellular carcinoma cell line, duck hepatitis B virus replication 565
Chicken herpesvirus(A), *see* Marek's disease virus(A)
Chicken influenza virus(A), defective-interfering particles 323
Chicken syncytial virus (CSV)(A), as insertional mutagens 937
Chickenpox, *see* Varicella
Chickenpox virus(A), *see* varicellazoster virus (VZV)(A)
Chickens
  avian leukosis viruses 69
  fowlpox 496
  infectious bursal disease, *see* Infectious bursal disease of chickens virus (IBDV)(A)
  Marek's disease (MD), *see* Marek's disease (MD)
  Newcastle disease virus 917
Chicory blotch virus (ChiBV)(P) 215
Chicory yellow mottle virus (CYMV)(P) 904
Chikungunya (CHIK) virus(A) **236–241**, 1270, 1615–1616
  epidemiology 239, 324
  evolution 238
  genetics and replication 238
  geographic and seasonal distribution 237
  history 236
  host range and virus propagation 237–238
  infections
    clinical features 239–240
    diagnosis 1616
    of eye 445
    of heart 965
    immune response 240–241
    pathology and histopathology 240
    prevention and control 241, 1616
  serologic relationship and variation 238–239

taxonomy and classification 237, 1615
transmission and pathogenicity 239, 324, 1615
vaccines 241
Children, cytomegalovirus infection 294, 295
Chilo iridescent virus (CIV)$^{(I)}$ 316, 1430
economic importance 1435
genome 1435
see also Iridoviruses$^{(I)}$; Tipula iridescent virus$^{(I)}$
Chimpanzee herpesvirus, *see* Herpesviruses$^{(A)}$; Herpesvirus pan$^{(A)}$
Chimpanzees
hepatitis C virus infection 571, 572, 574
human respiratory syncytial virus (HRSV) infection 1215
as model, difficulties 613
polio and poliovirus infections 634, 635, 636
Chinese hamster ovary (CHO) cells, cowpox virus growth 262
Chironomus luridus entomopoxvirus$^{(I)}$ 393, 396
Chlamydia, isolation, centrifugation culture 338
Chloramphenicol acetyltransferase (CAT) 692
mouse mammary tumor virus (MMTV) T-cell tumors 858
Chlorella NC64A viruses$^{(P)}$ 36, 38
ecology 39
Chlorella Pbi viruses$^{(P)}$ 38
Chlorella viruses$^{(P)}$ 36
taxonomy and classification 36
Chlorella-like algae, viruses 36
Chlorine 273
Chloriridovirus$^{(I)}$ 1430, 1433
evolution 1435
serology and members 1433
Chloris striae mosaic virus (CSMV)$^{(P)}$, structure 517–518
Chloroplasts
furoviruses affecting 514
vesicles, tymoviruses in 1501, 1502
Chordopoxviridae 497, 1037
Orthopoxvirus 1507
Chordopoxvirinae 392, 848, 1339, 1507
characteristics and structure 1339
fowlpox 497
Molluscipoxvirus 848
Orthopoxvirus, *see Orthopoxvirus*
see also Molluscum contagiosum virus (MCV)$^{(A)}$
Chorioallantoic membrane (CAM)
cultivation of viruses, history 630
fowlpox virus culture 500
inoculation method 338
Marek's disease virus isolation/propagation 834
virus isolation 338
Choristoneura biennis entomopoxvirus (CbEPV)$^{(I)}$ 394
Choristoneura fumiferana entomopoxvirus (CfEPB)$^{(I)}$ 398
infection 395
Choroid plexus 908
Chromatin 232
Chromosomal translocations
EBV infection (Burkitt's lymphoma) 406
proto-oncogene activation 938
Chromosome(s) 232
Chromosome 9, interferon genes 734
Chromosome 12, interferon-γ gene 734
Chromosome 21, influenza virus susceptibility 665
Chronic fatigue syndrome, human herpesvirus 6 (HHV-6) and 625
Chronic infections 787, 1076, 1089
Chronic paralysis virus (CPV)$^{(I)}$ 655–657
pathology and epizootiology 655–656
properties and genome 655, 656
transmission 657
Chronic paralysis virus associate (CVPA)$^{(I)}$ 655–657
Chronic persistent viral infections, *see* Persistent viral infections
Chrysanthemum stunt viroid$^{(P)}$ 1564
Chrysanthemum virus B (CVB)$^{(P)}$ 215
α-Chymotrypsin 1032
cI repressor
coliphages P2 and 186 1005
lambda, *see* Lambda bacteriophage$^{(B)}$
P22 bacteriophage 1012
see also Repressors
Cichlid virus$^{(A)}$ 475
Cirrhosis 557, 573, 579
hepatitis B virus and 555
Cis-acting proteins, bacteriophage ϕX174 995
cis-acting signals 1205
Cisternae 230
Citrus bent leaf virus$^{(P)}$ 1564
Citrus exocortis viroid (CEVd)$^{(P)}$ 1564
pathogenicity 1569
Citrus leaf rugose virus (CLRV)$^{(P)}$ 32, 33
host range and geographic distribution 34
Citrus tatter leaf virus (CTLV)$^{(P)}$ 242
Citrus tristeza virus (CTV)$^{(P)}$ 244, 245
control 248
disease and economic significance 247
transmission 246
Citrus variegation virus (CVV)$^{(P)}$ 32, 33
Classification, *see* Taxonomy and classification
Clathrin cages 229
Clathrin-coated pits 1204
Clearance of viruses, *see* Host defenses
Climatic changes, zoonoses and 1621
Clitoria yellow virus$^{(P)}$ 1500
Clone 525
Cloning, *see* Gene cloning
Clostero-like viruses 242, 243
Closterovirus$^{(P)}$ 1399
Closteroviruses$^{(P)}$ **242–248**
cytopathic effects 246
diseases and economic significance 247–248
epidemiology and control 248
future perspectives 248
genome organization and expression 245
genome structure 245
history 242
host range and geographic distribution 245–246
host relationship 246
structure and composition 244–245, 1582
coat proteins 244
taxonomy and classification 242, 243–244
subgroups 244
transmission 246–247
whitefly-transmitted 244, 247
Clostridium botulinum 105
neurotoxins, *see* Botulinum toxins
Cloudy wing virus (CWV)$^{(I)}$ 656, 659
Clover yellow mosaic virus (ClYMV)$^{(P)}$ 1143
genome 1144
Clover yellows virus (CYV)$^{(P)}$ 243
CMV, *see* Human cytomegalovirus (HCMV)$^{(A)}$; Cytomegaloviruses$^{(A)}$
Cmv-1, mouse cytomegalovirus (MCMV) resistance and 668
Co-infection, interference and 728
Coagulation abnormalities, Marburg and Ebola viruses 831
Cocksfoot mild mosaic virus (CMMV)$^{(P)}$ 1346
see also Sobemovirus$^{(P)}$
Cocksfoot mottle virus (CfMV)$^{(P)}$ 1346
see also Sobemovirus$^{(P)}$
Cocoa necrosis virus$^{(P)}$ 904
Coconut cadang-cadang 1111
Coconut cadang-cadang viroid$^{(P)}$ 1564, 1567
Coconut palm rhinoceros beetle, baculovirus in 137, 138
Coconut tinangaja viroid$^{(P)}$ 1564
Coconut trees, economic significance of viruses 1111, 1567
Coffee ring spot virus$^{(P)}$ 1246
Coggins assay 435
Cohesive sites 773
Cohort studies 399
Cold sores 590
Cold-adapted mutants 526
Colds, *see* Common cold
Cole latent virus (CoLV)$^{(P)}$ 215
Coleus blumei viroid$^{(P)}$ 1564, 1565
ColIb plasmid
T5 abortive infection and 1387
T7 phage inhibition 1395
Colicins, group A 467
Coliphage
F-specific isometric RNA 82
as indicator of contamination 79
RNA, *see* RNA bacteriophage$^{(B)}$
Coliphage 186$^{(B)}$ **1003–1009**
apl gene in lytic state 1005
B gene product 1006
DNA replication 1007
genome 1004
life cycle, control 1003–1006
lysogeny 1005
promoters and repressors 1005
prophage induction 1005
Coliphage λ, *see* Lambda bacteriophage$^{(B)}$
Coliphage Mu$^{(B)}$, *see* Mu bacteriophage$^{(B)}$
Coliphage N4$^{(B)}$, *see* N4 bacteriophage$^{(B)}$
Coliphage P2$^{(B)}$, *see* P2 coliphage$^{(B)}$
Coliphage P4$^{(B)}$, *see* P4 coliphage$^{(B)}$
Coliphage T2$^{(B)}$, *see* T2 bacteriophage$^{(B)}$
Coliphage T4$^{(B)}$, *see* T4 bacteriophage$^{(B)}$
Coliphage T6$^{(B)}$, *see* T6 bacteriophage$^{(B)}$
Colitis, hemorrhagic 103
Color deviation, plant viruses causing 1109, 1110
Colorado tick fever 945, 953, 1617
pathology 953, 954
protection against 955
Colorado tick fever virus (CTFV)$^{(A)}$ 942, 1617
eye infections 445
geographic distribution 944, 1617
replication 953
taxonomy and classification 1617
transmission 944, 1617
Colorectal carcinoma, interferon therapy 744
Coltivirus$^{(A)}$ 942
serogroup$^{(A),(B)}$ 943
Coltiviruses$^{(A)}$ **941–964**
economic significance 950
gene reassortment 946
import/export restrictions 942, 950, 956
replication 962–963
vaccines 954, 955
see also Orbiviruses$^{(A)}$
Columbia-SK virus$^{(A)}$ 205
Columnea latent viroid (CLVd)$^{(P)}$ 1564
host range 1570
Commelina yellow mottle virus (CoYMV)$^{(P)}$ 139, 1398
host range and host relationship 141
multiplication 140
structure and composition 140
transmission 142

Common cold 1221, 1223
causative agents 1223, 1224
clinical features 1258
incidence 1257
rhinoviruses causing 1253, 1257, 1258
virus transmission 1219, 1257
see also Coronaviruses(A); Rhinoviruses(A)
Comoviridae 902
Comovirus(P) 1400
Comoviruses(P) **249–254**
biological properties 249
components (top/middle/bottom) 249, 250
economic importance 254
epidemiology 254
fabaviruses similarity 454
genome expression 251, 252
genome structure 251, 1585
RNA1/RNA2 249, 251, 252
VPg linked to 5′ termini 250, 253
geographic distribution 253–254
host range 254
nepoviruses relationship 902
physical properties 249–250
proteins
functions 251, 253
polyprotein cleavage 251, 1575
relationships with other virus groups 253, 902
replication 253
serology 249
structure 250–251, 1580
picornavirus comparison 1578, 1585
symptoms and cytopathology 254
taxonomy and classification 249
transmission 254
Complement
alternate pathway 697
C3b receptor, HSV glycoproteins 599, 609
fixation, rabies-like viruses 1188
Complementary DNA (cDNA), *see* cDNA
Complementation 527–528
groups 529
inhibition by dominant-negative mutants 731
Complementation tests, transduction application 112
Complex nucleocapsids 1209
Concatemer 1010
see also under DNA
Concentration, of virus preparations 1172
Conditional lethal mutants 526
Condylomata acuminata 1019, 1024
Congenital disease, definition 401
Conjunctiva, entry of viruses 1077–1078
Conjunctivitis
acute hemorrhagic (AHC) 449–450, 1078
causative agents 379, 444, 449
clinical features 382
pandemic and epidemics 380, 381, 383
pathology and histopathology 382
treatment 383
coxsackievirus infection 271
palpebral 444
etiology 449
viral etiologies 448, 449
viruses causing 442
see also Eye infections
Consensus sequences, intergenic, coronaviruses 257
Contagium vivum fluidum 628
Corn, diseases, *see entries beginning* Maize
Cornell N line, resistance to Marek's disease virus 836
Coronaviridae 255, 1399, 1409
characteristics 443
eye infections 445
replication/transcription strategy 1206–1207
structure 443, 1583
toroviruses 1453
see also Coronaviruses(A); *individual viruses (page 255)*
Coronavirus-like particles(A) 376–377
Coronavirus-like superfamily 1456
Coronaviruses(A) **255–260**
assembly 257
cDNA clone 258
cell damage mechanism 911
classification 255
defective-interfering 258
epidemiology 258
future perspectives 260
genetic analysis 258, 260
geographic distribution 258
glycoproteins (S and M) 255, 256
HCV-229E 259
HE glycoprotein 256
host range and tissue tropism 255–256
human strains 255, 258, 259, 376–377
infections 258, 259, 376–377
common cold 1223
frequency 1223
gastroenteritis 376–377
immune responses 260
pathology 258–260
prevention and control 260
respiratory disease 1223
LDV and EAV similarity 763, 768
membrane proteins 1556
mutation frequency 257–258
receptors 256
replication and genetics 257–258, 1205
serogroups 256
HCV-229E group 256
serologic and evolutionary relationships 256–257
structure 255
toroviruses relationship 1453, 1456
virus propagation 255–256
Corticoviridae 1398, 1405
characteristics and taxonomy 97
structure 1583
Corynebacterium diphtheriae 101, 102
Corynebacterium pseudotuberculosis 101
Corynebacterium ulcerans 101
Coryza, *see* Common cold
Cosmids 115
Costelytra zealandica iridescent virus (CzIV)(I) 1432
genome 1435
Cot deaths 1224
Cotton-top marmosets, baboon herpesvirus infection 612, 613
Cottontail rabbit, herpesvirus sylvilagus (HVS) infection, *see* Herpesvirus sylvilagus (HVS)(A)
Cottontail rabbit papillomavirus (CRPV) (Shope papillomavirus)(A) 1013, 1026, 1305–1311, 1498
future perspectives 1311
genetics 1307–1308
geographic and seasonal distribution 1306
history 1305
host range and viral propagation 1306
infections
clinical features 1309
immune response 1310
lesions 1306
pathology and histopathology 1310
prevention and control 1310–1311
malignant transformation 1309
transmission and tissue tropism 1308–1309
see also Bovine papillomaviruses (BPV)(A)
Cowpea chlorotic mottle virus (CCMV)(P) 181, 1538
host range 185
infectivity of cDNA 1540
pathology 185
transmission 184
Cowpea mild mottle virus (CpMMV)(P) 215
Cowpea mosaic virus (CPMV)(P) 249, 250, 251, 254, 1538
infectivity of cDNA 1540
Cowpea mottle virus (CMeV)(P) 218, 219
Cowpea severe mosaic virus (CPSMV)(P) 249, 254
Cowpox 261
Cowpox virus(A) **261–267**, 1341, 1618
ATI inclusion protein gene 397
Brighton reference strain 261, 263
C (capsule) form 262
'd' antigen 263
epidemiology 264
evolution 263–264
extracellular enveloped (EEV) 261, 266
future perspectives 267
genetics 263
genome structure 262
genomic stability and mutations 262
geographic and seasonal distribution 263, 1618
hemagglutinin gene sequence 262
history 261, 1503, 1507
host range and virus propagation 263
infections
clinical features 265
human 264, 265, 267
immune response 266–267
pathology and histopathology 265–266
prevention and control 267
intracellular naked (INV) 261
M (mulberry) form 262
pathogenicity 265
physical properties 262
proteins, properties 262
replication 263, 264
serologic relationships and variability 264
smallpox virus relationship 1340
structure and properties 261–262
taxonomy and classification 261
tissue tropism 264–265
transmission 264–265, 1618
vaccination 267, 1503, 1507
vaccinia virus relationship
differences 261, 263
similarity 261, 262
'white pock' mutants 262, 263, 266, 1344
Cowpox-like viruses 261
Coxsackievirus A(A) 268, 447
CA24 variant, *see* Coxsackievirus A24 variant (CA24v)(A)
common cold 1223
diarrhea due to 376
isolation 338
pharyngitis due to 1223–1224
prototype 637
Coxsackievirus A21(A) 1223
Coxsackievirus A23(A) 641
Coxsackievirus A24 variant (CA24v)(A) 269, 379, 449, 641
common cold 1223
enterovirus (EV) serotype 70 (EV70) homology 381
Coxsackievirus B(A) 268
acute myopericarditis 966
B3 prototype, *see* Coxsackievirus B3 (CVB3)(A)
chronic infections 268, 271

enterovirus (EV) serotype 70 (EV70)
relationship 380
experimental diabetes due to 970
mutation rate 967
pancreatitis 969
receptors 269, 1081
serotypes 639
swine vesicular disease virus (SVDV)
similarity 387, 388
Coxsackievirus B3 (CVB3)^(A) 639
inappropriate immune response to 668
susceptibility to chronic myocarditis in
mice 668
Coxsackieviruses^(A) **268–274**
cross-antigenicity 273
defective particles 269
epidemiology 269–270
evolution and serologic relationships 269
future perspectives 273–274
genetics 269
genome structure 269
geographic and seasonal distribution 268
history 268, 637, 638–640
discovery 638
tissue culture 639–640
host range and virus propagation 268–269
host susceptibility 270
infections
clinical features 270–271, 447
frequency and respiratory disease 1223
immune response 272–273, 639
pathology and histopathology 271–272
persistent 273
prevention and control 273
pathogenicity 270
poliovirus infection with 639
polypeptides 269
replication 270
taxonomy and classification 268
tissue tropism 270
transmission 270
CR2, *see* CD21
Crandell feline kidney cells 455
Creutzfeldt–Jakob disease (CJD) 912, **1361–1369**
Alzheimer's disease and 1367, 1368
atypical 1367
bovine spongiform encephalopathy (BSE)
relationship 1358, 1364
clinical features 1367
corneal transplantation and 447, 1365
cytological and molecular pathology 1367–1369
infection route effect 1368
diagnosis 1368
experimental 1364, 1365, 1366
pathology 1368
eye infections 447
history 1361–1362
host range and viral propagation 1364–1365
host susceptibility and genetic factors for
1364, 1367
incidence and distribution 1366–1367
incubation period 1365, 1366
infection without disease 1366
infectious agent 1361
adaptation to new host 1364
inactivation 1362
intracerebral injection 1365
physical properties 1363
properties 1362–1364
retroviral hypothesis 1363–1364
routes of inoculation 1365
taxonomy and classification 1362
tissue tropism and replication 1365–1366
models 1361–1362
scrapie relationship 1361–1361, 1364
spontaneous case 759
tissue culture studies 1369
transmission 1364
maternal and vertical 1366
by transfusion? 1365
treatment 1367
see also Spongiform encephalopathies
Cricket paralysis virus (CrPV)^(I) 1100
ecology and transmission 1102
economic importance 1103
encephalomyocarditis virus (ECMV) cross
reactivity 1102
host range 1101
morphogenesis 1101
serological relationships 1102
taxonomy and classification 1100
see also Picornaviruses^(I), insect
Cricket paralysis virus (CrPV)^(I)$_{ark}$ 1102
Cricket paralysis virus (CrPV)^(I)$_{brk}$ 1102, 1103
Crimean-Congo hemorrhagic fever (CCHF)
191, 1616
Crimean-Congo hemorrhagic fever (CCHF)
virus^(A) 190, 191, 1616
distribution and transmission 1616
Crimson clover latent virus^(P) 904
crk oncogene 71, 936, 940
Crop failures and reductions 1110, 1111
see also Plant viruses, economic
significance
'Cross-protection' 729
of plants 729, 1108
Cross-reacting immune responses 64
Cross-reactivation 527
Croup 1223
causative agents 1224
human parainfluenza viruses (PIV-1, PIV-2) 1027, 1030
Cryoglobulinemia, essential mixed, interferon
therapy 743
Crypt cells, hyperplasia 377
Cryptoviridae 275, 276
Cryptovirus^(P) 1398
Cryptoviruses^(P) **274–278**
economic significance 277
future perspectives 277
history 274–275
host relationships 276
molecular biology 277
serology 276–277
structure and genome 275, 1580
subgroups 275–276, 277
taxonomy and nomenclature 275–276
transmission and host range 275
Cucumber green mottle mosaic tobamovirus
(CGMMV)^(P) 1436, 1437
Cucumber mosaic virus (CMV)^(P) 278, 1538
cross-protection 285
cyclic concentrations 281
host range and economic significance 279
pathology and pathogenesis 283
strains (groups I/II) 282
see also Cucumoviruses^(P)
Cucumber necrosis virus (CNV)^(P) 1447, 1538
21 and 20kD proteins 1449
defective interfering RNA 1452
genome 1448–1449
mutants 1452
tobacco necrosis virus similarity 899
transmission 1451
Cucumber pale fruit viroid^(P) 1564
Cucumber soil-borne virus (CSBV)^(P) 218,
219
Cucumber virus 1 (CV1)^(P) 278
see also Cucumber mosaic virus (CMV)^(P)
Cucumovirus^(P) 1400
Cucumoviruses^(P) 30, **278–285**
antigenic determinants 281
bromoviruses relationship 181
coat protein 281
diagnostic hosts 283
ecology and epidemiology 283
economic significance 279
genetics 281–282
genome structure 280–281
RNAs (RNA1,RNA2,RNA3,RNA4)
280, 281
geographic distribution 279
history 278
host range 279, 283
host relationships 282–283
in vitro construction 281
pathology and pathogenesis 282–283
prevention and control 284–285
cross-protection 285
genetically engineered 285
resistant cultivars 284–285
transmission prevention 284
virus elimination 284
pseudorecombinants 281, 285
replication 281
satellite RNAs 282
serologic relationships and variability 282
structure and composition 280, 1580
taxonomy and classification 278–279
transmission 283–284
virus propagation and purification 279–280
see also Cucumber mosaic virus (CMV)^(P)
Cucurbits, cucumovirus infection 279
Culicoides 944, 951
orbivirus transmission 944, 945, 950
Cultivar resistance, *see* Plant resistance (to
viruses)
Cultivation of viruses
history 630
see also Cell culture; *specific viruses*
Cutaneous T-cell lymphoma, interferon
therapy 743
Cyanobacteria 285–286
filamentous 288
metabolism 290–292
nitrogen-fixing filamentous 291
photoautotrophic metabolism 290
taxa 286
unicellular 287
Cyanobacteria bacteriophage^(B) **285–292**
AN group, genetics 289
AS-1, metabolism 291
filamentous cyanobacteria 288
metabolism 290–291
genetics 289–290
LPP-1, multiplication 291
metabolism and 290–292
N-1 290
effect on metabolism 291
NP-1T 290
replication, light requirement 291
types 286–289
unicellular cyanobacteria 287
Cyanophage, *see* Cyanobacteria
bacteriophage^(B)
Cycas necrotic stunt virus^(P) 904
Cyclic AMP response elements (CRE) 169
cytomegaloviruses 306
in HTLV-1 683
Cyclic AMP-responsive element binding
protein (CREB)
in HTLV-1 683
in HTLV-2 692
Cyclin A 20

Cyclin A *continued*
  E2F transcription factor complex 20, 21
Cyclobutyl-G (R-BHCG) 44, 46
Cydia pomonella granulosis virus(I), 130, 128
Cymbidium ringspot virus (CyRSV)(P) 1447, 1538
  defective interfering RNA 1452
  genome 1448
Cynodon mosaic virus (CynMV)(P) 215
Cynosurus mottle virus (CyMV)(P) 1346
  see also Sobemovirus(P)
Cypoviruses(I) 1194
Cystic fibrosis transmembrane conductance regulator (CFTR) 1530
Cystitis
  acute hemorrhagic, adenovirus infections 6, 973
  BK virus infection 752
Cystosori 514, 515
Cystoviridae 978, 1398, 1405
  characteristics and taxonomy 100
  structure 1583
Cytokines
  in autoimmunity mechanisms 64, 65
  interferon therapy combined with 744
  resistance, immune evasion by 701
  viral molecules interfering with, immune evasion 1094
  see also Interferon; *specific cytokines and interleukins*
Cytomegalic inclusion disease (CID) 296, 297
Cytomegaloviruses (CMV)(A) **292–312**
  animal **304–307**
  antibody titers in herpesvirus papio infections 613
  classification 293, 304, 307
  cytopathic effects 337
  distinguishing characteristics 293
  DNA-binding proteins 299, 300, 305
  epidemiology 305
  equine (ECMV), *see* Equine herpesviruses(A), EHV-2
  evolution 294, 304, 309
  gene expression control 306, 309
  genome structure 304, 305, 307
    major immediate-early region (MIE) 305, 306
  history 304, 307
  human, *see* Human cytomegalovirus (HCMV)(A)
  inclusions 339
  isolation, centrifugation culture 338
  murine, *see* Murine cytomegalovirus (MCMV)(A)
  physical properties 305, 308
  proteins 307
    IE1 305, 306
    IE2 *trans*-activators 305, 306
  receptor, β_2-microglobulin 667
  replication strategies 305, 308
  specific nuclear antigen 976
  structure 304, 307
  transcription 305, 306, 308
  see also Human cytomegalovirus (HCMV)(A); Mouse cytomegalovirus (MCMV)(A)
Cytomegaly 297, 307
Cytopathic effects (CPE) 1169–1170, 1588, 1590
  animal enteroviruses 387
  definition 1169
  historical aspects 639, 640
  plaque assays 1172
  in tube culture 336–337
  see also under individual viruses
Cytopathic virus, bovine immunodeficiency virus (BIV) 162
Cytoplasm 231
Cytoplasmic compartment 230
Cytoplasmic inclusion bodies, *see* Inclusion bodies (inclusions), cytoplasmic
Cytoplasmic polyhedrosis viruses (CPV)(I) **312–319**
  abortive replication cycle 316
  amplification cycles 316
  assembly 317
  in biological control programmes 318
  cell culture 316, 318
  cell tropism 313
  economic significance 318
  genome structure 312–313
  host interactions 313, 318
  inclusion bodies, *see* Polyhedra
  latency and persistence 313
  mixed infections 316
  morphogenesis 316–318
  pathology 315
  protein and antigen synthesis, kinetics 315–316
  reassortants 312, 318
  replication 315–316, 316
  serology 313–314
  structure and composition 312, 313
  taxonomy and classification 314
  transcription 315, 316
  transmission 318
Cytoplasmic vesicles
  pea enation mosaic virus (PEMV) infection 1087–1088
  varicella–zoster virus (VZV) infection 1515–1526
  see also Vesicles
Cytoskeleton 231
  host cell and virus-induced changes 1589–1590
    frog virus 3 (FV3) 507
    see also Herpes simplex viruses (HSV)(A); Polioviruses(A); Reoviruses(A); Vesicular stomatitis virus (VSV)(A)
  virus assembly on 1590
Cytotoxic agents, interferons with 742–743, 743, 744
Cytotoxic T lymphocytes (CTL) 698, 700, 703–704, 1092
  adverse immune reactions 702
  antigen presentation 704, 705
  antigen processing/recognition 726
  class I restriction 698, 703, 708, 726, 727
  endogenous processing pathway 704, 705
  killing mechanism 1092
  numbers 706, 707
  perforins and granzymes produced 704
  precursors (TLCp) 706, 707
  in viral infections 706
    bluetongue virus (BTV) 954
    coronavirus 260
    cytomegalovirus, after transplantation 1478
    ectromelia (mousepox) virus 865
    Epstein–Barr virus (EBV) 409
    hepatitis A 553
    hepatitis B 557
    herpes simplex virus (HSV) 592
    human respiratory syncytial virus (HRSV) 1217
    increase during 706, 707
    influenza viruses 714, 719
    JC and BK viruses 754, 756
    lymphocytic choriomeningitis virus (LCMV) 706–707, 801, 804, 805, 810
    measles virus 846
    murine cytomegalovirus (MCMV) 311
    respiratory infections 1222
    rotavirus 1279
    Sendai virus 1304
    SV40 1329
    Theiler's virus 1429
    varicella–zoster virus (VZV) 1517
  see also T cells, $CD8^+$

D
DAI protein 1094
Dairy foods, fermented, phage problems 117, 118
Dairy industry, phage-resistant starter strains 120, 121
Dam methylase 874
Dandelion latent virus (DLV)(P) 215
Dandelion yellow mosaic virus (DYMV)(P) 1042, **1042–1046**
  cytopathology 1045
  epidemiology and control 1045
  genome 1043
    properties/sequence 1042, 1043
  geographic distribution 1043
  history 1042
  host range 1044
  physical properties 1042, 1043
  serology 1043
  symptoms 1044
  taxonomy and classification 1042
  'top' and 'bottom' particles 1043
  transmission 1044, 1045
  virus-like particles (VLPs) 1044, 1045
Dane particles 555
  see also Hepatitis B virus (HBV)(A)
Daphne virus S (DaVS)(P) 215
Dark green epinasty virus(P) 198
Darna trima 1417, 1422
Darna trima virus (DtV)(I) 1417
  antibodies 1422
Dasheen mosaic virus(P), economic significance 1111, 1113
Dasychira pudibunda virus (DpV)(I) 1417
DBA/2 mice, susceptibility to Theiler's mouse encephalomyelitis virus (TMEV) 667–668
DD motif, L proteins of hantaviruses 540
De Quervain subacute thyroiditis 487
Deafness, in Lassa fever 783
Deer papillomavirus 1305
Defective viruses 320, 321
  abortive infections by 1170
  noninterfering 321
Defective-interfering (DI) viruses **320–323**, 526, 729
  amplification 321, 322
  assay 322
  biological effects 322
  carmoviruses 222–223
  in cell cultures 1171
  cyclic variations 321–322
  defective viruses *vs* 321
  defectiveness and need for helper virus 321
  equine herpesviruses 426
  evolutionary role 322
  in experimental animals 322
  feline leukemia virus (FeLV) 323, 461
  future perspectives 323
  generation 320–321
  genetic recombination 320, 461
  hepatitis A virus 323, 548
  hepatitis B virus 323
  history 320
  influenza viruses 526, 715
  interference 321

lymphocytic choriomeningitis virus (LCMV) 809, 810
mutations 526
in natural infections 323
difficulties in detecting 323
parvoviruses 1171
phytoreoviruses 1097–1098
properties 526
rhabdoviruses (plant) 1245
RNA viruses 526
Semliki Forest virus (SFV) 1333
simian sarcoma virus 536
Sindbis (SIN) virus 1333
Sonchus yellow net virus 1245
structure 320
tomato bushy stunt virus 1452
tombusviruses 1447, 1450, 1451–1452
toroviruses 1455
vesicular stomatitis virus (VSV) 320, 731
wound tumor virus 1097–1098
Definition, viruses 8, 42, 1203, 1563
Deformed wing virus (DWV)(I) 656, 658
Degenerative brain disease 1366
Dehydration, morbillivirus infections 1266
Delayed-type hypersensitivity (DTH) 697, 702, 706, 1222
herpes simplex virus infections 592
Theiler's murine encephalomyelitis virus (TMEV)
demyelination mechanism 1428–1429
susceptibility 668
Delta antigen 574
see also under Hepatitis delta virus (HDV)(A)
Deltaherpesvirus(A) 1514
Dementia, viruses causing 912
Demyelinating peripheral neuropathy 912
Demyelination
in JC virus infection 756
Marek's disease virus infection 835
Theiler's virus causing 1423, 1426, 1427, 1428
immune-mediated mechanism 1428–1429
visnamaedi virus infections 1592
Dendritic cells 697, 699
Dendritic follicular cells (DFCs) 699
Dendrobium vein necrosis (DVNV)(P) 243
Dengue 1 (DEN-1) virus(A) 324
Dengue 2 (DEN-2) virus(A) 324
topotypes 326
Dengue 3 (DEN-3) virus(A) 324, 327
Dengue 4 (DEN-4) virus(A) 324, 327
Dengue fever 324, 329
clinical features 329
importance 324
see also Dengue hemorrhagic fever/dengue shock syndrome (DHF/DSS); Dengue viruses(A)
Dengue hemorrhagic fever/dengue shock syndrome (DHF/DSS) 324
clinical features 328, 329
epidemics 326, 327, 328
grades 329
incubation period 329
pathogenesis 328–329
pathology and histopathology 329–330
prevention and control 330–331
spread and factors involved 327, 331
treatment 329
WHO diagnostic criteria 329
see also Dengue viruses(A)
Dengue viruses(A) **324–331**
antigenic and biologic variation 326, 327
antigenic determinants 327
detection 327
epidemiology 327
primitive forest cycle 327
rural cycle 327
urban cycle 327
evolution 326
future perspectives 331
genetic drift 326
genetics 326
geographic and seasonal distribution 325
history 324
host range 325
infections
heart 965
immune response 330
prevention and control 330–331
secondary 330
viremia 325, 328
see also Dengue fever; Dengue hemorrhagic fever
oligonucleotide fingerprinting 326
pathogenicity 328–329
proteins 325
replication, site 328
serologic relationships and variability 325, 326–327
structure 324–325
taxonomy and classification 324–325
transmission 327, 328, 401
vaccines 330
virulence 328
virus propagation 325–326
see also Dengue hemorrhagic fever/dengue shock syndrome (DHF/DSS)
Dense granules, tombusvirus infections 1451
Densonucleosis viruses(I) **331–335**
accumulation in nuclei 334
classification 331
disease and economic significance 334
ecology 334
genome structure 332–333
hairpin structures 332
inverted terminal repeat 332
host range 331, 333–334
replication 334–335
structure and composition 332
proteins 332
tissue culture difficulties 333
virus propagation 333–334
Densovirus(I) 331, 1052, 1068
in *Parvoviridae* 331, 1052
see also Densonucleosis viruses(I)
Deoxypyrimidine kinase (dPyK), varicella–zoster virus (VZV) 1524
Dependovirus(A) 331, 1052, 1068
see also Adeno-associated viruses (AAV)(A)
Dermatitis, infectious, HTLV-1 and 694
Dermis, virus inoculation 1076
Derzsy's disease 1068
Desmodium yellow mottle virus(P) 1500
Detection of early antigen fluorescent foci (DEAFF) 976
'Devil's grip' 639
d'Herelle, F. 628, 629, 644–645
DI particles, *see* Defective-interfering (DI) viruses
Diabetes insipidus, Duvenhage virus (rabies-like) association 1189
Diabetes mellitus
HLA-DQ loci sequences 65
type 1 (insulin-dependent IDDM)
autoimmunity 970, 971
coxsackievirus infections 271
viral infections 64–65
virus role in pathogenesis 969–971
viruses preventing, in animal models 65
Diagnostic techniques **335–348**
antibodies (viral) detection **345–346**, 1620
ELISA 346
immunofluorescence 346
radioimmunoassay 346
Western blotting 346
antigens (viral) detection **343–345**, 1620
ELISA 344–345
fluorescence 344
immuno-enzyme staining 344
latex agglutination 345
radioimmunoassay 345
direct observation
by EM 342
virus-induced changes by light microscopy 339, 341–342
flow chart 336
future perspectives 342–343, 348
identification 339–341
methods (summaries) 335, 336, 343, 344
microscopy **335–343**
electron (EM) 342
identification methods 339–341
light 339, 341–342
nucleic acids (viral) detection **346–348**, 1620
amplification methods 347–348
dot hybridization 347
in situ hybridization 347
sandwich hybridization 347
primary identification 339–341
immunofluorescence tests 339, 341
immunoperoxidase tests 341
neutralization tests 341
reverse transcriptase 341
primary isolation 335–338, 1620
cell cultures, *see* Cell cultures
embryonated eggs 338, 343
infant mice 338
specimens for 343
zoonoses 1620
see also individual techniques
Dianthovirus(P) 1400
structure 1580
Dianthoviruses(P) **349–353**
cytopathology and serology 353
deletion mutants 351
economic significance 352
epidemiology and control 352
gene functions 351
genetics 351
genome structure 349–351
RNA1 349–350, 350–351
RNA2 350
geographic distribution 353
host range and symptomatology 352
proteins
capsid protein 351
capsid protein domains 349
cell-to-cell movement proteins 351
pseudorecombination studies 351
replicase 351
replication 351
polypeptides required 351
structure and composition 349, 1580
taxonomy and classification 349
tombusviruses relationship 1447
transmission 352–353
Diapedesis 1079
Diarrhea
acute 373
adenovirus ('enteric') infections 5, 6, 375
astroviruses causing 376
bovine viral 179
caliciviruses causing 375
coronavirus infection 259
infantile, adenoviruses causing 5

Diarrhea *continued*
Norwalk agent causing 926, 930
rotaviruses causing 1274, 1278
group A 374
symptoms and clinical features 377
toroviruses causing 1456, 1457
viruses causing 377
see also Gastroenteritis
Diarrheal illnesses 925–926
3-(3,4-Dichlorophenyl)-1,1-dimethyl urea (DCMU) 290, 291
Dieffenbachia, virus infections 1111, 1113
Digitaria streak virus (DSV)[(P)] 518, 1538
gene expression 519
Dihydrofolate reductase (DHFR), gene small (U-RNA) 616
Dilated cardiomyopathy (DCM) 966
clinical features 968
pathogenesis 967
Diodia vein chlorosis virus (DVCV)[(P)] 243
Dioscorea bulbifera bacilliform virus (DbBV)[(P)] 139
Diphtheria 101–102
toxin 101–102
fragments A and B 102
phage β encoding 101–102
production 102
toxin repressor (DTxR) 102
toxoid 101
Diploid genomes 525, 529
Direct fluorescence antibody procedure (DFA) 339
'Direct transfection' 77
Disinfectants, coxsackievirus resistance to 273
Distamycin A 985
Distemper viruses[(A)] **1260–1268**
evolution 1263
see also Canine distemper virus (CDV)[(A)]; Phocid distemper virus (PDV)[(A)]
DN19 virus[(A)] 641
DNA
amplification 347–348
modified 'plaque' assays 1172–1173
of tumors/cell lines 938
see also Gene amplification; Polymerase chain reaction (PCR)
cloning system
phage P1 1002
see also Gene cloning; Vectors
complementary, *see* cDNA
concatemeric 1010
T1 phage 1372
T7 phage 1392, 1393
vaccinia virus 1509
degradation in host cell, *see under* Host
discontinuities, in caulimoviruses 224
double-strand break (DSB), *see under* Genetic recombination, bacteriophage
double-stranded (dsDNA)
adenovirus 8, 15
African swine fever virus[(A)] 24
algal viruses 36
ascoviruses 58
bacteriophage ϕ29 981
badnaviruses 140
bovine herpesviruses 155, 156
caulimoviruses 223, 224
cowpox virus 262
cyanobacteria bacteriophage 286
cytomegalovirus 299, 305
entomopoxviruses (EPVs) 392, 394
Epstein–Barr virus (EBV) 410
equine herpesviruses 424
fowlpox virus (FPV) 497
frog virus-4 (FV4) 42
granulosis viruses 128
herpesvirus saimiri 615, 616
herpesvirus sylvilagus (HVS) 621
herpesviruses 299, 305
HIV reverse transcription process 675
human herpesvirus 6 (HHV-6) 625
Lucké tumor herpesvirus (LTHV) 41
Marek's disease virus 833
mouse mammary tumor virus replication 857
nonoccluded baculovirus 137
nuclear polyhedrosis viruses (NPVs) 131, 132
parapoxviruses 1037–1038
partial, avian hepatitis B viruses 564
partial, hepadnaviruses 564
partial, hepatitis B virus 556, 560, 562
poxviruses 497, 1155
retrotransposons (formation in) 1234
transposable phage 868
undivided in totiviruses 1465
vaccinia virus 1508
double-stranded (dsDNA) circular
animal papillomaviruses 1305
herpesvirus saimiri transformed cells 616
JC and BK viruses 752
Marek's disease virus 834
murine polyomaviruses 1135, 1136, 1138, 1140
polydnaviruses 1133
SV40 1322, 1323
double-stranded (dsDNA) circular permuted and terminally redundant
fish lymphocystis disease virus 798
frog virus 3 (FV3) 504, 507, 508
iridoviruses 1434
phage T4 504, 507, 508
see also Terminal redundancy
double-stranded (dsDNA) linear
ascoviruses 59
bacteriophage P1 997, 998
fish lymphocystis disease virus (FLDV)[(A)] 798
frog virus 3 (FV3) 503, 504
herpes simplex viruses 587, 594
iridoviruses 1430
molluscum contagiosum virus (MCV) 849
N4 bacteriophage 891, 892
P22 bacteriophage 1010
parapoxviruses 1037–1038
PRD1 bacteriophage 1165, 1166
pseudorabies virus 1174
smallpox virus 1339
SPO1 phage 1352, 1353
T1 bacteriophage 1371
T4 phage 1377
T5 phage 1384
T7 phage 1389, 1390
tree shrew herpesviruses 1489–1490
varicella–zoster virus (VZV) 1518, 1519
yabapox and tanapox viruses 1597
double-stranded provirus, reticuloendotheliosis viruses (REV) 1229
double-stranded replicative form (RF)
bacteriophage ϕ29 983
bacteriophage ϕX174 993
densonucleosis viruses 334
filamentous bacteriophage 467
geminiviruses 521
polioviruses 1130
foreign, destruction by M/R systems 671
host, *see* Host
integration into host genome 527
adenovirus 13–14
avian leukosis virus 68
bovine leukemia virus 167, 169, 172
Campbell model 1005
caprine arthritis encephalitis virus 201
equine infectious anemia virus (EIAV) 431
feline leukemia virus 460
hepadnaviruses 556
human immunodeficiency virus (HIV) 675–676
human papillomaviruses (HPV) 1024, 1497
lambda phage, *see* Lambda bacteriophage
latent infections 787
Marek's disease virus 834
murine leukemia viruses (MuLVs) 886
murine polyomaviruses 1137
phage, *see individual bacteriophage*; Bacteriophage, temperate
reticuloendotheliosis viruses (REV) 1229, 1231
retroviruses 527, 556, 935, 1208, 1473
sequencing 530
at transcriptionally active sites 14
transformation and 1493
transformation by retroviruses 1473
transposable phage, *see* Mu bacteriophage[(B)]
inversion, transposable phage for study of 876
isolation, cyanobacteria bacteriophage 289
methylation, *see* Host-controlled modification and restriction; Methylation
modification, *see* DNA modification/restriction (M/R) systems; Host-controlled modification and restriction; Methylation
multicopy single-stranded (msDNA) 1009
packaging into bacteriophage 108–109, 999
see also specific phage; Transduction
positively supercoiled, in SSV1 virus 56
proviral (integrated)
foamy viruses 485
see also DNA, integration into host genome *(above)*
repair, Dam methylation 673
replication 232, 1209
animal cytomegaloviruses 305, 306, 308
bidirectional 85, 302
coliphage P4 1007
inhibitors 474
initiation via DNA methylation 673
'leaping polymerase' 320–321
Okazaki fragments 996
protein-primed in *E. coli* 1167, 1168
by reverse transcription, in caulimoviruses 224–225
role of bacteriophage ϕX174 in 996
rolling circle model 993
bacteriophage 85, 86
bacteriophage ϕX174 993
cytomegaloviruses 302
equine herpesviruses 426
filamentous bacteriophage 468
geminiviruses 521
lambda bacteriophage 773
P2 bacteriophage 1007
P22 bacteriophage 1011
T5 phage 1387
varicella–zoster virus 1523
self-priming' mechanism 1209
single-stranded DNA phage 84, 85
in transposable phage (Mu) 873

two stage in nucleus and cytoplasm, frog virus 3 (FV3) 504, 507
see also DNA, synthesis; *under specific viruses*
single-stranded binding protein 11
Escherichia coli 892
N4 bacteriophage 891, 892
single-stranded (ssDNA)
annealing, *see under* Genetic recombination
circular, bacteriophage ϕX174 989
densonucleosis viruses 331, 332
filamentous bacteriophage 464
geminiviruses 517, 518
generation in single-stranded DNA phage 84, 85
parvoviruses 1053, 1057, 1068, 1069
regions, in hepatitis B virus (HBV) 555
terminal palindromic structures 331
parvoviruses 1057
supercoiled, in caulimoviruses 224
synthesis
inhibition, avian hepatitis B control 568
retroviruses, reticuloendotheliosis viruses (REV) 1229
T1 phage 1372–1373
see also DNA, replication
terminally redundant
in phage 108, 110
see also DNA, double-stranded (dsDNA) circular permuted
transposition 868
mechanism 872–873
transposable phage for study of 876
see also Mu bacteriophage(B)
in viruses, discovery 629
DNA bacteriophage
double-stranded (dsDNA) 1405
Bacillus subtilis 72–78
lambda 772, 773
T1 phage 1371, 1372
evolution 82
halobacterial phage 50, 52
single-stranded (ssDNA) 1405
classification and families 1405
recombination 84–85
replication 84, 85
taxonomy and classification 95
of thermophiles 56
see also DNA, double-stranded (dsDNA); DNA, single-stranded (ssDNA)
DNA gyrase, effect on T7 phage 1395
DNA methyltransferases 76
algal viruses encoding 38, 39
T1 phage 1372
DNA modification/restriction (M/R) systems 670
bacteriophage Mu and D108 873–874
bacteriophage P1 997, 1001
function 671
phage typing and 80
see also Host-controlled modification and restriction; Methylation
DNA polymerase 1209
α-like 984
bacteriophage ϕ29 983, 984, 985
hepatitis B virus 555
host 1205
human cytomegalovirus 300
'leaping' mechanism 320–321
N4 bacteriophage 894
RNA-dependent 1208
in caulimoviruses 225
see also Reverse transcriptase
varicellazoster virus (VZV) 1521
virus-encoded 1204–1205
DNA probe 347
DNA site-specific endonucleases, *see* Restriction endonucleases
DNA viruses 440
animal viruses 525, 1409
atomic structures 1572
characteristics 442
classification 1405–1409
double-stranded 1208, 1398
bacterial 1405
classification 1398
invertebrate 1408
plant 1407
vertebrates 1409
electron microscopy 340, 342
enveloped
bacterial 1405
invertebrate 1408
vertebrates 1409
eye infections, *see* Eye infections
families
algal, fungal and protozoal 1406
bacterial 1405
invertebrates 1408
plants 1407
vertebrates 525, 1409
gene regulation 1205, 1209
genetic content 1208
intramolecular recombination 527
latency 787
mutation rate 438, 525
non-enveloped
algal, fungal and protozoal 1406
bacterial 1405
invertebrate 1408
plant 1407
vertebrates 1409
nuclear-replicating 1169
oncogenes 631
replication 232, 1208–1209
reverse transcription during replication (pararetroviruses), caulimoviruses 223, 224–225
sequencing of genome 530
single-stranded 1208, 1398
bacterial 1405
classification 1398
invertebrate 1408
plant 1407
as vectors (plant viruses) 1536, 1539–1540
vertebrates 1409
transcription 1204
strategy 1208–1209
tumor 1476–1477, 1590
integration 527
JC and BK viruses 752
see also under Transformation
as vectors 1528, 1536, 1539–1540
see also Adenovirus(A); Poxviruses(A)
viral and host gene relationship 440
yeast and smut fungus, *see* Totiviruses
Dominant-negative mutants 729, 729–732
as antiviral agents 731–732
attenuated, as antivirals 732
complementation and reassortment inhibition 731
'genotypic dominance' 730
live virus vaccines 732
mechanisms of interference 730–731
'all things in moderation' hypothesis 731
'attractive genome' hypothesis 731
'direct competition' hypothesis 731
'road block' hypothesis 730–731
'rotten apple' hypothesis 730
persistent infections 730
'phenotypic dominance' 730
selection technique 730
transfection of genes 731
Dot hybridization 347
Double minutes 938
DpnI restriction enzyme 672
Drosophila
carbon dioxide sensitivity 1311, 1312, 1313, 1315
cells, nodavirus culture 924
oncogenes studied in 940, 941
Sigma virus, *see* Sigma rhabdoviruses(I)
Drosophila C virus (DCV)(I) 1100
ecology and transmission 1102
host range 1101
morphogenesis 1101
serological relationships 1102
taxonomy and classification 1100
see also Picornaviruses, insect
Drosophila line virus (DLIV)(I) 919
Drosophila melanogaster 1313–1314
host-range mutants of Piry virus 235
sigma rhabdovirus infection 1311–1315
loss-of-function alleles 1314
permissive alleles 1314
restrictive alleles 1313–1314
viruses 1100
Drosophila X virus (DXV)(I) 144
Drug abusers, intravenous (IVDA), HTLV-2 in 687, 688
Ds element 1542
Duck(s)
Newcastle disease virus 917
reticuloendotheliosis viruses 1230
Duck adenovirus(A) 16–17
Duck hepatitis B virus (DHBV)(A) 387, 390, 555
geographic and seasonal distribution 555
host range and viral propagation 556
as model for chemotherapy of hepatitis B 568, 569
vaccines 391
see also Hepatitis B viruses (HBV)(A), avian
Duck infectious anemia virus (DIAV)(A) 1227
Dulcamara carlavirus A (DuCVA)(P) 215
Dulcamara carlavirus B (DuCVB)(P) 215
Dulcamara mottle virus (P) 1500
Duncan syndrome, *see* X-linked lymphoproliferative syndrome (XLPS)
Duvenhage virus(A) 1186
epidemiology 1188
geographic and seasonal distribution 1187
host range and viral propagation 1187
prevention and control 1189
transmission and tissue tropism 1189
vaccine 1189
see also Rabies-like viruses(A)
Dysentery 103
causative agent, bacteriophage discovery 644, 645

E
E2F transcription factor 20
E4F transcription factor 21
E-36 virus(P) 198
Ear infections, murine cytomegalovirus 310
Eastern equine encephalitis (EEE) virus(A) 416, 1616
clinical features of infection 421, 445
diagnosis 1616
epidemiology 419–420
geographic distribution 1616

Eastern equine encephalitis (EEE) virus(A) *continued*
  pathogenicity 421
  pathology and histopathology 422
  prevention and control 1616
  transmission 1616
  see also Equine encephalitis viruses(A)
EAV, *see* Equine arteritis virus (EAV)(A)
Ebola hemorrhagic fever 827
Ebola virus(A) 446, **827–832**, 1619
  diagnosis and cell cultures 832, 1619
  epidemiology 830
  evolution 830
  future perspectives 832
  genome structure and properties 828
  geographic distribution 829–830, 1619
  history 827
  host range and viral propagation 830
  infections
    clinical features 831
    eye 446
    immune response 831
    pancreatic 969
    pathology and histopathology 831
    prevention and control 831–832, 1619
  pathogenicity 831
  physical properties 828
  properties 828
  proteins
    L protein and GP glycoprotein 829
    NP nucleoprotein 829
    properties 828–829
  replication 829
  Reston virus relationship 828, 830
  serologic relationship 830
  taxonomy and classification 827–828, 1619
  transcription 829
  transmission and tissue tropism 831, 1619
  see also Marburg virus(A)
Ebola-Zaire virus(A) 830, 831
EBV, *see* Epstein–Barr virus (EBV)(A)
ECE (equine coital exanthema) virus, *see* Equine herpesviruses(A), EHV-3
Echinochola ragged stunt virus (ERSV)(P) 1096
Echovirus type 9(A) 641
Echoviruses(A) **354–360**
  antigenic variants 355, 356
  atypical echovirus type 22 355
  cell-specific receptor for 357
  common cold due to 1223
  cytopathic effects 355
  detection, in environmental samples 357
  future perspectives 359–360
  genetics 355–356
  geographic and seasonal distribution 355
  history 354, 640–641
  host range and viral propagation 355
    failure in newborn mice 354, 355, 358
  infections
    clinical features 358
    common cold 1223
    echovirus type 11 356
    echovirus type 25 358
    epidemiology 356–357
    immune response 359
    outbreaks 356–357
    pathology and histopathology 358–359
    viremia 357
  JV-4 prototype strain 356, 358
  pathogenicity 357
  proteins 355, 357
  resistance in environment 357, 360
  serological relationships 356, 359
  taxonomy and classification 354
  transmission and tissue tropism 357
  types 354
  vaccine development 360
  variability and rapid evolution 355, 356
Ecological changes, zoonoses transmission and 1621–1622
Economic significance
  insect viruses, *see* Insect viruses
  killer toxins (totiviruses) 1467, 1471
  plant viruses, *see* Plant viruses
*Eco*P1 1001
*Eco*P15, bacteriophage protection from 671
*Eco*RII, bacteriophage protection from 671
Ectromelia virus(A) **861–866**
  epidemiology 862–863
    enzootic mousepox 862
    experimental 863
    susceptibility of mouse strains 862–863
  future perspectives 866
  geographic and seasonal distribution 862
  history 861–862
  host range and viral propagation 862
  infections
    chronic, clinically inapparent 862
    clinical features 863
    immune response 865
    pathology and histopathology 863–865
    prevention and control 865–866
  intranasal inoculation 864–865
  intraperitoneal inoculation 864
  laboratory diagnosis 865
  pathogenesis 863
  resistance to 862
  screening tests 865
  susceptibility, *H-2*-linked genes 668
  taxonomy and classification 862
  see also Rabbitpox virus(A)
Eczema vaccination, smallpox vaccination 1513
Edema
  Argentine/Bolivian hemorrhagic fevers 782
  Lassa fever 782, 783
Edema disease 103
Edge Hill virus(A) 325
Eel virus European (EVE) (*Birnaviridae*)(A) 475
Eggplant mild mottle virus (EMMV)(P) 215
Eggplant mosaic virus(P) 1500
Eggplant mottled crinkle virus(P) 1447
Eggplant mottled dwarf virus (EMDV)(P) 1243, 1246
Egtved disease 478
Egtved virus(A), *see* Viral hemorrhagic septicemia virus (VHSV) (Egtved virus)(A)
Egypt bee virus(I) 656, 658
Eimeria stiedae virus(A) (ESV) 532, 534
Elderberry carlavirus (ECV)(P) 215
Elderberry latent virus (ELV)(P) 218
Electron density map (EDM), three-dimensional 1571
Electron microscopy
  historical aspects 640
  negative staining 340, 342
  particle counting 1173
  thin-sectioning method 340, 342
  virus observation 340, 342
ELISA, *see* Enzyme-linked immunosorbent assay (ELISA)
Elm mottle virus (EMtV)(P) 33, 34
Embryogenesis, oncogenes in 940–941
Embryonated eggs
  virus isolation 338, 343
  virus propagation in 1172
Embryonic fibroblasts, human 336
Encephalitis 911
  acute, Japanese encephalitis virus causing 746, 749
  acute post-measles 64, 845
  bovine, herpesviruses causing 157
  Bunyavirus causing 190
  coxsackievirus infections 272
  enterovirus (EV) serotype 71 (EV71) causing 383
  equine encephalitis viruses causing 421, 422
  herpes simplex virus 589, 590
  in mumps 882
  neonatal echovirus 359
  old dog 1265
  Theiler's murine encephalomyelitis virus (TMEV) 1423
  tick-borne 383
  type A 746
  type B 746
Encephalitis viruses(A) **361–372**, 1615, 1616–1617
  antigens 365
  epidemiology 365
  evolution 364
  features/characteristics 363
  genetics 364
  genome structure 362
  geographic and seasonal distribution 362–363, 1615, 1616, 1617
  history 361–362
  host range and viral propagation 363–364, 1616, 1617
  infections
    clinical features 365–366
    diagnosis/detection 365, 1615, 1616, 1617
    immune response 366, 1616
    pathology and histopathology 366
    prevention and control 366–367, 1615, 1616, 1617
  pathogenicity 365
  serologic relationships and variability 364–365
  structure and properties 362
  taxonomy and classification 362
  tissue tropism 364, 365, 1615
  transmission 364, 365, 1615, 1616, 1617
  see also Tick-borne encephalitis virus (TBEV)(A); *other specific viruses (pages 363, 367)*
Encephalomyelitis 911
  acute disseminated 911
  caprine arthritis encephalitis syndrome 199, 203
  postinfectious 911
Encephalomyocarditis virus (EMCV)(A) 205, 1101
  cross-reactivity with CrPV 1102
  EMC(D) variant 212
  experimental diabetes due to 970
  genome, Mengo virus comparison 211, 212
  genome structure and translation 207, 209
  geographic distribution and host range 211
  pathogenicity in mice 212
  secondary structure 1124
  Theiler's murine encephalomyelitis virus (TMEV) relationship 1423
  translation 1123, 1125
  see also Cardioviruses(A)
Encephalopathy, in dengue hemorrhagic fever/dengue shock syndrome (DHF/DSS) 328, 329
Endemic infection 399
Endocytosis 229, 1080
  HSV penetration 606
  pit-mediated 1169, 1204
  poliovirus 1123
  receptor-mediated 229, 1169, 1204

polioviruses 1123
reovirus infection 1192
rotaviruses 1287
see also Pathogenesis of viral infections; Receptors for viruses
Endogenous proviruses, of avian leukosis viruses 69, 71
'Endogenous viruses' 1170
Endomyocarditis, echoviruses causing 355
Endonucleases, *see* Restriction endonucleases
Endoplasmic reticulum (ER) 230, 1559
functions 230
in viral infections 230
HSV glycoprotein synthesis 604
rotavirus replication 1283, 1286, 1289
rough 230
hemagglutinin attachment/ modification 720
measles virus protein processing 842
poliovirus infections 1130
Sendai virus 1301
smooth 230
vesicular stomatitis virus (VSV) protein synthesis 1551, 1558–1559
Endosomes 1204, 1287
Endothelial cells
damage in equine arteritis virus infection 770
dysfunction in Lassa fever 782
hog cholera virus infection 652
infections and invasion pathway via 909, 1079
in measles 846
orbivirus and coltivirus infection 953
Enhancers 1081
Enhancing factor, granulosis virus 128
Enhancins 128
genes 128
Enrichment procedures, bacteriophage detection in soil samples 122
'Enteric' adenoviruses 1, 2, 5, 374–375
Enteric cytopathogenic human orphan viruses, *see* Echoviruses
Enteric infections
coronaviruses 258–259
transmission pattern 401–402
Enteric viruses **373–378**
adenoviruses 1, 2, 5, 374–375
astroviruses 376
caliciviruses 375–376
coronavirus/coronavirus-like agents 376–377
enteroviruses 376
history 373
infections
clinical features 377
entry route 1077
immune responses 378
pathology and histopathology 377
prevention and control 378
treatment 377–378
Norwalk virus 375
parvovirus/parvovirus-like agents 376
rotaviruses
group A 373–374
nongroup A 374
'silent' infections 373
toroviruses 377
see also individual viruses (as above); Norwalk virus[A]
Enteritis
acute 373
hemorrhagic, 'enteric' adenovirus causing 375
see also Diarrhea; Enteric viruses; Gastroenteritis
'Enteroviral' syndrome 640
Enterovirus[A] 268, 384, 1077
coxsackieviruses 268, 269
echoviruses 354
eye infections 447
poliovirus 637–638, 1115
see also Enteroviruses[A]
Enterovirus (EV) serotype 68 (EV68)[A] 378, 379, 641
California and Rhyne strains 379
Enterovirus (EV) serotype 69 (EV69)[A] 378, 379
Enterovirus (EV) serotype 70 (EV70)[A] 379–383, 447, 641
coxsackievirus A24 variant (CA24v) homology 381
epidemiology 381
genetics and evolution 381
genome structure/properties 380
geographic and seasonal distribution 380
history 379–380, 641
host range and viral propagation 381
infections
immune response to 382–383
prevention and control 383
route 1070
pandemics/epidemics 380, 381, 383
pathogenicity 382
physical properties 380
proteins, properties 380
prototype (J670/71) strain 380
relationship with other viruses 380, 381
replication 381
serologic relationship and variability 381
structure and properties 380
transmission and tissue tropism 381–382, 1078
see also Conjunctivitis, acute hemorrhagic (AHC)
Enterovirus (EV) serotype 71 (EV71)[A] 383–384, 641
epidemiology, transmission and tissue tropism 384, 641
geographic and seasonal distribution 383
host range and viral propagation 383
infections
clinical features 383–384, 641
immune response 384
pathology and histopathology 384
pathogenicity 384
prototype (BrCr/1970) strain 383
serologic relationships 383
Enterovirus (EV) serotype 72[A], *see* Hepatitis A virus (HAV)[A]
Enteroviruses[A] 376, **378–391**
animal **384–391**
assembly, uptake, release and cytopathology 387
clinical features of infections 389–390
epidemiology 388–389
genetics and evolution 388
genome structure/properties 385
geographic and seasonal distribution 387
history 384, 641–642
host range and viral propagation 387–388
immune responses to 390, 968
pathogenicity 389
pathology and histopathology 390
physical properties 385–386
post-translational processing 386
prevention and control of disease 391
properties 385
proteins 385
replications 386
serologic relationships and variability 388
taxonomy and classification 384–385, 641–642
transcription/translation 386
transmission and tissue tropism 389
antigenic variation 356
diseases associated 640
in feces 373
heart infections 965, 965–966
pathogenesis 967
history 638, 641
host protein synthesis shutdown 491
human serotypes **378–384**
see also Enterovirus (EV) serotypes 68/ 69/70/71 (above)
identification 341
laboratory diagnosis 968
neonatal disease 965
pancreatitis 969
poliovirus as member 638
RNA phage as index organisms 1336
secondary structure 1124
structure and genomic organization 359
taxonomy and classification, history 641–642
vaccines 359, 391
see also Enterovirus[A]
Entomopoxvirinae 392, 1507
Entomopoxviruses (EPVs)[I] **392–398**
antigenic relationships 392
characteristics/pathology of infections 394–395
gene regulation 397
genome structure 393–394
group A (Coleopteran) 392, 393, 395–396
group B (Lepidopteran) 392, 393, 395
group B (Orthopteran) 396
group C (Dipteran) 392, 393, 396
host range and viral propagation 392
hybridization studies 394
occlusion bodies 393, 396
replication 396–397
shuttle vectors for 397
spheroidin 392, 393, 394, 398
structure and composition 393
taxonomy and classification 392–393
Entry of viruses, into host cells 229–230, 400, 1168–1169
see also Pathogenesis of viral infections; Penetration, by viruses
env gene
avian leukosis viruses 67
bovine immunodeficiency virus 160
bovine leukemia virus 167
c-*sis* fusion and transformation 1473
caprine arthritis encephalitis virus (CAEV) 201
equine infectious anemia virus (EIAV) 430
feline leukemia virus (FeLV) 460, 461
foamy viruses 482, 484
HTLV-1 683
HTLV-2 690, 691
human immunodeficiency virus (HIV) 675
lymphoproliferative disease virus (LPDV) of turkeys 813
murine leukemia viruses (MuLVs) 884
recombinants (MCF) 885, 887–888
origin 439
reticuloendotheliosis viruses (REV) 1229, 1230
simian immunodeficiency virus (SIV) 1318
transcription/translation 485
transmembrane anchor domain (TM) 482, 484

env gene *continued*
type D retroviruses 1239, 1240
Env protein 68
Enveloped viruses 1209, 1556, 1586
DNA, *see* DNA viruses, enveloped
nonspheroidal 1584
RNA, *see* RNA viruses, enveloped
spheroidal 1583–1584
uncoating 1204
see also Membranes
Environment
changes and zoonoses transmission 1621–1622
coxsackievirus spread 269
Enviroxime 44, 47
Enzyme(s)
elevation in LDV infection 769, 770
genes, origins 439–440
Enzyme immunoassay (EIA)
badnaviruses 141, 142
respiratory viruses 1225
see also Enzyme-linked immunosorbent assay (ELISA)
Enzyme-linked immunosorbent assay (ELISA)
adenoviruses 4
advantages 344
avidity assays 346
capilloviruses 198
carlavirus 217
Chikungunya, O'nyong nyong and Mayaro viruses 240
cytomegalovirus (CMV) 1479
encephalitis viruses 1616
equine infectious anemia virus (EIAV) 435
feline immunodeficiency virus 458
fowlpox virus (FPV)[(A)] 501
hantavirus 544
hepatitis E virus 586
IgM antibody capture ('μ-capture') 346, 1620
iridoviruses 1432
lactate dehydrogenase-elevating virus (LDV) 768
maize chlorotic dwarf virus (MCDV)[(P)] 826
orbiviruses and coltiviruses 943, 955
pseudorabies virus 1178
Ross River virus (RRV) 1273
toroviruses 1452
viral antibody assay 346
viral antigen assay 344–345
zoonoses diagnosis 1620
Ependymal cells 908
Epidemic, definition 399
Epidemic keratoconjunctivitis, *see* Keratoconjunctivitis, epidemic
Epidemic myalgia 271, 639
'Epidemic polyarthritis' 1268, 1269, 1270, 1272
see also Ross River virus (RRV)[(A)]
'Epidemic polyarthritis with rash' 1268
Epidemiology of viral diseases **398–404**
assessment of disease occurrence/outcome 399
cause–effect relationships, studies 399
cycles 447, 449
of eye 447
history 398–399
implications for disease prevention 403–404
mathematical modeling 403
modes of transmission 400–401
molecular studies 399–400
perpetuation of viruses in nature 402–403
role 399
seasonal infections 447
types of epidemiological studies 399–400
virus transmission 400
Epidermal growth factor (EGF) 938, 1474
receptor 1474
vaccinia virus binding 1560
Epidermodysplasia verruciformis (EV) 1014, 1022, 1497–1498
clinical features 1019
HPV type causing 1018, 1024, 1497
Episomal DNA 1587
Episome, Epstein–Barr virus (EBV) 410
Epistasis 903
Epithelial cells
bovine papillomaviruses (BPV) infection 1308, 1309
EBV infection 407, 411
torovirus infections 1458
'Epithelial giant cells', in measles 846
Epithelial hyperplasia, fowlpox virus infection 499, 502
Epitope, *see* Antigenic determinants
Epizootic hemorrhagic disease (of deer) 942, 953
Epizootic hemorrhagic disease (of deer) virus (EHDV)[(A)] 941, 943
clinical features of infection 953
genetics 946
pathogenicity 952
Epstein–Barr virus (EBV)[(A)] **404–416**
apoptosis inhibition 1591
B cell infection, as immune evasion mechanism 702
baboon and chimpanzee herpesviruses relationship 609, 611, 613
biological heterogeneity 410
Burkitt's lymphoma, *see* Burkitt's lymphoma
capsid antigen (VCA), antibodies to 406, 611
carriers 406
cell transformation, genes involved 413, 1477
defective virions 323, 415
DNA replication, origins 410, 411, 790
early antigens (EA), antibodies to 406
EBER1 and EBER2 411, 414, 1094
adenovirus VA RNAs similarity 413
EBNA, *see nuclear antigens (EBNAs) (below)*
'EBNA type' 406
evolution 405
experimental infections and lymphomas in 1495
frequency and respiratory disease types 1223
future perspectives 410
genetics 405
deletions 405
genome structure/properties 410–411, 1494
geographic and seasonal distribution 405
herpesvirus saimiri and herpesvirus ateles relationship 614, 617
herpesvirus sylvilagus (HVS) relationship 624
het DNA 323
history 404
HLA-DQ sequence similarity 65
host range and viral propagation 405
immune evasion 702, 1094
in immunodeficient patients 1495
infection, *see* Epstein–Barr virus (EBV) infection *(below)*
IR1 410, 413
latency 790–791, 1090, 1092, 1591
type I 415
type II in nasopharyngeal carcinoma 415
type III 415
latency period for tumors 1499
latency-associated genes 411, 790
in Burkitt's lymphoma and nasopharyngeal carcinoma 414–145
expression 408, 411–414
functions of 412–416
latent membrane proteins (LMP) 408, 412, 790, 1494
expression in nasopharyngeal carcinoma 415
functions 413, 414
genes 412, 413–414
in tumors 1494, 1495
latent and productive infections 1090
lytic cycle genes 411
BCRF/vIL-10 416
BZLF1 415
early and late 415
expression 415
gp350/220 410, 415–416
membrane antigen (MA) 407, 611
vaccines 409
molecular biology **410–416**
nasopharyngeal carcinoma and, *see* Nasopharyngeal carcinoma (NPC)
nuclear antigens (EBNAs) 405, 408, 611, 790, 1092
EBNA-1 412, 790, 1482, 1494
EBNA-1 and HCMV phosphoprotein similarity 300
EBNA-1 promoter 415
EBNA-2 412–413, 1494
EBNA-3 413
functions 412, 413
glycine–alanine repeats in EBNA-1 412
homology with baboon herpesviruses 611
leader protein (EBNA-LP) 411–412, 413
sequences and transcription 411–412
pathogenicity 407–408
as persistent infection 1090
polyclonal activation of B cells 63–64, 408, 1482
properties and structure 410
prototype (B95.8) strain 405
re-activation 790–791
receptor 1561
replication 407, 410
serologic relationships and variability 405–406
taxonomy and classification 404
transcription, *trans*-activation 412, 413
transformation by 413, 1477, 1494–1495
role in Burkitt's lymphoma 1494–1495
see also Burkitt's lymphoma; Nasopharyngeal carcinoma (NPC)
transmission and tissue tropism 407
tree shrew herpesvirus relationship 1490
types 1/2 405
vaccines 409
varicella–zoster virus (VZV) comparison 1515
X-linked lymphoproliferative syndrome 1495
ZEBRA protein 790–791
Epstein–Barr virus (EBV) infection
after transplantation 1481–1483
bone marrow 1482
clinical features 1482
future perspectives 1483
incidence 1481–1482

pathogenesis and diagnosis 1482
treatment 1482–1483
B cell lymphomas and 1495
clinical features 408
epidemiology 406–407
eye 444
heart 965
immune response 406, 409
latent 1092, 1591
see also Epstein–Barr virus (EBV), latency *(above)*
lymphoproliferative syndromes associated 1495
pancreatic 969
pathology and histopathology 408–409
prevention and control 409–410, 1483
T cell lymphomas and 1495
see also Burkitt's lymphoma; Infectious mononucleosis (IM); Nasopharyngeal carcinoma (NPC)
Equilibrium centrifugation 1172
Equine abortion virus[A], *see* Equine herpesviruses[A], EHV-1
Equine arteritis virus (EAV)[A] 256, **763–771**
antibodies 770–771
assembly, uptake and cytopathology 766
coronavirus similarity 763, 768
evolution 768
genetics 767
genome structure and properties 764–765, 766, 768
geographic and seasonal distribution 767
history 763
host range and viral propagation 767
infection
clinical features 769
immune response 770–771
pathology and histopathology 770
prevention and control 771
pathogenicity 769
physical properties 765
proteins 765
S protein 768
replication 765–766
serologic relationships and variability 768
structure and properties 764
taxonomy and classification 763
tissue tropism 769
transmission 769
vaccine 771
virulent and avirulent mutants 767, 769
see also Lactate dehydrogenase-elevating virus (LDV), of mice[A]
Equine coital exanthema virus (ECE virus), *see* Equine herpesviruses[A], EHV-3
Equine cytomegalovirus (ECMV)[A], *see* Equine herpesviruses[A], EHV-2
Equine encephalitis viruses[A] **416–423**
attachment and entry 418
epidemiology 419–420
epizootics 420
evolution 419
future perspectives 423
genetics 419
genome structure/properties 417
geographic and seasonal distribution 418
history 415–417
host range and viral propagation 418–419
infections
clinical features 421–422
eye infections 445
immune responses 422
pathology and histopathology 422
prevention and control 422
pathogenicity 421
physical properties 418
proteins, properties 417–418
replication 418
sites 420
structure and properties 417, 443, 445
taxonomy and classification 417
transcription and translation 418
transmission and tissue tropism 419, 420
vaccines 422
Equine encephalosis virus (EEV)[A] 943
Equine herpesviruses[A] **423–429**
EHV-1 423
'abortion storms' 424
capsid species 426
clinical features 427
defective interfering particles 426
genome structure 425
Kentucky A strain 424, 425
vaccines to 428–429
EHV-2 304, 423
geographic and seasonal distribution 424
EHV-3 423
clinical features 428
EHV-4 423–424
genome structure 425
epidemiology 427
evolution 426
future perspectives 429
genetics and DNA homology 424–425
genome structure 424–425
IR gene sequences 425
geographic and seasonal distribution 424
history 423
host range and viral propagation 424
infections
clinical features 427–428
immune response 428
pathology and histopathology 428
persistent 426
prevention and control 428–429
pathogenicity 427
replication 425–426
serologic relationships and variability 427
taxonomy and classification 424
transcription 425
transmission and tissue tropism 427
vaccines 428–429
Equine infectious anemia 430, 434
Equine infectious anemia virus (EIAV)[A] **430–436**
assembly 431–432
carriers 434
cross-reaction with bovine immunodeficiency virus 164
cytopathicity 432
epidemiology 433
evolution 433
future perspectives 435–436
genetics 432–433
genome structure/properties 430–431
gag, *pol*, *env* genes 430
*S*1 and *S*2 genes 430
geographic and seasonal distribution 432
host range and viral propagation 432
immune response 435
control over viral replication 435
infections
clinical features 434
diagnostic assays 435
pathology and histopathology 434
mutation rate 432
pathogenicity 433
proteins 431
gp90 and gp45 431
provirus DNA integration 431
replication 431–432
reverse transcriptase 431, 432
serologic relationships and variability 433
structure 430
taxonomy and classification 430
transcription and translation 431
transmission and tissue tropism 433
tricistronic messenger RNA 431
vaccine development 430, 435
viremia 433, 434
Equine rhino-pneumonitis virus[A], *see* Equine herpesviruses[A], EHV-1
erbA oncogene 936, 941, 1475
erbB oncogene 936, 941, 1474, 1475
Erysimum latent virus[P] 1500
Erythema infectiosum 1054
Erythrocytes, reovirus receptor 1200
Erythrogenic toxins 103
Erythropoietin, in leukemogenesis by murine leukemia viruses 888
Escherichia coli
attB site 109
bacteriophage ϕX174 989
coliphage P2 discovery 1003
CRPcAMP system, mutations inactivating 817
D108 phage 868
DNA replication inhibition by N4 bacteriophage 893
dnaF gene 893, 894
enterohemorrhagic (EHEC) 102
F plasmids 464
filamentous bacteriophage 464
Fis factor 818, 822
gyrase 892
gyrB gene 893, 894
hfl mutants 817
him genes 817
mutants 818
hip genes 817
mutants 818
Hu proteins 872
insulin production, phage problems 120–121
integrated suppression, by phage P1 1002
interferons expressed in 733
λ prophage 80, 81
see also Lambda bacteriophage
lig 893, 894
McrA and McrBC systems 672–673
methylases 673
methylation-dependent restriction 672–673
multicopy single-stranded DNA (msDNA) 1009
nfrA and *nfrB* genes, N4 phage infection and 895
nfrC 895
P1 phage 997
PRD1 bacteriophage 1165–1168
recA gene mutation 92
RNA phage 1334, 1336, 1338
rnc (RNaseIII) gene, mutations 817
Shiga-like toxins 102–103
single-stranded-DNA-binding protein (SSB) 467, 892
T7 phage 1389
T-even phage 1376
see also T4 bacteriophage; T-even phage
vector systems, *see* Gene cloning; Lambda bacteriophage
Escherichia coli B
T1 bacteriophage 1371
T7 phage 1389
Escherichia coli K-12
isolation 772

Escherichia coli K-12 *continued*
lambda bacteriophage discovery 772
Mu bacteriophage adsorption 874
N4 bacteriophage 891
Esocid lymphosarcoma virus^(A)^ 475
ets oncogene 936
Eucaryotes, virusmediated gene transfer 112
Europe, zoonoses in 1618
European eel virus (EV)^(A)^ 144
European wheat striate mosaic virus (EWSMV)^(P)^ 1410, 1411
distribution and importance 1414
Euxoa scandens, cytoplasmic polyhedrosis viruses (CPV) 312, 317
Euyarchaeota 50
Evolution, microbial, phage as promoters of 112
Evolution of viruses **436–441**
bacteriophage 82, 437, 440
defective-interfering viruses role 322
genera/groups evolution 438–439
'modular' 440
molecular evidence 437
origins of genera/groups 439–440
origins of viruses 440–441
premolecular evidence 436–437
rates 440–441
species evolution 437–438
co-evolution with host 439, 440
patterns of relatedness 438
sequence variation 438
transduction role 112–113
Evolutionary variant, definition 2
Exanthem subitum 625
Exanthema
echovirus infections 358
equine 423, 428
Excisionase 774
Exocytosis, HSV 607
Exons 232
Expression vectors 114–115
see also Gene cloning
Extracellular matrix 228
Eyach virus^(A)^ 945, 1617
Eye infections **441–450**
DNA viruses 442–444
Adenoviridae 444
Herpesviridae 310, 442–444
murine cytomegalovirus 310
Papovaviridae 444
Poxviridae 442
etiological diagnosis 449–450
epidemiological information 447
RNA viruses 443
Arenaviridae 447
Bunyaviridae 446–447
Coronaviridae 445
coxsackievirus 271
Filoviridae 446
Flaviviridae 445
Orthomyxoviridae 446
Paramyxoviridae 445–446
Picornaviridae 271, 447
Reoviridae 445
Retroviridae 447
Rhabdoviridae 446
Togaviridae 445
see also Conjunctivitis; *individual viruses*
Eyelids, infections/lesions 449

F
Fabavirus^(P)^ 1400
Fabaviruses^(P)^ **451–454**
comovirus relationship 253
cytopathology 452
genome structure 452
host range and symptomatology 452–453
nepoviruses relationship 902
relationships 253, 454, 902
serology 453
structure and composition 451–452, 1580
T, M and B components 451–452
taxonomy and classification 451
transmission 453
Families, viral 94–95, 1398–1400, 1403
definition 94, 1403
descriptive characters used 1404–1405
nomenclature 1402
Fat body, entomopoxviruses (EPVs) infections 394, 395
Fatal familial insomnia 1367, 1368
Fatigue, Ross River virus (RRV) 1273
FBJ murine osteogenic murine sarcoma virus (FBJ-MSV)^(A)^ 884
oncogenes 936
Fc receptor, varicellazoster virus (VZV) 1521
Feces
virus transmission, hepatitis A virus (HAV) 551
viruses excreted in 373
Feline calicivirus (FCV)^(A)^
antigenic variation 1545
clinical features of infection 1546
epidemiology 1545
genomic organization 929
geographic and seasonal distribution 1545
history 1544
immunity 1546
pathogenesis 1545–1546
prevention and control 1546
vaccination 1546
see also Caliciviruses^(A)^; Vesicular exanthema virus of swine (VES)^(A)^
Feline distemper 1061
Feline enteric coronavirus (FECV)^(A)^ 255, 256, 259
Feline immunodeficiency virus (FIV)^(A)^ **454–459**
epidemiology 456
evolution 455
future perspectives 459
genetics 455
geographic distribution 455
history 454–455
host range and virus propagation 455
infections
clinical features 456–457
diagnosis 457–458
immune response 458
pathology and histopathology 458
prevention and control 458–459
neoplastic diseases associated 457
pathogenicity 456
recombination with host genes 461
serologic relationships and variability 456
taxonomy and classification 455
transmission and tissue tropism 456
vaccines 459
Feline infectious enteritis 1061
Feline infectious peritonitis virus (FIPV)^(A)^ 255, 256, 259
Feline leukemia virus (FeLV)^(A)^ 454, **459–463**
carriers, feline immunodeficiency virus (FIV) infection 457
defective-interfering particles 323, 461
epidemiology 462
future perspectives 463
genetics and pathogenicity 461
geographic distribution 461
history 459–460
host range and virus propagation 461
infections
clinical features and pathology 462
immune response 462
prevention and control 463
as insertional mutagens 937
replication and assembly 460
reticuloendotheliosis viruses relationship 1229
structure and genome structure 460
subgroups 460
taxonomy and classification 460
transmission 461–462
tumor-specific changes 461
vaccines 463
variants 461
highly immunosuppressive (FAIDS) 462
Feline panleucopenia virus (FPV)^(A)^
clinical features of infection 1055
see also Feline parvovirus (FPV)^(A)^; Parvoviruses^(A)^
Feline parvovirus (FPV)^(A)^ 1052, **1061–1067**
epidemiology 1053, 1065
evolution 1064
future perspectives 1067
genetics and genome 1053, 1064
geographic and seasonal distribution 1061–1062
history 1061
host range and viral propagation 1062
infections
clinical features 1055, 1065–1066
immune response 1067
pathology and histopathology 1056, 1066
prevention and control 1067
pathogenicity 1065
serologic relationships and variability 1064
structure 1063
taxonomy and classification 1061
vaccines 1056
Feline sarcoma viruses (FeSVs)^(A)^ 459–460
genetics and pathogenicity 461
Feline syncytium-forming virus (FeSFV)^(A)^ 481
geographic and seasonal distribution 486
Fermentation, industrial, *see* Industrial fermentations
Fermon enterovirus^(A)^ 379, 641
see also Enterovirus (EV) serotype 68 (EV68)^(A)^
Ferredoxin 291
Ferrets
influenza viruses in 710
kuru in 759
fes oncogene 936
Festuca leaf streak virus (FLSV)^(P)^ 1246, 1247
Festuca necrosis virus (FNV)^(P)^ 246
Fetus
exposure to mumps virus 882
human parvovirus B19 effect 1054
Feulgen reaction 342
Fever blisters 590
fgr oncogene 936
Fibroblast growth factor (FGF)
HSV receptor 606–607
oncogenesis by mouse mammary tumor virus (MMTV) 860
Fibroblasts
human, human cytomegalovirus culture 293, 305
human diploid, *see* Human diploid fibroblasts (HDF)
transformation
by reticuloendotheliosis viruses 1231–1232
sis gene 536

Fibromas, rabbit 1158
Fibropapillomas, bovine papillomaviruses (BPV) 1306, 1308, 1309, 1310
Fibrosarcoma, feline 459
Fifth disease 1054
Fig virus S (FVS)(P) 215
Figwort mosaic virus (FMV)(P) 225
Fiji disease virus (FDV)(P) 1096, 1111
Fijivirus(P) 1096, 1098, 1195
Filamentous bacteriophage(B) **464–469**
  classification 464
  gene expression and proteins 465–467
    capsid proteins 465
    for membrane-associated assembly 466
    pI and pI* 466
  genome structure 464–465
    intergenic region 465, 466
  infection and life cycle 467–469
    assembly 468
    DNA replication 468, 993
  polyphage 469
  structural classes 464
  structure 464–465
  uses 469
Filamentous ribonucleoprotein particles (RNPs), tenuiviruses 1410
Filamentous virus (FV)(I) 656, 659–660
Filamentous viruses 1571
  see also Helical symmetry
Filoviridae 830, 1399, 1409, 1619
  antigenic cross-reactivity 1215
  eye infections 446
  gene map 1212
  Marburg and Ebola viruses 827
  replication/transcription strategy 1207
  structure and characteristics 443, 446, 1584
Filoviruses(A) 827
'Filterable virus' 627
Fingerprints 529
Fis factor 818, 822
Fish
  neoplasms 474
  see also Fish viruses
Fish herpesviruses(A) **470–474**, 477, 479, 480
  evolution 470, 471–473
  genome structure 471, 472
  geographic and seasonal distribution 471
  history and classification 470–471
  host range and virus propagation 471
  infections
    clinical features and pathology 473
    immune response 473–474
    prevention and control 473–474
  pathogenesis 473
  properties and gene expression 471–473
  salmonid 470
  see also Channel catfish virus (CCV)(A)
Fish lymphocystis disease virus (FLDV)(A) 479–480, **798–801**
  enzymes 799
  epidemiology 800
  evolution 800
  future perspective 800–801
  genetics 800
  genome structure and properties 798–799
    sequences 799
  geographic and seasonal distribution 799
  host range and viral propagation 800
  infections
    clinical features 800
    pathology and histopathology 800
  life cycle 800
  proteins 799
  structure and properties 798
  taxonomy and classification 798
  transcription 799
Fish pox virus(A) 479
Fish rhabdoviruses(A) 1249–1253
  culture 1250
Fish viruses **474–480**
  classification 474
  low virulence 479–480
  of moderate–high virulence 475–479
  specific-pathogen-free (SPF) stock 475
  transmission 474
  see also Fish herpesviruses(A); Infectious pancreatic necrosis virus of fish (IPNV)(A); *other viruses above*
Flaviviridae 1399, 1409
  characteristics and structure 443, 1583
  evolution 364, 370
  eye infections 445
  Flavivirus, see Flavivirus(A)
  hepatitis C virus 570
  Japanese encephalitis virus 746
  Pestivirus 176
    hog cholera virus 649
    see also Bovine viral diarrhea virus (BVDV)(A)
  replication/transcription strategy 1206–1207
  see also Flaviviruses(A)
Flavivirus(A) 362, 1606
  dengue viruses 324, 327
  encephalitis viruses 362, 367, 368
  yellow fever virus 1606
  see also Dengue viruses(A); Encephalitis viruses(A); Flaviviruses(A); Yellow fever virus(A)
Flaviviruses(A)
  dengue antibody cross-reaction 330
  dengue virus relationship 325
  epidemiology 365, 371
  features 363
  genetics and evolution 364, 370
  genome structure 362, 368
  geographic and seasonal distribution 362–363, 369
  hepatitis C virus relationship 570, 571
  host range and viral propagation 363–364, 370
  infections
    clinical features 365–366, 372
    encephalitis 911
    eye infections 445
    immune response to 366, 372
    pathology and histopathology 366, 372
  isolation 338
  Japanese encephalitis virus cross-reactions 748
  pathogenicity 365, 371
  physical properties 368–369
  proteins 362, 368
  replication 369
  serologic relationships and variability 364, 370
  transmission and tissue tropism 365, 371
  see also Encephalitis viruses(A); Tick-borne encephalitis virus (TBEV)(A)
Flexal virus(A) 777
Flock house virus (FHV)(I) 919, 924
Fluorescein isothiocyanate (FITC) 344, 1251
Fluorescence, *see* Immunofluorescence
5-Fluorouracil, interferon therapy combined with 744
fms oncogene 936
Foamy viruses(A) **480–488**
  adsorption and attachment 484
  antibodies to 487
    titers in herpesvirus papio infections 613
  assembly, release and uptake 486
  bovine syncytial virus (BSV) 159
  cytopathology 480, 481, 486
  evolution 487
  future perspectives 488
  genetics 486
  genome structure/properties 482–483
  geographic and seasonal distribution 486
  history 480
  host range 486
  in vitro cell cultures 484–485
  isolates 480, 481
  pathogenicity 487
  physical properties 484
  post-translational processing 486
  proteins 483–484
    Env proteins 484
    matrix (MA) 483, 486
    properties 483–484
  replication 486
    strategy 480, 485
  serologic relationships and variability 487
  structure and properties 482
    glycoprotein spikes 482
  taxonomy and classification 480
  transcription 485
    transactivation (*taf* gene) 485, 487, 488
  translation 485–486
  transmission and tissue tropism 487
Food industry, bacteriophage problems, *see* Industrial fermentations
Foot and mouth disease, *see* Foot and mouth disease viruses (FMDVs)(A), infections
Foot and mouth disease viruses (FMDVs)(A) **488–496**
  antigenic sites 489, 495
  assembly and release 492
  attachment to cell receptors 492, 1081
  capsid structure 207
  epidemiology 494–495
  genetics and evolution 494
  genome structure/properties 489–490
    untranslated region (UTR) 490
    VPg 490, 491
  geographic distribution 492–494
  history 488, 627, 628
  host range and viral propagation 488, 494
  infections 627
    clinical features 495
    immune responses to 495
    prevention and control 495–496
  pathogenicity 495
  post-translational processing 492
  protein products 490–491
    leader protein 490
    P1, P2 and P3 regions 490
    structural proteins (VP1-4) 489, 490
    VPg roles 490, 491
  RNA replication 491
  serologic relationships and variability 494
  serotypes 492, 494
  structure and properties 489
    GH loop of VP1 489
    unique 489
  swine vesicular disease virus differentiation 390
  taxonomy and classification 489
  translation 491, 1125
  transmission and tissue tropism 495
  vaccines 495–496
  virus-infection-associated (VIA) antigen 495
Fort Morgan virus(A) 417, 418
fos oncogene 936, 1475
Foscarnet 44, 46
  avian hepatitis B control 568
  human cytomegalovirus infections 46, 298

Fowl, Marek's disease (MD), *see* Marek's disease (MD)
Fowl paralysis 833, 835
Fowl plague 710
Fowl plague orthomyxovirus(A), mutation and evolution 437
Fowlpox virus (FPV)(A) **496–503**
assembly site and release 500
carriers and latent infection 501
epidemiology 501
future perspectives 502
genes, essential and nonessential 500–501
genetics and evolution 500–501
genome structure/properties 497–498
hybridization studies 497–498
sequencing 498
geographic and seasonal distribution 500
history 496
host range and viral propagation 500
infections
clinical features 501–502
immune response 502
pathology and histopathology 502
prevention and control 502
pathogenicity 501
physical properties 498
plaque mutants 500
properties 497
proteins 498
39kD immunodominant 498
'reactivation' phenomenon 499
recombinants 501, 502
replication 498–499
adsorption and uncoating 498–499
DNA 499
serologic relationships and variability 501
taxonomy and classification 497
transcription characterization 499–500
translation characterization 500
transmission and tissue tropism 501
vaccines 502
vaccinia virus comparisons 497, 498, 499
Foxtail mosaic virus (FoMV)(P) 1143
genome 1144
fps oncogene 936
Frameshifting
bovine immunodeficiency virus 160
bovine leukemia virus 170
luteoviruses 794
lymphoproliferative disease virus (LPDV) of turkeys 812, 813
measles virus 839
mouse mammary tumor virus (MMTV) 857
Saccharomyces cerevisiae L-A 1601–1602
simultaneous slippage model 350
T7 phage 1389
translation of Ty elements 1234
'slippery site' mechanism 1234
Ty1 element 1234
type D retroviruses 1238
Frangipani mosaic tobamovirus (FrMV)(P) 1436, 1440
Friend murine leukemia virus (MuLV)(A) 884, 887
Friend spleen focus-forming virus (SFFV)(A) 887, 888
Frog virus 3 (FV3)(A) 24, **503–508**
assembly, assembly sites and release 506–507
bacteriophage shared features 505, 507, 508
chlorella viruses differences 36
cytoskeletal changes induced by 507
evolution 507–508
future perspectives 508
genetics 507
genome structure/properties 504
concatemeric DNA, functions 505
methylation 504, 505, 506
history 503
host mRNA synthesis shutdown 504, 506
pathogenicity 507
proteins, properties 504
recombination 504
frequencies 507
replication 504, 1434
DNA 504–505
structure and composition 1432
structure and properties 503–504, 1432
taxonomy and classification 503
transcription 505–506
IE/DE and L mRNAs 506
translation 506
Frog virus 4 (FV4)(A) 40, 41–42
Fuchsia latent virus (FuLV)(P) 215
Fujinami sarcoma virus (FuSV)(A), oncogenes 936
Fungal infections, susceptibility change due to plant viruses 1112
Fungal protoplasts 1047, 1049
Fungal transmission
Bymovirus 515
furoviruses 508
spores, of partitiviruses 1050
tobacco necrosis virus 896, 897
tombusviruses (cucumber necrosis virus (CNV)) 1451
Fungal viruses, *see* Mycoviruses(P)
Fungi, filamentous, totiviruses infecting 1464
Furcraea necrotic streak virus (FNSV)(P) 349
Furovirus(P) 1400
Furoviruses(P) **508–516**
cell-to-cell movement proteins 512, 513
deletion mutants 513, 514
economic significance 515
epidemiology and control 515–516
genetic engineering 516
genetics and replication 513–514
genome structure and expression 510–513
beet necrotic yellow vein virus (BNYVV) 512–513
peanut clump virus (PCV) 512
potato mop-top virus (PMTV) 512
RNA1 and RNA2 510, 511
RNA3, RNA4 513
soil-borne wheat mosaic virus (SBWMV) 510–512
geographic distribution 509, 516
host interactions 514
host range 509
inclusion bodies 514
serology 510
structure and composition 509–510, 1582
taxonomy and classification 508–509
translation 510, 512
transmission and vectors 514–515
Fusion
viral and cellular membranes 1557–1558
inhibition 1557
pH-dependence 1557
see also Cell fusion; Syncytium formation
Fusion proteins 1557, 1558
see also Hemagglutinin
Fusogenic proteins 228, 229–230

G
Gaeumannomyces graminis virus 019/6-A(P) 1047
Gaeumannomyces graminis virus 87-1-H (GgV-87-1-H)(P) 1464
Gaeumannomyces graminis virus T1-A(P) 1047
gag gene
avian leukosis viruses 67
bovine immunodeficiency virus (BIV) 160
bovine leukemia virus 167
caprine arthritis encephalitis virus (CAEV) 201
equine infectious anemia virus (EIAV) 430
feline immunodeficiency virus (FIV) 455
feline leukemia virus (FeLV) 460
foamy viruses 482, 483
HTLV-1 683
HTLV-2 690, 691
human immunodeficiency virus (HIV) 675
mouse mammary tumor virus (MMTV) 855, 857
murine leukemia viruses (MuLVs) 884
reticuloendotheliosis viruses (REV) 1229, 1230
in retrotransposons 1233
simian immunodeficiency virus (SIV) 1318
type D retroviruses 1237, 1238, 1240
Gag protein
avian leukosis viruses 67
bovine leukemia virus 167, 170
human immunodeficiency virus (HIV) 676, 677–678
lymphoproliferative disease virus (LPDV) of turkeys 812
murine leukemia viruses (MuLVs) 884
Saccharomyces cerevisiae L-A 1601
gag-pol fusion protein
reticuloendotheliosis viruses (REV) 1230
Saccharomyces cerevisiae L-A 1601
gag-pro product
avian leukosis viruses 67, 68
type D retroviruses 1238
gag-pro-pol-env genes, bovine leukemia virus 167
GagProPol precursor, avian leukosis viruses 68
Galinsoga mosaic virus (GMV)(P) 218, 219
Galleria mellonella densonucleosis virus (GmDNV)(I) 331, 332, 333, 334
histopathology 334
replication 335
in tissue culture 334
Gamma-globulins
coxsackievirus infections 273
human cytomegalovirus infections 298
see also Immunoglobulin(s)
Gammaherpesvirinae 155, 156, 157
amphibian herpesviruses 40–42
see also under Herpesviridae
Ganciclovir 44, 46
herpes simplex virus 44, 46
human cytomegalovirus infection 46, 298
infections after cardiac transplantation 966
oral bioavailability problem 46
Ganglia, herpes simplex virus in 589, 590, 789, 790
Garbage, vesicular exanthema virus of swine (VES) transmission 1544, 1545
GardnerRasheed feline sarcoma virus (GR-FeSV)(A), oncogenes 936
Garlic mosaic virus (GaMV)(P) 215
Gastric acidity 1077
Gastroenteritis
acute nonbacterial 926, 930
astrovirus causing 376
coronaviruses causing 376–377
coxsackievirus infection 270
Norwalk agent causing 375, 926, 930
discovery of 926
parvoviruses and parvovirus-like particles causing 376

rotavirus 1278, 1279
history 1274
seasonal distribution 1275
viral 926
see also Diarrhea; Enteritis
Gastrointestinal bleeding, in dengue hemorrhagic fever/dengue shock syndrome (DHF/DSS) 329
Gastrointestinal disease, reovirus infection 1194
Gastrointestinal tract
entry of viruses 1077
host defenses 1077
'GDD' box 1443
Geminivirus(P) 1398
Geminivirus I, II, III 1407
Geminiviruses(P) **517–524**
coat protein mutants 1539
deletion mutants 519, 520, 523
DNA, subgroups I/III relationship 519, 520, 524
DNA replication 521, 1539
evolution 523–524
subgroup III 523
gene amplification vectors based on 522–523
genes and functions of 519
genome organization and expression 518–521, 1539
large intergenic region (LIR) 518–519
subgenomic DNA 519, 520
subgroup I 518–519
subgroup II 520
subgroup III 519–520
geographic distribution 522
history 517
pathogenesis 523
structure and composition 517–518, 1539, 1582
subgroups 517
taxonomy and classification 517
tomato yellow leaf curl virus (TYLCV) 521
transcription 518, 519, 520
transmission and host range 517, 521–522
as vectors 522–523, 1539–1540
instability 1539
monopartite 1539
Gene(s)
acquisition from hosts 439
in cell growth 938
coat proteins and resistance to viruses 1108
de novo, evolution of species and 439
dosage 731
mapping, *see* Genome(s), mapping
overlapping, discovery 990
phage-resistance 120
shared between genera 440
viral disease resistance 665, 666, 668
plant viruses 1106–1107, 1147, 1440–1441
see also Host genetic resistance; Plant resistance
Gene amplification
vectors based on geminiviruses 522–523
see also DNA, amplification; Polymerase chain reaction (PCR)
Gene cloning 530
animal viruses as vectors 530
bacteriophage
coliphages P2 and P4 1008
T5 bacteriophage 1388
bacteriophage as vectors **113–116**
B. subtilis phage 77
bacteria (non-*E. coli*) 116
E. coli 113–116
expression vectors 114–115
filamentous bacteriophage 115, 469
insertion vectors 114
λ, *see* Lambda bacteriophage
M13 vectors 115, 469
mini-Mu derivatives 876
P1 phage 115
replacement vectors 114
requirements of 113
SPO1 bacteriophage 1356
hybrid systems
cosmids (plasmids/phage) 115
multi-component phage vectors 116
phagemids 115, 116, 469
plasmids as vectors, *see* Plasmids
vectors 530
see also Genetic engineering; Recombinant DNA technology; Vectors
Gene expression, viral, restriction and immune evasion 1092, 1093
Gene inactivation, DNA methylation and 13
Gene regulation (viruses) 1205
Gene therapy 731
adeno-associated virus (AAV) system 1060, 1535
adenoviruses as vectors 1530
retrovirus vector 1533, 1534
Genera, viral 94, 1403
definition 1403
evolution 438–439
genes shared between 440
nomenclature 1402
origins 439–440
Genetic diversity 664
Genetic drift 437
dengue viruses(A) 326
Genetic engineering 530
animal viruses, uses 530–531
see also Gene cloning; Recombinant DNA technology
Genetic factors, determining infection outcome 1082
Genetic reassortment 524, 527, 528, 701, 1221
coltiviruses 946
influenza viruses 711–712
inhibition by dominant-negative mutants 731
lymphocytic choriomeningitis virus (LCMV) 802
mapping 529
orbiviruses 946
reoviruses 1192, 1199, 1202
rotaviruses 1277
Genetic recombination 527
bacteriophage, *see* Genetic recombination, bacteriophage *(below)*
bacteriophage and host DNA 108, 109
bromoviruses 184
defective-interfering viruses generation 320, 461
host cell genes and viral genes, feline leukemia virus 461
illegitimate 109
intramolecular 524, 527, 528
lymphocytic choriomeningitis virus (LCMV) 802
mapping 529
N4 bacteriophage 893
origin of viral species 439
orthopoxviruses 263
P1 bacteriophage 1002
P2 coliphage 1008
P22 bacteriophage 1013
plant viruses as vectors 1542
concerns 1543
polioviruses 1132
Sindbis (SIN) virus 1333
SPO1 phage 1356
T1 phage 1372
T4 phage 1382
transducing genome and host DNA 110
Genetic recombination, bacteriophage **83–92**, 108, 109
bacteriophage ϕX174 996
double-strand break (DSB) 91–92
λ phage 86
single-stranded DNA phage 84
T4 phage 91
T7 phage 89
λ phage 85–87, 774–775
parallels between mechanisms/species 91–92
replication association
ssDNA phage 85
T4 phage 90, 91
T7 phage 87, 88
signals stimulating 91
single-stranded DNA phage 84–85
characteristics of 84
frequency and negative interference 84
RecA-dependent 84
RecA-independent 84–85
ssDNA annealing 91, 92
λ phage 87
single-stranded DNA phage 85
T4 phage 91
T7 phage 89
T4 phage 90–91
T7 phage 87–89
see also specific bacteriophage
Genetic resistance, host, *see* Host genetic resistance
Genetic susceptibility
to infections 1082
see also Host genetic resistance
Genetic transmission
reticuloendotheliosis viruses 1230
sigma rhabdoviruses 1313
see also Transovarial transmission
Genetic variation 524
Genetics of animal viruses **524–531**
genomes 525
mapping 529
history 524–525
interactions between gene products 527–529
molecular genetics 530
mutagenesis 526
mutations 525–526
see also Mutations
recombination, *see* Genetic recombination
transgenic mice 531
uses of genetic engineering 530–531
Genital cancer
HPV and 1496–1497
see also Cervical carcinoma
Genital warts 1017–1018
in HIV infections and AIDS 1018
HPV pathogenicity 1023–1024
HPV types causing 1018, 1021
risk 1017
see also Human papillomaviruses (HPV)(A); Warts
Genitourinary tract, entry of viruses 1077
Genome(s)
activation 30
animal viruses 525
DNA 525
history 630, 631
RNA 525
mapping 529
complementation groups 529

Genome(s) *continued*
    marker rescue by transfection 531
    recombination maps 529
    restriction maps 529
  'master copy' 437
  rescue phenomenon 1154
  sequencing 530
  structure and organization 1140
Genomic cluster, definition 2
Genomic libraries, screening 114
Gentiana carlavirus (GeCV)(P) 215
Genus, definition 94
Gerstmann–Sträussler–Scheinker disease (GSS) 1361
  clinical features 1367
  cytological and molecular pathology 1368
  inheritance 1367
  maternal and vertical transmission 1366
  transmission and infection route 1365
  see also Spongiform encephalopathies
Giant cells, in measles 846
Giardia lamblia 532
  transfection 1467
Giardiasis 532, 533
Giardiavirus(A) 1464
Giardiavirus (GLV) **532–535**, 1464
  antisera 533
  dsRNA transfection 1467
  evolution 534
  history 532
  host range and geographic distribution 532
  infection and replication 533–534
  physical and biochemical characteristics 532–533
  ssRNA and polypeptide in infected cells 534
  taxonomy and classification 532
Gibbon ape leukemia virus (GaLV)(A) **535–536**
  cell culture 535
  clinical features of infections 535–536
  GaLV-Br strain 535
  GaLV-H strain 535–536
  immune response 536
  as insertional mutagens 937
  pathogenicity 536
  San Francisco strain (GaLV-SF) 535
  SEATO strain (GaLV-SEATO) 535
  serologic relationships and variability 536
  transmission 535
Ginger chlorotic fleck virus (GCFV)(P) 1346
  see also Sobemovirus(P)
Gingivostomatitis 589
Gladiolus carlavirus (GCV)(P) 215
Glandular fever 1224
Glial fibrillary protein (GFAP) 1368
Global warming, orbiviruses and coltiviruses 955
Globotetraosylceramide 103
Globotriaosylceramide 103
Glomerulonephritis 973
  equine arteritis virus infection 770
Glomerulosclerosis, focal segmental 973
Glucocorticoid-response elements (GRE) 858
Glutamine synthetase 291
Glycine mosaic virus (GMV)(P) 249
Glycine mottle virus (GMeV)(P) 218, 219
Glycocalyx 228
Glycophorin, reovirus receptor 1200
Glycoproteins
  cell attachment proteins 1081, 1557
  plasma membrane 228
  virus-induced 231
Glycosylation 26, 303
  membrane proteins 1557
  N-linked
    hepatitis B virus (HBV) 563
    measles virus proteins 842
    toroviruses 1455
  Newcastle disease virus 916
  varicella–zoster virus (VZV) 1525
  see also Oligosaccharides
Glycosyltransferases 230
Goatpox 1160
  see also Poxviruses(A), sheep and goats
Goats, caprine arthritis encephalitis virus infection 201
Golden shiner virus disease(A) 479
Golgi apparatus 230–231, 1559
  bunyavirus maturation 196, 1556
  'cis' and 'trans' cisternae 230
  coronaviruses localization 255
  cowpox virus envelope from 261
  HSV glycoprotein transport to 604, 607
  torovirus assembly 1455
Gonometa virus (GoV)(I) 1100
  economic importance 1103
  serological relationships 1102
  taxonomy and classification 1100
  see also Picornaviruses(I), insect
Goose hepatitis virus (GPV)(A) 1052
  clinical features of infection 1055
Goose parvovirus (GPV)(A) **1067–1075**
  clinical features of infection 1074–1075
  geographic and seasonal distribution 1071
  history 1068
  host range and viral propagation 1071–1072, 1072
  properties and structure 1068
  serologic relationships and variability 1072, 1073
  taxonomy and classification 1068
  transmission and epidemiology 1073
  see also Parvoviruses(A)
Gossas virus(A) 1249
Gp34 glycoprotein 1361, 1362, 1364, 1367
  prion protein from 1361
Graft transmissibility, cryptoviruses 276
Graft versus host disease (GVHD) 1477, 1488
  cytomegalovirus (CMV) disease risk 1478, 1479
Graminella nigrifrons 824, 825
Granulin 127, 131
  gene 127, 128
Granulocytemacrophage colony-stimulating factor (GM-CSF) 740–741
Granulosis viruses (GVs)(I) **127–130**
  capsid proteins 128
  capsule, *see occlusion body structure (below)*
  cell lines susceptible to 130
  ecology 130
  genes 128–129
    enhancin 128
    granulin 127, 128
    iap 128–129
  genetics 128–129
  infection mechanism 129
  occlusion body structure 127–128, 131
    enhancing factor 128
    granulin 127, 131
    synergistic factor 128
  pathology and cytopathology 129, 129–130
  replication 129–130
  structure 128
  taxonomy and classification 127
  tissue tropism 129
  use as insecticides 130
Granzymes 704
Grapevine Algerian latent virus(P) 1447
Grapevine Bulgarian latent virus(P) 904
Grapevine chrome mosaic virus (GCMV)(P) 904, 905
Grapevine corky bark associated virus (GCBAV)(P) 243
Grapevine fanleaf virus (GFLV)(P) 904, 906
  coat proteins 905
  economic significance 906
  epidemiology 906
Grapevine leafroll disease virus (GLRAVs)(P) 244, 247, 248
Grapevine Tunisian ringspot virus(P) 904
Grapevine virus A (GVA)(P) 244, 247
Grapevine yellow speckle viroid(P) 1564
Grass carp reovirus (GCRV)(A) 475–476
GroEL 1378
Gross virus(A) 883, 884
Ground squirrel hepatitis virus (GSHV)(A) 555
  genome structure/properties 560
  geographic and seasonal distribution 555
  host range and viral propagation 556
  pathogenicity 558
  properties 560
Groundnut bud necrosis virus (GBNV)(P) 1459, 1460, 1463
Groundnut crinkle virus (GrCV)(P) 215
Groundnut ringspot virus (GRSV)(P) 1459, 1460
Groundnut rosette assistor luteovirus (GRAV)(P) 792
  economic significance 797
  transmission 795
Groundnut rosette disease 797
Groups of viruses 1398–1400, 1403
Growth factors
  hematopoietic 938
  Johne's bacillus 643
  oncogene action 938–939, 1473–1474
  receptors, oncogene action 939, 1474
  see also Platelet-derived growth factor (PDGF)
Growth hormone, reduced synthesis, in LCMV infections 804
GTP-binding proteins 1474, 1475
GTPase-activating proteins (GAP) 1475
Guanarito virus(A) **776–786**
  characteristics and structure 776, 777
  geographic and seasonal distribution 777, 778
  history 776
  host range and viral propagation 777, 778
  serologic relationships and variability 779
  taxonomy and classification 776
  see also Lassa virus(A); Venezuelan hemorrhagic fever (VHF)
Guanidine hydrochloride, poliovirus RNA synthesis inhibition 1131
Guanosine triphosphate (GTP)-binding proteins 1474–1475
Guanylyltransferase, reovirus 1192, 1197
Guarnieri bodies 496, 867
  vaccinia virus infection 1510
  see also Inclusion bodies, B-type
Guillain-Barré syndrome 912
  Duvenhage virus (rabies-like) association 1189
  echovirus association 359
Guinea-pig, foot and mouth disease virus infection 494
Gumboro disease 143
GUS gene 1541
Gynura latent virus (GyLV)(P) 215
Gypsy moth virus (GMV)(I) 919

H
H-1 virus(A) 1068, 1069
  clinical features of infection 1074
  host range and viral propagation 1072

tissue tropism and pathogenicity 1074
H-2D gene, mouse cytomegalovirus (MCMV) resistance 667
H-2D^{d} allele, susceptibility to Theiler's mouse encephalomyelitis virus (TMEV) 668
H-2D^{k} haplotype 667
Ha-*ras* oncogene 936
HADEN (hemadsorbing enteric virus)(A), *see* Bovine parvovirus (BPV)(A)
Hairpin loop
filamentous bacteriophage 464
parvoviruses 1058
satellite tobacco necrosis virus 900
Hairy cell leukemia
HTLV-2 and 682, 686–687, 687, 1496
interferon action 738, 741–742
clinical resistance 742
treatment schedule 742
Halobacterial phage(B) 50–54
host relationship and ecology 52–53
see also Bacteriophage ϕH(B)
Halobacterium halobium, ϕH bacteriophage, *see* Bacteriophage ϕH(B)
Halobacterium salinarium, ϕH bacteriophage, *see* Bacteriophage ϕH(B)
Halophiles 50
Hamster(s), encephalitis 421
Hamster cells, abortive infection with adenovirus (Ad12) 13
Hamster polyomavirus (HaPV)(A) 1135
Hamster syncytium-forming virus (HaSFV)(A) 481
Hand-foot-and-mouth disease (HFMD) 270, 642
enterovirus (EV) serotype 71 (EV71) causing 383
Hantaan virus(A) 538, 974
genome structure/properties 539
see also Hantaviruses(A); Korean hemorrhagic fever (KHF)
Hantavirus(A) 186, 187, 538, 974, 1614
cell culture 188
genome, sequences and coding strategies 193, 194
human infections/viruses 189–190
structure and genome 188
transmission and ecology 189
see also Bunyaviruses(A); Hantaviruses(A)
Hantaviruses(A) **538–545**, 974, 1614
cytopathology 541
detection 542, 544–545, 1614
epidemiology 543
future perspectives 545
genetics and evolution 542
genome structure/properties 539
coding properties 540
consensus sequences at 3′ termini 539
L, M and S segments 539
geographic and seasonal distribution 541–542, 1614
history 538, 1613
host range and viral propagation 542
infections
clinical features 543–544
immune complexes 543, 545
immune response 544–545
kidney 974
pathology and histopathology 544
prevention and control 545
see also Korean hemorrhagic fever (KHF)
pathogenicity 543
physical properties 539
plaque assays 542
post-translational processing 541
proteins 539–540
G1 and G2 539, 541
L protein 540
N (nucleocapsid) 539
properties 539–540
replication 540–541
assembly and transport 541
serologic relationships and variability 542–543
structure and properties 538–539
taxonomy and classification 538, 1614
transcription 540–541
translation 541
transmission and tissue tropism 543, 1614
vaccines 545
Haploid genome 525, 529
Hardy–Zucherman feline sarcoma virus (HZ-FeSV)(A), oncogenes 936
Hare fibroma virus (HFV)(A) 1154
see also Poxviruses(A)
Hare sarcoma 1154
Harrisina brillians granulosis virus(I) 129
Harvey murine sarcoma virus (Ha-MSV)(A) 884, 1474–1475
oncogenes 936
Hashimoto's thyroiditis 976
Hawaii agent(A) 926
antigenic relationships 928
see also Norwalk virus(A)
Head disease, fish 478
'Headful packaging', bacteriophage 108–109, 999
'Headful replication' 1601
Health care workers, hepatitis C in 573
Hearing loss, in cytomegalovirus infection 296
Heart
transplantation, infections after 966, 969
virus infections 964–969
coxsackievirus 271
etiology 964–967
murine cytomegalovirus 310
reovirus 1193
Heat shock proteins 1000
immune response to 64
virus-infected cells expressing 64
HeLa cells, HPV in 1025
Helenium virus S (HelVS)(P) 215
Helical symmetry 1209, 1571, 1574
atomic structure 1571
Helicase, plant viruses lacking 1447, 1448–1449
Helicoverpa ascovirus(I) 59, 62
Helicoverpa zea 59, 62
Heliothis armigera, cytoplasmic polyhedrosis viruses (CPV) 312
Heliothis armigera stunt virus (HaSV)(I) 1417
Helleborus carlavirus (HeCV)(P) 215
Helminthosporium victoriae, degenerative disease 1051
Helminthosporium victoriae 145S virus (HvV-145S)(P) 1047, 1051
Helminthosporium victoriae 190S virus (HvV-190S)(P) 1051, 1464, 1466
Helper viruses 321, 1587
adeno-associated viruses (AAVs) 1052, 1060
carmovirus satellite RNA 223
cucumoviruses 281
pea enation mosaic virus 1087
reticuloendotheliosis virus (REV-T) 1231
simian sarcoma associated virus as 536
spleen focus-forming virus 884
totiviruses 1464
Ustilago maydis virus 1468, 1470
Hemadsorption inhibition reaction, African swine fever virus 28
Hemadsorption test, virus detection 337
Hemadsorption virus type 2 (HA2)(A) 1300
Hemagglutinating units (HAU) 1173
Hemagglutinating virus of Japan (HVJ)(A) 1300
Hemagglutination-inhibition (HAI) tests 1173
alphavirus classification by 417
bovine parainfluenza virus type 3 (PIV-3) 1031, 1032–1033
cowpox virus 266
Japanese encephalitis virus 748, 750
reoviruses 1190, 1194
Ross River virus (RRV) 1272
Hemagglutinin (HA)
antigen 1557
bluetongue virus 959
bovine parainfluenza virus type 3 (PIV-3) 1034
measles virus 838, 839, 840, 841
reoviruses 1191
rotavirus 1286
Sendai virus 1300, 1301, 1304
see also Hemagglutininneuraminidase (HN)
Hemagglutinin (HA), influenza viruses 710, 715, 722, 1081, 1557
antigenic determinants 716, 722
antigenic drift and shift 711
atomic structure 1572
conformation, neutralizing antibodies 699, 722
detection 337
epitopes on 725
functions 716
fusion activity and cleavage 716, 1557–1558
pathogenicity and 713
post-translational processing 720
properties and structure 715–716, 722
synthesis 722, 1558, 1559
three-dimensional structure 716, 722, 1081
Hemagglutinin-esterase envelope (HE) glycoprotein 715, 717
coronaviruses 255, 256
Hemagglutinin–neuraminidase (HN)
human parainfluenza viruses (PIV) 1027, 1028
mumps virus (MuV) 877, 878
Newcastle disease virus 914, 915
Hematogenous spread 1078–1079
of viruses to nervous system 909, 910, 1079
Hematologic malignancies
interferon action in 738, 741–743
see also Leukemia; Lymphomas
Hematopoietic growth factors 938
Hematoporphyrin derivative (HPD), in cottontail rabbit papillomavirus (CRPV) infections 1310
Hematoxylin and eosin (H&E) technique 342
Hemobartonellosis 457
Hemolysis, equine infectious anemia virus (EIAV) infection 434
Hemolysis inhibiting antibodies (HLI), measles virus 841
Hemolytic uremic syndrome 103
Hemophiliacs
hepatitis C in 573
hepatitis D virus infection 578
Hemorrhages
hog cholera virus infection 652
subconjunctival 382
see also Bleeding
Hemorrhagic disease
Crimean-Congo hemorrhagic fever (CCHF) 191, 1616

Hemorrhagic disease *continued*
dengue hemorrhagic fever/dengue shock syndrome (DHF/DSS) 329
encephalitis viruses causing 362
Hemorrhagic fever 975, 1613, 1614
arenaviruses causing 776, 779
filoviruses (Marburg and Ebola viruses) 446, 827, 831
Hemorrhagic fever with renal syndrome (HFRS) 538, 974, 1613, 1614
clinical features 543–544
epidemiological forms 543
geographic and seasonal distribution 541
pathology and histopathology 544
Hemorrhagic fever viruses
Kyasanur Forest disease virus (KFDV) 362, 363, 366, 1619
Omsk hemorrhagic fever virus (OHFV) 362, 363, 366, 1618
Simian (SHFV), *see* Simian hemorrhagic fever virus (SHFV)[A]
see also Junin virus[A]; Lassa virus[A]; Machupo virus[A]
Henderson–Paterson bodies 851
Hepadnaviridae 564, 1398, 1409
Avihepadnaviruses 564
characteristics/features 555, 564, 1209
hepatitis B virus 555
latent infections 788
Orthohepadnaviruses 564
retroviruses homology 556, 566
structure 1583
see also Duck hepatitis B virus (DHBV)[A]; Hepadnaviruses[A]; Hepatitis B virus (HBV)[A]
Hepadnaviruses[A]
HBcAg as conserved polypeptide 560
hepatitis D virus differences 575
prerequisite for hepatitis D virus 575, 578, 579
see also specific hepatitis viruses
Heparan sulfate proteoglycan (HSPG), HSV receptor 606–607
Hepatic injury/necrosis, *see under* Liver
Hepatitis
acute infectious, *see* Hepatitis A
adenovirus infections 6
coronavirus infections 259
coxsackievirus infection 270
cytomegalovirus (CMV) 1479
fulminant, hepatitis D virus infection 579
'long incubation' 557
non-A, non-B, *see* Hepatitis C
serum, *see* Hepatitis B
transfusion-associated 572
see also Hepatitis B; Hepatitis C
Hepatitis A 546, 551, 554, 1484
clinical features 552
diagnosis 552
epidemiology 550–551
immune response 553
incubation period 552
needle-borne 551
outbreaks 551
pathology and histopathology 552–553
prevention and control 553–554
transfusion-associated 551
Hepatitis A virus (HAV)[A] **546–554**
antigenic determinants/sites 547, 553
antigenic variability absent 550
attenuation 548, 553
cDNA clone 546
cell culture-adapted strain 549, 550, 553
defective-interfering particles 323, 548
epidemiology 550–551
evolution 550
genetics 549–550
genome structure/properties 546, 547
VPg linked 547
genotypes 549–550
geographic and seasonal distribution 549
history 546, 642
host range and viral propagation 549
mutations 548, 550
pathogenicity 551
physical properties 548
resistance to thermal denaturation 548
proteins 547–548
2A, 2B, 2C 548
3B, 3C, 3D 548
properties 547–548
pX 547
VP0 546–547
VP1, VP2, VP3 546
VP4 and myristylation 546
receptors 548
replication 546, 548–549
downregulation of RNA replication 548
encapsidation of positive-strand RNA 548
RR/CPE+ phenotype 549, 550
serologic relationships and variability 550
structure and properties 546–547
taxonomy and classification 546
translation 549
transmission and tissue tropism 401, 551
vaccines 553–554
attenuated 549, 553–554
formalin-inactivated 553
Hepatitis B
acute 557
after transplantation, *see under* Hepatitis B virus (HBV)[A]
chronic 557, 558
clinical features 558, 1485–1486
epidemiology 557
hepatitis delta virus infection associated 575
immune response 558–559
pancreatic infections 969
pathogenic mechanisms 557–558
pathology and histopathology 558
persistent infections 555, 557
prevention and control 559, 568–569
in control of hepatitis D 579
interferon therapy 568
see also Hepatitis B virus (HBV)[A], vaccines
Hepatitis B virus (HBV)[A] **554–569**
antigenic variation 556
antigens 556
suppressed in chronic delta hepatitis 579
assembly site, release and uptake 563–564, 569
avian, *see* Hepatitis B viruses (HBV)[A], avian
carriers 557, 558, 1083, 1091, 1498
superinfection with hepatitis D virus 579
Dane particles 555, 560
defective genome and disease exacerbation 323
discovery 554
DNA sequences in transgenic mouse 531
electron microscopy 342, 560
envelope 555, 560
evasion of interferon action 737
evolution 556, 566
future perspectives 559
general features **554–559**
genetics 556
genome structure/properties 555, 560, 561
negative (complete) DNA strand 560
geographic and seasonal distribution 555
HBcAg (core antigen) 556, 560
antibodies to 558
conserved polypeptide 560
delta antigen similarity 574
translation 562
HBeAg 556, 557, 560
antibodies to 558
properties 560, 561
HBsAg (surface antigen; S) 554, 556, 560, 1485
antibodies to 558
carriers 557, 558, 1498
in HBV vaccines 559
hepatitis D virus use 575, 577
properties 561
structure 565
subtypes 556
hepatocellular carcinoma and, *see* Hepatocellular carcinoma (HCC)
history 554–555
host range and viral propagation 555–556
incomplete virus particles 554
infections after transplantation 1484–1486
clinical features and treatment 1485–1486
epidemiology and pathogenesis 1485
incidence 1485
prevention 1486
interferon clinical use/action 738
kidney infections 973
latency period for tumors 1499
molecular biology **560–564**
packaging of pregenomic RNA 562, 569
pathogenicity 557–558
as persistent infection 1089, 1090
physical properties 562
post-translational processing 563
$preS_2$/S gene 1498
proteins 560–562
L and M 560, 561
long-term expression of L protein 561
properties of 560–562
S, *see* Hepatitis B virus (HBV)[A], HBsAg
X protein 561–562, 1498
re-activation 1485, 1486
receptors 1081
replication 555, 562
CCC DNA 562, 569
CCC DNA copy number control 569
positive-strand DNA synthesis 562
pregenomic RNA (C-mRNA) 561, 562
reverse transcription of negative-strand DNA 562
reverse transcriptase 561, 569
serologic relationships and variability 556–557
structure and properties 555, 560
subviral particles concentration 560
surface antigen, *see* HBsAg *(above)*
taxonomy and classification 555, 564
titer in blood 560
transcription 562–563
CCC DNA template 562
L and M proteins 563
S and X proteins 563, 1498
transformation mechanism 1477, 1497
see also Hepatocellular carcinoma (HCC)
translation 561, 563
X protein 563
transmission and tissue tropism 557

vaccines 559, 568
adenovirus recombinants as 1529
Hepatitis B viruses (HBV)(A), avian 555, **564–569**
assembly, packaging 565
characteristics 564–565
epidemiology 566–567
evolution 566
future perspectives and importance 568–569
genetics and mutation rate 566
geographic and seasonal distribution 565
hepatitis 566, 567
history 564
host range and viral propagation 565–566
immune response 568
pathogenicity and clinical features 567
pathology and histobiology 567–568
prevention and control 568
replication 564–565
reverse transcriptase 569
serologic relationships and variability 566
surface antigen 565, 566
taxonomy and classification 564–565
tissue tropism 567
transmission
congenital 565, 566, 567
horizontal 565
as vectors for therapy of inherited diseases 569
see also Duck hepatitis B virus (DHBV)
Hepatitis C
acute 573
chronic 573
clinical features 573
community-acquired 573
detection tests 574
immune response 573–574
incubation period 569, 573
pathology and histopathology 573
prevention and control 574
Hepatitis C virus (HCV)(A) **569–574**
animal model (chimpanzees) 571, 572, 574
assay 1484
epidemiology 572–573
evolution 572
future perspectives 574
genetics 571–572
genome properties 570
subgenomic RNA 570, 571
geographic and seasonal distribution 571
hepatoma and 1498
history 569–570
host range and viral propagation 571
infections after transplantation 1484–1485
clinical features and treatment 1486
incidence 1485
pathogenesis 1485
prevention 1486
mutations
rate 572
variable/hypervariable sites 572
pathogenicity 573
persistent infections 1090, 1498
physical properties 571
proteins 570–571
envelope proteins 570
NS2, NS3, NS4, NS5 570
prototype strain 572
replication 571
serologic relationships and variability 572
structure and properties 570
taxonomy and classification 570
transmission and tissue tropism 573
vaccines 574
Hepatitis D (delta hepatitis)
acute 579
chronic 579
clinical features 578–579
diagnosis 579
epidemiology 578
immune response 579
incubation period 579
pathology and histopathology 579
prevention and control 579
Hepatitis delta virus (HDV)(A) **574–580**
antibodies (anti-HD) 578, 579
assembly and packaging 577
use of HBsAg 575, 577
cytopathogenicity and immune-induced cytopathology 578
delta antigen (HDAg) 574
cell death due to 578
functional domains 576–577
large and small species 576
ORF for 576
properties 576
translation 577
evolution 577–578
future perspectives 580
genetics 577
genome structure/properties 576
circular RNA 575, 576
double-stranded rod RNA 576
geographic distribution 575
hepadnaviruses as prerequisite 575, 578, 579
history 574–575
host range and viral propagation 575
mutations 577
pathogenicity 578
as persistent infection 1090
proteins 576–577
HBsAg from hepatitis B virus 577
replication 577
antigenomic sense mRNA 577
RNA 577
ribozyme activity 576
serologic relationships and variability 578
structure and properties 575–576
taxonomy and classification 575
transmission 578
viroid-like domain 1565
Hepatitis E
clinical features 584–585
detection, tests 585–586
epidemics and outbreaks 580, 583
epidemiology 583–584
gamma globulin in 586
immune response 585–586
lack of neutralizing antibody 585
incubation period 584
mortality rate 581, 585
pathology and histopathology 585
prevention and control 586
viremia 584
Hepatitis E virus (HEV)(A) **580–586**
adaptation 583
antigen (HEVAg), immunofluorescent staining 584
Burma isolate 582, 583
cross-protective immunity 586
degradation by trypsin 584
excretion 584
future perspectives 586
genetics and evolution 583
genome properties 581, 582
geographic and seasonal distribution 581–582
history 580
host range and viral propagation 582–583
immune electron microscopy (IEM) 581
Mexico isolate 583
origin 584
pathogenicity 584
re-infection with 585
serologic relationships and variability 583
structure and properties 581
taxonomy and classification 580–581
transmission and tissue tropism 583, 584
vaccine development 586
Hepatitis viruses
antivirals 45
infections after transplantation 1484–1486
see also individual hepatitis viruses (above)
Hepatocellular carcinoma (HCC) 557, 1477, 1498
carcinogenic mechanism 557–558, 562, 1498
duck hepatitis B virus and 564
hepatitis B virus and 555, 1498
X protein role 562, 1498
hepatitis C virus (HCV) and 1498
long-term expression of L protein 561
see also Hepatitis B virus (HBV)(A)
Hepatocytes
degeneration, in hepatitis E 585
'ground glass cells' 557
hepadnaviruses in 557
avian hepatitis B viruses 567
hepatitis B virus 555, 557
in hepatitis A 552
in hepatitis D virus infection 579
'microtubular structure' 579
programmed senescence and avian hepatitis B 568
Hepatoencephalomyelitis virus (HEV)(A) 1190
Hepatoma, *see* Hepatocellular carcinoma (HCC)
Hepatoma cell lines, hepatitis B virus replication 562
HER agent 1074
Heracleum latent virus (HLV)(P) 244, 246
Heracleum virus 6 (HV6)(P) 246
Herd immunity 402
Heron hepatitis B virus(A)
genomic sequence 566
tissue tropism 567
transmission 566, 567
see also Hepatitis B viruses (HBV)(A), avian
Herpangina 270, 1223
Herpes labialis 590
'Herpes pac homology' sequence 302
Herpes simplex viruses (HSV)(A) **587–609**, 907
acyclovir-resistance 1484
anterograde transport in nerves 908, 1079
antibodies to IE175 592
antibodies neutralizing during attachment 606
as antiviral chemotherapy target 603
assembly and release 602–603, 607
attachment 605–606
capsid 593
cellular receptors for 606–607
cytopathology 603, 911
proteins involved 599–600, 608
cytoplasmic de-envelopment 607, 608
deletion mutants 604, 606, 607
DNA polymerase 599, 1521
acyclovir triphosphate inhibiting 43, 592
DNA replication 600
enzymes with repair function 599, 600
ORI binding protein (U_L9) 599, 600
proteins involved 599

Herpes simplex viruses (HSV)^(A) *continued*
  entry 602, 908, 1079
  envelope 587, 593–594
    see also glycoproteins (below)
  epidemiology 588–589
  future perspectives 603
  general features **587–593**
  genetic map 596
  genetics 588
  genome structure 587, 594–597, 604, 1521
    'a' sequence 594, 602
    CMV similarity 299
    general arrangement 594, 595
    long/short segments DNA 594, 595
    mRNAs 594
    promoters 595–597
    replication origins 597
    sequence inversions 594
    U_L and U_s sequences 587, 1519
  glycoproteins 594, **603–609**, 1519
    antibodies to 591, 606
    attachment role 602, 605–606
    biosynthesis 604–605
    cell–cell fusion role 599, 607–608
    dispensable in cell culture 604
    domains 604
    envelopment role 607
    essential for cell infection 604
    function as receptors 599, 608–609
    gB 594, 603, 604
    gB role in penetration 602, 606
    gB sequence conserved 594, 604
    gC 603, 604, 1521
    gC binding C3b 599, 609
    gC role in attachment 606
    gC-minus mutants 606
    gD 603, 604
    gD, in cell membranes 608
    gD, cell receptor for 607
    gD, cell–cell fusion role 608
    gD, penetration role 602, 606
    gD, restriction to superinfection 608
    gE binding IgG 608
    general characteristics 603–604
    gG as type-specific 594
    gH role in cell–cell fusion 608
    gH role in penetration 602, 606
    gH sequence 604
    gI binding IgG 608
    HCMV envelope glycoprotein homologs 301
    maturation rate 605
    'nonessential' 1527
    penetration role 602, 606
    processing of and hypothesis on 607
    recombinant 605
    in related species 604
    restriction to superinfection 608
    role 599, 602, 605–608
    role in host response 599, 608
    structure 604–605
    U_L20 role in transit to extracellular space 607
    VZV homology 1519, 1521, 1527
  Golgi apparatus interaction 604, 607
  α herpesviruses, relationship 600
  history 587
  host range and viral propagation 588
  HSV-1 587
    clinical features of infections 589–590
    dominant-negative mutants as antivirals 731
    eye infections 443
    genital infections 589
    genome 604
    homology between strains 594
    HSV-2 homology 594
    latency 601
    rhinitis 590
    seroprevalence 588
    skin infection 588
  HSV-2 587
    cervical cancer and 592
    clinical features of infections 590
    eye infections 443
    genital infections 588
    genome 604
    latency 601
    prevalence 588
    transforming DNA sequences 592
  ICP4 negative mutants 789
  identification 339
  immune evasion mechanism 1092, 1093
  infection, *see* Herpes simplex viruses (HSV)^(A) infection *(below)*
  isolation 338
  latency 590–591, 789–790, 1526
    experiments 789
    mechanisms for establishment 601, 789
    as plasmids 789
    viral gene products controlling 590, 789
  latency associated transcripts (LATs) 595, 600, 789, 790, 1526, 1527
    expression and processing 602
  lytic cycle 789
    proteins expressed 597–598
    transcription regulation 598, 1524
  maturation 607
  membrane proteins 603–604
    see also glycoproteins (above)
  molecular biology **593–603**
  neoplastic transformation 592
  neuroinvasiveness 1080
  olfactory pathway of entry 909
  oligosaccharides linked 604, 605
  pathogenicity 589
  penetration 606
  persistence at peripheral sites 591
  as persistent infection 1090
  physical properties 600
  post-translational processing 602
  proteins 597–600, 604
    α0 protein 598
    α4 transcript 598
    α27 protein 598
    βDNA-binding protein (U_L29) 599, 601, 1521
    β_γ proteins (U_L46, 47) 598–599
    capsid assembly 599
    in cytopathology 599–600, 608
    DNA replication 599
    functional classification 598–600
    general properties 597–598
    in host response 599, 608
    ICP4 1522
    ICP8 1521
    nomenclature 597
    scaffolding (U_L26) 602
    structural 599
    transcriptional regulators 598–599
    VZV homology 1519, 1521, 1522
  reactivation of latent virus 587, 589, 590, 790
    LATs role 790
    mechanism 602
  as research model, advantages 603
  restriction endonuclease cleavage patterns 588, 594
  serologic relationships 588
  shedding 588, 589
  structure and properties 593–594
  syn- mutants 608
  taxonomy and classification 587
  tegument 587, 593
  thymidine kinase 789
    acyclovir phosphorylation 43, 592
    cytopathogenic role 599
  transcription 600–602
    delayed early (β) transcripts 601
    early transcripts 600–601, 1522
    late transcripts 601
    latent phase transcripts 601–602
    properties of transcripts 594–595
    regulatory proteins 598–599, 1522
    splicing of transcripts 595, 1524
    transcript encoding U_L15 ORF 595
    VZV homology 1524
  transcription factors 597, 598, 1522
    αTIF 597, 598, 601, 1524
    Oct-1 and CFF role 597, 598
  transcription map 596
  transcription unit 595, 597
  transcriptional activators 598, 1524
    αTIF (U_L49) 597, 598, 601
  translation 594, 602, 1525
  transmission and tissue tropism 401, 589
  tree shrew herpesvirus relationship 1490
  type 1, gene in recombinant VZV 1515
  unrestricted mutants 608
  as vaccine vector 603
  vaccines 592, 593
    vaccinia recombinants 592, 593
Herpes simplex viruses (HSV) infection
  after transplantation 1484
    clinical features and treatment 1484
    epidemiology and pathogenesis 1484
  antiviral efficacy
    acyclovir 43, 592–593
    ganciclovir 44, 46
  clinical features 589–590
  complications 590
  disseminated disease 1484
  during pregnancy 588, 591
  encephalitis 589, 590, 911, 1484
  eye infections 443, 588, 590
  of heart 965
  hepatitis 1484
  immune responses 591–592
  neonatal 591
  pathology 591
  pharyngitis 1223
  prevention and therapy 592–593
  recurrent, prostaglandin role 592
Herpes zoster, *see* Zoster
Herpesviridae 155, 156, 404, 1398, 1409, 1514
  Alphaherpesvirinae 155, 156, 1514
    Simplexvirus, herpes simplex viruses 587
    see also Alphaherpesvirinae
  Betaherpesvirinae 304, 307
    cytomegaloviruses 293
    equine herpesviruses 424
    features 293
  bovine 156, 157
  characteristics 442, 587, 609, 614
  equine herpesviruses 424
  evolution 470, 473
    mammalian/birds relationship 473
  eye infections 442–444
  Gammaherpesvirinae 155, 156, 157
    amphibian 40–42
    baboon and chimpanzee herpesviruses 609, 610
    features 614
    gamma-1, features 614, 615
    gamma-2, features 614, 615
    herpesvirus saimiri and herpesvirus

    - teles 614
  - herpesvirus sylvilagus (HVS) 621
  - *Lymphocryptovirus* 610
    - Epstein–Barr virus (EBV) 404
  - lymphocyte invasion and persistence in 609
- latent infections 788
- Marek's disease virus 833
- structure 442–443, 1583
- tree shrew herpesviruses 1489
- *see also* Herpesviruses(A); *individual viruses*

Herpesvirus 4(A), *see* Epstein–Barr virus (EBV)(A)

Herpesvirus 5(A), *see* Human cytomegalovirus (HCMV)(A)

Herpesvirus 6(A), *see* Human herpesvirus 6 (HHV-6)(A)

Herpesvirus 7(A) 625

Herpesvirus ateles(A) **614–621**
- EBV relationship 614, 617
- epidemiology 617
- evolution 617
- genome structure 616
- geographic and seasonal distribution 615
- history 614
- host range and viral propagation 615
- *in vitro* immortalization of lymphocytes 615, 619
- infection
  - clinical features 619, 620
  - immune response 620
  - pathology and histopathology 620
  - prevention and control 620
- pathogenicity 619
- serologic relationships and variability 617
- structure 614
- taxonomy and classification 614
- transmission and tissue tropism 619

Herpesvirus pan(A) 610–614
- *see also* Herpesviruses(A), baboon and chimpanzee

Herpesvirus papio(A) 610–614
- *see also* Herpesviruses(A), baboon and chimpanzee

Herpesvirus saimiri(A) **614–621**
- antibodies 620
- EBV relationship 614, 617
- epidemiology 617, 619
- evolution 617
- future perspectives 621
- genome structure/properties 615–617
  - H-genome 615
  - herpesvirus ateles differences 616
  - immortalization *STP-A11* ORF 616–617, 619
  - immortalization/oncogenicity sequence 616, 618
  - M-genome 615–616
  - sequencing 616
  - STP-C488 617, 619, 621
  - in tumor and transformed cells 616
- geographic and seasonal distribution 614–615
- history 614
- host range and viral propagation 615
- *in vitro* immortalization of lymphocytes 615, 619
- infection
  - clinical features 619–620
  - immune response 620
  - pathology and histopathology 620
  - prevention and control 620
- nononcogenic variants 616
- pathogenicity 619
- protein, STP-C488 617
- serologic relationships and variability 617
- structure 614
- taxonomy and classification 614
- transmission and tissue tropism 619
- vaccines 620
- as vector for gene expression 621

Herpesvirus salmonis(A) 470, 476

Herpesvirus sylvilagus (HVS)(A) **621–624**
- epidemiology and transmission 623
- future perspectives 624
- genetics and evolution 623
- genome structure and properties 621–622
  - partial restriction map 622
- geographic and seasonal distribution 623
- history 621
- host range and viral propagation 623
- infection
  - clinical features 623–624
  - immune response 624
  - pathology and histopathology 624
  - prevention and control 624
- pathogenicity 623
- physical properties 622–623
- proteins, properties 622
- replication 623
- serologic relationships and variability 623
- structure and properties 621
- taxonomy and classification 621

Herpesvirus of turkeys (HVT)(A) 833
- distribution and epizootiology 834–835
- neoplastic transformation absent 834
- properties and classification 833
- in vaccine against Marek's disease 833
- *see also* Marek's disease virus (MDV)(A)

Herpesviruses(A)
- amphibian, *see* Amphibian herpesviruses
- antivirals 43, 44, 47
- baboon and chimpanzee **609–614**, 1514
  - antigens 611, 613
  - cell cultivation 610
  - clinical features of infection 612
  - EBV relationship 609, 611, 613
  - epidemiology 611–612
  - evolution 611
  - future perspectives 613–614
  - genetics 611
  - geographic and seasonal distribution 610
  - history 609–610
  - host range and viral propagation 610–611
  - immune response to 613
  - pathogenicity 612
  - pathology and histopathology 612–613
  - prevention and control 613
  - serologic relationships and variability 611
  - Simian T-lymphoma virus-1 (STLV-1) association 611, 612, 613
  - taxonomy and classification 610
  - transmission and tissue tropism 611, 612
- fish, *see* Fish herpesviruses(A)
- genes 304
- infections after transplantation 1478–1484
- inside nuclei 233
- kidney infections 972
- latent, interferon clinical use/action 738
- transformation mechanism 1493
- tree shrew, *see* Tree shrew herpesviruses(A)
- as vectors 1535

Herpetic whitlow 590

Heteropolyploidy 525, 529

Hexons 15

Hibiscus chlorotic ringspot virus (HCRSV)(P) 218, 219

Hibiscus latent ringspot virus(P) 904

High frequency recombination 524
- *see also* Genetic reassortment

Highlands J virus(A) 418, 419
- epidemiology 420

*Hind*III restriction enzyme 262, 290

Hirame rhabdovirus(A) 476

Histones
- murine polyomavirus 1135, 1140
- SV40 1323

History of immunology 696–697
- cell-mediated immunity 708

History of virology **627–648**
- animal virology 630–632
  - biochemistry 630–631
  - cultivation 630
  - impact on immunology 631–632, 696–697
  - structural studies 631
  - tumors 631
  - vaccines and disease control 632
- bacteriophage, *see* Bacteriophage
- chemical composition of viruses 628–629
- earliest record of virus diseases 1115
- early investigations 627–628
- enteroviruses, *see* Enteroviruses
- insect viruses 632
- origins of term 'virus' 627
- physical studies of viruses 628
- plant virology 632–633
  - tobacco mosaic virus (TMV) 632–633
  - transmission 633
  - types of viruses 633
- poliovirus, *see* Poliovirus
- taxonomy and nomenclature 633
- *see also specific viruses*

HIV-1, *see* Human immunodeficiency virus (HIV)(A)

HLA antigen
- in diabetes 970, 971
- HPV infection association 1020, 1026

HLA-DQ loci, sequence variation in diabetes 65

Hodgkin's disease
- clinical features 408
- EBV association 1495
  - EBV gene expression 408
  - role 407
- varicella–zoster virus infection 1483

Hodgkin's-like disease, tree shrew herpesvirus isolation 1491, 1492

Hog cholera virus (HCV)(A) 175, 178, **649–654**
- antibodies 653
- antigens 651, 652
- economic significance 653, 654
- epidemiology 651
- future perspectives 654
- genetics and evolution 651
- genome structure and properties 649
- geographic and seasonal distribution 650
- glycoprotein expressed in pseudorabies virus 1179
- history 649
- host range and viral propagation 650–651
- infection
  - clinical features 652
  - congenital 651, 652
  - diagnosis 653, 654
  - immune response 653
  - pathology and histopathology 652–653
  - prevention and control 653–654, 654
- pathogenicity and virulence 652, 654
- physical properties 649–650
- proteins 649
- serologic relationships and variability 651
- structure and properties 649

Hog cholera virus (HCV)(A) *continued*
  taxonomy and classification 649
  transmission and tissue tropism 651–652
  vaccines and vaccination programs 653, 654
Holliday junctions 88–89
Homogeneously staining regions (HSRs) 938
Honey bee viruses(I) **654–660**
  acute paralysis virus (APV) 657–658
  Apis iridescent virus 660
  Arkansas bee virus (ABV) 660
  bee viruses X and Y 659
  Berkeley bee virus 660
  black queen cell virus (BQCV) 658–659
  chronic paralysis virus (CPV) and chronic paralysis virus associate 655–657
  cloudy wing virus 659
  deformed wing virus (DWV) 658
  Egypt bee virus 658
  filamentous virus 659–660
  future perspectives 660
  history and general characteristics 654–655
  Kashmir bee virus 656, 658
  physicochemical properties 655, 656
  sacbrood virus (SBV) 657
  slow paralysis virus 660
Hop latent viroid(P) 1564
Hop latent virus (HopLV)(P) 215
Hop mosaic virus (HopMV)(P) 215
Hop stunt viroid (HSVd)(P) 1564, 1566
  host range and transmission 1566
Hop trefoil cryptic virus 1 (HTCVi)(P) 277
Hordeivirus(P) 661, 1400
Hordeiviruses(P) **661–664**
  cytopathology 661
  economic importance 663
  genome structure and organization 661–662
  host range 663
  molecular biology 662–663
  pathogenicity 663
  physical properties 661
  structure 1582
  taxonomy and classification 661
  as vectors 663–664
  see also Barley stripe mosaic virus (BSMV)(P)
Horizontal transmission, *see* Transmission of viruses
Hormone-response elements (HRE) 858
  mouse mammary tumor virus (MMTV) 858, 859
Hormones
  effect on infection outcome 1083
  transcription initiation stimulation 859
Horse inoculation test 435
Horseradish peroxidase 341, 344
Horses
  abortions 423, 427, 763, 769
  Borna disease virus (BDV) infection 149, 151, 152
  encephalitis, *see* Equine encephalitis viruses(A)
  Japanese encephalitis virus infection 747
  orbivirus infections 942, 950, 952
  toroviruses (Berne virus) 1452, 1457
Host
  adaptation, in speciation 438
  cell killing 1590–1591
  cell structure, *see* Cell structure/function in virus infections
  chromosomal genes effect on viral replication 1467
  cytoskeleton
    viral infection effect on 1589–1590
    see also Cytoskeleton
  defensive response 1591
    see also Host defenses
  DNA 232
    in transduction 108, 109, 1375
  DNA degradation 1588
    T4 phage 1377
    T5 phage 1386
  DNA synthesis inhibition by viruses 1205, 1588
    N4 bacteriophage 893
    reoviruses 1191, 1203
    SPO1 bacteriophage 1356
    T1 bacteriophage 1372, 1373
  effects of virus infection on 1588–1590
    DNA and RNA synthesis 1588
    host cell morphology and viability 1588
    inclusion bodies, *see* Inclusion bodies
    macromolecular synthesis 1588
    membranes and cytoskeleton 1589–1590
    protein synthesis 1588–1589
    transformation, *see* Transformation
    see also Cytopathic effects (CPE)
  factors, determining infection outcome 1082–1083
  functions, in replication of plant viruses 1105
  gene acquisition from, evolution and 439
  interactions with virus **1587–1591**
    polydnaviruses significance 1134
    types of infections 1587–1588
  lipid metabolism change 1559
  macromolecular shutdown, hepatitis A virus not associated 549
  membranes, viral infection effect on 1589–1590
  mRNA synthesis shutdown, frog virus 3 (FV3) 504, 506
  'phylogenetically contained' virus groups 441
  protein, p220 491
  protein synthesis, failure to shut down, nonoccluded baculoviruses 138
  protein synthesis inhibition 1205, 1588–1589
    cardioviruses 210, 491
    entero- and rhinoviruses 491
    foot and mouth disease viruses 491
    in interference 728
    polioviruses 1119, 1123
    SPO1 bacteriophage 1356
    T1 bacteriophage 1371
    T4 phage 1379
    viroids 1567
  resistance to viruses 1591
    plant viruses, *see* Plant resistance
    see also Host genetic resistance
  response, viral infections 665
  severity of viral diseases relationship 665
  suicide of cell? 1590–1591
  viral gene relationship 440
  viral infection beneficial, yeast killer system 1467
  viral species co-evolution 439, 440
  virus spread in 1078–1080
  viruses as tools for probing 1591
Host defenses 697, 1076, 1077, 1591
  clearance of virus particles 1078
    factors affecting 1079
  DNA methylation role 13
  respiratory 1077
  skin 1076
  virus in bloodstream 1078
Host genetic resistance **664–669**, 1082
  future perspectives on 669
  history 664–665
  organismal level, antiviral effector mechanisms 665, 667–668
  polymorphous resistance genes 665, 666, 668
    classification 665
  target cell level
    constitutive expression at 665, 667
    inducible expression 665, 667
  unknown mechanisms 668
  see also Gene(s); Plant resistance
Host range
  plant viruses 1106
  see also individual viruses
Host-controlled modification and restriction 111, **669–673**
  antirestriction mechanisms by phage 671
  biological function 671
  classification of systems 670–671
    class I enzymes 670
    class II restriction enzymes 670
    class III 670–671
  cofactor requirements 670
  DNA methylation 669–670
  Hsd system 670, 671
    Rgl system differences 672
  methylation-dependent restriction 672–673
  nonclassical restriction/modification 673
  restriction 669
    of nonglucosylated HMC-containing DNA 671–672
  Rgl system 672
  T-even phage 1383
    T1 phage 1375
Hostvector system, transduction application 112
HPMPC (antiviral agent) 44, 46
Hpochoeris mosaic virus (HMV)(P) 509
HPV, *see* Human papillomaviruses (HPV)(A)
hsd genes 669
Hsd systems 670, 671
HTLV-1, *see* Human T-cell leukemia virus type 1 (HTLV-1)(A)
Human caliciviruses (HCV)(A) 375
Human carcinoma cell line (HEp-2) 337
  cytopathic effects of viruses 337
Human cytomegalovirus (HCMV)(A) **292–304**
  AD169 strain 294, 299
  in AIDS 702, 1479
  dense bodies 303–304
  detection/diagnosis 1479
    shell vial technique 1479
  DNA replication 299, 302
  DNA-binding phosphoproteins 300
  DNA-binding protein (DB140; UL57) 299
  epidemiology 294–295
  evolution 294
  future perspectives 298
  genetics 294
  genome structure 294, 299
    cellular gene homologs 299
    sequencing 294, 299
  geographic and seasonal distribution 293
  glycoproteins 301–302
  history 292–293
  host range and virus propagation 293–294
  human herpesvirus 6 (HHV-6) relationship 625
  infection, *see* Human cytomegalovirus (HCMV) infection *(below)*
  latent infection 295, 1091
  molecular biology **299–304**
  noninfectious enveloped particles (NIEPs) 303

pathogenicity 296
as persistent infection 1090, 1091
physical properties 302
proteins 294, 299–300
basic phosphoprotein (BPP) 300
capsid and tegument 300–301
early (β) 299, 300, 302
envelope 301–302
IE2 300
immediate-early (α) 299, 300, 302
integral membrane protein (IMP) 302
late (γ) 299, 302–303
MIEP (IE1: UL123) 300
nonvirion 300
virion 300–301
reactivation 295, 296
replication 302–304
animal cytomegaloviruses and 305, 306
assembly and release 303
DNA 299, 302
post-translational modification 303–304
transcription 302–303
translation 303
strains 293, 294
structure and composition 299
taxonomy and classification 293
tissue tropism 295
transmission 294, 295
blood transfusions 293, 295, 296
neonatal/congenital 294, 295
organ transplantation 295, 1478, 1479
vaccines 298
see also Cytomegaloviruses (CMV)[A]
Human cytomegalovirus (HCMV) infection
after transplantation 1478–1481
cardiac 966, 969
clinical aspects 1479
diagnosis 969, 1479
immunomodulation 1481
incidence 1478
pathogenesis 1478–1479
pathology 1480
prophylaxis 1479, 1481
re-activation 1478
risk of CMV disease 1478
treatment 1479
antivirals 46, 298, 1479, 1481
clinical features 296–297
congenital 444
diagnosis 975–976
diarrhea in 377
eye infections 444
frequency and respiratory disease types 1223
of heart 965
hepatitis 1479
immune response 296, 297, 1478
incubation period 296
intravenous immunoglobulin 1479, 1481
of kidney 972–973 972
metabolism of infected cells 303
models of 310–311
mortality and morbidity 296
pancreatic 969
pathology and histopathology 297, 303
pneumonia 1225
pneumonitis 1479
prevention and control 297–298, 1481
primary 296
secondary 296
serological diagnosis 296, 297
subclinical 293
viremia 295
Human diploid fibroblasts (HDF) 337
cytopathic effects of viruses 337
Human enteric coronaviruses (HECV)[A] 255
Human foamy virus (HFV)[A] 480, 481
pathogenicity 487
simian foamy virus (SFV) relationship 484
Human herpesvirus 4[A], *see* Epstein–Barr virus (EBV)[A]
Human herpesvirus 5[A], *see* Human cytomegalovirus (HCMV)[A]
Human herpesvirus 6 (HHV-6)[A] **624–626**
in AIDS patients and HIV carriers 624, 626
epidemiology and pathogenicity 626
genome structure and properties 305, 625
history 624–625
host range and viral propagation 625
human cytomegalovirus relationship 625
infection
clinical features 625–626
prevention and control 626
as persistent infection 1090
proteins, properties 625
structure and properties 625
transmission 626
Human herpesvirus 7 (HHV-7)[A] 625
Human immunodeficiency virus 1 (HIV-1)[A]
antivirals 48–49
cardiomyopathy in infancy 965
DNA integration mechanisms, similar to Mu 876
dominant-negative mutants as antivirals 731
genetic diversity 679
genome sequence, bovine immunodeficiency virus (BIV) similarity 160
gp41 protein, astrocyte cross-reaction 66
HTLV-1 infection consequences 694
origin and geographic distribution 679
prevalence in blood donors 688
prototype (LAV LAI) 680
transmission 165
vpu gene 439
see also Human immunodeficiency virus (HIV)[A]
Human immunodeficiency virus 2 (HIV-2)[A] 674, 1317
AIDS and progression to 679
genome sequence 679
host range and viral propagation 680
origin and geographic distribution 679
serologic relationships and variability 679, 680
vpx gene 439, 679
Human immunodeficiency virus (HIV)[A] **674–681**
antigenic variation 679, 701, 707
attachment and receptor 1557
CD4 molecule binding 48, 678, 707, 1557, 1560
cell damage mechanism 911
cross-reaction with bovine immunodeficiency virus 164
DNA synthesis and integration 675–676
envelope glycoproteins 678
variability 679, 701, 707
evolution
genes 439
patterns of relatedness 438
gene expression regulation 676–677
shift from early to late 676–677
genetic variability 679
HTLV-1 comparison 690
genetics 679
genome structure 675, 1318
gag, *env* and *pol* genes 675
gp41 66, 674, 675
gp120 1557, 1560
gp160, immunization with 681, 707
history 674
host range and viral propagation 679–680
immunosuppression 701, 708
infection, *see* Human immunodeficiency virus (HIV) infection *(below)*
long terminal repeat (LTR) 1081
morphogenesis and assembly 675
mouse models 48
mRNA processing and Rev system 676–677, 1208
origin and geographic distribution 679
p24 (CA antigen) 164, 675, 677
physical properties 675
protease
atomic structure 1572
selective inhibitors 49
'quasispecies' 679
receptors 1557, 1560, 1561
reverse transcriptase 678
error rate 679
reverse transcription 675
serologic relationships 680
simian immunodeficiency virus (SIV) relationship 1318, 1319
structural proteins 677–678
env proteins 678
Gag and Pol 49, 675, 676, 677–678
Nef protein 678
nucleocapsid protein (p9) 677–678
transmembrane protein 678
vif gene product 678
vpr gene product 678
vpu gene product 439, 678
vpx gene product 439, 678
structure and composition 674–675
spikes 674
susceptibility and sexual differences in 1083
tat protein 676
trans-activation of JC virus 755
taxonomy and classification 674
thyroid hormones 976
tissue tropism 1081
$CD4^+$ cells 674, 680
transcription 675, 676
negative regulatory element (NRE) 676
regulation and Tat system 676
transactivation HIV-2 676
transcriptional activators 676
transmission 401, 680, 681
types
see Human immunodeficiency virus 1 (HIV-1)[A]
see Human immunodeficiency virus 2 (HIV-2)[A]
vaccine
baculovirus expression system 1534
development and animal models 1506
vectors based on 1532
visna–maedi viruses as models 1594
see also AIDS
Human immunodeficiency virus (HIV) infection
antivirals 45, 48–49, 681
evaluation 48
strategies 48
autoimmune component to pathology 65
clinical features 912
enhancing antibodies in 702
enteropathy 377
eye in 447
heart in 965, 966
human papillomavirus infections 1018
immune response 680, 701

Human Immunodeficiency virus (HIV) *continued*
cell-mediated 680, 707–708
interferon-α action 740
in vitro data 741
kidney in 973
myocardial damage 966
opportunistic infections 680–681, 708
prevention and control 681
see also AIDS
Human leucocyte antigen (HLA), *see* HLA antigen
Human lymphotropic virus(A), *see* Human herpesvirus 6 (HHV-6)(A)
Human monkeypox virus(A), eye infections 442
Human papillomavirus type 1 (HPV1)(A) 1016, 1018
Human papillomavirus type 2 (HPV2)(A) 1018
Human papillomavirus type 3 (HPV3)(A) 1018
Human papillomavirus type 5 (HPV5)(A) 1014, 1015
skin carcinomas 1014, 1018, 1024, 1497
Human papillomavirus type 6 (HPV6)(A) 1018
genital warts 1018, 1021
tissue tropism 1018, 1023
variants 1015
Human papillomavirus type 8 (HPV8)(A) 1497
Human papillomavirus type 11 (HPV11)(A) 1014, 1015, 1017, 1018
genital warts 1018, 1021
tissue tropism 1018, 1023
Human papillomavirus type 16 (HPV16)(A) 1014, 1015
cervical cancer and 1014, 1017, 1019, 1021, 1024, 1496
incidence 1024, 1496
evolution 1016, 1022
immune response to 1020
variants 1015, 1016
Human papillomavirus type 18 (HPV18)(A) 1014, 1019, 1024
cervical cancer and 1021, 1496
immune response to 1020
Human papillomaviruses (HPV)(A) **1013–1026**
carcinogenicity 1023–1025, 1476, 1497
host cell control 1025
p53 and pRb association 1477, 1497
transformation mechanism 1493, 1497, 1498
see also Human papillomaviruses (HPV)(A), infection *(below)*
DNA
homology 1014
integration into host 1024, 1497
structure 1022
E6 and E7 genes/proteins 1019, 1024, 1025, 1476, 1497
in carcinogenesis 1476, 1497
epidemiology 1017–1018
anogenital types 1023
evolution 1015–1016, 1022
factors affecting oncogenic potential 1018, 1023
future perspectives 1021, 1026
general features **1013–1021**
genetics 1015, 1022
genotypes 1014, 1017
geographic and seasonal distribution 1014, 1022
history 1013–1014, 1021
host cell regulation of genes 1025
host range and viral propagation 1014–1015, 1022
immortalization 1025
life cycle 1018, 1476
as persistent infection 1090, 1497
phylogenetic tree 1014, 1016
plurality 1014
replication 1020, 1022
in vitro systems needed 1021, 1022, 1023
serologic relationships and variability 1016–1017
taxonomy and classification 1014, 1021–1022
tissue tropism 1016, 1018
anogenital types 1023
transmission 1018, 1023
Human papillomaviruses (HPV)(A) infection
anogenital 1019, **1021–1026**, 1496–1497
high-risk 1022
HPV types associated 1014, 1017, 1021, 1022, 1496
low-risk 1022
pathology and histopathology 1020, 1025
risk 1017
warts 1018, 1021, 1023–1024, 1025
see also cervical cancer (below)
antivirals for 45
cancers associated 1015, 1018, 1019, 1021
see also under Human papillomaviruses (HPV)(A) *(above)*
cervical cancer 1014, 1017, 1018, 1019, 1476
development 1024–1025
clinical features 1019
cutaneous 1016, 1017, 1497–1498
evolution 1016, 1022
transmission 1018
diagnosis 1022–1023
eye 444
HLA association 1020, 1026
immune response 1020, 1026
interferon clinical use/action 738, 1021
laryngeal carcinoma and 1497
latency before cancer development 1025, 1499
latent infections 788
mucosal 1016
oral disease 1019, 1497
transmission 1018
pathogenicity 1015, 1018–1019
anogenital types 1023–1025
pathology and histopathology 1019–1020, 1022
anogenital lesions 1025
prevention and control 1020–1021, 1026
treatment 738, 1021
types associated with cancers 1018–1019, 1021, 1022
Human parainfluenza viruses(A), *see* Parainfluenza viruses (PIV)(A), human
Human parvovirus B14(A), fetal infection 964
Human parvovirus B19(A)
epidemiology 1053
gene expression 1058
geographic and seasonal distribution 1052
infection
clinical features 1054–1055
pathology and histopathology 1056
in pregnancy 1054–1055
latent infections 1060
as persistent infection 1090
transmission 1054
see also Parvoviruses(A)
Human respiratory coronaviruses (HCV)(A) 255
Human respiratory syncytial virus (HRSV)(A) 1210
antigen
cross-reactivity 1214–1215
mapping 1214
subgroups 1214
antigenic variation 1214, 1216
budding 1211
cDNA-encoded analog of vRNA 1214, 1218
cellular receptor 1215
gene map 1212, 1213
genome 1213–1214
hemagglutinin absent 1211
infection
after transplantation 1487
bronchiolitis 1222
clinical features 1216
diagnosis 1216
epidemics 1220
frequency and respiratory disease types 1223
immune response 1216, 1217
immunoprophylaxis and treatment 1217–1218
incubation period 1216
mortality from 1216
pathogenesis 1216
pneumonia 1224
repeated 1216
resistance to reinfection 1217
in young infants 1216
infectious cycle 1215
neutralizing antibodies 1214, 1217, 1218
RNA replication 1213–1214
structure and proteins 1211–1213
F protein 1211, 1212, 1214
G protein 1211, 1212–1213, 1214
matrix proteins 1213
SH protein 1213
transmembrane proteins 1211–1212
taxonomy and classification 1210–1211
tissue tropism 1215
transcription 1213
transmission and spread 1216
vaccines 1217, 1218
animal model needed 1506
disease enhancement 1216, 1217, 1218
live attenuated strains 1218
recombinant/subunit 1218
see also Respiratory syncytial virus(A)
Human rhinovirus type 14 (HRV14)(A)
antiviral agents and structure of 1573
'canyon' hypothesis 1121, 1123, 1255, 1573
Human rhinovirus type 39 (HRV39)(A) 1131
Human rhinovirus type 87 (HRV87)(A) 1254
Human rhinoviruses (HRV)(A) 1253
see also Rhinoviruses(A)
Human spumavirus (HS$_R$V)(A), *see* Human foamy virus (HFV)(A)
Human T-cell leukemia virus (HTLV)(A) **682–695**
Human T-cell leukemia virus type 1 (HTLV-1)(A) 167, **682–686**
adult T-cell leukemia association 682, 685, 1496
antibodies 683, 685
bovine leukemia virus
as model 174
relationship 167, 170, 171, 682
carriers 685
detection 694–695
HTLV-2 discrimination 695
diseases associated 685–686
epidemiology 683–684
future perspectives 686
gene expression regulation 683, 691–692
trans-activating effect of Tax 683, 685, 691–692

genome structure 682–683, 1496
tax and rex genes 683, 692–693, 694, 1496
history 682, 686
host range 684
HTLV-2 serologic cross-reactivity 690
integration 685
latency period for tumors 1499
leukemogenesis 167, 685, 692, 694, 1496
molecular biology 682–683
pathogenesis *in vivo* 694
pathogenicity 684–686
prevalence in blood donors 688
prevention and control 686
pX sequence 682, 683
receptor 689
related viruses 682, 686, 687
reverse transcriptase fidelity 690
taxonomy and classification 682, 687
transformation mechanism 167, 685, 692, 694, 1476, 1496
transmission 685, 686
tropical spastic paraparesis 685, 686, 912–913
variability, HIV comparison 690
vesicular stomatitis virus (VSV) pseudotypes 689
see also Adult T-cell leukemia (ATL)
Human T-cell leukemia virus type 1 (HTLV-1)-associated myelopathy (HAM) 685, 686
Human T-cell leukemia virus type 2 (HTLV-2)[(A)] 682, **686–695**, 1496
in cell lines 689, 694
Mo T-cell line 686, 687
clinical features of infections 693–694
detection and control 694–695
HTLV-1 discrimination 695
epidemiology 687–689
gene expression regulation 691–693
Rex protein action 691, 693
tax gene 691–692
trans activating effect of Tax protein 691–692
genome structure 690–691, 690–693
env gene 690, 691
gag gene 691
long terminal repeat 689, 692
pol gene 690, 691
protease gene 691
rex gene 691, 692–693, 694
tax gene 690–691, 694
history 686–687
host range 689
integration 693
isolate (HTLV-2_{NRA}) 690, 693
leukemia and diseases associated 687, 1496
low genetic variability 689–690
molecular genetics 689–690
pathogenesis *in vivo* 694
PCR determination 688, 694
prevalence in blood donors 688
proteins 691
p19 691
Rex 692–693
Tax 691–692
receptor 689
related viruses 687
reverse transcriptase 691
serologic cross-reactivity with HTLV-1 690
T-cell line infection 689
taxonomy and classification 687
transcription 692
transcription factors 692
transforming mechanism 692, 694
vesicular stomatitis virus (VSV) pseudotypes 689
Human T-cell leukemia virus type III (HTLV-III)[(A)], *see* Human immunodeficiency virus (HIV)[(A)]
Human T-cell leukemia virus type III (HTLV-IIIB)[(A)] 680
Human wart virus, *see* Human papillomaviruses (HPV)[(A)]
Humoral response 1091
dengue viruses 330
encephalitis viruses[(A)] 366
Epstein-Barr virus (EBV) infections 409
equine encephalitis virus infection 422
herpes simplex virus infections 591
human cytomegalovirus infections 297
human papillomaviruses 1020, 1026
human parainfluenza viruses 1030
human respiratory syncytial virus (HRSV) infections 1217
influenza viruses 714
lactate dehydrogenase-elevating virus (LDV) infection 769, 770
Marek's disease virus 836
measles virus 846
mumps virus (MuV) 882
parvoviruses 1056
polioviruses 1118
polyclonal, in LDV infection 770
rotaviruses 1280
Sendai virus 1304
Theiler's viruses 1428
varicellazoster virus (VZV) 1517
vesicular stomatitis virus (VSV) 1555
see also Antibodies; Immune response; Immunoglobulin(s)
Hv-2 locus, mouse hepatitis virus type 4 (MHV4) susceptibility 667
Hybridization methods 347, 348
dot hybridization 347
in situ 347
rinderpest and distemper virus detection 1266
sandwich 347
viroid detection 1568
Hydra viridis, chlorella symbiotic with, viruses in 36
Hydrophobia, rabies 1184
Hydrops fetalis 1054, 1056
Hydroxymethylcytosine (HMC) 671, 672
in T-even phage 1376, 1383
Hydroxymethyluracil 1352, 1353, 1356
in SPO1 bacteriophage 72
(S)-1-(3-Hydroxy-2-phosphonylmethoxypropyl)cytosine (HPMPC) 44, 46
Hygiene
coxsackievirus infection prevention 273
Norwalk virus transmission prevention 932
Hypereosinophilic syndrome, interferon therapy 743
Hypergammaglobulinemia, in Aleutian disease 1056, 1066
Hyperplastic alveolar nodules (HAN) 859–860
Hypersensitive response (HR), tobamoviruses 1440
Hypertrophy, in ascovirus infection 61
Hz-1 baculovirus[(I)] 136, 137
genome 137
infection cycle 137–138
protein synthesis 138

I
Ibaraki virus[(A)] 950, 953
ICAM-1
down regulation 1094
rhinovirus receptor 1254, 1255
Ichnoviruses[(I)] 1133
see also Polydnaviruses[(I)]
ICI 130685 44, 47
Icosahedral cytoplasmic deoxyriboviruses (ICDV)[(A)] 798
Icosahedral symmetry 1210, 1571
assembly 1574–1577
atomic structure 1571–1572
capsid structure (principles) 1574–1579
capsid symmetry of individual viruses (summary) 1580, 1581
Crick and Watson hypothesis 1574
P=3 1578, 1579
viruses with 1580, 1581
quasiequivalent shells 1577–1579
T=1, viruses with 1580, 1581
T=3 250, 1576, 1578, 1579
viruses with 1580, 1581
T=4 1416, 1418, 1576
T=7 1576, 1578
viruses with 1580, 1581
T=9, viruses with 1580
T=13 1576, 1581
Icterus 552, 554
ICTV, *see* International Committee for Taxonomy of Viruses (ICTV)
Identification of viruses, *see under* Diagnostic techniques
Idiotypicantiidiotypic antibody network 699
Idoxuridine, in herpes simplex virus infections 592
Ilarvirus[(P)] 30, 1400
Ilarviruses[(P)] **30–35**
economic significance 34
epidemiology and control 34
genome structure 31, 32
host range and geographic distribution 34–35
host relationship 33
replication 32
structure and composition 31, 1580
transmission 34
see also Alfalfa mosaic virus (AlMV)[(P)]
Immune complexes 699
equine arteritis virus (EAV) infections 769, 770
formation, adverse effects 702
hantavirus infections 543, 545
hepatitis B 558
lactate dehydrogenase-elevating virus (LDV) 768, 769, 770
Immune electron microscopy (IEM) 342
Norwalk virus 926
Immune enhancement 702
dengue hemorrhagic fever/dengue shock syndrome (DHF/DSS) 328
Immune evasion, *see* Immune response, evasion
Immune response **696–709**, 931, 1091, 1092
adaptive (specific) 697–699, 1091, 1092
cytotoxic 698, 1092
delayed-type hypersensitivity 697
MHC restriction 697, 698, 703–704, 1092
see also Antibodies; Cell-mediated immunity; Humoral response
adenovirus infections 6–7, 17
adverse reactions to infections 702–703
African swine fever virus infection 29
animal enteroviruses 390
avian hepatitis B viruses 568
avian leukosis viruses 70–71
baboon and chimpanzee herpesviruses 613
birnaviruses[(A)] 147–148

Immune response *continued*
BK virus 756, 1486
Borna disease virus (BDV) infection 153–154
bovine herpesviruses 157
bovine leukemia virus 171, 173
bovine papillomaviruses (BPV) 1310
bovine parainfluenza virus type 3 (PIV-3) 1034
bovine viral diarrhea virus (BVDV) 180
caprine arthritis encephalitis (CAE) syndrome 204
capripoxviruses 1164
cardioviruses 213
cloning 929, 932
coronaviruses[(A)] 260
cottontail rabbit papillomavirus (CRPV) 1310
coxsackieviruses 639
cross-reacting 64
in diabetes 65
dengue viruses 330
diagnosis of infection 932
distemper viruses 1266
echovirus infections 359
ectromelia (mousepox) virus 865
encephalitis viruses[(A)] 366
enteric viruses 378
enterovirus (EV) serotype 70 (EV70) 382–383
enterovirus (EV) serotype 71 (EV71) 384
Epstein–Barr virus (EBV) infection 406, 409
equine arteritis virus (EAV) 770–771
equine encephalitis virus infection 422
equine herpesviruses 428
equine infectious anemia virus (EIAV) 435
evasion 700–702, 1091–1094
antibody response 700–701
antigenic variation 1093
cautions and doubts on mechanisms 1094
cell-mediated immunity 701–702
chronic infections 787
cytokine function interference 1094
down regulation of cell-surface adhesion molecules 1094
down regulation of MHC class I antigens 1094
infectious virus sanctuaries 701, 1092–1093
interferon inhibition 1094
latency 787, 1092
leporipoxviruses 1156, 1157, 1159
restricted viral gene expression 1092, 1093
sites for 701, 1092–1093
Theiler's murine encephalomyelitis virus (TMEV) 1428
see also Antigenic variation
feline immunodeficiency virus infections 458
feline leukemia virus (FeLV) 462
flaviviruses 372
foot and mouth disease viruses (FMDVs) 495
fowlpox virus (FPV) 502
future perspectives 932
gibbon ape leukemia virus (GaLV) 536
hantaviruses[(A)] 544–545
to heat shock proteins 64
hepatitis A 553
hepatitis B 558–559
hepatitis D (delta hepatitis) 579
hepatitis E 585–586
herpes simplex viruses (HSV) 591–592
herpesvirus saimiri[(A)] 620
herpesvirus sylvilagus (HVS) 624
history 696–697
animal virology impact on 631–632, 696–697
hog cholera virus (HCV) 653
human cytomegalovirus (HCMV) 296, 297, 1478
human immunodeficiency virus (HIV) 680, 701
human papillomaviruses (HPV) 1020, 1026
human parainfluenza viruses 1030
human respiratory syncytial virus (HRSV) 1216, 1217
influenza viruses 699–700, 714, 727
Japanese encephalitis virus 750
JC and BK viruses 756
Junin and Machupo viruses 785
in kuru 761
lactate dehydrogenase-elevating virus (LDV) 770
Lassa virus 784–785
lentiviruses 204
lymphocytic choriomeningitis virus (LCMV)[(A)] 804–805
Marburg and Ebola viruses 831
Marek's disease virus[(A)] 836
measles virus 846
molluscum contagiosum virus (MCV) 852
monkeypox virus 1345
morbilliviruses 1266
mumps virus (MuV) 882
murine cytomegalovirus (MCMV) 311
Newcastle disease virus 918
nonadaptive (nonspecific) 697
orbiviruses and coltiviruses 954
parapoxviruses 1041
parvoviruses 1056, 1067
polioviruses 1117–1118
poxviruses of rabbits, hares, squirrels and swine 1159
prevention and control 932
primary 1222
rabbitpox virus 867
rabies virus 1185
respiratory viruses 1222–1223
retroviruses, type D 1242
rhinoviruses 1258
rinderpest virus 1266
Ross River virus (RRV) 1273–1274
rotavirus 1279–1280
rubella virus 1297, 1298
simian immunodeficiency viruses (SIV) 1320, 1321
simian virus 40 (SV40) 1329
smallpox (variola) virus 1342
spongiform encephalopathy agents 1360
tanapox virus 1599–1600
Theiler's murine encephalomyelitis virus (TMEV) 1428
tick-borne encephalitis virus (TBEV) 372
toroviruses 1458
treatment of infection 932
vaccines, *see* Vaccines
vaccinia virus 1513
varicella–zoster virus (VZV) 1517
vesicular exanthema virus of swine (VES) 1546
vesicular stomatitis virus (VSV) 1555
to viral infections 699–700, 706
infection route effect 700
primary 699–700
secondary 700
yabapox virus 1599
yellow fever 1611–1612
see also Cell-mediated immunity; Humoral response; Immunity
Immune serum globulin, administration in hepatitis A 553
Immune spleen cells, in ectromelia (mousepox) virus infection 865
Immune surveillance 704, 708
evasion, EBV 409, 415
see also Immune response, evasion
Immune system cells, virus infections, coronaviruses 260
Immune-mediated tissue injury
Theiler's murine encephalomyelitis virus (TMEV) 1428–1429
see also Delayed-type hypersensitivity (DTH)
Immunity
bacteriophage as agent of (Bordet's theory) 646, 648
effect on survival of viruses 402
'herd' 402
nonhost (in plants), *see* Plant resistance (to viruses)
see also Immune response
Immunization
after transplantation 1488
'intracellular' 731
live vaccines, avoidance in transplant patients 1488
passive, transplant patients 1488
principles and Jenner's work 261
response in transplant patients 1488
routes, protection after 1505–1506
transplant patients 1488
in zoonoses control 1621
see also Vaccine vectors; Vaccines
Immunization challenge test, rabies-like viruses 1188
Immuno-enzyme staining, viral antigen assay 344
Immunoassays, viral antigen detection 343–344
Immunocompromised patients 702
adenovirus infections 6, 7
CMV causing eye infections 444
cutaneous and genital warts in 1017
herpes simplex virus infections 590
influenza in 713
kidney infections 972, 973, 974
VZV infections 1516, 1517
Immunodeficiency
echovirus infection 359
'enteric' adenovirus infections in 375
lymphomas in 406–407, 408
severe combined (SCID) 1506
see also T cells, deficiency
Immunodominant alleles 705
Immunofluorescence 339, 341
antibody-staining procedure, for hepatitis E virus antigen 584
direct (DIF) 339, 344
indirect (IIF), *see* Indirect immunofluorescent antibody procedure (IFA)
respiratory viruses 1225
viral antibody assay 346
hepatitis E virus 585
viral antigen assay 344
Immunoglobulin(s)
administration in hepatitis A 553
passive, polioviruses 1118
super gene family, cellular receptor for poliovirus 1122
varicella–zoster virus (VZIG) 1516, 1517, 1517–1518
see also Antibodies

Immunoglobulin A (IgA)
  coxsackievirus infections 273
  hepatitis A 553
  influenza viruses 714
  measles virus 846
  mumps virus (MuV) 882
  in nasopharyngeal carcinoma 1496
  production after immunization 1504, 1505
  rotavirus infections 1280
  secretory 700, 1505
Immunoglobulin G (IgG) 345
  avidity assays 346
  in central nervous system (CNS) 908
  Chikungunya, O'nyong nyong and Mayaro viruses 240
  dengue virus infection 330
  Fc receptors, HSV glycoproteins 599, 608–609
  hepatitis A 553
  hepatitis E 585
  influenza viruses 714
  Lassa virus 785
  rubella virus 1296, 1297
  Theiler's viruses 1428
  varicella–zoster virus (VZV) 1517
  viral, detection 345–346
  yellow fever virus 1611
Immunoglobulin M (IgM)
  antibody capture ('μ-capture') 346, 1621
  bovine leukemia virus infection 173
  in central nervous system (CNS) 908
  Chikungunya, O'nyong nyong and Mayaro viruses 240
  coxsackievirus infections 273
  dengue virus infection 330
  hantavirus-specific 544, 545
  hepatitis A 553
  hepatitis E 585
  herpes simplex viruses 592
  Japanese encephalitis virus 750
  Lassa virus 785
  rubella virus 1296, 1297
  viral 345
    detection 346
  yellow fever virus 1611
Immunological memory 698, 707
Immunological tolerance 632, 701, 703, 708
  LCMV role 801
Immunopathology
  Borna disease virus (BDV) infection 153, 154
  T cell-mediated 706–707
Immunoperoxidase (IP) tests 341
Immunosuppression
  baboon and chimpanzee herpesviruses causing 612, 613
  birnaviruses causing 148
  cytomegalovirus infection after 295, 296
  evasion of immune response by 701, 702
  feline leukemia virus infection 462
  hog cholera virus (HCV) causing 653
  lactate dehydrogenase-elevating virus (LDV) inducing 769, 770
  lymphocytic choriomeningitis virus (LCMV) 804
  reticuloendotheliosis viruses causing 1231
  simian immunodeficiency virus (SIV) causing 1320
  Theiler's murine encephalomyelitis virus (TMEV) demyelination reversal 1428
  in transplantation 1477–1478
    CMV infections 1478–1481
    EBV infections 1482
    see also Transplantation
  type D retroviruses 1236, 1241, 1242
Immunosuppressive therapy
  JC and BK infections in 755
  vaccination and 1488
Impatiens necrotic spot virus (INSV)(P) 1459, 1460
  economic significance 1463
  genome 1461
  see also Tospoviruses(P)
In situ hybridization 347
Incidence rate, definition 399
Inclusion bodies (inclusions) 1590
  acidophilic (A-type ATI) 262, 263, 265
    ectromelia (mousepox) virus 864
  adenovirus infections 6
  amorphous cytoplasmic (AI), potyviruses 1152
  amorphous X bodies 452
  B-type (Guarnieri bodies) 1156
    rabbitpox 867
    vaccinia virus infection 1510
  beet yellows virus (BYV)-type 246
  caulimoviruses 223, 225
  closterovirus infections 246
  cytoplasmic
    carlavirus infections 217
    ectromelia (mousepox) virus 863–864
    fowlpox 496, 497, 502
    nepoviruses 906
    potyviruses 1151
    reoviruses 1202
  dense granular (DGI), maize chlorotic dwarf virus 826
  dianthovirus infections 353
  direct observations by light microscopy 339, 341–342
  eosinophilic 265
    poxviruses 1156
  furoviruses 514
  honey bee viruses 655
  human cytomegalovirus infections 296, 297
  iridoviruses 1433
  laminated (LIC) 1146
  molluscum contagiosum virus (MCV) infection 851
  nuclear, potyviruses 1152
  polyhedra, *see* Polyhedra
  potexviruses 1146
  potyviruses 1151, 1152
  proteinaceous, *see* Proteinaceous inclusion bodies
  tenuiviruses 1415
  tobamoviruses 1439
  varicellazoster virus (VZV) infections 1526
  yabapox virus infections 1598
Inclusion body myositis, mumps virus (MuV) 883
Indian bunchy top viroid(P) 1564
Indian cassava mosaic virus(P) 520
Indian goat dermatitis 1160
Indirect immunofluorescent antibody procedure (IFA) 339, 344
  dengue virus detection 327
  hantavirus detection 544–545
Indoleamine 2, 3-dioxygenase 736
Indonesian soybean dwarf virus(P) 792
Industrial fermentations, bacteriophage in **116–121**
  acetone and butanol synthesis 119
  contamination sources 117
  control of phage 119–120
    antiphage chemicals 119–120
    judicial starter use 119
    phage-inhibitory media 119–120
    sanitation 120
  dairy industry 117, 118
  detection of phage 117–118, 120
  eucaryotic protein synthesis 118–119
  factors predisposing to phage problems 116–117
  future perspectives 121
  industries affected by problems 118–119
  phage mutation 118
  phage-resistant strains 119, 120–121
  scale-up of starter volume difficulties 117
  starter failure 117, 118
Infection routes 1076, 1077
  see also Pathogenesis of viral infections
Infectious bovine rhinotracheitis (IBR) 155, 157
Infectious bulbar paralysis 1173
  see also Pseudorabies virus (PRV)(A)
Infectious bursal disease of chickens virus (IBDV)(A) 143
  antigenic structure 147
  cell tropism 145
  cytopathology 148
  gene products 146–147
  genome segment B, product 147
  genome structure 146–147
  host range and epidemiology 144
  in vitro replication 146
  incomplete particles 146
  infection
    clinical features and pathology 145
    immune response 147–148
    laboratory diagnosis 148
    prevention and control 148
  pathogenic properties 145
  serotypes 147
  stability 144–145
  taxonomy and classification 144
  VP2, VP3 and VP1 proteins 147
  VP4 protein 146, 147
Infectious canine hepatitis 17
Infectious ectromelia 861
  see also Ectromelia virus(A)
Infectious hematopoietic necrosis virus (IHNV)(A) 476, 1250, 1251–1252
  clinical features and pathology 1252
  history 1251
  molecular aspects 1252
  transmission 1252
Infectious mononucleosis (IM) 404, 1224
  clinical features 408
  EBV gene expression in 408
  epidemiology 406
  fatal 1495
  immune response 409
  see also Epstein–Barr virus (EBV)(A)
Infectious pancreatic necrosis virus of fish (IPNV)(A) 143, 475, 476–477
  antigenic structure 147
  clinical features of infection 145
  host range and epidemiology 144
  in vitro replication 145, 146
  laboratory diagnosis 148
  pathogenic properties 145
  prophylaxis and control 148
  serotypes 147
  stability 144
  taxonomy and classification 144
Infectious pustular balanoposthitis 155
Infectious pustular vulvovaginitis (IPV) 155, 157
Infectivity assays 1172
  totiviruses 1467
Inflammation, bloodbrain barrier disruption 908
Inflammatory reaction 697
  Borna disease virus (BDV) infection 153
  caprine arthritis encephalitis syndrome 203

Inflammatory reaction *continued*
herpes simplex virus 591
Influenza 446, 713–714, 1223
antivirals 47, 714
resistance 47
clinical features 713–714
epidemics and pandemics 709, 711, 714–715
influenza A 1220
influenza A infection 446
influenza B infection 446
pathology and histopathology 714
prevention and control 714
see also Influenza viruses[(A)]
Influenza A virus[(A)]
antivirals for 47, 714
attenuated dominant-negative mutant as antiviral 732
avian 713
cell-mediated immunity, cytotoxic T cells 708
evolution 711
patterns of relatedness 438
pseudorecombination 439
frequency and respiratory disease types 1223
genome 1207
heart infections 965, 967
history 709–710
host range and viral propagation 710, 1587
inhibition by vesicular stomatitis virus (VSV) 728
M2 protein 719, 720
mutation rate 711
pathogenesis of myocarditis 967
productive infection 1587
schematic representation 716
subtypes 710, 711
Asian (H2N2) 711
Hong Kong (H3N2) 711, 712, 722, 724
Russian flu (H1N1) 712
transcription and replication strategy 1207
Influenza B virus[(A)]
BM2 protein 719, 720
cytopathic effects 337
frequency and respiratory disease types 1223
heart infections 965
history 710
host range and viral propagation 710
NB and NA proteins 719, 720
vaccine 714
Influenza C virus[(A)], proteins 715, 717
Influenza viruses[(A)] 446, **709–727**
antigen structure **722–727**
see also Influenza viruses, proteins
antigenic determinants 714, 716, 722
B- and T-cell 725–726
on hemagglutinin and neuraminidase 725
structures 724–725
T-cell, characteristics 726–727, 727
antigenic drift 1221
antivirals 47, 714
resistance and M2 protein 47, 719
assembly 228
avian 710, 713
cultivation 711
defective-interfering 526, 715
effect on host mRNA 1588
epidemiology 711–713
epitopes 724
escape mutants 725
evasion of interferon action 737
evolution 711, 712
pandemic strains 712, 713
general features **709–715**
genetic manipulation 721
using cDNA-derived RNA 721
genome structure and properties 715, 1207
geographic and seasonal distribution 710
history 630, 709–710
host range and viral propagation 710–711
identification 339
immune response to 699–700, 708, 714, 727
infectivity 721
isolation 338
molecular biology **715–722**
monoclonal antibodies 725
mortality 713
mRNA 232, 719, 720
mutations 711, 725
Mx1 gene regulating resistance 667, 697, 711, 736
pathogenicity 713
physical properties 719
proteins
BM2 protein 719, 720
fusion, *see* Hemagglutinin
hemagglutinin-esterase (HEF) 715, 717
M1 protein 715, 717–718, 720
M2 protein 715, 719, 720
NA and NB proteins 719, 720
NP protein 719
NS1 and NS2 proteins 719, 720
polymerase proteins (PB1, PB2, PA) 718–719
properties 715–719
RNA coding for 720
see also Hemagglutinin (HA); Neuraminidase (NA)
replication cycle 232, 716, 718, 719–721, 1207
M1 protein role 718, 721
packaging, assembly and release 721
post-translational processing 720–721
random packaging hypothesis 721
RNA replication 719–720, 1207
transcription 720, 1207
translation 720, 1207
RNA
complementary (cRNA) 719
viral (vRNA) 232, 719, 720
serologic relationships and variability 711
antigenic shift and antigenic drift 711
structure and properties 715
susceptibility, chromosome 21 665
swine 710
taxonomy and classification 710
transformant 721
transmission and tissue tropism 713
types 710
vaccination 714
vaccines 714, 1226, 1507
'Inhibitor of virus replication' (*N* gene product) 1107, 1147
Initiation factors (eIF-2), inactivation 1205, 1589
Inoviridae 464, 1398, 1405
characteristics and taxonomy 99
structure 1582
Inovirus virions[(B)], characteristics and taxonomy 99
Insect cell lines 1587
Insect larvae, cytoplasmic polyhedrosis virus infection, *see* Cytoplasmic polyhedrosis viruses (CPV)[(I)]
'Insect picornavirus', *see* Picornaviruses[(I)], insect
Insect viruses 58
economic significance
ascoviruses 63
cytoplasmic polyhedrosis viruses (CPV) 318
densonucleosis viruses 334
entomopoxviruses 397–398
nuclear polyhedrosis viruses (NPVs) 131, 136
tetraviruses 1422
see also Biological control; Insecticides
history 632
taxonomy and classification 1408
vector transmission 62
as vectors, *see* Baculoviruses
see also individual families (page 1408)
Insecticides, biological
cytoplasmic polyhedrosis viruses (CPV) 318
entomopoxviruses as 397–398
granulosis viruses as 130
nuclear polyhedrosis viruses (NPVs) as 136
ultra low volume (ULV), mosquito control 330
see also Biological control
Inserted sequence (IS) elements 108, 110
transposons mediated by 110
see also Transposition; Transposons
see also Terminal inverted repeats
Insertion vectors 114
see also Gene cloning
Insertional mutagenesis, adenoviral DNA integration 14
Insulin
deficiency in duck hepatitis B virus infections 568
Escherichia coli producing 120–121
Intasome 774
Integrase 68
bovine leukemia virus (BLV) 168
lambda bacteriophage 773, 774
Integration of DNA into host genome, *see under* DNA
Integration into host genome, lambda bacteriophage 109, 773–774
Integrative host factor (IHF) 774, 817, 822
bacteriophage Mu regulation 872
for coliphages P2 and 186 1005
Integrative suppression 1002
Integrin family, cell receptors for, foot and mouth disease viruses 492
Interference 526, **728–732**
defective viruses 321
see also Defective-interfering (DI) viruses
dominant-negative mutants, *see* Dominant-negative mutants
feline leukemia virus (FeLV) 460
history 728, 733
'intrinsic' 729
types 728–729
incompatibility of heterologous viruses 728–729
interferons 728
superinfection exclusion 729
see also Defective-interfering (DI) viruses; Interferon
Interferon **733–745**, 1591
action 735
anti-antiviral mechanisms 737
in antiviral state 665, 666, 735–737
dsRNA-dependent protein kinase 736
MX proteins 736
2′-5′ oligoadenylate synthetases synthesis 736
proteins induced by 665, 736–737

binding to receptors 735, 737
bovine papillomavirus infections 1310
clinical uses 737–738, **739–745**
in AIDS 739–741
in cancer 738, 741–745
cytotoxic agents with 742–743, 743, 744
5-fluorouracil with 744
in hematologic malignancies 738, 741–743
in hepatitis B 568, 738
low oral doses 739
maintenance therapy in myeloma 743
neutralizing antibodies development with 744
prognosis inversely related to titer 738
in solid tumors 743–744
synergistic with specific antibodies 738
toxicity 740, 744–745
see also specific interferons
DAI protein 1094
future perspectives 738–739, 745
general features **733–739**
genes, molecules and receptors 733–735
herpes simplex virus infections 591
history 733
human papillomavirus infections 1021
hybrid 733
immune (type II), *see* Interferon-γ
inducers, polyribonucleotides 738
inducible resistance by 665, 667
induction 735
by bacterial infections 738
by protozoa 738
by viruses 735
induction receptor 735
ineffective in ectromelia (mousepox) virus infection 865
inhibition 1094
lymphocytic choriomeningitis virus (LCMV) infections 804–805
mechanism of action 737
mRNA 735
murine cytomegalovirus (MCMV) infection 311
nonadaptive immune response 697
production
in *E. coli* 733
in viral hemorrhagic septicemia virus infection 1251
purification attempts 733
respiratory infections 1222
rhinovirus infections 1258
transcription of genes 735
trophoblast 734–735
type II (IFN-γ), *see* Interferon-γ
varicella–zoster virus infections 1518
virus interference by 728
Interferon stimulated response elements (ISRE) 736, 737
Interferon-α 45, 733–734
acid-lability 734
in AIDS 739–741
cytotoxic agents with 739–740
in vitro data 741
mechanism of action 740
production 734
binding to receptor 735, 737
biological response modifiers with 741
cells producing 734
in hairy cell leukemia 738, 741–742
in malignant melanoma 744
receptors 734
recombinant 738
side-effects 740
toxicity 740, 744–745
attempts to decrease 740–741
Interferon-β 734
binding to receptor 735, 737
cells producing 734
in HIV infection 741
receptors 734
recombinant 738
sequence homology with IFN-α 734
Interferon-γ 734
cells producing 734
in herpes simplex virus infections 592
in HIV infection 741
immune evasion of leporipoxviruses 1159
receptor 734
release, in hepatitis A 553
resistance to protozoal infections 738
T cell secretion 705
Interferon-ω (interferon-α^{II}) 734
Interleukin(s)
function in immune response 697
secretion by antigen-presenting cells 697
Interleukin-2 (IL-2)
EBV infection prevention after transplantation 1483
herpes simplex virus infections 592
HTLV-1 transformation mechanism 1476, 1496
T cell secretion 705
Interleukin-6 (IL-6) 734
Interleukin-10 (IL-10), EBV homolog 416
Intermediate filaments 231
adenovirus E1B protein (176R) association 21
changes due to frog virus 3 (FV3) 507
functions in viral infections 231
Internal Ribosomal Entry Site (IRES) 207
hepatitis A virus (HAV) 547
polioviruses 1123, 1125
Theiler's viruses 1424
Internal terminal repeats
parapoxviruses 1038
see also Terminal repeats
International Committee on Nomenclature of Viruses (ICNV) 633, 637, 1397, 1402
meeting 1397
International Committee on Taxonomy of Viruses (ICTV) 93, 1397, 1402
database project 1401
description of new phages 95
Fifth Report 1397, 1401, 1402
nepovirus classification 901, 902
nomenclature of taxa 1402
operation 1397, 1401
phytoreoviruses 1096
plant rhabdoviruses 1243
rules and regulations 95
species definition 1403
subcommittees 95
universal system 93–95
Intestinal villi
animal parvovirus infections 1056
blunting, rotavirus infection 1279
Intestine, rotavirus replication 1278, 1280
'Intracellular immunization' 731
Intracisternal A-type particles (IAPs) 1240
as insertional mutagens 937
Intracytoplasmic A-type particles (ICAPs) 1237, 1239
Intracytoplasmic inclusion bodies, *see* Inclusion bodies (inclusions)
Intraepithelial neoplasia (IN)
HPV-induced lesions 1020
see also Cervical intraepithelial neoplasia (CIN)
Intramolecular recombination, *see under* Genetic recombination
Intrauterine infections, of heart 964–965
Introns 232
'escaped' 1570
gram-positive bacteria 1352
viroid relationship 1570
Invertebrate viruses
vector transmission, in ascoviruses 62
see also Insect viruses
Inverted terminal repeat (ITR), *see specific viruses*; Terminal inverted repeats
Ippy virus(A) 777
Iridescence 1433
Iridocyclitis 446
Iridoviridae 1398, 1408, 1409, 1430
fish 477, 479
fish lymphocystis disease virus (FLDV) 798
frog virus 3 (FV3) 503
structure 1583
Iridovirus(I) 1430, 1433
evolution 1435
members and serology 1432–1433
tipula iridescent virus (TIV) 1430
Iridoviruses(I) **1430–1436**
African swine fever virus classification 24
characteristics 503, 1430
composition 503, 1432
lipid bilayer 1432
cytopathology 1434
DNA 1431–1432
components 1432, 1435
packaging 1434
ecology 1435
economic importance 1434–1435
evolution 1435
gene expression and control 1209, 1434
genetic modification 1435
genome structure 1431–1432, 1434, 1435
inclusion bodies 1433
insect hosts 1431
morphogenesis 1433
pathology 1433–1434
polypeptides and enzymes 1432
protein synthesis 1434
replication 1209, 1434
sites 1434
serology 1432–1433
structure 1431–1432
taxonomy and classification 1430–1431, 1435
transmission 1432, 1435
viroplasmic centers 1433, 1434
virus–host interaction 1432
see also Tipula iridescent virus (TIV)(I)
Iron, toxin production 102
Isfahan (ISF) virus(A) **233–236**
genome structure 234
host range, transmission and distribution 234–235
pathogenicity and genetics 235
proteins 234
'Isle of Wight disease' 655
Islet cells
antibodies 971
cross-reacting antibodies to, in diabetes 65
virus role in diabetes pathogenesis 970, 971
Isolation of viruses, *see under* Diagnostic techniques

J
Jamestown Canyon virus 1617
Japanese eel iridovirus(A) 477
Japanese encephalitis 746
clinical features 749
diagnosis 750, 1615
pathology and histopathology 749
prevention and control 750

Japanese encephalitis virus (JEV)(A) 445, **746–751**, 1615
diagnosis 750, 1615
epidemiology 748–749
evolution 370, 748
future perspectives 751
genetics 748
genome properties 746
history 746
host range and viral propagation 747–748
immune response to 750, 1615
infection, *see* Japanese encephalitis *(above)*
overwintering in temperate climatic areas 750–751, 1615
pathogenicity 749
physical properties 747
proteins 746–747
E protein epitopes 747
NS1 747
properties 746–747
RNA replication 747
serologic relationships and variability 748
slowly sedimenting hemagglutinin (SHA) 747
soluble complement fixing antigen (SCF) 747
structure and properties 746
subtypes 748
Beijing-1 (JaGAr-01) 748, 750
Nakayama 748
taxonomy and classification 746, 1615
tissue tropism 749
transmission 747, 748, 749, 1615
vaccination 748, 750
vaccine 750, 1615
bivalent 750, 1615
recombinant 751
Jaundice, hepatitis E 584
JC virus(A) **752–757**, 1617
antibodies to 754, 756
antigenic determinant 754
antigenic variants 754
archetype 753, 755
classification 752
DNA replication 752
epidemiology 754–755
evolution 754
future perspectives 757
genetics 753–754
genome
hypervariability 753, 754
promoterenhancer regions 753, 754
properties 752–753
geographic and seasonal distribution 753
history 752
host range and viral propagation 753
infection
after transplantation 486–1487
clinical features 756, 912
diagnosis 753, 757
immune response 756
of kidneys 974
oligodendrocytes 910
pathology and histopathology 756
prevention and control 756–757
progressive multifocal leukoencephalopathy 912
Mad-1 strain 752
Mad-11 variant 754
pathogenicity 755–756
persistent infections 755, 1090
reactivation 755
properties 752–753
reactivation 1486, 1487
serologic relationships and variability 754
T antigens 752, 753, 1081
titer measurement 753
transmission and tissue tropism 755, 1081
see also Progressive multifocal leukoencephalopathy (PML)
JE virus(A), *see* Japanese encephalitis virus (JEV)(A)
'Jellyroll β barrel' structures 349, 1575
tombusviruses 1448
Joest-Degen inclusion bodies 153
jun oncogene 71, 936, 1475
Juncopox virus (JPV)(A), genome structure 497
Jungle yellow fever 1610, 1615
see also Yellow fever
Junin virus(A) 447, **776–786**, 1614
antibody response 785
characteristics and structure 776, 777
diagnosis 785
geographic and seasonal distribution 777, 778
history 776
host range and viral propagation 777, 778
kidney infections 975
serologic relationships and variability 779
taxonomy and classification 776
transmission 780, 781
vaccines 785
see also Argentine hemorrhagic fever (AHF)
Junonia coenia densonucleosis virus (JcDNV)(I) 331, 332, 333
cell lines for 333

K
Kalanchoe carlavirus (KaCV)(P) 215
Kalanchoe latent virus (KLV)(P) 215
Kalanchoe top-spotting virus (KTSV)(P) 139
Kaolin, effect on phage stability 123
Kaposi's sarcoma 1498–1499
in AIDS 681
interferon-α in 739–740
cytotoxic agents with 739–740
orbital/conjunctival 447
Kashmir bee virus (KBV)(I) 656, 658
KBSH 1072, 1073
Kennedya yellow mosaic virus (KYMV)(P) 1500
see also Tymoviruses(P)
'Kennel cough' syndrome 17
Keratinocyte, molluscum contagiosum virus (MCV) infection 851, 852
Keratitis, etiology 449
Keratoconjunctivitis
epidemic 1
adenovirus (Ad8) causing 2, 6
clinical features 6
etiology 449
follicular 443
kex1 and *kex2* mutations 1603
Ki-*ras* oncogene 936
Kidney
JC and BK viruses tropism 755
transplantation
CMV infections 972, 973
hepatitis virus infections 1485, 1486
virus infections 971–976
categorization 972
cytomegalovirus 295, 297, 972
hepatitis virus 1485, 1486
latent infections 972
mechanism of damage 972
SV40 1328
virus persistence in 1093
in yellow fever 1611
Killer toxins 1464, 1468, 1600
M_1 1600, 1601, 1603
processing 1603
Ustilago maydis virus (UmV)(P) 1468, 1470
see also Saccharomyces cerevisiae virus L-A (ScV-L-A)(I); Totiviruses(P)
Kirsten murine sarcoma virus (Ki-MSV)(A) 884
oncogenes 936
kit oncogene 936
Klamath virus(A) 1249
Koilocytosis, HPV infections 1020, 1025
Koplik's spots 845
Korean hemorrhagic fever (KHF) 538, 974, 1614
geographic and seasonal distribution 541
pathology and histopathology 544
see also Hantaan virus(A); Hantaviruses(A)
Kotonkan virus(A) 1186
Kozak rule 1201
Kupffer cells, reduction in LDV infection 769, 770
Kuru **758–761**, 912, 1361
cell lines 760
clinical features 760–761
epidemiology 759
evolution 759
eye infections 447
genetics 759
geographic and seasonal distribution 758
history 758
host range and viral propagation 758–759
immune response 761
maternal and vertical transmission 1366
pathogenicity 760
pathology and histopathology 761
persistence 760
prevention and control 761
serologic relationships and variability 759
strain adaptation and variable behaviors 759
taxonomy and classification 758
tissue tropism 760
transmission 760, 1361, 1365
'Kuru' plaques 761
KV (polyomavirus)(A) 1135
Kyasanur Forest disease 366, 1619
Kyasanur Forest disease virus (KFDV)(A) 362, 363, 1619
geographic distribution and transmission 1619
see also Encephalitis viruses(A)

L
L cells
reovirus infection 1191
reovirus receptor 1200
vesicular stomatitis virus (VSV) infection 1587
L-A, *see under Saccharomyces cerevisiae*
L-*myc* gene 938
La Crosse virus(A) 189, 190, 446, 1617
genome structure 1413
geographic distribution and transmission 1617
transmission, aircraft transport and 1622
LA-1 oncogene 592
Laboratory animals, virus propagation in 1171–1172
Lactate dehydrogenase, elevation in LDV infections 763, 769, 770
Lactate dehydrogenase-elevating virus (LDV)(A), of mice **763–771**
antibodies 770
assembly, uptake and cytopathology 766
classification 764
CNS lesions 769, 770
contamination sources 768

coronavirus similarity 763, 768
evolution 768
future perspectives 771
genetics 767
genome structure and properties 763, 764–765, 768
RI RNA and subgenomic RNA 765
geographic and seasonal distribution 766–767
history 763
host range and viral propagation 767
infection
clinical features 769
immune response 770
pathology and histopathology 769–770
prevention and control 771
LDV-C strain 767, 769
pathogenicity 769
persistent infection 770
physical properties 765
proteins 765
S protein 768
receptor 1561
replication 765–766
serologic relationships and variability 768
structure and properties 764
taxonomy and classification 763–764
tissue tropism 769
transmission 768–769
Lactobacillus casei, phage-resistant strain development 120
Lactococci, phage-resistant strain development 120
Lactose, malabsorption 378
lacZ gene
cloning vectors and 114, 115
insertion vectors and 114
Lagos bat virus[(A)] 1186
epidemiology 1188–1189
host range and viral propagation 1187
transmission and tissue tropism 1189
see also Rabies-like viruses[(A)]
Lambda bacteriophage (λ)[(B)] **772–776**, 814–823
abortive lysogens 815
antitermination 819
att mutants 818
attB sites/recombination 109, 815, 822
attP sites/recombination 109, 815, 822
attR and *attL* sites 823
capsid proteins 772
characteristics and taxonomy 98, 772
cI repressor 820–821
action 816
amino and carboxyl domains 821
binding to oR 816, 819–820, 821
binding sites 816
initial synthesis and *c*II/*c*III 816
regulatory loops 817
structure 820–821
*c*I repressor gene 114, 115, 773, 774, 816
*c*I p^c mutations 821
λ*c*I857 mutant 816
mutations 816, 821
positive regulation 816
*c*II and *c*III genes 773
mutants 816
mutations inhibiting integration 818
regulation 817
transcription 819
translation in *rnc* hosts 817
cII protein 819
clear (λc) mutants 815, 816–817
complementation patterns/rationale 816–817
*c*Y mutants 816
λ*c*I857 mutant 816
λpRM mutant 816
oL and oR inactivation 817
as cloning vehicle 113–115, 775
expression vectors 114–115
factors influencing 113
insertion vectors 114
replacement vectors 114
restrictions on 775
cos sites 109, 114
in cosmids 115
distance between 114
Cro protein 773, 817, 821–822
binding to oR3, oR2, oR1 820
function 817
λ*c*I857*cro* mutant 817
structure 821–822
*c*Y mutants 816
derepression 774
DNA metabolism 86
DNA packaging in T1 bacteriophage 1373
DNA replication 85, 773
Fis binding site 823
gal and *bio* transducing phage 109
Gam protein 85
gam$^-$ mutant 85, 86
gene regulation 817, 819
genetic map 815, 818, 819
functionsite specificity 815
genome size 114, 772, 814
genome structure 772–773, 814, 815
base composition 772–773
head assembly 773
history 772
in vitro replication 775
ind gene, mutations (induction) 821
int gene 773, 817
expression control 819
mutants 817
Int protein
binding sites 822
function in integration/excision 817, 822
integration into host genome 109, 773–774, 815, 822–823
attB site 109
attP and *attB* recombination 815, 822
host mutations blocking 818
Int protein action 817, 822
phage mutations affecting 817–818
λ*gal* 109, 111
fate after infection 111
λZAP vector 116
lysogeny 109, 773–774, **814–823**
advantages 815
derepression mechanism 774
establishment 815, 818–819
frequency 815
genes involved 815
maintenance 815, 819–822
lytic cycle 773
*c*I repressor action 816
transcription 773, 815, 816
maintenance of prophage 819–822
oR region 819–820
mutants, *see specific mutants above/below*
mutations 815, 816–818
failure to lysogenize 816
host, affecting repression 817
host, blocking integration/excision 818
infrequent lysogenization 816
phage, affecting repression 816–817
phage defective in integration/excision 817–818
see also individual mutants (above)
N protein, function 819
nut sequence 819
operators 816
mutants inactivating 817
oR region 819–820
cI repressor interaction 821
control 819–820
Cro interaction 821–822
mutations 820
oR1, oR2, oR3 820
Orf protein 86
origin of replication (*ori*) 815
packaging of DNA into 85, 109, 114, 773
phage 434 recombination 817
plasmid formation 775
pRM mutant 816
promoters
bacteriophage ϕ29 cloning 983
pI 819
pL 816, 818–819
pR 816, 818–819, 819
pRE and pQ 819
pRM 819
prophage excision 109, 774, 815
host mutations blocking 818
phage mutations affecting 817–818
prophage induction 815, 820, 821
Q protein function 819
ral function 671, 775
recombination 85–87, 774–775
gal and *bio* genes 109
Rec-dependent 86, 87, 774
Rec-independent 86, 774
recA gene mutation 86
RecBCD-promoted, features 87
Red proteins role 85–86, 775
red genes 85, 775
red$^-$ *gam*$^-$ mutant 86
Redβ protein 87
regulation 773–774
restriction/modification 775
retroregulation 819
sib site 819
structure 772
superinfection immunity 816
ter cleavage site 815
transcription 818–819
transduction by 109
turbid plaques 815
xis mutants 817–818
Xis protein binding site 823
Lambdoid phage 772, 815
Lameness, vesicular exanthema virus of swine (VES) 1544, 1546
Lamium mild mosaic virus (LMMV)[(P)] 451
host range 452
serology 453
transmission 453
Langerhans cells, in herpes simplex virus infections 591
Lapine parvovirus (LPV)[(A)] 1052
Larvae, viruses, *see* Cytoplasmic polyhedrosis viruses (CPV)[(I)]
Laryngeal carcinoma, HPV and 1497
Laryngeal papillomatosis 1019
Laryngitis 1221
Laryngotracheobronchitis 1224
causative agents 1224
Lassa fever
clinical features 783
diagnosis 1614
drug prophylaxis 785
epidemiology 779–780
immune response 784–785
incubation 781, 783
kidney infections 975
mortality 784

Lassa fever *continued*
nosocomial outbreaks 780, 782
passive antibody protection 785
pathogenesis 781–782
pathology and histopathology 784
prevention and control 785
ribavirin in 785, 786
viremia 782, 783
see also Lassa virus(A)
Lassa virus(A) 447, **776–786**, 1614
characteristics and structure 776, 777
evolution 779
genetics 778–779
geographic and seasonal distribution 777, 778
history 776
host range and viral propagation 777, 778
LCMV relationship 803
persistent infections 779
serologic relationships and variability 779
taxonomy and classification 776
transmission 780, 781, 1614
vaccine 785
viremia 786
Latency associated transcript (LAT) 590
herpes simplex virus (HSV) 1527
HSV-1, equine herpesvirus homology 426
pseudorabies virus 1174
Latency/latent infections **787–791**, 1076, 1089–1090, 1170
alphaherpesviruses and gammaherpesviruses 157
amphibian herpesviruses 41
consequences of 787
cytoplasmic polyhedrosis viruses (CPV)(I) 313
definition 787, 1076, 1089–1090, 1170, 1587
EBV infection, *see* Epstein–Barr virus (EBV)
extrachromosomal state of HSV 590
features 787
feline immunodeficiency virus 458
feline leukemia virus 462
HSV, *see* Herpes simplex viruses (HSV)(A)
immune evasion by 701
of neurons 908
re-activation 787
viruses establishing 787, 788
see also DNA, integration into host genome; *specific viruses*
Lateral bodies
entomopoxviruses (EPVs) 393
fowlpox virus 500
Latex agglutination, viral antigen detection 345
Latex particles, in virus titration 1173
Latino virus(A) 777
LCMV, *see* Lymphocytic choriomeningitis virus (LCMV)(A)
LDV, *see* Lactate dehydrogenase-elevating virus (LDV)(A)
Leaf, morphology changes, plant viruses causing 1110
Leafhopper transmission 633
badnaviruses 142
geminiviruses 521
maize chlorotic dwarf virus (MCDV) 825, 825–826
phytoreoviruses 1098, 1099
rhabdoviruses 1248
'Leaping polymerase' mechanism 320–321
Leishmania braziliensis virus(A) (LBV) 532, 534
Lelystad virus (LV)(A) **763–771**
evolution 768
genome structure and properties 764–765
geographic and seasonal distribution 767
history 763
host range and viral propagation 767
properties 764
proteins 765
taxonomy and classification 763
see also Lactate dehydrogenase-elevating virus (LDV)(A), of mice
Lentiviridae 199
caprine arthritis encephalitis virus (CAEV) 199
members 199
oncoviruses comparison 199
see also Lentiviruses(A)
Lentivirinae 159
Lentiviruses(A) 1593
CA antigens 161, 164, 200
characteristics and structure 1316
equine infectious anemia virus (EIAV) 430, 433
evolution 163–164
feline immunodeficiency virus (FIV) 455
genes 1318
genetic variation 202, 1319
genomes 160, 200, 1317
groups 1316, 1317, 1319
guanine to adenosine hypermutation 202
history 1593
immune response 204
regulatory genes, origin 439
simian immunodeficiency virus (SIV) 1316
transmission and tissue tropism 165
see also Bovine immunodeficiency virus (BIV)(A); Equine infectious anemia virus (EIAV)(A); Human immunodeficiency virus (HIV)(A); Simian immunodeficiency virus (SIV)(A); Visnamaedi viruses(A)
Lepidoptera
ascoviruses of 58
granulosis viruses (GVs) 127
Leporipoxvirus(A)
members of 1154, 1155
see also Myxoma virus (MYX)(A); Poxviruses(A); Shope fibroma virus (SFV)(A)
Lettuce big vein virus(P), transmission 515
Lettuce infectious yellows virus (LIYV)(P) 247
Lettuce necrotic yellow vein virus (LNYV)(P) 1246
disease 1247
Leukemia
acute myelogenous, interferon therapy 743
chronic lymphocytic, interferon therapy 743
chronic myelogenous, interferon therapy 742
hairy cell, *see* Hairy cell leukemia
murine leukemia viruses causing, *see* Murine leukemia viruses (MuLVs)(A)
varicella vaccine and 1517
Leukemia viruses
transmission 401
see also individual leukemia viruses
Leukopenia, in LCMV infections 804, 805
Leukosis, sporadic bovine 173
Leviviridae 1334, 1399, 1405
Alloleviviruses (Group B) 1334, 1338
characteristics and taxonomy 100
Leviviruses (Group A) 1334, 1338
phylogeny 1338–1339
structure 1580
Leviviruses(B) 1334, 1338
characteristics and taxonomy 100
LexA protein 774
LFA-3 adhesion molecule, down regulation 1094
Lilac chlorotic leafspot virus (LCLSV)(P) 243
Lilac mottle virus (LMV)(P) 215
Lily symptomless virus (LSV)(P) 215
Lily virus X (LVX)(P) 1143, 1144
Lipid-containing viruses 1556
fowlpox virus 497
iridoviruses 1432
membrane bilayer 1556
membrane synthesis 1559
Lipopolysaccharide, T7 phage receptor 1389
Lipothrixviridae 1398, 1405
characteristics and members 96
Liver
cancer, *see* Hepatocellular carcinoma (HCC)
cholestatic changes 585
disease, chronic, hepatitis B 557, 558
in hepatitis A 551, 552
in hepatitis E 585
injury
mechanism in hepatitis B 557
reovirus infection 1193
necrosis
ectromelia (mousepox) virus infection 864
equine infectious anemia virus (EIAV) infection 434
reovirus infection 1193
transplantation
CMV infections after 1479
hepatitis virus infections after 1485, 1486
Locust coccobacillus 645
Lolium enation viruses (LEV)(P) 1096
Long terminal repeat (LTR) (retroviruses) 1208, 1318
avian leukosis virus 67, 68
bovine immunodeficiency virus (BIV) 160
bovine leukemia virus 167
deletion in retrovirus-based vectors 1533
equine infectious anemia virus (EIAV) 430–431
feline leukemia virus (FeLV) 460, 461
foamy viruses 482
gene expression 1208
HIV 1081
HTLV-2 689
lymphoproliferative disease virus (LPDV) of turkeys 812
MMTV, *see* Mouse mammary tumor virus (MMTV)(A)
murine leukemia viruses (MuLVs), *see* Murine leukemia viruses (MuLVs)(A)
reticuloendotheliosis viruses (REV) 1229
in retrotransposons, *see* Retrotransposons
simian immunodeficiency virus 1318
visnamaedi viruses 1594
Lonicera latent virus (LonLV)(P) 215
Louping ill 367, 370
Lox-cre recombination system 998, 1002
LPV (polyomavirus)(A) 1135
Lucerne Australian latent virus(P) 904
Lucerne Australian symptomless virus(P) 905
Lucerne transient streak virus (LTSV)(P) 1346
structure 1348
see also Sobemovirus(P)
Luciferase gene 1541
Lucké tumor 41, 503
Lucké tumor herpesvirus (LTHV)(A) 40–41
oncogenicity 41

properties 41
replication and gene expression 41
Lumpy skin disease (LSD) of cattle 1160
clinical features 1163
epidemiology 1162
geographic and seasonal distribution 1161
host range and viral propagation 1161
see also Poxviruses$^{(A)}$, sheep and goats
Lumpy skin disease virus, *see* Poxviruses$^{(A)}$
Lung
murine cytomegalovirus infection 310
Sendai virus infection 1303
Luteovirus 1399
Luteoviruses$^{(P)}$ 218, **792–797**
cytopathology 795
economic significance 797
epidemiology and disease management 796–797
evolution 439, 795–796
future perspectives 797
genome organization 793–794
gene expression 794
gene functions 793
geographic distribution 796
heterologous encapsidation in mixed infections 795
host range 796
host relationships 794–795
location in plant 795
pea enation mosaic virus (PEMV) relationship 1083, 1085, 1086
satellite RNAs 794
serologic relationships 794
structure and composition 793, 1580
subgenomic RNA, translation 794
subgroups 792
symptomatology 794
taxonomy and classification 792–793
translation 794
transmission 795
Lutzomyia sandflies, vesicular stomatitis virus (VSV) transmission 1552, 1554
'Luxury' functions 911
loss in LCMV infections 804, 805
Lwoff, A. 647
LY 217896 44, 47
Ly-4 locus 836
Lychnis ringspot virus (LRSV)$^{(P)}$ 661
economic importance 663
host range 663
Lymantria dispar nuclear polyhedrosis virus$^{(I)}$ 131, 132
Lymph nodes 705–706
ectromelia (mousepox) virus infection 864
feline immunodeficiency virus in 458
Lymphadenopathy
feline immunodeficiency virus (FIV) infection 457
progressive, *see* Progressive generalized lymphadenopathy (PGL)
Lymphatics, virus growth in 909
Lymphoblastoid cell lines (LCLs)
Burkitt's lymphoma derived 1494
Epstein–Barr virus (EBV) 405, 411
Lymphocryptovirus$^{(A)}$ 404
see also Epstein–Barr virus (EBV)
Lymphocystis cell 798
Lymphocystis disease (LD) 798, 799–800
clinical features 800
pathology and histopathology 800
Lymphocystis disease virus$^{(A)}$, *see* Fish lymphocystis disease virus (FLDV)$^{(A)}$
Lymphocytes 64, 697
depression by rubella virus 1297
in hepatitis E 585
myxoma virus replication in 1156, 1157, 1159
polyclonal activation, bovine leukemia virus transcription 172
see also B cells; T cells
Lymphocytic choriomeningitis virus (LCMV)$^{(A)}$ 776, **801–811**, 802, 806
adverse immune reactions, Tc cells 702, 804, 805
Armstrong strain 801, 802, 803
Biosafety Levels 803
cell damage mechanism 911
cytopathic effects 778, 911
defective-interfering (DI) 809, 810
diabetes prevention in animal models 65
enzymes 806
polymerase activity 809
epidemiology 803
evolution 803
future perspectives 805
gene expression 807, 808–809
in novel environments 811
general features **801–805**, 806
genetics 802–803
genome 807
small S RNA and large L RNA 802, 807
Z protein sequences 807
geographic and seasonal distribution 802
glucose intolerance and 970
history 776, 801–802
host range and viral propagation 777, 778, 802
immune responses to 804–805
anti-viral antibodies 801, 805
cytotoxic T cells 805
nonspecific 804–805
immunological tolerance induced by 701, 801, 805
immunosuppressive variants 804
importance 708, 801, 805
infection
acute, progression to persistent 809
clinical features 804
eye infections 447
immunopathology 706–707, 708, 804
pathology and histopathology 804
prevention and control 805
infection of mice 801
docile variants 804
immune response 804–805
pathogenicity 804
Lassa virus relationship 803
loss of 'luxury' cell functions 804
molecular biology **806–811**
mutation rate 802
pathogenicity 803–804
persistent infections 779, 801, 803, 806, 809, 810
in situ hybridization and 810–811
proteins 807–808
GPC and GP-1/GP-2 807–808
L protein 807, 808
NP 807
Z protein 808
purification 806
receptor 809
replication 802, 808
spleen-derived (Clone 13) 810
split-tolerance to 802
strains and reassortant viruses 809–810
structure and properties 806
T cell-mediated immunopathology 706–707, 804, 805
taxonomy and classification 802
transcription 808–809
L-protein and Z-protein 809
NP mRNA and GPC mRNA 808
translation 807
transmission and tissue tropism 803
Traub strain 801
variants 810
virulence, mapping of 809
WE strain 801, 803
Lymphocytosis, bovine leukemia virus 166, 172
Lymphoid cells, viral infection 64
Lymphomas 173
baboons with 610, 612, 613
see also Herpesviruses$^{(A)}$, baboon and chimpanzee
Burkitt's, *see* Burkitt's lymphoma
in cats 458, 459
EBV causing 1482, 1495
feline immunodeficiency virus infections 458
herpesvirus ateles causing 620
herpesvirus causing 833, 835
prevention by vaccination 833, 836–837
Hodgkin's, *see* Hodgkin's disease
in immunodeficiency diseases 406–407, 408
Marek's disease virus causing 833, 834, 835
murine leukemia viruses causing, *see* Murine leukemia viruses (MuLVs)$^{(A)}$
non-Hodgkin's
in AIDS patients 407, 408, 410
interferon therapy 742
reticuloendotheliosis viruses 1231
tree shrew herpesviruses and 1491
Lymphoproliferative disease (LPD) of turkeys 811
clinical features 813–814
diagnosis 814
etiology 811–812
Lymphoproliferative disease virus (LPDV) of turkeys$^{(A)}$ **811–814**
features and classification 812
genome organization 812–813
sequences 812
similarity with avian sarcoma-leukemia virus (ASLV) 812
history 811
host range and viral propagation 813
isolation 814
leukemogenic potential 813
pathogenicity 813
proteins 812
NC and MA 812
replication and sites of 813
serology 814
Lymphoproliferative disorder
herpesvirus saimiri causing 620
human herpesvirus 6 (HHV-6) infection 625
Lymphosarcoma 173
feline 461, 462
fish 475
lymphoblastic 535
Lysis, *see* Cell lysis
'Lysis from without' 79
Lysogenic bacteria, immunity 629, 815
Lysogenic conversion 101
see also Bacteriophage, toxins and phage conversion
Lysogeny **814–823**
advantages 815
algal viruses 39
Bacillus subtilis bacteriophages 75
bacteriophage λ, *see* Lambda bacteriophage

Lysogeny *continued*
  cyanobacteria bacteriophage 290
  discovery 629–630
  frequency 815
  halobacterial phage 52, 53
  history 629–630, 647
  immunity of lysogens 629, 815
  phage problems in industrial fermentations 117
  phage in soil 124
  see also Bacteriophage, temperate; Lambda bacteriophage[B]; Prophage
Lysosomal membranes, damage 1589
Lysosomes 229, 705
Lysosomotropic agents 1287
Lyssavirus[A] 1186, 1614
  evolution 1188
  see also Rabies-like viruses[A]

M
M/R systems, *see* DNA modification/restriction (M/R) systems; Host-controlled modification and restriction
MAC-1 group viruses 1227
Machlovirus[P], structure 1580
Machupo virus[A] 447, **776–786**, 1614
  antibody response 785
  characteristics and structure 776, 777
  geographic and seasonal distribution 777, 778
  history 776
  host range and viral propagation 777, 778
  kidney infections 975
  prevention and control 1621
  serologic relationships and variability 779
  taxonomy and classification 776
  transmission 781
McrA and McrB restriction systems 672–673, 1383
Macrophage
  antigen presentation 704
  antigen processing 697
  clearance of virus particle 1079
  infectious subviral reovirus particles in 1193
  resistance, herpes simplex virus 591
  virus infections 702
    caprine arthritis encephalitis virus (CAEV) 201, 202, 203
    coronavirus 260
    equine infectious anemia virus (EIAV) 433
    feline immunodeficiency virus (FIV) 456, 457
    lactate dehydrogenase-elevating virus (LDV) 766, 769
    simian immunodeficiency virus (SIV) 1320
  virus uptake 1079
  visna–maedi virus replication 1594
Maculopapules, in rubella 1296
Mad-itch agent 1173
Mad-itch disease, of cattle 1173
Madin Darby canine kidney (MDCK) cells
  HSV glycoprotein gC role in attachment 606
  influenza A virus infection 1587
MadinDarby bovine kidney (MDBK) cells, bovine parainfluenza virus type 3 (PIV-3) in 1032, 1034
Maedi 1592
  see also Visnamaedi viruses[A]
Magnaporthe grisea 1233
Magnesium, in DNA transposition in bacteriophage Mu 872
Maize, transposable elements 1542
Maize chlorotic dwarf virus group 1399
Maize chlorotic dwarf virus (MCDV)[P] 824, **824–827**
  control 825
  economic importance 824, 825
  epidemiology 825
  genome structure 825
  geographic distribution 825
  helper factor encoded by 826
  host range 826
  host relationships 826
  identification/diagnosis 826–827
  MCDV-M1 825
  MCDV-white stripe (-WS) 825
  overwintering host 825, 826
  strains 825
  structure and composition 824
  symptoms of infection 826
  synergistic interaction of strains 825
  taxonomy and classification 824
  transmission 825, 825–826
    specific binding to mouthparts of leafhopper 826
Maize chlorotic mottle virus (MCMV)[P] 218
  tobacco necrosis virus similarity 899
Maize mosaic virus (MMV)[P] 1247
  disease and pathology 1247
Maize rough dwarf virus (MRDV)[P] 1096
  genome 1097
Maize streak virus (MSV)[P] 517, 1538
  gene expression 519
  inoculation 1537
  as vector 1539
Maize stripe virus (MstV)[P] 1410, 1411, 1414
  gene expression strategies 1413
  genome 1412
  host relationships 1415
Major histocompatibility complex (MHC)
  antigen expression, downregulation in immune evasion 701
  B^{21} allele, resistance to Marek's disease virus 836
  class I antigens 725
    antigen processing 726, 1092
    antigen-binding site 726
    cells expressing 726
    down regulation and immune evasion strategy 1094
    endogenous processing pathway 704, 705
    increase, in autoimmunity development 64
    increased expression in LCMV infections 805
    β_2-microglobulin interaction 704
    structure 726
    upregulated, in autoimmunity 64, 65
  class I genes 705
    resistance to mouse cytomegalovirus and 667
    transcription, adenovirus gene E1A effects 21
  class I restriction 698, 703, 708, 726, 727, 1504–1505
  class II antigens 725
    cells expressing 704, 726
    in exogenous processing pathway 704, 705
    expression in bovine leukemia virus infection 173
    in immune response 697
    pyrogenic toxins binding 104–105
    structure 726
    upregulation, in autoimmunity 64
  class II restriction 697, 704–705, 726, 727, 1504–1505
    in Borna disease virus (BDV) infection 154
  complex formation, with adenovirus E3 gene 19, 23
  H-2 locus, resistance to mouse mammary tumor virus (MMTV) 856
  history 632
  severity of viral diseases relationship 665
  susceptibility to viral diseases and 668
  T-cell epitope production for vaccines 1505
mak mutants 1602
Malaria, falciparum, resistance 665
'Malarial catarrhal fever' 942
Malignancy 1472
  see also Cancer; Carcinogenesis; Tumor
Malignant rabbit fibroma virus (MRV)[A] 1154
  pathogenesis, mechanisms 1156
  replication in lymphocytes 1156, 1157, 1159
  see also Myxoma virus (MYX)[A]
Mammalian hepadnaviruses, *see* Hepadnaviridae; Hepatitis B virus (HBV)[A]
Mammary tumors, mice 855
  see also Mouse mammary tumor virus (MMTV)[A]
Mammillitis 155, 157
Manawatu virus (MwV)[I] 919
Maracuja mosaic tobamovirus (MaMV)[P] 1436
Marafivirus[P] 1399
  structure 1580
'Marble spleen' disease, of chickens 17
Marburg virus[A] 446, **827–832**, 1619
  evolution 830
  future perspectives 832
  genome structure and properties 828
  geographic distribution 829–830, 1619
  history 827
  host range and viral propagation 830, 1619
  infection
    clinical features 831
    diagnosis and cell cultures 832, 1619
    epidemiology 830
    immune response 831
    pancreatic 969
    pathology and histopathology 831
    prevention and control 831–832, 1619
  pathogenicity 831
  physical properties 828
  proteins
    L protein and GP glycoprotein 829
    NP nucleoprotein 829
    properties 828–829
  replication 829
  serologic relationship 830
  structure and properties 828
  taxonomy and classification 827–828, 1619
  transcription 829
  transmission and tissue tropism 830, 831, 1619
Marek's disease (MD) 832
  acute 833, 835
  clinical features 835
  lymphoma 833, 834, 835
  neural form ('fowl paralysis') 833, 835
  pathology and pathogenesis 835–836
    A-, B- and C- type lesions 835
  transient paralysis 835
  see also Marek's disease virus (MDV)[A]
Marek's disease tumor-associated surface antigen (MATSA) 835
Marek's disease virus (MDV)[A] **832–837**
  A, B and C antigens 833

  atherosclerosis in chickens 967–968
  cell interactions 833
  cell-free preparations 834
  distribution and epizootiology 834–835
  future perspectives 837
  genetic resistance 836
    loci 836
  genome structure 833
  history 832–833
  immune responses to 836
  infection, *see* Marek's disease (MD)
  isolation and propagation 834
  lymphoma 834
  mixed infections 834
  nonproductive infection 833, 834
  prevention and control 836–837
  productive infection 833
  proteins 833
  semiproductive infection 833
  serotypes (3) 833
  structure and properties 833–834
  taxonomy and classification 833
  transmission 834
  vaccination, herpesvirus of turkeys (HVT) 833
  vaccines 836–837
    failure 837
    serotype 1 836–837
    serotype 2 837
    serotype 3 837
  see also Marek's disease (MD)
Marin county agent 376
Marker rescue technique 527, 531, 1530
Mason-Pfizer monkey virus (M-PMV)(A) 812, 1236
  assembly and release 1239
  genome 1237
  post-translational processing 1239
  proteins 1238
  see also Retroviruses, type D
Mastadenovirus(A) 1, 2, 14, 16
Mastitis, in caprine arthritis encephalitis syndrome 203
Mastocytosis, interferon therapy 743
Mastomys 780
Mastomys natalensis 778
Mathematical modeling 403
'Maulgrind' 1037
Maus-Elberfeld (ME) virus(A) 206
Mayaro virus (MAY)(A) **236–241**, 1618
  epidemiology 239
  evolution 238
  genetics and replication 238
  geographic and seasonal distribution 237, 1618
  history 237
  host range and virus propagation 237–238
  infection
    clinical features 239–240
    immune response 240–241
    pathology and histopathology 240
    prevention and control 241
  serologic relationship and variability 238–239
  taxonomy and classification 237
  transmission and pathogenicity 239, 1618
MCF recombinant viruses, *see under* Murine leukemia viruses (MuLVs)(A)
Mealybug vectors
  badnaviruses 142
  closteroviruses 247
Measles 446, 838
  clinical features 838, 845
  complications 845
  diarrhea in 377
  epidemics 838, 844
  heart infection in 965
  immune response 846
  incubation period 845
  pathology and histopathology 845–846
  pneumonia 1225
  population size importance in infections 402, 844
  prevention and control 846–847
    see also Measles virus, vaccines
  subclinical infections 844
  viremia 846
  see also Measles virus(A)
Measles, mumps and rubella (MMR) vaccine 846, 883
Measles virus(A) **838–847**
  attenuation 846
  canine distemper virus relationship 1262
  defective-interfering particles 323
  epidemiology 844
  evolution 844, 1263
  future perspectives 847
  genetics 843
  genome structure and properties 838, 839, 1262
    gene order 839, 843
    intergenic sequences 839, 842
  geographic and seasonal distribution 843
  hemagglutinating activity 841
  hemolytic activity 841
  history 838
  host range and viral propagation 843
  immune response to 846
  mutation rates 843
  P gene 843
  pathogenicity 845
  persistent infection 402, 701, 843, 844, 845, 1090
    mutation rates 843
  physical properties 841
  post-infectious encephalomyelitis 64, 845
  post-translational processing 842
  proteins 838
    C and V proteins 843
    F protein signal peptide 841
    fusion (F) 839, 840, 841
    fusion (F) action 841
    fusion (F) glycosylation 842
    H (hemagglutinin) 838, 839, 840, 841
    H (hemagglutinin) glycosylation 842
    L and P 839, 840, 841
    matrix (M) 839, 840, 841
    matrix (M), transport 842
    N protein retention in nuclei 842
    properties 839–841
  replication 841–843
    assembly, release and cytopathology 842–843
    sites 846
    strategy 841
  serologic relationships and variability 844
  structure and properties 839
  subacute sclerosing panencephalitis (SSPE) 909, 912
    strains from 845
  taxonomy and classification 838–839, 1211
  transcription 839, 841–842, 843
  translation 842
  transmission and tissue tropism 844–845
  vaccines
    administration route 1506
    attenuated live 844, 846–847
    complications after 847
    inactivated killed 846
    reaction to 847
Media, phage-inhibitory 119–120
Medical Lake macaque virus(A) 1514
Melanoma, malignant, interferon therapy 743–744
Melolontha entomopoxvirus(I) 392, 395
Melon necrotic spot virus (MNSV)(P) 218
  genome structure 220, 221
  tobacco necrosis virus relationship 900
    RNA comparison 898, 899
  transmission 219
Membranes **1556–1560**
  acquisition by budding 1556
  fusion 228, 1557–1558
  host cell, viral infection effect on 1589–1590
  lipid bilayer 228, 1556
  proteins 1556–1557
    integral 1556–1557
    peripheral 1556, 1557
  synthesis 1558–1559
  see also Plasma membrane
Membranous glomerulonephropathy 973
Mengo virus(A) 205
  attenuated strain 213
  capsid structure 206, 207
  cDNA clones 213
  cell culture 211
  genome structure 207–209
    encephalomyocarditis (EMC) virus comparison 211, 212
  Theiler's viruses relationship 1423, 1424
  translation 209
  see also Cardioviruses(A)
Meningitis
  aseptic
    coxsackievirus infection 270, 638
    echovirus infection 356, 358
    enterovirus (EV) serotype 71 (EV71) causing 383, 384
    LCMV causing 801, 804
  lymphocytic, echoviruses 358
  viral (acute aseptic) 801, 911
Meningoencephalitis
  chronic, echovirus infection 359
  lymphocytic choriomeningitis virus (LCMV) infection 804
  simian immunodeficiency virus (SIV) infections 1320
Meristem tip culture, carlaviruses 217
'Mesenchymal giant cells', in measles 846
Methanobacterium thermoautotrophicum ΨM phage(B) 54–55
Methanobacterium voltae, virus-like particles (VLP J2) 51, 55
Methanogens, viruses of 54–55
Methionine codons 530
5-Methyl deoxycytidine (5-mC) 12
Methyl transferase gene, *de novo* origin 439
Methylases 670, 673
  M.dam and M.dcm 673
Methylated bases, *Paramecium bursaria Chlorella* virus 1 (PBCV-1) in 37
Methylation
  DNA 12–13, 669–670
    adenovirus as model system 12
    cyanophage tolerance 289
    frog virus 3 (FV3) 504, 506
    functions 673
    herpesvirus saimiri transformed cells 616
    T1 phage 1375
    T2 and T4 phage 1383
    see also Host-controlled modification and restriction
  mRNA, frog virus 3 (FV3) 505
  protection from restriction enzymes 671, 672
  restriction dependent on 672–673

5-Methylcytosine
  in fish lymphocystis disease virus (FLDV) 798
  in ϕN bacteriophage 52
Methyltransferases 671
Mice
  C3Hf strain 855–856
  C56BL/6 strain, mousepox resistance 862–863
  C57BL/6 strain
    immune response to MMTV 860
    mousepox resistance 863
    resistance to mouse mammary tumor virus (MMTV) 856
  I strain
    immune response to MMTV 860
    resistance to mouse mammary tumor virus (MMTV) 856
  infant, virus isolation 338
  LCMV infection, *see* Lymphocytic choriomeningitis virus (LCMV)(A)
  LDV infection, *see* Lactate dehydrogenase-elevating virus (LDV)(A), of mice
  transgenic, *see* Transgenic mice
  see also viruses beginning Mouse, Murine
Microfilaments
  actin 231
  changes due to frog virus 3 (FV3) 507
  see also Cytoskeleton
β_2-Microglobulin
  class I MHC interaction 704, 705
  cytomegalovirus receptor 667
  cytomegalovirus replication 303
Microscopy
  direct observations of inclusions 339, 341–342
  identification of viruses by 339–341
  stains, for virus identification 342
  see also Diagnostic techniques; Electron microscopy (EM)
Microtubules 231
  functions in viral infections 231–232
  reovirus infections 1202–1203
  virus-induced changes 1590
    frog virus 3 (FV3) 507
Microvilli 228
Microvilli-like projections, frog virus 3 (FV3) 506
Microviridae 1398, 1405
  characteristics and taxonomy 99
  structure 1580
Microvirus virions(B), characteristics and taxonomy 99, 1580
mil oncogene 936
Miliary recruits, adenovirus infections 4
Milker's modules 1036, 1039, 1040
Mill Hill-2 virus (MH2)(A)
  oncogenes 936
  transduction of oncogenes 935
Mini-Mu, *see* Mu bacteriophage(B)
Mink, kuru in 759
Mink cell focus-forming murine leukemia virus (MCF-MLF)(A), as insertional mutagens 937
Mink enteritis 1055
  clinical features 1065–1066
  pathology and histopathology 1056
  prevention and control 1067
Mink enteritis virus (MEV)(A) 1052
  epidemiology 1053, 1065
  evolution 1064
  future perspectives 1067
  genetics 1064
  genome and genetics 1053
  geographic and seasonal distribution 1061–1062
  history 1061
  host range and viral propagation 1062
  immune response 1067
  serologic relationships and variability 1064
  taxonomy and classification 1061
  see also Parvoviruses(A)
Minor lymphocyte-stimulating (Mls) antigens 860
Minute virus of canines (MVC)(A) 1057
  history 1061
  host range and viral propagation 1062
Minute virus of mice (MVM)(A) 1052, **1067–1075**
  gene expression 1058, 1059
  genome and replication 1069, 1170
  history 1068
  host range and viral propagation 1071, 1072
  infection
    clinical features 1055, 1074
    pathology and histopathology 1056
  morphogenesis 1070
  properties and structure 1068–1069
  proteins 1059
  tissue tropism and pathogenicity 1073, 1074
  transcription and translation 1069–1070
  variants 1068
Miscanthus streak virus(P) 518
Mitochondria 230
Mitosis
  'closed' 1234
  oncogenes stimulating 940
Mitotic (M)-phase-promoting factor (MPF) 940
Mo T-cell line 686–687
Mo T-cell virus(A) 686
Mobala virus(A) 777
Modeling, in epidemiology 403
Modification
  host-controlled, *see* DNA modification/restriction (M/R) systems; Host-controlled modification and restriction
  of proteins, *see* Post-translational processing
Mokola virus(A) 1181, 1182, 1186
  epidemiology 1188–1189
  evolution 1182, 1188
  genetics 1181, 1187–1188
  host range and viral propagation 1187
  prevention and control 1189
  transmission and tissue tropism 1189
  vaccine 1189
  see also Rabies-like viruses(A)
Mole crickets, iridoviruses of 1434
Molecular biology 627, 633
  history, bacterial viruses and 629
  see also Gene cloning; Recombinant DNA technology
Molecular cloning 530
  see also Gene cloning
Molecular genetics 530
Molecular mimicry 64–65, 66, 703
  cardiomyopathy pathogenesis 967
  diabetes pathogenesis 971
Molluscipoxvirus(A) 848
Molluscum bodies 848, 851
Molluscum contagiosum
  clinical features 851
  incubation period 851
  papules 851
  pathology and histopathology 851–852
  prevention and control 852
Molluscum contagiosum virus (MCV)(A) **848–853**
  assembly 850
  cytopathic effects 850
  epidemiology 851
  eye infections 442
  genetics and evolution 850–851
  genome structure and properties 849
  geographic and seasonal distribution 850
  history 848
  host range and viral propagation 850
  immune response to 852
  inability to culture 850
  p43K gene 850
  pathogenicity 851
  proteins
    p43K homology to vaccinia virus protein 849, 850
    properties 849
  replication 849, 851
  restriction digests 848, 849, 851
  serologic relationships and variability 850–851
  structure and properties 848–849
  subtypes 849
  taxonomy and classification 848
  transcription and translation 849–850
  transmission and tissue tropism 851
  vaccinia virus similarity 850
Moloney murine leukemia virus (Mo-MuLV)(A) 886, 887
  as insertional mutagens 937
Moloney murine sarcoma virus (Mo-MSV)(A) 884
  oncogenes 936
Monkeypox 1343
  clinical features 1345
    in humans 1345
  mortality 1345
  outbreaks 1343
  pathology and histopathology 1345
  prevention and control 1345
Monkeypox virus(A) 267, 1341, **1343–1345**
  epidemiology and transmission 1344
  evolution 1344
  future perspectives 1345
  genetics 1344
  geographic and seasonal distribution 1343
  history 1343
  host range and virus propagation 1343
  immune response to 1345
  properties and replication 1343
  restriction maps 1344
  smallpox virus relationship 1340
  taxonomy and classification 1343
  vaccination 1345
  'white pock' mutants 1344
  see also Smallpox
Monoclonal antibodies 344, 348
  African swine fever virus 28
  diagnostic techniques using, *see specific techniques*
  flaviviruses 370
  hepatitis A virus neutralization 553
  hepatitis E virus 583
  hog cholera virus 653
  in immunofluorescence tests for virus identification 339
  lactate dehydrogenase-elevating virus (LDV) 768, 770
  reovirus type 3 receptor (Reo3R) 1561, 1562
  virus identification 399
  zoonoses diagnosis 1620
Monocytes
  dengue virus replication 328
  feline immunodeficiency virus (FIV) infection 456, 457

  VZV infection 1516, 1517
  see also Macrophage
Mononegavirales 830, 1397, 1406
Mononuclear cells, VZV infection 1516, 1517
Montgomery County virus(A) 926
  antigenic relationships 928
  see also Norwalk virus(A)
Montmorillonite, effect on phage stability 123
Mopeia virus(A) 777
Morbillivirus(A) 445–446, 914, 1027, 1211, 1260–1261
  evolution 1263
  genome structure 1261, 1262
  identification and diagnosis 1266, 1267
  measles virus 838
  pathogenicity 1265
  pathology and histopathology 1265–1266
  prevention and control 1266–1267
  proteins 1262–1263
  rinderpest virus as archevirus 844
  transmission 1264–1265
  vaccines 1266, 1267
  viruses included 914
  see also Canine distemper virus (CDV)(A); Measles virus(A); Peste des petits ruminants (PPR) virus(A); Phocid distemper virus (PDV)(A); Rinderpest virus(A)
Morison's cell inclusions 655
Moroccan pepper virus(P) 1447
mos oncogene 884, 936, 941
Mosquito transmission
  bunyavirus 189
  Chandipura, Piry and Isfahan viruses 234, 235
  Chikungunya (CHIK), O'nyong nyong and Mayaro virus 237, 239
  control, in Chikungunya, O'nyong nyong and Mayaro virus infection prevention 241
  dengue virus 324, 325, 328
  encephalitis virus 364, 365
  equine encephalitis virus 419, 420
  fowlpox virus (FPV) 501
  Japanese encephalitis virus 747, 748, 749
  Ross River virus (RRV) 1272
  spread and transport methods 1622
  Wesselsbron virus (WSLV) 369, 370
  yellow fever 1606, 1609, 1610
  see also Aedes aegypti
Motor neurons, Theiler's viruses infection 1427
Mottle virus group 1398
Mouse cytomegalovirus (MCMV)(A) 304
  propagation 305
Mouse encephalomyelitis virus(A) 206
  see also Theiler's murine encephalomyelitis virus (TMEV)(A)
Mouse hepatitis virus (MHV)(A) 255
  cytopathology 256
  epidemiology 258
  pathology 259
Mouse hepatitis virus type 4 (MHV4)(A), susceptibility, *Hv-2* locus 667
Mouse mammary tumor virus (MMTV)(A) **853–861**, 1237
  cloning 856
  DMBA-LV 858
  endogenous (germline) 855
    immune response 860
  exogenous (milk-borne) 855
  gag gene 855
    transcription 857
  genetic experiments difficulties 856
  genetics 855–856
  genome structure 853, 854, 855
  glycoproteins 856
    gp52 856
  high-mammary-cancer-incidence mice 855, 858, 859, 860
  history 853
  hormone-response elements (HRE) 858, 859
  host range 855
  hst gene 860
  immune response 860–861
  incidence of tumors 855, 856
    genetic factors influencing 855
  as insertional mutagens 937
  int-1 integration locus 860
  int-2 and *int-3* integration loci 860
  integration 857, 860
    loci, characteristics of 860
    site (*int-41*) 860
  long terminal repeat (LTR) 854, 855, 856
    deletions and transcription regulation 859
    deletions/insertions 858
  low-mammary-cancer-incidence mice 855, 858, 859, 860
  Mtv-6 855, 861
  Mtv-8 855, 861
  Mtv-9 861
  Mtv-17 855
  oncogenesis 859–860
    mapping of integration loci 860
    Wnt-1 (*int-1*) 860
  pro and *pol* translation 857
  proteins, IN protein 857
  proto-oncogenes 938
  proviruses 855, 857
    LTR deletions/insertions 858
    structure 854
  receptor (*MTVR-1*) 856, 857
  replication 856–857
    unintegrated DNA types 857
  resistance to 856
  sag open reading frame 855, 857
  Sag protein 861
  structure and composition 853–855
    primer-binding site (PBS) 853, 855
  T-cell proliferation by 860, 861
  T-cell receptor depletion by 861
  taxonomy and classification 853
  tissue tropism 857–858
    kidney 858
    mammary gland 857–858
    T-cell tumors 858
  transcription 854, 857
  transcriptional regulation 858–589
    control elements 859
    control regions 854
    hormone-response element (HRE) 859
    positive regulators 859
  translation 857
    block 856
  transmission 855
  tumor types 859
  Wnt-1 integration locus 860
Mouse-pathogenic viruses 637, 638
  taxonomy and nomenclature 641
  see also Coxsackieviruses(A)
Mousepox 862
Mousepox virus(A), *see* Ectromelia virus(A)
Movement protein, genes 1541
mRNA 232
  adenovirus 10, 12
  antisense, in control of phage 121
  capping, discovery 631
  'caps' 232
  cleavage sites 530
  discovery 630
  host, degradation 1588
  host synthesis, shutdown by frog virus 3 (FV3) 504, 506
  influenza virus effect on 1588
  methylation, frog virus 3 (FV3) 505
  polyadenylation 232
  processing, human immunodeficiency virus (HIV) 676–677
  transcapping 232
  viral 1205
    degradation by 2′-5′ oligoadenylate synthetases 736
    influenza virus, capping 232
    nodaviruses 922
    orbiviruses and coltiviruses 957–958
    selective translation 1589
  see also under transcription of specific viruses
mRNA 2′-0-methyltransferase 1097
Mu bacteriophage(B) **868–876**
  attachment 868–870
  c gene product (repressor) 870, 872, 873
    binding to operators 872
    expression 872
  chromosomal DNA linked 868, 870
  com gene 874
  DNA modification 873–874
  DNA replication 873
  DNA transposition 868, 870
    frequency 872
    mechanism 872–873
    requirements for 872
  encapsidation and packaging 873
  genetic map 868, 869, 871
  as genetic tool 875–876
  genome 'inserted' in chromosomal DNA in virus 871
  genome propagation mode 871
  gin gene 874
  host range mutants 876
  IAS (internal activating sequence) element 872
  invertible G-loop 874
  late gene expression and DNA maturation 873
  left end regulatory regions 871
  life cycle 868–871
    lytic/lysogenic decision 871–872
  lysogenic response 870, 872
    c gene product (repressor) 870, 872, 873
  lytic cycle 870, 872
    early gene expression regulation 871–872
  mini-Mu derivatives 875–876
  mom gene 873–874
  Mu-1 868
  N gene product 870
  ner gene 870, 872
  Ner protein 872, 873
  pac and pacase 873
  promoter (P_c) 872
  prophage formation for lytic/lysogenic response 871
  random insertion into host genome 868, 870, 871
  S and U genes 874
  transposase (*A* gene product) 870, 872
    binding site 872
    expression 872
Muchupo virus(A)
  transmission 781
  see also Bolivian hemorrhagic fever (BHF)
Mucosal administration of vaccines 1505
Mucosal disease 179
  clinical features 180
  HPV infections 1016, 1019, 1497

- Mucosal disease (MD) virus$^{(A)}$ 175
- Mucosal surfaces, herpes simplex virus infection 589
- Mulberry latent virus (MuLV)$^{(P)}$ 215
- Mulberry ringspot virus$^{(P)}$ 904
- Multifocal leucoencephalopathy 1487
- Multinucleated giant cells, in measles 846
- Multiple myeloma, interferon therapy 743
- Multiple sclerosis
  - animal models 1423
  - coronavirus infections 259
- Multiplication cycle
  - blocking and abortive infections 1170
  - viruses 1168–1169
  - *see also* Replication of viruses
- Multiplicity of infection (MOI) 1171
- Multiplicity reactivation 527
- Multivesicular bodies, tombusvirus infections 1451
- Mumps 446, 876
  - clinical features 882
  - history 876
  - incidence 883
  - incubation period 882
  - neurological complications 1083
  - pathology and histopathology 882
  - prevention and control 882–883
- Mumps virus (MuV)$^{(A)}$ 446, **876–883**, 907
  - cDNA clone 883
  - cytopathic effects 881
  - diabetes and 971
  - epidemiology 881
  - evolution 881
  - in fetus 882
  - future perspectives 883
  - genetics 881
  - genome structure and properties 877
    - gene order 877, 878
    - intergenic sequences 879
  - geographic and seasonal distribution 880
  - growth in choroid plexus epithelium 909
  - heart infections 965
  - hemagglutinin, amino acid sequence 1034
  - hemagglutininneuraminidase (HN) 877, 878
    - processing 880
  - history 876
  - host range and viral propagation 881
  - immune response to 882
  - inclusion body myositis and 883
  - isolation 338
  - mRNAs 878, 880
  - mutants 881
  - pancreatitis 969
  - pathogenicity 882
  - physical properties 878
  - proteins
    - fusion (F) 877, 880
    - fusion (F), role 880
    - properties 877–878, 879
    - SH protein 878
  - receptor 878
  - replication 878–880
    - budding 880
  - RNA replication 880
  - serologic relationships and variability 881
  - structure and properties 877
  - taxonomy and classification 876–877
  - transcription 878, 878–879
    - consensus sequence 879
    - cotranscriptional editing 880
  - translation 880
  - transmission and tissue tropism 881–882
  - vaccines 883
    - live attenuated 883
- Mung bean yellow mosaic virus$^{(P)}$ 520
- Murine acquired immunodeficiency syndrome (MAIDS) 888
- Murine cytomegalovirus (MCMV)$^{(A)}$ **307–312**
  - assembly, release and cytopathology 308
  - evolution 309
  - future perspectives 311–312
  - genetics 309
  - genome structure 307
  - geographic and seasonal distribution 308–309
  - history 307
  - host range 309
  - infection
    - clinical features 310
    - immune response 311
    - pathology and histopathology 311
    - prevention and control 311
  - latency 311
  - as models of human infection 310–311
  - pathogenicity 309–310
  - physical properties 308
  - proteins 307–308
  - replication 308
  - resistance to 310
    - *Cmv-1* and 668
    - *H-2D* gene 667
  - structure 307
  - taxonomy and classification 307
  - transcription and translation 308
  - transmission and tissue tropism 309
  - virus propagation 309, 310
  - *see also* Cytomegaloviruses (CMV)$^{(A)}$
- Murine leukemia viruses (MuLVs)$^{(A)}$ **883–890**, 1135, 1473
  - amphotropic 885
  - classification 884, 885
  - distribution 885
  - ecotropic 885
  - endogenous 885, 887
  - Friend–Moloney–Rauscher (FMR) group 885
  - *gag*⁻ mutants 884
  - genome structure 884–885
  - germ line infections 885
  - Gibbon ape leukemia virus (GaLV) relationship 536
  - Gross group 885
  - history 883–884
  - host range 885–886
    - envelope glycoprotein interaction with receptor 885
    - Fv-1 restriction (N/B tropism) 885–886
  - immunodeficiency (MAIDS) 888
  - infection sites 886
  - as insertional mutagens 937
  - latency of disease 887
  - leukemogenesis 886–888
    - DNA integration sites 886, 887
    - enhancer activation 886
    - MCF recombinant action 888
    - preleukemic changes 888
    - promoter insertion 886
  - long terminal repeats (LTRs) 884
    - activation of proto-oncogenes 886–887
    - enhancers controlling tissue specificity of disease 887
  - MCF recombinant (*env* recombinants) 885, 887–888
  - neuropathic 888
  - polytropic 885
  - reticuloendotheliosis viruses relationship 1227, 1229
  - retroviral vectors based on 888–890
  - structure and replication 884–885
  - translation 884–885
  - type-specific antigens 885
  - variants 884
  - xenotropic 885
- Murine parvovirus$^{(A)}$, *see* Parvoviruses$^{(A)}$
- Murine poliovirus$^{(A)}$, *see* Theiler's murine encephalomyelitis virus (TMEV)$^{(A)}$
- Murray Valley encephalitis 366, 1269, 1619
- Murray Valley encephalitis virus (MVEV)$^{(A)}$ 361, 363, 1619–1620
  - genetics 364
  - geographic and seasonal distribution 362, 1620
  - Japanese encephalitis virus relationship 748
  - transmission 1620
- *Mus musculus*
  - LCMV infection 802
  - mousepox 862
- Muscles, striated, coxsackievirus infection 271
- Muscular disability, echovirus infections 358
- Muskmelon vein necrosis virus (MmVNV)$^{(P)}$ 215
- Mutagenesis 526
  - insertional, *see* Insertional mutagenesis
  - localized random, transduction application 112
  - phage resistance 120, 121
  - site-directed 527
  - SPO1 phage 1356
- Mutagens 526
  - insertional
    - retroviruses as 935, 937, 937–938
    - transposable bacteriophage as 875
  - viruses acting as 1493
- 'Mutant spectrum' 437
- Mutations 438, 525
  - affecting antigenic determinants 526
  - classification by phenotype 526
  - conditional lethal mutants 526
  - defective-interfering (DI) mutants 526
  - deletion 526
  - immune evasion and 1093
  - influenza viruses 711
  - non-lethal 526
  - point 525, 526
    - in evolution of virus species 437
  - polioviruses 1116
  - rate 525
    - adenoviruses 2
    - DNA genomes 438, 525
    - foot and mouth disease viruses 494
    - reoviruses 1192
    - RNA viruses 438, 441, 525, 1220
  - spontaneous, frequency 120
  - suppressor 526
  - types 526
  - *see also individual genes/viruses*
- *Mx1* gene, resistance to influenza viruses 667, 697, 711, 736
- *Mx* gene, cells lacking and influenza susceptibility 736
- Mx proteins 736
  - sequence homology with GTP-binding proteins 736
- Myalgia, coxsackievirus infection 271
- Myalgic encephalomyelitis syndrome 271
- *myb* oncogene 936
- *myc* oncogene 936, 937, 938
- *Mycobacterium avium-intracellulare*, interferon-γ in 738, 741
- *Mycogone perniciosa* virus (MpV)$^{(P)}$ 1464
- Mycosis fungoides, interferon therapy 743
- Mycoviruses (fungal viruses) 1047
  - cryptovirus affinities 276
  - dsRNA 1464

see also Totiviruses(P)
families 1047, 1406
infectivity assays 1467
mixed infections 1050
see also Partitiviruses(P)
Myelin breakdown, *see* Demyelination
Myeloblastosis associated virus (MAV)(A), as insertional mutagens 937
Myelocytomatosis virus 29(MC29)(A), oncogenes 936
Myelodysplastic syndromes, interferon therapy 743
Myeloma, refractory, interferon therapy 743
Myelopathy
HTLV-1 associated 685, 686
progressive spastic and HTLV-2 infection 693–694
Myeloproliferative disorders, interferon therapy 742
Myeloproliferative sarcoma virus (MPSV)(A), oncogenes 936
Myocarditis
asymptomatic 964
canine parvovirus (CPV) infection 1066
chronic, coxsackievirus B3 (CVB3) infections 668
echovirus infection 359
subclinical 964
see also Heart, virus infections
Myocardium, virus infections 964–969
Myoclonus, in Creutzfeldt-Jakob disease (CJD) 1367
Myopericarditis
acute 965
clinical features 968
chronic relapsing 966
laboratory diagnosis 968
Myositis
coxsackievirus infection
inclusion body 883
Myoviridae 95, 96, 286, 1007, 1398, 1405
characteristics and taxonomy 97
structure 1582
Myristic acid, VP4 of poliovirus 1115, 1120
Myristylation 27, 386
animal enteroviruses 386
hepatitis A virus VP4 546
Myrobalan latent ringspot virus(P) 904
Mystery swine disease (MSD) 763
Myxoma 1158
Myxoma leporipoxvirus(A), evolution studies 436–437
Myxoma virus (MYX)(A) 402, 1153, 1154
California strain 1157, 1158
clinical features of infections 1158
cytopathology 1156
genetic resistance 665
immune response 1159
pathogenesis, mechanisms 1156
pathogenicity 1157–1158
properties, DNA and proteins 1155
release and virulence 1157
replication in lymphocytes 1156, 1157, 1159
transmission and tissue tropism 1157
vectors 1156
see also Poxviruses(A)
Myxomatosis 402, 1153, 1156, 1157
Myxoviruses, antivirals 44, 47

N

N4 bacteriophage(B) **891–895**
DNA polymerase 894
effect on host cells 893
future perspectives 895
gene expression 892–893
genetic and physical map 893
genome structure 891
3′ extensions 893, 894
growth cycle/morphogenesis 891–892
host range 894–895
infection process 895
joint fragments 893, 894
mutants 893
promoters 892
receptor 895
recombination 893
replication 893–894
in vitro 894
phage-encoded functions 894
RNA polymerase 891, 892, 893
in initiation of replication 894
single-stranded-DNA-binding proteins 891, 892
structure 891
taxonomy and classification 891
transcription 891, 892–893
early 892
in vitro 892–893
late 893
middle 893
N-*myc* proto-oncogene 938
Nairobi sheep disease 191
Nairovirus(A) 187, 191
genome 188
sequences and coding strategies 193, 194
structure 188
transmission 189
see also Bunyaviruses(A)
Nandina stem pitting virus (NSPV)(P) 243
Narcissus mosaic virus (NaMV)(P) 1143, 1146
Narcissus tip necrosis virus (NTNV)(P) 218, 219
Nasopharyngeal carcinoma (NPC) 404, 1495–1496
clinical features 408
EBV gene expression 408, 414–415, 1495
epidemiology 406
evasion of immune surveillance 409
foamy virus isolation 480, 487
IgA antibodies 1496
pathology and histopathology 409
see also Epstein–Barr virus (EBV)(A)
Nasturtium mosaic virus (NasMV)(P) 215
Nasturtium ringspot virus (NRSV)(P) 451
serology 453
Natural killer (NK) cells 1091
function 1092
in cells immortalized by herpesvirus saimiri 615
herpes simplex virus infections 591
lymphocytic choriomeningitis virus (LCMV) infections 805
Marek's disease virus 836
murine cytomegalovirus (MCMV) infection 311
nonadaptive immune response 697, 1091
pseudorabies virus infection 1177
respiratory infections 1222
Navarro virus(A) 1249
Neckar river virus(P) 1447
Necrotizing enterocolitis 259
Necrovirus(P) 1399
Necroviruses(P) **896–901**
structure 1580
see also Tobacco necrosis virus (TNV)(P)
nef gene 1318
Negative regulatory element (NRE), HIV 676
Negri bodies 1181, 1184, 1590
Nematode transmission 633
dianthoviruses(P) 352
distribution 1446
nepoviruses 901, 902, 906
tobraviruses 1445, 1446
Neonatal infections
cytomegalovirus 294, 295
immunological tolerance induced 701
Neoplastic alterations, in latent infections 787
Neoplastic disease
feline immunodeficiency virus (FIV) infection 457
see also Cancer; Tumor
Nephropathia epidemica (NE)
geographic and seasonal distribution 541
pathology 543, 544
Nephrosis 973
Nepovirus(P) 1400
Nepoviruses(P) **901–907**
coat protein 902, 905
sequence comparisons 903, 904
comovirus relationship 253, 902
components (top/middle/bottom) 902
control of infections 906
cytopathology 906–907
definitive and tentative members 904–905
effects on host plants 906
epidemiology 906
fabaviruses similarity 454
genetics 903
genome structure/properties 902–903
history 901–902
host range 906
molecular properties 902–903
physical properties 902, 904–905
purification 902
RNA (RNA-1, RNA-2) 902, 903
satellites 905
structure 1581
taxonomy and classification 901, 902
transmission 901, 902, 906
VPg linked to RNA 902, 903
Nerine latent virus (NeLV)(P) 215
Nervous system, viruses(A) **907–913**
anatomic considerations 907–908
cell damage mechanisms 911
clinical features of infections 911–913
acute 911–912
slow infections 912–913
coxsackievirus infection 270
hematogenous spread 909, 910, 1079
neural cell infection 909–911, 1079–1080
pathways of CNS invasion 908–909, 1079–1080
viral proteins affecting 1080
trans-neuronal and trans-synaptic transport 1080
see also Central nervous system (CNS); Neurological disease; Neurons
neu oncogene 938
Neuraminidase (NA)
bovine parainfluenza virus type 3 (PIV-3) 1031
discovery 630
influenza viruses 710, 715, 722–724
antigenic drift and shift 711
atomic structure 1572
epitopes on 725
function 723
orientation 723
properties and structure 716–717, 722–723
receptor-destroying activity 717
three-dimensional structure 723–724
post-translational processing 720
see also Hemagglutininneuraminidase (HN)
Neuroblastoma 938

Neurofibrillary tangles 1368
Neurofibromatosis 1475
Neuroinvasive, definition 907
Neurological disease
  coronavirus infections 259
  cytomegalovirus infection 296
  dengue hemorrhagic fever/dengue shock syndrome (DHF/DSS) 329
  feline immunodeficiency virus (FIV) infection 457
  see also Central nervous system (CNS); Nervous system; Neurons
Neurolymphomatosis, fowl, *see* Marek's disease (MD)
Neuronophagia 1427
  Japanese encephalitis virus infection 749
Neuronotropic, definition 907
Neurons 907, 908
  acidophilic inclusions 153
  control of HSV latent infections 789
  latent infections 789, 908
  lytic infections 911
  noncytopathic infections 911
  satellite cells around, VZV latency 1516, 1526
  surface markers and latent infections 789
  Theiler's murine encephalomyelitis virus (TMEV) in 1426
  virus infections 909–911, 1079
  virus receptors 908
Neuropil, compact 909
Neurospora crassa, TAD element 1233
Neurotransmitter
  enzymes reduced, in LCMV infections 804
  receptors, for viruses 1080
Neurotropic, definition 907
Neurotropism
  Borna disease virus (BDV) 152, 153
  enterovirus (EV) serotype 70 (EV70) 381–382
  kuru agent 760
Neurovaccinia virus 866
Neurovirulent
  definition 907
  Theiler's viruses 1425–1426, 1426, 1427
Neutralization tests
  rabies-like viruses 1188
  virus identification 341
Neutropenia, interferon-α causing 740
NeVTA virus(A) 470
New Zealand black beetles virus (BBV)(I) 919
Newbury agent 375
Newcastle disease 916
  clinical features 918
  history 914
  immune response 918
  pathology and histopathology 918
  prevention and control 918
Newcastle disease virus (NDV)(A) 446, **914–919**, 1027
  antigenic variation 917
  assembly 916
  cell cultivation in hog cholera virus infected cells 651
  effect on host cell 1589
  epidemiology 917
  evolution 917
  future perspectives 918
  genetics 917
  genome structure/properties 915
  genomic sequences 1029
  geographic and seasonal distribution 916
  hemagglutinin, amino acid sequence 1034
  history 914
  host range and virus propagation 917
  importation to US linked to smuggled birds 1622
  infection, *see* Newcastle disease
  'intrinsic interference' 729
  nomenclature (PMV-1, PMV-2, PMV-3) 917
  nucleocapsid protein (NP) 914, 915
  pathogenicity 918
  post-translational processing 916
  proteins
    fusion glycoprotein 915
    large (L) protein 915
    matrix protein (M) 915
    phosphoprotein (P) 915
    properties 915
  replication 915–916, 917
  serologic relationships and variability 917
  structure and properties 914–915
  taxonomy and classification 914
  transcription and translation 916
  transmission and tissue tropism 918
  vaccination 916, 918, 919
Nicotiana glutinosa, tobacco mosaic virus infection 1436
Nicotiana sylvestris, tobamovirus resistance 1440
Nicotiana velutina mosaic virus (NVMV)(P) 509, 515
Nitrogen metabolism, enzymes 291
NOD mouse 65
Nodamura virus (NOV)(I) 919, 925
Nodaviridae 919, 1400, 1408
  structure 1581
Nodavirus(I) 919
Nodaviruses(I) **919–925**
  assembly 923–924
  cDNA clones 923
  coat protein 920
  composition and structure 920
  culture and propagation 924
  epidemiology and pathogenesis 924
  gene expression regulation 922
  genome structure and coding potential 921–922
  host range 924
  mRNA 922
  nucleotide sequences 921
  proteins 920
    synthesis 922
  RNA replication 923
  RNA (RNA1 and RNA2) 919, 921–922
  RNA-dependent RNA polymerase 921, 923
  serological relationship 919
  taxonomy and classification 919
  tetraviruses similarity 1419
  three-dimensional structure 920–921
  translation 922
  virus-host cell interactions 924
Nomenclature of viruses 1396, 1402–1403
  recommendations to editors 1402–1403
  suffixes 1402
  see also Taxonomy and classification
Non-A, non-B hepatitis
  enterically-transmitted (ET-NANBH) 580
    see also Hepatitis E
  see also Hepatitis C virus (HCV)
Noncytopathic infections, neural cells 911
Nonhost resistance, *see* Plant resistance (to viruses)
Nonoccluded baculoviruses, *see under* Baculoviruses
Nonproductive infections, types 1170–1171
North American, zoonoses in 1616–1617
  see also Colorado tick fever virus (CTFV)(A), ; Encephalitis viruses(A)
Northern cereal mosaic virus(P) 1246
Norwalk virus(A) 375, **925–933**, 926
  antibody 930
  antigenic relationships 927, 928
  antigenic types 375
  in *Caliciviridae* 928, 929, 933
  calicivirus cross-reactions 375, 928
  electron microscopy 342
  genome structure/organisation 928, 929, 933
  hepatitis E virus similarities 581, 585
  history 373, 925–936, 1274
  host range and propagation 928
  infection
    clinical features 930–931
    epidemiology 930
    gastroenteritis 375
    immune response 378, 931
    outbreaks 926, 930
  infectivity studies in volunteers 926, 927
  Montgomery Country virus 929
  pathogenicity 930
  RNA 927, 928–929
  serologic relationships 929–930
  serotypes 929
  structure 927
  Taunton agent 929
  taxonomy and classification 926–928, 929
  transmission 401, 930
Nosema apis 659, 660
Nosocomial infections 1613
Nostoc, cyanophage 290
Nostoc muscorum 291
Nuclear polyhedrosis viruses (NPVs)(I) 127, **130–136**
  budded virus (BV, extracellular virus [ECV]) 131, 133
  DNA binding protein 135
  ecology 135–136
  economic importance 131, 136
  envelope 134, 135
  gene expression regulation 132–133
    trans-activating factors 132, 133
  genome structure 131, 132, 135
  gp64 protein 135
  history 130–131
  homology absent between cytoplasmic polyhedrosis viruses 315
  host range 132
  hr-enhancer sequences 132
  importance 131
  life cycle and pathogenesis 133, 134
  multiple (MNPV) nucleocapsid 131
  p10 protein 133, 134–135
  p74 gene 135
  polyhedra-derived virion (PDV, occluded virus [OV]) 131, 133
  recombinant and genetic engineering 136
  replication 131, 132
  single (SNPV) nucleocapsid 131
  structure and composition 133–134
    nucleocapsids 131, 135
  taxonomy and host distribution 131–132
  transposable elements 132
Nuclearcytoplasmic exchanges 232–233
Nucleic acid
  amplification methods 347–348
    see also Polymerase chain reaction (PCR)
  purification 347
  subgenomic 322
  viral, detection methods 347–348
    see also Hybridization methods
  see also DNA; RNA
Nucleic acid sequence based amplification

(NASBA) 348
Nucleic acidprotein interactions 1572
Nucleocapsid forms 1209
  see also Helical symmetry; Icosahedral symmetry
Nucleoside analogs 568
  fish herpesvirus control 473–474
  herpes simplex virus infections 592
  HIV inhibition 48
Nucleoside triphosphate phosphohydrolase (NPH I) 394
Nucleosomes 232
'Nucleotype' 949
Nucleus 232
  hypertrophy, in ascovirus infection 61
  protein movement into 232, 233
Nudaurelia beta virus (NβV)(I) 1416, 1417
  ecology 1422
  genome 1418
  replication 1421
  see also Tetraviruses(I)
Nudaurelia cytherea capensis 1417, 1421, 1422
Nudaurelia omega virus (NωV)(I) 1417
  ecology 1422
  genome 1418, 1420

O
Oat golden stripe virus (OGSV)(P) 509
Oat sterile dwarf virus (OSDV)(P) 1096
Obodhiang virus(A) 1186
Occlusion body (OB) 127
  entomopoxviruses (EPVs) 393, 396
  nuclear polyhedrosis viruses (NPVs) 131
  proteins 127–128
  structure 127–128
Odontoglossum ringspot tobamovirus (ORSV)(P) 1436, 1538
Oily hair, reovirus infections 1193
Okazaki fragments 996, 1326
Okra mosaic virus(P) 1500
Olfactory nerve system, pseudorabies virus 1176
Olfactory rods 908
Olfactory spread, of viruses 908–909
2′-5′ Oligoadenylate synthetases 736
Oligodendrocytes
  altered in JC virus infection 756
  infection 910
  reovirus type 3 receptor (Reo3R) function 1562
  Theiler's viruses infection 1427, 1429
Oligosaccharides, N-linked
  HSV glycoproteins 604, 605
  modification 230, 231
  see also Glycosylation
Oliguria 544
Olive latent ringspot virus(P) 904
Olive latent-1 virus (OLV-1)(P) 1346
  see also Sobemovirus(P)
Olpidium brassicae 515
  tobacco necrosis virus transmission 896, 897
Olpidium radicale, tombusvirus transmission 1451
Omsk hemorrhagic fever 366, 1618
Omsk hemorrhagic fever virus (OHFV)(A) 362, 363, 1618
  epidemiology 365
  see also Encephalitis viruses(A)
Oncogenes 439, 527, **934–941**
  avian leukosis viruses 69, 1473
  cell proliferation control 938, 940
  cellular (c-*onc*) 934
    see also genes beginning c-; Proto-oncogenes
  cooperation 1475
  definition 934
  in development and embryogenesis 940–941
  discovery 631, 934–935, 935
  future perspective 941
  identification 934, 935
  LA-1 592
  retroviral 67, 631, 934, 935, 936
  signal transduction and 938–940, 1473
    growth factor action 938–939, 1473–1474
    growth factor receptor 939, 1474
    signal transducer 939–940, 1473, 1474
    transcriptional regulators 940, 1475
  spontaneous activation of proto-oncogenes 938
  transduction 934, 935
    multiple 935
  transmembrane receptors 938
  viral (v-*onc*) 71, 934, 935–941
    in avian leukosis virus infections 69, 70
    transformation-associated genes (*STP*) of herpesvirus saimiri 616–617, 621
    see also individual viral oncogenes beginning v-
  see also Oncogenesis; Retroviruses(A); Transformation; Tumor formation
Oncogenesis 1076, 1493
  adenoviruses in 1, 8, 10, 14, 18
    see also Adenovirus(A)
  discovery 631
  herpesvirus saimiri and herpesvirus ateles 616–617, 619
  mouse mammary tumor virus (MMTV) 859–860
  murine leukemia viruses 887
  see also Carcinogenesis; Oncogenes; Transformation; Tumor formation
Oncogenic viruses 631, 1472–1476, 1476–1477
  see also Oncoviruses; Retroviruses(A); Transformation
Oncorhynchus masou virus (OMV)(A) 470, 477
Oncovirinae 159, 1227
Oncoviruses(A)
  bovine leukemia virus (BLV) 159, 167
  Lentiviridae comparison 199
  type C, feline leukemia virus 460
Ononis yellow mosaic virus(P) 1500, 1501
O'Nyong Nyong (ONN) virus(A) **236–241**, 1619
  epidemiology 239
  evolution 238
  genetics and replication 238
  geographic and seasonal distribution 237, 1619
  history 236–237
  host range and virus propagation 237–238
  infection
    clinical features 239–240
    immune response 240–241
    pathology and histopathology 240
    prevention and control 241
  serologic relationship and variation 238–239
  taxonomy and classification 237
  transmission and pathogenicity 239, 1619
Open reading frames (ORFs)
  redundancy and mutations, in evolution 438
Opportunistic infections
  in HIV infections 680–681, 708
  in SIV infections 1320
Oral mucosa, HPV affecting 1019
Oral rehydration solutions, rotavirus infections 1280
Orbivirus(A) 942
  characteristics 956
  serogroup(a),(b) 943, 946
Orbiviruses(A) **941–964**, 1194
  Culicoides-transmission 942, 944, 950
  economic significance 950
  epidemiology 949–950
  evolution 946, 958
  future perspectives 955–956, 963
  gene reassortment 946
  general features **941–956**
  genetics 945–946
  genome
    profiles 946
    sequence 956
    structure/organization 957–958
  geographic and seasonal distribution 944–945
  history 941
  host range and viral propagation 943, 945
  import/export restrictions 942, 950, 956
  infection
    clinical features 952–953
    diagnostic tests 955
    immune response 954
    pathology and histopathology 953–954
    prevention and control 942, 950, 954–955
  molecular biology **956–964**
  mRNA 957, 962, 963
  'nucleotype' 949
  outbreaks of disease 942, 950
  'overwintering' 950, 951, 955
  pathogenicity 952
  phylogenetic associations 947
  protein synthesis 963
  proteins 948, 958–962
  replication 945, 951, 953, 962–963
  RNA 957
  serogroups 942–943, 943, 946, 948, 955
  serologic relationships and variability 948–949
  serotypes 946, 948, 955
  structure and properties 956, 957
  taxonomy and classification 942–944
  tests to distinguish vaccinated animals 956
  tick-borne 944, 950
  'topotype' 947, 948
  transmission and tissue tropism 950, 950–952
  vaccines 952, 954, 955
  vector distribution 950
  virulence factors 946, 952
  see also Bluetongue virus (BTV)(A)
Orchitis, in mumps 882
Orders (virus) 1398–1400, 1403, 1406
  definition 95
  nomenclature 1402
Oregon sockeye salmon disease virus (OSDV)(A) 1251
Orf 1037, 1040–1041
  see also Parapoxviruses(A)
ORFs, *see* Open reading frames
Orfviruses(A) 1037
  deletions 1039
  genome 1039
  importance 1041
  see also Parapoxviruses(A)
Organ system infections **964–977**
  acute myopericarditis 965
  adenoviruses 973
  arenaviruses 975
  atherosclerosis 966–967
  chronic cardiac disease 966
  clinical features 968

Organ system infections *continued*
cytomegalovirus (CMV) 969, 972–973
diabetes mellitus 969–971
diagnosis 975
enteroviruses 965, 969
epidemiology 967
etiology 964–967
hantaviruses 974
hepatitis B 973
HIV 973
human polyomaviruses 974
infection after cardiac transplantation 966
intrauterine and perinatal infections 964
laboratory diagnosis 968
mumps 969
myocarditis 965
pathogenesis 967, 969–971
see also Heart; Kidney; Pancreas
Organ transplantation, *see* Transplantation
Orgyria pseudotsugata multiple nuclear polyhedrosis virus(I) 131, 132, 135
Origins of viruses, *see* Evolution of viruses
Oropouche disease 190
Oropouche virus(A) 1618
'Orphan' viruses 640
reoviruses 1193
Orthohepadnaviruses 564
see also Hepadnaviridae; Hepatitis B virus (HBV)(A)
Orthomyxoviridae 1399, 1409
eye infections 446
influenza viruses 710
replication/transcription strategy 1207
structure and characteristics 443, 446, 1583
see also Influenza viruses(A); Orthomyxoviruses(A)
Orthomyxoviruses(A)
evolution 437, 439
mutation 437
Orthopoxvirus(A) 261, 1339, 1341, 1343, 1507
antigenic cross-reactivity 264
cowpox virus as model 267
genetic recombination in 264
inactivated, failure, explanation 867
mousepox (ectromelia) virus 861, 862
see also Cowpox virus(A); Vaccinia virus(A)
Orthoreovirus(A) 1194
Oryctes baculovirus(I) 136, 137
biological control using 138–139
genome 137
infection cycle 137–138
protein synthesis 138
Oryctes rhinoceros 138
Otitis media 1223, 1487
in measles 845
Overwintering
Japanese encephalitis virus 750–751, 1615
maize chlorotic dwarf virus 825, 826
orbiviruses 950, 951, 955
pea enation mosaic virus (PEMV) 1087
Ovine herpesviruses(A) 156
Ovine respiratory syncytial virus (ORSV)(A) 1210
'Owl eye' appearance, human cytomegalovirus 294, 297
Owl monkey kidney (OMK) cells, herpesvirus saimiri growth 615
Owl monkeys, herpesvirus saimiri infection 620
Oyster virus (OV)(A) 144

P
P1 bacteriophage(B) **997–1003**
antirestriction mechanisms 671, 1001
applications in genetic engineering 1002–1003
bof gene 999
*c*1 repressor 999, 1001
as cloning vehicle 115
control of host range by DNA inversion 1001–1002
DNA 997, 998
circularization 998
DNA-restriction-modification 997, 1001
*Eco*P1 1001
fate of transferred donor DNA 110–111
gene expression 1001
genome structure and organization 998
history 997
host cell lysis 999
importance 997
incompatibility determinants 1000
integration into host chromosome 1002
lox-cre site-specific recombination system 998, 1002
lysogeny and immunity 999
lytic cycle 998
mod and *res* genes 1001
morphogenetic pathways 999
mutants/mutations 997, 998, 999
pac site 999
packaging 999, 1002
headful of DNA 108, 999
phage coats, sizes 108
plasmid replication and partition 1000–1001
prophage 997, 1000
recombination 1002
replication
cycle 998–999
origins 998
replication and partition region 1000
structure and proteins 997–998
transposon insertion 110
P2 coliphage(B) **1003–1009**
capsid, genes for 1007, 1008
cox gene in lytic state 1005
DNA replication 1007
genome 1004
as helper to coliphage P4 1003, 1007
life cycle, control 1003–1006
lysogeny 1005
lytic cycle 1005–1006
ogr gene 1006
promoters and repressors 1005
recombinant DNA technology 1008
recombination 1008
structure and morphogenesis 1007–1008
P4 coliphage(B) **1003–1009**
capsid 1008
cosmids 1008
DNA replication 1007
genome 1003, 1004
interactions with helper phage 1003, 1007
life cycle control 1006–1007
lysogenic state 1006
as parasite of coliphages 186 and P2 1003
plasmid (lytic) state 1006, 1008
prophage induction 1007
recombinant DNA technology 1008
sid gene and *sir* mutants 1008
structure and morphogenesis 1008
P22 bacteriophage(B) **1009–1013**
antirepressor 1012–1013
assembly 1011
discovery 1009
DNA 1010
replication 1011
genetic structure and organisation 1010–1011
lysogeny 1011–1012
lytic cycle 1011
packaging 1010, 1011
headful of DNA 108, 1010
research 1012
recombination 1013
research 1012–1013
structure 1009–1010
tailspike protein 1010, 1013
p34 protein 20
p53 gene 938
adenovirus E1B protein 496R association 22
mutations, in hepatocellular carcinoma 558
SV40 transformation mechanism 1476
p53 protein, HPV E6 protein binding 1024, 1477, 1496
p220 protein 491
P elements 1232
pac gene 76, 108, 873
P22 bacteriophage 1010, 1012
T1 phage 1373, 1374
Palindromic sequences, *see* DNA, single-stranded (ssDNA)
Pancreas
avian hepatitis B viruses in 567
injury, reovirus infection 1193
tumors, interferon therapy 744
virus infections 969–971
Pandemic, definition 399
Pangola stunt virus (PSV)(P) 1096
Panicum mosaic virus (PMV)(P) 1346
see also Sobemovirus(P)
Papaya mosaic virus (PaPMV)(P) 1143
genome 1144
structure and assembly 1143
Papillomas 1305
see also Warts
Papillomaviruses(A) 1305
animal **1305–1311**
physical properties 1306
taxonomy and classification 1305–1306
types 1306
see also Bovine papillomaviruses (BPV)(A); Cottontail rabbit papillomavirus (CRPV)(A)
antivirals 45
eye infections 444
human, *see* Human papillomaviruses (HPV)(A)
latent infections 788
Papovaviridae 1014, 1135, 1398, 1409
characteristics 442, 1014, 1305, 1322
eye infections 444
murine polyomaviruses 1135
papillomaviruses 1014, 1305
see also Human papillomaviruses (HPV)(A); Papillomaviruses(A)
Polyomavirus
JC and BK viruses 752
SV40 1322
see also BK virus(A); JC virus(A); Simian virus 40 (SV40)(A)
replication 232
T antigen protein 232
structure 1581
vectors 1536
Papules
capripoxvirus infections 1163
molluscum contagiosum 851
Paraflumorbillivirinae 1211
antigenic cross-reactivity 1215
genome 1213
members 1211
proteins 1212
see also Canine distemper virus (CDV)(A); Measles virus(A); Mumps virus

(MuV)[A]; Parainfluenza viruses (PIV)[A]; Sendai virus[A]
Parainfluenza viruses (PIV)[A], animal 1027, **1031–1036**
avian 1036
canine PIV-5 1035
PIV-3, *see* Bovine parainfluenza virus type 3 (PIV-3)[A]
simian virus 41 (SV41) 1035
see also Bovine parainfluenza virus type 3 (PIV-3)[A]; Simian virus 5 (SV5)[A]
Parainfluenza viruses (PIV)[A], human **1027–1036**
epidemiology and transmission 1029–1030
future perspectives 1030
genetics and evolution 1029
genome organization 1027–1028
genomic sequences 1029
geographic and seasonal distribution 1029
hemagglutinin, amino acid sequence 1034
hemagglutininneuraminidase (HN) 1027, 1028
host range and viral propagation 1029
infections
bronchiolitis 1224
clinical features 446, 1030
frequency and respiratory disease types 1223
immune response 1030
pathology 1030
pneumonia 1224
prevention and control 1030
mRNA 1028
murine subtype, *see* Sendai virus
pathogenicity 1030
proteins 1027–1028
attachment 1027, 1028
fusion (F) 1027, 1028
M protein 1027, 1028
replication 1028–1029
serologic relationships and variability 1029
structure 1027–1028
taxonomy and classification 1027, 1211
transcription and translation 1028–1029
types 1027
vaccines 1030
Parainfluenzavirus[A] 914
Paralysis
bees 654, 655
coxsackievirus infection 270
echovirus infections 358, 359
flaccid motor, enterovirus (EV) serotype 70 (EV70) infection 382
posterior, in caprine arthritis encephalitis syndrome 203
viruses causing 912
Paralytic poliomyelitis, *see* Poliomyelitis
Paramecium bursaria, chlorella symbiotic with, viruses in 36
Paramecium bursaria Chlorella virus 1 (PBCV-1)[P] 36
attachment 37
ecology 39–40
genome structure 37
host cell changes after infection 37
proteins (Vp54) 36
replication 37–38
structure and composition 36–37
taxonomy and classification 36
transcription 37, 38
Paramural bodies 1566
Paramyxoviridae 1027, 1210, 1399, 1409
characteristics and structure 443, 914, 1583
eye infections 445–446
gene map 1212
in *Mononegavirales* 830
Morbillivirus 1027, 1260–1261
measles virus 838
Paraflumorbillivirinae 1211
Paramyxovirus 1027, 1300
mumps virus 876
Newcastle disease virus 914
Pneumovirinae 1210–1211
see also Human respiratory syncytial virus (HRSV)[A]
replication/transcription strategy 1207
viruses included and genera 914, 1210, 1211
see also individual genera and viruses (page 1027)
Paramyxovirus 445–446, 914, 1027
mumps virus (MuV) 876
viruses included 914
see also Newcastle disease virus (NDV)[A]; *Parainfluenzavirus*
Paramyxoviruses 1211
antivirals 44, 47
avian 1036
features 876
fusion protein 1557, 1558
M protein 842
mutation rates 843
see also Measles virus[A]; *Morbillivirus*[A]; Parainfluenza viruses (PIV)[A]
Parana virus[A] 777
Parapoxvirus[A] 1037
Parapoxviruses[A] **1036–1041**
epidemiology 1040
evolution 1039–1040
future perspectives 1041
genetics 1039
genome structure and properties 1037
geographic and seasonal distribution 1039
history 1036–1037
host range and virus propagation 1039
infection
clinical features 1040
immune response 1041
pathology and histopathology 1041
prevention and control 1041
pathogenicity 1040
physical properties 1038
proteins 1038
replication 1038–1039
serologic relationships and variability 1040
structure and properties 1037
taxonomy and classification 1037
transmission and tissue tropism 1039, 1040
see also Orfviruses[A]
Pararetroviruses[P]
badnaviruses as 140
plant 223, 1537
replication 1537
as vectors 1537–1539
see also Caulimoviruses[P]
Pararotaviruses[A], *see* Rotaviruses[A], nongroup A
Parasitoid wasps, luteovirus control 796–797
Parasitoids, polydnaviruses 1133, 1134
Parotid virus[A] 1135
Parotitis, in mumps 882
'Parsnip yellow fleck virus group' 1042, 1399
Parsnip yellow fleck virus (PYFV)[P] **1042–1046**
anthriscus yellows virus (AYV) relationship 1042, 1044
cytopathology 1045
epidemiology and control 1045
genome properties/sequence 1042, 1043
geographic distribution 1043
history 1042
host range 1044
physical properties 1042
serology 1043
structure 1581
symptoms 1044
taxonomy and classification 1042
'top' and 'bottom' particles 1043
transmission 1044
virus-like particles (VLPs) 1044, 1045
Partial sequencing 399
Particle assays 1173
Partitiviridae 1047, 1398, 1406
cryptovirus polymerase similarity 277
Partitivirus[P] 1047
Partitiviruses[P] **1047–1051**
antigenic properties 1048
dsRNA segments 1047, 1048
history 1047
host range 1050
multiplication 1049
particle properties 1047
H1 and H2 particles 1048, 1049
L1 and L2 1047, 1049
M1 and M2 particles 1047–1048, 1049
structure and composition 1048
taxonomy and classification 1047
transcription/translation 1048–1049
transmission 1049–1050
virus-host relationships 1050–1051
Parvoviridae 1052, 1061, 1068, 1398, 1408, 1409
Densoviruses 331, 1052
see also Densonucleosis viruses[I]
Dependoviruses 331, 1052
see also Adeno-associated viruses (AAV)[A]
genera 1052, 1068
Parvovirus, *see Parvovirus*
structure 1581
Parvovirus 1052, 1068
see also Parvoviruses[A]
Parvovirus-like particles, gastroenteritis 376
Parvoviruses[A] 376, 1052, **1052–1075**
'autonomous' 1052
cats, dogs and mink **1061–1067**
classification 331, 1052
coat proteins 1059–1060
defective particles 1171
DNA replication 1057–1058, 1069, 1070–1071
evolution 1053, 1064, 1072
future perspectives 1057, 1060, 1067
gene expression 1058–1059
general features **1052–1057**
genetics 1053, 1064, 1072
genome structure/organization 1057–1058, 1064, 1069
geographic and seasonal distribution 1052–1053, 1061–1062, 1071
history 1052, 1061, 1068
host range and viral propagation 1053, 1062, 1071–1072
human, *see* Human parvovirus B19[A]
human fecal 1057
infection
clinical features 1054–1055, 1065–1066, 1074–1075
epidemiology 1053, 1065, 1073
gastroenteritis 376, 1055
immune responses 1056, 1067, 1075
pathology and histopathology 1056, 1066, 1074–1075
prevention and control 1056–1057, 1067, 1075
infection steps (early) 1070

Parvoviruses$^{(A)}$ *continued*
  latent infections 788, 1060, 1170
  lytic cycle 1060
  molecular biology **1057–1060**
  morphogenesis 1070–1071
  NS proteins 1069–1070
  pathogenesis 1054
  pathogenicity 1065, 1073–1074
  properties 1057, 1068–1069
  proteins 1059–1060
  replication 1054, 1073
    strategy 1060
  rodents, pigs, cattle and geese **1067–1075**
  serologic relationships and variability 1064–1065, 1072
  structure 1057, 1062–1063
  taxonomy and classification 1052, 1061, 1068
  tissue tropism 1053–1054, 1073–1074
  transcription and translation 1069–1070
  transmission 1053–1054, 1073
  vaccines 1057, 1067, 1075
  viremia 1054, 1056
  see also individual viruses (page 1052)
Paspalum striate mosaic virus$^{(P)}$ 518
Passiflora latent virus (PaLV)$^{(P)}$ 215
Passiflora yellow mosaic virus$^{(P)}$ 1500
Pathogenesis of viral infections **1076–1083**
  definition 1076
  entry route 1076–1078
    conjunctiva 1077–1078
    gastrointestinal tract 1077
    genitourinary system 1077
    receptors 1080–1081
    respiratory tract 1076–1077
    skin 1076
    tropism and 1080, 1082
    see also Receptors for viruses
  entry of viruses 229–230, 1168–1169
  host factors affecting 1082–1083
  local infections 1078
  spread in host 1078–1080
    bloodstream 909, 910, 1078–1079
    nerves 908–909, 1079–1080
    tropism and 1082
  stages 1076
  tropism 1080–1082
  virulence 402
  see also Neurotropism; Replication of viruses; Tropism, viral
Pathogenesis-related proteins (PR) 1107, 1108
Pbi viruses (*Chlorella*)$^{(P)}$ 38
Pea early browning virus (PEBV)$^{(P)}$ 1442, 1443, 1444, 1538
Pea enation mosaic virus (PEMV)$^{(P)}$ 795, **1083–1089**, 1400
  aphid-nontransmissible 1084
  as chimeric virus 1083
  cytopathology 1087–1088
  distribution in aphid tissues 1088
  epidemiology 1087
  evolution 795
  future perspectives 1088–1089
  genome structure 1084, 1085, 1086
  host range 1086
  luteoviruses relationship 1083, 1085, 1086
  molecular biology 1084–1086
  overwintering 1087
  prevention and control 1087
  replication 1088
  resistance/tolerance 1087
  RNA1 1084–1085
    relationship to other viruses 1085
  RNA2 1084, 1085–1086
    infective 1084, 1089
  RNA3 1084, 1085, 1089
  RNA interactions 1089
  satellite RNA (RNA3) 1084, 1085, 1089
  serology 1084
  structure and composition 1084, 1581
    top and bottom components 1084
  symptomatology 1087
  taxonomy and classification 1083–1084
  transmission 1086–1087
Pea enation mosaic virus (PEMV) group 1400
Pea leaf roll virus$^{(P)}$ 792
Pea mild mosaic virus (PMiMV)$^{(P)}$ 249
Pea streak virus (PeSV)$^{(P)}$ 215
Peach latent mosaic viroid$^{(P)}$ 1564
Peach rosette mosaic virus$^{(P)}$ 904
Peanut chlorotic streak virus$^{(P)}$ 225
Peanut clump virus (PCV)$^{(P)}$ 509
  genome structure and expression 512
Peanut stunt virus (PSV)$^{(P)}$ 278
  pathology and pathogenesis 283
  strains 282
  see also Cucumoviruses$^{(P)}$
Peanut yellow spot virus (PYSV)$^{(P)}$ 1459, 1460, 1463
Pear blister canker viroid$^{(P)}$ 1564
Pelargonium flower break virus (PFBV)$^{(P)}$ 218, 219
Pelargonium leaf curl virus (PLCV)$^{(P)}$ 1447
Pelargonium line pattern virus (PLPV)$^{(P)}$ 218
Penetration, by viruses 228–229, 1169, 1204, 1558
Penicillium brevicompactum virus (PbV)$^{(P)}$ 1047
Penicillium chrysogenum virus (PcV)$^{(P)}$ 1047
Penicillium chrysogenum virus (PcV) group$^{(P)}$ 1047
Penicillium cyaneo-fulvum virus (Pc-fV)$^{(P)}$ 1047
Penicillium stoloniferum virus F (PsV-F)$^{(P)}$ 1047
Penicillium stoloniferum virus S (PsV-S)$^{(P)}$ 1047
Pentons 15
Pepino latent virus (PepLV)$^{(P)}$ 215
Pepper mild mottle tobamovirus (PMMV)$^{(P)}$ 1436
Pepper ringspot virus (PRV)$^{(P)}$ 1442
Peptides
  antibodies to 699
  enrichment, filamentous bacteriophage as vehicles for 469
Perch iridovirus$^{(A)}$ 477
Perch rhabdovirus disease 477
Perforins 704
Pericarditis
  in AIDS, features 966, 968
  echovirus infection 359
Perinatal infections, of heart 964–965
Peripheral blood lymphocytes (PBL)
  immortalized by herpesvirus saimiri and herpesvirus ateles 615, 619
  vesicular stomatitis virus (VSV) infection 1555
Peripheral nerves, Marek's disease virus infection 835
Permissiveness 1170
Persistent viral infections 402, 1076, **1089–1095**, 1170, 1587
  Borna disease virus (BDV) 153, 154
  bovine viral diarrhea virus (BVDV) 178, 179
  bunyaviruses 196
  cautions on and challenges 1094–1095
  chronic (productive) 1089
  consequences 1090–1091
  definition 1170
  dominant-negative mutants 730
  equine herpesviruses 426
  equine infectious anemia virus (EIAV) 434, 435
  extracellular matrix role 228
  factors leading to 701–702, 1091–1094
  human cytomegalovirus 295, 296
  immune evasion, *see* Immune response, evasion
  immune responses 435
  immunological tolerance induced 701
  latent 1089–1090
  vaccine development 1506
  viruses causing 1089, 1090
Peste des petits ruminants (PPR) virus$^{(A)}$ 838, 1260
  clinical features of infections 1265
  diagnosis 1266
  epizootiology 1264
  evolution 1263
  outbreaks 1264
  pathology and histopathology 1266
  transmission 1264
  vaccination and control 1267
  see also Distemper viruses$^{(A)}$; Rinderpest virus$^{(A)}$
Pestivirus$^{(A)}$ 176
  hog cholera virus 649
  see also Bovine viral diarrhea virus (BVDV)$^{(A)}$; *Flaviviridae*
Pestiviruses, hepatitis C virus relationship 570, 571
Petunia asteroid mosaic virus$^{(P)}$ 1447
Petunia ringspot virus$^{(P)}$ 451
  serology 453
Peyer's patches 1077
  poliovirus infections 1117, 1122
  reovirus infection 1193
pH, viral fusion and 1557
Phage, *see* Bacteriophage
Phage conversion 101
Phage Group 627, 629
Phage typing **78–81**
  advantages 78
  applications 80–81
  automation 79, 80
  history 79
  host characteristics 80–81
  isolation of phages for 79–80
  phage characteristics/properties 79–80
  propagation of phage 80
  routine test dilution (RTD) concept 79
  scheme 80
  technique 79
Phagemid 115, 116, 469
Phagocytes, virus uptake 1079
Phagocytosis 704, 705
Pharyngitis 1221, 1223, 1223–1224
  causative agents 1223, 1224
  in Lassa fever 783
Pharyngoconjunctival fever (PCF) 449
  adenoviruses causing 444
Phenotypic mixing 528–529
ϕX174 bacteriophage, *see* Bacteriophage ϕX174$^{(B)}$
Phialophora radicicola virus 2-2-A (PrV-2-2-A)$^{(P)}$ 1047
Philosamia X virus (PXV)$^{(I)}$ 1417
Phlebotomus fever 191
Phlebotomus fever virus$^{(A)}$ 190, 191
Phlebovirus$^{(A)}$ 186, 187, 190–191, 446
  genome 188
    sequences and coding strategies 193, 194
  structure 188
  transmission 189, 191
  see also Bunyaviruses$^{(A)}$

Phloem necrosis, closterovirus infections 247
Phocid distemper virus (PDV)(A) 838, 844, 1260
  evolution 1263
  host range and viral propagation 1261
  pathology and histopathology 1266
  serologic relationships and variability 1263–1264
  see also Distemper viruses(A); Rinderpest virus(A)
Phosphatidylinositol 1432
Phosphonoacetate 474
Phosphorylation 26, 27, 303
  antivirals 43, 46
  capsid protein in totiviruses 1466
  human cytomegalovirus 303
  murine polyomavirus proteins 1142
  varicella–zoster virus (VZV) 1525
Photosynthesis
  cyanobacteria 290
  inhibitors 290–291
Photosystem II, inhibition 291
Phycodnaviridae 36, 1398, 1406
Physalis mosaic virus (P) 1500
Phytoreovirus(P) 1096, 1195, 1407
Phytoreovirus I(P)
  genome structure 1097
  members and distribution 1096, 1098
Phytoreovirus II (*Fijivirus*)(P) 1195
  genome structure 1097
  members and distribution 1096, 1098
Phytoreovirus III(P)
  genome structure 1097
  members and distribution 1096, 1098
Phytoreoviruses(P) **1095–1100**
  culture 1098
  defective-interfering (DI) RNAs 1097
  economic significance 1098
  evolution 1097
  future perspectives 1099
  gene expression 1098
  genome structure and organisation 1097–1098
  geographic distribution 1098
  history 1095–1096
  plant host range 1099
  replication 1098
  RNA (dsRNA) 1095, 1096, 1097
  structure 1096, 1581
  symptom expression 1099
  taxonomy and classification 1096
  transmissibility, loss of 1099
  transmission 1099
  tumor induction 1099
Phytorhabdovirus(P), structure 1584
Phytovirus(P) 1195
Pichinde virus(A) 777, 806
'Picobirnaviruses'(A) 144
'Picorna-like viruses', insect 1102
Picorna-like viruses supergroup 253
Picornaviridae 1115, 1399, 1408, 1409, 1423
  Aphthovirus 489, 1115, 1423
  cardioviruses 206, 1423
  Enterovirus 1115, 1423
    animal enteroviruses 384
    coxsackieviruses 268
    echoviruses 354
    polioviruses 1115
    see also Enterovirus(A); Poliovirus(A)
  eye infections 447
  groups and members 1115, 1423
  Hepatovirus 1115
    hepatitis A virus 546
  insect 1100
  replication/transcription strategy 1206–1207
  Rhinovirus 1115, 1253
  RNA virus core (RVC) motif 207
  structure and characteristics 443, 447, 1115, 1423, 1581
  subdivisions and members 489, 1115, 1423
  see also individual groups/families and viruses (page 1115, 1423); Picornaviruses(A); Picornaviruses(I), insect
Picornaviruses(A)
  antivirals 44–45, 47–48
  assembly 231, 1578
  capsid-binding agents 47, 48
  classification, history 637–638
  cytopathology 212–213
  diabetes due to 970
  genome 1585
    organization 929
  hepatitis A virus 546, 549
  infection mechanism, sequestration into endosomes 230
  initiation of protein synthesis 1123
  insect picornavirus relationship 1101
  myristylated proteins 386
  polypyrimidine tract 1125
  receptor-binding sites on 1081
  replication, potyviruses relationship 1151
  rhinoviruses 1254
  structure 250
    capsids 47, 1578
    comovirus comparison 1578, 1585
  translation 386
  see also individual virus groups (pages 1115, 1423); *Picornaviridae*
Picornaviruses(I), insect **1101–1103**
  antibodies 1103
  cytopathic effects 1102
  ecology and transmission 1102–1103
  economic importance 1103
  history 1100
  host range and virus propagation 1101
  morphogenesis 1101
  pathology 1101–1102
  serological relationships 1102
  taxonomy and classification 1100–1101
'Picornaviruses, plant' 1042
pif gene 1395
Pigs
  African swine fever virus, *see* African swine fever virus(A)
  cardiovirus infections 211, 213
  feeding of garbage and swine fever 651
  foot and mouth disease 494, 495
  hog cholera virus infection, *see* Hog cholera virus (HCV)(A)
  lelystad virus infection 763, 767
  management of industry and pseudorabies 1175
  pseudorabies 1173, 1175
    see also Pseudorabies virus (PRV)(A)
Pike fry rhabdovirus disease 477–478
Pike fry rhabdovirus (PFV)(A) 477–478, 1250, 1253
Pili
  F conjugative pili 464, 467
  filamentous bacteriophage receptor 464, 467
  N pili 464
  RNA phage infection via 1334, 1335
pim-1 proto-oncogene 887
Pineapple virus (PV)(P) 243
Pinnipeds, caliciviruses **1544–1547**
  see also Caliciviruses(A)
Pinocytosis 229
Piper yellow mottle virus (PYMV)(P) 139, 141
Piry virus(A) **233–236**
  cross-reactions 234
  genome structure 234
  host range 234–235
    mutants (agD) 235
  infection features 235
  pathogenicity and genetics 235
  proteins 234
  transmission and distribution 234–235
'Pits', in capsids, cardioviruses 207
pl gene, type D retroviruses 1240
Plant(s)
  subliminal infection 1106
  'susceptibility factors' 1106
  transgenic, *see* Transgenic plants
Plant breeding, use of plant resistance 1108–1109
'Plant picornaviruses' 1042
Plant resistance (to viruses) **1104–1109**
  acquired 1104, 1108, 1114
    coat protein genes 1108
    cross-protection 1108
    local and systemic 1108
    'molecular spanners' 1108
    pathogenesis-related (PR) proteins 1107, 1108
  to alfalfa mosaic virus (AlMV) 34
  to barley yellow dwarf virus 797
  to cucumoviruses 284–285
  cultivar 1104, 1106–1107
    dominant and recessive alleles 1106, 1107
    dosage dependence 1107
    localization resistance 1107
    necrotic local lesions 1107
    negative resistance 1107
    resistance genes 1106–1107, 1147, 1440–1441
    virus movement protein 1107
  importance of 1104, 1114
  induced, *see* Plant resistance (to viruses), acquired
  nonhost 1104, 1105–1106
    inhibitors 1106
    as negative resistance 1106
    subliminal infection 1106
  to pea enation mosaic virus 1087
  to potyviruses 1152
  targets of mechanisms for 1104–1105, 1441
  to tenuiviruses 1416
  to tobacco mosaic virus, *see* Tobacco mosaic virus (TMV)(P)
  to tobamoviruses, *see* Tobamoviruses(P)
  to tomato mosaic virus, *see* Tomato mosaic virus (ToMV)(P)
  to tospoviruses 1463
  transgenic plants 1108, 1109, 1114, 1441, 1463
  types 1104
  use in plant breeding 1108–1109, 1114
  viroid disease 1567
Plant virology, history, *see* History of virology
Plant viruses
  atomic structures 1571, 1572
  bacilliform 139
  cell-to-cell spread 1105
    nonhost resistance by preventing 1106
  control strategies 1104
  damage from, *see* Plant viruses, economic significance
  dominant-negative mutants 729
  economic significance 1104, **1109–1114**
    Alfalfa mosaic virus (AlMV) 34
    assessment of damage 1113–1114
    carlaviruses 217
    carmoviruses 219

Plant viruses *continued*
closteroviruses 247–248
comoviruses 254
control 1113, 1114
cost of maintaining healthy plants 1113
crop reduction 1110
cryptoviruses 277
cucumoviruses 279
dianthoviruses 352
direct damage 1110–1113
economies affected 1110
fabaviruses 452
furoviruses 515
hordeiviruses 663
ilarviruses 34
indirect effects 1113
killer toxins (totiviruses) 1467, 1471
luteoviruses 797
maize chlorotic dwarf virus (MCDV) 824, 825
necroviruses 901
nepoviruses 906
potexviruses 1147
potyviruses 1148, 1152
reductions in growth 1111–1112
of resistance mechanisms 1108–1109
rhabdoviruses 1247
sobemoviruses 1351
tenuiviruses 1414
tobamoviruses and TMV 1437
tobraviruses 1444
tospoviruses 1462–1463
viroids 1567
see also Plant resistance
hepatitis D virus RNA relationship 578, 580
incidence in crops, surveys 1114
inoculation of plants with 1537, 1566
pararetroviruses 223
'pathogenicity factors' 1106
replicative cycle 1105
resistance to, *see* Plant resistance
spreading 1537
symptoms caused by 1105, 1110
color deviation 1109, 1110
transmission 633
types, discovery 633
as vectors, *see* Vectors
see also individual families (as above)
Plantago mottle virus(P) 1500
Plantain virus 6 (PlV6)(P) 218, 219
Plantain virus 8 (PlV8)(P) 215
Planthopper transmission
rhabdoviruses 1248
tenuiviruses 1414
Plaque
counting 81
formation 338
lytic phage 80
turbid, temperate phage 80
Plaque assays 1172
algal viruses 37, 39
bacteriophage 80, 81
in soil samples 122
modified for noncytopathogenic viruses 1172–1173
Plaque forming units (PFU) 338
Plasma cell 698
Plasma membrane 228–230
bilayer organization 228
channels in 228
contribution to virus infections 228–230
membrane fusion 228
signal transduction 229
see also Membranes
Plasmalemmasomes 1566
Plasmaviridae 1398, 1405
characteristics and members 96
Plasmids
bacterial appendages encoded 80
bacteriophage P1 1000
bromovirus cDNA in 182
co-transfection, adenovirus recombinants 1529
ColIb, *see* ColIb plasmid
cosmids 115
effect on typing phage 80
F plasmid 80, 464, 1395
pCV1 77
phage-resistance genes 120
recombinant nuclear polyhedrosis viruses 136
replicons, in mini-Mu used as cloning vehicles 876
T7 phage DNA replication and 1392
transduced by T1 phage 1375
transfer, transduction application 112
as vectors 115, 116
Agrobacterium Ti-plasmid 522, 1537, 1566
vaccinia-based systems 1530–1531
Plasminogen, expression in SF cells 1535
Plasmodesmata
tobamovirus modification of 1440
transformation, caulimoviruses 225
Platelet-derived growth factor (PDGF) 536, 939, 1308
simian sarcoma virus (SIV) oncogenic action 1473–1474
Platelets, aggregation inhibition, in Lassa fever 782
Plectonema, cyanophage 290
Pleiotropy 903
Pleurodynia 271, 639
Plodia interpunctella granulosis virus(I) 128
Plum pox virus (PPV)(P) 1149, 1538
PMS1 phage(B) 51, 54
Pneumabort-K vaccine 428–429
Pneumoencephalitis 914
Pneumonia 1222, 1224–1225
after transplantations 1487
causative agents 1224
adenovirus 6
human parainfluenza virus (PIV-3) 1027, 1030
chronic interstitial, in caprine arthritis encephalitis syndrome 203
interstitial
cowpox virus infection 265
in SIV infections 1321
in measles 845
Pneumonia virus of mice (PVM)(A) 1211
antigenic cross-reactivity 1214–1215
Pneumonitis
acute interstitial, Aleutian mink disease virus 1055, 1061
cytomegalovirus (CMV) 297, 1479
herpes simplex virus (HSV) 590
Pneumotropism, Sendai virus 1302–1303
Pneumovirinae 1210–1211
see also Respiratory syncytial virus (RSV)(A)
Pneumovirus(A) 445–446, 1027, 1210
viruses included 914
see also Respiratory syncytial virus (RSV)(A)
PO-1-Lu virus(A) 1237, 1239, 1240
Poa semilatent virus (PSLV)(P) 661
economic importance 663
host range 663
Podoviridae 95, 96, 286, 1398, 1405
characteristics and taxonomy 98
structure 1582
Poinsettia mosaic virus(P) 1500
pol gene
bovine immunodeficiency virus 160, 161, 164
bovine leukemia virus 167
caprine arthritis encephalitis virus (CAEV) 201
conserved segment 164
equine infectious anemia virus (EIAV) 430
feline immunodeficiency virus (FIV) 455
feline leukemia virus (FeLV) 460
foamy viruses 482, 483
HTLV-1 683
HTLV-2 690, 691
human immunodeficiency virus (HIV) 675
lymphoproliferative disease virus (LPDV) of turkeys 812, 813
mouse mammary tumor virus (MMTV) 857
murine leukemia viruses (MuLVs) 884
reticuloendotheliosis viruses (REV) 1229
in retrotransposons 1233
simian immunodeficiency virus (SIV) 1318
type D retroviruses 1238
Pol protein
human immunodeficiency virus (HIV) 676, 677–678
murine leukemia viruses (MuLVs) 884
see also Reverse transcriptase
polehole gene 940
polII promoter 595, 600
Polioencephalitis, pig, *see* Teschen disease
Poliomyelitis 634, 1115, 1117
abortive 1117
clinical features 912, 1117
eradication by vaccination 360
in LDV infection of immunosuppressed mice 769
nonparalytic 638, 640, 1117
see also Coxsackieviruses(A); Meningitis, aseptic
'provoking effect' 1082
see also Polioviruses(A)
Polioviruses(A) **1115–1132**
3C protease 209, 1127, 1131
'A-particle' 1123
adaptation, cardiovirus discovery 205
adsorption and infection 1077, 1122, 1123, 1204
antibodies
development 636, 1118
monoclonal 1122
patterns in outbreaks 636–637
antigenic variation 1116
capsid structure and antigenicity 1119–1122
'beta annulus' ('beta cylinder') 1120
'canyon' 1120–1121, 1122
myristic acid on VP4 1115, 1120
neutralization antigenic sites 1121
VP1, VP2, VP3 and VP4 proteins 1115, 1119
cell-free synthesis 1132
cellular receptor 1117, 1122–1123
absence and resistance to infection 1591
binding site for 1121, 1123
cDNA 1122
hybrid 1122
modification 1122
chimeric 1535
EV70 expression 383
coxsackievirus infection with 639
cytopathology 337, 1117, 1130

  detection, history 640, 1115
  dicistronic 1123, 1128
  effect on host cell 1119, 1123, 1588, 1589
  enterovirus (EV) serotype 70 (EV70) relationship 380
  epidemiology 1116
  evasion of interferon action 737
  future perspectives 1118–1119
  general features **1115–1119**
  genetic recombination 1132
  genetics 1116
  genome organization and processing 929, 1126, 1127–1128, 1206
  geographic and seasonal distribution 1115–1116
  history 634–636, 1115, 1503
    chimpanzee infections 635, 636
    classification 637–638
    cultivation 630
    growth in tissue culture 635, 1503
    laboratory findings application 636–637
    olfactory lesions discovery 634, 635
    oral–alimentary route of infection 634–635
    research using rhesus monkeys 634
    RNA splicing 631
    rodent-adapted strains 635, 1116
    transmission and fecal contamination aspects 635
  host range and viral propagation 1116
  hybrid (PV1 and PV2) 1122
  immune response 636, 1117–1118
    see also Polioviruses(A), antibodies
  infection
    clinical features 1117
    outbreak, pre- and post-epidemic specimens 636
    pathology and histopathology 1117
    prevention and control 636, 1118
    see also Poliomyelitis
  infectious cycle 1122, 1129
    initiation of infection 209, 1122
    internalization and endocytosis 1123
    membranous vesicles 1129, 1130
    site of entry 1082, 1122
  inhibition of host cap-dependent translation 1119, 1123, 1589
  internal ribosome entry site (IRES) 386, 1123, 1125
  isolation 343
  Lansing strain 635, 1122
  molecular biology **1119–1132**
  monoclonal antibodies 1122
  morphogenesis 1132
  mouse model 635, 1122
  mRNA 1123
  mutants 1130
    complementation 1131
    host cell protein synthesis inhibition defect 1128
  mutations
    NTR 1125, 1131
    rate 1116
  neutralization 'escape mutations' 1121
  non-translated region (NTR) 1123, 1125
    mutations 1125, 1131
  p220 1128
  passive immunoglobulin 1118
  pathogenicity 1117
  protein 2A 209, 1127, 1128
  proteinases 1127, 1128, 1131
  proteolytic processing of polyprotein 1119, 1126, 1127–1128, 1206
  resistance to inhibitors 1131
  RNA 1115, 1119
    forms 1130
  RNA polymerase 1131, 1206
  RNA replication 1128–1132, 1206–1207
    VPg role 1131, 1206–1207
  serologic relationships and variability 1116
  serotypes and strains 1115, 1119
  structure 642, 1115, 1119
  taxonomy and classification 637–638, 1115
  terminology, acceptance 638
  tissue tropism 1082, 1117, 1122
  transgenic mice 1122
  translation 1123–1126, 1206–1207
    attenuation loop 1125
    cap-independent 1119, 1123
    cis-acting sequences 1125
    defective 1125
    initiation 1123, 1206
    polypyrimidine tract-binding protein (PTB) 1125
    Ribosome Landing Pad (RLP; IRES) 386, 1123, 1125
    trans-acting factors (p52, p57) 1125
  transmission 401, 1117
  type I 1117
    Mahoney strain, VP3 relationship to hepatitis A virus VP3 550
    properties 1120
  vaccines 1118
    antigen chimeras in 1535
    'Cutter incident' 1082, 1118
    formalin-inactivated (Salk) 1115, 1118, 1503
    history 640, 1115, 1503
    live attenuated (Sabin) 1115, 1117, 1118, 1503
    live *vs* inactivated, protection after 1505–1506
    SV40 contamination 1322
    type 3 virus in 1506
  vesicular stomatitis virus (VSV) incompatibility 728
  virulence attenuation 1125
  VPg 1131, 1206–1207
  WHO declaration 1115, 1118, 1503
Pollen transmission
  capilloviruses 198
  cryptoviruses 276
  ilarviruses 34
  nepoviruses 906
Polyadenylation sequence/signal
  bovine leukemia virus 169
  potexviruses 1144, 1146
Polyamines, in comoviruses(P) 250
Polyarthritis with rash 1268
Polyclonal antisera 344
Polycythemia vera, interferon therapy 742
Polydnaviridae 1398, 1408
Polydnaviruses(I) **1133–1135**
  DNA (linear and circular) 1133, 1134
  DNA packaging 1133
  genome 1133
  life cycle 1134
  significance 1134
  structure and morphogenesis 1133
  taxonomy and classification 1133
Polyhedra 133, 313, 314–315
  degradation after infection 316
  envelope 134
  formation 317
  shape 317
  transmission 318–319
  see also Cytoplasmic polyhedrosis viruses (CPV)(I)
Polyhedrin 131, 133, 315, 1534
  crystallization 317
  gene 315
    in *Oryctes* baculovirus 137
  hyperexpression and importance of 133, 136
  promoter 1534
  synthesis 315–316
Polymerase chain reaction (PCR) 81, 83, 347–348, 531
  badnaviruses 142
  bovine leukemia virus 173
  CMV detection 1479
  hantaviruses 542, 543, 545
  hog cholera virus detection 654
  HTLV-1 and HTLV-2 688, 694
  lymphoproliferative disease virus (LPDV) of turkeys 814
  rabies virus evolution and 1182
  respiratory viruses 1225
  zoonoses diagnosis 1620
Polymorphonuclear leukocytes (PMN) 697
Polymyositis, coxsackievirus infections 272
Polymyxa betae 514
Polymyxa graminis 514
Polyomavirus(A) 1322
  JC and BK viruses 752
  properties 1322
  SV40 1322
  see also BK virus(A); JC virus(A); Polyomaviruses(A); Simian virus 40 (SV40)(A)
Polyomavirus hominis 1 752
Polyomavirus hominis 2 752
Polyomaviruses(A)
  infections after transplantation 1486–1487
  kidney infection 974
  latent infections 788, 1170
  see also BK virus(A); JC virus(A)
Polyomaviruses(A), murine **1135–1142**
  assembly and release 1142
  cell culture 1135
  cytopathology 1142
  DNA 1135, 1137, 1140
    integration 1137
    replication 1136, 1137, 1141
  general features **1135–1139**
  genetics 1138–1139
  genome organization 1137–1138
    early region 1137–1138
    late region 1138
    ORFs 1138
    regulatory region 1137
  history 1135
  importance of 1139
  lytic infection 1135–1136, 1141
  molecular biology **1139–1142**
  mutants 1138–1139
  persistent infection 1136
  physical properties 1140–1141
  post-translational processing 1142
  productive infections 1135–1136, 1141
  proteins
    early proteins 1136, 1137–1138
    late region 1136
    properties 1140
    roles 1138
    VP1 1140
  receptor 1136
  relationship to other polyomaviruses 1139
  semipermissive and nonpermissive cells for 1136
  structure and properties 1135, 1139–1140
  T antigens 1136, 1137, 1140
    LT 1136, 1137, 1138, 1140
    MT 1136, 1138, 1140
    ST 1137, 1138, 1140
  taxonomy and classification 1135
  temperature-sensitive mutants 1138–1139

Polyomaviruses$^{(A)}$, murine *continued*
transcription 1137–1138, 1141
transformation 1136–1137
translation 1141
Polyploidy 528, 529
Polypurine tract (PPT), in HIV genome 675
Polypyrimidine tract, picornaviruses 1125
Polypyrimidine tract-binding protein (PTB) 1125
Polyribosomes 231, 1590
Pongine herpesvirus$^{(A)}$, *see* Herpesvirus pan$^{(A)}$
Poplar mosaic virus (PMV)$^{(P)}$ 215
Population dynamics 403
Population size
critical 402
effect on survival of viruses 402–403
Porcine epidemic abortion and respiratory syndrome (PEARS) 763
Porcine hemagglutinating encephalomyelitis virus (HEV)$^{(A)}$ 255
Porcine parvovirus (PPV)$^{(A)}$ 1052, **1067–1075**
genome and replication 1069
geographic and seasonal distribution 1071
history 1068
host range 1053, 1071, 1072
infection
clinical features 1055, 1074
epidemiology 1053
immune response 1075
prevention and control 1075
NS proteins 1069–1070
properties and structure 1068–1069
serologic relationships and variability 1072, 1073
tissue tropism and pathogenicity 1073, 1074
transcription and translation 1069–1070
transmission and epidemiology 1073
vaccines 1075
viral propagation 1071, 1072
Porcine respiratory coronavirus (PRCV)$^{(A)}$ 255, 256, 258
Porcine respiratory syndrome (PRRS) 763
Pork, export and hog cholera virus infection 653, 654
Post-translational processing
African swine fever virus proteins 26–27
animal enteroviruses 386
foamy viruses 486
foot and mouth disease viruses (FMDVs) 492
glycosylation, *see* Glycosylation
hantaviruses 541
hepatitis B virus (HBV) 563
herpes simplex viruses (HSV) 602
human cytomegalovirus (HCMV) 303–304
influenza viruses 720–721
measles virus 842
mumps virus (MuV) 880
murine polyomavirus 1142
myristylation 27, 386, 546
Newcastle disease virus 916
phosphorylation, *see* Phosphorylation
prion hypothesis of spongiform encephalopathy aetiology 1362, 1363
proteolytic cleavage 26, 303
reticuloendotheliosis viruses (REV) 1230
retroviruses, type D 1239
rhinoviruses 1255
rubella virus 1293
Semliki Forest virus (SFV) 1332
Sendai virus 1301
simian virus 40 (SV40) 1326
Sindbis (SIN) virus 1332
Theiler's viruses 1424–1425
toroviruses 1455
totiviruses 1465, 1466
vaccinia virus 1510
varicella–zoster virus (VZV) 1525
Post-viral fatigue syndrome 271
after echovirus infections 358
Postherpetic neuralgia 1517
Postinfectious encephalomyelitis 911
Postmeasles encephalomyelitis
cell damage mechanism 911
see also Subacute sclerosing panencephalitis (SSPE)
'Postviral syndrome', Ross River virus (RRV) 1273
Potassium, in T1 phage infection 1371
Potato, crop value 1111
Potato aucuba mosaic virus (PoAMV)$^{(P)}$ 1143
Potato black ringspot virus$^{(P)}$ 904
Potato leaf roll virus (PLRV)$^{(P)}$ 792
control 797
economic significance 797
host range 796
pea enation mosaic virus (PEMV) relationship 1085
symptomatology 794
Potato mop-top virus (PMTV)$^{(P)}$ 509
epidemiology and control 515
genome structure and expression 512
tobacco mosaic virus relationship 510
Potato spindle tuber disease 1563
Potato spindle tuber viroid (PSTVd)$^{(P)}$ 1563, 1564
economic significance 1567
genome structure 1564, 1565
history 633
host range and transmission 1566
pathogenicity 1569
replication 1568
symptomatology 1567
see also Viroids
Potato virus M (PVM)$^{(P)}$ 214, 215, 1111
economic significance 217
genome structure 215
translation 216
Potato virus S (PVS)$^{(P)}$ 214, 215
translation 216
Potato virus T (PVT)$^{(P)}$ 197, 198, 242, 244
Potato virus U (PVU)$^{(P)}$, 904
Potato virus X (PVX)$^{(P)}$ 1111, 1143, 1538
classification of strains 1147
cytopathology and symptoms 1146
resistance and immunity 1147
virulence determinants 1147
Potato virus Y group 1148
Potato viruses
control 1113
economic significance 1111, 1147
reduction of quality and market value 1112–1113
reduction in vigor 1112
Potato yellow dwarf virus (PYDV)$^{(P)}$ 1245, 1246
disease 1247
Potato yellow mosaic virus (PYMV)$^{(P)}$ 520
'Potentially leukemic cells' (PLC) 887
Potexvirus$^{(P)}$ 1399
Potexviruses$^{(P)}$ **1142–1147**
assembly 1143–1144
characteristics 1142–1143
classification 1142–1143
cytopathology 1146–1147
economic significance 1147
genome structure and organization 216, 1144–1145
genomic RNA (gRNA) 1142, 1143, 1144
conserved sequences and promoters 1145
replication 1145
mutations 1146
reconstitution *in vitro* 1143
replicative strategy and molecular biology 1145–1146
resistance and protection 1147
structure and composition 1143–1144, 1582
subgenomic RNA (sgRNA) 1144, 1145
symptoms due to 1146
translation 1145
'triple gene block' 1144, 1146
as vectors 1146
Potyviridae 1148
characteristics 1148
subgroups 1148
Potyvirus$^{(P)}$ 1399
Potyviruses$^{(P)}$ **1148–1153**
control schemes 1152
economic significance 1148, 1152
future perspectives 1152–1153
gene expression 1150–1151
regulation 1151
genome structure and organization 1148–1150
HC-PRO protein 1151, 1152
hepatitis C virus relationship 570
host range 1148
inclusion bodies 1151, 1152
NIa proteinase 1150, 1151
proteolytic processing 1149, 1150, 1151
replication 1151, 1153
RNA 1148, 1149
structure 1582
symptoms of infections 1152
taxonomy and classification 1148
transgenic plants 1152, 1153
transmission 1148, 1152
viral-encoded proteins in 1152
virus–host relationship 1151–1152
see also Tobacco etch virus (TEV)$^{(P)}$
Poultry, *see* Chickens; Duck(s)
Poultry industry 918
Newcastle disease virus, problems 916, 918
Powassan (POW) virus$^{(A)}$ 1616–1617
Poxviridae 261, 1037, 1398, 1408, 1409, 1507
Capripoxvirus 954, 1160
characteristics 442, 1037
Chordopoxvirinae, *see Chordopoxvirinae*
classification 392
detection methods 442
Entomopoxvirinae 392, 1507
eye infections 442
Leporipoxvirus 1153–1160, 1155
Parapoxvirus 1037
replication 1209
structure 1584
Suipoxvirus 1154–1160
Yatapoxviruses 1597
see also individual genera (as above) and viruses; Poxviruses$^{(A)}$
Poxviruses$^{(A)}$ **1153–1168**
African swine fever virus similarity 24
DNA 1155
structure 497, 849
entomopoxviruses (EPVs) comparison 392
gene expression and control 1209
genome structure 497, 849, 1155
isolation 338
nucleus role in infection 499
rabbit, hare, squirrel and swine **1153–1160**
clinical features 1158
DNA 1155

DNA replication 1155–1156
evolution 1157
future perspectives 1159–1160
genetic variability 1157
geographic and seasonal distribution 1156
history 1153–1155
host range and viral propagation 1156–1157, 1159
immune evasion mechanism 1156, 1157, 1159
immune response 1159
pathogenesis, mechanisms 1156
pathogenicity 1157–1158
pathology and histopathology 1158–1159
prevention and control 1159
proteins 1155
replication 1155
structure and properties 1155
taxonomy and classification 1155
transcription and translation 1155–1156
transmission and tissue tropism 1157
tumors in hyperproliferative state 1158
'virokines' and 'viroreceptors' 1156
see also Malignant rabbit fibroma virus (MRV)[(A)]; Myxoma virus (MYX)[(A)]; Rabbitpox virus[(A)]; Shope fibroma virus (SFV)[(A)]; Swinepox virus (SPV)[(A)]
'reactivation' phenomenon 499
replication 499, 1169, 1204
uncoating 499
sheep and goats **1160–1165**
clinical features of infections 1163–1164
epidemiology 1162
evolution 1162
future perspectives 1165
genetics 1161
geographic and seasonal distribution 1160–1161
history 1160
host range and viral propagation 1161
immune response 1164
outbreaks 1162
pathogenicity 1163
pathology and histopathology 1164
prevention and control 1164–1165
serologic relationships and variability 1162
taxonomy and classification 1160
transmission and tissue tropism 1163, 1164
vaccines 1161, 1162, 1165
viability 1163
transcription 499, 1204
uncoating 1169
as vectors 1530–1532
applications 1531
history and development 1530–1531
viruses used 1531
see also Vaccinia virus[(A)]
see also other specific poxviruses
PR3 bacteriophage[(B)] 1165
PR4 bacteriophage[(B)] 1165
PR5 bacteriophage[(B)] 1165
PR772 bacteriophage[(B)] 1165
PRD1 bacteriophage[(B)] **1165–1168**
characteristics 97
discovery 1165
DNA and proteins 1165–1166
DNA replication 1166–1167
protein-primed 1167, 1168
genome and genetics 1166
lie-cycle 1166–1167
membrane 1166, 1168
structure 1165–1166
taxonomy and classification 97, 1165
Predatorprey interactions 126
Pregnancy
coxsackievirus infections 271
human parvovirus B19 infection 1054–1055
rubella in, *see* Rubella, congenital
rubella vaccine in 1298
Prelysosomes 229
'Prestige' vaccine 429
Prevalence rate, definition 399
Primary human fetal glial (PHFG) cells, JC and BK virus growth 753
Primosome 991, 993
Prion diseases, *see* Spongiform encephalopathies
Prion hypothesis 1357, 1362–1363
Prion protein (PrP) 760, 761, 1359, 1361
changes 1367
tissue culture studies 1369
Western blotting 1368
Prions 912, 913, 1361
discovery 631
see also Spongiform encephalopathies
pro gene
mouse mammary tumor virus (MMTV) 857
type D retroviruses 1240
Pro-polyhedron 317
Productive infections, definition 1170, 1587
Progressive generalized lymphadenopathy (PGL)
feline immunodeficiency virus (FIV) infection 457
type D retroviruses 1241
Progressive multifocal leukoencephalopathy (PML) 752, 755, 1091
clinical features 756
JC virus causing 912
pathology and histopathology 756
prevention and control 756
see also JC virus[(A)]
Prohormone processing 1603, 1605
Promoters
adenovirus 11
βHSV thymidine kinase (U_L23) 595
bovine papillomaviruses 1307
CAT box 595
coliphages P2 and 186 1005
cyanophage N-1 290
filamentous bacteriophage 466–467
herpes simplex viruses 595–597
HTLV-2 692
human cytomegalovirus 302, 306
influenza viruses 719–720
λ
in plasmid expression vectors 115
see also Lambda bacteriophage[(B)]
major late of adenovirus Ad12 13
methylation and gene inactivation 13, 673
Mu bacteriophage 872
murine cytomegalovirus 308
N4 bacteriophage 892
P22 bacteriophage 1011
parvoviruses 1058
polII 595, 600
polyhedrin 1534
specificities of coliphage 1395
SPO1 bacteriophage 73, 1354
T7 phage 1391, 1392, 1394
TATA box, *see* TATA box
vaccinia virus 1509
Propagation of viruses **1168–1173**
in cell culture 1171
cytopathic effects, *see* Cytopathic effects
host range 1170–1171
multiplication cycle 1168–1169
productive and nonproductive infections 1170–1171
purification of viruses 1172
virus titration 1172–1173
in whole organisms 1171–1172
see also Cell cultures
Prophage 81, 815
activation/induction 815
ϵ-prophage of *Salmonella* 81
history 630
incompatibility 1000
λ, *see* Lambda bacteriophage[(B)]
transposon insertion 110
see also Bacteriophage, temperate; Lambda bacteriophage[(B)]; Lysogeny
'Prophage transformation', cloning DNA in *B. subtilis* phage 77
Prospect Hill virus[(A)] 974, 975
genome structure/properties 539
proteins 540
Prostaglandin E_2 (PGE_2), in recurrent HSV infections 592
Proteases, animal enteroviruses 386
Protein
folding, P22 tailspike protein 1013
'free' 231
host, shutdown, *see under* Host
membrane 1556–1557
modification
endoplasmic reticulum function 230
see also Post-translational processing
movement into nucleus 232, 233
nucleic acid interactions 1572
post-translational processing, *see* Post-translational processing
production by genetic engineering techniques 531
sequences 530
synthesis
birnaviruses 146
endoplasmic reticulum function 230
nonoccluded baculoviruses 138
T1 bacteriophage 1372
viral 1205
virus effect on host's, *see under* Host
virus-specified 231
transport 232
endoplasmic reticulum function 230
viral, gene evolution 439–440
Protein 2′-5′ A synthetase 1591
Protein kinase 1591
dsRNA dependent 736
granulosis viruses 128
nuclear polyhedrosis viruses 135
Proteinaceous inclusion bodies
absence from badnaviruses 141
caulimoviruses 141
maize chlorotic dwarf virus inclusions similarity 826
see also Viroplasm
Proteinases, polioviruses 1127, 1128
Proteinuria 544
in Lassa fever 783
Proteoglycans 228
Proto-oncogenes 934
activation 938
murine leukemia viruses (MuLVs) 886–887
affected by retroviral integration 935, 937
alteration/activation 938
avian leukosis viruses 69
cellular, avian leukosis viruses causing insertional activation 70

- Proto-oncogenes *continued*
  - common proviral insertion sites 887
  - groups 938–940
  - spontaneous activation 938
  - transformation mechanism of retroviruses 1473
  - *see also* Oncogenes
- Proton motive force (PMF) 1371
- Protoplasts
  - Alfalfa mosaic virus (AlMV) replication 32
  - *Bacillus subtilis* 77
  - bromovirus replication 182
  - plant rhabdovirus replication 1246–1247
- Protozoa, families of viruses 1406
- 'Provoking effect', tissue tropism 1082
- Prune dwarf virus (PDV)(P) 30, 32, 33
- Prunus necrotic ringspot virus (PNRV)(P) 30, 32, 33
- 'Pseudocholera infantum' 373
- Pseudocowpox 1036, 1040, 1041
- Pseudocowpox virus(A) 1039
- Pseudoknot structures
  - foot and mouth disease viruses (FMDVs) 490
  - satellite tobacco necrosis virus (STNV) 900
- *Pseudoletia separata*, virus isolated from 397
- *Pseudoletia unipuncta* granulosis virus(I) 128
- Pseudolumpy skin disease 155
- Pseudolysogeny 40, 74
- *Pseudomonas*
  - filamentous bacteriophage 464
  - transposable phage 868, 874–875
    - *see also* Bacteriophage D3112(B)
- *Pseudomonas aeruginosa*
  - exotoxin A 102
  - PRD1 bacteriophage 1165
  - RNA phage 1335
- *Pseudomonas pseudoalcaligenes* 979
- Pseudorabies 1173, 1176
- Pseudorabies virus (PRV)(A) **1173–1179**
  - Bucharest strain 1178
  - deletion mutations 1174, 1177
  - DNA 1173, 1174
    - packaging 1174
    - replication 1174
  - epidemiology 1175
  - future perspectives 1179
  - genome 1174
    - immediate early, early and late genes 1174
  - geographic distribution 1174–1175
  - history 1173
  - hog cholera virus DNA expression 1179
  - host range 1175
  - infection
    - clinical features 1176
    - diagnosis 1178–1179
    - factors influencing 1175
    - immune response 1177
    - pathology and histopathology 1176–1177
    - prevention and control 1177–1178, 1179
  - inhibition by vesicular stomatitis virus (VSV) 728
  - latent infections 1175, 1176, 1177
  - molecular biology 1174
  - monoclonal antibodies 1177
  - outbreaks 1175, 1176
    - cost 1175
  - physical properties 1176
  - replication 1176
  - structural glycoproteins 1177
  - taxonomy and classification 1174
  - thymidine kinase 1177, 1178
  - tissue tropism 1176
  - transmission and pathogenesis 1175–1176
  - U_s region 1174
  - vaccines 1177, 1178, 1179
    - gene-deleted 1178, 1179
    - modified-live (MLV) 1178
    - OMNIMARK 1178
- Pseudorecombination 439
- Pseudotemperate phage, of *Bacillus subtilis* 73, 74–75
- Pseudotypes 528
- *Pseuduoplusia includens* virus (PiV)(I) 1417
- ΨM phage(B) 54–55
  - host range 54–55
  - structure 54
- Pulmonary edema, in Lassa fever 782
- Purification of viruses 1172
- Puumala virus(A) 974, 975, 1614
  - genome structure/properties 539
  - proteins 540
- Pyrogenic toxins 103–105
  - as superantigens 104–105

Q

- $Q\beta$-replicase amplification system 348
- Quail, Marek's disease virus infection 832, 834
- Quail adeno-associated viruses (AAAVs)(A) 1052
- Quail pea mosaic virus (QPMV)(P) 249
- Quailpox virus(A), genome structure 497
- Quasiequivalence theory 1577–1579
- 'Quasispecies' 437, 438

R

- R-BHCG 44, 46
- RA-1 virus 1057
- Rabbit(s)
  - eradication and myxomatosis 1153, 1157
  - myxoma leporipoxvirus evolution studies 436–437
- Rabbit coronavirus (RbCV)(A) 255
- Rabbit fibroma virus(A), *see* Poxviruses(A); Shope fibroma virus (SFV)(A)
- Rabbit hemorrhagic disease (RHD) virus(A) 1544
  - classification and properties 1544
  - epidemiology 1545
  - genome 1547
  - geographic distribution 1545
  - history 1544
  - host range 1545
  - immunity 1546
  - infection
    - clinical features 1546
    - prevention and control 1546–1547
  - Norwalk virus similarity 929
  - pathogenesis 1546
  - *see also* Caliciviruses(A); Vesicular exanthema virus(A)
- Rabbit myxomatosis, *see* Myxomatosis
- Rabbit papilloma virus(A), history 631
- Rabbitpox 866, 866–867, 1511, 1512
- Rabbitpox virus(A) **866–867**, 1511
  - epidemiology and genetics 866
  - history and classification 866
  - immune response 867
  - infection
    - clinical features 866–867
    - pathology and histopathology 867
    - prevention and control 867
  - pathogenesis 866, 1512
  - 'pockless' 867
  - *see also under* Poxviruses(A)
- Rabbitpox-Utrecht 866
- Rabies 446, 1184, 1185, 1614
  - clinical features 912, 1184
  - 'dumb' (passive) 912
  - 'furious' 1184
  - heart infections 965
  - history 1180–1181, 1613
  - immune response 1185
  - outbreaks 1181
  - 'paralytic' 1184
  - pathology and histopathology 1184
  - prevention and control 1185, 1614
  - resistance and survival 1183–1184, 1614
  - treatment 1614
- Rabies virus(A) 446, **1180–1185**, 1186, 1613–1614
  - detection 1614
  - epidemiology 1183
  - evolution 1182, 1188
  - future perspectives 1185
  - genetics 1181–1182
  - genome and regulatory sequences 1246
  - geographic and seasonal distribution 1181, 1184, 1614
  - glycoprotein, fowlpox virus recombinant 501
  - GL intergene 1182
  - history 1180–1181
    - cultivation 630
  - host range and viral propagation 1184, 1614
  - isolates 1182
  - laboratory adapted ('fixed') 1183
  - movement and crossing of international boundaries 1621–1622
  - neural spread 1080, 1184
  - pathogenicity 1183–1184
  - proteins 1181, 1182
  - ψ pseudogene 1182
  - receptor 1561
  - replication 1181, 1182
  - serologic relationships and variability 1182–1183
  - 'street' 1182, 1183, 1184, 1185
  - structure 1181
  - taxonomy and classification 1181, 1614
  - tissue tropism 1181, 1183
  - transcription 1182
  - transmission 401, 1181, 1183, 1184, 1614
  - vaccination
    - bat exposure and 1189
    - of wild animals 1185
  - vaccines 632, 1181, 1183, 1184, 1621
    - G glycoprotein in vaccinia virus 1531
    - preparation 1184, 1621
    - recombinant 708, 1621
- Rabies-like viruses(A) 1181, 1183, **1186–1190**
  - epidemiology 1188–1189
  - evolution 1182, 1188
  - future perspectives 1189–1190
  - genetics 1181, 1187–1188
  - genome 1186, 1188
  - geographic and seasonal distribution 1187
  - history 1186
  - host range and viral propagation 1187
  - infection
    - clinical features 1189
    - immune response 1189
    - pathology and histopathology 1189
    - prevention and control 1189
  - monoclonal antibodies 1188
  - pathogenicity 1189
  - serologic relationships and variability 1188
  - serotypes 1187
  - structure and characteristics 1186
  - taxonomy and classification 1186–1187
  - transmission and tissue tropism 1189
  - *see also* Duvenhage virus(A); Lagos bat virus(A); Mokola virus(A)

Raccoon parvovirus (RPV)[(A)] 1061
clinical features of infection 1065–1066
host range and viral propagation 1062
taxonomy and classification 1061
Radiculomyelitis, enterovirus (EV) serotype 70 (EV70) infection 382
Radioimmunoassay (RIA)
viral antibody assay 346
viral antigen detection 345
Radish mosaic virus (RaMV)[(P)] 249
raf oncogene 936, 940
Ramsay–Hunt syndrome 1517
Rana pipiens
frog virus 3 infection 503
frog virus-4 (FV4) infection 42
kidney tumors and virus causing 40, 41
Rana silvatica 41
Range paralysis, *see* Marek's disease (MD)
Ras family of proteins 1474–1475
GTPase activity 1474, 1475
Rash
coxsackievirus infection 270
human parvovirus B19 infection 1054
measles 845
mousepox 862, 864
rabbitpox 867
Ross River virus (RRV) 1273
rubella 1295, 1296
smallpox 1342
varicella (chickenpox) 1516
Raspberry ringspot virus (RRSV)[(P)] 904, 906
coat proteins 905
Rat(s), borna disease virus infection 151
Rat coronavirus (RCV)[(A)] 259
Rat virus (RV)[(A)] 1052
history 1068
host range and viral propagation 1071, 1072
infection
clinical features 1055, 1074
pathology and histopathology 1056
serologic relationships and variability 1072, 1073
tissue tropism and pathogenicity 1074
Rauscher murine leukemia virus (MuLV) 740
RAV-0 virus[(A)] 70
RB gene (retinoblastoma)
adenovirus E1A protein association 20
HPV E7 binding to product of 1024
transformation mechanism 1476, 1477, 1496
SV40 1476
Reactivation 527
Reassortment, *see* Genetic reassortment
RecA protein 774
RecA synaptase, host, single-stranded DNA phage recombination 84
RecBCD enzyme
host
in λ phage recombination 85
single-stranded DNA phage recombination 84, 85
recombination pathway 817
RecE, in λ phage recombination 86
Receptor-mediated endocytosis, *see* Endocytosis
Receptors for viruses 1080–1081, 1081, 1169, 1203, 1557, **1560–1563**
absence and resistance to infection 1591
definition 1203
dual 227, 229
'pits' in cardioviruses 207
recognition site on viral surface 1081, 1573
reovirus type 3 (Reo3R) 1561–1563
cells expressing 1562
cellular function 1562
CNS 1562–1563
functional studies 1562
inhibition 1562
monoclonal antibodies 1561, 1562
structure and sequence 1561, 1562
structure 228
types and viruses using 229, 1081, 1169, 1204, 1560, 1561
viral attachment proteins 1081
see also Endocytosis; Pathogenesis of viral infections
RecF, in λ phage recombination 86
Recombinant, definition 2
Recombinant DNA technology 530
coliphage P2 and P4 1008
Newcastle disease virus and 918
plant resistance to viruses 1108, 1109, 1114
viral antigen source, for serological diagnosis 345
see also Genetic engineering
Recombination, *see* Genetic recombination
Recombination maps 529
Red clover mottle virus (RCMV)[(P)] 249, 254
Red clover necrotic mosaic virus (RCNMV)[(P)] 349, 1538
genome structure 350
host range 352
serotypes 353
structure and composition 349
Red clover vein mosaic virus (RCVMV)[(P)] 215, 217
Red disease, fish 478
red genes 85, 775
Red system, λ phage recombination 86–87
Redundancies, molecular 437
Reed–Sternberg cells 407, 409
rel oncogene 936, 940, 1227, 1230, 1231, 1475
reticuloendotheliosis viruses (REV) 1230, 1475
Release of viruses 1078, 1169, 1559–1560
see also Budding
Renal allografts, BK virus infection 752
Renal cell carcinoma (adenocarcinoma)
frog virus 3 (FV3) 503
interferon therapy 744
Renal failure 973
Reo3R, *see under* Receptors for viruses
Reoviridae 1190, 1194, 1398, 1407, 1408, 1409
characteristics 443, 1095, 1190, 1207–1208, 1275, 1581
Cypoviruses 1194
eye infections 445
Fijivirus 1096, 1195
fish 475, 479
Orbiviruses 1194
Orthoreovirus 1194
see also Reoviruses
Phytorevirus 1096, 1195, 1407
Rotavirus 1194, 1275
structure 1581
transcription/replication strategy 1207–1208
see also individual genera/viruses
Reovirus type 1 Lang (T1L)[(A)]
hemagglutinin 1561
inclusions 339, 1202
spread 1080, 1193
Reoviruses[(A)] **1190–1203**
adsorption to receptors 1077, 1191, 1200
assembly 231
attachment protein (hemagglutinin) 1191, 1199
avian 1195
capping importance to 1201
cell lysis by 1192, 1202
cellular receptors 1561–1563
see also under Receptors for viruses
cytopathic effects 1191, 1202–1203
dsRNA (segmented) 1190, 1196, 1196–1197
replication 1192
effect in host cells 339, 1191, 1202–1203
enzymes 1193, 1197
capping role 1192, 1201
transcriptase 1192, 1197, 1201
epidemiology 1194
in feces 373
gene segment reassortment 1192, 1202
protein σ3 1199
general features **1190–1194**
genetics 1192
genome 1196–11197
classes (L, M, S) 1196
segments and proteins encoded 1196
hemagglutinin (σ1) 1190, 1191, 1200, 1561
history 1190
host range and viral propagation 1191
infection cycle 1077, 1193
spread 1080, 1193
infections
clinical features 1193–1194
diagnosis and treatment 1194
pancreatic damage 970
pathology 1193
infectious RNA 1202
infectious subviral particles (ISVPs) 1077, 1191, 1200
difference with reovirus particles 1191, 1200
transcription role 1192, 1201
L cell infection 1191, 1200
molecular biology **1194–1203**
oligonucleotides 1197
pathogenicity 1193
physical properties 1195
protein-synthesizing system 1202
proteins 1190–1191, 1197–1199
dsRNA segment encoding 1196
functions (summary) 1198
λ2 (spike) 1190, 1192, 1197
λ3 and μ2 (minor core) 1190, 1197
λ 1190, 1197
μ1C 1198
μNS and μNSC 1199
nonstructural 1199
σ1 1190, 1191, 1200, 1561
structure and function 1199
σ2 1190, 1197
σ3 1190, 1191
functions 1198–1199
σNS and σ1S 1199
receptors for 1200–1201
replication 1191–1192, 1199–1202, 1207–1208
strategy 1192, 1199–1200
transcriptase activation 1192, 1201
serotypes 1190, 1191, 1195
structure 1190–1191, 1195–1199
core shell 1190, 1195, 1197
outer capsid 1190, 1191, 1195–1196, 1197–1199
spikes 1190, 1195
taxonomy and classification 1190
tissue tropism 1193
transcription 1192, 1193, 1201–1202
abortive reiterative 1197
frequencies 1201
ISVPs role 1192, 1201
translation 1192, 1200, 1201–1202
dsRNA-containing complexes (dsRCCs) 1200
ts mutants 1192, 1202

Replacement vectors 114
Replicase, tymoviruses 1500–1501
Replication
DNA, *see* DNA, replication
RNA, *see* RNA, replication
Replication of viruses 42, 1169, **1203–1210**
assembly, *see* Assembly of viruses
blocking and abortive infections 1170, 1587
DNA viruses 1208–1209
early events 1203–1204
attachment 228, 229, 1081, 1203–1204
penetration and uncoating 1169, 1204, 1558
host chromosomal genes effect (totiviruses) 1467
RNA viruses 1205–1208
see also RNA viruses
stages 1203
synthesis of virus-specific macromolecules 1204–1205
cytoskeleton 1205
gene regulation 1205
genome replication 1204–1205
protein synthesis 1205
transcription, *see* Transcription
Replicative form (RF), *see under* DNA; RNA
Replicative intermediate (RI)
animal enteroviruses 386
bacteriophage ϕ29 983
bunyaviruses 196
cardioviruses 209
foot and mouth disease viruses 491
polioviruses 1130
Repressors
interferon synthesis 735
lambda bacteriophage 114, 115, 773, 774
see also cI repressor
Resistance
antivirals 47, 719, 1484
cytokines 701
disinfectants 273
host (animal), *see* Host genetic resistance
insect viruses 1313–1314
plant viruses, *see* Plant resistance
Resistance genes
plant viruses 1106–1107, 1147, 1440–1441
see also Plant resistance
Respiratory distress, in Lassa fever 783
Respiratory, enteric, orphan viruses, *see* Reoviruses$^{(A)}$
Respiratory syncytial virus (RSV)$^{(A)}$ **1210–1219**
antigenic and sequence relatedness to other viruses 1214–1215
antigenic subgroups 1214
cytopathic effects 337, 1215, 1216
epidemiology 1216
future perspectives 1218
genome organization 1213–1214
history 1210
immunity 1217
infections
after transplantation 1487
antivirals 47, 1218
clinical features and pathogenesis 1216
diagnosis 1216
of eye 446
immunopathology 702, 1216
immunoprophylaxis 1217–1218
in tissue culture and animals 1215–1216
treatment 1217–1218
proteins 1211–1213
replication 1213–1214
structure and characteristics 446, 1211–1213
taxonomy and classification 1210–1211
tissue tropism 1215
transcription 1213–1214
see also Human respiratory syncytial virus (HRSV)$^{(A)}$
Respiratory tract
entry of viruses 1076–1077
host defences 1077
Respiratory tract infections 1219
adenovirus 4, 6, 1223
animals 1225
bacterial infections with 1222
cattle, *see* Bovine parainfluenza virus type 3 (PIV-3)$^{(A)}$
chemotherapy 1226
coronaviruses 258–259, 1223
coxsackieviruses 271, 1223
deaths from 1219
echoviruses 357, 358, 1223
epidemiology 1219–1220
equine herpesviruses 423, 427
herpes simplex virus 589, 1223
human parainfluenza viruses 1027, 1029, 1223, 1224
human respiratory syncytial virus (HRSV) 1216, 1223, 1224
incubation period 1220
influenza 713, 1223, 1224
laboratory diagnosis 1225–1226
measles 846, 1225
outbreaks 1219
pathogenesis and immunity 1222–1223
pathways 1221
recovery 1222
reoviruses 1194
rhinoviruses 1223, 1258
Sendai virus 1303
transmission 402, 1219, 1220
types 1223–1225
upper tract 1223
vaccines 1226
problems 1226
see also individual viruses; Respiratory viruses
Respiratory viruses 444, **1219–1226**
culture 1225
evolution 1220
families 1219
immune response to 1222
infection route 1219
infections after transplantation 1487–1488
pathogenesis of infection 1222–1223
spread of 1219, 1220
transmission 1219
see also Respiratory tract infections
Reston virus$^{(A)}$ 827
Ebola virus relationship 828, 830
host range and viral propagation 830
pathology and histopathology 831
transmission and tissue tropism 831
see also Ebola virus$^{(A)}$; Marburg virus$^{(A)}$
Restriction, host-controlled, *see* DNA modification/restriction (M/R) systems; Host-controlled modification and restriction
Restriction endonucleases 529, 669, 670
algal viruses encoding 38–39
analysis, cyanobacteria bacteriophage 289
bacteriophage proteins inactivating 671, 1386
classification 670–671
*Cvi*JI 39
*Cvi*RI and *Cvi*RII 39
*Eco*P1 1001
*Eco*P15 671
*Eco*R11 671
fragment patterns 529
host, inactivation by T5 phage 1386
N4 bacteriophage resistance 894
as part of restrictionmodification (RM) systems 111
profiles, cowpox virus 262
*Rsa*I 39
Restriction fragment length polymorphisms (RFLP) 529
Restriction maps 529
Restrictive infections, definition 1171, 1204
'Reticuloendothelial giant cells', in measles 846
Reticuloendotheliosis 1231
Reticuloendotheliosis virus strain T (REV-T)$^{(A)}$ 1227
disease due to 1231
genome 1228
helper viruses 1231
oncogenes 936, 1227, 1230
rel gene 936, 1227, 1230
replication deficient 1227, 1228, 1230, 1231
transformation by 1230, 1231, 1475
cells *in vitro* 1231–1232
target cells 1232
Reticuloendotheliosis viruses (REV)$^{(A)}$ **1227–1232**
budding 1227
DNA synthesis and integration 1229, 1231
genome structure and properties 1228
helper viruses for REV-T 1231
history 1227
host range and viral propagation 1230
murine leukemia virus (MLV) relationship 1227, 1229
pathogenicity 1231–1232
in vitro 1231–1232
in vivo 1231
post-translational processing 1230
proteins 1229
rel gene 1227, 1230
replication 1229–1230
reverse transcriptase 1229
structure and properties 1227
syndromes due to 1231
taxonomy and classification 1227
transcription 1229–1230
transformation by 1230, 1231, 1232
transmission 1230–1231
$tRNA^{Pro}$ 1227, 1229
Reticuloendotheliosis-associated virus (REV-A)$^{(A)}$ 1227
genome 1228
as helper virus to REV-T 1231
Retina, Borna disease virus (BDV) infection 153
Retinal ischemia, 'cotton-wool spots' 447
Retinitis, CMV causing 444
Retinoblastoma, susceptibility locus, *see* RB gene (retinoblastoma)
Retinoic acid 305, 306
HPV transcription and 1025
receptors 1025
Retinoids, cottontail rabbit papillomavirus (CRPV) infections 1310
Retrograde axon transport 908
'Retroid elements' 440, 441
Retron 1009
Retronphage 1009
Retronphage ϕR67 and ϕR86 1009
Retronphage ϕR73 1009
Retroperitoneal fibromatosis (RF), type D retroviruses 1236, 1241
Retroregulation 819
xis gene, expression control 819

Raccoon parvovirus (RPV)(A) 1061
- clinical features of infection 1065–1066
- host range and viral propagation 1062
- taxonomy and classification 1061

Radiculomyelitis, enterovirus (EV) serotype 70 (EV70) infection 382
Radioimmunoassay (RIA)
- viral antibody assay 346
- viral antigen detection 345

Radish mosaic virus (RaMV)(P) 249
raf oncogene 936, 940
Ramsay–Hunt syndrome 1517
Rana pipiens
- frog virus 3 infection 503
- frog virus-4 (FV4) infection 42
- kidney tumors and virus causing 40, 41

Rana silvatica 41
Range paralysis, *see* Marek's disease (MD)
Ras family of proteins 1474–1475
- GTPase activity 1474, 1475

Rash
- coxsackievirus infection 270
- human parvovirus B19 infection 1054
- measles 845
- mousepox 862, 864
- rabbitpox 867
- Ross River virus (RRV) 1273
- rubella 1295, 1296
- smallpox 1342
- varicella (chickenpox) 1516

Raspberry ringspot virus (RRSV)(P) 904, 906
- coat proteins 905

Rat(s), borna disease virus infection 151
Rat coronavirus (RCV)(A) 259
Rat virus (RV)(A) 1052
- history 1068
- host range and viral propagation 1071, 1072
- infection
  - clinical features 1055, 1074
  - pathology and histopathology 1056
- serologic relationships and variability 1072, 1073
- tissue tropism and pathogenicity 1074

Rauscher murine leukemia virus (MuLV) 740
RAV-0 virus(A) 70
RB gene (retinoblastoma)
- adenovirus E1A protein association 20
- HPV E7 binding to product of 1024
- transformation mechanism 1476, 1477, 1496
  - SV40 1476

Reactivation 527
Reassortment, *see* Genetic reassortment
RecA protein 774
RecA synaptase, host, single-stranded DNA phage recombination 84
RecBCD enzyme
- host
  - in λ phage recombination 85
  - single-stranded DNA phage recombination 84, 85
- recombination pathway 817

RecE, in λ phage recombination 86
Receptor-mediated endocytosis, *see* Endocytosis
Receptors for viruses 1080–1081, 1081, 1169, 1203, 1557, **1560–1563**
- absence and resistance to infection 1591
- definition 1203
- dual 227, 229
- 'pits' in cardioviruses 207
- recognition site on viral surface 1081, 1573
- reovirus type 3 (Reo3R) 1561–1563
  - cells expressing 1562
  - cellular function 1562
  - CNS 1562–1563
  - functional studies 1562
  - inhibition 1562
  - monoclonal antibodies 1561, 1562
  - structure and sequence 1561, 1562
- structure 228
- types and viruses using 229, 1081, 1169, 1204, 1560, 1561
- viral attachment proteins 1081
- *see also* Endocytosis; Pathogenesis of viral infections

RecF, in λ phage recombination 86
Recombinant, definition 2
Recombinant DNA technology 530
- coliphage P2 and P4 1008
- Newcastle disease virus and 918
- plant resistance to viruses 1108, 1109, 1114
- viral antigen source, for serological diagnosis 345
- *see also* Genetic engineering

Recombination, *see* Genetic recombination
Recombination maps 529
Red clover mottle virus (RCMV)(P) 249, 254
Red clover necrotic mosaic virus (RCNMV)(P) 349, 1538
- genome structure 350
- host range 352
- serotypes 353
- structure and composition 349

Red clover vein mosaic virus (RCVMV)(P) 215, 217
Red disease, fish 478
red genes 85, 775
Red system, λ phage recombination 86–87
Redundancies, molecular 437
Reed–Sternberg cells 407, 409
rel oncogene 936, 940, 1227, 1230, 1231, 1475
- reticuloendotheliosis viruses (REV) 1230, 1475

Release of viruses 1078, 1169, 1559–1560
- *see also* Budding

Renal allografts, BK virus infection 752
Renal cell carcinoma (adenocarcinoma)
- frog virus 3 (FV3) 503
- interferon therapy 744

Renal failure 973
Reo3R, *see under* Receptors for viruses
Reoviridae 1190, 1194, 1398, 1407, 1408, 1409
- characteristics 443, 1095, 1190, 1207–1208, 1275, 1581
- *Cypoviruses* 1194
- eye infections 445
- *Fijivirus* 1096, 1195
- fish 475, 479
- *Orbiviruses* 1194
- *Orthoreovirus* 1194
  - *see also* Reoviruses
- *Phytorevirus* 1096, 1195, 1407
- *Rotavirus* 1194, 1275
- structure 1581
- transcription/replication strategy 1207–1208
- *see also individual genera/viruses*

Reovirus type 1 Lang (T1L)(A)
- hemagglutinin 1561
- inclusions 339, 1202
- spread 1080, 1193

Reoviruses(A) **1190–1203**
- adsorption to receptors 1077, 1191, 1200
- assembly 231
- attachment protein (hemagglutinin) 1191, 1199
- avian 1195
- capping importance to 1201
- cell lysis by 1192, 1202
- cellular receptors 1561–1563
  - *see also under* Receptors for viruses
- cytopathic effects 1191, 1202–1203
- dsRNA (segmented) 1190, 1196, 1196–1197
  - replication 1192
- effect in host cells 339, 1191, 1202–1203
- enzymes 1193, 1197
  - capping role 1192, 1201
  - transcriptase 1192, 1197, 1201
- epidemiology 1194
- in feces 373
- gene segment reassortment 1192, 1202
  - protein σ3 1199
- general features **1190–1194**
- genetics 1192
- genome 1196–11197
  - classes (L, M, S) 1196
  - segments and proteins encoded 1196
- hemagglutinin (σ1) 1190, 1191, 1200, 1561
- history 1190
- host range and viral propagation 1191
- infection cycle 1077, 1193
  - spread 1080, 1193
- infections
  - clinical features 1193–1194
  - diagnosis and treatment 1194
  - pancreatic damage 970
  - pathology 1193
- infectious RNA 1202
- infectious subviral particles (ISVPs) 1077, 1191, 1200
  - difference with reovirus particles 1191, 1200
  - transcription role 1192, 1201
- L cell infection 1191, 1200
- molecular biology **1194–1203**
- oligonucleotides 1197
- pathogenicity 1193
- physical properties 1195
- protein-synthesizing system 1202
- proteins 1190–1191, 1197–1199
  - dsRNA segment encoding 1196
  - functions (summary) 1198
  - λ2 (spike) 1190, 1192, 1197
  - λ3 and μ2 (minor core) 1190, 1197
  - λ 1190, 1197
  - μ1C 1198
  - μNS and μNSC 1199
  - nonstructural 1199
  - σ1 1190, 1191, 1200, 1561
    - structure and function 1199
  - σ2 1190, 1197
  - σ3 1190, 1191
    - functions 1198–1199
  - σNS and σ1S 1199
- receptors for 1200–1201
- replication 1191–1192, 1199–1202, 1207–1208
  - strategy 1192, 1199–1200
  - transcriptase activation 1192, 1201
- serotypes 1190, 1191, 1195
- structure 1190–1191, 1195–1199
  - core shell 1190, 1195, 1197
  - outer capsid 1190, 1191, 1195–1196, 1197–1199
  - spikes 1190, 1195
- taxonomy and classification 1190
- tissue tropism 1193
- transcription 1192, 1193, 1201–1202
  - abortive reiterative 1197
  - frequencies 1201
  - ISVPs role 1192, 1201
- translation 1192, 1200, 1201–1202
  - dsRNA-containing complexes (dsRCCs) 1200
- *ts* mutants 1192, 1202

Replacement vectors 114
Replicase, tymoviruses 1500–1501
Replication
 DNA, *see* DNA, replication
 RNA, *see* RNA, replication
Replication of viruses 42, 1169, **1203–1210**
 assembly, *see* Assembly of viruses
 blocking and abortive infections 1170, 1587
 DNA viruses 1208–1209
 early events 1203–1204
  attachment 228, 229, 1081, 1203–1204
  penetration and uncoating 1169, 1204, 1558
 host chromosomal genes effect (totiviruses) 1467
 RNA viruses 1205–1208
  see also RNA viruses
 stages 1203
 synthesis of virus-specific macromolecules 1204–1205
  cytoskeleton 1205
  gene regulation 1205
  genome replication 1204–1205
  protein synthesis 1205
  transcription, *see* Transcription
Replicative form (RF), *see under* DNA; RNA
Replicative intermediate (RI)
 animal enteroviruses 386
 bacteriophage ϕ29 983
 bunyaviruses 196
 cardioviruses 209
 foot and mouth disease viruses 491
 polioviruses 1130
Repressors
 interferon synthesis 735
 lambda bacteriophage 114, 115, 773, 774
 see also cI repressor
Resistance
 antivirals 47, 719, 1484
 cytokines 701
 disinfectants 273
 host (animal), *see* Host genetic resistance
 insect viruses 1313–1314
 plant viruses, *see* Plant resistance
Resistance genes
 plant viruses 1106–1107, 1147, 1440–1441
 see also Plant resistance
Respiratory distress, in Lassa fever 783
Respiratory, enteric, orphan viruses, *see* Reoviruses^(A)^
Respiratory syncytial virus (RSV)^(A)^ **1210–1219**
 antigenic and sequence relatedness to other viruses 1214–1215
 antigenic subgroups 1214
 cytopathic effects 337, 1215, 1216
 epidemiology 1216
 future perspectives 1218
 genome organization 1213–1214
 history 1210
 immunity 1217
 infections
  after transplantation 1487
  antivirals 47, 1218
  clinical features and pathogenesis 1216
  diagnosis 1216
  of eye 446
  immunopathology 702, 1216
  immunoprophylaxis 1217–1218
  in tissue culture and animals 1215–1216
  treatment 1217–1218
 proteins 1211–1213
 replication 1213–1214
 structure and characteristics 446, 1211–1213
 taxonomy and classification 1210–1211
 tissue tropism 1215
 transcription 1213–1214
 see also Human respiratory syncytial virus (HRSV)^(A)^
Respiratory tract
 entry of viruses 1076–1077
 host defences 1077
Respiratory tract infections 1219
 adenovirus 4, 6, 1223
 animals 1225
 bacterial infections with 1222
 cattle, *see* Bovine parainfluenza virus type 3 (PIV-3)^(A)^
 chemotherapy 1226
 coronaviruses 258–259, 1223
 coxsackieviruses 271, 1223
 deaths from 1219
 echoviruses 357, 358, 1223
 epidemiology 1219–1220
 equine herpesviruses 423, 427
 herpes simplex virus 589, 1223
 human parainfluenza viruses 1027, 1029, 1223, 1224
 human respiratory syncytial virus (HRSV) 1216, 1223, 1224
 incubation period 1220
 influenza 713, 1223, 1224
 laboratory diagnosis 1225–1226
 measles 846, 1225
 outbreaks 1219
 pathogenesis and immunity 1222–1223
 pathways 1221
 recovery 1222
 reoviruses 1194
 rhinoviruses 1223, 1258
 Sendai virus 1303
 transmission 402, 1219, 1220
 types 1223–1225
 upper tract 1223
 vaccines 1226
  problems 1226
 see also individual viruses; Respiratory viruses
Respiratory viruses 444, **1219–1226**
 culture 1225
 evolution 1220
 families 1219
 immune response to 1222
 infection route 1219
 infections after transplantation 1487–1488
 pathogenesis of infection 1222–1223
 spread of 1219, 1220
 transmission 1219
 see also Respiratory tract infections
Reston virus^(A)^ 827
 Ebola virus relationship 828, 830
 host range and viral propagation 830
 pathology and histopathology 831
 transmission and tissue tropism 831
 see also Ebola virus^(A)^; Marburg virus^(A)^
Restriction, host-controlled, *see* DNA modification/restriction (M/R) systems; Host-controlled modification and restriction
Restriction endonucleases 529, 669, 670
 algal viruses encoding 38–39
 analysis, cyanobacteria bacteriophage 289
 bacteriophage proteins inactivating 671, 1386
 classification 670–671
 *Cvi*JI 39
 *Cvi*RI and *Cvi*RII 39
 *Eco*P1 1001
 *Eco*P15 671
 *Eco*R11 671
 fragment patterns 529
 host, inactivation by T5 phage 1386
 N4 bacteriophage resistance 894
 as part of restrictionmodification (RM) systems 111
 profiles, cowpox virus 262
 *Rsa*I 39
Restriction fragment length polymorphisms (RFLP) 529
Restriction maps 529
Restrictive infections, definition 1171, 1204
'Reticuloendothelial giant cells', in measles 846
Reticuloendotheliosis 1231
Reticuloendotheliosis virus strain T (REV-T)^(A)^ 1227
 disease due to 1231
 genome 1228
 helper viruses 1231
 oncogenes 936, 1227, 1230
 rel gene 936, 1227, 1230
 replication deficient 1227, 1228, 1230, 1231
 transformation by 1230, 1231, 1475
  cells *in vitro* 1231–1232
  target cells 1232
Reticuloendotheliosis viruses (REV)^(A)^ **1227–1232**
 budding 1227
 DNA synthesis and integration 1229, 1231
 genome structure and properties 1228
 helper viruses for REV-T 1231
 history 1227
 host range and viral propagation 1230
 murine leukemia virus (MLV) relationship 1227, 1229
 pathogenicity 1231–1232
  in vitro 1231–1232
  in vivo 1231
 post-translational processing 1230
 proteins 1229
 rel gene 1227, 1230
 replication 1229–1230
 reverse transcriptase 1229
 structure and properties 1227
 syndromes due to 1231
 taxonomy and classification 1227
 transcription 1229–1230
 transformation by 1230, 1231, 1232
 transmission 1230–1231
 $tRNA^{Pro}$ 1227, 1229
Reticuloendotheliosis-associated virus (REV-A)^(A)^ 1227
 genome 1228
 as helper virus to REV-T 1231
Retina, Borna disease virus (BDV) infection 153
Retinal ischemia, 'cotton-wool spots' 447
Retinitis, CMV causing 444
Retinoblastoma, susceptibility locus, *see* RB gene (retinoblastoma)
Retinoic acid 305, 306
 HPV transcription and 1025
 receptors 1025
Retinoids, cottontail rabbit papillomavirus (CRPV) infections 1310
Retrograde axon transport 908
'Retroid elements' 440, 441
Retron 1009
Retronphage 1009
Retronphage ϕR67 and ϕR86 1009
Retronphage ϕR73 1009
Retroperitoneal fibromatosis (RF), type D retroviruses 1236, 1241
Retroregulation 819
 xis gene, expression control 819

Retrosternal chest pain, in Lassa fever 783
Retrotransposons 1232, 1318
of fungi **1232–1236**
assembly 1234
DNA integration 1234–1235
host genes affecting 1235, 1236
long terminal repeat (LTR)-containing 1232–1233, 1235
reverse transcriptase 1232, 1234
structural features 1232–1233
TAD element 1233
translation 1234
transposition mechanism 1233–1235
types 1233
virus-like particles 1234, 1235–1236
long terminal repeat (LTR)-containing types 1232, 1233, 1235
in nuclear polyhedrosis viruses 132
poly(A) types 1232
Retrovir 45, 48
Retroviral model, spongiform encephalopathy infectious agent 1363–1364
Retroviral vectors, *see also* Retroviruses$^{(A)}$, as vectors
Retroviridae 67, 159, 167, 1227, 1399, 1409
B-type 1237
mouse mammary tumor virus (MMTV) 853
C-type
murine leukemia viruses 884
reticuloendotheliosis viruses 1227
see also under Retroviruses$^{(A)}$
D-type 1237
see also Retroviruses$^{(A)}$, type D
eye infections 447
fish 475
Hepadnaviridae relationship 556, 566
Lentivirus 159, 455, 1316, 1593
feline immunodeficiency virus (FIV) 455
human immunodeficiency virus (HIV) 674
Oncovirinae 159, 167, 1227
HTLV-1 and HTLV-2 682, 687
reticuloendotheliosis viruses 1227
Spumavirinae 159, 1593
structure and characteristics 159, 443, 447, 682, 687, 1227, 1316, 1583
subfamilies 159, 1227
see also Retroviruses$^{(A)}$; *specific genera and viruses*
Retroviruses$^{(A)}$ 158, 159, 1590
acute leukemia viruses 682
avian, lymphoproliferative disease virus (LPDV) of turkeys 812
avian leukosis viruses **66–71**
bovine immunodeficiency virus (BIV) **158–166**
bovine leukemia virus 166, 167, 682
C-type
bovine leukemia virus differences 170
EBV-like viruses relationship 613
Gibbon ape leukemia virus (GaLV) 535
morphology 67
see also Retroviridae
chronic leukemia viruses 682
defective, phenotypic mixing 528
diploid genome 525
discovery 631
DNA 1208
DNA integration and transformation mechanism 527, 556, 935, 1208, 1473
see also Transformation
evolution 70
antiquity 440
foamy viruses 480
genes, order 67
genome 67
HTLV-1/HTLV-2 682, 687
as insertional mutagens 935, 937–938
isolation 341
latent infections 788
lenti 1593
mammalian, reticuloendotheliosis viruses relationship 1227
oncogenes 67, 631, 934, 935, 1473
list/characteristics 936
primates, *see* Retroviruses$^{(A)}$, type D, *(below)*
provirus 1208
replication 68, 431–432, 556, 884, 935
strategy 485, 935, 1169, 1204, 1208
replication-defective, and acute transformation by 1472–1473
species origin, recombination 439
spuma, *see* Spumaviruses$^{(A)}$
transformation by, *see* Transformation
translation 1208
transmission 935
as vectors 1532–1534
advantages/disadvantages of types 890, 1533
applications and future prospects 1534
cancer therapy 1534
double expression (DE) 1532, 1533
gene therapy 1533, 1534
general features 1532–1534
limitations and concerns 1533–1534
principles 888–890, 1533
self-inactivating (SIN) vectors 1532, 1533
vectors with internal promoters (VIP) 1532, 1533
see also Lentiviruses$^{(A)}$; Oncoviruses$^{(A)}$; *specific viruses*; Spumaviruses
Retroviruses$^{(A)}$, type D **1236–1242**
in AIDS patient 1242
assembly site and release 1239
cell fusion *in vitro* 1239
cytopathology 1239
endogenous viruses 1237, 1240
enzymes 1238
epidemiology 1241
evolution 1237, 1240
future perspectives 1242
genetics 1240
genome structure and properties 1237, 1238
geographic distribution 1239
history 1236–1237
host range and viral propagation 1240
infection
clinical features 1241
human infections 1242
immune response 1242
pathology and histopathology 1241–1242
prevention and control 1242
intracytoplasmic A-type particles (ICAPs) 1237, 1239
pathogenicity 1241
physical properties 1238
post-translational processing 1239
proteins 1238
receptors 1240
interference 1240
replication strategy 1238
reverse transcriptase 1237, 1238
serologic relationships and variability 1240–1241
structure and properties 1237
taxonomy and classification 1237
tissue tropism 1241, 1242
transcription 1238
translation 1238–1239
transmission 1241
$tRNA^{Lys}$ 1237
vaccines 1242
rev gene 1318
caprine arthritis encephalitis virus 200
HIV 676
origin 439
Rev protein 676, 1208, 1319
HIV 676–677
rev-responsive element (RRE) 1319
Reverse genetics 665, 1528, 1535
Reverse gyrase 56
Reverse transcriptase (RT) 158, 159, 935, 1204, 1208
antivirals inhibiting 48, 49
avian hepatitis B virus 564
avian leukosis viruses 67–68
badnaviruses 140
bovine immunodeficiency virus (BIV) 158, 159
bovine leukemia virus (BLV) 167
cellular gene 1235
detection 341
discovery 631
equine infectious anemia virus (EIAV) 431, 432
feline leukemia virus (FeLV) 460
hepatitis B virus (HBV) 561
HTLV-1, fidelity of 690
HTLV-2 691
human immunodeficiency virus (HIV) 678
lymphoproliferative disease virus (LPDV) of turkeys 814
origins of viruses and 441
plant viruses, caulimoviruses 225
reticuloendotheliosis viruses (REV) 1229
retrotransposons 1232, 1234
in transposons 1232, 1234
type D retroviruses 1237, 1238
Reverse transcription 68, 1208
caprine arthritis encephalitis virus (CAEV) 201
DNA genome replication by, caulimoviruses 223, 224–225
hepadnaviruses (hepatitis B virus) 556
HIV 675
'jumping' mechanism 68, 70
murine leukemia viruses (MuLVs) 884
rex gene
bovine leukemia virus 167
HTLV-1 683, 693–693, 1496
HTLV-2 687, 691, 692–693
Rex protein
HTLV-2 692–693
mechanism of action 683, 693
transformation and 694
Rex-responsive elements (RxRE) 693
Reye's syndrome 713, 1516
Rgl system 672
in order *Mononegavirales* 830
Rhabdoviridae 1210, 1399, 1408, 1409, 1548
animal, and bird 1249, 1409
antigenic cross-reactivity 1215
characteristics and structure 443, 446, 1181, 1186, 1548, 1584
evolution 441
eye infections 446
fish 476, 477–479, 1249
gene map 1212
Lyssavirus (rabies-like viruses) 1186, 1614
plant viruses 1243, 1407
rabies virus 1181

Rhabdoviridae continued
replication/transcription strategy 1207
subgroups A and B 1243
Vesiculovirus 233, 1243, 1548
see also Rabies virus(A); Rabies-like viruses(A); Vesicular stomatitis virus (VSV)(A)
Rhabdoviruses(A), animal **1249–1253**
avian 1249
filoviruses similarity 828
fish 1249–1253
plant rhabdovirus comparison 1244, 1245
rabies-like viruses 1186
see also individual viruses
Rhabdoviruses(I), insect, *see* Sigma rhabdoviruses(I)
Rhabdoviruses(P), plant 139, **1243–1249**, 1407
animal rhabdovirus comparison 1244, 1245
bullet-shaped nucleocapsid 1244
defective-interfering (DI) particles 1245
ecology and pathology 1247
economic significance 1247
genomic structure and organization 1244, 1245
geographic distribution and host range 1247
morphology 1244–1245
proteins 1244
size 1244–1245
taxonomy and classification 1243
transcription 1245
transmission 1243, 1248
kinetics 1248
ultrastructure and replication 1245–1246
Rhesus monkey kidney cells (RhMK) 337
cytopathic effects of viruses 337
Rhesus monkeys, poliovirus research using 634
Rhinitis 1221, 1223
causative agents 1223, 1224
Rhinomune vaccine 429
Rhinopneumonitis, equine 423, 427
see also Equine herpesviruses(A), EHV-1
Rhinovirus 1115, 1253
Rhinoviruses(A) **1253–1259**
anti-viral drugs 1258, 1259
assembly, release and cytopathology 1256
'canyon' hypothesis 1121, 1123, 1255
capsid proteins 1254
cell receptor 1254, 1255
cytopathic effects 337
epidemiology 1257
evolution 1256
future perspectives 1259
in gastrointestinal tract 1077
genetics 1256
genome structure and properties 1254
geographic and seasonal distribution 1256
history 1253
host protein synthesis shutdown 491
host range and virus propagation 1256
hydrophobic pocket in capsid 1259
identification 341
infection
clinical features 1258
common colds 1223, 1253, 1257, 1258
of eye 447
frequency and respiratory disease types 1223
immune response 1258
incidence of common colds 1257
initiation 209
pathology and histopathology 1258
prevention and control 1258
interferon clinical use/action 738, 1258
pathogenicity 1258
physical properties 1255
post-translational processing 1255
properties 642, 1254
proteins 1254
nonstructural 1254–1255
structural 1254
receptors 1081, 1561
replication strategy 1255
early events 1255
RNA synthesis 1255–1256
serological relationships and variability 1256–1257
serotypes 701, 1254, 1256, 1259
structure 642, 1254
taxonomy and classification 1253–1254
three-dimensional structures 1254, 1259
tissue tropism 1257
translation 1255
transmission 401, 1257
vaccine possibilities 1259
see also Human rhinoviruses(A)
Rhizidiovirus(P) 1398, 1406
Rhizoctonia solani virus 717 (RsV-717)(P) 1047
Rhizomania 509, 514
rho factor 1380
Rhodamine B 344
Ribavirin 44, 47
human respiratory syncytial virus (HRSV) infection 1218, 1487
Lassa fever 785, 786
Ribgrass mosaic tobamovirus (RMV)(P) 1436
Ribosomal frameshifting, *see* Frameshifting
Ribosomal pausing 1234
Ribosome Landing Pad (RLP) 1123
Ribosomes, host, in arenaviruses 776, 806
Ribozyme activity, hepatitis D virus 576
Rice black streaked dwarf virus (RBSDV)(P) 1096
genome 1097
Rice dwarf virus (RDV)(P) 1096
Rice gall dwarf virus (RGDV)(P) 1096
Rice grassy stunt virus (RGSV)(P) 1410, 1411
Rice hoja blanca virus (RHBV)(P) 1410, 1411
distribution and importance 1414
'transmitters' and 'nontransmitters' 1415
Rice ragged stunt virus (RRSV)(P) 1096
Rice stripe necrosis virus (RSNV)(P) 509
Rice stripe virus (RStV)(P) 1410, 1411
distribution and importance 1414
gene expression strategies 1414
genome 1412
Rice transitory yellowing virus(P) 1247
Rice tungro bacilliform virus (RTBV)(P) 139, 1538
distribution and economic importance 141
epidemiology and control 142
genome 140
host relationship 141
structure and multiplication 140
transmission 142
Rice tungro spherical virus (RTSV)(P) 142, 824
Rice yellow mottle virus (RYMV)(P) 1346
see also Sobemovirus(P)
Rift Valley fever 191, 1618
diagnosis and prevention 1619
epidemics 1621
Rift Valley fever virus(A) 190, 191, 446, 1618–1619
assembly and budding 196
geographic distribution 1618
host range 1618
transmission 1619
Rimantadine 44, 47
Rinderpest 1260
clinical features 1265
diagnosis 1266–1267
eradication campaign 1267
geographic and seasonal distribution 1261
immune response 1266
outbreaks 1260, 1264
pathology and histopathology 1265–1266
prevention and control 1266
Rinderpest virus (RV)(A) 838, **1260–1268**
as archevirus of *Morbillivirus* 844
epizootiology 1264
evolution 1263
future perspectives 1267–1268
genetics 1261–1263
geographic distribution 1261
history 1260
host range and viral propagation 1261
pathogenicity 1265
serologic relationships and variability 1263–1264
taxonomy and classification 1260–1261
transmission and tissue tropism 1264, 1264–1265
vaccination 1267
vaccine 1264, 1266, 1267
tissue culture attenuated 1267
Ring-necked pheasant virus (RPV)(A), as insertional mutagens 937
Ringspot disease, nepoviruses causing 904, 906
Rio Grande cichlid rhabdovirus(A) 478
RNA
ambisense single-stranded
arenaviruses 776, 802
bunyaviruses 1207
lymphocytic choriomeningitis virus (LCMV) 802, 807
tenuiviruses 1413, 1414
tospoviruses 1461
amplification methods 347, 348
double-stranded (dsRNA)
bacteriophage ϕ6 978
birnaviruses 143, 144, 1207
carlavirus replication 216
carmovirus replication 222
coltiviruses 942, 957–958
comovirus replication 253
cryptoviruses 275
cytoplasmic polyhedrosis viruses (CPV) 312
families 1207
foot and mouth disease viruses 491
giardiavirus replication 533
interferon action/induction 735, 737
orbiviruses 942, 957–958
partitiviruses 1047, 1048
phytoreoviruses 1095, 1096, 1097
protein kinase dependent on 736
Reoviridae, see Reoviridae
replication/transcription strategy 1207–1208
rotaviruses 1275, 1281
segmented in reoviruses, *see* Reoviruses
tobacco necrosis virus replication 898
editing mechanism, in measles virus 839
fingerprinting, zoonoses diagnosis 1620
host, inhibition by cardioviruses 210
messenger, *see* mRNA; RNA, positive-strand
messenger-sense, *see* RNA, positive-strand
mutation rate 438, 441, 525, 1220
negative-strand
bunyaviruses 186, 192, 193
Chandipura, Piry and Isfahan viruses 234

complementary terminal sequence 1462
effect of defective-interfering viruses 322
evolution 1182
ϕH bacteriophage immunity 54
hantaviruses 538, 539
human parainfluenza viruses 1027
influenza viruses 710, 715
measles virus 838, 839
Newcastle disease virus 915
nonsegmented 1207, 1210
nonsegmented in Marburg and Ebola viruses 827, 828
nonsegmented in mumps virus 876, 877
rabies virus 1181
rabies-like viruses 1186
replication strategy 1207
respiratory syncytial virus 1210, 1213
rinderpest and distemper viruses 1261, 1262
segmented 1207
Sendai virus 1300
sigma rhabdovirus 1312
tospoviruses (L segment) 1461
vesicular stomatitis virus (VSV) 1548, 1552
virus families 1207
see also RNA, single-stranded (ssRNA); RNA viruses
positive-strand
animal enteroviruses 385
astroviruses 376
bipartite, in nodaviruses 919, 921–922
bovine viral diarrhea virus (BVDv) 176
bromoviruses 181, 182
bunyaviruses 196
caprine arthritis encephalitis virus (CAEV) 199, 200
cardioviruses 206, 207
carlaviruses 214
carmoviruses 218, 220
Chikungunya (CHIK), O'Nyong Nyong (ONN) and Mayaro viruses 238
closteroviruses 242, 244, 245
comoviruses 251
coronaviruses 255
coxsackieviruses 268
cucumoviruses 280
dianthoviruses 349–351
echoviruses 355
encephalitis viruses 362
enterovirus (EV) serotype 70 (EV70) 380
equine arteritis virus (EAV) 764–765
equine encephalitis viruses 417
equine infectious anemia virus (EIAV) 430
feline leukemia virus (FeLV) 460
flaviviruses 362, 368, 369
foamy viruses 480
foot and mouth disease viruses (FMDVs) 489, 490
hepatitis A virus 547
hepatitis C virus 570
hepatitis E virus 581
hog cholera virus 649
hordeiviruses 661
Japanese encephalitis virus 746
lactate dehydrogenase-elevating virus (LDV) 764–765
luteoviruses 792, 793
lymphoproliferative disease virus (LPDV) of turkeys 812
maize chlorotic dwarf virus (MCDV) 824
mechanism of interferon action 737
murine leukemia viruses (MuLVs) 884
Norwalk virus 927, 928–929
parsnip yellow fleck viruses 1042, 1043
pea enation mosaic virus (PEMV) 1084
polioviruses 1115, 1119
poly(A) tail 482, 490, 581
poly(C) tract 490
potexviruses 1142, 1143, 1144
potyviruses 1148, 1149
replication strategy 1205, 1206–1207
reticuloendotheliosis viruses (REV) 1227, 1228
rhinoviruses 1254
RNA phage 1334
Ross River virus (RRV) 1269
rubella virus 1291
Sindbis and Semliki Forest viruses 1330
sobemovirus 1346, 1349
tetraviruses 1417
Theiler's viruses 1424
tick-borne encephalitis virus (TBEV) 368, 369
tobacco necrosis virus 896, 899
tobamoviruses 1438
tobraviruses 1442
tombusviruses 1447, 1449
toroviruses 1453
tymoviruses 1500, 1501
type D retroviruses 1237
vesicular exanthema virus and caliciviruses 1544
virus families with 1206
yellow fever virus 16–7
see also RNA, single-stranded (ssRNA)
primer 195
retroviruses, reticuloendotheliosis viruses (REV) 1227, 1229
probe 347
replication
birnaviruses 146, 147
bluetongue virus 962–963
bromovirus 182
bunyaviruses 192, 195, 196
cardioviruses 209
carlaviruses 216
cytoplasmic polyhedrosis viruses (CPV) 315
foot and mouth disease viruses 491
hepatitis A virus (HAV) 548–549
hepatitis D virus 577
influenza viruses 719–720
mumps virus (MuV) 880
nodaviruses 923
plant viruses, vectors 1541
rolling circle mechanism, hepatitis D virus 577
rotaviruses 1288–1289
viroids 1568
see also other specific viruses
replicative form, foot and mouth disease viruses 491
replicative intermediate, *see* Replicative intermediate (RI)
'reporter' molecule 348
satellite 1570
carmoviruses 222–223
evolutionary link with viroids 1565, 1566
single-stranded (ssRNA)
Borna disease virus (BDV) 149, 150
bovine immunodeficiency virus (BIV) 160
bovine leukemia virus 167, 169
circular, hepatitis D virus 575, 576
fabaviruses[P] 452
HIV 675
insect picornaviruses 1101
tenuiviruses 1412
viroids 1563, 1564
see also RNA, negative-strand; RNA, positive-strand
small nuclear (snRNA) 12
small (U-RNA), in herpesvirus saimiri 616
splicing
adenovirus transcription 12
discovery 631
plant viruses as vectors for studying 1542
subgenomic (sgRNA)
carlaviruses 214
carmoviruses 220
transcription, *see* Transcription
tRNA, *see* tRNA
VA ('virus associated') in adenoviruses 10, 15
RNA bacteriophage[B] 1405
discovery 630
double-stranded 1405
taxonomy and classification 95
RNA bacteriophage[B], single-stranded **1334–1339**, 1405
6S RNA 1338
coliphage 1334–1335
diversity 1338
ecology 1336
gene expression 1337
genetic map 1336
Group A 1334
Group B 1334
lysing protein 1336–1337
host range 1335–1336
as index organisms for pathogenic enteroviruses 1336
infection process 1334–1335
maturation (A) protein 1334, 1335
noncoliphage 1335
phage and groupings 1334
phylogeny 1338–1339
replicase 1337
replication 1337–1338
RNA variants (Q_β) 1338
structure 1334
taxonomy and classification 1334
RNA polymerase
DNA-dependent, discovery 630
Escherichia coli
N4 bacteriophage late transcription 893
T7 phage transcription 1394, 1395
filamentous bacteriophage infection 467
host
modification by PBS1 bacteriophage 75
modification by T4 phage 1377
modification by T5 phage 1386–1387
T4 bacteriophage transcription 1380
virus infection effect 1588
N4 bacteriophage 891, 892, 893, 894
partitiviruses 1047, 1048, 1049
phage
promoter specificities 1395
see also Promoters; *specific phage*
reovirus 1192, 1197
RNA-dependent 1204
alfalfa mosaic virus (AlMV) 35
coronaviruses 257
cryptic viruses 275, 276, 277
foot and mouth disease viruses 491

RNA polymerase *continued*
    GDD motif 221, 253
    giardiavirus 532
    nodaviruses 921, 923
    phytoreoviruses 1097
    rotaviruses 1276, 1286, 1287, 1288
    sobemoviruses 1350
    totiviruses 1465, 1466
  RNA-directed, hepatitis E virus 581
  SPO1 phage 1353, 1354
RNA polymerase II
  African swine fever virus 26
  cellular
    frog virus 3 (FV3) transcription 506
    transcription of silent methylated genes 506
  DNA-dependent 1204
  hepatitis D virus 577
  parvovirus transcription 1058
  subunit TFIID 21
  varicella–zoster virus transcription 1524
RNA polymerase III
  adenovirus 15
  consensus motifs 183
RNA replicase, carlaviruses 216
RNA virus supergroup, plant virus groups in 218
RNA viruses 440
  animal viruses 525
    smallest 649
  assembly 1573
  atomic structures 1572
  avian leukosis viruses 67
  cDNA clones
    infectivity 1540
    isolation 1528
  characteristics 443
  classification 1405–1409
  complementary termini in 188, 192, 1412, 1413, 1462
  defective-interfering mutants 526
  double-stranded 532, 534, 633, 1207, 1207–1208, 1399
    bacterial 1405
    classification 1398
    invertebrate 1408
    plant 1407
    properties 534
    vertebrates 1409
  electron microscopy 340, 342
  enveloped
    bacterial 1405
    invertebrate 1408
    plant 1407
    vertebrates 1409
  envelopes and spikes 444
  evolution 441
  eye infections, *see* Eye infections
  families
    algal, fungal and protozoal 1406
    bacterial 1405
    invertebrates 1408
    plants 1407
    vertebrates 1409
  intramolecular recombination 527
  mutation rate 438, 441, 525, 1220
  negative-strand
    evolution 1182
    nonsegmented 830, 1207, 1210
    segmented 1207
    see also RNA, negative-strand
  non-enveloped
    algal, fungal and protozoal 1406
    bacterial 1405
    invertebrate 1408
    plant 1407
    vertebrates 1409
  plant
    acquired resistance mechanism 1108
    ilarviruses 30
    as vectors, *see* Vectors
  positive-strand
    replication 1205, 1206–1207
    see also RNA, positive-strand
  replication 1169, 1205–1208
    double-stranded RNA 1207–1208
    negative-strand RNA 1207
    positive-strand RNA 1206–1207
    reverse transcription strategy 1208
    segmented negative-strand 1207
    unsegmented negative-strand 1207
  single-stranded 633, 1399–1400
    bacterial 1405
    classification 1399–1400
    invertebrate 1408
    plant 1407
    ribosomal frameshifting 1601
    Saccharomyces cerevisiae satellite viruses 1601
    vertebrates 1409
  structures 444
  transcription 1204
  transcriptional strategies 1205–1208
    see also RNA viruses, replication
  uncoating 1169
  as vectors 1528
  yeast, *see* Yeast RNA viruses
  see also individual RNA viruses; RNA
RNase H, in badnaviruses 140
Rocio virus(A) 1618
Rodent parvoviruses(A) 1055, 1056, **1067–1075**
  geographic and seasonal distribution 1071
  NS proteins 1069
  transcription and translation 1069–1070
  see also H-1 virus(A); Minute virus of mice (MVM)(A); Rat virus (RV)(A)
Rodents
  arenavirus hosts 778, 781
    as reservoir 779, 780
  control, arenavirus prevention 785
  hantavirus infections and transmission 542, 543
  poliovirus strains 635
ros oncogene 936
Roseola infantum 625
'Ross River fever' 1268
Ross River virus (RRV)(A) **1268–1274**, 1620
  assembly 1270
  cDNA clone 1269
  epidemiology 1272
  Fijian strain 1271
  genetics and evolution 1271–1272
  genome structure and properties 1269
  geographic and seasonal distribution 1270–1271, 1620
  hemagglutinin (E1) 1269, 1270
  history 1268
  host range and virus propagation 1271
  infection
    chronic form 1272
    clinical features 1273
    of eye 445
    immune response 1273–1274
    pathology and histopathology 1273
    prevention and control 1274
  mutation 1271, 1272
  pathogenicity 1272–1273
  proteins 1270
    E1 1269, 1270
    E2 1270
    non-structural 1269, 1270, 1271
  replication 1270
  serological relationships and variability 1272
  structure and properties 1269
  taxonomy and classification 1269
  Townsville T48 strain 1269, 1271, 1272
  transmission 1620
  vaccination 1274
  variants 1271
Rotaviruses(A) 926, **1274–1290**
  animal 1277
  antigen processing 1280
  antigenic determinants 1277, 1280
  cell lysis due to 1290
  cellular receptor 1286
  cytopathic effects 1290
  double-shelled 1281, 1286
  electron microscopy 1275, 1276, 1281
  electropherotypes 1276, 1277
  epidemiology 1278
  evolution 1277
  future perspectives 1281, 1290
  gene reassortment 1277
  general features **1274–1281**
  genetics 1275–1277
  genome structure and properties 1277, 1281, 1282, 1284–1285
    protein products 1277, 1284–1285
    terminal consensus sequences 1282
  geographic and seasonal distribution 1275
  Group A 373–374, 1275
    cross-species infections 374
    G-types and P-types 1277–1278, 1283
    genome structure 374, 1282
    geographic distribution 374
    infection 374, 1278, 1279
    serologic relationships and variability 1277
    serotypes 1277, 1278
  Group B 374, 1275, 1277
    infection 1277, 1279
  Group C 374, 1275, 1277
    infection 1277
  Groups D, F, and G 374, 1277
  hemagglutinin 1277, 1286
    see also Rotaviruses, proteins, VP4
  history 373, 1274–1275
  host range and viral propagation 1275
    plaque forming units 1275
  infections
    clinical features 1279
    gastroenteritis 1274, 1275, 1278, 1279
    immune response 1279–1280
    pathology and histopathology 1279
    prevention and control 1280–1281
    route of 1077, 1278
    treatment 1280
  monoclonal antibodies 1279–1280
  mRNA 1286
  neutralizing antibodies 1277, 1280, 1283
  nongroup A (B-F) 374, 377, 1277
    see also specific groups above
  'novel' (atypical), *see* Rotaviruses(A), nongroup A
  pathogenicity 1278–1279
  physical properties 1283
  proteins 1281–1282
    G-types and P-types 1277–1278, 1283
    nonstructural 1283
    NS28 1277, 1283, 1286, 1289
    properties 1283
    structural 1281, 1283
    subcore 1281, 1283
    synthesis and assembly 1289–1290
    VP1 1288
    VP2 shells (core particles) 1281, 1283

VP4 1277, 1280, 1283
VP4 cleavage 1275, 1282, 1287
VP4 serotypic diversity 1280, 1283
VP4 spikes 1282, 1283
VP5* and VP* 1282, 1287
VP6 1277, 1282, 1283
VP7 1277, 1280, 1283, 1289–1290
VP7 serotypes 1277, 1283
replication 1286–1290
adsorption, penetration and uncoating 1286–1287
assembly 1289–1290
general features 1286
internalization 1286
maturation 1290
release 1290
transcriptase activation 1288
transcription 1276–1277, 1287–1288
rhesus 1287
RNA replication 1288–1289
cell-free system 1288–1289
serologic classification 1277
serologic relationships and variability 1277–1278
serotypes (P-type and G-type) 374, 1277–1278, 1283
single-shelled particles 1277, 1286
structure and properties 1275, 1281–1282
spikes 1282, 1283
subviral particles 1289
assembly *in vitro* 1289
taxonomy and classification 1194, 1275
tissue tropism 1278
transcriptase 1276, 1287, 1288
transcription 1276–1277, 1287–1288
transmission 401, 1278
vaccination 1280, 1281
vaccine candidates 1278, 1281
Rous sarcoma virus (RSV)(A) 66, 69, 934, 935
oncogenes 936, 1474
quantitative assays 630
transformation mechanism 1474
RRE (Rev responsive element) 676
Rubella 1291
clinical features 1295–1296
congenital 1294–1295, 1296
diabetes and 970
consequences 1294–1295
diagnosis 1296
epidemiology 1294–1295
gestational 1294–1295, 1296
heart infections 965
immune response 1297, 1298
infants 1297
pandemic 1291, 1294
pathology and histopathology 1296–1297
prevention and control 1297–1298
rash 1295, 1296
vaccines 1291, 1297
antibodies 1296, 1297
E1 subunit 1298
in pregnancy 1298
RA27/3 strain 1297
reactions to 1297–1298
see also Measles, mumps and rubella (MMR) vaccine
see also Rubella virus(A)
Rubella virus(A) **1291–1298**
adsorption and penetration 1292
African green monkey kidney cells (AGMK), rubella virus culture 1293
antibodies to 1296, 1297
vaccine-induced 1296, 1297
assembly site and release 1293
clinical features of infection, *see* Rubella
cytopathology 1293
epidemiology 1294–1295
evolution 1294
future perspectives 1298
genome structure and properties 1291–1292
geographic and seasonal distribution 1293
glycoproteins (E1 and E2) 1291, 1293
history 1291
host range and virus propagation 1293
HPV77 strain 1294
Judith strain 1294
M_{33} strain 1294
pathogenicity 1295
persistent infection of cell cultures 1294
physical properties 1292
placental infection 1295
post-translational processing 1293
proteins 1292
C protein 1292
nonstructural 1292
re-infection 1296
replication strategy 1292–1293
RNA 1291
serologic relationships and variability 1294
structure and properties 1291
systemic spread 1295
taxonomy and classification 1291
Therien strain 1294
thyroiditis 976
translation 1292
transmission and tissue tropism 1295
Rubivirus(A) 1291
Rubus Chinese seedborne virus(P) 905
Russian spring summer encephalitis virus (RSSEV)(A) 361
Ryegrass cryptic virus (RGCV)(P) 275
Ryegrass spherical virus (RGSV)(P) 275

S
SA11 (simian rotavirus)(A) 1282
see also Rotaviruses(A)
SA12 polyomavirus(A) 1135
Sacbrood 654, 657
Sacbrood virus (SBV)(I) 656, 657
history 654
pathology and epizootiology 657
Thailand (TSBV) 657
Saccharomyces cerevisiae
20S RNA replicon 1604–1605
future perspectives 1605
chromosome genes affecting dsRNA viral replication 1602–1603, 1604
killer toxins, *see below*
mak mutants 1602
T dsRNA 1604–1605
Ty elements 1233
W dsRNA 1604–1605
Saccharomyces cerevisiae virus L-A (ScV-L-A)(P) 532, 534, 1464, 1468, 1600, 1601
applications 1605
expression and ribosomal frameshifting 1601–1602
future perspectives 1605
host functions affecting 1602–1603, 1604
killer toxins 1464, 1468, 1600
economic significance 1467, 1471
K_{28} 1604
M_1 1600, 1601, 1603, 1604
M_1 action 1603–1604
processing 1603
packaging 1601
replication cycle 1601
satellite viruses 1600
M_1, *see* M_1 *(above)*
replication 1601
see also Totiviruses(P)
Saccharomyces cerevisiae virus L-BC (ScV-L-BC)(P) 1465, 1600, 1601
host functions affecting 1604
Sacramento River Chinook disease virus (SRCDV)(A) 1251
Saguaro cactus virus (SCV)(P) 218, 219
SAIDS, *see* Simian AIDS (SAIDS)
St Louis encephalitis 366
St Louis Encephalitis virus (SLEV)(A) 361, 363, 445, 1616
epidemiology 365
geographic and seasonal distribution 362, 1616
Japanese encephalitis virus relationship 748
topotypes 364
see also Encephalitis viruses(A)
Saliva
feline leukemia virus transmission 461
rabies virus transmission 1183
type D retrovirus transmission 1241
Salivary gland
human cytomegalovirus infection 295, 297
mumps virus (MuV) in 881, 882
Salivary gland viruses 293
see also Human cytomegalovirus (HCMV)(A)
Salmon, virus infections 144
Salmonella
ϵ-prophage 81
filamentous bacteriophage 464
Salmonella Lille 1452
Salmonella typhi, Vi antigen 79
Salmonella typhimurium
P22 bacteriophage 1009
PRD1 bacteriophage 1165
RNA phage 1335
transduction discovery 107
Salmonid herpesviruses(A) 470
Salmonids, virus infections 470, 476
Salt concentrations, halobacterial phage and 52
Sammon's Opuntia tobamovirus (SOV)(P) 1436
San Miguel sea lion virus(A) 584, 1544, 1545
see also Caliciviruses(A); Vesicular exanthema virus of swine (VES)(A)
Sandflies, phlebotomus fever virus transmission 191
Sandfly fever viruses(A) 191, 1616
Sandwich hybridization 347
Satellite tobacco mosaic virus (STMV)(P) 1441
Satellite tobacco necrosis viruses (STNV)(P), *see* Tobacco necrosis virus (TNV)(P)
Satellite viruses, structure 1581
Satsuma dwarf virus(P) 905
Scarlatinal toxins 103
Schefflera ringspot virus (SRV)(P) 139
Schistocerca gregaria, control 397
Schizosaccharomyces cerevisiae, Tf1 elements 1233
Scrapie 912, **1361–1369**
Creutzfeldt–Jakob disease (CJD) relationship 1361–1361, 1364
cytological and molecular pathology 1368
history 1361
infectious agent 1361
bovine spongiform encephalopathy (BSE) and 1358, 1360
inactivation 1362
routes of inoculation 1365
taxonomy and classification 1362
susceptibility and incubation period 1367
susceptibility (*sinc*) 1361
tissue culture studies 1369

Scrapie *continued*
see also Creutzfeldt-Jakob disease (CJD); Spongiform encephalopathies
Scrapie-associated fibrils (SAF) 1360, 1361
Scrophularia mottle virus(P) 1500
Sea lion foamy virus(A) 481
sea oncogene 936
Seal morbillivirus infections 1260, 1261
clinical features 1265
epizootiology 1264
evolution of viruses 1263
Seasons, respiratory virus infections 1219
Seed transmission
bromoviruses 184
capilloviruses 198
cucumoviruses 284
ilarviruses 34
tobraviruses 1445
Self antigens, altered, expression in autoimmunity 64
Self-reactive antibodies, *see* Autoantibodies
'Self-sufficient' virus 277
Semliki Forest virus (SFV)(A) **1330–1333**
assembly 1332
cDNA 1330
defective-interfering (DI) particles 1333
evolution 1333
genetic recombination 1333
genetics 1333
genome 1330
geographic and seasonal distribution 1330
Group A 238
history 1330
host range and propagation 1330
infection
immunological response 1333
pathology and histopathology 1333
persistent 1333
viremia but failure to invade CNS 1079
molecular biology 1330–1333
mRNA 1331, 1332
mutation frequency 1333
post-translational modifications 1332
proteins 1330, 1331
receptors 1331
relationships 1333
Ross River virus (RRV) 1269, 1270, 1271
replication 1330
RNA replication 1332
structure 1330, 1418
taxonomy and classification 1330
transcription 1332
translation 1330, 1331, 1332
transmission 1330
as vector 1535–1536
Sendai virus(A) 1027, **1299–1304**
activating proteases 1301, 1302, 1303
adsorption and penetration 1300
antigen 1302
assembly and release 1301
bovine parainfluenza virus type 3 (PIV-3) homology 1033
cell fusion 1302
cytopathology 1301
dominant-negative mutant mechanism of action 731
epidemiology 1302
F (fusion) protein 1028, 1300, 1301, 1303
genome structure and properties 1300
genomic sequences 1029
geographic distribution 1302
hemagglutinin (HN) 1028, 1300, 1301, 1304
amino acid sequence 1034
history 1299–1300
host range and virus propagation 1302
infection
clinical features 1303
immune response 1303–1304
pathology and histopathology 1303
prevention and control 1304
mutant 1303
neuraminidase 1028
pathogenicity 1302–1303
post-translational processing 1301
proteins 1300
receptor 1561
replication 1300–1301
RNA replication 1301
serological relationship and variability 1302
structure and properties 1300
taxonomy and classification 1211, 1300
transcription 1300–1301
translation 1301
transmission and tissue tropism 1302
vaccines 1304
Sensory ganglia, varicella–zoster virus (VZV) latency 1516, 1517
Sentinel studies 400
Seoul virus(A) 974, 975, 1614
genome structure/properties 539
Sepik virus(A) 370
Sequence analysis 530
'Sequiviruses'(P) 1042
Sericesthis iridescent virus (SIV)(I) 1433
Seroconversion 345
Seroepidemiologic studies 400
Serological tests
badnaviruses 141
kidney infections 976
respiratory viruses 1226
viral antibody detection 345–346
viral antigen assays 343–345
Seroprevalence rate, definition 399
Serpl protein 1156
Serum neutralization (SN) tests, pseudorabies virus diagnosis 1178
Severe combined immunodeficiency (SCID), mice 1506
Sewage
echovirus transmission 357
poliovirus transmission 635
samples, RNA phage as index organisms 1336
T1 bacteriophage 1371
Sex, outcome of infection and 1082–1083
Sexual transmission 1077
hepatitis A virus 551
hepatitis B virus 557
hepatitis C virus 573
hepatitis D virus 578
herpes simplex viruses 589
HIV infection and AIDS 681
HTLV-1 684
human papillomaviruses (HPV) 1018, 1020, 1023
Marburg virus 830, 831
molluscum contagiosum virus (MCV) 851
SF21AE cells, baculovirus recombinants 1535
Shallot latent virus (SLV)(P) 215
economic significance 217
Shedding of viruses 400
Sheep
Borna disease virus (BDV) infection 149, 151, 152
orbivirus infection 952
see also Bluetongue virus (BTV)(A)
visna and maedi 1592
see also Visna–maedi viruses(A)
Sheeppox 1160
see also Poxviruses(A), sheep and goats
Shellfish, hepatitis A outbreaks 551
Shiga-like toxins
classification (SLT-1/II) 102
Escherichia coli 102–103
mechanism of action 103
phage encoding genes 103
protein structure 103
receptors 103
role in dysentery 103
structural genes 102
synthesis 102–103
Shigella dysenteriae 102, 103
T1 bacteriophage 1371
Shingles, *see* Zoster
Shipping fever 1033, 1210
Shock
Argentine/Bolivian hemorrhagic fevers 782
Lassa fever 782
Shope fibroma virus (SFV)(A) 1154
pathogenicity 1157
virulence and variability 1157
see also Poxviruses(A)
Shope papillomavirus(A), *see* Cottontail rabbit papillomavirus (CRPV)(A)
Shrews, rabies-like viruses 1187
Shuttle vectors, entomopoxviruses(I) 397
Shuttle vesicles 228
Sialic acid 724
reovirus receptor 1200
rotavirus attachment and 1286
Sialidase, *see* Neuraminidase (NA)
Sialodacryadenitis virus of rats (SDAV)(A) 255
Sibine fusca, densonucleosis virus 334
Sickle cell gene, malaria resistance 665
σ^A (σ^{43}) protein 73
Sigma factor
SPO1 phage 1353, 1354
T4 bacteriophage transcription 1380
Sigma rhabdoviruses(I) **1311–1315**
carbon dioxide sensitivity 1311, 1312, 1313, 1315
ecology 1315
genome structure 1312
hereditary transmission 1313
history 1311
intergenic sequences 1312
pathology 1312
proteins 1312, 1315
replication 1314–1315
resistance to 1313–1314
serology 1312–1313
stabilized maternal lines 1313
structure and properties 1312
taxonomy 1312–1313
temperature-sensitive mutants (*hap* mutants) 1314
virus-host interactions 1313–1314
Signal peptides 135, 841
Signal transduction 229, 1473, 1474–1475
oncogenes and, *see* Oncogenes
Silkworm, densonucleosis virus infection 334
Silkworm disease 130
history 632
Silkworm industry 318
Simian AIDS (SAIDS) 1236
clinical features 1241
epidemiology 1241
pathogenicity 1241
pathology and histopathology 1241–1242
viruses 1236
see also Retroviruses(A), type D
Simian cytomegalovirus (SCMV)(A) 304
propagation 305

Simian foamy virus (SFV)[A] 481
genome structure 482, 483
geographic and seasonal distribution 486
proteins 483
relationship with human foamy virus (HFV) 484
Simian hemorrhagic fever virus (SHFV)[A] **367–372**, 764
epidemiology 371
genetics and evolution 370
geographic and seasonal distribution 369
history 367
host range and viral propagation 370
infection, clinical features 372
properties and genome structure 368
serologic relationships and variability 370
taxonomy and classification 368
transmission and tissue tropism 371
see also Encephalitis viruses[A]
Simian hepatitis A virus (HAV)[A] 549, 550
Simian herpesvirus[A], *see* Herpesvirus ateles[A]; Herpesvirus saimiri[A]
Simian immunodeficiency virus (SIV)[A] 1236, **1316–1321**
in AIDS research 1321
antigenic variations 1319
bovine immunodeficiency virus cross-reaction 164
cell lines 1318
epidemiology 1320
evolution 1319
future perspectives 1321
genes 1318
genetics 1318–1319
genomic organization 1318
geographic and seasonal distribution 1316–1317
history 1316
HIV-1 relationship 680, 1318, 1319
HIV-2 relationship 679, 680
host range and virus propagation 1317–1318
infection
clinical features 1320
immune response 1320, 1321
pathology and histopathology 1320–1321
prevention and control 1321
mutations 1319
pathogenicity 1320
precautions with 1321
proteins 1319
serologic relationships and variability 680, 1319–1320
taxonomy and classification 1316
transmission and tissue tropism 1320
Simian retrovirus type 1 (SRV-1)[A] 1237
genetics 1240
see also Retroviruses[A], type D
Simian retrovirus type 2 (SRV-2)[A] 1237
Simian retrovirus type 3 (SRV-3)[A] 1237
Simian retrovirus type 4 (SRV-4)[A] 1237
Simian retrovirus type 5 (SRV-5)[A] 1237
Simian rotavirus (SA11)[A] 1282
Simian rotavirus (SA11), *see also* Rotaviruses[A]
Simian sarcoma associated virus (SSAV)[A]
Gibbon ape leukemia virus (GaLV) relationship 536
as helper virus 536
Simian sarcoma virus (SSV)[A]
as defective virus 536
Gibbon ape leukemia virus (GaLV) relationship 536
oncogenes 936, 1473
transformation mechanism 1473–1474
Simian T-cell leukemia virus (STLV)[A], taxonomy and classification 682
Simian T-cell leukemia virus type III (STLV-III)[A] 1316
Simian T-lymphoma virus-1 (STLV-1)[A] 611
herpesvirus papio association 611, 612, 613
transmission 612
Simian virus 5 (SV5)[A] 1027, 1035
genome sequence 1029, 1035
hemagglutinin, amino acid sequence 1034
Simian virus 40 (SV40)[A] **1322–1329**
antigenic determinant 1328
cell lysis by 1327
defective-interfering particles 1327
DNA, forms 1323
DNA replication 1325
enhancer elements 1326
epidemiology 1328
evolution 1327–1328
future perspectives 1329
genetic map 1323
genetics 1327
genome structure and properties 1323–1324
homology with JC and BK viruses 752
geographic and seasonal distribution 1327
histones associated 1323
history 629, 1322
host range and virus propagation 1327
infection
human exposure 1328
immune response 1329
pathogenicity and pathology 1328–1329
prevention and control 1329
intramolecular recombination with adenovirus 527
oligodendrocyte infection 910
physical properties 1323
post-translational processing 1326
properties 1322
proteins 1324–1325
agnoprotein LP1 1325, 1326
'late' structural 1324, 1325
nonstructural 1324
T-ag 1324, 1325
T-ag properties and functions 1324
replication 1323, 1325–1327
origin of 1324, 1325
overview of strategy 1325
T-ag role 1324, 1325, 1326
SELP protein 753
serologic relationships and variability 1328
structure and properties 1323
T antigens 752, 753, 1324–1325, 1476
tumor suppressor genes interaction 1476
taxonomy and classification 1322
transcription 1326
T-ag control of 1326
transformation mechanism 1476
transgenic mice 1328
translation 1326
transmission and tissue tropism 1328
tumorigenic potential 1327
tumors 1327, 1328, 1329
uptake and release 1326–1327
vaccine contaminations 1322, 1328
Simian virus 41 (SV41)[A] 1035
Simplexvirus[A] 587
Simultaneous slippage model 350
Sindbis (SIN) virus[A] 417, 418, **1330–1333**, 1616
assembly 1332
cDNA 1330
defective-interfering (DI) particles 1333
evolution 238, 1333
genetic recombination 1333
genetics 1333
genome 1292, 1330
geographic and seasonal distribution 1330, 1616
history 1330
host range and propagation 1330, 1616
infection
diagnosis 1616
of eyes 445
immune response 422, 1333
pathology and histopathology 1333
persistent 1333
molecular biology 1330–1333
mRNA 1331, 1332
mutation frequency 1333
nsp4 product, beet necrotic yellow vein virus homology 512
post-translational modifications 1332
proteins 1330, 1331
receptors 1331
relationship 1333
Ross River virus (RRV) 1269, 1270, 1271
replication 1330
RNA replication 1332
structure 1330, 1418, 1586
taxonomy and classification 1330
transcription 1332
translation 1330, 1331, 1332
transmission 1330, 1616
as vector 1535–1536
see also Western equine encephalitis (WEE) virus[A]
Sindbis (SIN) virus supergroup 1436, 1442
Sindbis-like phytoviruses[P] 218
Sindbis-like superfamily, cucumoviruses in 278
Single-stranded RNA bacteriophage[B], *see* RNA bacteriophage
Sinusitis 1223, 1487
Siphoviridae 95, 96, 1398, 1405
characteristics and taxonomy 98
structure 1582
sis oncogene 536, 936, 1473–1474
SIVmac virus 1316, 1318
SJL mouse strain 667
SKI genes (superkiller) 1467, 1602
ski oncogene 936
Skin
barrier to viruses 1076
cancer
causative agents (HPV types) 1018
human papillomaviruses causing 1014, 1018, 1024, 1497–1498
see also Epidermodysplasia verruciformis (EV)
entry of viruses 1076
infections
coxsackieviruses 270
murine cytomegalovirus 311
lesions
cowpox 264, 265
ectromelia (mousepox) virus infection 864
papilloma 1013
rash, *see* Rash
warts, *see* Warts
Slapped-cheek disease 1054
Slow paralysis virus[I] 656, 660
'Slow viruses'
eye infections 447
history 1593
see also Lentiviruses[A]

Small cell lung cancer, interferon therapy 744
Small intestine
  epithelial cells
    torovirus replication 1457–1458
    virus replication in 377
  see also Villi
Small nuclear RNAs (snRNAs), adenoviral 12
Smallpox
  clinical features 1342
  epidemiology 1341
  geographic and seasonal distribution 1340
  global eradication campaign 1340, 1342, 1503, 1507
  hemorrhagic 1342
  history 1339, 1503, 1507
  immune response 1342
  incubation period 1341, 1342
  last case 632
  outbreaks 1341, 1342
  pathology and histopathology 1342
  prevention and control 1342
  vaccination 261, 1342, 1503
    clinical features after 1512–1513
    complications 1512, 1512–1513
  vaccine 708, 1507
    Berne strain 1512
    jet injectors 1512
    vaccinia virus 1507, 1512, 1531
    see also Vaccinia virus(A)
Smallpox (variola) virus(A) 442, **1339–1343**
  DNA sequence 1343
  epidemiology 1341
  evolution 1340–1341
  future perspectives 1343
  genetics 1340
  geographic and seasonal distribution 1340
  history 1339, 1503, 1507
  host range and virus propagation 1340
  reference collections 1340, 1343
  serologic relationships and variability 1341
  stocks maintained 1340, 1343
  structure and properties 1340
  taxonomy and classification 1339–1340
  transmission 1340, 1341–1342
SMEDI enteroviruses(A) 390
SMEDI (stillbirths, mummification, embryonic death and infertility) syndrome 389, 1068, 1071
Smut fungus
  killer strains 1464
  viruses, *see also* Totiviruses(P); *Ustilago maydis* virus-H1 (UmV-H1)(P)
Snakehead rhabdovirus(A) 478
Sneezes, virus transmission 1219
Snow Mountain agent(A) 926
  antigenic relationships 928
  see also Norwalk virus(A)
Snowshoe hare (SSH) virus(A) 1617
Snyder-Theilin feline sarcoma virus (ST-FeSV)(A), oncogenes 936
Sobemovirus(P) 1399
Sobemovirus(P) **1346–1352**
  capsid organization and stability 1348
  cytopathology 1347
  encapsidated subgenomic and satellite RNAs 1349–1350
  epidemiology and control 1351
  experimental host range and symptomatology 1346–1347
  genome 1349
  geographical distribution 1346
  infection process 1350
  insect vectors 1347, 1351
  messenger-sense RNA 1346, 1349
  natural hosts 1346
  physical properties 1347–1348
  proteins 1351
  replication and gene expression 1350–1351, 1351
  serology 1348–1349, 1351
  structure and properties 1347–1348, 1581
  taxonomy and classification 1346
  translation 1350, 1351
  transmission 1347
  VpG 1349
Social changes, zoonoses control and 1622
Soil
  bacterial survival and numbers 121
  bacteriophage in, *see* Bacteriophage
  community dynamics 126
  as microbial environment 121–122
  pH 123–124
  temperature and moisture content 124
Soil transmission
  dianthoviruses 352–353
  furoviruses 509, 514
  tombusviruses 1450
Soil-borne wheat mosaic virus (SBWMV)(P) 509
  deletion mutants 513
  epidemiology and control 515, 516
  genome structure and expression 510–512
  tobacco mosaic virus relationship 510, 512
Solanaceous tymoviruses(P) 1501
Solanum nodiflorum virus (SNMV)(P) 1346
  see also Sobemovirus(P)
Solanum yellows virus(P) 792
Sonchus yellow net virus (SYNV)(P) 1245, 1246
  defective particles of 1245
  genome 1245, 1246
    regulatory sequences 1246
  replication 1246–1247
  see also Rhabdoviruses, plant(P)
'Sore muzzle' disease 942
Sorghum chlorotic spot virus (SCSV)(P) 509
SOS response 774, 1005
South African ovine maedi-visna virus (SA-OMVV)(A) 199
South America, zoonoses in 1617–1618
Southampton virus(A) 933
  see also Norwalk virus(A)
Southern bean mosaic virus (SBMV)(P) 1346
  capsid stability 1348
  coat protein 899, 900
  genome and gene expression 1350
  geographical distribution 1346
  structure and properties 1347–1348, 1348
  transmission 1347
Southern bean sobemovirus(P), tombusviruses relationship 1448
Southern potato virus (SPV)(P) 215
Sowbane mosaic virus (SoMV)(P) 1346
  see also Sobemovirus(P)
Sowthistle yellow vein virus (SYVV)(P) 1246
Soybean
  bean pod mottle virus (BPMV) infection 254
  peanut stunt virus (PSV) infection 283
Soybean dwarf virus(P) 792, 793
speA gene 104
Species, viral 94, 1403
  co-evolution with host 439, 440
  definition 82, 94, 1403
    evolutionary basis 437
  evolution 437–438
    patterns of relatedness 438
    see also Evolution of viruses
  genomic sequences differences 438, 439
  'type' 94
Spermidine 250
Spherical viruses 1571
  see also Icosahedral symmetry
Spheroidin 392, 393, 394, 398
  gene sequence 394
  similarity between species 398
Spider monkeys, herpesvirus of, *see* Herpesvirus ateles(A)
Spinal cord
  Borna disease virus (BDV) infection 153
  Theiler's viruses infection 1427
Spleen
  infarction 652
  lesions, ectromelia (mousepox) virus infection 864
  necrosis 668
Spleen focus-forming virus (SFFV)(A) 884, 887, 888
  helper virus for 884
Splenomegaly
  lactate dehydrogenase-elevating virus (LDV) infection 770
  lymphoproliferative disease (LPD) of turkeys 813
SPO1 bacteriophage(B) 72–73, **1352–1356**
  cloning vehicle 1356
  conditional lethal mutations 1353, 1355
  DNA replication 1355
    concatemers 1353, 1354, 1355
    origins 1355
  effect on the host cell 1355–1356
  evolution 1352
  gene regulation 1353–1355
  genome structure and gene function 1352, 1353
  growth cycle 1353
  history 1352
  host DNA distinguishing 1356
  morphogenesis 1355
  mutagenesis 1356
  mutations 1354
  promoters 1354
  recombination 1356
  structure and proteins 1352
  taxonomy 1352
  terminal redundancies 1355
  transcription 72–73, 1354
    early 1354
    late 1354, 1355
    middle 1354
    TF1 1354
  transfection 1356
Spodoptera ascovirus(I) 59, 61, 62
Spodoptera litura granulosis virus(I) 128
Spongiform encephalopathies **1357–1369**
  clinical features and treatment 1367
  Creutzfeldt–Jakob disease (CJD) **1361–1369**
    see also Creutzfeldt–Jakob disease (CJD)
  cytological and molecular pathology 1367–1369
  future perspectives 1369
  Gp34 glycoprotein 1361, 1362, 1364, 1367
  human 912
    see also Creutzfeldt-Jakob disease (CJD); Kuru
  incidence and distribution 1366
  infectious agents
    hypotheses 1357, 1362–1363
    inactivation 1362
    physical properties 1363
    properties 1357, 1363–1364
  mule deer, elk and bovine **1357–1360**
    clinical features of infection 1359
    epidemiology 1358
    evolution 1358
    future perspectives 1360

genetics 1358
geographic and seasonal distribution 1357
history 1357
host range and viral propagation 1357–1358
immune response 1360
pathogenicity 1359
pathology and histopathology 1359–1360
prevention and control 1360
serologic relationships and variability 1358
taxonomy and classification 1357
transmission and tissue tropism 1359
see also Bovine spongiform encephalopathy (BSE)
scrapie **1361–1369**
subacute
kuru **758–761**, 912
see also Kuru
tissue culture studies 1369
transmissible **1361–1369**
vertical transmission 1366
see also Creutzfeldt–Jakob disease (CJD); Gerstmann–Sträussler–Scheinker (GSS) disease; Prions; Scrapie
Spongospora subterranea 514
Spores, fungal, partitivirus transmission 1050
Sporoptera frugiperda$^{(I)}$ 59, 62
Sporulation, bacterial, phage in soil and 124, 125
Spraing disease 514
Spread of viruses 1078–1080, 1082
Spring viremia of carp virus (SVCV)$^{(A)}$ 478, 1245, 1250, 1252–1253
genome and regulatory sequences 1246
SPT genes 1235, 1236
Spuma retroviruses$^{(A)}$, *see* Spumaviruses$^{(A)}$
Spumavirinae 159
Spumaviruses$^{(A)}$ 159, 480, 1593
bovine syncytial virus (BSV) 159
see also Foamy viruses$^{(A)}$
Squash, yellow summer, economic significance of viruses 1112
Squash leaf curl virus (SqLCV)$^{(P)}$ 520, 522
Squash mosaic virus (SqMV)$^{(P)}$ 249, 254
Squirrel fibroma virus (SqFV)$^{(A)}$ 1154
see also Poxviruses$^{(A)}$
Squirrel monkey retrovirus (SMRV)$^{(A)}$ 1237, 1240
Squirrel monkeys, herpesvirus of, *see* Herpesvirus saimiri$^{(A)}$
Src homology region 2 (SH2 domains) 940
src oncogene 936, 940, 1474
SSV1 virus$^{(B)}$ 50, 51, 52, 55, 56–57
gene expression model 56
genetic map 57
structure and genome 56, 57
transcription 57
SSV1-type phage 1398
characteristics and members 96, 1405
classification 1405
Stains, light microscopy 342
Staphylococcus aureus, pyrogenic toxins 103–105
enterotoxins and genes for 104
Starter cultures, phage infection 118, 119
Stomatitis papulosa virus$^{(A)}$ 1037
Stratum Malpighii, virus entry 1076
Strawberry latent ringspot virus (SLRSV)$^{(P)}$ 905, 906
Strawberry mild yellow edge-associated virus (SMYEAV)$^{(P)}$ 1143
genome 1144
Strawberry pseudomild yellow edge virus (SPMYEV)$^{(P)}$ 215
Streptococcus pneumoniae, *DpnI* action 672
Streptococcus pyogenes, pyrogenic toxins 103–105
phage conversion 104
SpeA 104
SpeC 104
Streptomyces, phage
adsorption 124
detection from soil 123
hostphage interactions 126
interactions and factors affecting 124
stability in soil 123
Striped jack nervous necrosis virus (SJNNV)$^{(I)}$ 919, 924, 925
Structure of viruses **1571–1586**
architecture 1571
atomic **1571–1573**
antigenic sites 1573
antiviral agents 1573
assembly of viruses 1573
evolution 1572–1573
helical viruses 1571
host receptor recognition site 1573
icosahedral viruses 1571–1572
nucleic acidprotein interaction 1572
of proteins 1572
enveloped nonspheroidal viruses 1584
enveloped spheroidal viruses 1583–1584
method of determination 1571
nonenveloped nonspheroidal viruses 1582–1583
nonenveloped spheroidal viruses 1580–1581
principles **1573–1586**
complex structures 1586
icosahedral capsids 1574–1578
quasiequivalence 1576, 1577–1578
subunit tertiary structure 1579, 1585
Structure/activity relationships (SARs) 43
Stuffer fragments 114
Stunted growth syndrome 1195
Styloviridae$^{(P)}$ 286
Subacute sclerosing panencephalitis (SSPE) 323, 402, 845, 909
clinical features 912
see also Measles virus$^{(A)}$
Subclinical infections 402
Subconjunctival hemorrhage, etiology 449
Subfamilies of viruses 1398–1400
nomenclature 1402
Sublethal viral infections, 'luxury' functions loss 804, 805
Subterranean clover mottle virus (SCMoV)$^{(P)}$ 1346
see also Sobemovirus$^{(P)}$
Subterranean clover redleaf virus$^{(P)}$ 792, 793
Subtilisphage$^{(B)}$, *see Bacillus subtilis* bacteriophages$^{(B)}$
Suckling mouse
coxsackievirus discovery 638
foot and mouth disease 494
Sugarbeet
economic significance of viruses 1111
yellows disease, control and prevention 1114
Sugarcane bacilliform virus (ScBV)$^{(P)}$ 139, 141, 142
Suipoxvirus$^{(A)}$
member 1154
see also Swinepox virus (SPV)$^{(A)}$
Sukhumi baboons, Simian T-lymphoma virus-1 (STLV-1) association with herpesvirus papio 611, 612, 613
Sulfation, varicellazoster virus (VZV) 1525
Sulfolobus shibatae, SSV1 bacteriophage, *see* SSV1 virus$^{(B)}$
'Summer grippe' 271
Sunn-hemp mosaic tobamovirus (SHMV)$^{(P)}$ 511, 1436, 1437
'Superantigens', endogenous 860
Suramin, avian hepatitis B control 568
Susan McDonough feline sarcoma virus (SM-FeSV)$^{(A)}$, oncogenes 936
Susceptibility to viral diseases 667, 668, 1170, 1591
see also Genetic susceptibility; Host genetic resistance
SV40, *see* Simian virus 40 (SV40)
Swamp fever, *see* Equine infectious anemia
Sweet clover necrotic mosaic virus (SCNMV)$^{(P)}$ 349
host range 352
serotypes 353
Swim bladder inflammation (SBI) disease 1252
'Swimming pool conjunctivitis' 1078
Swine
abortion, Japanese encephalitis virus infection 751
vesicular exanthema, *see* Vesicular exanthema virus of swine (VES)$^{(A)}$
Swine fever 649
see also Hog cholera virus (HCV)$^{(A)}$
Swine herpesvirus-1$^{(A)}$, *see* Pseudorabies virus$^{(A)}$
Swine infertility and respiratory syndrome (SIRS) 763
Swine influenza viruses 710
Swine vesicular disease 387, 388
clinical features 389–390
pathology and histopathology 390
vaccines 391
Swine vesicular disease virus (SVDV)$^{(A)}$ 385
host range and propagation 387
transmission and tissue tropism 389
see also Enteroviruses$^{(A)}$, animal; Vesicular exanthema virus$^{(A)}$
Swinepox virus (SPV)$^{(A)}$ 1154–1155
clinical features of infections 1158
geographic and seasonal distribution 1156
pathogenicity 1158
virulence and variability 1157
see also Poxviruses$^{(A)}$
Syncytium formation 1589
bovine immunodeficiency virus (BIV) 162
bovine parainfluenza virus type 3 (PIV-3) substrains 1034–1035
HSV strains 599
Sendai virus 1301
varicellazoster virus (VZV) 1526
see also Cell fusion
Syncytium-forming viruses, *see* Foamy viruses$^{(A)}$
Synovitis, in visnamaedi virus infections 1592
Syrian hamster, LCMV infection 803

T
T1 bacteriophage$^{(B)}$ **1371–1376**
adsorption 1371
amber mutations 1373
DNA packaging 1373
DNA synthesis 1372–1373
early steps in infection 1371–1372
esp gene 1375
genetic map 1373–1374
genome structure 1371
host range 1371
hr mutation 1374, 1375
morphogenesis 1373
mutants 1372, 1373–1374
mutation 1373

T1 Bacteriophage$^{(B)}$ *continued*
pac gene 1373, 1374, 1375
pip mutant 1373
protein synthesis 1372
recombination (Grn) 1372
restriction and modification 1375
structure and properties 1371
transcription 1372
transduction 1375–1376
T2 bacteriophage$^{(B)}$ 1376
characteristics and taxonomy 97
tail fibres 1379
see also T4 bacteriophage$^{(B)}$
T3 bacteriophage$^{(B)}$ 1389
antirestriction mechanisms 671
cDNA of RNA viruses used as vectors 1540
nonclassical restriction/modification 673
Ocr protein 671
promoters 1395
see also T7 bacteriophage$^{(B)}$
T4 bacteriophage$^{(B)}$ **1376–1384**
arn function 671
characteristics and taxonomy 97
concatemeric DNA functions 505
DNA packaging 90, 1377, 1379, 1383
DNA replication 90, 91, 1377, 1382
gene expression integration 1380, 1381–1382
in vitro 1383
origins 1382
recombination-dependent initation 1382
frog virus 3 (FV3) DNA shared features 505, 508
genes
gene 59 90
immediate early 1379
late 1382
nonessential 1379
prereplicative 1382
genome structure and map 629, 1377
'ghosts' 1379
history 629, 1376
host range 1379
HSV replication similarity 600
infection cycle 1377–1379
interconnecting processes 1376–1377
mutations 1377
overview 1376–1377
phenotypic mixing with T2 phage 1379
plasmid integration 91
recombination 90–91, 1382
double-strand break-promoted 91
importance 90
mutations affecting 90, 92
replication dependence 90, 91, 92
restrictionmodification 1383
SPO1 phage relationship 1352
structure and assembly 1377–1379
head 1378–1379
scaffolding proteins 1378
tail fibers 1379
tails 1379
temporal control of gene expression 1379–1382
transcription 1380
initiation 1380
post-initiation effects 1380
premature termination 1380
promoters (early, middle, late) 1380
self-splicing introns 1380
translational control 1377, 1380–1382
UvsX and UvsY proteins 90, 91
T5 bacteriophage$^{(B)}$ **1384–1388**
abortive infection in Collb hosts 1387
cell receptor 1385
cloning genes 1388
DNA polymerase gene 1388
DNA replication 1387
DNA transfer 1385
early genes 1386
effect on host cell metabolism 1385–1386
future perspectives 1388
genetic maps 1385
genome structure 1384–1385
deletions 1384
nicks 1384
sequence 1385
terminal repetition 1384
head 1384
inactivation of host restriction endonucleases 1386
infection process 1385
late genes 1386
morphogenesis 1387
morphology 1384
pre-early genes 1385
restriction maps 1385
tail 1384
transcription regulation 1386–1387
transfection 1387–1388
T6 bacteriophage$^{(B)}$ 1376
characteristics and taxonomy 97
see also T4 bacteriophage$^{(B)}$
T7 bacteriophage$^{(B)}$ **1388–1396**
antirestriction mechanisms 671
capsid assembly and DNA packaging 1392–1393
cDNA of RNA viruses used as vectors 1540
cell lysis by 1393
characteristics and taxonomy 98
combination 1389
DNA metabolism 87
DNA packaging 88
DNA replication 87, 88, 1391–1392
bidirectional 1392
origin 1391, 1392
recombinogenic intermediates 88
terminal repeat synthesis 1393, 1394
early proteins 1391
effect on host cells 1391, 1393
genetic map 1389–1390
genome and terminal repeats 88
growth inhibition 1395
head 1388, 1390
host functions in development of 1394, 1395
host range 1389, 1395
infection cycle 1391, 1392
class I genes 1391
class II genes 1391
class III genes 1392
mRNA 1392, 1393
nucleases encoded 88–89
Ocr protein 671
promoters 1391, 1392, 1394
sequences 1389, 1395
properties, ecology and evolution 1388–1389
receptor 1389
recombination 87–89
frequency and DNA copy number 88
gene 3 endonuclease 89
gene 6 protein 88–89
in vitro systems 89
replication coupling 87, 88
RNA polymerase 348, 1389, 1394, 1395–1396
expression systems based on 1395–1396
ssDNA-binding protein 89
structure 1388–1389, 1390
superinfection exclusion 1389, 1395
tail 1388, 1390
transcription 1391, 1394, 1395
processing of primary transcripts 1394
termination 1395
T12 bacteriophage$^{(B)}$, pyrogenic toxins of *S. pyogenes* 104
T=3 viruses, *see under* Icosahedral symmetry
T cell receptors (TCR) 698, 703, 726, 1092, 1504
$\alpha\beta$ and $\gamma\delta$ 703, 1504
autoimmune diseases and 668
deletion, by mouse mammary tumor virus (MMTV) 861
heterodimers 703
loci near, susceptibility to Theiler's viruses 668
pyrogenic toxins binding 105
T cells 697
activated, target for neoplastic transformation by Marek's disease virus 835
activation 697–698
polyclonal, by viruses 63
antigen presentation 704, 725–726
bluetongue virus (BTV)-specific 954
categories 703, 726
$CD4^+$ helper (Th) 697, 726, 1091, 1092, 1504
antigen presentation 704, 705, 725–726, 1092
antigen processing/recognition 727
in cell-mediated response 706
class II restriction 697, 704–705, 726
classes 705, 1504
cytotoxic clones 1505
decline in feline immunodeficiency virus (FIV) infection 457
decline in HIV infection and mechanism 680, 707–708
exogenous processing pathway 704, 705
function 726, 1091, 1092, 1504
HTLV-1 infection 684, 685
human herpesvirus 7 (HHV-7) infection 625
human immunodeficiency virus (HIV) tropism 674, 680
lymphokine secretion 705
respiratory infections 1222
target for neoplastic transformation by Marek's disease virus 835
TH1 and TH2 subsets 705, 1504
$CD8^+$ 703–704, 726, 1091, 1092
class I MHC restriction 698, 703, 726, 1092, 1505
function 726, 1091, 1092, 1504
in HIV infections 680, 681
in immune response in diabetes 65
see also Cytotoxic T lymphocytes (CTL); T cells, suppressor
congenital deficiency, respiratory infections 1222
cytotoxic ($CD8^+$), *see* Cytotoxic T lymphocytes (CTL)
deficiency
cytomegalovirus infections 296
lactate dehydrogenase-elevating virus (LDV) infection 770
respiratory infections 1222
depressed response, in recurrent HSV infections 592
effector 697
epitopes 724

difference from B-cell epitopes 725–726
from influenza virus proteins 726–727, 727
production for vaccines 1505
functions 1504
gamma-2 herpes viruses replication 617, 619
hepatitis B 557, 558–559
herpes simplex virus (HSV) infections 592
herpesvirus saimiri and herpesvirus ateles tropism 617, 619
human herpesvirus 6 (HHV-6) infection 625
immunopathology mediated by 706–707
influenza virus infections 726–727
leporipoxvirus replication 1156, 1157, 1159
lymphomas in Marek's disease 835
memory cells 699, 707
mumps virus in 882
myxoma virus replication in 1156, 1157, 1159
primary response 705–706
proliferation, by mouse mammary tumor virus (MMTV) 860, 861
recirculation shutdown in infections 706
reovirus type 3 receptor (Reo3R) 1562
suppression, by reticuloendotheliosis viruses 1231
suppressor (CD8) 697, 1504
see also T cells, $CD8^+$ (above)
surveillance by 704
Theiler's virus infection 1428
in vaccination, roles 1504–1505
viral infection, as immune evasion mechanism 702
see also Cell-mediated immunity
T-cell leukemia 1496
adult (ATL), *see* Adult T-cell leukemia (ATL); Human T-cell leukemia virus type 1 (HTLV-1)[(A)]
T-cell malignancy
HTLV-2 infection and 693
mouse mammary tumor virus (MMTV) 858
T-cell variant of hairy-cell leukemia 693
T-even phage 1376
biological concepts formulated from 1376
homology 1376
interconnecting processes 1376–1377
restriction of nonglucosylated HMC-containing DNA 671–672
see also other specific bacteriophage; T4 bacteriophage[(B)]
TAAATG motif 397
TAATATTAC sequence 519, 520, 523
TAATG motif, molluscum contagiosum virus (MCV) 850
Tacaribe complex, LCMV cross-reactions 803
Tacaribe virus[(A)] 777, 779, 806
TAD element 1233
taf gene, foamy viruses 485, 487, 488
Tahyna (TAH) virus[(A)] 1618
Talfan disease 389
Tamarin marmosets, herpesvirus saimiri and herpesvirus ateles infection 619
Tamiami virus[(A)] 777
Tanapox virus[(A)] **1597–1600**
epidemiology 1599
evolution 1597
future perspectives 1600
genome structure 1597
history 1597
infection
clinical features and pathology 1599
immune response 1599–1600
prevention and control 1600
pathogenicity 1599
physical properties 1598
proteins 1597
serologic relationships and variability 1598–1599
structure and properties 1597
taxonomy and classification 1597
transmission and tissue tropism 1599
see also Yabapox virus[(A)]
tat gene 1318
caprine arthritis encephalitis virus 200
defective 323
HIV 676, 1208
origin 439
TAT protein 1208
HIV, *trans*-activation of JC virus 755
Kaposi's sarcoma and 1499
TATA box 169, 332, 485, 530
herpes simplex viruses 595
lymphoproliferative disease virus (LPDV) of turkeys 812
mouse mammary tumor virus (MMTV) 859
TATGARAT sequence, HSV 596, 598
Taunton agent[(A)]
antigenic relationships 928
see also Norwalk virus[(A)]
tax gene
bovine leukemia virus 167
HTLV-1 683, 685, 686, 691, 1476, 1496
HTLV-2 687, 690–691, 691
Tax proteins
HTLV-1 and HTLV-2 691–692
functions 683, 691–692
transforming capacity of HTLV-1 685, 692, 694
Tax-responsive element binding protein (TREBs) 683
Taxonomy and classification **1396–1410**
animal viruses 525
bacteriophage **93–100**
see also under Bacteriophage
clusters of viruses 440
descriptive characters used at family level 1404–1405, 1409
families 94–95, 1398–1400, 1403
genera 94, 1403
groups 1403
hierarchical system 1397
history 633, 1396–1397
ICTV, *see* International Committee on Taxonomy of Viruses (ICTV)
nomenclature of taxa 1396, 1402–1403
order of presentation of families and groups 1398–1400, 1401–1402
criteria 1401–1402
orders 95, 1403, 1406
species 94, 1403
systems 1401
Adansonian 1401
family and genus levels 1401
Linnean 1401
taxa descriptions 1407–1410
universal classification system 93–95, 1403–1407
see also Evolution of viruses; Families; Genera; International Committee on Taxonomy of Viruses (ICTV); Species
Tectiviridae 1165, 1398, 1405
characteristics and taxonomy 97
structure 1581
Tegument, Epstein–Barr virus (EBV) 410
Tellina virus (TV)[(I)] 144
Temperate phage 772
Temperature, Lucké tumor herpesvirus (LTHV) replication 41
Temperature-sensitive mutants 526
'Tenangaja' 1111
Tenuivirus[(P)] 1400
Tenuiviruses[(P)] **1410–1416**
ambisense (gene expression) 1413, 1414
antibodies 1415
biological properties 1410
economic importance 1414
epidemiology and control 1415–1416
filamentous ribonucleoprotein particles (RNPs) 1410–1411
gene expression strategies 1413–1414
genome composition and structure 1412–1413
complementary termini 1412, 1413
RNA components 1411, 1412
geographic distribution 1414
geographic incidence 1411
host range 1411
noncapsid protein (NCP) 1415
nucleoprotein (N protein) 1415
resistance to 1416
serological and nucleic acid-based relationships 1415
structure and composition 1410–1412, 1582
taxonomy and classification 1410
transmission 1410, 1411, 1414–1415
virus-host relationships 1415
virus-vector interactions 1414–1415
Tephrosia symptomless virus (TSV)[(P)] 218, 219
Terminal inverted repeats 15
adenoviruses 10
African swine fever virus 24
bacteriophage ϕ29 981
densonucleosis viruses 332
fowlpox virus 498
molluscum contagiosum virus (MCV) 849
Paramecium bursaria Chlorella virus 1 (PBCV-1) 36, 37
poxviruses 1155
smallpox virus 1340
vaccinia virus 1508
see also Inserted sequence (IS) elements; Terminal repeats
Terminal redundancy
badnaviruses 140
iridoviruses 1432, 1434
SPOI bacteriophage 1353, 1354, 1355
T1 phage 1371
T5 bacteriophage 1384
see also DNA, double-stranded (dsDNA) circular permuted and terminally redundant
Terminal repeats
in Epstein–Barr virus (EBV) 410
internal, parapoxviruses 1038
long, *see* Long terminal repeat (LTR)
T7 phage, synthesis of 1393, 1394
see also Terminal inverted repeats
Terminology, viral 627, 1168
Teschen disease (pig polioencephalitis) 384, 389
clinical features 389
pathology and histopathology 390
vaccines 391
Teschen disease virus[(A)] 384, 387
see also Enteroviruses[(A)], animal
Tetrahydroimidazobenzodiazepinone (TIBO) compounds 48–49
Tetraviridae 1399, 1408, 1416
members 1417

Tetraviridae continued
  structure 1581
Tetraviruses(I) **1416–1422**
  antibodies 1422
  capsid structures 1418–1419, 1420
  coat proteins 1418
  ecology 1422
  economic importance 1422
  genome structure 1417, 1417–1420
  pathology 1420–1421
  physical properties 1417, 1418
  replication 1421
    site of 1421
  serology 1420
  structure and composition 1417
  symptoms 1421
  taxonomy and classification 1416
  translation 1421
  transmission 1420
  virus-host interaction 1420–1421
Tf1 elements 1233
TGTTCT consensus sequence 859
Thailand sacbrood virus (TSBV)(I) 657
Theiler's murine encephalomyelitis virus (TMEV)(A) 206, 667, **1423–1430**
  chimeric viruses 1426
  classification 1423
  genome determinants important in pathogenesis 1425–1426
  geographic distribution 1423
  history 1423
  host range 1423
  immune evasion 1428
  immune-mediated demyelination mechanism 1428–1429
  infection
    clinical features 1427
    delayed hypersensitivity mechanism 1428–1429
    demyelinating disease 1423, 1426, 1427, 1428
    immune response 1428
    pathology and histopathology 1427, 1428–1429
    poliomyelitis 1426, 1427
  internal ribosome entry site (IRES) 1424
  neurovirulence groups 1425–1426
  pathogenesis 1426–1427
  persistence, sites of 1427–1428
  physical properties 1425
  polyprotein and post-translational processing 1424–1425
  protein cleavage system 1424, 1425
  receptor 1424
  replication 1425
  RNA genome 1424
  serologic relationships 1423
  strains 1423, 1426
  three-dimensional structure 1423–1424
  TO(B) strain 1426
  transmission and tissue tropism 1426
Theiler's viruses(A) 206, **1423–1430**
  see also Theiler's murine encephalomyelitis virus (TMEV)(A)
Thermophiles, viruses of 51, 55–57
Thermoproteus tenax, viruses 51, 55–56
Thioredoxin 291, 468
Thosea asigna virus (TaV)(I) 1417
Thottapalayam viruses(A) 542
'Three (3-)day' fever 212
Thrips transmission
  tomato spotted wilt virus (TSWV) 192
  tospoviruses 1462, 1463
Thrombocythemia, essential, interferon therapy 742
Thrombocytopenia, equine infectious anemia virus (EIAV) infection 434
Thymidine kinase (TK)
  African swine fever virus 25
  capripoxviruses 1161
  fish lymphocystis disease virus (FLDV) 799
  fowlpox virus 499
  pseudorabies virus (PRV) 1177, 1178
  vaccinia virus 1510
  viral, antiviral phosphorylation 43, 46
  viral and host gene relationship 440
Thymus gland atrophy 458, 462
  hog cholera virus infection 653
Thyroid gland, virus infections 976
Thyroid hormone receptor, c-*erbA* encoding 941
Thyroid-binding globulin (TBG) 976
Thyroid-releasing hormone (TRH) 976
Thyroiditis 976
TIBO compounds 48–49
Tick-borne encephalitis 371–372
  enterovirus (EV) serotype 71 (EV71) causing 383
  flaviviruses causing 445
Tick-borne encephalitis virus (TBEV)(A) 361–362, **367–372**, 1615, 1618
  antigenic complex, viruses in 370
  cytopathic effects 369
  Eastern subtype 361, 367
  epidemiology 371
  genetics and evolution 370
  geographic and seasonal distribution 369, 1615, 1618
  history 367
  host range and viral propagation 370
  infection
    clinical features 371–372
    immune response 372
    prevention and control 372, 1615
  pathogenicity 371
  physical properties 368–369
  properties and genome structure 368
  proteins 368
  replication 369
  serologic relationships and variability 370
  subtypes 370
  taxonomy and classification 367–368
  transmission and tissue tropism 371, 1615, 1618
  Western subtype 361–362, 367
    see also Central European encephalitis virus (CEEV)(A)
Tick-borne infections
  Crimean-Congo hemorrhagic fever (CCHF) (bunyavirus) 191, 1616
  encephalitis virus 361–362, 364, 365
    see also Tick-borne encephalitis virus (TBEV)(A)
  orbivirus and coltivirus 944, 950
Ticks, African swine fever virus propagation in 27
Tight junctions 227, 908
Tipula iridescent virus (TIV)(I) **1430–1436**
  DNA components 1432, 1435
  taxonomy and classification 1430, 1435
  see also Iridoviruses(I)
Tissue, organization 226–228
Tissue culture
  coxsackieviruses, historical aspects 639–640
  history 630, 635
  poliovirus growth 635
Tissue tropism, *see* Tropism, viral
Tissues (paper), impregnated with virucidal agents 1259
Titration of viruses 1172–1173
Tn917 (transposon) 76, 77
Tobacco black ring virus (TBRV)(P), genetic map 903
Tobacco etch virus (TEV)(P) 1149
  gene expression 1150–1151
  genome organization 1149, 1150
  symptoms 1110
  see also Potyviruses(P)
Tobacco mild green mosaic tobamovirus (TMGMV)(P) 1436
Tobacco mosaic disease 627
  see also Tobacco mosaic virus (TMV)(P)
Tobacco mosaic virus (TMV)(P) 1538
  assembly, discovery 633
  atomic structure 1571
  core polymerase sequence motif (GDD) 512
  economic significance 1111
  effect on fungal infections 1112
  furoviruses relationship 510, 512
  history 627–628
    chemical composition 629
    infectivity assay 632
    RNA infectious 632
    structural studies 628
    structure 632
  host range 1106
  importance 1436
  mutants 632
  N gene for resistance 1107, 1147, 1440
  resistance mechanisms 1107, 1147
  satellite (STMV) 1441
  structure 1574
  tobraviruses relationship 1442
  see also Tobamoviruses(P)
Tobacco necrosis virus (TNV)(P) 218, **896–901**
  classification 900
  coat protein 899, 900
  economic importance 901
  future perspectives 901
  genome expression 898–899
  genome structure/properties 897–898
    subgenomic 899
  German TNV isolate 897, 898, 899
  history 633
  melon necrotic spot virus (MNSV) comparison 898, 899
  properties 897
  purification 897
  relationships
    between strains 899–900
    with other plant viruses 900
  replication 898–899, 1575
  RNA 897
  satellite viruses (STNV) 896, 897, 1574
    genome organization 900–901
    serotypes 897
    STNV-1 900
    STNV-2 900
    structure and assembly 1574–1575
  Southern bean mosaic virus (SBMV) similarities 1351
  symptoms, distribution and diseases 896
  TNV-A strain 897
    relationships 899
    RNA 897–898, 899
  TNV-D strain 897
    relationships 899
  TNV-WF strain (South African) 896, 897, 901
  translation 899, 900
  transmission 896–897
Tobacco necrotic dwarf virus (TNDV)(P) 792, 794
Tobacco rattle virus (TRV)(P) 1442, 1538

CAM strain 1442
furoviruses relationship 511, 512
history 633
SYM strain 1442, 1443
TCM strain 1443
see also Tobraviruses[(P)]
Tobacco ringspot virus (TRSV)[(P)] 904, 906
coat proteins 905
economic significance 1111
Tobacco streak virus (TSV)[(P)] 30
host range and geographic distribution 34
host relationship 33
Tobacco stunt virus[(P)], transmission 515
Tobacco vein mottling virus (TVMV)[(P)] 1149, 1538
Tobamovirus[(P)] 1399
furoviruses moved from 508
Tobamoviruses[(P)] **1436–1441**
'A-protein' 1437
assembly 1437, 1438
coat protein 1439
cytology 1439
economic significance 1437
epidemiology 1437
evolution 439
genetics 1439
genome structure and molecular biology 1438–1439
history 1436
host range 1436–1437
multiplication and spread 1439–1440
prevention and control 1437
proteins, host factor interaction 1440
proteins (126/183kD) 1439
resistance 1440–1441
hypersensitive response (HR) 1440
N gene 1107, 1147, 1440
N' gene 1440
Tm-1, *Tm2*, *Tm-2*^{2} genes 1106, 1107, 1108–1109, 1440–1441
transgenic plants 1441
RNA 1437, 1438
subgenomic 1437, 1438, 1439
tRNA-like terminus 1438
RNA replication 1439–1440
satellite viruses 1441
structure and composition 1437–1438, 1583
symptoms 1440
taxonomy and classification 1436
transmission 1437
see also Tobacco mosaic virus (TMV)[(P)]
Tobravirus[(P)] 1400
Tobraviruses[(P)] **1442–1446**
'anomalous' isolates 1444
cDNA 1442
detection and identification 1445
diseases caused 1444–1445
economic importance 1444
epidemiology and control 1446
'GDD' box 1443
genetics 1443–1444
genome organization and molecular biology 1442–1443
homologous termini 1443, 1444
geographic distribution 1445
host range 1444
L and S particles 1442
non-multiplying infections (NM) 1442
pseudo-recombinants 1444
RNA 1442
tRNA-like structure 1442
RNA1 1442, 1443
homology with other RNA1s 1443, 1444
RNA2 1442, 1443
serological relationships 1443, 1444
structure and properties 1442, 1583
symptoms of infections 1444–1445
taxonomy and classification 1442
tobamoviruses relationship 1442
transmission 1445
nematodes 1445, 1446
see also Tobacco rattle virus (TRV)[(P)]
Togaviridae 1399, 1408, 1409
Alphavirus 237, 1269, 1330, 1615
Arterivirus
equine arteritis virus (EAV) 763
lactate dehydrogenase-elevating virus (LDV) 763
characteristics and structure 443, 1291, 1583
eye infections 445
Flaviviridae before re-classification 362
Japanese encephalitis virus 746, 1615
see also Flaviviridae
genome 1292
Pestivirus, before re-classification 176
replication/transcription strategy 1206–1207
Rubivirus 1291
see also individual genera and viruses; Togaviruses[(A)]
Togaviruses[(A)]
assembly 1560
fusion 1558
nonarthropod-borne 175
Tol proteins 467
Toluca-1 virus[(A)] 379
antiserum 379
Tom-P[(P)] 1243
Tomato apical stunt viroid[(P)] 1564
Tomato aspermy virus (TAV)[(P)] 278
pathology and pathogenesis 283
strains 282
see also Cucumoviruses[(P)]
Tomato black ring virus (TBRV)[(P)] 904, 906
coat proteins 905
Tomato bushy stunt virus (TBSV)[(P)] 1447, 1538
cherry strain 1447
defective interfering RNA 1452
genome 1448
structure 1575, 1578
see also Tombusviruses[(P)]
Tomato chlorotic spot virus (TCSV)[(P)] 1459, 1460
Tomato golden mosaic virus (TGMV)[(P)] 520, 1538
host range 522
as vector 1539
Tomato mosaic virus (ToMV)[(P)] 1436
cultivar resistance mechanism 1106–1107
host range 1106
resistance
Tm2 and *Tm-2*^{2} genes 1106, 1107, 1108–1109, 1440–1441
Tm-1 gene 1107, 1440–1441
use in plant breeding 1108–1109
Tm-2^2 gene, economic importance 1109
Tomato planta macho viroid[(P)] 1564
Tomato ringspot virus (ToRSV)[(P)] 904
coat proteins 905
Tomato spotted wilt virus (TSWV)[(P)] 192, 1459, 1460
economic significance 1462–1463
genome 1461, 1462
host range 192
resistance to 1463
structure 1584
transgenic plants resistant 1463
transmission 192
see also Tospoviruses[(P)]
Tomato top necrosis virus[(P)] 905
Tomato vein yellowing virus[(P)] 1243
Tomato yellow leaf curl virus (TYLCV)[(P)] 517
evolution 524
genome organization and expression 521
Tomato yellow top virus[(P)] 792
Tomatoes, spotted wilt 1459
Tombusvirus[(P)] 1399
Tombusviruses[(P)] 219, **1447–1452**
capsid stability and calcium 1448
cDNA clones 1449
coat protein 1447, 1450
cytopathology 1451
defective-interfering RNA 1447, 1450, 1451–1452
formation 1450, 1452
deletion mutants 1450, 1452
evolutionary relationships 1447
genome organization and molecular biology 1448–1450
21 and 20kD proteins 1449–1450
coat protein 1450
geographic distribution 1450
helicase motif lacking 1447, 1448–1449
host range 1450
polymerase 1447
RNA 1447, 1448
subgenomic 1449
satellite RNA 1447, 1451–1452
structure and properties 1447–1448, 1581
P domain 1447, 1448
R domain 1448
S domain 1448
symptomatology 1450
taxonomy and classification 1447
transmission 1450–1451
see also Tomato bushy stunt virus (TBSV)[(P)]
Tonsillar carcinomas 1019
Topoisomerase II, African swine fever virus 25, 26, 28
'Topotype' 947
Toroviridae 377
Torovirus[(A)] 1409
Torovirus-like particles 1453
Toroviruses[(A)] **1452–1458**
antibodies 1457, 1458
assembly 1455
cDNA hybridization studies 1453, 1456
defective-interfering particles 1455
envelope 1454
epidemiology 1457
evolution 1456, 1458
future perspectives 1458
genome structure and properties 1453
geographic and seasonal distribution 1455–1456
history 1452–1453
host range and viral propagation 1456
infection
clinical features 1457
immune response 1458
pathology and histopathology 1457–1458
N and P (peplomer) proteins 1454
pathogenicity 1457
physical properties 1454
post-translational processing 1455
proteins 1454
replication 1454
serologic relationships and variability 1456–1457
structure and properties 1453
taxonomy and classification 1453

Toroviruses(A) *continued*
  transcription 1454–1455
  translation 1455
  transmission and tissue tropism 1457
  see also Berne virus (BEV)(A)
Toscana virus(A) 191
Tospovirus(P) 186, 187, 191–192, 1459
  genome sequences and coding strategies 193, 194
  structure and genome 188
  transmission 189
  see also Bunyaviruses(P); Tomato spotted wilt virus (TSWV)(P)
Tospoviruses(P) **1459–1464**
  assembly 1462
  detection 1463
  disease management 1463
  economic significance 1462–1463
  genome structure
    coding capacity 1459, 1461–1462
    complementary terminal sequences 1462
  geographic distribution 1462–1463
  history 1459
  host range 1462–1463
  molecular events during infection cycle 1462
  monoclonal antibodies 1463
  proteins 1459
    G proteins 1459
  resistance of plants to 1463
  RNA 1459, 1461–1462
    ambisense S and M segments 1461
    negative-strand L segment 1461
  structure and properties 1459
    Bunyaviridae members comparison 1459, 1460
  taxonomy and classification 1459, 1460
  transgenic plants resistant to 1463
  transmission 1463
  vectors 1462
Totiviridae 1051, 1398, 1406, 1464, 1468
  giardiavirus 532, 1464
  members 532, 1464
  structure 1581
Totiviruses(P) **1464–1471**
  capsid protein heterogeneity 1465–1466
  dsRNA 1464, 1465
    replication 1466
    replication *in vitro* 1466
  general features **1464–1468**
  helper viruses 1464
  host range 1467
  infectivity assays 1467
  M-dsRNA (satellite) 1464, 1465
    genes required for maintenance of 1467
    replication 1466
    SKI (superkiller) genes effect 1467
  mixed infections 1467
  post-translational modification 1465, 1466
  properties 1465
  RNA polymerase 1465, 1466
  structure and composition 1465
  taxonomy and classification 1464–1465
  transmission 1467
  virus–host relationships 1467
  see also Saccharomyces cerevisiae virus L-A (ScV-L-A)(P); *Ustilago maydis* virus-H1 (UmV-H1)(P)
tox gene 101–102
Toxic shock syndrome 103, 104
Toxic shock-like syndrome 104
Toxins
  bacterial protein 101
  bacteriophage-encoded, *see under* Bacteriophage
  victorin 1051
Toxoplasma gondii, interferon-γ in 738, 741
Tracheitis 1221
Trager duck spleen necrosis virus (SNV)(A) 1227
Trans-acting signals 1205
Trans-activation
  adenovirus early gene (E1A) 11
  see also Rex protein; Tax protein; Transcription
Trans-neuronal transport 1080
Trans-synaptic transport 1080
Transactivational factor (*taf*), foamy viruses 485, 487, 488
Transcapsidation 528, 529
Transcriptase
  bunyavirus 195
  cytoplasmic polyhedrosis viruses (CPV)(I) 315
  see also Reverse transcriptase
Transcription 232, 1137–1138, 1138, 1141, 1204–1205
  adenoviral genes 11–12
  African swine fever virus 25–26
  animal enteroviruses 386
  bacteriophage ϕ6 979
  bacteriophage ϕ29, *see* Bacteriophage ϕ29
  bacteriophage ϕH 53–54
  bacteriophage ϕX174 995
  biphasic, in fowlpox virus 499
  birnaviruses 146, 1207–1208
  bovine immunodeficiency virus (BIV) 160–162
  bovine leukemia virus 169
  bovine papillomaviruses 1307
  bromoviruses 183–184
  bunyaviruses 195, 1207
  caliciviruses 1206
  caprine arthritis encephalitis virus 201
  'cascade' mechanism, filamentous bacteriophage 467
  coliphages P2 and 186 1006
  comoviruses 253
  coronaviruses 257, 1206–1207
  cyanobacteria bacteriophage 290
  cytomegaloviruses 302–303, 305, 306
  DNA viruses 1204–1205, 1209
  EBV latency-associated genes 411–412
  equine herpesviruses 425
  equine infectious anemia virus (EIAV) 431
  feline leukemia virus (FeLV) 460
  filamentous bacteriophage 466–467
  Filoviridae 1207
  fish lymphocystis disease virus (FLDV) 799
  flaviviruses 1206
  foamy viruses 485
  fowlpox virus (FPV) 499–500
  frog virus 3 (FV3) 505–506
  geminiviruses 518, 519, 520
  hantaviruses 540–541
  hepatitis B virus (HBV) 562, 562–563
  herpes simplex virus (HSV), *see* Herpes simplex viruses (HSV)
  herpesviruses, cytomegaloviruses 302–303, 305, 306
  host cell genes, bovine leukemia virus Tax protein effect 172
  HTLV-2 691–693
  human immunodeficiency virus (HIV) 676
  human parainfluenza viruses 1028–1029
  human respiratory syncytial virus (HRSV) 1213
  influenza viruses 720, 1207
  initiation sites 530
  interferon genes 735
  lambda bacteriophage 773, 818–819
  lymphocytic choriomeningitis virus (LCMV) 808–809
  Marburg and Ebola viruses 829
  measles virus 839, 841–842, 843
  MMTV, *see* Mouse mammary tumor virus (MMTV)
  mumps virus (MuV) 878, 878–879
  murine cytomegalovirus 308
  murine polyomavirus 1137–1138, 1141
  N4 bacteriophage, *see* N4 bacteriophage
  Newcastle disease virus 916
  nuclear polyhedrosis viruses (NPVs) 132, 133
  Orthomyxoviridae 1207
  Paramecium bursaria Chlorella virus 1 (PBCV-1) 37, 38
  Paramyxoviridae 1207
  partitiviruses 1048
  parvoviruses 1058
  primary 195
  primers
    bunyaviruses 195
    'cap-snatch' mechanism 195
    reticuloendotheliosis viruses (REV) 1227, 1229
  provirus, avian leukosis viruses 68
  pseudorabies virus 1174
  RB (retinoblastoma) protein inhibiting 20
  reoviruses, *see* Reoviruses
  repression, by adenovirus E1A proteins 18, 19, 20
  reticuloendotheliosis viruses (REV) 1227, 1229–1230
  retroviruses, type D 1238
  reverse, *see* Reverse transcription
  Rhabdoviridae 1207
  RNA viruses 1204
  rotaviruses 1276–1277, 1287–1288
  secondary 195
  Semliki Forest virus (SFV) 1332
  Sendai virus 1300–1301
  simian virus 40 (SV40) 1326
  Sindbis (SIN) virus 1332
  SPO1 phage, *see* SPO1 bacteriophage(B)
  subgenomic minus-strand templates 257
  T1 bacteriophage 1372
  T4 bacteriophage, *see* T4 bacteriophage
  T7 bacteriophage 1391, 1394, 1395
  Tax protein effect 683, 692
  terminators for, filamentous bacteriophage 466
  toroviruses 1454–1455
  trans-activation 11
    in foamy viruses 485, 487, 488
    Tax protein 683, 692
    see also Rex protein; Tax protein
  transfer of products after 232
  translation requirement, bunyaviruses 195
  Ustilago maydis virus (UmV) 1471
  vaccinia virus 1509
  vesicular stomatitis virus (VSV) 1207, 1549–1551
  VZV, *see* Varicella–zoster virus (VZV)
Transcription factors
  activation, in adenovirus transformation 21
  AP1 1475
  E2F 20
  E4F 21
  herpes simplex virus (HSV) 597, 598
  human cytomegalovirus 302
  IRF-1 735
  NFkB 676, 683, 1230, 1475
  Oct-1 597, 598
  Oct-1 and CFF role 597, 598

oncogenes action/as regulators 940, 1475
TF1, SPO1 phage 1354, 1355
transformation mechanism 1475
vaccinia virus transcription 1509
see also Transcriptional activators
Transcriptional activators 1081
herpes simplex virus (HSV) 597, 598
single-stranded-DNA-binding proteins of N4 bacteriophage 891, 892, 893
SP1 676
see also Transcription factors
Transcriptional control elements
hypervariability in JC virus 753, 754
mouse mammary tumor virus (MMTV) 859
Transcriptional enhancer 1081
human interferon-γ 734
Transcriptional factor, NF-1, hormone receptors and 859
Transcriptional regulators, oncogenes action 940
Transcriptional termination motif (TTTTTNT) 38
Transducing phage 997
of *Bacillus subtilis* 76
generalized 76, 107, 109
Bacillus subtilis phage (PBS1) 76
see also P1 bacteriophage[(B)]; P22 bacteriophage[(B)]; Transduction
specialized 76, 109, 110, 112
Bacillus subtilis phage (SPβ) 76
see also Lambda bacteriophage[(B)]; Transduction
T1 phage 1375–1376
Transduction 82, **107–113**
abortive 108, 111
applications 112
barriers against acquisition of foreign genes 111
definition 107
detection 111
discovery 107
experimental evidence 107
fate of donor DNA after infection 110–111
autonomous replication 110
recombination 110, 111
gene transfer efficiency 112
generalized 107
fate of transferred DNA 110
phage, *see* Transducing phage
oncogenes 934, 935
role in biological evolution 112–113
specialized 76, 107–108
'addition' 76
fate of transferred DNA 110–111
insertion length 110
recombination as source of phage genome 109
'replacement' 76
transducing phage, *see* Transducing phage
transposition as source of phage genome 110
T1 bacteriophage 1375–1376
frequency and sites 1375
uptake of cellular DNA into phage 108–109
packaging of discreet genomes 109
packaging of headful of DNA 108–109
Transfection 1204
Bacillus subtilis 76–77
baculovirus recombinants 1535
'direct', cloning DNA in *B. subtilis* phage 77
dominant-negative genes 731
Giardia lamblia 1467
retroviral vector generation 889–890
SPO1 phage DNA 1356
T5 bacteriophage 1387–1388
Transformation **1472–1477**, 1493, 1590
adenoviruses, *see* Adenovirus
avian leukosis viruses 69
baboon herpesviruses 610
Bacillus subtilis 77
bovine papillomaviruses (BPV) 1308, 1309
DNA tumor viruses 1476–1477, 1493
adenoviruses 1476
Epstein–Barr virus (EBV) 1477, 1494, 1495
hepatitis B virus 1477
human papilloma virus 1476–1477, 1497
SV40 1476
see also DNA viruses; *individual viruses*
features of viruses causing 1472
future perspectives 1477
heritable nature of transformed phenotype 1472, 1493
'hit and run' mechanism 1493
murine polyomaviruses 1136–1137
'prophage', cloning DNA in *B. subtilis* phage 77
reticuloendotheliosis viruses 1231–1232
life span extension 1232
retroviruses 1208, 1472–1476, 1473
acute (polyclonal tumors) 1472–1473, 1475
cell signaling changes 1473
cell signaling through transcription factors 1475
growth factors and receptors 1473–1474
HTLV-1 167, 685, 692, 694, 1476
human 1475–1476
intracellular signal transduction molecules 1474–1475
long latency (monoclonal/oligoclonal tumors) 1472, 1473
oncogene cooperation 1475
viral oncogenes 1473
see also Growth factors
yabapox virus 1598
see also Oncogenesis; Tumor formation
Transforming infections 1587
'Transforming proteins' 787
Transgenic animals
mouse mammary tumor virus (MMTV) transcription 859
transfection of dominant-negative genes 732
Transgenic mice 531
bovine papillomaviruses (BPV) 1306
foamy virus pathogenic potential 487
LCMV genes expression 811
poliovirus cDNA 1122
SV40 DNA in 1328
transmissible agent of spongiform encephalopathies 1368
vaccine development 1506
Wnt-1 gene of mouse mammary tumor virus in 860
Transgenic plants
alfalfa mosaic virus (AlMV) 35
resistance to 34
geminivirus DNA 522
potyviruses 1152, 1153
resistance genes 1108, 1109, 1114, 1441, 1463
Transient digestive system disorder (TDSD) 1195
Translation 1205
animal enteroviruses 386
avian leukosis viruses RNA 68
bacteriophage ϕX174 995
bovine immunodeficiency virus (BIV) 160–162
bovine leukemia virus 169–170
bunyaviruses 195
caprine arthritis encephalitis virus 201
cardioviruses 207, 209
carlaviruses 216
caulimoviruses 224
comoviruses 251
cucumoviruses 281
dianthoviruses 351
endoplasmic reticulum function 230
equine infectious anemia virus (EIAV) 431
feline leukemia virus (FeLV) 460
filamentous bacteriophage 467
foamy viruses 485–486
foot and mouth disease viruses (FMDVs) 491
fowlpox virus (FPV) 500
frog virus 3 (FV3) 506
hantaviruses 541
hepatitis A virus (HAV) 549
hepatitis B virus (HBV) 561, 563
hepatitis D virus 577
herpes simplex viruses (HSV) 594, 602
human cytomegalovirus 302
human immunodeficiency virus (HIV) 676, 677
human parainfluenza viruses 1028–1029
influenza viruses 720
initiation factor, poliovirus effect on 1119, 1123, 1589
internal ribosome entry site (IRES) 386, 491
Japanese encephalitis virus 747
luteoviruses 794
lymphocytic choriomeningitis virus (LCMV) 807
measles virus 842
mouse mammary tumor virus (MMTV) 857
mumps virus (MuV) 880
murine cytomegalovirus 308
murine leukemia viruses (MuLVs) 884–885
murine polyomavirus 1137–1138, 1138, 1141
nepoviruses 902
Newcastle disease virus 916
nodaviruses 922
partitiviruses 1048
phytoreoviruses 1098
polioviruses, *see* Polioviruses[(A)]
potexviruses 1145
retroviruses, type D 1238–1239
rhinoviruses 1255
rinderpest and distemper viruses 1262
rubella virus 1292
Semliki Forest virus (SFV) 1330, 1331, 1332
Sendai virus 1301
simian virus 40 (SV40) 1326
Sindbis (SIN) virus 1330, 1331, 1332
sobemoviruses 1350, 1351
tetraviruses 1421
tobacco necrosis virus 899, 900
tobamoviruses 1439
tobraviruses 1443
toroviruses 1455
Ty elements 1234
tymoviruses 1501
Ustilago maydis virus (UmV) 1471
vaccinia virus 1509–1510
varicella–zoster virus (VZV) 1525
yabapox virus 1598

Translational control, T4 phage 1377, 1380–1382
Transmissible gastroenteritis virus of swine (TGEV)(A) 255, 256
Transmissible spongiform encephalopathy (TSE) of mule deer and elk(A) 1357
clinical signs 1359
infectious agent 1357
intracerebral injection 1358
pathology 1359–1360
transmission 1359
see also Spongiform encephalopathies
Transmission of viruses 400
aerosol 1613
arthropod, *see* Arboviruses(A); Arthropod transmission
assisted, fabaviruses 453
common patterns 401–402
fecaloral cycle 401–402
respiratory cycle 402
direct contact 1613
horizontal 400–401
airborne 401, 1613
common vehicle 400
direct contact 400
indirect contact 400, 1613
nosocomial 401
vector-borne 401
in utero, feline immunodeficiency virus (FIV) 456
modes of 400–401
nosocomial 1613
plant viruses 633
propagative 1099
transovarial, *see* Transovarial transmission
vector- and arthropod-borne 1613
vertical, *see* Vertical transmission
zoonoses 1613
Transovarial transmission
dengue viruses 328
encephalitis viruses 365
iridoviruses 1432
phytoreoviruses 1099
plant rhabdoviruses 1248
tetraviruses 1420
vesicular stomatitis virus (VSV) 1554
Transplacental transmission, *see* Transmission of viruses; Vertical transmission of viruses
Transplantation **1477–1489**
bone marrow, *see* Bone marrow transplantation (BMT)
immunization after 1488
immunosuppression 1477–1478
viral infections after 1478–1488
cytomegalovirus 295, 296
hepatitis viruses 1484–1486
herpesviruses 1478–1484
polyomaviruses 1486–1487
respiratory viruses 1487–1488
see also individual viruses/groups of viruses
see also Allograft recipients
Transposable bacteriophage 868
use as genetic tools 875–876
see also Mu bacteriophage
Transposable elements, *see also* Transposons
Transposase, bacteriophage Mu 870, 872
Transposition
plant viruses as vectors for studying 1542
as source of specialized transducing phage genome 110
see also Genetic recombination; Inserted sequence (IS) elements; Retrotransposons; Transposons
Transposons 110
classes 1232
composite 110
'hitchhiking' 104
insertion into prophage 110
in nuclear polyhedrosis viruses 132
Tn917 76, 77
see also Retrotransposons; Ty elements
Tree shrew herpesviruses(A) **1489–1492**
enzymes 1490
epidemiology 1491
evolution 1491
future perspectives 1492
genetics 1490–1491
genome structure and properties 1489–1490
geographic and seasonal distribution 1490
history 1489
host range and viral propagation 1490
infection
clinical features 1492
malignant lymphomas 1491, 1492
pathology and histopathology 1492
isolation from lymphomas 1491, 1492
pathogenicity 1491
proteins 1490
rabbits response 1491
serologic relationships and variability 1491
structure and properties 1489
taxonomy and classification 1489
transmission and tissue tropism 1491
Trichomonas vaginalis virus (TVV) 532, 534
Trichoplusia ascovirus(I) 59, 61, 62
Trichoplusia ni 59, 62
cytoplasmic polyhedrosis viruses (CPV) 317
Trichoplusia ni granulosis virus(I) 128
Trichoplusia ni virus (TnV)(I) 1417
discovery 1422
Tricornaviruses(P)
conserved motif with dianthoviruses(P) 351
genome structure and organization 661–662
Trifluridine (Viroptic) 44
'Triple gene block'
carlaviruses 216
potexviruses 1144, 1146
Trisodium phosphonoformate, *see* Foscarnet
tRNA 232
mimicry, in bromoviruses 183
reticuloendotheliosis viruses (REV) 1227, 1229
terminal in
tobamoviruses 1438
tobraviruses 1442
tymoviruses 1500
$tRNA^{met}$, priming for DNA replication, in caulimoviruses 225
type D retroviruses 1237
in virus-like particle of transposons 1234
Trophoblast interferons 734–735
Tropical spastic paraparesis (TSP) 685, 686, 912–913
Tropism, viral 1080–1082, 1557
definition 1080
enhancers and transcriptional activators 1081
receptors 1080–1081, 1557
site of entry and pathway of spread 1082
tissue-specific promoters 1081
viral cell attachment proteins 1081
see also individual viruses
Trout
infectious pancreatic necrosis virus (IPNV) infections 476–477
rainbow, herpesviruses 470, 476
viral hemorrhagic septicemia virus infection 1250
virus infections 144
Trypsin 584
rotavirus uptake and 1287
Sendai virus activation 1301, 1302, 1303
TTTTT motif, molluscum contagiosum virus (MCV) 850
Tube culture, cytopathic effects in 336–337
Tubules
in nepovirus infections 906–907
in orbivirus infections 961, 963
Tulare apple mosaic virus (TAMV)(P) 30
host range and geographic distribution 34
host relationship 33
Tulip breaking virus(P) 1111
economic significance 1109, 1112
Tulip virus X(P) 1146
Tumor 1472
DNA amplification 938
monoclonal 1493
non-immortalized, leporipoxviruses 1158
solid, murine leukemia viruses causing 884
see also Cancer; *individual tumors*
Tumor formation 1477
oncogenes 934, 935
activation 938
phytoreoviruses 1099
retroviruses 935
see also Carcinogenesis; Transformation
Tumor necrosis factor (TNF)
adenovirus gene products modulating response 7, 1094
immune evasion of leporipoxviruses 1159
molecules protecting from lysis by, immune evasion 1094
sensitivity induced by adenovirus E1A gene 18, 22
Tumor suppressor genes
p53, *see* p53 gene
retinoblastoma (RB) gene 20, 1024, 1476, 1477, 1496
Tumor viruses (human) **1493–1500**, 1590
criteria for causal relationship with cancer 1493–1494
DNA viruses, *see* DNA viruses; Transformation
future perspectives 1499
latency periods 1499
retroviruses, *see* Retroviruses; Transformation
risk definition methods 1493
see also Carcinogenesis; Transformation
Tunicamycin 486, 1247
Tupaia herpesvirus (THV-1)(A) 1489
Tupaia herpesvirus (THV-2)(A) 1489, 1490
Tupaia herpesvirus (THV-3)(A) 1489
Turbinates, air passage 1221
Turbot herpesvirus(A) 470
Turkey(s)
birnavirus infection 144
Marek's disease virus infection 832, 834
reticuloendotheliosis viruses 1230
Turkey bluecomb coronavirus (TCV)(A) 255
Turkey herpesvirus(A), *see* Marek's disease virus (MDV)(A)
Turkey rhinotracheitis virus (TRTV)(A) 1211
antigenic cross-reactivity 1215
genome 1213
Turkey virus hepatitis 387, 390
Turnip crinkle carmovirus (TCV)(P) 218, 1538
coat-protein-binding sites 220, 222
genome structure 220, 221
in vitro assembly 220
infectivity of cDNA 1540
mutagenesis studies 222
tobacco necrosis virus relationship 900

  tombusviruses relationship 1448
  transmission 219
Turnip rosette virus (TRoSV)(P) 1346
  see also Sobemovirus(P)
Turnip yellow mosaic virus (TYMV)(P) 511, 1500, 1538
  see also Tymoviruses(P)
Twort, F.W. 628, 643–644, 645
Twortd'Herelle phenomenon 648
Ty elements 1233
  assembly 1234
  host genes affecting 1235, 1236
  translation 1234
  Ty1 1234–1235
  Ty3 1234, 1235
  virus-like particles 1235–1236
  see also Retrotransposons, of fungi
'Tymobox' 1501
Tymovirus(P) 1399
Tymoviruses(P) **1500–1502**
  distinguishing features 1500
  evolution 439
  genome structure and proteins 1501
  host interactions 1502
  host ranges 1502
  overlapping (OP) genes 1501
  replicase 1500–1501
  replication strategy 1501–1502
  structure and composition 1501, 1581
  symptoms 1502
  taxonomy and classification 1500–1501
  transmission 1502
'Type species' 94
Tyrosine kinase
  oncogene activity 938, 939
  SH2 domains on substrates 1474
  v-*src* oncogene action 1474

U
UCD-144 cell line 535
UDP-glycosyltransferase 133
Ullucus C virus (UCV)(P) 249
Ullucus mild mottle tobamovirus (UMMV)(P) 1436
Unclassified viruses, Borna disease virus (BDV) 149
Uncoating of virus 1169, 1204, 1558
Upper respiratory tract infections (URTI; URI) 1223
  see also Respiratory tract infections
Urbanization, dengue virus spread 324, 327, 331
Urethritis 590
Ustilago maydis 1470
Ustilago maydis virus (UmV)(P) **1468–1471**
  capsid cross-reactivity 1469
  dsRNA 1468
    classes (H, M, L) 1468
  economic significance 1471
  exclusion phenomenon 1469
  genetics 1469
  genome structure 1468–1469
  geographic distribution 1471
  helper virus 1468, 1470
  killer toxins 1468, 1470
    KP1 1468, 1470
    KP4 1468, 1470, 1470–1471
    KP6 1468, 1470
      structure and effect 1470
    physical characteristics 1470
    resistance to 1470
  multiplication 1469
  satellite dsRNA 1468
  structure and composition 1468
  subtypes 1468, 1469
  taxonomy and classification 1468
  transcription and translation 1471
  transformation vectors 1470
  transmission 1470
  virushost relationship 1469–1470
  see also Totiviruses(P)
Ustilago maydis virus-H1 (UmV-H1)(P) 1464
Ustilago maydis virus-P1 (UmV-P1)(P) 534
Uukuviruses(A) 190–191

V
v-*crk* oncogene 1474
v-*erbA* 941, 1475
v-*erbB* 941, 1474, 1475
v-*fes* 935
v-*fos* 1475
v-*jun* 1475
v-*mos* 940
v-*myc* 935
v-*onc* 936
  viruses and diseases 936
  see also Oncogenes, viral
v-*ras* 1474
v-*rel* 1230, 1232, 1475
v-*sis* 1473
v-*src* 935, 1474
Vaccine vectors
  adenoviruses, *see* Adenovirus(A)
  fowlpox virus (FPV) as 502
  herpes simplex virus (HSV) as 603
  see also Vaccinia virus(A)
Vaccines **1503–1507**
  adenovirus 7, 17, 1506
    recombinants 1529
  adverse reactions to 1297–1298, 1513
  African horse sickness virus (AHSV) 954, 955
  animal models 1506
  antibodies induced by 1504
  antigenic determinants 1504, 1505
  avian encephalomyelitis 391
  birnaviruses 148
  bluetongue virus (BTV) 954, 955
  bovine herpesviruses 157
  bovine immunodeficiency virus (BIV) 165
  bovine leukemia viruses 173
  bovine parvoviruses 1311
  canine distemper virus (CDV) 1266, 1267
  canine parvovirus 1057
  cell-mediated immunity and 708, 1504
  Central European encephalitis virus 367
  channel catfish virus 473–474
  Chikungunya, O'nyong nyong and Mayaro viruses 241
  coltiviruses 954, 955
  coronaviruses 260
  cowpox 267, 1503, 1507
  current types 1503
  dengue viruses 330
  development, factors affecting feasibility 1506–1507
    animal models 1506
  dominant-negative mutants 732
  duck hepatitis B virus 391
  Duvenhage virus 1189
  echoviruses 360
  enteroviruses 359, 391
  equine arteritis virus (EAV) 771
  equine encephalitis virus 422
  equine herpesviruses 428–429
  equine infectious anemia virus 430, 435
  feline calicivirus 1546
  feline immunodeficiency virus (FIV) 459
  feline leukemia virus (FeLV) 463
  feline parvoviruses (FPV) 1056
  foot and mouth disease virus 495–496
  fowlpox 497, 498, 499, 502
  hantaviruses 545
  hepatitis A virus, *see* Hepatitis A virus (HAV)(A)
  hepatitis B virus, *see* Hepatitis B virus (HBV)(A)
  hepatitis C virus 574
  hepatitis E virus 586
  herpes simplex viruses (HSV) 592, 593
  herpesvirus saimiri 620
  herpesvirus of turkeys 833
  history 632, 708, 1503
  human cytomegalovirus (HCMV) infections 298
  human immunodeficiency virus, *see* Human immunodeficiency virus (HIV)(A)
  human parainfluenza viruses 1030
  immune response
    antibodies role 1504
    cell-mediated 1504
    effector T cells, role 1504–1505
    epitopes recognized by antibodies 1504
  immunological requirements 1506
  influenza viruses 714, 1226, 1507
  Japanese encephalitis virus, *see* Japanese encephalitis virus (JEV)(A)
  lassa virus 785
  live
    avoidance in transplant patients 1488
    dominant-negative mutants in 732
  Marek's disease virus, *see* Marek's disease virus (MDV)(A)
  measles, mumps and rubella (MMR) 846, 883
  measles virus, *see* Measles virus(A)
  Mokola virus 1189
  molluscum contagiosum virus, *see* Molluscum contagiosum virus (MCV)(A)
  Morbillivirus 1266, 1267
  mumps virus 883
  Newcastle disease virus 916, 918, 919
  orbiviruses 952, 955
  parainfluenza viruses 1030
  parvoviruses 1057, 1067, 1075
  poliovirus, *see* Polioviruses(A)
  porcine parvovirus 1075
  poxviruses, of sheep/goats 1161, 1162, 1165
  'Prestige' 429
  pseudorabies, *see* Pseudorabies virus (PRV)(A)
  rabies, *see* Rabies virus(A)
  recombinant vaccinia virus, *see* Vaccinia virus(A)
  requirements (safety and efficacy) 1504
  retroviruses, type D 1242
  rhinoviruses 1259
  Rift Valley fever virus 191
  rinderpest, *see* Rinderpest virus (RV)(A)
  rotaviruses 1278, 1281
  rubella, *see* Rubella
  Sendai virus 1304
  smallpox, *see* Smallpox
  swine vesicular disease 391
  T-cell epitope production 1505
  trials 400
  under trial 1503
  varicella–zoster virus, *see* Varicella–zoster virus (VZV)(A)
  vesicular stomatitis virus (VSV) 1555, 1618
  yellow fever virus 1608, 1612, 1615
  see also other specific vaccines
Vaccinia
  progressive 1512–1513
  see also Smallpox; *under* Vaccinia virus(A)

Vaccinia growth factor 530
Vaccinia virus(A) **1507–1513**
accidental infections 1512, 1513
assembly, uptake and release 1510
Copenhagen strain 1531
cowpox virus
differences 261, 263
similarity 261, 262
cultivation 630, 1511
cytopathology 1510
DNA packaging, lack of constraints 1530
DNA replication 499, 1509
early/intermediate and late genes 1509
entomopoxviruses (EPVs) comparison 393, 394, 395
epidemiology 1512
evolution 1511
fowlpox virus comparisons 497, 498
future perspectives 1513
genetics 1511
genome structure and properties 1508
geographic and seasonal distribution 1511
history 1507
host range and viral propagation 1511
infection 267, 1512–1513
clinical features 1512–1513
of eye 442, 1513
heart infections 965
immune response 1513
pathology and histopathology 1513
prevention and control 1513
leporipoxvirus replication similarity 1155
molluscum contagiosum virus (MCV)
gene expression similarity 850
mutants 1511
neurovaccinia variants, rabbitpox 866
orfvirus genome similarities 1038
Paramecium bursaria Chlorella virus 1 (PBCV-1) similarity 36
pathogenicity 1512
physical properties 1509
post-translational processing 1510
proteins 1508–1509
p37K homology to molluscum contagiosum virus (MCV) protein 849, 850
VGF protein 1560
receptor 1560, 1561
recombinant virus as vaccines 1510, 1531
advantages 1509–1510
concerns 1511, 1512
rabies 1531, 1621
see also Vaccinia virus(A), vaccines *(below)*
recombinants 1530
recovery and screening 1530
serologic relationships and variability 1511–1512
structure and properties 1508
surface tubules 1508
taxonomy and classification 1507
transcription 1509
transfected DNA 1530
translation 1509–1510
transmission and tissue tropism 1512
vaccines
EBV membrane antigen (MA) 409
monkeypox 1345
safety concerns 1531
smallpox 1507, 1512, 1513, 1531
see also Vaccinia virus(A), recombinant virus
as vectors 1510–1511, 1530
ALVAC 1531
applications 1531
LCMV genes 811
methodology 1530
NYVAC 1531, 1532
recombinant plasmids 1530–1531
recombinants, *see* Vaccinia virus(A), recombinants
white pock lesions 1511
'Vacuolating virus', *see* Simian virus 40 (SV40)(A)
Vacuolation, cucumoviruses 282
Vacuolization, in Creutzfeldt-Jakob disease (CJD) 1367–1368
Varicella
antiviral treatment 46, 1518
clinical features 1516–1517
complications 1516–1517
epidemiology 1516
history 1514
incubation period 1516
transmission and tissue tropism 1516
see also Varicella–zoster virus (VZV)(A)
Varicella-like illnesses, simian 1514
Varicella–zoster virus (VZV)(A) **1514–1527**
acyclovir-resistance 1518
animal models 1515
antivirals 46, 1518
assembly, uptake and release 1515–1526
αTIF (trans-inducing factor) 1519, 1524
HSV homology 600, 1524
cell-associated nature 1515, 1522, 1526
cytopathology 1515–1526
DNA polymerase 1521
DNA replication 1523–1524
'Challberg' genes analogues 1524
genes involved 1521
epidemiology 1516
Epstein–Barr virus (EBV) comparison 1515
evolution 1515
Fc receptor 1521
future perspectives 1518, 1526–1527
gene products
gene *4* 1522
gene *29* 1521, 1527
gene *47* 1519, 1525
gene *61* 1522
gene *62* 1524
gene *63* 1522
IE62 1522, 1525, 1527
immediate-early 1522
general features **1514–1518**
genetics 1515
genome structure and properties 1515, 1519, 1520–1521
gene mapping 1524
isomeric forms 1519
L and S segments 1519
geographic and seasonal distribution 1514
glycoproteins 1519, 1521
gpI 1521
gpII 1525
gpIV 1521, 1525
gpV 1525
HSV homology 1519, 1521, 1527
post-translational processing 1525
transcription 1524
history 1514
host range and viral propagation 1514–1515, 1527
in immunocompromised and AIDS 1516, 1517
immunoglobulin (VZIG) 1483, 1516, 1517, 1517–1518
infection
clinical features 1516–1517
of eyes 444
heart infections 965
immune response 1517
pathology and histopathology 1517
prevention and control 1517–1518
see also Varicella; Zoster
infection after transplantation 1483
features and treatment 1483
incidence and pathogenesis 1483
latency
genes expressed 1527
sites 1516, 1517, 1527
molecular biology **1518–1527**
neural spread 1079, 1516, 1526
ORFS 1519–1520–1521
pathogenicity 1516
physical properties 1522–1523
post-translational processing 1525
proteins 1518–1519, 1519, 1520–1521, 1521–1522
genes coding 1520–1521
HSV homology 1519, 1521, 1522
tegument proteins 1519, 1522
re-activation 1483
model 1527
see also Zoster
recombinant 1515
replication (lytic cycle) 1523–1524
sites 1516
sensitivity to environment 1522–1523
serologic relationships and variability 1515–1516
serological tests 1515
structure and properties 1518–1519
taxonomy and classification 1514
tissue culture 1515, 1526
tissue tropism 1527
transcription 1524–1525
homology with HSV 1524
immediate-early, early and late 1524
low abundance transcripts 1524
modulatory proteins 1522, 1524, 1526
splicing 1524
transactivators 1522
transformation possibility 1527
translation 1525
transmission and tissue tropism 1516
transrepressor 1522
UV irradiation sensitivity 1523
vaccination 1518
vaccine 1483
concerns 1527
immune response 1518
Oka strain 1516, 1518, 1527
zoster after 1518, 1527
variability of strains 1515
Variola 1341
see also Smallpox
Variola major 1341, 1342
Variola minor 1341, 1342
Variola virus(A), *see* Smallpox (Variola) virus(A)
Variolae vaccinae 261
Variolation 1341
Varroa jacobsoni 657, 658, 659
Vascular endothelial cells 909
Vascular permeability
increase, in dengue hemorrhagic fever/dengue shock syndrome (DHF/DSS) 328, 329–330
invasion of tissues by viruses 1079
Vectors **1528–1543**
adeno-associated viruses (AAV) 1060
advances and development of 1528
animal viruses **1528–1536**
adenoviruses 1528–1530
baculovirus 1534–1535
control 1621

future perspectives 1536
miscellaneous 1535–1536
movement and spread of zoonoses 1622
poxvirus-based systems 1530–1532
retrovirus 1532–1534
see also under individual viruses
barley stripe mosaic virus (BSMV) as 663–664
cloning, *see* Gene cloning
gene amplification, based on geminiviruses 522–523
for gene expression, herpesvirus saimiri 621
history 1528
importance and uses 1528, 1536
in inherited diseases therapy, avian hepatitis B viruses 569
plant RNA viruses 1540–1542
cDNA infectivity 1540
coat protein genes 1541
hordeiviruses 663–664
movement protein genes 1541
proteins in RNA replication 1541
plant viruses **1536–1543**
advantages 1543
applications 1542–1543
concerns 1543
future perspectives 1543
geminivirus 1539–1540
independent *vs* 'disarmed' 1542, 1543
inoculation of plants with 1537
instability 1536–1537
pararetrovirus 1537–1539
recombination risk concern 1543
splicing and recombination 1542
spread in host 1537
for studying DNA and RNA rearrangements 1542
virus types 1536
plasmid, *see* Plasmids
retroviral 888–890
RNA viruses as 1528
vaccine, *see* Vaccine vectors; Vaccinia virus[(A)]
see also Gene cloning; *individual viruses (as above)*
Velvet tobacco mottle virus (VTMoV)[(P)] 1346
see also Sobemovirus[(P)]
Velvet-bean caterpillar 136
Venezuelan equine encephalitis (VEE) virus[(A)] 417, 1617–1618
diagnosis 1617
epidemiology 420
geographic distribution 1617
IAB variant 1617
infection
clinical features 421, 445
immune response 422
pathology and histopathology 422
prevention and control 1617–1618
mutant TRD strains 421
pathogenicity 421
taxonomy and classification 1617
transmission 1613, 1617
see also Equine encephalitis viruses[(A)]
Venezuelan hemorrhagic fever (VHF) 776
clinical features 784
epidemiology 781
pathology and histopathology 784
see also Guanarito virus[(A)]
Vero cells
African swine fever virus infection 25, 26–27
arenavirus cultivation 778
Vero toxins, *see* Shiga-like toxins
Vertical transmission of viruses 400, 401
bovine leukemia virus 171
Creutzfeldt-Jakob disease (CJD) 1366
foamy viruses 487
HIV infection 681
HTLV-1 infection 684, 686
mouse mammary tumor virus (MMTV) 855
reticuloendotheliosis viruses 1230
see also Transmission of viruses
Vesicles 228
cell, occluded virions in 129
membranous
closterovirus infections 246
in poliovirus infections 1129, 1130
pea enation mosaic virus (PEMV) in 1087–1088
shuttle 228
see also Cytoplasmic vesicles
Vesicular exanthema virus of swine (VES)[(A)] **1544–1547**
classification and properties 1544
epidemiology and transmission 1544, 1545
geographic distribution 1544, 1545
history 1544
host range and viral propagation 1545
infection
clinical features 1546
immune response 1546
prevention and control 1546
outbreaks 1544
pathogenesis 1545–1546
serologic relationships and variability 1545
see also Caliciviridae
Vesicular fluid, in herpes simplex virus infections 591
Vesicular lesions
foot and mouth disease 495
vesicular exanthema virus of swine (VES) 1546
vesicular stomatitis virus (VSV) infection 1554
Vesicular stomatitis virus (VSV)[(A)] 446, **1547–1556**, 1618
assembly and release 228, 1551
attachment 1557
Chandipura, Piry and Isfahan viruses comparisons 234
cross-reactions 234
defective-interfering 320, 731
effect on host cell 1588, 1589
epidemiology 1553
evolution 1552
experimental hosts 1552
future perspectives 1555
genetics 1552
genome and regulatory sequences 1246
genome structure and properties 1548
geographic and seasonal distribution 1551–1552, 1618
history 1547–1548
host range and viral propagation 1552, 1587, 1618
HTLV-1 pseudotypes 689
incompatibility with heterologous viruses 728
Indiana serotype 1548, 1553
infection
clinical features 1554–1555
clinical features in humans 1554–1555
immune response 1555
outbreaks 1547, 1551, 1553
pathology and histopathology 1555
prevention and control 1555, 1618
membrane proteins 1558
mutants 1589
mutation rate 525, 1552
neutralization epitopes 1555
New Jersey serotype 1548, 1553, 1555
pathogenicity 1554
physical properties 1549
plant rhabdovirus size comparison 1244
protein synthesis 1551
proteins 1548–1549
glycoprotein (G) 1549
large (L) polymerase protein 1548, 1549
matrix (M) protein 1548, 1549
nucleocapsid (N) 1548, 1548–1549, 1551
polymerase-associated phosphoprotein (P) 1548, 1549
replication 1549–1551
strategy 1207
resistance, Mx protein 736
RNA packaging 1551
RNA polymerase mutant, mechanism of interference by dominant-negative mutant 730
RNA replication 1207, 1549–1551, 1550–1551
serologic relationships and variability 1553
serotypes 1548
structure and properties 1548
superinfection exclusion 729
taxonomy and classification 1548, 1618
tissue tropism 1554
transcription 1207, 1549–1551
attenuated mechanism of regulation 1550
transcriptionreplication switch mechanism 1550–1551
transmission 1552, 1554
to humans 1554, 1618
ts mutant, mechanism of interference by dominant-negative mutant 730
types 1, 2 and 3 viruses 1548
uptake and infection 1549
vaccines 1555, 1618
Vesicularization of cells, in poliovirus infections 1130
Vesiculation
carmoviruses 219
cucumoviruses 282
Vesiculovirus[(A)] 233, 1243, 1548
transmission 234, 235
see also Vesicular stomatitis virus (VSV)[(A)]
Vi antigen 79
Vibrio cholerae
RNA phage 1335
transposable phage (VcA1) 875
Vicia cryptic virus (VCV)[(P)] 275
Victorin 1051
Vidarabine
herpes simplex virus infections 592
herpesvirus inhibition 43, 44
varicellazoster virus infections 1518
vif gene 1318
Villi
torovirus replication 1457–1458
virus replication in 377
see also Small intestine
Vimentin 507
Viral arthritis syndrome (VAS) 1195
Viral hemorrhagic septicemia virus (VHSV) (Egtved virus)[(A)] 478–479, 1250–1251
clinical features of infection 1250
history and host range 1250
immunology and diagnosis 1250
interferon production 1251
molecular aspects 1251
pathology and histopathology 1250
transmission and prophylaxis 1251

- Viral Infections, types 1587–1588
- Viral membranes, *see* Membranes
- Viral oncogenes, *see* Oncogenes
- Viral structure, *see* Structure of viruses
- Virazole, *see* Ribavirin
- Viremia
  - persistent, nervous system viruses 909
  - primary 1078
  - secondary 1078
- Virgin-soil epidemic 403
- Virino hypothesis 1363
- Virion 1573
  - definition 1168
- Virion hypothesis 1357
- Viroids **1563–1570**
  - 'captured', introns as 1570
  - chimeras 1569
  - control 1567
  - cytopathic effects 1566
  - definition 1563
  - detection methods 1567–1568
  - diseases 1563, 1567
  - economic significance 1567
  - epidemiology 1567
  - evolution, satellite RNA link 1565, 1566
  - genome structure 1563–1565, 1570
    - domains 1564–1565
  - geographic distribution 1567
  - history 1563
  - host range 1565–1566, 1570
  - as living fossils 1570
  - molecular biology 1568–1570
  - origin and evolution 1570
  - pathogenicity 1569–1570
  - replication 1568–1569
    - cleavage 1569, 1570
  - RNA, hepatitis D virus RNA relationship 578
  - RNA replication 1568
  - self-clearing 1564
  - species 1564
  - symptomatology 1566–1567
  - taxonomy and classification 1565
  - thermal denaturation 1569
  - transmission 1565–1566
- Viropexis 10
- Viroplasm 1279
  - rotavirus 1279, 1286
  - tospovirus infections 1459
  - *see also* Proteinaceous inclusion bodies
- Viroptic, *see* Trifluridine (Viroptic)
- Virosomes 1156
- Virulence genes, bacteriophage 106
- Virulence of viruses 402
- Virus 1203
  - attachment and entry, *see* Attachment of viruses; Entry of viruses; Pathogenesis of viral infections
  - definition 8, 42, 1203, 1563
  - lysis by antibodies 1504
  - origin of term 627
  - replication, *see* Replication of viruses
  - shedding 400
  - terminology 1168
  - transmission, *see* Transmission of viruses
  - yield 400
  - *see also specific topics*
- Virus diarrhea (VD) virus(A) 175
- Virus kingdoms 1403
- Virus-like particles (VLPs) 50
  - algal 35
  - cryptoviruses 274
  - dsRNA association 275
  - J2 virus-like particle 51, 55
  - *Methanobacterium voltae* 55
- Virus–host cell interactions, *see under* Host
- Visceral lymphomatosis 70
- Visna disease 912, 1366, 1592
  - *see also* Visnamaedi viruses(A)
- Visna virus(A) 199
  - caprine arthritis encephalitis virus (CAEV) comparison 199, 1592, 1593
  - feline immunodeficiency virus (FIV) similarity 455
  - *see also* Visnamaedi viruses(A)
- Visna–maedi viruses(A) **1592–1595**
  - cell biology and pathogenesis 1594
  - epizootiology 1593
  - future perspectives 1594
  - genome 1593, 1594
  - history 1592
  - host range 1593
  - infection
    - clinical and pathological criteria 1592
    - immune responses 1594
    - pathology 1592–1593
    - prevention and control 1594
  - as models for HIV studies 1594
  - nomenclature for disease 1592
  - replication 1594
  - taxonomy and classification 1593
  - transmission 1593
  - variability 1593–1594
- VLP, *see also* Virus-like particles (VLPs)
- Voandzeia necrotic mosaic virus(P) 1500
- Voandzeia mosaic virus (VoMV)(P), 215
- Vomiting
  - Norwalk agent causing 926, 930
  - viral infections 377
  - *see also* Gastroenteritis
- Von Economo's encephalitis 746
- *vpu* gene 1318
  - origin 439
- *vpx* gene 1318
  - origin 439
- VZV, *see* Varicella–zoster virus (VZV)(A)

W

- Walleye herpesvirus(A) 470, 480
- Warts 1013, 1014, 1305
  - clinical features 1019
  - epidemiology 1017
  - HPV types causing 1018
  - immune response to 1310
  - types 1019
  - *see also* Genital warts; Human papillomaviruses (HPV)(A)
- Wasps, ascovirus transmission 61–62
- Water, drinking, contamination with hepatitis E virus 581, 582, 583, 584
- Watermelon silver mottle virus (WSMV)(P) 1463
- Weed infestation, tobraviruses 1446
- Weight loss, in feline immunodeficiency virus (FIV) infection 457
- Wesselsbron virus (WSLV)(A) **367–372**, 1619
  - clinical features of infections 372
  - epidemiology 371
  - genetics and evolution 370
  - geographic and seasonal distribution 369, 1619
  - history 367
  - host range and viral propagation 370
  - taxonomy and classification 368
  - transmission 1619
  - *see also* Encephalitis viruses(A)
- West Nile virus (WNV)(A) 362, 363, 1615
  - clinical features of infections 365–366
  - genetics 364
  - geographic and seasonal distribution 362, 1615
  - Japanese encephalitis virus relationship 748
  - transmission 1615
  - *see also* Encephalitis viruses(A)
- Western blot assay
  - hepatitis E virus 586
  - parapoxviruses 1041
  - viral antibody detection 346
- Western equine encephalitis (WEE) virus(A) 416, 1616
  - cytopathic effects 639
  - epidemiology 420
  - evolution 439
  - geographic distribution 1616
  - history 639
  - infection
    - clinical features 421, 445
    - diagnosis 1616
    - pathology and histopathology 422
    - prevention and control 1616
  - pathogenicity 421
  - subtypes 417, 418
  - transmission 1616
  - *see also* Equine encephalitis viruses(A)
- Wheat dwarf virus (WDV)(P) 518, 1538
  - gene expression 519
  - as vector 1539
- Wheat yellow leaf virus (WYLV)(P) 243
- White bryony mosaic virus (WBMV)(P) 215
- White clover, infections 1112
- White clover cryptic virus I (WCCVI)(P) 275–276
- White clover cryptic virus II (WCCVII)(P) 276
- White clover mosaic virus (WClMV)(P) 215, 1143, 1538
  - genome 1144
  - RNA transcripts 1146
- White clovergrass pastures 1112
- Whitefly transmission
  - closteroviruses 244, 247
  - geminiviruses 521
- Wild cucumber mosaic virus(P) 1500
- Wilting, plant viruses causing 1110
- WIN 51711 45, 48
- 'WIN compounds' 1121
- Wineberry latent virus(P) 243
- 'Winter Vomiting Disease' 926
- *Wiseana* iridescent virus (WIV)(I) 1434
  - economic importance 1435
  - genome and ecology 1435
- WiskottAldrich syndrome 1017
- *Wnt-1* 860
- Woodchuck hepatitis virus (WHV)(A) 555
  - genome structure/properties 560
  - geographic and seasonal distribution 555
  - for hepatitis D virus infections 575
  - host range and viral propagation 555–556
  - pathogenicity 558
  - properties 560
- World Health Organization (WHO), poliovirus declaration 1115, 1118, 1503
- Wound tumor virus (WTV)(P) 1096
  - defective-interfering (DI) RNA 1097–1098
  - genome 1097

X

- X ray diffraction 1571
- X-linked lymphoproliferative syndrome (XLPS) 406, 407, 408, 1495
- X-ray crystallography 1571
  - animal virology 631
  - influenza virus antigenic determinants 724–725
  - nodaviruses 920
- *Xanthomonas*, filamentous bacteriophage 464
- Xenodiagnosis 1620

Y
Y62-33 virus(A) 417, 418
Yabapox virus(A) **1597–1600**
epidemiology 1599
evolution 1597
genome structure 1597
history 1597
infection
clinical features and pathology 1599
immune response 1599
pathogenicity 1599
physical properties 1598
proteins 1597
replication 1598
serologic relationships and variability 1598–1599
structure and properties 1597
taxonomy and classification 1597
transformation 1598
transmission and tissue tropism 1599
see also Tanapox virus(A)
Yamagushi 73 sarcoma virus(A), oncogenes 936
Yamame tumor virus(A) 470
Yarrowia lipolytica virus (YLV)(P) 1465
Yatapoxviruses(A) 1597
Yeast
20S RNA 1600, 1601
killer phenomenon 1464, 1467, 1600
RNA replicons 1600, 1601
suppressive sensitive strains (S-dsRNA) 1465
T dsRNA 1600
viruses 1464
see also Totiviruses(P)
Yeast dsRNA virus (ScV-L)(P), *see Saccharomyces cerevisiae* virus L-A (ScV-L-A(P); Totiviruses(P)
Yeast RNA viruses **1600–1606**
see also entries beginning Saccharomyces cerevisiae
Yellow fever
clinical features 1611
diagnosis 1615
epidemics 1621
epidemiology and transmission 1609–1611
genetic resistance in mice 664
heart infections 965
history 1606, 1614
immune response 1611–1612
incidence 1609
jungle 1610, 1615
outbreaks, concern over future 1615
pathogenesis and pathology 1611
prevention and control 1612, 1615
Yellow fever virus(A) **1606–1612**, 1614–1615
genome structure and properties 1607
genes 1608
geographic and seasonal distribution 1609, 1615
history 628, 1606
host range and viral propagation 1608–1609, 1615
infection, *see* Yellow fever
isolation and assay 1609
morphogenesis 1606–1607
proteins 1607
replication 1607
sites 1608–1609, 1611
strains and 'topotypes' 1608
structure and properties 1606–1607
taxonomy and classification 1606
transmission 1606, 1609, 1615
movement of vectors and 1622
urban and forest cycles 1610
vaccine 1612, 1615
17D strain 1608, 1612
variation and evolution 1607–1608
viscerotropic 1611
yes oncogene 936

Z
Zenker's fixing solution 342
Zenker's necrosis 379
Zidovudine (AZT), interferon-α with 739–740
Zinc sulfate 634
Zoonoses 403, **1613–1622**
in Africa 1618–1619
in Asia 1619
in Australia 1619–1620
in Central/South America 1617–1618
control 1620–1621
ecological change 1621–1622
problems 1621–1622
social changes and 1622
definition 1613
diagnosis 1620
in Europe 1618
geographic distribution 1613
history 1613
in North American 1616–1617
see also Colorado tick fever virus(A); Encephalitis viruses(A)
spread 1621–1622
transmission 1613
cycles 403
see also Transmission of viruses
widely distributed 1613–1616
Crimean-Congo hemorrhagic fever virus (CCHF) 1616
sandfly fever viruses 1616
see also Chikungunya (CHIK) virus(A); Encephalitis viruses(A); Hantaviruses(A); Rabies virus(A); Sindbis (SIN) virus(A); Yellow fever virus(A)
Zoster
clinical features 1517
eye infections 444
history 1514
re-activation of VZV 1516
recurrent 1516, 1517
transmission and tissue tropism 1516
in transplant patients 1483
see also Varicella–zoster virus (VZV)(A)
Zovirax, *see* Acyclovir
Zwiegerzuejte 1592